Medical and Nutritional Complications of Alcoholism

Mechanisms and Management

With Contributions by

Siamak A. Adibi, M.D., Ph.D., *Montefiore Hospital and University of Pittsburgh School of Medicine, Pittsburgh, Pennsylvania*

Enrique Baraona, M.D., *Mt. Sinai School of Medicine of the City University of New York and Bronx Veterans Affairs Medical Center, New York, New York*

Murray Epstein, M.D., F.A.C.P., *University of Miami School of Medicine and Veterans Affairs Medical Center, Miami, Florida*

Lawrence Feinman, M.D., *Mt. Sinai School of Medicine of the City University of New York and Bronx Veterans Affairs Medical Center, New York, New York*

Howard S. Friedman, M.D., *Downstate Medical Center of the State University of New York, The Brooklyn Hospital, New York, New York*

Anthony J. Garro, Ph.D., *University of Medicine and Dentistry of New Jersey, New Jersey Medical School, Newark, New Jersey*

Barbara H. J. Gordon, Ph.D., *Hunter College of the City University of New York, New York, New York*

Gary G. Gordon, M.D., *New York Medical College, Valhalla, New York*

Mark A. Korsten, M.D., *Mt. Sinai School of Medicine of the City University of New York, and Bronx Veterans Affairs Medical Center, New York, New York*

Maria A. Leo, M.D., *Mt. Sinai School of Medicine of the City University of New York, New York, New York*

John Lindenbaum, M.D., *Columbia University College of Physicians and Surgeons, New York, New York*

Finbarr C. Martin, M.D., *Department of Medicine for the Elderly, St. Thomas' Hospital, London, England*

Fiorenzo Paronetto, M.D., *Mt. Sinai School of Medicine of the City University of New York and Bronx Veterans Affairs Medical Center, New York, New York*

Timothy J. Peters, M.D., *King's College School of Medicine and Dentistry University of London, King's College Hospital, London, England*

Romano C. Pirola, *Department of Medicine, Prince Henry Hospital, Little Bay, Australia*

Alan S. Rosman, M.D., *Mt. Sinai School of Medicine of the City University of New York and Bronx Veterans Affairs Medical Center, New York, New York*

Maurice Victor, M.D., *Veterans Affairs Medical and Regional Office Center, White River Junction, Vermont, and Dartmouth Medical School, Hanover, New Hampshire*

Medical and Nutritional Complications of Alcoholism

Mechanisms and Management

Charles S. Lieber

Bronx Veterans Affairs Medical Center and
Mount Sinai School of Medicine of the City University of New York
New York, New York

Plenum Medical Book Company • New York and London

Library of Congress Cataloging-in-Publication Data

Medical and nutritional complications of alcoholism : mechanisms and
  management / [edited by] Charles S. Lieber
        p.   cm.
    Includes bibliographical references and index.
    ISBN 0-306-43558-6
    1. Alcohol--Physiological effect. 2. Alcoholism--Pathophysiology.
I. Lieber, Charles S., 1931-
    [DNLM: 1. Alcohol, Ethyl--adverse effects. 2. Alcoholism-
-complications. 3. Nutrition--drug effects. 4. Nutrition
Disorders--etiology. 5. Nutrition Disorders--therapy.   WM 274
M4872]
QP801.A3M35 1992
616.86'1--dc20
DNLM/DLC
for Library of Congress   92-49893
                                                                      CIP

ISBN 0-306-43558-6

© 1992 Plenum Publishing Corporation
233 Spring Street, New York, N.Y. 10013

Plenum Medical Book Company is an imprint of Plenum Publishing Corporation

Printed in the United States of America

To my children
Colette, Daniel, Leah, Samuel, and Sarah
in appreciation for their patience and understanding
during the writing of this book and the ones that preceded it

Preface

In the Western world, alcohol is the most abused drug. For all the attention being directed toward heroin, cocaine, and marijuana, the favorite mood-altering drug in the United States, as in almost every human society, is alcohol. In nature, the fermentation of sugars is the major source of ethanol, but how humans first encountered it is unclear. It most likely occurred in either fermented fruit juices (wine), fermented grain (beer), or fermented honey (mead). Whether the Paleolithic Stone Age man knew of ethanol is undetermined, but it is abundantly clear that his Neolithic descendants were familiar with the product of fermentation. With the exception of the original inhabitants of Tierra del Fuego, the Australian aborigines, and some polar tribes, all known human groups (even among the surviving Stone Age cultures) are familiar with alcoholic beverages. The purpose of this volume is not to discuss the effects of ethanol on mood and behavior that have led to its widespread use but, rather, to focus on the consequences, sometimes disastrous, of the metabolism of ethanol in the body and how, as a result, alcoholics differ from nonalcoholics biochemically and pathologically.

A few decades ago, the medical issues relating to the disease of chronic alcoholism were not widely studied, because the intrinsic toxicity of alcohol was not fully appreciated and alcoholism was considered primarily a social or behavioral problem. However, the prevalence of just one medical problem, cirrhosis of the liver, has now reached a magnitude such that this complication of alcoholism represents, in and of itself, a major public health problem. We now recognize that 75% of all medical deaths attributable to alcoholism are the result of cirrhosis of the liver; in large urban areas, it has become the fourth leading cause of death in the active age group of 25 to 65 years. Although not all cirrhotic subjects are alcoholics, it is now generally recognized that a majority of patients with cirrhosis do admit to excessive alcohol consumption. Other tissues can also be severely affected, including brain, gut, heart, endocrine systems, bone, blood, and muscle. A question often raised is, "In what way does an alcoholic differ from a nonalcoholic?" Inquiries have focused on psychological make-up, behavioral differences, and socioeconomic factors. More recently, however, physical differences have been delineated. Prior to the development of various disease entities, chronic ethanol exposure results in profound biochemical and morphological changes. Consequently, an alcoholic does not respond normally to alcohol, other drugs, or even other toxic agents. Some of these persistent changes are consequences of the injurious effects of ethanol and associated nutritional disorders, whereas others may represent adaptive responses to the profound changes in intermediary metabolism that are a direct and immediate consequence of the oxidation of ethanol itself.

In this volume we describe these general effects of alcohol. More specifically, we summarize the symptomatology, pathogenesis, and available treatment for various medical complications of alcoholism. These complications involve virtually the entire field of internal medicine, much of pathology and biochemistry, and some pharmacology. Because of its wide impact, syphilis was called "the great imitator"; this can now be said of alcoholism, since its complications can mimic a great many other diseases. Indeed, there is hardly any tissue in the body that is spared by alcohol. To know the pathophysiology of alcohol-induced lesions is to understand the pathological responses of most organs.

Because of their wide impact, the medical disorders associated with alcoholism represent one of the most important public health problems confronting our society. The significant progress made in developing tools for diagnosis and elucidating concepts of patho-genesis may open up new avenues for prevention. In-deed, the significant progress made in our understand-ing of the metabolism of ethanol and its effects on the body allows for some cautious optimism. Although we may not understand why alcoholism develops in only some ethanol consumers and why major complications such as liver disease occur in only a minority of heavy drinkers, we may nevertheless be able to intervene successfully once we acquire a better understanding of how ethanol affects the chemistry and structure of the body. A public health approach at three levels of inter-vention can now be conceptualized: (1) biological markers of alcoholism are being developed that, when perfected, may allow for early detection of heavy drink-ing; (2) in these individuals at risk, improved blood tests may eventually permit detection of alcohol-induced liver injury and other medical disorders at reversible,

early stages; and (3) recognition of preexisting enzyme defects or other preexisting abnormalities, such as pre-cirrhotic lesions, or the detection of contributory genetic factors, may permit screening for alcoholics predis-posed to the development of major complications of alcoholism such as cirrhosis. A concentrated therapeu-tic effort on this more manageable subgroup might arrest the disease at a potentially reversible stage, prior to the social and medical disintegration of the individ-ual. At our present state of knowledge, full implementa-tion of such a public health approach is not yet possible, but I hope that this volume, by summarizing the ad-vances made to date, may kindle the start of such an approach and bring closer the day when full implemen-tation will be feasible.

Charles S. Lieber

Bronx, New York

Acknowledgments

No single individual's expertise could be sufficient to span the vast area of medicine involved in the medical disorders associated with alcoholism. I was fortunate in the writing of this book to benefit from considerable help from a large number of colleagues. Those specifically involved in the preparation of some segments of the book are identified in the appropriate chapters. Others have helped in more general ways. I am particularly indebted to Ms. Leonore DeCarli, my long-standing associate, whose unceasing dedication to our research projects has allowed me to collect many of the data reported in this book. Dr. C. Kim has also been extremely helpful in proofreading and contributing many constructive criticisms concerning several of the chapters. I am grateful to my office staff, Ms. Renee Cabell and Ms. Patricia Walker, who very skillfully typed a large part of this monograph. Furthermore, I wish to thank Ms. Gloria Spivacek and her staff at the Medical Media Department of the Bronx VA Medical Center for much of the skillfully executed artwork and photography, and Ms. Margaret Kinney and her staff at the medical library of the Bronx VA Medical Center for their assistance in collecting the bibliographic citations in this volume.

Contents

Chapter 4.   **Ethanol and Lipid Disorders, Including Fatty Liver, Hyperlipemia, and Atherosclerosis**

Charles S. Lieber

Chapter 5.   **Effects of Ethanol on Amino Acid and Protein Metabolism**

Siamak A. Adibi, Enrique Baraona, and Charles S. Lieber

Chapter 6.   **Interaction of Ethanol with Other Drugs**

Charles S. Lieber

Chapter 11. . **Ethanol and the Pancreas**

Mark A. Korsten, Romano C. Pirola, and Charles S. Lieber

Chapter 12. **Cardiovascular Effects of Ethanol**

Howard S. Friedman

Chapter 13. **Effects of Alcohol Abuse on Skeletal Muscle**

Finbarr C. Martin and Timothy J. Peters

Chapter 14. **The Effects of Alcohol on the Nervous System: Clinical Features, Pathogenesis, and Treatment**

Maurice Victor

Chapter 15. **Alcohol Abuse: Carcinogenic Effects and Fetal Alcohol Syndrome: Description, Diagnosis, and Prevention**

Anthony J. Garro, Barbara H. J. Gordon, and Charles S. Lieber

Chapter 16. **Alcohol and the Kidney**

Murray Epstein

Chapter 17.   **Nutrition: Medical Problems of Alcoholism**

Lawrence Feinman and Charles S. Lieber

Chapter 18.   **Biological Markers of Alcoholism**

Alan S. Rosman and Charles S. Lieber

1

Metabolism of Ethanol

Charles S. Lieber

1.1. Pathways of Ethanol Oxidation

Ethanol is not only produced by yeast but can also be found in mammals in trace amounts (McManus *et al.*, 1966). Bacterial fermentation in the gut is one way in which it is produced (Krebs and Perkins, 1970). However, it is primarily an exogenous compound that is readily absorbed from the gastrointestinal tract. Only 2–10% of ethanol absorbed is eliminated through the kidneys and lungs; the rest is oxidized in the body, principally in the liver. The rate of ethanol removal from the blood is, indeed, remarkably decreased or halted by hepatectomy or procedures damaging the liver (Thompson, 1956). Moreover, the predominant role of liver for ethanol metabolism was shown directly in individuals with portacaval shunts undergoing hepatic vein catheterization (Winkler *et al.*, 1969). Extrahepatic metabolism of ethanol is relatively small (Forsander and Raiha Niels, 1960; Larsen, 1959), except for the stomach (*vide infra*). This relative organ specificity probably explains why, despite the existence of intracellular mechanisms to maintain homeostasis, ethanol disposal produces striking metabolic imbalances in the liver. These effects are aggravated by the lack of a feedback mechanism to adjust the rate of ethanol oxidation to the metabolic state of the hepatocyte and the inability of ethanol, unlike other major sources of calories, to be stored in the liver or to be metabolized or stored to a significant degree in peripheral tissues (Table 1.1). The hepatocyte contains three main pathways for ethanol metabolism, each located in a different subcellular compartment: the alcohol dehydrogenase pathway of the cytosol (or soluble frac-

tion of the cell), the microsomal ethanol-oxidizing system located in the endoplasmic reticulum, and catalase located in the peroxisomes (Fig. 1.1).

1.1.1. The Alcohol Dehydrogenase Pathway

1.1.1.1. Chemical Characterization of ADH-Mediated Ethanol Oxidation

A major pathway for ethanol disposition involves alcohol dehydrogenase (ADH), an enzyme of the cell sap (cytosol) that catalyzes the conversion of ethanol to acetaldehyde. There is some debate concerning the lobular distribution of ADH in the liver. Early histochemical studies carried out on rat liver either showed periportal maxima of ADH activity (Greenberger *et al.*, 1965) or were unable to demonstrate any uneven distribution pattern (Berres *et al.*, 1970), contrasting with the microquantitative measurements of Morrison and Brock (1967), which had originally revealed that the activity of ADH in the perivenular (pericentral) are of the liver lobule in humans and in female rats was about 1.7 times higher than in the periportal area. By means of modern immunohistochemical techniques, human ADH was now demonstrated mainly in hepatocytes around the terminal hepatic (central) vein (Buehler *et al.*, 1982). Other approaches to separate periportal and perivenous hepatocyte populations, employing anterograde and retrograde collagenase perfusion in the male rat liver, led to the conclusion that there were no regional differences of ADH activity (Väänänen *et al.*, 1984). Mare and castrated male horses showed high hepatic ADH activity,

TABLE 1.1.   Characteristics of Ethanol Metabolism

Large caloric load, sometimes in excess of all other nutrients
Almost no renal or pulmonary excretion
No storage mechanism in the body
Oxidation predominantly in the liver
No feedback control of rates of ethanol oxidation

which was evenly distributed in the liver acinus (Maly and Sasse, 1985). It was also observed that the high activity of liver ADH in adult females relative to adult males in some rodent strains does not constitute a general phenomenon but depends on the species and strains of animals studied (Maly and Sasse, 1985).

The *raison d'être* of ADH enzyme might be to rid the body of the small amount of alcohol produced by fermentation in the gut. Another possible explanation for presence of alcohol dehydrogenase in the liver is the fact that this enzyme has a broad substrate specificity, which includes dehydrogenation of steroids (Okuda and Takigawa, 1970), oxidation of glycols in the metabolism of norepinephrine (Mårdh *et al.*, 1985), and omega oxidation of fatty acids (Bjorkhem, 1972). These compounds may represent the "physiological" substrates for ADH, although the small amount of endogenous ethanol could play such a role.

Multiple forms of ADH exist. Studies over the past

15 years on human liver ADH molecular forms have revealed a degree of complexity in ADH isoenzymes that has no counterpart in lower animal species (Li, 1977; Vallee and Bazzone, 1983). Human alcohol dehydrogenase is a dimeric zinc metalloenzyme for which three classes, I, II, and III, have been distinguished (Jörnvall *et al.*, 1987). Subunits hybridize within but not between classes. Human liver ADH exists in multiple molecular forms that arise from the association of eight different types of subunits, α, β_1, β_2, β_3, γ_1, γ_2, π, and χ, into active dimeric molecules. A genetic model accounts for this multiplicity as products of five gene loci, ADH1 through ADH5 (Bosron and Li, 1987). Polymorphism occurs at two loci, ADH2 and ADH3, which encode the β and γ subunits. All of the known homodimeric and heterodimeric isoenzymes have been isolated and purified to homogeneity.

Class I isozymes migrate cathodically on starch gel electrophoresis (Bosron and Li, 1987). There are three types of subunits in class I: α, β, and γ. The primary structures of all three forms have been established, as have the overall properties and the effects of the amino acid substitutions between the various forms. Each subunit has 374 residues, of which 35 exhibit differences among the α, β, and γ chains. Corresponding cDNA structures are also known, as are the genetic organization and details of the gene structures. Allelic variants occur at the β and γ loci. Corresponding amino acid substitutions have been characterized, and enzymatic differences between the allelic forms are explained by

A
$$CH_3CH_2CH + NAD^+ \xrightarrow{\text{ADH}} CH_3CHO + NADH + H^+$$

B
$$CH_3CH_2OH + NADPH + H^+ + O_2 \xrightarrow{\text{MEOS}} CH_3CHO + NADP^+ + 2H_2O$$

C
$$NADPH + H^+ + O_2 \xrightarrow[\substack{\text{NADPH} \\ \text{Oxidase}}]{} NADP^+ + H_2O_2$$
$$+$$
$$H_2O_2 + CH_3CH_2OH \xrightarrow{\text{Catalase}} 2H_2O + CH_3CHO$$

D
$$HYPOXANTHINE + H_2O + O_2 \xrightarrow[\substack{\text{Xanthine} \\ \text{Oxidase}}]{} XANTHINE + H_2O_2$$
$$+$$
$$H_2O_2 + CH_3CH_2OH \xrightarrow{\text{Catalase}} 2H_2O + CH_3CHO$$

FIGURE 1.1. Ethanol oxidation by (A) alcohol dehydrogenase (ADH) and nicotinamide adenine dinucleotide (NAD); (B) the hepatic MEOS and NADPH; (C) a combination of NADPH oxidase and catalase; (D) xanthine oxidase and catalase.

defined residue exchanges. The subunits of class I are derived from at least three genetic loci (Smith *et al.*, 1971; Bahr-Lindström *et al.*, 1986; Duester *et al.*, 1986) and constitute the α subunit (the major form expressed in fetal liver), different β subunits, distributed nonidentically in various populations (β_1 common in Caucasian populations, β_2 common in Oriental populations, and β-Indianapolis found at least in some African populations (von Wartburg *et al.*, 1965; Jörnvall *et al.*, 1984; Bosron *et al.*, 1983), and the two allelic types of γ subunit (γ_1 and γ_2, both of high frequency). Von Wartburg *et al.* (1965) first differentiated the normal ADH of human liver (pH optimum 10.5) from the so-called atypical type (pH optimum 8.8), which shows a several-fold higher activity. The normal human main liver ADH has only the β_1 subunit, whereas the atypical type also has the β_2 subunit. Both subunits are controlled by the ADH2 locus. The atypical form of ADH occurs in frequencies between 5 and 20% in European populations: English, 10% (Smith *et al.*, 1971); Swiss, 20% (von Wartburg and Schurch, 1968); and Germans, 9% (Harada *et al.*, 1978). In Mongoloid populations, the frequency is up to 90% (Stamatoyannopoulos *et al.*, 1975). The K_i values for 4-methylpyrazole inhibition of the class I isozymes are in the micromolar range (Bosron and Li, 1987).

Other data also reveal the genetic organization and the gene structures (Duester *et al.*, 1986) as well as the protein structures of subunits of the class II type and subunits of the class III type. Class II isozymes migrate more anodically than class I isozymes and, unlike the latter, which generally have low K_m values for ethanol, class II (or π-) ADH has a high K_m (34 mM) and an insensitivity to 4-methylpyrazole inhibition with a K_i of 2 mM at pH 7.5 (Li and Magnes, 1975; Bosron *et al.*, 1979). Class III (χ-ADH) does not participate in the oxidation of ethanol in the liver because of its very low affinity for that substrate; it is not inhibited by 12 mM 4-methylpyrazole (Pares and Vallee, 1981).

In the rat (Julia *et al.*, 1987) and the baboon (Holmes and VandeBerg, 1987), isoenzymes of ADH have been characterized that exhibit many analogies with human ADH classes I–III.

In ADH-mediated oxidation of ethanol, hydrogen is transferred from the substrate to the cofactor nicotinamide adenine dinucleotide (NAD), converting it to its reduced form (NADH) (see Fig. 1.1A), and acetaldehyde is produced. The dissociation of the NADH–enzyme complex has been shown to be a rate-limiting step in this reaction (Theorell and Chance, 1951). As a net result, the first step in the oxidation of ethanol

generates an excess of reducing equivalents in the cytosol, primarily as NADH. Thus, in normal rats given ethanol, there is a marked shift in the redox potential of the cytosol, as measured by changes in the lactate and pyruvate ratio (Veech *et al.*, 1972; Domschke *et al.*, 1974). The altered redox state, in turn, is responsible for a variety of metabolic abnormalities. Some of these, such as hyperlactacidemia, are linked to the utilization of the excess NADH in the cytosol (Fig. 1.2). The reducing equivalents can also be transferred to NADPH, and the increased NADPH can be utilized for synthetic pathways in the cytosol and microsomal functions, as illustrated in Fig. 1.2 and discussed in more detail later in this chapter.

Some of the hydrogen equivalents formed in this reaction are transferred from the cytosol into the mitochondria. Since the mitochondrial membrane is impermeable to NADH, it is generally believed that the reducing equivalents enter the mitochondria via shuttle mechanisms. Several shuttle systems have been proposed (three of which are illustrated in Fig. 1.2), namely, the malate cycle (Chappell, 1968) (quantitatively, probably the most important one), the fatty acid elongation cycle (Whereat *et al.*, 1969; Grunnet, 1970), and the α-glycerophosphate cycle (Bucher and Klingenberg, 1958). In these cycles, NADH reduces the oxidized partner of the shuttle pair (that is, oxaloacetate) in the presence of the cytoplasmic enzyme (malate dehydrogenase), thereby regenerating NAD. The reduced component (that is, malate) now traverses the mitochondrial membrane where it reacts with the mitochondrial enzyme and NAD (or FAD in the case of α-glycerophosphate) to generate NADH or FADH plus the oxidized partner. FADH and NADH are oxidized by the respiratory chain, and the oxidized partner enters the cytoplasm, where it is available for another round of the shuttle cycle. Whereas the α-glycerophosphate shuttle seems to play an important role in some tissues (for example, in insect flight muscle), its significance in the liver has been questioned, because the activity of hepatic α-glycerophosphate dehydrogenase is low, except in special situations such as hyperthyroidism (Hassinen, 1967) or treatment with clofibrate (Kahonen *et al.*, 1971). Normally, fatty acids are oxidized via β-oxidation and the citric acid cycle of the mitochondria, which serves as "hydrogen donor" for the mitochondrial electron transport chain. When ethanol is oxidized, however, the generated hydrogen equivalents, which are shuttled into the mitochondria, supplant the citric acid cycle as source of hydrogen. Following the administration of ethanol, the mitochondria are shifted to a more

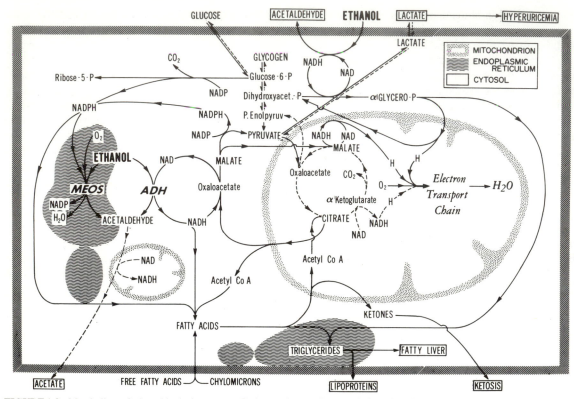

FIGURE 1.2. Metabolism of ethanol in the hepatocyte. Pathways that are decreased after ethanol abuse are represented by dashed lines.

reduced redox state as measured by changes in the ratio of β-hydroxybutyrate over acetoacetate (Domschke *et al.*, 1974).

1.1.1.2. Metabolic Changes Associated with the Reduced State

As emphasized before (Lieber and Davidson, 1962; Lieber, 1968), the metabolic effects of alcohol that can be attributed to the generation of NADH include hypoglycemia, hyperlactacidemia, and interference with galactose, serotonin, and other amine metabolism. The increased availability of NADH also results in alteration of hepatic steroid metabolism in favor of the reduced compounds (Cronholm and Sjövall, 1970; Admirand *et al.*, 1970). It is noteworthy that ADH activity was found in testis (Van Thiel *et al.*, 1974), which makes it possible that ethanol ingestion is associated with increased testicular concentrations of NADH, possibly altering testicular sex steroidogenesis, as discussed in Chapter 3.

1.1.1.3. Rate-Limiting Factors in ADH-Mediated Ethanol Metabolism

According to classic concepts, the major rate-limiting factor in the metabolism of ethanol by the ADH pathway is the capacity of the liver to reoxidize the NADH produced from the reduction of NAD by the hydrogen from the ethanol. Some results, however, show that ADH is not necessarily present in excess and that, under a variety of circumstances, it is the level of the enzyme itself that actually becomes rate limiting (Crow *et al.*, 1977; Lieber, 1984). The concept that rates of ethanol metabolism may be determined by the level of ADH activity has been the subject of controversy for many years. It has attracted attention because ADH was found to be heterogeneous (see Section 1.1.1.1). The "atypical" ADH isolated by von Wartburg *et al.* (1965) has, *in vitro*, a much higher activity at physiological pH than the normal variety. Although those individuals with "atypical" ADH had enzyme activities several times higher than normal *in vitro*, this was not accompanied by an acceleration of the metabolism of ethanol *in vivo* (Edwards and Price-Evans, 1967). This discrepancy supports the view that in the process of ethanol oxidation, ADH itself may not be the major rate-limiting factor, provided at least a normal amount of ADH is present. Under these conditions, velocities may depend on availability of the cofactor NAD and the cell's capacity to dissociate the ADH–NADH complex and reoxidize NADH. This concept is supported by the observation that in patients with uremia, blood ethanol clearance is not accelerated despite a severalfold increase of liver ADH activity (Mezey *et al.*, 1975).

Although ADH activity above normal may not increase the rate of ethanol oxidation, diminution of ADH activity can reduce it. An experimental example of the effect of decreased ADH level on the rate of ethanol metabolism is furnished by the effect of low-protein diets, which have been shown to diminish hepatic ADH levels in rats (Horn and Manthei, 1965; Bode *et al.*, 1970a; Wilson *et al.*, 1986) and to slow considerably the metabolism of ethanol both in rats (Bode *et al.*, 1970a; Pekkanen *et al.*, 1978; Wilson *et al.*, 1986) and in human beings (Bode *et al.*, 1971). Thus, compared with normal subjects, in malnourished alcoholics, the brain effects of ethanol may be greater because of slower ethanol metabolism causing higher and more sustained blood ethanol levels; conversely, effects of ethanol secondary to its metabolism may be less striking. It was also shown (Bode and Thiele, 1975) that prolonged fasting markedly slows the metabolism of ethanol. The mechanism of the effect of fasting has been clarified by the definition of some rate-limiting factors in the oxidation of ethanol. Reoxidation of cytosolic NADH, generated by ethanol oxidation in liver via the ADH pathway, requires transport of reducing equivalents into the mitochondria via substrate shuttles and reoxidation of the reducing equivalents by the mitochondrial respiratory chain (see Section 1.1.1.1). Data from Cederbaum *et al.* (1977a) now suggest that each of these factors may be rate limiting in ethanol oxidation, depending on whether rats were fed or starved prior to the experiment.

In liver cells from starved animals, the addition of the components of the malate–asparate shuttle, for example, glutamate or malate, resulted in a marked stimulation of ethanol oxidation; this effect was sensitive to inhibition by the transaminase inhibitor cycloserine. Addition of dihydroxyacetone stimulated ethanol uptake by 45% and caused a three-fold increase in α-glycerophosphate content. The increase of ethanol oxidation by the addition of components of the malate–aspartate cycle (and its inhibition by transaminase inhibitors) or of the α-glycerophosphate cycle suggests that in the fasting state the oxidation of ethanol by isolated liver cells is limited by the rate of transfer of reducing equivalents from the cytosol (where they are generated by ADH) to the mitochondria and that this transport itself is regulated by the intracellular concentrations of the intermediates of the shuttles. Stimulation of ethanol oxidation by lactate may involve a similar mechanism (Crow *et al.*, 1978). Inhibition of the electron transport chain by the addition of amobarbital resulted in a significant reduction of ethanol oxidation, suggesting that the flux through the respiratory chain also regulates rates of ethanol metabolism. However, the actual reoxidation of reducing equivalents by the respiratory chain is not limiting under these conditions, since uncoupling agents, which stimulate oxygen consumption, did not stimulate ethanol oxidation.

In liver cells isolated from fed rats, the rate of ethanol oxidation was about twice that in the fasting state. Since the concentrations of malate, aspartate, and α-glycerophosphate in liver cells from fed rats were higher than in the cells from starved rats, it seems likely that the higher rate of ethanol oxidation in the liver cells from fed rats was caused by increased activities of the hydrogen transport cycles. In contrast to starved rats, malate addition was rather ineffective in stimulating ethanol oxidation in the liver cells of fed rats, indicating that the hydrogen transport cycles were not rate limiting. Under these conditions, it might be expected that reoxidation of reducing equivalents would be rate limiting in ethanol oxidation. Indeed, addition of uncoupling agents caused a large stimulation of ethanol oxidation, suggesting that in livers of fed rats the mitochondrial reoxidation of NADH, whether transported from the cytosol or generated directly in the mitochondria by acetaldehyde dehydrogenase, is a rate-limiting step in ethanol oxidation. Similarly, treatment of rats with $3,3',5$-triiodo-L-thyronine (T_3) for a period of 6 days led to a 45% decrease in total liver alcohol dehydrogenase, but the rate of ethanol elimination *in vivo* was the same in T_3-treated and control animals (Smith and Dawson, 1985). These results do not support the notion that ethanol elimination *in vivo* is normally governed primarily by the level of ADH under these conditions.

With regard to all these experiments, a word of caution is needed concerning the possible variability caused by species differences. For instance, in spontaneously hypertensive rats, rates of ethanol metabolism appear to be modulated by ADH activity, which in turn is strikingly affected by sex hormones, with inhibition of testosterone and stimulation by estradiol (Rachmamin *et al.*, 1980). In conventional rats as well, a number of studies have shown hormonal influences, as discussed elsewhere (Lieber, 1984, 1987). Interaction between sex hormones and ethanol-metabolizing enzymes was tested in detail by Teschke *et al.* (1986). Estradiol increased the hepatic activities of ADH and catalase in both ovariec-tomized and sham-operated female rats on the control diet, whereas this enhancing property was virtually lost in animals on the alcohol diet. The hepatic activities of the microsomal ethanol-oxidizing system (MEOS) re-mained unaffected under these experimental conditions irrespective of the diet used. Testosterone increased the hepatic activities of MEOS and decreased the ADH activity in female rats on the control diet, but these changes were either not clear-cut or markedly reduced in similarly treated female rats fed the alcohol diet. Ac-cording to Lumeng and Crabb (1984), changes in alco-hol elimination rates produced by fasting and castration mainly reflected changes in the V_{max} of liver ADH. It has been shown that a decrease in the rate of degradation is the principal cause for the increase in liver ADH follow-ing castration (Mezey and Potter, 1985). As discussed in Chapter 3, chronic ethanol consumption has a profound interaction with testosterone metabolism, including a castrationlike effect.

It is also important to note that a substantial frac-tion of ethanol metabolism persists even in the presence of pyrazole (a potent inhibitor of ADH) in isolated perfused liver (Papenberg *et al.*, 1970), liver slices (Lieber and DeCarli, 1975), and isolated liver cells (Thieden, 1971; Grunnet *et al.*, 1973; Matsuzaki and Lieber, 1975). Furthermore, in the presence of pyrazole, glucose labeling from $1R$-[^{3}H]ethanol was nearly abol-ished, whereas H^{3}HO production was inhibited less than 50%. In view of the stereospecificity of ADH for $1R$[^{3}H]ethanol, these findings again suggest "the pres-ence of a significant pathway not mediated by cytosolic ADH" (Rognstad and Clark, 1974). The rate of this non-ADH-mediated oxidation varied, depending on the con-centration of ethanol used, from 20 to 25% (Lieber and DeCarli, 1968, 1972; Papenberg *et al.*, 1970) to half or more (Thieden, 1971; Grunnet *et al.*, 1973; Matsuzaki and Lieber, 1975; Matsuzaki *et al.*, 1981) of the total ethanol metabolism. The fact that the cytosolic redox state was unaffected is consistent with the conclusion that this pyrazole-insensitive residual ethanol metabo-lism is not ADH mediated (Grunnet and Thieden, 1972). Similarly, in glycogen storage disease, the strikingly accelerated ethanol metabolism without an associated redox change (Lowe and Mosovich, 1965) raised the possibility of a non-ADH pathway. Moreover, a strain of

mice has been described that lacks the low-K_m ADH but is nevertheless capable of significant ethanol oxidation up to two-thirds of normal (Burnett and Felder, 1980; Shigeta *et al.*, 1984). Thus, it is now obvious that non-ADH pathways play a significant role in ethanol oxida-tion. Theoretically, two enzyme systems could account for the ADH-independent pathway, namely, the micro-somal ethanol-oxidizing system and catalase.

1.1.2. Microsomal Ethanol-Oxidizing System

1.1.2.1. Chemical Characterization of MEOS-Mediated Ethanol Oxidation

The first indication of an interaction of ethanol with the microsomal fraction of the hepatocyte was provided by the morphological observation that ethanol feeding results in a proliferation of the smooth endoplasmic reticulum (SER) (Iseri *et al.*, 1964, 1966; Lane and Lieber, 1966). This increase in SER resembles that seen after the administration of a wide variety of xenobiotic compounds including known hepatotoxins (Meldolesi, 1967), numerous therapeutic agents (Conney, 1967), and food additives (Lane and Lieber, 1967). Most of these substances that induce a proliferation of the SER are metabolized, at least in part, in the microsomal fraction of the hepatocyte that comprises the SER. The observa-tion that ethanol produced proliferation of the SER raised the possibility that, in addition to its oxidation by ADH in the cytosol, ethanol may also be metabolized by the microsomes. A microsomal system capable of meth-anol oxidation had been described (Orme-Johnson and Ziegler, 1965), but its capacity for ethanol oxidation was extremely low. Furthermore, this system could not oxid-ize long-chain aliphatic alcohols such as butanol and was exquisitely sensitive to the catalase inhibitors azide and cyanide. Therefore, Ziegler (1972) concluded that this system is clearly different from the cytochrome P-450-dependent system and involves the H_2O_2-mediated ethanol peroxidation by catalase. However, a micro-somal ethanol-oxidizing system with a rate of ethanol oxidation ten times higher than that reported by Orme-Johnson and Ziegler (1965) was described (Lieber and DeCarli, 1968, 1970a; Lieber *et al.*, 1974). The system required NADPH and O_2 and was relatively insensitive to catalase inhibition. Furthermore, the MEOS was differentiated from the system reported by Orme-Johnson and Ziegler (1965) and from catalase by its ability to oxidize long-chain aliphatic alcohols (Teschke *et al.*, 1975a), which are not substrates for catalase (Chance and Oshino, 1971).

1.1.2.2. Differentiation of MEOS from ADH and Catalase

Differentiation of MEOS in total rat liver microsomes from alcohol dehydrogenase was achieved by subcellular localization, pH optimum *in vitro* (7.4 versus 10), cofactor requirements (Fig. 1.1A,B), and effects of inhibitors such as pyrazole (Lieber and DeCarli, 1973; Lieber *et al.*, 1970). Studies with inhibitors have also indicated that a major fraction of the ethanol-oxidizing activity in microsomes is independent of catalase (Lieber and DeCarli, 1970a, 1973; Lieber *et al.*, 1970).

A clear dissociation of the NADPH-dependent from an H_2O_2-mediated ethanol oxidation in microsomes was observed with pyrazole, which inactivates catalase *in vivo* (Lieber *et al.*, 1970), and azide, which inhibits catalase *in vitro* (Lieber and DeCarli, 1970a). This dissociation has also been found in preparations with the same control rates of H_2O_2-mediated and NADPH-mediated ethanol metabolism, to obviate the objection that effectiveness of inhibition might differ if

the ratio of H_2O_2 generated and the amount of catalase varies (Lieber and DeCarli, 1973). Thus, under experimental conditions with complete abolition of the peroxidatic activity of catalase, the NADPH-dependent ethanol oxidation still proceeded at a significant rate; this again dissociates the NADPH-dependent MEOS activity from a process involving catalase–H_2O_2. Subsequently, MEOS was solubilized and separated from alcohol dehydrogenase and catalase activities by diethylaminoethylcellulose column chromatography (Fig. 1.3) (Teschke *et al.*, 1972, 1974a; Mezey *et al.*, 1973). Differentiation of MEOS from ADH in the column fractions was shown by the failure of NAD to promote ethanol oxidation at pH 9.6, by cofactor requirements (NADPH and O_2), by the apparent K_m for ethanol (7 to 9 mM), and by the insensitivity of the MEOS to the ADH inhibitor pyrazole. The MEOS was also distinguished from a process involving catalase-H_2O_2 by the lack of catalytic activity, by the apparent K_m for oxygen (8.3 μM), by the insensitivity to the catalase inhibitors azide and cyanide, and by the inability of an H_2O_2-generating system (glucose–glucose oxidase) to sustain

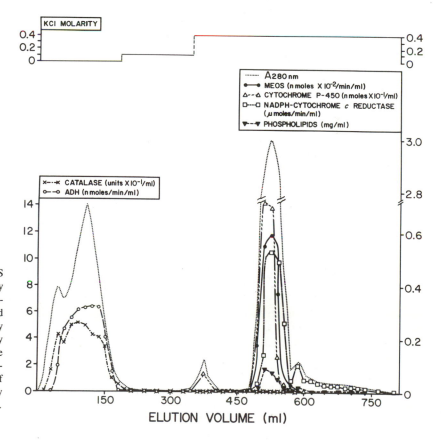

FIGURE 1.3. Separation of MEOS from ADH and catalase activities by ion-exchange column chromatography on DEAE–cellulose. Sonicated microsomes from rats (fed laboratory chow) were further solubilized by treatment with sodium deoxycholate and put onto a DEAE–cellulose column (2.5 × 45 cm). The separation of the enzyme activities was achieved by a stepwise increase of the gradient. (From Teschke *et al.*, 1974a.)

ethanol oxidation in the isolated fraction (Teschke *et al.*, 1974a). Thus, with specific and sensitive methods, MEOS activity could be clearly differentiated from an enzymatic process involving peroxidatic activity of catalase. These results, as well as other reports (Hildebrandt and Speck, 1973; Mezey *et al.*, 1973; Hildebrandt *et al.*, 1974), therefore fail to support the concept that a catalase–H_2O_2 system accounts for the oxidation of ethanol in the microsomes. In addition, MEOS activity could be dissociated from microsomal NADPH oxidase activity (Hasumura *et al.*, 1975), which generates H_2O_2 in microsomes (Gillette *et al.*, 1957).

It has also been reported that microsomes from acatalasemic mice fail to oxidize ethanol (Vatsis and Schulman, 1973), but this claim has been subsequently retracted (Vatsis and Schulman, 1974). Indeed, hepatic microsomes of acatalasemic mice subjected to heat inactivation displayed decreased catalytic activity, but NADPH-dependent MEOS remained active and unaffected (Lieber and DeCarli, 1974; Teschke *et al.*, 1975b). Even without heat inactivation, in the acatalasemic strain, the NADPH-dependent metabolism was much more active than the H_2O_2-mediated one, whereas microsomes of control mice displayed equal rates of H_2O_2- and NADPH-dependent ethanol oxidation (Vatsis and Schulman, 1974). These results therefore support the conclusion that hepatic microsomes of normal and acatalasemic mice contain a NADPH-mediated ethanol-oxidizing system that is catalase independent.

1.1.2.3. Nature of MEOS Activity

Of particular interest regarding the nature of MEOS are studies with different alcohols as substrates. Previously, a NADPH-dependent oxidation of methanol and ethanol, but not of longer-chain alcohols, was reported (Orme-Johnson and Ziegler, 1965). This was considered as evidence for an obligatory role of catalase in microsomal alcohol oxidation (Thurman *et al.*, 1972), since catalase reacts peroxidatically primarily with methanol and ethanol but not with alcohols with longer aliphatic chains (Chance, 1947). Subsequently, however, the NADPH-dependent microsomal alcohol-oxidizing system was found capable of metabolizing methanol, ethanol, *n*-propanol, and *n*-butanol to their respective aldehydes in hepatic microsomal preparations as well as in column fractions that contained the microsomal components cytochrome P-450, NADPH–cytochrome *c* reductase, and phospholipids and exhibited no ADH or catalase activity (Teschke *et al.*, 1974b, 1975a). In hepatic microsomal preparations, the oxidation rate of

ethanol is approximately twice that of butanol in the presence of NADPH and O_2. With a H_2O_2-generating system, rates similar to the NADPH-dependent oxidation are achieved with ethanol, whereas *n*-butanol is a substrate only for the NADPH-dependent microsomal system but, unlike ethanol, is not a substrate for catalase–H_2O_2. The latter finding is in excellent agreement with previous reports regarding the substrate specificity of catalase (Chance, 1947; Chance and Oshino, 1971; Keilin and Hartree, 1945). The system of Orme-Johnson and Ziegler (1965) did not oxidize *n*-propanol and *n*-butanol, which suggests that its low activity may well have been caused by contaminating catalase.

Reconstitution of the ethanol-oxidizing activity with the three microsomal components cytochrome P-450, NADPH–cytochrome *c* reductase, and lecithin has been demonstrated (Fig. 1.4) Ohnishi and Lieber, 1977a). In these experiments cytochrome P-450 was partially purified by protease treatment and subsequently by column chromatography on DEAE–cellulose using a stepwise KCl gradient (Comai and Gaylor, 1973). NADPH–cytochrome *c* reductase was partially purified, essentially by the method of Levin *et al.* (1974). Successful reconstitution of the MEOS was confirmed using highly purified microsomal cytochrome P-450 from phenobarbital-treated rats (Miwa *et al.*, 1978), ethanol-treated rabbits (Koop *et al.*, 1982), and isoniazid- and ethanol-treated rats (Ryan *et al.*, 1985, 1986), respectively. An "ethanol-specific" P-450 has now been identified. This was first suggested by the increase of a

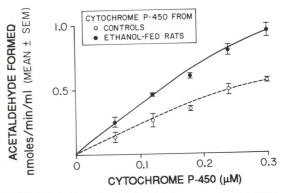

FIGURE 1.4. Ethanol oxidation by the reconstituted MEOS. Active oxidation of alcohol is achieved by the combination of three components: cytochrome P-450, NADPH-cytochrome *c* reductase, and L-α-dioleoyl lecithin. P-450 extracted from microsomes of ethanol-fed rats was more active than that of controls. (From Ohnishi and Lieber, 1977a.)

cytochrome P-450 species showing high affinity for cyanide after ethanol administration (Joly *et al.*, 1972; Comai and Gaylor, 1973; Hasumura *et al.*, 1975). Evidence in favor of an increase of a special species of cytochrome P-450 after ethanol treatment was also derived from inhibitor studies (Ullrich, 1975). More direct proof was obtained from studies of microsomal proteins (Ohnishi and Lieber, 1977b): the rise in cytochrome P-450 involves a hemeprotein different from those induced by phenobarbital or 3-methylcholanthrene treatment. Moreover, SDS–polyacrylamide gel electrophoresis showed induction of a microsomal protein of 53,400 da (Fig. 1.5). The partially purified cytochrome P-450 from ethanol-fed rats was more active for alcohol oxidation than the control preparation in the presence of an excess of NADPH–cytochrome c reductase and 1-α-dioleoyl lecithin. There was no significant difference in the capacity of partially purified NADPH–cytochrome c reductase from either ethanol-fed rats or controls to promote ethanol oxidation in the presence of cytochrome P-450 and L-dioleoyl lecithin (Ohnishi and Lieber, 1977b).

Studies by Joly *et al.* (1976, 1977) also showed that chronic ethanol administration to rats is associated with the appearance of a form of cytochrome P-450 with spectral and catalytic properties different from those of the cytochrome P-450 of control, phenobarbital-treated, and methylcholanthrene-treated rats. An ethanol-inducible form of P-450 (LM3a), purified from rabbit liver microsomes (Koop *et al.*, 1982; Ingelman-Sundberg and Johansson, 1984) catalyzed ethanol oxidation at rates much higher than other P-450 isoenzymes and also had an enhanced capacity to oxidize 1-butanol, 1-pentanol, and aniline (Morgan *et al.*, 1982), acetaminophen (Morgan *et al.*, 1983), CCl_4 (Ingelman-Sundberg and Johansson, 1984), acetone (Koop and Cassaza, 1985), and N-nitrosodimethylamine (NDMA) (Yang *et al.*, 1985). Similar results have been obtained with cytochrome P-450j, a major hepatic P-450 isoenzyme purified from ethanol- or isoniazid-treated rats (Ryan *et al.*, 1985, 1986). Evidence was also provided for the existence of a P-450j-like isoenzyme in humans (HLj) (Wrighton *et al.*, 1986, 1987; Song *et al.*, 1986). Wrighton *et al.* (1986) isolated DNAs complementary to human P-450j. The amino acid sequence of human P-450j, deduced by sequencing of the complementary DNA (cDNA) inserts, was reported to be 94% homologous to the published N termini for HLj (Wrighton *et al.*, 1986) over the first 18 amino acid residues.

We now have succeeded in obtaining the purified human protein (Fig. 1.6) in a catalytically active form,

with a high turnover rate for ethanol and other specific substrates (Lasker *et al.*, 1987a). The N-terminal sequences of cytochrome P-450ALC, P-450-B, and P-450-C were determined as shown in Fig. 1.7. The N-terminal sequences of HLj (Wrighton *et al.*, 1986) and human P-450j (Song *et al.*, 1986) are given for comparison, and those residues identical to P-450ALC are boxed. In a new nomenclature for cytochromes P-450, it was proposed that the ethanol-inducible form be designated as P450IIE1 (Nebert *et al.*, 1987). The designation P450IIE1 should be reserved for this specific P-450 alcohol oxygenase. However, other microsomal cytochrome P-450 isoenzymes can also contribute to ethanol oxidation (Lasker *et al.*, 1987b). Thus, the term MEOS should be maintained when one refers to the overall capacity of the microsomes to oxidize ethanol rather than to the fraction of the activity that is specifically catalyzed by P450IIE1.

The rat P450IIE1 gene was isolated, characterized, and localized to chromosome 7 (Umeno *et al.*, 1988a), and the human gene to chromosome 10 (Umeno *et al.*, 1988b). In rabbits, two genes may be involved (Khani *et al.*, 1988). Preliminary evidence for genetic variability of MEOS activity has been provided (Petersen and Atkinson, 1980).

The reconstituted MEOS showed a dependency on cytochrome P-450 and the reductase and required synthetic phospholipids (such as lecithin) for its maximal activity. The K_m of the reconstituted MEOS for ethanol was 10 mM, which is similar to the K_m measured in crude microsomes and the MEOS fraction isolated by column chromatography (Lieber and DeCarli, 1968, 1970a; Teschke *et al.*, 1974a). This reconstituted system required NADPH as a cofactor, did not react to an H_2O_2-generating system, and was insensitive to the catalase inhibitor azide. These characteristics were also similar to those observed in crude microsomes. The activity was dependent on the amount of cytochrome P-450 present. The involvement of reduced cytochrome P-450 in MEOS was measured in the presence and absence of carbon monoxide, both in the dark and with illumination by white light and by lights of varying wavelength between 200 and 500 nm. Carbon monoxide inhibition was observed between 430 and 460 nm (Fabry and Lieber, 1979). These action spectra show the involvement of reduced cytochrome P-450 in the activity of MEOS.

How cytochrome P-450 mediates ethanol oxidation has not been fully clarified. According to Morgan *et al.* (1982), a reconstituted system containing P-450 (LM3a oxidizes ethanol by a direct, classic monooxygenase-

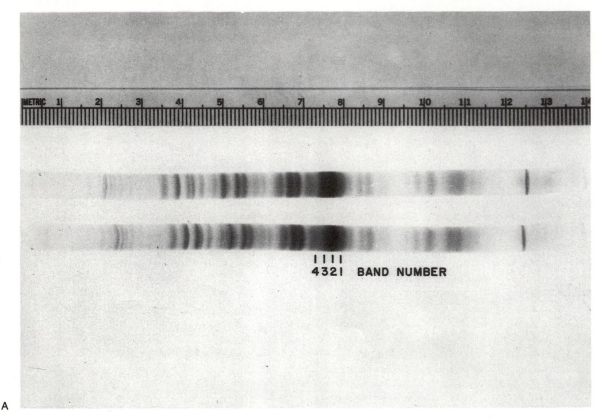

A

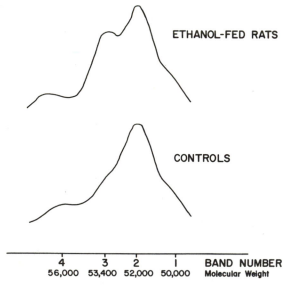

B

ETHANOL-FED RATS

CONTROLS

| 4 | 3 | 2 | 1 | BAND NUMBER
56,000 53,400 52,000 50,000 Molecular Weight

FIGURE 1.5. Coomassie blue protein profiles obtained by electrophoresis of liver microsomes from ethanol-fed (top line) and control (bottom line) rats. (A) Band 3 indicates the position to which a protein of molecular weight 53,400 would migrate. The area of bands 1 to 4 was also scanned. (B) Scans of the molecular weight region of the gels shown in A. Relative density of each band at 550 nm is plotted on the ordinate. Peak heights of band 2 in both gels are automatically set to the same height. Migration distance is on the abscissa. Bands 1, 2, 3, and 4 have the apparent molecular weights shown beneath the band number. (From Ohnishi and Lieber, 1977a.)

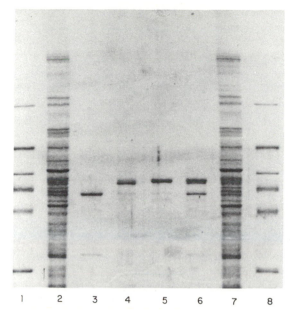

FIGURE 1.6. SDS–PAGE of human liver microsomes and purified cytochrome P-450. Samples were analyzed on a slab gel 0.75 mm thick containing 7.5% acrylamide using the discontinuous buffer system. Migration proceeds from top to bottom. Lanes 2 and 7, microsomes (10 μg); lanes 3, 4, and 5, cytochrome P-450-B, P-450-ALC, and P-450-C, respectively (0.5 μg); lane 6, mix of all three P-450s (0.25 μg each); lanes 1 and 8, protein standards with molecular weights of 89,000, 68,000, 58,000, 53,000, 43,000, and 29,000 (0.5 μg each). (From Lasker *et al.*, 1987a.)

type reaction. In contrast, Ingelman-Sundberg and Johansson (1984), employing an ethanol- or benzene-inducible form of rabbit liver P-450 (P-450 LMeb), favored the role of OH·formed in an iron-promoted Haber–Weiss reaction. In a more recent study, however, Ekström *et al.* (1986) found that liver microsomes from ethanol-treated rats contain P-450 isozyme(s) specifically effective in ethanol and NADPH oxidation but not in OH·production. Thus, only part of the MEOS mechanism may involve generation of hydroxyl radicals, which in turn might be "scavenged" by ethanol. Indeed, compounds that interact with hydroxyl radicals inhibit MEOS activity (Cederbaum *et al.*, 1977b; Ohnishi and Lieber, 1978). Furthermore, deprivation of redox-active iron through desferrioxamine inhibits by about 50% the microsomal oxidation of ethanol *in vitro* and reduces very significantly *in vivo* the overall ethanol elimination rate in rats (Nordmann *et al.*, 1987). Conversely, chronic overloading of rats with iron results in an increased rate of ethanol elimination, although alcohol dehydrogenase and catalase activities are reduced and cytochrome P-450 is depleted in the livers of such iron-overloaded animals. Thus, both a monooxygenase (P450IIE1-mediated) and a hydroxyl radical (iron-promoted) mechanism appear to play a role in microsomal ethanol oxidation.

1.1.3. Role of Catalase

The hepatocyte contains catalase primarily in the peroxisomes and in the mitochondria. Small amounts are also found in the isolated microsomes, but in the latter fraction catalase is considered to be a contaminant added during isolation rather than a component of the

CYTOCHROME											Residue								
		1	2	3	4	5	6	7	8	9	10	11	12	13	14	15	16	17	18	19
P-450-B		Met	Glu	Pro	Phe	Val	Val	Leu	Val	Leu	–	Leu	Ser	Ser	Met	Leu	Leu	Ser		
P-450-C		Met	–	Ser	Leu	Val	Val	Leu	Val	Leu	Leu	Leu	Ser	–	Leu	Leu	Leu			
P-450-ALC		Met	Ala	Leu	Gly	Val	Thr	Val	Ala	Leu	Leu	Val	Trp	Ala	Ala	Phe	Leu	Leu	Leu	Val
HLj		Ala	Ala	Leu	Gly	Val	Thr	Val	Ala	Leu	Leu	Val	Trp	Ala	Ala	Phe	Leu	Leu	Leu	
Human P-450j (Met)		Phe	Ala	Leu	Gly	Val	Thr	Val	Ala	Leu	Leu	Val	Trp	Ala	Ala	Phe	Leu	Leu	Leu	Val

FIGURE 1.7. N-Terminal amino acid sequences of human P-450-ALC, P-450-B, P-450-C, and related proteins. The N-terminal sequences of HLj and human P-450j are given for comparison, and those residues identical to P-450-ALC are boxed. Whether the N-terminal Met of human P-450j is present in the native protein is not known, since the sequence was deduced from a cDNA clone. Underlined P-450-B and P-450-C residues indicate those common to P-450-ALC. (From Lasker *et al.*, 1987a.)

membrane of the endoplasmic reticulum itself (Redman *et al.*, 1972).

It has been shown that catalase is capable of oxidizing ethanol *in vitro* in the presence of an H_2O_2-generating system (Keilin and Hartree, 1945) (see Fig. 1.1C,D). However, the idea that catalase plays a significant role in ethanol metabolism has been rejected by many (Bartlett, 1952; Lester and Benson, 1970; Papenberg *et al.*, 1970; Feytmans and Leighton, 1973). It is generally accepted that the H_2O_2-mediated ethanol peroxidation by catalase is limited by the rate of H_2O_2 generation rather than the amount of catalase itself. Thus, an indirect answer to the question of the extent of the role of catalase in ethanol metabolism can be derived from the rate of H_2O_2 production, which has been estimated to be 3.0 to 3.6 μmol/hr per g liver (Boveris *et al.*, 1972; Sies, 1974), which represents 2% of the *in vivo* rate of ethanol oxidation of 178 μmol/hr per g liver (Lieber and DeCarli, 1972). Actually, this rate of ethanol oxidation (2%), possibly occurring through a catalase- and H_2O_2-mediated mechanism, is probably an overestimate, since not all of the H_2O_2 generated in the liver can be utilized by the peroxidatic reaction of catalase (Oshino *et al.*, 1973). Finally, the hepatic oxygen concentration, which was estimated to be less than 50 μmol, may be far too low for significant rates of peroxisomal H_2O_2 generation, which has an apparent K_m of 100 μmol for oxygen (Boveris *et al.*, 1972). To evaluate a possible role of catalase in ethanol metabolism, aminotriazole (a catalase inhibitor) has been widely used. Except for Wendell and Thurman (1979), who found that in ethanol-pretreated rats aminotriazole decreased rates of ethanol oxidation by approximately 25%, no significant change in the rate of ethanol oxidation was observed either *in vivo* (Smith, 1961; Feytmans and Leighton, 1973; Roach *et al.*, 1972) or *in vitro*, in liver slices (Smith, 1961) or in perfused liver (Papenberg *et al.*, 1970). Thus, under physiological conditions, catalase appears to play no major role and cannot account quantitatively for the ADH-independent pathway of ethanol metabolism.

More recently, however, it has been proposed that the catalase contribution might be enhanced if significant amounts of H_2O_2 became available through β-oxidation of fatty acids such as octanoate, palmitate, and oleate in peroxisomes (Handler and Thurman, 1985). However, it should be pointed out that enzymes do not oxidize short-chain fatty acids such as octanoate. It should be noted that this phenomenon was observed only in the absence of ADH activity, achieved either through inhibition by 4-methylpyrazole or the use of ADH-negative deermice. Indeed, unless 4-methyl-

pyrazole is present, the rate of ethanol metabolism is reduced by adding fatty acids (Williamson *et al.*, 1969), and β-oxidation of fatty acids is inhibited by NADH produced from ethanol metabolism via ADH (Williamson *et al.*, 1969). Thus, under physiological conditions with ADH present, it is unlikely that β-oxidation of fatty acids stimulates H_2O_2 generation or that peroxidation of ethanol by catalase becomes quantitatively significant. In fact, in perfused livers of both fed and fasted rats, uptake of butanol was decreased by the addition of oleate, whereas that of methanol was increased, but only in the fasted state (Handler and Thurman, 1988); no data on ethanol uptake were provided in that study.

More recently the contribution of peroxisomal fatty acid β-oxidation to ethanol metabolism was examined in deermice hepatocytes. Addition of 1 mM oleate to hepatocytes isolated from fasted ADH-positive deermice in the presence of 4-methylpyrazole or to hepatocytes from fasted or fed ADH-negative deermice produced only a slight and statistically not significant increase in ethanol oxidation. Lactate (10 mM), which is not a peroxisomal substrate, showed a greater effect on ethanol oxidation. There was also a lack of oleate effect on the oxidation of ethanol by hepatocytes of ADH-positive deermice. Furthermore, in ADH-negative deermice, the catalase inhibitor azide (0.1 mM) did not inhibit the increase in ethanol oxidation by oleate and lactate. The rate of oleate oxidation by hepatocytes from fasted ADH-negative deermice was much lower than that of ethanol. These results indicated that in deermice hepatocytes, peroxisomal fatty acid oxidation does not play a major role in ethanol metabolism (Inatomi *et al.*, 1989).

In deermice lacking low-K_m ADH, Handler *et al.* (1987) reported a high H_2O_2 generation rate comparable to the rate of ethanol metabolism. This high value was partly based on an unusually high content of catalase heme (79 nmol/g liver) measured by the method described by Oshino *et al.* (1975), who reported 13 nmol/g liver in rats. Using the same method, we found similar values of catalase heme in rats (16.5 ± 1.4) and deermice (19.8 ± 2.1) (Alderman *et al.*, 1989a). These results are consistent with a low H_2O_2 generation rate and a minimal role for the catalase pathway. Furthermore, it must be pointed out that when fatty acids were used by Handler and Thurman (1985) to stimulate ethanol oxidation, this effect was very sensitive to inhibition by aminotriazole (AT), a catalase inhibitor. Therefore, if this mechanism were to play an important role *in vivo*, one would expect a significant inhibition of ethanol metabolism after AT administration *in vivo*, when physiological amounts of fatty acids and other substrates for H_2O_2 generation are present. A number of studies, however, have shown that AT treatment *in vivo* has little, if any, effect on ethanol oxidation *in vivo* (Bartlett, 1952; Smith, 1961; Kinard *et al.*, 1956; Feytmans and Leighton, 1973; Roach *et al.*, 1972). More recent studies by Takagi *et al.* (1986) and Kato *et al.* (1987a, 1988) have confirmed this relative lack of effect of AT on ethanol metabolism *in vivo* while verifying its inhibitory effect on catalase-mediated ethanol peroxidation *in vitro*. In a report by Glassman *et al.* (1985) it was speculated that when H_2O_2 generation limits ethanol oxidation by catalase, the small amount of remaining catalase activity after AT could metabolize ethanol at near-normal rates. However, Takagi *et al.* (1986) found that AT effectively inhibits peroxidation of ethanol by catalase under the treatment conditions used, with H_2O production rates comparable to those estimated *in vivo* (Levin *et al.*, 1986). In addition, all MEOS assays contained 1 mM azide, effectively precluding catalase participation. Furthermore, Kato *et al.* (1987a,b) found that urate (1 mM), a substrate for H_2O_2 production, increased ethanol oxidation fourfold in control but not in AT-treated cells, confirming effective catalase inhibition by AT in this live model system. Moreover, 1-butanol (10 mM), a competitive inhibitor of MEOS that is not oxidized by catalase, halved basal ethanol oxidation in both control and AT-treated cells. It was concluded that in animals lacking ADH, MEOS mediates virtually all of the remaining ethanol oxidation. Reduction of ethanol metabolism by 1-butanol was also demonstrated *in vivo* in deermice (Kato *et al.*, 1988).

It was suggested by Glassman *et al.* (1985) that a significant fraction of *in vivo* ethanol elimination in ADH$^-$ deermice could be through excretion and expiration rather than oxidation. However, rates of ethanol metabolism measured *in vitro* (in the absence of any losses) could account for a large fraction of the metabolism observed *in vivo* (Takagi *et al.*, 1986). Furthermore, an earlier study that directly measured loss of ethanol in urine and breath of deermice given ethanol indicated that nonoxidative elimination amounted to only 15% of the total (Shigeta *et al.*, 1984).

Similar conclusions were reached using another approach, namely, the differential isotope effects on the velocity of ethanol oxidation: when deuterated substrate is used, the reaction is slowed differently, according to the pathway involved. The two deermouse strains (ADH$^+$ and ADH$^-$) were used to calculate flux through the known ethanol-oxidizing pathways (ADH, MEOS, and catalase) *in vivo* (Takagi *et al.*, 1985; Alderman *et al.*, 1987). In ADH$^-$ deermice, the isotope effects at 7.0 and 58 mM blood ethanol were consistent with ethanol oxidation principally by MEOS in each case. Pretreatment

of ADH$^-$ animals with the catalase inhibitor AT did not significantly affect the values. These conclusions were challenged by Handler *et al.* (1988) because of the use of 1R-[^{3}H]ethanol in a reaction that is not stereospecific (Ekström *et al.*, 1987). However, 1R-[^{2}H,1,2-^{14}C]-, not 1R-[^{3}H]ethanol was employed. Since the same substrate was used by Alderman *et al.* (1987) *in vitro* and *in vivo*, the conclusions regarding the contribution of MEOS to the overall rate of metabolism are perfectly valid. Extrapolating from *in vitro* to *in vivo* data, Handler *et al.* (1988) also claim that in alcohol-fed ADH$^-$ deermice MEOS can only account for 28% of *in vivo* ethanol metabolism. It should be pointed out, however, that from Handler's own data, the MEOS can account for the bulk of ethanol metabolism in the perfused liver. Subsequently, Norsten *et al.* (1989) concluded that at least 50% of ethanol elimination of ADH$^-$ deermice was caused by a mitochondrial dehydrogenase. However, the activity was measured at pH 10, and no indication was given of whether any such activity was observed in intact mitochondria or at physiological pH, nor was this activity well characterized. Using the same strain of ADH$^-$ deermice, Inatomi *et al.* (1990) could not find any activity at physiological pH. In either solubilized or intact mitochondria the oxidation found at pH 10 was minimal and did not appear to reflect alcohol dehydrogenase activity. The conclusions of Norsten *et al.* (1989) were also based, in part, on the results of *in vivo* isotope exchange, which was equated with metabolism, an assumption that is not necessarily correct, since *in vivo* systems exist that can catalyze exchange but no net metabolism, and vice versa.

1.1.4. Relative Importance of ADH and Non-ADH Pathways

Despite the considerable controversy that originally surrounded this issue, and irrespective of original claims to the contrary (Thurman *et al.*, 1972), it is now agreed by the principal contenders involved that catalase cannot account for microsomal ethanol oxidation (Thurman and Brentzel, 1977; Teschke *et al.*, 1977). The questions now focus on the relative importance of the ADH and non-ADH pathways for the metabolism of ethanol in the liver. There are now several lines of evidence indicating that a non-ADH pathway significantly contributes to ethanol oxidation in the liver. This includes the incomplete inhibition of ethanol metabolism by ADH inhibitors and the pattern of labeling of acetaldehyde derived from stereospecifically labeled ethanol (both previously discussed under ADH) (Rognstad and Clark, 1974) and the increased rate of ethanol metabolism at high ethanol concentrations well above those needed to saturate alcohol dehydrogenase (Lieber and DeCarli, 1972; Matsuzaki *et al.*, 1981).

Actually, estimates of the magnitude of the non-ADH pathway obtained by measurements of residual ethanol metabolism after inhibition with the ADH inhibitor pyrazole or 4-methylpyrazole are underestimations, in view of the fact that these inhibitors also reduce the activity of the microsomal ethanol-oxidizing system (Teschke *et al.*, 1977).

Increasing ethanol metabolism with rising ethanol concentrations was found not only in the presence of an ADH inhibitor but also in its absence in isolated hepatocytes (Grunnet *et al.*, 1973; Matsuzaki *et al.*, 1981), in isolated perfused livers (Gordon, 1968), and *in vivo* in humans (Lereboullet *et al.*, 1976; Feinman *et al.*, 1978; Salaspuro and Lieber, 1978), rats (Feinman *et al.*, 1978), and baboons (Salaspuro and Lieber, 1978; Pikkarainen and Lieber, 1980).

The fact that ethanol metabolism increases with rising ethanol concentrations well above the level needed to fully saturate the low-K_m ADH suggests the involvement of a non-ADH pathway, at least in a species such as the rat devoid of the anodic high-K_m ADH active with ethanol. Moreover, the acceleration of ethanol metabolism at higher ethanol concentrations explains sporadic observations, which now have been confirmed (Lereboullet *et al.*, 1976; Feinman *et al.*, 1978; Salaspuro and Lieber, 1978), that ethanol disappearance from the blood is not linear at high ethanol concentrations that fully saturate the ADH pathway. Further, the persistence of a substantial rate of ethanol metabolism in mice congenitally devoid of ADH, in association with a high MEOS activity (Burnett and Felder, 1980; Shigeta *et al.*, 1984), illustrates most elegantly the importance *in vivo* of non-ADH ethanol metabolism. Finally, the contribution of cytochrome P-450 to ethanol metabolism has now also been determined in the presence of ADH.

To quantitate contributions of the various metabolic pathways, ethanol oxidation rates were initially examined in isolated deermouse hepatocytes with or without added 4-methylpyrazole (Takagi *et al.*, 1985, 1986); 4-methylpyrazole significantly reduced rates of ethanol oxidation in both ADH$^+$ and ADH$^-$ hepatocytes. This decrease seen in the ADH$^-$ cells was used to correct for the inhibitory effect of 4-methylpyrazole on cytochrome P-450-catalyzed ethanol oxidation in ADH$^+$ deermouse hepatocytes. After such correction, cytochrome P-450-dependent oxidation was found to catalyze 42% of the total ethanol metabolism at 10 mM substrate and 62% at 50 mM substrate. Similar results were derived *in vivo*

with the ADH inhibitor 4-methylpyrazole and with the catalase inhibitor aminotriazole, the latter used at a dose that inhibited H_2O_2-dependent ethanol peroxidation (Takagi *et al.*, 1986). By a different approach (involving measurement of isotope effects), cytochrome P-450 was calculated to account for 35% of the total substrate oxidation at low ethanol concentrations and about 70% at high ethanol concentrations (Takagi *et al.*, 1985; Alderman *et al.*, 1987). With a third approach, namely, by determining the fate of ^3H from [2-^3H]ethanol, the contribution of non-ADH pathways to ethanol oxidation was found in rats to approach 50% (Vind and Grunnet, 1985). Thus, very different experimental approaches yielded similar results, namely, that cytochrome P-450 plays a significant role in hepatic ethanol oxidation even in the presence of ADH.

1.2. Alteration in the Metabolism of Ethanol after Chronic Ethanol Consumption

Regular drinkers tolerate large amounts of alcoholic beverages, mainly because of central nervous system adaptation. In addition, alcoholics develop increased rates of blood ethanol clearance, that is, metabolic tolerance (Kater *et al.*, 1969; Ugarte *et al.*, 1972). Experimental ethanol administration also results in an increased rate of ethanol metabolism (Lieber and DeCarli, 1970a; Tobon and Mezey, 1971; Misra *et al.*, 1971; Feinman *et al.*, 1978; Pikkarainen and Lieber, 1980; Nomura *et al.*, 1983; Shigeta *et al.*, 1984). The progressive acceleration of ethanol metabolism after chronic ethanol consumption is not to be confounded with the rise that occurs after an acute dose of ethanol (the so-called "swift increase" in alcohol metabolism), which appears to result from a stress-associated adrenalin discharge (Yuki and Thurman, 1980). The mechanism of the chronic acceleration is still the subject of discussion.

1.2.1. ADH-Related Ethanol Metabolism

Although some discrepant results have been published before (Hawkins *et al.*, 1966; Hawkins and Kalant, 1972), it is now agreed (Videla *et al.*, 1973; Kalant *et al.*, 1975) that ADH activity does not increase after chronic ethanol feeding, a finding consistent with the observation of other groups (Singlevich and Barboriak, 1971; Raskin and Sokoloff, 1972; de Saint-Blanquat *et al.*, 1972). In some studies, there was actually a decrease of ADH activity in the liver after chronic ethanol consumption (Lieber and DeCarli, 1970a; Brighenti and Pancaldi, 1970; Salaspuro et al., 1981), a finding in keeping with the observation that alcoholics display decreased hepatic ADH activity in the presence (Zorzano et al., 1989) or even in the absence (Ugarte et al., 1967) of liver damage. In view of the observation of Crow et al. (1977) that, under certain circumstances, the level of activity of ADH may become a major rate-limiting factor for the metabolism of ethanol, the decrease of ADH after chronic ethanol consumption may on occasion acquire functional significance.

One of the mechanisms that could contribute to the acceleration of ADH-dependent ethanol metabolism after ethanol consumption (based on increased NADH reoxidation) involves enhanced ATPase activity (susceptible to ouabain inhibition) Bernstein et al., 1973) and the creation of a hypermetabolic state akin to hyperthyroidism (Israel et al., 1975; Bernstein et al., 1975). Israel et al. (1975) reported that in liver slices, ouabain, an inhibitor of the Na^+,K^+-activated ATPase, can completely block the extra ethanol metabolism elicited by chronic ethanol treatment. Dinitrophenol increases the rate of ethanol metabolism in the livers of the treated animals only in the presence of ouabain. Oxygen consumption was also found to be increased in the livers of animals chronically treated with ethanol (Israel et al., 1973; Bernstein et al., 1973; Thurman et al., 1976), mimicking the effects of thyroxine. However, not all investigators found such an increase in oxygen consumption (Cederbaum et al., 1977c; Gordon, 1977). In alcohol-fed baboons, decreased O_2 consumption was in fact observed at high blood ethanol levels (Lieber et al., 1988). In the studies of Israel et al. (1975), a maximal effect was produced after 18 to 21 days of treatment. For both ethanol- and thyroxine-treated animals, an increased rate of oxygen consumption occurred with a concomitant loss of dinitrophenol effect. Mitochondrial α-glycerophosphate oxidase was found to be increased in the livers of animals treated with ethanol or with thyroxine. Thus, it was proposed that the hypermetabolic condition that occurs in the livers of animals chronically treated with ethanol may in some respects be similar to that found in the livers of animals treated with thyroid hormones, in which the hypermetabolic state also appears to be associated with an increased hydrolysis of ATP by the Na^+, K^+-ATPase system (Ismail-Beigi and Edelman, 1970, 1971) and a resulting lowering of the phosphorylation potential.

Controversy is evident in the literature concerning the effects of thyroid treatment on rates of alcohol metabolism. In some studies in which the animals were pretreated with thyroxine for several days (Ylikahri and Maenpaa, 1968; Ylikahri, 1970), the rate of ethanol metabolism was found to be increased. Increases of up to 100% have also been reported more recently in humans with hyperthyroidism (Ugarte and Persea, 1978). In contrast, other workers have found no effect or slight inhibitions after thyroxine pretreatment (Stokes and Lasley, 1967; Rawat and Lundquist, 1968; Hillbom and Pikkarainen, 1970; Hillbom, 1971; Lindros, 1972). Some of these discrepancies were attributed to the fact that higher levels of thyroxine inhibit alcohol dehydrogenase activity. In particular, Israel et al. (1975) attribute the lack of increased ethanol metabolism after large doses of thyroxine under their experimental conditions to this ADH inhibition. It must be pointed out, however, that in another study carried out by the same group (Videla et al., 1975), ethanol disappearance rate in vivo was increased by cold acclimation, even though liver alcohol dehydrogenase activity was reduced to the same extent as that after thyroxine (Israel et al., 1975). It is also noteworthy that chronic alcohol consumption was found to be associated with decreased incorporation of ^{131}I into the thyroid in rats (Ramakrishnan et al., 1976) and with low T_3 values in humans (Kalant et al., 1977).

Other discrepancies concern the changes in α-glycerophosphate level, which has been found to be increased in the liver after ethanol but reduced or unchanged after thyroid hormone treatment (Bode et al., 1970b; Ylikahri, 1970). Furthermore, mitochondrial α-glycerophosphate oxidase was found by Israel et al. (1975) to be increased after chronic ethanol feeding, whereas others found no change (Pilstrom and Kiessling, 1972) or even decreases (Rubin et al., 1972). It must be pointed out, however, that the experimental conditions were not identical: whereas Israel et al. (1975) used a low-fat diet in association with ethanol administration, other studies were conducted with ethanol and a diet containing an amount of fat comparable to that of human consumption. As a consequence, under the conditions used by Israel et al. (1975), ethanol consumption did not result in liver changes comparable to those seen in human alcoholic liver injury—for example, no fatty liver was observed. In contrast, under conditions that mimic the clinical situation with development of fatty liver, chronic ethanol consumption was not found to be associated with increased ATPase activity (Gordon, 1977), and the increase in the rates of ethanol metabolism after chronic ethanol consumption could not be abolished by ouabain (Cederbaum et al., 1977c), which indicates that the theory of enhanced

ATPase activity may not be applicable to the situation that normally prevails after chronic alcohol consumption. Indeed, it has now been acknowledged that the ouabain effect is nonspecific and does not imply involvement of ATPase (Yuki *et al.*, 1980).

Among other mechanisms that enhance metabolic activity, one should mention the increase in the release of epinephrine from the adrenal glands after ethanol administration in both animals and humans (Klingman and Goodall, 1957; Perman, 1958, 1960, 1961; Anton, 1965; Ogata *et al.*, 1971). This may indirectly accelerate ethanol metabolism. Epinephrine is known to produce a calorigenic effect that appears to be similar to the one produced by thyroid hormones (Harrison, 1964; Himms-Hagen, 1967; Gale, 1973). The calorigenic effect produced by ethanol in the liver, as measured in liver slices, could be reproduced by a single large dose of epinephrine (Bernstein *et al.*, 1975). Oxygen consumption by liver slices of animals given a 2 mg/kg dose of epinephrine bitartrate increased by 40 to 50%. In those livers, all the extra oxygen consumption, but not the basal respiration, could be abolished by ouabain, an inhibitor of the sodium pump. Dinitrophenol did not affect the respiratory rate in the liver of epinephrine-treated animals but markedly increased that in controls. However, the effect of 2,4-dinitrophenol, an uncoupler of oxidative phosphorylation, on ethanol metabolism is controversial: increases (Eriksson *et al.*, 1974; Seiden *et al.*, 1974) and decreases (Krarup and Olsen, 1974) of ethanol metabolism have been reported. If, following chronic ethanol consumption, changes affecting the ADH pathway (such as ATPase activity) are exclusively responsible for the acceleration of ethanol metabolism, the decreases should be fully abolished by pyrazole treatment; however, this was not the case (Lieber and DeCarli, 1970b, 1972; Salaspuro *et al.*, 1975). This raised the possibility of the involvement of non-ADH pathways, especially at higher ethanol concentrations at which ADH inhibitors only partially inhibited the accelerated ethanol concentrations. Since all these experiments were carried out in rat liver, the high-K_m ADH forms do not have to be taken into account in the concentration range studied.

1.2.2. Non-ADH-Related Acceleration of Ethanol Metabolism

Following chronic ethanol consumption, MEOS significantly increases in activity (Lieber and DeCarli, 1968, 1970a). This is associated with an increase in various constituents of the smooth fraction of the membranes involved in drug metabolism, such as phospholipids, cytochrome P-450 reductase, and cytochrome P-450 (Ishii *et al.*, 1973; Joly *et al.*, 1973; Sato *et al.*, 1978). The increase of cytochrome P-450 was associated with the appearance of a distinct form of cytochrome P-450 now called P450IIE1 (see Section 1.1.2.3). Increased levels of P450IIE1 were found in microsomes obtained from recently drinking alcoholics (Lasker *et al.*, 1986). Mechanisms involved in P450IIE1 regulation are under active investigation. Research with a P450IIE1 cDNA probe indicated that P450IIE1 protein induction by "ethanollike" agents may be regulated by posttranslational events (Song *et al.*, 1986), whereas other studies showed an increase in translatable P450IIe1 mRNA following treatment with ethanol (Kubota *et al.*, 1988), possibly reflecting species differences.

That chronic ethanol feeding results in an increased activity in liver tissue of a non-ADH and noncatalase pathway was also shown not only in liver microsomes but also in slices and in isolated hepatocytes. Ethanol oxidation was enhanced in isolated liver tissue by increasing the ethanol concentration employed *in vitro* from 10 to 30 mM. Of particular interest was the observation that this phenomenon was more pronounced in ethanol-fed rats than in their pair-fed controls (Teschke *et al.*, 1977). To test whether or not MEOS is involved in this adaptive increase, ADH and catalase activities were inhibited by pyrazole and sodium azide, respectively. The activity of the non-ADH and noncatalase pathway, which is most likely through MEOS, was significantly higher in ethanol-fed rats than in controls. In addition, the difference between the two groups was more striking at 30 mM than at 10 mM. Similarly, when a relatively constant blood ethanol level was maintained through continuous infusion in the baboon, the acceleration of ethanol metabolism with higher blood levels was more pronounced in alcohol-fed than in control animals (Pikkarainen and Lieber, 1980). All these data indicate that a non-ADH pathway, most likely MEOS, represents a major mechanism for the acceleration of ethanol metabolism at high ethanol concentrations. A similar change was shown in humans: in volunteers, alcohol consumption resulted in a progressive acceleration of blood ethanol clearance, particularly at high ethanol concentrations (Salaspuro and Lieber, 1978) (Fig. 1.8).

Calculations show that, when corrected for microsomal losses during the preparative procedure, the rise in MEOS activity might account for all of the increases in ethanol metabolism of hepatocytes isolated from ADH⁻ deermice (Alderman *et al.*, 1989b) and for a major fraction of the increase in blood ethanol clearance

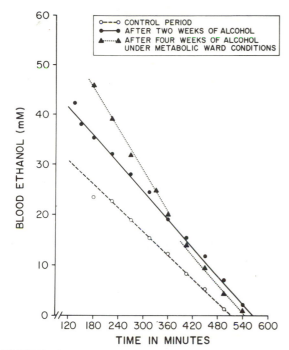

FIGURE 1.8. The effect of chronic alcohol consumption on the blood ethanol elimination curve in a human volunteer. The dose of ethanol administered was 1.0 g/kg in the first experiment, 1.3 g/kg after 2 weeks of alcohol, and 1.5 g/kg after 4 weeks of alcohol. (From Salaspuro and Lieber, 1978.)

in vivo (Lieber and DeCarli, 1972). Some of the difference that is not accounted for may actually result from a secondary increase in oxidation via ADH, a pathway limited by the rate of NADH reoxidation. This, indeed, could be accelerated by an increase in MEOS activity, since the latter is associated with enhanced NADPH utilization, and the NADPH : NADP and NADH : NAD systems are linked (Veech *et al.*, 1969; Cronholm and Fors, 1976). For instance, NADH could reduce oxaloacetate to malate, which could be oxidized to pyruvate with generation of NADPH. Pyruvate could regenerate the oxaloacetate in an ADP-requiring reaction. The hypothesis of the link of the oxidation of ethanol with the reduction of oxaloacetate was proposed two decades ago (Lieber, 1968) (see Fig. 1.2). Experimental evidence indirectly supporting this concept has been provided subsequently (Damgaard *et al.*, 1972; Selmer and Grunnet, 1976). Moreover, there is evidence that NADH may serve as partial electron donor for microsomal drug-detoxifying systems (Cohen and Estabrook, 1971). Consistent with the concept of the interaction of the NAD

and NADP systems is the observation that acute administration of ethanol leads to a smaller shift of the reduced state in livers from animals adapted to ethanol than in those from control rats, both in cytosolic and mitochondrial redox levels (Domschke *et al.*, 1974), although such an effect may also be secondary to decreased ADH (Salaspuro *et al.*, 1981).

In other studies, galactose elimination was used to assess availability of cytosolic NAD. In malnourished alcoholics, the redox change produced by an acute dose of alcohol was attenuated, which was at the time attributed to malnutrition (Salaspuro and Kesaniemi, 1973). Subsequently, such a phenomenon was observed after chronic alcohol feeding even in well-nourished baboons. In the animals fed ethanol, the capacity to clear ethanol from the blood increased progressively, whereas the associated redox change decreased (Salaspuro and Lieber, 1978; Salaspuro *et al.*, 1981).

Increases of MEOS activity in alcoholics or after chronic alcohol feeding were also found in humans (Mezey and Tobon, 1971; Kostelnik and Iber, 1973), an effect that decreased after discontinuation of alcohol. There was no strict parallelism between the increases in blood ethanol clearances and enzyme activity, and this was conceivably related to changes in blood flow or liver mass or both (these were not assessed). Furthermore, in the case of a membrane-bound enzyme system such as MEOS, it is difficult to extrapolate from an *in vitro* condition to *in vivo* activities.

There is also some debate about whether ethanol feeding enhances *catalase* activity in rats. Both an increase (Carter and Isselbacher, 1971) and no change (von Wartburg *et al.*, 1961; Hawkins *et al.*, 1966; Lieber and DeCarli, 1970a) have been reported. In humans there was no increase (Ugarte *et al.*, 1972). This question, however, may not be fully relevant to the rate of ethanol metabolism, since, as previously discussed, peroxidative metabolism of ethanol in the liver is probably limited by the rate of hydrogen peroxide formation rather than by the amount of available catalase (Boveris *et al.*, 1972). Ethanol consumption, however, does enhance the activity of hepatic NADPH oxidase (Lieber and DeCarli, 1970a; Carter and Isselbacher, 1971; Thurman, 1973), which can participate in H_2O_2 generation. It is conceivable that this mechanism contributes to ethanol metabolism *in vivo* (and to its increase after chronic ethanol consumption) by furnishing the H_2O_2 needed for peroxidative oxidation of ethanol. Such a mechanism must not necessarily involve catalase, particularly in the case of H_2O_2 generation outside the peroxisomes. Catalase appears to participate primarily in the oxidation of methanol, at least in the rat, whereas in the monkey alcohol dehydrogenase may play a greater role (Makar *et al.*, 1968).

In addition to the enhancement of MEOS activity, chronic ethanol feeding results in a striking increase of microsomal NADPH-dependent oxidations of methanol, *n*-propanol, and *n*-butanol (Teschke *et al.*, 1974a). These experiments show, therefore, that the adaptive response observed after chronic ethanol consumption can be ascribed predominantly to a catalase- and H_2O_2-independent mechanism, since the response was also demonstrated with *n*-propanol and *n*-butanol, substrates that virtually fail to react peroxidatically with catalase–H_2O_2. Because of all the experimental evidence obtained (including the substrates involved, the insensitivity to ADH and catalase inhibitors), and because the activating effect of high ethanol concentrations is consistent with the saturation of MEOS ($K_m = 8$ to 10 mM for ethanol), the adaptive increase of ethanol oxidation in the liver after chronic ethanol feeding most likely involves primarily the activity of MEOS. This was most conclusively demonstrated in deermice. Chronic ethanol treatment increased *in vivo* ethanol metabolism in ADH⁻ deermice (Shigeta *et al.*, 1984), suggesting that the metabolic tolerance to ethanol is mediated by non-ADH pathways. Similarly, after chronic ethanol feeding, ethanol (50 mM) metabolism by ADH⁻ hepatocytes doubled (Kato *et al.*, 1987, 1988). 1-Butanol, a competitive inhibitor of MEOS but neither a substrate for nor inhibitor of catalase-mediated peroxidation, reduced ethanol metabolism by 43, 50, and 57% at 10, 15, and 50 mM butanol, respectively, in controls and by 67, 75, and 78% at the same respective butanol levels after ethanol treatment. A potent catalase inhibitor, azide (0.2 mM), decreased ethanol metabolism only slightly (by less than 15%). Effective inhibition of intracellular peroxidation by 0.1 mM azide was documented by finding that urate-promoted ethanol metabolism was abolished under these conditions. These results indicate that MEOS, not catalase, accounts for most of the ethanol metabolism in ADH⁻ hepatocytes, including the portion induced by chronic ethanol consumption.

1.2.3. Energy Cost Related to Stimulated Microsomal Function and Its Contribution to a "Hypermetabolic" State and Altered Phosphorylation Potential

Oxidation of foreign compounds by the microsomes requires NADPH and O_2. As discussed before,

this is also the case for the microsomal ethanol-oxidizing system (see Fig. 1.1B), whereas ethanol oxidation by ADH pathway requires NAD and generates NADH. The reoxidation of NADH can be coupled with oxidative phosphorylation. In contrast, MEOS activity generates only heat without apparent conservation of chemical energy. Conceivably, this could be responsible for the fact that ethanol produces a greater increase in oxygen consumption in alcoholics than in normal individuals (Trémolières and Carré, 1961), since MEOS is "induced" in the former but not in the latter. Moreover, this may also contribute to the lesser growth (Saville and Lieber, 1965; Lieber *et al.*, 1965) seen in animals that are fed ethanol compared to those receiving isocaloric carbohydrate, since heat produced in excess of the needs for body temperature regulation represents energy wastage. Consistent with this view is the observed rise in temperature of rats given ethanol (Hosein and Bexton, 1975). Energy wastage may explain, at least in part, the weight loss of volunteers on isocaloric replacement of food with ethanol and the relative failure to gain weight on addition of ethanol to the diet compared with the effect of supplementation of other calories (Pirola and Lieber, 1972) (see Figs. 17.2 and 17.3). If chronic alcohol intake increases the energy requirements of the body, this should be reflected in a higher rate of oxygen consumption, which was indeed verified experimentally in rats fed ethanol chronically (Pirola and Lieber, 1975, 1976). These findings are in keeping with other animal studies in which metabolic rates were increased by the administration of ethanol and barbiturates in doses known to induce hepatic microsomal enzymes. Thus, pretreatment with barbiturates enhanced oxygen consumption in rats tested under various conditions: in the absence of drugs, during hexobarbital anesthesia, and after the administration of aminopyrine (Pirola and Lieber, 1975, 1976).

In addition to the energy-wasteful pathway of ethanol metabolism, there are, of course, a number of other mechanisms whereby ethanol might affect the efficient disposal of ingested energy, such as interference with digestion and absorption and physiological changes requiring increased expenditure of energy (Pirola and Lieber, 1972). As discussed before, increased ATPase activity, to the extent that it occurs, could also contribute to energy wastage. Ethanol might also enhance other catabolic pathways that are not effectively coupled with the formation of high-energy phosphate bonds, such as the initial steps in amino acid degradation. Indeed, it is possible that ethanol could increase protein catabolism as suggested by changes in urinary nitrogen (Rodrigo *et al.*, 1971; see also Chapter 6). Decreased energy utilization from dietary fat may also be contributory, since the lack of weight gain was not observed in rats fed ethanol with a low-fat diet (Lieber *et al.*, 1988). In any event, ethanol calories do not fully "count," at least at a relatively high ethanol intake. This applies especially to alcoholics in whom ethanol represents a large fraction of the total calories.

Participation of the microsomal pathway in ethanol oxidation may also explain some otherwise paradoxical alterations in the phosphorylation potential of the liver. Indeed, whereas the potential is unchanged at low doses of ethanol (Pösö and Forsander, 1976; Guynn and Pieklik, 1975), there is a decrease at high doses of ethanol (French, 1966; Ammon, 1967). Furthermore, after chronic ethanol treatment, although no effect was observed in the early stages (Gordon, 1977) a decrease was later noted (Israel *et al.*, 1973; Gordon, 1973, 1977). In many animals the difference in phosphorylation potential between high and low doses of ethanol could be explained by the greater participation of MEOS with the higher concentration of ethanol, which may actually somewhat inhibit ADH (substrate inhibition). After chronic ethanol feeding, the induction of the MEOS system may accentuate the effect; moreover, mitochondrial injury (discussed in Chapter 7) contributes to a relative lack of phosphorylation.

1.2.4. Metabolic and Forensic Implications of the Altered Ethanol Metabolism in the Alcoholic

Increased tolerance to alcohol is a key feature of chronic alcohol abuse. It is mainly a result of increased central nervous system tolerance, that is, decreased susceptibility of the brain to the effects of alcohol. Metabolic tolerance, which is enhanced degradation of ethanol, also contributes to the tolerance, particularly in the early stages of alcoholism. At late stages, the development of severe liver disease may offset the adaptive increase in ethanol metabolism. Moreover, malnutrition, through the decrease in the activities of the ADH pathway, may also counteract the metabolic tolerance. The demonstration that the metabolic tolerance involves increased activity of the microsomal ethanol-oxidating system has a number of practical implications. As discussed in subsequent chapters, the "induction" of liver microsomes has profound consequences on hormone and drug metabolism. It may also promote liver injury through enhanced activation of potentially hazardous or toxic compounds or the enhanced generation of hepatotoxic metabolites of ethanol such as acetaldehyde.

The energy wastage related to the activity of microsomal systems has been discussed in this chapter. Another consequence pertains to forensic medicine. Because of the activity of the microsomal system, which has a higher K_m (8 to 10 mM or 37 to 46 mg/100 ml) than ADH (K_m of 0.5 to 1.0 mM), the rate of ethanol disappearance in the bloodstream is significantly greater at high than at low ethanol concentrations. This is particularly evident after chronic ethanol consumption (Fig. 1.8), which results in "induction" of the MEOS system (Salaspuro and Lieber, 1978, 1980). The observation of an acceleration of ethanol disappearance in the blood at high ethanol concentrations is of particular significance for the medicolegal application of blood ethanol measurements. Heretofore, a common procedure to determine retrospectively the blood ethanol concentrations at a given time was to extrapolate linearly from a subsequent determination with the assumption of a standard rate of metabolism. In view of the findings corroborating the nonlinear disappearance of ethanol and the adaptive increase after chronic consumption, conventional calculations should be interpreted with caution.

1.3. Effects of Liver Disease, Blood Flow, Circadian Rhythm, Gender, and Other Factors on Hepatic Alcohol Metabolism

As discussed before, chronic ethanol consumption is associated with an increased rate of ethanol disappearance in the blood. When alcohol abuse results in severe liver disease, this acceleration disappears, and on occasion there may be an actual reduction in blood alcohol clearance. This occurs, however, only with very severe liver disease shown to be associated with reduced liver ADH (Dow *et al.*, 1975). In patients with moderate cirrhosis, the rate of alcohol metabolism may be normal (Dacruz *et al.*, 1975). In addition to the activity of the enzymes of the pathways discussed previously, total hepatic mass is an important parameter, not often measurable. Other factors that play a key role in ethanol metabolism *in vivo* include availability of cofactors and the capacity of the liver to dispose of their products (NADH and acetaldehyde). In addition to activation of MEOS, high ethanol concentrations inhibit ADH activity (substrate inhibition) (Theorell *et al.*, 1955; von Wartburg *et al.*, 1964).

Another factor that is difficult to assess is blood flow. Chronic ethanol consumption tends to increase hepatic blood flow in the baboon (Lieber *et al.*, 1988).

Acutely, the results depend on the dose used: some investigations show no effect (Smythe *et al.*, 1953; Castenfors *et al.*, 1960; Edwards *et al.*, 1987) or even a decrease (Lundquist *et al.*, 1962), whereas most studies report an increase (Mendeloff, 1954; Stein *et al.*, 1963; Shaw *et al.*, 1977; Jauhonen *et al.*, 1982; Bredfeldt *et al.*, 1985; McKaigney *et al.*, 1986; Bendtsen *et al.*, 1987; Carmichael *et al.*, 1987; Lieber *et al.*, 1988). The increase in portal blood flow after ethanol administration was attributed to a preportal vasodilatory effect of adenosine formed from acetate metabolism in extrahepatic tissues (Carmichael *et al.*, 1988). In general, when unchanged or decreased flow was observed, there was an association with low blood ethanol levels. Experiments are difficult to control, especially because very large doses of ethanol may also produce hypothermia (Nikki *et al.*, 1971). Hypothermia, in turn, has been found to result in both decrease of liver blood flow (Brauer *et al.*, 1959) and slowing of ethanol metabolism (Larsen, 1971; Krarup and Larsen, 1972).

Most studies also fail to mention the time of day that the experiments were carried out. Such information is important in view of the circadian variation of ethanol metabolism (Wilson *et al.*, 1956; Sturtevant *et al.*, 1976; Pinkston and Soniman, 1979), which is particularly manifested and perhaps even altered in the alcoholic (Jones and Paredes, 1974). Stress associated with the experimental conditions is rarely described, but it may significantly affect ethanol metabolism (Mezey *et al.*, 1979). Gender may also be of importance, with women possibly having a faster overall ethanol metabolism than men (Cole-Harding and Wilson, 1987), although their gastric metabolism may be lower (see Section 1.6). Furthermore, the volume of distribution of ethanol in females in less than that in males (Marshall *et al.*, 1983), and generally their body weight is lower. As a net result, women are more affected by ethanol than men, with, on the average, an equivalent effect for half the total dose.

1.4. Effects of Other Drugs on Ethanol Metabolism

The action of drugs such as phenobarbital on ethanol metabolism is complex. In addition to affecting liver weight and liver blood flow, they may also interfere with the pathways of ethanol metabolism. Indeed, barbiturates increase total hepatic MEOS activity (Lieber and DeCarli, 1970c), and they were also found to enhance rates of blood ethanol clearance (Fischer and Oelssner,

1961; Lieber and DeCarli, 1970c; Mezey and Robles, 1974; Ruebner *et al.*, 1975). Other drugs may exert a similar effect. Indeed, asthmatics were found to exhibit an accelerated clearance of ethanol from the blood, possibly as a result of long-standing drug use (Sotaniemi *et al.*, 1972). Similarly, diabetics were reported to display accelerated ethanol metabolism in association with tolbutamide treatment (Carulli *et al.*, 1971). Some researchers failed to verify the barbiturate effect (Tephly *et al.*, 1969; Klaassen, 1969). In the latter investigations, however, long-acting barbiturates were used, and ethanol clearance was tested in close association with barbiturate administration, at a time when blood barbiturate levels were probably elevated. Under these conditions, it was found that barbiturates interfere with blood ethanol clearance (Lieber and DeCarli, 1972). A similar mechanism may also have played a role in a study by Khanna *et al.* (1972), who used a large dose of barbiturates. Indeed, inhibition of ethanol oxidation by microsomes *in vitro* has been demonstrated for some drugs (Hildebrandt *et al.*, 1974). Thus, whereas pretreatment with these drugs may accelerate ethanol metabolism, the presence of the drug may have an inhibitory effect. Psychotropic medications such as chlorpromazine also influence serum ethanol concentrations by inhibiting ADH (Messiha, 1980). Isoniazid may also have a similar effect (Whitehouse *et al.*, 1980).

Possible effects of cimetidine on the metabolism of ethanol have been the subject of debate. The influence of a 1-week regimen of either cimetidine (1000 mg per day) or ranitidine (300 mg per day) on serum ethanol concentrations after a single oral dose of ethanol (0.8 g/kg body weight) was investigated in a randomized, placebo-controlled study in eight male volunteers by Seitz *et al.* (1984). Compared with the placebo, cimetidine, but not ranitidine, produced a significant increase in the peak serum ethanol concentration. However, the ethanol elimination rate was not affected by cimetidine. When ethanol (1.0 g/kg body weight) was administered intravenously, cimetidine failed to induce a change in ethanol metabolism. The effect of H_2-receptor antagonists on ethanol clearance was also studied in animals. Female Sprague–Dawley rats received a single dose of ethanol together with an intraperitoneal injection of either cimetidine, ranitidine, or isotonic saline. After absorption, ethanol elimination was significantly inhibited by both cimetidine and ranitidine at high ethanol concentrations (20 to 60 mM) but not at blood ethanol levels below 20 mM. These results suggested an increase in ethanol absorption caused by cimetidine (but not by ranitidine) in humans at ethanol serum concentrations below 20 mM. Alternatively, the observed effect may have been caused, at least in part, by decreased gastric metabolism of ethanol because of ADH inhibition by cimetidine (Caballeria *et al.*, 1989a) (see Section 1.6). At ethanol serum concentration above 20 mM, both H_2-receptor antagonists inhibited ethanol elimination in the rat, suggesting inhibition of microsomal ethanol oxidation. Patel *et al.* (1986) found both isoniazid and cimetidine to inhibit the elimination of ethanol in rabbits, and Feely *et al.* (1982) found a corresponding effect of cimetidine in humans. However, Dobrilla *et al.* (1984) found no such effect, and Mitchell *et al.* (1985) found that cimetidine did not alter the *in vitro* activities of either hepatic alcohol dehydrogenase or MEOS at concentrations as high as 5 mM. *In vivo*, cimetidine did not alter the rate of ethanol elimination at high ethanol concentrations (20 to 60 mM) when compared to that obtained in animals given isovolumetric amounts of saline, and cimetidine was found to have no effect on ethanol metabolism when ethanol was given intravenously (Couzigou *et al.*, 1984; Caballeria *et al.*, 1989a). The discrepancy between cimetidine effects when ethanol is given orally and intravenously is again consistent with a primary gastric effect of the drug (see Section 1.6).

In general, the drug effects (whether increases or decreases of ethanol metabolism) are rather moderate. The only compound that was shown to produce a significant acceleration of ethanol metabolism is fructose (Lundquist and Wolthers, 1958; Brown *et al.*, 1972). Fructose accelerates ethanol metabolism via the ADH pathway. Indeed, its effect is completely abolished by pyrazole (Berry, 1971) and may be caused by a speeding up of the reoxidation of NADH (Holzer and Schneider, 1955), either directly or indirectly. The metabolism of fructose results in rapid production of α-glyceraldehyde and pyruvate. Both can serve to reoxidize NADH, but they also stimulate the shuttle mechanisms (Thieden *et al.*, 1972; Damgaard *et al.*, 1972, 1973). To be effective, however, large amounts of fructose are required (1 to 2 g/kg body weight). Moreover, fructose was shown to reduce hepatic ATP and inorganic phosphate (P_i) (Ylikahri *et al.*, 1971; Lindros and Hillbom, 1971; Bode *et al.*, 1973a; Hultman *et al.*, 1975). In view of the moderate nature of the fructose effect and the potential hepatotoxicity of the compound (Bode *et al.*, 1973b), its use is not recommended.

Fructose solutions were also not found to be superior to placebo in alcoholism detoxification (Iber, 1987). Experimentally, oxygenation of the drinking water has been reported to accelerate ethanol elimination by 60%

(Hyvärinen *et al.*, 1978). By contrast, oxygen breathing was ineffective (Kinard *et al.*, 1951). According to others, oxygen breathing (Larsen, 1968) and increasing oxygen pressure (Mattie, 1963) even slightly depressed rates of ethanol oxidation.

In summary, whereas acute and chronic ethanol administration has remarkable effects on drug metabolism (see Chapter 6), the converse action, namely, the action of drugs and other agents on ethanol metabolism, is much less striking.

1.5. Ethnic and Genetic Factors in Ethanol Metabolism

In addition to the environmental factors discussed previously, individual differences in rates of ethanol metabolism appear also to be in part genetically controlled (Vessel *et al.*, 1971), and the possible role of heredity in the development of alcoholism in humans has also been emphasized (Goodwin, 1971; Goodwin *et al.*, 1977). Studies of alcohol elimination rates, however, in adoptees with and without alcoholic parents failed to reveal any significant difference (Utne *et al.*, 1977). Differences in rates of ethanol metabolism (Fenna *et al.*, 1971) and sensitivity to alcohol (Wolff, 1972) according to racial background were reported. Fenna *et al.* (1971) found that blood ethanol clearance is significantly faster in Caucasians than in Eskimos and Native Americans, whereas Native Americans have a greater capacity for acceleration of blood ethanol clearance after chronic ethanol consumption. The differences could not be correlated with previous ethanol intake or dietary habits, suggesting that a genetic mechanism may be implicated, although the influence of environmental factors was not fully excluded. Moreover, the results cannot be generalized to all Eskimos and Native Americans. Indeed, Ewing *et al.* (1974) found that Caucasians and Orientals did not differ significantly in their rate of ethanol metabolism, and Bennion and Li (1976) reported that Native Americans from Arizona metabolized ethanol at the same rate as Caucasians, and the activity of their liver ADH was the same. Specifically, no atypical form of ADH was found. In contrast, Fukui and Wakasugi (1972) observed a strikingly increased incidence of atypical ADH in the Japanese population compared with the reported results in Caucasians (von Wartburg and Schurch, 1968; Smith *et al.*, 1971). Furthermore, studies carried out by Reed *et al.* (1976) in Ojibwa Indians revealed a high rate of metabolism.

The same investigation and another study (Hanna, 1978) showed that Chinese subjects metabolize ethanol more rapidly than Caucasians. The apparent acceleration in the Chinese subjects could possibly be explained on the basis of their reactive leanness compared with the Caucasians under investigation. Indeed, a difference in adipose tissue may strikingly affect calculation of ethanol metabolism as carried out by the Widmark technique. However, it has also been shown that Orientals have a high frequency of an ADH isozyme (B_2-B_2) with a high capacity to metabolize ethanol (Thomasson et al., 1991). The variable results among American Indians may be understood on the basis of the considerable difference in ethnic background of the various tribes studied. In a review of the literature on racial comparisons of alcohol metabolism, Reed (1978) concluded that variations between races are not different from the variations among individuals in a race. On the other hand, interpretation of the results is complicated by variable and not-well-defined degrees of alcoholism in the population studied. As discussed before, chronic alcohol consumption is known to affect rates of alcohol metabolism. Moreover, the techniques used are not uniform (oral versus intravenous alcohol administration) and do not carefully distinguish rates of ethanol metabolism at high and low blood levels. Some studies also used breath alcohol measurements, the accuracy and reproducibility of which have been seriously questioned (Alobaidi et al., 1976). In any event, although it is likely that ethnic differences exist not only in terms of ethanol metabolism but also with respect to physiological responses to ethanol, additional studies are needed to define the respective roles of racial background, habitual alcohol consumption, and associated nutritional factors (for example, protein malnutrition and obesity).

1.6. Extrahepatic Ethanol Metabolism

1.6.1. Gastric ADH

Although it is generally recognized that the liver is the main site of ethanol metabolism, extrahepatic metabolism occurs: ethanol oxidation in the digestive tract of the rat has been previously reported (Lamboeuf et al., 1981, 1983) and has been related to the ADH present in this tissue (Hempel and Pietruszko, 1979; Pestalozzi et al., 1983). In the rat, the ability of the stomach to oxidize ethanol has been demonstrated in vitro (Carter and Isselbacher, 1971) and in vivo (Lamboeuf et al., 1983). However, the magnitude of gastrointestinal etha-

nol metabolism was assumed to be small (Lamboeuf et al., 1981, 1983). Some authors (Lin and Lester, 1980) could not demonstrate any significant gastrointestinal ethanol oxidation when they gave an acute high dose to rats, and they, as well as more recently Wagner (1986), concluded that this process was of negligible quantitative significance. The issue was reopened when it was shown that a significant fraction of alcohol ingested in doses in keeping with usual "social drinking" does not enter the systemic circulation in the rat and is oxidized mainly in the stomach (Julkunen et al., 1985a,b). This process was also shown to occur in humans (Fig. 1.9) (Julkunen et al., 1985a; DiPadova et al., 1987). Furthermore, Julkunen et al. (1985a) found as great an increase of acetate in peripheral blood after oral as after i.v. administration of ethanol, which clearly showed that ethanol must have been oxidized and therefore that some first-pass metabolism had occurred locally, a process that has now been confirmed by the data of Wagner et al. (1989) obtained in human subjects. Moreover, gastrectomy was associated with an abolition of the first-pass

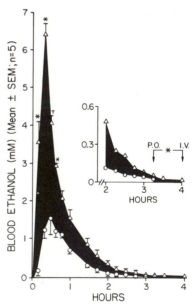

FIGURE 1.9. First-pass metabolism of ethanol in five nonalcoholics after 0.15 g/kg body weight ethanol. Ethanol was administered in a 5% dextrose solution (5 g/100 ml) perorally or intravenously 1 hr after a standard breakfast. Drinking time was 10 min, and that of the intravenous infusion 20 min. The black area represents the amount of ingested alcohol that did not enter the systemic circulation and expresses the magnitude of the first-pass metabolism of ethanol. ○, p.o. ethanol; △, i.v. ethanol. (From DiPadova et al., 1987.)

metabolism (Caballeria *et al.*, 1989b). Thus, the main site of the first-pass metabolism appears to be the upper G.I. tract both in rodents (Julkunen *et al.*, 1985a,b) and in humans (Caballeria *et al.*, 1989b).

In the rat stomach, an alcohol dehydrogenase (ADH) isoenzyme has been described (Cederbaum *et al.*, 1975; Julia *et al.*, 1987) that, contrary to the main isoenzymes of the liver, is not inhibited by high ethanol concentrations. Although this gastric ADH has a high K_m for ethanol, it is effective in the oxidation of ethanol at the high concentrations prevailing in the gastric lumen during alcohol consumption (Halsted *et al.*, 1973). A similar isoenzyme has been identified in the baboon stomach (Holmes *et al.*, 1986). Previously, only isoenzymes with low K_m for ethanol have been reported in the human stomach (Hempel and Pietruzko, 1979). However, the latter studies were done in autopsy material and at low ethanol concentrations; when we reassessed gastric ADH activity in fresh surgical specimens using ethanol concentrations similar to those prevailing in the stomach during the drinking of alcoholic beverages, two types of alcohol dehydrogenase isoenzymes (differing in their affinity for ethanol, sensitivity to 4-methylpyrazole, and electrophoretic migration) were identified in the human stomach. At the high concentrations prevailing in the gastric lumen during alcohol consumption, the sum of their activities could account for considerable oxidation of ethanol (Hernandez-Munoz *et al.*, 1990). The magnitude of this process was assessed in the rat to amount to about 20% of the ethanol administered when given at a low dose (Caballeria *et al.*, 1987). This process determines, in part, the bioavailability of alcohol and thus modulates its potential toxicity. This gastric barrier is low in females (DiPadova *et al.*, 1988; Frezza *et al.*, 1990), thereby contributing to their increased susceptibility to ethanol. Commonly used H_2 antagonists, such as cimetidine, decrease the activity of gastric ADH (Caballeria *et al.*, 1989a), thereby enhancing peripheral blood levels of ethanol, a potentially hazardous complication if it occurs in an unsuspecting driver. Therefore, in general drinkers who require treatment with H_2 blockers, drugs such as famotidine that do not affect gastric ADH and consequently do not increase blood ethanol levels should be used in preference to equipotent H_2 blockers such as cimetidine and ranitidine that do have these side effects (Caballeria *et al.*, 1989c).

The decrease of the first-pass metabolism in the fasting state has no obvious explanation but could be tentatively attributed, in part, to the accelerated gastric emptying and increased intestinal absorption (Welling

and Tse, 1983). By contrast, when alcohol is ingested with a meal, it is mainly and rapidly absorbed in the stomach (Cortot *et al.*, 1986), where its oxidation is also maximal (Julkunen *et al.*, 1985a,b), at least in rodents. It is noteworthy that in humans, alcohol induced a marked delay of gastric emptying of solids, with a smaller slowing effect on gastric emptying of the liquid phase of the solid–liquid meal and of the homogenized meal (Jian *et al.*, 1986). In any event, experimentally, the contribution of the small intestine below the ligament of Treitz to alcohol absorption appears negligible in the fed but not in the fasted state.

In alcoholics, first-pass metabolism was much smaller, and after fasting it had virtually disappeared (DiPadova *et al.*, 1987). The reduced first-pass metabolism in alcoholics may in part reflect diminished gastric ADH activity, since (1) a significant decrease of gastric ADH activity has been documented in rats fed alcohol-containing diets chronically (Julkunen *et al.*, 1985a,b), and (2) alterations in the rate of gastric emptying do not appear to be contributory (Keshavarizian *et al.*, 1985), at least in humans. In other species, e.g., dogs (Tzamaloukas *et al.*, 1988) or rats (Gentry *et al.*, 1990), and different experimental conditions, gastric retention may play a role.

1.6.2. Ethanol Metabolism in Other Tissues

Tissues other than liver and stomach are capable of ethanol oxidation. For instance, pulmonary slices and microsomes can convert ethanol to acetaldehyde (Pikkarainen *et al.*, 1981). The quantitative role of such extrahepatic oxidative pathways is unsettled and may vary from species to species. Qualitatively, however, such extrahepatic metabolism, when it occurs at crucial sites, could acquire pathological significance, for instance, in the brain (Cohen *et al.*, 1983) and also the intestine (Seitz *et al.*, 1978) and esophagus (Farinati *et al.*, 1985), where it may contribute to carcinogenesis (see Chapter 15).

1.6.3. Nonoxidative Metabolism

The possible pathogenic role of a nonoxidative pathway of ethanol to form fatty acid ethyl esters was raised by Laposata and Lange (1986). The capacity of ethanol to form ethyl esters *in vivo* had been demonstrated by Goodman and Deykin (1963) and also by Lange (1982), who purified the enzyme (Mogelson and Lange, 1984). Laposata and Lange (1986) found that in acutely intoxicated subjects, concentrations of fatty acid ethyl esters were significantly higher than in controls in pancreas, liver, heart, and adipose tissue. Since this nonoxidative ethanol metabolism occurs in humans in the organs most commonly injured by alcohol abuse, and since some of these organs lack oxidative ethanol metabolism, Laposata and Lange (1986) postulated that fatty acid ethyl esters may have a role in the production of alcohol-induced injury. Further experiments are needed to verify this interesting hypothesis.

1.7. Summary

Ethanol oxidation, once thought to be a simple, one-enzyme-mediated reaction, has now been shown to be a complex process affected by various enzyme systems, nutritional status, liver disease, genetic factors, and prior history of alcohol and drug use. Recent advances in our knowledge of hepatic metabolism of ethanol enable us to understand a number of metabolic alterations associated with the oxidation of ethanol that develop in the alcoholic. We have also gained better insight into various consequences of chronic ethanol consumption including the metabolic tolerance to ethanol that develops in the alcoholic. The acceleration of blood ethanol clearance after chronic ethanol consumption is highest at high ethanol blood levels, which may have some forensic importance. The acute and chronic interactions between ethanol and drug metabolism are now better understood.

REFERENCES

Admirand, W. H., Cronholm, T., and Sjövall, J.: Reduction of dehydroepiandrosterone sulfate in the liver during ethanol metabolism. *Biochim. Biophys. Acta* **202:**343–348, 1970.

Alderman, J., Takagi, T., and Lieber, C. S.: Ethanol metabolizing pathways in deermice: Estimation of flux calculated from isotope effects. *J. Biol. Chem.* **262:**7497–7503, 1987.

Alderman, J. A., Kato, S., and Lieber, C. S.: Characteristics of butanol metabolism in alcohol dehydrogenase-deficient deermice. *Biochem. J.* **257:**615–619, 1989a.

Alderman, J., Kato, S., and Lieber, C. S.: The microsomal ethanol oxidizing system mediates metabolic tolerance to ethanol in deermice lacking alcohol dehydrogenase. *Arch. Biochem. Biophys.* **271:**33–39, 1989b.

Alobaidi, T. A. A., Hill, D. W., and Payne, J. P.: Significance of variation in blood: Breath partition coefficient of alcohol. *Br. Med. J.* **2:**1479–1481, 1976.

Ammon, H. P. T., and Estler, C. J.: Influence of acute and chronic administration of alcohol on carbohydrate breakdown and energy metabolism in the liver. *Nature* **216:**158–159, 1967.

Anton, A. H.: Ethanol and urinary catecholamines in man. *Clin. Pharmacol. Ther.* **6:**462–469, 1965.

Bahr-Lindström, H. von, Hoog, J.-O., Heden, L.-O., Kaiser, R., Fleetwood, L., Larsson, K., Lake, M., Holmquist, B., Holmgren, A., and Hempel, T.: cDNA and protein structure for the α subunit of human liver alcohol dehydrogenase. *Biochemistry* **25:**2465–2470, 1986.

Bartlett, G. R.: Does catalase participate in the physiologic oxidation of alcohols? *Q. J. Stud. Alcohol* **13:**583–589, 1952.

Bendtsen, F., Henriksen, J. H., Widding, A., and Winkler, K.: Hepatic venous oxygen content in alcoholic cirrhosis and non-cirrhotic alcoholic liver disease. *Liver* **7:**176–181, 1987.

Bennion, L. J., and Li, T.-K.: Alcohol metabolism in American Indians and whites. Lack of racial differences in metabolic rate and liver alcohol dehydrogenase. *N. Engl. J. Med.* **294:**9–13, 1976.

Bernstein, J., Videla, L., and Israel, Y.: Metabolic alterations produced in liver by chronic ethanol administration. Changes related to energetic parameters of the cell. *Biochem. J.* **134:**515–522, 1973.

Bernstein, J., Videla, L., and Israel, Y.: Hormonal influences in the development of the hypermetabolic state of the liver produced by chronic administration of ethanol. *J. Pharmacol. Exp. Ther.* **192:**583–591, 1975.

Berres, H. H., Goslar, H. G., and Jaeger, K. H.: Morphologische und histochemische Veranderungen in der Leber nach einmaliger Alkoholbelastung und nach Versuchen ihrer. *Acta Histochem.* **35:**173–185, 1970.

Berry, M. N.: Effects of microsomal-metabolized drugs (MMD) on oxidation of ethanol, sorbitol or glycerol in rat. *Clin. Res.* **19:**471, 1971 (abstract).

Bjorkhem, I.: On the role of alcohol dehydrogenase in ω-oxidation of fatty acids. *Eur. J. Biochem.* **30:**441–451, 1972.

Bode, Ch., Goebell, H., and Stahler, M.: Anderungen der Alkoholdehydrogenase Aktivität in der Rattenleber durch Eiweissmangel und Athanol. *Z. Gesampte Exp. Med.* **152:**111–124, 1970a.

Bode, Ch., Bode, C., Goebell, H., Godbersen, H., and Strohmeyer, G.: On the pathogenesis of ethanol-induced lipid accumulation in the liver. II. Effect of stimulation of the α-glycerophosphate cycle on metabolic changes in the liver caused by ethanol. *Horm. Metab. Res.* **2:**282–286, 1970b.

Bode, Ch., Buchwald, B., and Goebell, H.: Inhibition of ethanol breakdown due to protein deficiency in man. *Germ. Med.* **1:**149–151, 1971.

Bode, J. C., and Thiele, D.: Hemmung des Athanolabbaus beim Menschen durch Fasten: Reversibilität durch Fructose-Infusion. *Dtsch. Med. Wochensch.* **100:**1849–1851, 1975.

Bode, J. C., Zelder, O., Rumpelt, H. J., and Wittkamp, U.: Depletion of liver adenosine phosphates and metabolic effects of intravenous infusion of fructose or sorbitol in man and in the rat. *Eur. J. Clin. Invest.* **3:**436–441, 1973a.

Bode, J. C., Bode, C., Rumpelt, H. J., and Zelder, O.: Loss of hepatic adenosine phosphates and metabolic consequences following fructose or sorbitol administration in man and in the rat. In: *Regulation of Hepatic Metabolism*, F. Lundquist and N. Tygstrup (eds.), New York, Academic Press, pp. 267–284, 1973b.

Bosron, W. F., and Li, T.-K.: Catalytic properties of human liver alcohol dehydrogenase isoenzymes. *Enzyme* **37**:19–28, 1987.

Bosron, W. F., Li, T.-K., Dafeldecker, W. P., and Vallee, B. L.: Human liver π-alcohol dehydrogenase: kinetic and molecular properties. *Biochemistry* **18**:1101–1105, 1979.

Bosron, W. F., Magnes, L. J., and Li, T.-K.: Human liver alcohol dehydrogenase: ADH Indianapolis results from genetic polymorphism at the ADH2 gene locus. *Biochem. Genet.* **21**:735–744, 1983.

Boveris, A., Oshino, H., and Chance, B.: The cellular production of hydrogen peroxide. *Biochem. J.* **128**:617–630, 1972.

Brauer, R. W., Holloway, R. J., Krebs, J. S., Leong, G. F., and Carroll, H. W.: The liver in hypothermia. *Ann. N.Y. Acad. Sci.* **80**:395–423, 1959.

Bredfeldt, J. E., Riley, E. M., and Groszmann, R. J.: Compensatory mechanisms in response to an elevated hepatic oxygen consumption in chronically ethanol-fed rats. *Am. J. Physiol.* **248**:G507–G511, 1985.

Brighenti, L., and Pancaldi, G.: Effetto della somministrazione di alcool etilico su alcune attivita enzimatiche del fegato di ratto. *Soc. Ital. Biol. Spermentale* **46**:1–5, 1970.

Brown, S. S., Forrest, J. A. H., and Roscoe, P.: A controlled trial of fructose in the treatment of acute alcoholic intoxication. *Lancet* **2**:898–900, 1972.

Bucher, T., and Klingenberg, M.: Wege des Wasserstoffs in der lebendigen Organisation. *Angew. Chem.* **70**:552–570, 1958.

Buehler, R., Hess, M., and von Wartburg, J.-P.: Immunohistochemical localization of human liver alcohol dehydrogenase in liver tissue, cultured fibroblasts and hela cells. *Am. J. Pathol.* **108**:89–99, 1982.

Burnett, K. G., and Felder, M. R.: Ethanol metabolism in peromyscus genetically deficient in alcohol dehydrogenase. *Biochem. Pharmacol.* **28**:1–8, 1980.

Caballeria, J., Baraona, E., and Lieber, C. S.: The contribution of the stomach to ethanol oxidation in the rat. *Life Sci.* **41**:1021–1027, 1987.

Caballeria, J., Baraona, E., Rodmilans, M., and Lieber, C. S.: Effects of cimetidine on gastric alcohol dehydrogenase activity and blood ethanol levels. *Gastroenterology* **96**:388–392, 1989a.

Caballeria, J., Frezza, M., Hernandez, R., DiPadova, C., Korsten, M. A., Baraona, E., and Lieber, C. S.: The gastric origin of the first pass metabolism of ethanol in man: effect of gastrectomy. *Gastroenterology* **97**:1205–1209, 1989b.

Caballeria, J., Baraona, E., Rodamilans, M., and Lieber, C. S.: Effects of cimetidine on gastric alcohol dehydrogenase activity and blood ethanol levels (Letter to the Editor) *Gastroenterology* **96**:1067–1068, 1989c.

Carmichael, F. J., Saldivia, V., Israel, Y., McKaigney, J. P., and Orrego, H.: Ethanol-induced increase in portal hepatic blood flow: Interference by anesthetic agents. *Hepatology* **7**:89–94, 1987.

Carmichael, F. J., Saldivida, V., Varghese, G. A., Israel, Y., and Orrego, H.: Ethanol-induced increase in portal blood flow role of acetate and A_1- and A_2-adenosine receptors. *Am. J. Physiol.* **225**:G417–G423, 1988.

Carter, E. A., and Isselbacher, K. J.: The role of microsomes in the hepatic metabolism of ethanol. *Ann. N.Y. Acad. Sci.* **179**:282–294, 1971.

Carulli, N., Manenti, F., Gallo, M., and Salviolli, G. F.: Alcohol-drugs interaction in man: Alcohol and tolbutamide. *Eur. J. Clin. Invest.* **1**:421–424, 1971.

Castenfors, H., Hultman, E., and Josephson, B.: Effect of intravenous infusions of ethyl alcohol on estimated hepatic blood flow in man. *J. Clin. Invest.* **39**:776–781, 1960.

Cederbaum, A. I., Pietrusko, R., Hempel, J. Becker, F. F., and Rubin, E.: Characterization of a nonhepatic alcoholic dehydrogenase from rat hepatocellular carcinoma and stomach. *Arch. Biochem. Biophys.* **171**:348–360, 1975.

Cederbaum, A. I., Dicker, E., and Rubin, E.: Transfer and reoxidation in reducing equivalents as the rate-limiting steps in the oxidation of ethanol by liver cells isolated from fed and fasted rats. *Arch. Biochem. Biophys.* **183**:638–646, 1977a.

Cederbaum, A. I., Dicker, E., Rubin, E., and Cohen, G.: The effect of dimethylsulfoxide and other hydroxyl radical scavengers on the oxidation of ethanol by rat liver microsomes. *Biochem. Biophys. Res. Commun.* **78**:1254–1262, 1977b.

Cederbaum, A. I., Dicker, E., Lieber, C. S., and Rubin, E: Factors contributing to the adaptive increase in ethanol metabolism due to chronic consumption of ethanol. *Alcoholism: Clin. Exp. Res.* **1**:27–31, 1977c.

Chance, B.: An intermediate compound in the catalase–hydrogen reaction. *Acta Chem. Scand.* **1**:236–267, 1947.

Chance, B., and Oshino, N.: Kinetics and mechanisms of catalase in peroxisomes of the mitochondrial fraction. *Biochem. J.* **122**:225–233, 1971.

Chappell, J. H.: Systems used for the transport of substrates into mitochondria. *Br. Med. Bull.* **24**:150–157, 1968.

Cohen, B. S., and Estabrook, R. W.: Microsomal electron transport reactions. III. Cooperative interactions between reduced diphosphopyridine nucleotide and reduced triphosphopyridine nucleotide linked reaction. *Arch. Biochem. Biophys.* **143**:54–65, 1971.

Cohen, G., Sinet, P. M., and Heikkila, R. E.: Ethanol oxidation by catalase in rat brain *in vivo*. In: *Biological Approach to Alcoholism* (C. S. Lieber, ed.), Washington, D.C., Research Monograph-11, DDHS Publication No. (ADM) 83-1261, Superintendent of Documents, U.S. Government Printing Office, pp. 311–315, 1983.

Cole-Harding, S., and Wilson, J. R.: Ethanol metabolism in men and women. *J. Stud. Alcohol* **48**:380–387, 1987.

Comai, K., and Gaylor, J. L.: Existence and separation of three forms of cytochrome P-450 from rat liver microsomes. *J. Biol. Chem.* **248**:4947–4955, 1973.

Conney, A. H.: Pharmacological implications of microsomal enzyme induction. *Pharmacol. Rev.* **19**:317–366, 1967.

Cortot, A., Jobin, G., Ducrot, F., Aymes, C., Giraudeaux, V., and Modigliani, R.: Gastric emptying and gastrointestinal absorption of alcohol ingested with a meal. *Dig. Dis. Sci.* **31**:343–348, 1986.

Couzigou, P., Fleury, B., Bourjac, M., Betbeder, A. U., Vincon, G., Richard-Molard, B., Albin, H., Amouretti, M., and Beraud, C.: Pharmacocinétique de l'alcool après perfusion intraveineuse de trois heures avec et sans cimétidine chez dix sujets sains non alcooliques. *Gastroenterol. Clin. Biol.* **8:** 103–108, 1984.

Cronholm, T., and Fors, C.: Transfer of the 1-pro-R and the 1-pro-S hydrogen atoms of ethanol in metabolic reduction *in vivo Eur. J. Biochem.* **70:**83–87, 1976.

Cronholm, T., and Sjövall, J.: Effect of ethanol metabolism on redox state of steroid sulphates in man. *Eur. J. Biochem.* **13:**124–131, 1970.

Crow, K. E., Cornell, N. W., and Veech, R. L.: The rate of ethanol metabolism in isolated rat hepatocytes. *Alcoholism: Clin. Exp. Res.* **1:**43–47, 1977.

Crow, K. E., Cornell, N. W., and Veech, R. L.: Lactate-stimulated ethanol oxidation in isolated rat hepatocytes. *Biochem. J.* **172:**29–36, 1978.

Dacruz, A. G., Correia, J. P., and Menezes, L.: Ethanol metabolism in liver cirrhosis and chronic alcoholism. *Acta Hepato-Gastroenterol.* **22:**369–374, 1975.

Damgaard, S. E., Sestoft, L., Lundquist, F., and Tygstrup, N.: The interrelationship between fructose and ethanol metabolism in the isolated perfused pig liver. *Acta Med. Scand.* **542:**(Suppl.) 131–140, 1972.

Damgaard, S. E., Lundquist, F., Tønnesen, K., Hansen, F. V., and Sestoft, D.: Metabolism of ethanol and fructose in the isolated perfused pig liver. *Eur. J. Biochem.* **33:**87–97, 1973.

de Saint-Blanquat, G., Fritsch, P., and Derache, R.: Activité alcool-déhydrogénasique de la muqueuse gastrique sous l'effet de différents traitements éthanoliques chez le rat. *Pathol. Biol.* **10:**249–253, 1972.

DiPadova, C., Worner, T. M., Julkunen, R. J. K., and Lieber, C. S.: Effects of fasting and chronic alcohol consumption on the first pass metabolism of ethanol. *Gastroenterology* **92:** 1169–1173, 1987.

DiPadova, C., Frezza, M., and Lieber, C. S.: Gastric metabolism of ethanol: implications for its bioavailability in men and women. In: *Biomedical and Social Aspects of Alcohol and Alcoholism* (K. Kuriyama, A. Takada, and H. Ishii, eds.), Amsterdam: Elsevier Science Publishers, B. V., pp. 81–84, 1988.

Dobrilla, G., de Pretis, G., Piazzi, L., Chilovi, R., Coberiato, M., Valentini, M., Pastorino, A., and Vallaperta, P.: Is ethanol metabolism affected by oral administration of cimetidine and ranitidine at therapeutic doses? *Acta Hepato-Gastroenterol.* **31:**35–37, 1984.

Domschke, S., Domschke, W., and Lieber, C. S.: Hepatic redox state: Attenuation of the acute effects of ethanol induced by chronic ethanol consumption. *Life Sci.* **15:**1327–1334, 1974.

Dow, J., Krasner, N., and Goldberg, A.: Relation between hepatic alcohol dehydrogenase activity and the ascorbic acid in leucocytes of patients with liver disease. *Clin. Sci. Mol. Med.* **49:**603–608, 1975.

Duester, G., Smith, M., Bilanchone, V., and Hatfield, G. W.: Molecular analysis of the human class I alcohol dehydrogenase gene family and nucleotide sequence of the gene encoding the β subunit. *J. Biol. Chem.* **261:**2027–2033, 1986.

Edwards, D. J., Babiak, L. M., and Beckman, H. B.: The effect of a single oral dose of ethanol on hepatic blood flow in man. *Eur. J. Clin. Pharmacol.* **32:**481–484, 1987.

Edwards, J. A., and Price-Evans, D. A.: Ethanol metabolism in subjects possessing typical and atypical liver alcohol dehydrogenase. *Clin. Pharmacol. Ther.* **8:**824–829, 1967.

Ekström, G., Cronholm, T., and Ingelman-Sundberg, M.: Hydroxyl-radical production and ethanol oxidation by liver microsomes isolated from ethanol-treated rats. *Biochem. J.* **233:**755–761, 1986.

Ekström, G., Norsten, C., Cronholm, T., and Ingelman-Sundberg, M.: Cytochrome P-450 dependent ethanol oxidation. Kinetic isotope effects and absence of steroselectivity. *Biochemistry* **26:**7346–7354, 1987.

Eriksson, C. J. P., Lindros, K. O., and Forsander, O. A.: 2,4-Dinitrophenol-induced increase in ethanol and acetaldehyde oxidation in the perfused rat liver. *Biochem. Pharmacol.* **23:** 2193–2195, 1974.

Ewing, J. A., Rouse, B. A., and Pellizzari, E. D.: Alcohol sensitivity and ethnic background. *Am. J. Psychiatry* **131:** 206–210, 1974.

Fabry, T. L., and Lieber, C. S.: The photochemical action spectrum of the microsomal ethanol oxidizing system. *Alcoholism: Clin. Exp. Res.* **3:**219–222, 1979.

Farinati, F., Zhou, Z. C., Bellah, J., Lieber, C. S., and Garro, A. J.: Effect of chronic ethanol consumption on activation of nitroso-pyrrolidine to a mutagen by rat upper alimentary tract, lung and hepatic tissue. *Drug. Metab. Dispos.* **13:**210–214, 1985.

Feely, J., and Wood, A. J. J.: Effects of cimetidine on the elimination actions of ethanol. *J.A.M.A.* **247:**2819–2821, 1982.

Feinman, L., Baraona, E., Matsuzaki, S., Korsten, M., and Lieber, C. S.: Concentration dependence of ethanol metabolism *in vivo* in rats and man. *Alcoholism: Clin. Exp. Res.* **2:**381–385, 1978.

Fenna, D., Mix, L., Schaefer, O., and Gilbert, J. A. L.: Ethanol metabolism in various racial groups. *Can. Med. Assoc. J.* **105:**472–475, 1971.

Feytmans, E., and Leighton, F.: Effects of pyrazole and 3-amino-1,2, or 4-triazole on methanol and ethanol metabolism by the rat. *Biochem. Pharmacol.* **22:**349–360, 1973.

Fischer, H. D., and Oelssner, W.: Der Einfluss von Barbituraten auf die Alkoholelimination bei Mausen. *Klin. Wochenschr.* **39:** 1265, 1961.

Forsander, O. A., and Raiha Niels, C. R.: Metabolites produced in the liver during alcohol oxidation. *J. Biol. Chem.* **235:**34–36, 1960.

French, S. W.: Effect of acute and chronic ethanol ingestion in rat liver ATP. *Proc. Soc. Exp. Biol. Med.* 121:681–685, 1966.

Frezza, M., Di Padova, C., Pozzato, G., Terpin, M., Baraona, E., and Lieber, C. S.: High blood alcohol levels in women: Role of decreased gastric alcohol dehydrogenase activity and first pass metabolism. *N. Engl. J. Med.* **322:**95–99, 1990.

Fukui, M., and Wakasugi, C.: Liver alcohol dehydrogenase in a Japanese population. *Jpn. J. Legal Med.* **26:**46–51, 1972.

Gale, C. C.: Neuroendocrine aspects of thermoregulation. *Annu. Rev. Physiol.* **35:**391–430, 1973.

Gentry, R. T., Baraona, E., and Lieber, C. S.: Chronic alcohol consumption accelerates gastric emptying and increases peak blood ethanol in rats. *Alcoholism: Clin. Exp. Res.* (abstract) **14:**291, 1990.

Gillette, J. R., Brodie, B. B., and La Du, B. N.: The oxidation of drugs by liver microsomes: On the role of TPNH and oxygen. *J. Pharmacol. Exp. Ther.* **119**:532–540, 1957.

Glassman, A. B., Bennett, C. E., and Randall, C. L.: Effects of ethyl alcohol on human peripheral lymphocytes. *Arch. Pathol. Lab. Med.* **109**:540–542, 1985.

Goodman, D. W., and Deykin, D.: Fatty acid ethyl ester formation during ethanol metabolism *in vivo*. *Proc. Soc. Exp. Biol. Med.* **113**:65–67, 1963.

Goodwin, D. W.: Is alcoholism hereditary? A review and critique. *Arch. Gen. Psychiatry* **25**:545–549, 1971.

Goodwin, D. W., Schulsinger, F., Knop, J., Mednick, S., and Guze, S. B.: Psychopathology in adopted and nonadopted daughters of alcoholics. *Arch. Gen. Psychiatry* **34**:1005–1009, 1977.

Gordon, E. R.: The utilization of ethanol by the isolated perfused rat liver. *Can. J. Physiol. Pharmacol.* **46**:609–616, 1968.

Gordon, E. R.: Mitochondrial functions in an ethanol-induced fatty liver. *J. Biol. Chem.* **248**:8271–8280, 1973.

Gordon, E. R.: ATP metabolism in an ethanol induced fatty liver. *Alcoholism: Clin. Exp. Res.* **1**:21–25, 1977.

Greenberger, N. J., Cohen, R. B., and Isselbacher, K. J.: The effect of chronic ethanol administration on liver alcohol dehydrogenase activity in the rat. *Lab. Invest.* **14**:264–271, 1965.

Grunnet, N.: Oxidation of extramitochondria NADPH by rat liver mitochondria: Possible role of acyl-CoA elongation enzymes. *Biochem. Biophys. Res. Commun.* **41**:909–917, 1970.

Grunnet, N., and Thieden, H. I. D.: The effect of ethanol concentrations upon *in vivo* metabolite levels of rat liver. *Life Sci.* (Part II) **11**:983, 1972.

Grunnet, N., Quistorff, B., and Thieden, H. I.D.: Rate-limiting factors in ethanol concentration, fructose, pyruvate and pyrazole. *Eur. J. Biochem.* **40**:275–282, 1973.

Guynn, R. W., and Pieklik, J. R.: Dependence on dose of the acute effects of ethanol on liver metabolism *in vivo*. *J. Clin. Invest.* **56**:1411–1419, 1975.

Halsted, C. H., Robles, E. A., and Mezey, E.: Distribution of ethanol in the human gastrointestinal tract. *Am. J. Clin. Nutr.* **26**:831–834, 1973.

Handler, J. A., and Thurman, R. G.: Fatty acid-dependent ethanol metabolism. *Biochem. Biophys. Res. Commun.* **133**:44–51, 1985.

Handler, J. A., and Thurman, R. G.: Hepatic ethanol metabolism is mediated predominantly by catalase-H_2O_2 in the fasted state. *FEBS Lett.* **238**:139–141, 1988.

Handler, J. A., Bradford, B. U., Glassman, E. B., Forman, D. T., and Thurman, R. G.: Inhibition of catalase-dependent ethanol metabolism in alcohol dehydrogenase-deficient deermice by fructose. *Biochem. J.* **284**:415–421, 1987.

Handler, J. A., Koop, D. R., Coon, M. J., Takei, Y., and Thurman, R. G.: Identification of P-450$_{ALC}$ in microsomes from alcohol dehydrogenase-deficient deermice: contribution to ethanol elimination *in vivo*. *Arch. Biochem. Biophys.* **264**:114–124, 1988.

Hanna, J. M.: Metabolic responses of Chinese, Japanese and Europeans to alcohol. *Alcoholism: Clin. Exp. Res.* **2**:89–92, 1978.

Harada, S., Agarwal, D. P., and Goedde, H. W.: Human liver alcohol dehydrogenase isoenzyme variations: Improved separation methods using prolonged high voltage starch gel electrophoresis and isoelectric focusing. *Hum. Genet.* **40**:215–220, 1978.

Harrison, T. S.: Adrenal medullary and thyroid relationships. *Physiol. Rev.* **44**:262–285, 1964.

Hassinen, I.: Hydrogen transfer into mitochondria in the metabolism of ethanol. *Ann. Med. Exp. Biol. Fenn* **45**:35–46, 1967.

Hasumura, Y., Teschke, R., and Lieber, C. S.: Hepatic microsomal ethanol oxidizing system (MEOS): Dissociation from reduced nicotinamide adenine dinucleotide phosphate-oxidase and possible role of form 1 of cytochrome P-450. *J. Pharmacol. Exp. Ther.* **194**:469–474, 1975.

Hawkins, R. D., and Kalant, H.: The metabolism of ethanol and its metabolic effects. *Pharmacol. Rev.* **24**:67–157, 1972.

Hawkins, R. D., Kalant, H., and Khanna, J. M.: Effects of chronic intake of ethanol on rate of ethanol metabolism. *Can. J. Physiol. Pharmacol.* **44**:241–257, 1966.

Hempel, J. D., and Pietruszko, R.: Human stomach alcohol dehydrogenase: Isoenzyme composition and catalytic properties. *Alcoholism: Clin. Exp. Res.* **3**:95–98, 1979.

Hernandez-Munoz, R., Caballeria, J., Baraona, E., Uppal, R., Greenstein, R., and Lieber, C. S.: Human gastric alcohol dehydrogenase: Its inhibition by H_2-receptor antagonists and its effect on bioavailability of ethanol. *Alcoholism: Clin. Exp. Res.* **14**:946–950, 1990.

Hildebrandt, A. G., and Speck, M.: Investigations on metabolism of ethanol in rat liver microsomes. *Arch. Pharmacol. Suppl.* **277**:165, 1973.

Hildebrandt, A. G., Speck, M., and Roots, I.: The effects of substrates of mixed function oxidase on ethanol oxidation in rat liver microsomes. *Naunyn-Schmiedebergs Arch. Pharmacol.* **281**:371–382, 1974.

Hillbom, M. E.: Thyroid state and voluntary alcohol consumption of albino rats. *Acta Pharmacol. Toxicol.* **29**:95–105, 1971.

Hillbom, M. E., and Pikkarainen, P. H.: Liver alcohol and sorbitol dehydrogenase activities in hypo- and hyperthyroid rats. *Biochem. Pharmacol.* **19**:2097–2103, 1970.

Himms-Hagen, J.: Sympathetic regulation of metabolism. *Pharmacol. Rev.* **19**:367–461, 1967.

Holmes, R. S., and VandeBerg, J. L.: Baboon alcohol dehydrogenase isozymes: phenotypic changes in liver following chronic consumption of alcohol. In: *Current Topics in Biological and Medical Research, Vol. 16: Agriculture, Physiology, and Medicine*, New York, Alan R. Liss, pp. 1–20, 1987.

Holmes, R., Courtney, Y. R., and VandeBerg, J. L.: Alcohol dehydrogenase isoenzymes in baboons: tissue distribution, catalytic, properties and variant phenotypes in liver, kidney stomach, and testis. *Alcoholism: Clin. Exp. Res.* **10**:623–630, 1986.

Holzer, H., and Schneider, S.: Zum Mechanismus der Beeinflussung der Alkohol-Oxydation in der Leber durch Fructose. *Klin. Wochenschr.* **33**:1006–1009, 1955.

Horn, R. S., and Manthei, R. W.: Ethanol metabolism in chronic protein deficiency. *J. Pharmacol. Exp. Ther.* **147**:385–390, 1965.

Hosein, E. A., and Bexton, B.: Protective action of carnitine on liver lipid metabolism after ethanol administration to rats. *Biochem. Pharmacol.* **14**:1859–1863, 1975.

Hultman, E., Nilsson, L. H., and Sahlin, K.: Adenine nucleotide content of human liver. *Scand. J. Clin. Lab. Invest.* **35**:245–251, 1975.

Hyvärinen, J., Leakso, M., Sippel, H., Roine, R., Fluopaniemi, T., Leinonen, L., and Flytomen, V.: Alcohol detoxification accelerated by oxygenated drinking water. *Life Sci.* **22**:553–560, 1978.

Iber, F. L.: Evaluation of an oral solution to accelerate alcoholism detoxification. *Alcoholism: Clin. Exp. Res.* **11**:305–308, 1987.

Inatomi, N., Kato, S., Ito, D., and Lieber, C. S.: Role of peroxisomal fatty acid beta-oxidation in ethanol metabolism. *Biochem. Biophys. Res. Commun.* **163**:418–423, 1989.

Inatomi, N., Ito, I., and Lieber, C. S.: Ethanol oxidation by deermice mitochondria under physiologic conditions. *Alcoholism: Clin. Exp. Res.* **14**:130–133, 1990.

Ingelman-Sundberg, M., and Johansson, I.: Mechanisms of hydroxyl radical formation and ethanol oxidation by ethanol-inducible and other forms of rabbit liver microsomal cytochromes P-450. *J. Biol. Chem.* **259**: 6447–6458, 1984.

Iseri, O. A., Gottlieb, L. S., and Lieber, C. S.: The ultrastructure of ethanol-induced fatty liver. *Fed. Proc.* **23**:579, 1964 (abstract).

Iseri, O. A., Lieber, C. S., and Gottlieb, L. S.: The ultrastructure of fatty liver induced by prolonged ethanol ingestion. *Am. J. Pathol.* **48**:535–555, 1966.

Ishii, H., Joly, J.-G., and Lieber, C. S.: Effect of ethanol on the amount and enzyme activities of hepatic rough and smooth microsomal membranes. *Biochim. Biophys. Acta* **291**:411–420, 1973.

Ismail-Beigi, F., and Edelman, I.S.: Mechanism of thyroid calorigenesis: Role of active sodium transport. *Proc. Natl. Acad. Sci. U.S.A.* **67**:1071–1078, 1970.

Ismail-Beigi, F., and Edelman, I.S.: The mechanism of the calorigenic action of thyroid hormone. *J. Gen. Physiol.* **57**:710–722, 1971.

Israel, Y., Videla, L., MacDonald, A., and Bernstein, J.: Metabolic alterations produced in the liver by chronic ethanol administration. Comparison between the effects produced by ethanol and by thyroid hormones. *Biochem. J.* **134**:523–529, 1973.

Israel, Y., Videla, L., Fernandes-Videla, V., and Bernstein, J.: Effects of chronic ethanol treatment and thyroxine administration on ethanol metabolism and liver oxidative capacity. *J. Pharmacol. Exp. Ther.* **192**:565–574, 1975.

Jauhonen, P., Baraona, E., Miyakawa, H., and Lieber, C. S.: Mechanism for selective perivenular hepatotoxicity of ethanol. *Alcoholism: Clin. Exp. Res.* **6**:350–357, 1982.

Jian, R., Cortot, A., Ducrot, F., Jobin, G., Chaybialle, J. A., and Modigliani, R.: Effect of ethanol ingestion and postprandial gastric emptying and secretion, biliopancreatic secretions, and duodenal absorption in man. *Dig. Dis. Sci.* **31**:604–614, 1986.

Joly, J. G., Ishii, H., and Lieber, C. S.: Microsomal cyanidebinding cytochrome: Its role in hepatic ethanol oxidation. *Gastroenterology* **62**:174, 1972 (abstract).

Joly, J. G., Ishii, H., Teschke, R., Hasumura, Y., and Lieber, C. S.: Effect of chronic ethanol feeding on the activities and submicrosomal distribution of reduced nicotinamide adenine dinucleotide phosphate (NADPH)-cytochrome P-450 reductase and the demethylases for amino-pyrine and ethylmorphine. *Biochem. Pharmacol.* **22**:1532–1535, 1973.

Joly, J. G., Hetu, C., Mavier, P., and Villeneuve, J. P.: Mechanisms of induction of hepatic drug-metabolizing enzymes by ethanol. I. Limited role of microsomal phospholipids. *Biochem. Pharmacol.* **25**:1995–2001, 1976.

Joly, J. G., Villeneuve, J. P., and Mavier, P.: Chronic ethanol administration induces a form of cytochrome P-450 with specific spectral and catalytic properties. *Alcoholism: Clin. Exp. Res.* **1**:17–20, 1977.

Jones, B. M., and Paredes, A.: Circadian variation of ethanol metabolism in alcoholics. *Br. J. Addict.* **69**:3–10, 1974.

Jörnvall, H., Hempel, J., Vallee, B. L., Bosron, W. F., and Li, T.-K.: Human liver alcohol dehydrogenase: Amino acid substitution in the β2β2 Oriental isozyme explains functional properties, establishes an active site structure, and parallels mutational exchanges in the yeast enzyme. *Proc. Natl. Acad. Sci. U.S.A.* **81**:3024–3028, 1984.

Jörnvall, H., Hoog, J.-O., Bahr-Lindstrom, H., and Vallee, B. L.: Mammalian alcohol dehydrogenases of separate classes: Intermediates between different enzymes and intraclass isozymes. *Proc. Natl. Acad. Sci. U.S.A.* **84**:2580–2584, 1987.

Julia, P., Farres, J., and Parès, X.: Characterization of three isoenzymes of rat alcohol dehydrogenase. Tissue distribution and physical and enzymatic properties. *Eur. J. Biochem.* **162**:179–189, 1987.

Julkunen, R. J. K., DiPadova, C., and Lieber, C. S. First pass metabolism of ethanol—a gastrointestinal barrier against the systemic toxicity of ethanol. *Life Sci.* **37**:567–573, 1985a.

Julkunen, R. J. K., Tannenbaum, L., Baraona, E., and Lieber, C. S.: First pass metabolism of ethanol: An important determinant of blood levels after alcohol consumption. *Alcohol* **2**:437–441, 1985b.

Kahonen, M. T., Ylikahri, R. H., and Hassinen, I.: Ethanol metabolism in rats treated with ethyl-α-p-chlorophenoxyisobutyrate (clofibrate). *Life Sci.* **10** (Part II):661–670, 1971.

Kalant, H., Khanna, J. M., and Endrenyi, L., Effect of pyrazole on ethanol metabolism in ethanol-tolerant rats. *Can. J. Physiol. Pharmacol.* **53**:416–422, 1975.

Kalant, H., Orrego, H., Israel, Y., Walfish, P. C., Rankin, J., and Findlay, J.: Serum triiodothyronine (T3) as an index of severity and prognosis in alcoholic liver disease (ALD). Effect of propylthiouracil (FIU). *Gastroenterology* **73**:A29, 1977 (abstract).

Kater, R. M. H., Carulli, N., and Iber, F. L.: Differences in the rate of ethanol metabolism in recent drinking alcoholic and non-drinking subjects. *Am. J. Clin. Nutr.* **22**:1608–1617, 1969.

Kato, S., Alderman, J., and Lieber, C. S.: Ethanol metabolism in alcohol dehydrogenase deficient deermice is mediated by the microsomal ethanol oxidizing system, not by catalase. *Alcohol Alcoholism*, (Suppl.) **1**:231–234, 1987a.

Kato, S., Alderman, J., and Lieber, C. S.: Respective roles of the microsomal ethanol oxidizing system (MEOS) and catalase in ethanol metabolism by deermice lacking alcohol dehydrogenase. *Arch. Biochem. Biophys.* **254**:586–591, 1987b.

Kato, S., Alderman, J., and Lieber, C. S.: *In vivo* role of the microsomal ethanol oxidizing system in ethanol metabolism by deermice lacking alcohol dehydrogenase. *Biochem. Pharmacol.* **37**:2706–2708, 1988.

Keilin, D., and Hartree, E. F.: Properties of catalase. Catalysis of coupled oxidation of alcohols. *Biochem. J.* **39**:293–301, 1945.

Keshavarizian, A., Wobbleton, J., Greer, P., and Iber, F. L.: Stomach emptying of solid meal in alcohol withdrawal. *Alcohol: Clin. Exp. Res.* **9**:88–92, 1985.

Khani, S. C., Porter, T. D., and Coon, M. J.: Isolation and partial characterization of the gene for cytochrome P-450 3a (P-450$_{ALC}$) and a second closely related gene. *Biochem. Biophys. Res. Commun.* **150**:10–17, 1988.

Khanna, J. M., Kalant, H., and Lin, G.: Significance *in vivo* of the increase in microsomal ethanol-oxidizing system after chronic administration of ethanol, phenobarbital, and chlorcyclizine. *Biochem. Pharmacol.* **21**:2215–2226, 1972.

Kinard, F. W., McCord, W. M., and Aull, J. C.: The failure of oxygen, oxygen-carbon dioxide, or pyruvate to alter alcohol metabolism. *Q. J. Stud. Alcohol* **12**:179–183, 1951.

Kinard, F. W., Nelson, G. H., and Hay, M. G.: Catalase activity and ethanol metabolism in rat. *Proc. Soc. Exp. Biol. Med.* **92**:772–773, 1956.

Klaassen, C. D.: Ethanol metabolism in rats after microsomal metabolizing enzyme induction. *Proc. Soc. Exp. Biol. Med.* **132**:1099–1102, 1969.

Klingman, G. I., and Goodall, McC.: Urinary epinephrine and levarterenol excretion during acute sublethial alcohol intoxication in dogs. *J. Pharmacol. Exp. Ther.* **121**:313–318, 1957.

Koop, D. R., and Casazza, J. P.: Identification of ethanol-inducible P-450 enzyme 3a as the acetone and acetol monooxygenase of rabbit microsomes. *J. Biol. Chem.* **260**:13607–13611, 1985.

Koop, D. R., Morgan, E. T., Tarr, G. E., and Coon, M. J.: Purification and characterization of a unique isozyme of cytochrome P-450 from liver microsomes of ethanol-treated rabbits. *J. Biol. Chem.* **257**:8772–8780, 1982.

Kostelnik, M. E., and Iber, F. L.: Correlation of alcohol and tolbutamide blood clearance rates with microsomal enzyme activity. *Am. J. Clin. Nutr.* **26**:161–164, 1973.

Krarup, N., and Larsen, J. A.: The effect of slight hypothermia on liver function as measured by the elimination rate of ethanol, the hepatic uptake and excretion of indocyanine green and bile formation. *Acta Physiol. Scand.* **84**:396–407, 1972.

Krarup, N., and Olsen, C.: Energy requirement of the transport of reducing equivalents from cytosol to mitochondria in perfused rat liver. *Life Sci.* **15**:65–72, 1974.

Krebs, H. A., and Perkins, J. R.: The physiological role of liver alcohol dehydrogenase. *Biochem. J.* **118**:635–644, 1970.

Kubota, S., Lasker, J. M., and Lieber, C. S.: Molecular regulation of an ethanol-inducible cytochrome P-450IIE1 in hamsters. *Biochem. Biophys. Res. Commun.* **150**:304–310, 1988.

Lamboeuf, Y., De Saint Blanquat, G., and Derache, R.: Mucosal alcohol dehydrogenase- and aldehyde dehydrogenase-mediated ethanol oxidation in the digestive tract of the rat. *Biochem. Pharmacol.* **30**:542–545, 1981.

Lamboeuf, Y., La Droitte, P., and De Saint Blanquat, G.: The gastrointestinal metabolism of ethanol in the rat. Effect of chronic alcohol intoxication. *Arch. Int. Pharmacodyn. Ther.* **261**:157–169, 1983.

Lane, B. P., and Lieber, C. S.: Ultrastructural alterations in human hepatocytes following ingestion of ethanol with adequate diets. *Am. J. Pathol.* **49**:593–603, 1966.

Lane, B. P., and Lieber, C. S.: Effects of butylated hydroxytoluene on the ultrastructure of rat hepatocytes. *Lab. Invest* **16**:341–348, 1967.

Lange, L. G.: Nonoxidative ethanol metabolism: Formation of fatty acid ethyl esters by cholesterol esterase. *Proc. Natl. Acad. Sci. U.S.A.* **79**:3954–3957, 1982.

Laposata, E. A., and Lange, L. G.: Presence of nonoxidative ethanol metabolism in human organs commonly damaged by ethanol abuse. *Science* **231**:497–499, 1986.

Larsen, J. A.: Extrahepatic metabolism of ethanol in man. *Nature* **184**:12346, 1959.

Larsen, J. A.: The effect of oxygen breathing at atmospheric pressure on the metabolism of glycerol and ethanol in cats. *Acta Physiol. Scand.* **73**:186–195, 1968.

Larsen, J. A.: The effect of cooling on liver function in cats. *Acta Physiol. Scand.* **81**:197–207, 1971.

Lasker, J. M., Bloswick, B. P., Ardies, C. M., and Lieber, C. S.: Evidence for an ethanol-inducible form of liver microsomal cytochrome P-450 in humans. *Hepatology* **4**:1108, 1986 (abstract).

Lasker, J. M., Raucy, J., Kubota, S., Bloswick, B. P., Black, M., and Lieber, C. S.: Purification and characterization of human liver cytochrome P-450-ALC. *Biochem. Biophys. Res. Commun.* **148**:232–238, 1987a.

Lasker, J. M., Tsutsumi, M., Bloswick, B. P., and Lieber, C. S.: Characterization of benzoflavone (BF)-inducible hamster liver cytochrome P-450 isozyme catalytically similar to cytochrome P-450-ALC. *Hepatology* **7**:432, 1987b (abstract).

Lereboullet, J., Barres, G., and Briard, J. P.: La courbe d'alcoolémie de Widmark est-elle toujours fiable? *Bull. Acad. Nat. Med.* **160**:312–315, 1976.

Lester, D., and Benson, G. D.: Alcohol oxidation in rats inhibited by pyrazole, oximes and amides. *Science* **169**:282–284, 1970.

Levin, W., Ryan, D., West, S., and Lu, A. Y. H.: Preparation of partially purified lipid-depleted cytochrome P-450 and reduced nicotinamide adenine dinucleotide phosphate-cytochrome c reductase from rat liver. *J. Biol. Chem.* **249**:1747–1754, 1974.

Levin, W., Thomas, P. E., Oldfield, N., and Ryan, D. E.: N-demethylation of N-nitrosodimethylamine catalyzed by purified rat hepatic microsomal cytochrome P-450: isozyme specificity and role of cytochrome b5. *Arch. Biochem. Biophys.* **248**:158–165, 1986.

Li, T.-K.: Enzymology of human alcohol metabolism. *Adv. Enzymol.* 45:427–283, 1977.

Li, T.-K., and Magnes, L. J.: Identification of a distinctive molecular form of alcohol dehydrogenase in human livers with high activity. *Biochem. Biophys. Res. Commun.* **63**:202–208, 1975.

Lieber, C. S.: Metabolic effects produced by alcohol in the liver and other tissues. *Adv. Intern. Med.* **14**:151–199, 1968.

Lieber, C. S.: Alcohol and the liver. In: *Liver Annual-IV* (I. M. Arias, M. S. Frenkel, and J. H. P. Wilson, eds.), Amsterdam, The Netherlands, Excerpta Medica, pp. 130–186, 1984.

Lieber, C. S.: Alcohol and the liver, In: *Liver Annual-VI* (I. M. Arias, M. S. Frenkel, and J. H. P. Wilson, eds.), Amsterdam, The Netherlands, Excerpta Medica, pp. 163–240, 1987.

Lieber, C. S., and Davidson, C. S.: Some metabolic effects of ethyl alcohol. *Am. J. Med.* **33**:327–329, 1962.

Lieber, C. S., and DeCarli, L. M.: Ethanol oxidation by hepatic microsomes: Adaptive increase after ethanol feeding. *Science* **162**:917–918, 1968.

Lieber, C. S., and DeCarli, L. M.: Hepatic microsomal ethanol oxidizing system: *In vitro* characteristics and adaptive properties *in vivo*. *J. Biol. Chem.* **245**:2505–2512, 1970a.

Lieber, C. S., and DeCarli, L. M.: Reduced nicotinamide-adenine dinucleotide phosphate oxidase: Activity enhanced by ethanol consumption. *Science* **170**:78–80, 1970b.

Lieber, C. S., and DeCarli, L. M.: Effect of drug administration on the activity of the hepatic microsomal ethanol oxidizing system. *Life Sci.* **9**:267–276, 1970c.

Lieber, C. S., and DeCarli, L. M.: The role of the hepatic microsomal ethanol oxidizing system (MEOS) for ethanol metabolism *in vivo*. *J. Pharmacol. Exp. Ther.* **181**:279–287, 1972.

Lieber, C. S., and DeCarli, L. M.: The significance and characterization of hepatic microsomal ethanol oxidation in the liver. *Drug Metab. Dispos.* **1**:428–440, 1973.

Lieber, C. S., and DeCarli, L. M.: Oxidation of ethanol by hepatic microsomes of acatalasemic mice. *Biochem. Biophys. Res. Commun.* **60**:1187–1192, 1974.

Lieber, C S., and DeCarli, L. M.: Alcoholic liver injury: Experimental models in rats and baboons. In: *Alcoholic Intoxication and Withdrawal: Advances in Experimental Medicine and Biology, Vol. 59*, (M. Gross, ed.), Plenum, New York, pp. 379–393, 1975.

Lieber, C. S., Jones, D. P., and DeCarli, L. M.: Effects of prolonged ethanol intake: Production of fatty liver despite adequate diets. *J. Clin. Invest.* **44**:1009–1021, 1965.

Lieber, C. S., Rubin, E., and DeCarli, L. M.: Hepatic microsomal ethanol oxdizing system (MEOS): Differentiation from alcohol dehydrogenase and NADPH oxidase. *Biochem. Biophys. Res. Commun.* **40**:858–865, 1970.

Lieber, C. S., Teschke, R., Hasumura, Y., and DeCarli, L. M.: Interaction of ethanol with liver microsomes: In: *Alcohol and Aldehyde Metabolizing Systems* (R. G. Thurman, T. Yonetani, J. R. Williamson, and B. Chance, eds.), New York, Academic Press, pp. 243–256, 1974.

Lieber, C. S., Lasker, J. M., DeCarli, L. M., Saeli, J., and Wojtowicz, T.: Role of acetone, dietary fat and total energy intake in induction of hepatic microsomal ethanol oxidizing system. *J. Pharmacol. Exp. Ther.* **247**:791–795, 1988.

Lin, G. W. J., and Lester, D.: Significance of the gastrointestinal tract in the *in vivo* metabolism of ethanol in the rat. *Adv. Exp. Med. Biol.* **132**:281–286, 1980.

Lindros, K. O.: Role of the redox state in ethanol-induced suppression of citrate-cycle flux in the perfused liver of normal, hyper- and hypothyroid rats. *Eur. J. Biochem.* **26**:338–346, 1972.

Lindros, K. O., and Hillbom, M. E.: Hepatic redox state and ketone body metabolism during oxidation of ethanol and fructose in normal, hyper- and hypothyroid rats. *Ann. Med. Exp. Biol. Fenn.* **49**:162–169, 1971.

Lowe, C. U., and Mosovich, L. L.: The paradoxical effect of alcohol on carbohydrate metabolism in four patients with liver glycogen disease. *Pediatrics* **35**:1005–1008, 1965.

Lumeng, L., and Crabb, D. W.: Rate determining factors for ethanol metabolism in fasted and castrated male rats. *Biochem. Pharmacol.* **33**:2623–2628, 1984.

Lundquist, F., and Wolthers, H.: The influence of fructose on the kinetics of alcohol elimination in man. *Acta Pharmacol.* **14**: 290–294, 1958.

Lundquist, F., Tygstrup, N., Winkler, K., Mellemgaard, K., and Munck-Peterson, S. T.: Ethanol metabolism and production of free acetate in the human liver. *J. Clin. Invest.* **41**:955–961, 1962.

Makar, A. B., Tephly, T. R., and Mannering, G. J.: Methanol metabolism in the monkey. *Mol. Pharmacol.* **4**:471–483, 1968.

Maly, I. P., and Sasse, D.: Microquantitative determination of the distribution patterns of alcohol dehydrogenase activity in the liver of rat, guinea pig and horse. *Histochemistry* **83**:431–436, 1985.

Mårdh, G., Luehr, C. A., and Vallee, B. L.: Human class I alcohol dehydrogenases catalyze the oxidation of glycols in the metabolism of norepinephrine. *Proc. Natl. Acad. Sci. U.S.A.* **82**:4979–4982, 1985.

Marshall, A. W., Kingstone, D., Boss, M., and Morgan, M. Y.: Ethanol elimination in males and females: Relationship to menstrual cycle and body composition. *Hepatology* **3**:701–706, 1983.

Matsuzaki, S., and Lieber, C. S.: ADH-independent ethanol oxidation in the liver and its increase by chronic ethanol consumption. *Gastroenterology* **69**:845 (abstract), 1975.

Matsuzaki, S., Gordon, E., and Lieber, C. S.: Increased ADH independent ethanol oxidation at high ethanol concentrations in isolated rat hepatocytes: The effect of chronic ethanol feeding. *J. Pharm. Exp. Ther.* **217**:133–137, 1981.

Mattie, H.: Elimination of ethanol in rats *in vitro* at different oxygen pressures. *Acta Physiol. Pharmacol.* **12**:1–11, 1963.

McKaigney, J. P., Carmichael, F. J., Saldivia, V., Israel, Y., and Orrego, H.: Role of ethanol metabolism in the ethanol-induced increase in splanchnic circulation. *Am. J. Physiol.* **250**:G519–G523, 1986.

McManus, I. R., Contag, A. O., and Olson, R. E.: Studies on the identification and origin of ethanol in mammalian tissues. *J. Biol. Chem.* **241**:349–356, 1966.

Meldolesi, J.: On the significance of the hypertrophy of the smooth endoplasmic reticulum in liver cells after administration of drugs. *Biochem. Pharmacol.* **16**:125–131, 1967.

Mendeloff, A. I.: Effect of intravenous infusions of ethanol upon estimated hepatic blood flow in man. *J. Clin. Invest.* **33**: 1298–1302, 1954.

Messiha, F. S.: Chlorpromazine and ethanol intoxication: An underlying mechanism. *Neurobehav. Toxicol. Teratol.* **7**:185, 1980.

Mezey, E., and Potter, J. J.: Effect of castration on the turnover of rat liver alcohol dehydrogenase. *Biochem. Pharmacol.* **34:** 369–371, 1985.

Mezey, E., and Robles, E. A.: Effects of phenobarbital administration on rates of ethanol clearance and on ethanol-oxidizing enzymes in man. *Gastroenterology* **66:**248–253, 1974.

Mezey, E., and Tobon, F.: Rates of ethanol clearance and activities of the ethanol-oxidizing enzymes in chronic alcoholic patients. *Gastroenterology* **61:**707–715, 1971.

Mezey, E., Potter, J. J., and Reed, W. D.: Ethanol oxidation by a component of liver microsomes rich in cytochrome P-450. *J. Biol. Chem.* **248:**1183–1187, 1973.

Mezey, E., Vestal, R., and Potter, J. J.: Effect of uremia on rate of ethanol disappearance from the blood and on the activities of the ethanol-oxidizing enzymes. *J. Lab. Clin. Med.* **86:**931–937, 1975.

Mezey, E., Potter, J. J., and Kvetransky, R.: Effect of stress by repeated immobilization on hepatic alcohol dehydrogenase and ethanol metabolism. *Biochem. Pharmacol.* **28:**657–663, 1979.

Misra, P. S., Lefevre, A., Ishii, H., Rubin, E., and Lieber, C. S.: Increase of ethanol, meprobamate and pentobarbital metabolism after chronic ethanol administration in man and in rats. *Am. J. Med.* **51:**346–351, 1971.

Mitchell, M. C., Potter, J. J., and Mezey, E.: Ethanol and H_2-receptor antagonists. *Hepatology* **5:**909–910, 1985.

Miwa, G. T., Levin, W., Thomas, P. E., and Lu, A. Y. H.: The direct oxidation of ethanol by a catalase- and alcohol dehydrogenase-free reconstituted system containing cytochrome P-450. *Arch Biochem. Biophys.* **187:**464–475, 1978.

Mogelson, S., and Lange, L. G.: Nonoxidative ethanol metabolism in rabbit myocardium: Purification to homogeneity of fatty acyl ethyl ester synthase. *Biochemistry* **23:**4075–4081, 1984.

Morgan, E. T., Koop, D. R., and Coon, M. J.: Catalytic activity of cytochrome P-450 isozyme 3a isolated from liver microsomes of ethanol-treated rabbits. *J. Biol. Chem.* **257:**13951–13957, 1982.

Morgan, E. T., Koop, D. R., and Coon, M. J.: Comparison of six rabbit liver cytochrome P-450 isozymes in formation of a reactive metabolite of acetaminophen. *Biochem. Biophys. Res. Commun.* **112:**8–13, 1983.

Morrison, G. R., and Brock, F. E.: Quantitative measurement of alcohol dehydrogenase activity within the liver lobule of rats after prolonged alcohol ingestion. *J. Nutr.* **92:**286–292, 1967.

Nebert, D. W., Adesnik, M., Coon, M. J., Estabrook, R. W., Gonzalez, F. J., Guengerich, P., Gunsalus, I. C., Johnson, E. F., Kemper, B., Levin, W., Phillips, I. R., Sato, R., and Waterman, M. R.: The P450 gene superfamily: Recommended nomenclature. *DNA* **6:**1–11, 1987.

Nikki, P., Vapaatalo, H., and Karppanen, H.: Effect of ethanol on body temperature, postanaesthetic shivering and tissue monoamines in halothane-anesthetized rats. *Ann. Med. Exp Biol. Fenn.* **49:**157–161, 1971.

Nomura, F., Pikkarainen, P. H., Jauhonen, P., Arai, M., Gordon, E. R., Baraona, E., and Lieber, C. S.: The effect of ethanol administration on the metabolism of ethanol in baboons. *J. Pharmacol. Exp. Ther.* **227:**78–83, 1983.

Nordmann, R., Ribière, C., and Rouach, H.: Involvement of iron and iron-catalyzed free radical production in ethanol metabolism and toxicity. *Enzyme* **37:**57–69, 1987.

Norsten, C., Cronholm, T., Ekström, G., Handler, J. A., Thurman, R. G., and Ingelman-Sundberg, M.: Dehydrogenase-dependent ethanol metabolism in deermice (penomyscus maniculatus) lacking cytosolic alcohol dehydrogenase. *J. Biol. Chem.* **264:**5593–5597, 1989.

Ogata, M., Mendelson, J. H., Mello, N. K., and Majchrowicz, E.: Adrenal function and alcoholism. II. Catecholamines. *Psychosom. Med.* **33:**159–180, 1971.

Ohnishi, K., and Lieber, C. S.: Reconstitution of the microsomal ethanol oxidizing system (MEOS): Qualitative and quantitative changes of cytochrome P-450 after chronic ethanol consumption. *J. Biol. Chem.* **252:**7124–7131, 1977a.

Ohnishi, K., and Lieber, C. S.: Reconstitution of the hepatic microsomal ethanol oxidizing system (MEOS) in control rats after ethanol feeding. In: *Alcohol and Aldehyde Metabolizing System, Vol. II*, (R. G. Thurman, J. R. Williamson, H. Drott, and B. Chance, eds.), New York, Academic Press, pp. 341–350, 1977b.

Ohnishi, K., and Lieber, C. S.: Respective role of superoxide and hydroxyl radical in the activity of the reconstituted microsomal ethanol-oxidizing system. *Arch. Biochem. Biophys.* **191:**798–803, 1978.

Okuda, K., and Takigawa, N.: Rat liver 5β-cholestane-3α,7α,12α, 26-tetrol dehydrogenase as a liver alcohol dehydrogenase. *Biochim. Biophys. Acta.* **22:**141–148, 1970.

Orme-Johnson, W. H., and Ziegler, D. M.: Alcohol mixed function oxidase activity of mammalian liver microsomes. *Biochem. Biophys. Res. Commun.* **21:**78–82, 1965.

Oshino, N., Chance, B., Sies, H., and Bucher, T.: The role of H_2O_2 generation in perfused rat liver and the reaction of catalase compound I and hydrogen donors. *Arch Biochem. Biophys.* **154:**117–131, 1973.

Oshino, N., Jamieson, D., Sugano, T., and Chance, B.: Optical measurement of the catalase–hydrogen peroxide intermediate (compound I) in the liver of anaesthetized rats and its implication to hydrogen peroxide production *in situ*. *Biochem. J.* **146:**67–77, 1975.

Papenberg, J., von Wartburg, J. P., and Aebi, H.: Metabolism of ethanol and fructose in the perfused rat liver. *Enz. Biol. Clin.* **11:**237–250, 1970.

Pares, X., and Vallee, B. L.: New human liver alcohol dehydrogenase forms with unique kinetic characteristics. *Biochem. Biophys. Res. Commun.* **98:**122–130, 1981.

Patel, S. R., Joshi, N. J., Bhavsar, V. H., and Kelkar, V. V.: The effect of isoniazid and cimetidine on ethanol and methanol elimination in rabbits. *Acta Pharmacol. Toxicol.* **58:**234–236, 1986.

Pekkanen, L., Eriksson, K., and Silivonen, M.-L.: Dietarily induced changes in voluntary ethanol consumption and ethanol metabolism in the rat. *Br. J. Nutr.* **40:**103–113, 1978.

Perman, E. S.: The effect of ethyl alcohol on the secretion from the adrenal medulla in man. *Acta Physiol. Scand.* **44:**241–247, 1958.

Perman, E. S.: The effect of ethyl alcohol on the secretion from the adrenal medulla of the cat. *Acta Physiol. Scand.* **48:**323–328, 1960.

Perman, E. S.: Observation on the effect of ethanol on the urinary excretion of histamine, 5-hydroxyindole acetic acid, catecholamines and 17-hydroxycorticosteroids in man. *Acta Physiol. Scand.* **51:**62–67, 1961.

Pestalozzi, D. M., Buhler, R., von Wartburg, J. P., and Hess, M.: Immunohistochemical localization of alcohol dehydrogenase in the human gastrointestinal tract. *Gastroenterology* **85:** 1011–1016, 1983.

Petersen, D. R., and Atkinson, N.: Evidence for a genetic basis of microsomal ethanol oxidation (MEOS) induction in mice. In: *Animal Model in Research* (E. Erikson, J. D. Sinclair, and J. D. Kiianmack, eds.), New York, Academic Press, pp. 71–79, 1980.

Pikkarainen, P., and Lieber, C. S.: Concentration dependency of ethanol elimination rates in baboons: Effects of chronic alcohol consumption. *Alcoholism: Clin. Exp. Res.* **4:**40–43, 1980.

Pikkarainen, P. H., Baraona, E., Jauhonen, P., Seitz, H., and Lieber, C. S.: Contribution of oropharynx microflora and of lung microsomes to acetaldehyde in expired air after alcohol ingestion. *J. Lab. Clin. Med.* **97:**631–638, 1981.

Pilstrom, L., and Kiessling, K. H.: A possible localization of α-glycerophosphate dehydrogenase to the inner boundary membrane of mitochondria in livers from rats fed with ethanol. *Histochemistry* **32:**329–334, 1972.

Pinkston, J. N., and Soniman, K. F. A.: Effect of light and fasting on the circadian variation of ethanol metabolism in the rat. *J. Interdisc. Cycle Res.* **10:**185–193, 1979.

Pirola, R. C., and Lieber, C. S.: The energy cost of the metabolism in drugs, including ethanol. *Pharmacology* **7:**185–196, 1972.

Pirola, R. C., and Lieber, C. S.: Energy wastage in rats given drugs that induce microsomal enzymes. *J. Nutr.* **105:**1544–1548, 1975.

Pirola, R. C., and Lieber, C. S.: Hypothesis: Energy wastage in alcoholism and drug abuse: Possible role of hepatic microsomal enzymes. *Am. J. Clin. Nutr.* **29:**90–93, 1976.

Pösö, A. R., and Forsander, O. A.: Influence of ethanol oxidation rate on the lactate/pyruvate ratio and phosphorylation state of the liver in fed rats. *Acta Chem. Scand. [B]* **30:**801–806, 1976.

Rachmamin, G., MacDonald, J. A., Wahid, S., Clapp, T. A., Khanna, T. M., and Israel, Y.: Modulation of alcohol dehydrogenase and ethanol metabolism by sex hormones in the spontaneously hypertensive rat. Effect of chronic ethanol administration. *Biochem. J.* **186:**483–490, 1980.

Ramakrishnan, S., Prasanna, C. V., and Balasubramanian, A.: Effect of alcohol intake on rat hepatic enzymes and thyroid function. *Int. J. Biochem. Biophys.* **13:**49–51, 1976.

Raskin, N. H., and Sokoloff, L.: Ethanol induced adaptation of alcohol dehydrogenase activity in rat brain. *Nature* **236:**138–140, 1972.

Rawat, A. K., and Lundquist, F.: Influence of thyroxine on the metabolism of ethanol and glycerol in rat liver slices. *Eur. J. Biochem.* **5:**13–17, 1968.

Redman, C. M., Grab, D. J., and Irukunla, R.: The intracellular pathway of newly formed rat liver catalase. *Arch. Biochem. Biophys.* **151:**496–501, 1972.

Reed, T. E.: Racial comparisons of alcohol metabolism: Background, problems and results. *Alcoholism: Clin. Exp. Res.* **2:**83–87, 1978.

Reed, T. E., Kalant, H., Gibbins, R. J., Kapur, B. M., Rankin, J. G.: Alcohol and acetaldehyde metabolism in Caucasians, Chinese and Americans. *Can. Med. Assoc. J.* **115:**851–855, 1976.

Roach, M. K., Khan, M., Knapp, M., and Reese, W. N.: Ethanol metabolism *in vivo* and the role of hepatic microsomal ethanol oxidation. *Q. J. Stud. Alcohol* **33:**751–755, 1972.

Rodrigo, C., Anpezana, C., and Baraona, E.: Fat and nitrogen balances in rats with alcohol-induced fatty liver. *J. Nutr.* **101:** 1307–1310, 1971.

Rognstad, R., and Clark, D. G.: Tritium as a tracer for reducing equivalents in isolated liver cells. *Eur. J. Biochem.* **42:**51–60, 1974.

Rubin, E., Beattie, D. S., Toth, A., and Lieber, C. S.: Structural and functional effects of ethanol on hepatitic mitochondria. *Fed. Proc.* **31:**131–140, 1972.

Ruebner, B. H., Krieger, R. I., Miller, J. L., Isao, M., and Rorvik, M.: Hepatic and metabolic effects of ethanol on rhesus monkeys. In: *Advances in Experimental Medicine and Biology, Vol. 59*, (M. M. Gross, ed.), New York, Plenum, pp. 395–405, 1975.

Ryan, D. E., Ramanthan, L., Iida, S., Thomas, P. E., Haniu, M., Shively, J. E., Lieber, C. S., and Levin, W.: Characterization of a major form of rat hepatic microsomal cytochrome P450 induced by isoniazid. *J. Biol. Chem.* **260:**6385–6393, 1985.

Ryan, D. E., Koop, D. R., Thomas, P. E., Coon, M. J., and Levin, W.: Evidence that isoniazid and ethanol induce the same microsomal cytochrome P-450 isozyme 3a. *Arch. Biochem. Biophys.* **246:**633–644, 1986.

Salaspuro, M. P., and Kesaniemi, Y. A.: Intravenous galactose elimination tests with and without ethanol loading in various clinical conditions. *Scand. J. Gastroenterol.* **8:**681–686, 1973.

Salaspuro, M. P., and Lieber, C. S.: Non-uniformity of blood ethanol elimination: Its exaggeration after chronic consumption. *Ann. Clin. Res.* **10:**294–297, 1978.

Salaspuro, M. P., and Lieber, C. S.: Comparison of the detrimental effects of chronic alcohol intake in humans and animals. In: *Animal Models in Alcohol Research* (K. Eriksson, J. D. Sinclair and K. Klianmaa, eds.), London, Academic Press, pp. 359–376, 1980.

Salaspuro, M. P., Lindros, K. O., and Pikkarainen, P.: Ethanol and galactose metabolism as influenced by 4-methylpyrazole in alcoholics with and without nutritional deficiencies. Preliminary report of a new approach to pathogenesis and treatment in alcoholic liver disease. *Ann. Clin. Res.* **7:**269–272, 1975.

Salaspuro, M. P., Shaw S., Jayatilleke, E., Ross, W. A., and Lieber, C. S.: Attenuation of the ethanol induced hepatic redox change after chronic alcohol consumption: Mechanism and metabolic consequences. *Hepatology* **1:**33–38, 1981.

Sato, N., Kamada, T., Shichiri, M., Hayashi, M., Matsumura, T., Abe, H., and Hayihara, B.: The levels of the mitochondrial and microsomal cytochromes in drinkers' livers. *Clin. Chem. Acta* **87:**347–351, 1978.

Saville, P. D., and Lieber, C. S.: Effect of alcohol on growth bone density and muscle magnesium in the rat. *J. Nutr.* **87:**477–488, 1965.

Seiden, H., Israel, Y., and Kalant, H.: Activation of ethanol metabolism by 2,4-dinitrophenol in the isolated perfused rat liver. *Biochem. Pharmacol.* **23**:2334–2337, 1974.

Seitz, H. K., Garro, A. J., and Lieber, C. S.: Effect of chronic ethanol ingestion on intestinal metabolism and mutagenicity of benzo(a)pyrene. *Biochem. Biophys. Res. Commun.* **85**: 1061–1066, 1978.

Seitz, H. K., Veith, S., Czygan, P., Bosche, J., Simon, B., Gugler, R., and Kommerell, B.: *In vivo* interactions between H_2-receptor antagonists and ethanol metabolism in man and in rats. *Hepatology* **4**:1231–1234, 1984.

Selmer, J., and Grunnet, N.: Ethanol metabolism and lipid synthesis by isolated liver cells from fed rats. *Biochem. Biophys. Acta* **428**:123–137, 1976.

Sies, H.: Biochemistry of the peroxisome in the liver cell. *Angew. Chem.* [Engl.] **13**:706–718, 1974.

Shaw, S., Heller, E., Friedman, H., Baraona, E., and Lieber, C. S.: Increased hepatic oxygenation following ethanol administration in the baboon. *Proc. Soc. Exp. Biol. Med.* **156**:509–513, 1977.

Shigeta, Y., Nomura, F., Iida, S., Leo, M. A., Felder, M. R., and Lieber, C. S.: Ethanol metabolism *in vivo* by the microsomal ethanol oxidizing system in deermice lacking alcohol dehydrogenase (ADH). *Biochem. Pharmacol.* **33**:807–814, 1984.

Singlevich, T. E., and Barboriak, J. J.: Liver lipds after ethanol, chlorcyclizine and DDT. *Fed. Proc.* **30**:581 (abstract), 1971.

Smith, M. E.: Interrelations in ethanol and methanol metabolism. *J. Pharmacol.* **134**:233–237, 1961.

Smith, M. M., and Dawson, A. G.: Effect of triiodothyronine on alcohol dehydrogenase and aldehyde dehydrogenase activities in rat liver. Implication for the control of ethanol metabolism. *Biochem. Pharmacol.* **34**:2291–2296, 1985.

Smith, M., Hopkinson, D. A., and Harris, H.: Developmental changes and polymorphism in human alcohol dehydrogenase. *Ann. Hum. Genet.* **34**:251–271, 1971.

Smythe, C. McC., Heinemann, H. P., and Bradley, S. E.: Estimated hepatic blood flow in the dog: effect of ethyl alcohol on it, renal blood flow, cardiac output and arterial pressure. *Am. J. Physiol.* **172**:737–741, 1953.

Song, B.-J., Gelboin, H. V., Park, S.-S., Yang, C. S., and Gonzalez, F. J. Complementary DNA and protein sequences of ethanol-inducible rat and human cytochrome P-450s, *J. Biol. Chem.* **261**:16689–16697, 1986.

Sotaniemi, E., Isoaho, R., Huhti, E., Huikko, M., and Koivisto, O.: Increased clearance of ethanol from the blood of asthmatic patients. *Ann. Allergy* **30**:254–257, 1972.

Stamatoyannopoulos, G., Chen, S., and Fukui, M.: Liver alcohol-dehydrogenase in Japanese: High population frequency of atypical form and its possible role in alcohol sensitivity. *Am. J. Hum. Genet.* **27**:789–796, 1975.

Stein, W. S., Lieber, C. S., Leevy, C. M., Cherrick, G. R., and Abelmann, W. H.: The effect of ethanol upon systemic and hepatic blood flow in man. *Am. J. Clin. Nutr.* **13**:68–74, 1963.

Stokes, P. E., and Lasley, B.: Further studies on blood alcohol kinetics in man as affected by thyroid hormones, insulin and D-glucose. In *Biochemical Factors in Alcoholism* (R. P. Maickel, ed.), New York, Pergamon Press, pp. 101–114, 1967.

Sturtevant, F. M., Sturtevant, R. P., Scheving, L. E., and Pauly, J. E.: Chronopharmacokinetics of ethanol. II. Circadian

rhythm in rate of blood level decline in a single subject. *Naunyn-Schmiedebergs. Arch. Pharmakol.* **293**:203–208, 1976.

Takagi, T., Alderman, J., and Lieber, C. S.: *In vivo* roles of alcohl dehydrogenase (ADH), catalase and the microsomal ethanol oxidizing system (MEOS) in deermice. *Alcohol* **2**:9–12, 1985.

Takagi, T., Alderman, J., Gellert, J., and Lieber, C. S.: Assessment of the role of non-ADH ethanol oxidation *in vivo* and in hepatocytes from deermice. *Biochem. Pharmacol.* **35**:3601–3606, 1986.

Tephly, T. R., Tinelli, F., and Watkins, W. D.: Alcohol metabolism: Role of microsomal oxidation *in vivo. Science* **166**:627–628, 1969.

Teschke, R., Hasumura, Y., Joly, J.-G., Ishii, H., and Lieber, C. S.: Microsomal ethanol-oxidizing system (MEOS): Purification and properties of a rat liver system free of catalase and alcohol dehydrogenase. *Biochem. Biophys. Res. Commun.* **49**: 1187–1193, 1972.

Teschke, R., Hasumura, Y., and Lieber, C. S.: Hepatic microsomal ethanol oxidizing system: Solubilization, isolation and characterization. *Arch. Biochem. Biophys.* **163**:404–415, 1974a.

Teschke, R., Hasumura, Y., and Lieber, C. S.: NADPH dependent oxidation of methanol, ethanol, propanol and butanol by hepatic microsomes. *Biochem. Biophys. Res. Commun.* **60**: 851–857, 1974b.

Teschke, R., Hasumura, Y., and Lieber, C. S.: Hepatic microsomal alcohol oxidizing system. Affinity for methanol, ethanol and propanol. *J. Biol. Chem.* **250**:7397–7404, 1975a.

Teschke, R., Hasumura, Y., and Lieber, C. S.: Hepatic microsomal alcohol oxidizing system in normal and acatalsemic mice: Its dissociation from the peroxidatic activity of catalase-H_2O_2. *Mol. Pharmacol.* **11**:841–849, 1975b.

Teschke, R., Matsuzaki, S., Ohnishi, K., DeCarli, L. M., and Lieber, C. S.: Microsomal ethanol oxidizing system (MEOS): Current status of its characterization and its role. *Alcoholism: Clin. Exp. Res.* **1**:7–15, 1977.

Teschke, R., Wannagat, F-J., Löwendorf, F., and Strohmeyer, G.: Hepatic alcohol-metabolizing enzymes after prolonged administration of sex hormones and alcohol in female rats. *Biochem. Pharmacol.* **35**:521–527, 1986.

Theorell, H., and Chance, B.: Studies on liver alcohol dehydrogenase. II. The kinetics of the compounds of horse liver alcohol dehydrogenase and reduced diphosphopyridine nucleotide. *Acta Chem. Scand.* **5**:1127–1144, 1951.

Theorell, H., Nygaard, A. P., and Bonnichsen, R.: Studies on liver alcohol dehydrogenase. III. The influence of pH and some anions on the reaction velocity constants. *Acta Chem. Scand.* **9**:1148–1165, 1955.

Thieden, H. I. D.: The effect of ethanol concentration on ethanol oxidation rate in rat liver slices. *Acta Chem. Scand.* **25**:3421–3427, 1971.

Thieden, H. I. D., Grunnet, N., Damgaard, S. E., and Sestoft, L.: Effect of fructose and glyceraldehyde on ethanol metabolism in human liver and in rat liver. *Eur. J. Biochem.* **30**:250–261, 1972.

Thomasson, H. R., Crabb, D. W., Edenberg, H. J., Ching, C., Ostrofsky, Y. M., Li, T-K.: Allele frequencies for alcohol and aldehyde dehydrogenases in Hawaiians and Byelorussians. *Alcoholism: Clin. Exp. Res.* **15**:358, 1991 (abstract).

Thompson, G. N.: *Alcoholism.* Springfield, IL, Charles C. Thomas, 1956.

Thurman, R. G.: Induction of hepatic microsomal reduced nicotinamide adenine dinucleotide phosphate-dependent production of hydrogen peroxide by chronic prior treatment with ethanol. *Mol. Pharmacol.* **9**:670–675, 1973.

Thurman, R. G., and Brentzel, H. J.: The role of alcohol dehydrogenase in microsomal ethanol oxidation and the adaptive increase in ethanol metabolism due to chronic treatment with ethanol. *Alcoholism: Clin. Exp. Res.* **1**:33–38, 1977.

Thurman, R. G., Ley, H. G., and Scholz, R.: Hepatic microsomal ethanol oxidation. Hydrogen peroxide formation and the role of catalase. *Eur. J. Biochem.* **25**:420–430, 1972.

Thurman, R. G., McKenna, W. R., and McCaffrey, T. B.: Pathways responsible for the adaptive increase in ethanol utilization following chronic treatment with ethanol: Inhibitor studies with the hemoglobin-free perfused rat liver. *Mol. Pharmacol.* **12**:156–166, 1976.

Tobon, F., and Mezey, E.: Effect of ethanol administration on hepatic ethanol and drug-metabolizing enzymes and on rates of ethanol degradation. *J. Lab. Clin. Med.* **77**:110–121, 1971.

Trémolières, J., and Carré, L.: Etudes sur les modalitiés d'oxydation de l'alcool chez l'homme normal et alcoolique. *Rev. Alcool.* **7**:202–227, 1961.

Tzamaloukas, A. H., Jackson, J. E., Hermann, J. J., Gallegos, J. C., Long, D. A., and McLane, M.M.: Ethanol elimination in the anesthetized dog: Intragastric versus intravenous administration. *Alcohol* **5**:111–116, 1988.

Ugarte, G., Pino, M.E., and Insunza, I.: Hepatic alcohol dehydrogenase in alcoholic addicts with and without hepatic damage. *Am. J. Dig. Dis.* **12**:589–592, 1967.

Ugarte, G., and Persea, T.: Influence of hyperthyroidism on the rate of ethanol metabolism in man. *Nutr. Metab.* **22**:113–118, 1978.

Ugarte, G., Pereda, I., Pino, M. E., and Iturriaga, H.: Influence of alcohol intake, length of abstinence and meprobamate on the rate of ethanol metabolism in man. *Q. J. Stud. Alcohol* **33**:698–705, 1972.

Ullrich, V., Weber, P., and Wollenberg, P.: Tetrahydrofurante—an inhibitor for ethanol-induced liver microsomal cytochrome P-450. *Biochem. Biophys. Res. Commun.* **64**:808–813, 1975.

Umeno, M., Song, B. J., Kozak, C., Gelboin, H. V., and Gonzalez, F. J.: The rat P450IIE1 gene: Complete intron and exon sequence, chromosome mapping, and correlation of developmental expression with specific 5′ cytosine demethylation. *J. Biol. Chem.* **263**:4956–4962, 1988a.

Umeno, M., McBride, O. W., Yang, C. S., Gelboin, H. V., and Gonzalez, F.J.: Human ethanol-inducible P450IIE1: Complete gene sequence, promoter characterization, chromosomes mapping, and cDNA-directed expression. *Biochemistry* **27**:9006–9013, 1988b.

Utne, H. E., Vallo Hansen, F., Winkler, K., and Schulsinger, F.: Alcohol elimination rates in adoptees with and without alcoholic parents. *J. Stud. Alcohol* **38**:1219–1223, 1977.

Väänänen, H., Salaspuro, M., and Lindros, K.: The effect of chronic ethanol ingestion on ethanol metabolizing enzymes in

isolated periportal and perivenous rat hepatocytes. *Hepatology* **4**:862–866, 1984.

Vallee, B. L., and Bazzone, T. J.: Isozymes of human liver alcohol dehydrogenase. Isozymes. *Curr. Top. Biol. Med. Res.* **8**:219–244, 1983.

Van Thiel, D. H., Gavaler, J., and Lester, R.: Ethanol inhibition of vitamin A metabolism in the testes: Possible mechanisms for sterility in alcoholics. *Science* **186**:941–942, 1974.

Vatsis, K. P., and Schulman, M. P.: Absence of ethanol metabolism in "acatalatic" hepatic microsomes that oxidize drugs. *Biochem. Biophys. Res. Commun.* **52**:588–594, 1973.

Vatsis, K. P., and Schulman, M. P.: An unidentified constituent of ethanol oxidation in hepatic microsomes. *Fed. Proc.* **33**:554 (abstract), 1974.

Veech, R. L., Eggleston, L. V., and Krebs, H. A.: The redox state of free nicotinamide-adenine dinucleotide phosphate in the cytoplasm of rat liver. *Biochem. J.* **115**:609–619, 1969.

Veech, R. L., Guynn, R., and Veloso, D.: The time-course of the effects of ethanol on the redox and phosphorylation states of rat liver. *Biochem. J.* **127**:387–397, 1972.

Vessell, E. S., Page, J. G., and Passananti, G. T.: Genetic and environmental factors affecting ethanol metabolism in man. *Clin. Pharmacol. Ther.* **12**:192–201, 1971.

Videla, L., Bernstein, J., and Israel, Y.: Metabolic alterations produced in the liver by chronic ethanol administration. Increased oxidative capacity. *Biochem. J.* **134**:507–514, 1973.

Videla, L., Flattery, K. V., Sellers, E. A., and Israel, Y.: Ethanol metabolism and liver oxidative capacity in cold acclimation. *J. Pharmacol. Exp. Ther.* **192**:575–582, 1975.

Vind, C., and Grunnet, N.: Contribution of non-ADH pathways to ethanol oxidation in hepatocytes from fed and hyperthyroid rats. *Biochem. Pharmacol.* **34**:655–661, 1985.

von Wartburg, J. P., and Schurch, P. M.: Atypical human liver alcohol dehydrogenase. *Ann. N.Y. Acad. Sci.* **151**:936–946, 1968.

von Wartburg, J. P., Rothlisberger, M., and Eppenberger, M. H.: Hemmung der Athylalkoholoxydation durch Fusselöle. *Helv. Med. Acta* **28**:696–704, 1961.

von Wartburg J. P., Bethune, J. L., and Vallee, B. L.: Human liver-alcohol dehydrogenase. Kinetic and physiochemical properties. *Biochemistry* **3**:1775–1782, 1964.

von Wartburg, J. P., Papenberg, J., and Aebi, H.: An atypical human alcohol dehydrogenase. *Can. J. Biochem.* **43**:889–898, 1965.

Wagner, J. G.: Lack of first-pass metabolism of ethanol at blood concentrations in the social drinking range. *Life Sci.* **39**:407–414, 1986.

Wagner, J. G., Wilkinson, P. K., and Ganes, D. A.: Parameters V_m and K_m for elimination of alcohol in young male subjects following low doses of alcohol. *Alcohol Alcoholism* **24**:555–564, 1989.

Welling, P. G., and Tse, F. L.: Food interactions affecting the absorption of analgesic and antiinflammatory agents. *Drug Nutr. Interact.* **2**:153–168, 1983.

Wendell, G., and Thurman, R. G.: Effect of ethanol concentration on rates of ethanol elimination in normal and alcohol-treated rats *in vivo. Biochem. Pharmacol.* **28**:273–279, 1979.

Whereat, A. F., Orishimo, M. W., and Nelson, J.: The location of different synthetic systems for fatty acids in inner and outer mitochondrial membranes from rabbit heart. *J. Biol. Chem.* **244:**6498–6506, 1969.

Whitehouse, L. W., Paul, C. J., and Thomas, B. H.: Isoniazid induced tolerance to ethanol in rabbits, guinea pig and rat. *Biopharm. Drug Dispos.* **1:**235–245, 1980.

Williamson, J. R., Scholz, R., Browning, E. T., Thurman, R. G., and Fukami, M. H.: Metabolic effects of ethanol in perfused rat liver. *J. Biol. Chem.* **25:**5044–5054, 1969.

Wilson, R. H.L., Newman, E. J., and Newman, H. W.: Diurnal variation in rate of alcohol metabolism. *J. Appl. Physiol.* **8:** 556–558, 1956.

Wilson, J. S., Korsent, M. A., and Lieber, C. S.: The combined effects of protein deficiency and chronic ethanol administration on rat ethanol metabolism. *Hepatology* **6:**823–829, 1986.

Winkler, K., Lundquist, F., and Tygstrup, N.: The hepatic metabolism of ethanol in patients with cirrhosis of the liver. *Scand. J. Clin. Lab. Invest.* **23:**59–69, 1969.

Wolff, P. H.: Ethnic differences in alcohol sensitivity. *Science* **275:**449–450, 1972.

Wrighton, S. A., Thomas, P. E., Molowa, D. T., Haniu, M., Shively, J. E., Maines, S. L., Watkins, P. B., Parker, G., Mendez-Picon, G., Levin, W., and Guzelian, P. S.: Characterization of ethanol-inducible human liver N-nitrosodimethylamine dimethylase. *Biochemistry* **25:**6731–6735, 1986.

Wrighton, S. A., Thomas, P. E., Ryan, D. E., and Levin, W.: Purification and characterization of ethanol-inducible human hepatic cytochrome P-450Hlj. *Arch. Biochem. Biophys.* **258:** 292–297, 1987.

Yang, C. S., Tu, Y. Y., Koop, D. R., and Coon, M. J. Metabolism of nitrosamines by purified rabbit liver cytochrome P-450 isozymes. *Cancer Res.* **45:**1140–1145, 1985.

Ylikahri, R. H.: Ethanol-induced changes in hepatic α-glycerophosphate and triglyceride concentration in normal and thyroxine-treated rats. *Metabolism* **19:**1036–1045, 1970.

Ylikahri, R. Y., and Maenpaa, P. H.: Rate of ethanol metabolism in fed and starved rats after thyroxine treatment. *Acta Chem. Scand.* **22:**1707–1709, 1968.

Ylikahri, R. H., Hassinen, I. E., and Kahonen, M. T.: Metabolic interactions of fructose and ethanol in perfused liver of normal and thyroxine-treated rats. *Metabolism* **20:**555–567, 1971.

Yuki, T., and Thurman, R. G.: The swift increase in alcohol metabolism. Time course for the increase in hepatic oxygen uptake and the involvement of glycolysis. *Biochem. J.* **186:** 119–126, 1980.

Yuki, T., Thruman, R. G., Schwabe, U., and Schonz, R.: Metabolic changes after prior treatment with ethanol. Evidence against an involvement of the Na^+-K^+-activated ATPase in the increase in ethanol metabolism. *Biochem. J.* **186:**997–1000, 1980.

Ziegler, D. M.: Discussion in Microsomes and Drug Oxidation. In: *Microsomes and Drug Oxidations* (R. W. Estabrook, R. J. Gilhette, and K. C. Liebman, eds.), Baltimore, Williams & Wilkins, pp. 458–460, 1972.

Zorzano, A., Ruiz del Arbol L., and Herrera, E.: Effect of liver disorders on ethanol elimination and alcohol and aldehyde dehydrogenase activities in liver and erythrocytes. *Clin. Sci.* **76:**51–57, 1989.

2

Acetaldehyde and Acetate

Charles S. Lieber

All known pathways of ethanol oxidation in the liver result in production of acetaldehyde, which in turn is metabolized to acetate.

2.1. Pathways of Acetaldehyde and Acetate Metabolism

It is now generally accepted that normally more than 90% of the acetaldehyde formed by the liver is oxidized in that organ (Lindros, 1974). Several enzyme systems responsible for acetaldehyde oxidation have been described. Racker (1949) was the first to report the presence of an NAD-linked aldehyde dehydrogenase (ALDH) in bovine liver. Subsequent investigators detected aldehyde dehydrogenase activity in mitochondria (Walkenstein and Weinhouse, 1953), cytoplasm (Büttner, 1965; Deitrich, 1966), and microsomes (Tietz *et al.*, 1964; Dietrich, 1966; Tottmar *et al.*, 1973) of rat liver. The various enzymes have been characterized primarily by their differential affinities for the aldehyde substrate (Lundquist *et al.*, 1962; Grunnet, 1973; Marjanen, 1973; Koivula and Koivusalo, 1975; Tottmar *et al.*, 1973).

More than one form of ALDH exists in both the cytoplasm and the mitochondria of most species. At least 17 distinct bands possessing NAD-dependent aldehyde-oxidizing activity have been observed with rat liver homogenates following isoelectric focusing (Weiner *et al.*, 1974). Smith and Packer (1972) and Tottmar *et al.* (1973) obtained evidence, using kinetic analysis, for two mitochondrial enzymes. One of these isozymes, designated enzyme I, had a K_m for acetaldehyde of approximately 10 μM; the other isozyme, enzyme II, had a K_m for acetaldehyde of 0.9 to 1.7 mM. In addition, these investigators noted that enzyme II could also function with NADP as a cofactor. Furthermore, evidence was obtained suggesting that enzyme I is confined to the mitochondrial matrix, whereas enzyme II is most likely in the intermembrane space (Siew *et al.*, 1976). Understanding of the number and properties of cytosolic ALDH isozymes has been facilitated by the use of various drugs that induce specific isozymes of ALDH. Redmond and Cohen (1971) noted that phenobarbital treatment results in a twofold increase in ALDH activity in a 700 *g* supernatant fraction of mouse liver homogenates. Deitrich (1971) studied the phenomenon more extensively in rats and observed that a tenfold induction is seen in the cytosol, but only in certain strains of rats. Mitochondrial ALDH activity is unaffected by phenobarbital treatment in these animals. The induction is controlled by a single gene displaying incomplete dominance. Subsequent studies of phenobarbital induction (Deitrich *et al.*, 1972) appeared to support the notion that the cytosol contains more than one ALDH isozyme. One is induced by phenobarbital, and the other is not; however, each exhibits different substrate specificity. More recent reports (Lindahl and Evces, 1984a,b) indicate that, apart from mitochondrial ALDH released as an artifact of subcellular fractionation procedures, there may not be a true constitutive cytosolic isozyme in the rat. Moreover, the phenobarbital-inducible activity in the cytosol is distinct from any normal form but does appear identical to an isozyme appearing during the promotion phase of hepatocarcinogenesis. There is now evidence for another inducible cytosolic ALDH isozyme, which is increased up to 100-fold after exposure to 2,3,7,8-tetrachlorodibenzo-p-dioxin (TCDD). Unlike the phenobarbital enzyme, induction occurs in all known rat strains (Roper $et\ al.$, 1976).

Number and properties of ALDH isozymes vary among species. In the horse liver, two isozymes have been purified, one with a high K_m for acetaldehyde and one with a low K_m; the latter was far more sensitive to the inhibitor effects of disulfiram (Eckfeldt and Yonetani, 1976). Multiple isozymes were also found in human liver (Koivula, 1975). It is further apparent that subcellular distribution of isozymes varies among species (Lindahl and Evces, 1984b), although the majority of low-K_m activity is invariably mitochondrial (Grunnet, 1973; Parrilla $et\ al.$, 1974; Corall $et\ al.$, 1976). Because acetaldehyde levels in the liver are very low following ethanol administration (Eriksson, 1973; Kesäniemi, 1974), Grunnet (1973) has suggested that the major portion of acetaldehyde oxidation within the mitochondria occurs in the matrix. This is logical, since enzyme I, which is found in the matrix, has a much lower K_m for acetaldehyde than does enzyme II or any of the cytoplasmic isozymes. These latter could play a minor role in acetaldehyde metabolism, as suggested by the observation that induction of some of these enzymes by phenobarbital in some susceptible strains of rats may be associated with a lowering of blood acetaldehyde (Petersen $et\ al.$, 1977). No such response, however, was observed by Eriksson $et\ al.$ (1975). On incubation of isolated hepatocytes with tritiated ethanol, it was calculated that 20% of acetaldehyde formed from ethanol is metabolized in the cytosol (Grunnet $et\ al.$, 1976).

Whether the results obtained in rodents and the derived conclusions apply to other species (including humans) is still the subject of debate. Only limited studies have been carried out in primates, primarily in human autopsy samples (Kraemer and Deitrich, 1968; Blair and Bodley, 1969; Greenfield and Pietruszko, 1977; Harada $et\ al.$, 1980). These studies yielded conflicting results and suggested the existence of one (Kraemer and Deitrich, 1968; Blair and Bodley, 1969), two (Greenfield and Pietruszko, 1977), or four (Harada $et\ al.$, 1980) forms of ALDH in human liver. More recent investigations have clarified this situation to some degree, indicating that two principal hepatic isozymes exist [originally designated E_1 for the low-K_m ($<10\ \mu$M) and E_2 for the high-K_m (100 μM) forms by Greenfield and Pietruszko, 1977)]. These correspond to the isozymes II and I, respectively, identified by Harada $et\ al.$

(1980). The latter isozyme is apparently similar to, and possibly identical with, ALDH found in erythrocytes. By immunostaining, these two main enzymes were localized to the mitochondria and cytosol, respectively (Maeda $et\ al.$, 1988). Two additional isozymes, III and IV, reported by this group have K_m values for acetaldehyde in the millimolar range and are probably not important in acetaldehyde catabolism. An active site for the human enzyme has been identified (Abriola $et\ al.$, 1987). In those studies employing biopsy specimens, the subcellular locus of human acetaldehyde metabolism has been variously reported to be cytosolic (Thomas $et\ al.$, 1982; Matthewson $et\ al.$, 1986), with about half the low-K_m ALDH activity in cytosol (Takada $et\ al.$, 1984), or mitochondrial, as in rodents (Tipton and Henehan, 1984). Among subhuman primates, baboon liver has detectable ALDH activity in the mitochondrial, microsomal, and soluble fractions; each fraction appears to contain two forms of ALDH, one with a high and another with a low K_m for acetaldehyde. In the microsomes, there is activity only with millimolar concentrations of acetaldehyde. In the baboon liver, more than 60% of total and 80% of the low-K_m ALDH activity resides in the mitochondria (Lebsack $et\ al.$, 1981). In the human liver, ALDH$_1$ and ALDH$_2$ are distributed equally in the periportal and centrilobular regions (Maeda $et\ al.$, 1988).

Hasumura $et\ al.$ (1975, 1976) found that isolated, intact rat liver mitochondria are capable of oxidizing acetaldehyde at the rate of approximately 15 nmol/min per mg protein at 30°C with 180 μM acetaldehyde as substrate. Concomitantly, the mitochondria consumed oxygen at the same rate. This reaction was markedly affected by several modifiers of the mitochondrial respiratory chain. ADP increased the rate of acetaldehyde oxidation by 88%. Also significant is the stimulation of oxidation observed on addition of 2,4-dinitrophenol, an uncoupler of the mitochondrial respiratory chain. A similar stimulatory effect of dinitrophenol on acetaldehyde oxidation was found in perfused livers (Eriksson, 1973). By contrast, rotenone, an inhibitor of mitochondrial respiration at site 1, virtually abolished the rate of acetaldehyde oxidation. As reported for isolated hepatocytes (Parrilla $et\ al.$, 1974), acetaldehyde oxidation in the isolated mitochondria was also diminished by the addition of substrates for mitochondrial NAD-linked dehydrogenases. The results clearly indicate that the oxidation of acetaldehyde in liver mitochondria is linked to the mitochondrial respiratory chain at the site of NAD-linked dehydrogenases. Since the ADP-to-O ratio with 180 μM acetaldehyde was 2.6, a ratio similar to that with glutamate as substrate, it is also likely that acetaldehyde oxidation is coupled with oxidative phosphorylation.

Dietary factors may also alter the rate of oxidation of acetaldehyde. Indeed, it has been noted that in some groups of rats, alcohol resulted in much higher acetaldehyde levels than ordinarily observed (Lindros et al., 1975; Marchner and Tottmar, 1976a). This puzzling finding has now been resolved. It was found that the activity of the mitochondrial enzyme was two to three times lower in rats fed the commercial diet in question, which explains why the acetaldehyde blood levels were two to four times higher in rats fed this diet. The dietary factor was shown to be located in the calcinated bone-meal fraction of the diet and was identified as cyanamide (Marchner and Tottmar, 1976b; Pettersson and Jenkins, 1977), which is known to inhibit aldehyde dehydrogenase. Low-protein diet can also reduce acetaldehyde dehydrogenase activity (Lindros et al., 1977), but it does not necessarily increase blood acetaldehyde (Lindros et al., 1979). Agents that trap acetaldehyde, such as penicillamine, can reduce its blood level (Nagasawa et al., 1977). A number of commonly used drugs can similarly increase blood acetaldehyde via ALDH inhibition. Among these are some hypoglycemic agents (Johnston et al., 1986) and cephalosporin antibiotics containing a l-methyltetrazole-5-thio side chain (Kitson, 1986). Intermediate- and high-K_m ALDH isozymes are sensitive to irreversible inhibition by disulfiram (Deitrich and Erwin, 1971; Kitson, 1975; Vallari and Pietruszko, 1982), whereas low-K_m isozymes are most strongly inhibited by disulfiram metabolites (Harada et al., 1982). This combination of properties makes disulfiram a potentially useful tool for maintenance of sobriety in motivated individuals (Sereny et al., 1986). The principal, though not the only, aversive component in the disulfiram–alcohol reaction appears to be elevated plasma acetaldehyde levels (Beyeler et al., 1985).

The exact fate of acetaldehyde is still the subject of debate. That acetyl-CoA is formed from acetaldehyde is indicated by the observation that [^{14}C]ethanol can be traced to a variety of metabolites of which acetyl-CoA is a precursor, such as fatty acids and cholesterol, as reviewed elsewhere (Lieber, 1967). It is noteworthy that a large fraction of the carbon skeleton of ethanol is incorporated in hepatic lipids after ethanol administration (Scheig, 1971; Brunengraber et al., 1974). The acetaldehyde that results from the oxidation of ethanol could be converted to acetyl-CoA via acetate. The reverse possibility, that acetaldehyde is converted directly to acetyl-CoA, which in turn could be either

incorporated into various metabolites or yield acetate, has not been ruled out. In any event, acetate has been found to increase markedly in the blood after ethanol administration (Lundquist et al., 1962; Crouse et al., 1968). Although in vitro the liver can readily utilize acetate, in vivo most of the acetate is metabolized in peripheral tissues (Katz and Chaikoff, 1955), particularly in the presence of ethanol, which inhibits the oxidation of acetate in the liver (Lindros, 1970).

2.2. Effect of Chronic Ethanol Consumption on Acetaldehyde and Acetate Metabolism

2.2.1. Alterations in Acetaldehyde Metabolism

Chronic ethanol consumption results in a significant reduction of the capacity of rat mitochondria to oxidize acetaldehyde, regardless of the presence of substrates for NAD-linked dehydrogenase (Hasumura et al., 1975). Since prolonged intake of ethanol depresses the oxidation of some NAD-dependent substrates (Cederbaum et al., 1974a), the observed reduction of acetaldehyde metabolism can be ascribed, at least in part, to the decreased ability of NADH reoxidation in mitochondria of ethanol-fed animals (Hasumura et al., 1975). Indeed, Koivula and Lindros (1975) and Hasumura et al. (1976) found that the activity of the NAD-dependent aldehyde dehydrogenase, characterized by high affinity for acetaldehyde, remained unchanged following chronic ethanol consumption, whereas others found it increased (Horton and Barrett, 1976; Greenfield et al., 1976) or even decreased (Lebsack et al., 1981). More recent reports increasingly indicate that ethanol depresses functionally important ALDH activity (Palmer and Jenkins, 1985), as do other hepatotoxins such as carbon tetrachloride (Hjelle and Petersen, 1981) and malondialdehyde (Hjelle et al., 1982). Inhibition by the latter, a frequently studied product of lipid peroxidation, may be important, since both ethanol and acetaldehyde are thought to stimulate hepatic lipid peroxidation. Discrepancies between the decreased rate of acetaldehyde oxidation in intact mitochondria and the inconsistently affected enzyme activity in disrupted organelles suggest that the rate-limiting step of acetaldehyde oxidation in hepatic mitochondria may be the mitochondrial capacity to reoxidize NADH rather than the aldehyde dehydrogenase activity itself (Hasumura et al., 1975). In other studies, the low-K_m aldehyde dehydrogenase was decreased after ethanol intake in some strains of rats (Koivula and Lindros, 1975). Furthermore, chronic ethanol consumption also decreased total and low-K_m mitochondrial ALDH activity in baboon liver (Lebsack et al., 1981). Liver disease, whether alcoholic or not, was associated with a decrease of hepatic aldehyde dehydrogenase activity (Panes et al., 1989).

The decreased capacity of mitochondria of alcohol-fed subjects to oxidize acetaldehyde, associated with unaltered or even enhanced rates of ethanol oxidation (and therefore acetaldehyde generation), may result in an imbalance between production and disposition of acetaldehyde. Such a mechanism may result in the elevated acetaldehyde levels observed after chronic ethanol consumption in rats (Koivula and Lindros, 1975), in humans (Korsten et al., 1975; DiPadova et al., 1987), and in baboons (Pikkarainen et al., 1981a). In the baboon, plasma free acetaldehyde correlated positively with the production rate of acetaldehyde ($r = 0.69$, $p < 0.001$) and negatively with liver mitochondrial ALDH activity ($r = 0.59$, $p < 0.01$). When these determinants of blood acetaldehyde were combined (acetaldehyde production rate/ALDH activity), a correlation coefficient of 0.81 ($p < 0.001$) resulted, suggesting that in addition to increased production, decreased catabolism may contribute to the higher acetaldehyde levels seen after chronic ethanol consumption (Pikkarainen et al., 1981a). One attractive hypothesis that might explain, at least in part, the increased level of acetaldehyde is a relative shift of the oxidation of acetaldehyde

from the low-K_m mitochondrial pathway to the high-K_m cytosolic one. Such a shift might be anticipated in view of the dependence of the mitochondrial pathway on mitochondrial integrity and the known alterations of mitochondrial functions after chronic ethanol consumption (see Chapter 7), including decreased acetaldehyde catabolism. If a greater proportion is processed via nonmitochondrial systems, one would expect a higher steady-state concentration of acetaldehyde to be required to maintain the same total flux of acetaldehyde in view of the much higher K_m of the nonmitochondrial systems. Enhancement of the effect might be expected if the total flux of acetaldehyde is increased (because of accelerated ethanol metabolism).

The higher liver acetaldehyde level might be expected to result in a higher hepatic vein acetaldehyde level. This indeed has been documented in baboons fed alcohol chronically and then challenged with ethanol acutely (Lieber et al., 1989) (Fig. 2.1). The latter in turn would elevate the peripheral blood acetaldehyde level. The blood acetaldehyde level also depends on peripheral metabolism, which can be altered (Lamboeuf et al., 1980). In any event, evidence of changes in blood acetaldehyde levels has been slow in coming. Truitt (1971) studied acetaldehyde levels after oral administration of ethanol to alcoholics and nonalcoholics but found no statistically significant differences. Freund and O'Hollaren (1965) found a plateau of acetaldehyde concentration in human alveolar air after ethanol administration, but the relationship to ethanol metabolism was not clarified. Majchrowicz and Mendelson (1970) also

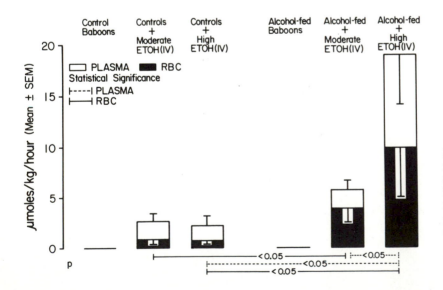

FIGURE 2.1. Acute and chronic effects of ethanol on splanchnic acetaldehyde output in the plasma and red blood cells (RBC). The increased release of both red blood cell and plasma acetaldehyde after an acute dose of ethanol was much greater in alcohol-fed baboons than in controls, particularly at the high ethanol dose. (From Lieber et al., 1989.)

found elevated blood acetaldehyde levels in alcoholics. However, because of lack of relationship between blood ethanol and acetaldehyde concentrations, they related this finding to the amount of acetaldehyde contained as a congener in alcoholic beverages. Magrinat *et al.* (1973) reached a similar conclusion.

Korsten *et al.* (1975) found a relationship between blood acetaldehyde and ethanol levels in humans after intravenous administration of ethanol that yielded concentrations high enough to saturate the intrahepatic ethanol-oxidizing systems, both ADH and MEOS. This study revealed that the acetaldehyde level remained relatively constant despite wide variations in blood ethanol above 150 mg/100 ml, but the acetaldehyde plateau abruptly terminated when ethanol concentration reached about 100 mg/100 ml. Moreover, the plateau level of acetaldehyde was significantly higher in alcoholics than in nonalcoholics. This difference in blood acetaldehyde between alcoholics and nonalcoholics was considered to be a consequence of chronic ethanol consumption. A more recent study using improved methodology for acetaldehyde measurement has confirmed this difference (DiPadova *et al.*, 1986, 1987), although at absolute acetaldehyde blood levels less than originally reported.

In addition to the above mentioned metabolic causes for the elevation of blood acetaldehyde in the alcoholic, a genetic influence has been postulated (Schuckit and Rayses, 1979). After a test dose of alcohol, blood acetaldehyde levels were significantly higher in relatives of alcoholics than in matched controls. The methodology of acetaldehyde measurement has been questioned (Eriksson, 1980a), but the study illustrates an actively debated issue, namely, whether a primary defect in ALDH plays a role in alcoholism. With such a defect, the resulting moderately elevated acetaldehyde levels after drinking could lead to inhibition of dehydrogenation of biogenic aldehydes (MacKerell *et al.*, 1986) or formation of addictive condensation products with biogenic amines (Myers, 1978). Consistent with this, Thomas *et al.* (1982) reported persistently decreased cytosolic ALDH after cessation of drinking and concurrent with the return of other hepatic cytosolic enzymes to control levels. More recent studies, however, have consistently indicated that depressed hepatic ALDH activity follows excessive drinking (Nuutinen *et al.*, 1983; Jenkins *et al.*, 1984). These studies are consistent with decreased ALDH activities seen following chronic alcohol consumption in subhuman primates and rats (Lebsack *et al.*, 1981). There are also indications that, besides quantitative changes in activity, qualitative alterations in ALDH isozymes can occur—baboons fed alcohol were found to possess extra cytosolic isozymes with a changed disulfiram resistance (Alderman *et al.*, 1982, 1985). A concomitant decrease in red cell ALDH of alcoholics has also been reported (Lin *et al.*, 1984; Maring *et al.*, 1983; Towell *et al.*, 1986). After withdrawal from alcohol, the red cell ALDH returns toward normal (Lin *et al.*, 1984).

2.2.2. Acetaldehyde Determinations in Blood and Expired Air

Previous methods for acetaldehyde determinations were plagued by two problems: rapid disappearance of acetaldehyde from the bloodstream, which tends to lower the increased level, and artifactual production of acetaldehyde from ethanol (Eriksson, 1980b). When proteins are precipitated from the blood, there is a major artifactual rise of acetaldehyde, particularly in the presence of red blood cells. Recent methodological advances partially alleviate these two problems.

The disappearance of acetaldehyde from whole blood could represent binding to red blood cells (Gaines *et al.*, 1977) or metabolism of acetaldehyde by red blood cells or plasma. It has been shown that acetaldehyde in the rat is unevenly distributed between blood cells and plasma, with about 75% in the cells (Eriksson *et al.*, 1977; Tottmar *et al.*, 1978). In human blood, although some found no binding to red blood cells (Eriksson *et al.*, 1977), some report considerable acetaldehyde in the red cells, particularly of the alcoholic (Baraona *et al.*, 1987a) (Fig. 2.2). Aldehyde dehydrogenase activity has also been found in erythrocytes (Inoue *et al.*, 1978, 1979; Pietruszko and Vallari, 1978), and this could be responsible for some acetaldehyde disappearance, at least during and immediately after the withdrawal of the blood, while it is still at 37°C. In addition, reversible binding of acetaldehyde to red cells can be an important mediator of systemic toxicity of acetaldehyde. It was found that acetaldehyde could be transported by red cells to extrahepatic tissues in amounts much larger than previously appreciated from plasma measurements. Furthermore, the magnitude of this transport is markedly enhanced in alcoholics (Baraona *et al.*, 1987a,b) (Fig. 2.2). Such an investigation was made possible by development of a method permitting acetaldehyde measurement in red cells with good recovery and small artifactual formation (DiPadova *et al.*, 1986).

The values of blood acetaldehyde thus found are far below the levels in the blood that have been found previously (Truitt, 1971; Kinoshita, 1974; Korsten *et al.*, 1975) and are similar to those found in cerebrospinal

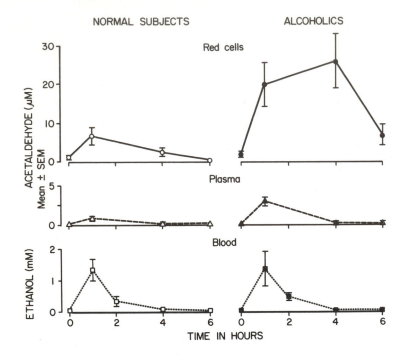

NORMAL SUBJECTS   ALCOHOLICS

FIGURE 1.2. Blood acetaldehyde after alcohol administration. The levels increased mainly in red blood cells, with much smaller changes in plasma acetaldehyde. The largest increase was observed in the red blood cells of the alcoholics. (From Baraona *et al.*, 1987a.)

fluid (Lindros and Hillbom, 1979). The high levels with the previous methods can be attributed, in part, to the fact that with these methods artifactual formation of acetaldehyde from ethanol was not fully eliminated or not totally compensated for by the correction factors used. It is noteworthy that even with improved procedures, although the absolute levels were low, an elevation of plasma acetaldehyde was found after chronic ethanol consumption (Pikkarainen *et al.*, 1979, 1981a; Lindros *et al.*, 1980; DiPadova *et al.*, 1987).

In view of the difficulties related to blood acetaldehyde measurement, in some studies, breath acetaldehyde has been used as an index of blood acetaldehyde (Freund and O'Hollaren, 1965; Fukui, 1969; Zeiner *et al.*, 1979). In order to study to what extent this is justified, Jauhonen *et al.* (1982) measured ethanol and acetaldehyde concentrations in right atrium blood of catheterized baboons (while awake) and in breath samples collected through an endotracheal tube or a facial mask. In the tube samples, breath acetaldehyde was 16% of that in the mask (2.4 ± 0.7 and 15.2 ± 3.7 nM, respectively), whereas ethanol was 77% higher in the tube than in the mask samples. Tube acetaldehyde levels correlated between free acetaldehyde in pulmonary blood and alveoli (148 ± 9) closely resembled the ones obtained *in vitro* between plasma and air (147 ± 8) at 37°C. A partition factor of 211 ± 13 was found between total blood acetaldehyde and alveolar air acetaldehyde.

This factor did not differ significantly from the value of 190 used by other authors (Stowell *et al.*, 1980). Unlike alveolar air collected through the endotracheal tube, mask acetaldehyde did not correlate with blood acetaldehyde levels and resulted in a great overestimation. Similarly, in human volunteers given ethanol, acetaldehyde in end-expiratory samples yielded a two- to sixfold overestimation of blood acetaldehyde concentrations, particularly in smokers; this overestimation may be explained by the active oxidation of ethanol to acetaldehyde by oropharyngeal microorganisms (Pikkarainen *et al.*, 1981b); this process increases the acetaldehyde concentration in areas through which end-expiratory air passes in the sampling process (Lieber *et al.*, 1983) (Figs. 2.3 and 2.4). Furthermore, rat lung microsomes produce acetaldehyde from ethanol, an effect enhanced by chronic alcohol consumption (Pikkarainen *et al.*, 1981b).

Thus, breath acetaldehyde originates at least in part from microbial oxidation of ethanol in the oropharynx. This contribution can be only slightly diminished by taking end-expiratory samples. The overestimation is almost completely abolished by rinsing the mouth with the ADH inhibitor pyrazole, at least in nonsmokers; it was much less inhibited in smokers (Jauhonen *et al.*, 1982). It is tempting to speculate that smoking increases lung acetaldehyde production by inducing microsomal ethanol oxidation, but direct evidence for this is still

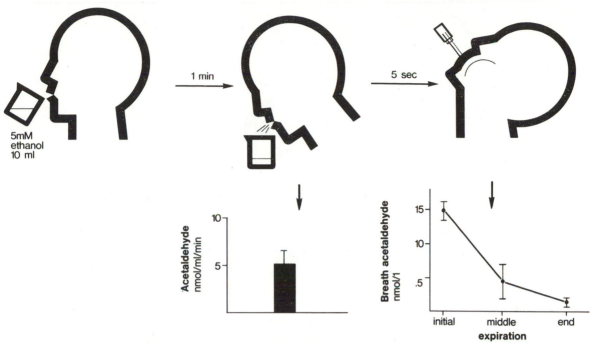

FIGURE 2.3. *In vivo* formation of acetaldehyde from ethanol in the oral cavity. Five volunteers rinsed their mouths with 10 ml of 5 mM ethanol solution in saline for 1 min. The solution was then collected into an ice-cold beaker containing 1 ml of 6 N perchloric acid. The acetaldehyde produced was measured by gas chromatography in 1.0-ml aliquots. Immediately after the mouth rinsing, breath samples (initial, middle, and end-expiratory) were taken. (From Lieber *et al.*, 1983.)

lacking. Increased acetaldehyde production because of bacterial overgrowth in smokers is also contributory (Miyakawa *et al.*, 1986). For all these reasons, except when blood acetaldehyde levels are very high, calculations based on measurements of total breath acetaldehyde usually overestimate blood acetaldehyde concentrations. These limitations of breath acetaldehyde as an index of blood levels have been confirmed in another investigation (Stowell *et al.*, 1984).

Whether chronic ethanol consumption alters acetate metabolism and blood levels has not been extensively studied. It has been reported that chronic alcohol consumption is associated with elevated blood acetate levels and may serve as a marker of drinking behavior (see Chapter 18).

2.3. Effect of Liver Injury on Acetaldehyde Metabolism

Mitochondrial alterations common to most forms of liver injury could decrease the capacity of this organelle to oxidize acetaldehyde derived from alcoholic and nonalcoholic sources. Indeed, impaired acetaldehyde metabolism in patients with nonalcoholic liver disorders has been described by Matthewson *et al.* (1986). Experimentally, chronic liver injury was found to be associated with increased levels of acetaldehyde derived from physiological sources, including the degradation of threonine (Ma *et al.*, 1989). Moreover, liver injury exaggerated the accumulation of acetaldehyde during alcohol consumption. The accumulation resulted primarily from a decreased capacity of the hepatic mitochondria to oxidize this metabolite. Mitochondrial injury is frequent in liver disease produced by a variety of agents, and the integrity of this organelle is required for the oxidation of acetaldehyde (Hasumura *et al.*, 1976). Such alterations were also documented in experimental models, and they included a significant decrease in the *in vitro* capacity to oxidize acetaldehyde (Ma *et al.*, 1989). Corresponding changes may be responsible for the accumulation of acetaldehyde regardless of the source. In addition, liver injury decreases the enzyme activities involved in the major degradative pathways of threonine, thereby possibly increasing the availability of this amino acid for the minor aldolase pathway, which results in production of acetaldehyde (Ma *et al.*, 1989).

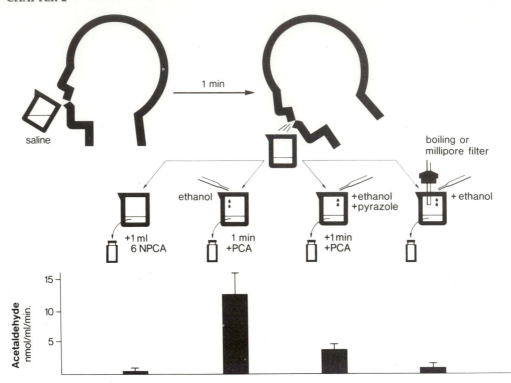

FIGURE 2.4. *In vitro* formation of acetaldehyde from ethanol in saline mouthwashes. Five volunteers kept saline in their mouths for 1 min. The reaction was stopped by addition of perchloric acid, and acetaldehyde was measured in 1 ml aliquots by gas chromatography. Acetaldehyde was produced from ethanol, and this was inhibited by 76% in the presence of 1 mM pyrazole and inhibited completely by prior boiling (for 10 min) or by Millipore filtration. (From Lieber *et al.*, 1983.)

These increased endogenous levels of acetaldehyde provide a likely explanation for the finding of increased titers of antibodies against acetaldehyde–protein adducts in patients with nonalcoholic liver injury (Hoerner *et al.*, 1988). In any event, the decreased capacity to remove acetaldehyde could potentiate the acetaldehyde-mediated toxicity of ethanol in patients with nonalcoholic liver disease. This may provide a rationale for the hitherto empirical recommendation made to these patients to avoid the use of alcohol.

2.4. Effects Attributable to Acetaldehyde and Acetate after Acute and Chronic Ethanol Consumption

The metabolic effects of acetaldehyde are still the subject of debate even though it has been speculated for a long time that this compound may contribute to the complications of alcoholism (Truitt and Duritz, 1966) and to the numerous toxic effects noted in this disease (Walsh, 1971). Acetaldehyde stimulates the release of catecholamines (Eade, 1959) and of prostacyclin (Guivernau *et al.*, 1987); these, as well as some direct effects of acetaldehyde on blood vessels, might be responsible for the flushing of the skin sometimes observed after alcohol ingestion (see also Chapter 4). In some Orientals, even small amounts of alcohol that have almost no effect on Caucasians can produce a rapid facial flush, frequently associated with tachycardia, headache, and nausea (Wolff, 1973). The propensity for flushing appears to be genetically determined and caused by an elevation of blood acetaldehyde secondary to a decrease in disposition secondary to the lack of a low-K_m aldehyde dehydrogenase activity (Goedde *et al.*, 1979; Harada *et al.*, 1979; Ijiri, 1974; Teng, 1981). It has been demonstrated that individuals who exhibit the flushing reaction possess an enzymatically inactive material that is immunologically identical with active E_2 (Ikawa *et* al., 1983; Impraim et al., 1982; Yoshida et al., 1984), indicating that a structural mutation has occurred and that the inactive allele is dominant (Crabb et al., 1989).

A study of ALDH isozymes among four ethnically different subpopulations in China revealed an E_2 deficiency among 25 to 50% (Goedde et al., 1984). In the Japanese population, the E_2 defect occurs in about half of the individuals (Agarwal et al., 1981). When E_2 is inactive, the major cytosolic isozyme (E_1) is responsible for acetaldehyde oxidation, and its activity probably determines individual variation in acetaldehyde-mediated responses (Inoue et al., 1984). Unlike Orientals, ALDH phenotypes among American Indians living in New Mexico were similar to those of Caucasian populations (Rex et al., 1985). Furthermore, Inoue et al. (1980) have shown an inverse relationship between blood acetaldehyde and the activity of ALDH in erythrocytes. In any event, among Orientals with flushing, there was a tendency toward higher blood acetaldehyde levels (Mizoi et al., 1979), but the difference was not statistically significant in all studies (Ewing et al., 1974). Ojibwa Indians were also found to have higher blood acetaldehyde levels than Caucasians. Chinese did not have higher blood acetaldehyde levels in one study (Reed et al., 1976), but breath acetaldehyde was increased in another (Zeiner et al., 1979). The aversive cardiovascular effects of acetaldehyde may contribute to the relatively lower incidence of cirrhosis in "flushers" (Yoshihara et al., 1983). The flushing phenotype may confer some resistance to the development of alcoholism (Harada et al., 1983). In fact, the flushing reaction seen in susceptible Orientals mimics to a lesser degree the disulfiram reaction caused by the elevation of acetaldehyde following aldehyde dehydrogenase inhibition. This reaction has gained widespread therapeutic use as a reinforcement for abstinence in alcoholism rehabilitation programs.

The metabolism of some amines (such as serotonin and norepinephrine) shifts from conversion to the acid to the production of the reduced (alcoholic) compound, partly because of the increased NADH-to-NAD ratio. The major reason for this change, however, appears to be the competitive inhibition of the corresponding aldehyde dehydrogenase by acetaldehyde (MacKerell et al., 1986). In addition, acetaldehyde has been shown to participate in and to favor condensation reactions of biogenic amines (Cohen and Collins, 1970; Davis and Walsh, 1970). The products of these interactions have been shown to appear in the urine of alcoholics (Collins et al., 1979). It has been postulated that these products have addictive properties if sufficient amounts are generated in vivo, as discussed in more detail in Chapter 14. Furthermore, some of the condensation products may be hepatotoxic (Moura et al., 1977). Alterations in CNS amine levels have also been reported in experimental animals given acetaldehyde to achieve concentrations comparable to those noted after intravenous alcohol administration. Moreover, withdrawal symptoms were observed when acetaldehyde administration was discontinued. These symptoms were similar to those that occur during alcohol withdrawal (Ortiz et al., 1974). On the other hand, Amir (1978) noted that brain aldehyde dehydrogenase increased following prolonged ethanol administration; this would tend to reduce the brain acetaldehyde level after chronic ethanol consumption.

Acetaldehyde can affect many tissues. It appears to interfere with pyridoxal phosphate metabolism and to displace pyridoxal phosphate from its protein binding, thereby promoting degradation of vitamin B_6, which may contribute to the low plasma levels of this vitamin in alcoholics (Veitch et al., 1975). Another effect of acetaldehyde is impaired myocardial protein synthesis (Schreiber et al., 1972, 1974); this effect might contribute to the development of alcoholic cardiomyopathy. One must add to this list stimulation of collagen synthesis, which has been reported in cultured baboon liver myofibroblasts (Savolainen et al., 1984) and Ito cells (Moshage et al., 1990).

The most startling changes caused by acetaldehyde involve the liver (see Chapter 7). Other hepatotoxic agents have been shown to act through an active metabolite that promotes lipid peroxidation or that covalently binds to proteins. Acetaldehyde, the active metabolite of ethanol, may do both. Enhanced lipid peroxidation by acetaldehyde has been demonstrated in isolated, perfused liver (Muller and Sies, 1982). Acetaldehyde was also found to "bind" to hepatic plasma membranes (Barry and McGivan, 1985) and to membranes of the endoplasmic reticulum (Nomura and Lieber, 1981) and specifically to the ethanol-inducible cytochrome P450IIE1 (Behrens et al., 1988). This binding was increased after chronic ethanol consumption (Nomura and Lieber, 1981) and may contribute to the role of non-ADH ethanol metabolism pathways in the development of liver injury (Takada et al., 1986), especially since acetaldehyde produced by this non-alcohol-dehydrogenase (ADH) pathway appeared to be degraded more slowly than that produced by the ADH pathway in the liver (Yasuhara et al., 1986). It is tempting to postulate that the resulting increased acetaldehyde–protein adduct formation may be responsible, at least in part, for the appearance of antibodies against such adducts. Indeed, using an animal model, Israel *et al.* (1986) demonstrated that acetaldehyde adducts may serve as neoantigens generating an immune response in mice. Our own studies (Hoerner *et al.*, 1986) have shown that antibodies against acetaldehyde adducts (produced *in vitro*) are present in the serum of most alcoholics (Fig. 2.5). Hepatic injury initially may promote the release of significant amount of acetaldehyde-altered proteins, and the severity of liver disease can play a role in the appearance of circulating antibodies against acetaldehyde adducts not only in alcoholics but also in some nonalcoholics with liver injury (Hoerner *et al.*, 1988). In turn, complement-binding acetaldehyde-adduct-containing immune complexes may contribute to the perpetuation or exaggeration of liver disease. This may represent one of the immune mechanisms—cell-mediated as well as humoral—that can play a role in the pathogenesis of alcoholic liver injury, as discussed in Chapters 7 and 9.

Of particular interest is the binding of acetaldehyde to tubulin, the constituent protein of microtubules. The resulting impairment of polymerization and the alterations in microtubule association have been incriminated in inhibition of protein, lipoprotein, and glycoprotein secretion (see Chapters 5 and 7). Binding to lipoproteins alters their catabolism (Savolainen *et al.*, 1987) (see Chapter 4).

Binding of acetaldehyde may also occur in other tissues. Gains *et al.* (1977), for instance, reported morphological changes in erythrocytes caused by acetaldehyde binding. Binding to hemoglobin (Peterson and Polizzi, 1987; Stevens *et al.*, 1981) and to the red cell membrane proteins spectrin and actin (Gaines *et al.*, 1977) has been demonstrated. The phenomenon of reversible binding of acetaldehyde to red cells (see Section 2.2.2; Baraona *et al.*, 1987a), by serving as a transport mechanism, may mediate extrahepatic acetaldehyde toxicity.

Ethanol clearance is either the same or faster in alcoholics than in nonalcoholics; this "induction" may be the source of greater production of the toxic metabolite acetaldehyde. On the other hand, low-K_m hepatic acetaldehyde dehydrogenase activity, as discussed before, is predominantly intramitochondrial (Marjanen, 1972; Grunnet *et al.*, 1973; Tottmar *et al.*, 1973; Lebsack *et al.*, 1981), and the reoxidation of the cofactor NADH to NAD is also, in part, dependent on mitochondrial integrity. Thus, both may be impaired by chronic alcohol consumption. As mentioned before, the resulting defective acetaldehyde dehydrogenation may contribute to higher acetaldehyde levels. The latter may in turn enhance the functional disturbance of the mitochondria by reducing the activity of various shuttles involved in the disposition of reducing equivalents and by inhibiting oxidative phosphorylation (Cederbaum *et al.*, 1974b).

Cederbaum *et al.* (1975a) have also shown that acetaldehyde depresses the capacity of liver mitochondria to oxidize fatty acids, thereby mimicking defects associated with chronic alcohol consumption (Cederbaum *et al.*, 1975b). The concentrations of acetaldehyde required to achieve the hepatic effects in mitochondria of normal animals were greater than those likely to occur *in vivo*. However, mitochondria of rats fed ethanol chronically were found to have an increased susceptibility to the effects of acetaldehyde; under these conditions, concentrations of acetaldehyde that could occur in the liver were found to depress mitochondrial functions, including lipid oxidation (Matsuzaki *et al.*, 1977) (Fig. 2.6). Thus, acetaldehyde causes mitochondrial dysfunction, which in turn promotes higher acetaldehyde levels; this "vicious cycle" (Hasumura *et al.*, 1975) (Fig. 2.7) could lead to a progressive increase of acetaldehyde levels and to damage in the liver, which are discussed in

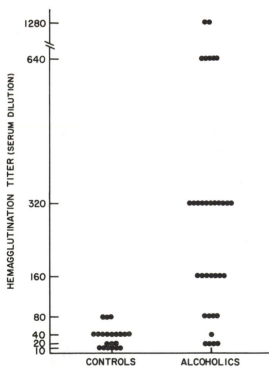

FIGURE 2.5. Hemagglutination titers of antibodies against acetaldehyde adducts in alcoholic patients and control subjects. (From Hoerner *et al.*, 1986.)

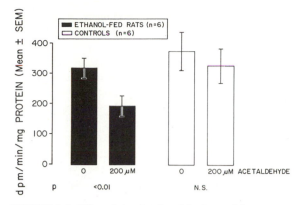

FIGURE 2.6. Effect of chronic ethanol feeding and the presence of acetaldehyde on fatty acid oxidation in hepatic mitochondria. Palmitic acid oxidation was studied by measuring CO_2 production at 37°C using $1-^{14}C$-labeled palmitic acid. (From Matsuzaki et al., 1977.)

Chapter 7. Acetaldehyde may also contribute to the fetal alcohol syndrome, since it has been shown that the rise in maternal blood acetaldehyde is markedly exaggerated at the end of pregnancy and even more so during lactation (Gordon et al., 1985) (see Chapter 15).

The role of acetate is less clear than that of acetaldehyde. Acetate was found (Liang and Lowenstein, 1978) to increase cardiac output, myocardial contractility, and coronary and portal blood flow; the latter has been attributed to adenosine formed (Carmichael et al., 1988). The effects of a rise of circulating acetate on intermediary metabolism in various tissues have not been well defined. In adipose tissue, acetate inhibits lipolysis (Nilsson and Belfrage, 1978), and it was found

to be responsible, at least in part, for the decreased release of free fatty acids (FFA) and the fall of circulating FFA (Crouse et al., 1968) (Fig. 2.8). A fall in FFA, a major fuel for peripheral tissues, may have significant metabolic consequences. In the liver, acetate was also shown to promote steatosis (Morgan and Mendenhall, 1977).

In summary, the metabolites of ethanol, namely, acetate and particularly acetaldehyde, are of great interest, since they are probably responsible for many of the intracellular metabolic effects of ethanol, not only in the liver but also in other tissues such as heart, adipose tissue, and probably the brain. Of particular significance is the observation that after a given dose of alcohol, the blood acetaldehyde level of alcoholics is higher than that of nonalcoholics because of metabolic differences, genetic factors, or both. Acetaldehyde not only is responsible for much of the tissue damage of the alcoholic but also has been incriminated in the development of dependence and some of the manifestations of intolerance to alcohol exhibited by some individuals. The latter phenomenon can be artificially induced through the use of acetaldehyde dehydrogenase inhibitors such as disulfiram. The adverse reaction thereby produced is useful in helping some alcoholics maintain sobriety.

2.5. Summary

Ethanol is converted in equimolar amounts in the liver to acetaldehyde and acetate. In view of the large amount of ethanol ingested by the alcoholic, it is not surprising that the correspondingly large quantities of

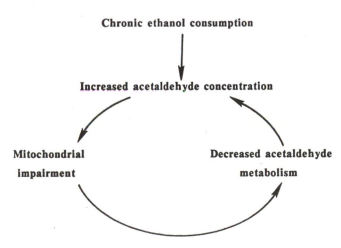

FIGURE 2.7. Possible relationship among ethanol consumption, altered acetaldehyde levels, and mitochondrial impairment through a "vicious cycle." (From Hasumura et al., 1975.)

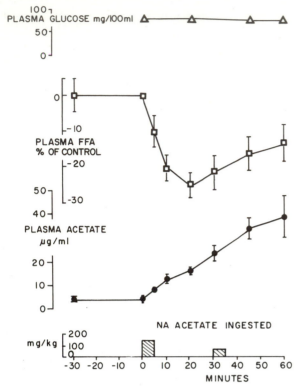

FIGURE 2.8. Effect of oral administration of sodium acetate on plasma glucose (△), FFA (□), and acetate (●). Points represent average values for five volunteers. Variation is expressed as S.E.M. Plasma FFA at zero time averaged 572 ± 91 μEq per liter. (From Crouse *et al.*, 1968.)

acetaldehyde and acetate produced have, in turn, an important metabolic impact. The deleterious effects of acetaldehyde (and possibly acetate) on the liver are discussed in Chapter 7. Furthermore, some of the acetaldehyde and most of the acetate are exported from the liver into the bloodstream for peripheral metabolism. A number of tissues other than the liver can thereby be affected, particularly the brain and fetus (by acetaldehyde), the heart and vascular tissue (by acetaldehyde), and adipose tissue (by acetate). Schemes have been elaborated to explain the development of physical dependence on the basis of the effects of small concentrations of acetaldehyde on the brain. On the other hand, some of the manifestations of alcohol intolerance exhibited by predisposed individuals can also be attributed to acetaldehyde, but in this case with high concentrations. Cardiomyopathy may be related, at least in part, to acetaldehyde-mediated impairment of protein synthesis in the heart. Acetaldehyde stimulates the adrenals and

impairs vitamin metabolism. Inhibition of lipolysis in adipose tissue is mediated through the action of acetate. In view of the reactivity of acetaldehyde and acetate, many more of the changes associated with chronic ethanol intake will undoubtedly be traced eventually to the effects of these two metabolites of ethanol.

REFERENCES

Abriola, D. P., Fields, R., Stein, S., MacKerell, D., Jr., and Pietruszko, R.: Active site of human liver aldehyde dehydrogenase. *Biochemistry* **26:**5679–5684, 1987.

Agarwal, D. P., Harada, S., and Goedde, H.W.: Racial differences in biological sensitivity to alcohol: The role of alcohol dehydrogenase and aldehyde dehydrogenase isozymes. *Alcoholism: Clin. Exp. Res.* **5:**12–16, 1981.

Alderman, J. A., Sanny, C. G., Gordon, E. R., and Lieber, C. S.: Partial characterization of hepatic aldehyde dehydrogenase from the baboon. In: *Enzymology of Carbonyl Metabolism: Aldehyde Dehydrogenase and Aldo/Keto Reductase* (H. Weiner and B. Wermuth, eds.), pp. 77–89, New York, Liss, 1982.

Alderman, J. A., Sanny, C., Gordon, E. R., and Lieber, C. S.: Ethanol feeding can produce secondary alterations in aldehyde dehydrogenase isozymes. *Alcohol* **2:**91–95, 1985.

Amir, S.: Brain aldehyde dehydrogenase: Adaptive increase following prolonged administration in rats. *Neuropharmacology* **17:**463–467, 1978.

Baraona, E., DiPadova, C., Tabasco, X., and Lieber, C. S.: Red blood cells: A new major modality for acetaldehyde transport from liver to other tissues. *Life Sci.* **40:**253–258, 1987a.

Baraona, E., DiPadova, C., Tabasco, X., and Lieber, C. S.: Transport of acetaldehyde in red blood cells. *Alcohol Alcohol.* [Suppl.] **1:**203–206, 1987b.

Barry, R. E., and McGivan, J. D.: Acetaldehyde alone may initiate hepatocellular damage in acute alcoholic liver disease. *Gut* **26:**1065–1069, 1985.

Behrens, U. H., Hoerner, M., Lasker, J. M., and Lieber, C. S.: Formation of acetaldehyde adducts with ethanol-inducible P450IIE1 *in vivo. Biochem. Biophys. Res. Commun.* **154:** 584–590, 1988.

Beyeler, C., Fisch, H. U., and Preisig, R.: The disulfiram–alcohol reaction: Factors determining and potential tests predicting severity. *Alcoholism: Clin. Exp. Res.* **9:**118–124, 1985.

Blair, A. H., and Bodley, F. H.: Human liver aldehyde dehydrogenase: Partial purification and properties. *Can J. Biochem.* **47:** 265–272, 1969.

Brunengraber, H., Boutry, M., Lowenstein, L., and Lowenstein, J. M.: The effect of ethanol on lipogenesis by the perfused liver. In: *Alcohol and Aldehyde Metabolizing Systems* (R. G. Thurman, T. Yonetani, J. R. Williamson, and B. Chance, eds.), New York, Academic Press, p. 329, 1974.

Büttner, H.: Aldehyd- und Alkoholdehydrogenase. Aktivität in Leber und Niere der Ratte. *Biochemistry* **341:**300–314, 1965.

Carmichael, F. J., Saldivia, V., Varghese, G. A., Israel, Y., and Orrego, H.: Ethanol-induced increase in portal blood flow:

role of acetate and A_1- and A_2-adenosine receptors. *Gastrointest. Liver Physiol.* **18**:G417–G423, 1988.

Cederbaum, A. I., Lieber, C. S., and Rubin, E.: Effects of chronic ethanol treatment on mitochondrial functions. *Arch. Biochem. Biophys.* **165**:560–569, 1974a.

Cederbaum, A. I., Lieber, C. S., and Rubin, E.: The effect of acetaldehyde on mitochondrial function. *Arch. Biochem. Biophys.* **161**:26–39, 1974b.

Cederbaum, A. I., Lieber, C. S., Beattie, D. S., and Rubin, E.: Effect of chronic ethanol ingestion on fatty acid oxidation by hepatic mitochondria. *J. Biol. Chem.* **250**:5122–5129, 1975a.

Cederbaum, A. I., Lieber, C. S., and Rubin, E.: Effect of acetaldehyde on fatty acid oxidation and ketogenesis by hepatic mitochondria. *Arch. Biochem. Biophys.* **169**:29–41, 1975b.

Cohen, G., and Collins, M.: Alkaloids from catecholamines in adrenal tissue: Possible role in alcoholism. *Science* **167**:1749–1751, 1970.

Collins, M. A., Nijm, W. P., Borge, F. G., Teas, G., and Goldfarb, C.: Dopamine-related tetrahydroisoquinolines: Significant urinary excretion by alcoholics after alcohol consumption. *Science* **206**:1184–1186, 1979.

Corall, R. J. M., Havre, P., Margolis, J., Kong, M., and Landau, B. R.: Subcellular site of acetaldehyde oxidation in rat liver. *Biochem. Pharmacol.* **25**:17–20, 1976.

Crabb, D. W., Edenberg, H. J., Bosron, W. F., and Li, T-K.: Genotypes for aldehyde dehydrogenase deficiency and alcohol sensitivity. The inactive *ALDH2*2* allele is dominant. *J. Clin. Invest.* **83**:314–316, 1989.

Crouse, J. R., Gerson, C. D., DeCarli, L. M., and Lieber, C. S.: Role of acetate in the reduction of plasma free fatty acids produced by ethanol in man. *J. Lipid Res.* **9**:509–512, 1968.

Davis, V. E., and Walsh, M. J.: Alcohol, amines and alkaloids: Possible biochemical basis for alcohol addiction. *Science* **167**:1005–1007, 1970.

Deitrich, R. A.:Tissue and subcellular distribution of mammalian aldehyde oxidizing capacity. *Biochem. Pharmacol.* **15**:1911–1922, 1966.

Deitrich, R. A.: Genetic aspects of increase in rat liver aldehyde dehydrogenase induced by phenobarbital. *Science* (Washington) **173**:334–336, 1971.

Deitrich, R. A.,and Erwin, V. G.: Mechanism of the inhibition of aldehyde dehydrogenase *in vivo* by disulfiram and diethyldithiocarbamate. *Mol Pharmacol* **7**:301–307, 1971.

Deitrich, R. A., Collins, A. C., and Erwin, V. G.: Genetic influence upon phenobarbital-induced increase in rat liver supernatant aldehyde dehydrogenase activity. *J. Biol.Chem* **247**:7232–7236, 1972.

DiPadova, C., Alderman, J., and Lieber, C. S.: Improved methods for the measurement of acetaldehyde concentrations in plasma and red blood cells. *Alcoholism: Clin. Exp. Res.* **10**:86–89, 1986.

DiPadova, C., Worner, T. M., and Lieber, C. S.: The effect of abstinence on the blood acetaldehyde response to a test dose of alcohol in alcoholics. *Alcoholism: Clin. Exp. Res.* **11**:559–561, 1987.

Eade, N. R.: Mechanisms of sympathomimetic action of aldehydes. *J. Pharmacol. Exp. Ther.* **127**:29–34, 1959.

Eckfeldt, J., and Yonetani, T.: Kinetics and mechanism of the F1

isozyme of horse liver aldehyde dehydrogenase. *Arch. Biochem. Biophys.* **173**:273–281, 1976.

Eriksson, C. J. P.: Ethanol and acetaldehyde metabolism in rat strains genetically selected from their ethanol preference. *Biochem. Pharmacol.* **22**:2283–2292, 1973.

Eriksson, C. J. P.: Elevated blood acetaldehyde levels in alcoholics and their relatives: A reevaluation. *Science* **207**:1383–1384, 1980a.

Eriksson, C. J. P.: Problems and pitfalls in acetaldehyde determinations. *Alcoholism: Clin. Exp. Res.* **4**:22–29, 1980b.

Eriksson, C. J. P., Marselos, M., and Koivula, T.: Role of cytosolic rat liver aldehyde dehydrogenase in the oxidation of acetaldehyde during ethanol metabolism *in vivo*. *Biochem. J.* **152**:709–712, 1975.

Eriksson, C. J. P., Sippel, H. W., and Forsander, O. A.: The occurrence of acetaldehyde binding in rat blood. *FEBS Lett.* **75**:205–208, 1977.

Ewing, J. A., Rouse, B. A., and Pellizzari, E. D.: Alcohol sensitivity and ethnic background. *Am. J. Psychiatry* **131**:206–210, 1974.

Freund, G., and O'Hollaren, P.: Acetaldehyde concentrations in alveolar air following a standard dose of ethanol in man. *Lipid Res.* **6**:471–477, 1965.

Fukui, Y.: Gas chromatograph determination of acetaldehyde in the expired air after ingestion of alcohol. *Jpn. J. Leg. Med.* **23**:24–29, 1969.

Gaines, K. C., Salhany, J. M., Tuma, D. J., and Sorrell, M. F.: Reaction of acetaldehyde with human erythrocyte membrane proteins. *FEBS Lett.* **75**:115–119, 1977.

Goedde, H. W., Harada, S., and Agarwal, D. P.: Racial differences in alcohol sensitivity: A new hypothesis. *Hum. Genet.* **51**:331–334, 1979.

Goedde, H. W., Bentemann, H. G., Kriese, L., Bogdanski, P., Agarwal, D. P., Ruofu, D., Liangzhong, C., Meiyuig, C., Yida, Y., Jiujiu, X., Shizhe, L., and Yongfa, W.: Aldehyde dehydrogenase isozyme deficiency and alcohol sensitivity in four different Chinese populations. *Hum. Hered.* **34**:183–186, 1984.

Gordon, B. H. J., Baraona, E., Miyakawa, H., Finkelman, F., and Lieber, C. S.: Exaggerated acetaldehyde response after ethanol administration during pregnancy and lactation in rats. *Alcoholism: Clin. Exp. Res.* **9**:17–22, 1985.

Greenfield, N. J., and Pietruszko, R.: Two aldehyde dehydrogenases from human liver. Isolation via affinity chromatography and characterization of the isozymes. *Biochim. Biophys. Acta* **483**:35–45, 1977.

Greenfield, J. J., Pietruszko, R., Lin, G., and Lester, D.: The effect of ethanol ingestion on the aldehyde dehydrogenases of rat liver. *Biochim. Biophys. Acta* **428**:627–632, 1976.

Grunnet, N.: Oxidation of acetaldehyde by rat-liver mitochondria membrane. *Eur. J. Biochem.* **35**:236–243, 1973.

Grunnet, N., Quistorff, B., and Thieden, H.I.D.: Rate-limiting factors in ethanol oxidation by isolated rat liver parenchymal cells. *Eur. J. Biochem.* **40**:275–282, 1973.

Grunnet, N., Thieden, H.I.D., and Quistorff, B.: Metabolism of 1-3H-ethanol by isolated liver cells. Time-course of the transfer of tritium from R, S-1-3H ethanol to lactate and β-hydroxybutyrate. *Acta Chem. Scand.* **B30**:345–352, 1976.

Guivernau, M., Baraona, E., and Lieber, C. S.: Acute and chronic effects of ethanol and its metabolites on vascular production of prostacyclin in rats. *J. Pharmacol. Exp. Ther.* **240:**59–64, 1987.

Harada, S., Misawa, S., Agarwal, D. P., and Goedde, H. W.: Studies on liver alcohol and acetaldehyde dehydrogenase variants in Japanese. *Hoppe-Seylers Z. Physiol. Chem.* **360:** 278, 1979.

Harada, S., Agarwal, D. P., and Goedde, H. W.: Electrophoretic and biochemical studies of human aldehyde dehydrogenase isozymes in various tissues. *Life Sci.* **26:**1773–1780, 1980.

Harada, S., Agarwal, D. P., and Goedde, H. W.: Mechanism of alcohol sensitivity and disulfiram-ethanol reaction. *Subst. Alcohol Actions/Misuse* **3:**107–115, 1982.

Harada, S., Agarwal, D. P., Goedde, H. W., and Ishikawa, B.: Aldehyde dehydrogenase isozyme variation and alcoholism in Japan. *Pharmacol. Biochem. Behav.* **18** (Suppl. 1):151–153, 1983.

Hasumura, Y., Teschke, R., and Lieber, C. S.: Acetaldehyde oxidation by hepatic mitochondria. Decrease after chronic ethanol consumption. *Science* **189:**727–729, 1975.

Hasumura, Y., Teschke, R., and Lieber, C. S.: Characteristics of acetaldehyde oxidation in rat liver mitochondria. *J. Biol. Chem.* **251:**4908–4913, 1976.

Hjelle, J. J., and Petersen, D. R.: Decreased *in vivo* acetaldehyde oxidation and hepatic aldehyde dehydrogenase inhibition in C57BL and DBA mice treated with carbon tetrachloride. *Toxicol. Appl. Pharmacol.* **59:**15–24, 1981.

Hjelle, J. J., Grubbs, J. H., and Petersen, D. R.: Inhibition of mitochondrial aldehyde dehydrogenase by malonidaldehyde. *Toxicol. Lett.* **14:**35–43, 1982.

Hoerner, M., Behrens, U. J., Worner, T., and Lieber, C. S.: Humoral immune responses to acetaldehyde adducts in alcoholic patients. *Res. Commun. Chem. Pathol. Pharmacol.* **54:** 3–12, 1986.

Hoerner, M., Behrens, U. J., Worner, T. M., Blacksberg, I., Braly, L. F., Schaffner, F., and Lieber, C.S.: The role of alcoholism and liver disease in the appearance of serum antibodies against acetaldehyde adducts. *Hepatology* **8:**569–574, 1988.

Horton, A. A., and Barrett, M. C.: Rates of induction of mitochondrial aldehyde dehydrogenase in rat liver. *Biochem. J.* **156:**177–179, 1976.

Ijiri, I.: Studies of the relationship between the concentrations of blood acetaldehyde and urinary catecholamine and the symptoms after drinking alcohol. *Jpn. J. Stud. Alcohol* **9:**35–59, 1974.

Ikawa, M., Impraim, C. C., Wang, G., and Yoshida, A.: Isolation and characterization of aldehyde dehydrogenase isozymes from usual and atypical human livers. *J. Biol. Chem.* **258:** 6282–6287, 1983.

Impraim, C., Wang, G., and Yoshida, A.: Structural mutation in a major human aldehyde dehydrogenase gene results in loss of enzyme activity. *Am. J. Hum. Genet.* **34:**837–841, 1982.

Inoue, K., Ohbora, Y., and Yamasawa, K.: Metabolism of acetaldehyde by human erythrocytes. *Life Sci.* **23:**179–184, 1978.

Inoue, K., Nishimukai, H., and Yamasawa, K.: Purification and partial characterization of aldehyde dehydrogenase from human erythrocytes. *Biochim. Biophys. Acta* **569:**117–123, 1979.

Inoue, K., Fukunaga, M., and Yamasawa, K.: Relationship between human erythrocyte aldehyde dehydrogenase and sensitivity to alcohol. *Drug Alcohol Depend.* **6:**31–32, 1980.

Inoue, K., Fukunaga, M., Kiriyama, T., and Kamura, S.: Accumulation of acetaldehyde in alcohol-sensitive Japanese: Relation to ethanol and acetaldehyde oxidizing capacity. *Alcoholism: Clin. Exp. Res.* **8:**319–322, 1984.

Israel, Y., Hurwitz, E., Niemela, O., and Arnon, R.: Monoclonal and polyclonal antibodies against acetaldehyde-containing epitopes in acetaldehyde–protein adducts. *Proc. Natl. Acad. Sci. U.S.A.* **83:**7923–7927, 1986.

Jauhonen, P., Baraona, E., Miyakawa, M., and Lieber, C. S.: Origin of breath acetaldehyde during ethanol oxidation. *J. Lab. Clin. Med.* **100:**908–916, 1982.

Jenkins, W. J., Cakebread, K., and Palmer, K. R.: Effect of alcohol consumption on hepatic aldehyde dehydrogenase activity in alcoholic patients. *Lancet* **2:**1048–1049, 1984.

Johnston, C., Saunders, J. B., Barnett, H., Ricciardi, B. R., Hopkinson, D. A., and Pyke, D. A.: Chlorpropamide-alcohol flush reaction and isoenzyme profiles of alcohol dehydrogenase and aldehyde dehydrogenase. *Clin. Sci.* **71:**513–517, 1986.

Katz, J., and Chaikoff, I. L.: Synthesis via the Kreb's cycle in the utilization of acetate by rat liver slices. *Biochim. Biophys. Acta* **18:**87–101, 1955.

Kesäniemi, Y. A.: Metabolism of ethanol and acetaldehyde in intact rats during pregnancy. *Biochem. Pharmacol.* **23:**1157–1162, 1974.

Kinoshita, M.: Gas chromatographic determination of blood acetaldehyde after ingestion of ethanol in man. *Jpn. J. Stud. Alcohol* **9:**1–9, 1974.

Kitson, T. M.: The effect of disulfiram on the aldehyde dehydrogenases of sheep liver. *Biochem. J.* **151:**407–412, 1975.

Kitson, T. M.: The effect of 5,5-dithiobis(1-methyltetrazole) on cytoplasmic aldehyde dehydrogenase and its implications for cephalosporin-alcohol reactions *Alcoholism: Clin. Exp. Res.* **10:**27–32, 1986.

Koivula, T.: Subcellular distribution and characterization of human liver aldehyde dehydrogenase fractions. *Life Sci.* **16:** 1563–1570, 1975.

Koivula, T., and Koivusalo, M.: Different forms of rat liver aldehyde dehydrogenase and their subcellular distribution. *Biochim. Biophys. Acta* **397:**9–23, 1975.

Koivula, T., and Lindros, K. O.: Effects of long term ethanol treatment on aldehyde and alcohol dehydrogenase activities in rat liver. *Biochem. Pharmacol.* **24:**1937–1942, 1975.

Korsten, M. A., Matsuzaki, S., Feinman, L., and Lieber, C. S.: High blood acetaldehyde levels after ethanol administration in alcoholics. Difference between alcoholic and non-alcoholic subjects. *N. Engl. J. Med.* **292:**386–389, 1975.

Kraemer, R. J., and Deitrich, R. A.: Isolation and characterization of human liver aldehyde dehydrogenase. *J. Biol. Chem.* **243:** 6402–6408, 1968.

Lamboeuf, Y., de Saint Blanquat, G., and Derache, R.: Effect of withdrawal from alcohol on extra-hepatic and alcohol and aldehyde-dehydrogenase activities in the rat. *Experientia* **36:** 286–288, 1980.

Lebsack, M. E., Gordon, E. R., and Lieber, C. S.: The effect of chronic ethanol consumption on aldehyde dehydrogenase activity in the baboon. *Biochem. Pharmacol.* **30:**2273–2277, 1981.

Liang, C. S., and Lowenstein, J. M.: Metabolic control of the circulation. *J. Clin. Invest.* **62**:1029–1038, 1978.

Lieber, C. S.: Chronic alcoholic hepatic injury in experimental animals and man: Biochemical pathways and nutritional factors. *Fed. Proc.* **26**:1443–1448, 1967.

Lieber, C. S., Baraona, E., Gordon, E., Jauhonen, P., Lebsack, M., and Pikkarainen, P.: Elevation of acetaldehyde after chronic alcohol consumption: Pathogenesis and pathologic consequences. In: *Biological and/or Genetic Factors in Alcoholism* (V. Hesselbrock, E. Shaskan, and R. Meyer, eds.), Research Monograph-9, DHHS Publication No. (ADM) 83-1199, Washington, D.C., Superintendent of Documents, U.S. Government Printing Office, pp. 67–91, 1983.

Lieber, C. S., Baraona, E., Hernandez-Munoz, R., Kubota, S., Sato, N., Kawano, S., Matsumura, T., and Inatomi, N.: Impaired oxygen utilization: A new mechanism for the hepatotoxicity of ethanol in sub-human primates. *J. Clin. Invest.* **83**:1682–1690, 1989.

Lin, C. C., Potter, J. J., and Mezey, E.: Erythrocyte aldehyde dehydrogenase activity in alcoholism. *Alcoholism: Clin. Exp. Res.* **8**:539–541, 1984.

Lindahl, R., and Evces, S.: Rat liver aldehyde dehydrogenase. II. Isolation and characterization of four inducible isozymes. *J. Biol. Chem.* **259**:11991–11996, 1984a.

Lindahl, R., and Evces, S.: Comparative subcellular distribution of aldehyde dehydrogenase in rat, mouse and rabbit liver. *Biochem. Pharmacol.* **33**:3383–3389, 1984b.

Lindros, K. O.: Suppression by ethanol of acetate and hexanoate oxidation studied by nonrecirculating liver perfusion. *Biochem. Biophys. Res. Commun.* **41**:635–642, 1970.

Lindros, K. O.: Acetaldehyde oxidation and its role in the overall metabolic effects of ethanol in the liver in regulation of hepatic metabolism. In: *Proceedings of the Alfred Benson Symposium VI*, Copenhagen 1973 (F. Lundquist and N. Tygstrup, eds.), Copenhagen, Munksgaard, 1974.

Lindros, K. O., and Hillbom, M. E.: Acetaldehyde in cerebrospinal fluid: Its near-absence in ethanol-intoxicated alcoholics. *Med. Biol.* **57**:246–247, 1979.

Lindros, K. O., Koivula, T., and Eriksson, C. J. P.: Acetaldehyde levels during ethanol oxidation: A diet-induced change and its relation to liver aldehyde dehydrogenases and redox states. *Life Sci.* **17**:1589–1598, 1975.

Lindros, K. O., Pekkanen, L., and Koivula, T.: Effect of a low-protein diet on acetaldehyde metabolism in rats. *Acta Pharmacol. Toxicol.* **40**:134–144, 1977.

Lindros, K. O., Pekkanen, L., and Koivula, T.: Enzymatic and metabolic modification of hepatic ethanol and acetaldehyde oxidation by the dietary protein level. *Biochem. Pharmacol.* **28**:2313–2320, 1979.

Lindros, K. O., Stowell, A., Pikkarainen, P., and Salaspuro, M.: Elevated blood acetaldehyde in alcoholic with accelerated ethanol elimination. *Pharmacol. Biochem. Behav.* **13**:119–124, 1980.

Lundquist, F., Tygstrup, N., Winkler, K., Mellemgaard, K., and Munck-Petersen, S.: Ethanol metabolism and production of free acetate in the human liver. *J. Clin. Invest.* **41**:955–961, 1962.

Ma, X-L., Baraona, E., Hernandez-Munoz, R., and Lieber, C. S.: High levels of acetaldehyde in non-alcoholic liver injury after threonine or alcohol administration. *Hepatology* **10**:933–940, 1989.

MacKerell, D., Erich, E., Blatter, B. S., and Pietruszko, R.: Human aldehyde dehydrogenase: Kinetic identification of the isozyme for which biogenic aldehydes and acetaldehyde compete. *Alcoholism: Clin. Exp. Res.* **10**:266–270, 1986.

Maeda, M., Hasumura, Y., and Takeuchi, J.: Localization of cytoplasmic and mitochondrial aldehyde dehydrogenase isozymes in human liver. *Lab. Invest.* **59**:75–81, 1988.

Magrinat, G., Dolan, J. P., Biddy, R. L., Miller, L. D., and Korol, B.: Ethanol and methanol metabolites in alcohol withdrawal. *Nature* **244**:234–235, 1973.

Majchrowicz, E., and Mendelson, J. H.: Blood concentration of acetaldehyde and ethanol in chronic alcoholics. *Science* **168**:1100–1102, 1970.

Marchner, H., and Tottmar, O.: Inhibition of the acetaldehyde dehydrogenases in rat liver by a cyanamide derivative present in a commercial standard diet for small animals. *Acta Pharmacol. Toxicol.* **39**:331–343, 1976a.

Marchner, H., and Tottmar, O.: Influence of the diet on the metabolism of acetaldehyde in rats. *Acta Pharmacol. Toxicol.* **38**:59–71, 1976b.

Maring, J.-A., Weigand, K., Brenner, H. D., and von Wartburg, J.-P.: Aldehyde oxidizing capacity of erythrocytes in normal and alcoholic individuals. *Pharmacol. Biochem. Behav.* **18** (Suppl. 1):135–138, 1983.

Marjanen, L.: Intracellular localization of aldehyde dehydrogenase in rat liver. *Biochem. J.* **127**:633–639, 1972.

Marjanen, L.: Comparison of aldehyde dehydrogenases from cytosol and mitochondria of rat liver. *Biochim. Biophys. Acta* **327**:238–246, 1973.

Matsuzaki, S., Teschke, R., Ohnishi, K., and Lieber, C. S.: Acceleration of ethanol metabolism by high ethanol concentrations and chronic ethanol consumption: Role of the microsomal ethanol oxidizing system (MEOS). In: *Alcohol and the Liver, Vol. 3*, (M. M. Fisher and J. G. Rankin, eds.), New York, Plenum, pp. 119–143, 1977.

Matthewson, K., Almardini, H., Bartlett, K., and Record, C. O.: Impaired acetaldehyde metabolism in patients with non-alcoholic liver disorders. *Gut* **27**:756–764, 1986.

Miyakawa, H., Baraona, E., Chang, J. C., Lesser, M. D., and Lieber, C. S.: Oxidation of ethanol to acetaldehyde by broncho-pulmonary washings. Role of bacteria. *Alcoholism: Clin. Exp. Res.* **10**:517–520, 1986.

Mizoi, Y., Ijiri, I., Tatsuno, Y., Kijima, T., Fujiwara, S., Adachi, T., and Hishida, S.: Relationship between facial flushing and blood acetaldehyde levels after alcohol intake. *Pharmacol. Biochem. Behav.* **10**:303–311, 1979.

Morgan, D. D., and Mendenhall, C. L.: The role of acetate in the pathogenesis of the acute ethanol-induced fatty liver in rats. *Gastroenterology* **73**:1235, 1977 (abstract).

Moshage, H., Casini, A., and Lieber, C. S.: Acetaldehyde selectively stimulates collagen production in cultured rat liver fat-storing cells but not in hepatocytes. *Hepatology* **12**:511–518, 1990.

Moura, D., Azevedo, I., and Osswald, W.: Hepatotoxicity of the condensation product of adrenaline with acetaldehyde. *J. Pharm. Pharamcol.* **29**:255–256, 1977.

Muller, A., and Sies, H.: Role of alcohol dehydrogenase activity and of acetaldehyde in ethanol-induced ethane and pentane production by isolated perfused rat liver. *Biochem. J.* **206:** 153–156, 1982.

Myers, R. D.: Tetrahydroisoquinolines in the brain: The basis of an animal model of alcoholism. *Alcoholism: Clin. Exp. Res.* **2:**145–154, 1978.

Nagasawa, H. T., Goon, D. J. W., DeMaster, E. G., and Alexander, C. S.: Lowering of ethanol-derived circulating blood acetaldehyde in rats by D-penicillamine. *Life Sci.* **20:**187–194, 1977.

Nilsson, N. O., and Belfrage, P.: Effects of acetate acetaldehyde and ethanol on lipolysis in isolated rat adipocytes. *J. Lipid Res.* **19:**737–741, 1978.

Nomura, F., and Lieber, C. S.: Binding of acetaldehyde to rat liver microsomes: Enhancement after chronic alcohol consumption. *Biochem. Biophys. Res. Commun.* **100:**131–137, 1981.

Nuutinen, H., Lindros, K., and Salaspuro, M.: Determinants of blood acetaldehyde level during ethanol oxidation in chronic alcoholics. *Alcoholism: Clin. Exp. Res.* **7:**163–168, 1983.

Ortiz, A., Griffiths, P. J., and Littleton, J. A.: A comparison of the effects of chronic administration of ethanol and acetaldehyde to mice: Evidence for a role of acetaldehyde in ethanol dependence. *J. Pharm. Pharmacol.* **26:**249–260, 1974.

Palmer, K. R., and Jenkins, W. J.: Aldehyde dehydrogenase in alcoholic subjects. *Hepatology* **5:**260–263, 1985.

Panes, J., Soler, X., Pares, Al., Caballeria, T., Farres, T., Rodes, T., and Pares, X.: Influence of liver disease on hepatic alcohol and aldehyde dehydrogenases. Gastroenterology **97:**708–714, 1989.

Parrilla, R., Ohkawa, K., Lindros, K. O., Zimmerman, U. T., Kobayashi, K., and Williamson, T. R.: Functional compartmentation of acetaldehyde oxidation in rat liver. *J. Biol. Chem.* **249:**4926–4933, 1974.

Petersen, D., Collins, A. C., and Deitrich, R. A.: Role of liver cytosolic aldehyde dehydrogenase isozymes in control of blood acetaldehyde concentrations. *J. Pharmacol. Exp. Ther.* **201:**471–481, 1977.

Peterson, C. M., and Polizzi, C. M.: Improved method for acetaldehyde in plasma and hemoglobin-associated acetaldehyde: Results in teetotalers and alcoholics reporting for treatment. *Alcoholism: Clin. Exp. Res.* **4:**477–480, 1987.

Pettersson, H., and Jenkins, W. J.: Acetaldehyde occurrence in cerebrospinal fluid during ethanol oxidation in rats and its dependence on the blood level and on dietary factors. *Biochem. Pharmacol.* **26:**237–240, 1977.

Pietruszko, R., and Vallari, R. C.: Aldehyde dehydrogenase in human blood. *FEBS Lett.* **92:**89–91, 1978.

Pikkarainen, P. H., Salaspuro, M. P., and Lieber, C. S.: A method for the determination of "free" acetaldehyde in plasma. *Alcoholism: Clin. Exp. Res.* **3:**259–261, 1979.

Pikkarainen, P. H., Gordon, E. R., Lebsack, M. E., and Lieber, C. S.: Determinants of plasma free acetaldehyde level during the steady state oxidation of ethanol: Effects of chronic ethanol feeding. *Biochem. Pharmacol.* **30:**799–802, 1981a.

Pikkarainen, P. H., Baraona, E., Jauhonen, P., Seitz, H. K., and Lieber, C. S.: Contribution of oropharynx microflora and of lung microsomes to acetaldehyde in expired air after alcohol ingestion. *J. Lab. Clin. Med.* **97:**631–638, 1981b.

Racker, E.: Aldehyde dehydrogenase, a diphosphopyridine nucleotide-linked enzyme. *J. Biol. Chem.* **177:**883–892, 1949.

Redmond, G., and Cohen, G.: Induction of liver acetaldehyde dehydrogenase: Possible role in ethanol tolerance after exposure to barbiturates. *Science (Washington)* **171:**387–389, 1971.

Reed, T. E., Kalant, H., Gibbins, R. J., Reed, B. M., and Rankin, T. G.: Alcohol and acetaldehyde metabolism in Caucasians, Chinese and Americans. *Can. Med. Assoc. J.* **115:**851–855, 1976.

Rex, D. K., Bostron, W. F., Smialek, J. E., and Li, T.: Alcohol and aldehyde dehydrogenase isoenzymes in North American Indians. *Alcoholism: Clin. Exp. Res.* **9:**147–152, 1985.

Roper, M., Stock, T., and Deitrich, R. A.: Phenobarbital and tetrachlorodibenzo-*p*-dioxin induce different isoenzymes of aldehyde dehydrogenase *Fed. Proc.* **35:**282 (abstract), 1976.

Savolainen, E-R., Leo, M. A., Timpl, R., and Lieber, C. S.: Acetaldehyde and lactate stimulate collagen synthesis of cultured baboon liver myofibroblasts. *Gastroenterology* **87:** 777–787, 1984.

Savolainen, M. J., Baraona, E., and Lieber, C. S.: Acetaldehyde binding increases the catabolism of rat serum low-density lipoproteins. *Life Sci.* **40:**841–846, 1987.

Scheig, R.: Lipid synthesis from ethanol in liver. *Gastroenterology* **60:**751, 1971 (abstract).

Schreiber, S. S., Briden, K., Oratz, M., and Rothschild, M. A.: Ethanol, acetaldehyde and myocardial protein synthesis. *J. Clin. Invest.* **51:**2808–2819, 1972.

Schreiber, S. S., Oratz, M., Rothschild, M. A., Reff, F., and Evan, C.: Alcoholic cardiomyopathy. II. The inhibition of cardiac microsomal protein synthesis by acetaldehyde. *J. Mol. Cell Cardiol.* **6:**207–213, 1974.

Schuckit, M. A., and Rayses, V.: Ethanol ingestion: Differences in blood acetaldehyde concentrations in relatives of alcoholics and controls. *Science* **203:**54–55, 1979.

Sereny, G., Sharma, V., Holt, J., and Gordis, E.: Mandatory supervised antabuse therapy in an outpatient alcoholism program: A pilot study. *Alcoholism: Clin. Exp. Res.* **10:**290–292, 1986.

Siew, C., Deitrich, R. A., and Erwin, V. G.: Localization and characteristics of rat liver mitochondrial aldehyde dehydrogenases. *Arch. Biochem. Biophys.* **176:**638–649, 1976.

Smith, L., and Packer, L.: Aldehyde oxidation in rat liver mitochondria. *Arch. Biochem. Biophys.* **148:**270–276, 1972.

Stevens, V. J., Fantl, W. J., Newman, C. B., Sims, R. V., Cerami, A., and Peterson, C. M.: Acetaldehyde adducts with hemoglobin. *J. Clin. Invest.* **67:**361–369, 1981.

Stowell, A. R., Lindros, K. O., and Salaspuro, M. P.: Breath and blood acetaldehyde concentrations and their correlation during normal calcium carbimide-modified ethanol oxidation in man. *Biochem. Pharmacol.* **29:**783–797, 1980.

Stowell, L., Johnsen, J., Aune, H., Vatne, K., Ripel, A., and Morland, J.: A reinvestigation of the usefulness of breath analysis in the determination of blood acetaldehyde concentrations. *Alcoholism: Clin. Exp. Res.* **8:**442–447, 1984.

Takada, A., Takase, S., Nei, J., and Matsuda, Y.: Subcellular distribution of ALDH isozymes in the human liver. *Alcoholism: Clin. Exp. Res.* **8:**123 (abstract), 1984.

Takada, A., Matsuda, Y., and Takase, S.: Effects of dietary fat on alcohol-pyrazole hepatitis in rats: The pathogenetic role of the nonalcohol dehydrogenase pathway in alcohol-induced hepatic cell injury. *Alcoholism: Clin. Exp. Res.* **10**:403–411, 1986.

Teng, Y.-S.: Human liver aldehyde dehydrogenase in Chinese and Asiatic Indians: Gene deletion and its possible implications in alcohol metabolism. *Biochem. Genet.* **19**:107–114, 1981.

Thomas, M., Halsall, S., and Peters, T. J.: Role of hepatic acetaldehyde dehydrogenase in alcoholism demonstration of persistent reduction of cytosolic activity in abstaining patients. *Lancet* **2**:1057–1059, 1982.

Tietz, A., Lindberg, M., and Kennedy, E. P.: A new pteridine-requiring enzyme system for the oxidation of glyceryl ethers. *J. Biol. Chem.* **239**:4081–4090, 1964.

Tipton, K. F., and Henehan, G. T. M.: Distribution of aldehyde dehydrogenase activities in human liver. *Alcoholism: Clin. Exp. Res.* **8**:131, 1984 (abstract).

Tottmar, S. O. C., Pettersson, H., and Kiessling, K.-H.: The subcellular distribution and properties of aldehyde dehydrogenases in rat liver. *J. Biochem. (Tokyo)* **135**:577–586, 1973.

Tottmar, S. O. C., Marchner, H., and Pettersson, H.: Determination of acetaldehyde in rat blood by the use of rat liver aldehyde dehydrogenase. *Ann. Biochem.* **91**:241–249, 1978.

Towell, J. F., Barboriak, J. J., Townsend, W. F., Kalbfleisch, J. H., and Wang, R. I. H.: Erythrocyte aldehyde dehydrogenase: Assay of a potential biochemical marker of alcohol abuse. *Clin. Chem.* **32**:734–738, 1986.

Truitt, E. B.: Blood acetaldehyde levels after alcohol consumption by alcoholics and nonalcoholic subjects. In: *Biological Aspects of Alcohol*. (M. K. Roach, W. M. McIsaac, and P. J. Creaven, eds.), Austin, University of Texas Press, pp. 212–232, 1971.

Truitt, E. B., and Duritz, G.: The role of acetaldehyde in the actions of ethanol. In: *Biochemical Factors in Alcoholism* (P. P. Maickel, ed.), New York, Pergamon Press, pp. 61–69, 1966.

Vallari, R. C., and Pietruszko, R.: Human aldehyde dehydrogenase: Mechanism of inhibition of disulfiram. *Science* **216**:637–639, 1982.

Veitch, R. L., Lumeng, L., and Li, T.-K.: Vitamin B6 metabolism in chronic alcohol abuse: The effect of ethanol oxidation on hepatic pyridoxal-5-phosphate metabolism. *J. Clin. Invest.* **55**:1026–1032, 1975.

Walkenstein, S. S., and Weinhouse, S.: Oxidation of aldehydes of mitochondria of rat tissues. *J. Biol. Chem.* **200**:515–523, 1953.

Walsh, M. J.: Role of acetaldehyde in the interactions of ethanol with neuroamines. In: *Biological Aspects of Alcohol* (M. K. Roach, W. M. McIsaac, and P. J. Creaven, eds.), Austin, University of Texas Press, p. 233, 1971.

Weiner, H., King, P., Hu, J. H. J., and Bensch, W. R.: Mechanistic and enzymatic properties of liver aldehyde dehydrogenases. In: *Alcohol and Aldehyde Metabolizing Systems* (R. G. Thurman, T. Yonetani, L. R. Williamson, and B. Chance, eds.), New York, Academic Press, pp. 101–125, 1974.

Wolff, P. H.: Vasomotor sensitivity to alcohol in diverse mongoloid populations. *Am. J. Hum. Genet.* **25**:193–199, 1973.

Yasuhara, M., Matsuda, Y., and Takada, A.: Degradation of acetaldehyde produced by the nonalcohol dehydrogenase pathway. *Alcoholism: Clin. Exp. Res.* **10**:545–549, 1986.

Yoshida, A., Huang, I-Y., and Ikawa, M.: Molecular abnormality of an inactive aldehyde dehydrogenase variant commonly found in Orientals. *Proc. Natl. Acad. Sci. U.S.A.* **81**:258–261, 1984.

Yoshihara, H., Sato, N., Kamada, T., and Abe, H.: Low K_m ALDH isozyme and alcoholic liver injury. *Pharmacol. Biochem. Behav.* **18**:425–428, 1983.

Zeiner, A. R., Paredes, A., and Christensen, H.D.: The role of acetaldehyde in mediating reactivity to an acute dose of ethanol among different racial groups. *Alcoholism: Clin. Exp. Res.* **3**:11–18, 1979.

3

Alcohol, Hormones, and Metabolism

Gary G. Gordon and Charles S. Lieber

There are some sluggish men who are improved by drinking, as there are fruits that are not good till they are rotten.
Samuel Johnson

Alcohol has distinct and well-known effects on sexual and reproductive function that have been extensively researched (Gordon *et al.*, 1982). In contrast, the effects of alcohol on other endocrine glands and hormonal axes are more subtle and, as they are generally unaccompanied by clinical change, have received much less attention. In consequence the literature on this subject is patchy, and the studies available are often difficult to interpret. Much of the early data were obtained using necessarily crude techniques for measurement and in ignorance of the fluctuant nature of hormonal secretion. In more recent years investigators have tended to study heterogeneous populations of chronic alcoholics, paying little, if any, attention to the presence of malnutrition or liver disease or to the effects of alcohol withdrawal. Although a large amount of information has been obtained from experiments involving acute or chronic administration of alcohol to animals *in vitro* and *in vivo*, this does not, in general, translate easily to the situation in humans. Therefore, many gaps in our knowledge of the effects of alcohol on endocrine function still exist.

3.1. Hypothalamic–Pituitary–Thyroidal Axis

In the 1940s, it was suggested that alcoholism might have an endocrinological basis (Gross, 1945; Goldfarb and Berman, 1949). Indeed, early studies in experimental animals suggested a possible relationship between thyroid functional status and preference for alcohol (Richter, 1956, 1957; Prieto *et al.*, 1958; Hillbom, 1971). A number of investigators have failed to find any etiological relationship between thyroid status and drinking behavior in humans (Satterfield and Guze, 1961; Selzer and Van Houten, 1964; Augustine, 1967; Stokes, 1971; Wright, 1978) or any beneficial effect of thyroid hormone in the treatment of alcohol abuse (Satterfield and Guze, 1961; Kalant *et al.*, 1962).

The majority of alcoholics are euthyroid (Augustine, 1967; Hollander *et al.*, 1967; Chopra *et al.*, 1974, 1975; Nomura *et al.*, 1975; Green *et al.*, 1977; Borzio *et al.*, 1983), but their clinical appearance and the results of their thyroid function tests might occasionally lead to an erroneous diagnosis of hyper- (Summerskill and Molnar, 1962; Augustine, 1967; Stuart and Schultz, 1978; Kallner, 1981) or hypothyroidism (Kydd and Mann, 1951; Augustine, 1967; Hanski *et al.*, 1979; Habnener, 1981).

3.1.1. Laboratory Findings

Abnormal laboratory indices of thyroid function have been observed in alcoholic individuals, especially those with significant liver disease, that resemble changes observed in other chronic illnesses (Chopra et al., 1974; Green et al., 1977; Van Thiel et al., 1979; Oppenheimer et al., 1982; Kabadi and Premachandra, 1983; Borzio et al., 1983). These changes include alterations in thyroid hormone metabolism, changes in thyroid hormone binding by plasma proteins, and changes in basal and stimulated thyroid-stimulating hormone (TSH) responses (Pamenter and Boyden, 1986).

In alcoholics free of liver disease, both total and free serum T_4 and T_3 levels are normal (Wright et al., 1976; Green et al., 1977; Modigliani et al., 1979), as are plasma concentrations of thyroxine-binding globulin (TBG) (Agner et al., 1986). Basal TSH levels are normal (Descos et al., 1974; Agner et al., 1986), although some patients show abnormal TSH responses to exogenous thyrotrophin-releasing hormone (TRH) (Wright et al., 1976; Green et al., 1977; Modigliani et al., 1979).

In alcoholics with significant liver disease, serum total T_4 levels are normal (Carter et al., 1974; Nomura et al., 1975; Green et al., 1977; Hepner and Chopra, 1979; Israel et al., 1979; Modigliani et al., 1979; Borzio et al., 1983; Hegedus, 1984; Agner et al., 1986), whereas serum free T_4 concentrations are normal or high (Chopra et al., 1974; Nomura et al., 1975; Green et al., 1977; Schlienger, 1979). Serum total T_3 levels are invariably reduced (Chopra et al., 1974; Nomura et al., 1975; Green et al., 1977; Hepner and Chopra, 1979; Israel et al., 1979; Modigliani et al., 1979; Borzio et al., 1983; Hegedus, 1984; Agner et al., 1986), but free T_3 levels may be normal (Chopra et al., 1974). Serum rT_3 values are usually raised (Chopra et al., 1974; Hepner and Chopra, 1979; Kabadi and Premachandra, 1983; Borzio et al., 1983). Plasma albumin and thyroxine-binding prealbumin (TBPA) concentrations are usually reduced in individuals with alcoholic cirrhosis, but plasma TBG values may be high (Agner et al., 1986), normal (Schlienger 1979), or low (Inada and Sterling, 1967). Serum TSH concentrations are normal or slightly elevated in alcoholics with cirrhosis (Chopra et al., 1974; Cuttelod et al., 1974; Pokroy et al., 1974; Green et al., 1977; Kabadi and Premachandra, 1983; Borzio et al., 1983; Teschke et al., 1983; Agner et al., 1986), and they may show an abnormal TSH response to exogenous TRH (Chopra et al., 1974; Pokroy et al., 1974; Green et al., 1977; Schlienger, 1979; Modigliani et al., 1979). The most common abnormalities observed are either an exaggerated response with a delayed peak height or an attenuated response (Green et al., 1977; Schlienger, 1979).

3.1.2. Mechanisms of Changes

3.1.2.1. Effects of Alcohol on Hypothalamic–Pituitary Control

In patients with alcoholic cirrhosis, serum TSH levels tend to be normal or slightly elevated despite the presence of low circulating T_3 concentrations (Chopra et al., 1974; Green et al., 1977; Kabadi and Premachandra, 1983; Teschke et al., 1983). These patients may also show exaggerated and delayed serum TSH responses to stimulation with exogenous TRH. The increased basal and prolonged TSH responses to TRH might reflect either a reduction in TSH clearance or else a decrease in its volume of distribution (Cuttelod et al., 1974) but also could indicate inadequate feedback control of TSH secretion as a result of either modifications in peripheral hormone metabolism or else defective hypothalamic–pituitary functioning (Green et al., 1977; Van Thiel et al., 1979).

It has been suggested (Röjdmark et al., 1984) that the blunted plasma TSH response to exogenous TRH observed in chronic alcoholics during acute alcohol withdrawal results from dopaminergic inhibition of the thyrotrope. This suggestion was based on the finding that metaclopramide, a specific dopamine D_2-receptor blocker, reversed the blunted plasma TSH response to TRH observed in a group of chronic alcoholic men on admission to hospital.

It is also possible that the abnormalities observed in TSH secretion in alcoholic men reflect changes in circulating estrogen concentrations. Estrogens are thought to sensitize the thyrotrope to the actions of the neurohormone (De Lean and Labrie, 1977) so that in the presence of hyperestrogenemia the plasma TSH response to exogenous TRH are exaggerated, although basal plasma TSH levels remain unaltered (De Lean and Labrie, 1977).

3.1.2.2. Changes in Hormone Binding and Peripheral Hormone Metabolism

It is unlikely that the changes observed in peripheral hormone concentrations in individuals abusing alcohol result from simple changes in hormone binding. Plasma albumin and TBPA concentrations may be reduced in alcoholics with cirrhosis, but plasma TBG levels may be high (Agner et al., 1986), normal (Schlienger, 1979), or low (Inada and Sterling, 1967). However, binding capacity and TBG avidity may be reduced (Inada and Sterling, 1967; McConnon et al., 1972). Recently, Geurts et al. (1981) reported transient but significant alterations in serum T_4 levels in alcoholic patients with little or no liver disease, which they attributed to changes in TBG synthesis and/or secretion by the liver. On admission to hospital the patients they studied had low plasma total T_4 and TSH levels, although free T_4 concentrations were normal. The patients were clinically euthyroid and had normal plasma TSH levels and normal plasma T_3 and TSH responses to exogenous TRH. During the first week of admission plasma total T_4 and TSH concentrations gradually increased, to become normal by the end of 2 weeks. Significant correlations existed between plasma total T_4 and TBG levels and between plasma TBG levels and the concentrations of serum aspartate and alanine transferases. Thus, the decrease in plasma total T_4 levels was attributed to the reduction in circulating TBG concentrations that resulted from an effect of alcohol on hepatic TBG synthesis and/or secretion.

A number of nonspecific changes may occur in alcoholic patients, such as increases in plasma free fatty acid levels and decreases in plasma pH, which might interfere with the binding of thyroid hormones to their carrier proteins (Chopra et al., 1982). In addition, an inhibitor has been identified in serum from alcoholics with severe liver disease that prevents adequate binding of transport proteins to the thyroid hormones (Oppenheimer et al., 1982).

In alcoholics with cirrhosis, thyroid production rates of T_3 are normal (Chopra et al., 1974), whereas T_4 production rates are slightly reduced (Chopra, 1976) and T_3 and T_4 metabolic clearance rates are unchanged (Chopra et al., 1974; Nomura et al., 1975). However, the peripheral conversion of T_4 to T_3 is significantly reduced in these patients (Nomura et al., 1975; Lumholtz et al., 1978). In contrast, the conversion of T_4 to rT_3 is normal in alcoholics with liver disease, but the metabolic clearance rate of rT_3 is reduced. When viewed in the context of the decreased T_4 production rate (Chopra, 1976), however, the production of rT_3 in patients with alcohol-related liver disease is either normal or increased, whereas T_3 production is clearly diminished (Chopra et al., 1974; Chopra, 1976; Lumholtz et al., 1978). These changes, which explain to a large extent the alterations observed in circulating thyroid hormone concentrations in patients with alcohol-related liver disease, result from a decrease in the activity of iodothyronine-5'-deiodinase, the enzyme that catalyzes the formation of T_3 from T_4 and the metabolism of rT_3. Individuals with the most severe liver disease tend to have the lowest serum T_3 and highest rT_3 levels (Nomura et al., 1975; Green et al., 1977; Hepner and Chopra, 1979; Israel et al., 1979), whereas clinical improvement is associated with a return of circulating T_3 concentrations toward normal (Carter et al., 1974; Chopra et al., 1974; Hepner and Chopra, 1979; Israel et al., 1979; Kabadi and Premachandra, 1983; Borzio et al., 1983).

3.1.3. Interpretation of Thyroid Function Tests in Alcoholic Individuals

The changes observed in circulating thyroid hormones in individuals abusing alcohol occur as a result of change in peripheral hormone metabolism, predominantly as a result of associated liver disease but also because of the effects of alcohol per se. Alcohol may, in addition, affect thyroidal uptake of iodine and may interfere with the TSH response to exogenous TRH because of an effect on the hypothalamus.

In consequence, thyroid function tests may be difficult to interpret in alcoholic individuals, especially if samples are obtained shortly after patients have been admitted to hospital, before withdrawal from alcohol is complete. In general, measurement of free T_3 and T_4 levels and, if necessary, of the T_3 and TSH responses to exogenous TRH will clarify the situation.

It is important to remember, however, that alcoholic individuals may develop incidental thyroid disease.

3.2. Parathormone and Calcitonin

Hypocalcemia is often noted in hospitalized alcoholics (Martin et al., 1959; Fankushen et al., 1964) and is often associated with the presence of pancreatitis, intestinal malabsorption of chronic liver disease, and severe hypoalbuminemia. Hypocalcemia may also be observed in hospitalized alcoholics who are free of complicating disease, but it usually resolves quite rapidly following cessation of alcohol and restoration of adequate nutrition. This reduction in serum calcium levels undoubtedly reflects the presence of hypomagnesemia. Estep et al. (1969) gave parathormone to a group of alcoholic individuals with hypocalcemia, but only one-third responded with an increase in serum calcium levels, suggesting that the majority of such patients were peripherally resistant or insensitive to the action of this hormone. However, when the associated hypomagnesemia was corrected, the response to para-thormone was restored. It is also possible that the hypo-magnesemia might have inhibited release of endoge-nous parathormone.

In humans, alcohol given acutely does not produce hypocalcemia but does cause an increase in circulating parathormone levels (Williams *et al.*, 1978).

Alcohol increases circulating calcitonin levels when given acutely to normal human adults (Williams *et al.*, 1978) and to patients with medullary carcinoma of the thyroid (Cohen *et al.*, 1973; Wells *et al.*, 1975). Wells *et al.* (1975) noted that circulating calcitonin levels were higher following oral than intravenous alco-hol. This they attributed to the fact that oral ethanol stimulates gastrin secretion, which in turn facilitates calcitonin release (Hennessy *et al.*, 1974). Gastrin levels were unchanged following intravenous ethanol, sug-gesting that ethanol must also have a direct stimulatory effect on the thyroid C cells. The significance of the effect of alcohol on calcitonin secretion in humans is unknown.

3.3. Adrenocortical Function: Glucocorticoid Secretion

It has been recognized for some years that chronic alcoholics may develop certain physical features remi-niscent of those observed in individuals with Cushing's syndrome, such as facial mooning, truncal obesity, and proximal muscle wasting. In a small number of alco-holics the clinical signs are much more florid, so that the clinical picture may be indistinguishable from that of true Cushing's syndrome; the term "pseudo-Cushing's syndrome" is used to describe this condition (Frajria and Angeli, 1977; Rees *et al.*, 1977; Smals and Klop-penborg, 1977; Smals *et al.*, 1977; Jordan *et al.*, 1979; Lamberts *et al.*, 1979; Fink *et al.*, 1981). Both syn-dromes are characterized by typical faces, a buffalo hump, proximal muscle wasting, abdominal striae, ex-cessive bruising, arterial hypertension, and vertebral collapse, though not all features need be present. The two conditions may also be biochemically indistinguish-able, since glucose intolerance, raised plasma cortisol levels, impaired diurnal cortisol rhythm, increased urin-ary steroid loss, and inadequate suppression of cortisol secretion by dexamethasone may be features of both (Frajria and Angeli, 1977; Rees *et al.*, 1977; Smals and Kloppenborg, 1977; Smals *et al.*, 1977; Oxenkrug,

1978; Jordan *et al.*, 1979; Lamberts *et al.*, 1979; Jenkins and Page, 1981; Fink *et al.*, 1981). However, in contrast to true Cushing's syndrome, the clinical and biochemi-cal abnormalities observed in "pseudo-Cushing's syn-drome" are reversible and disappear within days or weeks of withdrawal from alcohol (Rees *et al.*, 1977; Smals *et al.*, 1977; Lamberts *et al.*, 1979).

Paradoxically, there is increasing evidence that al-coholics may develop a characteristic syndrome of hypothalamic–pituitary–adrenocortical insufficiency that also reverses with abstinence from alcohol (Merry and Marks, 1973). Thus, in about 25% of chronic alco-holics, the cortisol response to insulin-induced hypo-glycemia may be either absent of severely attenuated (Merry and Marks, 1973; Wright *et al.*, 1976; Chalmers *et al.*, 1977), and there may be no objective or subjective clinical response (Wright *et al.*, 1976). Similarly, alco-holic individuals may show a poor adrenocortical re-sponse to surgical stress (Margraf *et al.*, 1967).

3.3.1. Laboratory Findings

Much of the information on glucocorticoid metab-olism in individuals abusing alcohol cannot be inter-preted accurately because patient populations are often too poorly defined. Thus, information on the patient's nutritional status, the nature and severity of any liver disease, and the recent drinking behavior is omitted in many of the studies available. Care should be taken, therefore, to distinguish where possible between basal laboratory findings in alcoholics with and without liver disease, in individuals withdrawing from alcohol, and in individuals exhibiting features of the pseudo-Cushing's syndrome.

In alcoholic individuals free of liver disease, basal plasma cortisol levels are normal (Kissin *et al.*, 1960; Shishov, 1973) or high (Margraf *et al.*, 1967; Stokes, 1973; Soulairac and Aymard, 1977), and the diurnal rhythm of secretion is preserved (Soulairac and Ay-mard, 1977; Stokes, 1973). Basal plasma ACTH concen-trations tend to be normal in these patients (Dobrzanski and Pieschl, 1976), and urinary 17-hydroxycortico-steroid (17-OHCS) excretion is unchanged (Kissin *et al.*, 1960; Margraf *et al.*, 1967).

In alcoholic individuals with significant liver dis-ease, basal plasma cortisol levels are normal (Brown *et al.*, 1954; Peterson *et al.*, 1955; Peterson, 1960; Bajpai *et al.*, 1973; McCann and Fulton, 1975; Kley *et al.*, 1975; Fanghänel *et al.*, 1976) or low (Perkoff *et al.*, 1959; Doe *et al.*, 1960; Sholiton *et al.*, 1961; Dioguardi *et al.*, 1974), and the diurnal rhythm of secretion is generally preserved (Perkoff *et al.*, 1959; Sholiton *et al.*, 1961; Dioguardi *et al.*, 1974), although it may be disturbed in individuals with decompensated cirrhosis. No information is available on basal plasma ACTH concentrations in these individuals; urinary 17-OHCS excretion is invariably reduced (Brown *et al.*, 1954; Klein *et al.*, 1955; Peterson *et al.*, 1955; Peterson, 1960; Lederer and Bataille, 1967; Zumoff *et al.*, 1967).

There appears to be little consensus as to the effects of alcohol withdrawal on plasma cortisol levels, although overall it appears that significant increases may be observed in approximately 50% of chronic alcoholics (Kissin *et al.*, 1960; Mendelson and Stein, 1966; Mendelson *et al.*, 1971; Merry and Marks, 1972; Shishov, 1973; Stokes, 1973; Oxenkrug, 1978; Fink *et al.*, 1981; Beevers *et al.*, 1982). Plasma ACTH concentrations may be normal (Pohorecky, 1974; Genazzani *et al.*, 1982) or elevated (Dobrzanski and Pieschl, 1976) in these individuals during alcohol withdrawal; cerebrospinal fluid (CSF) ACTH concentrations are elevated (Genazzani *et al.*, 1982).

Laboratory abnormalities are not well documented in individuals with the pseudo-Cushing's syndrome. The most consistent and convincing abnormalities recorded to date are increased plasma cortisol levels and impaired cortisol suppression following low-dose dexamethasone (Frajria and Angeli, 1977; Rees *et al.*, 1977; Smals and Kloppenborg, 1977; Smals *et al.*, 1977; Oxenkrug, 1978; Jordan *et al.*, 1979; Lamberts *et al.*, 1979; Fink *et al.*, 1981; Jenkins and Page, 1981). Abnormal diurnal variation of cortisol secretion has been noted in some of these patients but is not well recorded (Pohorecky, 1974; Frajria and Angeli, 1977; Rees *et al.*, 1977). Plasma ACTH concentrations are raised in the few subjects in whom they have been measured (Lamberts *et al.*, 1979; Rees *et al.*, 1977; Jenkins and Page, 1981). Low plasma cortisol-binding globulin (CBG) concentrations have been noted in a small number of these patients (Frajria and Angeli, 1977), and hypersensitivity to metyrapone and an increase in the cortisol secretory rate have also been recorded (Rees *et al.*, 1977; Lamberts *et al.*, 1979).

3.3.2. Mechanisms of the Changes

No single hypothesis is available to explain the effects of alcohol on plasma cortisol concentrations. Alcohol may have direct effects on the adrenal cortex and/or the hypothalamic–pituitary axis and may in addition alter peripheral hormone metabolism. To a large extent our understanding of these effects of alcohol has been hampered because too little account has been taken of the "nonspecific" stress effects of alcohol intoxication and withdrawal.

3.3.2.1. "Nonspecific" Stress Effect

The earliest animal studies of the effects of alcohol on adrenocortical function were based on indirect measurements of adrenocortical activity.

Ellis (1966) was the first to measure plasma corticosteroid levels in animals receiving ethanol and showed that plasma 17-OHCS levels increased in a dose-related manner. Ellis (1966) concluded that alcohol activated the adrenal cortex via its action on the hypothalamus. Direct evidence that the adrenocortical response to acute alcohol exposure is mediated via ACTH was provided by Noble *et al.* (1971), who documented a fall in ACTH concentrations in the mouse pituitary within 10 min of alcohol administration.

In general, the responses to acute alcohol exposure are attenuated in animals chronically exposed to alcohol. Crossland and Ratcliffe (1968) showed that the plasma corticosterone response to a single intraperitoneal injection of alcohol was reduced in rats fed alcohol for several weeks. They also noted that the corticosteroid response to a number of other stressful stimuli were reduced in these animals. Similarly, mice chronically exposed to alcohol no longer show diurnal variation in plasma corticosterone levels (Kakihana *et al.*, 1968, 1971; Kakihana and Moore, 1976; Tabakoff *et al.*, 1978). However, cessation of chronic alcohol administration to mice is followed by an abstinence syndrome characterized by hyperactivity, seizures, and increases in plasma corticosterone levels (Bongiovanni and Eisenmenger, 1951; Chart and Shipley, 1953; Peterson, 1960; Laragh *et al.*, 1962; Coppage *et al.*, 1972; Goldstein, 1972; Kakihana *et al.*, 1971; Wright *et al.*, 1976; Tabakoff *et al.*, 1978).

Early studies in normal human subjects failed to detect any consistent change in urinary steroid output following acute alcohol ingestion (Kissin *et al.*, 1960; Perman, 1961a). However, Fazekas (1966) was able to show a clear correlation between blood alcohol levels in nonalcoholic volunteers given alcohol acutely and resultant plasma 17-OHCS increments in that the higher the blood alcohol level the greater the magnitude and the duration of the plasma 17-OHCS response. In general, the blood alcohol level needed to approach 100 mg/100 ml (22 mM) before an effect of alcohol on blood corticosteroid level became noticeable.

Merry and Marks (1969) gave 284 ml of whiskey (≈90 g of alcohol), in water, to three normal volunteers and noted progressive increments in plasma 17-OHCS levels, with a maximum at 90 to 120 min after ingestion. However, their subjects were clearly distressed, with nausea, vomiting, and features of intoxication, and it seems likely that the corticosteroid response was caused, at least in part, by the "nonspecific" stress of drunkenness.

Other workers have found no evidence to suggest that large doses of ethanol given acutely to normal volunteers produce increases in plasma cortisol concentrations. Thus, Jeffcoate et al. (1979) and Davis and Jeffcoate (1983) have studied the effects of both oral and intravenous ethanol, in doses of 0.5 to 0.75 g/kg, in 22 different subjects on 28 separate occasions at varying times of day and noted no significant increase in plasma cortisol levels.

In general, it can be concluded that the cortisol response seen in some healthy volunteers in reaction to acute intoxicating amounts of alcohol probably reflects an acute hypothalamic stress reaction resulting in pituitary release of ACTH. This conclusion is supported by the findings of a study by Jenkins and Connolly (1968), who gave intravenous alcohol to four patients with hypothalamic or pituitary disorders but intact adrenal responses. Significant increases were observed in plasma cortisol levels in the individuals with hypothalamic lesions but not in the patients with pituitary disease.

Alcoholic individuals appear to tolerate acute alcohol, even in intoxicating doses, without evidence of adrenocortical activation (Merry and Marks, 1969, 1973). However, more prolonged administration of alcohol to chronic alcoholics is accompanied by increases in plasma cortisol concentrations (Mendelson et al., 1971; Stokes, 1973). In chronic alcoholics, withdrawal from alcohol is generally associated with increased plasma cortisol levels (Mendelson and Stein, 1966; Mendelson et al., 1971; Stokes, 1973). In humans there is no evidence, however, that alcohol given acutely or chronically either to normal volunteers or to chronic alcoholics affects the circadian rhythm of cortisol secretion (Prinz et al., 1980; Bertello et al., 1982; Rosman et al., 1982).

It therefore appears that alcohol may produce "nonspecific" stimulatory effects on the hypothalamic–pituitary–adrenocortical axis, both in animals and in normal volunteers, that are mediated via the hypothalamus. In animals fed alcohol chronically, and in habitual drinkers, tolerance to this effect develops. The increases in plasma cortisol concentrations following chronic exposure to alcohol must result from separate and distinct effects either on the adrenal gland itself or on the hypothalamic–pituitary axis.

3.3.2.2. Direct Effects of Alcohol on the Adrenal Cortex

The existence of an effect of alcohol on the adrenal cortex separate from an effect mediated via the hypothalamic–pituitary axis would be difficult to demonstrate in intact humans. There an be no doubt, however, that in chronic alcoholics, plasma ACTH and cortisol levels are concordant, and the adrenal cortex is normally responsive to ACTH (Bongiovanni and Eisenmenger, 1951; Margraf et al., 1967; Dioguardi et al., 1974).

Cobb et al. (1979) and Cobb and Van Thiel (1982) have shown that ethanol and acetaldehyde, in concentrations similar to these observed in the plasma of intoxicated individuals, increase corticosterone production by the isolated, perfused rat adrenal gland to the same degree as physiological concentrations of ACTH. It remains to be seen whether this effect occurs in intact humans.

3.3.2.3. Effects of Alcohol on the Hypothalamic–Pituitary Axis

Merry and Marks (1972) studied 15 alcoholics admitted to hospital for withdrawal from alcohol. During the early withdrawal phase plasma cortisol levels were raised, probably as a result of increased ACTH release mediated by heightened activity in the higher neural centers and the hypothalamus. If alcohol or amobarbital were given during this period to abolish withdrawal symptoms, plasma cortisol levels returned to normal. However, diazepam, given in a dose of 30 to 40 mg, suppressed withdrawal symptoms equally effectively but had no effect on plasma cortisol levels (Merry and Marks, 1972). These findings suggest that alcohol might have a specific effect on the hypothalamic–pituitary axis separate from its "nonspecific" stress effect.

Further support has been gained for this suggestion from the results of studies of dexamethasone suppression in chronic alcoholics with normal plasma cortisol levels (Oxenkrug, 1978; Fink et al., 1981; Hasselbalch *et al.*, 1982). Oxenkrug (1978) gave oral dexamethasone to 12 chronic alcoholics with normal basal plasma cortisol levels and found that suppression of plasma cortisol levels with impaired. Hasselbalch *et al.* (1982) studied 15 middle-aged brewery workers who were consuming approximately 100 g of ethanol daily. Four of these individuals showed facial mooning, two had a buffalo hump and truncal obesity, and one had hypertension. All had normal basal urinary free cortisol levels, and in only one individual did plasma cortisol levels fail to suppress with oral dexamethasone; this subject was not cushingoid. In this study the possibility that the subjects may not have absorbed the administered dexamethasone was not considered. In order to eliminate this possibility, Fink *et al.* (1981) undertook intravenous dexamethasone suppression tests in ten chronic alcoholics on the day following their admission to hospital for detoxification. Three of the patients were clinically cushingoid, but the remaining seven showed no evidence of pseudo-Cushing's syndrome. All subjects had normal basal plasma cortisol levels, which was attributed to the fact that they had received either chlomethiazole or epanutin for control of withdrawal symptoms. However, four patients showed an impaired plasma cortisol response to dexamethasone, although only one of these was clinically cushingoid. The abnormal response to dexamethasone shown by alcoholic individuals without features of the pseudo-Cushing's syndrome and with normal basal plasma cortisol levels suggests a specific effect of alcohol on the hypothalamus or pituitary.

Likewise, in about 25% of chronic alcoholics, the cortisol response to insulin-induced hypoglycemia may be absent or severely attenuated, suggesting that alcohol may blunt the sensitivity of the hypothalamic–pituitary axis to the stress of the hypoglycemia (Wright *et al.*, 1976; Chalmers *et al.*, 1977).

Ethanol intoxication and withdrawal are associated with increased adrenergic activity in the brain, a factor known to inhibit hypothalamic release of CRF and thereby impair the normal ACTH stress response (Littleton, 1980; Tabakoff and Hoffman, 1980).

3.3.2.4. Changes in Hormone Binding and Peripheral Hormone Metabolism

The liver is the major organ of catabolism of adrenal corticosteroids, so that it might be expected that abnormalities of clearance would occur in patients with alcohol-related liver injury. Evidence suggests that, in cirrhotic patients, the cortisol production rate may be reduced (Cope and Black, 1958; Peterson, 1960; Zumoff *et al.*, 1967; Scavo *et al.*, 1974; McCann and Fulton, 1975), the clearance of radiolabeled cortisol, infused unlabeled cortisol, and ACTH may be impaired (Englert *et al.*, 1957; Peterson, 1960; Scavo *et al.*, 1974; Del Prete *et al.*, 1975; Andreani *et al.*, 1976; Rosman *et al.*, 1982), and the cortisol turnover rate is decreased (Peterson, 1959, 1960; Peterson *et al.*, 1955; Scavo *et al.*, 1974).

In patients with liver disease, the activities of the steroid-catabolizing enzymes are impaired, so that the pathway of cortisol breakdown is shifted in favor of hydroxylated rather than A-ring-reduced metabolites (Brown *et al.*, 1954; Klein *et al.*, 1955; Peterson, 1960; Zumoff *et al.*, 1967). The hydroxylated metabolites are not measured in the routine procedure for urinary 17-OHCS, which explains their low levels in patients with liver disease despite normal plasma cortisol levels (Peterson *et al.*, 1955; Bajpai *et al.*, 1973). These hydroxylated metabolites may have glucocorticoid activity and hence feedback effects on the hypothalamic–pituitary axis, in contrast to the common reduced metabolites, which are without hormonal effect (Gordon *et al.*, 1968). Additionally, the formation of glucuronic conjugates is reduced in patients with significant liver disease (Peterson *et al.*, 1955).

In individuals with liver disease the transport and distribution of cortisol to its metabolic sites is altered because of changes in circulating CBG and albumin concentrations. Plasma CBG concentrations are significantly reduced in individuals with cirrhosis, and plasma albumin levels may be at the lower limits of the reference range or else are reduced (Daughaday, 1958; Fanghänel *et al.*, 1976). If plasma CBG concentrations are low, the percentage of unbound cortisol increases, so that even if total plasma cortisol levels are reduced plasma ACTH values will remain normal (Dioguardi *et al.*, 1974).

Changes in hormone-metabolizing enzymes and in hormone binding can also occur in individuals abusing alcohol who do not have liver disease. Thus, alcohol may alter the activity of enzymes involved in the peripheral metabolism of steroid hormones (Lieber, 1973), and low CBG values have been described in two alcoholics with pseudo-Cushing's syndrome, but with no evidence of alcohol-related liver injury (Frajria and Angeli, 1977). If in these patients the unbound fraction of cortisol is increased and the feedback response diminished, then tissue exposure to cortisol will increase. In certain susceptible individuals this could lead to the development of cushingoid features.

3.4. Adrenocortical Function: Mineralocorticoid Secretion

Alcohol has a marked effect on the renal excretion of sodium. Nicholson and Taylor (1938) gave 180 to 212 ml of 95% ethanol to a group of normal volunteers over an 8-hr period and found that the initial diuresis was followed by a period of fluid retention even though the subjects continued to drink alcohol. Marked sodium retention occurred both during and after the diuretic phase. These findings have since been confirmed by other investigators (Rubini et al., 1955). During more prolonged periods of drinking, sodium retention persists despite the development of hypernatremia and an increase in serum osmolality not solely attributable to the increase in serum sodium levels (Ogata et al., 1968). In dogs, prolonged ingestion of 2 to 4 g/kg body weight of ethanol daily for 8 weeks results in sodium and water retention with consequent increases in plasma volume, extracellular fluid volume, and total body water (Beard et al., 1965).

In view of the known effects of alcohol on the adrenocortical secretion of cortisol, it was logical to suppose that alcohol might stimulate aldosterone secretion. Alcohol in intoxicating doses has little or no effect on plasma aldosterone levels in normal subjects, although withdrawal from alcohol is associated with a significant increase in plasma aldosterone concentrations, probably reflecting a nonspecific stress effect (Farmer and Fabre, 1975; Linkola et al., 1976, 1979). If alcohol is given to previously abstinent alcoholics, the aldosterone production rate increases during the first few days but then falls to basal levels or below as drinking continues (Farmer and Fabre, 1975).

At present, therefore, there is no convincing evidence that aldosterone plays a significant role in the sodium retention associated with prolonged drinking.

In patients with liver disease, the aldosterone secretion rate is increased (Dobrzanski and Pieschl, 1976), in part because of increased activity in the renin–angiotensin system. In addition, the aldosterone clearance rate is decreased because its hepatic metabolism is impaired (Laragh et al., 1962; Mendelson et al., 1971). In consequence, plasma aldosterone levels are elevated in these patients (Bongiovanni and Eisenmenger, 1951; Chart and Shipley, 1953; Luetscher and Johnson, 1954; Laragh et al., 1962; Wolff et al., 1962; Lommer et al., 1968; Coppage et al., 1972; Wright et al., 1976). However, in cirrhotic patients with normal renal function, hyperaldosteronism is unlikely to be of major importance in the genesis of sodium retention.

3.5. Adrenal Medullary Function

3.5.1. Description

Since the original observation made by Matunga (1942) that cutting the splanchnic neural outflow in experimental animals prevented the hyperglycemic response to alcohol administration, considerable evidence has accumulated, in both experimental animals and in humans, to show that alcohol stimulates adrenal medullary secretion of catecholamines.

In normal human volunteers doses of alcohol in the range 0.27 to 0.71 g/kg body weight increase plasma and/or urinary catecholamine concentrations (Kinzius, 1950; Abelin et al., 1958; Perman, 1958, 1961b; Anton, 1965). However, in a group of alcoholic individuals who had been abstinent from alcohol for a week, acute alcohol ingestion had no effect on urinary catecholamine excretion even though blood alcohol levels of 330 mg/100 ml (72 mM) were attained. This may reflect a degree of tolerance to the effects of acute alcohol ingestion on catecholamine secretion (Carlsson and Häggendal, 1967).

Ogata et al. (1971) studied the effects of more prolonged drinking on catecholamine excretion in four abstinent male alcoholics allowed limited access to alcohol for 20 days and subsequently free access to alcohol for a further 20 days. Increases were observed in urinary epinephrine, norepinephrine, metanephrine, and normetanephrine excretion during both drinking periods and were more marked during the period of free drinking, when blood alcohol levels were significantly higher. It is of interest that catecholamine excretion did not diminish during the period of alcohol ingestion, suggesting that the sympathetic system and the adrenal medulla might not develop tolerance to chronic alcohol ingestion.

3.5.2. Mechanisms of the Changes

Studies to date suggest that the anxiety and hyperactivity associated with alcohol intoxication and withdrawal activate catecholamine release from the adrenal medulla and/or neural tissue (Giacobini et al., 1960a,b; Ogata et al., 1971). However, there is also evidence to suggest that alcohol specifically affects adrenal medullary activity and the peripheral metabolism of the released catecholamines (Anton, 1965; Davis et al., 1967a,b; Ogata et al., 1971).

In four abstinent male alcoholics allowed access to alcohol over a period of 40 days, there was a marked decrease in urinary 3-methoxy-4-hydroxymandelic acid (vanillylmandelic acid; VMA) excretion but a corresponding increase in urinary 3-methoxy-4-hydroxyphenylglycol (MHPG) excretion. This represents a shift in catecholamine metabolism from the oxidative to the reductive pathway. Tracer studies using radiolabeled norepinephrine confirm this shift in peripheral amine metabolism following alcohol (Davis et al., 1967a,b), which could result from (1) the change in hepatic redox potential associated with alcohol consumption, (2) competition between catecholamines and aldehyde for aldehyde dehydrogenase, which is responsible for the conversion of 3-methoxy-4-hydroxymandelic aldehyde to VMA, or (3) increased activity of alcohol dehydrogenase induced by alcohol (Smith and Gitlow, 1967).

The clinical significance of altered catecholamine secretion and metabolism in humans is not known. However, it would seem sensible to obtain details of alcohol consumption from patients suspected of having a pheochromocytoma or other disorder of the adrenal medulla.

3.6. Growth Hormone

Changes occur in plasma growth hormone levels and in somatomedin activity in individuals abusing alcohol that appear to be without clinical consequence, although an alcohol-related defect in growth hormone secretion may be present in some individuals who develop alcohol-induced hypoglycemia.

3.6.1. Laboratory Findings

In alcoholic individuals free of liver disease, basal plasma growth hormone levels are normal (Ganda et al., 1978) or high (Kuska et al., 1978). These individuals show normal growth hormone responses to hypoglycemia (Riesco et al., 1974), but some show abnormal growth hormone responses to TRH (Van Thiel et al., 1978), L-dopa (Riesco et al., 1974), glucagon (Riesco et al., 1974; Kuska et al., 1978), and glucagon and propranolol (Ganda et al., 1978). In alcoholics with significant liver disease, basal plasma growth hormone levels are usually raised (Hernandez et al., 1969; Samaan et al., 1969; Conn and Daughaday, 1970; Toccafondi et al., 1970; Cameron et al., 1972; Greco et al., 1974; De Portugal-Alvarez et al., 1974; Wu et al., 1974; Pimstone et al., 1975; Panerai et al., 1977; Zanoboni and Zanobini-Muciaccia, 1977; Kuska et al., 1978; Yamaguchi et al., 1979). Abnormal or paradoxical increases

in plasma growth hormone levels are seen after an oral or intravenous glucose load (Samaan et al., 1969; Conn and Daughaday, 1970; Toccafondi et al., 1970; Descos et al., 1974; Greco et al., 1974) and after arginine (Greco et al., 1974; Ogura, 1975; Kasperska-Czyzykowa and Rogala, 1978), glucagon (Ghirlanda et al., 1976; Kuska et al., 1978), TRH (Phillips and Safrit, 1971; Panerai et al., 1977), and L-dopa (Greco et al., 1974). Plasma growth hormone responses to insulin-induced hypoglycemia are variable in these patients (Arky and Freinkel, 1964; De Portugal-Alvarez et al., 1974; Andreani et al., 1976; Tamburrano et al., 1976; Priem et al., 1976; Wright et al., 1976; Chalmers et al., 1977; Ganda et al., 1978). Somatomedin activity is reduced in the presence of liver disease (Marek et al., 1974; Wu et al., 1974; Takano et al., 1977).

3.6.2. Mechanisms of the Changes

3.6.2.1. Hypothalamic–Pituitary Dysfunction

The plasma growth hormone response to acute alcohol in normal humans is variable and may depend, in part, on the period of fasting before investigation (Arky and Freinkel, 1964; Bellet et al., 1971; Bagdade et al., 1972; Toro et al., 1973; Andreani et al., 1976). However, in normal subjects, alcohol suppresses nonsleep-related bursts of growth hormone secretion (Leppäluoto et al., 1975) and causes attenuation of the plasma growth hormone responses to insulin-induced hypoglycemia (Priem et al., 1976) and arginine (Andreani et al., 1976; Tamburrano et al., 1976). This suggests that alcohol has an acute effect on growth hormone secretion that is mediated either directly via the hypothalamic–pituitary axis or indirectly via the cerebral neurotransmitters that govern growth hormone release.

In alcoholic individuals, basal plasma growth hormone levels may be elevated, and the dynamic responses to various stimuli may be paradoxical or attenuated (Arky and Freinkel, 1964; Samaan et al., 1969; Conn and Daughaday, 1970; Greco et al., 1974; De Portugal-Alvarez et al., 1974; Priem et al., 1976; Chalmers et al., 1977; Panerai et al., 1977; Zanoboni and Zanobini-Muciaccia, 1977; Kuska et al., 1978), probably because of an effect of alcohol on the hypothalamic–pituitary axis or on central neurotransmitter release. The hyperestrogenemia often observed in male alcoholics may also contribute to the altered hypothalamic–pituitary control of growth hormone secretion (Frantz and Rabkin, 1965). Estrogens are known to increase growth hormone responses to various stimuli, and basal plasma growth hormone levels are raised in males given exogenous estrogen (Carlson *et al.*, 1973).

3.6.2.2. Metabolic Changes

Changes in circulating free fatty acid levels and protein concentrations and changes in glucose homeostasis all affect growth hormone secretion.

Bellet *et al.* (1971) gave alcohol acutely to healthy volunteers after an overnight fast and observed increases in plasma growth hormone levels associated with a fall in circulating free fatty acid concentrations. If normal subjects are given alcohol acutely, either orally or intravenously, after a prolonged fast, they develop hypoglycemia and a marked increase in plasma free fatty acid concentrations (Arky and Freinkel, 1964; Bagdade *et al.*, 1972; Andreani *et al.*, 1976). Under these circumstances plasma growth hormone levels either fall (Arky and Freinkel, 1964) or remain essentially unchanged (Bagdade *et al.*, 1972; Andreani *et al.*, 1976). Increased plasma free fatty acid concentrations are known to impair the plasma growth hormone response to insulin-induced hypoglycemia (Blackard *et al.*, 1971). Thus, the impairment observed in plasma growth hormone response to alcohol and alcohol-induced hypoglycemia in alcoholic individuals may reflect concomitant increases in circulating free fatty acid levels. However, plasma free fatty acid levels and the plasma growth hormone responses to various stimuli do not correlate significantly (Bellet *et al.*, 1971; Chalmers *et al.*, 1977). In addition, in a study by Andreani *et al.* (1976) in which alcohol was given by infusion to normal volunteers after an overnight fast, plasma free fatty acid and growth hormone levels both fell, and the subsequent plasma growth hormone response to an arginine infusion was impaired.

The abnormalities observed in plasma growth hormone responses in alcoholic individuals have been ascribed to changes in glucose homeostasis, but there is little evidence to support this contention. Thus, Samaan *et al.* (1969) found no significant correlation between plasma growth hormone response to an oral glucose load and the degree of impairment of glucose tolerance in alcoholic individuals, whereas Yamaguchi *et al.* (1979) showed that the paradoxical rise in plasma growth hormone levels following exogenous TRH could not be abolished by hyperglycemia. In addition, no correlation exists between the magnitude of the plasma growth hormone response to various stimuli and plasma glucose, insulin, or glucagon levels (Conn and Daughaday,

1970; Toccafondi *et al.*, 1970; De Portugal-Alvarez *et al.*, 1974; Kuska *et al.*, 1978).

3.6.2.3. Changes in Peripheral Hormone Metabolism

Basal plasma growth hormone concentrations tend to be raised in patients with cirrhosis; the highest levels are frequently observed in patients with the most severe disease (Toccafondi *et al.*, 1970). This finding has been attributed to a decrease in the metabolic clearance of the hormone, but the evidence for this is inconclusive (Cameron *et al.*, 1972; Taylor *et al.*, 1972; Owens *et al.*, 1973; Pimstone *et al.*, 1975).

In cirrhotic patients the expected gradient of somatomedin activity across the liver is absent, suggesting a decrease in hepatic production, probably as a result of hepatocellular dysfunction (Schimpff *et al.*, 1977, 1978). It is also possible that this reduction in somatomedin production results from the hyperestrogenemia commonly observed in these patients, since estrogens are known to decrease the somatomedin response to growth hormone in humans (Wiedemann and Schwartz, 1972; Wiedemann *et al.*, 1976).

3.7. Antidiuretic Hormone and Oxytocin

Ethanol produces a water diuresis and increase urine flow in humans, most likely because it inhibits antidiuretic hormone (ADH) secretion. It has also been suggested that ADH may play a significant role in the development of tolerance to and physical dependence on alcohol (Hoffman *et al.*, 1978). However, further study is required to assess this contention. Ethanol inhibits oxytocin secretion and has been used to inhibit the onset of premature labor (Ganda *et al.*, 1978).

3.7.1. Mechanisms of the Changes

In 1932, Murray suggested that alcohol was a diuretic and demonstrated that its diuretic effect could be abolished by administration of pituitary extract. In 1951, Van Dyke and Ames (1951) showed that alcohol did not increase urine output further in dogs with posterior pituitary insufficiency and diabetes insipidus. Thus, the concept arose that alcohol produced its diuretic effect by inhibiting secretion of ADH from the posterior pituitary. In recent years this has been confirmed by direct measurement of the effects of alcohol on ADH levels in blood (Helderman *et al.*, 1978; Linkola *et al.*, 1978) and urine (Marquis *et al.*, 1975).

The diuretic effect of ethanol is only apparent in normally hydrated individuals at a time when the blood alcohol levels are rising, and it can be overcome by administration of ADH (Haggard *et al.*, 1941; Eggleton, 1942; Kleeman, 1972). Ethanol will not induce a diuresis if the extracellular fluid is hyperosmolar, as this stimulus to ADH release overrides the inhibitory effect of alcohol.

Very little information is available on the effects of ethanol on oxytocin secretion. Ethanol inhibits oxytocin release in human postpartum nursing mothers (Wagner and Fuchs, 1968). In pregnant women oxytocin infusions have been used to induce labor. Conversely, ethanol infusions are used to arrest the onset of premature labor (Fuchs *et al.*, 1967). It is generally assumed, therefore, that the effects of ethanol on oxytocin secretion are similar to its effects on ADH secretion.

3.7.2. Alcohol and Adenosine 3′,5′-Cyclic Monophosphate (cAMP)

Short-chain alcohols have been shown to enhance adenylate cyclase activity *in vitro* and to stimulate hormone-mediated adenylate cyclase activity. There are, however, no systematic studies of the effects of alcohol on cAMP *per se*.

Gorman and Bitensky (1970) showed that in mammalian liver, short-chain alcohols increased basal adenylate cyclase activity. These alcohols also produced a marked but reversible increase in glucagon-responsive adenylate cyclase activity but had no effect on the enzyme response to epinephrine (Gorman and Bitensky, 1970). In these experiments phosphodiesterase activity was unchanged, and there was no evidence that the activation of the adenylate cyclase resulted from increased magnesium binding. In contrast, Birnbaumer *et al.* (1969) showed that hormonal activation in adipocytes did result from enhanced magnesium binding.

Many workers have suggested that the effects of alcohol on adenylate cyclase may be secondary to changes in membrane fluidity. However, the effect of membrane fluidity on adenylate cyclase activity is disputed. Thus, Hanski *et al.* (1979) reported an increase in membrane fluidity and in adenylate cyclase activity in turkey erythrocytes following addition of *cis*-vaccenic acid, whereas Sinensky *et al.* (1979) found that a decrease in membrane fluidity in Chinese hamster ovary cells was accompanied by an increase in adenylate cyclase activity.

A major problem with many of the studies performed to date is that high concentrations of alcohol have been employed. It is unclear, therefore, whether any of the *in vivo* effects of ethanol on hormonal functions can be attributed to changes in adenylate cyclase activity. More studies are obviously needed.

3.7.3. Insulin and Glucagon

The changes that occur in plasma insulin and glucagon concentrations in alcoholic individuals, whether or not they have associated liver disease, are usually without clinical consequence. Changes in these two hormones do not play a major role in the pathogenesis of alcohol-related hypo- and hyperglycemia.

Alcohol has no effect on basal plasma insulin levels in normal subjects (Bellet *et al.*, 1971; Nikkilä and Taskinen, 1975) or in chronic alcoholics free of liver disease (Farmer *et al.*, 1971). However, alcohol given orally or intravenously to normal subjects and to noninsulin-dependent diabetics enhances the plasma insulin response to a glucose load (Metz *et al.*, 1969; Dornhorst and Ouyang, 1971; Friedenberg *et al.*, 1971; Phillips and Safrit, 1971; McMonagle and Felig, 1975; Nikkilä and Taskinen, 1975; O'Keefe and Marks, 1977) and similarly influences the responses to tolbutamide (Kühl and Andersen, 1974) and arginine (Andreani *et al.*, 1976). Chronic alcoholics, on the other hand, show enhanced plasma insulin responses to glucose loading without alcohol pretreatment (Farmer *et al.*, 1971). The mechanism of these enhanced insulin responses is not known.

Glucose intolerance is often observed in alcoholic cirrhotic subjects and indeed in individuals with chronic liver disease of other etiologies (Samols and Holdsworth, 1968). It is generally without clinical consequence and probably reflects the inability of the damaged liver to store glycogen. In patients with alcoholic cirrhosis, plasma insulin levels, both in the fasting state and in response to oral or intravenous glucose, amino acids, and tolbutamide, are consistently raised (Megyesi *et al.*, 1967; Conn and Daughaday, 1970; Greco *et al.*, 1974; De Portugal-Alvarez *et al.*, 1974; Johnston and Alberti, 1976; Stewart *et al.*, 1983). The hyperinsulinemia results from impaired breakdown of insulin as a result of liver cell damage (Johnston and Alberti, 1976; Johnson *et al.*, 1978; Smith-Laing *et al.*, 1979) rather than as a consequence of hypersecretion and portal–systemic shunting of blood (Megyesi *et al.*, 1967). This combination of glucose intolerance and hyperinsulinemia suggests the presence of peripheral insulin resistance.

The effects of ethanol on fasting plasma glucagon levels and on the plasma glucagon responses to glucose loading in normal subjects and in alcoholic individuals free of liver disease have not been explored. Experimentally, ethanol appears to stimulate glucagon secretion in rats (Jauhonen, 1978). Fasting total plasma immunoreactive glucagon values are increased in patients with cirrhosis (Marco *et al.*, 1973; Greco *et al.*, 1974; Sherwin *et al.*, 1974; Smith-Laing *et al.*, 1979); they do not suppress following glucose (Smith-Laing *et al.*, 1979) and increase abnormally after alanine and arginine stimulation (Greco *et al.*, 1974).

Controversy exists over the etiology of this hyperglucagonemia. Alford *et al.* (1979) measured total plasma immunoreactive glucagon in a group of cirrhotic patients both before and after they underwent portal–systemic shunt surgery. Before surgery, basal plasma glucagon levels wee raised, and the metabolic clearance rate of glucagon was normal. Following surgery, the basal plasma glucagon levels remained elevated, but the metabolic clearance rate decreased. The authors concluded that the hyperglucagonemia observed before shunting reflected hypersecretion, whereas following surgery the raised plasma glucagon levels reflected both hypersecretion and impaired degradation. Sherwin *et al.* (1978), however, measured the 3485-da glucagon in plasma in cirrhotic patients with and without demonstrable portal–systemic shunting of blood and found that basal plasma glucagon levels were elevated only in the patients with portal venous bypass but that the metabolic clearance rate was normal in all patients studied. They suggested that the hyperglucagonemia observed in cirrhotic patients with portal–systemic shunting resulted from hypersecretion of glucagon. The stimulus for glucagon hypersecretion in these patients remains unknown.

3.8. Pancreatic and Gastrointestinal Hormones

It has been postulated that alcohol consumption modifies pancreatic and gastrointestinal hormone release. However, few systematic studies have been carried out in humans, and those available provide conflicting results. More studies are needed to establish the role of acute and chronic alcohol intake on the hormonal regulation of the gut and pancreas.

3.8.1. Secretin and Pancreatic Polypeptide

Secretin is released from the duodenum and jejunum in response to gastric acid and stimulates the secretion of pancreatic juice rich in water and electrolytes. Straus *et al.* (1975) showed that oral ingestion of alcohol in amounts normally consumed during social drinking resulted in a rapid and striking increase in plasma immunoreactive secretin levels in normal humans. The rapidity of the response and the fact that serum gastrin levels did not change led these authors to suggest that alcohol directly stimulated secretin release. Conversely, Llanos *et al.* (1977) showed that although oral alcohol stimulated secretin release in normal individuals, this response was not rapid but delayed. In addition, they showed that oral alcohol also stimulated gastrin release, whereas intraduodenal alcohol had no effect on the secretion of either hormone. Thus, these authors suggested that alcohol stimulates secretin release indirectly by liberating gastric acid (Llanos *et al.*, 1977). Henry *et al.* (1981) found, however, that oral ingestion of 60 ml of vodka (19 g of ethanol) had no effect on secretin release in normal individuals. Obviously the question of whether alcohol stimulates secretin release remains to be answered.

Pancreatic polypeptide is secreted by the F cells of the pancreas; secretion is inhibited by pancreatic somatostatin and appears to be under vagal control. This hormone exhibits a variety of actions, including stimulation of glycogenolysis and the secretion of gastric and pancreatic enzymes. Fink *et al.* (1983) measured plasma pancreatic polypeptide concentrations in nine chronic alcoholic men clinically free of significant complicating illness both before and 3 hr after a test meal. Mean plasma pancreatic polypeptide concentrations were consistently higher in alcoholics than in healthy controls, not only on admission to hospital but also 2 weeks later. The significance of this finding is not yet clear.

3.8.2. Gastrin

The term gastrin is used for a family of polypeptides present in cells of the antral pyloric glands. Gastrin is the most potent known stimulant of gastric acid secretion. The major physiological stimulus for gastrin secretion is the presence of food, most notably protein, in the stomach. Ethanol ingestion stimulates gastric acid production, but controversy exists as to whether it does so directly by stimulating the parietal cells or indirectly via release of gastrin. Two groups of workers (Becker *et al.*, 1974; Llanos *et al.*, 1977) have shown that ingestion of moderate amounts of alcohol will cause significant increases in plasma gastrin levels in healthy individuals, although lower doses were without effect (Becker *et al.*, 1974). Equally, two other groups of workers (Straus *et al.*, 1975; Henry *et al.*, 1981) have shown that alcohol ingestion by healthy control subjects was unaccompanied by an increase in plasma gastrin levels. However, the latter two groups of workers used only very small amounts of alcohol.

Little is known about the effects of chronic alcohol ingestion on gastrin release in humans. Treffort *et al.* (1975), however, showed that the gastrin response to meals was greater in dogs fed alcohol chronically than in control animals.

3.9. Alteration of Carbohydrate Metabolism by Alcohol

The interactions of alcohol and carbohydrate metabolism are complex. In order to simplify the analysis, the changes that occur in the "fed state" or anabolic phase will be distinguished from those produced in the "fasted state" or catabolic phase. The influence of ethanol on the blood glucose concentrations varies greatly with the nutritional state. Indeed, by and large in the well-nourished state, the blood glucose level increases after ethanol ingestion, whereas in the starved state, in which hepatic glycogen reserve is diminished, the blood concentration does not increase after ethanol administration (Tennent, 1941) and may even fall.

3.9.1. Interaction of Ethanol and Carbohydrate in the Anabolic Phase of Carbohydrate Metabolism

After their ingestion, carbohydrates undergo gastrointestinal digestion and intestinal absorption. These processes are little affected by ethanol, as discussed in detail in Chapter 10.

The effects of ethanol on glucose metabolism during this anabolic phase are not well defined in adipose tissue and muscle. At high concentrations, ethanol inhibits glucose uptake in peripheral tissues (Lochner *et al.*, 1967), possibly as a consequence of the acetate generated by ethanol, as discussed in Chapter 2.

During the anabolic phase of glucose metabolism, the liver plays a major role, since it retains approximately two-thirds of a glucose load administered orally. Insulin, the release of which is stimulated by the glu-

cose, favors conversion of glucose to glycogen in the liver. Thus, insulin suppresses hepatic glucose output, and it is therefore not surprising that in humans splanchnic glucose output diminishes during the administration of alcohol (Lundquist *et al.*, 1962). After a large dose of ethanol, however, there can be an acute discharge of epinephrine by the adrenals, possibly mediated by acetaldehyde resulting from the oxidation of ethanol (see Chapter 2). The ensuing activation of hepatic phosphorylase may account for the rise in plasma glucose sometimes encountered after ingestion of alcohol.

In addition to these actions on glucose metabolism, ethanol also affects metabolism of other sugars such as galactose and fructose. After its absorption, galactose is converted to glucose in the liver as follows:

$$\text{Galactose} + \text{ATP} \rightarrow \text{galactose-1-P} + \text{ADP} \qquad (1)$$

$$\text{Galactose-1-P} + \text{UDP-glucose} \leftrightarrow \text{UDP-galactose} + \text{glucose-1-P} \qquad (2)$$

$$\text{UDP-Galactose} \longleftrightarrow \text{UDP-glucose} \qquad (3)$$

The last reaction involves UDP-galactose-4-epimerase and oxidized diphosphopyridine nucleotide (NAD) and is inhibited by the reduced cofactor (NADH).

Alcohol decreases galactose utilization and increases its excretion in the urine (Bauer and Wozasek, 1934; Tygstrup and Lundquist, 1962). The inhibition occurs at reaction 3, caused by the rise in NADH/NAD (Isselbacher and McCarthy, 1960; Isselbacher and Krane, 1961). In the alcoholic with liver disease, galactose utilization is less inhibited by ethanol than in normals (Tygstrup, 1964). Similarly, in some alcoholics, the redox change produced by an acute dose of alcohol was attenuated. At the time, this was attributed to malnutrition (Salaspuro and Kesaniemi, 1973). More recently, such a phenomenon was observed after chronic alcohol feeding even in well-nourished baboons. In the animals fed ethanol, their capacity to remove ethanol from the blood increased progressively, whereas the associated redox change (as measured by galactose clearance) decreased (Salaspuro *et al.*, 1981). It has now become clear that, independent of possible effects of liver disease, chronic ethanol consumption itself results in a progressive attenuation of the redox change secondary to acute ethanol administration. Indeed, in rats fed alcohol as part of a nutritionally adequate liquid diet, after only 3 or 4 weeks the striking redox changes produced by the acute dose of ethanol in control animals is markedly attenuated (Domschke *et al.*, 1974). Such an attenuation of the redox changes after chronic ethanol consumption was confirmed by Khanna *et al.* (1975) in the rat and by Salaspuro *et al.* (1981) in the baboon. In the baboon, however, the attentuation of the redox change produced by an acute dose of ethanol occurs more slowly than in the rat. After 3 months of ethanol consumption, the acute redox change was still apparent and disappeared only after a longer period of time, whereas in rats, this effect was already seen after about 1 month (Domschke *et al.*, 1974).

The mechanism involved appears to be, at least in part, a decrease in cytosolic alcohol dehydrogenase, documented in rats (Lieber and DeCarli, 1970) and in baboons (Salaspuro *et al.*, 1981). As discussed in Chapter 1, when alcohol dehydrogenase decreases, it may become rate limiting for the metabolism of ethanol via this pathway. As a consequence, the amount of NADH produced is diminished, and no redox change will ensue. Overall ethanol metabolism is nevertheless increased under these conditions because of the adaptive rise in the activity of the microsomal ethanol-oxidizing system. Thus, the lack of inhibition of galactose metabolism observed after chronic alcohol consumption is caused not only by the much later development of liver disease but also by the shift in ethanol metabolism from the alcohol dehydrogenase to the microsomal ethanol-oxidizing system (MEOS) pathway. As discussed in Chapter 1, increased reoxidation of NADH by the mitochondria might also play role in this attenuation of the redox change.

Another sugar of special metabolic importance for ethanol is fructose. Its effects on ethanol metabolism have been discussed in Chapter 1. Not only does fructose accelerate ethanol metabolism but, conversely, in the presence of ethanol, a greater portion of the fructose is converted to glycerol than when fructose is ingested alone. Indeed, fructose, after phosphorylation to fructose-1-phosphate, is split into dihydroxyacetone phosphate, which, under the influence of NADH generated by alcohol metabolism, is converted to α-glycerol phosphate. In addition, the D-glyceraldehyde resulting from the fructose-1-phosphate can also be converted to glycerol under the influence of the excess NADH generated by ethanol metabolism.

3.9.2. Alteration in the Metabolism of Carbohydrate in the Catabolic Phase

Hepatic glycogenolysis is augmented after the ingestion of alcohol. This results from the release of epinephrine by the adrenal glands and the subsequent activation of hepatic adenylate cyclase. Enhanced glycogenolysis results in increments in the plasma glucose.

Indeed, release of epinephrine from the adrenal medulla is increased by ethanol (Perman, 1961b, 1962), and its effect can be blunted by the administration of ganglionic blockers (Klingman *et al.*, 1958). In addition, glucagon can also be affected by ethanol. Ethanol infusion increases plasma glucagon concentrations in fasting pigs (Tiengo *et al.*, 1974) and humans (Palmer and Ensinck, 1975). These studies demonstrated that at least moderate hypoglycemia is required for glucagon release. In rats, plasma immunoreactive glucagon doubled after an acute dose of ethanol (Jauhonen, 1978) without changes in insulin and with an increase in cAMP. On the other hand, Forsander *et al.* (1965) also reported an increased glycogenolysis in the absence of hormones in the ethanol-perfused rat liver. The magnitude of the increase in liver glucose output caused by alcohol depends on the amount of glycogen present and the enzymatic mechanism available. Liver glycogen, however, is reduced after acute (Jauhonen, 1978) and chronic (Mirone, 1965; Lefevre *et al.*, 1970) alcohol administration. Moreover, the activation of the adenylate cyclase system by alcohol may not be striking: the enzyme lacks sufficient substrate because hepatic ATP is often decreased after chronic alcohol consumption (French, 1966; see also related discussion in Chapter 1). In contrast with chronic alcohol consumption, hepatic ATP was reported to be increased after an acute dose (French, 1966; Ourea *et al.*, 1967), although this has been a subject of controversy (Hyams and Isselbacher, 1964; Gajdos *et al.*, 1967).

The pathways of gluconeogenesis are altered by ethanol and its metabolites. Alcohol affects amount and uptake of precursors by the liver and the kidney. Alcohol decreases the blood level of alanine, an important glucogenic amino acid (Felig *et al.*, 1969), because of both reduced delivery from muscle and enhanced liver uptake (Pozefsky *et al.*, 1969). Kreisberg *et al.* (1972) reported that in normal volunteers in the postabsorptive state, acute ethanol (19.6 g ethanol, followed 1 hr later by 9.8 g ethanol) inhibited gluconeogenesis from alanine and reduced plasma alanine levels by 25%, although clearance did not change. There was an increased recovery of [14C]alanine in lactate. It was suggested that ethanol might inhibit gluconeogenesis from alanine by diminishing the production and/or release of alanine from skeletal muscle.

The increased hepatic NADH during the oxidation of alcohol (see Chapter 1) also decreases the incorporation of amino acids into glucose by inhibiting the conversion of glutamic acid to α-ketoglutarate through interference with the activity of glutamic dehydrogenase (Friden, 1959). Indeed, in liver slices, alcohol depresses the formation of [^{14}C]glucose from ^{14}C-labeled alanine, glutamic acid, lactic acid, and glycerol (Freinkel *et al.*, 1969).

Glycerol is also an important building block for glucose production. It is released from adipose tissue during the lipolysis that is associated with the fasting state. As discussed in Chapter 4, however, an acute dose of ethanol decreases glycerol release from adipose tissue (Feinman and Lieber, 1967). This is a short-term effect; thereafter, mobilization of glycerol returns to normal or even supernormal levels. However, ethanol decreases the uptake of glycerol by the splanchnic area. The first stage in glycerol metabolism is phosphorylation by glycerokinase to α-glycerophosphate. In some way this process must be inhibited by excess NADH produced during ethanol oxidation. The second stage is oxidation of glycerophosphate to dihydroxyacetone phosphate by glycerophosphate dehydrogenase. The dihydroxyacetone phosphate formed in this reaction may be transformed to glucose and glycogen or to pyruvate (and lactate), or it may be reduced again to α-glycerophosphate by glycerophosphate dehydrogenase. This last process is favored by the increased concentration of NADH found during ethanol oxidation. In this way, gluconeogenesis from glycerol is inhibited by ethanol, and a considerable concentration of α-glycerophosphate might be building up, since the possibility of removing dihydroxyacetone phosphate by other routes is limited (Nikkä and Ojala, 1963a,b; Lundquist *et al.*, 1965). As discussed in Chapters 1 and 4, increased α-glycerophosphate favors hepatic triglyceride accumulation by trapping fatty acids (Lundsgaard, 1946; Lundquist *et al.*, 1965). The inhibition of glycerol metabolism by alcohol may result, in part, from the accumulation of α-glycerophosphate secondary to the increase in the NADH-to-NAD ratio. Furthermore, enhanced formation of α-glycerophosphate from glucose may contribute to the alcohol-induced hypoglycemia (Wolfe *et al.*, 1976).

Another important precursor of glucose is lactate. Lactate, which is produced in muscles and erythrocytes, is reconverted in the liver to pyruvate. Pyruvate, in turn, is utilized in part as a glucose precursor. However, the NADH/NAD changes associated with alcohol metabolism favor the conversion of pyruvate to lactate rather than the reverse reaction. Lactate is thereby shifted away from gluconeogenesis. Krebs *et al.* (1969) showed that gluconeogenesis from lactate in the perfused liver of starved rats was inhibited by ethanol and explained this finding by a decreased concentration of pyruvate, which

is one of the factors limiting the rate of the pyruvate carboxylase reaction. The pyruvate carboxylase reaction is a key step in gluconeogenesis from lactate and precursors such as alanine and serine. Decreased conversion of lactate to glucose was also demonstrated in humans (Kreisberg *et al.*, 1971b).

As discussed elsewhere in the book (Chapters 4 and 7), the metabolism of ethanol is associated with decreased activity in the citric acid cycle, which results particularly in the depression of fatty acid oxidation (Lieber and Schmid, 1961). Furthermore, fatty acid oxidation is one of the regulators of gluconeogenesis (Williamson *et al.*, 1966). Therefore, alcohol also impairs the formation of glucose by inhibiting mitochondrial fatty acid oxidation. Moreover, as a first step to proceed to the gluconeogenic pathway, most amino acids (except alanine) are converted to α-ketoglutarate and oxaloacetate, which are tricarboxylic acid (TCA) cycle intermediates, after deamination prior to oxidation. As discussed before, generation of NADH induced by ethanol oxidation suppresses TCA cycle activity, as evidenced by reduction of $^{14}CO_2$ production from [^{14}C]pyruvate (Field *et al.*, 1963), [^{14}C]alanine (Freinkel *et al.*, 1965), [^{14}C]acetate and [^{14}C]palmitate (Lieber and Schmid, 1961; Rebouças and Isselbacher, 1961), and gluconeogenesis from amino acids is therefore inhibited during ethanol oxidation. Of interest are the studies of Lochner *et al.* (1967) on hepatic gluconeogenesis in dogs with chronic end-to-side portacaval shunts starved for 2 to 3 days. Under these circumstances, ethanol produced not only a 60% decrease in hepatic glucose output but also a 23% inhibition of peripheral glucose utilization. Furthermore, the infusion of glutamate and α-ketoglutarate, both NAD-dependent precursors of glucose, failed to augment the depressed rate of hepatic gluconeogenesis induced by ethanol. Methylene blue, a redox dye that favors oxidation of NADH to NAD, prevented the fall in hepatic glucose output, supporting the concept that the increased NADH-to-NAD ratio resulting from ethanol oxidation causes a block in the gluconeogenic pathway.

Another possible contributory mechanism to the hypoglycemia is the reported depression by ethanol of two key gluconeogenic enzyme activities, pyruvate carboxylase and fructose diphosphatase, effects that, interestingly, are preventable by folate (Stifel *et al.*, 1976). On the other hand, experimental data by Madison *et al.* (1967) indicate that the gluconeogenic pathway from fructose-1,6-diphosphate to glucose is not blocked during ethanol-induced suppression of hepatic gluconeogenesis. Furthermore, Ishii *et al.* (1973) showed that chronic ethanol administration to rats increases hepatic microsomal glucose-6-phosphatase activity, an effect that could be glucocorticoid mediated. Indeed, as discussed in Section 3.3, under some circumstances ethanol feeding increases the level of glucocorticoids (Ellis, 1966; Margraf *et al.*, 1967), which in turn increases glucose-6-phosphatase activity (Ashmore and Weber, 1968).

3.9.3. Alcoholic Hypoglycemia

A dramatic but uncommon complication of acute alcohol abuse is severe hypoglycemia, which may be responsible for some of the unexplained sudden deaths in acute alcoholic intoxication. The best known form of alcoholic hypoglycemia is the one that occurs in fasted or severely malnourished individuals who drink large amounts of ethanol, so-called fasting alcoholic hypoglycemia.

3.9.3.1. Fasting Alcoholic Hypoglycemia

3.9.3.1a. Clinical Presentation. The clinical presentation of alcoholic hypoglycemia follows a well-known pattern of hypoglycemia of any etiology and therefore is not described here. It usually becomes manifest 6–36 hr after the consumption of an alcoholic beverage in substantial amounts by a fasting or poorly nourished subject, particularly in children. The most commonly encountered circumstances and clinical presentation have been reviewed elsewhere (Marks and Medd, 1964; Marks, 1978). The most common findings are conjugate deviation of the eyes, extensor rigidity of the extremities, unilateral or bilateral Babinski's reflexes, and convulsions (Madison, 1968). Transient or permanent hemiparesis may occur. Trismus is not an uncommon finding (Brown and Harvey, 1941; De Moura *et al.*, 1967; Neame and Jourbet, 1961), and the combination of trismus and coma in the alcoholic should suggest to the physician the diagnosis of hypoglycemia, particularly if hypothermia, which is common in this condition (Kedes and Field, 1964), is also present (De Moura *et al.*, 1967).

Blood analysis will confirm the hypoglycemia and usually also reveals low plasma insulin levels and therefore makes it possible to rule out hypoglycemia secondary to insulinoma (Kahil *et al.*, 1964). Of clinical relevance is the fact that the diagnosis of hypoglycemia may be easily overlooked because of the associated alcoholic intoxication; under those conditions the coma is often attributed to the alcohol rather than to hypoglycemia.

The attacks occur either as a single event or recurrently (Tucker and Porter, 1941; Fredericks and Lazor, 1963; Freinkel *et al.*, 1963). Fatal outcomes are not uncommon (Madison, 1968).

3.9.3.1b. Pathogenesis and Treatment. Alcohol-induced hypoglycemia, although observed by 1941 by Brown and Harvey (1941), has been clearly attributed to alcohol (rather than to hepatotoxicity from congeners) only since 1963 (Freinkel *et al.*, 1963; Field *et al.*, 1963; Arky and Freinkel, 1966). The pioneering studies of the latter investigators clearly established, under metabolic ward conditions, that administration of pure alcohol to fasting individuals can reproduce the syndrome. The main mechanism appears to be multifactorial inhibition of gluconeogenesis. Many workers have proposed that the inhibition of gluconeogenesis is secondary to the alcohol-induced redox change, as discussed previously. However, it was noted that chronic alcohol consumption is associated with an attenuation of the redox changes produced by an acute dose of ethanol (Domschke *et al.*, 1974; Salaspuro *et al.*, 1981), which helps to explain the observed refractoriness of chronic malnourished alcoholics to alcohol hypoglycemia (Hed and Nygren, 1968; Salaspuro, 1970, 1971). Children, in whom such an attenuation has obviously not occurred, might be expected to have maximal redox changes after ethanol, and, indeed, as mentioned before, alcoholic hypoglycemia is unusually frequent in children: it can even occur after sponge bathing (Moss, 1970).

That hypoglycemia produced after a prolonged (at least 36 hr) fast is related to alcohol metabolism has been confirmed by the demonstration that it can be inhibited in humans by 4-methylpyrazole, an inhibitor of ethanol metabolism (Salaspuro *et al.*, 1977). However, it should be pointed out that the hepatic glycogen concentration in rats fasting for 24 hr decreased to very low values after a single ethanol load, but even so the blood glucose concentration remained constant and within normal values (Jauhonen *et al.*, 1975; Jauhonen, 1978). This means that gluconeogenesis was still keeping pace with peripheral glucose utilization in spite of the known inhibition of gluconeogenesis by ethanol. As discussed before, glucagon is enhanced after ethanol, and the hepatic cAMP is increased, reflecting either the change in glucagon or catecholamine release or both. These normal effects, as expected, do counteract any hypoglycemia tendency, especially since ethanol inhibits the cAMP-induced insulin release (Colwell *et al.*, 1973). Arky and Freinkel (1964) have noted that disorders of other factors, especially glucocorticoids, may contribute to the syndrome. Deficiency of growth hormone secretion has also been incriminated (Andreani *et al.*, 1976; Priem *et al.*, 1976).

Treatment is simple: glucose (20–30 g) should be given either orally or, if that is not possible, intravenously. Glucagon has been shown not to be useful under these circumstances (Andreani *et al.*, 1976; Freinkel *et al.*, 1963).

3.9.3.2. Other Types of Hypoglycemia

It has been pointed out that alcohol might provoke reactive hypoglycemia, which can occur secondary to hyperglycemia after carbohydrate-rich meals. As discussed previously, alcohol may increase plasma insulin secretion under these conditions, with associated enhanced peripheral glucose utilization (Metz *et al.*, 1969; Friedenberg *et al.*, 1971; Nikkila and Taskinen, 1975), thereby favoring reactive hypoglycemia.

The hypoglycemic effect of alcohol can also occur in sulfonylurea-treated diabetics, sometimes leading to hypoglycemic coma and even death (Arky *et al.*, 1968). Alcoholic hypoglycemia can also be potentiated by strenuous exercise (Haight and Keatinge, 1973). Even ingestion of wine, which contains some carbohydrates, will lower blood glucose (on the average by 10%) (Murdock, 1971).

3.9.4. Hyperglycemia

Depending on the conditions, ethanol may accelerate rather than inhibit gluconeogenesis (Krebs *et al.*, 1973). In fact, hyperglycemia has sometimes been described in association with alcoholism, but its mechanism is still obscure. Alcoholic pancreatitis and an increase in circulating catecholamines could be contributory. It must be pointed out, however, that after chronic ethanol ingestion, subsensitivity of the liver to norepinephrine may develop (French *et al.*, 1976).

Glucose intolerance (Phillips and Safrit, 1971) may also be caused, at least in part, by decreased peripheral glucose utilization (Lochner *et al.*, 1967; Kreisberg *et al.*, 1971b), possibly secondary to ethanol-induced ketosis (Lefevre *et al.*, 1970). On the other hand, ethanol has been shown to increase the rate of removal of an intravenous glucose load (Metz *et al.*, 1969) and to improve and accelerate body glucose turnover (Searle *et al.*, 1974). There is so far no adequate explanation for the discrepancy between these findings.

3.9.5. Abnormalities of Carbohydrate Metabolism in Alcoholic Liver Disease

For the past 50 years it has been recognized that carbohydrate intolerance frequently accompanies chronic liver disease, whether or not it is associated with alcoholism (Debry and Charles, 1965). Not only is there an increased frequency of overt diabetes in cirrhotic patients, but there is also impairment of glucose tolerance, and increased insulin resistance can occur without clinical manifestations of diabetes (Creutzfeld *et al.*, 1962). Abnormal glucose tolerance is associated not only with cirrhosis but also with fatty liver (Rehfeld *et al.*, 1973).

Glucose intolerance and frank diabetes mellitus are present in 45% to 70% of patients with liver disease secondary to alcoholism (Samols and Holdsworth, 1968), whereas the incidence of diabetes mellitus in the general population is only 2% to 3%. The standard intravenous glucose tolerance test is less frequently abnormal than the oral test in alcoholic cirrhosis (Moyer and Womack, 1948); some cirrhotics even have a normal intravenous glucose tolerance tests (Amatuzio *et al.*, 1952). The degree and duration of tolbutamide-induced hypoglycemia is greater in patients with alcoholic liver disease than in normal subjects (Kahil *et al.*, 1964). Because of the exaggerated insulin response to oral glucose seen in many cirrhotics, a deficiency in β-cell function is not a likely cause of glucose intolerance (Megyesi *et al.*, 1967; Collins and Crofford, 1969; Samaan *et al.*, 1969). The exaggerated insulin response has also been interpreted as indicative of peripheral resistance to endogenous insulin, as previously recognized in cirrhosis (Danowski *et al.*, 1956). Increased levels of growth hormone have been incriminated (Hernandez *et al.*, 1969; Samaan and Stone, 1967). However, the effect of growth hormone is thought to be mediated by somatomedins (Hall and Luft, 1974), which are found to be decreased in cirrhotics who have elevated growth hormone levels (Wu *et al.*, 1974). Thus, it is questionable whether growth hormone plays a key role in insulin resistance. Ishii *et al.* (1975) showed that the exaggerated insulin response to glucose in the cirrhotic patient is accompanied by an increased diameter of islet cells as well as an increased islet cell volume in autopsied cases of liver cirrhosis when compared with controls and diabetics. On the other hand, Felber *et al.* (1967) suggested that this could result from the elevation of free fatty acids. Thus, the cause of the hyperinsulinemia observed in the cirrhotic patient is complex. Since the liver can remove up to 25% of insulin passing through it, it is possible that hepatic uptake of insulin is impaired by the liver diseases and by the portal–systemic shunting often seen in liver cirrhosis. Many other factors obviously play a role, since these defects do not explain impaired assimilation of rapidly intravenously injected glucose or the absence of evidence that surgical portacaval anastomosis further impairs glucose tolerance in patients with liver disease (Megyesi et al., 1967).

The insulin–glucose ratio is also increased during glucagon stimulation, suggesting increased β-cell sensitivity (Rehfeld et al., 1973). Glucagon levels appear to be affected by the presence of ethanol (Section 3.9.2). In normal human subjects, basal glucagon levels tend to be higher after ethanol ingestion (McMonagle and Felig, 1975). Cirrhosis is associated with hyperglucagonemia (Marco et al., 1973; Sherwin et al., 1974). In addition, cirrhosis is associated with hypoalaninemia (Johnston and Alberti, 1976). It may be that the elevated glucagon levels seen in cirrhosis are responsible for the hypoalaninemia of cirrhosis by a stimulatory effect of glucagon on the hepatic uptake of alanine.

Fasting hypoglycemia is uncommon in alcoholic liver disease (Zimmerman et al., 1953) unless massive hepatocellular destruction is present.

3.9.6. Abnormalities of Carbohydrate Metabolism Secondary to Pancreatic Disease

Abnormal glucose tolerance tests have been observed in 60 to 75% of patients with chronic pancreatitis secondary to alcoholism. In approximately 30% of patients with chronic pancreatitis, diabetes mellitus requires hypoglycemic agents (Chey et al., 1967).

3.10. Effects of Ethanol on Ketone Metabolism

Until the demonstration of a direct effect of alcohol on ketone metabolism (Lefevre et al., 1970), the interrelation between alcohol and ketone metabolism offered a confusing picture. To assess the "antiketogenic" effect of alcohol, ethanol had been given to both diabetic and "normal" subjects, in whom ketonuria was induced by fever or low-carbohydrate diets; alcohol was found to decrease (Benedict and Torok, 1906; Staubli, 1908) or not to affect (Higgins et al., 1916) ketonuria. Most of these studies were brief (Benedict and Torok, 1906) or

involved rather nonspecific chemical assay methods. Ketonemia (Schlierf et al., 1964) after ethanol ingestion has also been noted, but five of the seven subjects were diabetic, and ethanol was not given isocalorically. These various factors may explain the conflicting results obtained.

Clinically, ketonuria has been observed occasionally in acutely intoxicated alcoholic patients and was generally attributed to poor nutrition or prolonged fasting (Dillon et al., 1940). However, our studies (illustrated in Fig. 3.1) revealed that chronic administration of ethanol in association with a calorically adequate fat-containing diet induces marked hyperketonemia and that, for an identical caloric and carbohydrate intake, ethanol is more ketogenic than isocaloric fat (Lefevre et al., 1970).

Theoretically, alcohol feeding could induce the observed hyperketonemia and ketonuria by enhancing hepatic ketone production, by reducing peripheral ketone utilization, or by a combination of both mechanisms. Increased ketone production from exogenous palmitate was indeed observed in liver slices obtained from rats fed alcohol chronically (Lefevre et al., 1970). In addition, theoretically there could be a reduction of peripheral utilization of ketones caused by the elevated plasma acetate levels after ethanol (Crouse et al., 1968). However, acetate added to diaphragm incubations did not alter acetoacetate consumption whether or not the animals were pretreated with alcohol. Moreover, in the fasting state, no ethanol was being metabolized. This makes it unlikely that the ketonemia observed was secondary to a block of muscle utilization of ketones by acetate. Alcohol pretreatment in vivo induced a small decrease (10%) of $^{14}CO_2$ production from labeled β-hydroxybutyrate in incubated diaphragms. No decreased utilization of acetoacetate by the diaphragm was observed either under the influence of alcohol in vitro or after pretreatment of the rat with an alcohol-containing diet. To the extent that the diaphragm can be considered as representing the whole muscular mass, these experiments indicate that the accumulation of blood ketones results to a greater extent from increased hepatic production of ketones (observed in the liver slices) than from decreased utilization in muscle.

It must be pointed out that the alcohol effect on ketogenesis was predominant in the fasting state, when alcohol had disappeared from the blood and when its metabolic effects on the redox potential of the cells had already regressed, as indicated by the normal ratio between β-hydroxybutyrate and acetoacetate. Furthermore, in rats, increased ketone production in liver slices

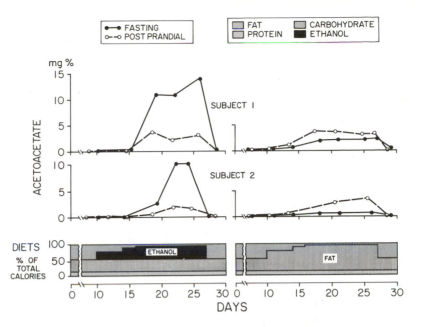

FIGURE 3.1. The effect of isocaloric replacement of dietary carbohydrate (by either alcohol or fat) on blood acetoacetate concentration in two subjects in both the fasting and the postprandial state. (From Lefevre *et al.*, 1970.)

occurred only after several days of ethanol pretreatment, and addition of ethanol *in vitro* was without effect. Similarly, splanchnic ketogenesis was unaffected when ethanol was given in acute experiments to volunteers (Lundquist *et al.*, 1962).

In other studies *in vitro* in the presence of ethanol, hepatic ketogenesis has been reported to be reduced by some researchers (Ontko, 1973) or moderately increased by others (Williamson *et al.*, 1969). Only a moderate increase (Lindros, 1970) was observed after an acute dose of ethanol *in vivo*. Thus, the striking action of ethanol on ketone metabolism is characterized by a delayed change, which suggests that the alcohol effect may not result primarily from the presence of the ethanol and its immediate biochemical consequences but rather from a progressive change in metabolic pathways in the liver, the nature of which has not yet been elucidated. Glycogen depletion may play a permissive role (Lefevre *et al.*, 1970). Furthermore, the more striking ketogenesis in our patients in the fasting state may also reflect increased susceptibility in the fasting liver to some persistent effect of alcohol, including perhaps alterations in mitochondrial function, discussed in more detail in Chapter 7. Although overall fatty acid oxidation is depressed because of inhibition of citric acid cycle activity (see Chapter 4), β-oxidation to the level of acetyl-CoA is actually increased (Cederbaum *et al.*,

1975). Contrasting with the enhanced β-oxidation of fatty acids in hepatic mitochondria of rats fed ethanol chronically, in the presence of ethanol, the opposite occurs (Ontko, 1973). In addition, impaired citric acid cycle activity may play a role in decreasing ketone metabolism.

The ethanol–fat diet contained only a minor proportion of carbohydrates; this factor, although important, is probably not sufficient by itself to account for the ketonemia, since a diet containing 4% of total calories as carbohydrates and 66% as alcohol but with only 5% as fat had no appreciable ketogenic effect (Table 3.1). Furthermore, it has been observed that hyperketonemia produced by starvation in rats can be abolished or reduced by feeding fat in adequate caloric amounts (Mayes, 1962). In humans, except under maximal starvation, a ketogenic diet furnishing over 92% of its calories in the form of fat and thus containing a low amount of carbohydrate may, if given early enough, delay or alleviate ketosis (Freund, 1965). Since our ketogenic regimen contained 15% of calories as protein, which is glucogenic, and 36% as fat (containing glycerol, also a precursor for gluconeogenesis), it seems unlikely that lack of exogenous carbohydrate itself is the sole explanation for the ketotic effect observed with the ethanol diet. The key role played by alcohol was confirmed by our finding that a diet consisting of alcohol

TABLE 3.1.   Effect of Ethanol and/or Dietary Fat on Blood Ketones[a]

Diet Composition (% of total calories)				Number of Subjects	β-Hydroxybutyrate		Acetoacetate	
Ethanol	Fat	Carbohydrate	Protein		Fasting	Postprandial	Fasting	Postprandial
—	36	49	15	4	1.8 ± 0.4	2.1 ± 0.2	0.27 ± 0.06	0.26 ± 0.05
46	36	3	15	4	49.9 ± 6.3	27.4 ± 4.9	7.6 ± 0.9	1.7 ± 0.2
—	82	3	15	4	15.8 ± 3.0	16.2 ± 2.1	2.2 ± 0.5	2.3 ± 0.3
—	5	80 / 70	15 / 25	6	1.2 ± 0.4	1.7 ± 0.6	0.26 ± 0.07	0.29 ± 0.08
46	5	34	15	3	2.9 ± 0.5	3.6 ± 0.5	0.27 ± 0.07	0.17 ± 0.02
60	5	10	25	3	5.2 ± 0.9	4.2 ± 0.8	0.44 ± 0.10	0.29 ± 0.12
66	5	4	25	1	9.5 ± 0.5	7.1 ± 1.9	1.0 ± 0.4	0.53 ± 0.18

[a]From Lefevre *et al.* (1970).

(46% of calories) and fat (36% of calories) is far more ketogenic than a high-fat diet alone (82% of total calories), although both had the same low carbohydrate content (see Table 3.1).

Although the biochemical mechanisms underlying hepatic ketogenesis are much debated (Johnson *et al.*, 1961; Williamson *et al.*, 1968) and may vary in different "ketogenic" conditions, one probably can still accept the classic theory that the formation of ketones is regulated by the combination of (1) an increased fat load on the liver, through increased fat mobilization (fasting) or excessive fat intake (high-fat, low-carbohydrate diets), associated with (2) a relative depletion of hepatic carbohydrate intermediaries necessary for the oxidation of fat through the citric acid cycle. These factors might ultimately result, as discussed by Bremer (1969), in an increase of the ratio of acetyl-CoA over CoA, which could be the determining factor regulating acetoacetate production.

Alcohol does not increase the intestinal absorption of fat (Lieber *et al.*, 1966), nor does it produce peripheral fat mobilization (Jones *et al.*, 1965; Lieber and Spritz, 1966; Feinman and Lieber, 1967) except with very large amounts of ethanol (Brodie *et al.*, 1961; Lieber *et al.*, 1966). The hepatic metabolism of alcohol itself results in the production of two-carbon fragments (Lieber, 1968). The excess of acetyl radicals originating from ethanol, however, cannot suffice by itself to explain the excessive ketone production, since a diet with a large amount of alcohol (but containing only 5% of fat as total calories) did not induce hyperketonemia (see Table 3.1). Moreover, most of the two-carbon fragments derived from ethanol are released from the liver as acetate (Lundquist *et al.*, 1962), which in turn is predominantly metabolized in peripheral tissues (Katz and

Chaikoff, 1955). The fact that a diet containing 46% ethanol and 36% fat is more ketogenic than one containing 82% fat also indicates that ethanol does not act solely as a source of two-carbon fragments. Moreover, it must be pointed out that the alcohol effect on ketogenesis was predominant in the fasting state, when alcohol had disappeared from the blood. Thus, an effect persisting beyond ethanol ingestion must be postulated. An explanation for increased ketone production during fasting was suggested by Bremer (1969), who observed a decreased concentration of propionylcarnitine in the liver of fasting rats; this may, in turn, result in a decrease of propionyl-CoA, which is an inhibitor of the acetoacetate-forming enzyme system.

Since oxaloacetate depletion has been considered to be one of the mechanisms responsible for ketone overproduction (Mayes, 1962), it is of interest to note that acute administration of alcohol to rats did not modify the absolute amount of oxaloacetate, although it increased the ratio of malate over oxaloacetate (Rawat, 1968). However, the effects of chronic administration of ethanol on these parameters and, more specifically, on the mitochondrial oxaloacetate are still unknown. Our findings of a delayed effect of ethanol on ketone metabolism and the role of associated dietary factors could tentatively be explained by the following hypothesis: The delay in the appearance of ketonemia after ethanol feeding may be caused by the fact that time is required for the depletion of the glycogen stores in the liver. Glycogen depletion most likely results from a combination of a low-carbohydrate diet and an alcohol-induced block of gluconeogenesis. If dietary fatty acids are then given with the ethanol, their oxidation is probably impaired, and ketogenesis markedly enhanced, both because of the lack of carbohydrate and because of the ethanol-induced inhibition of the activity of the citric acid cycle. Much less ketonemia appears in the absence of ethanol, even with a high-fat diet, probably because gluconeogenesis from glycerol and proteins can proceed normally under these conditions. Moreover, the difference in carbohydrate intake between the ethanol and high-fat diets (caused by the glyceride glycerol), although small, could contribute to higher hepatic glycogen levels. Also, less ketonemia appeared with a high dose of ethanol alone (in the virtual absence of dietary fat) because the two-carbon fragments derived from ethanol are exported from the liver as acetate (which is utilized primarily in peripheral tissues (Katz and Chaikoff, 1955) instead of being converted to ketone bodies in the liver, as is the case for fatty acids. An additional mechanism that may favor the ketosis is the mitochondrial liver injury that develops after chronic ethanol consumption, as discussed in detail in Chapter 7.

Whatever its mechanism, the effect of alcohol on ketones involves not only acetoacetate (see Fig 3.1) but also β-hydroxybutyrate (Fig. 3.2). Actually, the effect on β-hydroxybutyrate was four to six times greater than that on acetoacetate. This selective effect on β-hydroxybutyrate may have some practical importance. Indeed, if blood ketone levels are assessed by the nitroprusside reaction (Acetest, Ketostix), which is insensitive to β-hydroxybutyrate and determines acetoacetate or acetone, an accumulation of ketones may be readily overlooked in the alcoholic. This may explain why alcoholic ketoacidosis have been relatively underreported. Although an elevation of ketones can be induced experimentally by the administration of alcohol- and fat-containing diets in any volunteer thus tested, clinically severe ketoacidosis only appears in some susceptible individuals; the term "alcoholic ketoacidosis" is reserved for this unusually severe complication.

3.11. Alcoholic Ketoacidosis

Although mild ketosis may accompany a variety of conditions, diabetes mellitus was usually the only recognized cause of severe ketoacidosis in adults. However, in 1940, Dillon *et al.* described alcoholics hospitalized with severe ketoacidosis and normal or only slightly elevated (or even decreased) blood glucose concentrations. This ketosis was attributed to poor nutritional status or fasting or both. However, since the publication by Lefevre *et al.* (1970) of the experimental production of ketosis by ethanol in volunteers, severe ketosis in alcoholics or "alcoholic ketosis" or "alcoholic ketoacidosis" has been reported by several groups (Jenkins *et al.*, 1971; Levy *et al.*, 1973; Cooperman *et al.*, 1974; Fulop and Hoberman, 1975).

The episodes of ketoacidosis may recur after subsequent alcoholic debauches. Although the studies of Lefevre *et al.* (1970) clearly showed that chronic consumption of alcohol with a fat-containing diet regularly

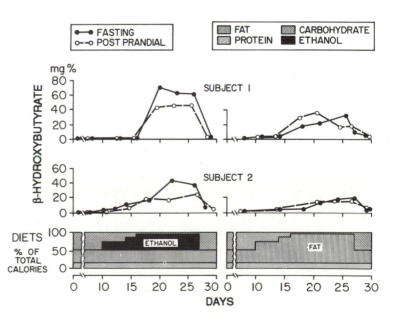

FIGURE 3.2. The effect of isocaloric replacement of dietary carbohydrate (by either alcohol or fat) on blood β-hydroxybutyrate concentration in two subjects in both the fasting and the postprandial state. (From Lefevre *et al.*, 1970.)

leads to ketosis, the pathogenesis of the severe keto-acidosis that may occur in some alcoholics but not in the common variety of heavy drinker has not been fully explained. It is probable that all the mechanisms discussed in the preceding paragraph are operative with, in addition, an unusually high free fatty acid load caused by fatty acid mobilization from adipose tissue subsequent to starvation and associated glycogen store depletion. Indeed, unusually high levels of free fatty acids have been reported in some of these patients. This clinical syndrome is more common than might be suggested by the relative paucity of the reports. This underreporting is probably a result of the fact, already mentioned, that the commonly used nitroprusside reaction does not react with β-hydroxybutyrate, which, as shown in Fig 3.2, is selectively elevated in the ketosis resulting from alcohol. The test may therefore be negative or only mildly or moderately positive even in the presence of severe alcoholic ketoacidosis.

Characteristically, the clinical syndrome occurs in nondiabetics, although alcoholic ketoacidosis may of course complicate the diabetic variety. Not uncommonly, it occurs in subjects with "biochemical" evidence of mild diabetes that does not require treatment with insulin or hypoglycemic agents or even dietary restrictions. Typically, prior to the admission for alcoholic ketoacidosis, the patients have had a period of increased alcohol consumption followed by a period of anorexia, hyperemesis, and cessation of food and alcohol intake. Starvation, progressive dehydration, and mild mental obstruction may follow. Ketoacidosis usually occurs 40 hr or more after the last intake of alcohol. On admission, the moderately decreased, normal, or moderately elevated blood glucose (Cooperman *et al.*, 1974; Miller *et al.*, 1978) contrasts with the severe ketosis. Most patients also have acidosis and plasma hydroxybutyrate levels that may exceed 20 mM, although in some cases no acidosis is present (Fulop and Hoberman, 1975) because of coexisting respiratory or metabolic alkalosis. Clinical improvement occurs rapidly following the intravenous administration of glucose and large amounts of fluid; insulin and bicarbonate are usually not necessary (*Nutrition Reviews*, 1978).

The ketosis usually disappears within 12 to 18 hr of intravenous administration of glucose and saline. Therefore, unless complicated by lactic acidosis (discussed in Section 3.12), the condition is relatively benign if appropriately treated.

A slightly to moderately abnormal glucose tolerance is often present, but it can rapidly disappear with rehabilitation of the patient and may be caused by associated liver damage produced by alcohol, as discussed in Section 3.9. In the presence of acidosis, bicarbonate is often administered but in fact is not really necessary (Miller *et al.*, 1978). The frequency of the syndrome has been estimated as one for every four diabetic ketoacidosis cases (Cooperman *et al.*, 1974).

3.12. Hyperlactacidemia and Lactic Acidosis

The accumulation of lactate in the blood has long been recognized as a normal response to vigorous exercise and as a manifestation of acute hypoxia secondary to shock or asphyxia. However, it was not until 1961 that Huckabee called attention to the occurrence of hyperlactatemia in a variety of clinical circumstances lacking evidence of either circulatory failure or hypoxemia (Huckabee, 1961a,b). Since then, a large and still-growing literature on this subject has appeared, and some experts now suspect that lactic acidosis may be the most frequent form of acute metabolic acidosis. It is also a not uncommon cause of congenital or acquired chronic acidosis (Alberti and Nattrass, 1977). The net rate of production of lactate in all cells is determined by the rate of formation of pyruvate (with which the lactate is in equilibrium) and by the cytoplasmic redox potential, which influences the equilibrium relationship between lactate and pyruvate. The formation of pyruvate, in turn, is determined mainly by the factors regulating glycolysis.

The concentration of lactate in cells and in the blood is normally about ten times that of pyruvate, but this ratio increases when the redox potential of the cell is reduced by hypoxia. Lactate circulates in the blood at a normal concentration of about 1 mM (9 mg/dl) and is removed from the circulation mainly by the liver. The liver converts the lactate back to glucose (or glycogen).

Lactic acidosis results whenever the rate of production of lactate exceeds the capacity of the lactate-utilizing reactions. The result is a rise in the blood level of lactate and pyruvate, a fall in plasma bicarbonate, and usually—though not always—some reduction in blood pH.

By far the most common clinical form of lactic acidosis is that accompanying severe acute circulatory or respiratory failure, in which the sudden rise in blood lactate is a biochemical sign of tissue hypoxia. Under ordinary circumstances, ethanol by itself does not reduce oxygen availability, since increased oxygen consumption associated with hepatic ethanol oxidation is offset by increased flow (see Chapter 1). Oxygen supply to tissues could decline for many reasons, including poor cardiac output (secondary to the reduced extracellular fluid volume or to an ethanol-induced cardiomyopathy), a pulmonary disorder (e.g., aspiration pneumonia, underlying lung disease), a decreased hemoglobin (blood loss, nutritional anemia), or an increased oxygen affinity for hemoglobin (low 2,3-diphosphoglycerol levels in erythrocytes secondary to ethanol-induced phosphate depletion).

As discussed in Chapter 1, the redox changes associated with the oxidation of ethanol also result in a shift of pyruvate to lactate. This leads to increased lactate levels, resulting from either increased hepatic lactate production (Jorfeldt and Juhlin-Dannfelt, 1978) or, depending on the metabolic state of the liver, decreased utilization by the liver of lactate derived from extrahepatic tissues (Krebs, 1967; Kreisberg et al., 1971a) or both. As a consequence, lactate rises in the blood. The levels usually achieved are moderate, as illustrated in Table 3.2, and are much lower than those described in so-called lactic acidosis. However, ethanol may strikingly exacerbate the hyperlactacidemia resulting from some other causes, for instance, essential chronic lactic acidosis (Sussman et al., 1970), and among diabetic patients (Daughaday et al., 1962), particularly those treated with phenformin, a drug known to cause lactic acidosis in its own right. Indeed, until it was ordered off the general market, the drug most commonly responsible for drug-induced lactic acidosis in the United States was phenformin, a biguanide oral hypoglycemic agent. Phenformin and its close congener, metformin, are still used in other countries. In addition to lactic acidosis, the hyperlactacidemia also has clinically significant consequences with regard to uric acid metabolism (Fig. 3.3).

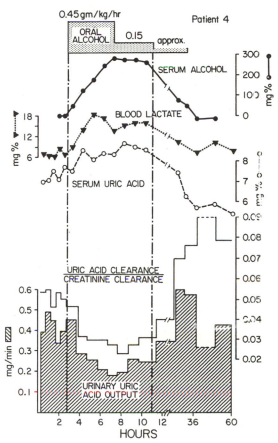

FIGURE 3.3. Blood and urine studies after oral ethanol in a volunteer. The rise in blood lactate was associated with a fall in urinary uric acid and a corresponding rise in the serum. (From Lieber et al., 1962.)

TABLE 3.2.   Blood Lactate in Alcoholics[a]

Subject	Serum Alcohol (mg/100 ml)	Blood Lactate (mg/100 ml) Drunk	Blood Lactate (mg/100 ml) Sober
4	325	36.0	7.5
6	320	24.5	8.0
8	265	16.5	
11	400	17.0	
18	290	21.5	
19	262	13.0	
20	350	34.5	
21	278	15.5	9.0
Mean	311	22.3	

[a]From Lieber et al. (1962).

If the patient is hypokalemic or has a reasonable urine output, potassium should be added in the first intravenous infusion. Since these patients are almost always phosphate depleted, and this might compromise oxygen delivery to tissues, some of the potassium should be given as the phosphate salt, and glucose should be given to prevent hypoglycemia. As for the treatment of alcoholic ketoacidosis (see Section 3.11), insulin and bicarbonate are usually not necessary.

3.13. Alcoholic Hyperuricemia and Its Relationship to Gouty Attacks

It is an ancient tradition among both the public and the medical profession to consider the consumption of alcoholic beverages as one of the most common predisposing or precipitating causes of acute gouty attacks. In the third edition of A. B. Garrod's *Treatise on Gout*, published in 1876, he wrote: "There is no truth in medicine better established than the fact that the use of fermented liquors is the most powerful of all the predisposing causes of gout." Alcohol, or some congener, that is, some nonalcoholic component of alcoholic beverages, was incriminated. Sweet wines, such as port wine, were considered to be especially harmful.

But in the 20th century this relationship between gout and alcohol has been underemphasized and even widely questioned. This declining acceptance of the relationship of alcohol to gout has resulted, in part, from the fact that the concept of the role of alcohol in gout was based only on uncontrolled clinical impressions and that, surprisingly, the relationship between gout and alcohol had not been carefully studied.

We became intrigued by this problem when one of our patients with alcoholic liver disease suffered from gouty attacks that were clearly precipitated by bouts of alcoholism and occurred with such regularity that he had learned spontaneously to take colchicine prophylactically after each alcoholic bout. This prompted a study (Lieber *et al.*, 1962) that we started simply by measuring the serum uric acid in patients hospitalized for acute alcoholic intoxication. As shown in Fig. 3.4, in each of these 12 individuals the initial serum uric acid level was higher when they were acutely intoxicated than thereafter. About half of our patients had a serum uric acid level above what we consider to be the upper limit of normal. All of these values returned to within the normal range on recovery from alcoholic intoxication. In some of them the drop was quite striking from about 9 to 11 mg/100 ml to 3 to 5 mg/100 ml. On the average, the fall was 40% which was highly significant on statistical analysis.

Of course, these observations did not prove a causal relationship between alcohol and the elevation of serum uric acid, since in such acutely intoxicated patients it is difficult to rule out a number of other factors. These patients had no history of gout, and they had no renal insufficiency and no hematological problem; their hematocrits did not change. It was, however, impossible to rule out, for instance, the theoretical possibility of an effect of some nonalcoholic component of the alcoholic beverages or the influence of starvation, which is known to affect the serum uric acid level.

In order to determine whether alcohol *per se* is capable of altering the level of serum uric acid, pure ethanol was given to subjects hospitalized on a metabolic ward, and they were maintained on a diet of

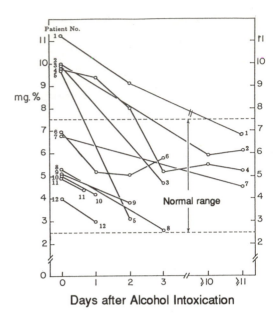

FIGURE 3.4. Serum uric acid concentrations in 12 subjects when acutely intoxicated with alcohol and at various intervals thereafter. The normal range is also indicated (mean ± 2 S.D.). (From Lieber *et al.*, 1962.)

constant purine content. As shown in Fig. 3.3, administration of ethanol was accompanied by a rise in serum uric acid, which returned to normal after discontinuation of the alcohol. The rise in serum uric acid produced by ethanol was accompanied by a decrease in urinary uric acid output without alteration in glomerular filtration rate, as indicated by endogenous creatinine clearance levels. In all patients in whom urinary studies were done, calculation of the deficit in urinary uric acid excretion showed that this could account for the observed rise in serum uric acid.

In various conditions accompanied by an elevated blood lactate, such as exercise (Quick, 1935; Nichols *et al.*, 1951), toxemia of pregnancy (Handler, 1960), and glycogen storage disease (Jeandet and Lestradet, 1961), decrease in urinary urate excretion and high serum uric acid levels have been described. These effects have been attributed to the rise in blood lactate, since lactate administration is known to decrease urinary uric acid excretion (Gibson and Doisy, 1923; Michael, 1977; Yu *et al.*, 1957; Burch and Kurke, 1968). Thus, it appears reasonable to assume that the rise in serum uric acid concentration and the concomitant decrease in urinary uric acid output that we observed in patients given ethanol resulted, at least in part, from the rise in blood lactate. This concept was supported by the observation that the ingestion of sodium lactate in subjects formerly given ethanol led to increases in blood lactate concentrations, decreases in urine urate excretion, and increases in serum urate similar to those that occur after ethanol ingestion (Fig. 3.5). Therefore, it is reasonable to assume that the hyperuricemia observed in patients spontaneously intoxicated with alcoholic beverages (see Fig. 3.4) results at least in part from the following sequence of events: oxidation of ethanol to acetaldehyde generates NADH, which reduces pyruvate to lactate. The resulting rise in blood lactate decreases urinary uric acid output, which leads to an increase in serum uric acid concentration.

The role of this effect on uric acid metabolism has been pursued by McLachlan and Rodnan (1967). These studies confirmed that after ingestion of large amounts of alcohol (112 to 135 g) there is an increase in serum

uric acid concentration and a concomitant decrease in urinary uric acid output. In addition, subjects who fasted for a period of 3 days demonstrated a uniform increase in serum uric acid concentrations. The mean increase was 2 mg/100 ml, and this increment was easily reversed by feeding the patient. During the fast, the blood lactate concentrations remained at control values, whereas β-hydroxybutyrate concentrations rose steadily over the 2-day period, reaching concentrations of 20 to 40 mg/100 ml in most patients at the end of the second day of fasting. At this time, the urinary uric acid excretion was decreased compared to that during control periods, presumably as the result of the tubular inhibition of uric acid secretion by β-hydroxybutyrate. Patients who had received less than 100 g of alcohol and were fasted for two sequential days demonstrated an increase in serum uric acid concentration greater than patients who had simply been fasted for 2 days. When serum concentrations of lactate and β-hydroxybutyrate under these experimental conditions were compared with those in which alcohol and food were given, both metabolites were found to be significantly higher in subjects receiving alcohol and fasting than they were in the fed–alcohol group. The rise in ketones after fasting, the associated reduction in uric acid excretion, and the rise in blood uric acid are well known (Lecocq and McPhaul, 1965). Indeed, experimentally, both acetoacetate and β-hydroxybutyrate infusion caused uric acid retention (Goldfinger et al., 1965). Therefore, one can assume that the fasting state (via ketosis) and ingestion of alcohol (via hyperlactacidemia and ketosis) could be additive and produce a significant increase in serum uric acid concomitant with a decrease in urinary uric acid.

Two additional factors are worth mentioning in the context of alcoholism and hyperuricemia. These factors are the influence of delirium tremens on urate metabolism and the probable occurrence of increased serum uric acid concentrations during and after a seizure. Delirium tremens, a confusional state associated with motor hyperactivity (see Chapter 14), might be expected to have an associated hyperuricemia on the basis of increased lactate production resulting from the increased muscular activity in the untreated patient with delirium tremens.

It has been demonstrated that grand mal seizures result in an increased concentration of uric acid in the serum (Crawford, 1941). Thus, on that basis, the alcoholic prone to so-called "rum fits" or "withdrawal" seizures may have an increased serum concentration of uric acid. Furthermore, various hepatotoxic agents result in an increased breakdown of liver nucleoproteins

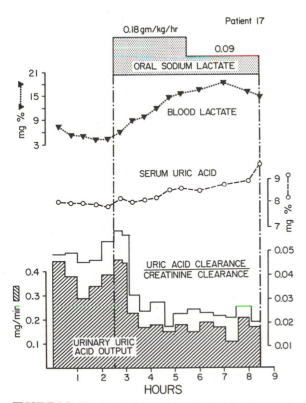

FIGURE 3.5. Blood and urine studies and oral sodium lactate. A rise in blood lactate produced changes in urinary and serum uric acid similar to those produced by ethanol. (From Lieber *et al.*, 1962.)

and enhanced release of uric acid into the bloodstream. It is possible that such a mechanism may contribute to the hyperuricemia of the alcoholic. Indeed, in gouty subjects with apparently higher production of uric acid, ethanol was found to exacerbate this process (Fuller and Fox, 1982), probably because of increased purine production secondary to enhanced ATP breakdown caused by the ethanol-induced rise in acetate.

Correction of nutritional deficiency in severely malnourished individuals could also be associated with a relative rise of uric acid (Olin *et al.*, 1976). After the initial drop of serum uric acid associated with alcohol withdrawal (see Fig. 3.4), there can be a secondary rise as illustrated by Case 6 of Fig. 3.4 and also as described more recently by Olin *et al.* (1976). The various mechanisms postulated for increased production of uric acid could account for this occasional persistence of hyperuricemia beyond cessation of alcohol intake.

Finally, the question arises as to what extent the hyperuricemia in the alcoholic promotes clinical gout. As indicated before, more than a century of clinical observations have documented the aggravating or exacerbating effects of alcohol abuse on gouty attacks in gouty individuals, and this has been verified experimentally (McLachlan and Rodnan, 1967). There is, however, no indication that because of the hyperuricemia, alcoholics may be more prone to develop gout. In a general population of alcoholics, though hyperuricemia was commonly present, fewer than 1% of the patients suffered from clinical gout (Olin *et al.*, 1976). Thus, secondary gout caused by alcoholism has not been documented and is unlikely in view of the intermittent nature of the alcohol-induced hyperuricemia, although in individuals with primary gout, alcohol can aggravate the attacks.

ACKNOWLEDGMENT. The authors wish to thank Juanita Ravikumar, M.D. and Michael Bergman, M.D. for their assistance in the preparation of the chapter.

REFERENCES

Abelin, I., Herren, C. L., and Berli, W.: Über die erregende Wirkung des Alkohols auf den Adrenalin und Noradrenalinhaushalt des menschlichen Organismus. *Helv. Med. Acta* **25:** 591–600, 1958.

Agner, T., Hagen, C., Nyboe Andersen, B., and Hegedus, L.: Pituitary–thyroid function and thyrotropin, prolactin and growth hormone responses to TRH in patients with chronic alcoholism. *Acta Med. Scand.* **220:**57–62, 1986.

Alberti, K. G. M. M., and Nattrass, M.: Lactic acidosis. *Lancet* **2:**25–29, 1977.

Alford, F. P., Dundley, F. J., Chisholm, D. J., and Findlay, D. M.: Glucagon metabolism in normal subjects and in cirrhotic patients before and after portasystemic venous shunt surgery. *Clin. Endocrinol.* **11:**413–424, 1979.

Amatuzio, D. S., Schriftner, N., Stutzmann, F. L., and Nesbit, S.: Blood pyruvic acid response to intravenous glucose or insulin in the normal and in patients with liver disease and with diabetes mellitus. *J. Clin. Invest.* **31:**751–756, 1952.

Andreani, D., Tamburrano, G., and Javicoli, M.: Alcohol hypoglycemia: Hormonal changes. In: *Hypoglycemia: Proceedings of the European Symposium* (D. Andreani, P. Lefebree, and V. Mark, eds.), Stuttgart, Georg Thieme, pp. 99–105, 1976.

Anton, A. H.: Ethanol and urinary catecholamines in man. *Clin. Pharmacol. Ther.* **6:**462–469, 1965.

Arky, R. A., and Freinkel, N.: The response of plasma human growth hormone to insulin and ethanol-induced hypoglycemia in two patients with "isolated andrenocorticotropic defect." *Metabolism* **13:**547–550, 1964.

Arky, R. A., and Freinkel, N.: Alcohol hypoglycemia: V. Alcohol infusion to test gluconeogenesis in starvation, with special reference to obesity. *N. Engl. J. Med.* **274:**426–433, 1966.

Arky, R. A., Veverbrants, E., and Abramson, E. A.: Irreversible hypoglycemia: A complication of alcohol and insulin. *J.A.M.A.* **206:**575–578, 1968.

Ashmore, J., and Weber, G.: Hormonal control of carbohydrate metabolism in liver. In: *Carbohydrate Metabolism and Its Disorders, Vol. 1* (F. Dickens, P. J. Randle, and W. J. Whelan, eds.), New York, Academic Press, p. 335, 1968.

Augustine, J. R.: Laboratory studies in acute alcoholics. *Can. Med. Assoc. J.* **96:**1367–1370, 1967.

Bagdade, J. D., Bierman, E. L., and Porte, D., Jr.: Counterregulation of basal insulin secretion during alcohol hypoglycemia in diabetic and normal subjects. *Diabetes* **21:**65–70, 1972.

Bajpai, H. S., Singh, R. K., Nair, K. S. S., Chansouria, J. P. M., and Uduga, K. M.: Observations on adrenocortical function in cirrhosis of the liver. *J. Assoc. Phys. India* **21:**761–766, 1973.

Bauer, R., and Wozasek, O.: Alcohol and liver function. *Rev. Gastroenterol.* **1:**95–103, 1934.

Beard, J. D., Barlow, G., and Overman, R. R.: Body fluids and blood electrolytes in dogs subjected to chronic ethanol administration. *J. Pharmacol. Exp. Ther.* **148:**348–355, 1965.

Becker, H. D., Reeder, D. D., and Thompson, J. C.: Gastrin release by ethanol in man and in dogs. *Ann. Surg.* **179:**906–909, 1974.

Beevers, D. G., Bann, L. T., Saunders, J. B., Paton, A., and Walters, J. R.: Alcohol and hypertension. *Contrib. Nephrol.* **30:**92–97, 1982.

Bellet, S., Yoshimine, N., DeCastor, O. A. P., Roman, L., Parmar, S., and Sandberg, H.: Effects of alcohol ingestion on growth hormone levels: Their relationship to 11-hydroxycorticoid levels and serum FFA. *Metabolism* **20:**762–769, 1971.

Benedict, H., and Torok, B.: Der Alcohol in der Ernährung der Zuckerkranken. *Z. Klin. Med.* **60:**329–348, 1906.

Bertello, P., Agrimonti, F., Gurioli, L., Frairia, R., Fornaro, D., and Angeli, A.: Circadian patterns of plasma cortisol and testosterone in chronic male alcoholics. *Alcoholism: Clin. Exp. Res.* **6**:475–481, 1982.

Birnbaumer, L., Pohl, S. L., and Rodbell, M.: Adenyl cyclase in fat cells. 1. Properties and the effects of adrenocorticotrophin and fluoride. *J. Biol. Chem.* **244**:3468–3476, 1969.

Blackard, W. G., Hull, E. W., and Lopez, S. A.: Effect of lipids on growth hormone secretion in humans. *J. Clin. Invest.* **50**:1439–1443, 1971.

Bongiovanni, A. M., and Eisenmenger, W. J.: Adrenal cortical metabolism in chronic liver disease. *J. Clin. Endocrinol.* **11**:152–172, 1951.

Borzio, M., Caldara, R., Borzio, F., Piepoli, V., Rampini, P., and Ferrari, C.: Thyroid function tests in chronic liver disease: Evidence for multiple abnormalities despite clinical euthyroidism. *Gut* **24**:631–636, 1983.

Bremer, J.: Pathogenesis of ketonemia. *Scan. J. Clin. Lab. Invest.* **23**:105–108, 1969.

Brodie, B. B., Butler, J. M., Jr., Horning, M. G., Maickel, R. P., and Maling, H. U.: Alcohol induced triglyceride disposition in liver through derangement of fat transport. *Am. J. Clin. Nutr.* **9**:432–435, 1961.

Brown, H., Willardson, D. G., Samuels, L. T., and Tyler, F. H.: 17-Hydroxycorticosteroid metabolism in liver disease. *J. Clin. Invest.* **33**:1524–1538, 1954.

Brown, T. M., and Harvey, A. M.: Spontaneous hypoglycemia in "smoke" drinkers. *J.A.M.A.* **117**:12–22, 1941.

Burch, R. E., and Kurke, N.: The effect of lactate infusion on serum uric acid. *Proc. Soc. Exp. Biol. Med.* **127**:17–20, 1968.

Cameron, D. P., Burger H. G., Catt, K. J., Gordon, E., and Watts, J. M.: Metabolic clearance of human growth hormone in patients with hepatic and renal failure, and in the isolated perfused pig liver. *Metabolism* **21**:895–904, 1972.

Carlson, H. E., Jacobs, L. S., and Daughaday, W. H.: Growth hormone, thyrotropin and prolactin responses to thyrotropin-releasing hormone following diethylstilbestrol pretreatment. *J. Clin. Endocrinol. Metab.* **37**:488–490, 1973.

Carlsson, C., and Häggendal, J.: Arterial noradrenaline levels after ethanol withdrawal. *Lancet* **2**:889, 1967 (letter).

Carter, J. N., Eastman, C. J., Corcoran, J. M., and Lazarus, L.: Effect of severe chronic illness on thyroid function. *Lancet* **2**:971–974, 1974.

Cederbaum, A. L., Lieber, C. S., Beattie, D. S., and Rubin, E.: Effect of chronic ethanol ingestion on fatty acid oxidation by hepatic mitochondria. *J. Biol. Chem.* **250**:5122–5129, 1975.

Chalmers, R. J., Bennie, E. H., Jonson, R. H., and Kinnell, H. G.: The growth hormone response to insulin induced hypoglycemia in alcoholics. *Psychol. Med.* **7**:607–611, 1977.

Chart, J. J., and Shipley, E. G.: The mechanism of sodium retention in cirrhosis of the liver, *J. Clin. Invest.* **32**:560 (abstract), 1953.

Chey, W. Y., Shay, H., Nielsen, O. F., and Lorber, S. H.: Evaluation of tests of pancreatic function in chronic pancreatic disease. *J.A.M.A.* **201**:347–350, 1967.

Chopra, I. J.: An assessment of daily production and significance of thyroidal secretion of 3,3′,5′-triiodothyronine (reverse T_3) in man. *J. Clin. Invest.* **58**:32–40, 1976.

Chopra, I. J., Solomon, D. H., Chopra, U., Young, R. T., and Teco, G. N. C.: Alterations in circulating thyroid hormones and thyrotrophin in hepatic cirrhosis: Evidence for euthyroidism despite subnormal serum triiodothyronine. *J. Clin. Endocrinol. Metab.* **39**:501–511, 1974.

Chopra, I. J., Chopra, J., Smith, S. R., Reza, M., and Solomon, D. H.: Reciprocal changes in serum concentrations of 3,3′,5′-triiodothyronine (reverse T_3) and 3,5,3′-triiodothyronine (T_3) in systemic illnesses. *J. Clin. Endocrinol. Metab.* **41**:1043–1049, 1975.

Chopra, I. J., Solomon, D. H., Chua-Teco, G. N., and Eisenberg, J. B.: An inhibitor of the binding of thyroid hormones to serum protein is present in extrathyroidal tissues. *Science* **215**:407–409, 1982.

Cobb, C. F., and Van Thiel, D. H.: Mechanism of ethanol-induced adrenal stimulation. *Alcoholism: Clin. Exp. Res.* **6**:202–206, 1982.

Cobb, C. F., Van Thiel, D. H., Ennis, M. F., Gavaler, J. S., and Lester, R.: Is acetaldehyde an adrenal stimulant? *Curr. Surg.* **36**:431–434, 1979.

Cohen, S. L., MacIntyre, I., Grahame-Smith, D., and Walter, J. G.: Alcohol stimulated calcitonin release in medullary carcinoma of the thyroid. *Lancet* **2**:1172–1174, 1973.

Collins, J. F., and Crofford, O. B.: Glucose intolerance and insulin resistance in patients with liver disease. *Arch. Intern. Med.* **124**:142–148, 1969.

Colwell, A. R., Feinzimer, M., Copper, D., and Zuckerman, L.: Alcohol inhibition of cyclic AMP-induced insulin release. *Diabetes* **22**:854–857, 1973.

Conn, H. O., and Daughaday, W. H.: Cirrhosis and diabetes. V. Serum growth hormone levels in Laennec's cirrhosis. *J. Lab. Clin. Med.* **76**:678–688, 1970.

Cooperman, M. T., Davidoff, F., Spark, R., and Pallotta, J.: Clinical studies of alcoholic ketoacidosis. *Diabetes* **23**:433–439, 1974.

Cope, C. L., and Black, E.: The production rate of cortisol in man. *Br. Med. J.* **1**:1020–1024, 1958.

Coppage, W. S., Jr., Island, D. P., Cooner, A. E., and Liddle, G. W.: The metabolism of aldosterone in normal subjects and in patients with hepatic cirrhosis. *J. Clin. Invest.* **41**:1672–1680, 1972.

Crawford, M. D.: Plasma uric acid and urea findings in eclampsia. *J. Obstet. Gyn. Brit. Emp.* **48**:60–72, 1941.

Creutzfeldt, W., Willie, D., and Kaup, H.: Intravenous injection of glucose, insulin, and tolbutamine in healthy persons, diabetics, cirrhotics and patients with insulinoma. *Dts. Med. Wochenschr.* **87**:2189–2194, 1962.

Crossland, J., and Ratcliffe, F.: Some effects of chronic alcohol administration in the rat. *Br. J. Pharmacol. Chemother.* **32**:413–414 (abstract), 1968.

Crouse, J. R., Gerson, C. D., DeCarli, L. M., and Lieber, C. S.: Role of acetate in the reduction of plasma free fatty acids produced by ethanol in man. *J. Lipid Res.* **9**:509–512, 1968.

Cuttelod, S., Le Marchand-Beraud, T., Magnenat, P., Perret, C., Poli, S., and Vanotti, A.: Effect of age and role of kidneys and liver on thyrotropin turnover in man. *Metabolism* **23**:101–113, 1974.

Danowski, T. S., Gillespie, H. K., Fergus, E. B., and Puntereri,

A. J.: Significance of blood sugar and serum electrolyte changes in cirrhosis following glucose, insulin, glucagon, or epinephrine. *Yale J. Biol. Med.* **29:**361–375, 1956.

Daughaday, W. H.: Binding of corticosteroids by plasma proteins. *Arch. Intern. Med.* **101:**286–290, 1958.

Daughaday, W. H., Lipicky, R. J., and Rasinski, D. C.: Lactic acidosis as a cause of nonketonic acidosis in diabetic patients. *N. Engl. J. Med.* **267:**1010–1014, 1962.

Davis, J. R. E., and Jeffcoate, W. J.: Lack of effect of ethanol on plasma cortisol in man. *Clin. Endocrinol.* **19:**461–466, 1983.

Davis, V. E., Brown, H., Huff, J. A., and Cashaw, J. L.: Ethanol-induced alterations of norepinephrine metabolism in man. *J. Lab. Clin. Med.* **69:**787–799, 1967a.

Davis, V. E., Cashaw, J. L., Huff, J. A., Brown, H., and Nicholas, N. L.: Alteration of endogenous catecholamine metabolism by ethanol ingestion. *Proc. Soc. Exp. Biol. Med.* **125:**1140–1143, 1967b.

Debry, G., and Charles, J.: Etude des perturbations du métabolisme des glucides chez les sujets atteints de cirrhose hépatique. *Diabetes* **13:**152–165, 1965.

De Lean, A., and Labier, F.: Sensitizing effect of treatment with estrogens on TSH response to TRH in male rats. *Am. J. Physiol.* **233:**E235–E239, 1977.

Del Prete, S., Malacco, E., Bonzi, G., Natelli, A., Roncoroni, S., and Jalanbo, H.: Significato clinico-prognostico del tempo di emivita del cortisolo plasmatico (dopo carico endovenoso) in corso di malatte epatiche acute e croniche. *Minerva Med.* **66:**1051–1057, 1975.

De Moura, M. C., Corrleia, J. P., and Madeira, F.: Clinical alcohol hypoglycemia. *Ann. Intern. Med.* **66:**893–905, 1967.

De Portugal-Alvarez, J., Perezagua-Clamagirand, C., Souto, J. M., Catalan, E., Vila, T., and Velasco-Martin, A.: Relationship between growth hormone, insulin, and urinary oestrogens in hepatic cirrhosis. *Acta Diabetol. Lat.* **11:**1–8, 1974.

Descos, L., M'Bendi, S., Sassolas, G., Audigier, J.-C., Sidigian, M., Faure, A., and Bizollon, C.-A.: Dosage de l'insuline et de l'hormone de croissance plasmatiques au cours de l'épreuve d'hyperglycémie provoquée par voie orale dans la cirrhose alcoolique. *Arch. Fr. des Mal. App. Dig.* **63:**17–24, 1974.

Dillon, E. S., Dyer, W. W., and Smelo, L. S.: Ketone acidosis in nondiabetic adults. *Med. Clin. North Am.* **24:**1813–1822, 1940.

Dioguardi, N., D'Alonzo, R., and Peromini, G.: Observations on plasma cortisol and ACTH in chronic liver disease and after corticosteroid treatment. *Chronobiological 1 Suppl.* **1:**449–468, 1974.

Dobrzanski, T., and Pieschl, D.: Untersuchungen uber den Gehalt des Plasmas an ACTH, an STH und an andersen Hormonen unter Belastung in Verschiedenen Gruppen mit Chlormethiazol, Haloperidol oder Reserpin bei Alkoholdelir, alkoholischen Halluzinosen und chronischen Alkoholikern. *Psychiatr. Neurol. Med. Psychol.* **28:**26–32, 1976.

Doe, R. P., Vennes, J. A., and Flink, E. B.: Diurnal variation of 17-hydroxycorticosteroids, sodium, potassium, magnesium and creatinine in normal subjects and in cases of treated adrenal insufficiency and Cushing's syndrome. *J. Clin. Endocrinol. Metab.* **20:**253–265, 1960.

Domschke, S., Domschke, W., and Lieber, C. S.: Hepatic redox state: Attenuation of the acute effects of ethanol induced by chronic ethanol consumption. *Life Sci.* **15:**1327–1334, 1974.

Dornhorst, A., and Ouyang, A.: Effect of alcohol on glucose tolerance. *Lancet* **2:**957–959, 1971.

Eggleton, M. G.: The diuretic action of alcohol in man. *J. Physiol. [Lond.]* **101:**172–191, 1942.

Ellis, F. W.: Effect of ethanol on plasma corticosterone levels. *J. Pharmacol. Exp. Therap.* **153:**121–127, 1966.

Englert, E., Jr., Brown, H., Wallach, S., and Simons, E. L.: Metabolism of free and conjugated 17-hydroxycorticosteroids in subjects with liver disease. *J. Clin. Endocrinol. Metab.* **17:**1395–1406, 1957.

Estep, H., Shaw, W. A., Watlington, C., Hobe, R., Holland, W., and Tucker, St. G.: Hypocalcemia due to hypomagnesemia on reversible parathyroid hormone unresponsiveness. *J. Clin. Endocrinol. Metab.* **29:**842–848, 1969.

Fabre, L. F., Jr., Farmer, R. W., Pellizzari, E. D., and Farrell, G.: Aldosterone secretion in pentobarbital-anesthetized ethanol-infused dogs. *Q. J. Stud. Alcohol* **33:**476–484, 1972.

Fanghänel, G., Blitz, I., Uribe, M., and Morato, T.: Determinacion de cortisol y globulina transportadora de corticoides (gtc) en normales, embarazades y pacientes con cirrosis hepatica y sindrome de malabsorcion intestinal. *Rev. Invest. Clin.* **28:**151–159, 1976.

Fankushen, D., Raskin, D., Dimich, A., and Wallach, S.: The significance of hypomagnesemia in alcoholic patients. *Am. J. Med.* **37:**802–812, 1964.

Farmer, R. W., and Fabre, L. F., Jr.: Some endocrine aspects of alcoholism. *Adv. Exp. Med. Biol.* **56:**277–289, 1975.

Farmer, R. W., Farrell, G., Pellizzari, E. D., and Fabre, L. F., Jr.: Serum insulin levels during oral GTT in chronic alcoholics. *Fed. Proc.* **30:**250 (abstract), 1971.

Fazekas, I. G.: Hydrocortisone content of human blood and alcohol content of blood and urine after wine consumption. *Q. J. Stud. Alcohol* **27:**439–446, 1966.

Feinman, L., and Lieber, C. S.: Effect of ethanol on plasma glycerol in man. *Am. J. Clin. Nutr.* **20:**400–403, 1967.

Felber, J. P., Magnenat, P., and Vannotti, A.: Tolerance au glucose diminuée et réponse insulinique élevée dans la cirrhose. *Schweiz. Med. Wochenschr.* **97:**1537–1539, 1967.

Felig, P., Owen, O. W., Wahren, J., and Cahill, G. F., Jr.: Amino acid metabolism during prolonged starvation. *J. Clin. Invest.* **48:**584–594, 1969.

Field, J. B., Williams, H. E., and Mortimore, G. E.: Studies on the mechanism of ethanol-induced hypoglycemia. *J. Clin. Invest.* **42:**497–506, 1963.

Fink, R. S., Short, F., Marjot, D. H., and James, V. H. T.: Abnormal suppression of plasma cortisol during the intravenous infusion of dexamethasone to alcoholic patients. *Clin. Endocrinol.* **15:**97–102, 1981.

Fink, R. S., Adrian, T. E., Marjot, D. H., and Bloom, S. R.: Increased plasma pancreatic polypeptide in chronic alcohol abuse. *Clin. Endocrinol.* **18:**417–421, 1983.

Forsander, O. A., Raiha, N., Salaspuro, M., and Maepaa, P.: Influence of ethanol on the liver metabolism of fed and starved rats. *Biochem. J.* **94:**259–265, 1965.

Frajria, R., and Angeli, A.: Alcohol-induced pseudo-Cushing's syndrome. *Lancet* **1:**1050–1051, 1977.

Frantz, A. G., and Rabkin, M. T.: Effects of estrogen and sex difference on secretion of human growth hormone, *J. Clin. Endocrinol. Metab.* **25:**1470–1480, 1965.

Fredericks, E. J., and Lazor, M. Z.: Recurrent hypoglycemia associated with acute alcoholism. *Ann. Intern. Med.* **59**:90–94, 1963.

Freinkel, N., Singer, D. L., Arky, R. A., Bleicher, S. J., Anderson, J. B., and Silbert, C. K.: Alcohol hypoglycemia. I. Carbohydrate metabolism of patients with clinical alcohol hypoglycemia and the experimental reproduction of the syndrome with pure ethanol. *J. Clin. Invest.* **42**:1112–1133, 1963.

Freinkel, H., Cohen, A. K., Arky, R. A., and Foster, A. E.: Alcohol hypoglycemia. II. A postulated mechanism of action based on experiments with rat liver slices. *J. Clin. Endocrinol. Metab.* **25**:76–94, 1965.

French, S. W.: Effect of acute and chronic ethanol ingestion on rat liver ATP. *Proc. Soc. Biol. Med.* **121**:681–685, 1966.

French, S. W., Palmer, D. S., and Narod, M. E.: Noradrenergic subsensitivity of rat liver homogenates during chronic ethanol ingestion. *Res. Commun. Chem. Pathol. Pharmacol.* **13**:283–295, 1976.

Freund, G.: The calorie deficiency hypothesis of ketogenesis tested in man. *Metabolism (Clin. Exp.)* **14**:985–990, 1965.

Frieden, C.: Glutamic dehydrogenase. I. The effect of coenzyme on the sedimentation velocity and kinetic behavior. *J. Biol. Chem.* **234**:809–814, 1959.

Friedenberg, R., Metz, R., Mako, M., and Surmaczynska, V.: Differential plasma insulin response to glucose and glucagon stimulation following ethanol priming. *Diabetes* **20**:397–403, 1971.

Fuchs, F., Fuchs, A-R., Poblete, V. F., Jr., and Risk, A.: Effects of alcohol on threatened premature labor. *Am. J. Obstet. Gynecol.* **99**:627–637, 1967.

Fuller, J., and Fox, I. H.: Ethanol-induced hyperuricemia: Evidence for increased urate production by activation of adenine nucleotide exposure. *N. Engl. J. Med.* **307**:1598–1602, 1982.

Fulop, M., and Hoberman, H. D.: Alcoholic ketosis. *Diabetes* **24**:785–790, 1975.

Gajdos, A., Gajdos-Torok, M., Palma-Carlos, A., and Palma-Carlos, L.: Surproduction des porphyrines et variations du taux hépatique et érthrocytaire de l'ATP au cours de l'intoxication alcoolique aigue du rat. *Nouv. Rev. Frane Hematol.* **7**:15–27, 1967.

Ganda, O. P., Sawin, C. T., Iber, F., Glennon, J. A., and Mithcell, M. L.: Transient suppression of growth hormone secretion after chronic ethanol intake. *Alcoholism: Clin. Exp. Res.* **2**:297–299, 1978.

Garrod, A. B.: A Treatise on Gout and Rheumatic Gout, 3rd ed., London, Longmans, Green, p. 217, 1876.

Genazzani, A. R., Nappi, G., Facchinetti, F., Mazzella, G. L., Parrini, D., Sinforiani, E., Petraglia, F., and Savoldi, F.: Central deficiency of β-endorphin in alcohol addicts. *J. Clin. Endocrinol. Metab.* **55**:583–586, 1982.

Geurts, J., Demeester-Mirkine, N., Glinoer, D., Prigogine, T., Fernandez-Deville, M., and Covilain, J.: Alterations in circulating thyroid hormones and thyroxine binding globulin in chronic alcoholism. *Clin. Endocrinol.* **14**:3–8, 1981.

Ghirlanda, G., Manna, R., Rebuzzi, A., Altomonte, L., and Greco, A. V.: Variazioni dell'insulina e del GH plasmatici dopo glucogone endovena nella cirrosi epatica. *Minerva Med.* **67**:3187–3192, 1976.

Giacobini, E., Izikowitz, S., and Wegmann, A.: A urinary excre-

tion of noradrenaline and adrenaline during acute alcohol intoxication in alcoholic addicts. *Experientia* **16**:467, 1960a.

Giacobini, E., Izikowitz, S., and Wegmann, A.: Urinary norepinephrine and epinephrine excretion in delirium tremens. *Arch. Gen. Psychiatry* **3**:289–296, 1960b.

Gibson, H. V., and Doisy, E. A.: A note on the effect of some organic acids upon the uric acid excretion of man. *J. Biol. Chem.* **55**:605–610, 1923.

Goldfarb, A. I., and Berman, S.: Alcoholism as a psychosomatic disorder; endocrine pathology of animals and man excessively exposed to alcohol; its possible relation to behavioral pathology *Q. J. Stud. Alcohol* **10**:415–429, 1949.

Goldfinger, S., Klinenberg, J. R., and Seegmiler, J. E.: Renal retention of uric acid induced by infusion of beta-hydroxybutyrate and acetoacetate. *N. Engl. J. Med.* **272**:351–355, 1965.

Goldstein, D. B.: An animal model for testing effects of drugs on alcohol withdrawal reactions. *J. Pharmacol. Exp. Ther.* **183**:14–22, 1972.

Gordon, G. G., Altman, K., and Southren, A. L.: Lack of inhibition by 6β-hydroxycortisol and *o,p′*-DDD on the cortisol mediated induction of rat hepatic tryptophan oxygenase; the corticoid effect of 6α-hydroxycortisol. *Endocrinology* **1968**:384–386, 1968.

Gordon, G. G., Southren, L., and Lieber, C. S.: The effects of alcohol and alcoholic liver disease on the endocrine system and intermediary metabolism. In: *Medical Disorders of Alcoholism, Pathogenesis and Treatment.* (C. S. Lieber, ed.), Philadelphia, W. B. Saunders Co., Chapter 3, pp. 65–140, 1982.

Gorman, R. E., and Bitensky, M. W.: Selective activation by short chain alcohols of glucagon responsive adenyl cyclase in the liver. *Endocrinology* **87**:1075–1081, 1970.

Greco, A. V., Ghirlanda, G., Patrono, C., Fedeli, G., and Manna, R.: Behavior of pancreatic glucagon, insulin and HGH in liver cirrhosis after arginine and I.V. glucose. *Acta Diabetol. Lat.* **11**:330–339, 1974.

Green, J. R. B., Snitcher, E. J., Mowat, N. A. G., Ekins, R. P., Rees, L. H., and Dawson, A. M.: Thyroid function and thyroid regulation in euthyroid men with chronic liver disease. Evidence of multiple abnormalities. *Clin. Endocrinol.* **7**:453–461, 1977.

Gross, M.: The relation of the pituitary gland to some symptoms of alcoholic intoxication and chronic alcoholism. *Q. J. Stud. Alcohol* **6**:25–35, 1945.

Habnener, J. F.: Hormone biosynthesis and secretion. In: *Endocrinology and Metabolism* (P. Felig, J. D. Baxter, A. E. Broadus, and L. A. Frohman, eds.), New York, McGraw-Hill, pp. 25–59, 1981.

Haggard, H. W., Greenberg, L. A., and Carroll, R. P.: Studies in the absorption, distribution and elimination of alcohol; diuresis from alcohol and its influence on the elimination of alcohol in the urine. *J. Pharmacol. Exp. Ther.* **71**:349–357, 1941.

Haight, J. S., and Keatinge, W. R.: Failure of thermoregulation in the cold during hypoglycaemia induced by exercise and ethanol. *J. Physiol. [Lond.]* **229**:87–97, 1973.

Hall, K., and Luft, A.: Growth hormone and somatomedin. In: *Advances in Metabolic Disorders* (R. Levin and A. Luft, eds.), Vol. 7, New York, Academic Press, pp. 1–36, 1974.

Handler, J. S.: The role of lactic acid in the reduced excretion of uric acid in toxemia of pregnancy. *J. Clin. Invest.* **39**:1526–1532, 1960.

Hanski, E., Rimon, G., and Levitzki, A.: Adenylate cyclase activation by the β-adrenergic receptors as a diffusion-controlled process. *Biochemistry* **18**:846–853, 1979.

Hasselbalch, H., Selmer, J., Sestoft, L., and Kehlet, H.: Hypothalamo–pituitary–adrenocortical function in chronic alcoholism. *Clin. Endocrinol.* **16**:73–76, 1982.

Hed, R., and Nygren, A.: Alcohol-induced hypoglycemia in chronic alcoholics with liver disease. *Acta Med. Scand.* **183**:507–510, 1968.

Hegedus, L.: Decreased thyroid gland volume in alcoholic cirrhosis of the liver. *J. Clin. Endocrinol. Metab.* **58**:930–933, 1984.

Helderman, J. H., Vestal, R. E., Rowe, J. W., Tobin, J. D., Andres, R., and Robertson, G. L.: The response of arginine vasopressin to intravenous ethanol and hypertonic saline in man: The impact of aging. *J. Gerontol.* **33**:39–49, 1978.

Hennessy, J. F., Wells, S. A., Jr., Ontjes, D. A., and Cooper, C. W.: A comparison of pentagastrin injection and calcium infusion as provocative agents for the detection of medullary carcinoma of the thyroid. *J. Clin. Endocrinol. Metab.* **39**:487–495, 1974.

Henry, R. W., Lavery, T., and Buchanan, K. D.: The effect of alcohol on gastro entero-pancreatic hormones. *Ulster Med. J.* **50**:120–122, 1981.

Hepner, G. W., and Chopra, I. J.: Serum thyroid hormone levels in patients with liver disease. *Arch. Intern. Med.* **139**:17–20, 1979.

Hernandez, A., Zorrilla, E., and Gershberg, H.: Decreased insulin production, elevated growth hormone levels, and glucose intolerance in liver disease. *J. Lab. Clin. Med.* **73**:25–33, 1969.

Higgins, H. L., Peabody, F. W., and Fritz, R.: A study of acidosis in three normal subjects, with incidental observations on the action of alcohol as an antiketogenic agent. *J. Med. Res.* **34**:263–272, 1916.

Hillbom, M. E.: Thyroid state and voluntary alcohol consumption of albino rats. *Acta Pharmacol. Toxicol.* **29**:95–105, 1971.

Hoffman, P. L., Ritzman, R. F., Walter, R., and Tabakoff, B.: Arginine vasopressin maintains ethanol tolerance. *Nature* **276**:614–616, 1978.

Hollander, D., Meek, J. C., and Manning, R. T.: Determination of free thyroxine in serum of patients with cirrhosis of the liver. *N. Engl. J. Med.* **276**:900–902, 1967.

Huckabee, W. E.: Abnormal resting blood lactate. I. The significance of hyperlactatemia in hospitalized patients. *Am. J. Med.* **30**:833–839, 1961a.

Huckabee, W. E.: Abnormal resting blood lactate. II. Lactic acidosis. *Am. J. Med.* **30**:840–848, 1961b.

Hyams, D. E., and Isselbacker, K. J.: Prevention of fatty liver by administration of adenosine triphosphate. *Nature* **204**:1196–1197, 1964.

Inada, M., and Sterling, K.: Thyroxine turnover and transport in Laennec's cirrhosis of the liver. *J. Clin. Invest.* **46**:1275–1282, 1967.

Ishii, H., Joly, J.-G., and Lieber, C. S.: Increase of microsomal glucose-6-phosphatase activity after chronic ethanol administration. *Metabolism* **22**:799–806, 1973.

Ishii, H., Mamori, H., Takahashi, H., Muraoka, U., Arai, U., and Shigeta, Y.: Klinische Untersuchungen zur Pathogenese des hepatogenen Diabetes. *Leber Magen Darm.* **5**:52–57, 1975.

Israel, Y., Walfish, P. G., Orrego, H., Blake, J., and Kalant, H.: Thyroid hormones in alcoholic liver disease. Effect of treatment with 6-n-propylthiouracil. *Gastroenterology* **76**:116–122, 1979.

Isselbacher, K. J., and Krane, S. M.: Studies on the mechanism of the inhibition of galactose oxidation by ethanol. *J. Biol. Chem.* **236**:2394–2398, 1961.

Isselbacher, K. J., and McCarthy, E. A.: Effects of alcohol on the liver: Mechanism of the impaired galactose utilization. *J. Clin. Invest.* **39**:999–1000, 1960.

Jauhonen, V. P.: Effect of acute ethanol load on plasma immunoreactive insulin and glucagon. *Horm. Metab. Res.* **10**:214–219, 1978.

Jauhonen, V. P., Savalainen, M. J., and Hassinen, I. E.: Cyclic AMP-linked mechanisms in ethanol-induced derangements of metabolism in rat liver and adipose tissue. *Biochem. Pharmacol.* **24**:1879–1883, 1975.

Jeandet, J., and Lestradet, H.: L'hyperlactacidemia, cause probable de l'hyperuricémie dans la glycogenose hépatique. *Rev. F. Clin. Biol.* **6**:71–72, 1961.

Jeffcoate, W. J., Herbert, M., Cullen, M. H., Hasting, A. G., and Walder, C. P.: Prevention of effects of alcohol intoxication by naloxone. *Lancet* **2**:1157–1159, 1979.

Jenkins, J. S., and Connolly, J.: Adrenocortical response to ethanol in man. *Br. Med. J.* **2**:804–805, 1968.

Jenkins, R. M., and Page, M. McB.: Atypical case of alcohol-induced Cushingoid syndrome. *Br. Med. J.* **282**:17–18, 1981.

Jenkins, D. W., Eckel, R. E., and Craig, J. W.: Alcoholic ketoacidosis. *J.A.M.A.* **217**:177–183, 1971.

Johnson, R. E., Passmore, R., and Sargent, F.: Multiple factors in experimental human ketosis. *Arch. Intern. Med.* **107**:111–118, 1961.

Johnston, D. G., and Alberti, K. G. M. M.: Carbohydrate metabolism in liver disease. *Clin. Endocrinol. Metab.* **5**:675–702, 1976.

Johnston, D. G., Alberti, K. G. M. M., Wright, R., Smith-Laing, G., Stewart, A. M., Sherlock, S., Faber, O., and Binder, C.: C-Peptide and insulin in liver disease. *Diabetes* **27**(Suppl 2):201–206, 1978.

Jones, D. P., Perman, E., and Lieber, C. S.: Free fatty acid turnover and triglyceride metabolism after ethanol ingestion in man. *J. Lab. Clin. Med.* **66**:804–813, 1965.

Jordan, R. M., Jacobson, J. M., and Young, R. L.: Alcohol-induced Cushingoid syndrome. *South. Med. J.* **72**:1347–1348, 1979.

Jorfeldt, L., and Juhlin-Dannfelt, A.: The influence of ethanol on splanchnic and skeletal muscle metabolism in man. *Metabolism* **27**:97–106, 1978.

Kabadi, U. M., and Premachandra, B. N.: Serum T_3 and reverse T_3 levels in hepatic cirrhosis: Relation to hepatocellular damages and normalization on improvement in liver dysfunction. *Am. J. Gastroenterol.* **78**:195–203, 1983.

Kahil, M. F., Brown, H., and Dobson, H. L.: Post-alcoholic hypoglycemia versus islet cell tumor. *Gastroenterology* **46**:467–470, 1964.

Kakihana, R., and Moore, J. A.: Circadian rhythm of corticosterone in mice: The effect of chronic consumption of alcohol. *Psychopharmacologia* **46**:301–305, 1976.

Kakihana, R., Noble, E. P., and Butte, J. C.: Corticosterone response to ethanol in inbred strains of mice. *Nature* **218**: 360–361, 1968.

Kakihana, R., Butte, J. C., Hathaway, A., and Noble, E. P.: Andrenocortical response to ethanol in mice: Modification by chronic ethanol consumption. *Acta Endocrinol.* **67**:653–664, 1971.

Kalant, H., Sereny, G., and Charlebois, R.: Evaluation of triiodothyronine in the treatment of acute alcoholic intoxication. *N. Engl. J. Med.* **267**:1–6, 1962.

Kallner, G.: Assessment of thyroid function in chronic alcoholics. *Acta Med. Scand.* **209**:93–96, 1981.

Kasperska-Czyzykowa, T., and Rogala, H.: Growth hormone in blood plasma of patients with liver cirrhosis. *Acta Med. Pol.* **19**:485–492, 1978.

Katz, J., and Chaikoff, I. L.: Synthesis via the Krebs cycle in the utilization of acetate by rat liver slices. *Biochim. Biophys. Acta* **18**:87–101, 1955.

Kedes, L. H., and Field, J. B.: Hypothermia: A clue to hypoglycemia. *N. Engl. J. Med.* **271**:785–786, 1964.

Khanna, J. M., Kalant, H., and Loth, J.: Effect of chronic intake of ethanol on lactate/pyruvate and β-hydroxybutyrate/acetoacetate ratios in rat liver. *Can. J. Physiol. Pharmacol.* **53**: 299–303, 1975.

Kinzius, H.: Adrenalin und Arbeit: die Beeinflussung des Blutadrenalinspiegels durch Pervitin, Luminal und Alkohol. *Arbeitsphysiologie* **14**:243–248, 1950.

Kissin, B., Schenker, V., and Schenker, A. C.: The acute effect of alcohol ingestion on plasma and urinary 17-hydroxycorticosteroids in man. *Am. J. Med. Sci.* **239**:690–705, 1960.

Kleeman, C. R.: Water metabolism. In: *Clinical Disorders of Fluid and Electrolyte Metabolism* (C. R. Kleeman, ed.), New York, McGraw Hill, pp. 243–257, 1972.

Klein, R., Papadatos, C., Fortunato, J., Byers, C., and Punterei, A.: Serum corticoids in liver disease. *J. Clin. Endocrinol. Metab.* **15**:943–952, 1955.

Kley, H. K., Nieschlag, E., Wiegelmann, W., Solbach, H. G., and Kurskemper, H. L.: Steroid hormones and their binding in plasma of male patients with fatty liver, chronic hepatitis and liver cirrhosis. *Acta Endocrinol.* **79**:275–285, 1975.

Klingman, G. I., Haag, H. B., and Bane, R.: Studies on severe alcohol intoxication in dogs. II. Mechanism involved in the production of hyperglycemia, hypokalemia and hemoconcentration. *Q. J. Stud. Alcohol* **19**:543–552, 1958.

Krebs, H. A.: Effects of ethanol on gluconeogenesis in the perfused rat liver. In: *Konferenz der Gesellschaft für biologische Chemie vom 27-29*. April, 1967, pp. 216–224, Berlin, Springer-Verlag, 1967.

Krebs, H. A., Freedlander, R. A., Hems, R., and Stubbs, M.: Inhibition of hepatic gluconeogenesis by ethanol. *Biochem. J.* **112**:117–124, 1969.

Krebs, H. A., Hems, R., and Lund, P.: Accumulation of amino acids by the perfused rat liver in the presence of ethanol. *Biochem. J.* **134**:697–705, 1973.

Kreisberg, R. A., Owen, W. C., and Siegal, A. M.: Ethanol-induced hyperlactacidemia: Inhibition of lactate utilization. *J. Clin. Invest.* **50**:166–174, 1971a.

Kreisberg, R. A., Siegal, A. M., and Owen, W. C.: Glucose-lactate interrelationships: Effect of ethanol. *J. Clin. Invest.* **50**:175–185, 1971b.

Kreisberg, R. A., Owen, W. C., and Siegal, A. M.: Hyperlactacidemia in man: Ethanol-phenformin synergism. *J. Clin. Endocrinol. Metab.* **34**:29–35, 1972.

Kühl, C., and Andersen, O.: Glucose and tolbutamide mediated insulin response after preinfusion with ethanol. *Diabetes* **23**:821–826, 1974.

Kuska, J., Król, Z., and Libera, T.: Studies on the mechanism of stimulation of growth hormone (IR-HGH) by glucagon in patients with chronic liver diseases. *Mater. Med. Pol.* **10**:3–8, 1978.

Kydd, D. M., and Man, E. B.: Precipitable iodine of serum (SPI) in disorders of the liver. *J. Clin. Invest.* **30**:874–878, 1951.

Lamberts, S. W. J., Klijn, J. G. M., de Jong, F. H., and Birkenhäger, J. C.: Hormone secretion in alcohol-induced pseudo-Cushing's syndrome. *J.A.M.A.* **242**:1640–1643, 1979.

Laragh, J. H., Cannon, P. J., Ames, R. P., Sicinski, A. M., Bentzel, C. J., and Meltzer, J. I.: Angiotensin II and renal sodium transport: Natriuresis and diuresis in patients with cirrhosis and ascites. *J. Clin. Invest.* **41**:1375–1376, 1962 (abstract).

Lecocq, F. R., and McPhaul, J. J., Jr.: The effects of starvation, high fat diets, and ketone infusions on uric acid balance. *Metabolism* **14**:186–197, 1965.

Lederer, J., and Bataille, J. P.: Mécanisms de l'hypogonadisme masculin au cours de la cirrhose. *Ann. Endocrinol.* **28**:821–826, 1967.

Lefevre, A., Adler, H., and Lieber, C. S.: Effect of ethanol on ketone metabolism. *J. Clin. Invest.* **49**:1775–1782, 1970.

Leppäluoto, J., Rapeli, M., Varis, R., and Ranta, T.: Secretion of anterior pituitary hormones in man: Effects of ethyl alcohol. *Acta Physiol. Scand.* **95**:400–406, 1975.

Levy, L. J., Duga, J., Girgis, M., and Gordon, E. E.: Ketoacidosis associated with alcoholism in nondiabetic subjects. *Ann. Intern. Med.* **78**:213–219, 1973.

Lieber, C. S.: Metabolic effects produced by alcohol in the liver and other tissues. *Adv. Intern. Med.* **14**:151–199, 1968.

Lieber, C. S.: Liver adaptation and injury in alcoholism. *N. Engl. J. Med.* **288**:356–362, 1973.

Lieber, C. S., and DeCarli, L. M.: Hepatic microsomal ethanol oxidizing system: *In vitro* characteristics and adaptive properties *in vivo*. *J. Biol. Chem.* **234**:2505–2512, 1970.

Lieber, C. S., and Schmid, R.: The effect of ethanol on fatty acid metabolism: Stimulation of hepatic fatty acid synthesis *in vitro*. *J. Clin. Invest.* **40**:394–399, 1961.

Lieber, C. S., and Spritz, N.: Effects of prolonged ethanol intake in man: Role of dietary, adipose, and endogenously synthesized fatty acids in the pathogenesis of the alcoholic fatty liver. *J. Clin. Invest.* **45**:1400–1411, 1966.

Lieber, C. S., Jones, D. P., Losowsky, M. S., and Davidson, C. S.: Interrelation of uric acid and ethanol metabolism in man. *J. Clin. Invest.* **41**:1863–1870, 1962.

Lieber, C. S., Spritz, N., and DeCarli, L. M.: Role of dietary, adipose, and endogenously synthesized fatty acids in the pathogenesis of the alcoholic fatty liver. *J. Clin. Invest.* **45:** 51–62, 1966.

Lindros, K. O.: Interference of ethanol and sorbitol with hepatic ketone body metabolism in normal hyper- and hypothyroid rats. *Eur. J. Biochem.* **13:**111–116, 1970.

Linkola, J., Fyhrquist, F., Nieminen, M. M., Weber, J. H., and Tontti, K.: Renin-aldosterone axis in ethanol intoxication and hangover. *Eur. J. Clin. Invest.* **6:**191–194, 1976.

Linkola, J., Ylikhari, R., Fyhrquist, F., and Wallenius, M.: Plasma vasopressin in ethanol intoxication and hangover. *Acta Physiol. Scand.* **104:**180–187, 1978.

Linkola, J., Fyhriquist, F., and Ylikahri, R.: Renin, aldosterone and cortisol during ethanol intoxication and hangover. *Acta Physiol. Scand.* **106:**75–82, 1979.

Littleton, J.: Alcohol and neurotransmitters. *Clin. Endocrinol. Metab.* **7:**369–384, 1980.

Llanos, O. L., Swierczek, J. S., Teichmann, R. K., Rayford, P. L., and Thompson, J. C.: Effect of alcohol on the release of secretin and pancreatic secretion. *Surgery* **81:**661–667, 1977.

Lochner, A., Wulff, J., and Madison, L. L.: Ethanol-induced hypoglycemia. I. Acute effects of ethanol on hepatic glucose output and peripheral glucose utilization in fasted dogs. *Metabolism* **16:**1–18, 1967.

Lommer, D., Dusterdieck, G., Jahnecke, J., Vecsei, P., and Wolff, H. P., Sekretin plasmakonzetration, Verteilung, Stoffwechsel und Ausscheidung von Aldosteron bei Gesunden und Kranken. *Klin. Wochenschr.* **46:**741–751, 1968.

Luetscher, J. A., Jr., and Johnson, B. B.: Observations on the sodium-retaining corticoid (aldosterone) in the urine of children and adults in relation to sodium balance and edema. *J. Clin. Invest.* **33:**1441–1446, 1954.

Lumholtz, I. B., Faber, J., Sørensen, M. B., Kirkegaard, C., Siersbeck-Nielsen, K., and Friis, T. H.: Peripheral metabolism of T_4, T^3, reverse T_3, $3',5'$-diiodothyronine and $3',3'$-diiodothyronine in liver cirrhosis. *Horm. Metab. Res.* **10:** 566–567, 1978.

Lundquist, F., Tygstrup, N., Winkler, K., Mallemgaard, K., and Munck-Petersen, S.: Ethanol metabolism and production of free acetate in the human liver. *J. Clin. Invest.* **41:**955–961, 1962.

Lundquist, F., Tygstrup, N., Winkler, K., and Jensen, K. B.: Glycerol metabolism in the human liver: Inhibition by ethanol. *Science* **150:**616–617, 1965.

Lundsgaard, E.: Glycerol oxidation and muscle exercise. *Acta Physiol. Scand.* **12:**27–33, 1946.

Madison, L. L.: Ethanol induced hypoglycemia. *Adv. Metab. Dis.* **3:**85–109, 1968.

Madison, L. L., Lochner, A., and Wulff, J.: Ethanol-induced hypoglycemia. II. Mechanism of suppression of hepatic gluconeogenesis. *Diabetes* **16:**252–259, 1967.

Marco, J., Diego, J., Villanueva, M. L., Diaz-Fierros, M., Valverde, I., and Sergovia, J. M.: Elevated plasma glucagon levels in cirrhosis. *N. Engl. J. Med.* **289:**1107–1111, 1973.

Marek, J., and Schüllerová, O.: Somatomedin in chronic liver disease. *Rev. Czech. Med.* **22:**194–200, 1974.

Margraf, H. W., Moyer, C. A., Ashford, L. E., and Lavalle, L. W.: Adrenocortical function in alcoholics. *J. Surg. Res.* **7:**55–62, 1967.

Marks, V.: Alcohol-induced hypoglycemia and endocrinopathy. In: *Alcoholism* (G. Edwards and M. Grant, eds.), Baltimore, University Park Press, pp. 208–217, 1978b.

Marks, V., and Medd, W. E.: Alcohol-induced hypoglycemia. *Br. J. Psychiatry* **110:**228–232, 1964.

Marquis, C., Marchetti, J., Burlet, C., and Boulange, M.: Sécrétion urinaire et hormone antidiurétique chez des rats soumis à une administration répétée d'ethanol. *C. R. Soc. Biol.* (Paris) **169:**154–161, 1975.

Martin, H. E., McCuskey, C., Jr., and Tupikova, N.: Electrolyte disturbance in acute alcoholism with particular reference to magnesium. *Am. J. Clin. Nutr.* **7:**191–196, 1959.

Matunga, H.: Experimentelle Untersuchunger üben den Einfluss des Alkohols auf den Kohlehydratstoffwechsel. I. Über die Wirkung des Alkohols auf den Blutzuckerspiegel und den Glykogengehalt der Leber mit besonderer Berucksichtigung seines Wirkungsmechanismus. *Tohoku J. Exp. Med.* **44:**130–157, 1942.

Mayes, P. A.: A calorie deficiency hypothesis of ketogenesis. *Metabolism (Clin. Exp.)* **11:**781–799, 1962.

McCann, V. J., and Fulton, T. T.: Cortisol metabolism in chronic liver disease. *J. Clin. Endocrinol. Metab.* **40:**1038–1044, 1975.

McConnon, J., Row, V. V., and Volpé, R.: The influence of liver damage in man on the distribution and disposal rates of thyroxine and triiodothyronine. *J. Clin. Endocrinol. Metab.* **34:**144–151, 1972.

McLachlan, M. J., and Rodnan, G. P.: Effects of food, fast, and alcohol on serum uric acid levels and acute attacks of gout. *Am. J. Med.* **42:**38–42, 1967.

McMonagel, J., and Felig, P.: Effects of ethanol ingestion on glucose tolerance and insulin secretion in normal and diabetic subjects. *Metabolism* **24:**625–632, 1975.

Megyesi, C., Samols, E., and Marks, V.: Glucose tolerance and diabetes in chronic liver disease. *Lancet* **2:**1051–1056, 1967.

Mendelson, J. H., and Stein, S.: Serum cortisol levels in alcoholic and nonalcoholic subjects experimentally induced ethanol intoxication. *Psychol. Med.* **28:**616–626, 1966.

Mendelson, J. H., Ogata, M., and Mello, N. K.: Adrenal function and alcoholism. I. Serum cortisol. *Psych. Med.* **33:**145–157, 1971.

Merry, J., and Marks, V.: Plasma hydrocortisone response to ethanol in chronic alcoholics. *Lancet* **1:**921–923, 1969.

Merry, J., and Marks, V.: The effect of alcohol, barbiturate and diazepam on hypothalamic/pituitary/adrenal function in chronic alcoholics. *Lancet* **2:**990–992, 1972.

Merry, J., and Marks, V.: Hypothalamic–pituitary–adrenal function in chronic alcoholics. *Adv. Exp. Med. Biol.* **35:**167–169, 1973.

Metz, R., Berger, S., and Mako, M.: Potentiation of the plasma insulin response to glucose by prior administration of alcohol: An apparent islet-priming effect. *Diabetes* **18:**517–522, 1969.

Michael, S. T.: The relation of uric acid excretion to blood lactic acid in man. *Am. J. Physiol.* **141:**71–74, 1977.

Miller, P. B., Heining, R. E., and Worterhouse, C.: Treatment of alcoholic acidosis: The role of dextrose and phosphorus. *Arch. Intern. Med.* **138:**67–72, 1978.

Mirone, L.: Effect of ethanol on growth and liver components in mice. *Life Sci.* **4**:1823–1830, 1965.

Modigliani, E., Periac, P., Perret G., Hugues, J. N., and Coste, T.: La réponse au TRH au cours de 53 hépatopathies alcooliques chroniques. *Ann. Med.* **130**:297–302, 1979.

Moss, M. H.: Alcohol-induced hypoglycemia and coma caused by alcohol sponging. *Pediatrics* **46**:445–447, 1970.

Moyer, J., and Womack, C.: Glucose tolerance. II. Evaluation of glucose tolerance in liver disease and comparison of the relative value of three types of tolerance tests. *Am. J. Med. Sci.* **216**:446–457, 1948.

Murdock, H. R., Jr.: Blood glucose and alcohol levels after administration of wine to human subjects. *Am. J. Clin. Nutr.* **24**:394–396, 1971.

Murray, M. M.: The diuretic action of alcohol and its relation to pituitrin. *J. Physiol. [Lond.]* **76**:379–386, 1932.

Neame, P. B., and Jourbet, S. M.: Post-alcoholic hypoglycemia and toxic hepatitis. *Lancet* **2**:893, 1961.

Nichols, J., Miller, A. T., Jr., and Hiatt, E. P.: Influence of muscular exercise on uric acid excretion in man. *J. Appl. Physiol.* **3**:501–507, 1951.

Nicholson, W. M., and Taylor, H. M.: The effect of alcohol on water and electrolyte balance in man. *J. Clin. Invest.* **17**:279–285, 1938.

Nikkilä, E. A., and Ojala, K.: Ethanol-induced alterations in the synthesis of hepatic and plasma lipids and hepatic glycogens from glycerol-C^{14}. *Life Sci.* **2**:717–721, 1963a.

Nikkilä, E. A., and Ojala, K.: Role of hepatic L-glycerophosphate and triglyceride synthesis in production of fatty liver by ethanol. *Proc. Soc. Exp. Biol. Med.* **113**:814–817, 1963b.

Nikkilä, E. A., and Taskinen, M. R.: Ethanol-induced alterations of glucose tolerance, postglucose hypoglycemia, and insulin secretion in normal, obese and diabetic subjects. *Diabetes* **24**:933–943, 1975.

Noble, E. P., Kakihana, R., and Butte, J. C.: Corticosterone metabolism in alcohol-adapted mice. In: *Biological Aspects of Alcohol* (M. K. Roach, W. M. Melsaac, and P. J. Creaven, eds.), University of Texas Press, Austin, pp. 389–417, 1971.

Nomura, S., Pittman, C. S., Chambers, J. B., Jr., Buck, M. W., and Shimizu, T.: Reduced peripheral conversion of thyroxine to triiodothyronine in patients with hepatic cirrhosis. *J. Clin. Invest.* **56**:643–652, 1975.

Nutritional significance of lactose intolerance. *Nutr. Rev.* **36**:133–134, 1978.

Ogata, M., Mendelson, J. H., and Mello, N. K.: Electrolyte and osmolality in alcoholics during experimentally induced intoxication. *Psychol. Med.* **30**:463–488, 1968.

Ogata, M., Mendelson, J. H., Mello, N. K., and Majchrowitz, E.: Adrenal function and alcoholism. II. Catecholamines. *Psychol. Med.* **33**:159–180, 1971.

Ogura, C.: Studies on glucose, amino acid and protein metabolism in patients with liver cirrhosis in relation to plasma levels of human growth hormone. *Hokkaido J. Med. Sci.* **50**:489–506, 1975.

O'Keefe, S. J. D., and Marks, V.: Lunchtime gin and tonic a cause of reactive hypoglycemia. *Lancet* **1**:1286–1288, 1977.

Olin, J. S., Devenyi, P., and Weldon, K. L.: Uric acid in alcoholics. *Q. J. Stud. Alcohol* **34**:1202–1207, 1976.

Ontko, J. A.: Effects of ethanol on the metabolism of free fatty acids in isolated liver cells. *J. Lipid Res.* **14**:78–86, 1973.

Oppenheimer, J. H., Schwartz, H. L., Mariash, C. N., and Kaiser, F. E.: Evidence for a factor in the serum of patients with non-thyroid disease which inhibits iodothyronine binding by solid matrices, serum proteins and rat hepatocytes. *J. Clin. Endocrinol. Metab.* **54**:757–766, 1982.

Ourea, E., Raiha, N. C. R., and Suomalainen, H.: Influence of some alcohols and narcotics on the adenosine phosphates in the liver of the mouse. *Ann. Med. Exp. Fenn.* **45**:57–62, 1967.

Owens, D., Srivastava, M. C., Tompkins, C. V., Nabarro, J. D. N., and Sönksen, P. H.: Studies on the metabolic clearance rate, apparent distribution space and plasma half-disappearance time of unlabeled human growth hormone in normal subjects and in patients with liver disease, renal disease, thyroid disease and diabetes mellitus. *Eur. J. Clin. Invest.* **3**:284–294, 1973.

Oxenkrug, F. G.: Dexamethasone test in alcoholics. *Lancet* **2**:795, 1978.

Palmer, J. P., and Ensinck, J. W.: Stimulation of glucagon secretion by ethanol-induced hypoglycemia in man. *Diabetes* **24**:295–300, 1975.

Pamenter, R. W., and Boyden, T. W.: Mini review: Interactions of ethanol with the hypothalamic–pituitary–thyroid axis. *Life Sci.* **34**:707–712, 1986.

Panerai, A. E., Salerno, F., Manneschi, M., Cocchi, D., and Müller, E. E.: Growth hormone and prolactin response to thyrotropin releasing hormone in patients with severe liver disease. *J. Clin. Endocrinol. Metab.* **45**:134–140, 1977.

Perkoff, G. T., Eik-Nes, K., Nugent, C. A., Fred, H. L., Nimer, R. A., Rush, L., Samuels, L. T., and Tyler, F. H.: Studies of the diurnal variations of plasma 17-hydroxycorticosteroids in man. *J. Clin. Endocrinol. Metab.* **19**:432–443, 1959.

Perman, E. S.: The effect of ethyl alcohol on the secretion from the adrenal medulla in man. *Acta Physiol. Scand.* **44**:241–247, 1958.

Perman, E. S.: Observations on the effect of ethanol on the urinary excretion of histamine, 5-hydroxyindole acetic acid, catecholamines and 17-hydroxycorticosteroids in man. *Acta Physiol. Scand.* **51**:62–67, 1961a.

Perman, E. S.: Effect of ethanol and hydration on the urinary excretion of adrenaline and noradrenaline and on the blood sugar of rats. *Acta Physiol. Scand.* **51**:68–74, 1961b.

Perman, E. S.: Effect of ethanol on oxygen uptake and on blood glucose concentration in anesthetized rabbits. *Acta Physiol. Scand.* **55**:189–202, 1962.

Peterson, R. E.: Metabolism of adrenocorticosteroids in man. *Ann. NY Acad. Sci.* **82**:846–853, 1959.

Peterson, R. E.: Adrenocortical steroid metabolism and adrenal cortical function in liver disease. *J. Clin. Invest.* **39**:320–331, 1960.

Peterson, R. E., Wyngaarden, J. B., Guerra, S. D., Brodie, B. B., and Bunim, J.: Physiological disposition and metabolic fate of hydrocortisone in man. *J. Clin. Invest.* **34**:1779–1794, 1955.

Phillips, G. B., and Safrit, H. F.: Alcoholic diabetes. Induction of glucose intolerance with alcohol. *J.A.M.A.* **217**:1513–1519, 1971.

Pimstone, B. L., Le Roith, D., Epstein, S., and Kronheim, S.: Disappearance rates of plasma growth hormone after intravenous somatostatin in renal and liver disease. *J. Clin. Endocrinol. Metab.* **41:**392–395, 1975.

Pohorecky, L. A.: Effects of ethanol on central and peripheral noradrenergic neurons. *J. Pharmacol. Exp. Ther.* **189:**380–391, 1974.

Pokroy, M., Epstein, S., Hendricks, S., and Pimstone, B.: Thyrotrophin response to intravenous thyrotrophin-releasing hormone in patients with hepatic and renal disease. *Horm. Metab. Res.* **6:**132–136, 1974.

Pozefsky, T., Felig, P., Tobon, J. D., Soeldner, J. S., and Cahill, G. F., Jr.: Amino acid balance across tissues of the forearm in postabsorptive man. Effects of insulin at two dose levels. *J. Clin. Invest.* **48:**2273–2282, 1969.

Priem, H. A., Shanley, B. C., and Malan, C.: Effect of alcohol administration on plasma growth hormone response to insulin-induced hypoglycemia. *Metabolism* **25:**397–403, 1976.

Prieto, R., Varela, A., and Maardones, J.: Influence of oral administration of thyroid powder on the voluntary alcohol intake by rats. *Acta Physiol. Lat.* **8:**203 (abstract), 1958.

Prinz, P. N., Roehrs, T. A., Vitaliano, P. P., Linnoila, M., and Weitzman, E. D.: Effect of alcohol on sleep and nighttime plasma growth hormone and cortisol concentrations. *J. Clin. Endocrinol. Metab.* **51:**759–764, 1980.

Quick, A. J.: The effect of exercise on the excretion of uric acid, with a note on the influence of benzoic acid on uric acid elimination in liver disease. *J. Biol. Chem.* **110:**107–112, 1935.

Rawat, A. K.: Effects of ethanol infusion on the redox state and metabolic levels in rat liver in vivo. *Eur. J. Biochem.* **6:**585–592, 1968.

Rebouças, G., and Isselbacher, K. J.: Studies on the pathogenesis of the ethanol induced fatty liver. I. Synthesis and oxidation of fatty acids by the liver. *J. Clin. Invest.* **40:**1355–1362, 1961.

Rees, L. H., Besser, G. M., and Jeffcote, W. J.: Alcohol-induced pseudo-Cushing's syndrome. *Lancet* **1:**726–728, 1977.

Rehfeld, J. F., Juhl, E., and Hilden, M.: Carbohydrate metabolism in alcohol-induced fatty liver. Evidence for an abnormal insulin response to glucagon in alcoholic liver disease. *Gastroenterology* **64:**445–451, 1973.

Richter, C. P.: Loss of appetite for alcohol and alcoholic beverages produced in rats by treatment with thyroid preparations. *Endocrinology* **59:**472–478, 1956.

Richter, C. P.: Production and control of alcoholic cravings in rats. In: *Neuropharmacology, Vol. 3* (H. A. Abramson, ed.), Princeton, N.J., Josiah Macy Foundation, pp. 39–146, 1957.

Riesco, J., Costamaillere, L., Litvak, J., and Zlatar, N.: Secrecion de somatotrofina en el etilismo cronico falta de respuesta a la estimalacion con L-dopa y glucagon. *Rev. Med. Chil.* **102:**443–446, 1974.

Röjdmark, S., Adner, N., Andersson, D. E. H., Austern, J., and Lamminpää, K.: Prolactin and thyrotropin responses to thyrotropin-releasing hormone and metaclopramide in men with chronic alcoholism. *J. Clin. Endocrinol. Metab.* **59:**595–600, 1984.

Rosman, P. M., Farag, A., Benn, R., Tito, J., Mishik, A., and

Wallace, E. Z.: Modulation of pituitary-adrenocortical function: Decreased secretory episodes and blunted circadian rhythmicity in patients with alcoholic liver disease. *J. Clin. Endocrinol. Metab.* **55:**709–717, 1982.

Rubini, M. E., Kleeman, C. R., and Lamdin, E.: Studies on alcohol diuresis. I. The effect of ethyl alcohol on water, electrolyte and acid-base metabolism. *J. Clin. Invest.* **34:**439–447, 1955.

Salaspuro, M. P.: Alcohol-induced hypoglycemia as related to the unchanged redox state of the fatty liver. *Scand. J. Clin. Lab. Invest.* **25:**73, 1970.

Salaspuro, M. P.: Influence of the unchanged redox state of liver during ethanol oxidation on galactose and glucose metabolism in protein deficiency. In: *Alkohol und Leber* (W. Gerok, F. Sickinger, and H. H. Hewarkewer, eds.), Stuttgart, Shattauer, pp. 59–62, 1971.

Salaspuro, M. P., and Kesaniemi, Y. A.: Intravenous galactose elimination test with and without ethanol loading in various clinical conditions. *Scand J. Gastroenterol.* **8:**681–686, 1973.

Salaspuro, M. P., Pikkarainen, P., and Lindros, K.: Ethanol-induced hypoglycemia in man: Its suppression by the alcohol dehydrogenase inhibitor 4-methylpyrazole. *Eur. J. Clin. Invest.* **7:**487–490, 1977.

Salaspuro, M. P., Shaw, S., Jayatilleke, E., Ross, W. A., and Lieber, C. S.: Attenuation of the ethanol induced hepatic redox change after chronic alcohol consumption in baboons: Metabolic consequences *in vivo* and *in vitro*. *Hepatology* **1:**33–38, 1981.

Samaan, N., and Stone, D.: Growth hormone and serum insulin in response to a glucose load in chronic hepatic cirrhosis. *J. Lab. Clin. Med.* **70:**986, 1967.

Samaan, N. A., Stone, D. B., and Eckhardt, R. D.: Serum glucose, insulin, and growth hormone in chronic hepatic cirrhosis. *Arch. Intern. Med.* **124:**149–152, 1969.

Samols, E., and Holdsworth, D.: Disturbances in carbohydrate metabolism: Liver disease. In: *Carbohydrate Metabolism and Its Disorders* (F. Dickens, P. J. Randle, and W. J. Whelan, eds.), Vol. 2, London, Academic Press, pp. 289–336, 1968.

Satterfield, H. J., and Guze, S. B.: Treatment of alcoholic patients with triiodothyronine. *Dis. Nerv. Syst.* **22:**227, 1961.

Scavo, D., Cugini, P., Di Lascio, G., Lupi, A., and Rossi, G.: Cortisol metabolism in liver cirrhosis. *Folia Endocrinol.* **27:**670–677, 1974.

Schimpff, R. M., Lebrec, D., and Donnadieu, M.: Somatomedin production in normal adults and cirrhotic patients. *Acta Endocrinol.* **86:**355–362, 1977.

Schimpff, R. M., Lebrec, D., and Donnadieu, M.: Serum somatomedin activity measured as sulphation factor in peripheral, hepatic and renal veins of patients with alcoholic cirrhosis. *Acta Endocrinol.* **88:**729–736, 1978.

Schlienger, J. L.: Thyroid status in fifty patients with alcoholic cirrhosis. *Z. Gastroenterol.* **17:**452–461, 1979.

Schlierf, G., Gunning, B., Uzawa, H., and Kinsell, L. W.: The effects of calorically equivalent amounts of ethanol and dry wine on plasma lipids, ketones, and blood sugar in diabetic and non-diabetic subjects. *Am. J. Clin. Nutr.* **15:**85–89, 1964.

Searle, G. L., Shames, D., Cavalieri, R. R., Bagdade, J. D., and

Porte, D. J.: Evaluation of ethanol hypoglycemia in man: Turnover studies with C-¹⁴C glucose. *Metabolism* **23**:1023–1035, 1974.

Selzer, M. L., and Van Houten, W. H.: Normal thyroid function in chronic alcoholism. *J. Clin. Endocrinol. Metab.* **24**:380–382, 1964.

Sherwin, R., Joshi, P., Hendler, R., Felig, P., and Conn, H. O.: Hyperglucagonemia in Laennec's cirrhosis: The role of portal-systemic shunting. *N. Engl. J. Med.* **290**:239–242, 1974.

Sherwin, R. S., Fisher, M., Bessoff, J., Snyder, N., Hendler, R., Conn, H. O., and Felig, P.: Hyperglucagonemia in cirrhosis: Altered secretion and sensitivity to glucagon. *Gastroenterology* **74**:1224–1228, 1978.

Shishov, V. I.: Sostoianie funkstii kory nadpochechnikov pri khronicheskom alkogolizme. *Zh. Nevropatol. Psikhiatr.* **73**: 235–238, 1973.

Sholiton, L. J., Werk, E. E., Jr., and Marnell, R. T.: Diurnal variation of adrenocortical function in non-endocrine disease states. *Metabolism* **10**:632–646, 1961.

Sinensky, M., Minneman, K. P., and Molinoff, P. B.: Increased membrane acyl chain ordering activates adenylate cyclase. *J. Biol. Chem.* **254**:9135–9141, 1979.

Smals, A., and Kloppenborg, P.: Alcohol-induced pseudo-Cushing's syndrome. *Lancet* **1**:1369, 1977.

Smals, A. G. H., Njo, K. T., Knoben, J. M., Ruland, C. M., and Kloppenborg, P. W. C.: Alcohol induced Cushingoid syndrome. *J. R. Coll. Physicians* **12**:36–41, 1977.

Smith, A. A., and Gitlow, S.: Effects of disulfiram and ethanol on the catabolism of norepinephrine in man. In: *Biochemical Factors in Alcoholism* (R. P. Maickel, ed.), Oxford, Pergamon, pp. 53–59, 1967.

Smith-Laing, G., Sherlock, S., and Faber, O. K.: Effect of spontaneous portal-systemic shunting on insulin metabolism. *Gastroenterology* **76**:685–690, 1979.

Soulairac, A., and Aymard, N.: Rythme circadien du cortisol plasmatique dans l'alcoolisme mental chronique. *Ann. Med. Psychol.* **2**:317–322, 1977.

Staubli, C.: Beiträge zur Pathologie und Therapie des Diabetes Mellitus. *Dtsch. Arch. Klin. Med.* **93**:107–160, 1908.

Stewart, A., Johnston, D. G., Alberti, K. G. M. M., Nattrass, M., and Wright, R.: Hormone and metabolite profiles in alcoholic liver disease. *Eur. J. Clin. Invest.* **13**:397–403, 1983.

Stifel, F. F., Greene, H. L., Lufkin, E. G., Wrensch, M. R., Hagler, L., and Herman, R. H.: Acute effects of oral and intravenous ethanol on rat hepatic enzyme activities. *Biochim. Biophys. Acta* **428**:633–638, 1976.

Stokes, P. E.: Alcohol–endocrine interrelationships. In: *The Biology of Alcoholism* (B. Kissin and H. Begleiter, eds.), Vol. I: Biochemistry, New York, Plenum, pp. 397–436, 1971.

Stokes, P. E.: Adrenocortical activity in alcoholics during chronic drinking. *Ann. NY Acad. Sci.* **215**:77–83, 1973.

Straus, E., Urbach, H-J., and Yalow, R. S.: Alcohol-stimulated secretion of immunoreactive secretin. *N. Engl. J. Med.* **293**:1031–1032, 1975.

Stuart, D. D., and Schultz, A. L.: Thyroid function tests simulating Graves' disease in alcoholic hepatitis. *Ann. Intern. Med.* **89**:514–515, 1978.

Summerskill, W. H. J., and Molnar, G. D.: Eye signs in hepatic cirrhosis. *N. Engl. J. Med.* **266**:1244–1248, 1962.

Sussman, K. E., Alfrey, A., Kirsch, W. M., Zweig, P., Felig, P., and Messner, F.: Chronic lactic acidosis in an adult. *Am. J. Med.* **48**:104–112, 1970.

Tabakoff, B., and Hoffman, P. L.: Alcohol and neurotransmitters. In: *Alcohol Tolerance and Dependence* (H. Rigter and J. Crabbe, eds.), Elsevier/North Holland, Amsterdam, pp. 201–226, 1980.

Tabakoff, B., Jafee, R. C., and Tizmann, R. F.: Corticosterone concentrations in mice during ethanol drinking and withdrawal. *J. Pharm. Pharmacol.* **30**:371–374, 1978.

Takano, K., Hizuka, M., Shizume, K., Hayashi, M., Motoike, Y., and Obata, H.: Serum somatomedin peptides measured by somatomedin A radio-receptor assay in chronic liver disease. *J. Clin. Endocrinol. Metab.* **45**:828–832, 1977.

Tamburrano, G., Tamburrano, S., Gambardella, S., and Andreani, D.: Effects of alcohol on growth hormone secretion in acromegaly. *J. Clin. Endocrinol. Metab.* **42**:193–196, 1976.

Taylor, A. L., Lipman, R. L., Salam, A., and Mintz, D. H.: Hepatic clearance of human growth hormone. *J. Clin. Endocrinol. Metab.* **34**:395–399, 1972.

Tennent, D. M.: Factors influencing the effects of alcohol on blood sugar and liver glycogen. *Q. J. Stud. Alcohol* **2**:263–270, 1941.

Teschke, R., Moreno, F., Heinen, E., Hermann, J., Krüskemper, H-L., and Strohmeyer, G.: Hepatic thyroid hormone levels following chronic alcohol consumption: Direct experimental evidence in rats against the existence of a hyperthyroid hepatic state. *Hepatology* **3**:469–474, 1983.

Tiengo, A., Fedele, D., Frasson, P., *et al.*: Ethanol effect on glucagon secretion in the pig. *Horm. Metab. Res.* **5**:245–246, 1974.

Toccafondi, R., Maioli, M., and Palmas, S.: Growth hormone levels and chronic liver cirrhosis. *Riv. Crit. Clin. Med.* **70**: 120–125, 1970.

Toro, G., Kolodny, R. C., Jacobs, L. S., Masters, W. H., and Daughaday, W. H.: Failure of alcohol to alter pituitary and target organ hormone levels. *Clin. Res.* **21**:505 (abstract), 1973.

Treffot, M. J., Tiscornia, O. M., Palasciano, G., Hage, G., and Sarles, H.: Chronic alcoholism and endogenous gastrin. *Am. J. Gastroenterol.* **63**:29–32, 1975.

Tucker, H. S. G., and Porter, W. B.: Hypoglycemia following alcoholic intoxication. *Am. J. Med. Sci.* **204**:559–566, 1942.

Tygstrup, N.: The galactose elimination capacity in control subjects and in patients with cirrhosis of the liver. *Acta Med. Scand.* **175**:281–289, 1964.

Tygstrup, N., and Lundquist, F.: The effect of ethanol on galactose elimination in man. *J. Lab. Clin. Med.* **59**:102–109, 1962.

Van Dyke, H. B., and Ames, R. G.: Alcohol diuresis. *Acta Endocrinol.* **7**:110–121, 1951.

Van Thiel, D. H., Gavaler, J. S., Wight, C. S., Smith, W. I., Jr., and Abuid, J.: Thyrotropin-releasing hormone (TRH)-induced growth hormone (hGH) responses in cirrhotic men. *Gastroenterology* **75**:66–70, 1978.

Van Thiel, D. H., Smith, W. I., Jr., Wight, C., and Abuid, J.: Elevated basal and abnormal thyrotropin releasing hormone-induced thyroid stimulating hormone secretion in chronic alcoholic men with liver disease. *Alcoholism: Clin. Exp. Res.* **3:**302–308, 1979.

Wagner, G., and Fuchs, A-R.: Effect of ethanol on uterine activity during suckling in postpartum women. *Acta Endocrinol.* **58:** 133–141, 1968.

Wells, S. A., Jr., Cooper, C. W., and Ontjes, D. A.: Stimulation of thyrocalcitonin secretion by ethanol in patients with medullary thyroid carcinoma—an effect apparently not mediated by gastrin. *Metabolism* **24:**1215–1219, 1975.

Wiedemann, E., and Schwartz, E.: Suppression of growth hormone-dependent human serum sulfation factor by estrogen. *J. Clin. Endocrinol. Metab.* **34:**51–58, 1972.

Wiedemann, E., Schwartz, E., and Frantz, A. G.: Acute and chronic estrogen effects upon serum somatomedin activity, growth hormone, and prolactin in man. *J. Clin. Endocrinol. Metab.* **42:**942–952, 1976.

Williams, G. A., Bowser, E. N., Hargis, G. K., Kukreja, S. C., Shah, J. H., Vora, N. M., and Henderson, W. J.: Effect of ethanol on parathyroid hormone and calcitonin secretion in man. *Proc. Soc. Exp. Biol. Med.* **159:**187–191, 1978.

Williamson, D. H., Bates, M. W., and Krebs, H. A.: Activity and intracellular distribution of enzymes of ketone-body metabolism in rat liver. *Biochem. J.* **108:** 353–361, 1968.

Williamson, J. R., Kreisberg, R. A., and Felts, P. W.: Mechanism for the stimulation of gluconeogenesis by fatty acids in perfused rat liver. *Proc. Natl. Acad. Sci. U.S.A.* **56:**247–254, 1966.

Williamson, J. R., Browning, E. T., and Scholz, R.: Control mechanisms of gluconeogenesis and ketogenesis. I. Effects of oleate on gluconeogenesis in perfused rat liver. *J. Biol. Chem.* **244:**4607–4616, 1969.

Wolfe, B. M., Havel, J. R., Marliss, E. B., Kane, J. P., Seymour, J., and Ahuja, S. P.: Effects of a 3-day fast and of ethanol on splanchnic metabolism of FFA, amino acids and carbohydrates in healthy young men. *J. Clin. Invest.* **57:**329–340, 1976.

Wolff, H. P., Lommer, D., and Torbica, M.: Utersuchungen uber das plasmaaddosteron und den Aldosteronstott wechsel bei Gesunden, Herz, Leber- und Nierenkranken. *Schweiz, Med. Wochenschr.* **95:**387–395, 1962.

Wright, J.: Endocrine effects of alcohol. *Clin. Endocrinol. Metab.* **7:**351–367, 1978.

Wright, J. W., Merry, J., Fry, D., Merry, J., and Marks, V.: Abnormal hypothalamic–pituitary gonadal function in chronic alcoholics. *Br. J. Addict.* **71:**211–215, 1976.

Wu., A., Grant, D. B., Humbley, J., and Levi, A. J.: Reduced serum somatomedin activity in patients with chronic liver disease. *Clin. Sci. Mol. Med.* **47:**359–366, 1974.

Yamaguchi, K., Fukushima, H., and Uzawa, H.: Response of human growth hormone, prolactin and thyrotropin to thyrotropin releasing hormone in liver cirrhosis and diabetes mellitus. *Endocrinol. Jpn.* **26:**81–88, 1979.

Yu, T. F., Sirota, J. H., Berger, L., Halpern, M., and Gutman, A. B.: Effect of sodium lactate infusion on urate clearance in man. *Proc. Soc. Exp. Biol. N.Y.)* **96:**809–813, 1957.

Zanoboni, A., and Zanobini-Muciaccia, W.: Elevated basal growth hormone levels and growth hormone response to TRH in alcoholic patients with cirrhosis. *J. Clin. Endocrinol. Metab.* **45:**576–578, 1977.

Zimmerman, H. J., Thomas, L. J., and Scherr, E. H.: Fasting blood sugar in hepatic disease with reference to infrequency of hypoglycemia. *Arch. Intern. Med.* **91:**577–584, 1953.

Zumoff, B., Bradlow, H. L., Gallagher, T. F., and Hellman, L.: Cortisol metabolism in cirrhosis. *J. Clin. Invest.* **46:**1735–1743, 1967.

4

Ethanol and Lipid Disorders, Including Fatty Liver, Hyperlipemia, and Atherosclerosis

Charles S. Lieber

4.1. Interaction of the Metabolism of Ethanol and Lipids

Ethanol interference with lipid metabolism may lead not only to the well-known deposition of neutral lipids in the liver (Fig. 4.1) and some other tissues but also to functional disturbances because of changes in the lipid environment of cellular membranes. Ethanol-induced lipid changes have been implicated in the pathogenesis of diseases such as pancreatitis, hemolytic anemia, cardiomyopathy, and atherosclerosis.

4.1.1. Effects of Excessive Hepatic NADH Generation

The oxidation of ethanol results in the transfer of hydrogen to NAD. During this process, the redox potential of the liver shifts to a more reduced level, indicating that the hepatocytes' capacity to handle reducing equivalents has been exceeded. The altered redox level is reflected in an enhanced NADH-to-NAD ratio, and, in turn, this excess NADH produces changes in the flux of other substrates and alters the ratio of those metabolites that are dependent for reduction on the NADH/NAD couple. It was therefore proposed that the altered NADH/NAD ratio is responsible for a number of metabolic abnormalities associated with alcohol abuse, including alterations of lipid metabolism (Lieber and Davidson, 1962).

In the absence of ethanol, the reducing equivalents utilized in the respiratory chain are provided mainly by the oxidation of fatty acids. During the oxidation of ethanol, the activity of the citric acid cycle is depressed (Forsander *et al.*, 1965; Lieber *et al.*, 1967), partly because of a slowing of the reactions of the cycle that require NAD. The mitochondria then use the hydrogen equivalents originating from ethanol rather than from oxidation through the citric acid cycle of two-carbon fragments derived from fatty acids. Thus, fatty acids that normally serve as the main energy source of the liver are supplanted by ethanol. Decreased fatty acid oxidation caused by ethanol has been demonstrated in liver slices (Blomstrand *et al.*, 1973; Lieber and Schmid, 1961), perfused liver (Lieber *et al.*, 1967), isolated hepatocytes (Ontko, 1973), and *in vivo* (Blomstrand and Kager, 1973). This decrease in fatty acid oxidation results in the deposition of dietary fat in the liver, when available, or fatty acids derived from endogenous synthesis in the absence of dietary fat (Lieber *et al.*, 1966a, 1969; Lieber and Spritz, 1966; Mendenhall, 1972).

The increased NADH-to-NAD ratio is also associ-

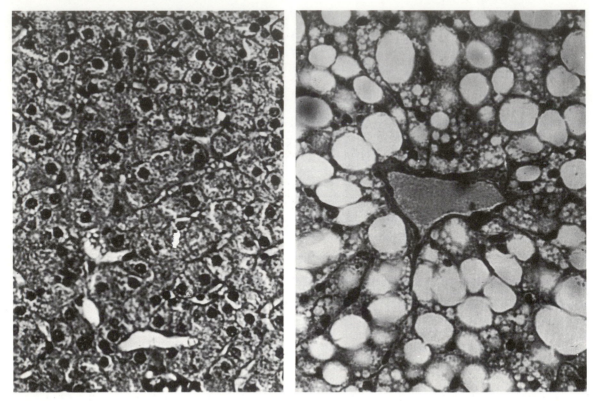

FIGURE 4.1. Sections of baboon liver show normal hepatic architecture during a control period (left) and conspicuous fatty metamorphosis after 15 months of administration of ethanol with a high-protein diet (right). Hematoxylin and eosin. ×400.

ated with a rise in the concentration of α-glycerophosphate (Nikkilä and Ojala, 1963), which favors hepatic triglyceride accumulation by trapping fatty acids. Theoretically, lipogenesis at all levels (enhanced synthesis of fatty acid, triacylglycerols, and cholesterol and/or its esters) can be considered as metabolic systems whereby excess reducing equivalents formed during the oxidation of ethanol can be utilized. *In vitro* studies in rats (Gordon, 1972; Lieber and Schmid, 1961) and in humans (Holmström, 1969) have demonstrated that in the presence of ethanol, lipogenesis is increased, possibly by the elongation pathway of transhydrogenation to nicotinamide adenine dinucleotide phosphate (NADPH) (Lieber, 1968). However, this pathway may only participate in removing excess NADH under limited conditions, for in *in vitro* experiments in naive rats following an acute dose of ethanol, no evidence of enhanced fatty acid synthesis was observed (Guynn *et al.*, 1973). In animals consuming ethanol for prolonged periods of time, conflicting results have been obtained. Arakawa *et al.* (1975) found an increase in lipogenesis, and Savolainen *et al.* (1977a) found no change. In monkeys,

hepatic steatosis after alcohol was accompanied by increased incorporation of [^{14}C]acetate (Vasdev *et al.*, 1974). *In vitro*, addition of ethanol caused a 20% decreased incorporation of tritium into fatty acids, probably because of a decrease in the specific activity of the NADPH (Selmer and Grunnet, 1976).

The effects observed following an acute dose of ethanol and those in animals consuming ethanol for prolonged periods of time differ (Salaspuro *et al.*, 1981). After chronic ethanol consumption, the acute inhibition of fatty acid oxidation fades (Salaspuro *et al.*, 1981) in keeping with the attenuation of the redox change discussed previously in Chapter 1.

4.1.2. Interaction of Ethanol with Hepatic Microsomal Lipid Metabolism

Most of the enzymes involved in the synthesis of complex lipids (such as triacylglycerols and phospholipids) and lipoproteins are bound to the membranes of the endoplasmic reticulum. Since ethanol is also metabolized by the microsomal enzyme system, it may interfere with lipid metabolism by mechanisms not related to changes in the cytoplasmic or mitochondrial redox change. The exact mechanism of the ethanol-induced alterations of the microsomal functions has not been clarified. As discussed in Chapter 1, chronic ethanol feeding results in proliferation of the membranes of the smooth endoplasmic reticulum, the morphological counterpart of the microsomal fraction obtained by ultracentrifugation. Therefore, the lipid-synthesizing capacity of the liver may be enhanced partly by the increase in the mass of the smooth endoplasmic reticulum. In addition, it has been demonstrated that both acute and chronic ethanol administration produce some specific alterations in the activities of enzymes involved in the synthesis of complex lipids and lipoproteins. Moreover, the microsomal ethanol oxidation interferes with lipid metabolism by generating oxygen radicals such as $\cdot O_2^-$ and OH$\cdot$ (Chance et al., 1979; Reitz, 1975), which in turn can initiate a cascade of autocatalytic lipoperoxidation reactions and damage cell membranes (Tappel, 1973). Microsomal "induction" produced by chronic ethanol ingestion increases activities of enzymes such as NADPH oxidase (Lieber and De-Carli, 1970b; Reitz, 1975), resulting in enhanced $\cdot O_2^-$ and H_2O_2 production, thereby also favoring lipid peroxidation.

One difficulty in assessing the effect of ethanol on lipid metabolism is the fact that [^{14}C]acetate, which is commonly used as a precursor for lipid studies, is heavily diluted by the unlabeled acetate that results from ethanol metabolism. Effects of ethanol on lipids are therefore easily overshadowed by isotopic dilution problems. This is one of the reasons why the literature is replete with conflicting assessments of the effects of ethanol on lipid metabolism. In addition, effects may vary depending on the dose and duration of ethanol administration. Despite the complexity of the possible mechanism involved, the resulting effect is a striking accumulation of triglycerides and cholesterol esters in the liver after chronic ethanol feeding. Moreover, as described in Chapter 1, a large percentage of administered labeled ethanol ends up as labeled lipids, which again underscores the importance of ethanol as a precursor for lipid synthesis.

4.1.3. Effect of Ethanol on Membrane Lipids and Related Membrane Properties

The important role of lipids in membrane structure has long been recognized. Bacteria and other prokaryotes can adapt to drugs in their environment by changing the lipid composition of the membranes. This

has been extrapolated to mammalian cells, and consequently, several studies have been made of ethanol effects on the lipid composition and functional properties of cellular membranes. It has been shown that the presence of a physiological concentration of ethanol (20–80 mM) increases the fluidity of the plasma membranes of mouse erythrocytes and synaptosomes in vitro (Chin and Goldstein, 1977; Chin et al., 1978; Johnson et al., 1979). Moreover, membranes isolated from animals chronically treated with ethanol were more resistant to the in vitro fluidizing effect of ethanol than membranes prepared from their respective pair-fed controls (Chin et al., 1978; Johnson et al., 1979). These results were confirmed in man by a study of alcoholic erythrocyte membranes (Beaugé et al., 1985). The changes in the physicochemical properties of the membranes have been linked to the development of tolerance at the level of the intact animal and to the disturbed metabolic function of cell organelles at the cellular level. It has been proposed that the changes in mitochondrial function produced as a consequence of prolonged ethanol consumption may be directly related to alteration in the properties of cellular membranes (Rottenberg et al., 1981; Waring et al., 1981), but others found that no such simple correlation explains the functional changes observed in the mitochondria of ethanol-fed animals (Gordon et al., 1982).

In contrast to other membranes, chronic ethanol feeding resulted in an increase in hepatic plasma membrane fluidity as assessed by fluorescence anisotropy (Polokoff et al., 1985; Schüller et al., 1985; Yamada and Lieber, 1984; Zysset et al., 1985). Such changes may relate to the specialized role of the liver in bile excretion, lipoprotein metabolism, and ethanol oxidation.

Plasma membrane glycoprotein assembly in the liver following acute ethanol administration is significantly impaired (Mailliard et al., 1984). Chronic ethanol feeding also alters the lipid composition of cellular membranes. The more constant changes in lipid composition are the increase in cholesterol, the decrease in phospholipids, or the increase in the total cholesterol/phospholipid ratio and the decrease of the arachidonate/linoleate ratio in the phospholipids of extraneural membranes (Arai et al., 1984a; Chin et al., 1978; Cunningham et al., 1982, 1983; John et al., 1980; Rouach et al., 1984; Sun and Sun, 1985). Differences in experimental conditions, especially dietary factors, could explain the discrepancies between some studies. In liver plasma membrane, the total and free cholesterol were found to be decreased (Polokoff et al., 1985; Yamada and Lieber, 1984) or normal, with increased cholesteryl esters (Zysset et al., 1985; Kim et al., 1988). Arai et al. (1984a) have also shown a decrease in mitochondrial membrane phospholipids from baboon liver associated with increased phospholipase A_2 activity.

Some inhibitory effects by alcohol on $\Delta 5$-, $\Delta 6$-, and $\Delta 9$-acyl-CoA desaturase activities have been reported in rat liver (Nervi *et al.*, 1980; Rao *et al.*, 1984; Wang and Reitz, 1983); these could explain the relative decrease in unsaturated fatty acyl chains. It is not clear whether these rather small alterations are associated with a biologically significant change in the physical state of the membranes (Taraschi and Rubin, 1985). The role of phospholipid composition (Sun and Sun, 1985) and cholesteryl esters (Dawidowicz, 1985; Kim *et al.*, 1988; Zysset *et al.*, 1985) in membrane properties and their modifications by ethanol have been pointed out. Other changes may be produced by the incorporation of ethanol *per se* into membrane lipids. Both *in vitro* and *in vivo*, the presence of ethanol leads to formation of abnormal lipids by incorporation of ethanol to fatty acids or phospholipids with an ester or ether bond (Alling *et al.*, 1983; Goodman and Deykin, 1963; Lange *et al.*, 1981; Polokoff and Bell, 1978). The same mechanism may be involved in both processes. It has been demonstrated that ethanol can replace water molecules in the reactions catalyzed by cholesterol esterase (Lange, 1982) and phospholipase D (Kovatchev and Eibl, 1978). In the former reaction ethyl esters of fatty acids are formed, whereas in the latter the product is an ethylated derivative of phosphatidic acid. These compounds do not exist under normal physiological conditions in the absence of ethanol. It has been postulated that this may help elucidate the mechanism of ethanol intoxication (Alling *et al.*, 1983).

4.1.4. Effect of Acetaldehyde and Acetate on Lipid Metabolism

As discussed in Chapter 2, acetaldehyde can depress the capacity of liver mitochondria to oxidize fatty acids, particularly in organelles derived from alcohol-fed animals. The necessary concentrations of acetaldehyde, although relatively high, do probably occur in some circumstances in the liver *in vivo* (Baraona and Lieber, 1982), at least in some subcompartments. The impairment of the mitochondrial function together with NADH produced by the oxidation of acetaldehyde results in an additional load of reducing equivalents. Furthermore, acetaldehyde reduces the capacity of the liver to export proteins (Baraona *et al.*, 1977; Matsuda *et al.*, 1979), possibly through interference with microtubules; this effect may be associated with a diminution

of lipoprotein secretion (discussed in Chapter 7). Acetaldehyde may also form adducts with lipoproteins, thereby altering their catabolism (Savolainen *et al.*, 1987).

Acetate inhibits lipogenesis in adipose tissue, as discussed in detail in Chapter 2.

4.1.5. Main Sites of the Ethanol-Induced Disturbances in Lipid Metabolism

The most common disturbance of lipid metabolism produced by ethanol ingestion is the fatty liver, characterized by accumulation of triacylglycerols as cytoplasmic lipid droplets in the hepatocytes. Fatty liver develops within a few hours after the ingestion of ethanol and is dose dependent (Horning *et al.*, 1960; Mallov and Bloch, 1956; Savolainen and Hassinen, 1980). The deposition of triacylglycerols develops progressively over the first month of alcohol administration (Fig. 4.2) and persists if the ingestion of ethanol continues, at least for 1 year in the rat (Lieber and DeCarli, 1970a) and for many years in the baboon (Lieber *et al.*, 1972; Popper and Lieber, 1980; Salaspuro *et al.*, 1981). Fatty liver may also be prominent in patients with alcoholic hepatitis or cirrhosis (Lieber, 1982). Eventually other structural damage, such as alterations of mitochondria, may develop (Arai *et al.*, 1984b; Iseri *et al.*, 1966). In rats, cholesteryl esters (Lefevre *et al.*, 1972; Lieber *et al.*, 1963) and fatty acyl ethyl esters (Goodman and Deykin, 1963) also accumulate in the liver.

Lipids also accumulate in tissues other than the liver, such as the heart (Kikuchi and Kako, 1970; Lange

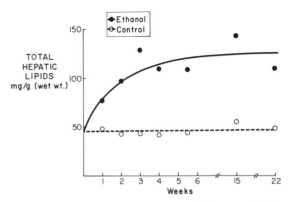

FIGURE 4.2. Time course of hepatic total lipid accumulation in rats fed ethanol with a diet containing normal amounts of protein (18% of calories) and fat (35% of calories). Lipid accumulation levels off after 1 month. (From Lieber and DeCarli, 1970a.)

et al., 1981; Lieber et al., 1966b; Vasdev et al., 1974), lungs (Liau et al., 1981), pancreas (Wilson et al., 1982), and in the blood (Lieber et al., 1963; Vasdev et al., 1974), partly because of the ethanol-induced derangement of hepatic lipid metabolism and also because of a direct action of ethanol and its oxidation on lipid metabolism of the peripheral tissues.

4.2. Pathogenesis of the Alcoholic Fatty Liver

4.2.1. Etiologic Role of Ethanol

Until two decades ago the concept prevailed that malnutrition was primarily responsible for the development of the alcoholic fatty liver. This notion was based largely on experimental work in rats given ethanol in drinking water (Best et al., 1949). With this technique, ethanol consumption usually does not exceed 10 to 25% of the total caloric intake of the animal. A comparable amount of alcohol, when given with an adequate diet, resulted in negligible ethanol levels in the blood (Lieber et al., 1965). When ethanol was incorporated in a totally liquid diet, the amount of ethanol consumed was increased to 36% of total calories, a proportion comparable to moderate alcohol intake in humans. With these nutritionally adequate diets, isocaloric replacement of sucrose or other carbohydrate by ethanol consistently produces five- to tenfold increase in hepatic triglycerides (DeCarli and Lieber, 1967; Lieber et al., 1963, 1965; Porta et al., 1965). Moreover, ethanol as well as insulin and dexamethasone were shown to promote accumulation of triacylglycerol in cultured hepatocytes (Dich et al., 1983). It is noteworthy, as shown in Fig. 4.3, that isocaloric replacement of carbohydrate by fat instead of ethanol did not produce steatosis (Lieber, 1968).

Volwiler et al. (1948) and Summerskill et al. (1957) failed to detect any deleterious effects from alcohol administration in patients recovering from alcoholic fatty liver. In these studies, however, the amounts of alcohol given were less than the usual intake of alcoholics. With larger amount of alcohol, Menghini (1960) found that the clearance of fat from the alcoholic fatty liver was prevented. Moreover, individuals with a morphologically normal liver (with or without a history of alcoholism) developed a fatty liver when given ethanol in a variety of nondeficient diets as an isocaloric substitution for carbohydrates (Lieber et al., 1963, 1965; Lieber and Rubin, 1968). This was evident both by

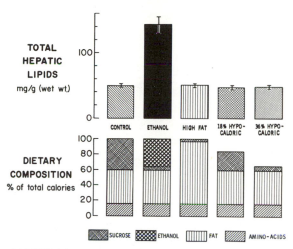

FIGURE 4.3. Effect on total hepatic lipids of five types of liquid diets fed to rats for 24 days. Isocaloric replacement of carbohydrate by ethanol produced an increase in hepatic lipids. Ethanol rather than the reduction in carbohydrate must be incriminated, since isocaloric substitution of carbohydrate with fat had no similar effect. (From Lieber, 1968.)

morphological examination and by direct measurement of the lipid content of the liver biopsies, which revealed up to a 25-fold rise in triglyceride concentration. Even with a high-protein, vitamin-supplemented diet, there was significant increase in hepatic triglycerides, as measured in precutaneous biopsies (Lieber et al., 1975) (Fig. 4.4). These studies therefore established the fact that even in the presence of an adequate diet ethanol can produce a fatty liver.

4.2.2. The Influence of Dietary Factors

4.2.2.1. Role of Dietary Fat

To investigate the role of the amount and kind of dietary fat in the pathogenesis of alcohol-induced liver injury, rats were given liquid diets containing a constant amount of ethanol (36% of energy) and an adequate amount of protein for rodents (18% of total calories), with varying amounts of fat (Lieber and DeCarli, 1970a) (Fig. 4.5). In the 2% fat diet, the only lipid given was linoleate to avoid essential fatty acid deficiency. Reduction in dietary fat to a lever of 25% (or less) of total calories was accompanied by a significant decrease in the steatosis induced by ethanol (Lieber and DeCarli, 1970a). The importance of dietary fat was confirmed in volunteers. For a given alcohol intake, much more steatosis developed with diets of normal fat content than

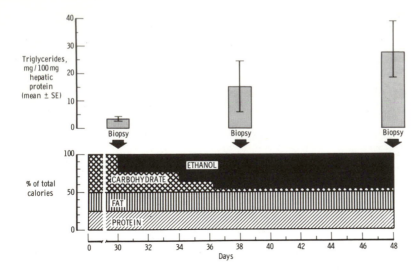

FIGURE 4.4. Effect of ethanol on hepatic triglycerides in five volunteers given a high-protein, low-fat diet. There was a striking increase in spite of the good diet. (From Lieber *et al.*, 1975.)

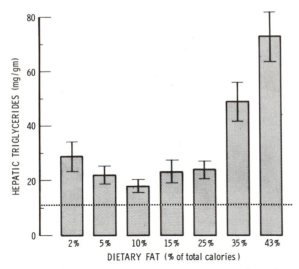

FIGURE 4.5. Hepatic triglycerides in seven groups of rats given ethanol (36% of calories) with a diet normal in protein (18% of calories) but varying in fat content. Average hepatic triglyceride concentration in the control animals is indicated by a dotted line. A reduction of dietary fat decreased the capacity of alcohol to produce steatosis. (From Lieber and DeCarli, 1970a.)

with low-fat diets (Lieber and Spritz, 1966). In addition to the amount, the chain length of the dietary fatty acid is also important for the degree of fat deposition in the liver. Replacement of dietary triglycerides containing long-chain fatty acids by fat containing medium-chain fatty acids reduces the capacity of alcohol to produce a fatty liver in rats (Lieber and DeCarli, 1966). The

propensity of medium-chain fatty acids to undergo oxidation rather than esterification probably explains this phenomenon (Lieber *et al.*, 1967).

4.2.2.2. Role of Protein and Lipotropic Factors (Choline and Methionine)

In a growing rat, deficiencies in dietary protein and lipotropic factors (choline and methionine) can produce a fatty liver (Best *et al.*, 1949), whereas primates are far less susceptible to protein and lipotrope deficiency than rodents (Hoffbauer and Zaki, 1965). Clinically, treatment with choline of patients suffering from alcoholic liver injury has been found to be ineffective in the face of continued alcohol abuse (Olson, 1964; Phillips and Davidson, 1954; Post *et al.*, 1952; Volwiler *et al.*, 1948), and, experimentally, massive supplementation with choline failed to prevent the fatty liver produced by alcohol in volunteer subjects (Rubin and Lieber, 1968). This is not surprising, since, unlike rat liver, human liver contains very little choline oxidase activity, which may explain the species difference with regard to choline deficiency. The phospholipid content of the liver represents another key difference between the fatty liver produced by ethanol and that caused by choline deficiency. After the administration of ethanol, hepatic phospholipids increase (Lieber *et al.*, 1965), whereas in the fatty liver produced by choline deficiency, they decrease (Ashworth *et al.*, 1961). Similarly, hepatic carnitine is decreased by choline deficiency (Corredor *et al.*, 1967) but increased after ethanol feeding (Kondrup and Grunnet, 1973). Ultrastructurally, the two types of fatty liver also differ (Iseri *et al.*, 1966). Furthermore, orotic acid, which reduces the choline-deficiency fatty liver, has no such effect on the ethanol variety (Edreira *et al.*, 1974). Finally, whereas choline deficiency is associated with a reduction in circulating lipoproteins, including high-density lipoproteins (HDL) (Chalvardjian, 1970), the opposite is true of alcohol (Baraona and Lieber, 1970): the increased incorporation of [^{3}H]palmitic acid into lipoproteins after alcohol contrasts with the decrease in choline deficiency (Haines, 1966). Thus, hepatic injury induced by choline deficiency appears to be primarily an experimental disease of rats with little, if any, relevance to human alcoholic liver injury. Even in rats, massive choline supplementation failed to fully prevent the ethanol-induced lesion, whether alcohol was administered acutely (Di Luzio, 1958) or chronically (Lieber and DeCarli, 1966).

The effect of protein deficiency has not yet been clearly delineated in human adults. In children, protein deficiency leads to hepatic steatosis, one of the manifestations of kwashiorkor. In adolescent baboons, protein restriction to 7% of total calories did not result in conspicuous liver injury even after 19 months either by biochemical analysis or by light and electron microscopic examination. Significant steatosis was observed only when the protein intake was reduced to 4% of total energy (Lieber *et al.*, 1972). On the other hand, an excess of protein (25% of total calories, or 2.5 times the recommended amount) did not prevent alcohol from producing fat accumulation in human volunteers (see Fig. 4.4) (Lieber *et al.*, 1975). Thus, in humans, ethanol is capable of producing striking changes in liver lipids even in the presence of a protein-enriched diet. When protein deficiency is present, the deficiency may potentiate the effect of ethanol. In the rat, a combination of ethanol and a diet deficient in both protein and lipotropic factors leads to more pronounced hepatic steatosis than with either factor alone (Klatskin *et al.*, 1954; Lieber *et al.*, 1969). This potentiation is not unexpected, since, as discussed subsequently, increased lipoprotein secretion offsets, in part, the alcohol-induced steatosis. Protein or choline deficiency or both result in impaired lipoprotein secretion, which can be expected to markedly potentiate hepatic lipid accumulation secondary to alcohol. In addition to its role as lipotrope, methionine may also act as a cysteine and glutathione precursor, thereby affecting lipid peroxidation, as discussed in Chapter 7.

4.2.3. Origin of Fatty Acids Deposited in the Alcoholic Fatty Liver and Mechanism of Accumulation

Lipids that accumulate in the liver originate from three main sources: dietary lipids, which reach the bloodstream as chylomicrons; adipose tissue lipids, which are transported to the liver as free fatty acids (FFA); and lipids synthesized in the liver itself (Fig. 4.6). These fatty acids can accumulate in the liver as a result of a variety of metabolic disturbances (Lieber, 1969). The five major mechanisms that have been pro-

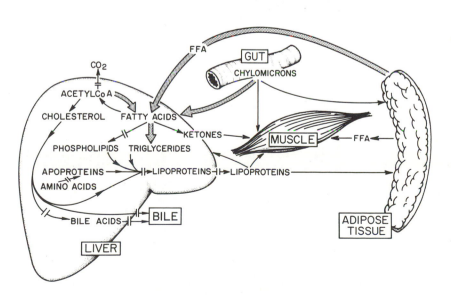

FIGURE 4.6. Possible mechanisms of fatty liver production through either increase (shaded arrows) or decrease (arrows) of lipid transport and metabolism.

posed are (1) increased availability of precursors for hepatic lipid synthesis, (2) increased hepatic lipogenesis because of ethanol-induced stimulation of enzymes, (3) decreased lipid breakdown, (4) decreased hepatic secretion of lipids, and (5) enhanced hepatic uptake of circulating lipids. Depending on the experimental conditions, any of the three sources and the various mechanisms can be implicated.

4.2.3.1. Increased Availability of Substrates for Hepatic Lipid Biosynthesis

Ethanol interferes in many ways with the supply of precursors of lipogenesis from both hepatic and extrahepatic sources. First, ethanol oxidation provides reducing equivalents and two-carbon units for lipid synthesis. Second, the reduced redox state inhibits the oxidation of fatty acids and diverts them into esterification, which is further enhanced by the increased concentration of *sn*-glycerol-3-phosphate. Third, ethanol affects the amount of fatty acids transported from the adipose tissue and the small intestine into the liver. As a net consequence, ethanol results in enhanced incorporation of fatty acids in hepatic lipids (Johnson, 1974).

4.2.3.1a. Increased Mobilization of Peripheral Fat. In rats given one large, sublethal dose of ethanol, it was observed that fatty acids resembling those of adipose tissue accumulate in the liver (Brodie *et al.*, 1961; Lieber *et al.*, 1966a). Experimental procedures or agents that reduce the normal rate of peripheral fat mobilization (that is, adrenalectomy, spinal cord transection, or ganglioplegic drugs) prevent or decrease this type of hepatic fat accumulation (Brodie *et al.*, 1961; Mallov, 1957; Rebouças and Isselbacher, 1961). More direct approaches, however, such as studies in rats with prelabeled epididymal fat pads, yielded conflicting information, with evidence for increased (Kessler and Yalovsky-Mishkin, 1966) or unchanged (Poggi and Di Luzio, 1964) fatty acid mobilization. Similarly, in rats, one single dose of ethanol has been reported to result in increased (Brodie *et al.*, 1961; Savolainen *et al.*, 1977b), unchanged (Elko *et al.*, 1961), or decreased (Jauhonen *et al.*, 1975) circulating levels of FFA depending on the dose and degree of fasting. In rats, only half of the triglyceride fatty acids accumulating in the liver after a large dose of ethanol were derived from plasma unesterified fatty acids (Abrams and Cooper, 1976). In humans, even with amounts of ethanol as large as 300 g/day, the concentration of circulating FFA did not increase; it rose only after ingestion of very large doses

of ethanol (400 g/day) (Lieber *et al.*, 1963) (Fig. 4.7). In short-term studies, ethanol administration produced a fall in the level of circulating FFA in humans (Jones *et al.*, 1963; Lieber *et al.*, 1962) with reduced peripheral venous–arterial differences in FFA (Lieber *et al.*, 1962), decreased FFA turnover (Jones *et al.*, 1965), and concomitant reduction in circulating glycerol (Feinman and Lieber, 1967) (Fig. 4.8). A similar fall in FFA was found in the rat (Jauhonen and Hassinen, 1978). This effect of ethanol on FFA mobilization from adipose tissue was found to be mediated by acetate (Crouse *et al.*, 1968) and acetaldehyde (Jauhonen and Hassinen, 1978). Since stressful doses of ethanol probably both stimulate fatty acid mobilization (via catecholamine release) and depress it (via the acetaldehyde and acetate produced), the net effect may depend on the particular experimental conditions, including the dose of ethanol and the nutritional status of the subject. This may account for some of the apparent contradictions of the literature.

4.2.3.1b. Decreased Hepatic Fatty Acid Oxidation. Decreased fatty acid oxidation caused by ethanol has been demonstrated in liver slices (Lieber and Schmid, 1961; Blomstrand *et al.*, 1973), perfused liver (Lieber *et al.*, 1967), isolated hepatocytes (Ontko, 1973), liver biopsies (Leung and Peters, 1986), and *in vivo* (Blomstrand and Kager, 1973). This decrease in fatty acid oxidation results in the deposition of dietary fat in the liver when available or fatty acids derived from endogenous synthesis in the absence of dietary fat (Lieber *et al.*, 1966a; Lieber and Spritz, 1966; Mendenhall, 1972). The inhibition can be attributed to the redox change and the decrease in the activity of the citric acid cycle. After chronic ethanol consumption, the acute inhibition of fatty acid oxidation was less pronounced, in keeping with the attenuation of the redox change (Salaspuro *et al.*, 1981). An exaggeration of this change takes place in the perivenular zones of the liver, where a greater redox shift is favored by low oxygen tensions prevailing in that zone (Baraona *et al.*, 1983a; Jauhonen *et al.*, 1982). These oxygen tensions may become lower with progression of the liver injury because of impairment of the sinusoidal circulation.

Structural changes of the mitochondria (Arai *et al.*, 1984b; Iseri *et al.*, 1966) could also be responsible for some alterations in lipid metabolism beyond those produced by the altered redox state. The mitochondria showed decreased oxidation of two-carbon fragments of fatty acids (Gordon, 1973; Hasumura *et al.*, 1976; Lieber and Schmid, 1961; Matsuzaki and Lieber, 1977; Rubin *et al.*, 1972) but no changes in the β-oxidation

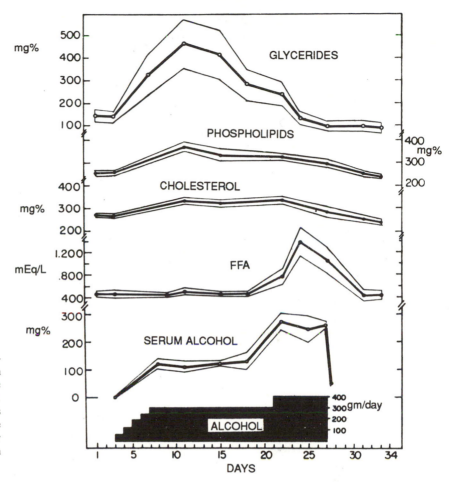

FIGURE 4.7. Effect of prolonged alcohol intake on serum lipids in seven chronic alcoholic individuals (mean ± S.E.M.). The higher triglyceridemia lasts for several weeks, whereas the rise in FFA only occurs with very large amounts of ethanol. (From Lieber *et al.*, 1963.)

(Cederbaum *et al.*, 1975). Carnitine palmitoyltransferase I (CPT-I), the overt form of carnitine palmitoyltransferase, is generally considered to catalyze the rate-limiting step in the transport of long-chain fatty acids into the mitochondrial matrix. Ethanol consumption was found to decrease CPT-I activity (Guzman and Geelen, 1988) and to increase enzyme sensitivity to inhibition by exogenously added malonyl-CoA. Furthermore, addition of ethanol or acetaldehyde to the incubation medium strongly depressed CPT-I activity and rates of fatty acid oxidation in hepatocytes from ethanol-treated rats. Thus, decreased fatty acid oxidation, whether a function of the reduced citric acid cycle activity (secondary to the altered redox potential) or of the permanent changes in mitochondrial structure, offers the most likely explanation for the deposition of fat in the liver, at least during the initial stage of chronic alcohol abuse.

4.2.3.1c. Increased Hepatic Lipogenesis. Earlier *in vitro* studies in rats (Lieber and Schmid, 1961) and in humans (Holmström, 1969) have demonstrated that, in the presence of ethanol, hepatic fatty acid synthesis can be increased. Subsequent studies have revealed a more complex picture. In experiments *in vitro* in naive rats following an acute dose of ethanol, no evidence of enhanced fatty acid synthesis was observed (Guynn *et al.*, 1973). In animals consuming ethanol for prolonged periods of time, conflicting results have been obtained too. An increase (Arakawa *et al.*, 1975; Vasdev *et al.*, 1974) in hepatic fatty acid synthesis or no change (Savolainen *et al.*, 1977a) was observed. In human liver biopsies, a decrease was reported (Venkatesan *et al.*, 1986).

It should be noted that assessing the effect of ethanol on hepatic fatty acid synthesis is extremely difficult, mainly for two reasons. First, ethanol, by its

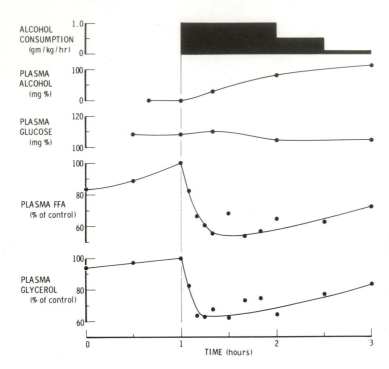

FIGURE 4.8. Effect of ethanol on plasma glucose, FFA, and glycerol concentrations. The FFA and glycerol concentrations are expressed as percentages of the control values immediately preceding alcohol administration. Each point represents the mean value in the five individuals studied. Ethanol ingestion produced a sharp temporary drop in FFA and glycerol. (From Feinman and Lieber, 1967.)

metabolism, causes heavy isotope dilution not only when [^{14}C]acetate incorporation into fatty acids is measured but also in the case of tritium incorporation from ^{3}H$_2$O into fatty acids because of a decrease in the specific activity of the NADPH (Selmer and Grunnet, 1976). Second, since the fatty acid synthesis rate *in vivo* is very sensitive to the amount of both carbohydrate and fat in the diet, it is very difficult to add a large amount of ethanol to the diet without drastically changing the carbohydrate/fat ratio (Savolainen *et al.*, 1977a). In any event, irrespective of any direct effect on fatty acid synthesis, ethanol interferes with the intermediary metabolism by providing a large part of the two-carbon units for lipogenesis (Brunengraber *et al.*, 1974). When the effects of ethanol on fatty acid synthesis and esterification were studied in hepatocytes isolated from fed and 24-hr-fasted rats, it was found that addition of ethanol markedly increased the incorporation of this label in triglycerides and phospholipids; the increase was higher in fasted rat hepatocytes in both the glycerol backbone and the acyl groups of glycerolipids. Ethanol increased [U-^{14}C]palmitate incorporation into triglycerides, but only in hepatocytes from fasted rats (Maquedano *et al.*, 1988).

4.2.3.1d. Increased Supply of *sn*-Glycerol-3-Phosphate and Esterification. It has been proposed that the partition of fatty acids between oxidation and esterification is determined on the outer membrane of mitochondria by the relative activities of carnitine acyltransferase and *sn*-glycerol-3-phosphate acyltransferase. Therefore, an increase in the concentration of *sn*-glycerol-3-phosphate because of the ethanol-induced redox shift would favor the esterification of fatty acids. Indeed, it has been shown that a correlation exists between the hepatic *sn*-glycerol-3-phosphate concentration and triacylglycerol accumulation both *in vivo* (Nikkilä and Ojala, 1963) and in perfused livers (Fellenius *et al.*, 1973). Both of these phenomena are prevented by thyroxine (Ylikahri, 1970) and clofibrate treatment (Savolainen *et al.*, 1977b) or by pyrazole or its derivatives (Morgan and Di Luzio, 1970), suggesting that the concentration of *sn*-glycerol-3-phosphate is an important factor in the pathogenesis of ethanol-induced fatty liver. However, opposite results were also obtained (Bustos *et al.*, 1970; Johnson *et al.*, 1971). The conflicting results are probably caused by differences in the *sn*-glycerol-3-phosphate levels in the various experimental conditions used. Because of *sn*-glycerol-3-phosphate concentration limits the rate of esterification only within a narrow range (Declercq *et al.*, 1982), and because there are probably several pools of *sn*-glycerol-3-phosphate in the hepatocyte, the measurement of the total concentration does not reveal the regulatory role of the pool of *sn*-glycerol-3-phosphate involved in glycerolipid synthesis.

A clinical study with some markers of lipolysis showed an increase of hepatic free fatty acids with progression of alcoholic liver injury (Mavrelis *et al.*, 1983). Cytosolic free fatty acids are potentially toxic (Shaw, 1985); this accumulation could play a primary role, but it also could be secondary to the liver injury. Usually the free fatty acids are bound to the fatty-acyl-binding protein (Ockner *et al.*, 1982), which increases after chronic alcohol consumption (Pignon *et al.*, 1987), but the effects of liver injury on the amount of this protein and its saturation are not known. Increased formation of fatty acyl esters, documented in liver biopsies (Leung and Peters, 1986), may tend to counteract the accumulation of free fatty acids.

4.2.3.2. Increased Activity of the Enzymes of Hepatic Lipid Synthesis

In addition to its effects of increasing the supply of precursors, ethanol may enhance the rate of triglyceride synthesis by stimulating the activities of the enzymes catalyzing triglyceride synthesis. These enzymes are bound to the membranes of the endoplasmic reticulum with the exception of glycerophosphate acyltransferase and phosphatidate phosphohydrolase, which are, in part, found also in the outer membrane of mitochondria and in the cytoplasm, respectively. Phosphatidate phosphohydrolase fulfills almost all criteria of a rate-limiting enzyme, and its activity changes in most situations parallel the rate of triglyceride synthesis (Fallon *et al.*, 1977). Both acute and chronic ethanol administration stimulate phosphatidate phosphohydrolase activity in the microsomal and soluble fractions (Lamb *et al.*, 1979; Pritchard *et al.*, 1977; Savolainen, 1977; Savolainen and Hassinen, 1980), and the increase in the enzyme activity precedes the accumulation of hepatic triglycerides (Savolainen, 1977). It should be noted, however, that on prolonged ethanol administration, the activity is slowly restored to control values (Tijburg *et al.*, 1988). The ethanol-induced stimulation of the soluble phosphatidate phosphohydrolase was shown to be caused partly by an increase in the amount of the enzyme, but also partly by activation of the existing phosphohydrolase (Sturton *et al.*, 1981). This dual action may be caused by

the mediation of the ethanol effect by two separate mechanisms: the former by glucocorticoids (Glenny and Brindley, 1978; Lehtonen *et al.*, 1979) and the latter by low-molecular-weight molecules, probably by *sn*-glycerol-3-phosphate (Savolainen and Hassinen, 1978) or acetaldehyde (Ide and Nakazawa, 1987).

The first enzyme of hepatic glycerolipid synthesis, glycerophosphate acyltransferase, is affected only by chronic ethanol ingestion (Joly *et al.*, 1973). Diacylglycerol transferase, the last enzyme of the triacylglycerol synthesis pathway, is the only one specific for the formation of triacylglycerols; the others share a common pathway with phospholipid synthesis. Therefore, a diacylglycerol acyltransferase would be a plausible site for the ethanol effect, since there is a much greater increase in the triacylglycerols than in the phospholipids. Previous studies showed no change after an acute dose of ethanol (Pritchard *et al.*, 1977). In baboons, at the early stages of fatty liver, there were increased microsomal diacylglycerol acyltransferase and cytosolic phosphatidate phosphohydrolase activities, which disappeared with progression of the liver injury (Savolainen *et al.*, 1984).

In keeping with the increased phospholipid content of the liver after chronic ethanol administration (Lieber *et al.*, 1965), increased activity of two enzymes involved in the synthesis of phosphatidylcholine, namely, choline phosphotransferase and phosphatidylethanolamine methyltransferase, was reported (Uthus *et al.*, 1976).

In alcohol-fed rats, cholesteryl esters accumulate in the liver (Lefevre *et al.*, 1972; Field *et al.*, 1985), and acyl coenzyme A : cholesterol acyltransferase and 3-hydroxy-3-methylglutaryl-coenzyme A (HMG-CoA) reductase activities increase (Field *et al.*, 1985). The latter enzyme was found unchanged in activity in rats given alcohol in drinking water (Lakshman and Veech, 1977). Moreover, an incomplete feedback inhibition of HMG-CoA reductase activity by chylomicron remnants from alcohol-fed rats was described (Lakshman and Ezekiel, 1982).

The increased activity of enzymes of hepatic lipid synthesis with the increased supply of free fatty acid could explain the hepatic lipoprotein overproduction associated with alcoholic hyperlipemia.

4.2.3.3. Decreased Lipid Breakdown

The accumulation of esterified lipids in the fatty liver could result from an impaired capacity of lysosomal lipases and esterases to hydrolyze the glycerol esters. However, chronic ethanol administration increases the appearance of lysosomes (Rubin and Lieber, 1967; Mezey *et al.*, 1980), and the activity of lysosomal enzymes is either increased or unchanged (Mezey *et al.*, 1976, 1980). Acutely, ethanol has no uniform effect on these enzymes (Berg and Mørland, 1973; Von Platt *et al.*, 1971), but it increases the activity of both microsomal and cytosolic phospholipase A (Sturton *et al.*, 1978). In liver biopsies of alcoholics, activity of hepatic lipases was either unchanged or increased (Leung *et al.*, 1985). Hydrolysis of cholesterol ester was found to be reduced after chronic ethanol consumption (Takeuchi *et al.*, 1974). Chronic feeding increases the phospholipase A_2 activity in baboons (Arai *et al.*, 1984b).

4.2.3.4. Decreased Secretion of Lipids from the Liver

Lipids may leave the hepatocyte either as lipoproteins secreted into the bloodstream or as constituents of bile. Ethanol increases the output of triacylglycerol-carrying lipoproteins from the liver (see Section 4.5.1), but this compensatory mechanism is relatively inefficient in preventing fat accumulation, probably because of impairment of protein secretion secondary to cytoskeletal lesions (Baraona *et al.*, 1977; Matsuda *et al.*, 1979: see Chapter 5).

For the fate of cholesterol synthesized in the hepatocyte or taken up from the circulation, the secretion into bile (with or without conversion into bile acids) is as important as the secretion into the blood as lipoproteins. Any disturbance in these functions leads to an imbalance in cholesterol content of the liver or other tissues because the carbon skeleton of cholesterol is not catabolized in the body but is eliminated almost exclusively by excretion through the bile into the feces. In the rat, ethanol increases cholesterogenesis in the liver (Lefevre *et al.*, 1972) and in the small intestine. Furthermore, ethanol feeding decreases degradation of cholesterol to bile acids and excretion of bile acid in feces (Lefevre *et al.*, 1972) (Fig. 4.9), whereas after cessation of ethanol administration, the opposite effect was observed in the bile (Boyer, 1972). In the presence of alcohol, however, biliary secretion of bile acids was depressed (Ideo *et al.*, 1978; Marin *et al.*, 1975). The reduction of bile acid excretion could be secondary to diminution in cholesterol 7α-hydroxylase activity (Lakshman and Veech, 1977). In contrast with the rat, the baboon does not accumulate cholesteryl ester in the liver after chronic ethanol consumption (Baraona *et al.*, 1983b),

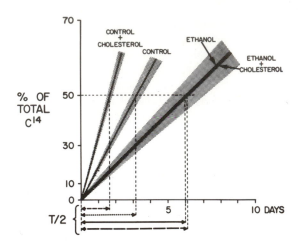

FIGURE 4.9. Cumulative excretion in feces of labeled bile acids following an intraperitoneal injection of [^{14}C]cholic acid. Ethanol suppressed the cholesterol-induced acceleration of bile acid turnover. (Shaded areas represent S.E.M.) (From Lefevre *et al.*, 1972.)

possibly because of increased fecal cholesterol elimination (Karsenty *et al.*, 1985).

The main event leading to the development of the alcoholic fatty liver can be summarized as follows: ethanol, which has an almost "obligatory" hepatic metabolism, replaces the fatty acids as a normal fuel for the hepatic mitochondria. This results in fatty acid accumulation directly, because of decreased lipid oxidation, and indirectly, because one way for the liver to dispose of excess hydrogen generated by ethanol oxidation is to synthesize more lipids. Fatty acids derived from adipose tissue accumulate in the liver only when very large amounts of ethanol are given. The lipids increase in the liver despite the fact that the transport mechanism via release of lipoproteins from the liver into the bloodstream is stimulated by ethanol, at least during the initial state of intoxication. At one point, an equilibrium is reached between a lessening in lipid oxidation (as a result of progressive attenuation of redox change) and enhanced lipoprotein secretion. Lipids, then, cease to accumulate further.

4.3. Agents and Procedures that Prevent the Alcoholic Fatty Liver

The role of dietary fat and lipotropic factors has already been discussed. Differences in dietary fat may explain some of the discrepancies in reports concerning the effect of antioxidants that reduced or prevented hepatic steatosis in some studies (Di Luzio, 1966; Hartman and Di Luzio, 1968) but not in others (Lieber and DeCarli, 1966). The negative results were obtained with diets containing 43% total calories as fat, whereas partial protection was observed with a relatively low-fat diet. Since dietary fat potentiates the steatogenic effect of ethanol, it is quite conceivable that antioxidants may be moderately active with low-fat diets but incapable of counteracting the much stronger effects of ethanol combined with dietary fat. Chlorophenoxyisobutyrate, a drug used to reduce hyperlipemia, partially protected against the alcoholic fatty liver (Brown, 1966; Spritz and Lieber, 1966), possibly through a reduction in glycerolipid formation (Adams et al., 1971). The protective action also could be related to the abolition of the redox change (Kähönen et al., 1972) or enhanced fatty acid metabolism (Savolainen et al., 1977b).

The fatty liver produced by one large dose of ethanol has been reported to be prevented by β-sympathicolytic agents (Estler and Ammon, 1967), pyridyncarbinol (Ammon and Zeller, 1965), cold exposure (Radomski and Wood, 1964), and nicotinic acid (Baker et al., 1973). By contrast, the effect of ethanol given chronically is exacerbated by nicotinic acid (Sorrell et al., 1976). In rats, hyperbaric oxygen was found to be protective (L'Huiller et al., 1967).

Adenine (Hernandez-Munoz et al., 1978) and adenosine triphosphate (ATP) (Hyams and Isselbacher, 1964) have been reported to protect against acute ethanol-induced fatty liver. When given in moderate amounts, however, ATP restored liver adenosine triphosphate levels to normal without protecting against the ethanol-induced fatty liver (Marchetti et al., 1968). The partial protection afforded by much larger doses (Hyams and Isselbacher, 1964) can possibly be attributed to nonspecific effects (hypothermia and impaired peripheral fat mobilization). Chlorpromazine, which inhibits ADH activity, failed to prevent the fatty liver produced by an acute large ethanol dose (Koff and Fitts, 1972). Controversy exists over whether pyrazole, another ADH inhibitor, prevents the fatty liver produced by a single large dose of ethanol. Some researchers found no reduction (Bustos et al., 1970), whereas others found prevention (Blomstrand and Forsell, 1971; Morgan and Di Luzio, 1970). The difference is perhaps caused by the dose of the drug (Nordmann et al., 1972) or the sex of the animal (Domanski et al., 1971) used. Although some acute effects of ethanol on lipid metabolism were prevented by pyrazole (Prancan and Nakano, 1972), the effects of pyrazole on the consequences of

chronic ethanol ingestion were inconclusive (Kalant et al., 1972). A derivative of pyrazole, 3,5-dimethylpyrazole, was shown to reduce the fatty liver resulting from a single large dose of ethanol by blocking free fatty acid mobilization from adipose tissue (Bizzi et al., 1966).

Anabolic steroids were reported to be ineffective by some (Fenster, 1966; Eberhardt et al., 1975) but not by others (Mendenhall, 1968) in accelerating the disappearance of fat from the alcoholic fatty liver. Experimentally, carnitine (Sachan et al., 1984) and a mixture of pyruvate, dihydroxyacetone, and riboflavin were protective (Stanko et al., 1978). These results are still awaiting clinical confirmation.

In summary, there is no practical modality to prevent the alcoholic fatty liver other than by the control of alcohol intake.

4.4. Effects of Ethanol on Blood Lipids: Characteristics and Pathogenesis of Alcoholic Hyperlipemia

Alcohol consumption is associated with changes in all lipid components of the plasma. Hypercholesteremia was first reported by Ducceshi in 1915 and has recently stimulated great interest because of a possible link between moderate alcohol consumption and the prevention of coronary heart disease. Serum lactescence and hypertriglyceridemia after bouts of excessive drinking were first reported by Feigl in 1918.

4.4.1. Blood Lipid Perturbation Induced by Alcohol

4.4.1.1. Plasma Triacylglycerols

The plasma triacylglycerols are the major determinants of the serum lactascence or turbidity found after bouts of excessive drinking. They increase in all lipoprotein fractions but proportionally more in those lipoproteins that are normally rich in triacylglycerols: very-low-density lipoproteins (VLDL), which migrate as pre-β-lipoproteins on electrophoresis, and chylomicrons or chylomicronlike particles, which stay at the origin of the electrophoretic strip (Baraona et al., 1983b). Thus, alcoholic hyperlipemia is usually classified as type IV or type V. However, the phenotypic pattern changes rapidly after alcohol withdrawal, from type V to type IV and to type II, because of the rapid clearance of chylomicrons, followed by VLDL and the slower clearance of cholesterol and phospholipids, which predominate in low-density lipoproteins (LDL) or β-lipoproteins. This striking change as it occurred with time in a given subject is illustrated in Fig. 4.10 (Losowsky *et al.*, 1963).

The incidence of hypertriglyceridemia varies with the population studied. Among hyperlipemic patients, alcohol constitutes the second major cause, following diabetes (Chait *et al.*, 1972). In epidemiologic studies, plasma triacylglycerol levels correlate with alcohol consumption (Castelli *et al.*, 1977; Ostrander *et al.*, 1971), but significant increases in plasma triacylglycerols are rather rarely encountered in patients hospitalized for alcoholism or its complications. Only 27–28% of hospitalized alcoholics had fasting triacylglycerol levels over normal limits (2 mM), and 17% over 3 mM, with phenotypes IV, II, or V, in that order of frequency (Böttiger *et al.*, 1976; Sirtori *et al.*, 1973). The withdrawal state favors hypertriglyceridemia by the reversal of the inhibition of extrahepatic lipoprotein lipase (Ekman *et al.*, 1981). This hypertriglyceridemia is frequently seen in patients with fatty livers but rarely among cirrhotics

(Cachera *et al.*, 1950; Marzo *et al.*, 1970; Patek and Earampamoorthy, 1976). A type III hyperlipoproteinemia can occur in severe liver disease (Muller *et al.*, 1974). Thus, it is apparent that classification of the hyperlipemia in the alcoholics is fraught with difficulties.

4.4.1.2. Plasma Cholesterol

In patients with marked hypertriglyceridemia, plasma cholesterol is also increased. Cholesterol is a component of all lipoprotein fractions, but most of it is transported in LDL and in HDL or β-lipoproteins. In hospitalized alcoholics, total plasma cholesterol was not significantly increased (Böttiger *et al.*, 1976). However, in about 30% of alcoholics seeking medical attention (Johansson and Laurell, 1969) and 86% of patients after a recent drinking bout (Johansson and Medhus, 1974), a prominent α-lipoprotein band was found on electrophoresis. This alteration normalized after approximately 2 weeks of abstinence (Devenyi *et al.*, 1980; Johansson and Medhus, 1974). By contrast, LDL tend to be reduced in alcoholics (Taskinen *et al.*, 1982). The formation of acetaldehyde adducts of LDL and their increased disappearance from the circulation (Savolainen *et al.*, 1987) may offer an explanation for the lack of increase, or the decrease, in serum LDL levels found in alcoholics. The changes in cholesterol-carrying lipoproteins are also shown in epidemiologic studies. There is a strong positive correlation between alcohol consumption and plasma HDL cholesterol and a weaker but significant negative correlation with LDL cholesterol (Castelli *et al.*, 1977; Hulley and Gordon, 1981).

Since only a few other factors (chronic administration of some drugs and vigorous physical exercise) are capable of increasing HDL cholesterol, this has been proposed as one of the useful biological markers of alcoholism (Barboriak *et al.*, 1980; Danielsson *et al.*, 1978; Sanchez-Graig and Annis, 1981). This increase occurs in women (who normally have higher HDL cholesterol) as well as in men, indicating that this is not merely a consequence of the feminization of the alcoholics.

By contrast, alcoholics with advanced liver disease failed to show the increase in HDL cholesterol shortly after their last drink, nor did they show a significant change after abstinence (Devenyi *et al.*, 1980). The main apolipoproteins of HDL, apolipoprotein AI and apolipoprotein AII, are increased by alcohol (Camargo *et al.*, 1985; Fraser *et al.*, 1983; Poynard *et al.*, 1985).

HDL is a heterogeneous family of lipoproteins of

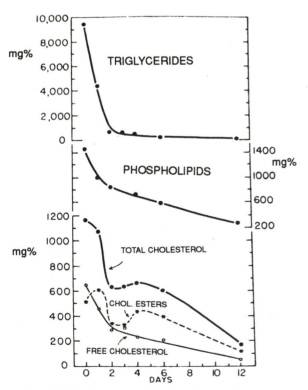

FIGURE 4.10. Changes in plasma lipid fractions during recovery from alcoholic hyperlipemia. (From Losowsky *et al.*, 1963.)

different sizes, composition, and metabolic characteristics (Kraus, 1982; Lieber, 1987) (Fig. 4.11). The HDL increase predominates in the light subfractions (HDL₂) or heavy subfractions (HDL₃), depending on the alcohol intake (see Section 4.5). A minor subfraction that floats as HDL, although it contains apolipoprotein B (with a unique antigenic determinant), is named Lp(a) and has been found to be markedly decreased in heavy alcohol consumers (Marth *et al.*, 1982).

4.4.1.3. Other Plasma Lipids

Plasma FFA concentration is increased or decreased depending on the alcohol intake and the nutritional conditions (see Section 4.2.3.1a). Alcohol consumption increases both total and HDL plasma phospholipid levels (Böttiger *et al.*, 1976; Cuvelier *et al.*, 1985; Lieber *et al.*, 1963; Puchois *et al.*, 1984).

In alcoholics, some modifications of the plasma fatty acid patterns have been described, including a relative increase in oleic acid and a relative decrease in linoleic acid in triacylglycerols and phospholipids (Alling *et al.*, 1979; Norbeck *et al.*, 1979); others have shown a low level of arachidonic acid (Holman and Johnson, 1981; Marzo *et al.*, 1970). Johnson *et al.* (1985) assessed the phospholipid fatty acyl pattern in a large group of patients. They concluded that excessive alcohol

consumption alone does not induce essential fatty acid deficiency and that cirrhosis, with or without alcoholism, is accompanied by a deficiency in arachidonic acid.

4.4.2. Etiologic Role of Ethanol

The administration of an ethanol dose that results in blood concentration over 100 mg/dl and mild intoxication increased triacylglycerols in the plasma of normal volunteers (Jones *et al.*, 1963; Verdy and Gattereau, 1967; Kaffarnik and Schneider, 1970; Belfrage *et al.*, 1973; Avogaro and Cazzolato, 1975; Taskinen and Nikkilä, 1977), whereas the administration of smaller doses did not (Friedman *et al.*, 1965). In about 25% of the subjects drinking 1 g/kg in the evening, the fasting serum samples obtained next morning reflect a type IV hyperlipemia (Taskinen and Nikkilä, 1977). Thus, prior alcohol ingestion has to be investigated in patients with hyperlipemia. Acute ethanol administration does not, however, produce significant changes in plasma cholesterol (Taskinene *et al.*, 1985). The effects of ethanol on plasma triacylglycerols and plasma cholesterol are exaggerated by chronic ethanol administration. Either as a supplement to a normal diet (Fig. 4.7) (Lieber *et al.*, 1963; Schapiro *et al.*, 1965; Belfrage *et al.*, 1977) or as a substitute for other foods (Losowsky *et al.*, 1963), ethanol produced a fourfold increase in plasma triacyl-

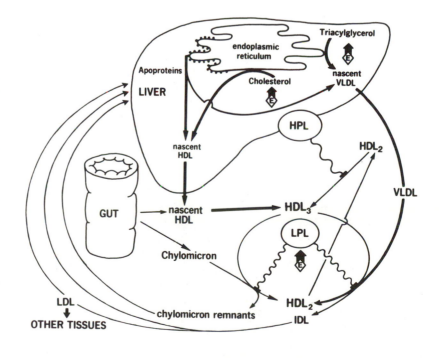

FIGURE 4.11. Interactions between lipoproteins and ethanol. An increased production of HDL by the liver will result in an increased HDL₃ concentration, observed in association with moderate ethanol consumption. The catabolism of VLDL and chylomicrons by extrahepatic lipoprotein lipase is associated with transformation of HDL₃ to HDL₂. An increased production of VLDL by the liver and an increased extrahepatic lipoprotein lipase activity will result in an increased HDL₂ concentration, observed in association with high alcohol intake. LPL, extrahepatic lipoprotein lipase; HPL, hepatic lipoprotein lipase; E→, stimulation by ethanol; ∿, Enzyme-mediated transformation. (From Lieber, 1987.)

glycerols and lesser increases in cholesterol and phospholipids. This required several days of administration and blood ethanol concentrations greater than 100 mg/dl. Administration of small doses to normal subjects did not produce significant hyperlipemia (Glueck *et al.*, 1980), but HDL cholesterol was elevated (Belfrage *et al.*, 1973, 1977; Fraser *et al.*, 1983; Hartung *et al.*, 1983).

The lipemic response to chronic ethanol consumption in human volunteers is usually transient. After reaching a maximal levels in several days or weeks, serum lipids decrease and even normalize despite continuation of the ethanol intake (Lieber *et al.*, 1963; Schapiro *et al.*, 1965; Belfrage *et al.*, 1977; Schneider *et al.*, 1983). A consistent lipemic response has been produced by acute administration of ethanol to rabbits (Bezman-Tarcher *et al.*, 1966; Warembourg *et al.*, 1970), but only after chronic administration of ethanol-containing diets (with or without an acute ethanol dose) was a significant postprandial hyperlipemia observed in rats (Baraona and Lieber, 1970; Baraona *et al.*, 1973). Lipids increased in all the fractions, including HDL. In nonhuman primates (which have a lipoprotein composition and metabolism more akin to those of humans), chronic ethanol administration produces a sustained increase in fasting triacylglycerol, cholesterol, and phospholipid levels in the plasma, at least during early stages of alcoholic liver injury (Baraona *et al.*, 1983b; Leathers *et al.*, 1981; Vasdev *et al.*, 1974).

4.4.3. Clinical Manifestations of Alcoholic Hyperlipemia

Usually, alcoholic hyperlipemia is first suspected because of the incidental observation of serum lactascence or of hypercholesterolemia. It is commonly associated with hepatic, gastrointestinal, or pancreatic complications of alcoholism, and on this basis a combination of varying symptoms is seen, most frequently including anorexia, nausea, vomiting, abdominal pain, fever, jaundice, and transient hemolysis. Indeed, transient hemolytic anemia can accompany episodes of hyperlipemia in alcoholics (the so-called Zieve's syndrome) (Zieve, 1958). In rats chronically fed with ethanol, reticulocytosis and increased osmotic fragility of the erythrocytes have been reported. These abnormalities correlated with the degree of hyperlipemia (Baraona and Lieber, 1969), and lipid abnormalities of the red cell membrane have been incriminated (Goebel *et al.*, 1979).

As discussed before, the degree of ethanol-induced hyperlipemia varies but is usually moderate. In a few

individuals, plasma lactescence can be severe because of potentiating factors such as underlying abnormality of lipid metabolism (that is, a *"forme fruste"* of essential hyperlipemia), pancreatitis (Cameron *et al.*, 1974), or diabetes. Massive hypertriacylglyceridemia may result in some clinical manifestations ("chylomicronemia syndrome"), as abdominal pain (Bloomfield and Shenson, 1947) with or without pancreatitis (Albrink and Klatskin, 1957; Greenberger *et al.*, 1966; Nestel, 1967), lipemia retinalis (Grossberg, 1966), and eruptive xanthoma (Chait *et al.*, 1972; Nestel, 1967). The diagnosis of pancreatitis may be difficult because serum lactescence interferes with the enzyme determinations (Fallat *et al.*, 1973; Greenberger *et al.*, 1966). Indeed, alcohol can produce pancreatitis and hyperlipemia, but pancreatitis was also reported in nonalcoholic patients with primary dyslipoproteinemia (Brunzell and Bierman, 1982; Farmer *et al.*, 1973).

4.4.4. Potentiating Factors of Alcoholic Hyperlipemia

4.4.4.1. Role of Nutritional Factors

The lipemic response to acute ethanol administration in volunteers is greatly enhanced by dietary fat (Talbot and Keating, 1962; Brewster *et al.*, 1966; Barboriak and Meade, 1968). The increase includes both VLDL and chylomicrons (Wilson *et al.*, 1970). In patients with alcoholic fatty liver, administration of a high-fat meal, even in the absence of ethanol, produced a striking increase in serum triacylglycerols (Borowsky *et al.*, 1980; Avgerinos *et al.*, 1983). Thus, dietary fat represents a striking potentiating factor of alcoholic hyperlipemia. This effect tends to disappear with the progression of the liver damage (Fig. 4.12) (Borowsky *et al.*, 1980). Contrasting with the potentiating effect of usual fat mixtures, ω_3 fatty-acid-rich fish oil lowers the ethanol-induced increase in blood lipids (Lakshman *et al.*, 1988).

4.4.4.2. Role of Underlying Alteration of Lipid Metabolism

The subclinical defects in lipid metabolism most commonly aggravated by alcohol consumption are reduced ability to remove serum lipids and overproduction of VLDL. The first type of alteration is found in subjects with primary (familial) or secondary (usually diabetic) decreased activity of lipoprotein lipase (Chait *et al.*, 1972; Losowsky *et al.*, 1963). In another group of

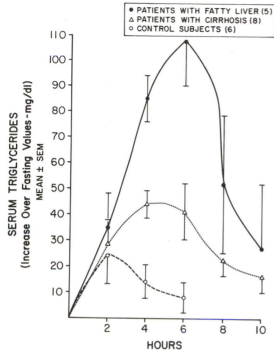

FIGURE 4.12. Serum triglyceride response of control subjects and alcoholic patients after a high-fat meal. Note the marked response of patients with fatty liver and the less striking but still significant response in patients with cirrhosis. (From Borowsky *et al.*, 1980.)

4.4.4.3. Role of Associated Liver Injury

An inverse relationship between the magnitude of alcoholic hyperlipemia and the severity of the liver injury has been observed (Cachera *et al.*, 1950; Marzo *et al.*, 1970). In baboons fed an alcohol-containing diet for several years, the hypertriglyceridemia and the hypercholesterolemia decreased or disappeared during the progression from alcoholic fatty liver to fibrotic stages (Savolainen *et al.*, 1984). In humans, no simple association was evident between a well-defined plasma lipid, lipoprotein, and/or apolipoprotein profile and chronic alcoholism in the presence or absence of liver injury (Duhamel *et al.*, 1984; Sabesin and Weidman, 1984).

At advanced stages, striking alteration of serum lipoprotein occurs, regardless of the etiology of the liver injury. Electrophoretically, these alterations are characterized by a marked decrease or disappearance of both pre -β- and α-lipoproteins with appearance of a broad β band (Papadopoulus and Charles, 1970) and, chemically, by a decreased cholesteryl ester/free cholesterol ratio. These changes are associated with decreased lecithin:cholesterol acyltransferase (LCAT) activity (Day *et al.*, 1979; Sabesin *et al.*, 1977). The apo-AI, major apoprotein of HDL and activator of LCAT, is markedly decreased (Poynard *et al.*, 1986; Sabesin *et al.*, 1977), and apo-E may be increased (Weidman *et al.*, 1982). Lipoprotein abnormalities associated with cholestasis (McIntyre, 1978), portacaval anastomosis (Balint, 1981), malnutrition, pancreatitis, or endocrine abnormalities may also complicate the clinical picture (Lieber, 1982).

4.4.5. Pathogenesis

Hyperlipemia results when the rate of entry of lipoprotein lipids into the plasma exceeds the rate of exit.

4.4.5.1. Hepatic Lipoprotein Production

The initial phase of hepatic lipid deposition after ethanol consumption is accompanied by an increased release of lipoproteins into the blood; this tends to counteract lipid accumulation in the liver. The effect is dose dependent, as discussed elsewhere (Lieber, 1982; Lieber and Pignon, 1989). In both humans (Jones *et al.*, 1963; Lieber *et al.*, 1963) and rats (Baraona and Lieber, 1970), ethanol administered in commonly used amounts produces hyperlipemia rather than hypolipemia, and the most striking change takes place in the very-low-density

individuals, alcohol aggravated either a primary or a secondary (obesity) overproduction of VLDL (type IV) or both VLDL and LDL (type IIb) (Debry *et al.*, 1979; Ginsberg *et al.*, 1974; Mendelson and Mellow, 1973; Taskinen and Nikkilä, 1977). Another possible mechanism could be an increased capacity to secrete serum lipoproteins on challenge with alcohol. This may account for the observation that some alcoholic patients appear to have an unusual sensitivity to the hyperlipemic effect of ethanol (DeGennes *et al.*, 1972; Kudzma and Schonfeld, 1971). A large percentage of patients with marked and sustained alcoholic hyperlipemia have relatives with hyperlipemia (DeGennes *et al.*, 1972). The participation of an underlying defect in lipid metabolism should be suspected in any alcoholic with severe hyperlipemia. Conversely, the lack of response to the usual treatment for hyperlipemia should lead to a search for alcoholism.

lipoprotein fraction. The alcohol-induced hyperlipemia can occur in the fasting state (Jones *et al.*, 1963); it is markedly exaggerated, however, when alcohol is given with a fat-containing diet (Barboriak and Meade, 1968; Brewster *et al.*, 1966; Wilson *et al.*, 1970).

The effect of alcohol depends not only on the dose but also on the duration of intake. Even moderate but regular use of alcohol raises blood lipids significantly (Barboriak and Hogan, 1976; Ostrander *et al.*, 1974). This alcohol effect does not result solely from caloric overload, since no comparable hyperlipemia was produced by isocaloric amounts of either carbohydrate or lipids (Losowsky *et al.*, 1963). Incorporation into lipoprotein of intragastrically administered [³H]palmitate and intravenously injected [¹⁴C]lysine was significantly increased by alcohol administration (Baraona and Lieber, 1970). These as well as other studies suggested that the ethanol-induced hyperlipemia results, at least in part, from enhanced lipoprotein production. This contrasts with choline (Lombardi *et al.*, 1968) and protein (Flores *et al.*, 1970; Seakins and Waterlow, 1972) deficiencies, which produce the opposite effect.

Direct evidence was provided of this increase in hepatic lipoprotein production (Savolainen *et al.*, 1986; Wolfe *et al.*, 1976). Savolainen *et al.* (1986) have shown by catheterization studies that the livers of alcohol-fed baboons secrete more VLDL-TG than the livers of the pair-fed controls, and the main part of the increased hepatic output of VLDL is caused by an enhanced production of abnormally large VLDL particles, which have flotation characteristics similar to chylomicrons (S_f > 400) (Fig. 4.13) (Savolainen *et al.*, 1986). This may account for the observation of chylomicronlike particles in the fasting blood of some alcoholics (Chait *et al.*, 1972). Moreover, Sane *et al.* (1984) have described an increased fractional catabolic rate and total turnover (production) rate of traicylgycerol VLDL in alcoholic men. After withdrawal both the synthetic and catabolic rate return to normal.

In addition to VLDL, the liver normally secretes an apo-E-rich HDL. In alcoholics, as well as in nonalcoholics after chronic administration of phenobarbital, a correlation between the plasma levels of HDL and hepatic microsomal activities has been pointed out (Cushman *et al.*, 1982; Luoma *et al.*, 1982). Malmendier and Delcroix (1985) have performed kinetic studies of HDL apo-AI in healthy volunteers before and after a 4-week period of alcohol intake. The synthetic rate of this apolipoprotein is increased, providing direct evidence in favor of increased HDL production (Lieber, 1987) (Figure 4.11).

In general, ethanol could act by enhancing the availability of fatty acids, which, in turn, can induce hepatic synthesis of lipoproteins (Alcindor *et al.*, 1970).

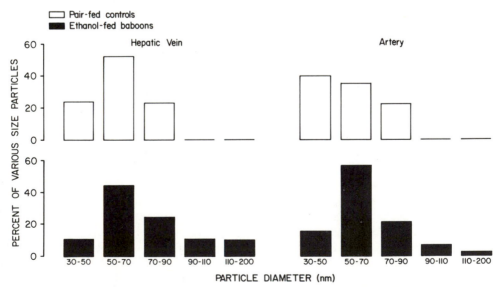

FIGURE 4.13. Frequency distribution of particle diameters in the hepatic venous and arterial S_f > 400 serum lipoproteins. In alcohol-fed baboons, large particles (with diameter greater than 90 nm) were released in the hepatic venous blood and found, although less frequently, in arterial blood during the fasting state. Such large particles were not found in the controls. (From Savolainen *et al.*, 1986.)

Furthermore, fatty acids are esterified (Stein and Schapiro, 1958) and lipoproteins are formed (Jones *et al.*, 1967) in the endoplasmic reticulum, the enzymes of which are "induced" after chronic ethanol consumption, as discussed in Chapter 1.

4.4.5.2. Lipid Absorption and Production by the Intestine

The suggestion has been made that, after alcohol feeding, the intestine releases more lipid into the lymph, either by decreasing oxidation of fatty acids or by increasing the synthesis of lipids from sources other than dietary fat (Ockner *et al.*, 1969; Windmueller *et al.*, 1973). However, as discussed before (Lieber, 1982), although an adequate supply of dietary lipids represents a permissive factor, changes in lymph lipid output do not seem to play a major role in the lipemic effect of ethanol, and the site of origin of increased production of serum lipoprotein is not intestinal. Similarly, the contribution of lymph lipids to the hepatic steatosis appears to be a minor one (Ockner *et al.*, 1973).

4.4.5.3. Removal of Intestinal Lipids by the Liver

The possibility that intestinal lipids could be poorly removed by the liver was raised when, after the injection of chylomicrons doubly labeled in the triacylglycerol and cholesteryl ester moieties to alcohol-fed rats, the clearance of chylomicron cholesteryl esters was impaired to a greater extent than the clearance of chylomicron triacylglycerol (Redgrave and Martin, 1977; Lakshman and Ezekiel, 1982). These observations are consistent with the possibility that after hydrolysis of the chylomicron triacylglycerol by extrahepatic lipoprotein lipases, the cholesteryl-ester-enriched remnants are poorly removed by the liver. This alteration could be caused by a perturbation of the apo-E-specific liver uptake or saturation of the system by catabolic products of endogenous VLDL, which share similar uptake mechanism. In the rat, Lakshman *et al.* (1986) showed that hepatocytes from alcohol-fed animals have a lower affinity and uptake of remnants, with a marked impairment of hepatic catabolism of the lipid moieties. In alcohol-fed baboons, the hepatic extraction of labeled chylomicrons and VLDL triacylglycerols during passage through the splanchnic vascular bed was increased rather than decreased (Baraona *et al.*, 1983b). This raises the possibility that an excessive supply of products of both endogenous and dietary origin saturates the hepatic uptake, leading to lipid accumulation in the plasma during the postprandial state.

4.4.5.4. Removal of Plasma Lipoproteins

The mechanisms of removal are similar for hepatic and intestinal lipoproteins. Most of the early studies focused on the total postheparin lipoprotein lipase activity and found no change after acute (Barboriak, 1966; Verdy and Gattereau, 1967; Wilson *et al.*, 1970) or chronic Kudzma and Schonfeld, 1971) ethanol administration. Recent studies have focused either on tissue measurements of lipoprotein lipase activity or on postheparin measurements after separation of the hepatic and extrahepatic components. A large dose of ethanol can produce a slight decrease in extrahepatic lipoprotein lipase activity and some delay in the clearance of serum triacylglycerols (Jaillard *et al.*, 1974; Nikkilä *et al.*, 1978; Nilsson-Ehle *et al.*, 1978; Schneider *et al.*, 1983). Most of the delayed clearance, however, appears to be secondary to the excessive supply of endogenous and exogenous lipids competing for the lipases (Schneider *et al.*, 1983). Even in those alcoholics with severe type V hyperlipemia (in whom lipoprotein lipase activities could be impaired), the turnover rates of VLDL apo-B and VLDL triacylglycerol are increased rather than decreased (Sigurdsson *et al.*, 1976). This reflects the predominant role of increased lipoprotein production.

By contrast, chronic alcohol consumption has been found to increase the activity of extrahepatic lipoprotein lipase in the majority of subjects (Belfrage *et al.*, 1977; Ekman *et al.*, 1981; Taskinen *et al.*, 1982), also according to a kinetic study (Sane *et al.*, 1984). This change may represent a compensatory adaptation to the increased supply of plasma lipids, which would increase their removal. This may account for the transient nature of hypertriglyceridemia during alcohol administration (Lieber *et al.*, 1963; Belfrage *et al.*, 1977). Mordasini *et al.* (1982) documented that the rise in plasma triacylglycerol (mainly VLDL) is associated with an initial decrease in lipoprotein lipase activity, whereas the return of triacylglycerol to normal levels (despite continuous ethanol intake) is associated with increased lipase activity. This adaptation may also account for the modest degree of hypertriglyceridemia found in the majority of alcoholics and the paradoxical increase in VLDL and LDL during the initial period of abstinence (Ekman *et al.*, 1981). In heavy drinkers with type V hyperlipemia, postheparin lipoprotein lipase activity was decreased (Breier *et al.*, 1984), confirming some observations made almost three decades ago (Losowsky *et al.*, 1963).

The enhanced production and metabolism of VLDL after chronic alcohol consumption may have important consequences for plasma cholesterol, since VLDL (as well as chylomicron) supplies free cholesterol to the HDL fraction of the plasma during the extrahepatic lipolysis. This process is associated with the conversion of HDL_3 into HDL_2 (Lieber, 1987) (Fig. 4.11). Thus, an increase of this lipase activity induces an increased HDL_2, as seen in heavy drinkers. The hepatic lipoprotein lipase is thought to regenerate HDL_3 from HDL_2 (Nikkilä et al., 1982). An increase in HDL_3 could result from increased production of nascent hepatic HDL or from an increase of the hepatic lipase activity (Lieber, 1987) (Fig. 4.11). In alcoholics with mild liver injury, this hepatic activity was found to be unchanged (Ekman et al., 1981) or increased (Taskinen et al., 1982), but it may decrease in more severe forms of liver impairment such as alcoholic hepatitis (Muller et al., 1974; Freeman et al., 1977). In normal subjects (Goldberg et al., 1984; Nikkilä et al., 1978; Schneider et al., 1983; Taskinen et al., 1985), an acute inhibition was described. Thus, the response of these enzymes to alcohol is time dependent and probably dose dependent, as for the extrahepatic lipase.

Although chronic ethanol administration increases hepatic cholesterol and bile acid synthesis in relation to serum high-density-lipoprotein cholesterol in rats (Maruyama et al., 1986), alcoholic hypercholesterolemia may have a predominantly extrahepatic origin (Karsenty et al., 1985). In alcohol-fed baboons, ethanol feeding increased free cholesterol in all plasma lipoprotein fractions and esterified cholesterol in very-low-density lipoprotein, intermediate-density lipoprotein (IDL), and high-density lipoprotein. The major increase occurred in HDL, mainly as esterified cholesterol. The latter was associated with decreased transfer of esterified cholesterol from HDL to low-density lipoprotein. By contrast, the smaller increase in HDL-free cholesterol was associated with increased turnover in the plasma, increased splanchnic uptake, and increased fecal excretion of plasma cholesterol, mainly as neutral steroids (Karsenty et al., 1985), with evidence of increased synthesis (Cluette et al., 1984). Cholesterol extraction predominated over release in the splanchnic vascular bed, suggesting that the excess of cholesterol excreted in the feces originated in extrasplanchnic tissues (Karsenty et al., 1985). Thus, these findings indicate that alcohol consumption favors mobilization of tissue free cholesterol for hepatic removal and excretion. By contrast, the increase in HDL cholesterol (mainly esterified) appears to be a poor indicator of cholesterol mobilization.

4.4.6. Treatment

The first step in the treatment is alcohol abstinence or at least a decrease in consumption. Associated factors should be treated: obesity by decreased caloric intake and diabetes by diet and drugs, if needed.

If the hyperlipemia persists after alcohol withdrawal, restriction in dietary fat and cholesterol (type V hyperlipemia) or restriction in simple carbohydrates (type IV hyperlipoproteinemia) is necessary. Exogenous estrogen should be avoided.

If after dietary change and alcohol withdrawal the triacylglycerol remains elevated, appropriate drug therapy can be instituted (Janus and Lewis, 1978; Schaefer and Levy, 1985).

4.5. Ethanol, Atherosclerosis, and High-Density Lipoproteins

For more than a century, the effect of alcoholism on atherosclerosis has been the subject of a lively debate, as discussed in detail elsewhere (Lieber and Pignon, 1989).

One common problem in many of the reported studies is the difficulty of establishing a proper control group. For example, some studies (Kittner et al., 1983; Klatsky et al., 1974) included in the group of teetotalers both lifetime teetotalers and ex-drinkers, who retain excess morbidity and mortality (Pell and D'Alonzo, 1973), including a higher incidence of coronary heart disease death (Dyer et al., 1981; Klatsky et al., 1981). However, removal of the past heavy drinker did not eliminate a statistically significant difference between teetotalers and light drinkers (Klatsky et al., 1981). In a comprehensive review, Eichner (1985) has pointed out the importance of taking into account the various factors associated with coronary heart disease and choosing an appropriate endpoint. Thus, a study of a random sample of all coronary events in Auckland, New Zealand, has shown that heavy drinkers are more likely to die immediately from a heart attack (Fraser and Upsdell, 1981). Obviously, a higher incidence of coronary heart disease in teetotalers than in moderate drinkers may be caused not only by a difference in alcohol intake but also by a variety of other factors that are present in the teetotalers (the light use of alcohol could be a marker of behavior only) (Eichner, 1985). Popham et al. (1983) have shown a negative correlation between consumption of alcohol and of dairy products, and the association between alcohol intake and cardiovascular mortality disappeared when the consumption of milk proteins was taken into account; thus, this association could be only indirect (Barboriak, 1984; Popham et al., 1983).

Although the relationship between alcohol and coronary heart disease is controversial (Eichner, 1985; Gordon and Doyle, 1985), the combined evidence from all the studies still suggests that moderate alcohol consumption may be associated with reduced risk of coronary heart disease. Such correlation is further strengthened if one takes into account the saturated fat and cholesterol content of the diet (Hegsted and Ausman, 1988). However, further studies are needed before we can accept a causal relationship, especially since, as pointed out by Eichner (1985), in the nonsmoking subpopulation of the Kaiser Permanente experience (Klatsky et al., 1981), the protective effect of light alcohol consumption disappeared, and the nonsmoking teetotalers had the lowest coronary heart disease mortality. It has been claimed that the alcohol-mortality relationships are produced by preexisting disease and by the movement of individuals with such disease into nondrinking or occasional-drinking categories (Shaper et al., 1988).

One mechanism that has been evoked in the protective effect of alcohol is its action on the blood lipids other than triacylglycerol. The rise in HDL associated with alcohol has been shown both experimentally (Baraona and Lieber, 1970; Baraona et al., 1973, 1983b) and in a number of clinical studies (Johansson and Medhus, 1974; Castelli et al., 1977; Ernest et al., 1980; Thornton et al., 1983); the increase includes not only cholesterol but also apolipoproteins AI and AII (Camargo et al., 1985; Fraser et al., 1983; Lakshman et al., 1988). It occurs in teen-agers (Glueck et al., 1981), in both older and younger men (Barrett-Connor and Sharez, 1982), inactive or active (Willet et al., 1980), although the latter has been questioned (Hartung et al., 1983). This alcohol-induced increase of HDL may be particularly significant with regard to the problem of atherosclerosis in view of the negative correlation between HDL blood levels and the development of coronary heart disease (Miller and Miller, 1975; Miller et al., 1977), either because HDL promotes the transport of cholesterol out of the cell (Miller et al., 1985) or because HDL competes with LDL for uptake, thereby suppressing the increment in cell sterol content induced by LDL. A recent study in baboons supports the former hypothesis: the turnover of HDL free cholesterol was higher in alcohol-fed baboons than in pair-fed controls, with increased uptake by the liver and excretion into feces, suggesting a more active transport of the cholesterol

from peripheral tissues to the liver and subsequently into the feces (Karsenty et al., 1985).

In another study in monkeys, the area of coronary arteries occupied by atherosclerotic lesions was inversely correlated with HDL cholesterol (Rudel et al., 1981). Thus, the concept that moderate drinking may exert protective cardiovascular effects was strengthened when elevated levels of circulating HDL emerged as a possible mechanism. When the role of HDL in cholesterol transport and its protective effect against atherosclerosis became apparent, it made sense to postulate that the decreased incidence of coronary heart disease in moderate drinkers might be a result of the ethanol-induced elevation of HDL. However, HDL is a heterogeneous group of lipoproteins with two major subclasses: the less-dense HDL_2, epidemiologically associated with reduction in coronary heart disease, and the denser HDL_3, with a weaker link to coronary heart disease (Ballantyne et al., 1982; Gofman et al., 1966; Miller et al., 1981). Moreover, agents or conditions that are thought to affect coronary heart disease through HDL (such as exercise and female sex) have been shown to increase HDL_2, not HDL_3 (Kraus, 1982). A negative correlation (Levy et al., 1984) was found between HDL_3 mass (but not HDL_2 mass) and coronary lesions. It was reported that the increase in HDL after alcohol consumption involved primarily HDL_2 (Ekman et al., 1981; Taskinen et al., 1982, 1985). However, these observations were made in individuals with a relatively high intake of alcohol. Similarly, after an acute dose of ethanol (40 g, equivalent to 4 oz of 86-proof beverage) there was a transient increase in both HDL_2 and HDL_3 (Goldberg et al., 1984).

It is now well recognized that large amounts of alcohol have adverse effects not only on the liver (see Chapter 7) but also on virtually all the tissues of the body, including the cardiovascular system (Friedman, 1984; Knochel, 1983; Lieber, 1982), and it is generally agreed that such high intakes are not associated with protection against coronary heart disease (Devenyi et al., 1980; Friedman, 1984). Furthermore, Haskell et al. (1984) reported that moderate doses of ethanol (about 0.5–2.2 oz or 12–51 g of absolute ethanol/day) raised levels of HDL_3 but not levels of HDL_2 and that on abstention from moderate consumption, levels of HDL_3, not of HDL_2, decreased. The relationship between alcohol and HDL_3 at moderate alcohol dose was confirmed by two other studies (Haffner et al., 1985; Williams et al., 1985). Williams et al. (1985) pointed out the correlation between nutritional components, such as alcohol and starch, and HDL_3 (but not HDL_2). Thus, we now must revise some of the previously derived implications (Lieber, 1984). For instance, the conclusion of Hartung *et al.* (1983) that "nonexercisers can maintain levels of HDL similar to those of individuals who jog regularly by ingesting three beers a day" must be reassessed in light of the fact that the HDL subfractions involved appear to be different in the two groups and that the relationship between some of these HDL subfractions (such as HDL_3) and coronary heart disease is not completely established. With the progress of alcoholic liver injury, the HDL fractions tend to decrease, and abnormal lipoproteins appear in the bloodstream (Borowsky *et al*, 1980; Devenyi *et al.*, 1980; Poynard *et al.*, 1985; Sabesin *et al.*, 1977).

The relationship among alcohol, coronary heart disease, and HDL could be explained by the protein moiety (Camargo *et al.*, 1985). The major apoproteins of HDL are apo-AI and apo-AII. Apo-AI is thought by some to be a better marker for risk of coronary heart disease than HDL cholesterol (Avogaro *et al.*, 1979; Maciejko *et al.*, 1983; Riesen *et al.*, 1980). The association between apo-AII concentration and coronary heart disease is less well established (Camargo *et al.*, 1985). The apo-AI is increased, and the apo-AII is increased or normal in alcoholics (Camargo *et al.*, 1985; Malmendier *et al.*, 1983) as well as in healthy volunteers after moderate consumption (Fraser *et al.*, 1983; Camargo *et al.*, 1985).

Lipoproteins other than HDL are, of course, also involved in the process of atherosclerosis. It is of interest that a strong positive correlation was observed between the particle size of LDL and the area of atherosclerotic plaques in the aorta and coronary arteries. Therefore, the decreased size of LDL particles observed in alcohol-fed animals may attenuate the atherosclerotic process (Rudel *et al.*, 1981).

The main route by which cholesterol is eliminated from the body is the excretion into feces. Chronic alcohol feeding increased fecal excretion of cholesterol and bile acids derived from it (Cohen and Raicht, 1981; Topping *et al.*, 1982; Karsenty *et al.*, 1985). The secretion of neutral steroids was increased more than the secretion of acidic sterols (bile acids). After an acute ethanol dose, the opposite effect may be found (Monroe *et al.*, 1981).

Mechanisms other than lipoproteins might also explain the protective effect of alcohol on cardiovascular disease. Inhibition of platelet aggregation and increased fibrinolytic activity have been suggested as alternative explanations (Ashley, 1982; Barboriak, 1984). One other mechanism of ethanol could be a block of the vaso-

constriction response to stress associated with a reduced effect of stress on blood pressure (Zeichner *et al.*, 1983).

Whatever the mechanism whereby ethanol affects atherosclerosis, it must be pointed out that the protective action was observed in a population comprising moderate rather than heavy drinkers (Barboriak *et al.*, 1979a; Hennekens *et al.*, 1978; Marmot *et al.*, 1981; Yano *et al.*, 1977). The overall decreased mortality associated with alcohol intake was shown with consumption less than 34 g (Marmot *et al.*, 1981) or 31 ml of pure ethanol (Blackwelder *et al.*, 1980) or two drinks (Klatsky *et al.*, 1981) per day. Large amounts of alcohol exert cardiotoxic effects and may promote hypertension and have been shown to increase total mortality (Blackwelder *et al.*, 1980; Dyer *et al.*, 1981; Klatsky *et al.*, 1981; Marmot *et al.*, 1981).

In conclusion, we can distinguish three partially overlapping stages in the relationship among alcohol, coronary heart disease, and HDL. (1) Moderate alcohol intake is associated with a decreased incidence of coronary heart disease and with an increased level of HDL_3 rather than HDL_2. The decreased incidence of coronary complications could be caused by a protector effect of HDL_3 or could be secondary to some mechanisms unrelated to HDL, such as prostaglandin-mediated vascular effects (see Section 4.6) or enhanced fibrinolysis. Moreover, there are some epidemiologic arguments that the effect of alcohol on coronary heart disease is unrelated or related only partially to the effect of alcohol on HDL (Barboriak *et al.*, 1979b). Furthermore, an association does not prove a cause-and-effect relationship; it is possible that teetotalers may differ from moderate drinkers in ways other than alcohol intake. (2) In the absence of severe liver injury, high alcohol intake results in increased levels of HDL, primarily HDL_2, but at that level of alcohol intake there is no evidence for "protection" against coronary heart disease. (3) When alcohol abuse is associated with severe liver disease, the HDL fraction decreases (Lieber, 1984).

4.6. Interaction of Ethanol with Vascular Prostacyclin Production

Acute and chronic alcohol consumption are associated with vascular effects that could be mediated by increased production of prostacyclin, a potent vasodilator and platelet antiaggregator (Moncada and Vane, 1979). Some of the acute effects of ethanol, such as cutaneous vasodilation, tachycardia, and hypotension, resemble those produced by the administration of pros- tacyclin (Lewis and Dollery, 1983). These changes are particularly prominent when the blood levels of acetal- dehyde produced by alcohol oxidation are increased by the presence of aldehyde dehydrogenase inhibitors such as disulfiram (Kitson, 1977) or chlorpropamide (Jern- torp *et al.*, 1981) or in Oriental individuals who lack some of the aldehyde dehydrogenase activity (Harada *et al.*, 1983). Vasodilation of a lesser degree is also com- mon in intoxicated alcoholics. The similarity of the flushing reaction with the changes produced by adminis- tration of prostacyclin (PGI$_2$) (Lewis and Dollery, 1983) suggested the possibility that acetaldehyde could stimu- late the release of prostaglandins with vasodilator prop- erties.

To test this possibility, Guivernau *et al.* (1987) investigated the effects of ethanol and its metabolites (acetaldehyde and acetate) on the vascular production of prostacyclin (one of the main vasodilator prostaglandins synthesized by the blood vessels) in aortic rings from rats fed chow *ad libitum* or pair-fed liquid diets contain- ing either ethanol (36% of energy) or isocaloric car- bohydrate for 4–5 weeks. Acetaldehyde produced a concentration-dependent stimulation of prostacyclin production in the aortic rings of rats, whereas acetate did not (Guivernau *et al.*, 1987) (Fig. 4.14). This effect was associated with increased conversion of arachidonate and prostaglandin endoperoxide H$_2$ to prostacyclin. Eth- anol did not affect prostacyclin release in control rats, but in aortas from alcohol-fed animals, ethanol did stimulate prostacyclin formation. It was concluded that

acetaldehyde is a potent stimulant of vascular pros- tacyclin production and that the effect results, at least in part, from enhanced activity of prostacyclin syn- thase. Ethanol acquired such a stimulator effect on prostacyclin formation after chronic alcohol consump- tion. The stimulatory effect of HDL was also markedly increased by chronic alcohol administration (Guivernau *et al.*, 1989). The aortic rings from alcohol-fed rats produced considerably more prostacyclin than those from controls in response to HDL, regardless of the source of these lipoproteins (Guivernau *et al.*, 1989) (Fig. 4.15).

Thus, the principal mechanism for the enhanced vascular production of prostacyclin in response to HDL was the increased reactivity of the vessel and, to a lesser extent, the increased plasma levels of these lipoproteins. These findings suggest that the changes in vascular reactivity induced by chronic alcohol consumption are caused at least in part by an increase in the activity of the prostacyclin synthetic pathway. Supporting this possi- bility is the finding that the chlorpropamide–alcohol flush, which is also associated with high acetaldehyde levels (Harada *et al.*, 1983), was prevented by aspirin (Strakosch *et al.*, 1980) and indomethacin (Barnett *et al.*, 1980), two cyclooxygenase inhibitors. The flushing observed in Orientals, who develop high acetaldehyde concentrations after alcohol ingestion, was also pre- vented by the administration of aspirin (Truitt *et al.*, 1987). Vasodilation is also prominent in many intoxi- cated alcoholics, who also develop higher blood acetal- dehyde levels after alcohol ingestion, as discussed in Chapter 2.

It also has been postulated that the stimulator effect of HDL could counteract the deficient synthesis of prostaglandin in the atherosclerotic vessel and inhibit platelet aggregation at the site of the endothelial injury (Karsenty *et al.*, 1985). In addition, prostacyclin may have a direct antiatherogenic effect by decreasing cho- lesteryl ester deposition through an enhancement of catabolism (Hajjar, 1985). Furthermore, HDL has been recently reported to decrease thromboxane formation by platelets (Beitz *et al.*, 1986). Also, ethanol inhibits platelet thromboxane production both *in vitro* (Mik- khailidis *et al.*, 1983) and *in vivo* (Kangasaho *et al.*, 1982; Kontula *et al.*, 1982). In alcoholics, the situation may differ from that in moderate drinkers. The benefit of the enhanced response of the vessels to the stimulatory effect of HDL could be lost by a progressive decrease in the capacity of HDL to stimulate prostacyclin forma- tion. Moreover, chronic alcohol consumption enhances the response of the vessel not only to vasodilator agents

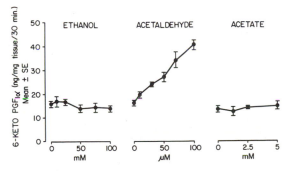

FIGURE 4.14. Effects of ethanol and its metabolites on prosta- cyclin production by aortic rings from chow-fed rats. Prostacyclin formation was estimated by the accumulation of 6-keto-PGF$_{1\alpha}$ in the medium. Contrasting with the lack of effect of ethanol and acetate, acetaldehyde increased prostacyclin formation in a concentration-dependent manner. (From Guivernau *et al.*, 1987.)

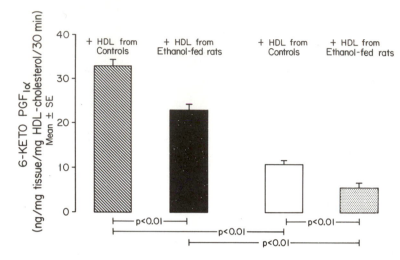

FIGURE 4.15. Contribution of HDL and aortic response to the stimulatory effect of chronic alcohol consumption on vascular prostacyclin production. The results indicate the increase (over basal production) of 6-keto-PGF$_{1\alpha}$. Two-way ANOVA revealed significant effects of both the source of the aorta and the source of HDL. (From Guivernau et al., 1989.)

but also to vasoconstrictors (Green and Claesson, 1986). An increased reactivity of the platelets to aggregant agents has also been found in severe alcoholics (Arai et al., 1986; Hillborn et al., 1985). Such a change might explain, in part, the opposite effects of moderate and large doses of alcohol on cardiovascular disease. Moreover, some changes may be related to the withdrawal from alcohol. It appears that in alcoholics during withdrawal, the ratio of the vasodilator prostanoids prostaglandin E and prostacyclin to the vasoconstrictor prostanoid thromboxane A$_2$ is lower than in normal subjects (Förstermann and Feuerstein, 1987).

4.7. Summary

The interaction of ethanol with lipid metabolism is complex. When ethanol is present, it becomes a preferred fuel for the liver and displaces fat as a source of energy. This favors fat accumulation. In addition, the altered redox state secondary to the oxidation of ethanol promotes lipogenesis, for instance, through an increase of α-glycerophosphate and enhanced formation of acylglycerols. The depressed oxidative capacity of mitochondria injured by chronic alcohol feeding also contributes to the development of the fatty liver. The accumulation of fat in the liver acts as a stimulus for the secretion of lipoprotein into the bloodstream and the development of hyperlipemia. Clinically, such heavy alcohol consumption is one of the most common (and

readily overlooked) factors that promote hyperlipemia. Hyperlipemia may also be caused by the proliferation of the endoplasmic reticulum after chronic ethanol consumption and the associated increase of enzymes involved in the assembly of triglycerides and lipoproteins. The propensity to enhance lipoprotein secretion is offset, at least in part, by a decrease in microtubules and the impairment of the secretory capacity of the liver resulting from liver damage following chronic ethanol consumption. The level of blood lipids depends on the balance between these two opposite changes. At the early stage of alcohol abuse, when liver damage is still small, hyperlipemia will prevail, whereas the opposite occurs with severe liver injury. When hyperlipemia occurs, it involves all lipoprotein classes, including HDL. The latter could play a role in opposing the development of atherosclerosis. Prostaglandins may also be involved in the latter process. Indeed, acetaldehyde (a metabolite of ethanol oxidation) is a potent stimulant of vascular production of prostacyclin, a powerful vasodilator and platelet antiaggregant agent. This effect may contribute to the flushing and other cardiovascular effects observed after alcohol. After chronic alcohol consumption, ethanol acquires an enhancing effect on prostacyclin production associated with a marked increase in the response of the vessel to the stimulatory effect of HDL.

ACKNOWLEDGMENT.   Grateful acknowledgment is made to Jean-Pierre Pignon, M.D., for updating some of the references in this chapter.

REFERENCES

Abrams, M. C., and Cooper, C.: Mechanism of increased hepatic uptake of unesterified fatty acid from serum of ethanol-treated rats. *Biochem. J.* **156**:47–54, 1976.

Adams, L. L., Webb, W. W., and Fallon, H. J.: Inhibition of hepatic triglyceride formation by clofibrate. *J. Clin. Invest.* **50**:2339–2346, 1971.

Albrink, M. J., and Klatskin, G.: Lactescence of serum following episodes of acute alcoholism and its probable relationship to acute pancreatitis. *Am. J. Med.* **23**:26–33, 1957.

Alcindor, L. G., Infante, R., Soler-Argilaga, C., Raisonnier, A., Polonovski, J., and Caroli, J.: Induction of the hepatic synthesis of β-lipoproteins by high concentrations of fatty acids: Effect of actinomycin D. *Biochim. Biophys. Acta.* **210**:483–486, 1970.

Alling, C., Aspenstom, G., Dencker, S. J., and Svennerholm, L.: Essential fatty acids in chronic alcoholism. *Acta Med. Scand. Suppl.* **631**:1–38, 1979.

Alling, C., Gustavsson, L., and Änggärd, E.: An abnormal phospholipid in rat organs after ethanol treatment. *FEBS Lett.* **152**:24–28, 1983.

Ammon, H. P. T., and Zeller, W.: Der Einfluss von α-Pyridylcarbinol auf die durch Alcohol erzeugte Fettleber der Ratte. *Arzneim. Forsch.* **15**:1369–1371, 1965.

Arai, M., Gordon, E. R., and Lieber, C. S.: Decreased cytochrome oxidase activity in hepatic mitochondria after chronic ethanol consumption and the possible role of decreased cytochrome aa₃ content and changes in phospholipids. *Biochim. Biophys. Acta* **797**:320–327, 1984a.

Arai, M., Leo, M. A., Nakano, M., Gordon, E. R., and Lieber, C. S.: Biochemical and morphological alterations of baboon hepatic mitochondria after chronic ethanol consumption. *Hepatology* **4**:165–174, 1984b.

Arai, M., Okuno, F., Nagata, S., Shigeta, Y., Takagi, S., Ebihara, Y., Kobayashi, T., Ishii, H., and Tsuchiya, M.: Platelet dysfunction and alteration of prostaglandin metabolism after chronic alcohol consumption. *Scand J. Gastroenterol.* **21**:1091 (abs), 1986.

Arakawa, M., Taketomi, S., Furuno, K., Matsuo, T., Iwatsuka, H., and Suzuoki, Z.: Metabolic studies on the development of ethanol-induced fatty liver in KK-Ay mice. *J. Nutr.* **105**:1500–1508, 1975.

Ashley, M. J.: Alcohol consumption, ischemic heart disease, and cerebrovascular disease: An epidemiological perspective. *J. Stud. Alcohol* **43**:869–887, 1982.

Ashworth, C. T., Wrightsman, F., and Buttram, V.: Hepatic lipids. *Arch. Pathol.* **72**:620–624, 1961.

Avgerinos, A., Chu, P., Greenfield, C., Harry, D. S., and McIntyre, N.: Plasma lipid and lipoprotein response to fat feeding in alcoholic liver disease. *Hepatology* **3**:349–355, 1983.

Avogaro, P., and Cazzolato, G: Changes in the composition and physico-chemical characteristics of serum lipoproteins during ethanol-induced lipemia in alcoholic subjects. *Metabolism* **24**:1231–1242, 1975.

Avogaro, P., Cazzolato, G., Bittolo Bon, G., and Quinci, G. B.: Are apolipoproteins better discriminators than lipids for atherosclerosis? *Lancet* **1**:901–903, 1979.

Baker, H., Luisada-Opper, A., Sorrell, M. F., Thomson, A. D., and Frank, O.: Inhibition by nicotinic acid of hepatic steatosis and alcohol dehydrogenase in ethanol-treated rats. *Exp. Mol. Pathol.* **19**:106–112, 1973.

Balint, J. A.: The Liver: Annual 1/1981 (I. M. Arias, M. Frenkel, and J. H. P. Wilson, eds.), Amsterdam, Excerpta Medica, pp. 31–48, 1981.

Ballantyne, F. C., Clark, R. S., Simpson, H. S., and Ballantyne, D.: High density and low density lipoprotein subfractions in survivors of myocardial infarction and in control subjects. *Metabolism* **31**:433–437, 1982.

Baraona, E., and Lieber, C. S.: Fatty liver, hyperlipemia and erythrocyte alterations produced by ethanol feeding in the rat. *Am. J. Clin. Nutr.* **22**:356–357, 1969.

Baraona, E., and Lieber, C. S.: Effects of chronic ethanol feeding on serum lipoprotein metabolism in the rat. *J. Clin. Invest.* **49**:769–778, 1970.

Baraona, E., and Lieber, C. S.: Effects of alcohol on hepatic transport of proteins. *Annu. Rev. Med.* **33**:281–292, 1982.

Baraona, E., Pirola, R. C., and Lieber, C. S.: The pathogenesis of postprandial hyperlipemia in rats fed ethanol-containing diets. *J. Clin. Invest.* **52**:296–303, 1973.

Baraona, E., Leo, M. A., Borowsky, S. A., and Lieber, C. S.: Pathogenesis of alcohol-induced accumulation of protein in the liver. *J. Clin. Invest.* **60**:546–554, 1977.

Baraona, E., Jauhonen, P., Miyakawa, H., and Lieber, C. S.: Zonal redox changes as a cause of selective perivenular hepatotoxicity of alcohol. *Pharmacol. Biochem. Behav.* **18**(suppl. 1): 449–454, 1983a.

Baraona, E., Savolainen, M., Karsenty, C., Leo, M. A., and Lieber, C. S.: Pathogenesis of alcoholic hypertriglyceridemia and hypercholesterolemia. *Trans. Assoc. Am. Physicians* **96**: 306–315, 1983b.

Barboriak, J. J.: Effect of ethanol on lipoprotein lipase activity. *Life Sci.* **5**:237–241, 1966.

Barboriak, J. J.: Alcohol, lipids and heart disease. *Alcohol* **1**:341–345, 1984.

Barboriak, J. J., and Hogan, W. J.: Preprandial drinking and plasma lipids in man. *Atherosclerosis* **4**:323–325, 1976.

Barboriak, J. J., and Meade, R. C.: Enhancement of alimentary lipemia by preprandial alcohol. *Am. J. Med. Sci.* **255**:245–251, 1968.

Barboriak, J. J., Anderson, A. J., Rimm, A. A., and Tristani, F. E.: Alcohol and coronary arteries. *Alcoholism: Clin. Exp. Res.* **3**:29–32, 1979a.

Barboriak, J. J., Anderson, A. J., and Hoffmann, R. G.: Interrelationship between coronary artery occlusion, high-density lipoprotein cholesterol, and alcohol intake. *J. Lab. Clin. Med.* **94**:348–353, 1979b.

Barboriak, J. J., Jacobson, G.R., Cushman, P., Herrington, R. E., Lipo, R. F., Daley, M. E., and Anderson, A. J.: Chronic alcohol abuse and high density lipoprotein cholesterol. *Alcoholism: Clin. Exp. Res.* **4**:346–349, 1980.

Barnett, A. H., Spiliopoulus, A. J., and Pyke, D. A.: Blockade of chlorpropamide-alcohol flush by indomethacin suggests an association between prostaglandins and diabetic vascular complications. *Lancet* **2**:164–166, 1980.

Barrett-Connor, E., and Sharez, L.: A community study of alcohol and other factors associated with the distribution of high density lipoprotein cholesterol in older vs younger men. *Am. J. Epidemiol.* **115:**888–893, 1982.

Beaugé, F., Stibler, H., and Borg, S.: Abnormal fluidity and surface carbohydrate content of the erythrocyte membrane in alcoholic patients. *Alcoholism: Clin. Exp. Res.* **9:**322–326, 1985.

Beitz, J., Block, H.-U., Beitz, A., Muller, G., Winkler, L., Dargel, R., and Mest, H. J.: Endogenous lipoproteins modify the thromboxane formation capacity of platelets. *Atherosclerosis* **60:**95–99, 1986.

Belfrage, P., Berg, B., Cronholm, T., Elmqvist, D., Hägerstrand, I., Johansson, B., Nilsson-Ehle, P., Nordén, G., Sjövall, J., and Wiebe, T.: Prolonged administration of ethanol to young, healthy volunteers: Effects on biochemical, morphological and neurophysiological parameters. *Acta Med. Sand. Suppl.* **552:**1–44, 1973.

Belfrange, P., Berg, B., Hägerstrand, I., Nilsson-Ehle, P., Tornquist, H., and Wiebe, T.: Alterations of lipid metabolism in healthy volunteers during long-term ethanol intake. *Eur. J. Clin. Invest.* **7:**127–131, 1977.

Berg, T., and Mørland, J.: Effects of chronic ethanol treatment on rat liver lysosomes. *Acta Pharmacol. Toxicol.* **33:**409–416, 1973.

Best, C. H., Hartroft, W. S., Lucas, C. S., and Ridout, J. H.: Liver damage produced by feeding alcohol or sugar and its prevention by choline. *Br. Med. J.* **II:**1001–1006, 1949.

Bezman-Tarcher, A., Nestel, P. J., Felts, J. M., and Havel, R. J.: Metabolism of hepatic and plasma triglycerides in rabbits given ethanol or ethionine. *J. Lipid Res.* **2:**248–257, 1966.

Bizzi, A., Tacconi, M. T., Veneroni, E., and Garattini, S.: Triglyceride accumulation in liver. *Nature* **209:**1025–1026, 1966.

Blackwelder, W. C., Yano, K., Rhoads, G. G., Kagan, A., Gordon, T., and Palesch, Y.: Alcohol and mortality: The Honolulu heart study. *Am. J. Med.* **68:**164–169, 1980.

Blomstrand, R., and Forsell, L.: Prevention of the acute ethanol-induced fatty liver by 4-methylpyrazole. *Life Sci.* **10**(Part II):523–530, 1971.

Blomstrand, R., and Kager, L.: The combustion of triolein-1-^{14}C and its inhibition by alcohol in man. *Life Sci.* **13:**113–123, 1973.

Blomstrand, R., Kager, L., and Lantto, O.: Status of studies on the ethanol-induced decrease of fatty acid oxidation in rat and human liver slices. *Life Sci.* **13:**1131–1141, 1973.

Bloomfield, A. L., and Shenson, B.: The syndrome of idiopathic hyperlipemia with crises of violent abdominal pain. *Stanford Med. Bull.* **5:**185–191, 1947.

Borowsky, S. A., Perlow, W., Baraona, E., and Lieber, C. S.: Relationship of alcoholic hypertriglyceridemia to stage of liver disease and dietary lipid. *Dig. Dis. Sci.* **25:**22–27, 1980.

Böttiger, L. E., Carlson, L. A., Hultman, E. M., and Romanus, V.: Serum lipids in alcoholics. *Acta Med. Scand.* **199:**357–361, 1976.

Boyer, J. L.: Effect of chronic ethanol feeding on bile formation and secretion of lipids in the rat. *Gastroenterology* **62:**294–301, 1972.

Breier, C., Lisch, H.-J., Drexel, H., and Braunsteiner, H.: Post-heparin lipolytic activities and alterations of the chemical composition of high density lipoproteins in alcohol-induced type V hyperlipidemia. *Atherosclerosis* **52:**317–328, 1984.

Brewster, A. C., Lankford, H. G., Schwartz, M. G., and Sullivan, J. F.: Ethanol and alimentary lipemia. *Am. J. Clin. Nutr.* **19:**255–259, 1966.

Brodie, B. B., Butler, W. M., Horning, M. G., Maickel, R. P., and Maling, H. M.: Alcohol-induced triglyceride deposition in liver through derangement of fat transport. *Am. J. Clin. Nutr.* **9:**432–435, 1961.

Brown, D. F.: The effect of ethyl α-p-chlorophenoxyisobutyrate on ethanol-induced hepatic steatosis in the rat. *Metabolism* **15:**868–873, 1966.

Brunengraber, H., Boutry, M., Lowenstein, L., and Lowenstein, J. M.: The effect of ethanol on lipogenesis by the perfused liver. in: *Alcohol and Aldehyde Metabolizing Systems* (R. G. Thurman, T. Yonetani, J. R. Williamson, and B. Chance, eds.), New York, Academic Press, pp. 329–337, 1974.

Brunzell, J. D., and Bierman, E. L.: Chylomicronemia syndrome: Interaction of genetic and acquired hypertriglyceridemia. *Med. Clin. North Am.* **66:**455–468, 1982.

Bustos, G. O., Kalant H., Khanna, J. M., and Loth, J.: Pyrazole and induction of fatty liver by a single dose of ethanol. *Science* **168:**1598–1599, 1970.

Cachera, R., Lamotte, M., and Lamotte-Barrillon, S.: Etude clinique, biologique et histologique des stéatoses du foie chez les alcooliques. *Sem. Hop.* **26:**3497–3514, 1950.

Camargo, C. A., Williams, P. T., Vranizan, K. M., Albers, J. J., and Wood, P. D.: The effect of moderate alcohol intake on serum apolipoproteins A-I and A-II. *J.A.M.A.* **253:**2854–2857, 1985.

Cameron, J. L., Capuzzi, D. M., Zuidema, G. D., and Margolis, S.: Acute pancreatitis with hyperlipemia: Evidence for a persistent defect in lipid metabolism. *Am. J. Med.* **56:**482–487, 1974.

Castelli, W. P., Gordon, T., Hjortland, M. C., Kagan, A., Doyle, J. T., Hames, C. G., Hulley, S. B., and Zukel, W. J.: Alcohol and blood lipids. *Lancet* **2:**153–155, 1977.

Cederbaum, A. L., Lieber, C. S., Beattie, D. S., and Rubin, E.: Effect of chronic ethanol ingestion on fatty acid oxidation by hepatic mitochondria. *J. Biol. Chem.* **250:**5122–5129, 1975.

Chait, A., February, A. E., Mancini, M., and Lewis, B. L.: Clinical and metabolic study of alcoholic hyperlipidemia. *Lancet* **2:**62–64, 1972.

Chalvardjian, A.: Mode of action of choline. V. Sequential changes in hepatic and serum lipids of choline-deficient rats. *Can. J. Biochem.* **48:**1234–1240, 1970.

Chance, B., Sies, H., and Boveris, A.: Hydroperoxide metabolism in mammalian organs. *Physiol. Rev.* **59:**527–605, 1979.

Chin, J. H., and Goldstein, D. B.: Effects of low concentration of ethanol on the fluidity of spin-labeled erythrocyte and brain membranes. *Mol. Pharmacol.* **13:**435–441, 1977.

Chin, J. H., Parson, L. M., and Goldstein, D. R.: Increased cholesterol content of erythrocyte and brain membranes in ethanol-tolerant mice. *Biochim. Biophys. Acta.* **513:**358–363, 1978.

Cohen, B. I., and Raicht, R. F.: Sterol metabolism in the rat: Effect of alcohol on sterol metabolism in two strains of rats. *Alcoholism: Clin. Exp. Res.* **5**:225–229, 1981.

Corredor, C., Mansbach, C., and Bressler, R.: Choline control of tissue carnitine content. *Fed. Proc.* **26**:278, 1967.

Crouse, J. R., Gerson, C. D., DeCarli, L. M., and Lieber, C. S.: Role of acetate in the reduction of plasma free fatty acids produced by ethanol in man. *J. Lipid Res.* **9**:509–512, 1968.

Cunningham, C. C., Filus, S., Bottenus, R. E., and Spach, P. I.: Effect of ethanol consumption on the phospholipid composition of rat liver microsomes and mitochondria. *Biochim. Biophys. Acta* **712**:225–233, 1982.

Cunningham, C. C., Bottenus, R. E., Spach, P. I., and Rudel, L. L.: Ethanol-related changes in liver microsomes and mitochondria from the monkey, Macaca fascicularis. *Alcoholism: Clin. Exp. Res.* **7**:424–430, 1983.

Cushman, P., Barboriak, J. J., Liao, A., and Hoffman, N. E.: Association between plasma high density lipoprotein cholesterol and antipyrine metabolism in alcoholics. *Life Sci.* **30**:1721–1724, 1982.

Cuvelier, I., Steinmetz, J., Mikstacki, T., and Siest, G.: Variations in total phospholipids and high-density lipoprotein phospholipids in plasma from a general population: Reference intervals and influence of xenobiotics. *Clin. Chemist.* **31**:763–766, 1985.

Danielsson, B., Ekman, R., Fex, G., Johansson, B. G., Kristensson, H., Nilsson-Ehle, P., and Wadstein, J.: Changes in plasma high density lipoproteins in chronic male alcoholics during and after abuse. *Scand. J. Clin. Lab. Invest.* **38**:113–119, 1978.

Dawidowicz, E. A.: The effect of ethanol on membranes. *Hepatology* **5**:697–699, 1985.

Day, R. C., Harry, D. S., Owen, J. S., Foo, A. Y., and McIntyre, N.: Lecithin-cholesterol acyltransferase and the lipoprotein abnormalities of parenchymal liver disease. *Clin. Sci.* **56**:575–583, 1979.

Debry, G., Mejean, L., Max, J. P., Lambert, D., Pointel, J.-P., and Drouin, P.: Effects of alcohol intake on several metabolic parameters in primary hyperlipoproteinemia. In: *Metabolic Effects of Alcohol* (P. Avogaro, C. R. Sirtori, and F. Tremoli, eds.), Amsterdam, The Netherlands, Elsevier/North Holland Biomedical Press, pp. 227–234, 1979.

DeCarli, L. M., and Lieber, C. S.: Fatty liver in the rat after prolonged intake of ethanol with a nutritionally adequate new liquid diet. *J. Nutr.* **91**:331–336, 1967.

Declercq, P. E., Debeer, L. J., and Mannaerts, G. P.: Glucagon inhibits triacylglycerol synthesis in isolated hepatocytes by lowering their glycerol 3-phosphate content. *Biochem. J.* **202**:803–806, 1982.

DeGennes, J. L., Thomopoulus, P., Truffert, J., and Labrousse de Tregomain, B.: Hyperlipemies dépendantes de l'alcool. *Nutr Metab.* **14**:141–158, 1972.

Devenyi, P., Robinson, G. M., and Roncari, D. A. K.: Alcohol and high-density lipoproteins. *Can. Med. Assoc. J.* **123**:981–984, 1980.

Dich, J., Bro, B., Grunnet, N., Jensen, F., and Kondrup, J.: Accumulation of traicylglycerol in cultured rat hepatocytes is increased by ethanol and by insulin and dexamethasone. *Biochem. J.* **212**:617–623, 1983.

Di Luzio, N. R.: Effect of acute ethanol intoxication on liver and plasma lipid fractions of the rat. *Am. J. Physiol.* **194**:453–456, 1958.

Di Luzio, N. R.: A mechanism of the acute ethanol-induced fatty liver and the modification of liver injury by antioxidant. *Lab. Invest.* **15**:50–63, 1966.

Domanski, R., Rifenberick, D., Stearns, F., Scorpio, R. M., Narrod, S. A., and Narrod, M. F.: Ethanol-induced fatty liver in rats: Effects of pyrazole and glucose. *Proc. Soc. Exp. Biol. Med.* **138**:18–20, 1971.

Ducceschi, V.: La colesterina del sangue nella intossieazione per alcool. *Arch. Fisiol.* **13**:147–153, 1915.

Duhamel, G., Nalpas, B., Goldstein, S., Laplaud, P. M., Berthelot, P., and Chapman, M. J.: Plasma lipoprotein and apolipoprotein profile in alcoholic patients with and without liver disease: On the relative roles of alcohol and liver injury. *Hepatology* **4**:577–585, 1984.

Dyer, A. R., Stamler, J., Paul, O., Berkson, D. M., Shekelle, R. B., Lepper, M. H., McKean, H., Lindberg, H. A., Garside, D., and Tokich, T.: Alcohol, cardiovascular risk factors, and mortality: The Chicago experience. *Circulation* **64**(Suppl III):20–27, 1981.

Eberhardt, G., Mo'hr, J., Oehlert, W., Schmidt, E., Schmidt, F. W., and Vondrasek, P.: Kontrollierte Studie zur therapeutischen Wirkung von B-Vitaminen und einem anabolen Steroid bei chronischer Hepatitis. *Dtsch. Med. Wochenschr.* **100**:2074–2082, 1975.

Edreira, J. G., Hirsch, R. L., and Kennedy, J. A.: Production of fatty liver with dietary ethanol despite orotic acid supplementation. *Q. J. Stud. Alcohol* **35**:20–25, 1974.

Eichner, E. R.: Alcohol versus exercise for coronary protection. *Am. J. Med.* **79**:231–240, 1985.

Ekman, R., Fex, G., Johansson, B. G., Nilsson-Ehle, P., and Wadstein, J.: Changes in plasma high density lipoproteins and lipolytic enzymes after long-term, heavy consumption. *Scand. J. Clin. Lab. Invest.* **41**:709–715, 1981.

Elko, E. E., Wooles, W. R., and Di Luzio, N. R.: Alterations and mobilization of lipids in acute ethanol-treated rats. *Am. J. Physiol.* **201**:923–926, 1961.

Ernest, N., Fisher, M., Smith, W., Gordon, T., Rifkind, B. M., Little, J. A., Mishkel, M. A., and Williams, O. D.: The association of plasma high-density lipoprotein cholesterol with dietary intake and alcohol consumption: The Lipid Research Clinics Prevalence study. *Circulation* **62**(Suppl. IV):41–52, 1980.

Estler, C. J., and Ammon, H. P. T.: The influence of beta adrenergic blockade on the ethanol-induced derangement of lipid transport. *Arch. Int. Pharmacodyn. Ther.* **166**:333–341, 1967.

Fallat, R. W., Vester, J. W., and Glueck, C. J.: Suppression of amylase activity by hypertriglyceridemia. *J.A.M.A.* **225**:1331–1334, 1973.

Fallon, H. J., Lamb, R. G., and Jamdar, S. C.: Phosphatidate phosphohydrolase and the regulation of glycerolipid biosynthesis. *Biochem. Soc. Trans.* **5**:37–40, 1977.

Farmer, R. G., Winkelman, E. I., Brown, H. B., and Lewis, L. A.: Hyperlipoproteinemia and pancreatitis. *Am. J. Med.* **54**:161–165, 1973.

Feigl, J.: Neue Untersuchungen zur Chemte des Biutes bei akuter Alkcohol Intoxikation und bei chronischem Alkoholismus mit besonderer Berucksichtigung der Fette und Lipoide. *Biochem. Z.* **92**:282–317, 1918.

Feinman, L., and Lieber, C. S.: Effect of ethanol on plasma glycerol in man. *Am. J. Clin. Nutr.* **20**:400–403, 1967.

Fellenius, E., Bengtsson, G., and Kiessling, K.-H.: The influence of ethanol-induced changes of the α-glycerophosphate level on hepatic triglyceride synthesis. *Acta Chem. Scand.* **27**:2893–2901, 1973.

Fenster, L. F.: The non-efficacy of short-term anabolic steroid therapy in alcoholic liver disease. *Ann. Intern. Med.* **65**:738–744, 1966.

Field, F. J., Boydstun, J. S., and LaBrecque, D. R.: Effect of chronic ethanol ingestion on hepatic and intestinal acyl coenzyme A: Cholesterol acyltransferase and 3-hydroxy-3-methylglutaryl coenzyme A reductase in the rat. *Hepatology* **5**:133–138, 1985.

Flores, H., Pak, N., Maccioni, A., and Monckeberg, F.: Lipid transport in kwashiorkor. *Br. J. Nutr.* **24**:1005–1011, 1970.

Forsander, O. A., Mäenpää, P. H., and Salaspuro, M. P.: Influence of ethanol on the lactate/pyruvate and β-hydroxybutyrate/acetoacetate ratios in rat liver experiments. *Acta Chem. Scand.* **19**:1770–1771, 1965.

Förstermann, U., and Feuerstein, T. J.: Decreased systemic formation of prostaglandin E and prostacyclin, and unchanged thromboxane formation, in alcoholics during withdrawal as estimated from metabolites in urine. *Clin. Sci.* **73**:277–283, 1987.

Fraser, G. E., and Upsdell, M.: Alcohol and other discriminants between cases of sudden death and myocardial infarction. *Am. J. Epidemiol.* **114**:462–476, 1981.

Fraser, G. E., Anderson, J. T., Foster, W., Goldberg, R., Jacobs, D., and Blackburn, H.: The effect of alcohol on serum high density lipoprotein (HDL). *Atherosclerosis* **46**:275–286, 1983.

Freeman, M., Kuiken, L., Ragland, J. B., and Sabesin, S. M.: Hepatic triglyceride lipase deficiency in liver disease. *Lipids* **12**:443–445, 1977.

Friedman, H. S.: Cardiovascular effects of alcohol with particular reference to the heart. *Alcohol* **1**:333–339, 1984.

Friedman, M., Rosenman, R. H., and Byers, S. O.: Effect of moderate ingestion of alcohol upon serum triglyceride responses of normo- and hyperlipemic subjects. *Proc. Soc. Exp. Biol. Med.* **120**:696–698, 1965.

Ginsberg, H., Olefsky, J., Farquhar, J. W., and Reaven, G. M.: Moderate ethanol ingestion and plasma triglyceride levels. *Ann. Intern. Med.* **80**:143–149, 1974.

Glenny, H. P., and Brindley, D. N.: The effect of cortisol, corticotropin and thyroxine on the synthesis of glycerolipids and on the phosphatidate phosphohydrolase activity in rat liver. *Biochem. J.* **176**:777–784, 1978.

Glueck, C. J., Hogg, E., Allen, C., and Gartside, P. S.: Effects of alcohol ingestion on lipids and lipoproteins in normal men: Isocaloric metabolic studies. *Am. J. Clin. Nutr.* **33**:2287–2293, 1980.

Glueck, C. J., Heiss, G., Morrison, J. A., Khoury, P., and Moore, M.: Alcohol intake, cigarette smoking and plasma lipids and lipoproteins in 12–19-year-old children. *Circulation* **64**:(Suppl. III):48–56, 1981.

Goebel, K. M., Goebel, F. D., Schubotz, R., and Schneider, J.: Hemolytic implications of alcoholism in liver disease. *J. Lab. Clin. Med.* **94**:123–132, 1979.

Gofman, J. W., Young, W., and Tandy, R.: Ischemic heart disease, atherosclerosis, and longevity. *Circulation* **34**:679–697, 1966.

Goldberg, C. S., Tall, A. R., and Krumholz, S.: Acuté inhibition of hepatic lipase and increase in plasma lipoproteins after alcohol intake. *J. Lipid Res.* **25**:714–720, 1984.

Goodman, D. S., and Deykin, D.: Fatty acid ethyl ester formation during ethanol metabolism *in vivo*. *Proc. Soc. Exp. Biol. Med.* **113**:65–67, 1963.

Gordon, E. R.: Effect of an intoxicating dose of ethanol on lipid metabolism in an isolated perfused rat liver. *Biochem. Pharmacol.* **21**:2991–3004, 1972.

Gordon, E. R.: Mitochondrial functions in an ethanol-induced fatty liver. *J. Biol. Chem.* **248**:8271–8280, 1973.

Gordon, E. R., Rochman, J., Arai, M., and Lieber, C. S.: Lack of correlation between hepatic mitochondrial membrane structure and functions in ethanol-fed rats. *Science* **216**:1319–1321, 1982.

Gordon, T., and Doyle, J. T.: Drinking and coronary heart disease: The Albany study. *Am. Heart J.* **110**:331–334, 1985.

Green, F. A., and Claesson, H.-E.: Ethanol-enhanced transmembrane penetration of arachidonic acid and activation of the 5-lipooxygenase pathway in human leukocytes. *Biochem. Biophys. Res. Commun.* **140**:782–788, 1986.

Greenberger, N. J., Hatch, F. T., Drummey, G. D., and Isselbacher, K. J.: Pancreatitis and hyperlipemia: A study of serum lipid alterations in 25 patients with acute pancreatitis. *Medicine* (Baltimore) **45**:161–174, 1966.

Grossberg, S. J.: Lipemia retinalis, hyperlipemia, and hepatosplenomegaly in chronic alcoholic patient. *N.Y. State J. Med.* **66**:2951–2953, 1966.

Guivernau, M., Baraona, E., and Lieber, C. S.: Acute and chronic effects of ethanol and its metabolites on vascular production of prostacyclin in rats. *J. Pharmacol. Exp. Ther.* **240**:59–64, 1987.

Guivernau, M., Baraona, E., Soong, J., and Lieber, C. S.: Enhanced stimulatory effect of HDL and other agonists on vascular prostacyclin production in rats fed alcohol-containing diets. *Biochem. Pharmacol.* **38**:503–508, 1989.

Guynn, R. W., Veloso, D., Harris, R. L., Lawson, J. W. R., and Veech, R. L.: Ethanol administration and the relationship of malonyl-coenzyme A concentrations to the rate of fatty acid synthesis in rat liver. *Biochem. J.* **136**:639–647, 1973.

Guzman, M., and Geelen, M. J. H.: Effects of ethanol feeding on the activity and regulation of hepatic carnitine palmitoyltransferase. *Arch. Biochem. Biophys.* **267**:580–588, 1988.

Haffner, S. M., Applebaum-Bowden, D., Wahl, P. W., Hoover, J. J., Warnick, G. R., Albers, J. J., and Hazzard, W. R.: Epidemiological correlates of high density lipoprotein subfractions, apolipoproteins A-I, A-II, and D, and lecithin cholesterol acyltransferase: Effects of smoking, alcohol, and adiposity. *Arteriosclerosis* **5**:169–177, 1985.

Haines, D. S. M.: The effects of choline deficiency and choline refeeding upon the metabolism of plasma and liver lipids. *Can. J. Biochem.* **44**:45–57, 1966.

Hajjar, D. P.: Prostaglandins and cyclic nucleotides: Modulators of arterial cholesterol metabolism. *Biochem. Pharmacol.* **34**:295–300, 1985.

Harada, S., Agarwal, D. P., Goedde, H. W., and Ishikawa, B.: Aldehyde dehydrogenase isoenzyme variation and alcoholism in Japan. *Pharmacol. Biochem. Behav.* **18**:151–153, 1983.

Hartman, A. D., and Di Luzio, N. R.: Inhibition of the chronic ethanol-induced fatty liver by antioxidant administration. *Proc. Soc. Exp. Biol. Med.* **127**:270–281, 1968.

Hartung, G. H., Foreyt, J. P., Mitchell, R. E., Mitchell, J. G., Reeves, R. S., and Gotto, A. M., Jr.: Effect of alcohol intake on high-density lipoprotein cholesterol levels in runners and inactive men. *J.A.M.A.* **249**:747–750, 1983.

Haskell, W. L., Camargo, C., Jr., Williams, P. T., Vranizan, K. M., Krauss, R. M., Lindgren, F. T., and Wood, P. D.: The effect of cessation and resumption of moderate alcohol intake on serum high-density-lipoprotein subfractions: A controlled study. *N. Engl. J. Med.* **310**:805–810, 1984.

Hasumura, Y., Teschke, R., and Lieber, C. S.: Characteristics of acetaldehyde oxidation in rat liver mitochondria. *J. Biol. Chem.* **251**:4908–4913, 1976.

Hegsted, M. D., and Ausman, L. M.: Diet, alcohol, and coronary heart disease in men. *J. Nutr.* **118**:1184–1189, 1988.

Hennekens, C. H., Rosner, B., and Cole, D. S.: Daily alcohol consumption and fatal coronary heart disease. *Am. J. Epidemiol.* **107**:196–200, 1978.

Hernandez-Munoz, R., Santamaria, A., Garcia-Sainz, J. A., Pina, E., and Chagoya-de-Sanchez, V.: On the mechanism of ethanol-induced fatty liver and its reversibility by adenosine. *Arch. Biochem. Biophys.* **190**:155–162, 1978.

Hillborn, M., Kangasaho, M., Lowbeer, C., Kaste, M., Muuronen, A., and Numminen, H.: Effects of ethanol on platelet function. *Alcohol* **2**:429–432, 1985.

Hoffbauer, F. W., and Zaki, F. G.: Choline deficiency in baboon and rat compared. *Arch. Pathol.* **79**:364–369, 1965.

Holman, R. T., and Johnson, S.: Changes in essential fatty acid profile of serum phospholipids in human disease. *Prog. Lipid Res.* **20**:67–73, 1981.

Holmström, B.: Studies on the metabolism of ^{14}C-labeled ethanol in man. V. Synthesis of liver fatty acids. *Ark. Kemi* **30**:333–345, 1969.

Horning, M. B., Williams, E. A., Maling, H. M., and Brodie, B. B.: Depot fat as source of increased liver triglycerides after ethanol. *Biochem. Biophys. Res. Commun.* **3**:635–640, 1960.

Hulley, S. B., and Gordon, S.: Alcohol and high-density lipoprotein cholesterol: Causal inference from diverse study designs. *Circulation* **64**(Suppl 3):57–63, 1981.

Hyams, D. E., and Isselbacher, K. J.: Prevention of fatty liver by administration of adenosine triphosphate. *Nature* **204**:1196–1197, 1964.

Ide, H., and Nakazawa, Y.: Role of acetaldehyde in ethanol-induced increase in the activity of phosphatidate phosphatase in rat liver. *Biochem. Pharmacol.* **36**:2443–2448, 1987.

Ideo, G., Bellobuono, A., and Bellati, G.: Bilirubin excretion in bile decreased by large amounts of ethanol in isolated and perfused rat liver. *Biomedicine* **29**:225–227, 1978.

Iseri, O. A., Lieber, C. S., and Gottlieb, L. S.: The ultrastructure of fatty liver induced by prolonged ethanol ingestion. *Am. J. Pathol.* **48**:535–555, 1966.

Jaillard, J., Dewailly, P., Sezille, G., Fruchart, J. C., Rault, C., and Scherpe-Reel, P.: Intérêt et signification du test d'hyperlipemie par voie veineuse dans l'exploration des hyperlipemies: Étude de ses variations sous l'effet de l'alcool. *Atherosclerosis* **19**:493–500, 1974.

Janus, E. D., and Lewis, B.: Alcohol and abnormalities of lipid metabolism. *Clin. Endocrinol. Metab.* **7**:321–332, 1978.

Jauhonen, V. P., and Hassinen, I. E.: Metabolic and hormonal changes during intravenous infusion of ethanol, acetaldehyde and acetate in normal and adrenalectomized rats. *Arch. Biochem. Biophys.* **191**:358–366, 1978.

Jauhonen, V. P., Savolainen, M. J., and Hassinen, I. E.: Cyclic AMP-linked mechanisms in ethanol-induced derangements of metabolism in rat liver and adipose tissue. *Biochem. Pharmacol.* **24**:1879–1883, 1975.

Jauhonen, P., Baraona, E., Miyakawa, H., and Lieber, C. S.: Mechanism for selective perivenular hepatotoxicity of ethanol. *Alcoholism: Clin. Exp. Res.* **6**:350–357, 1982.

Jerntorp, P. Ohlin, H., Bergstrom, B., and Almer, L.: Increase of plasma acetaldehyde: An objective indicator of the chlorpropamide alcohol flush. *Diabetes* **30**:788–791, 1981.

Johansson, B. G., and Laurell, C. B.: Disorders of serum α-lipoproteins after alcoholic intoxication. *Scand. J. Clin. Lab. Invest.* **23**:231–233, 1969.

Johansson, B. C., and Medhus, A.: Increase in plasma α-lipoproteins in chronic alcoholics after acute abuse. *Acta. Med. Scand.* **195**:273–277, 1974.

John, G. R., Littleton, J. M., and Jones, P. A.: Membrane lipids and ethanol tolerance in the mouse: The influence of dietary fatty acid composition. *Life Sci.* **27**:545–555, 1980.

Johnson, O.: Influence of blood ethanol concentration on the acute ethanol-induced liver triglyceride accumulation in rats. *Scand. J. Clin. Lab. Invest.* **33**:207–213, 1974.

Johnson, D. A., Lee, N. M., Cooke, R., and Loh, H. H.: Ethanol-induced fluidization of main lipid bilayers required presence of cholesterol in membranes for the expression of tolerance. *Mol. Pharmacol.* **15**:739–746, 1979.

Johnson, O., Hernell, O., Fex, G., and Olivecrona, T.: Ethanol-induced fatty liver: The requirement of ethanol metabolism for the development of fatty liver. *Life Sci.* **10**:553–559, 1971.

Johnson, S. B., Gordon, E., McClain, C., Low, G., and Holman, R. T.: Abnormal polyunsaturated fatty acid patterns of serum lipids in alcoholism and cirrhosis: Arachidonic acid deficiency in cirrhosis. *Proc. Natl. Acad. Sci. USA* **82**:1815–1818, 1985.

Joly, J. G., Feinman, L., Ishii, H., and Lieber, C. S.: Effect of chronic ethanol feeding on hepatic microsomal glycerophosphate acyltransferase activity. *J. Lipid Res.* **14**:337–343, 1973.

Jones, A. L., Ruderman, N. B., and Herrera, M. G.: Electron microscopic and biochemical study of lipoprotein synthesis in the isolated perfused rat liver. *J. Lipid Res.* **8**:429–446, 1967.

Jones, D. P., Losowsky, M. S., Davidson, C. S., and Lieber, C. S.:

Effects of ethanol on plasma lipids in man. *J. Lab. Clin. Med.* **62:**675–682, 1963.

Jones, D. P., Perman, E. S., and Lieber, C. S.: Free fatty acid turnover and triglyceride metabolism after ethanol ingestion in man. *J. Lab. Clin. Med.* **66:**804–813, 1965.

Kaffarnik, H., and Schneider, J.: Short-term effect of ethyl alcohol on serum lipids of healthy fasting persons. *Z. Gesamte Exp. Med.* **152:**187–200, 1970.

Kähönen, M. T., Ylikahri, R. H., and Hassinen, I.: Studies on the mechanism of inhibition of acute alcoholic fatty liver by clofibrate. *Metabolism* **21:**1021–1028, 1972.

Kalant, H., Khanna, J. M., and Bustos, G. O.: Effect of pyrazole on the induction of fatty liver by chronic administration of ethanol. *Biochem. Pharmacol.* **21:**811–819, 1972.

Kangasaho, M., Hillbom, M., Kaste, M., and Vapaatalo, H.: Effects of ethanol intoxication and hangover on plasma levels of thromboxane B_2 and 6-keto prostaglandin F1α and on thromboxane B_2 formation by platelets in man. *Throm. Haemostasis* **49:**232–234, 1982.

Karsenty, C., Baraona, E., Savolainen, M. J., and Lieber, C. S.: Effects of chronic ethanol intake on mobilization and excretion of cholesterol in baboons. *J. Clin. Invest.* **75:**976–986, 1985.

Kessler, J. I., and Yalovsky-Mishkin, S.: Effect of ingestion of saline, glucose, and ethanol on mobilization and hepatic incorporation of epididymal pad palmitate-1-^{14}C in rats. *J. Lipid Res.* **7:**772–777, 1966.

Kikuchi, T., and Kako, K. J.: Metabolic effects of ethanol on the rabbit heart. *Circ. Res.* **26:**625–634, 1970.

Kim, C. I., Leo, M. A., Lowe, N., and Lieber, C. S.: Effects of vitamin A and ethanol on liver plasma membrane fluidity. *Hepatology* **8:**735–741, 1988.

Kitson, T. M.: The disulfiram-ethanol reaction. *J. Stud. Alcohol* **38:**96–113, 1977.

Kittner, S. J., Garcia-Palmieri, M. R., Costas, R., Jr., Cruz-Vidal, M., Abbott, R. D., and Havlik, R. J.: Alcohol and coronary heart disease in Puerto Rico. *Am. J. Epidemiol.* **117:**538–550, 1983.

Klatskin, G., Krehl, W. A., and Conn, H. O.: The effect of alcohol on the choline requirement. I. Changes in the rat's liver following prolonged ingestion of alcohol. *J. Exp. Med.* **100:**605–614, 1954.

Klatsky, A. L., Friedman, G. D., and Siegelaub, A. B.: Alcohol consumption before myocardial infarction: Results from the Kaiser-Permanente epidemiologic study of myocardial infarction. *Ann. Intern. Med.* **81:**294–301, 1974.

Klatsky, A. L., Friedman, G. D., and Siegelaud, A. B.: Alcohol and mortality: A ten-year Kaiser-Permanente experience. *Ann. Intern. Med.* **95:**139–145, 1981.

Knochel, J. P.: Cardiovascular effects of alcohol. *Ann. Intern. Med.* **98** (Part 2):849–854, 1983.

Koff, R. S., and Fitts, J. J.: Chlorpromazine inhibition of ethanol metabolism without prevention of fatty liver. *Biochem. Med.* **6:**77–81, 1972.

Kondrup, J., and Grunnet, N.: The effect of acute and prolonged ethanol treatment on the contents of coenzyme A, carnitine and their derivatives in rat liver. *Biochem. J.* **132:**373–379, 1973.

Kontula, K., Vilnikka, L., Ylikorkala, O., and Ylikahri, R.: Effect of acute ethanol intake on thromboxane and prostacyclin in human. *Life Sci.* **31:**261–264, 1982.

Kovatchev, S., and Eibl, H.: The preparation of phospholipids by phospholipase D. *Adv. Exp. Med. Biol.* **101:**221–226, 1978.

Kraus, R. M.: Regulation of high density lipoprotein levels. *Med. Clin. North Am.* **66:**403–429, 1982.

Kudzma, D. J., and Schonfeld, G.: Alcoholic hyperlipidemia: Induction by alcohol but not by carbohydrate. *J. Lab. Clin. Med.* **77:**384–389, 1971.

Lakshman, M. R., and Ezekiel, M.: Relationship of alcoholic hyperlipidemia to the feed-back regulation of hepatic cholesterol synthesis by chylomicron remnants. *Alcoholism: Clin. Exp. Res.* **6:**482–486, 1982.

Lakshman, M. R., and Veech, R. L.: Short-term and long-term effects of ethanol administration *in vivo* on rat liver HMG-CoA reductase and cholesterol 7α-hydroxylase activities. *J. Lipid Res.* **18:**325–330, 1977.

Lakshman, M. R., Ezekiel, M., Campbell, B. S., and Muesing, R. A.: Binding, uptake, and metabolism of chylomicron remnants by hepatocytes from control and chronic ethanol-fed rats. *Alcoholism: Clin. Exp. Res.* **10:**412–418, 1986.

Lakshman, M. R., Chirtel, S. J., and Chambers, L. L.: Roles of ω3 fatty acids and chronic ethanol in the regulation of plasma and liver lipids and plasma apoproteins A_1 and E in rats. *J. Nutr.* **118:**1299–1303, 1988.

Lamb, R. G., Wood, C. K., and Fallon, H. J.: The effect of acute and chronic ethanol intake on hepatic glycerolipid biosynthesis in the hamster. *J. Clin. Invest.* **63:**14–20, 1979.

Lange, L. G.: Nonoxidative ethanol metabolism: Formation of fatty acid ethyl esters by cholesterol esterase. *Proc. Natl. Acad. Sci. USA* **79:**3954–3957, 1982.

Lange, L. G., Bergmann, S. R., and Sobel, B. E.: Identification of fatty acid ethyl esters as products of rabbit myocardial ethanol metabolism. *J. Biol. Chem.* **256:**12968–12973, 1981.

Leathers, C. W., Bond, M. G., Bullock, B. C., and Rudel, L. L.: Dietary ethanol and cholesterol in Macaca nemestrina serum lipid and hepatic changes. *Exp. Mol. Pathol.* **35:**285–299, 1981.

Lefevre, A. F., DeCarli, L. M., and Lieber, C. S.: Effect of ethanol on cholesterol and bile acid metabolism. *J. Lipid Res.* **13:**48–55, 1972.

Lehtonen, M. A., Savolainen, M. J., and Hassinen, I. E.: Hormonal regulation of hepatic soluble phosphatidate phosphohydrolase. *FEBS Lett.* **99:**162–166, 1979.

Leung, N. N. Y., and Peters, T. J.: Palmitic acid oxidation and incorporation into triglyceride by needle liver biopsy specimens from control subjects and patients with alcoholic fatty liver disease. *J. Clin. Sci.* **71:**253–260, 1986.

Leung, N. N. Y., Cairns, S. T., and Peters, T. J.: Activities and subcellular distributions of hepatic lipases in control subjects and in patients with alcoholic fatty liver. *J. Clin. Sci.* **69:**517–523, 1985.

Levy, R. I., Brensike, J. F., Epstein, S. E., Kelsey, S. F., Passamani, E. R., Richardson, J. M., Loh, I. K., Stone, N. J., Aldrich, R. F., Battaglini, J. W., Moriarty, D. J., Fisher, M. L., Friedman, L., Friedewald, W., and Detre, K. M.: The influence of changes in lipid values induced by cholestyramine and diet on progression of coronary artery disease: Results of NHLBI Type II Coronary Intervention Study. *Circulation* **69**:325–337, 1984.

Lewis, P. J., and Dollery, C. T.: Clinical pharmacology and potential of prostacyclin. *Br. Med. Bull.* **38**:281–284, 1983.

L'Huiller, J. R., Roudier, R., and Thuiller, J.: Evidence histologique de l'efficacité et de l'innocuité de, l'oxygène hyperbare dans le traitement de l'intoxication éthylique expérimentale. *Med. Pharmacol. Exp.* **16**:513–519, 1967.

Liau, D. F., Hashim, S. A., Pierson, R. N., and Ryan, S. F.: Alcohol-induced lipid change in the lung. *J. Lipid Res.* **22**:680–686, 1981.

Lieber, C. S.: Metabolic effects produced by alcohol in the liver and other tissues. *Adv. Intern. Med.* **14**:151–199, 1968.

Lieber, C. S.: Alcohol and the liver. In: *The Biological Basis of Medicine, Vol. 5* (E. E. Bittar, ed.), New York, Academic Press, pp. 317–344, 1969.

Lieber, C. S.: Alcohol, protein metabolism, and liver injury. *Gastroenterology* **79**:373–390, 1980.

Lieber, C. S. (ed.): *Medical Disorders of Alcoholism, Pathogenesis and Treatment*. Philadelphia, W. B. Saunders Co., 1982.

Lieber, C. S.: To drink (moderately) or not to drink? *N. Engl. J. Med.* **310**:846–848, 1984.

Lieber, C. S.: Alcohol and the Liver. In: *Liver Annual-VI* (J. L. Boyer and J. H. P. Wilson, eds.), Amsterdam, The Netherlands, Excerpta Medica, pp. 163–240, 1987.

Lieber, C. S., and Davidson, C. S.: Some metabolic effects of ethyl alcohol. *Am. J. Med.* **33**:319–327, 1962.

Lieber, C. S., and DeCarli, L. M.: Study of agents for the prevention of the fatty liver produced by prolonged alcohol intake. *Gastroenterology* **50**:316–322, 1966.

Lieber, C. S., and DeCarli, L. M.: Quantitative relationship between the amount of dietary fat and the severity of the alcoholic fatty liver. *Am. J. Clin. Nutr.* **23**:474–478, 1970a.

Lieber, C. S., and DeCarli, L. M.: Reduced nicotinamide-adenine dinucleotide phosphate oxidase: Activity enhanced by ethanol consumption. *Science* **170**:78–80, 1970b.

Lieber, C. S., and Pignon, J. P.: Ethanol and lipids. In: *Human Plasma Lipoproteins: Chemistry, Physiology and Pathology* (J. C. Fruchart, and J. Shepherd, eds.), Berlin, Walter De Gruyter and Co., pp. 245–280, 1989.

Lieber, C. S., and Rubin, E.: Alcoholic fatty liver in man on a high protein and low fat diet. *Am. J. Med.* **44**:200–207, 1968.

Lieber, C. S., and Schmid, R.: The effect of ethanol on fatty acid metabolism: Stimulation of hepatic fatty acid synthesis *in vitro. J. Clin. Invest.* **40**:394–399, 1961.

Lieber, C. S., and Spritz, N.: Effect of prolonged ethanol intake in man: Role of dietary, adipose, and endogenously synthesized fatty acids in the pathogenesis of the alcoholic fatty liver. *J. Clin. Invest.* **45**:1400–1411, 1966.

Lieber, C. S., Leevy, C. M., Stein, S. W., George, W. S., Cherrick, G. R., Abelmann, W. H., and Davidson, C. S.: Effect of ethanol on plasma free fatty acids in man. *J. Lab. Clin. Med.* **59**:826–832, 1962.

Lieber, C. S., Jones, D. P., Mendelson, J., and DeCarli, L. M.: Fatty liver, hyperlipemia and hyperuricemia produced by prolonged alcohol consumption, despite adequate dietary intake. *Trans. Assoc. Am. Physicians* **76**:289–300, 1963.

Lieber, C. S., Jones, D. P., and DeCarli, L. M.: Effects of prolonged ethanol intake: Production of fatty liver despite adequate diets. *J. Clin. Invest.* **44**:1009–1021, 1965.

Lieber, C. S., Spritz, N., and DeCarli, L. M.: Role of dietary, adipose and endogenously synthesized fatty acids in the pathogenesis of the alcoholic fatty liver. *J. Clin. Invest.* **45**:51–62, 1966a.

Lieber, C. S., Spritz, N., and DeCarli, L. M.: Accumulation of triglycerides in heart and kidney after alcohol ingestion. *J. Clin. Invest.* **45**:1041 (abstract), 1966b.

Lieber, C. S., Lefevre, A., Spritz, N., Feinman, L., and DeCarli, L. M.: Difference in hepatic metabolism of long- and medium-chain fatty acids: The role of fatty acid chain length in the production of the alcoholic fatty liver. *J. Clin. Invest.* **46**:1451–1460, 1967.

Lieber, C. S., Spritz, N., and DeCarli, L. M.: Fatty liver produced by dietary deficiencies: Its pathogenesis and potentiation by ethanol. *J. Lipid Res.* **10**:283–287, 1969.

Lieber, C. S., DeCarli, L. M., Gang, H., Walker, G., and Rubin, E.: Hepatic effects of long-term ethanol consumption in primates. In: *Medical Primatology, Part III* (E. I. Goldsmith and J. Moor-Jankowski, eds.), Basel, Karger, pp. 270–278, 1972.

Lieber, C. S., Teschke, R., Hasumura, Y., and DeCarli, L. M.: Differences in hepatic and metabolic changes after acute and chronic alcohol consumption. *Fed. Proc.* **34**:2060–2074, 1975.

Lombardi, B. P., Pani, P., and Schlunk, F. F.: Choline-deficiency fatty liver: Impaired release of hepatic triglycerides. *J. Lipid Res.* **9**:437–446, 1968.

Losowsky, M. S., Jones, D. P., Davidson, C. S., and Lieber, C. S.: Studies of alcoholic hyperlipemia and its mechanism. *Am. J. Med.* **35**:794–803, 1963.

Luoma, P. V., Sotaniemi, E. A., Pelkonen, R. O., and Ehnholm, C.: High-density lipoproteins and hepatic microsomal enzyme induction in alcohol consumers. *Res. Commun. Chem. Pathol. Pharmacol.* **37**:91–96, 1982.

Maciejko, J. J., Holmes, D. R., Kottke, B. A., Zinsmeister, A. R., Dinh, D. M., and Mao, S. J. T.: Apolipoprotein A-I as a marker of angiographically assessed coronary-artery disease. *N. Engl. J. Med.* **309**:385–389, 1983.

Maillard, M. E., Sorrell, M. F., Volentine, G. D., and Tuma, D. J.: Impaired plasma membrane glycoprotein assembly in the liver following acute ethanol administration. *Biochem. Biophys. Res. Commun.* **123**:951–958, 1984.

Mallov, S.: Effect of adrenalectomy on ethanol and fat metabolism in the rat. *Am. J. Physiol.* **189**:428–438, 1957.

Mallov, S., and Bloch, J. L.: Role of hypophysis and adrenals in fatty infiltration of liver resulting from acute ethanol intoxication. *Am. J. Physiol.* **184**:29–34, 1956.

Malmendier, C. L., and Delcroix, C.: Effect of alcohol intake on high and low density lipoprotein metabolism in healthy volunteers. *Clin. Chim. Acta* **152**:281–288, 1985.

Malmendier, C. L., Mailier, E. L., Amerijckx, J. P., and Fischer, M. L.: Plasma levels of apolipoproteins A-I, A-II in alcoholism relation to the degree of histological liver damage, and to liver function tests. *Acta Hepato-gastroenterol.* **30**:236–239, 1983.

Maquedano, A., Guzman, M., and Castro, J.: Dependence with nutritional status of the ethanol effects on fatty acid metabolism in rat hepatocytes. *Int J. Biochem.* **20**:937–941, 1988.

Marchetti, M., Ottani, V., Zanetti, P., and Puddu, P.: Aspects of lipid metabolism in ethanol-induced fatty liver. *J. Nutr.* **95**:607–611, 1968.

Marin, G. A., Karjoo, M., Ward, N., and Rosato, E.: Effects of alcohol on biliary lipids in the presence of a chronic biliary fistula. *Surg. Gynecol. Obstet.* **141**:352–356, 1975.

Marmot, M. G., Rose, G., Shipley, M. J., and Thomas, B. J.: Alcohol and mortality: A u-shaped curve. *Lancet* **1**:580–583, 1981.

Marth, E., Cazzolato, G., Bon, B., Avogaro, P., and Kostner, G. M.: Serum concentrations of Lp(a) and other lipoprotein parameters in heavy alcohol consumers. *Ann. Nutr. Metab.* **26**:56–62, 1982.

Maruyama, S., Murawaki, Y., and Hirayama, C.: Effects of chronic ethanol administration on hepatic cholesterol and bile acid synthesis in relation to serum high density lipoprotein cholesterol in rats. *Res. Commun. Chem. Pathol. Pharmacol.* **53**:3–21, 1986.

Marzo, S., Ghirardi, P., Sardini, D., Prandini, B. D., and Albertini, A.: Serum lipids and total fatty acids in chronic alcoholic liver disease at different states of cell damage. *Klin. Wochenschr.* (Berlin) **48**:949–950, 1970.

Matsuda, Y., Baraona, E., Salaspuro, M., and Lieber, C. S.: Effects of ethanol on liver microtubules and Golgi apparatus: Possible role in altered hepatic secretion of plasma proteins. *Lab. Invest.* **41**:455–463, 1979.

Matsuzaki, S., and Lieber, C. S.: Increased susceptibility of hepatic mitochondria to the toxicity of acetaldehyde after chronic ethanol consumption. *Biochem. Biophys. Res. Commun.* **75**:1059–1065, 1977.

Mavrelis, P. G., Ammon, H. V., Gleysteen, J. J., Komorowski, R. A., and Charaf, U. K.: Hepatic free fatty acids in alcoholic liver disease and morbid obesity. *Hepatology* **3**:226–231, 1983.

McIntyre, N.: Plasma lipids and lipoproteins in liver disease. *Gut* **19**:526–530, 1978.

Mendelson, J. H., and Mellow, N. K.: Alcohol-induced hyperlipidemia and beta lipoproteins. *Science* **180**:1372–1374, 1973.

Mendenhall, C. L.: Anabolic steroid therapy as an adjunct to diet in alcoholic hepatic steatosis. *Am. J. Dig. Dis.* **13**:783–791, 1968.

Mendenhall, C. L.: Origin of hepatic triglyceride fatty acids: Quantitative estimation of the relative contribution of linoleic acid by diet and adipose tissue in normal and ethanol-fed rats. *J. Lipid Res.* **13**:177–183, 1972.

Menghini, G.: L'aspect morpho-bioptique du foie de l'alcooholique (non cirrhotique) et son évolution. *Bull. Acad. Suisse Sci. Med.* **16**:36–52, 1960.

Mezey, E., Potter, J. J., and Ammon, R. A.: Effect of ethanol administration on the activity of the hepatic lysosomal enzymes. *Biochem. Pharmacol.* **25**:2263–2267, 1976.

Mezey, E., Potter, J. J., Slusser, R. J., Brandes, D., Romero, J., Tamura, T., and Halsted, C. H.: Effect of ethanol feeding on hepatic lysosomes in the monkey. *Lab. Invest.* **43**:88–93, 1980.

Mikkhailidis, D. P., Jeremy, J. Y., Barradas, M. A., Green, N., and Dandona, P.: Effect of ethanol on vascular prostacyclin (prostaglandin 12) synthesis, platelet aggregation, and platelet thromboxane release. *Br. Med. J.* **287**:1495–1498, 1983.

Miller, G. J., and Miller, N. E.: Plasma high-density-lipoprotein concentration and development of ischaemic heart disease. *Lancet* **1**:16–19, 1975.

Miller, N. E., Førde, O. H., Thelle, D. S., and Mjøs, O. D.: The Tromsø heart study. High-density lipoprotein and coronary heart disease: A prospective case-control study. *Lancet* **1**: 965–967, 1977.

Miller, N. E., Hammett, F., Saltissi, S., Rao, S., Van Zeller, H., Coltart, J., and Lewis, B.: Relation of angiographically defined coronary artery disease to plasma lipoprotein subfractions and apolipoproteins. *Br. Med. J.* **282**:1741–1744, 1981.

Miller, N. E., La Ville, A., and Crook, D.: Direct evidence that reverse cholesterol transport is mediated by high-density lipoprotein in rabbit. *Nature* **314**:109–111, 1985.

Moncada, S., and Vane, J. R.: Pharmacology and endogenous roles of prostacyclin endoperoxides, thromboxane A2 and prostacyclin. *Pharmacol. Rev.* **30**:293–297, 1979.

Monroe, P., Viahcevic, Z. R., and Swell, L.: Effects of acute and chronic ethanol intake on bile acid metabolism. *Alcoholism: Clin. Exp. Res.* **5**:92–100, 1981.

Mordasini, R., Kaffarnik, H., Schneider, J., and Riesen, W.: Alkohol und Plasmalipoproteine im akut und Langzeitversuch. *Schweiz. Med. Wochenschr.* **112**:1928–1931, 1982.

Morgan, J. C., and Di Luzio, N. R.: Inhibition of the acute ethanol-induced fatty liver by pyrazole. *Proc. Soc. Exp. Biol. Med.* **134**:462–468, 1970.

Muller, P., Fellin, R., Lamprecht, J., Agostini, B., Wieland, H., Rost, W., and Seidel, D.: Hypertriglyceridaemia secondary to liver disease. *Eur. J. Clin. Invest.* **4**:419–428, 1974.

Nervi, A. M., Peluffo, R. O., Brenner, R. R., and Leikin, A. I.: Effect of ethanol administration on fatty acid desaturation. *Lipids* **15**:263–268, 1980.

Nestel, P. J.: Some clinical and chemical manifestations of alcohol-induced hyperlipemia. *Australas Ann. Med.* **16**:139–143, 1967.

Nikkilä, E. A., and Ojala, K.: Role of hepatic L-α-glycerolphosphate and triglyceride synthesis in production of fatty liver by ethanol. *Proc. Soc. Exp. Biol. Med.* **113**:814–817, 1963.

Nikkilä, E. A., Taskinen, M.-R., and Huttunen, J. K.: Effect of acute ethanol load on postheparin plasma lipoprotein lipase and hepatic lipase activities and intravenous fat tolerance. *Horm. Metab. Res.* **10**:220–223, 1978.

Nikkilä, E. A., Kuusi, T., and Taskinen, M. R.: Role of lipoprotein lipase and hepatic endothelial lipase in the metabolism of high density lipoproteins: A novel concept on cholesterol transport in HDL cycle. In: *Metabolic Risk Factors in Ischemic Cardiovascular Disease* (L. A. Carlson and B. Pernow, eds.), New York, Raven, pp. 205–215, 1982.

Nilsson-Ehle, P., Carlstrom, S., and Belfrage, P.: Effects of ethanol intake on lipoprotein lipase activity in adipose tissue of fasting subjects. *Lipids* **13**:433–437, 1978.

Norbeck, H.-E., Walldius, G., Carlson, L. A., and Ideström, C.-M.: Serum lipoprotein abnormalities and adipose tissue metabolism in severe alcoholism. In: *Metabolic Effects of Alco-*

hol (P. Avogaro, C. R. Sirtori, and E. Tremoli, eds.), The Netherlands, Elsevier/North-Holland Biomedical Press, pp. 215–225, 1979.

Nordmann, R., Ribiere, C., Rouach, H., and Nordmann, J.: Paradoxical effects of pyrazole on acute ethanol-induced fatty liver. *Rev. Eur. Etud. Clin. Biol.* **17**:592–596, 1972.

Ockner, R. K., Hughes, F. B., and Isselbacher, K. J.: Very low density lipoproteins in intestinal lymph: Origin, composition and role in lipid transport in the fasting state. *J. Clin. Invest.* **48**:2079–2088, 1969.

Ockner, R. K., Mistilis, S. P., Poppenhausen, R. B., and Stiehl, A. F.: Ethanol induced fatty liver: Effect of intestinal lymph fistula. *Gastroenterology* **64**:603–609, 1973.

Ockner, R. K., Manning, J. A., and Kane, J. P.: Fatty acid binding protein: Isolation from rat liver, characterization, and immuno-chemical quantification. *J. Biol. Chem.* **257**:7872–7878, 1982.

Olson, R. E.: In: *Modern Nutrition in Health and Disease* (M. G. Wohl and R. S. Goodhart, eds.), Philadelphia, Lea and Febiger, pp. 1037–1050, 1964.

Ontko, J. A.: Effects of ethanol on the metabolism of free fatty acids in isolated liver cells. *J. Lipid Res.* **14**:78–86, 1973.

Ostrander, L. D., Jr., Johnson, B. C., and Block, W. D.: Relation-ship of alcohol intake to lipid levels in men from a general population. *Circulation* **44**:11–59, 1971.

Ostrander, L. D., Jr., Lamphiear, D. E., Block, W. D., Johnson, B. C., Ravenscroft, C., and Epstein, F. H.: Relationship of serum lipid concentrations to alcohol consumption. *Arch. Intern. Med.* **134**:451–456, 1974.

Papadopoulus, N. M., and Charles, M. A.: Serum lipoprotein patterns in liver disease. *Proc. Soc. Exp. Biol. Med.* **134**:797–799, 1970.

Patek, A. J., and Earampamoorthy, S.: Serum lipids in alcoholic patients with and without cirrhosis of the liver, with particular reference to endogenous familial hypertriglyceridemia. *Am. J. Clin. Nutr.* **29**:1122–1126, 1976.

Pell, S., and D'Alonzo, C. A.: A five-year mortality study of alcoholics. *J. Occup. Med.* **15**:120–125, 1973.

Phillips, G. B., and Davidson, C. S.: Nutritional aspects of cirrhosis in alcoholism: Effect of a purified diet supplemented with choline. *Ann. N.Y. Acad. Sci.* **57**:812–830, 1954.

Pignon, J. P., Bailey, N. C., Baraona, E., and Lieber, C. S.: Fatty acid binding protein: A major contributor to the ethanol-induced increase in liver cytosolic proteins in the rat. *Hepatology* **7**:865–871, 1987.

Poggi, M., and Di Luzio, N. R.: The role of liver and adipose tissue in the pathogenesis of the ethanol induced fatty liver. *J. Lipid Res.* **5**:437–441, 1964.

Polokoff, M. A., and Bell, R. M.: Limited palmitoyl-CoA pene-tration into microsomal vesicles as evidenced by a highly latent ethanol acyltransferase activity. *J. Biol. Chem.* **253**:7173–7178, 1978.

Polokoff, M. A., Simon, T. J., Harris, R. A., Simon, F. R., and Iwahashi, M.: Chronic ethanol increases liver plasma mem-brane fluidity. *Biochemistry* **24**:3114–3120, 1985.

Popham, R. E., Schmidt, W., and Israel, Y.: Variation in mortality from ischemic heart disease in relation to alcohol and milk consumption. *Med. Hypotheses* **12**:321–329, 1983.

Popper, H., and Lieber, C. S.: Histogenesis of alcoholic fibrosis and cirrhosis in the baboon. *Am. J. Pathol.* **98**:695–716, 1980.

Porta, E. A., Hartroft, W. S., and de la Iglesia, F. A.: Hepatic changes associated with chronic alcoholism in rats. *Lab. Invest.* **14**:1437–1455, 1965.

Post, J., Benton, J. G., Breakstone, R., and Hoffman, J.: Effects of diet and choline on fatty infiltration of the human liver. *Gastroenterology* **20**:403–410, 1952.

Poynard, T., Abella, A., Pignon, J. P., Naveau, S., Poitrine, A., Agostini, H., Zourabichvili, O., Leluc, R., and Chaput, J. C.: Apolipoprotein A-I (APO-A-I) and high-density-lipoprotein cholesterol (DLC) concentrations are elevated in alcoholic patients with liver steatosis. *Gastroenterology* **88**:1685 (ab-stract), 1985.

Poynard, T., Abella, A., Pignon, J.-P., Naveau, S., Leluc, R., and Chaput, J.-C.: Apolipoprotein AI and alcoholic liver disease. *Hepatology* **6**:1391–1395, 1986.

Prancan, A. V., and Nakano, J.: Effect of pyrazole on conversion of ethanol and acetate into lipids in rat liver. *Res. Commun. Chem. Pathol. Pharmacol.* **4**:181–191, 1972.

Pritchard, P. H., Bowley, M., Burditt, S. D., Cooling, J., Glenny, H. D., Lawson, N., Sturton, R. G., and Brindley, D. N.: The effects of acute ethanol feeding and of chronic benfluorex administration on the activities of some enzymes of glycero-lipid synthesis in rat liver and adipose tissue. *Biochem. J.* **166**:639–642, 1977.

Puchois, P., Fontan, M., Gentilini, J.-L., Gelez, P., and Fruchard, J. C.: Serum apolipoprotein A-II, a biochemical indicator of alcohol abuse. *Clin. Chim. Acta* **185**:185–189, 1984.

Radomski, M. W., and Wood, J. D.: The lipotropic action of cold. I. The influence of cold and choline deficiency on liver lipids of rats at different intakes of dietary methionine. *Can. J. Physiol. Pharmacol.* **42**:679–777, 1964.

Rao, G. A., Lew, G., and Larkin, E. C.: Alcohol ingestion and levels of hepatic fatty acid synthetase and stearoyl-CoA de-saturase activities in rats. *Lipids* **19**:151–153, 1984.

Rebouças, G., and Isselbacher, K. J.: Studies on the pathogenesis of the ethanol-induced fatty liver. 1. Synthesis and oxidation of fatty acids by the liver. *J. Clin. Invest.* **40**:1355–1362, 1961.

Redgrave, T. G., and Martin, G.: Effects of chronic ethanol consumption on the catabolism of chylomicron triacyl-glycerol and cholesteryl ester in the rat. *Atherosclerosis* **28**:69–80, 1977.

Riesen, W. F., Mordasini, R., Salzmann, C., Theler, A., and Gurtner, H. P.: Apoproteins and lipids as discriminators of severity of coronary heart disease. *Atherosclerosis* **37**:157–162, 1980.

Reitz, R. C.: A possible mechanism for the peroxidation of lipids due to chronic ethanol ingestion. *Biochim. Biophys. Acta* **380**:145–154, 1975.

Rottenberg, H., Waring, A., and Rubin, E.: Tolerance and cross-tolerance in chronic alcoholics: Reduced membrane binding of ethanol and other drugs. *Science* **213**:583–585, 1981.

Rouach, H., Clement, M., Orfanelli, M. T., Janvier, B., and Nordmann, R.: Fatty acid composition of rat liver mito-chondrial phospholipids during ethanol inhalation. *Biochim. Biophys. Acta* **795**:125–129, 1984.

Rubin, E., and Lieber, C. S.: Early fine structural changes in the human liver induced by alcohol. *Gastroenterology* **52:**1–13, 1967.

Rubin, E., and Lieber, C. S.: Alcohol induced hepatic injury in non-alcoholic volunteers. *N. Engl. J. Med.* **278:**869–876, 1968.

Rubin, E., Beattie, D. S., Toth, A., and Lieber, C. S.: Structural and functional effects of ethanol on hepatic mitochondria. *Fed. Proc.* **31:**131–140, 1972.

Rudel, L. L., Leathers, C. W., Bond, M. G., and Bullock, B. C.: Dietary ethanol-induced modifications in hyperlipoproteinemia and atherosclerosis in non-human primates (Macaca nemestrina). *Atherosclerosis* **1:**144–155, 1981.

Sabesin, S. M., and Weidman, S. W.: Lipoprotein profiles in chronic alcoholics: Use of high-density lipoprotein subspecies levels to differentiate subpopulations. *Hepatology* **4:**737–738, 1984.

Sabesin, S. M., Hawkins, H. L., Kuiken, L., and Ragland, J. B.: Abnormal plasma lipoproteins and lecithin-cholesterol acyltransferase deficiency in alcoholic liver disease. *Gastroenterology* **72:**510–518, 1977.

Sachan, D. S., Rhew, T. H., and Ruark, R. A.: Ameliorating effects of carnitine and its precursors on alcohol-induced fatty liver. *Am. J. Clin. Nutr.* **39:**738–744, 1984.

Salaspuro, M. P., Ross, W. A., Jayatilleke, E., Shaw, S., and Lieber, C. S.: Attenuation of the ethanol induced hepatic redox change after chronic alcohol consumption in baboons: Metabolic consequences *in vivo* and *in vitro*. *Hepatology* **1:**33–38, 1981.

Sanchez-Graig, M., and Annis, H. M.: Gamma-glutamyl transpeptidase and high-density lipoprotein cholesterol in male problem drinkers: Advantages of a composite index for predicting alcohol consumption. *Alcoholism. Clin. Exp. Res.* **5:**540–544, 1981.

Sane, T., Nikkilä, E. A., Taskinen, M.-R., Välimäki, M., and Ylikahri, R.: Accelerated turnover of very low density lipoprotein triglycerides in chronic alcohol users: A possible mechanism for the up-regulation of high density lipoprotein by ethanol. *Atherosclerosis* **53:**185–193, 1984.

Savolainen, M. J.: Stimulation of hepatic phosphatidate phosphohydrolase activity by a single dose of ethanol. *Biochem. Biophys. Res. Commun.* **75:**511–518, 1977.

Savolainen, M. J., and Hassinen, I. E.: Mechanisms for the effects of ethanol on hepatic phosphatidate phosphohydrolase. *Biochem. J.* **176:**885–892, 1978.

Savolainen, M. J., and Hassinen, I. E.: Effect of ethanol on hepatic phosphatidate phosphohydrolase: dose-dependent enzyme induction and its abolition by adrenalectomy and pyrazole treatment. *Arch. Biochem. Biophys.* **201:**640–645, 1980.

Savolainen, M. J., Hiltunen, J. K., and Hassinen, I. E.: Effect of prolonged ethanol ingestion on hepatic lipogenesis and related enzyme activities. *Biochem. J.* **164:**169–177, 1977a.

Savolainen, M. J., Jauhonen, V. P., and Hassinen, I. E.: Effects of clofibrate on ethanol-induced modifications in liver and adipose tissue metabolism: Role of hepatic redox state and hormonal mechanisms. *Biochem. Pharmacol.* **26:**425–431, 1977b.

Savolainen, M. J., Baraona, E., Pikkarainen, P., and Lieber, C. S.:

Hepatic triacylglycerol synthesizing activity during progression of alcoholic liver injury in the baboon. *J. Lipid Res.* **25:**813–820, 1984.

Savolainen, M., Baraona, E., Leo, M. A., and Lieber, C. S.: Pathogenesis of the hypertriglyceridemia at early stages of alcoholic liver injury in the baboon. *J. Lipid Res.* **27:**1073–1083, 1986.

Savolainen, M. J., Baraona, E., and Lieber, C. S.: Acetaldehyde binding increases the catabolism of rat serum low-density lipoproteins. *Life Sci.* **40:**841–846, 1987.

Schaefer, E. J., and Levy, R. I.: Pathogenesis and management of lipoprotein disorders. *N. Engl. J. Med.* **312:**1300–1310, 1985.

Schapiro, R. H., Scheig, R. L., Drummey, G. D., Mendelson, J. H., and Isselbacher, K. J.: Effect of prolonged ethanol ingestion on transport and metabolism of lipids in man. *N. Engl. J. Med.* **272:**610–615, 1965.

Schneider, J., Panne, E., Braun, H., Mordasi, R., and Kaffarnik, H.: Ethanol-induced hyperlipoproteinemia. *J. Lab. Clin. Med.* **101:**114–122, 1983.

Schüller, A., Moscat, J., Diez, E., Fernandez-Checa, J. C., Gavilanes, F. C., and Municio, A. M.: Functional properties of isolated hepatocytes from ethanol-treated rat liver. *Hepatology* **5:**677–682, 1985.

Seakins, A., and Waterlow, J. C.: Effect of a low-protein diet on the incorporation of amino acids into rat serum lipoproteins. *Biochem. J.* **129:**793–795, 1972.

Selmer, J., and Grunnet, N.: Ethanol metabolism and lipid synthesis by isolated liver cells from fed rats. *Biochim. Biophys. Acta* **428:**123–137, 1976.

Shaper, A. G., Wannamethee, G., and Walker, M.: Alcohol and mortality in British men: Explaining the U-shaped curve. *Lancet* **2:**1267–1273, 1988.

Shaw, W.: Possible role of lysolecithins and nonesterified fatty acids in the pathogenesis of Reye's syndrome, sudden infant death syndrome, acute pancreatitis, and diabetic ketoacidosis. *Clin. Chem.* **31:**1109–1115, 1985.

Sigurdsson, G., Nicoll, A., and Lewis, B.: The metabolism of low density lipoprotein in endogenous hypertriglyceridaemia. *Eur. J. Clin. Invest.* **6:**167–177, 1976.

Sirtori, C. R., Agradi, E., and Mariani, C.: Hyperlipoproteinemia in alcoholic subjects. *Pharmacol. Res. Commun.* **5:**81–85, 1973.

Sorrell, M. F., Baker, H., Tuma, D. J., Frank, O. M., and Barak, A. J.: Potentiation of ethanol fatty liver in rats by chronic administration of nicotinic acid. *Biochim. Biophys. Acta* **450:**231–238, 1976.

Spritz, N., and Lieber, C. S.: Decrease of ethanol-induced fatty liver by ethyl-α-p-chlorophenoxyisobutyrate. *Proc. Soc. Exp. Biol. Med.* **121:**147–149, 1966.

Stanko, R. T., Mendelow, H., Shinozuka, H., and Adibi, S. A.: Prevention of alcohol-induced fatty liver by natural metabolites and riboflavin. *J. Lab. Clin. Med.* **91:**228–235, 1978.

Stein, Y., and Schapiro, B.: Glyceride synthesis by microsome fractions of rat liver. *Biochim. Biophys. Acta* **30:**271–277, 1958.

Strakosch, C. R., Jeffreys, D. B., and Keen, H.: Blockade of chlorpropamide alcohol flush by aspirin. *Lancet* **1:**394–396, 1980.

Sturton, R. G., Pritchard, P. H., Han, L.-Y., and Brindley, D. N.: The involvement of phosphatidate phosphohydrolase and phospholipase A activities in the control of hepatic glycerolipid synthesis. *Biochem. J.* **174**:667–670, 1978.

Sturton, R. G., Butterwith, S. C., Burditt, S. L., and Brindley, D. N.: Effects of starvation, corticotropin injection and ethanol feeding on the activity and amount of phosphatidate phosphohydrolase in rat liver. *FEBS Lett.* **126**:297–300, 1981.

Summerskill, W. H. J., Wolfe, S. J., and Davidson, C. S.: Response to alcohol in chronic alcoholics with liver disease: Clinical, pathological and metabolic changes. *Lancet* **1**:335–343, 1957.

Sun, G. Y., and Sun, A. Y.: Ethanol and membrane lipids. *Alcoholism: Clin. Exp. Res.* **9**:164–180, 1985.

Takeuchi, N., Ito, M., and Yamamura, Y.: Esterification of cholesterol and hydrolysis of cholesteryl ester in alcohol induced fatty liver of rats. *Lipids* **9**:353–357, 1974.

Talbot, G. D., and Keating, B. M.: Effects of preprandial whiskey on postalimentary lipemia. *Geriatrics* **17**:802–808, 1962.

Tappel, A. L.: Lipid peroxidation damage to cell components. *Fed. Proc.* **33**:1870–1874, 1973.

Taraschi, T. F., and Rubin, E.: Biology of disease. *Lab. Invest.* **52**:120–131, 1985.

Taskinen, M.-R., and Nikkilä, E. A.: Nocturnal hypertriglyceridemia and hyperinsulinemia following moderate evening intake of alcohol. *Acta Med. Scand.* **202**:173–177, 1977.

Taskinen, M.-R., Välimäki, M., Nikkilä, E. A., Kuusi, T., Ehnholm, C., and Ylikahri, R.: High density lipoprotein subfractions and postheparin plasma lipases in alcoholic men before and after ethanol withdrawal. *Metabolism* **31**:1168–1174, 1982.

Taskinen, M.-R., Välimäki, M., Nikkilä, E. A., Kuusi, T., and Ylikahri, R.: Sequence of alcohol-induced initial changes in plasma lipoproteins. *Metabolism* **34**:112–119, 1985.

Thornton, J., Symes, C., and Heaton, K.: Moderate alcohol intake reduces bile cholesterol saturation and raises HDL cholesterol. *Lancet* **2**:819–822, 1983.

Tijburg, L. B. M., Maquedano, A., Bijleveld, C., Guzman, M., and Geelen, M. J. H.: Effects of ethanol feeding on hepatic lipid synthesis. *Arch. Biochem. Biophys.* **267**:568–579, 1988.

Topping, D. L., Weller, R. A., Nader, C. J., Calvert, G. D., and Illman, R. J.: Adaptive effects of dietary ethanol in the pig: Changes in plasma high-density lipoproteins and fecal steroid excretion and mutagenicity. *Am. J. Clin. Nutr.* **36**:245–250, 1982.

Truitt, E. B., Gaynor, C. R., and Mehl, D. L.: Aspirin attenuation of alcohol-induced flushing and intoxication in Oriental and Occidental subjects. *Alcohol Alcohol. Suppl.* **1**:595–599, 1987.

Uthus, E. O., Skurdal, D. N., and Cornatzer, W. E.: Effect of ethanol ingestion on choline phosphotransferase and phosphatidyl ethanolamine methyltransferase activities in liver microsomes. *Lipids* **11**:641–644, 1976.

Vasdev, S. C., Subrahmanyam, D., Chakravarti, R. N., and Wahi, P. L.: Effect of chronic ethanol feeding on the major lipids of red blood cells, liver, and heart of rhesus monkey. *Biochim. Biophys. Acta* **369**:323–330, 1974.

Venkatesan, S., Leung, N. W. Y., and Peters, T. J.: Fatty acid synthesis *in vitro* by liver tissue from control subjects and patients with alcoholic liver disease. *Clin. Sci.* **71**:723–728, 1986.

Verdy, M., and Gattereau, A.: Ethanol, lipase activity, serum-lipid level. *Am. J. Clin. Nutr.* **20**:997–1004, 1967.

Volwiler, W., Jones, C. M., and Mallory, T. B.: Criteria for the measurement of result of treatment in fatty cirrhosis. *Gastroenterology* **11**:164–182, 1948.

Von Platt, U., Stein, U., and Heissmeyer, H.: Catabolic mucopolysaccharide-protein metabolism in rat liver following administration of ethanol. *Z. Klin. Chem. Klin. Biochem.* **9**:126–129, 1971.

Wang, D. L., and Reitz, R. C.: Ethanol ingestion and polyunsaturated fatty acids: Effects on the acyl-CoA desaturases. *Alcoholism: Clin. Exp. Res.* **7**:220–226, 1983.

Warembourg, H., Biserte, G., Jaillard, J., Sezille, G., and Scherpereel, P.: Changes in serum and hepatic lipids induced by alcohol during chronic ethyl alcohol poisoning in rabbits. *Rev. Eur. Etud. Clin. Biol.* **15**:649–653, 1970.

Waring, A. J., Rottenberg, H., Ohnishi, T., and Rubin, E.: Membranes and phospholipids of liver mitochondria from chronic alcoholic rats are resistant to membrane disordering by alcohol. *Proc. Natl. Acad. Sci. U.S.A.* **78**:2582–2586, 1981.

Weidman, S. W., Ragland, J. B., and Sabesin, S. M.: Plasma lipoprotein composition in alcoholic hepatitis: Accumulation of apolipoprotein E-rich high density lipoprotein and preferential reappearance of "light"-HDL during partial recovery. *J. Lipid Res.* **23**:556–569, 1982.

Willet, W., Hennekens, C. H., Siegel, A. J., Adner, M. M., and Castelli, W. P.: Alcohol consumption and high density lipoprotein cholesterol in marathon runners. *N. Engl. J. Med.* **303**:1159–1161, 1980.

Williams, P. T., Kraus, R. M., Wood, P. D., Albers, J. J., Dreon, D., and Ellsworth, N.: Associations of diet and alcohol intake with high-density lipoprotein subclasses. *Metabolism* **34**:524–530, 1985.

Wilson, D. E., Schreibman, P. H., Brewster, A. C., and Arky, R. A.: The enhancement of alimentary lipemia by ethanol in man. *J. Lab. Clin. Med.* **75**:264–274, 1970.

Wilson, J. S., Colley, P. W., Sasula, L., Pirola, R. C., Chapman, B. A., and Somer, J. B.: Alcohol causes a fatty pancreas: A rat model of ethanol-induced pancreatic steatosis. *Alcoholism: Clin. Exp. Res.* **6**:117–121, 1982.

Windmueller, H. G., Herbert, P. N., and Levy, R. I.: Biosynthesis of lymph and plasma lipoprotein apoproteins by isolated perfused rat liver and intestine. *J. Lipid Res.* **14**:215–223, 1973.

Wolfe, B. M., Havel, J. R., Marliss, E. B., Kane, J. P., Seymour, J., and Ahuja, S. P.: Effects of a 3-day fast and of ethanol on splanchnic metabolism of FFA, amino acids, and carbohydrates in healthy young men. *J. Clin. Invest.* **57**:329–340, 1976.

Yamada, S., and Lieber, C. S.: Decrease in microviscosity and cholesterol content of rat liver plasma membranes after chronic ethanol feeding. *J. Clin. Invest.* **74**:2285–2289, 1984.

Yano, K., Rhoads, G. G., and Kagan, A.: Coffee, alcohol, and risk of coronary heart disease among Japanese men living in Hawaii. *N. Engl. J. Med.* **297:**405–409, 1977.

Ylikahri, R.: Ethanol-induced changes in hepatic alpha-glycerophosphate and triglyceride concentrations in normal and thyroxine-treated rats. *Metabolism* **19:**1036–1045, 1970.

Zeichner, A., Feuerstein, M., Swartzman, L., and Reznick, E.: Acute effects of alcohol on cardiovascular reactivity to stress in type A (coronary prone) businessmen. In: *Stress and Alcohol Use* (L. A. Pohorecky, and J. Brick, eds.), New York, Elsevier Science, pp. 353–368, 1983.

Zieve, L.: Jaundice, hyperlipemia, and hemolytic anemia: A heretofore unrecognized syndrome associated with alcoholic fatty liver and cirrhosis. *Ann. Intern. Med.* **48:**471–496, 1958.

Zysset, T., Polokoff, M. A., and Simon, F. R.: Effect of chronic ethanol administration on enzyme and lipid properties of liver plasma membranes in long and short sleep mice. *Hepatology* **5:**531–537, 1985.

5

Effects of Ethanol on Amino Acid and Protein Metabolism

Siamak A. Adibi, Enrique Baraona, and Charles S. Lieber

Alcohol interferes with amino acid and protein metabolism at various levels in different tissues. Effects of ethanol on the digestion of exogenous proteins and the absorption of their constituent amino acids is discussed in Chapter 10. In this chapter we focus primarily on the effects of ethanol on amino acid metabolism, the endogenous synthesis of proteins (primarily in the liver), their degradation, and their secretion from the liver into the bloodstream. Hepatic consequences of the alterations in these processes are emphasized.

5.1. Effects of Ethanol on Nitrogen Balance and Body Protein Composition

Urinary nitrogen excretion and nitrogen balance have been commonly used as parameters of whole-body protein metabolism. Klatskin (1961) and Rodrigo *et al.* (1971) have shown that ethanol feeding increases urinary nitrogen excretion and impairs nitrogen balance in rats. The clinical relevance of these observations has been investigated by McDonald and Margen (1976). In this investigation, six healthy male volunteers who were maintained in metabolic ward conditions were served 250 g of a California wine (9.3 g% alcohol) with each meal for a period of 18 days. The formula diet provided adequate amounts of calories, protein, and other nutri-

ents. The total dose of alcohol was 1.28–1.48 g/kg of body weight and 22–25% of total calories, depending on the subject's size. The caloric allowances were adjusted to keep body weight constant by adding to the basic formula another formula containing carbohydrate and fat but no protein. For the control, the same wine, but dealcoholized, was given in the same amount of 250 g with each meal to the same subjects, again for a period of 18 days. During the control period, an increased amount of the carbohydrate–fat formula was substituted for the calories contributed by alcohol. The conditions of the experiment were otherwise similar to those of the alcohol period.

The wine drinking significantly ($P < 0.01$) increased urinary excretion (grams/day) of total nitrogen (11.16 ± 1.34 versus 10.08 ± 0.84), urea nitrogen (9.03 ± 1.25 versus 7.51 ± 0.49), and uric acid (0.59 ± 0.11 versus 0.51 ± 0.10). Furthermore, the subjects were in positive nitrogen balance during the control period but in negative nitrogen balance during the wine period. The fact that ethanol consumption has a deleterious effect on nitrogen balance in humans has recently been confirmed by the results of the studies of Reinus *et al.* (1989). These investigators, using a nasogastric formula infusion method, evaluated the effect of ethanol relative to that of glucose on nitrogen balance in a group of alcoholic men. Before inclusion of ethanol in the diet, each subject was maintained in neutral or slightly positive energy and protein balance by continuous infusion of a complete liquid formula diet. When the glucose content of a complete liquid formula diet was replaced with ethanol (50% of total energy), the nitrogen balance became negative. Removal of ethanol from the diet resulted in return of nitrogen balance to a positive value.

The above studies establish that ethanol consumption increases protein catabolism in the body but does not show which tissues participate in this response. This problem was investigated in rats (Stanko *et al.*, 1979). The control group was fed for 30 days a liquid diet with 28% fat, 15% protein, and 57% carbohydrate. The experimental group was fed the same diet except with partial isocaloric replacement of the carbohydrate content of the diet with 95% ethanol, which resulted in the following caloric composition: 28% fat, 15% protein, 23% carbohydrate, and 34% ethanol. Despite identical caloric intake by control and ethanol-fed rats, chronic ethanol ingestion for 30 days resulted in a significantly smaller gain in body weight (Stanko *et al.*, 1979). Although there was no significant difference between either the liver or kidney weights of control and ethanol-fed rats, the muscle and intestinal mucosa weights of ethanol-fed rats were significantly smaller (Stanko *et al.*, 1979). The protein concentrations of liver, muscle and kidney were not significantly affected by ethanol but the protein concentration in intestinal mucosa was significantly reduced (Table 5.1). The total protein content of liver and kidney was not significantly affected by

ethanol, but the total protein content of the muscle and intestinal mucosa was significantly reduced (Table 5.1).

The above results indicate that chronic ethanol ingestion in rats has a varied effect on protein metabolism in tissues. Although tissue growth, as assessed by total protein content, is not affected in the liver and kidney, it is reduced in the muscle and intestinal mucosa. In view of the fact that muscle mass accounts for 45% of body weight, the impairment in muscle growth appears to be a significant component of the smaller body weight gain by ethanol-fed rats (Lieber *et al.*, 1965; Stanko *et al.*, 1979). Preedy and Peters (1988) have recently confirmed that chronic ethanol consumption decreases muscle protein content and have also shown that ethanol reduces muscle protein synthesis. The relevance of these observations in rats to the clinical situation is suggested by the muscle atrophy that is a feature of alcoholic myopathy (Perkoff *et al.*, 1966; Rubin, 1979). Although quantitatively the effect of ethanol appears to be greatest on the skeletal muscle, qualitatively the effect appears to be greatest on the intestinal mucosa. As shown in Table 5.1, intestinal mucosa was the only tissue that displayed a reduction in protein concentration with ethanol feeding. The study of amino acid pools in tissues showed that ethanol has a greater catabolic effect on intestinal mucosa than on other tissues. Two factors may account for the greater sensitivity of intestinal mucosa than liver, muscle, or kidney to the effect of ethanol. First, since ethanol was administered orally, intestinal mucosa was exposed to a greater concentration of ethanol than other tissues. Second, the protein turnover is more rapid in intestinal mucosa than other tissues.

5.2. Effects of Ethanol on Amino Acids

5.2.1. Amino Acid Pools

Alterations in amino acid and protein metabolism are usually reflected in the composition of amino acid pools in tissues (Adibi, 1971; Adibi *et al.*, 1973; Vazquez *et al.*, 1988). Amino acid concentrations were investigated in plasma, liver, intestine, kidney, and muscle of rats fed ethanol for 30 days. The design of the experiment was similar to the one described in Section 5.1 for the studies of effect of ethanol consumption on concentration of protein in rat tissues. Among the 15 individual amino acids studied (aspartic acid, threonine, serine, asparagine, glutamine, glutamic acid, glycine, alanine, α-amino-*n*-butyric acid, valine, methionine, isoleucine,

TABLE 5.1. Protein Composition of Organs[a]

Organ	Control Group	Ethanol Group
Liver		
Protein concentration (mg/g)	241 ± 6 (7)	239 ± 4 (11)
Protein content (g)	2.64 ± 0.09	2.56 ± 0.16
Gastrocnemius muscle		
Protein concentration (mg/g)	221 ± 5 (10)	223 ± 5 (12)
Protein content (g)	0.38 ± 0.01	0.31 ± 0.02[c]
Kidney		
Protein concentration (mg/g)	184 ± 6 (8)	199 ± 5 (15)
Protein content (g)	0.20 ± 0.01	0.20 ± 0.01
Intestinal mucosa		
Protein concentration (mg/g)	145 ± 3 (5)	132 ± 4[b] (6)
Protein content (g)	0.71 ± 0.03	0.52 ± 0.03[b]

[a]Rats were fed for 30 days either the control liquid diet or the ethanol diet before protein composition of organs was determined. The ethanol diet was similar to the control diet except for partial isocaloric replacement of carbohydrate content with ethanol. The calorie intake was maintained similar for both groups. Numbers in parentheses denote numbers of animals in each group.
[b]$p < 0.05$.
[c]$p < 0.02$.

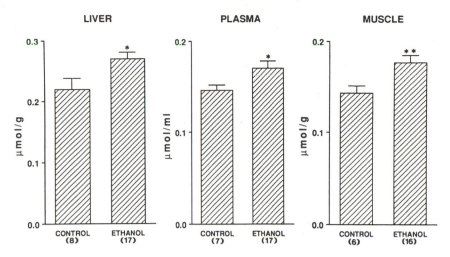

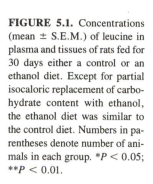

FIGURE 5.1. Concentrations (mean ± S.E.M.) of leucine in plasma and tissues of rats fed for 30 days either a control or an ethanol diet. Except for partial isocaloric replacement of carbohydrate content with ethanol, the ethanol diet was similar to the control diet. Numbers in parentheses denote number of animals in each group. *P < 0.05; **P < 0.01.

leucine, tyrosine, and phenylalanine), chronic ethanol ingestion had the most consistent and significant effect on plasma and tissue concentrations of three amino acids: leucine, alanine, and α-amino-*n*-butyric acid. In general, changes in plasma amino acid concentrations more closely reflected changes in amino acid concentrations of liver and muscle than those of kidney and intestinal mucosa. Ethanol feeding significantly increased plasma, muscle, and liver concentrations of leucine (Fig. 5.1). There were similar trends in plasma and tissue concentrations of the other two branched-chain amino acids (isoleucine and valine), but the increases reached statistical significance only in the muscle for valine and in plasma for isoleucine. In contrast to leucine, plasma and liver concentrations of alanine were

markedly decreased while the muscle concentration remained unaffected in the ethanol-fed rats (Fig. 5.2). Ethanol ingestion significantly increased concentration of α-amino-*n*-butyric acid in both plasma and muscle without affecting the concentration of this amino acid in the liver (Fig. 5.3). In addition to the changes described above, chronic ethanol ingestion caused significant increases in concentrations of glycine and phenylalanine in the muscle and significant increases in concentrations of glutamic acid, glutamine, asparagine, and phenylalanine in the liver (Stanko *et al.*, 1979).

Ethanol caused complex changes in amino acid concentrations in the kidney (Stanko *et al.*, 1979): some were increased (aspartate, threonine, asparagine, glutamine, α-amino-*n*-butyrate); some were decreased

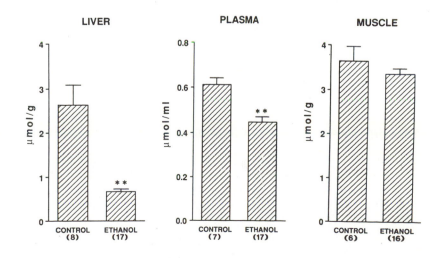

FIGURE 5.2. Concentration (mean ± S.E.M.) of alanine in plasma and tissues of rats fed for 30 days either a control or an ethanol diet. Except for partial isocaloric replacement of carbohydrate content with ethanol, the ethanol diet was similar to the control diet. Numbers in parentheses denote number of animals in each group. *P < 0.05; **P < 0.01.

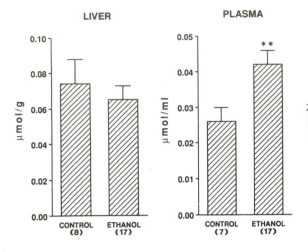

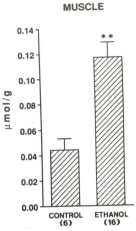

FIGURE 5.3. Concentration (mean ± S.E.M.) of α-amino-*n*-butyrate in plasma and tissues of rats fed for 30 days either a control or an ethanol diet. Except for partial isocaloric replacement of carbohydrate content with ethanol, the ethanol diet was similar to the control diet. Numbers in parentheses denote number of animals in each group. *$P < 0.05$; **$P < 0.01$.

(serine, alanine, methionine, and phenylalanine); and some remained unchanged (glutamate, glycine, valine, isoleucine, leucine, and tyrosine). Ethanol significantly increased concentrations of most amino acids in intestinal mucosa (Stanko *et al.*, 1979).

No information is available on the effect of ethanol consumption on concentrations of amino acids in human tissues, but some data are available on plasma that support the observations in rats. Shaw and Lieber (1978) studied the effect of chronic alcohol consumption on plasma concentrations of branched-chain amino acids and α-amino-*n*-butyric acid in humans. They found significant increases in plasma concentrations of branched-chain amino acids and α-amino-*n*-butyric acid (Shaw and Lieber, 1978). Avogaro *et al.* (1986) investigated the effects of two isocaloric diets, one including alcohol, on concentrations of branched-chain amino acids and alanine in alcoholics. They found that the diet with alcohol induced sustained increases in concentrations of branched-chain amino acids, persisting even after the postprandial phase, and a decrease in alanine concentration.

In summary, the most prominent finding after chronic ethanol consumption is the increased concentration of branched-chain amino acids in the muscle and decreased concentration of alanine in the liver.

5.2.2. Amino Acid Transport

The first step in cellular metabolism of amino acids is the transport of amino acids across the cell membrane. Chambers *et al.* (1966) found that addition of ethanol to the perfusion medium inhibits the uptake of α-aminoisobutyric acid (AIB) by the isolated perfused rat liver. Using primary cultures of adult rat hepatocytes, Dorio *et al.* (1984) investigated the effect of 100 mM ethanol on various neutral amino acid transport systems. The neutral amino acid transport systems characterized in rat hepatocytes include systems designated as A, L, ASC, and N. Although these systems have overlapping substrate specificities, they have preference for certain amino acids. The A system transports short-chain amino acids including alanine and glycine and the synthetic model amino acids AIB and MeAIB (methyl-α-amino-isobutyric acid). The L system transports branched-chain amino acids such as leucine, isoleucine, and valine. The amino acids transported by the ASC system include cysteine, and those transported by the N system include glutamine. Dorio *et al.* (1984) found that ethanol exposure for 20 hr selectively decreases amino acid uptake by the A and the N system by 40–70% but had no significant effect on the ASC and the L systems. The decrease in the A system was significant after 3 hr of ethanol exposure, and the activity was not affected by the presence or absence of ethanol during the uptake measurement. Kinetic analysis showed that ethanol treatment affected predominantly the high-affinity component of the A system activity by decreasing the apparent V_{max} without significantly changing the apparent K_m. Ethanol treatment did not prevent the cells from increasing A system activity in response to insulin and glucagon, but the magnitude of hormone-stimulated uptake was reduced. In contrast to these results, Heitman *et al.* (1987) found that exposure to ethanol (37 and 85 mM) for 48 hr stimulated AIB uptake by fetal hepatocytes. The onset of this stimulation of net uptake was progressive and required in excess of 6 hr of contact with ethanol. This ethanol stimulation of AIB uptake persisted for at least 24 hr following ethanol withdrawal. Although the conclusion of Heitman *et al.* (1987) appears contradictory to that of Dorio *et al.* (1984), actually both studies show similar findings if the results are expressed similarly. Dorio *et al.* expressed transport per cell protein, whereas Heitman *et al.* expressed transport per cell number. Heitman *et al.* (1987) found that ethanol exposure increased water and protein contents of hepatocytes. Consequently, when they calculated uptake per cell protein, they also found that ethanol inhibits amino acid transport.

The above studies were performed in intact cells, yet ethanol may also affect intracellular metabolism. Therefore, the effect of ethanol on amino acid transport may be a consequence of altered intracellular metabolism. Recently, Moseley and Murphy (1989) investigated the effect of ethanol on amino acid transport in basolateral liver plasma membrane vesicles. They found that ethanol produces a concentration-dependent and reversible inhibition of Na^+-independent leucine transport by system L. Initial rates of Na^+-dependent cysteine uptake were also inhibited in a concentration-dependent manner by ethanol treatment. No effect on alanine transport was observed under Na^+- equilibrated conditions, whereas initial rates of Na^+ flux were enhanced by ethanol treatment. Analysis of kinetic parameters demonstrated that ethanol decreases the V_{max} of the Na^+-dependent alanine uptake into basolateral plasma membrane vesicles without altering the K_m for alanine. From these results, Moseley and Murphy concluded that ethanol treatment indirectly decreases Na^+-dependent amino acid transport via dissipation of the electrochemical Na^+ gradient. In other words, the effect of ethanol appears to be a consequence of enhanced Na^+ flux rather than a direct effect on amino acid transport.

The above study examined the effect of acute ethanol exposure on membrane transport, which may be different from the effect of chronic ethanol consumption. To resolve this problem, Smith *et al.* (1990) investigated the effect of chronic ethanol consumption on amino acid uptake by basolateral liver plasma membrane vesicles. Rats were pair-fed for 6 weeks a diet containing 36% of calories as ethanol or a control diet in which ethanol was isocalorically replaced with carbohydrates. The chronic ethanol consumption reduced basolateral liver plasma membrane Na^+-dependent alanine transport activity by 36%. This reduction was caused primarily by impaired activity of amino acid transport system A. The response of system A to glucagon was reduced in the ethanol-fed rats, suggesting that impaired hormonal regulation is partially responsible for the lower system A activity. Kinetic analysis showed that ethanol consumption reduces the V_{max} of Na^+-dependent alanine transport without affecting K_m. The unexpected finding was that ethanol, in contrast to its effect on alanine uptake, increased Na^+-dependent glutamine uptake by 28%. In the absence of sodium, there was no difference in glutamine uptake by basolateral liver plasma membrane vesicles in control and ethanol-fed animals. The mechanism of differential effects of ethanol feeding on the two amino acid transport systems, namely, A and N, remains uncertain. However, Smith *et al.* speculated that the ethanol-fed animal compensates for the lower flux of amino acids through the A system by increased use of system N substrates, e.g., glutamine, to meet metabolic requirements.

A number of investigators using *in vitro* preparations of small intestine have reported that ethanol inhibits amino acid transport (e.g., Chang *et al.*, 1967). Green *et al.* (1981), studying the acute effect of ethanol on amino acid absorption by an intestinal perfusion technique in rats, have found a different result. When ethanol was perfused in concentrations of 0.5 or 1.0 M, concentrations that may be found in the gut lumen during drinking, there was no change in absorption of leucine, glycine, alanine, phenylalanine, lysine, or methionine (Green *et al.*, 1981). Even higher ethanol concentrations (1.3 and 1.6 M) did not inhibit absorption of leucine or methionine. However, when the concentration of ethanol in the perfusion solution was increased to 2 M or higher, amino acid absorption was reduced. This inhibition in absorption appeared to result from deleterious effects of unphysiological concentrations of ethanol on viability of intestinal mucosa.

Hajjar *et al.* (1981b), studying the effect of chronic ethanol consumption, also did not find an inhibition of amino acid absorption in rat intestine. They found that 7 weeks of treatment with a liquid diet in which 36% of the carbohydrate content was replaced isocalorically with ethanol results in shortening and thinning of the intestine. However, leucine absorption by the entire intestine was not significantly altered. In fact, leucine absorption, expressed per weight or per protein content of intestinal mucosa, was actually increased. Again, the explanation for these contradictory results remains uncertain, but differences in treatment and methods of investigation are probably involved.

Studies by Matthews and Adibi (1976) have shown that peptide absorption is a more important mechanism than amino acid absorption for intestinal assimilation of dietary proteins. Therefore, Hajjar *et al.* (1981a) also investigated the effect of ethanol on intestinal absorption of a dipeptide, glycylsarcosine, in rat intestine. Ethanol was administered either acutely (luminal perfusion of 0.7 M ethanol) or chronically (inclusion of ethanol in drinking water). The amount of ethanol consumed approximated 32% of daily caloric intake. Both total absorption per entire small intestine and specific absorption per milligram dry weight of mucosa were unaffected by ethanol. However, luminal appearance of glycine, an index of intracellular hydrolysis of glycylsarcosine, was significantly reduced. This was attributed to inhibition of cytosolic peptide hydrolases by ethanol (Hajjar *et al.*, 1981a). The fact that ethanol inhibits dipeptide hydrolysis has been shown by studies of Dinda and Beck (1982): the kinetic analysis indicated decreases in the V_{max} but no significant change in the K_m of peptide hydrolases. All of the above studies were performed in intact enterocytes. Beesley (1986), however, investigated the effect of acute ethanol exposure on amino acid transport by brush border membrane vesicles isolated from hamster jejunum. He found that ethanol, at concentrations found in the jejumun after moderate drinking, reduces Na$^+$-dependent uptake of alanine and phenylalanine. The inhibitory effect of ethanol was dose dependent and reversible. Furthermore, ethanol had no effect on the rate of uptake when vesicles were incubated with a KCl gradient or when NaCl was equilibrated across the vesicle membrane.

Finally, the fact that an alteration in amino acid transport by ethanol may have clinical relevance has been suggested by the results of the studies on the effect of ethanol in pregnant animals. Chronic ethanol consumption by pregnant women results in impairment of fetal growth and development of fetal alcohol syndrome. A number of investigators have found that chronic ethanol consumption in pregnant animals results in impaired amino acid uptake by placenta and in impaired maternal–fetal transfer of amino acids (Lin, 1981; Patwardhan *et al.*, 1981; Fisher *et al.*, 1981). These impairments in amino acid transport have been suggested to play a significant role in the development of fetal alcohol syndrome (Lin, 1981; Patwardhan *et al.*, 1981; Fisher *et al.*, 1981), as discussed in greater detail in Chapter 15.

However, a recent study (Schenker *et al.*, 1989) has questioned the results of an earlier study (Fisher *et al.*, 1981) that showed that exposure of human placenta to ethanol *in vitro* inhibits amino acid transport. Schenker *et al.* (1989) investigated the effect of brief exposure to ethanol on human placental transport of model amino

acids (α-aminoisobutyrate and cycloleucine) by two techniques—the perfused human placental cotyledon and human placental vesicle systems. They found no impairment of α-aminoisobutyrate transfer by ethanol using these techniques. However, ethanol caused a very small decrease of cycloleucine clearance by the perfused human placenta but not by the vesicles. From these studies Schenker *et al.* (1989) concluded that the transport of amino acids by the human placenta is relatively resistant to the inhibitory effect of ethanol. They attributed the previously demonstrated inhibitory effect of ethanol on amino acid transport by human placenta (Fisher *et al.*, 1981) to the use of pharmacological concentrations of ethanol. It is pertinent to note that the studies in experimental animals that have shown that ethanol inhibits placental transport of amino acids have employed chronic rather than acute ethanol exposure. Therefore, such an approach might be necessary to establish whether or not ethanol also has an adverse effect on placental transport of amino acids in humans.

5.2.3. Branched-Chain Amino Acids

Ward *et al.* (1985a) found in rats that the rate of leucine oxidation was not significantly altered after chronic administration of ethanol (20% v/v solution as drinking water for 28 days), but it was significantly decreased by a single acute dose of ethanol (9 g/kg body weight). They then investigated the effect of these treatments on the activity of key enzymes involved in leucine oxidation, namely, the activity of branched-chain amino acid transaminase and branched-chain ketoacid dehydrogenase. They found that the effects of acute and chronic ethanol administration on the activity of these enzymes are opposite to the effects of the same treatments on the rates of leucine oxidation described above. The acute administration of ethanol was without an effect on the activities of these enzymes in all tissues studied, but chronic administration of ethanol resulted in decreased activity of these enzymes in muscle and kidney. No explanation was offered for lack of correlation between effect of ethanol on rates of leucine oxidation and the activity of key enzymes involved in leucine oxidation.

Although the effect of ethanol consumption on oxidation of branched-chain amino acid is of great interest, the validity of results of the above studies remains uncertain. First, the method of ethanol administration was inclusion in drinking water. Previous experience has indicated that this method is not satisfactory; for example, rats ingest inadequate amounts of food and consequently become protein–calorie malnourished. In fact, this was observed by Ward *et al.* (1985a) in rats they studied. Second, the method of leucine oxidation was based on the determination of $^{14}CO_2$ production after a bolus injection of [^{14}C]leucine (Ward *et al.*, 1985a). This method of study of leucine oxidation is no longer considered adequate, since it does not take into account alteration in specific activity of leucine by treatment. Third, the methods used to assay enzyme activity were not optimal. For example, leucine was used to measure the activity of branched-chain ketoacid dehydrogenase (Ward *et al.*, 1985b). The substrate for this enzyme is not leucine but one of the ketoacids of branched-chain amino acids. Finally, no precautions were taken to prevent artificial activation of branched-chain ketoacid dehydrogenase *in vitro* (Ward *et al.*, 1985b). It is well established that branched-chain keto-acid dehydrogenase exists in both active (dephosphory-lated) and inactive (phosphorylated) forms (Paul and Adibi, 1982). Furthermore *in vitro* conditions of enzyme assay could influence the proportion of this enzyme in active and inactive forms. More recently, Baranyai and Blum (1989) investigated the effect of acute ethanol exposure (10 mM) on decarboxylation of α-ketoiso-caproate (the transamination product of leucine) by isolated hepatocytes from fed rats. They found $^{14}CO_2$ production from both [2-^{14}C]- and [U-^{14}C]ketoisocapro-ate to be decreased in ethanol-treated cells. However, the effect was not unique to ketoisocaproate, since $^{14}CO_2$ production from other substrates such as alanine, glu-tamine, and glucose was also decreased. Baranyai and Blum suggested an inhibition of the citric acid cycle by ethanol as the mechanism of reduced decarboxylation.

5.2.4. Tryptophan

Since the mid-1950s, a number of investigators have reported alterations in tryptophan oxidation in alcoholics or in experimental animals fed ethanol (for a previous review see Baraona *et al.*, 1982). There are two pathways in the oxidation of tryptophan. The major pathway is the oxidation of tryptophan to kynurenine, which is converted to 3-hydroxyanthranilic acid, nico-tinic acid, and other compounds. The liver tryptophan pyrrolase (tryptophan oxygenase) is believed to be the rate-limiting enzyme for this pathway. The minor path-way involves oxidation of tryptophan to 5-hydroxytryp-tophan and decarboxylation of this amino acid to 5-hy-droxytryptamine (serotonin).

A consequence of increased oxidation could be lowered plasma tryptophan concentration. In view of previous controversy on the effect of ethanol on plasma concentration of tryptophan, Branchey *et al.* (1981) investigated plasma tryptophan level in alcoholics and also in baboons and rats whose alcohol dietary intake was carefully monitored. They found that alcoholics have a lower plasma tryptophan concentration than con-trol subjects (56 ± 2 versus 78 ± 3 μM). However, their studies in baboons and rats showed no significant alter-ation in plasma tryptophan level. More recently, Buydens-Branchey *et al.* (1988) investigated the basal plasma tryptophan level of alcoholics after 3 days and after 1 month of abstinence. Plasma tryptophan was deter-mined both as free and as bound to plasma protein. They found basal level of total tryptophan to be 62 ± 9 μM after 3 days of sobriety versus 72 ± 2 μM 1 month later; free tryptophan values were 11 ± 1 μM versus 11 ± 3 μM. They also investigated plasma tryptophan levels after an oral tryptophan load of 50 mg/kg in these subjects. In comparison to the first oral tryptophan tolerance test, the second one showed no significant difference between plasma levels of total tryptophan but a significant difference between plasma levels of free tryptophan. Free tryptophan levels were higher after a few days than after 1 month of abstinence. The most striking finding of the study of Buydens-Branchey *et al.* (1988) was a significantly greater urinary excretion of kynurenine after 3 days than after 1 month of cessation of drinking. This greater urinary excretion of kynu-renine was attributed to stimulation of tryptophan oxy-genase activity by ethanol (Branchey *et al.*, 1981). In-deed, an earlier study by Branchey and Lieber (1982) showed that chronic ethanol feeding results in increased activity of tryptophan pyrrolase in rat liver.

A consequence of lowered plasma tryptophan con-centration may be reduced tryptophan entry into brain and consequently decreased serotonin synthesis. Al-though further studies may be necessary to establish this, ethanol may affect brain tryptophan without alter-ing the plasma level. For example, ethanol consumption in rats had no significant effect on plasma tryptophan concentration but reduced concentrations of tryptophan and serotonin in the brain (Branchey *et al.*, 1981). Since serotonin is a brain neurotransmitter and is believed to be a regulator of behavior, it is possible that abnormal behaviors in alcoholics may be related to alteration in serotonin metabolism.

5.2.5. Alanine

A major pathway for amino acid catabolism is conversion to glucose. Among the amino acids, alanine is the key substrate for hepatic gluconeogenesis. There is evidence that ethanol inhibits hepatic gluconeogenesis from alanine (Krebs *et al.*, 1969; Kreisberg *et al.*, 1972). The mechanism appears to require the action of alcohol dehydrogenase. Inhibition of alcohol dehydrogenase by pyrazole prevents inhibition of gluconeogenesis by ethanol (Krebs *et al.*, 1969).

The effect of ethanol on gluconeogenesis may have important clinical relevance. Alcohol abuse combined with fasting or poor nutrition may result in hypoglycemia (Freinkel *et al.*, 1965). This hypoglycemia has been attributed to an inhibition of hepatic gluconeogenesis by ethanol (Freinkel *et al.*, 1965) (see Chapter 3).

Despite inhibition of gluconeogenesis there is a sharp decrease in hepatic alanine concentration in rats fed ethanol (Fig. 5.2). The explanation appears to include an increased conversion of alanine to lactate (Kreisberg *et al.*, 1972). Transamination of alanine results in formation of pyruvate. In the face of increasing conversion of NAD to NADH by ethanol, there is increased reduction of pyruvate, resulting in an increased concentration of lactate.

5.2.6. Glutathione

Ethanol consumption appears to alter metabolism of glutathione, a tripeptide composed of glutamic acid, cysteine, and glycine (Speisky *et al.*, 1985; Videla *et al.*, 1980). Its cellular distribution has been studied in the liver, where it is largely (85–90%) localized in the cytoplasm; the remainder is compartmentalized in the mitochondria (Wahlander *et al.*, 1975). Pathways for hepatic glutathione metabolism include efflux from the liver, apparently by a carrier-mediated process (Ookhtens *et al.*, 1985), followed by assimilation by other tissues as a source of its constituent amino acids. Chronic ethanol feeding causes a modest fall (24%) in the cytosolic concentration of glutathione and a marked decrease (65%) in mitochondrial concentration of glutathione in rat liver (Fernandez-Checa *et al.*, 1987). The mechanism of this effect of ethanol appears to be increased hepatic efflux of glutathione (Fernandez-Checa *et al.*, 1987; Pierson and Mitchell, 1986). The hepatic depletion of glutathione has important implications in drug metabolism (see Chapter 6), hepatitis, and toxicity (see Chapter 7), but the metabolic significance of increased movement of glutathione from the liver to extrahepatic tissues remains uncertain.

The above studies were performed in experimental animals, but there is recent evidence that ethanol also affects metabolism of glutathione in humans. Vendemiale *et al.* (1989) investigated the effect of acute ethanol consumption on the plasma concentration of glutathione in healthy volunteers. The experimental group received an acute dose of ethanol (1.5 g/kg orally over a period of 3 hr). The control group received an isocaloric amount of carbohydrates. Blood samples were obtained every hour for 6 hr. Significant increases in plasma concentration of glutathione were observed in subjects receiving ethanol. Although the cause of increased plasma concentration was not studied, Vendemiale *et al.* suggested that an increased hepatic efflux of glutathione, as shown in experimental animals, was the mechanism.

5.2.7. Summary

Chronic ethanol consumption appears to have a catabolic effect on body protein metabolism as evidenced by increased urinary excretion of nitrogen in experimental animals and in humans. The metabolic basis of this increased nitrogen excretion appears to be increased protein catabolism in muscle, as studied in rats. In contrast to muscle, liver protein is preserved or even increased during chronic ethanol consumption. Quantitatively, the protein catabolic effect of ethanol is greatest in muscle. Qualitatively, the effect is greatest in intestinal mucosa, which has exposure to a higher concentration of ethanol and also has a higher rate of protein turnover. The most dramatic effects of ethanol on amino acid pools in the body are increases in concentration of branched-chain amino acids in muscle and a decrease in concentration of alanine in the liver. Inhibition of hepatic gluconeogenesis, as studied in rat liver, provides a metabolic basis for alcoholic hypoglycemia. Alcohol abuse during pregnancy is associated with the fetal alcohol syndrome, which includes retarded growth and development of the fetus. The inhibition of placental transport of amino acids and maternal–fetal transfer of amino acids by ethanol could contribute to the development of this syndrome (see Chapter 15).

5.3. Effects of Ethanol on Hepatic Protein Synthesis

Protein synthesis has been most often assessed by the incorporation of radioactive amino acids into either exportable or constituent proteins of the liver. Most investigators have used as tracers branched-chain amino acids, which are poorly oxidized or transaminated in the liver (Mortimore and Mondon, 1970). Except when stringent conditions of fasting and deprivation of substrates (Girbes *et al.*, 1983) were used, the acute exposure to ethanol has not changed the intracellular specific activity of the tracer amino acids in hepatocytes from rats either fed or fasted for 10–14 hr (Morland and Bessesen, 1977; Morland *et al.*, 1979a; Baraona *et al.*, 1980). Moreover, under these conditions, the rate of protein degradation was not affected by ethanol (Morland and Bessesen, 1977; Girbes *et al.*, 1983). Therefore, the changes in the incorporation of radioactive branched-chain amino acids have been equated with changes in protein synthesis. This is not the case after chronic administration of ethanol. More recently, the effects of ethanol on protein synthesis have been reassessed with methods that are independent of the specific activity of precursor amino acid pools, such as the measurement of specific mRNAs (Zern *et al.*, 1983, 1985) or the measurement of nascent polypeptides with labeled puromycin (Donohue *et al.*, 1985).

5.3.1. Effects of Ethanol *In Vitro*

There is a consensus that ethanol, added *in vitro*, can inhibit the synthesis of both constituent and export proteins in perfused livers (Rothschild *et al.*, 1971; Kirsch *et al.*, 1973; Chambers and Piccirillo, 1973; Morland, 1975; Oratz *et al.*, 1976, 1978), liver slices (Perin *et al.*, 1974; Perin and Sessa, 1975; Sorrell *et al.*, 1977a), isolated hepatocytes (Jeejeebhoy *et al.*, 1975; Morland and Bessesen, 1977; Baraona *et al.*, 1980; Dich and Tönnesen, 1980; Morland *et al.*, 1980; Girbes *et al.*, 1983), hepatocyte primary cultures (Voci *et al.*, 1988), microsomes (Renis *et al.*, 1975), ribosomes (Kuriyama *et al.*, 1971; Perin and Sessa, 1975), and even in mitochondria (Rubin *et al.*, 1970; Burke *et al.*, 1975; Renis *et al.*, 1975), an organelle capable of independent protein synthesis. The addition of approximately 50 mM ethanol to the perfusate of livers isolated from fed rabbits produced disaggregation of both free and bound polyribosomes, detachment of ribosomes from the rough endoplasmic reticulum, decreased RNA/DNA ratio, and inhibition of albumin production (Rothschild *et al.*, 1971, 1974). The disaggregation of polyribosomes after ethanol was greater in those attached to the endoplasmic membranes (Rothschild *et al.*, 1974).

5.3.1.1. The Role of Nutritional Factors

The effects of ethanol on liver protein synthesis are markedly influenced by the nutritional status of the animals. Although fasting by itself produced disaggregation of membrane-bound polyribosomes and decreased albumin production (Rothschild *et al.*, 1974), the addi-

tion of ethanol resulted in a further decrease in albumin production and in increased disaggregation of both free and membrane-bound polysomes. Similar results have been found in liver slices (Perin *et al.*, 1974) and in hepatocytes (Dich and Tönnesen, 1980): the inhibitory effects of ethanol on albumin production and protein synthesis were greater in 24-hr-fasted than in fed rats despite the lower basal levels of protein synthesis in the fasted animals. Thus, the combination of fasting and ethanol produced greater alterations than either factor alone (Rothschild *et al.*, 1983).

5.3.1.2. The Role of Ethanol Oxidation

To assess the extent to which inhibition of protein synthesis and associated alterations in metabolism are linked to the oxidation of ethanol, two major criteria have been used: the dependence of these effects on the concentration of ethanol and the effects of inhibitors of alcohol dehydrogenase (such as pyrazole or 4-methylpyrazole).

In liver slices from fed rats (Perin *et al.*, 1974; Perin and Sessa, 1975) and in isolated hepatocytes from fed (Baraona *et al.*, 1980) and 24-hr-fasted (Dich and Tönnesen, 1980) rats, the inhibitory effects of ethanol have been found to be near maximal at ethanol concentrations of 5–10 mM, which are sufficient to saturate alcohol dehydrogenase. With greater ethanol concentrations the inhibitory effects increased relatively little in these studies. However, in hepatocytes isolated from fasted rats, Morland *et al.* (1980) found that the inhibition was almost linearly dependent on the concentration of ethanol within the range of 10–105 mM.

Either pyrazole or 4-methylpyrazole almost completely prevented the inhibition of protein synthesis induced by 10 mM ethanol in liver slices from fed rats (Perin *et al.*, 1974; Sorrell *et al.*, 1977a) and by 50 mM ethanol in hepatocytes isolated from fed rats (Baraona *et al.*, 1980). In hepatocytes isolated from fasted rats, incubation with 80 mM ethanol produced a similar shift of polysomes toward smaller sizes, associated with decreases in protein synthesis and in the formation of initiation complexes: 4-Methylpyrazole fully prevented the ethanol-induced polysomal changes, suggesting that the decreased initiation of protein synthesis is a consequence of ethanol metabolism. However, in hepatocytes isolated from fasted rats, Morland *et al.* (1980) found that 4-methylpyrazole prevented most of the inhibition produced by 12–18 mM ethanol but very little of that produced by 57–82 mM.

Thus, there seems to be a consensus that the inhibitory effect of ethanol is at least in part dependent on its oxidation through the alcohol dehydrogenase pathway and in part independent, especially at high ethanol concentrations. Whether the latter is a direct effect of ethanol or mediated by oxidation through other pathways remains unknown.

5.3.1.3. Direct Effects of Alcohols at High Concentrations

A direct effect has been suggested by Morland *et al.* (1980), based on analogy with other alcohols such as propanol and tertiary butanol, which also inhibited protein synthesis in hepatocytes at 4–8 mM concentrations. Tertiary butanol is not metabolized by alcohol dehydrogenase. It appears that tissues display variable sensitivity to those nonoxidative effects of ethanol. For instance, 50 mM ethanol did not inhibit protein synthesis by nonparenchymal cells of the liver, which synthesize proteins at a rate comparable to that of hepatocytes but do not oxidize ethanol substantially (Morland *et al.*, 1979a; Bengtsson *et al.*, 1984). A similar lack of inhibitory effect of ethanol on proteins synthesis has been shown in a variety of extrahepatic tissues (Perin and Sessa, 1975). On the other hand, in reticulocytes, ethanol decreased heme and protein synthesis (Freedman *et al.*, 1975), an effect that has been attributed to a block in the initiation of translation (Wu, 1981). This has been related to the premature appearance of a protein repressor of the initiation of translation, which would be controlled by hemin (Freedman and Rosman, 1976).

Ethanol added *in vitro* to liver ribosomes (Kuriyama *et al.*, 1971) or to polyribosomes (Perin and Sessa, 1975) did not inhibit protein synthesis unless extremely high concentrations were reached (217–435, mM). At 200 mM ethanol concentrations, a 50% inhibition of the translation of both endogenous and exogenous mRNAs was observed in cell-free systems from Chinese hamster ovary cells; this was in part because of an inhibitory effect of ethanol and other alcohols on leucyl-tRNA synthetase activity (David *et al.*, 1983). Thus, ethanol shares with other alcohols the capacity to interfere directly with protein synthesis provided that high concentrations are achieved.

5.3.1.4. The Role of Metabolites of Ethanol

The first oxidation product of ethanol is acetaldehyde, a substance that can react with amino acids and proteins. The addition of a single dose of acetaldehyde to liver slices inhibited protein synthesis (Perin *et al.*, 1974; Perin and Sessa, 1975), but at concentrations greater than 1 mM, which are unlikely to occur *in vivo*.

Infusion of acetaldehyde into the incubation medium to maintain concentrations more akin to those likely to occur in the liver *in vivo* inhibited glycoprotein synthesis by rat liver slices, whereas the product of acetaldehyde oxidation, acetate, had no effect (Sorrell *et al.*, 1977a). However, the inhibition of protein and glycoprotein synthesis produced in slices by acetaldehyde concentrations between 320 and 660 μM was shown to be irreversible (Sorrell *et al.*, 1977b), whereas the inhibition produced by ethanol has been shown to be reversible (Perin *et al.*, 1974). Moreover, comparable concentrations of acetaldehyde inhibit protein synthesis in perfused guinea pig heart (Schreiber *et al.*, 1974) and in various rat tissues (Perin and Sessa, 1975) where ethanol is devoid of inhibitory activity. Other dissociations between the effects of ethanol and those of acetaldehyde on protein synthesis have been found in rabbit perfused liver (Oratz *et al.*, 1978). Acetaldehyde inhibits protein synthesis by liver ribosomes *in vitro*, but only at concentrations of approximately 2 mM, which are unlikely to occur *in vivo* (Kuriyama *et al.*, 1971). Even higher concentrations had to be employed to inhibit amino acyl-tRNA synthetase activities (David *et al.*, 1983).

Although acetaldehyde in concentrations higher than those occurring *in vivo* is capable of inhibiting protein synthesis, this does not necessarily indicate that acetaldehyde is the mediator of the ethanol effect. In hepatocytes isolated from fed rats, the rate of ethanol oxidation was doubled by removing reducing equivalents from the cytosol either with methylene blue (an artificial scavenger of reducing equivalents) or by facilitation of their transport into the mitochondria (with substrates for the malic–aspartic acid shuttle) (Baraona *et al.*, 1980). Under these conditions, the concentration of acetaldehyde increased from 7 to over 100 μM. However, the ethanol-induced inhibition of protein synthesis, instead of being exaggerated, was in fact prevented. Also, the trapping of acetaldehyde with penicillamine did not prevent the inhibitory effect of ethanol on protein synthesis by hepatocytes (Morland *et al.*, 1980). Thus, acetaldehyde is not a likely mediator of the inhibitory effect of ethanol on hepatic protein synthesis.

5.3.1.5. The Role of the Redox State

The inhibition of protein synthesis produced by ethanol in isolated hepatocytes was associated with a marked shift toward reduction of the cytosolic redox state, as judged by the increase in the lactate/pyruvate ratio. The addition of either methylene blue or substrates for the malic–aspartic acid shuttle decreased the lactate/pyruvate ratio and prevented the ethanol-induced inhibition of protein synthesis (Baraona *et al.*, 1980). In liver slices (Sorrell *et al.*, 1977a) and in isolated perfused livers (Rothschild *et al.*, 1971), ethanol oxidation was also associated with a marked increase in the lactate/pyruvate ratio, but the mechanism of this alteration has not been as well defined as in isolated hepatocytes. Methylene blue failed to prevent the ethanol-induced inhibition of glycoprotein synthesis in liver slices (Sorrell *et al.*, 1977a). Attempts to reproduce the redox changes produced by ethanol by administration of substances such as sorbitol or xylitol have resulted in either inhibition (Perin and Sessa, 1975; Sorrell *et al.*, 1977a) or no changes (Morland *et al.*, 1980; Dich and Tönnesen, 1980) in protein synthesis. Conversely, increasing the oxidized state of the cytosol by massive supplementation with pyruvate (5–10 mM) prevented the ethanol-induced inhibition of protein synthesis (Perin and Sessa, 1975; Dich and Tönnesen, 1980; Baraona and Lieber, 1983) but not that of glycoproteins (Sorrell *et al.*, 1977a). However, the addition of a large amount of lactate (10 mM), which actually increased the reduced state of the cytosol, also prevented the ethanol-induced inhibition of protein synthesis (Dich and Tönnesen, 1980; Baraona and Lieber, 1983). These dissociations make it very unlikely that the shift in redox state by itself could be the cause of the inhibition of protein synthesis induced by ethanol, but suggest that some other metabolic alteration secondary to the reduced state is probably involved.

5.3.1.6. The Role of Other Metabolic Disturbances Induced by Ethanol Oxidation

The increase in NADH/NAD ratio leads to a small increase in hydrogen concentration. Since a decrease in

pH from 7.4 to 7 inhibited hepatocyte protein synthesis by approximately 20%, Wallin *et al.* (1981) raised the possibility that the ethanol effect could be mediated by its ability to lower the pH; however, the change of approximately 0.1 unit of pH in their experiments is insufficient to explain the ethanol effect.

The oxidation of ethanol by hepatocytes and other *in vitro* liver preparation (Rothschild *et al.*, 1971; Sorrell *et al.*, 1977a; Baraona *et al.*, 1980) is associated not only with a marked increase of the lactate/pyruvate ratio but also with a decrease in the sum of lactate plus pyruvate, the latter reaching very low values. All factors that prevented the ethanol-induced inhibition of protein synthesis (4-methylpyrazole, methylene blue, substrates for the malic–aspartic acid shuttle or pyruvate, and lactate) have in common the property of sparing pyruvate depletion (Baraona and Lieber, 1983) (Fig. 5.4). The depletion of pyruvate is even more prominent in cells derived from fasted rats, in which the concentration of pyruvate is decreased even in the absence of ethanol (Veech *et al.*, 1969, 1972). In hepatocytes isolated from 24-hr-fasted rats, supplementation with pyruvate (or its precursor, lactate) increased the rate of protein synthesis and prevented the inhibitory effect of ethanol (Dich and Tönnesen, 1980). Furthermore, Girbes *et al.* (1983) have shown that in hepatocytes isolated from 48-hr-fasted rats (where ethanol, in the absence of exogenous substrates, does not inhibit protein synthesis beyond the effects of the prolonged fasting), 10 mM ethanol inhibited the stimulatory effects of substrates such as glucose, pyruvate, and lactate on protein synthesis. This suggests that ethanol oxidation interferes with the availability of substrates necessary for the synthesis of proteins. In the fed state, pyruvate results mainly from glycolysis, whereas in the fasted state, pyruvate is derived mainly from lactate that reaches the liver from peripheral tissues and

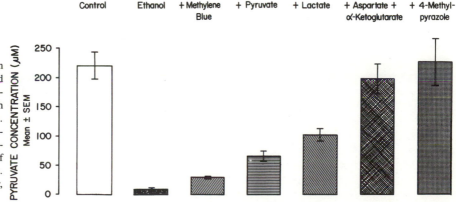

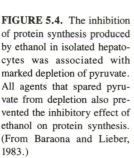

FIGURE 5.4. The inhibition of protein synthesis produced by ethanol in isolated hepatocytes was associated with marked depletion of pyruvate. All agents that spared pyruvate from depletion also prevented the inhibitory effect of ethanol on protein synthesis. (From Baraona and Lieber, 1983.)

from gluconeogenic amino acids. Ethanol oxidation interferes with both glycolysis and gluconeogenesis from lactate and some amino acids (Krebs *et al.*, 1969; Fellenius *et al.*, 1973). The mechanism by which the depletion of pyruvate is connected with the inhibition of protein synthesis is unknown. It is likely that under these conditions either the synthesis or utilization of amino acids may be altered.

5.3.1.7. The Role of Amino Availability for Protein Synthesis

The observation by Rothschild *et al.* (1971) that 10 mM tryptophan reversed the ethanol-induced inhibition of albumin production in perfused rabbit livers has been expanded by the demonstration that various amino acids given in large amounts can reverse not only the inhibition of albumin production (Jeejeebhoy *et al.*, 1972; Kirsch *et al.*, 1973; Rothschild *et al.*, 1974) but also that of the synthesis of other proteins (Perin and Sessa, 1975). The reversal of the inhibition of protein synthesis by supplementation with amino acids was found in perfused livers (Rothschild *et al.*, 1974) and liver slices (Perin and Sessa, 1975) obtained from fed animals. These observations are consistent with the possibility that ethanol may alter the utilization of amino acids for protein synthesis.

In the fasting state, the supplementation with amino acids increased albumin and protein synthesis, but it did not completely prevent the inhibitory effects of ethanol (Rothschild *et al.*, 1974; Morland and Bessesen, 1977; Dich and Tönnesen, 1980; Girbes *et al.*, 1983). However, if the perfused livers from fasted donors were supplemented with spermine, which reaggregated polyribosomes, then supplementation with arginine restored albumin production even in the presence of ethanol (Oratz *et al.*, 1976). These observations suggest that ethanol especially in the fasting state, may decrease not only the availability of amino acids for protein synthesis but also that of polyamines required for the organization and activity of RNA and other nucleic acids. These effects are probably linked to inhibition of the urea cycle by ethanol (Meijer *et al.*, 1975) and decreased synthesis of ornithine (Oratz *et al.*, 1983), the precursor of the polyamines putrescine, spermidine, and spermine.

In addition, as discussed before, ethanol can inhibit the uptake of some amino acids (Chambers and Piccirillo, 1973; Piccirillo and Chambers, 1976; Rosa and Rubin, 1980; Dorio *et al.*, 1984), the inhibitory effects being modest at ethanol concentrations below 100 mM. A similar inhibition has been reported *in vivo* using a multiple dilution technique through the splanchnic circulation in dogs treated with ethanol for 2 days (Cruz *et al.*, 1985). It is likely that the inhibition of protein synthesis observed after acute exposure to high ethanol concentrations may be mediated, at least in part, by inhibition of amino acid uptake.

5.3.1.8. The Role of Oxygen Tensions

Jeejeebhoy *et al.* (1975) noted that hypoxia markedly enhanced the inhibitory effects of ethanol on hepatocyte protein synthesis. *In vivo*, hepatocytes synthesize proteins at oxygen tensions considerably lower than those generally used *in vitro*. The hepatocytes of perivenular zones of the liver are normally exposed to lower oxygen tensions than periportal cells. When hepatocytes were exposed to oxygen tensions similar to those prevailing in perivenular zones, the inhibitory effect of ethanol on protein synthesis was markedly enhanced (Baraona *et al.*, 1983) (Fig. 5.5). This was associated with greater redox shift and more severe depletion of pyruvate.

5.3.1.9. Summary

In conclusion, the *in vitro* effects described above indicate that ethanol at high concentrations that are grossly inebriating *in vivo* may share with other alcohols the capacity to directly alter both amino acid uptake and protein synthesis not only in the liver but also in other tissues. At the lower concentrations usually found during development of experimental liver injury, the oxidation of ethanol is also capable of interfering with hepatic protein synthesis in *in vitro* liver preparations. This is associated with a severe shift in redox state and depletion of pyruvate. It is possible that, under these conditions, in which gluconeogenesis is maintained at the expense of pyruvate and some nonessential amino acids, the availability of these amino acids may become limiting for protein synthesis. These effects are markedly aggravated by fasting and by low oxygen tensions.

5.3.2. Acute Effects of Ethanol *in Vivo*

In contrast to the consistent inhibitory effects of ethanol *in vitro*, its administration to achieve a similar concentration *in vivo* has produced variable effects on hepatic protein synthesis. The disparities seem to depend greatly on the dose of ethanol used.

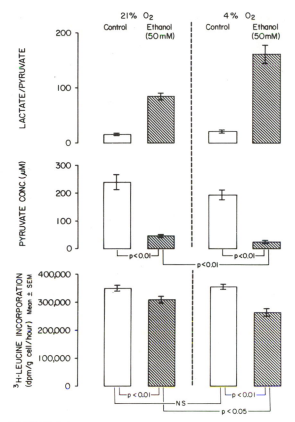

FIGURE 5.5. Effect of oxygenation on ethanol-induced redox shift, pyruvate depletion, and inhibition of protein synthesis in isolated rat hepatocytes. At normal barometric pressure, 4% O_2 produces oxygen tensions similar to those prevailing in perivenular zones, whereas 25% O_2 produces PO_2 equivalent to that in arterial blood. The incubated medium was supplemented with 3mM lactate and 0.3 mM pyruvate.

5.3.2.1. Effects on Constituent Proteins

Under very drastic conditions, namely, after intragastric administration of a sublethal dose of ethanol (7.5 g/kg body weight as a 50% solution) to fasted rats, the incorporation of intraperitoneally injected [^{14}C]leucine into total liver proteins and serum albumin and globulin was significantly decreased (Murty *et al.*, 1980). In pregnant rats, acute intragastric administration of 4 g of ethanol per kilogram body weight, as a 25% solution, also decreased protein synthesis in liver, brain, and other organs (Henderson *et al.*, 1980). Since acute ethanol administration has been reported to decrease the intrahepatic concentrations of some amino acids (Cer-

dan *et al.*, 1980), thereby increasing the specific activity of the tracer, protein synthesis was assessed from the incorporation of [^{14}C]valine, after the valine pool had been flooded with a large dose of the amino acid in order to minimize variations in the specific activity of free valine pools (Dunlop *et al.*, 1975). However, the administration of ethanol resulted in development of hypothermia (from 38 to 35.8°C), which by itself can inhibit protein synthesis (Dunlop *et al.*, 1975). Maintenance of body temperature at 38°C by heating the animal abolished the acute inhibitory effect of ethanol on protein synthesis (Henderson *et al.*, 1980).

Using the time for polypeptide chain completion (another technique for assessing protein synthesis independently of the specific activity of preceding amino acid pools), Cerdan *et al.* (1982) found no effect of moderate doses of ethanol on liver protein synthesis. This is consistent with other studies indicating no changes in the synthesis of total liver proteins after administration of ethanol (3–6 g/kg body weight) to naive rats (Seakins and Robinson, 1964; Ashworth *et al.*, 1965; Moojerkea and Chow, 1969; Morland, 1975; Baraona *et al.*, 1980; Cerdan *et al.*, 1980) and to other species (Raatikainen and Mäenpää, 1980), even though these doses of ethanol result in blood concentrations similar to those that inhibit protein synthesis *in vitro* (Baraona *et al.*, 1980). Thus, there is some question whether the effects on hepatic protein synthesis produced by acute administration of a large dose of ethanol are caused by ethanol itself or result from secondary changes such as hypothermia.

5.3.2.2. Effects on Secretory Proteins

Decreased hepatic production of exportable proteins (such as albumin and transferrin) has been demonstrated after acute administration of ethanol (3 g/kg) to the rat *in vivo* (Jeejeebhoy *et al.*, 1972). Similarly, inhibition of the hepatic production of serum lipoproteins has been reported after acute ethanol administration, especially when high ethanol concentrations were achieved either *in vitro* (Schapiro *et al.*, 1964) or *in vivo* (Dajani and Kouyoumjian, 1967; Madsen, 1969). However, no change or even an increase in lipoprotein production was observed with more moderate doses both *in vitro* (Gordon, 1972) and *in vivo* (Elko *et al.*, 1961; Seakins and Robinson, 1964; Wooles, 1966; Hirayama and Hiroshige, 1970; Baraona *et al.*, 1970, 1973; Abrams and Cooper, 1976). Hypo- and hyperlipemic effects of ethanol have been produced by varying the dose of alcohol administered (Estler, 1975). It is not clear whether the decreased production of plasma proteins is caused solely by a defect in synthesis or by a concomitant impairment of secretion (see following section).

5.3.2.3. Effects on the *Ex-Vivo* Capacity of Ribosomes to Synthesize Proteins

Acute ethanol administration has also produced controversial results on *in vitro* ribosomal protein synthesis. Administration of large doses of ethanol (4–7.5 g/kg body weight) to rats decreased the ability of ribosomes (Kuriyama *et al.*, 1971), membrane-bound polyribosomes (Murty *et al.*, 1980), or complete microsomes (Renis *et al.*, 1975) to incorporate labeled amino acids when assayed *in vitro* with the pH 5 cytosolic fraction or in the presence of the synthetic messenger poly(U) (Murty *et al.*, 1980). Cell sap from ethanol-treated rats was less active than that of control rats in promoting protein synthesis by various polyribosomal fractions obtained from either ethanol-treated or control animals. The decline in protein synthetic activity of the membrane-bound polyribosomes after ethanol was associated with alterations of the microsomal membranes such as decreased lactoperoxidase-catalyzed radio-iodination of microsomal protein and decreased incorporation of [^{14}C]choline into microsomal phospholipids (Murty *et al.*, 1980). By contrast, Princen *et al.* (1981) found that the capacity of either free or membrane-bound polyribosomes from ethanol-treated rats to synthesize albumin and total liver proteins was similar to those of controls. Moreover, cell sap from ethanol-treated rats stimulated protein synthesis more than that from controls. The apparent contradiction between the latter two studies may be ascribed to the different methodology used. Several investigators have found that acute ethanol administration to either rats or mice does not cause changes in yield, size, or state of aggregation of either free or membrane-bound polyribosomes and dose not affect initiation factors (Plapp *et al.*, 1971; Murty *et al.*, 1980; Princen *et al.*, 1981). In mice, the acute administration of a large ethanol dose decreased the incorporation of tritiated uridine in RNA of both hepatocytes and hepatic mesenchymal cells (Eletsky and Merkulov, 1969). The lack of consistent effects of *in vivo* administration of an acute ethanol dose on *in vitro* measured hepatic protein synthesis also contrasts with the direct inhibitory effects of ethanol *in vitro*.

The remarkable differences between the *in vivo* and *in vitro* effects of ethanol on hepatic protein synthesis have been attributed to profound differences in ethanol metabolism between these two conditions (Baraona *et*

al., 1980): the rate of ethanol oxidation is much lower in *in vitro* preparations than *in vivo*, probably as a result of the presence of excessive amounts of reducing equivalents as NADH, which inhibits ADH activity. *In vitro* addition of ethanol is associated with a more severe shift in cytosolic redox state than *in vivo* (Lindros and Aro, 1969; Baraona *et al.*, 1980; Cerdan *et al.*, 1982), suggesting a relative incapacity of *in vitro* preparations to transport reducing equivalents from the cytosol into the mitochondria (Meijer and Williamson, 1974; Crow *et al.*, 1977). Whereas *in vitro* the exaggerated shift in cytosolic redox state resulted in marked depletion of pyruvate, the concentration of pyruvate was only moderately decreased in freeze-clamped livers during ethanol oxidation *in vivo* (Domschke *et al.*, 1974; Baraona *et al.*, 1980).

Supporting the role of the state of ethanol oxidation, there is an interesting finding by Sidransky *et al.* (1980), who showed that administration of a large dose of ethanol (7.5 g/kg body weight) to Buffalo rats with Morris hepatoma 5123 produced polyribosomal disaggregation (especially of the membrane-bound polyribosomes) and decreased the capacity of the ribosomes to synthesize proteins in the hepatoma but not in the host liver. Many neoplastic tissues have reduced activity of the hydrogen translocation shuttles (both the malic–aspartic acid and the glycerophosphate shuttles) (Criss, 1971, 1973), a metabolic characteristic that is reminiscent of that found *in vitro* liver preparations. The possibility remains that the inhibitory effects of ethanol so readily demonstrable *in vitro* might require a very particular metabolic setting characterized by exaggerated redox shift and depletion of pyruvate. Such a setting could occur in perivenular areas of the hepatic acini, which have a preexisting low oxygen tension and develop an exaggerated redox shift after ethanol (Quistorff *et al.*, 1978; Jauhonen *et al.*, 1982, 1985; Baraona *et al.*, 1983). *In vivo* studies indicate zonal heterogeneity in hepatocyte protein synthesis (LeBouton, 1968, 1969; Courtoy *et al.*, 1981), with more controversial results when periportal and perivenular hepatocytes have been partially separated by their difference in buoyant density (Weigand *et al.*, 1974, 1977; Smith-Kielland *et al.*, 1982). Ethanol had similar inhibitory effects on protein synthesis in all these cell populations (Bengtsson *et al.*, 1984) but possible effects of the O_2 gradient were not excluded.

5.3.2.4. Summary

In conclusion, contrasting with the readily demonstrable inhibitory effects of ethanol on *in vitro* liver preparations, acute ethanol administration *in vivo* produces no consistent effects on hepatic protein synthesis.

5.3.3. Chronic Effects of Ethanol

In vivo studies of protein synthesis after chronic ethanol administration have also yielded conflicting results. The disparities seem to depend greatly on the technique of chronic alcohol administration and the presence of associated alterations that can affect protein metabolism.

5.3.3.1. Effects of Techniques of Chronic Administration of Ethanol

Decreased incorporation of labeled amino acids into liver proteins was found by Banks *et al.* (1970) in rats fed the ethanol-containing diets described by Porta *et al.* (1968); by Rawat (1976), using ethanol in a Metrecal® diet; and by Morland (1974) and Morland and Sjetnan (1976) after several weeks of administration of a mixture of solid and liquid diets containing ethanol. Furthermore, Morland (1974) showed that livers isolated from alcohol-fed rats have a decreased ability to incorporate amino acids into protein after stimulation with dexamethasone and have reduced activity of enzymes (tryptophan oxygenase, tyrosine aminotransferase) involved in protein metabolism.

These techniques of alcohol administration result in a relatively low ethanol intake (7.9–8.5 g ethanol per kilogram body weight), accounting for 25% of total energy (Smith-Kielland and Morland, 1979), and produce average blood ethanol concentrations of 5–6 mM (Morland and Sjetnan, 1976), It is noteworthy that in all these studies, fatty liver and hepatomegaly were minimal or absent, and the weight gain was less than 2 g/day (Smith-Kielland *et al.*, 1983a), suggesting that ethanol administration may have been complicated by some degree of undernutrition. Under these conditions, no differences in hepatic amino acid composition were found between ethanol-fed and control rats (Morland *et al.*, 1979b). In a more recent study (Smith-Kielland and Morland, 1981), increased hepatic concentration of branched-chain amino acids was found. However, after correction for the decreased specific activity of valine, net protein synthesis was still decreased in the ethanol-fed rats, although the magnitude of the inhibition was less than previously found. Moreover, the inhibition was prevented by seemingly trivial manipulations, such as taking the animals out of the cage for an intraperitoneal injection of saline.

The effects of chronic alcohol administration on protein synthesis were reassessed using rats pair-fed the liquid diets described by Lieber and DeCarli (1970). Under these conditions, rats gained weight at a rate of 3–4 g/day, and protein synthesis was found to be either unaffected (Donohue *et al.*, 1987) or slightly increased (Baraona *et al.*, 1977). The weight gain was slightly less in the alcohol-fed rat than in the control despite isocaloric feeding. The inefficiency of ethanol to promote growth and weight gain is discussed in Chapter 1.

5.3.3.2. Effects of Specific Activity of Amino Acid Precursors on Synthesized Protein

Despite a 9.4% difference in body weight between ethanol-fed and control rats, the former had hepatomegaly and fatty liver. In contrast with some of the previous reports, the incorporation of leucine into protein in these rats was not decreased.

In rats fed liquid diets, as also reported in humans (Avogaro *et al.*, 1986) and in baboons (Shaw and Lieber, 1978), alcohol feeding increased the concentration of leucine (and other branched-chain amino acids) in the plasma and even more in the liver. This increase diluted the specific activity of the tracer in the amino acid precursor pool. Therefore, the time course of the changes in the specific activity of the leucine bound to tRNA (thus immediately preceding its incorporation into the nascent peptide chain) was determined according to Wallyn *et al.* (1974). After correction for the dilution of the amino acid pool and the increased liver size of the alcohol-fed animals, the rate of protein synthesis (calculated according to the procedures reported by Morgan and Peters, 1971a,b) was found to be enhanced by chronic ethanol administration. These animals were studied in the fed state with blood ethanol concentrations of approximately 30 mM. Acute administration of similar amounts of ethanol-containing diet to naive rats produced no changes in hepatic protein synthesis (Baraona *et al.*, 1980).

An alternative to the laborious measurements of amino-acyl-tRNA specific activities has been to increase the size of the tracer amino acid pool in order to minimize variations in amino-acyl specific activities (Dunlop *et al.*, 1975). Using this method and the same animal model as Baraona *et al.* (1977), Wasterlain *et al.* (1983) found similar incorporation of either valine or lysine into liver proteins of alcohol-fed and pair-fed control rats, when expressed per gram of liver. This represents an increase in liver protein synthesis when the hepatomegaly of the alcohol-fed animals is taken into account. This contrasted with the inhibitory effects of chronic alcohol consumption on brain protein synthesis, similar to those previously shown in other models (Tewari and Noble, 1971). In pregnant rats fed similar liquid diets and using a similar method, no significant inhibition of hepatic protein synthesis was found unless the blood alcohol levels were raised to approximately 70 mM by an acute ethanol administration that resulted in hypothermia (Henderson *et al.*, 1980). Hepatic protein synthesis was not completely restored by heating the animal to normalize body temperature. Thus, there is evidence that indicates that under certain experimental conditions, high alcohol levels can inhibit the synthesis of liver proteins, but there is no evidence that those conditions are required for the production of alcoholic liver injury. Using rats pair-fed similar liquid diets but considerably more drastic procedures (such as fasting, laparotomy, and multiple surgical liver biopsies), Morland *et al.* (1983) and Smith-Kielland *et al.* (1983b) found a 21% decrease of hepatic protein synthesis in the alcohol-fed rats. Although the relevance of some of the latter experiments to the conditions normally prevailing in the animals during development of alcoholic liver injury is difficult to appreciate, these findings raise the possibility of a greater susceptibility of the alcohol-fed animals to conditions that may affect hepatic protein synthesis.

5.3.3.3. Effects on Hepatocytes Isolated from Alcohol-Fed Animals

Hepatocytes (Baraona *et al.*, 1980) as well as liver slices (Sorrell *et al.*, 1983; Donohue *et al.*, 1985) obtained from ethanol-fed rats display decreased incorporation of amino acids into proteins when incubated *in vitro* as compared to those from pair-fed controls (Fig. 5.6). This effect was associated with decreased formation of peptidyl-[³H]puromycin (Donohue *et al.*, 1985), an indirect measurement of the number of ribosomes actively undergoing protein synthesis at the time of the puromycin block (Wool and Kurihara, 1967; Nakano and Hara, 1979), suggesting that the decreased incorporation was indeed a result of decreased protein synthesis. By contrast, application of this methodology to *in vivo* assessment of liver protein synthesis revealed no significant differences between ethanol-fed rats and pair-fed controls (Donohue *et al.*, 1987). The reason for this discrepancy between *in vivo* and *in vitro* studies is unknown, although it raises again the possibility that chronic alcohol consumption may increase the suscep-

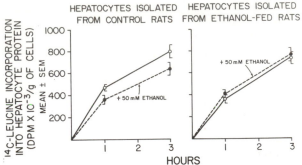

FIGURE 5.6. Disappearance of the inhibitory effect of ethanol on protein synthesis after chronic alcohol consumption: 50 mM ethanol did not alter the incorporation of [¹⁴C]leucine into cell protein of hepatocytes isolated from ethanol-fed rats, whereas it produced significant ($P < 0.01$) inhibition in hepatocytes isolated from naive rats (pair-fed controls). In the absence of ethanol, the hepatocytes from ethanol-fed rats displayed decreased leucine incorporation compared to those from control rats. (From Baraona *et al.*, 1980.)

tibility of the liver to adverse *in vitro* conditions for protein synthesis.

5.3.3.4. Acute Effects of Ethanol in Animals Chronically Fed Alcohol

Chronic alcohol consumption also changes the response of the liver to the acute *in vitro* effects of ethanol on protein synthesis. Indeed, ethanol failed to inhibit amino acid incorporation into cell protein when hepatocytes were obtained from rats fed alcohol-containing diets for several weeks, in contrast with the hepatocytes from pair-fed controls (Fig. 5.6). The hepatocytes from alcohol-fed rats also displayed smaller changes of the lactate/pyruvate ratio in the presence of ethanol and maintained higher pyruvate concentrations than the hepatocytes isolated from pair-fed controls (Baraona *et al.*, 1980). The decreased effect of ethanol in these hepatocytes compared to those in naive pair-fed controls can be attributed to the attenuation of the ethanol-induced redox change following chronic alcohol consumption [as reported in humans (Salaspuro and Kesanienni, 1973), in rats (Domschke *et al.*, 1974), and in baboons (Salaspuro *et al.*, 1981)], increased availability of shuttle substrates, or both. Similarly, the ethanol-induced inhibition of albumin synthesis that had been observed in naive animals was not reproduced when ethanol was administered to alcoholic subjects (Jeejeebhoy *et al.*, 1975). However, both in hepatocytes isolated from rats fed a mixture of solid and liquid diets (Morland *et al.*, 1980; Bengtsson *et al.*, 1984) and in liver slices from rats pair-fed liquid diets (Sorrell *et al.*, 1983), the addition of ethanol induced similar inhibition of *in vitro* protein synthesis in alcohol-fed as in control rats. There was no evidence, however, that either the alcohol feeding technique or the incubation conditions resulted in sufficient attenuation of the ethanol-induced redox shift.

5.3.3.5. Chronic Effects of Ethanol on *Ex-Vivo* Capacity of Ribosomes to Synthesize Proteins

A dissociation between the acute and chronic effects of ethanol on protein synthesis has also been reported *in vitro* in isolated mouse ribosomes (Kuriyama *et al.*, 1971) and rat microsomes (Renis *et al.*, 1975): after an acute dose of alcohol, these organelles exhibited an enhanced capacity to incorporate amino acids into protein when obtained from animals fed ethanol chronically, whereas the opposite effect is observed in naive animals.

Electron microscopic (Iseri *et al.*, 1966; Lane and Lieber, 1966; Rubin and Lieber, 1967a,b; Lieber and Rubin, 1968a,b) and subcellular fractionation (Ishii *et al.*, 1973) studies indicate altered morphology and paucity of rough endoplasmic reticulum in ethanol-fed rats, suggesting a decrease in the number of membrane-bound polysomes. However, there were no differences in liver RNA contents when expressed per total liver (Zern *et al.*, 1983). After prolonged administration of 20% ethanol in drinking solution, which tends to produce some degree of undernutrition, the amount of ribosomal RNA has been found decreased (Albertini *et al.*, 1970). This was associated with decreased turnover of ribosomal RNA. Conversely, autoradiographic studies (Combescot *et al.*, 1969) have found increased incorporation of labeled uridine into the ribosomes, which was interpreted as acceleration of their turnover.

In another study (Khawaja and Lindholm, 1978), chronic alcohol feeding to rats more than doubled the capacity of free ribosomes to incorporate labeled amino acids. This enhancement was observed in the presence of homologous pH 5 cytosolic fraction (but not with a heterologous one) and was associated with increased amino acid incorporation in tRNA of the pH 5 fraction. There are apparent discrepancies with regard to the effects of chronic alcohol administration on the capacity of bound ribosomes for protein synthesis. Whereas Renis *et al.* (1975) found increased leucine incorporation by microsomes (which contain bound ribosomes) of

ethanol-fed rats, Khawaja and Lindholm (1978) found a 25% decrease in the incorporation of a labeled amino acid mixture (but not of labeled phenylalanine) into bound ribosomes. Only when these ribosomes were detached from the membranes after treatment with Triton X-100 was an increased capacity to synthesize protein (similar to that of free ribosomes) demonstrated. In rats intoxicated by multiple gastric intubations with alcohol, Peters and Steele (1982) found a 20% decrease in leucine incorporation by membrane-bound polysomes. These polysomes were enlarged, suggesting a retardation of the polypeptide elongation step. By contrast, using rats chronically fed alcohol in nutritionally adequate liquid diets, Zern *et al.* (1983) found a 65% increase in protein synthesis by membrane-bound polysomes compared to either pair-fed or *ad libitum* controls. To further characterize the proteins produced, these investigators extracted and translated the RNA in a rabbit reticulocyte mRNA-dependent system, showing a marked increase in the synthesis of albumin by RNA from the ethanol-fed rats. This was associated with an increase in albumin mRNA in alcohol-fed rats, as shown by hybridization with recombinant cDNA. Similar findings have been more recently obtained in baboons fed ethanol in liquid diets for 2–10 years that had developed various degrees of hepatic fibrosis (Zern *et al.*, 1985): there was an almost twofold increase in the synthesis of secretory proteins, such as albumin and type I procollagen, associated with an increase in their specific mRNAs, whereas the levels of a representative constitutive protein, such as β-actin, remained unchanged.

5.3.3.6. Summary

Ethanol has the potential to inhibit protein synthesis when administered in high doses and when the metabolic derangements associated with ethanol oxidation are experimentally exaggerated. Whether such conditions are relevant to clinical situations that aggravate the effects of alcohol on the liver (such as malnutrition or hypoxia) remains to be documented. Selected areas, such as the perivenular zones, may be adversely affected. Thus far, however, the experimental evidence indicates that alcohol consumption can lead to the development of at least the initial stages of liver injury with no apparent impairment in the synthesis of total liver proteins or even with some increase. There is scarce information on the synthesis of specific proteins. Also, there is little controversy with regard to inhibitory effects of alcohol consumption on the synthesis of protein by other organs, such as the brain and the heart, and by the fetus (as discussed in other chapters). With progression of alcoholic injury to more severe stages, one would expect various liver functions, including protein synthesis, to be adversely affected. Serum levels of albumin are characteristically decreased in alcoholics with cirrhosis. Surprisingly, however, albumin production was not always decreased but often was normal or elevated, especially in patients with ascites (Rothschild *et al.*, 1969, 1987).

5.4. Effects of Ethanol on Hepatic Protein Secretion

5.4.1. *In Vivo* Studies

As previously mentioned, delayed appearance of newly labeled albumin and fibrinogen was first observed by Jeejeebhoy *et al.* (1972) after an acute dose of ethanol. Whereas total protein concentration in the liver remained unchanged after chronic ethanol consumption, the concentration of export proteins (such as albumin and transferrin) increased, suggesting an impairment of protein secretion (Baraona *et al.*, 1975, 1977). The secretory defect was documented in rats fed alcohol-containing diets, with the observation of delayed appearance of newly labeled albumin and transferrin in the serum and a corresponding retention of these newly labeled proteins in the liver (Baraona and Lieber, 1977; Baraona *et al.*, 1977) (Fig. 5.7). Acute alcohol administration (3 g/kg) *in vivo* to naive rats was also associated with delayed appearance of newly labeled albumin in the serum, retention of labeled albumin in the liver, and increased hepatic content of immunoreactive albumin (Baraona *et al.*, 1980). After acute ethanol administration, the degree of albumin retention was considerably smaller, and the secretory rate less reduced, than that of rats chronically fed an alcohol-containing diet and given the same acute ethanol dose. By contrast, withdrawal from ethanol for 20 hr restored the secretory rate to normal levels (Baraona *et al.*, 1981a). These findings indicated that the defect in secretion was linked to the oxidation of ethanol, the rate of which was greatest in the rats given ethanol chronically. Similarly effects of acute ethanol administration were found by Volentine *et al.* (1984) on the secretion of fucosylated glycoproteins. In addition, the latter investigators showed that the acute effect was prevented by inhibiting ethanol oxidation by pretreatment of the animals with pyrazole. Wallin *et al.* (1984) claimed that acute ethanol administration does not affect either liver protein synthesis or secretion.

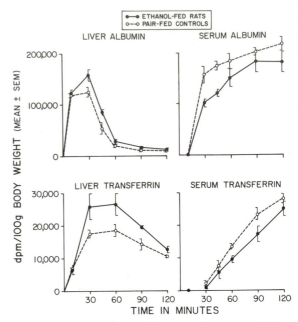

FIGURE 5.7. Incorporation of intravenously injected [^{14}C]leucine into liver and serum albumin and transferrin at various time intervals in rats pair-fed either ethanol-containing or control diets. Delayed appearance of newly labeled proteins in the serum coincided with retention in the liver. (From Baraona *et al.*, 1977.)

However, this study actually measured the synthesis of both constituent and secretory proteins of the liver rather than the secretory process.

5.4.2. *In Vitro* Studies

Sorrell *et al.* (1977a) reported that ethanol inhibits the incorporation of leucine or glucosamine into proteins released by liver slices more than into tissue proteins, indicating either a preferential inhibitory effect of ethanol on the synthesis of exportable protein or a secretory defect. Further studies by Sorrell and Tuma (1978), Tuma *et al.* (1981), Tuma and Sorrell (1981), Sorrell *et al.* (1983) favored the possibility that the acute inhibitory effects of ethanol are exerted not only on protein synthesis but on secretion as well. Under these conditions, 10 mM ethanol inhibited the release of pre-labeled protein and glycoproteins from liver slices of fed rats when synthesis was blocked with cycloheximide or puromycin. These effects were reproduced by infusion of acetaldehyde to achieve concentrations of 50–70 μM, which are likely to occur *in vivo* (Sorrell and Tuma, 1978; Tuma *et al.*, 1980), raising the possibility that acetaldehyde could be the mediator of these effects. At high concentrations of acetaldehyde (320–660 μM), the impairment of glycoprotein metabolism became irreversible (Sorrell *et al.*, 1977b). When secretory glycoproteins were separated from structural glycoproteins, a retention of the secretory fraction became evident (Tuma and Sorrell, 1981). The secretory defect produced by ethanol in rat liver slices affected albumin as well as glycoproteins (Tuma *et al.*, 1981). Except for the *in vitro* inhibitory effects on protein synthesis, the effects were similar to those produced by colchicine, a well-known inhibitor of microtubule assembly. A similar secretory defect was also documented in slices from rats fed alcohol-containing liquid diets (Sorrell *et al.*, 1983). A failure to demonstrate inhibitory effects of ethanol on protein secretion by isolated hepatocytes reported by Morland *et al.* (1981) is probably a result of lack of acetaldehyde (Morland *et al.*, 1980), probably as a consequence of a slow rate of ethanol oxidation (Sjoblom and Morland, 1979), under these conditions.

5.4.3. Mechanism of the Secretory Defect Induced by Ethanol

During intracellular transport, secretory proteins undergo a series of posttranslational changes. Some proteins, such as albumin, are discharged into the cisterna with an additional peptide or "prosegment," which is removed prior to secretion. Inhibition of the involved protease interferes with the secretion of albumin (Edwards *et al.*, 1979; Algramati and Sabatini, 1979). It is unlikely that alcohol acts at this site because both proalbumin and albumin accumulate in alcohol-fed animals (Baraona *et al.*, 1977), an effect similar to that produced by colchicine (Redman *et al.*, 1978).

5.4.3.1. The Role of Alterations in Glycosylation

The majority of plasma proteins (with the exception of albumin) undergo posttranslational addition of carbohydrate and are exported from the liver in the form of glycoproteins. The incorporation of glucosamine into glycoproteins is inhibited by acute ethanol administration both *in vivo* (Moojerkea and Chow, 1969) and *in vitro* (Sorrell *et al.*, 1977a), which suggests a possible site for the ethanol effect on protein secretion. *In vivo*, this alteration was attributed to decreased formation of the UDP-*N*-acetalhexosamine intermediate (Moojerkea and Chow, 1969). Ethanol may also inhibit subsequent

steps in the glycosylation process, including the final attachment to the protein moiety (Sorrell *et al.*, 1977a). It appears paradoxical, however, that the activity of the glycosyltransferases responsible for the carbohydrate binding to protein at the Golgi apparatus increases within 16 hr after an acute ethanol dose and even more after chronic ethanol feeding (Gang *et al.*, 1973). In liver slices, the decreased incorporation of glucosamine was associated with ethanol-induced inhibition of protein synthesis (Sorrell *et al.*, 1977a), to which the synthesis and attachment of the "core" polysaccharides is closely tied (Schmitt and Elbein, 1979). However, after complete inhibition of protein synthesis with cycloheximide, ethanol produced decreased release of glycoproteins prelabeled with fucose into the medium with retention in the tissue, indicating a dual effect of ethanol on synthesis and secretion of glycoproteins (Tuma and Sorrell, 1981; Tuma *et al.*, 1981). Also, the incorporation of fucose and a precursor of sialic acid into glycoproteins of the plasma membrane was inhibited by acute administration of a large ethanol dose (Maillard *et al.*, 1984). Thus, alcohol impairs terminal glycosylation of both secretory glycoproteins and plasma membrane glycoproteins, which share common pathways. However, inhibition of protein glycosylation with tunicamycin or 2-deoxy-D-glucose results in the formation of unglycosylated proteins that are secreted at normal rates by the hepatocytes (Struck *et al.*, 1978; Edwards *et al.*, 1979). Thus, the alcohol effects on glycosylation not only do not explain the altered secretion of albumin (which is not a mucoprotein), they also are not necessarily the cause of the secretory defect.

5.4.3.2. The Role of Microtubular Alterations

Ethanol must interfere with a secretory step that affects both glycoproteins and nonglycoproteins. Microtubules represent such a likely site. Indeed, drugs (such as colchicine and Vinca alkaloids) that alter microtubules have been reported to impair secretion of macromolecules from various organs, including the liver (LeMarchand *et al.*, 1973, 1974; Stein and Stein, 1973; Stein *et al.*, 1974; Redman *et al.*, 1975, 1978; Feldman *et al.*, 1975; Reaven and Reaven, 1980). This effect is associated with retention of secretory proteins in the Golgi and, to a lesser extent, in microsomal fractions of the liver (Redman *et al.*, 1978). An additional feature is the abundance of lysosomes and autophagic vacuoles (Redman *et al.*, 1978; Reaven and Reaven, 1980), which suggests increased degradation of the retained proteins.

By a variety of methods, it was shown that the export defect of alcohol-fed rats was associated with a significant decrease in the hepatic content of polymerized tubulin, the major chemical component of microtubules (Baraona *et al.*, 1977, 1984). Similar findings were obtained in alcohol-fed baboons (Matsuda *et al.*, 1978) and in alcoholics (Matsuda *et al.*, 1983). The binding with high affinity of 1 mole of colchicine per mole of tubulin dimer has permitted the use of [³H]colchicine binding as an assay for quantitating tubulin in cell extracts (Borisy, 1972; Sherline *et al.*, 1974). When the difference between free and total tubulin (Baraona *et al.*, 1977; Matsuda *et al.*, 1979) was assessed, or when the microtubule-derived tubulin was measured directly (Matsuda *et al.*, 1983; Baraona *et al.*, 1984), there was a significant decrease in both the concentration and the total hepatic content of polymerized tubulin in both alcoholics and alcohol-fed rats. The concentration of total tubulin was significantly decreased in the hepatic cytosol (Baraona *et al.*, 1975, 1977, 1984), whereas that of nonpolymerized tubulin remained unaffected (Baraona *et al.*, 1977, 1984; Matsuda *et al.*, 1983), revealing a defect in the state of polymerization of the microtubular protein and suggesting a secondary decrease in the mass of microtubules.

That this biochemical change was indeed caused by a decrease in microtubules was documented morphologically by Matsuda *et al.* (1979) and by Okanoue *et al.* (1984). Both studies found (using different morphometric parameters) that chronic alcohol feeding decreased rat liver microtubules by approximately 60%. The morphometric decrease was twice as large as the biochemical decrease in polymerized tubulin, suggesting that the microtubular injury is more severe than estimated by the measurement of polymerized tubulin. The most likely explanation for this discrepancy with the biochemical measurements is that the microtubules of alcohol-fed rats were structurally altered, and small fragments of this organelle may have not been recognized as microtubules by electron microscopy. The microtubules of alcohol-fed rats were not only decreased in number, they were significantly shorter and thicker than those of controls (Matsuda *et al.*, 1979) (Fig. 5.8). By contrast, Berman *et al.* (1983) failed to detect a morphometric decrease in liver microtubules in rats fed alcohol-containing diets. However, in the latter study, the normal volume density of microtubules appeared to be overestimated by 15-fold or more, which could obscure the effect produced by ethanol. The application of morphometric techniques to liver biopsies from alcoholics has shown that the microtubular alter-

ation is particularly prominent in the perivenular or centrolobular areas of the liver (Matsuda *et al.*, 1987).

The question of whether the microtubular alterations observed in alcohol-fed rats are secondary to other structural changes produced by ethanol in the liver or whether this is an effect directly linked to ethanol or its metabolites was first addressed in hepatocytes isolated from normal rats. Incubation of hepatocytes with 50 mM ethanol in a nutritionally enriched medium reproduced the decreases in both polymerized tubulin and volume density of microtubules observed *in vivo* (Matsuda *et al.*, 1979). These acute effects were prevented by pyrazole, indicating that they were linked to derangements generated by the oxidation of ethanol rather than a direct effect of ethanol.

5.4.3.3. The Role of Acetaldehyde

In a nutritionally enriched medium, ethanol oxidation by the hepatocytes resulted in accumulation of 130 μM acetaldehyde. Serial additions of acetaldehyde to maintain similar concentrations reproduced the effects of ethanol on both polymerized tubulin and visible microtubules (Matsuda *et al.*, 1979). The addition of 1 mM acetate (the product of acetaldehyde oxidation) also decreased polymerized tubulin with a tendency for a concomitant decrease in total tubulin. Similar effects have been observed by Kawahara *et al.* (1987) using cultured hepatocytes and documenting the changes in microtubules both by indirect immunohistochemistry and by scanning electronmicroscopy.

In vivo, the inhibitory effects of ethanol on either microtubules or polymerized tubulin were markedly exaggerated by pretreatment of the animals with disulfiram (an inhibitor of aldehyde dehydrogenase), incriminating acetaldehyde as the major agent responsible for these effects of ethanol oxidation (Baraona *et al.*, 1981a; Baraona and Lieber, 1982, 1983) (Fig. 5.9). The inhibitory effects of acute ethanol administration on microtubules were markedly enhanced in rats previously fed alcohol-containing diets for 4–6 weeks. On the other hand, there was rapid restoration of the microtubules after 20 hr of alcohol withdrawal in spite of the persistence of other morphological alterations. Conversely, a condition that reduces the rate of ethanol oxidation, such as fasting, decreased polymerized tubulin (Pipeleers *et al.*, 1977) and abolished the inhibitory effects of ethanol on polymerized tubulin in alcohol-fed rats (Baraona *et al.*, 1984). This was associated with smaller acetaldehyde concentrations in the liver than in fed animals. All these observations incriminate acetaldehyde as the most

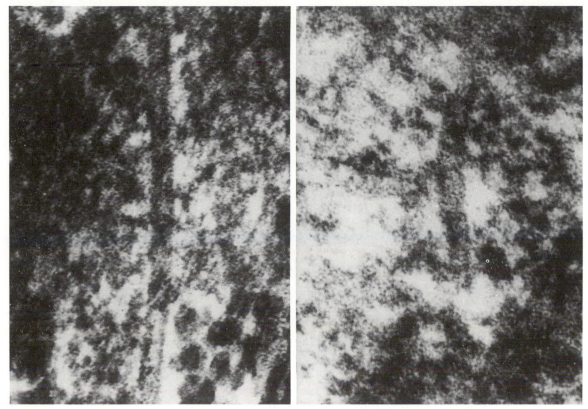

A   B

FIGURE 5.8. Electron micrographs (×174,000) of liver microtubules from rats pair-fed diets containing 36% of calories as either ethanol or additional carbohydrates for 4 weeks. Compared to the thin and long microtubules of pair-fed controls (A), those from the ethanol-fed rats (B) were shorter and thicker as well as decreased in number. (From Matsuda *et al.*, 1979.)

FIGURE 5.9. Acute effects of ethanol (with and without disulfiram) on hepatic acetaldehyde and microtubules in ethanol-fed rats and pair-fed controls. Disulfiram increased the ethanol-induced accumulation of acetaldehyde and the decrease in liver microtubules, especially in alcohol-fed rats*. Ethanol-containing diet was given for 4–6 weeks and replaced by control diet 20 hr before i.v. administration of ethanol. (From Baraona *et al.*, 1981a.)

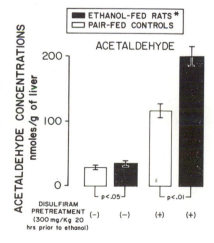

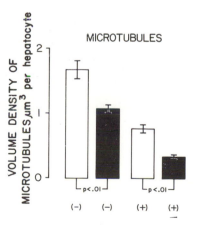

likely mediator of the inhibitory effects of ethanol on microtubules but do not exclude the possibility of an additional role of acetate.

5.4.3.4. The Role of Microsomal Ethanol-Oxidizing System (MEOS)

The enhancement of the microtubular alteration after chronic ethanol administration was associated with significantly higher levels of acetaldehyde in the liver as a consequence of the increased rates of ethanol oxidation (Baraona *et al.*, 1981a; Baraona and Lieber, 1982, 1983). Moreover, 4-methylpyrazole did not fully prevent the ethanol-induced increase in hepatic acetaldehyde concentrations and the decrease in microtubules in alcohol-fed animals. This suggests that the enhanced toxicity of ethanol on liver microtubules is caused by the development of alternative ethanol-oxidizing pathways such as MEOS, which is much less sensitive to the inhibitory effects of 4-methylpyrazole (Teschke *et al.*, 1977). Marked exaggeration of the microtubular alteration, appearance of striking ballooning of the hepatocytes, and retention of transferrin have been observed in rat fed high-fat diets containing both ethanol and pyrazole (Takada *et al.*, 1986). This has been attributed to enhanced microsomal oxidation of ethanol to acetaldehyde because of ADH inhibition and a stimulatory effect of

high-fat diets on MEOS induction (Kanayama *et al.*, 1984). Although the alterations are similar to those found in alcoholics with ballooning (Matsuda *et al.*, 1985a), the interpretation of the combined effects of ethanol and pyrazole is complicated by the inherent toxicity of pyrazole and its possible potentiation by the interactions of both drugs at the microsomal level (Lieber *et al.*, 1970). Chronic alcohol consumption not only enhanced the toxicity of ethanol on microtubules but also seemed to exacerbate the toxicity of acetaldehyde itself. The disruption of microtubules correlated with the level of acetaldehyde in alcohol-fed as well as in naive rats given alcohol acutely; however, at similar acetaldehyde concentrations in the liver, the decrease in microtubules was greater in alcohol-fed rats (Fig. 5.10). A possible mechanism for this enhanced toxicity of acetaldehyde in alcohol-fed rats has been proposed by Yasuhara *et al.* (1986), namely, that the oxidation of the acetaldehyde produced in the microsomes by MEOS proceeds at a slower rate than that produced by alcohol dehydrogenase in the cytosol. Indeed, the ratio between ethanol oxidation rate and hepatic acetaldehyde concentration, which can be considered an index of hepatic acetaldehyde degradation, decreased significantly after inhibition of alcohol dehydrogenase with 4-methylpyrazole. Therefore, microtubules could be exposed to more acetaldehyde when ethanol is oxidized by MEOS than by ADH.

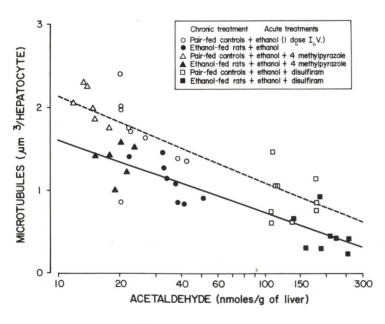

FIGURE 5.10. Correlation between the decrease in volume density of microtubules and the concentrations of acetaldehyde achieved in the liver by administration of ethanol with and without 4-methylpyrazole or disulfiram pretreatment. For similar levels of acetaldehyde, the microtubular disease was more intense in rats chronically fed alcohol-containing diets than in the controls. (From Baraona *et al.*, 1981a.)

5.4.3.5. Mechanism of the Inhibitory Effect of Acetaldehyde on Microtubules

Acetaldehyde (but not ethanol or acetate) binds to tubulin, competes with colchicine for a similar binding site, and inhibits *in vitro* polymerization of tubulin isolated from brain. Inhibitory effects of acetaldehyde on colchicine binding to liver tubulin were reported by Gabriel *et al.* (1977), although at high concentrations. Since liver cytosol rapidly converts acetaldehyde into ethanol through alcohol dehydrogenase and to acetate through cytosolic aldehyde dehydrogenases, the acting concentrations could have been much smaller. After inhibition of alcohol dehydrogenase, concentrations of acetaldehyde as low as 200 μM could be maintained by sequential additions of acetaldehyde to the liver cytosol. Under these conditions, acetaldehyde competitively inhibited the binding of colchicine to tubulin (Baraona *et al.*, 1981a; Baraona and Lieber, 1983). This suggests that acetaldehyde and colchicine may compete for a similar binding site to tubulin.

Jennett *et al.* (1980) showed that acetaldehyde was capable of inhibiting the *in vitro* polymerization of brain tubulin in a concentration-dependent manner, whereas ethanol lacked any effect and acetate favored tubulin polymerization. Acetaldehyde did not affect the rate of depolymerization (Jennett *et al.*, 1980). However, acetaldehyde augmented the depolymerizing effects of Ca^{2+} on preassembled microtubules (McKinnon *et al.*, 1987). The concentration of acetaldehyde required to demonstrate the inhibitory effects was at least 0.5–1 mM, which is not likely to occur *in vivo* during ethanol oxidation. However, the concentrations of tubulin used *in vitro* were also considerably greater than those existing *in vivo*. When *in vitro* polymerization was carried out using tubulin concentrations more in the range of those existing *in vivo*, inhibition could be detected with concentrations as low as 200 μM acetaldehyde (Baraona *et al.*, 1981a; Baraona and Lieber, 1983). Moreover, the capacity of acetaldehyde to inhibit tubulin polymerization increased with the time of exposure of tubulin to this metabolite (Tuma *et al.*, 1987).

Subsequently, it was clearly established that acetaldehyde binds with high affinity to the α subunit of either the brain (Jennett *et al.*, 1987) or the hepatic (Jennett *et al.*, 1989) tubulin dimer. This subunit contains a very reactive lysine in position 394, which become inaccessible for reaction in the assembled microtubules. Methylation of this lysine, after treatment of tubulin with formaldehyde, markedly reduces the ca-

pacity of tubulin to polymerize *in vitro* and, to a lesser extent, the binding of colchicine (Blank *et al.*, 1986). By contrast, methylation of assembled microtubules does not prevent repolymerization of the tubulin. Similar observations have been made with acetaldehyde (Smith *et al.*, 1989). Moreover, by using tubulin in which other binding sites have been adduced during the polymerized state, the stoichiometry of the acetaldehyde reaction with the highly reactive lysine required for polymerization could be determined: it took only 0.08 mol of acetaldehyde per mole of tubulin for complete inhibition of assembly to occur. This lack of stoichiometry is shared by other inhibitors of tubulin polymerization, such as colchicine. The fact that only a small number of tubulin molecules need be adduced in order to prevent the assembly of microtubules explains how such an alteration may be produced by the small acetaldehyde concentrations actually measured in the liver during ethanol oxidation (Baraona *et al.*, 1981a).

5.4.3.6. Summary

In conclusion, alcohol consumption inhibits the secretion of proteins from the liver. The secretory alteration induced by alcohol parallels that on microtubules under all experimental conditions thus far studied. Moreover, the development of the secretory and the microtubular alterations seems to share a similar pathogenesis: both are linked to the oxidation of ethanol and appear to be mediated by acetaldehyde. *In vitro*, the disruption of liver microtubules produced by ethanol was associated with engorgement of the secretory vesicles of the Golgi complex (Matsuda *et al.*, 1979), suggesting a concomitant impairment of secretion. Chronic alcohol consumption, which potentiated the disruptive effect of ethanol on microtubules, also exaggerated the engorgement of the Golgi. Conversely, 4-methylpyrazole, which decreased the acute effect of ethanol on microtubules, also decreased the Golgi engorgement. Moreover, the quantitative relationship between the microtubular decrease produced by ethanol and its effects on albumin and transferrin secretion (studied *in vivo*) are comparable to those obtained with small doses of colchicine on microtubules and secretion of very-low-density lipoproteins in the perfused rat liver (Reaven and Reaven, 1980). These various observations support the concept that integrity of microtubules is required for normal protein secretion and that the acetaldehyde-mediated alteration of microtubules is responsible, at least in part, for the impaired protein secretion after ethanol. The findings in alcoholics withdrawn from alcohol for several days (Matsuda *et al.*, 1983, 1985a, 1987) suggest that, after prolonged alcohol abuse, these alterations may persist even in the absence of active ethanol oxidation.

5.5. Effects of Ethanol on Hepatic Protein Catabolism

In the liver, the rate of proteolysis is commonly assessed by the release of branched-chain amino acids (such as valine) when reincorporation into protein can be suppressed or accounted for (Khairallah and Mortimore, 1976), since these amino acids are not metabolized significantly in the liver (Mortimore and Mondon, 1970). Adapting this method to isolated hepatocytes from normal rats, Morland and Bessesen (1977) found no changes in the rate of proteolysis after the addition of 50 mM ethanol. This observation was confirmed by Girbes *et al.* (1983). A similar lack of effect was found in cultured hepatocytes using 100 mM ethanol (Voci *et al.* 1988). However, in perfused livers from naive rats, the same concentration of ethanol (50 mM) inhibited protein degradation both in the absence of amino acids and in the presence of a mixture of 20 amino acids at a concentration 10 times higher than in normal plasma (Pösö *et al.*, 1987). The maximal inhibitory effect of ethanol was achieved with a 5 mM concentration, which is sufficient to saturate ADH, thereby suggesting that this effect might be a consequence of ethanol oxidation. The mechanism of the inhibition was not entirely clear, and two possible mechanisms have been postulated: (1) inhibition by a slight accumulation of ammonia, probably as a result of an inhibitory effect of ethanol on ureogenesis (Meijer *et al.*, 1975); and (2) decreased formation of autophagic vacuoles. The difference between the results in hepatocytes and in perfused liver has been attributed to a lesser sensitivity of the former preparation (Mortimore and Pösö, 1984).

There is a paucity of information with regard to hepatic protein degradation after chronic alcohol consumption. As in the case of protein synthesis, it is likely that the results will be greatly influenced by the nutritional state of the animals. Pösö (1987) reported that protein degradation was also decreased in rats fed ethanol for 3 to 12 weeks, but the details of the methods used, and particularly those to control associated nutritional changes, were not given. The rate of proteolysis

can also be estimated from independent measurements of the rate of turnover and the rate of synthesis. Using this approach, Donohue *et al.* (1987, 1989a) recently reported that the effects of chronic ethanol administration on the turnover of liver proteins [measured by a modification of the method of Swick (1958)] vary with the nutritional status, the duration of ethanol administration, and the subcellular fraction and the type of proteins involved. During the first 12 days of alcohol administration, the turnover of liver proteins in ethanol-fed rats was slower than that in pair-fed controls but more rapid than in rats fed chow diet *ad libitum*. This difference was not corrected by feeding the diets three times daily; thus, it appears unlikely that this difference could be solely related to periods of fasting in the food-restricted pair-fed controls but probably resulted from other differences in composition between the chow diet given *ad libitum* and the liquid diets used for the pair feeding. The inhibitory effect on proteolysis that occurred exclusively in the microsomal fraction was associated with a rapid increase in liver weight and disappeared with prolongation of alcohol administration. By contrast, the turnover of cytosolic proteins was not affected at any time. Previous results by others (Hofmann and Hosein, 1978) indicate that chronic ethanol consumption markedly accelerates mitochondrial protein degradation. With regard to the degradation of retained secretory proteins (crinophagy), it can be calculated from the data of Baraona *et al.* (1977) that the difference between the rate of synthesis of proalbumin and the rate of release of albumin into the serum should result in an accumulation of albumin far greater than that actually measured. Thus, it is likely that most of the retained proteins undergo degradation. After acute administration of ethanol, increased lipolysis (but not proteolysis) of retained VLDL was documented in the secretory vesicle fraction of the Golgi complex isolated from rat liver, with no significant changes in lysosomal activities (Falk *et al.*, 1985). Similarly, a preliminary report by Donohue *et al.* (1989b) indicates that the addition of 25 mM ethanol to liver slices obtained from rats, the proteins of which were prelabeled *in vivo* with radioactive leucine, increased the rate of degradation of retained plasma proteins (recognizable with anti-rat-serum antibodies).

A primary inhibition of protein catabolism could contribute to the increased protein content of the liver induced by alcohol consumption but does not account for the marked changes in protein composition and subcellular distribution discussed in greater detail in the next section.

5.6. Hepatic Accumulation of Protein after Chronic Ethanol Feeding

Two of the earliest and most conspicuous features of the hepatic damage produced by alcohol are the deposition of fat and the enlargement of the liver. The hepatomegaly has been traditionally attributed to the accumulation of lipids. However, in animals fed alcohol-containing diets, it was shown that lipids account for only half of the increase in liver dry weight (Lieber *et al.*, 1965); the other half is almost totally accounted for by an increase in proteins (Baraona *et al.*, 1975) (Fig. 5.11). The alcohol-induced increase in liver proteins (compared to pair-fed controls) occurred whether ethanol was substituted for carbohydrate or for lipid in the control diet (Baraona *et al.*, 1981a; Baraona and Lieber, 1983). This indicates that the effect is caused by ethanol and not by the associated manipulations of the other energy sources. The increase in protein was not associated with changes in concentration, indicating that water was retained in proportion to the increase in protein. The mechanism of water retention is not fully elucidated; it has been argued that the oncotic pressure of retained proteins is insufficient to account for the increase in cell water (Israel *et al.*, 1979), but the rise in both protein and amino acids (Baraona *et al.*, 1979) is probably accompanied by an increase in associated small ions, which could account osmotically for a large fraction of the

water increase. The increases in lipid, protein, amino acid, water, and electrolytes are also associated with increased size of the hepatocytes. The number of hepatocytes and the hepatic content of deoxyribonucleic acid (DNA) did not change significantly after alcohol treatment, and thus the hepatomegaly was mainly accounted for by the increased cell volume (Baraona *et al.*, 1975, 1977). Microspectrophotometric methods of hepatocytes isolated from ethanol-fed rats (stained with Feulgen and naphthol yellow S for DNA and protein, respectively) showed that, in addition to a 30% increase in protein, there was a 6% increase in DNA (Gaub *et al.*, 1981). The increased ploidy was caused by increased percentage of mono- or binuclear polyploid cells. This difference became particularly prominent after hepatectomy. This suggests that ethanol impairs cytokinesis in the telophase, perhaps as a consequence of impaired microtubular function. This is in keeping with the observation that in rapidly dividing hematopoietic cells, chronic ethanol administration also produces karyokinetic alterations with increased frequency of micronuclei (Baraona *et al.*, 1981b), in which alterations of the spindle microtubules could be involved. Polyploidy by itself could contribute to a small extent to the increase in cell proteins and hepatomegaly. There is also an increase in the number of hepatic mesenchymal cells after ethanol feeding (Baraona *et al.*, 1975), but this increase does not significantly contribute to the hepatomegaly because of the small contribution of these cells to the liver volume.

5.6.1. Subcellular Site of Protein Accumulation in the Liver

As discussed before, alcohol consumption results in proliferation of the smooth membranes of the endoplasmic reticulum both in rat and in humans (Iseri *et al.*, 1966; Lane and Lieber, 1966). This finding was subsequently confirmed (Rubin *et al.*, 1968; Carulli *et al.*, 1971) and established on a biochemical basis by the demonstration of an increase in both phospholipids and total protein content of the smooth membranes (Ishii *et al.*, 1973). The increase in microsomal protein, however, accounted for only 30% of the total protein increase. Mitochondria are also grossly altered after alcohol administration (Iseri *et al.*, 1966; Lane and Lieber, 1966; Rubin *et al.*, 1972). Swollen and giant mitochondria are commonly observed, but total mitochondrial proteins did not contribute significantly to the total

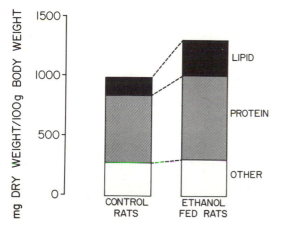

FIGURE 5.11. Effects of ethanol feeding on rat liver dry weight and protein and lipid contents. In ethanol-fed rats, the accumulation of lipids accounts for approximately half of the increase in liver dry weight; the other half is almost totally accounted for by proteins. (From Baraona *et al.*, 1975.)

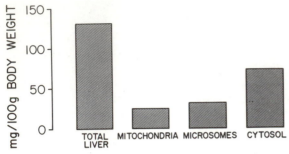

FIGURE 5.12. Distribution among subcellular fractions of the increase in liver protein induced by ethanol feeding. Most of the protein accumulation occurs in the cytosol. There is also significant accumulation in the microsomes. The trend toward an increase in mitochondrial proteins was not significant. (From Baraona *et al.*, 1975.)

increase in liver protein. More than half of the total increase in liver protein was actually due to increased soluble proteins of the cytosol (Baraona *et al.*, 1975) (Fig. 5.12).

5.6.2. Type of Proteins that Accumulate in the Liver after Chronic Alcohol Consumption

Despite the lack of changes in the concentration of total hepatic proteins, the concentration of some proteins that are primarily destined for export into the plasma (such as albumin and transferrin) was found significantly increased in the liver of ethanol-fed rats, whereas the concentration of soluble constituent proteins (such as ferritin) decreased (Baraona *et al.*, 1975, 1977). The increased concentration of export proteins reflects an even greater increase in the amount per total liver, since alcohol administration produced hepatomegaly. Conversely, decreased concentration of constituent proteins after alcohol feeding may merely reflect dilution in enlarged protein and water pools.

5.6.2.1. Contribution of Export Proteins

It was shown, in the case of albumin, that the increase involves precursor proteins (such as proalbumin) as well as mature serum albumin (Baraona *et al.*, 1977). Proteins destined for export are synthesized by bound ribosomes, discharged into the cisterna of the rough endoplasmic reticulum, and then transported to the smooth endoplasmic reticulum and Golgi apparatus (Pe-

ters, 1962; Glaumann, 1970; Glaumann and Ericsson, 1970; Peters *et al.*, 1971; Morgan and Peters, 1971a,b,; Redman and Cherian, 1972). Movement of secretory vesicles from the endoplasmic reticulum to the Golgi and from the Golgi to the plasma membrane involves vesicular transport. After acute administration of ethanol, about half of the retained secretory protein accumulated in the Golgi, while the other half distributed in the RER, SER, and cytosol (Volentine *et al.*, 1986). Though after chronic ethanol consumption immunoreactive albumin accumulated preferentially in the cytosol and transferrin in the microsomal fraction, the concentration of these proteins increased in both compartments (Baraona *et al.*, 1977). Since no evidence of leakage from the microsomes was obtained, the possibility that the cytosol might serve as a storage site of retained and partially digested proteins has to be considered. By immunoelectron microscopy, albumin was found to accumulate in the endoplasmic reticulum but not the cytosol of alcoholics (Feldman and Maurice, 1977). In the latter compartment, however, the concentration of albumin may have been too small to be detectable by the technique used.

More recently, a significant increase in the concentration of transferrin (but not albumin) was found in liver biopsies from 29 alcoholic patients with hepatitis or various degrees of liver fibrosis after several days of alcohol withdrawal (Matsuda *et al.*, 1983). There were very good correlations among the degree of ballooning, the hepatic retention of transferrin, and the decrease in polymerized tubulin (Matsuda *et al.*, 1983, 1985a). Immunohistochemically, striking retention of transferrin was demonstrated in the ballooned hepatocytes of alcoholics, whereas other liver diseases with some degree of ballooning did not show such an accumulation (Matsuda *et al.*, 1985a). The authors suggested that the presence of these swollen hepatocytes loaded with immunorecognizable transferrin could be an important differential feature between alcoholic and nonalcoholic ballooning. Furthermore, the authors produced striking centrolobular ballooning with retention of transferrin and microtubular disruption by feeding rats with alcohol-containing diets during chronic ADH inhibition with pyrazole (Matsuda *et al.*, 1985b; Takada *et al.*, 1986). This experimental lesion, so-called "alcohol–pyrazole hepatitis," mimics some of the features of alcoholic hepatitis, and its development is accelerated by high-fat diets (Takada *et al.*, 1986). These investigators suggested that the lesions are a consequence of the enhanced oxidation of ethanol by the MEOS pathway

(Takada *et al.*, 1986) and probably are mediated by increased exposure to acetaldehyde as a result of slower oxidation of this metabolite when it originates in the microsomes rather than in the cytosol (Yasuhara *et al.*, 1986).

The retention of transferrin in the liver was associated with the appearance of transferrin with reduced sialic acid content in the serum (Matsuda *et al.*, 1983, 1985b), an alteration previously described in alcoholics and proposed as a good marker of alcohol abuse (Stibler *et al.*, 1979, 1980, 1986; Behrens *et al.*, 1988a,b). At the initial stages of alcoholic liver injury, the serum levels of total transferrin are normal or only slightly decreased. Turnover studies in patients with alcoholic fatty liver revealed increased synthesis and turnover of transferrin (Potter *et al.*, 1985). As the liver injury progresses toward cirrhosis, the serum levels fall, and the turnover decreases. In the alcoholics with various degrees of fibrosis studied by Matsuda *et al.* (1983, 1985a), the retention in the liver was associated with decreased levels of transferrin in the serum. It is not clear at present whether the alterations of transferrin are primary effects of ethanol on glycosylation or secondary to either alterations in the secretion or in the recycling of transferrin through the liver by receptor-mediated endocytosis.

It must be pointed out that the increases in these two export proteins (albumin and transferrin) account for only a small fraction of the total increase in soluble proteins. Thus, the major contributors to the ethanol-induced accumulation of liver protein remain to be identified. It is likely that export proteins other than albumin and transferrin are also retained. In fact, accumulation of α-fetoprotein has been found in association with decreased levels in the plasma (Weesner *et al.*, 1980) in rats fed ethanol-containing diets. The degree of retention may be particularly important for those export proteins (such as lipoproteins) the production of which is greatly enhanced by ethanol feeding (Baraona and Lieber, 1970: Baraona *et al.*, 1973; Savolainen *et al.*, 1986). That the intracellular transport of lipoproteins is indeed hampered by ethanol was shown by Marinari *et al.* (1978). Increased lipolysis of lipoproteins retained in secretory vesicles after ethanol has also been documented (Falk *et al.*, 1985). Other findings (Burnett, 1979) also suggest that the effect of alcohol on lipoprotein output results from two opposing actions of ethanol, namely, a stimulatory effect on the synthesis of lipoprotein triglycerides and an inhibitory effect on their secretion.

5.6.2.2. Contribution of Constituent Proteins

Until recently, all the constituent proteins of the cytosol that were measured in ethanol-fed animals, such as ferritin and tubulin (Baraona *et al.*, 1977), actin (Zern *et al.*, 1985), ligandin (Reyes *et al.*, 1971), and alcohol (Nomura *et al.*, 1983) and lactic (Pignon *et al.*, 1987) dehydrogenases, have been found either unchanged or decreased in concentration. A remarkable exception has been the recent report (Pignon *et al.*, 1987) that chronic alcohol administration markedly increases the hepatic concentration and content of fatty-acid-binding protein (Fig. 5.13). This low-molecular-weight protein (or group of proteins) is very abundant in the cytosol of the hepatocyte, normally accounting for about 5% of the total cytosolic proteins. Thus, the ethanol-induced increase in this single protein accounted for almost one-third of the total increase in cytosolic proteins, thereby becoming the largest single contributor to this increase.

5.6.3. Possible Consequences of the Ethanol-Induced Effects on Liver Proteins

Some of these changes, such as the increase in fatty-acid-binding protein, may represent a compensatory response to the inhibitory effects of ethanol on mitochondrial fatty acid oxidation. This protein seems to play a key role in maintaining small concentrations of

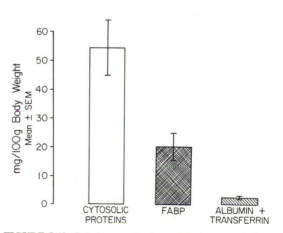

FIGURE 5.13. Relative contributions of the increases in immunoreactive fatty-acid-binding protein and albumin plus transferrin to the total increase in cytosolic proteins induced by chronic ethanol administration to rats. (Data from Pignon *et al.*, 1987; Baraona *et al.*, 1977.)

potentially deleterious free fatty acids by binding and stimulating their esterification (Bass, 1985; Glatz and Veerkamp, 1985). The alterations of transferrin may impair the recycling of iron and be linked to the siderosis frequently observed in the liver of the alcoholics. The swelling of the hepatocytes after chronic alcohol administration was found to be associated with a reduction of the intercellular space and with portal hypertension (Israel et al., 1979; Israel and Orrego, 1983). The interesting feature of this factor of portal hypertension is its potential reversibility. The relative contribution of this and other factors to the development of portal hypertension in the alcoholic is discussed in Chapter 7. The "ballooning" of the hepatocyte is commonly associated with necrosis and inflammation, as discussed in Chapter 7. It is conceivable that the striking architectural disorganization accompanying this alteration may impair the survival of these cells. Severe alcoholic liver damage is also frequently associated with the formation of Mallory bodies (see Chapter 7), clusters of fibrillar material that resemble intermediary filaments normally present in various tissues. These fibrils are morphologically and chemically different from microtubules and from microfilaments (actin) (French et al., 1976) and may represent prekeratin (Denk et al., 1981). However, because antitubulins such as griseofulvin and colchicine have the ability to induce the formation of Mallory bodies in mice (Denk et al., 1975; Denk and Eckerstorfer, 1977), it is possible that the appearance of these bodies in alcoholic hepatitis may be linked to the antitubulin effects of chronic alcohol ingestion. Acetaldehyde also binds with high-affinity reactive lysines of other components of the cytoskeleton, such as actin (Xu et al., 1989). Thus, whether it acts solely as an antitubular agent or whether other mechanisms are involved, chronic ethanol consumption promotes the retention of protein in the liver with possible associated deleterious effects. Accordingly, one may wonder whether protein restriction might be useful in patients in whom hepatocyte enlargement and ballooning are present.

5.7. Dietary Requirements for Protein in the Alcoholic

That dietary protein can modify the effect of alcohol on the liver has been clearly shown in rodents. Severe protein deficiency (Lieber et al., 1969) potentiates the effects of alcohol. The role of nutritional factors is less clear with regard to liver injury in primates, as discussed in detail in Chapter 7. At present, the optimal diet for the alcoholic is not established, particularly with regard to proteins. In rats, carbon-tetrachloride-induced cirrhosis can be prevented by a low-protein diet (Bhuyan et al., 1965). Early studies relating beneficial effects of high-protein diets (Patek et al., 1948) lacked proper controls. Subsequently the risk in cirrhotic patients of dietary-induced encephalopathy has become more apparent, as discussed in Chapter 14. Therefore, in the absence of experimental data to the contrary, high-protein diets do not seem indicated at the present time, and until this issue is resolved it may be prudent to settle for an intake of proteins that does not exceed the individual protein tolerance of the cirrhotic or the recommended amount (Munro, 1978), whichever is lower. In patients with liver disease, tolerance to dietary protein may be related not only to the amount but also to its amino acid content, with better tolerance for milk than for meat or blood proteins. Vegetable proteins were tolerated best of all (Greenberger et al., 1977), but their assimilation under these conditions has been questioned (see Chapter 17).

5.8. Summary

Alcohol consumption affects protein metabolism in various tissues. Most of the studies have focused on brain, intestine, muscle, and liver protein metabolism. Although there is agreement that alcohol consumption decreases brain protein synthesis, an alteration that could lead to the cerebral atrophy found in alcoholics, there is controversy with regard to the effects of ethanol on hepatic protein metabolism. Although inhibitory effects of ethanol on isolated liver preparations are readily demonstrable, no change or even stimulation of protein synthesis by alcohol consumption has been found in vivo under conditions that lead to alcoholic liver damage. In fact, at least during the initial steps, alcohol-induced liver injury is associated with accumulation of protein in the liver and hepatomegaly. The proteins that accumulate include proteins destined for export into the plasma. Experiments in vivo and in vitro indicate that, regardless of the effects of ethanol on protein synthesis, alcohol also impairs secretion of proteins from the liver. The mechanism of the secretory alteration is not fully elucidated, but that it may be linked to a decrease in liver microtubules, a cytoskeletal organelle required for intracellular transport, secretion, and other functions that depend on a high degree of architectural organization. Moreover, the degree of microtubular disruption correlates with the capacity of ethanol to increase intrahepatic acetaldehyde concentrations. This metabolite can bind to tubulin, the microtubule protein, and inhibit its polymerization during microtubule assembly. Acetaldehyde has also been incriminated as a mediator of the secretory defect. Accumulation of protein and water in the hepatocyte, along with alterations of the cytoskeleton, may be the basis for the "ballooning" of the hepatocytes, a lesion frequently associated with cell necrosis in alcoholics (see Chapter 7). The answer to the important question of whether dietary protein supplementation is beneficial or deleterious for the development of alcoholic liver injury requires further studies. In any event, in the patient with severe liver disease, protein tolerance is decreased.

REFERENCES

Abrams, M. A., and Cooper, C.: Quantitative analysis of metabolism of hepatic triglyceride in ethanol treated rats. *Biochem. J.* **156**:33–46, 1976.

Adibi, S. A.: Interrelationships between levels of amino acids in plasma and tissues during starvation. *Am. J. Physiol.* **221**:829–838, 1971.

Adibi, S. A., Modesto, T. A., Morse, E. L., and Amin, P. M.: Amino acid levels in plasma liver and skeletal muscle during protein deprivation. *Am. J. Physiol.* **225**:408–414, 1973.

Albertini, A., Fiaccavento, S., and Bonera, B.: Turnover of liver ribosomes in ethanol-intoxicated rats. *Chem. Biol. Interact.* **2**:57–60, 1970.

Algramati, I. D., and Sabatini, D. D.: Effect of protease inhibitors on albumin secretion in hepatoma cells. *Biochem. Biophys. Res. Commun.* **90**:220–226, 1979.

Ashworth, C. T., Johnson, C. F., and Wrightman, F. J.: Hepatic composition and morphologic correlations of hepatic protein synthesis in acute ethanol intoxication in rats. *Am. J. Pathol.* **46**:757–773, 1965.

Avogaro, A., Cibin, M., Croatto, T., Rizzo, A., Gallimbert, L., and Tiengo, A.: Alcohol intake and withdrawal: Effects on branched-chain amino acids and alanine. *Alcoholism: Clin. Exp. Res.* **10**:300–304, 1986.

Banks, W. L., Kline, E. S., and Higgins, E. S.: Hepatic composition and metabolism after ethanol consumption in rats fed liquid purified diets. *J. Nutr.* **100**:581–593, 1970.

Baranyai, J. M., and Blum, J. J.: Quantitative analysis of intermediary metabolism in rat hepatocytes incubated in the presence and absence of ethanol with a substrate mixture including ketoleucine. *Biochem. J.* **258**:121–140, 1989.

Baraona, E., and Lieber, C. S.: Effects of chronic ethanol feeding on serum lipoprotein metabolism in the rat. *J. Clin. Invest.* **49**:769–778, 1970.

Baraona, E., and Lieber, C. S.: Effects of ethanol on hepatic protein synthesis and secretion. In: *Currents in Alcoholism,*

Vol. 1 (F. A. Seixas, ed.), New York, Grune & Stratton, pp. 7–15, 1977.

Baraona, E., and Lieber, C. S.: Effects of alcohol on hepatic transport of proteins. *Annu. Rev. Med.* **33**:281–292, 1982.

Baraona, E., and Lieber, C. S.: Effects of ethanol on hepatic protein metabolism. In: *Alcohol and Protein Synthesis: Ethanol, Nucleic Acid, and Protein Synthesis in the Brain and Other Organs*, NIAAA Research Monograph-10, DHHS publication No. (ADM) 83-1198, pp. 75–95, 1983.

Baraona, E., Pirola, R. C., and Lieber, C. S.: Pathogenesis of postprandial hyperlipemia in rats fed ethanol-containing diets. *J. Clin. Invest.* **52**:296–303, 1973.

Baraona, E., Leo, M. A., Borowsky, S. A., and Lieber, C. S.: Alcoholic hepatomegaly: Accumulation of protein in the liver. *Science* **190**:794–795, 1975.

Baraona, E., Leo, M. A., Borowsky, S. A., and Lieber, C. S.: Pathogenesis of alcohol-induced accumulation of protein in the liver. *J. Clin. Invest.* **60**:546–554, 1977.

Baraona, E., Matsuda, Y., Pikkarainen, P., Finkelman, F., and Lieber, C. S.: Exaggeration of the ethanol-induced decrease in liver microtubules after chronic alcohol consumption. *Gastroenterology* **76**:1274, 1979.

Baraona, E., Pikkarainen, P., Salaspuro, M., Finkelman, F., and Lieber, C. S.: Acute effects of ethanol on hepatic protein synthesis and secretion in the rat. *Gastroenterology* **79**:104–111, 1980.

Baraona, E., Matsuda, Y., Pikkarainen, P., Finkelman, F., and Lieber, C. S.: Effects of ethanol on hepatic protein secretion and microtubules. Possible mediation by acetaldehyde. *Curr. Alcohol.* **8**:421–434, 1981a.

Baraona, E., Guerra, M., and Lieber, C. S.: Cytogenetic damage of bone marrow cells produced by chronic alcohol consumption. *Life Sci.* **29**:1797–1802, 1981b.

Baraona, E., Shaw, S., and Lieber, C. S.: Alcohol-induced changes of amino acid and protein metabolism. In: *Medical Disorders of Alcoholism: Pathogenesis and Treatment* (C. S. Lieber, ed.), Philadelphia, Saunders, pp. 178–236, 1982.

Baraona, E., Jauhonen, P., Miyakawa, H., and Lieber, C. S.: Zonal redox changes as a cause of selective perivenular hepatotoxicity of alcohol. *Pharmacol. Biochem. Behav.* **18**:449–454, 1983.

Baraona, E., Finkelman, F., and Lieber, C. S.: Reevaluation of the effects of alcohol consumption on rat liver microtubules: Effect of feeding status. *Res. Commun. Chem. Path. Pharmacol.* **44**:265–278, 1984.

Bass, N. M.: Function and regulation of hepatic and intestinal fatty acid binding protein. *Chem. Phys. Lipids.* **38**:95–114, 1985.

Beesley, R. C.: Ethanol inhibits Na^+-gradient-dependent uptake of L-amino acids into intestinal brush border membrane vesicles. *Dig. Dis. Sci.* **31**:987–992, 1986.

Behrens, U. J., Worner, T. M., Braly, L. F., Schaffner, F., and Lieber, C. S.: Carbohydrate-deficient transferrin (CDT), a marker for chronic alcohol consumption in different ethnic populations. *Alcoholism: Clin. Exp. Res.* **12**:427–432, 1988a.

Behrens, U. J., Worner, T. M., and Lieber, C. S.: Changes in carbohydrate deficient transferrin (CDT) levels after alcohol withdrawal. *Alcoholism: Clin. Exp. Res.* **12**:539–544, 1988b.

Bengtsson, G., Smith-Kielland, A., and Morland, J.: Ethanol

effects on protein synthesis in nonparenchymal liver cells, hepatocytes, and density populations of hepatocytes. *Exp. Mol. Pathol.* **41**:44–57, 1984.

Berman, W. J., Gil, J., Jennett, R. B., Tuma, D., Sorrell, M. F., and Rubin, E.: Ethanol, hepatocellular organelles, and microtubules. A morphometric study *in vivo* and *in vitro*. *Lab. Invest.* **48**:760–767, 1983.

Bhuyan, U. N., Nayak, N. C., Deo, M. G., and Ramalingaswami, V.: Effect of dietary protein on carbon tetrachloride-induced hepatic fibrogenesis in albino rats. *Lab. Invest.* **14**:184–190, 1965.

Blank, G. S., Yaffe, M. B., Szasz, J., George, E., Rosenberry, T. L., and Sternlicht, H.: The role of Lys 394 in microtubule assembly. *Ann. NY Acad. Sci.* **466**:467–481, 1986.

Borisy, G. G.: A rapid method for quantitative determination of microtubule protein using DEAE-cellulose filters. *Anal. Biochem.* **50**:373–385, 1972.

Branchey, L., and Lieber, C. S.: Activation of tryptophan pyrrolase after chronic alcohol administration. *Subst. and Alcohol Actions/Misuse* **2**:225–229, 1982.

Branchey, L., Shaw, S., Lieber, C. S.: Ethanol impairs tryptophan transport into the brain and depresses serotonin. *Life. Sci.* **29**:2751–2755, 1981.

Burke, J. P., Tumbelson, M. E., Hicklin, K. W., and Wilson, R. B.: Effect of chronic ethanol ingestion on mitochondrial protein synthesis in Sinclair (s-1) miniature swine. *Proc. Soc. Exp. Biol. Med.* **148**:1051–1056, 1974.

Burnett, D. A.: Paradoxical effect of ethanol on hepatic triglyceride secretion. *Gastroenterology* **77**:A–6, 1979.

Buydens-Branchey, L., Branchey, M., Worner, T. M., Zucker, D., Aramsobatdee, E., and Lieber, C. S.: Increase in tryptophan oxygenase activity in alcoholic patients. *Alcoholism: Clin. Exp. Res.* **12**:163–168, 1988.

Carulli, N., Manenti, F., Gallo, M., and Salvioli, G. F.: Alcohol-drugs interactions in man: Alcohol and tolbutamide. *Eur. J. Clin. Invest.* **1**:421–424, 1971.

Cerdan, S., Robles, S. S., Ayuso, M. S., and Parrilla, R.: Acute effects of ethanol on hepatic protein synthesis in the rat *in vivo*. *IRCS Med. Sci.* **8**:229, 1980.

Cerdan, S., Sanchez, S., Martin-Requero, A., Ayuso-Parrilla, M. S., and Parrilla, R.: Role of the state of reduction of the NAD system on the regulation of hepatic protein synthesis in the rat *in vivo*. *Intl. J. Biochem.* **14**:615–620, 1982.

Chambers, J. W., and Piccirillo, V. J.: Effects of ethanol on amino-acid uptake and utilization by the liver and other organs of rats. *Q. J. Stud. Alcohol* **34**:707–717, 1973.

Chambers, J. W., Georg, R. H., and Bass, A. D.: The effect of ethanol on the uptake of α-aminoisobutyric acid by the isolated perfused rat liver. *Life Sci.* **5**:2293–2300, 1966.

Chang, T., Lewis, J., and Glazko, A. J.: Effect of ethanol and other alcohols on the transport of amino acids and glucose by everted sacs of rat small intestine. *Biochim. Biophys. Acta* **135**:1000–1007, 1967.

Combescot, C., Courte, M.-T., Reynouard, F., and Weill, J.: Modifications de l'incorporation de l'uridine tritée dans l'acide ribonucléique hépatique du rat Wistar après l'administration prologée per os d'éthanol. *C. R. Soc. Biol.* **163**:2410–2412, 1969.

Courtoy, P. J., Lombart, C., Feldman, G., Rogier, E., and Mo-guilevsky, N.: Synchronous increase in four acute phase proteins synthesized by the same hepatocytes during the inflammatory reaction. A combined biochemical and morphologic kinetics study in the rat. *Lab. Invest.* **44**:105–115, 1981.

Criss, W. E.: A review of isozymes in cancer. *Cancer Res.* **31**:1523–1542, 1971.

Criss, W. E.: Control of the adenylate charge in the Morris "minimal-deviation" hepatomas. *Cancer Res.* **33**:51–56, 1973.

Crow, K. E., Cornell, N. W., Veech, R. L.: The rate of ethanol metabolism in isolated rat hepatocytes. *Alcoholism: Clin. Exp. Res.* **1**:43–47, 1977.

Cruz, M. A., Bravo, I., Rojas, S., and Gallardo, V.: Effects of ethanol ingestion on amino acid uptake in the dog liver *in vivo*. *Pharmacology* **30**:12–19, 1985.

Dajani, R. M., and Kouyoumjian, R.: A probable direct role of ethanol in the pathogenesis of fat infiltration in the rat. *J. Nutr.* **91**:535–539, 1967.

David, E. T., Fischer, I., and Moldave, K.: Studies on the effect of ethanol on eukaryotic protein synthesis *in vitro*. *J. Biol. Chem.* **258**:7702–7706, 1983.

Denk, H., and Eckerstorfer, R.: Colchicine-induced Mallory body formation in the mouse. *Lab. Invest.* **36**:563–565, 1977.

Denk, H., Gschnait, F., and Wolff, K.: Hepatocellular hyalin (Mallory bodies) in long term griseofulvin-treated mice: A new experimental model for study of hyaline formation. *Lab. Invest.* **32**:773–776, 1975.

Denk, H., Franke, W. W., Dragosics, B., and Zeiler, I.: Pathology of cytoskeleton of liver cells: Demonstration of Mallory bodies (alcoholic hyalin) in murine and human hepatocytes by immunofluorescent microscopy using antibodies to cytokeratin polypeptides from hepatocytes. *Hepatology* **1**:9–20, 1981.

Dich, J., and Tönnesen, I. C.: Effects of ethanol, nutritional status, and composition of the incubation medium on protein synthesis in isolated rat liver parenchymal cells. *Arch. Biochem. Biophys.* **204**:640–647, 1980.

Dinda, P. K., and Beck, I. T.: Effect of ethanol on peptidases of hamster jejunal brush-border membrane. *Am. J. Physiol.* **242**:G442–G447, 1982.

Domschke, S., Domschke, W., and Lieber, C. S.: Hepatic redox state: Attenuation of the acute effects of ethanol induced by chronic ethanol consumption. *Life Sci.* **15**:1327–1334, 1974.

Donohue, T. M., Sorrell, J. H., Sorrell, M. F., and Tuma, D. J.: Measurement of protein synthetic activity by determination of peptidyl[³H]puromycin formation in liver slices after ethanol administration. *Alcoholism: Clin. Exp. Res.* **9**:526–530, 1985.

Donohue, T. M., Sorrell, M. F., and Tuma, D. J.: Hepatic protein synthetic activity *in vivo* after ethanol administration. *Alcoholism: Clin. Exp. Res.* **11**:80–86, 1987.

Donohue, T. M., Zetterman, R. K., and Tuma, D. J.: Effect of chronic ethanol administration on protein catabolism in rat liver. *Alcoholism: Clin. Exp. Res.* **13**:49–57, 1989a.

Donohue, T. M., Chaisson, M. L., Zetterman, D., Krako, J. A., and Zetterman, R. K.: Effects of ethanol on catabolism of plasma proteins. *Alcoholism: Clin. Exp. Res.* **13**:326, 1989b.

Dorio, R. J., Hoek, J. B., and Rubin, E.: Ethanol treatment selectively decreases neutral amino acid transport in cultured hepatocytes. *J. Biol. Chem.* **259**:11430–11435, 1984.

Dunlop, D. S., Van Elden, W., and Lajtha, A.: A method for measuring brain protein synthesis rates in young adult rats. *J. Neurochem.* **24**:337–344, 1975.

Edwards, K., Nagashima, M., Dryburgh, H., Wykes, A., and Schreiber, G.: Secretion of protein from liver cells is suppressed by the proteinase inhibitor N-alpha-tosyl-L-lysyl chloromethane, but not by tunycamicin, an inhibitor of glycosylation. *FEBS Lett.* **100**:269–272, 1979.

Eletsky, Y. K., and Merkulov, M. F.: The effect of ethanol on the ³H-uridine incorporation in RNA of murine hepatic cells. *Farmakol. Toksikol.* (Moskva) **32**:561–564, 1969.

Elko, E. E., Wooles, W. R., and DiLuzio, N. R.: Alterations and mobilization of lipids in acute ethanol-treated rats. *Am. J. Physiol.* **201**:923–926, 1961.

Estler, C. J.: Comparison of hepatic triglyceride content and hepatic lipid secretion after various doses of ethanol. *Biochem. Pharmacol.* **24**:1871–1873, 1975.

Falk, M., Ahlberg, J., and Glaumann, H.: Ethanol intoxication stimulates lipolysis in isolated Golgi complex secretory vesicle fraction from rat liver. *Virchows Arch. (Cell Pathol.)* **49**:231–239, 1985.

Feldman, G., and Maurice, M.: Morphological findings of liver protein synthesis and secretion. In: *Membrane Alterations or Basis of Liver Injury* (H. Popper, L. Bianchi, and W. Reutter, eds.), Lancaster, MTP Press, p. 61, 1977.

Feldman, G., Koziner, B., Talamo, R., and Bloch, K. J.: Familial variable immunodeficiency: Autosomal dominant pattern as inheritance with variable expression of the defect(s). *J. Pediatr.* **87**:534–539, 1975.

Fellenius, E., Carlgren, H., and Kiessling, H.-H.: Inhibition of glucogenesis from proline by ethanol oxidation. *Life Sci.* **13**:595–599, 1973.

Fernandez-Checa, J. C., Ookhtens, M., and Kaplowitz, N.: Effect of chronic ethanol feeding on rat hepatocytic glutathione. Compartmentation, efflux, and response to incubation with ethanol. *J. Clin. Invest.* **80**:57–62, 1987.

Fisher, S. E., Atkinson, M., Van Thiel, D. H., Rosenblum, E., David, R., and Holtzman, I.: Selective fetal malnutrition: Effect of ethanol and acetaldehyde upon in vitro uptake of α-aminoisobutyric by human placenta. *Life Sci.* **29**:1283–1288, 1981.

Fisher, S. E., Barnicle, M. A., Steis, B., Holzman, I., and Van Thiel, D. H.: Effects of acute ethanol exposure upon *in vivo* leucine uptake and protein synthesis in the fetal rat. *Pediatr. Res.* **15**:335–339, 1981.

Freedman, M. L., and Rosman, J.: A rabbit reticulocyte model for the role of hemin-controlled repressor in hypochromic anemias. *J. Clin. Invest.* **37**:594–603, 1976.

Freedman, M. L., Cohen, H. S., Rosman, J., and Forte, F. J.: Ethanol inhibition of reticulocyte protein synthesis: The role of Haem. *Br. J. Haematol.* **30**:351–363, 1975.

Freinkel, N., Arky, R. A., Singer, D. L., Cohen, A. K., Bleicher, S. J., Anderson, J. B., Silbert, C. K., and Foster, A. E.: Alcohol hypoglycemia IV. Current concepts of its pathogenesis. *Diabetes* **14**:350–361, 1965.

French, S. W., Sim, J. S., Franks, K. E., Burbige, E. J., Denton, T., and Caldwell, M. G.: Alcoholic hepatitis. In: Alcoholic Hepatitis (M. M. Fisher and J. G. Rankin, eds.), New York, Plenum, 1977.

Gabriel, L., Bonelli, G., and Dianzani, M. U.: Inhibition of colchine binding to rat liver tubulin by aldehydes and by linoleic acid hydroperoxide. *Chem. Biol. Interact.* **19**:101–109, 1977.

Gang, H., Lieber, C. S., and Rubin, E.: Ethanol increases glycosyl transferase activity in the hepatic Golgi apparatus. *Nature [New Biol.]* **243**:123–125, 1973.

Gaub, J., Fauerholdt, L., Keidin, S., Kondrup, J., Petersen, P., and Wantzin, G. L.: Cytomorphometry of liver cells from ethanol-fed rats: ethanol causes increased polyploidization and protein accumulation. *Eur. J. Clin. Invest.* **11**:235–237, 1981.

Girbes, T., Susin, A., Ayuso, M. S., and Parrilla, R. P.: Acute effects of ethanol in the control protein synthesis in isolated rat liver cells. *Arch. Biochem. Biophys.* **226**:37–49, 1983.

Glatz, J. F. C., and Veerkamp, J. H.: Intracellular fatty acid-binding proteins. *Int. J. Biochem.* **17**:13–22, 1985.

Glaumann, H.: Studies on the synthesis and transport of albumin in microsomal subfractions from rat liver. *Biochim. Biophys. Acta* **224**:206–218, 1970.

Glaumann, H., and Ericsson, J. L. E.: Evidence for the participation of the Golgi apparatus in the intracellular transport of nascent albumin in the liver cell. *J. Cell Biol.* **47**:555–567, 1970.

Gordon, E. R.: Effect of an intoxication dose of ethanol on lipid metabolism in an isolated perfused rat liver. *Biochem. Pharmacol.* **21**:2991–3004, 1972.

Green, R. S., MacDermid, R. G., Scheig, R. L., and Hajjar, J. J.: Effect of ethanol on amino acid absorption across *in vivo* rat intestine. *Am. J. Physiol.* **241**:G176–G181, 1981.

Greenberger, N. J., Carley, J., Schenker, S., Bettinger, I., Stamnes, C., and Beyer, P.: Effects of vegetable and animal protein diets in chronic hepatic encephalopathy. *Dig. Dis.* **22**:845–855, 1977.

Hajjar, J. J., Krasner, K., Van Linda, B., and Scheig, R.: Effect of ethanol on glycylsarcosine absorption. *Life Sci.* **14**:1403–1407, 1981a.

Hajjar, J. J. Tomicic, T., Scheig, R. L.: Effect of chronic ethanol consumption on leucine absorption in the rat small intestine. *Digestion* **22**:170–176, 1981b.

Harbitz, I., Wallin, B., Hauge, J. G., and Morland, J.: Effect of ethanol metabolism on initiation of protein synthesis in rat hepatocytes. *Biochem. Pharmacol.* **33**:3465–3470, 1984.

Heitman, D. W., Frosto, T. A., Schenker, S., and Henderson, G. I.: Stimulatory effects of ethanol on amino acid transport by rat fetal hepatocytes. *Hepatology* **7**:307–314, 1987.

Henderson, G. I., Hoyumpa, A. M., Rothschild, M. A., and Schenker, S.: Effect of ethanol and ethanol-induced hypothermia on protein synthesis in pregnant and fetal rats. *Alcoholism: Clin. Exp. Res.* **4**:165–177, 1980.

Hirayama, C., and Hiroshige, K.: Effect of alpha-tocopherol and tocopheronolactone on ethanol induced fatty liver and triglyceridemia. *Experientia* **26**:1306–1308, 1970.

Hofmann, I., and Hosein, E. A.: Effects of chronic ethanol consumption on the rate of rat liver mitochondrial protein turnover and synthesis. *Biochem. Pharmacol.* **27**:457–463, 1978.

Iseri, O. A., Lieber, C. S., and Gottfried, L. S.: The ultrastructure of fatty liver induced by prolonged ethanol ingestion. *Am. J. Pathol.* **48**:535–555, 1966.

Ishii, H., Joly, J.-G., and Lieber, C. S.: Effect of ethanol on the amount and enzyme activities of hepatic rough and smooth microsomal membranes. *Biochem. Biophys. Acta* **291**:411–420, 1973.

Israel, Y., and Orrego, H.: On the characteristics of alcohol-induced liver enlargement and its possible hemodynamic consequences. *Pharmacol. Biochem. Behav.* **18**:433–437, 1983.

Israel, Y., Khanna, J. M., Orrego, H., Rachmin, G., Wahid, S., Britton, R., Macdonald, A., and Kalant, H.: Studies on metabolic tolerance to alcohol, hepatomegaly and alcoholic liver disease. *Drug. Alcohol. Dep.* **4**:109–118, 1979.

Jauhonen, P., Baraona, E., Miyakawa, H., and Lieber, C. S.: Mechanism for selective perivenular hepatotoxicity of ethanol. *Alcoholism: Clin. Exp. Res.* **6**:350–357, 1982.

Jauhonen, P., Baraona, E., Lieber, C. S., and Hassinen, I. E.: Dependence of ethanol-induced redox shift on hepatic oxygen tensions prevailing *in vivo*. *Alcohol* **2**:163–167, 1985.

Jeejeebhoy, K. N., Phillips, M. J., Bruce-Robertson, A., Ho, J., and Sodtke, U.: The acute effect of ethanol on albumin, fibrinogen and transferrin synthesis in the rat. *Biochem. J.* **126**:1111–1126, 1972.

Jeejeebhoy, K. N., Bruce-Robertson, A., Ho, J., and Sodtke, U.: The effect of ethanol on albumin and fibrinogen synthesis *in vivo* and in hepatocyte suspensions. In: *Alcohol and Abnormal Protein Synthesis* (M. A. Rothschild, M. Oratz, and S. S. Schrieber, eds.), New York, Pergamon, pp. 373–391, 1975.

Jennett, R. B., Tuma, D. J., and Sorrell, M. F.: Effect of ethanol and its metabolites on microtubule formation. *Pharmacology* **21**:363–368, 1980.

Jennett, R. B., Sorrell, M. F., Johnson, E. L., and Tuma, D. J.: Covalent binding of acetaldehyde to tubulin: Evidence for preferential binding to the α-chain. *Arch. Biochem. Biophys.* **256**:10–18, 1987.

Jennett, R. B., Sorrell, M. F., Saffari-Fard, A., Ockner, J. L., and Tuma, D. J.: Preferential covalent binding of acetaldehyde to the α-chain of purified rat liver tubulin. *Hepatology* **9**:57–62, 1989.

Kanayama, R., Takase, S., Matsuda, Y., and Takada, A.: Effect of dietary fat upon ethanol metabolism in rats. *Biochem. Pharmacol.* **33**:3283–3287, 1984.

Kawahara, H., Matsuda, Y., and Takada, A.: Effects of ethanol on the microtubules of cultured rat hepatocytes. *Alcohol Alcoholism* (**Suppl. 1**):307–311, 1987.

Khairallah, E. K., and Mortimore, G. E.: Assessment of protein turnover in perfused rat liver. Evidence for amino acid compartmentation from differential labeling of free and t-RNA bound valine. *J. Biol. Chem.* **251**:1375–1384, 1976.

Khawaja, J. A., and Lindholm, D. B.: Differential effect of ethanol ingestion on the protein synthetic activities of free and membrane-bound ribosomes from liver of the weanling rat. *Res. Commun. Chem. Pathol. Pharmacol.* **19**:129–139, 1978.

Kirsch, R. E., Frith, L. O., Stead, R. H., and Saunders, S. J.: Effect of alcohol on albumin synthesis by the isolated perfused rat liver. *Am. J. Clin. Nutr.* **26**:1191–1194, 1973.

Klatskin, G.: The effect of ethyl alcohol on nitrogen excretion in the rat. *Yale J. Biol. Med.* **34**:124–143, 1961.

Krebs, H. A., Freedland, A., Hems, R., and Stubbs, M.: Inhibition of hepatic gluconeogenesis by ethanol. *Biochem. J.* **112**:117–124, 1969.

Kreisberg, R. A., Siegal, A. M., and Owen, W. C.: Alanine and gluconeogenesis in man: Effect of ethanol. *J. Clin. Endocrinol.* **34**:876–883, 1972.

Kuriyama, K., Sze, P. Y., and Rauscher, G. E.: Effects of acute and chronic ethanol administration on ribosomal protein synthesis in mouse brain and liver. *Life Sci.* **10**:181–189, 1971.

Lane, B. P. and Lieber, C. S.: Ultrastructural alterations in human hepatocytes following ingestion of ethanol with adequate diets. *Am. J. Pathol.* **49**:594–603, 1966.

LeBouton, A. V.: Heterogeneity of protein metabolism between liver cells as studied by radioautography. *Curr. Mod. Biol.* **2**:111–114, 1968.

LeBouton, A. V.: Relations and extent of the zone of intensified protein metabolism in the liver acinus. *Curr. Mod. Biol.* **3**:4–8, 1969.

LeMarchand, Y., Singh, A., Assimacopoulos-Jeannet, F., Orci, L., Rouiller, C., and Jeanrenaud, B.: A role for the microtubular system in the release of very low density lipoproteins by perfused mouse livers. *J. Biol. Chem.* **248**:6862–6870, 1973.

LeMarchand, Y., Patzelt, C., Assimacopoulos-Jeannet, F., Loten, E. G., and Jeanrenaud, B.: Evidence for a role of the microtubular system in the secretion of newly synthesized albumin and other proteins by the liver. *J. Clin. Invest.* **53**:1512–1517, 1974.

Lieber, C. S., and DeCarli, L. M.: Quantitative relationship between amount of dietary fat and severity of alcoholic fatty liver. *Am. J. Clin. Nutr.* **23**:474–478, 1970.

Lieber, C. S., and Rubin, E.: Ethanol, a hepatotoxic drug. *Gastroenterology* **54**:642–646, 1968a.

Lieber, C. S., Rubin, E.: Alcoholic fatty liver in man on a high protein and low fat diet. *Am. J. Med.* **44**:200–206, 1968b.

Lieber, C. S., Jones, D. P., and DeCarli, L. M.: Effects of prolonged ethanol intake: Production of fatty liver despite adequate diets. *J. Clin. Invest.* **44**:1009–1021, 1965.

Lieber, C. S., Spritz, N., and DeCarli, L. M.: Fatty liver produced by dietary deficiencies: Its pathogenesis and potentiation by ethanol. *J. Lipid Res.* **10**:283–287, 1969.

Lieber, C. S., Rubin, E., DeCarli, L. M., Misra, P. S., and Gang, H.: Effects of pyrazole on hepatic function and structure. *Lab. Invest.* **22**:615–621, 1970.

Lin, G. W.-J.: Effect of ethanol feeding during pregnancy on placental transfer of alpha-aminoisobutyric acid in the rat. *Life Sci.* **28**:595–601, 1981.

Lindros, K. O., and Aro, H.: Ethanol-induced changes in levels of metabolites related to the redox state and ketogenesis in rat liver. *Ann. Med. Exp. Fenn.* **47**:39–42, 1969.

Madsen, N. P.: Reduced serum low-density lipoprotein levels after acute ethanol administration. *Biochem. Pharmacol.* **18**:261–262, 1969.

Mailliard, M. E., Sorrell, M. F., Volentine, G. D., and Tuma, D. J.: Impaired plasma membrane glycoprotein assembly in the liver following acute ethanol administration. *Biochem. Biophys. Res. Commun.* **123**:951–958, 1984.

Marinari, U. M., Cottalasso, D., Gambella, G. R., Averame, M. M., Pronzato, M. A., and Nanni, G.: Effects of acute ethanol intoxication of ^{3}H-palmitic acid transport through hepatocyte Golgi apparatus. *FEBS Lett.* **86**:53–56, 1978.

Matsuda, Y., Baraona, E., Salaspuro, M., and Lieber, C. S.: Pathogenesis and role of microtubular alterations in alcohol-induced liver injury. *Fed. Proc.* **37:**402, 1978.

Matsuda, Y., Baraona, E., Salaspuro, M., and Lieber, C. S.: Effects of ethanol on liver microtubules and Golgi apparatus. Possible role in altered hepatic secretion of plasma proteins. *Lab. Invest.* **41:**455–463, 1979.

Matsuda, Y., Takada, A., Kanayama, R., and Takase, S.: Changes of hepatic microtubules and secretory proteins in human alcoholic liver disease. *Alcoholism: Phmarmacol. Biochem. Behav.* **18 (Suppl. 1):**479–482, 1983.

Matsuda, Y., Takada, A., Sato, H., Yasuhara, M., and Takase, S.: Comparison between ballooned hepatocytes occurring in human alcoholic and non-alcoholic liver diseases. *Alcoholism: Clin. Exp. Res.* **9:**366–370, 1985a.

Matsuda, Y., Takase, S., Takada, A., Sato, H., and Yasuhara, M.: Comparison of ballooned hepatocytes in alcoholic and non-alcoholic liver injury in rats. *Alcohol* **2:**303–308, 1985b.

Matsuda, Y., Yasuhara, M., and Takada, A.: Subcellular differences in hepatic microtubular changes of alcoholic liver disease: morphometrical analysis. *Alcohol Alcohol.* **Suppl. 1:**503–507, 1987.

Matthews, D. M., and Adibi, S. A.: Peptide absorption. *Gastroenterology* **71:**151–161, 1976.

McDonald, J. T., and Margen, S.: Wine versus ethanol in human nutrition. I. Nitrogen and calorie balance. *Am. J. Clin. Nutr.* **29:**1093–1103, 1976.

McKinnon, G., Davidson, M., De Jersey, J. D., Shanley, B., and Ward, L.: Effects of acetaldehyde on polymerization of microtubule proteins. *Brain Res.* **416:**90–99, 1987.

Meijer, A. J., and Williamson, J. R.: Transfer of reducing equivalents across the mitochondrial membrane. I. Hydrogen transfer mechanisms involved in the reduction of pyruvate to lactate in isolated liver cells. *Biochim. Biophys. Acta* **333:**1–11, 1974.

Meijer, A. J., Gimpel, J. A., Deleeuw, G. A., Tager, J. M., and Williamson, J. R.: Role of anion translocation across the mitochondrial membrane in the regulation of urea synthesis from ammonia by isolated rat hepatocytes. *J. Biol. Chem.* **250:**7728–7738, 1975.

Moojerkea, S., and Chow, A: Impairment of glycoprotein synthesis in acute ethanol intoxication in rats. *Biochim. Biophys. Acta* **184:**83–92, 1969.

Morgan, E. H., and Peters, T.: The biosynthesis of rat serum albumin. V. Effect of protein depletion and refeeding on albumin and transferrin synthesis. *J. Biol. Chem.* **246:**3500, 1971a.

Morgan, E. H., and Peters, T.: Intracellular aspects of transferrin synthesis and secretion in the rat. *J. Biol. Chem.* **246:**3508–3511, 1971b.

Morland, J.: Effects of chronic ethanol treatment on tryptophan oxygenase, tyrosine aminotransferase and general protein metabolism in the intact and perfused rat liver. *Biochem. Pharmacol.* **73:**71–75, 1974.

Morland, J.: Incorporation of labelled amino acids into liver protein after acute ethanol administration. *Biochem. Pharmacol.* **24:**439–442, 1975.

Morland, J., and Bessessen, A.: Inhibition of protein synthesis by

ethanol in isolated rat liver parenchymal cells. *Biochim. Biophys. Acta.* **474:**312–320, 1977.

Morland, J., and Sjetnan, A. E.: Effect of ethanol intake on the incorporation of labeled amino acids into liver protein. *Biochem. Pharmacol.* **25:**2125–2130, 1976.

Morland, J., Bessesen, A., and Svendsen, L.: Incorporation of labeled amino acids into proteins of isolated parenchymal and non-parenchymal rat liver cells in the absence and presence of ethanol. *Biochim. Biophys. Acta* **561:**464–474, 1979a.

Morland, J., Flengsrud, R., Prydz, H., and Svendsen, L.: Hepatic amino acid levels in rats after long-term ethanol feeding. *Biochem. Pharmacol.* **28:**423–427, 1979b.

Morland, J., Bessesen, A., and Svendsen, L.: The role of alcohol metabolism in the effect of ethanol on protein synthesis in isolated rat hepatocytes. *Alcoholism: Clin. Exp. Res.* **4:**313–321, 1980.

Morland, J., Rothschild, M. A., Oratz, M., Mongelli, J., Donor, D., and Schreiber, S. S.: Protein secretion in suspensions of isolated rat hepatocytes: No influence of acute ethanol administration. *Gastroenterology* **80:**159–165, 1981.

Morland, J., Bessesen, A., Smith-Kielland, A., and Wallin, B.: Ethanol and protein metabolism in the liver. *Pharmacol. Biochem. Behav.* **18 (Suppl 1):**252–256, 1983.

Mortimore, G. E., and Mondon, C. E.: Inhibition by insulin of valine turnover in the liver. Evidence for a general control of proteolysis. *J. Biol. Chem.* **245:**2375–2383, 1970.

Mortimore, G. E., and Pösö, A. R.: Mechanism and control of deprivation-induced protein degradation in liver: Role of glucogenic amino acids. In: *Glutamine Metabolism in Mammalian Tissues* (D. Häussinger and H. Sies, eds.). Springer-Verlag, Berlin, pp. 138–157, 1984.

Moseley, R. H., and Murphy, S. M.: Effects of ethanol on amino acid transport in basolateral liver plasma membrane vesicles. *Am. J. Physiol.* **256:**G458–465, 1989.

Munro, H. N.: *Committee on Dietary Allowances*. Washington, D. C., The National Research Council, National Academy of Sciences, 1978.

Murty, N., Verney, E., and Sidransky, H.: Acute effects of ethanol on membranes of the endoplasmic reticulum and on protein synthesis in the rat liver. *Alcoholism: Clin. Exp. Res.* **4:**93–103, 1980.

Nakano, K., and Hara, H.: Measurement of the protein-synthetic activity *in vivo* of various tissues in rats by using [^{3}H]puromycin. *Biochem. J.* **184:**663–668, 1979.

Nonura, F., Pikkarainen, P. H., Jauhonen, P., Arai, M., Gordon, E. R., Baraona, E., and Lieber, C. S.: The effect of ethanol administration on the metabolism of ethanol in baboons. *J. Pharmacol. Exp. Ther.* **227:**78–83, 1983.

Okanoue, T., Ou, O., Ohta, M., Yoshida, J., Horishi, M. Y. T., Okuno, T., and Takino, T.: Effect of chronic ethanol administration on the cytoskeleton of rat hepatocytes—including morphometric analysis. *Acta Hepat. Jpn.* **25:**210–213, 1984.

Ookhtens, M., Hobdy, K., Corvasce, M. C., Aw, T. Y., and Kaplowitz, N.: Sinusoidal efflux of glutathione in the perfused rat liver: Evidence for carrier-mediated process. *J. Clin. Invest.* **75:**258–265, 1985.

Oratz, M., Rothschild, M. A., and Schreiber, S. S.: Alcohol, amino acids, and albumin synthesis. II. Alcohol inhibition of albumin synthesis reversed by arginine and spermine. *Gastroenterology* **71:**123–127, 1976.

Oratz, M., Rothschild, M. A., and Schreiber, S. S.: Alcohol, amino acids and albumin synthesis. III. Effects of ethanol, acetaldehyde and 4-methylpyrazole. *Gastroenterology* **74:** 672–676, 1978.

Oratz, M., Rothschild, M. A., Schreiber, S. S., Burks, A., Mongelli, J., and Matarese, B.: The role of the urea cycle and polyamines in albumin synthesis. *Hepatology* **3:**567–571, 1983.

Patek, A. J., Jr., Post, J., Ratnoff, O. D., Mankin, H., and Hillman, R. W.: Dietary treatment of cirrhosis of the liver. *J.A.M.A.* **138:**543–549, 1948.

Patwardhan, R. V., Schenker, S., Henderson, G. I., Abou-Mourad, N. M., and Hoyumpa, A. M., Jr.: Short-term and long-term ethanol administration inhibits the placental uptake and transport of valine in rats. *J. Lab. Clin. Med.* **98:**251–262, 1981.

Paul, H. S., and Adibi, S. A.: Role of ATP in regulation of branched-chain α-keto acid dehydrogenase activity in liver and muscle mitochondria of fed, fasted, and diabetic rats. *J. Biol. Chem.* **257:**4875–4881, 1982.

Perin, A., and Sessa, A.: *In vitro* effects of ethanol and acetaldehyde on tissue protein synthesis. In: *The Role of Acetaldehyde in the Action of Ethanol* (K. O. Lindros, and C. J. P. Eriksson, eds.), Satellite Symp. 6th Intl. Congr. Pharmacol. Helsinki, The Finnish Foundation for Alcohol Studies, pp. 105–122, 1975.

Perin, A., Scalabrino, G., Sessa, A., and Arnaboldi, A.: *In vitro* inhibition of protein synthesis in rat liver as consequence of ethanol metabolism. *Biochim. Biophys. Acta* 366:101–108, 1974.

Perkoff, G. Y., Hardy, P., and Velez-Garcia, E.: Reversible acute muscular syndrome in chronic alcoholism. *N. Engl. J. Med.* **274:**1277–1285, 1966.

Peters, T., Jr.: The biosynthesis of rat serum albumin. II. Intracellular phenomena in the secretion of newly formed albumin. *J. Biol. Chem.* **237:**1186–1189, 1962.

Peters, T., Jr., Fleischer, B., and Fleischer, S.: The biosynthesis of rat serum albumin. IV. Apparent passage of albumin through the Golgi apparatus during secretion. *J. Biol. Chem.* **246:** 240–244, 1971.

Piccirillo, V. J., and Chambers, J. W.: Inhibition of hepatic uptake of alpha amino-isobutyric acid by ethanol: effects of pyrazole and metabolites of ethanol. *Res. Commun. Chem. Pathol. Pharmacol.* **13:**297–308, 1976.

Pierson, J. L., and Mitchell, M. C.: Increased hepatic afflux of glutathione after chronic ethanol feeding. *Biochem. Pharmacol.* **35:**1533–1537, 1986.

Pignon, J.-P., Bailey, N. C., Baraona, E., and Lieber, C. S.: Fatty acid-binding protein: a major contributor to the ethanol-induced increase in liver cytosolic proteins in the rat. *Hepatology* **7:**865–871, 1987.

Pipeleers, D. G., Pipeleers-Marichal, M. A., and Kipnis, D. M.: Physiological regulation of total tubulin and polymerized tubulin in tissues. *J. Cell Biol.* **74:**351–357, 1977.

Plapp, F. V., Updike, R. D., and Chiga, M.: Effects of ethanol on mouse liver polysomal disaggregation by dimethylnitrosamine and lasiocarpine. *FEBS Lett.* **18:**121–123, 1971.

Porta, E. A., Koch, O. R., Gomez-Dumm, C. L. A. and Hartroft, W. S.: Effects of dietary protein on the liver of rats in experimental chronic alcoholism. *J. Nutr.* **94:**437–446, 1968.

Pösö, A. R.: Ethanol and hepatic protein turnover. *Alcohol Alcohol.* **Suppl. 1:**83–90, 1987.

Pösö, A. R., Surmacz, C. A., and Mortimore, G. E.: Inhibition of intracellular protein degradation by ethanol in perfused rat liver. *Biochem. J* **242:**459–464, 1987.

Potter, B. J., Chapman, R. W. G., Nunes, R. M., Sorrentino, D., and Sherlock, S.: Transferrin metabolism in alcoholic liver disease. *Hepatology* **5:**714–721, 1985.

Preedy, V. R., and Peters, T. J.: The effect of chronic ethanol ingestion on protein metabolism in Type-I and Type-II-fiber-rich skeletal muscles of the rat. *Biochem. J.* **254:**631–639, 1988.

Princen, J. M. G., Mol-Backx, G. P. B. M., and Yap, S. H.: Acute effects of ethanol intake on albumin and total protein synthesis in free and membrane-bound polyribosomes of rat liver. *Biochim. Biophys. Acta.* **655:**119–127, 1981.

Quistorff, B., Chance, B., and Takeda, H.: Two- and three-dimensional redox heterogeneity of rat liver. Effects of anoxia and alcohol on the lobular redox pattern. In: *Frontiers of Biological Energetics: Electrons to Tissues* (L. T. Dutton, L. S. Leigh, and A. Scarpa, eds.), New York, Academic Press, pp. 1487–1497, 1978.

Raatikainen, O. J., and Mäenpää, P. H.: Ethanol and liver protein synthesis *in vivo. Experientia* **36:**527, 1980.

Rawat, A. K.: Effect of maternal ethanol consumption on foetal and neonatal hepatic protein synthesis. *Biochem. J.* **160:**653–661, 1976.

Reaven, E. P., and Reaven, G. M.: Evidence that microtubules play a permissive role in hepatocyte very low density lipoprotein secretion. *J. Cell Biol.* **84:**28–29, 1980.

Redman, C. M., and Cherian, G.: The secretory pathways of rat serum glycoproteins and albumin. Localization of newly formed proteins with the endoplasmic reticulum. *J. Cell Biol.* **52:**231–245, 1972.

Redman, C. M., Banerjee, D., Howell, K., and Palade, G. E., Colchicine inhibition of plasma protein release from rat hepatocytes. *J. Cell Biol.* **66:**42–59, 1975.

Redman, C. M., Banerjee, D., Manning, C., Huang, C. Y., and Green, K.: *In vivo* effects of colchicine on hepatic protein synthesis and on the conversion of proalbumin to serum albumin. *J. Cell Biol.* **77:**400–416, 1978.

Reinus, J. F., Heymsfield, S. B., Wiskind, R., Casper, K., and Galambos, J. T.: Ethanol: relative fuel value and metabolic effects *in vivo. Metabolism* **38:**125–135, 1989.

Renis, M., Giovine, A., and Bertolino, A.: Protein synthesis in mitochondrial and microsomal fractions from rat brain and liver after acute and chronic ethanol administration. *Life Sci.* **16:**1447–1458, 1975.

Reyes, H., Levi, A. J., Gatmaitan, Z., and Arias, I. M.: Studies of Y and Z, two hepatic cytoplasmic organic anion-binding proteins: Effect of drugs, chemicals, hormones, and cholestasis. *J. Clin. Invest.* **50:**2242–2252, 1971.

Rodrigo, C., Antezana, C., and Baraona, E.: Fat and nitrogen balance in rats with alcohol-induced fatty liver. *J. Nutr.* **101:** 1307–1310, 1971.

Rosa, Y., and Rubin, E.: Effects of ethanol on amino acid uptake by rat liver cells. *Lab. Invest.* **43**:366–372, 1980.

Rothschild, M. A., Oratz, M., Zimmon, D., Schreiber, S. S., Winer, I., and Van Caneghem, A.: Albumin synthesis in cirrhotic subjects with ascitis studied with carbonate-^{14}C. *J. Clin. Invest.* **48**:344–350, 1969.

Rothschild, M. A., Oratz, M., Mongelli, J., and Schreiber, S. S.: Alcohol induced depression of albumin synthesis: Reversal by tryptophan. *J. Clin. Invest.* **50**:1812–1818, 1971.

Rothschild, M. A., Oratz, M., and Schreiber, S. S.: Alcohol, amino acids, and albumin synthesis. *Gasteroenterology* **67**: 1200–1213, 1974.

Rothschild, M. A., Oratz, M., and Schreiber, S. S.: Effects of nutrition and alcohol on albumin synthesis: *Alcoholism: Clin. Exp. Res.* **7**:28–30, 1983.

Rothschild, M. A., Oratz, M., and Schrieber, S. S.: Effects of ethanol on protein synthesis. *Ann. NY Acad. Sci.* **492**:233–244, 1987.

Rubin, E.: Alcoholic myopathy in heart and skeletal muscle. *N. Engl. J. Med.* **301**:28–33, 1979.

Rubin, E., and Lieber, C. S.: Early fine structural changes in human liver induced by alcohol. *Gastroenterology* **52**:1–13, 1967a.

Rubin, E., and Lieber, C. S.: Experimental alcoholic hepatic injury in man: Ultrastructural changes. *Fed. Proc.* **26**:1458–1467, 1967b.

Rubin, E., Hutterer, F., and Lieber, C. S.: Ethanol increases hepatic smooth endoplasmic reticulum and drug metabolizing enzymes. *Science* **159**:1469–1470, 1968.

Rubin, E., Beattie, D. S., and Lieber, C. S.: Effects of ethanol on the biogenesis of mitochondrial functions. *Lab. Invest.* **23**: 620–627, 1970.

Rubin, E., Beattie, D. S., Toth, A., and Lieber, C. S.: Structural and functional effects of ethanol on hepatic mitochondria. *Fed. Proc.* **31**:131–140, 1972.

Salaspuro, M. P., and Kesanienni, U. A.: Intravenous galactose elimination tests with and without ethanol loading in various clinical conditions. *Scand. J. Gastroenterol.* **8**:681–686, 1973.

Salaspuro, M. P., Shaw, S., Jayatilleke, E., Ross, W. A., and Lieber, C. S.: Attenuation of the ethanol induced hepatic redox change after chronic alcohol consumption in baboons: Metabolic consequences *in vivo* and *in vitro*. *Hepatology* **1**: 33–38, 1981.

Savolainen, M. J., Baraona, E., Leo, M. A., and Lieber, C. S.: Pathogenesis of the hypertriglyceridemia at early stages of alcoholic liver injury in the baboon. *J. Lipid Res.* **27**:1073–1083, 1986.

Schapiro, R. H., Drummey, G. D., Shimizu, Y., and Isselbacher, K. J.: Studies on the pathogenesis of ethanol-induced fatty liver. II. Effect of ethanol on palmitate-1-C^{14} metabolism by isolated perfused rat liver. *J. Clin. Invest.* **43**:1338–1347, 1964.

Schenker, S., Dicke, J. M., Johnson, R. F., Hays, S. E., and Henderson, G. I.: Effect of ethanol on human placental transport of model amino acids and glucose. *Alcoholism: Clin. Exp. Res.* **13**:112–119, 1989.

Schmitt, J. W., and Elbein, A. D.: Inhibition of protein synthesis also inhibits synthesis of lipid-linked oligosaccharides. *J. Biol. Chem.* **254**:12291–12294, 1979.

Schreiber, S. S., Oratz, M., Rothschild, M. A., Reff, F., and Evans, C.: Alcoholic cardiomyopathy. II. The inhibition of cardiac microsomal protein synthesis by acetaldehyde. *J. Mol. Cell Cardiol.* **6**:207–213, 1974.

Seakins, A., and Robinson, D. S.: Changes associated with the production of fatty livers by white phosphorus and by ethanol in the rat. *Biochem. J.* **92**:308–312, 1964.

Shaw, S., and Lieber, C. S.: Plasma amino acid abnormalities in the alcoholic: Respective role of alcohol, nutrition, and liver injury. *Gastroenterology* **74**:677–682, 1978.

Sherline, P., Bodwin, C. K., and Kipnis, D. M.: A new colchicine binding assay for tubulin. *Anal. Biochem.* **62**:400–407, 1974.

Sidransky, H., Verney, E., and Murty, C. N.: Effect of ethanol on polyribosomes and protein synthesis of transplantable hepatomas and host livers of rat. *Proc. Soc. Exp. Biol. Med.* **163**: 335–339, 1980.

Sjoblom, M., and Morland, J.: Metabolism of ethanol and acetaldehyde in parenchymal and nonparenchymal rat liver cells. *Biochem. Pharmacol.* **28**:3417–3423, 1979.

Smith, D. J., Da Silva Daley, G., and Polch, S. A.: Ethanol consumption decreases alanine uptake by rat basolateral liver plasma membrane vesicles. *Gastroenterology* **98**:429–436, 1990.

Smith, S. L., Jennett, R. B., Sorrell, M. F., and Tuma, D. J.: Acetaldehyde substoichiometrically inhibits bovine neurotubulin polymerization. *J. Clin. Invest.* **84**:337–341, 1989.

Smith-Kielland, A., and Morland, J.: Changes in protein, RNA and DNA content in various rat organs after long-term intake of ethanol. *Acta Pharmacol. Toxicol.* **45**:122–130, 1979.

Smith-Kielland, A., and Morland, J.: Reduced hepatic protein synthesis after long term ethanol treatment in fasted rats. Dependence on animal handling before measurement. *Biochem. Pharmacol.* **30**:2377–2379, 1981.

Smith-Kielland, A., Bengtsson, G., Svendsen, L., and Morland, J.: Protein synthesis in different populations of rat hepatocytes separated according to density. *J. Cell Physiol.* **110**:262–266, 1982.

Smith-Kielland, A., Blom, G. P., Svendsen, L., Bessesen, A., Morland, J.: A study of hepatic protein synthesis, three subcellular enzymes, and liver morphology in chronically ethanol fed rats. *Acta Pharmacol. Toxicol.* **53**:113–120, 1983a.

Smith-Kielland, A., Svendsen, L., Bessesen, A., and Morland, J.: Effect of chronic ethanol consumption on *in vivo* protein synthesis in livers from female and male rats fed two different diet regimens. *Alcohol Alcohol.* **18**:285–292, 1983b.

Sorrell, M. F., and Tuma, D. J.: Selective impairment of glycoprotein metabolism by ethanol and acetaldehyde in rat liver slices. *Gastroenterology* **75**:200–205, 1978.

Sorrell, M. F., Tuma, D. J., Schafter, E. C., and Barak, A. J.: Role of acetaldehyde in the ethanol-induced impairment of glycoprotein metabolism in rat liver slices. *Gastroenterology* **73**: 137–144, 1977a.

Sorrell, M. F., Tuma, D. J., and Barak, A. J.: Evidence that acetaldehyde irreversibly impairs glycoprotein metabolism in liver slices. *Gastroenterology* **73**:1138–1141, 1977b.

Sorrell, M. F., Nauss, J. M., Donohue, T. M., and Tuma, D. J.: Effects of chronic ethanol administration of hepatic glycoprotein secretion in the rat. *Gastroenterology* **84**:580–586, 1983.

Speisky, H., MacDonald, A., Giles, G., Orrego, H., and Israel, Y.: Increased loss and decrease synthesis of hepatic glutathione after acute ethanol administration. *Biochem. J.* **225**: 565–572, 1985.

Stanko, R. T., Morse, E. L., and Adibi, S. A.: Prevention of effects of ethanol on amino acid concentrations in plasma and tissues by hepatic lipotropic factors in rats. *Gastroenterology* **76**:132–138, 1979.

Stein, O., and Stein, Y.: Colchicine-induced inhibition of very low density lipoprotein release by rat liver *in vivo*. *Biochim. Biophys. Acta* **306**:142–147, 1973.

Stein, O., Sanger, L., and Stein, Y.: Colchicine-induced inhibition of lipoprotein and protein secretion into the serum and lack of interference with secretion of biliary phospholipids and cholesterol by rat liver *in vivo*. *J. Cell Biol.* **69**:90–103, 1974.

Stibler, H., Borg, S., and Allgulander, C.: Clinical significance of abnormal heterogeneity of transferrin in relation to alcohol consumption. *Acta. Med. Scand.* **206**:275–281, 1979.

Stibler, H., Borg, S., and Allgulander, C.: Abnormal microheterogeneity of transferrin—a new marker of alcoholism. *Subst. Alcohol Action/Misuse* **1**:247–252, 1980.

Stibler, H., Borg, S., and Joustra, M.: Micro anion exchange chromatography of carbohydrate-deficient transferrin in serum in relation to alcohol consumption. *Alcoholism* **10**:535–544, 1986.

Struck, D. K., Siuta, P. B., Lane, M. D., and Lennarz, W. J.: Effect of tunicamycin on the secretion of serum proteins on primary cultures of rat and chick hepatocytes. Studies on transferrin, very low density lipoproteins and serum albumin. *J. Biol. Chem.* **253**:5332–5337, 1978.

Swick, R. W.: Measurement of protein turnover in rat liver. *J. Biol. Chem.* **231**:751–764, 1958.

Takada, A., Matsuda, Y., and Takase, S.: Effects of dietary fat on alcohol–pyrazole hepatitis in rats: the pathogenetic role of the nonalcohol dehydrogenase pathway in alcohol-induced hepatic cell injury. *Alcoholism: Clin. Exp. Res.* **10**:403–411, 1986.

Teschke, R., Matsuzaki, S., Ohniski, K., DeCarli, L. M., and Lieber, C. S.: Microsomal ethanol oxidizing system (MEOS): Current status of its characterization and its role. *Alcoholism: Clin. Exp. Res.* **1**:7–15, 1977.

Tewari, S., and Noble, E. P.: Ethanol and brain protein synthesis. *Brain Res.* **26**:469–474, 1971.

Tuma, D. J., and Sorrell, M. F.: Effects of ethanol on the secretion of glycoproteins in rat liver slices. *Gastroenterology* **80**:273–278, 1981.

Tuma, D. J. Zetterman, R. K., and Sorrell, M. F.: Inhibition of glycoprotein secretion by ethanol and acetaldehyde in rat liver slices. *Biochem. Pharmacol.* **29**:35–38, 1980.

Tuma, D. J., Jennett, R. B.,and Sorrell, M. F.: Effect of ethanol on the synthesis and secretion of hepatic secretory glycoproteins and albumin. *Hepatology* **1**:590–598, 1981.

Tuma, D. J., Jennett, R. B., and Sorrell, M. F.: The interaction of acetaldehyde with tubulin. *Ann. NY Acad. Sci.* **492**:277–286, 1987.

Vazquez, J. A., Paul, H. S., and Adibi, S. A.: Regulation of leucine catabolism by caloric sources: Role of glucose and lipid in nitrogen sparing during nitrogen deprivation. *J. Clin. Invest.* **82**:1606–1613, 1988.

Veech, R. L., Eggleston, L. V., and Krebs, H. A.: The redox state of free nicotinamide–adenine dinucleotide phosphate in cytoplasm of rat liver. *Biochem. J.* **115**:609–619, 1969.

Veech, R. L., Guynn, R., and Veloso, D.: The time-course of the effects of ethanol on the redox and phosphorylation states of rat liver. *Biochem. J.* **127**:387–397, 1972.

Vendemiale, G., Altomare, E., Grattagliano, I., and Albano, O.: Increased plasma levels of glutathione and malondialdehyde after acute ethanol ingestion in humans. *J. Hepatol.* **9**:359–365, 1989.

Videla, L. A., Fernandez, V., Ugarte, G., and Valenzuela, A.: Effect of acute ethanol intoxication on the content of reduced glutathione of the liver in relation to its lipoperoxidative capacity in the rat. *FEBS Lett.* **111**:6–10, 1980.

Voci, A., Gallo, G., Balestrero, F., and Rugassa, E.: Effects of ethanol on protein metabolism in hepatocyte primary cultures from adult rat. *Cell. Biol. Int. Rep.* **12**:647–660, 1988.

Volentine, G. D., Tuma, D. J., and Sorrell, M. F.: Acute effects of ethanol on hepatic glycoprotein secretion in the rat *in vivo*. *Gastroenterology* **86**:225–229, 1984.

Volentine, G. D., Tuma, D. J., and Sorrell, M. F.: Subcellular location of secretory proteins retained in the liver during the ethanol-induced inhibition of hepatic protein secretion in the rat. *Gastroenterology* **90**:158–165, 1986.

Wahlander, A., Soboll, A., and Sies, H.: Hepatic mitochondrial and cytosolic glutathione content and the subcellular distribution of GSH-S transferases. *FEBS Lett.* **97**:138–140, 1975.

Wallin, B., Morland, J., and Fikke, A.-M.: Combined effects of ethanol and pH-change on protein synthesis in isolated rat hepatocytes. *Acta Pharmacol. Toxicol.* **49**:134–140, 1981.

Wallin, B., Bessesen, A., Fikke, A. M., Aabakke, J., and Morland, J.: No effect of acute ethanol administration on hepatic protein synthesis and export in the rat *in vivo*. *Alcoholism* **8**: 191–195, 1984.

Wallyn, C. S., Vidrich, A., Airhart, J., and Khairallah, E. A.: Analysis of the specific radioactivity of valine isolated from amino-acyl-transfer ribonucleic acid of rat liver. *Biochem. J.* **140**:545–548, 1974.

Ward, L. C., Carrington, L. E., and Daly, R.: Ethanol and leucine oxidation—I. Leucine oxidation by the rat *in vivo*. *Int. J. Biochem.* **17**:187–193, 1985a.

Ward, L. C., Ramm, G. A., Mason, S., and Daly, R.: Ethanol and leucine oxidation—II. Leucine oxidation by rat tissue *in vitro*. *Int. J. Biochem.* **17**:195–201, 1985b.

Wasterlain, C. G., Dwyer, B. E., Fando, J., Salinas-Fando, M., and Lieber, C. S.: The effects of ethanol and acetaldehyde on brain proteins and DNA synthesis are independent of nutritional effects. In: *Alcohol and Protein-Synthesis: Ethanol, Nucleic Acid, and Protein Synthesis in the Brain and Other Organs*, Washington, D.C., NIAAA, DHHS Public No. (ADM) 83-1198, pp. 199–218, 1983.

Weesner, R. E., Mendenhall, C. L., Morgan, D. D., Kessler, V., Kromme, C.: Serum-fetoprotein: Changes associated with acute and chronic ethanol ingestion in the resting and regenerating liver. *J. Lab. Clin. Med.* **95**:725–736, 1980.

Weigand, K., Otto, I., and Schopf, R.: Ficoll density separation of enzymatically isolated rat liver cells. *Acta Hepato-Gastroenterol.* **21:**245–253, 1974.

Weigand, K., Richter, E., and Esperer, H.-D.: Biochemical studies of isolated rat hepatocytes from normal and phenobarbital-treated liver as obtained by rate zonal centrifugation. *Acta Hepato-Gastroenterol.* **24:**170–174, 1977.

Wool, I. G., and Kurihara, K.: Determination of the number of active muscle ribosomes: Effects of diabetes and insulin. *Proc. NY Acad. Sci.* **58:**2401–2407, 1967.

Wooles, W. R.: Depressed fatty acid oxidation as a factor in the etiology of acute ethanol-induced fatty liver. *Life Sci.* **5:**267–276, 1966.

Wu, J. M.: Control of protein synthesis in rabbit reticulocytes. Inhibition of polypeptide synthesis by ethanol. *J. Biol. Chem.* **256:**4164–4167, 1981.

Xu, D. S., Jennett, R. B., Smith, S. L., Sorrell, M. F., and Tuma, D. J.: Covalent interactions of acetaldehyde with the actin/microfilament system. *Alcohol Alcohol.* **24:**281–289, 1989.

Yasuhara, M., Matsuda, Y., and Takada, A: Degradation of acetaldehyde produced by the nonalcohol dehydrogenase pathway. *Alcoholism: Clin. Exp. Res.* **10:**545–549, 1986.

Zern, M. A., Chakraborty, P. R., Ruiz-Opazo, N., Yap, S. H., and Shafritz, D. A.: Development and use of a rat albumin cDNA clone to evaluate the effect of chronic ethanol administration on hepatic protein synthesis. *Hepatology* **3:**317–322, 1983.

Zern, M. A., Leo, M. A., Giambrone, M.-A., and Lieber, C. S.: Increased type I procollagen mRNA levels and *in vitro* protein synthesis in the baboon model of chronic alcoholic liver disease. *Gastroenterology* **89:**1123–1131, 1985.

6

Interaction of Ethanol with Other Drugs

Charles S. Lieber

An "interaction" between alcohol and a drug is any alteration in the pharmacological properties of either because of the presence of the other. Interactions may be (1) *antagonistic*—the effect of either or both agents is blocked or reduced; (2) *additive*—the net effect of the combination is the sum of the effects of the individual agents; or (3) *supraadditive* (synergistic or potentiating)—the effect of the two agents in combination is greater than it would be if they were merely additive. Hypersensitivity is another type of drug interaction. It can occur in alcoholics or heavy drinkers whose alcohol-related pathologies may make them especially sensitive to other drugs. Chronic alcohol use can also produce *tolerance*; that is, a fixed amount of alcohol in the chronic alcohol abuser has less of an effect than it would in the nonchronic user. In *cross-tolerance*, a fixed amount of another drug has less of an effect in the alcohol user than in a nonuser. Indeed, chronic alcoholic patients have an increased tolerance to a variety of drugs when sober but a paradoxically increased susceptibility to them when intoxicated. These changes in the susceptibility to drugs are a result of changes in their rates of metabolism as well as of adaptive and synergistic effects of the drug on the primary site of action. Conversely, some effects are caused by the drugs' inhibition of alcohol metabolism. The latter effects of drugs on ethanol metabolism have been discussed in Chapter 1.

The present chapter focuses primarily on some interactions of acute and chronic ethanol consumption with the metabolism and/or toxicity of other drugs. Knowledge acquired over the last decade includes the recognition of an accessory pathway of ethanol metabolism in the liver, namely, the microsomal ethanol-oxidizing system (MEOS) (see Chapter 1). MEOS shares many properties with other microsomal drug-metabolizing enzymes, including utilization of cytochrome P-450, NADPH, and O_2 (Lieber and DeCarli, 1970). This sharing explains many of the alcohol–drug interactions. Furthermore, after chronic alcohol consumption, there is an increase in the activity of MEOS, and this induction "spills over" to various other drug-metabolizing systems in the microsomes, thereby affecting drug metabolism.

This new knowledge offers us better insight into altered drug metabolism in the alcoholic and possible pathological consequences. Theoretically, other mechanisms of ethanol–drug interactions are possible, but their role is largely speculative. Some of these are shown schematically in Fig. 6.1.

6.1. Interaction with Drug Absorption

Theoretically, alcohol consumption might alter drug absorption from the alimentary tract in a number of ways, notably in interfering with gastric emptying. There also have been reports of enhanced absorption of some drugs such as diazepam in the presence of ethanol

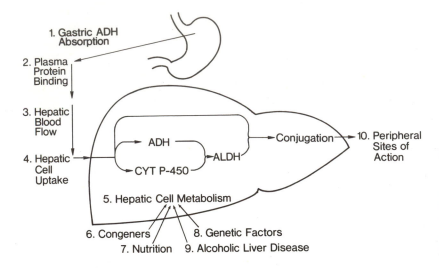

FIGURE 6.1. Schematic illustration of some of the sites of ethanol–drug interactions. Metabolic interactions may affect conjugations, cytochrome P-450-dependent microsomal pathways (CYT P-450), alcohol dehydrogenase (ADH), and acetaldehyde dehydrogenase (ALDH).

(Hayes *et al.*, 1977). Interactions of drugs with blood ethanol level through inhibition of gastric ADH have been reviewed in Chapter 1.

6.2. Interaction with Plasma Protein Binding

The effective therapeutic level of a drug depends not only on the total blood concentration but also on the degree of plasma protein binding. Thus, cirrhosis is associated with a doubling of the mean elimination half-life of lorazepam because of reduced plasma binding (Kraus *et al.*, 1978). In another study, patients with alcoholic cirrhosis had reduced protein binding of three organic bases (quinidine, dapsone, and triamterene) and of one organic acid (fluorescein), although another organic acid (diphenylhydantoin) was little affected (Affrime and Reidenberg, 1975). The basis of this reduced plasma protein binding appeared to be a nonspecific effect of hypoproteinemia. Ethanol may also directly alter plasma protein binding of drugs (Thiessen *et al.*, 1976). Whatever the mechanism, it seems that patients with alcoholic liver disease may develop toxicity from drugs such as quinidine even when the plasma concentrations are in the usual "therapeutic" range.

6.3. Interaction with Hepatic Blood Flow

A drug that is rapidly metabolized by the liver will be rapidly cleared from the hepatic circulation (high

"first-pass effect"). The rate of hepatic blood flow is then the dominant influence in the rate of clearance of that drug. In such cases, an effect of ethanol on hepatic blood flow could be important. There has been some controversy concerning the effects of ethanol on splanchnic blood flow, as discussed in detail in Chapter 1. In general, ethanol does not markedly affect the hepatic circulation.

6.4. Interaction with Hepatic Cell Uptake

Theoretically, ethanol might alter the plasma clearance of drugs by changing the hepatic concentrations of carrier proteins. However, there is as yet no experimental evidence for such an effect (Reyes *et al.*, 1971) except for the increase in the intrahepatic concentration of the fatty acid binding protein (Pignon *et al.*, 1987) (see Chapter 5).

6.5. Interaction with Hepatic Metabolism

6.5.1. Interaction of Ethanol with Cytochrome P-450-Dependent and Other Microsomal Drug Metabolism

The microsomal ethanol-oxidizing system (MEOS) shares with other microsomal drug-metabolizing systems many properties, including utilization of cytochrome P-450, NADPH, and O_2. This sharing explains many alcohol–drug interactions (Fig. 6.2). Furthermore, the induction of MEOS by chronic alcohol consumption is associated with corresponding increases in activity of other microsomal drug-metabolizing systems, with important clinical sequelae.

6.5.1.1. Acute Interaction

The main effect of the acute presence of ethanol is inhibition of drug metabolism (Fig. 6.2B and Fig. 6.3). The main mechanism involved appears to be competition for a partially common detoxification process, including competition for binding with cytochrome P-450 (Rubin *et al.*, 1971). Other mechanisms could include the release of steroid hormones that may inhibit some microsomal drug-metabolizing enzymes (Chung and Brown, 1976). In some instances, the NADH generated by the oxidation of ethanol might also inhibit citric acid cycle activity (Lieber, 1977) and possibly deplete intermediates that might be necessary for the generation of cytosolic NADPH (Reinke *et al.*, 1980).

The complexity of ethanol–drug interactions at the microsomal level is exemplified by the fact that for some drug-metabolizing systems (such as aniline hydroxylase), inhibition occurs at low ethanol concentrations, whereas for others (such as aminopyrine demethylase) high ethanol concentrations are required (Rubin *et al.*, 1970b). Furthermore, in the latter case, low ethanol concentrations were even stimulatory (Cinti *et al.*,

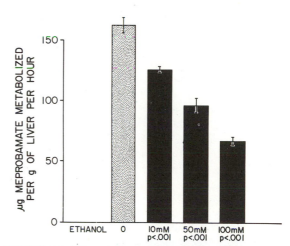

FIGURE 6.3. Effect of ethanol on metabolism of meprobamate by rat liver slices. Each flask contained 200 mg liver and was incubated for 120 min. Meprobamate concentration was 0.3 mM, including 1 μCi [14C]meprobamate. (From Rubin *et al.*, 1970b.)

1973), possibly because of enhanced NADH production and the likelihood that NADH may serve as a partial electron donor for microsomal drug-detoxifying systems (Cohen and Estabrook, 1971). However, the usual effect of the presence of ethanol is an inhibition of drug metabolism. The inhibitory effects may explain the observation that *in vivo*, simultaneous administra-

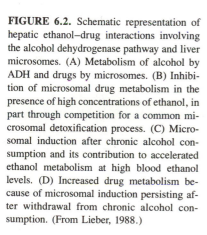

FIGURE 6.2. Schematic representation of hepatic ethanol–drug interactions involving the alcohol dehydrogenase pathway and liver microsomes. (A) Metabolism of alcohol by ADH and drugs by microsomes. (B) Inhibition of microsomal drug metabolism in the presence of high concentrations of ethanol, in part through competition for a common microsomal detoxification process. (C) Microsomal induction after chronic alcohol consumption and its contribution to accelerated ethanol metabolism at high blood ethanol levels. (D) Increased drug metabolism because of microsomal induction persisting after withdrawal from chronic alcohol consumption. (From Lieber, 1988.)

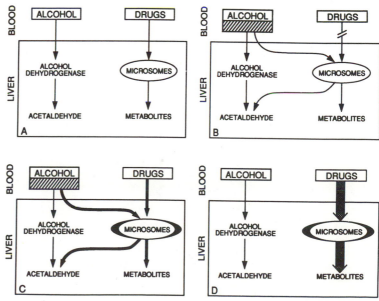

tion of ethanol and drugs slows the rate of drug metabolism (Rubin *et al.*, 1970b; Whitehouse *et al.*, 1975) (Fig. 6.3).

The acute interaction of ethanol with drug metabolism may have some important practical consequences. Indeed, in the United States more than 50% of all fatal road accidents are associated with an elevated blood alcohol level (Voas, 1973). One may wonder to what extent the loss of control on the road may be caused not only by ethanol itself but also by an ethanol–drug interaction, since a large segment of the population is given sedatives and tranquilizers. Indeed, many patients use drugs from this class in combination with alcohol because they are generally unaware that tranquilizers are CNS depressants that can increase the adverse effects of alcohol on performance skills and alertness. They are also unaware that ethanol, by competing with some of these drugs for an at least partially common microsomal detoxification pathway, may prolong and enhance the effects of these drugs. Indeed, the acute oral administration of ethanol results in a prolongation of the half-life of pentobarbital and meprobamate (see Fig. 6.4) in the blood in humans (Rubin *et al.*, 1970b). Alcohol acts synergistically with meprobamate (Equanil®, Miltown®) to depress performance tasks and driving-related skills such as time

estimation, attention, reaction time, body steadiness, oculomotor control, and alertness (Forney and Hughes, 1964). In the dog, the acute administration of ethanol reduced the clearance of benzodiazepines of low intrinsic clearance (demoxepan and desmethyldiazepam). Ethanol does not affect the clearance of benzodiazepines of higher intrinsic clearance (chlordiazepoxide and diazepam) when these drugs are given intravenously, probably because ethanol does not markedly affect hepatic blood flow, which in this case is rate limiting. However, ethanol does inhibit the metabolism (decreases the first-pass effect) of high-intrinsic-clearance drugs such as diazepam when those drugs are given orally (Hoyumpa *et al.*, 1979). In this latter case, hepatic blood flow is less of a rate-limiting factor, since the drug reaches the liver at relatively high concentrations directly via the portal vein. Thus, ethanol has been shown to inhibit the metabolism by microsomes of both high- and low-clearance drugs when they are given orally. These effects have some significant clinical implications.

The benzodiazepines (Librium®, Valium®, Dalmane®) are the most frequently used minor tranquilizers. Interaction between ethanol and benzodiazepines has been reviewed in detail by Hollister (1990). Benzodiazepines of low intrinsic clearance (demoxepam and desmethyldiazepam) have a reduced clearance in dogs given ethanol acutely. As discussed earlier, ethanol impairs the rate of disappearance from the plasma of benzodiazepines that have a high intrinsic clearance (diazepam, chlordiazepoxide) when these are given orally but not intravenously (Hoyumpa *et al.*, 1979; Desmond *et al.*, 1980; Whiting and Lawrence, 1979). Pharmacokinetic parameters of benzodiazepines metabolized via N-demethylation and/or hydroxylation tend to be more affected by ethanol consumption than those of benzodiazepines that are conjugated. In general, the longer-acting benzodiazepines such as diazepam, chlordiazepoxide, clobazam, and clorazepate are N-demethylated and/or hydroxylated, whereas shorter-acting compounds such as oxazepam and lorazepam are only conjugated. The 8-hr or longer dosing of ethanol with a single dose of diazepam, chlordiazepoxide, or lorazepam (Desmond *et al.*, 1980; Hoyumpa *et al.*, 1981; Sellers *et al.*, 1980) demonstrated that the presence of ethanol causes a decrease in the clearance of each of these benzodiazepines. At the doses used, the effect was greater for the oxidative metabolism of diazepam and chlordiazepoxide than for the conjugation of lorazepam. The results from other studies in which a single dose of ethanol was administered with a benzo-

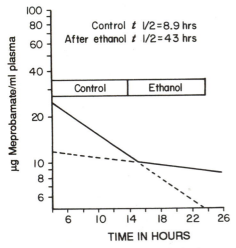

FIGURE 6.4. Effect of acute ethanol intoxication on disappearance of meprobamate from the blood in volunteer subjects. Subjects were given meprobamate, 12 to 15 mg/kg orally. Fourteen hours later, ethanol 1g/kg, was given and followed by 24 g every 2 hr. Solid lines are plotted from experimental points. Dashed lines are extrapolated. (From Rubin *et al.*, 1970c.)

diazepine are conflicting (Lane *et al.*, 1985). Clinically significant potentiation has been reported (Mørland *et al.*, 1974; Bo *et al.*, 1975).

The phenothiazines are also metabolized by microsomal pathways, and the acute administration of ethanol in conjunction with phenothiazines may result in competitive inhibition of hepatic metabolism with decreased clearance and enhanced sedative effects. Forrest *et al.* (1971) investigated the effects of short-term ethanol administration when combined with chlorpromazine in 12 patients taking long-term chlorpromazine therapy (400 to 900 mg daily). Up to a 33% decrease in urinary metabolites (conjugated plus unconjugated) occurred in 54% of the patient samples. Whether this was caused by decreased metabolism or decreased oral absorption of the drug cannot be determined from the reported data. In combination with alcohol there may be impaired coordination, and severe, potentially fatal respiratory depression (Zirkle *et al.*, 1959; Milner and Landauer, 1971). The use of phenothiazines to control symptoms of alcohol withdrawal may thus be hazardous. In addition, hypotension—a side effect of these drugs—may be exacerbated by alcohol.

Tricyclic antidepressants also increase susceptibility to convulsions and should be administered cautiously in alcohol withdrawal. Kissin (1974) found that the tricyclic antidepressants were either synergistic or antagonistic to ethanol according to their ratio of sedative to stimulant activity. Since these drugs produce hypotension, they should be prescribed only for alcoholics who can be carefully monitored. The combination of these drugs with alcohol adversely affects motor skills, particularly driving. Amitriptyline undergoes a 30 to 40% first-pass effect. Significant increases in both the amitriptyline total $AUC_{0-8 \, hr}$ (44%) and the free $AUC_{0-8 \, hr}$ resulted when it was combined with ethanol (Dorian *et al.*, 1983).

The interaction of ethanol with barbiturates represents a particular danger. In rats, the pentobarbital $t_{1/2}$ increased from 70 to 150 min, and in human subjects it was approximately doubled after ethanol administration (30 versus 59 hr) (Rubin *et al.*, 1970b). Chung and Brown (1976) demonstrated a decrease in hexobarbital hydroxylase activity of 45 to 50% in rats pretreated with a single dose of ethanol. The lethal dose for barbiturates is nearly 50% lower in the presence of alcohol than it is when the drug is used alone (Bogan and Smith, 1967). Blood levels of secobarbital or pentobarbital as low as 0.5 mg per 100 ml combined with blood alcohol levels of 0.1 g per 100 ml can cause death from respiratory depression (Bogan and Smith, 1967). A major source of

this potentiation is the inhibition by alcohol of the enzymes that metabolize barbiturates, so that the drugs are present in the bloodstream in higher than normal amounts. Synergism and tolerance may occur directly in brain tissue as well. Thus, barbiturates should not be used in the treatment of alcoholism.

In 1977, Horder reported a series of four deaths resulting from suicide attempts employing chlormethiazole in conjunction with alcohol. Others observed a similar phenomenon (Robinson and McDowall, 1979). A pharmacokinetic basis for this interaction was reported by Neuvonen *et al.* (1981); the oral chlormethiazole $AUC_{0-8 \, hr}$ was significantly increased, as was the chlormethiazole bioavailability after ethanol.

Ethanol also interacts with narcotics: combined use of morphine and alcohol potentiates the effects of both drugs and increases the probability of death (Kissin, 1974). Opiates have been reported to be involved frequently in deaths caused by alcohol–drug combinations. In suspensions of isolated hepatocytes from rats, the rate constants for codeine and morphine elimination were reduced by approximately one-third and one-fourth, respectively, in the presence of 60 mM ethanol; inhibition of codeine metabolism was dose dependent, extending from approximately 15% at 10 mM ethanol to 40 to 50% at 100 mM. A threefold increase in the ratio of the morphine concentration (formed from codeine) to the amount of codeine metabolized was observed in the presence of ethanol as compared to control cells. The mean morphine concentration was 170% higher in the ethanol-treated than in the control cells. The inhibition of morphine metabolism was accompanied by a similar reduction of morphine-3-glucuronide formation. The accumulation of morphine observed in the cell medium in the presence of ethanol might result from inhibition of other metabolic pathways for codeine; thus, shunting codeine to morphine formation, combined with the inhibitory effect of ethanol on morphine metabolism *per se* (Bodd *et al.*, 1985), may enhance morphine accumulation. Kissin (1974) also states that users of illicit methadone often use alcohol to experience a "high" that methadone alone does not provide. Experimentally, acute administration of ethanol resulted in increased brain and liver concentrations of methadone and decreased biliary output of pharmacologically active methadone (Borowski and Lieber, 1978). *In vitro*, ethanol inhibited *N*-demethylation of methadone by microsomes from livers of rats. Also, methadone metabolism was inhibited by carbon monoxide, indicating cytochrome P-450 dependence, as is the case for ethanol. Therefore, acute ethanol administration may enhance cerebral effects of methadone as a consequence of its inhibitory effect at liver microsomal sites.

Acute intoxication does not solely affect drugs acting on the CNS. For instance, it reduces the metabolism of warfarin, leading to increased anticoagulant effects and the danger of hemorrhage. The combination causes blood warfarin levels to be higher than expected from a given dose (Kissin, 1974; Coleman and Evans, 1975). Chronic alcohol abuse, however, can enhance enzyme activity (see Section 6.5.1.2.), leading to decreased anticoagulant effects. Because the effects on prothrombin time can change with varying intakes of alcohol, physicians need to monitor closely the prothrombin times in patients who drink. Alcohol also appears to enhance the bioavailability of propanol (Grabowski *et al.*, 1980).

Ethanol also has acute effects on antimicrobials. Ethanol was found to cause a 30% decrease in isoniazid $t_{1/2}$ in two subjects (one slow and one fast acetylator) when the blood alcohol level was maintained at 0.02 to 0.04% (Lester, 1964). Isoniazid is acetylated via *N*-acetyltransferase, and these results are consistent with findings on the effects of ethanol on drug acetylation. Following ethanol, the sulfadimidine elimination $t_{1/2}$ was decreased by 20% in both rapid and slow acetylators. Acute ethanol administration also significantly prolonged the tolbutamide $t_{1/2}$ (Carulli *et al.*, 1971).

Ethanol also interacts with industrial solvents. Occupational exposure to xylene is widespread in the manufacture and application of chemicals that contain xylene, notably paints, glues, printing inks, and pesticides. Alcohol beverages are extensively consumed by some populations, occasionally in the course of the work day, during lunchtime, and particularly after work. Since the excretion of xylene is delayed by its high solubility and storage in lipid-rich tissues, the simultaneous presence of xylene and ethanol in the body is not uncommon. Indeed, after ethanol intake blood xylene levels were found to increase by about one-and-one-half- to twofold while urinary methylhippuric acid excretion fell by about 50%, suggesting that ethanol decreases the metabolic clearance of xylene by about one-half during xylene inhalation (Riihimaki *et al.*, 1982). Such an alteration in xylene pharmacokinetics is likely to be caused primarily by ethanol-mediated inhibition of hepatic xylene metabolism. Acute interactions of this type also have been described for other industrial solvents.

Besides the enhancement of the toxicity of the parent compound, the presence of ethanol may also de-

crease the toxicity of certain solvents by the inhibition of the hepatic metabolism of a relatively innocuous parent compound to a toxic product. This is the case for carbon tetrachloride (CCl_4) (Teschke *et al.*, 1983).

6.5.1.2. Chronic Interaction

6.5.1.2a. Enhanced Drug Metabolism: Metabolic Drug Tolerance.
In addition to tolerance to ethanol (Fig. 6.2C and Chapter 1), alcoholics also tend to display tolerance to various other drugs (Fig. 6.2D). The tolerance of the alcoholic to various drugs has been generally attributed to central nervous system adaptation (Kalant *et al.*, 1970). However, there is sometimes a dissociation in the time course of the decreased drug sensitivity of the animals and the occurrence of central nervous system tolerance: the decreased drug sensitivity was found to precede the central nervous system tolerance (Ratcliffe, 1969). Thus, in addition to central nervous system adaptation, metabolic adaptation must be considered. Indeed, it has been shown that the rate of drug clearance from the blood is enhanced in alcoholics (Kater *et al.*, 1969a). Of course, this could be caused by a variety of factors other than ethanol, such as the congeners and the use of other drugs so commonly associated with alcoholism. Controlled studies showed, however, that administration of pure ethanol with nondeficient diets to either rats or humans (under metabolic ward conditions) resulted in a striking increase in the rate of blood clearance of meprobamate and pentobarbital (Misra *et al.*, 1971) (Fig. 6.5) and propranolol (Sotaniemi

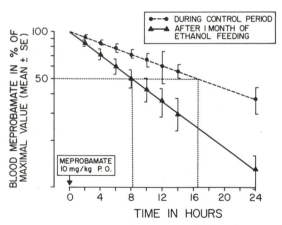

FIGURE 6.5. Effect of ethanol consumption on clearance of meprobamate from blood. Four volunteer alcoholics were tested before and after 1 month of ethanol ingestion. Half-lives are shown by the dotted lines on the *x* and *y* axes. (From Misra *et al.*, 1971.)

et al., 1981). Similarly, increases in the metabolism of antipyrine (Vessell *et al.*, 1971), tolbutamide (Kater *et al.*, 1969a,b; Carulli *et al.*, 1971), warfarin (Kater *et al.*, 1969a), propranolol (Pritchard and Schneck, 1977), diazepam (Sellman *et al.*, 1975), and rifamycin (Grassi and Grassi, 1975) were found. Furthermore, the capacity of liver slices from animals fed ethanol to metabolize meprobamate was increased (Misra *et al.*, 1971), which clearly showed that ethanol consumption affects drug metabolism in the liver itself, independent of changes in drug excretion or distribution or hepatic blood flow. In one study, failure to verify such an effect (Ioannides *et al.*, 1975) was probably caused by the very low dosage of ethanol administered.

The induction in the activity of MEOS after chronic alcohol consumption "spills over" to various other drug-metabolizing systems in liver microsomes, thereby accelerating drug metabolism in general (Fig. 6.2D). Repeated ethanol administration results in increased activities of a variety of microsomal drug-detoxifying enzymes (Rubin and Lieber, 1968; Ariyoshi *et al.*, 1970; Carulli *et al.*, 1971; Misra *et al.*, 1971; Joly *et al.*, 1973). On some occasions, some effects are observed after a single ethanol dose (Powis, 1975). As discussed in Chapter 1, ethanol consumption also increases the content of microsomal cytochrome P-450 and the activity of NADPH cytochrome P-450 reductase (Rubin *et al.*, 1968; Joly *et al.*, 1973; Luoma and Vorne, 1973). These increases occur predominantly in the smooth membranes (Ishii *et al.*, 1973; Joly *et al.*, 1973). Moreover, it has been shown that microsomal cytochrome P-450s, a reductase, and phospholipids play a key role in the microsomal hydroxylation of various drugs (Lu *et al.*, 1969). Therefore, the increase in the activity of hepatic microsomal drug-detoxifying enzymes and in the content of cytochrome P-450 induced by ethanol ingestion offers a likely explanation for the observation that ethanol consumption enhances the rate of drug clearance *in vivo*. Thus, a recovering alcoholic's previous history of alcohol abuse can be a key factor in prescribing decisions, because even after withdrawal alcoholics need doses different from those required by nondrinkers to achieve therapeutic levels of certain drugs, such as warfarin, diphenylhydantoin, tolbutamide, and isoniazid. The half-life of each of these drugs is 50% shorter in abstaining alcoholics than in nondrinkers (Kater *et al.*, 1969a,b; Martin, 1971). Experimentally, this effect of chronic ethanol consumption may be modulated, in part, by the dietary content in carbohydrates (Teschke *et al.*, 1981), lipids (Joly and Hetu, 1975), and proteins (Mitchell *et al.*, 1981). The metabolic tolerance

will persist several days to weeks after cessation of alcohol abuse, and the duration of recovery varies depending on the drug considered (Hetu and Joly, 1985).

6.5.1.2b. Increased Xenobiotic Toxicity and Carcinogenicity in Alcoholics.

The increased activity of hepatic microsomal enzymes resulting from prolonged alcohol consumption augments the toxicity of agents that are converted to toxic metabolites in the microsomes. This pertains particularly to those substrates for which the alcohol-inducible cytochrome P-450, when compared with other P-450s, displays an enhanced capacity for conversion to hepatotoxic metabolites. This pertains to acetaminophen and carbon tetrachloride (Morgan *et al.*, 1983; Johansson and Ingelman-Sundberg, 1985), and to the bioactivation of procarcinogens such as *N*-nitrosodimethylamine to ultimate carcinogens (Yang *et al.*, 1985; Levin *et al.*, 1986). It is known that carbon tetrachloride (CCl_4) exerts its toxicity after conversion to an active compound in the microsomes, and we found that alcohol pretreatment remarkably stimulates the toxicity of CCl_4 (Hasumura *et al.*, 1974). These experiments were carried out at a time when the ethanol had disappeared from the blood, which rules out the possibility of the increase of the toxicity of CCl_4 resulting from the presence of ethanol (Traiger and Plaa, 1972).

The potentiation of the CCl_4 toxicity by ethanol pretreatment may be accounted for by the increased production of toxic compounds of CCl_4, since the conversion of $^{14}CCl_4$ to $^{14}CO_2$ and covalent binding of CCl_4 metabolites to protein were significantly accelerated in microsomes of ethanol-pretreated rats (Hasumura *et al.*, 1974). Furthermore, pretreatment of rats with ethanol or rabbits with either imidazole or pyrazole, agents known to induce an ethanol-specific form of liver microsomal cytochrome P-450 (P450IIE1), caused, compared to controls, three- to 25-fold enhanced rates of CCl_4-dependent lipid peroxidation or chloroform production in isolated liver microsomes (Johansson and Ingelman-Sundberg, 1985). Thus, the clinical observation of the enhanced susceptibility of alcoholics to the hepatotoxic effect of CCl_4 (Moon, 1950) may be, at least in part, caused by an increased activation and biotransformation of CCl_4. A number of other mechanisms that have been invoked are less likely to play a role, as discussed by Zimmerman (1986). The increased vulnerability of the alcoholic is apparent in toxicity resulting from either ingestion or inhalation, as reviewed before (Lieber, 1982). Indeed, the increased vulnerability has been evident in instances of severe poisoning of alcoholics exposed to atmospheric concentrations of CCl_4 that led to no overt hepatic injury in simultaneously exposed non-alcoholic individuals (Guild et al., 1958), and the instances of toxicity among patients who received CCl_4 as a vermifuge involved mainly alcoholics (Rosenthal, 1930; Lamson et al., 1928; Smillie and Pessoa, 1923; Hall, 1921; Lambert, 1922; Gardner et al., 1925).

It is likely that a larger number of other toxic agents will be found to display a selective injurious action in the alcoholic. Indeed ethanol treatment aggravated hemopoietic toxicity (Nakajima et al., 1985) and markedly increased the activity of microsomal low-K_m benzene-metabolizing enzymes (Nakajima et al., 1987); benzene and ethanol apparently induce the same form of cytochrome P-450 in rabbits (Ingelman-Sundberg and Johansson, 1984), and P450IIE1 also was shown to be a benzene hydroxylase (Johansson and Ingelman-Sundberg, 1988; Koop et al., 1988), present and induced in bone marrow (Schnier et al., 1989). Hepatotoxicity of bromobenzene was also increased after chronic ethanol treatment (Hetu et al., 1983). Hepatic microsomal enflurane defluorination increased 10.5-fold 1 hr following cessation of chronic treatment (Pantuck et al., 1985). Under similar conditions, enhanced hepatotoxicity was demonstrated (Tsutsumi et al., 1990). Animals pretreated with alcohol develop potentiated centrizonal necrosis on exposure to halothane, especially when metabolism is rendered reductive (Takagi et al., 1983). It is noteworthy that halothane can be metabolized by various microsomal cytochrome P-450s, including P450IIE1 (Gruenke et al., 1988), and that P-450-mediated halothane metabolism is associated with lipid peroxidation and liver injury (Akita et al., 1988). Combined exposures to ethanol and xylene were accompanied by a marked proliferation of smooth endoplasmic reticulum (a subcellular site of cytochrome P-450 monooxygenases in the hepatocytes) and an increased NADPH-Fe^{2+}- and ascorbate-Fe^{2+}-driven lipid peroxidation in microsomal membranes—a potential toxic mechanism (Wiśniewska-Knypl et al., 1989).

Enhanced metabolism (and toxicity) pertains not only to industrial solvents and anesthetics but also to a variety of prescribed drugs. For instance, the observed increased hepatotoxicity of isoniazid in alcoholics (Mitchell and Jollows, 1975) may well be a result of increased production by the microsomes of an active metabolite of the acetyl derivative of the drug. Similarly, chronic ethanol administration enhances phenylbutazone hepatotoxicity, possibly because of increased biotransformation (Beskid et al., 1980).

The same mechanism of hepatotoxicity also pertains to some "over-the-counter" medications. Acetaminophen (paracetamol, N-acetyl-p-aminophenol), widely used as an analgesic and an antipyretic, is generally safe when taken in recommended doses. However, acetaminophen in large doses (that is, in cases of suicide attempts) has been shown to produce fulminant hepatic failure (Davidson and Eastham, 1966; Thomson and Prescott, 1966). A similar lesion can also be observed in experimental animals (Boyd and Bereczky, 1966; Dixon et al., 1971; Mitchell et al., 1973). Acetaminophen intake in doses much smaller than those taken for suicidal attempts has been shown to produce hepatic injury (Prescott and Wright, 1973). Following the studies of Mitchell et al. (1973) and others, it is now clear that in addition to glucuronate and sulfate conjugation, acetaminophen is also metabolized in the liver by the microsomal cytochrome P-450-dependent drug-metabolizing system; the latter biotransformation yields an active metabolite highly toxic to the liver. Because alcohol can also induce microsomal drug-metabolizing systems, it is to be expected that a history of alcohol consumption might favor the hepatotoxicity of acetaminophen. This has been suggested in various case reports. Scott and Stewart (1975) noted that "the majority of cases of paracetamol overdose are accompanied by some alcohol." Two fatal cases were described by Emby and Fraser (1977). Vilstrup et al. (1977) reported a case of severe liver damage in an alcoholic after only 5.4 g of acetaminophen.

A pattern of use among alcoholic patients presenting with acetaminophen-associated hepatic injury has appeared to be repetitive intake for headaches, including those associated with withdrawal symptoms, dental pain, or the pain of pancreatitis, leading in some to high daily doses. Although some of these (10 g) have been within the range of suicidal hepatotoxic doses, even smaller amounts well within the accepted tolerable range (2.5–6 g) have been incriminated as the cause of hepatic injury in alcoholic patients (Black, 1984; Seef et al., 1986), although others have challenged the low-dose toxicity (Lauterburg and Velez, 1988). Recognition of hepatotoxic effects in alcoholics taking therapeutic doses of acetaminophen is helped by the appearance of extremely high serum transaminase levels. Experimentally, after chronic ethanol feeding, enhanced urinary excretion and covalent binding of reactive metabolite(s) of acetaminophen to microsomes from ethanol-fed rats was observed by Sato et al. (1981) and confirmed by Prasad et al. (1985). The enhanced activation of acetaminophen to toxic compounds after chronic ethanol consumption can most likely be attributed to the ethanol-inducible form of cytochrome P-450, which has a high affinity for acetaminophen, as discussed above, and also reviewed by Black and Raucy (1986). These alterations were associated with the enhanced hepatotoxicity in ethanol-fed rats as evidenced by light (Fig. 6.6) and electron microscopy (Fig. 6.7) as well as increased serum glutamic oxaloacetic transaminase (SGOT) levels and glutamic dehydrogenase (GDH) levels in the blood (Fig. 6.8). Therefore, it is likely that the enhanced hepatotoxicity of acetaminophen after chronic ethanol consumption is caused, at least in part, by an increased microsomal production of reactive metabolite(s) of acetaminophen.

Consistent with this view if the observation that, in animals fed ethanol chronically, the potentiation of acetaminophen hepatotoxicity occurred after ethanol withdrawal (Sato et al., 1981). Enhanced microsomal production of reactive metabolites(s) may be responsible for this effect. In addition, depletion of reduced glu-

tathione (GSH) after both ethanol and acetaminophen may contribute to the toxicity of each compound, as discussed in greater detail in Chapter 7. Chronic use of acetaminophen may also be associated with toxicity, particularly in alcoholics (Barker et al., 1977). However, to what extent ethanol, when still present, competes with microsomal metabolism of acetaminophen remains to be evaluated clinically. Indeed, experimentally, unlike pretreatment with alcohol, which accentuates toxicity, the presence of ethanol in fact prevented the acute acetaminophen-induced hepatotoxicity (Fig. 9), most likely because of the inhibition of the biotransformation of acetaminophen to reactive metabolites (Sato and Lieber, 1981; Altomare et al., 1984a,b). In view of the fact that a large segment of the population regularly uses alcohol or drugs (or both) that are capable of inducing the microsomal system of the liver, it is apparent that a warning is justified concerning the enhanced risk of acetaminophen toxicity in alcohol users.

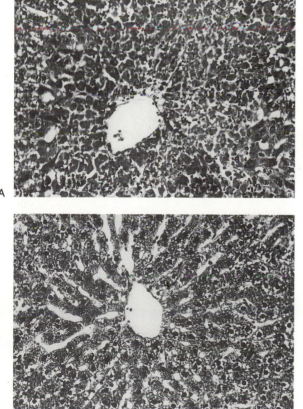

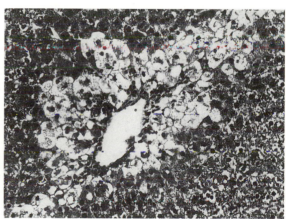

FIGURE 6.6. Liver damage 36 hr after the injection of acetaminophen (0.5 g/kg, i.p.). (Hematoxylin–eosin stain, ×100) (A) Massive hepatic centrilobular coagulative necrosis in ethanol-fed rat. (B) Prominent hydropic changes in ethanol-fed rat. (C) No apparent necrosis was observed in the pair-fed control littermate. (From Sato et al., 1981.)

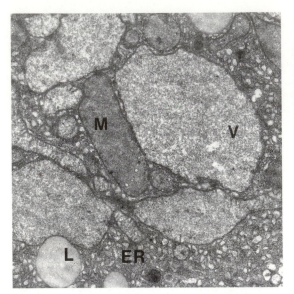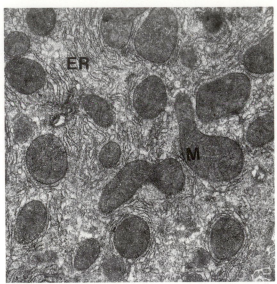

A   B

FIGURE 6.7. Electron micrographs of centrilobular hepatocytes 6 hr after the injection of acetaminophen (0.5 g/kg, i.p.). (×20,000) (A) Ethanol-fed rat: lipid droplet (L), vesiculated endoplasmic reticulum (ER), mitochondrial swelling with disoriented cristae (M), and vacuolated mitochondria with loss of cristae and loss of density of ground substance (V). (B) Pair-fed control with minimum changes: mitochondria (M) and endoplasmic reticulum (ER). (From Sato *et al.*, 1981.)

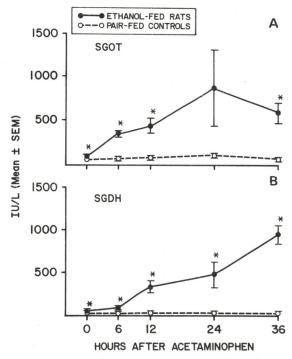

FIGURE 6.8. Time course of serum enzyme activities after the injection of acetaminophen (0.5 g/kg, i.p.). (A) Serum glutamic oxaloacetic transaminase (SGOT) activity. (B) Serum glutamic dehydrogenase (SGDH) activity. Each point represents the mean value for at least five animals. *$P > 0.05$. (From Sato *et al.*, 1981.)

It is also likely that various other compounds may exhibit enhanced toxicity in the alcoholic. Acute administration of aflatoxin B_1 (AFB_1) can lead to necrosis and steatosis in experimental animals (Zimmerman, 1978). Pretreatment of rats with four oral doses of ethanol over a 48-hr period has been reported to lead to a potentiation of the necrogenic effect of a single dose of AFB_1. Glinsukon *et al.* (1978) have speculated that the ethanol may act by enhancing conversion of AFB_1 to its toxic metabolite. Pretreatment with ethanol also enhances the necrogenic effects. of cocaine (Strubelt, 1982) and the hepatotoxic effects of *N*-nitrosodimethylamine (Maling *et al.*, 1975; Lorr *et al.*, 1984) and thioacetamide (Maling *et al.*, 1975). Pretreatment of rats with a 5% ethanol solution (oral) for 7 days led to enhanced toxicity of vinylidene chloride (Siegers *et al.*, 1983). Possible implication regarding carcinogenesis are discussed in Chapter 15.

6.5.2. Alcohol-Dehydrogenase-Dependent Reactions between Ethanol and Drugs

6.5.2.1. Direct Competition

The first oxidation step in the inactivation of the active principles of digitalis is catalyzed by alcohol dehydrogenase (ADH). Direct competition was found to occur *in vitro* between ethanol and the 20 pharmaco-logically active constituents of digoxin, digitoxin, and gitoxin (Frey and Vallee, 1979, 1980). The clinical relevance of this effect remains to be established, but it may have an important bearing on the individual variations commonly seen in response to cardiac glyco-sides. Ethylene glycol monobutyl ether (2-butoxyethanol), used in aerosols, can cause hematotoxicity via its metab-olite, butoxyacetic acid; production of the latter can be inhibited by the ADH inhibitor pyrazole or the aldehyde dehydrogenase inhibitor cyanamide (Ghanayem *et al.*, 1987).

A number of alcohols other than ethanol can serve as substrates for ADH. This pathway acquires particular importance when the product formed is highly toxic, as in the case of ethylene glycol. This compound, com-monly used as a solvent and in antifreeze solutions, is oxidized by ADH to its corresponding aldehyde (von Wartburg *et al.*, 1964) and then to the highly toxic oxalic acid. Ethanol competitively inhibits ethylene glycol oxidation by ADH, so that the compound is gradually excreted unchanged. Ethanol therapy has been found to be effective in cases of ethylene glycol poisoning (Wacker *et al.*, 1965).

Methanol poisoning can also be treated with etha-nol. The conversion of methanol to formaldehyde can be accomplished *in vitro* by ADH (Blair and Vallee, 1966; Makar *et al.*, 1968), by the microsomal ethanol-oxidizing system (Lieber and DeCarli, 1970), and by catalase (Tephly *et al.*, 1964). The relative importance of the three pathways varies with the species involved. In any event, ethanol can inhibit methanol metabolism and serve for the treatment of methanol poisoning. The therapeutic value of ethanol was first noted by an anony-mous U.S. army surgeon and reported by Bullar and Wood (1904). It was subsequently rediscovered by Röe (1943). Ethanol competitively inhibits methanol metab-olism. A dose sufficient to keep the level of ethanol in the blood between 100 and 150 mg per 100 ml is recom-mended. 4-Methylpyrazole, which inhibits both ADH and MEOS (see Section 1.1.2.2.), has also been pro-posed for therapeutic use (McMartin *et al.*, 1975).

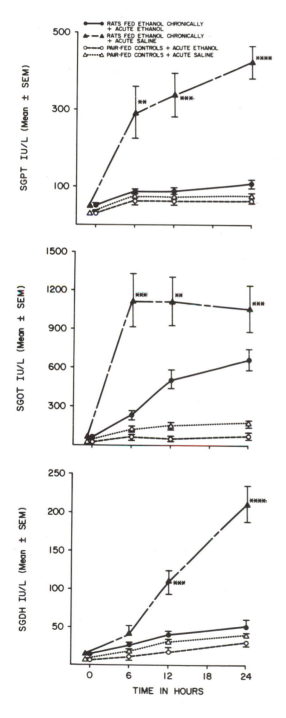

FIGURE 6.9. Effects of acute ethanol administration on the rise of serum GPT, GOT, and GDH activities induced by acetamino-phen. Rats pair-fed chronically ethanol-containing or isocaloric control diets were fasted for 18 hr and were given ethanol (3 g/kg, *per os*) acutely or saline and, concomitantly, acetaminophen (1 g/kg, intraperitoneally). Each point represents the mean value ± S.E.M. of at least five animals. **$P < 0.25$; ***$P < 0.005$; ****$P < 0.001$ when compared to the other three groups. (From Altomare *et al.*, 1984a.)

Trichloroethylene is a widely used solvent with a low vapor pressure. When inhaled, 70 to 80% of trichloroethylene is metabolized to trichloroethanol and trichloroacetic acid via chloral hydrate. These same metabolites result from the oral ingestion of chloral hydrate. Ingestion of ethanol (0.5 g/kg) 30 min after chloral hydrate resulted in a significantly increased trichloroethanol in human volunteers (Sellers *et al.*, 1972). The authors suggested that this resulted from (1) enhanced formation of trichloroethanol via alcohol dehydrogenase and (2) a decrease in the formation of trichloroethanol glucuronide, which requires NAD (nicotinamide-adenine dinucleotide).

6.5.2.2. Change in Redox State

Ethanol oxidation by the ADH pathway increases the intracellular NADH-to-NAD ratio. A similar change in redox state is produced by lactate and by sorbitol. Each of these three compounds was found to inhibit the glucuronidation of harmol, 2-naphthol, 4-methylumbelliferone, and phenolphthalein in isolated hepatocytes (Moldeus *et al.*, 1978). In contrast, other drug detoxification processes such as sulfation (Sundheimer and Brendel, 1984) are unaltered by ethanol. The inhibitory effect of ethanol or glucuronidation was reversed by 4-methylpyrazole, an inhibitor of ADH. The synthesis of UDP-glucuronic acid was also inhibited by ethanol, in both the presence and absence of a substance undergoing glucuronidation. Heretofore, UDP-glucuronic acid synthesis had not been considered to be rate limiting. However, these results suggest that inhibition of glucuronidation by ethanol is caused by a decreased UDP-glucuronic acid synthesis as a result of the increased NADH-to-NAD ratio that results from ADH-dependent oxidation of ethanol. Such a mechanism may also prove to be important in the interference, by ethanol, with the hepatic glucuronidation and excretion of other compounds, such as morphine (Bodd *et al.*, 1985, 1986).

6.5.3. Aldehyde-Dehydrogenase-Dependent Reactions between Ethanol and Drugs

Acetaldehyde is produced from ethanol in both the alcohol dehydrogenase (ADH) and MEOS pathways. It is highly toxic and is rapidly metabolized by aldehyde dehydrogenase (ALDH). Competitive inhibition of this enzyme by disulfiram (Antabuse®) has clinical application in the treatment of some alcoholic patients (Sereny *et al.*, 1986). Disulfiram is partially reduced *in vivo* to

diethyldithiocarbamate, which is an efficient inhibitor of Cu,Zn-containing superoxide dismutase (SOD) both *in vitro* and *in vivo*. Ohman and Marklund (1986) found that plasma extracellular SOD was moderately reduced (about 20%) in disulfiram-treated alcoholics as compared with nontreated alcoholics and healthy controls. No effect of disulfiram treatment on erythrocyte Cu,Zn-SOD activity was demonstrated. The full disulfiram reaction is partly caused by acetaldehyde retention and partly by an inhibition of dopamine oxidase, which results in lowered tissue levels of norepinephrine (Truitt and Walsh, 1971). To assess factors determining and potential tests predicting severity of the disulfiram-alcohol reaction (DAR), Beyeler *et al.* (1985) subjected 13 ambulatory alcoholics who consented to alcohol-aversive treatment with disulfiram to detailed investigations before, during, and after the DAR. It was concluded that the plasma acetaldehyde level appears to be the single most important factor determining severity of the cardiovascular response in the DAR and that acetaldehyde-oxidizing capacity measurements clearly define subjects under treatment with disulfiram.

A number of drugs when taken with alcohol also cause an acute disulfiramlike reaction with nausea, headache, vomiting, and possibly convulsions. These reactions are generally less severe than those produced by disulfiram. Some of these effects are caused by the inhibition of acetaldehyde metabolism. Such agents include the sulfonylureas and the antimicrobials chloramphenicol, furazolidine (Furoxone®), griseofulvin, metronidazole (Flagyl®), and quinacrine (Atabrine®). Cephalosporin antibiotics with a 1-methyltetrazole-5-thio side chain also have the ability to cause an unpleasant flushing reaction if they are taken some time before the drinking of alcohol. It was proposed that the explanation for this is that the side chain becomes liberated *in vivo* and oxidized to 5,5′-Dithiobis (1-methyltetrazole) or to a mixed disulfide analogue, which then inactivates aldehyde dehydrogenase. Support for this proposal was given by results concerning the interaction *in vitro* between the disulfides and sheep liver cytoplasmic aldehyde dehydrogenase. 5,5′-dithiobis (1-methyltetrazole) or to a mixed disulfide analogue, which then inactivates aldehyde dehydrogenase. Support for this proposal was given by results concerning the interaction *in vitro* between the disulfides and sheep liver cytoplasmic aldehyde dehydrogenase. 5,5′-Dithiobis (1-methyltetrazole) had a rapid and pronounced inhibitory effect, very similar in many ways (though not identical) to that of disulfiram, to which it has a structural similarity (Kitson, 1986).

6.5.4. Drug Conjugations Affected by Ethanol

Some alcohol–drug interactions involving changes in conjugation have already been discussed. These include the role of GSH in acetaminophen toxicity and the changes in redox state affecting glucuronidation.

6.5.4.1. Effects on Acetylation

As discussed in Chapters 1 and 2, ethanol is metabolized to yield acetate and may increase the concentration of acetyl coenzyme A (CoA), the acetate donor in the *N*-acetyltransferase reaction leading to acetylation of drugs. If a low concentration of acetyl-CoA is rate limiting in this reaction, ethanol metabolism may increase the velocity by enhancing the formation of acetyl-CoA. Indeed, when 16 healthy volunteers were given sulfadimidine, in both rapid and slow acetylators, the apparent half-life of the drug decreased by about 20% after ethanol, and the amount of drug acetylated, measured in blood and urine, increased (Olsen and Mørland, 1978).

6.5.4.2. Glycine Conjugation

Ethanol alone does not affect salicylate clearance (Levy, 1971). However, the formation of the glycine conjugate, salicylurate, is inhibited by benzoate, and the latter is sometimes marketed with a salicylate to prolong its action. Ethanol reduces the conjugation of benzoate and may thus influence salicylate disposition indirectly (Levy, 1971).

6.5.5. Congeners

Congeners are the many compounds in alcoholic drinks, other than ethanol and water, that contribute so much to the characteristics of each beverage. They undoubtedly influence the responses to drugs, as reviewed before (Lieber, 1982). A most dramatic example is the hypertensive crisis resulting from interaction between some congeners and monoamine oxidase (MAO) inhibitors. Some beers and wines contain tyramine, a pressor amine that releases norepinephrine from tissues. The MAO inhibitors block the degradation of tyramine and thus accentuate norepinephrine release. Similarly, some beers contain dihydroxyphenylalanine (dopa), a precursor of dopamine, whose accumulation has effects similar to those of tyramine.

6.5.6. Nutrition

The state of nutrition is one of many physiological factors that influence the activities of drug-metabolizing systems (Furner and Feller, 1971). The role of protein, carbohydrate, and lipid in the inducibility of the microsomes has been discussed before. Thus, malnutrition in the alcoholic might be expected to influence ethanol–drug interactions. The general impression seems to be that such an influence is often minor, but there is a dearth of relevant studies. Theoretically, substrate availability for conjugation reactions could be important in some situations such as acetaminophen toxicity.

6.5.7. Genetic Factors

Some ethanol–drug interactions may achieve clinical significance only in the presence of certain genetically determined diseases. An example is acute intermittent porphyria, the symptoms of which can be exacerbated by alcohol. This side effect of alcohol can probably be explained by an induction of the enzyme δ-aminolevulinic acid synthetase (ALA synthetase) in the liver (Shanley *et al.*, 1968; Rubin *et al.*, 1970a; Held, 1977). In contrast, the ALA dehydratase in the liver was not influenced by alcohol (Held, 1977).

In comparison with other porphyrinogenic agents, the effect of alcohol on the porphyrin concentration in the liver of starving rats is small (Held, 1977). Phenobarbital strikingly potentiated the alcohol effect on ALA synthetase (Held, 1977). The observed sensitization of the porphyrinogenic effect of alcohol by phenobarbital may be of clinical significance. Alcohol is often ingested in combination with drugs that may induce the drug-metabolizing enzyme system. In patients with acute intermittent porphyria, which may be exacerbated after alcohol or drug ingestion (Fillipini, 1968), such a combination may be much more dangerous than the ingestion of alcohol or another drug alone.

6.5.8. Alcoholic Liver Disease

Hepatic dysfunction from alcoholic liver disease could interfere with drug metabolism independently of the changes in plasma protein binding discussed earlier. Evidence for impaired metabolism of diazepam, antipyrine, and chlordiazepoxide, but not of lorazepam, was obtained in patients with cirrhosis (Kraus *et al.*, 1978; Klotz *et al.*, 1975). Fortunately, this does not appear to have a major clinical implication in stable compensated liver disease, but there is an obvious need for caution in prescribing for these patients, especially since there are no convenient objective markers of hepatic drug metabolism. In addition to these considerations, severe hepatic dysfunction alters the sensitivity of the brain to central nervous system depressants, thus increasing the risk of developing coma, as discussed in Chapter 14.

6.6. Pharmacological Interactions

6.6.1. Acute

Interactions between ethanol and other drugs at their primary site of action are numerous and well documented (Forney and Hughes, 1964) and are not discussed in detail here. Obvious and important examples of acute pharmacological interactions are the additive effects of ethanol with sedatives, tranquilizers, and narcotics. The prominent sedative side effect of antihistamines is increased to such an extent by alcohol that it is dangerous to perform any hazardous task while taking the combination. Alcohol also has an additive hypotensive effect in combination with reserpine, methyldopa (Aldomet®), hydralazine (Apresoline®, Dralzine®), guanethidine (Esimil®, Ismelin®), ganglionic blockers, nitroglycerin, and peripheral vasodilators. In addition, propranolol may mask the signs and symptoms (rapid heart beat, profuse sweating) of alcohol-caused hypoglycemia (Coleman and Evans, 1975).

6.6.2. Chronic

The enhanced pharmacological tolerance of alcoholics to central nervous system depressants is well known (Kalant *et al.*, 1970). Together with the metabolic tolerance discussed earlier, it has important clinical implications. The mechanisms of this pharmacological tolerance remain to be elucidated. Contributory factors could include changes in synaptosomal structure and fluidity that are produced by chronic ethanol intake (Littleton *et al.*, 1979; Johnson *et al.*, 1980).

6.7. Summary and Therapeutic Guidelines

Although there is no substitute for a detailed understanding of the needs of the individual patient and of the metabolic basis of alcohol–drug interactions, the following therapeutic guidelines can be suggested.

6.7.1. Avoidance of Alcohol

This should be absolute in any patient receiving MAO inhibitors or disulfiram and in patients engaged in hazardous motor tasks (including driving) while taking sedatives, tranquilizers, or narcotics. In addition, it is our practice to advise that alcohol intake be kept at a minimum level (one or two drinks with meals) while patients are receiving drugs capable of producing minor disulfiramlike reactions, such as metronidazole (Flagyl®), chloramphenicol, griseofulvin, furazolidine (Furoxone®), quinacrine (Atabrine®), and the sulfonylureas. Interactions with H_2 blockers (such as cimetidine and ranitidine, but not famotidine) are discussed in chapter 1.

6.7.2. Monitoring Drug Dosage

There may be an increased need for laboratory tests that monitor drug dosage or efficacy. This would apply especially to drugs whose hepatic metabolism or plasma protein binding may be affected by alcoholism, particularly if there is already a narrow margin between therapeutic and toxic blood levels. Examples include quinidine and warfarin.

6.7.3. Increased Drug Doses

Alcohol produces pharmacological and metabolic tolerance to drugs such as diphenylhydantoin, barbiturates, benzodiazepines, tolbutamide, propranolol, and rifamycin. Such agents may have to be prescribed in increased amounts for patients in the recovery phase of alcoholism.

6.7.4. Reduced Drug Dosage

The presence of alcoholic liver disease may require reduced dosage. As a general rule, altered doses of prescribed drugs are not indicated in patients with compensated alcoholic liver disease. However, a reduction in dosage is frequently indicated in patients with decompensated hepatic disease (for example, clinically evident alcoholic hepatitis or moderately advanced cirrhosis). Furthermore, such patients have an enhanced susceptibility to coma when they are taking central nervous system depressants or diuretics.

6.7.5. Treatment of Toxicity and Overdoses

There are two crucial considerations. One is whether the drug or its metabolite is responsible for the toxicity. When the metabolite is responsible, as in the case of acetaminophen or carbon tetrachloride, then the alcoholic will usually require vigorous treatment and supportive measures. If the drug is responsible, as in the case of an overdose of a sedative, then a more favorable outcome may be anticipated. The other consideration is the acute presence of alcohol. Significant blood alcohol levels may have a protective role in acetaminophen toxicity and a deleterious effect in the case of overdoses of sedatives.

REFERENCES

Affrime, M., and Reidenberg, M. M.: The protein binding of some drugs in plasma from patients with alcoholic liver disease. *Eur. J. Clin. Pharmacol.* **8**:267–269, 1975.

Akita, S., Morio, M., Kawahara, M., Takeshita, T., Fujii, K., and Yamamoto, M.: Halothane-induced liver injury as a consequence of enhanced microsomal lipid peroxidation in guinea pigs. *Res. Commun. Chem. Pathol. Pharmacol.* **61**:227–243, 1988.

Altomare, E., Leo, M. A., and Lieber, C. S.: Interaction of acute ethanol administration with acetaminophen metabolism and toxicity in rats fed alcohol chronically. *Alcoholism: Clin. Exp. Res.* **8**:405–408, 1984a.

Altomare, E., Leo, M. A., Sato, C., Vendemiale, G., and Lieber, C. S.: Interaction of ethanol with acetaminophen metabolism in the baboon. *Biochem. Pharmacol.* **33**:2207–2212, 1984b.

Ariyoshi, T., Takabatake, E., and Remmer, H.: Drug metabolism in ethanol-induced fatty liver. *Life Sci.* (Part II) **9**:361–369, 1970.

Barker, J. D., de Carle, D. J., and Anuras, S.: Chronic excessive acetaminophen use and liver damage. *Ann. Intern. Med.* **87**:299–301, 1977.

Beskid, M., Bialek, J., Dzieniszewski, J., Sadowski, J., and Tlalka, J.: Effect of combined phenylbutazone and ethanol administration on rat liver. *Exp. Pathol.* **18**:487–491, 1980.

Beyeler, C., Fisch, H. U., and Preisig, R.: The disulfiram–alcohol reactions: Factors determining and potential tests predicting severity. *Alcoholism: Clin. Exp. Res.* **9**:118–124, 1985.

Black, M.: Acetaminophen hepatotoxicity. *Annu. Rev. Med.* **35**:577–593, 1984.

Black, M., and Raucy, J.: Acetaminophen, alcohol and cytochrome P-450. *Ann. Intern. Med.* **104**:427–429, 1986.

Blair, A. H., and Vallee, B. L.: Some catalytic properties of human liver alcohol dehydrogenase. *Biochemistry* **5**:2026–2034, 1966.

Bo, O., Haffner, J. F. W., Langard, O., and Solheim, K.: Ethanol and diazepam as causative agents in road traffic accidents. In: *Alcohol, Drugs and Traffic Safety* (S. Israelstam and S. Lam-

bert, eds.), Toronto, Addiction Research Foundation of Ontario, pp. 439–448, 1975.

Bodd, E., Drevon, C. A., Kveseth, N., Olsen, H., and Morland, J.: Ethanol inhibition of codeine and morphine metabolism in isolated rat hepatocytes. *J. Pharmacol. Exp. Ther.* **237**:260–237, 1985.

Bodd, E., Gadenholt, G., Christensson, P. I., and Morland, J.: Mechanisms behind the inhibitory effect of ethanol on the conjugation of morphine in rat hepatocytes. *J. Pharmacol. Exp. Ther.* **239**:887–890, 1986.

Bogan, J., and Smith, H.: Analytical investigation of barbiturate poisoning—description of methods and a survey of results. *J. Forensic Sci.* **7**:37–45, 1967.

Borowsky, S. A., and Lieber, C. S.: Interaction of methadone and ethanol metabolism. *J. Pharmacol. Exp. Ther.* **207**:123–129, 1978.

Boyd, E. M., and Bereczky, G. M.: Liver necrosis from paracetamol. *Br. J. Pharamcol.* **26**:606–614, 1966.

Bullar, F., and Wood, C. A.: Poisoning by wood alcohol: Cases of death and blindness from Columbian spirits and other methylated preparations. *J.A.M.A.* **43**:1289–1296, 1904.

Carulli, N., Manenti, I., Gallo, M., and Salvioli, G. F.: Alcohol–drugs interaction in man: Alcohol and tolbutamide. *Eur. J. Clin. Invest.* **1**:421–424, 1971.

Chung, H., and Brown, D. R.: Mechanism of the effect of acute ethanol on hexobarbital metabolism. *Biochem. Pharmacol.* **25**:1613–1616, 1976.

Cinti, D. L., Grundin, R., and Orrenius, S.: The effect of ethanol on drug oxidations *in vitro* and the significance of ethanol–cytochrome P-450 interaction. *Biochem. J.* **134**:367–375, 1973.

Cohen, B. S., and Estabrook, R. W.: Microsomal electron transport reactions. III. Cooperative interactions between reduced diphosphopyridine nucleotide and reduced triphosphopyridine nucleotide-linked reactions. *Arch. Biochem. Biophys.* **143**:54–65, 1971.

Coleman, J. H., and Evans, W. E.: Drug interactions with alcohol. Reprint: *Alcohol Health Res. World Winter*, pp. 16–19, 1975.

Davidson, D. G. D., and Eastham, W. N.: Liver necrosis following overdose of paracetamol. *Br. Med. J.* **2**:497–499, 1966.

Desmond, P. V., Patwardhan, R. V., Schenker, S., and Hoyumpa, A. M.: Short-term ethanol administration impairs the elimination of chlordiazepoxide (Librium) in man. *Eur. J. Clin. Pharmacol.* **18**:275–278, 1980.

Dixon, M. F., Nimmo, J., and Prescott, L. F.: Experimental paracetamol-induced hepatic necrosis. A histopathological study. *J. Pathol. Bacteriol.* **103**:225–229, 1971.

Dorian, P., Sellers, E. M., Reed, K. L., Warsh, J. J., Hamilton, C., Kaplan, H. L., and Fan. T.: Amitriptyline and ethanol: Pharmacokinetic and pharmacodynamic interaction. *Eur. J. Clin. Pharmacol.* **25**:325–331, 1983.

Emby, D. J., and Fraser, B. N.: Hepatotoxicity of paracetamol enhanced by ingestion of alcohol. *S. Afr. Med. J.* **51**:208–209, 1977.

Fillipini, L.: Die akut intermittierende Porphyrie. *Intern. Praxis* **8**:575–580, 1968.

Forney, R. B., and Hughes, F. W.: Meprobamate, ethanol or meprobamate–ethanol combinations on performance of hu man subjects under delayed audiofeedback (DAF). *J. Psychol.* **57:**431–436, 1964.

Forrest, F. M., Forrest, I. S., and Finkle, B. S.: Alcohol–chlorpromazine interaction in psychiatric patients. *Agressologie* **13:**67–74, 1971.

Frey, W. A., and Vallee, B. L.: Human liver alcohol dehydrogenase—an enzyme essential to the metabolism of digitalis. *Biochem. Biophys. Res. Commun.* **91:**1543–1548, 1979.

Frey, W. A., and Vallee, B. L.: Digitalis metabolism and human liver alcohol dehydrogenase. *Proc. Natl. Acad. Sci. U.S.A.* **77:**924–927, 1980.

Furner, R. L., and Feller, D. D.: The influence of starvation upon hepatic drug metabolism in rats, mice and guinea pigs. *Proc. Soc. Exp Biol. Med.* **137:**816–819, 1971.

Gardner, G. H., Grove, R. C., Gustafson, R. K., Mauri, E. D., Thompson, M. J., Wells, H. S., and Lamson, P. D.: Studies on the pathological histology of experimental carbon tetrachloride poisoning. *Bull. Johns Hopkins Hosp.* **36:**107–133, 1925.

Ghanayem, B. I., Burka, L. T., and Matthews, H. B.: Metabolic basis of ethylene glycol monobutyl ether (2-butoxyethanol) toxicity: Role of alcohol and aldehyde dehydrogenases. *J. Pharmacol. Exp. Ther.* **242:**222–231, 1987.

Glinsukon, T., Taycharpipranai, S., and Tosulkao, C.: Aflatoxin B_1 hepatotoxicity in rats pretreated with ethanol. *Experientia* **34:**869–870, 1978.

Grabowski, B. S., Cady, W. J., Young, W. W., and Emery, J. F.: Effects of acute alcohol administration on propranolol absorption. *Int. J. Clin. Pharmacol. Ther. Toxicol.* **18:**317–319, 1980.

Grassi, G. G., and Grassi, C.: Ethanol–antibiotic interactions at hepatic level. *J. Clin. Pharmacol. Biopharmacol.* **11:**216–225, 1975.

Gruenke, L. D., Konopka, K., Koop, D. R., and Waskell, L. A.: Characterization of halothane oxidation by hepatic microsomes and purified cytochromes P-450 using a gas chromatographic mass spectrometric assay. *J. Pharmacol. Exp. Ther* **246:**454–459, 1988.

Guild, W. R., Young, J. V., and Merrill, J. P.: Anuria due to carbon tetrachloride intoxication. *Ann. Int. Med.* **48:**1221–1227, 1958.

Hall, M. C.: The use of carbon tetrachloride for the removal of hookworm disease. *J.A.M.A.* **77:**1641–1643, 1921.

Hasumura, Y., Teschke, R., and Lieber, C. S.: Increased carbon tetrachloride hepatotoxicity, and its mechanism, after chronic ethanol consumption. *Gastroenterology* **66:**415–422, 1974.

Hayes, S. L., Pablo, G., Radomski, T., and Palmer, R. F.: Ethanol and oral diazepam absorption. *N. Engl. J. Med.* **296:**186–189, 1977.

Held, H.: Effect of alcohol on the heme and porphyrin synthesis interaction with phenobarbital and pyrazole. *Digestion* **15:**136–146, 1977.

Hetu, C., and Joly, J-G.: Differences in the duration of the enhancement of liver mixed-function oxidase activities in ethanol-fed rats after withdrawal. *Biochem. Pharmacol.* **34:**1211–1216, 1985.

Hetu, C., Dumont, A., and Joly, J-G.: Effect of chronic ethanol administration on bromobenzene liver toxicity in the rat. *Toxicol. Appl. Pharmacol.* **67:**166–167, 1983.

Hollister, L. E.: Interactions between alcohol and benzodiazepines. In: *Recent Developments in Alcoholism, Vol. 8* (M. Galanter, ed.), New York, Plenum, pp. 233–239, 1990.

Horder, J. M.: Fatal chlormethiazole poisoning in chronic alcoholics. *Br. Med. J.* **2:**614, 1977.

Hoyumpa, A., Desmond, P., Roberts, R., Avant, G. R., Roberts, R. K., and Schenker, S.: Effect of ethanol on benzodiazepine disposition in dogs. *Clin. Res.* **27:**454, 1979 (abstract).

Hoyumpa, A. M., Patwardhan, R., Maples, M., Desmond, P. V., Johnson, R. F., Sinclair, A. P., and Schenker, S.: Effect of short-term ethanol administration on lorazepam clearance. *Hepatology* **1:**47–53, 1981.

Ingelman-Sundberg, M., and Johansson, I.: Mechanisms of hydroxyl radical formation and ethanol oxidation by ethanol-inducible and other forms of rabbit liver microsomal cytochrome P-450. *J. Biol. Chem.* **259:**6447–6458, 1984.

Ioannides, C., Lake, B. G., and Parke, D. V.: Enhancement of hepatic microsomal drug metabolism *in vitro* following ethanol administration. *Xenobiotica* **5:**665–676, 1975.

Ishii, H., Joly, J.-G., and Lieber, C. S.: Effect of ethanol on the amount and enzyme activities of hepatic rough and smooth microsomal membranes. *Biochim. Biophys. Acta* **291:**411–420, 1973.

Johansson, I., and Ingelman-Sundberg, M.: Carbon tetrachloride-induced lipid peroxidation dependent on an ethanol-inducible form of rabbit liver microsomal cytochrome P-450. *FEBS Lett.* **183:**265–269, 1985.

Johansson, I., and Ingelman-Sundberg, M.: Benzene metabolism by ethanol-, acetone-, and benzene-inducible cytochrome P-450 (IIE1) in rat and rabbit liver microsomes. *Cancer Res.* **48:**5387–5390, 1988.

Johnson, D. A., Lee, N. M., Cooke, R., and Loh, H.: Adaptation to ethanol-induced fluidization of brain lipid bilayers: Cross tolerance and reversibility. *Mol. Pharmacol.* **17:**52–55, 1980.

Joly, J-G., and Hetu, C.: Effects of chronic ethanol administration in the rat: Relative dependency on dietary lipids. *Biochem. Pharmacol.* **24:**1475–1480, 1975.

Joly, J.-G., Ishii, H., Teschke, R., Hasumura, Y., and Lieber, C. S.: Effect of chronic ethanol feeding on the activities and submicrosomal distribution of reduced nicotinamide adenine dinucleotide phosphate (NADPH)-cytochrome P-450 reductase and the dimethylase for aminopyrine and ethylmorphine. *Biochem. Pharmacol.* **22:**1532–1535, 1973.

Kalant, H., Khanna, J. M., and Marshman, J.: Effect of chronic intake of ethanol on pentobarbital metabolism. *J. Pharmacol. Exp. Ther.* **5:**318–324, 1970.

Kater, R. M. H., Roggin, G., Tobon, F., Zieve, P., and Iber, F. L.: Increased rate of clearance of drugs from the circulation of alcoholics. *Am. J. Med. Sci.* **258:**35–39, 1969a.

Kater, R. M. H., Tobon, F., and Iber, F. L.: Increased rate of tolbutamide metabolism in alcoholic patients. *J.A.M.A.* **207:**363–365, 1969b.

Kissin, B.: Interactions of ethyl alcohol and other drugs. In: *The Biology of Alcoholism, Vol. 3: Clinical Pathology* (B. Kissin and H. Begleiter, eds.), New York, Plenum, pp. 109–161, 1974.

Kitson, T. M.: The effect of 5,5′-Dithiobis(1-methyltetrazole) on cytoplasmic aldehyde dehydrogenase and its implications for cephalosporin–alcohol reactions. *Alcoholism: Clin. Exp. Res.* **10**:27–32, 1986.

Klotz, U., Avant, G. R., Hoyumpa, A., Schenker, S., and Wilkinson, G. R.: The effects of age and liver disease on the disposition and elimination of diazepam in adult man. *J. Clin. Invest.* **55**:347–359, 1975.

Koop, D. R., Orlandi, C. L., and Schnier, G. G.: Ethanol-inducible P-450 isozyme 3a is a benzene hydroxylase. *FASEB J.* **2**:A1012, 1988 (abstract).

Kraus, J. W., Desmond, P. V., Marshall, J. P., Johnson, R. F., Schenker, S., and Wilkinson, G. R.: Effects of aging and liver disease on disposition of lorazepam. *Clin. Pharmacol. Ther.* **24**:411–419, 1978.

Lambert, S. M.: Carbon tetrachloride in the treatment of hookworm disease: Observations in twenty thousand cases. *J.A.M.A.* **79**:2055–2057, 1922.

Lamson, P. D., Minot, A. S., and Robbins, B. H.: The prevention and treatment of carbon tetrachloride intoxication. *J.A.M.A.* **90**:345–349, 1928.

Lane, E. A., Guthrie and Linnoila, M.: Effects of ethanol on drug and metabolite pharmacokinetics. *Clin. Pharmacokin.* **10**:228–247, 1985.

Lauterburg, B. H., and Velez, M. E.: Glutathione deficiency in alcoholics: Risk factor for paracetamol hepatotoxicity. *Gut* **29**:1153–1157, 1988.

Lester, D.: The acetylation of isoniazid in alcoholics. *Quart. J. Stud. Alcohol* **25**:541–543, 1964.

Levin, W., Thomas, P. E, Oldfield, N., and Ryan, D. E.: *N*-demethylation of *N*-nitrosodimethylamine catalyzed by purified rat hepatic microsomal cytochrome P-450: Isozyme specificity and role of cytochrome b_5. *Arch. Biochem. Biophys.* **248**:158–165, 1986.

Levy, G.: Drug biotransformation interactions in man: Nonnarcotic analgesics. *Ann. N.Y. Acad. Sci.* **9**:32–42, 1971.

Lieber, C. S.: Metabolism of ethanol. In: *Metabolic Aspects of Alcoholism* (C. S. Lieber, ed.), Baltimore, Maryland, University Park Press, pp. 1–29, 1977.

Lieber, C. S; *Medical Disorders of Alcoholism: Pathogenesis and Treatment*, Philadelphia, W. B. Saunders, 1982.

Lieber, C. S., and DeCarli, L. M.: Hepatic microsomal ethanol-oxidizing system: *In Vitro* characteristics and adaptive properties *in vivo*. *J. Biol. Chem.* **245**:2505–2512, 1970.

Littleton, J. M., Geryk, R. J., and Grieve, S. J.: Alterations in phospholipid composition in ethanol tolerance and dependence. *Alcoholism: Clin. Exp. Res.* **3**:50–56, 1979.

Lorr, N. A., Miller, K. W., Chung, H. R., and Yand, C. S.: Potentiation of the hepatotoxicity of *N*-nitrosodimethylamine by fasting, diabetes, acetone, and isopropanol. *Toxicol. Appl. Pharmacol.* **73**:423–431, 1984.

Lu, A. Y. H., Junk, K. W., and Coon, M. J.: Resolution of the cytochrome P-450-containing omega-hydroxylation system of live microsomes into three components. *J. Biol. Chem.* **244**:3714–3721, 1969.

Luoma, P., and Vorne, M.: The combined effect of ethanol and phenobarbital on the activities of hepatic drug metabolizing enzymes in rats. *Acta Pharmacol. Toxicol.* **33**:442–448, 1973.

Makar, A. B., Tephly, T. R., and Mannering, G. J.: Methanol metabolism in the monkey. *Mol. Pharmacol.* **4**:471–483, 1968.

Maling, H. M., Stripp, B., Sipes, I. G., Highman, B., Waul, W., and Williams, M. A.: Enhanced hepatotoxicity of carbon tetrachloride thioacetamide and dimethylnitrosamine and some comparisons with potentiation by isopropanol. *Toxicol. Appl. Pharmacol.* **33**:291–308, 1975.

Martin, E. W.: *Hazards of Medication*. Philadelphia, J. B. Lippincott Company, p. 435, 1971.

McMartin, K. E., Makar, A. B., Martin, A. G., Palese M., and Tephly, T. R.: Methanol poisoning: The role of formic acid in the development of metabolic acidosis in the monkey and the reversal by 4-methylpyrazole. *Biochem. Med.* **13**:319–333, 1975.

Milner, G., and Landauer, A. A.: Alcohol thioridazine and chlorpromazine effects on skills related to driving behavior. *Br. J. Psychiatry* **118**:351–352, 1971.

Misra, P. S., Lefevre, A., Ishii, H., Rubin, E., and Lieber, C. S.: Increase of ethanol meprobamate and pentobarbital metabolism after chronic ethanol administration in man and in rats. *Am. J. Med.* **51**:346–351, 1971.

Mitchell, J. R., and Jollows, D. J.: Metabolic activation of drugs to toxic substances. *Gastroenterology* **68**:392–410, 1975.

Mitchell, J. R., Jollows, D. J., Potter, W. Z., Gillette, J. R., and Brodie, B. B.: Acetaminophen induced hepatic necrosis. I. Role of drug metabolism. *J. Pharmacol. Exp. Ther.* **187**:185–194, 1973.

Mitchell, J. R., Mack, C., Mezey, E., and Maddrey, W. C.: The effects of variation in dietary protein and ethanol on hepatic microsomal drug metabolism in the rat. *Hepatology* **1**:336–340, 1981.

Moldeus, P., Andersson, B., and Norling, A.: Interaction of ethanol oxidation with glucuronidation in isolated hepatocytes. *Biochem. Pharmacol.* **27**:2583–2588, 1978.

Moon, H. D.: The pathology of fatal carbon tetrachloride poisoning with special reference to the histogenesis of the hepatic and renal lesions. *Am. J. Pathol.* **26**:1041–1057, 1950.

Morgan, E. T., Koop, D. R., and Coon, M. J.: Comparison of six rabbit liver cytochrome P-450 isozymes in formation of a reactive metabolite of acetaminophen. *Biochem. Biophys. Res. Commun.* **112**:8–13, 1983.

Mørland, J., Setekliev, J., Haffner, J. F. W., Stromsaether, C. E., Danielsen, A., and Wethe, G. M.: Combined effects of diazepam and ethanol on mental and psychomotor functions. *Acta Pharmacol. Toxicol.* **34**:5–15, 1974.

Nakajima, T., Okuyama, S., Yonekura, I, and Sato, A.: Effects of ethanol and phenobarbital administration on the metabolism and toxicity of benzene. *Chem. Biol. Interact.* **55**:23–38, 1985.

Nakajima, T., Okino, T. and Sato, A.: Kinetic studies on benzene metabolism in rat liver—possible presence of three forms of benzene metabolizing enzymes in the liver. *Biochem. Pharmacol.* **36**:2799–2804, 1987.

Neuvonen, P. J., Pentikainen, P. J., Jostell, K. G., and Syvalahti, E.: Effects of ethanol on the pharmacokinetics of chlormethiazole in humans. *Int. J. Clin. Pharmacol. Ther. Toxicol.* **19**:552–560, 1981.

Ohman, M., and Marklund, S. L.: Plasma extracellular superoxide dismutase and erythrocyte Cu,Zn-containing superoxide dismutase in alcoholics treated with disulfiram. *Clin. Sci.* **70:**365–369, 1986.

Olsen, H., and Mørland, J.: Ethanol-induced increase in drug acetylation in man. *Br. Med. J.* **2:**1260–1262, 1978.

Pantuck, E. J., Pantuck, C. B., Ryan, D. E., and Conney, A. H.: Inhibition and stimulation of enflurane metabolism in the rat following a single dose or chronic administration of ethanol. *Anesthesiology* **62:**255–262, 1985.

Pignon, J. P., Bailey, N. C., Baraona, E., and Lieber, C. S.: Fatty acid-binding protein: A major contributor to the ethanol-induced increase in liver cytosolic proteins in the rat. *Hepatology* **7:**865–871, 1987.

Powis, G.: Effect of a single oral dose of methanol, ethanol and propan-2-ol, on the hepatic microsomal metabolism of foreign compounds in the rat. *Biochem. J.* **148:**.269–277, 1975.

Prasad, J. S., Crankshaw, D. L., Erickson, R. R., Elliot, C. E., Husby, A. D., and Holtzman, J. L.: Studies on the effect of chronic consumption of moderate amounts of ethanol on male rat hepatic microsomal drug-metabolizing activity. *Biochem. Pharmacol.* **34:**3427–3431, 1985.

Prescott, L. F., and Wright, N.: The effects of hepatic and renal damage on paracetamol metabolism and excretion following overdosage: A pharmacokinetic study. *Br. J. Pharmacol.* **49:**602–613, 1973.

Pritchard, J. F., and Schneck, D. W.: Effects of ethanol and phenobarbital on the metabolism of propanolol by 9000 *g* rat liver supernatant. *Biochem. Pharmacol.* **26:**2453–2454, 1977.

Ratcliffe, F.: The effect of chronic ethanol administration on the responses to amylobarbitone sodium in the rat. *Life Sci.* **8:** 1051–1061, 1969.

Reinke, L. A., Kauffman, F. C., Belinsky, S. A., and Thurman, R. G.: Interactions between ethanol metabolism and mixed-function oxidation in perfused rat liver: Inhibition of *p*-nitroanisole *o*-demethylation. *J. Pharmacol. Exp. Ther.* **213:**70–78, 1980.

Reyes, H., Levi, A. J., Gatmaitan, Z., and Arias, I. M.: Studies of Y and Z, two hepatic cytoplasmic organic anion-binding proteins: Effect of drugs, chemicals, hormones and cholestasis. *J. Clin. Invest.* **50:**2242–2252, 1971.

Riihimaki, V., Savolainen, K., Pfaffli, P., Pekari, K., Sippel, H. W., and Laine, A.: Metabolic interaction between *n*-xylene and ethanol. *Arch. Toxicol.* **49:**253–263, 1982.

Robinson, A. E., and McDowall, R. D.: Toxicological investigations of six chlormethiazole-related deaths. *Forensic Sci. Int.* **14:**49–55, 1979.

Röe, O.: Clinical investigations of methyl alcohol poisoning with special reference to pathogenesis and treatment of amblyopia. *Acta. Med. Scand.* **13:**558–608, 1943.

Rosenthal, S. M.: Some effects of alcohol upon the normal and damaged liver. *J. Pharmacol. Exp. Ther.* **38:**291–301, 1930.

Rubin, E., and Lieber, C. S.: Hepatic microsomal enzymes in man and rat: Induction and inhibition by ethanol. *Science* **162:** 690–691, 1968.

Rubin, E., Hutterer, F., and Lieber, C. S.: Ethanol increases hepatic smooth endoplasmic reticulum and drug-metabolizing enzymes. *Science* **159:**1469–1470, 1968.

Rubin, E., Bacchin, P., Gang, H., and Lieber, C. S.: Inhibition of hepatic microsomal and mitochondrial enzymes by ethanol. *Lab. Invest.* **22:**569–580, 1970a.

Rubin, E., Gang, H., Misra, P. S., and Lieber, C. S.: Inhibition of drug metabolism by acute ethanol intoxication: A hepatic microsomal mechanism. *Am. J. Med.* **49:**801–806, 1970b.

Rubin, E., Lieber, C. S., Alvares, A. P., Levin, W., and Kuntzman, R.: Ethanol binding to hepatic microsomes: Its increase by ethanol consumption. *Biochem. Pharmacol.* **20:**229–231, 1971.

Sato, C., and Lieber, C. S.: Mechanism of preventive effect of ethanol on acetaminophen-induced hepatotoxicity. *J. Pharmacol. Exp. Ther.* **218:**811–815, 1981.

Sato, C., Nakano, M., and Lieber, C. S.: Increased hepatotoxicity of acetaminophen after chronic ethanol consumption in the rat. *Gastroenterology* **80:**140–148, 1981.

Schnier, G. G., Laethem, C. L., and Koop, D. R.: Identification and induction of cytochromes P450, P450IIE1 and P450IA1 in rabbit bone marrow. *J. Pharmacol. Exp. Ther.* **251:**790–796, 1989.

Scott, C. R., and Stewart, M. J.: Letter: Cysteamine treatment in paracetamol overdose. *Lancet* **1:**452–453, 1975.

Seef, L. B., Cuccherini, B. A., Zimmerman, H. J., Alder, E., and Benjamin, S. B.: Acetaminophen hepatotoxicity in alcoholics (Clinical review). *Ann. Intern. Med.* **104:**399–404, 1986.

Sellers, E. M., Lang, M., Koch-Weser, J., LeBlanc, E., and Kalant, H.: Interaction of chloral hydrate and ethanol in man. I. Metabolism. *Clin. Pharmacol. Ther.* **13:**37–40, 1972.

Sellers, E. M., Naranjo, C. A., Giles, H. G., Frecker, R. C., and Beeching, M.: Intravenous diazepam and oral ethanol interaction. *Clin. Pharmacol. Ther.* **28:**638–645, 1980.

Sellman, R., Kanto, J., Raijola, E., and Pekkarinen, A.: Human and animal study on elimination from plasma and metabolism of diazepam after chronic alcohol intake. *Acta Pharmacol. Toxicol.* **36:**33–38, 1975.

Sereny, G., Sharama, V., Holt, J., and Gordis, E.: Mandatory supervised antabuse therapy in an outpatient alcoholism program: A pilot study. *Alcoholism: Clin. Exp. Res.* **10:**290–292, 1986.

Shanley, B. C., Zail, S. S., and Joubert, S. M.: Effects of ethanol on liver δ-aminolevulinate synthetase in rat. *Lancet* **1:**70–71, 1968.

Siegers, C. P., Heidbuchel, K., and Younes, M.: Influence of alcohol, dithiocarb and (+)-catechin on the hepatotoxicity and metabolism of vinylidene chloride in rats. *J. Appl. Toxicol.* **3:**90–95, 1983.

Smillie, W. G., and Pessoa, S. B.: Treatment of hookworm disease with carbon tetrachloride. *Am. J. Hyg.* **3:**35–45, 1923.

Sotaniemi, E.A., Anttila, M., Rautio, A., Stengard, J., Saukko, P., and Jarvensivu, P.: Propranolol and sotalol metabolism after a drinking party. *Clin. Pharmacol. Ther.* **29:**705–710, 1981.

Strubelt, O.: Ethanol potentiation of liver injury. *Adv. Pharmacol.* **5:**55–64, 1982.

Sundheimer, D. W., and Brendel, K.: Factors influencing sulfation in isolated rat hepatocytes. *Life Sci.* **34:**23, 1984.

Takagi, T., Ishii, H., Takahashi, H., Kato, S., Okuno, F., Ebihara, Y., Yamauchi, H., Nagata, S., Tashiro, M., and Tsuchiya, M.: Potentiation of halothane hepatotoxicity by chronic ethanol administration in rat: An animal model of halothane hepatitis. *Pharmacol. Biochem. Behav.* **18**(Suppl 1):461–465, 1983.

Tephly, T. R., Parks, R. E., and Mannering, G. J.: Methanol metabolism in the rat. *J. Pharmacol. Exp. Ther.* **143**:292–300, 1964.

Teschke, R., Moreno, F., and Petrides, A. S.: Hepatic microsomal ethanol oxidizing system (MEOS): Respective role of ethanol and carbohydrates for the enhanced activity after chronic alcohol consumption. *Biochem. Pharmacol.* **30**:45–51, 1981.

Teschke, R., Hauptmeier, K.-H., and Frenzel, H.: Effect of an acute dose of ethanol on the hepatotoxicity due to carbon tetrachloride. *Liver* **3**:100–109, 1983.

Thiessen, J. J., Sellers, E. M., Denbeigh, P., and Dolman, L.: Plasma protein binding of diazepam and tolbutamide in chronic alcoholics. *J. Clin. Pharmacol.* **16**:345–351, 1976.

Thomson, J. S., and Prescott, L. F.: Liver damage impaired glucose tolerance after paracetamol overdosage. *Br. Med. J.* **2**:506–507, 1966.

Traiger, G. J., and Plaa, G. L.: Relationship of alcohol metabolism to the potentiation of CCl$_4$ hepatotoxicity induced by aliphatic alcohols. *J. Pharmacol. Exp. Ther.* **183**:481–488, 1972.

Truitt, E. B., and Walsh, M. J.: The role of acetaldehyde in the actions of ethanol. In: *The Biology of Alcoholism, Vol. 1. Biochemistry* (B. Kissin and H. Begleiter, eds.), New York, Plenum, 1971.

Tsutsumi, R., Leo, M. A., Kim, C., Tsutsumi, M., Lasker, J., Lowe, N., and Lieber, C. S.: Interaction of ethanol with enflurane metabolism and toxicity: Role of P450IIE1. *Alcoholism: Clin. Exp. Res.* **14**:174–179, 1990.

Vessell, E. S., Page, J. G., and Passananti, G. T.: Genetic environmental factors affecting ethanol metabolism in man. *Clin. Pharmacol. Ther.* **12**:192–201, 1971.

Vilstrup, H., Henningsen, N. C., and Hansen, L. F.: Lever-beskadigeles efter paracetamol. *Ugeskr. Laeg.* **139**:831–834, 1977.

Voas, R. B.: Alcohol as an underlying factor in behavior leading to fatal highway crashes. In: *Proceedings of the First Annual Alcoholism Conference of the National Institute on Alcohol Abuse and Alcoholism.* (M. E. Chafetz, ed.), Washington, D. C., DHEW Publ. (NIH) 74-675, U.S. Government Printing Office, p. 324, 1973.

von Wartburg, J.-P., Bethune, J. L., and Valle, B. L.: Human liver alcohol dehydrogenase. Kinetic and physico-chemical properties. *Biochemistry* **3**:75–82, 1964.

Wacker, W. E. C., Haynes, H., Druyan, R., Fisher, W., and Coleman, J. E.: Treatment of ethylene glycol poisoning with ethyl alcohol. *J.A.M.A.* **194**:1231–1235, 1965.

Whitehouse, L. W., Paul, C. J., Coldwell, B. B., and Thomas, B. H.: Effect of ethanol on diazepam distribution in rat. *Res. Commun. Chem. Pathol. Pharmacol.* **12**:221–241, 1975.

Whiting, B., and Lawrence, J. R.: Effect of acute alcohol intoxication on the metabolism and plasma kinetics of chlordiazepoxide *Br. J. Clin. Pharmacol.* **7**:95–100, 1979.

Wiśniewska-Knypl, J. M., Wronska-Nofer, T., Jajte, J., and Jedlinska, U.: The effect of combined exposures to ethanol and xylene on rat hepatic microsomal monooxygenase activities. *Alcohol* **6**:347–352, 1989.

Yang, C. S., Koop, D. R., Wang, T., and Coon, M. J.: Immunochemical studies on the metabolism of nitrosamines by ethanol-inducible cytochrome P-450. *Biochem. Biophys. Res. Commun.* **128**:1007–1013, 1985.

Zimmerman, H. J.: *The Adverse Effects of Drugs and Other Chemicals on the Liver.* New York, Appleton–Century–Crofts, 1978.

Zimmerman, H. J.: Effects of alcohol on other hepatotoxins. *Alcoholism: Clin. Exp. Res.* **10**:3–15, 1986.

Zirkle, G. A., King, P. D., McAlee, D. B., and Van Dyke, R.: Effects of chlorpromazine and alcohol on coordination and judgement. *J.A.M.A.* **1**:1496–1499, 1959.

7

Alcohol and the Liver

Charles S. Lieber and Maria A. Leo

7.1. Epidemiology of Alcoholic Liver Disease

7.1.1. Mortality and Prevalence

The mortality rates for cirrhosis vary greatly from country to country. For example, in 1972 the World Health Organization reported mortality rates of 7.5/100,000 in Finland and 57.2/100,000 in France. The corresponding per-capita alcohol consumptions were 5.1 and 16.8 liters of absolute alcohol (International Statistics on Alcoholic Beverages: Production Trade and Consumption 1950–1972, 1977). Whereas deaths from other causes such as cardiovascular diseases decreased in the period from 1950 to 1974 by 2% in the United States, deaths from cirrhosis climbed 71.7%. Since then, mortality from cirrhosis has leveled off or even decreased in the United States. The reduction, however, pertains primarily to nonalcoholic cirrhosis (Liver Cirrhosis Mortality in the United States, 1988). In Canada, cirrhosis has been the most rapidly increasing cause of death in the population over 25 years of age, followed by lung and bronchial cancer and suicide (Schmidt, 1977), and it is now the fifth major cause of death for men in the productive years from 25 to 64 (Rankin, 1977). The concentration of mortality of alcoholic cirrhosis to a rather young population group also occurs in the United States. In Baltimore, it was the fifth leading cause of death for persons 25 to 44 years of age (Kuller *et al.*, 1969) and the fourth leading cause of death for those 25 to 64 years of age in New York City (New York City; Summary of Vital Statistics, 1984).

The male-to-female ratio of mortality is more than 2 : 1 for all cases of cirrhosis, and it is even greater when restricted to alcoholic cirrhosis (Hällen and Krook, 1963; Jolliffe and Jellinek, 1941). However, there is evidence that the progression to more severe liver injury is accelerated in women (Rankin, 1977). Indeed, Wilkinson *et al.* (1969) found women to be more susceptible than men to the development of alcoholic cirrhosis. Other studies also found the incidence of chronic advanced liver disease to be higher among women than among men for a similar history of alcohol abuse (Morgan and Sherlock, 1977; Maier *et al.*, 1979; Nakamura *et al.*, 1979). Pequignot *et al.* (1974, 1978) have also shown that daily intake of alcohol as low as 40 g in men and 20 g in women resulted in a statistically significant increase in the incidence of cirrhosis in a well-nourished population. For both men and women mortality from cirrhosis in the nonwhite population is at least double that of the white population in urban areas of the United States (Kuller *et al.*, 1969).

The incidence of liver cirrhosis has been calculated in 16 areas of the United States (Garagliano *et al.*, 1977). In the Baltimore area during 1973, the age-adjusted incidence rates (first hospitalizations) for all alcoholic liver diseases per 100,000 population over 20 years were: 36.3 for white males, 19.8 for white females, 60.0 for nonwhite males, and 25.4 for nonwhite females. An increasing incidence of alcoholic liver disease has also been demonstrated in Great Britain (Morgan and Sherlock, 1977), where it has become a major cause of liver damage. The morbidity of liver cirrhosis has been reported to be 28 times higher among the problem drinkers than among the nondrinkers in a survey of factory workers (Pell and D'Alonzo, 1968).

7.1.2. Relationship of Liver Disease to Alcohol Consumption

Despite the high incidence of alcoholism and cirrhosis throughout the world, the prevalence of cirrhosis in alcoholics is relatively low. According to autopsy series the prevalence of cirrhosis is about 18% and in series based on liver biopsy it ranges from 17 to 30.8% (Lelbach, 1966, 1967; Leevy, 1968; Von Olderhausen, 1970). The development of alcoholic liver disease correlates with both the magnitude and duration of alcohol consumption, as illustrated in Fig. 7.1. Pequignot (1958) and Pequignot *et al.* (1974) estimated that the average cirrhogenic dose is 180 g ethanol/day consumed regularly for approximately 25 years; the risk is increased five times with a consumption between 80 and 160 g/day and 25 times if daily ethanol consumption exceeds 160

g. As mentioned before, average cirrhogenic doses as well as threshold doses are lower in females than males.

A close correlation exists between per-capita alcohol consumption and cirrhosis mortality (Jolliffe and Jellinek, 1941; Schmidt, 1975; Lelbach, 1976; Schmidt, 1977). Both in Finland and in Canada liver cirrhosis mortality has been used as a means to measure the prevalence of alcoholism (Brunn *et al.*, 1960; Popham, 1970). The availability of alcoholic beverages has been shown to correlate with the prevalence of liver disease in various countries. During the First and Second World Wars in Europe and during the time of Prohibition in the United States, there was a clear decrease in mortality from cirrhosis (Jolliffe and Jellinek, 1941). Licensing laws, economic factors, and environmental conditions have also been shown to influence both per-capita alcohol consumption and mortality from cirrhosis. High alcohol taxes were associated with decreased death rates from cirrhosis in England and Wales from 12/100,000 in 1914 to 2/100,000 in 1945 (Stone *et al.*, 1968). Simi-

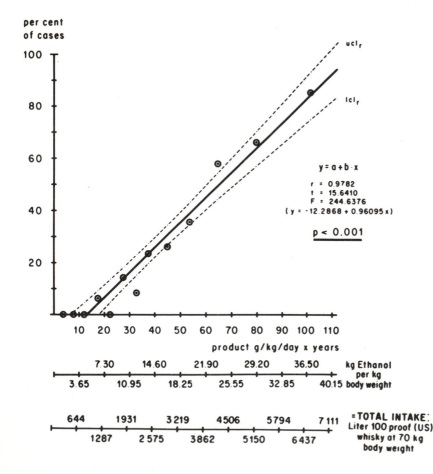

FIGURE 7.1. Correlation between total amounts of ethanol consumed per kilogram of body weight during drinking life and incidence of cirrhosis of the liver in 265 alcoholics (ucl$_r$, lcl$_r$ = upper and lower confidence limits of regression). (Reproduced by permission from Lelbach, 1975.)

larly, the termination of alcohol rationing in Sweden in 1955 was followed by an increase in deaths attributable to cirrhosis (Hällen and Krook, 1963). Wine, as an inexpensive source of ethanol, plays a dominant role in alcoholic liver diseases in some economically disadvantaged countries such as Chile and Portugal, where mortality from cirrhosis is particularly high. On the other hand, economic growth in such countries as West Germany resulted in an increase in problem drinkers and cirrhosis mortality (Martini and Bode, 1970). The incidence of liver damage in Austrian brewery workers with free daily allowance of beer (144 g ethanol) was 26% as opposed to only 9.3% of a group of metal workers (Frank et al., 1967).

It is now generally accepted that both the consumption and rates of alcohol problems can be regulated by means of various restrictions and tax policies. In both Canada and Finland the potential usefulness of the many diverse restrictions through which control can be exercised has been examined in great detail (Bruun et al., 1975; Popham et al., 1976). There exists no valid epidemiologic or clinical evidence that would suggest that drinking patterns—continuous versus periodic—or the type of alcoholic beverage influences mortality from cirrhosis (Schmidt, 1975).

7.2. Pathology and Symptomatology of the Various Stages of Alcoholic Liver Injury

The spectrum of alcoholic liver injury involves hepatic steatosis (fatty liver), early fibrosis (perivenular, perisinusoidal, and pericellular fibrosis), alcoholic hepatitis, and cirrhosis.

7.2.1. Fatty Liver

Fat accumulation in liver cells is the earliest and most common response to alcohol, and alcohol is its most common etiological factor (Leevy, 1962). The normal liver weighs about 1.5 kg, whereas the alcoholic fatty liver weighs 2.0 to 2.5 kg. In rare cases massive accumulation of fat may occur, resulting in a liver weighing from 4.0 to 6.0 kg. On hematoxylin- and eosin-stained sections, parenchymal accumulation of lipid is seen as clear intracytoplasmic vacuoles or as black or red globules with various Sudan stains. In massive steatosis the hepatocytes are uniformly filled with large fat droplets; the nucleus may be eccentrically

placed (Fig. 7.2), and when the cell membranes between adjacent hepatocytes rupture, numerous fatty cysts are formed. Increased hepatic lipid content may be demonstrated by biochemical measurements before it becomes histologically apparent. Evidence of necrosis is usually sparse, but sometimes hepatocytes are surrounded by mononuclear cells, indicating a mild inflammatory response. When pronounced, these formations are called "lipogranulomas." Characteristic ultrastructural changes reveal enlarged and distorted mitochondria with shortened cristae containing crystalline inclusions (Svoboda and Manning, 1964). The endoplasmic reticulum shows vacuolar dilation and proliferation. Despite adequate nutrition, alcohol can induce these alterations in baboons (Lieber et al., 1972) (Fig. 7.3) and in human volunteers (Lane and Lieber, 1966).

The clinical spectrum of alcoholic fatty liver may extend from silent, nonsymptomatic hepatomegaly to severe hepatocellular failure with cholestasis and portal hypertension. Most of the patients with pure fatty liver are virtually asymptomatic. Hepatomegaly is the most common clinical sign, present on physical examination in 75% of patients (Leevy, 1962). In more advanced cases, hepatic tenderness, anorexia, nausea, emesis, jaundice, fluid accumulation, and even spider angiomas are present. In many of these cases an associated alcoholic hepatitis may contribute to the clinical picture and laboratory findings (see below). The typical patient with more severe alcoholic fatty liver is young or middle-aged with a history of drinking alcohol heavily for weeks or months. The liver edge may be a few centimeters below the costal margin or may extend to the iliac crest. The liver is usually solid, and its surface is smooth. Mild tenderness is common, sometimes associated with severe epigastric or right upper quadrant pain (Goldberg and Thompson, 1961). Severe forms of fatty liver may present with a clinical picture mimicking extrahepatic obstructive jaundice, especially if associated with dark urine and alcoholic stools (Phillips and Davidson, 1957; Ballard et al., 1961).

Typical abnormalities in laboratory tests are slightly or moderately elevated serum transaminases (AST and ALT) and slight retention of bromsulfthalein (BSP) (Leevy, 1962; Devenyi et al., 1970). However, the correlation between the pathological and laboratory findings and the degree of hepatic steatosis is rather poor (Bradus et al., 1963). Serum prothrombin concentrations are frequently low, even in mild cases of alcoholic fatty liver. In cases of associated malnutrition or malabsorption, the prothrombin time can be corrected by administration of vitamin K. Both decreased and normal

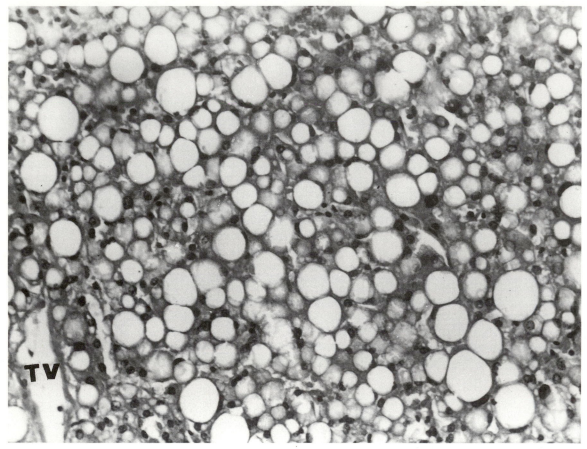

FIGURE 7.2. Severe fatty liver of an alcoholic. The hepatocytes are uniformly filled by large fat droplets, and nuclei are eccentrically placed. T.V., terminal venule (hematoxylin and eosin staining, ×250).

serum albumin values have been reported in alcoholic fatty liver (Leevy, 1962; Devenyi *et al.*, 1970). Serum protein values should be interpreted cautiously, however, since they may be influenced by direct metabolic effects of alcohol as well as by nutritional factors. In contrast to more advanced alcoholic liver injury, all the abnormalities in laboratory tests tend to return to normal rapidly within the first days of hospitalization. Transient anemia and albuminuria can be detected in about 25% of patients with fatty liver at the time of initial hospitalization. Albuminuria disappears within 24 to 72 hr, and its mechanism has not been resolved (Leevy, 1962). Anemia is usually caused by associated nutritional deficiencies or gastrointestinal hemorrhage (see Chapters 8 and 17). Rarely, alcoholic fatty liver may be complicated by hyperlipemia, hemolytic anemia, and jaundice (Zieve's syndrome: Zieve, 1958; see Chapters 4 and 8). The

possible occurrence of various metabolic abnormalities (such as hypoglycemia, alcoholic ketoacidosis, and lactic acidosis) is discussed in Chapter 3.

7.2.2. Early Fibrosis: Pericellular, Perisinusoidal, and Perivenular (or Pericentral) Fibrosis or Sclerosis

Although it can occur anywhere in the hepatic lobule, the earliest deposition of fibrous tissue is generally seen around the central veins and venules (also now called terminal hepatic venules). Fibrosis or sclerosis around central veins has been described in alcoholic hepatitis (see Section 7.2.3.) and is often associated with a necrotizing process called "sclerosing hyalin necrosis" (Edmondson *et al.*, 1967). When intensive, it may obliterate central veins and lead to postsinusoidal

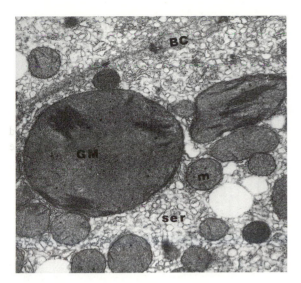

FIGURE 7.3. Ultrastructural changes after prolonged ethanol ingestion in a baboon. Note the prominence of the smooth endoplasmic reticulum (ser). Mitochondrial changes include swelling, shortening of the cristae, and some giant forms (GM). Some normal-sized mitochondria (m) are still present. Disruptions of mitochondrial membranes are common. (Uranyl acetate and lead staining, ×9,500.)

portal hypertension with ascites prior to the development of cirrhosis (Reynolds *et al.*, 1969). It is important to note, however, that pericentral (or perivenular) fibrosis can be seen in the absence of widespread inflammation and necrosis, in association with what most pathologists would label as "simple" fatty liver. Of 75 hospitalized alcoholic subjects in whom the diagnosis of simple fatty liver had been made on the basis of liver biopsy, 40% were found to have perivenular fibrosis on the basis of a review of the histology (Van Waes and Lieber, 1977a) (Fig. 7.4B). There was no evidence of diffuse alcoholic hepatitis. Since some previous alcoholic hepatitis may have been missed in these patients, sequential changes were studied in the baboon model of alcoholic liver injury. Perivenular fibrosis was again observed as early as the "simple" fatty-liver stage, before more severe liver lesions developed. The degree of perivenular fibrosis correlated with an increase in portal blood pressure (Lieber *et al.*, 1976; Miyakawa *et al.*, 1985). Perivenular fibrosis is commonly associated with pericellular fibrosis (Van Waes and Lieber, 1977a), which in the early stages again prevails in the perivenular areas but may occur anywhere in the lobule. Perivenular and perisinusoidal fibrosis usually parallel each other in extent, but dissociation can occur.

An interesting clinical and morphological pattern of acute alcoholic liver disease was described by Uchida *et al.* (1983). Histopathologically, it is characterized by striking microvesicular fatty change in the perivenular hepatocytes with little or no alcoholic hyalin. This entity was called alcoholic foamy degeneration (AFD) of the liver. According to Uchida *et al.* (1983), the lobular architecture is generally intact. The perivenular areas had a disrupted cord pattern with the hepatocytes strikingly swollen three to six times the normal size and laden with innumerable microvacuoles. The bloated hepatocytes had no nuclear displacement by the fat vacuoles, and the nucleus was often pyknotic, crenated, or lysed. Megamitochondria were found frequently in the affected hepatocytes. Delicate intrasinusoidal creeping collagenosis was noted. The foamy change coexisted with macrovesicular fat to a varying degree in the same or different hepatocytes. Those hepatocytes having both types of fatty change were swollen. In all cases, the foamy fatty change was more conspicuous around the terminal hepatic venules, involving the midzonal hepatocytes to a lesser degree. The periportal hepatocytes were intact or had only macrovesicular fat. In the follow-up biopsies of four patients, AFD showed resolution with little or no foamy fatty change. However, some degree of macrovesicular fatty change remained.

Glucose-6-phosphatase (G6P) activity was markedly decreased or absent. Succinic dehydrogenase activity was also decreased perivenularly, though not as strikingly as G6P activity. There was marked elevation of serum aminotransferases at the time of admission. However, serological markers for acute viral hepatitis A or B were absent. Leukocytosis (WBC in excess of 10,000) was found in only 24% of AFD cases. The total serum bilirubin was greater than 5 mg/dl in 81% of cases with AFD. Thus, this AFD is a further illustration of the fact that, at the so-called fatty liver stage and in the absence of florid alcoholic hepatitis, severe liver changes can be seen in the alcoholic. The merit of the paper by Uchida *et al.* (1983) is to point out that many of the changes described before in isolation can occur together. The striking swelling of hepatocytes after chronic alcohol consumption has been described before in the rat (Baraona *et al.*, 1975) and has also been found in baboons and humans, as discussed in Chapter 5. This swelling is secondary, at least in part, to a defect in protein secretion (see Chapter 5). The ultrastructural changes of AFD (megamitochondria and so on) have been described before in alcoholic liver injury, but, as pointed out by Uchida *et al.* (1983), they seem to occur predominantly in those cells with microvesicular steatosis in the case of

AFD. Furthermore, microvesicular fat is not uncommon in steatosis of the alcoholic (see, for instance, Fig. 7.4). A characteristic, although not pathognomonic, feature of AFD is the jaundice with bile pigment deposition in the cytoplasm as well as bile plugs in the dilated bile canaliculi. Jaundice, of course, is not unusual in the alcoholic. Over 65% of subjects with a history of excessive alcohol intake exhibit jaundice (Klatskin and Yesner, 1949; Summerskill *et al.*, 1960), and yet the mechanism of this disturbance in bilirubin metabolism is not fully understood. It may be akin, in part, to idiosyncratic jaundice that occurs occasionally and unpredictably after administration of many drugs. In alcoholic liver injury, cholestasis may occur under a variety of circumstances (Popper *et al.*, 1981): cholestasis associated with massive steatosis is present throughout the acinus, and extensive microvesicular steatosis (with centrally placed nucleic) may be associated with this cholestatasis (Ballard *et al.*, 1961; Morgan *et al.*, 1978), as also seen in AFD.

7.2.3. Alcoholic Hepatitis

This stage is characterized by the appearance of necrosis with an inflammatory reaction, including polymorphonuclear cells. Other terms have been *florid cirrhosis*, *acute hepatic insufficiency*, *progressive alcoholic cirrhosis*, or steatonecrosis—Mallory body type. The term *alcoholic hepatitis* used by Beckett, Livingston, and Hill (1961) is nowadays widely accepted by most pathologists and clinicians. Although fatty liver in alcoholics is very common, alcoholic hepatitis develops in only a portion of heavy drinkers, even after decades of abuse (Lelbach, 1975; Kyösola and Salorinne, 1975).

Histological characteristics of alcoholic hepatitis are ballooning and a great disarray of hepatocytes (Fig. 7.5), parenchymal and portal infiltration with polymorphonuclear leukocytes, and varying degrees of steatosis, necrosis, fibrosis, and cholestasis. The changes are seen predominantly in perivenular areas (Gerber and Popper, 1972). These has been some semantic confusion regarding the definition of alcoholic hepatitis. Contrasting with the relatively uncommon involvement of polymorphonuclear leukocytes, an increased number of mononuclear leukotyes is often found in alcoholic liver injury. When these are abundant, some hepatologists use the term *alcoholic hepatitis*, a practice that is inconsistent with the more commonly held view, codified by the International Association for the Study of Liver Diseases (Leevy *et al.*, 1976), that polymorphonuclear inflammation is the *sine qua non* of "alcoholic" hepatitis.

One way to avoid the confusion would be to call the classic cases *polymorphonuclear alcoholic hepatitis* and to use the term *mononuclear alcoholic hepatitis* for those with predominant mononuclear inflammation.

Ultrastructural changes in alcoholic hepatitis are similar to but more severe than the ones seen in fatty liver (Svoboda and Manning, 1964). Collagenous material may be found in the space of Disse and may be associated with loss of microvilli that normally project from the hepatocyte (Edmondson *et al.*, 1967; Klion and Schaffner, 1968). Mallory's alcoholic hyalin (irregular cytoplasmic bodies) can be considered as a diagnostic hallmark (Hall and Ophuls, 1925; Phillips and Davidson, 1954; Popper *et al.*, 1955; Edmondson *et al.*, 1963; Lischner *et al.*, 1971; Harinasuta and Zimmerman, 1971) but is not always present (Birschbach *et al.*, 1974). These bodies were first defined by Mallory (1911) as an irregular, coarse, hyaline meshwork that stains deeply with eosin. Cells containing these bodies may be surrounded by numerous polymorphonuclear leukocytes (Fig. 7.5). Electron microscopically, alcoholic hyalin is fibrillar, is not surrounded by a limiting membrane, and is readily distinguished from giant mitochondria (Biava, 1964; Flax and Tisdale, 1964; Iseri and Gottlieb, 1971; Yokoo *et al.*, 1972; French and Davies, 1975; Petersen, 1977). Three morphological forms have been described (Yokoo *et al.*, 1972) (see Fig. 7.6). The exact nature and pathogenesis of alcoholic hyalin has not been elucidated. At first, actinlike proteins were incriminated (Nenci, 1975). It was recognized, however, that these fibrils are chemically heterogeneous (Okamura *et al.*, 1975), morphologically and chemically different from microtubules and from microfilaments (actin) and vimentin (French *et al.*, 1976; Franke *et al.*, 1979), and may represent prekeratin (Franke *et al.*, 1979; Denk *et al.*, 1979; Borenfreund *et al.*, 1980). However, because antitubulins (such as griseofulvin and colchicine) have the ability to induce the formation of Mallory bodies in mice (Denk and Eckerstorfer, 1977; Denk *et al.*, 1979), it is possible that the appearance of these bodies in alcoholic hepatitis may be linked to the antitubulin effects of alcohol (see Chapter 5). The nearest morphological cytosolic counterpart of Mallory bodies might be the intermediate filament, as suggested by Sim *et al.* (1977). Microfilaments, though distinct from alcoholic hyalin, nevertheless show a related increase (Petersen, 1977).

Although alcoholic hyalin is a characteristic of alcoholic hepatitis, it has also been demonstrated in primary biliary cirrhosis (Monroe *et al.*, 1973), in cholestasis and cirrhosis (Gerber *et al.*, 1973), in hepato-

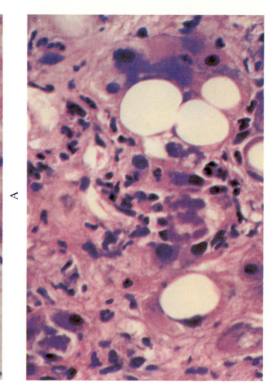

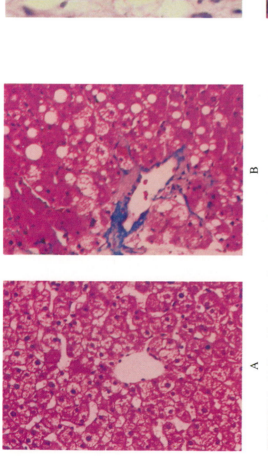

A

B

FIGURE 7.4. (A) Control liver biopsy of a nonalcoholic patient (primary cancer of the colon without evidence of liver involvement); the liver architecture is normal, and there is virtually no fibrous rim surrounding the terminal hepatic venule (trichrome stain, ×400). (B) Liver biopsy of a patient with alcoholic fatty liver without evidence of alcoholic hepatitis. Note the fibrous rim around the terminal hepatic venule (perivenular fibrosis). Some fibrosis strands surround adjacent sinusoids (perisinusoidal fibrosis) and hepatocytes (pericellular fibrosis). (Trichrome stain, ×250.) (Van Waes and Lieber, 1977a.)

FIGURE 7.5. (A) Patient with alcoholic hepatitis: large Mallory bodies around the nucleus of a swollen hepatocyte; no inflammation present. (B) Another patient with alcoholic hepatitis: hepatocytes with large Mallory bodies, surrounded by inflammatory cells, including polymorphonuclear cells (hematoxylin and eosin staining, ×600).

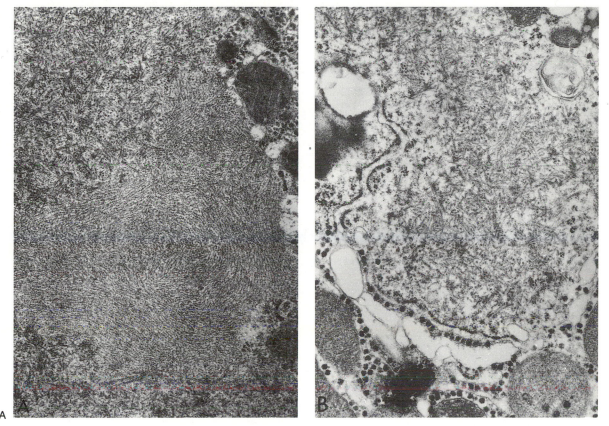

FIGURE 7.6. Electron micrographs of alcoholic hyalin demonstrating the three types of filaments as classified by Yokoo *et al.* (1972). (A) Light, sharp (young) filaments arranged in a parallel manner, type I (×12,700). (B) Randomly oriented darker filaments (older), type II (×19,000). (*Continued*)

cellular carcinoma (Keeley *et al.*, 1972), in tumor cells produced by diethyl nitrosamine (Borenfreund *et al.*, 1980), after other drug administration (Pessayre *et al.*, 1979), in Wilson's disease (Popper and Schaffner, 1961), in abetalipoproteinemia (Partin *et al.*, 1974), in Indian childhood cirrhosis (Roy *et al.*, 1971), in Weber–Christian disease (Kimura *et al.*, 1980), and in diabetes, but in the latter condition, the lesions may prevail in the periportal rather than in the perivenular zones (Nagore and Scheuer, 1988). Furthermore, hyalin is not required for the diagnosis of alcoholic hepatitis (Christoffersen and Juhl, 1971; Christoffersen and Nielsen, 1971). The diagnosis of alcoholic hepatitis is based on histological changes of parenchymal inflammation with polymorphonuclear leukocytes. In such cases alcoholic hyalin is found in approximately 30% of patients (Lischner *et al.*, 1971). The appearance of Mallory bodies does not correlate with mortality rates either retrospectively (Harina-

suta and Zimmerman, 1971) or prospectively (Birschbach *et al.*, 1974), but they are accompanied by more severe degrees of necrosis, cholestasis, and steatosis. The possible pathogenetic role of alcoholic hyalin is discussed in Section 7.3.

On rare occasions, the liver pathology resembles that of chronic active hepatitis (Goldberg *et al.*, 1977). Whether the latter lesion is related to alcohol or merely represents the coincidental appearance of another disease process in an alcoholic has not been settled. The fulminant and fatal nature of alcoholic hepatitis was stressed by early studies based on autopsy materials. In those cases patients were acutely ill, with fever, anorexia, nausea, abdominal pain, jaundice, tender hepatomegaly, ascites, edema of lower extremities, or even liver coma (Phillips and Davidson, 1954; Edmondson *et al.*, 1963; Hardison and Lee, 1966). In other studies, when the diagnosis was based on histological criteria

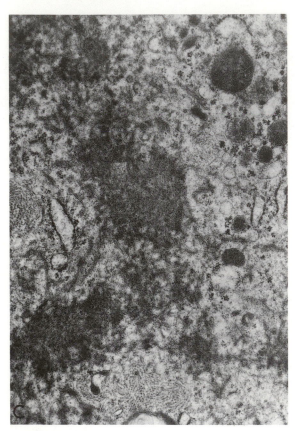

C

FIGURE 7.6. (*Continued*) (C) Homogeneous amorphous osmophilic material (oldest), type III (×12,700). Electron micrographs kindly provided by Dr. Oscar A. Iseri.

features of alcoholic hepatitis (hepatocellular necrosis, leukocytic infiltration, frequency of alcoholic hyalin) and the degree of hepatomegaly, leukocytosis, and in the intensity of symptoms (Harinasuta and Zimmerman, 1971; Christoffersen *et al.*, 1970). Patients with central hyalin fibrosis may have portal hypertension, splenomegaly, ascites, and esophageal and gastric varices, even in the absence of cirrhosis (Reynolds *et al.*, 1969).

Characteristic laboratory findings include the elevations of serum enzymes. Glutamyl transpeptidase (GTP), ornithine carbamyl transferase (OCT), aspartate aminotransferase (AST) (also called glutamic–oxaloacetic transaminase or GOT), and alanine aminotransferase (ALT) (also called glutamic–pyruvic transaminase or GPT) are elevated in all cases, and alkaline phosphatases (ALP) in 80% (Skrede *et al.*, 1976). In contrast to nonalcoholic liver disease, GOT/GPT (Cohen and Kaplan, 1979) or, even more clearly, GTP/ALP ratios are increased (Skrede *et al.*, 1976), and this has been used in the differential diagnosis. The role of these ratios as quantitative reflectors of liver cell damage can be questioned, however, since the origin of GOT is at least in part extrahepatic (Konttinen *et al.*, 1970), and the increased serum activity of GTP may be partly a result of enzyme induction (Teschke *et al.*, 1977; Ishii *et al.*, 1980). Such an induction, at first questioned (Mørland *et al.*, 1977), has now been confirmed (Gadeholt *et al.*, 1980). A better correlation between the degree of liver necrosis and inflammation was observed when the serum level of a mitochondrial enzyme, glutamic dehydrogenase (GDH), was measured (Van Waes and Lieber, 1977b). However, there is a rapid fall in the plasma GDH following cessation of the alcohol insult; early blood sampling of the patient is necessary in order to relate GDH to sustained liver injury. Under these conditions, a low value excludes alcoholic hepatitis; a high value indicates a liver lesion more severe than simple fatty liver. SGOT, SGPT, and GGPT are not as accurate as GDH in predicting the degree of injury (Worner and Lieber, 1980). Serum protein changes include a decrease in serum albumin in about half of the patients; haptoglobin and complement 3 are usually increased, whereas serum transferrin concentration is relatively low (Skrede *et al.*, 1975). Serum immunoglobulin A (IgA) is usually elevated and seems to correlate with the degree of liver damage as assessed from histological liver specimens (Iturriaga *et al.*, 1977). In addition, the depression of serum albumin and prolongation of the prothrombin time have been shown to correlate with the severity of the histological lesion (Helman *et al.*, 1971).

the clinical picture was characterized by milder symptomatology and a less dismal prognosis (Beckett *et al.*, 1961, 1962; Green *et al.*, 1963; Harinasuta *et al.*, 1967; Christoffersen *et al.*, 1970; Alexander *et al.*, 1971).

The spectrum of alcoholic hepatitis may thus range from mild anicteric hepatomegaly to fatal disease with jaundice, ascites, liver coma, gastrointestinal hemorrhage, and with all the complications of severe hepatic insufficiency. The only clinical difference between the mild and severe forms of alcoholic hepatitis is in the intensity and frequency of various symptoms (Gregory and Levi, 1972; Harinasuta *et al.*, 1967; Helman *et al.*, 1971; Lischner *et al.*, 1971). The most frequent symptoms of the milder cases include anorexia, fatigue, intermittent fever, right epigastric pain, and hepatomegaly. In general, patients with alcoholic hepatitis are more ill than those with a simple fatty liver, and there is a correlation between the severity of the histological

Other common laboratory findings include serum bilirubin elevations (90% of patients). It should be noted however, that serum bilirubin may also be increased in the absence of alcoholic hepatitis, for instance, in association with the steatosis (see Section 7.2.1) or because of hemolysis or alcoholic pancreatitis. Even more sensitive indicators of alcoholic liver injury are serum bile acids. The concentration of chenodeoxycholic acid, especially, is increased, but the correlation to the degree of liver cell necrosis is poor (Milstein *et al.*, 1976). Although BSP, galactose, indocyanine green, and antipyrine tests have been reported to be abnormal in alcoholic liver damage, they lack practical discriminative value in the evaluation of initial alcoholic liver injury, fatty liver, perivenular fibrosis, or alcoholic hepatitis. On occasion α-fetoprotein may appear in the serum of patients with alcoholic hepatitis (Miller *et al.*, 1975).

7.2.4. Cirrhosis

Macroscopically the early cirrhotic liver is golden yellow with fine uniform nodules (from 1 to 5 mm) on the surface. Traditionally this type of cirrhosis is called *micronodular* or Laennec cirrhosis (Fig. 7.7). The size of the liver varies, depending on the degree of fibrosis, inflammation, and steatosis, from a small, shrunken, and hard liver to a large organ weighing up to 4 kg. In later stages the size of the nodules increases, ranging from 5 to 50 mm, and deep scars appear. This macronodular cirrhosis resembles the postnecrotic cirrhosis. Microscopically, scar tissue distorts the normal architecture of the liver by forming bands of connective tissue joining portal and central zones. At first, nodules are regular in size and shape. In more advanced cirrhosis nodules become larger and frequently irregular in

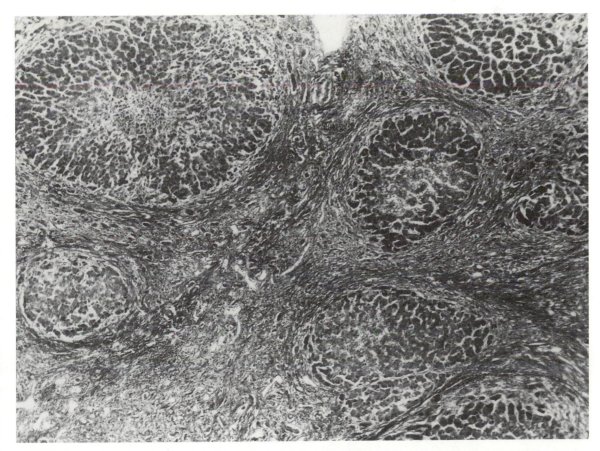

FIGURE 7.7. Section of liver showing cirrhosis in an alcoholic patient. The nodules are of regular shape and are monolobular; i.e., they do not contain portal tracts or central veins (hematoxylin and eosin staining, ×50).

size and shape. The nodules eventually become "re-lobulized" through formation of small portal tracts and efferent venous channels.

Among the most characteristic pathological features of cirrhosis are the changes in the hepatic blood circulation. The regenerative nodules may compress the hepatic veins, preventing the flow of blood out of the liver and resulting in postsinusoidal portal hypertension. With the progression of the fibrosis the outflow is further depressed (Rappaport, 1975). In addition, the total as well as the effective blood flow in the liver decreases in cirrhosis (Bradley *et al.*, 1952; Conn *et al.*, 1972). The total blood flow is reduced by extrahepatic portasystemic shunts, and the effective circulation is limited by neovascularization of connective tissue septa, which form anastomoses between the afferent branches of the portal vein and hepatic artery and the efferent tributaries of the hepatic veins (Popper, 1977). Collagen deposits in the space of Disse may further isolate the hepatocyte from its blood supply (Edmondson *et al.*, 1967; Klion and Schaffner, 1968).

The relationship between liver morphology and portal pressure in alcoholic liver disease has been the subject of long-standing debate. Some have postulated that swelling of the hepatocyte is the primary event (Fig. 7.8A,1), which in turn causes increased portal pressure (Fig. 7.8A,2) and secondary fibrosis (Fig. 7.8A,3) as a result of the increased pressure (Orrego *et al.*, 1981; Blendis *et al.*, 1982). If this were the case, fibrosis should develop first "upstream" (in the portal area), where pressure would be greatest. However, sequential liver biopsy specimens revealed that fibrosis starts and predominates around the terminal hepatic venules (Popper and Lieber, 1980; Miyakawa *et al.*, 1985). Because fibrosis started "downstream" (centrally) rather than "upstream" (periportally) from the swollen hepatocytes, it is most likely that, as shown in Fig. 7.8B,1, perivenular fibrosis and associated perisinusoidal fibrosis are the primary events and lead to secondary portal hypertension (Fig. 7.8B,2). Indeed, in the baboon, a significant increase in portal pressure was observed at the precirrhotic stage; it correlated significantly with the degree of perivenular fibrosis (PVF) but not with swelling of the hepatocyte (Miyakawa *et al.*, 1985).

These results in baboons are consistent with the findings of Krogsgaard *et al.* (1984) in humans. To determine whether an increase in hepatocyte volume compresses the vascular structures and causes portal hypertension, Krogsgaard *et al.* (1984) related the ratio of relative sinusoidal vascular volume to mean hepato-

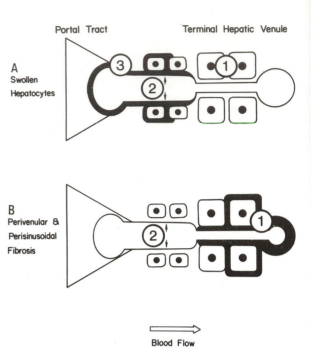

FIGURE 7.8. The two main theories of anatomic factors leading to portal hypertension. (A) Swelling of the hepatocytes (1) leads to increased pressure (2) and secondary fibrosis (3). (B) Perivenular and perisinusoidal fibrosis (1) leads to secondary portal hypertension (2), with swollen hepatocytes playing only a contributory role. Note that according to A, fibrosis appears first and foremost in the periportal areas, whereas under B, it first appears in the perivenular zone. (From Miyakawa *et al.*, 1985.)

cyte volume (which is a reflection of the vascular compression from enlargement of hepatocytes) to portal pressure. No significant correlation was found. Furthermore, mean hepatocyte volume did not significantly correlate with relative sinusoidal vascular volume. These findings are in accordance with the hypothesis that elevated hepatic vascular resistance and portal pressure in alcoholic liver disease are in part determined by the severity of the hepatic architectural destruction and subsequent distortion of the efferent venous system. A significant positive correlation was found between fibrosis and hepatic resistance. Necrosis, fatty change, occurrence of Mallory bodies, and inflammation showed no significant correlation with hepatic resistance (Krogsgaard *et al.*, 1985). The obstruction to the blood flow can be offset, at least in part, by an acute increase in splanchnic blood flow produced by ethanol. This had been shown originally in humans by Mendeloff (1954) and Stein *et al.* (1963) and has been confirmed in the baboon (Shaw *et al.*, 1977). Collagen deposition is also associated with an increase in diameter but a concomitant greater reduction in the number of patent fenestrations between sinusoids and the Disse space (Fig. 7.9) (Mak and Lieber, 1984), thereby resulting in a significant decrease in the total surface of the opening. These alterations isolate the hepatocyte from its blood supply and are most likely associated with a disturbance in the exchanges between the sinusoidal bloodstream and the hepatic parenchyma. Theoretically, an increase in mean diameter might favor uptake of large particles, such as chylomicrons. This, however, was not observed: the fractional hepatic uptake of chylomicrons was in fact reduced under these observations (Savolainen *et al.*, 1986).

In addition to fibrosis, various admixtures of fat, inflammatory reactions, and cholestasis may be seen. Histologically stainable iron deposits are common in cirrhotic livers, especially in alcoholic subjects (Bell, 1955; Powell, 1975) and in patients who have undergone portacaval shunts (Tisdale, 1961). The exact mechanism leading to iron accumulation is, however, unclear (Grace and Powell, 1974). Quantitative measurements of hepatic iron content in these disorders have demonstrated that iron stores are very seldom increased and never exceed 179 mmol (10 g). In some cases the finding can possibly be related to the high iron content of alcoholic beverages, especially red wines (Perman, 1967) and Kaffir beer used by the Bantu in South Africa (Isaacson *et al.*, 1961). Copper accumulations can sometimes be demonstrated with orcein or copper-staining methods (Salaspuro and Sipponen, 1976), particularly in those cases of alcoholic cirrhosis that are characterized by long-standing cholestasis (Salaspuro *et al.*, 1976).

Primary hepatocellular carcinoma is an increasingly frequent finding in liver cirrhosis, presumably because of improved management and prolonged survival (Gall, 1960; Lee, 1966). Pathogenetically, it more commonly relates to cirrhosis rather than to the effect of alcohol itself, although possible carcinogen properties of alcohol are now being considered (see Chapter 15).

Alcoholic cirrhosis becomes symptomatic at an average age of 50 years, especially when it is not complicated by acute fatty liver or alcoholic hepatitis. Most patients are detected when they seek treatment for complications. However, according to autopsy studies, cirrhosis may remain unrecognized antemortem in 40% of cirrhotic patients, and in about 20% it is discovered fortuitously on routine examination or during the evaluation of some other, unrelated disease. The most common signs of cirrhosis are weight loss (unless masked by ascites), weakness, and anorexia (Ratnoff and Patek, 1942). Jaundice can be detected in about two-thirds of patients when first admitted to the hospital and is often the main reason for more extensive examination. Low-grade and continuous fever is also common in decompensated cirrhosis (Tisdale and Klatskin, 1960). Jaundice and hepatomegaly are typical primary physical signs of cirrhosis. Secondary phenomena include portal hypertension with splenomegaly and ascites, encephalopathy with asterixis and other neurological signs, gastrointestinal hemorrhage from esophageal or gastric varices, caput medusae, edema, and bleeding tendencies resulting from clotting factor abnormalities. Tertiary complications include spontaneous peritonitis caused by anaerobic bacteria (Targan *et al.*, 1977) and, in males, gynecomastia and a female pattern of pubic hair distribution resulting from testicular atrophy and hormonal imbalance. Other manifestations include spider angiomas, palmar erythema, parotid enlargement, and Dupuytren's contracture. Abnormalities found on laboratory examination range from mild BSP retention to profound hyperbilirubinemia and hyperglobulinemia. Typically, serum electrophoresis reveals a decrease in serum albumin and prealbumin and a broad elevation of the β- and γ-globulins (Skrede *et al.*, 1976). Frequently, IgA and also IgG and IgM are elevated (Feizi, 1968; Agostini *et al.*, 1969; Skrede *et al.*, 1975; Iturriaga *et al.*, 1977). With more advanced liver injury a depression or even absence of α- and pre-β-lipoprotein bands can be demonstrated (Sabesin *et al.*, 1977; Borowsky *et al.*, 1980) (see Chapter 4). This is associated with a decrease of the esterified fraction of plasma cholesterol secondary to depressed lecithin : cholesterol acyltransferase (LCAT) activity (Simon and Scheig, 1970; Jones *et al.*, 1971; Blomhoff *et al.*, 1974). Serum transaminases and alkaline phosphatases are usually only mildly elevated. An impaired glucose tolerance and an increased incidence of diabetes have been described in patients with Laennec's cirrhosis (Conn *et al.*, 1971).

7.3. Pathogenesis of Alcoholic Liver Injury

7.3.1. Nutritional Factors

The pathogenesis of the alcoholic fatty liver is discussed in detail in Chapter 4, including the respective roles of malnutrition and metabolic effects of ethanol. A similar debate has been ongoing concerning the progression of liver injury beyond the fatty liver stage. Specifically, the question has been raised as to whether lesions more severe than steatosis, particularly cirrhosis, can be produced by alcohol in the absence of dietary deficiencies. Studies in rodents had been unsatisfactory because even when it is given with a liquid diet, the intake of alcohol in rodents does not reach the level of average consumption in the alcoholic, namely, 50% of total calories (Patek et al., 1975b). Such a consumption was achieved in the baboon through incorporation of ethanol in totally liquid diets (Lieber and DeCarli, 1974). The nonalcohol calories of the diet were provided by protein (36% of total nonalcohol calories), fat (42%), and carbohydrate (22%). All of the other nutrients were calculated to exceed the normal requirements of the baboon (National Academy of Sciences, 1978). Choline was given at twice the level recommended by Foy et al. (1964) for the baboon. This diet was also liberally supplemented with minerals and vitamins. The fat and carbohydrate composition of the baboon diet was calculated to mimic an optimal clinical situation in which the alcoholic may be trying to achieve such a high-protein diet with available natural foods while drinking. In fact, even if the alcoholic tried hard, it would be difficult for him to consume a diet richer in protein than the one administered to the baboons (36% of calories). Nevertheless, the baboons not only developed fatty liver but, after 2 to 5 years, one third also had progression of liver damage to cirrhosis (Lieber et al., 1975). Therefore, it was concluded that in addition to dietary factors alcohol itself plays a key etiological role in the development of liver injury.

Ainley et al. (1988) failed to reproduce alcohol-induced cirrhosis in baboons. However, only two animals received the regular ethanol-containing diet for more than 18 months. Since only one-third of our ani-

mals (Popper and Lieber, 1980) developed cirrhosis (with an average ethanol treatment of 3 years), the negative result of Ainley et al. (1988) is not surprising. Moreover, the amount of alcohol consumed by the baboons of Ainley et al. (1988) is not clear. A daily administration of 25 g/kg is reported; if the baboons had consumed that amount, they would not have survived, since it is a lethal dose. Actually, the reported postprandial blood levels were, if anything, lower than the one observed with a fivefold lower ethanol administration (Lieber et al., 1975).

If one can extrapolate from our baboon data to the human situation, it would seem that heavy drinkers, even if they make an extra effort to maintain a high-protein diet, will not necessarily succeed in preventing the development of cirrhosis unless they also control their alcohol intake. Some questions, however, still remained unresolved, namely, to what extent does malnutrition promote the toxic effect of alcohol, and at what level of intake does alcohol become toxic to the liver? That dietary factors can modify the effect of alcohol on the liver has been clearly shown in rodents. Although with regard to the kidney, alcohol does not raise choline requirements (Koch et al., 1977b), there is an increased need concerning the liver (Klatskin et al., 1954), possibly because of enhanced choline oxidation (Thompson and Reitz, 1976). The role of nutritional factors is less clear with regard to liver injury in primates, which are much less susceptible to protein and choline deficiencies than rodents (Hoffbauer and Zaki, 1965). In primates some fibrosis but, except possibly for a single animal (Patek et al., 1975a), no cirrhosis was ever produced with protein deficiency alone (de la Iglesia et al., 1967). To cause cirrhosis, one had to create dietary conditions not normally achievable with natural food (such as an

$\longrightarrow$

FIGURE 7.9. (A–C) Scanning electron micrographs of the luminal surface of the hepatic sinusoidal endothelium of control baboons. (A) Endothelial surface showing a bulging nuclear and perinuclear region (N) and flattened cytoplasmic extensions (C). Fenestrations (F) are evident in the cytoplasmic extensions but not in the perinuclear region. ×1000. (B) Large fenestrations (LF) with diameters greater than 500 nm are seen at the periphery of clusters of small fenestrations (F). Perisinusoidal cell processes (arrow) in the Disse space can be seen through the large fenestration. Some small fenestrations contain protruding hepatocyte microvilli (arrowheads). The cell border of neighboring endothelial cells is marked CB. ×10,000. (C) Luminal view of cytoplasmic extensions showing fenestrations arranged in clusters (F), forming sieve platelike structures separated by nonfenestrated cytoplasmic ridges (NF). ×10,000. (D–F) Scanning electron micrographs of the luminal surface of the sinusoidal endothelium of alcohol-fed baboons. (D) Alcohol feeding for 24 months. Clustering of the fenestrations (F) is evident, but the clusters appear to be smaller than those of control baboons. Surface view of a perisinusoidal cell (arrow) in the Disse space can be seen through a fenestration (greater than 500 nm in diameter). The overlapping of cytoplasmic extensions of the two endothelial cells at the cell border is marked (CB). ×10,000. (E) Alcohol feeding for 61 months. Two clusters of fenestrations are seen. Within the clusters, the fenestrations are of different sizes ranging from 50 nm (arrowheads) to up to 500 nm (arrows) in diameter. Note the extensive nonfenestrated endothelial surface. P, perisinusoidal cell process in the Disse space; CB, endothelial cell border. ×10,000. (F) Alcohol feeding for 104 months. There are distinctly fewer fenestrations in the endothelium here than in control baboons. ×10,000. (Mak and Lieber, 1984.)

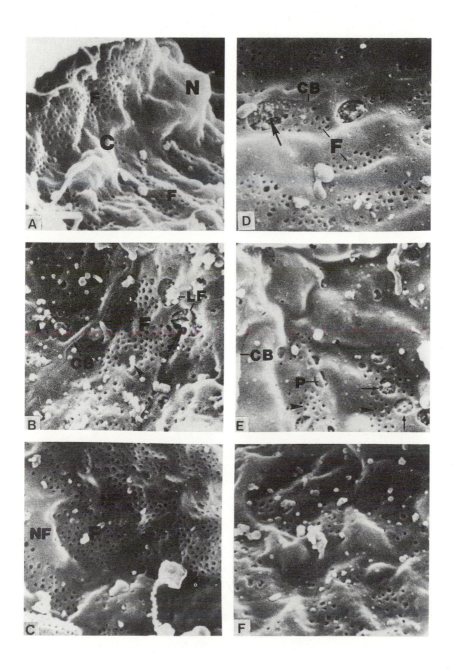

potentiate the alcohol-induced liver damage in the primate has not yet been determined. In fact, a low-protein diet was found to greatly reduce the fibrous responses of monkey livers to implantation of foreign material (Lancet, 1972). Experimentally, massive choline supplementation (ten times the recommended amount) failed to prevent the development of cirrhosis, even when the dose of choline was so high that some toxicity was produced (Lieber et al., 1985).

Another dietary factor that may contribute to fibrosis is excess of vitamin A (Leo and Lieber, 1988). Its role in promoting fibrosis has been questioned: vitamin A suppressed induction of experimental hepatic fibrosis by CCl_4 and pig serum (Senoo and Wake, 1985). However, it has been demonstrated that excessive vitamin A, when combined with ethanol, results in a striking reciprocal potentiation of the hepatotoxic effects, including the development of fibrosis (Leo and Lieber, 1983). Traditionally, vitamin A toxicity was reported following consumption of doses of the order of 100,000 IU per day for periods ranging from weeks to months (Jeghers and Marrao, 1958; Korner and Vollm, 1975). More recently, however, it was noted that a vitamin supplement of two capsules of 25,000 IU vitamin A daily for a little more than 2 years was associated with severe vitamin A toxicity and a rise of the plasma vitamin A level to ten times normal (Farris and Erdman, 1982). Moreover, liver damage from massive vitamin A deposition was reported in association with a serum vitamin A level that was actually below normal (perhaps because of protein deficiency and/or liver damage) (Weber et al., 1982).

There are at present no readily available measurements to detect early signs of toxicity. It is noteworthy that vitamin A levels in the serum may be quite normal in the presence of high liver vitamin A and associated hepatotoxicity, especially when there is concomitant liver disease. It has been proposed that toxicity may result when the capacity of retinol-binding protein to transport vitamin A is exceeded and excessive vitamin A is presented to cell membranes in a form other than bound to retinol-binding protein (Smith and Goodman, 1976), but such mechanisms may not pertain to the alcoholic. Increased serum levels of enzymes such as GDH and transaminase may reflect hepatotoxicity, but their practical use may be limited by the fact that in alcoholics or other drug users, it may be difficult to differentiate between an enzyme increase in the blood that reflects vitamin A toxicity and one that results from the direct hepatotoxic effects of ethanol. Similarly, measurements of vitamin A in liver biopsies may not be revealing, since the actual level found may again result from two opposite changes, namely, the decrease of

vitamin A associated with chronic ethanol and/or drug use (Leo and Lieber, 1982; Leo et al., 1984a) and the rise of vitamin A associated with vitamin A therapy. As a net result, the liver vitamin A content of the vitamin-A-supplemented, ethanol-fed rats was not greater than that of the control (normal-A) group, suggesting that metabolites of vitamin A rather than vitamin A itself may have been responsible for the potentiation of vitamin A toxicity by ethanol (Leo et al., 1982).

Furthermore, the theoretical possibility exists that the production of metabolites might be increased by chronic ethanol and/or drug use. Indeed, degradative pathways have been described in liver microsomes. Microsomes from hamsters (Roberts et al., 1979, 1980) and rats (Leo et al., 1984b; Sato and Lieber, 1982) were shown to metabolize retinoic acid to more polar metabolites, but the rate of this activity was extremely low, in the picomole range (per milligram microsomal protein); thus, although this system was shown to be inducible by phenobarbital (Roberts et al., 1979), ethanol (Sato and Lieber, 1982), and vitamin A (Leo et al., 1984a), it is doubtful whether it played a major role in the overall economy of retinol in the body. By contrast, new microsomal systems were discovered (Leo and Lieber, 1985; Leo et al., 1987) that have rates of activities in the nanomole range (per milligram microsomal protein) and therefore may potentially be of much greater quantitative significance for the production of some of the postulated toxic metabolites.

In addition to vitamin A excess, alcohol-induced fibrosis can probably also be potentiated by concomitant abuse of other drugs, as suggested by the increased incidence of hepatic cirrhosis in young adults in associated with adolescent onset of alcohol and parenteral heroin abuse (Novick et al., 1985). In some patients subjected to intestinal bypass for obesity, lesions similar to those present in some alcoholics—hepatitis and fibrosis—have occurred (Peters et al., 1975), and it has been concluded, therefore, that in both of the above situations, a similar pathogenesis, malnutrition, may pertain. However, such reasoning by analogy is hazardous, since pathogenesis of the liver lesion after bypass is still unsettled. Development of the lesion has been described despite nutritional correction with massive parenteral alimentation (Craig et al., 1980). Even if malnutrition were shown to be the culprit in some cases, extrapolation to the alcoholic is not necessarily warranted. It could be argued, for instance, that similar liver lesions have also been engendered by chemical agents in humans (Pessayre et al., 1979), not to mention the experimental evidence, referred to before, of the production of liver damage after alcohol and an adequate diet. It is possible, of course, that future research may unravel specific dietary factor(s) that may be lacking in the alcoholic and therefore would warrant selective repletion. At the present time, however, we must accept the fact that independent of dietary factors, alcohol exerts some direct toxicity in the liver, resulting not only in fatty liver development (see Chapter 4) but also in further progression of liver disease, as discussed in the next section.

7.3.2. Hepatotoxic Effects of Ethanol and Its Metabolites

Only 2 to 10% of the ethanol absorbed is eliminated through the kidneys and lungs. The rest must be oxidized in the body, principally in the liver, which contains the bulk of the body's enzymes capable of ethanol oxidation. This relative organ specificity probably explains why ethanol oxidation produces striking metabolic imbalances in the liver. These effects are aggravated by the lack of a feedback mechanism to adjust the rate of ethanol oxidation to the metabolic state of the hepatocyte and the inability of ethanol, unlike other major sources of calories, to be stored or metabolized to a significant degree in peripheral tissues. When ethanol is present, it becomes the preferred fuel for the liver. By displacing up to 90% of all other substrates normally utilized by the liver (Lundquist *et al.*, 1962), ethanol literally takes over the intermediary metabolism of the liver. Ethanol metabolism in the liver results in the production of hydrogen and acetaldehyde (Fig. 7.10; Lieber 1989). Each of these two products is directly

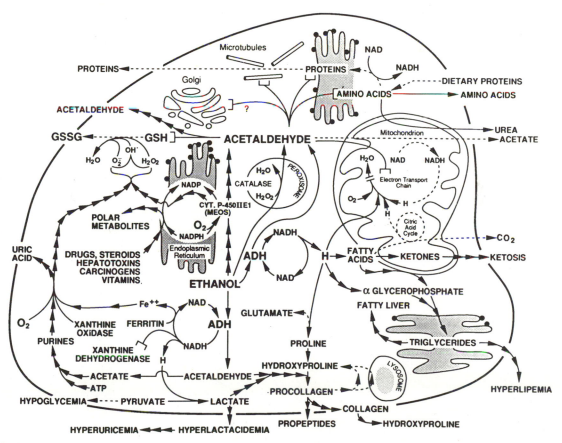

FIGURE 7.10. Oxidation of ethanol in the hepatocyte. Many disturbances in intermediary metabolism and toxic effects can be linked to the generation of NADH mediated by alcohol dehydrogenase (ADH); the induction of microsomal enzymes, especially P450IIE1; and acetaldehyde, the product of ethanol oxidation. NAD denotes nicotinamide adenine dinucleotide; NADH, reduced NAD; NADP, nicotinamide adenine dinucleotide phosphate; NADPH, reduced NADP; MEOS, the microsomal ethanol-oxidizing system; and ADH, alcohol dehydrogenase. The broken lines indicate pathways that are depressed, whereas repeating arrows indicate stimulation or activation. The bracket symbol denotes interference or binding.

responsible for a variety of metabolic alterations that play a role in the development of liver injury. In that regard, it is noteworthy that a relationship has been described between the severity of liver damage and the rate of ethanol metabolism (Ugarte *et al.*, 1977) and, hence, the production of metabolites of ethanol.

A link between hepatotoxicity of ethanol and its metabolism could also explain some of the zonal changes in alcoholic liver disease. Both in rats and baboons fed alcohol and in alcoholic patients, the swelling of the hepatocytes is most conspicuous in the centrolobular (also now called perivenular) area. This, of course, is the zone with the lowest oxygen tension, and, as discussed subsequently, lack of oxygen may contribute to liver damage. On the other hand, the perivenular area is the zone with the greatest proliferation of the endoplasmic reticulum after chronic ethanol consumption (Iseri *et al.*, 1966) and where MEOS activity was shown to be highest (Tsutsumi *et al.*, 1989). Furthermore, the activity of ADH may also be highest in this zone (Kato *et al.*, 1990a; see also Chapter 1). The preponderance of ethanol-induced lesions in the perivenular area could be explained by increased ethanol metabolism in this zone if the lesions were secondary to the oxidation of ethanol and/or the metabolites formed.

7.3.2.1. Effects of Excessive Hepatic NADH Generation and Toxicity Associated with a "Hypermetabolic State"

Major metabolic consequences of hydrogen generation and the associated altered redox state (with a shift in the NADH/NAD ratios) have been discussed before, particularly in Chapters 1, 3, and 4. Regeneration of NAD from NADH ultimately requires oxygen. Trémolières and Carré (1961) showed that oxygen consumption after an acute dose of alcohol increases more in alcoholics than in nonalcoholics. The cause of this is unknown. Two mechanisms have been postulated: (1) enhanced role of the induced MEOS (Pirola and Lieber, 1976) and (2) relative uncoupling of reoxidation of NADH in liver mitochondria (Israel *et al.*, 1975), possibly associated with increased catecholamine release or a hyperthyroidlike state. Some studies, however, have challenged the concept that chronic ethanol ingestion results in hyperthyroidism or liver hypermetabolism (Schaffer *et al.*, 1981; Teschke *et al.*, 1983). In any event, *in vivo*, increased oxygen consumption can be compensated for by enhanced blood flow (Shaw *et al.*, 1977; Villeneuve *et al.*, 1981; Iturriaga *et al.*, 1980; Carmichael *et al.*, 1987), an effect attributed to in-

creased adenosine (Orrego *et al.*, 1988). In baboons, defective O_2 utilization rather than lack of O_2 blood supply characterized liver injury produced by high concentrations of ethanol (Lieber *et al.*, 1989). Indeed, in baboons fed control diets, hepatic vein catheterization revealed that a moderate ethanol dose resulted in increased splanchnic O_2 consumption, offset by an increase in flow sufficient to prevent lack of O_2, as judged from normal hepatic venous P_{O_2}, tissue hemoglobin (Hb) concentration, and O_2 saturation of Hb (S_{O_2}). Flow-independent tissue O_2 consumption (V_{O_2}) also increased. The Hb, S_{O_2}, and V_{O_2} were measured by hepatic surface reflectance spectrophotometry through peritoneoscopy. After a high ethanol dose in controls and both a moderate and high dose in animals fed alcohol chronically, V_{O_2} failed to increase (or even decreased) without any decrease in S_{O_2} or in hepatic venous P_{O_2}, indicating impaired capacity of the hepatocytes to take up O_2, even in the presence of an ample O_2 supply.

A characteristic feature of the liver injury of the alcoholic is the predominance of the lesions in perivenular (also called centrolobular) zone or area 3 of the hepatic acinus (Edmondson *et al.*, 1967). The mechanism for this zonal selectivity of the effects of ethanol remains unknown. Two distinct but not mutually exclusive hypotheses have been raised: one claims that ethanol can produce hypoxic damage to perivenular hepatocytes, whereas the other postulates that conditions normally prevailing in perivenular zones enhance the metabolic toxicity of ethanol. The hypoxia hypothesis originated from the observation that liver slices from rats of fed alcohol chronically consume more oxygen than those of controls (Videla and Israel, 1970). It was then postulated that the increased consumption of oxygen would increase the gradient of oxygen tensions along the sinusoids to the extent of producing anoxic injury in perivenular hepatocytes (Israel *et al.*, 1975). Such a mechanism was illustrated experimentally when centrolobular liver necrosis was induced by hypoxia in chronic ethanol-fed rats (French *et al.*, 1984). Furthermore, both in alcoholics (Kessler *et al.*, 1954) and in animals fed alcohol chronically (Jauhonen *et al.*, 1982; Sato *et al.*, 1983), decreases in either hepatic venous oxygen saturation (Kessler *et al.*, 1954) or P_{O_2} (Jauhonen *et al.*, 1982) and in tissue oxygen tensions (Sato *et al.*, 1983) have been found during the withdrawal state. However, this decrease is within the range of values found in normal subjects. Moreover, the differences in hepatic oxygenation found during the withdrawal state disappeared (Jauhonen *et al.*, 1982; Shaw *et al.*, 1977) or decreased (Sato *et al.*, 1983) when alcohol was present in the blood. It is noteworthy, however, that with progression of fibrosis and fat accumulation in the liver, estimated regional hepatic tissue hemoglobin concentration decreases and this decreased oxygen supply to the liver may have an important role in the progression of alcoholic liver disease (Hayashi *et al.*, 1985).

An alternative hypothesis to explain the selective perivenular hepatotoxicity of ethanol postulates that the low oxygen tensions normally prevailing in perivenular zones could exaggerate the redox shift produced by ethanol (Jauhonen *et al.*, 1982). This was illustrated *in vivo* and in the isolated perfused liver, when the oxygen supply was varied to reproduce the oxygen tensions prevailing *in vivo* along the sinusoid (Jauhonen *et al.*, 1985). Varying the oxygen tensions within the physiological range produced a redox gradient of both cytochrome oxidase and NAD, with a more reduced state at tensions normally prevailing in perivenular zones. The degree of reduction of cytochrome oxidase at these physiological oxygen tensions was not associated with impairment in the ability of the liver to consume oxygen and to produce ATP, suggesting a lack of cellular anoxia. Ethanol at 25 mM increased hepatic oxygen consumption but had no direct effect on the state of reduction of cytochrome oxidase. The effects of ethanol and oxygen tensions on NADH fluorescence were additive, indicating that a greater redox shift should occur when ethanol is oxidized at oxygen tensions similar to those normally prevailing in perivenular zones than at those in periportal zones. This dependence of the ethanol-induced redox shift on oxygen tensions may contribute to the selective perivenular hepatotoxicity of alcohol (Jauhonen *et al.*, 1982). In addition, selective induction of the MEOS and highest ADH levels in zone 3 also contribute to the selective hepatotoxicity (see Chapter 6).

7.3.2.2. Hepatotoxic Effects of Acetaldehyde

Pathways of acetaldehyde metabolism and general effects of acetaldehyde are discussed in detail in Chapter 2, and its adverse effect on protein synthesis and secretion in Chapter 5. We focus here on possible sequelae of protein retention and additional aspects of the hepatotoxicity of acetaldehyde.

7.3.2.2a. Swelling of the Hepatocyte and Possible Relation to Necrosis. Two of the earliest and most conspicuous features of the hepatic damage produced by alcohol are the deposition of fat and the enlargement of the liver. This hepatomegaly was traditionally attributed

to the accumulation of lipids. However, in animals fed alcohol-containing diets, it was shown that lipids account for only half the increase in liver dry weight (Lieber *et al.*, 1965); the other half is almost totally accounted for by an increase in proteins (Baraona *et al.*, 1975) (see Chapter 5), possibly secondary to acetaldehyde-induced impairment of microtubule-mediated protein secretion. However, the measured increase in export proteins such as albumin and transferrin accounted for only a small fraction of the total increase in cytosolic protein. This raised the possibility that other export or even constituent proteins of the cytosol could also contribute to the increase. Indeed, Pignon *et al.* (1987) found that an increase in fatty-acid-binding protein (FABP) accounts for one-sixth to one-third of the increase in cytosolic protein induced by chronic ethanol feeding, thus becoming the largest known contributor to the ethanol-induced increase in these proteins. Pignon *et al.* (1987) also showed a striking discrepancy between the alcohol-induced increase in esterified fatty acids in the liver (mainly as triglycerides) and the modest increase in nonesterified fatty acids. In addition to the ethanol-induced increase in the synthesizing activities (Joly *et al.*, 1974; M. J. Savolainen *et al.*, 1984), the increase of FABP could also favor the esterification of fatty acids, thereby playing a possible role in preventing potentially deleterious accumulation of nonesterified fatty acids (Brenner, 1984) and fatty acid–CoA esters (Powell *et al.*, 1986) secondary to the ethanol-induced inhibition of their oxidation. The female gender was associated with a lesser response of FABP (Shevchuk *et al.*, 1991) thereby possibly contributing to a greater vulnerability to ethanol. It is noteworthy that impairment of microtubules was also found to affect hepatocellular transport of biliary lipids (Gregory *et al.*, 1978) and to be associated with engorgement of the Golgi (Matsuda *et al.*, 1979).

The increase in hepatic protein observed after ethanol was not associated with changes in concentration (Baraona *et al.*, 1977), indicating that water was retained in proportion to the increase in protein. The mechanism of the water retention is not fully elucidated, but the rise in both protein and amino acids, plus a likely increase in associated small ions, could account osmotically for a large fraction of the water increase. The increases in lipid, protein, amino acid, water, and electrolytes result in increased size of the hepatocytes. In the rat, the swelling affects both perivenular and periportal hepatocytes. Similar changes have been observed in baboons fed 50% of energy as ethanol, although in this species the accumulation of fat greatly exceeds that of protein (M. J. Savolainen *et al.*, 1984), and the enlargement of the hepatocytes has, as in humans, a clear centrolobular distribution (Miyakawa et al., 1985). The number of hepatocytes and the hepatic content of deoxyribonucleic acid (DNA) did not change after alcohol treatment, and thus the hepatomegaly is entirely accounted for by the increased cell volume (Baraona et al., 1975, 1977). There is also an increase in the number of hepatic mesenchymal cells after ethanol feeding (Baraona et al., 1975), but this increase does not significantly contribute to the hepatomegaly. The swelling of the hepatocytes after chronic alcohol administration was found to be associated with a reduction of the intercellular space and with portal hypertension (Orrego et al., 1981).

One suspects that ballooning and associated gross distortion of the volume of the hepatocytes may result in severe impairment of key cellular functions. In alcoholic liver disease some cells not uncommonly have a diameter that is increased two to three times, and thereby volume is increased tenfold or more. One may wonder to what extent this type of cellular disorganization, with protein retention and ballooning, may promote progression of the liver injury in the alcoholic. Indeed, there are causes of protein retention in the liver, such as α_1-antitrypsin deficiency, associated with progression to fibrosis and cirrhosis. By analogy, one can assume that protein retention may in some way also favor progression of liver disease in the alcoholic.

7.3.2.2b. Structural and Functional Alterations of the Mitochondria and Plasma Membranes.

As discussed before, studies with the electron microscope have revealed striking morphological alterations, including swelling and abnormal cristae, in the liver mitochondria of patients with alcoholism. Controlled studies in animals and humans (Iseri et al., 1966; Lane and Lieber, 1966; Rubin and Lieber, 1967a,b; Lieber and Rubin, 1968) have shown that these changes are caused by alcohol itself rather than by other factors such as a poor diet. These structural abnormalities are associated with functional impairments, especially decreased oxidation of fatty acids and of a variety of other substrates, including acetaldehyde (Hasumura et al., 1976).

Mitochondria of alcohol-fed animals have a reduction in cytochrome a and b content (Rubin et al., 1970; Koch et al., 1977a). The respiratory capacity of the mitochondria was found to be depressed (Kiessling and Pilstrom, 1968; Rubin et al., 1972; Gordon, 1973; Hasumura et al., 1975) using pyruvate, succinate, and acetaldehyde as substrates. As pointed out before, in human volunteers given ethanol, mitochondrial lesions developed even in the presence of a high-protein, low-fat diet (Lieber and Rubin, 1968). Oxidative phosphorylation was found to be altered (Cederbaum et al., 1974a). It is noteworthy that high concentrations of acetaldehyde mimic the defects produced by chronic ethanol consumption on oxidative phosphorylation (Cederbaum et al., 1974b). One may wonder to what extent chronic exposure to acetaldehyde is the cause of the defect observed after chronic ethanol consumption. Alterations of acetaldehyde metabolism associated with alcoholism have been discussed in Chapter 2. As has been pointed out, alcoholics may exhibit higher acetaldehyde levels than nonalcoholics for a given ethanol load and blood level. It is therefore reasonable to speculate that exposure to high acetaldehyde levels may in turn affect mitochondrial function and result in the vicious cycle depicted in Fig. 2.7 (Lieber, 1989). In normal liver, higher acetaldehyde concentrations than those usually achieved after alcohol ingestion are required to produce toxicity. However, after chronic alcohol consumption, the liver mitochondria are unusually susceptible to the toxic effects of acetaldehyde, and a variety of important mitochondrial functions, such as fatty acid oxidation, are depressed, even in the presence of relatively low acetaldehyde concentrations (Matsuzaki and Lieber, 1977). It has been proposed that the functional changes in mitochondria may also be related to altered cellular membranes produced as a consequence of prolonged ethanol consumption (Waring et al., 1981, 1982). Membrane binding of ethanol, anesthetics, and hydrophobic molecules in brain synaptosomes and liver mitochondria of rats is reduced after long-term consumption of ethanol. Membranes became resistant to structural disordering by ethanol and halothane. It has been suggested that the increased rigidity impairs normal membrane function, although such a correlation has not always been verified (Gordon et al., 1982).

Ethanol also strikingly alters other liver membranes, as reviewed by Sun and Sun (1985). Plasma membrane glycoprotein assembly in the liver following acute ethanol administration was significantly impaired (Mailliard et al., 1984). Furthermore, hepatocytes isolated from ethanol-fed animals exhibit pronounced morphological alterations of their plasma membranes by scanning electron microscopy (Fig. 7.11) and a reduced content of alkaline phosphatase (despite an increase in total liver alkaline phosphatase) during incubation of the cells with 50 mM ethanol. Moreover, chronic ethanol feeding resulted in reduced recovery of alkaline phosphatase in hepatic plasma membranes. These findings suggest that the increased serum alkaline phosphatase

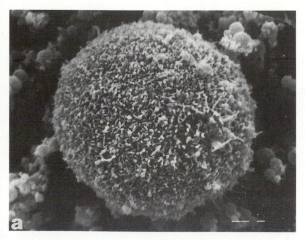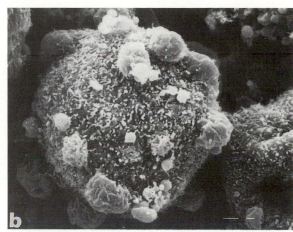

FIGURE 7.11. Scanning electron micrographs of isolated hepatocytes. (a) Typical hepatocyte from control rat showing abundant microvilli uniformly covering the cell surface. ×5000. (b) Hepatocyte from ethanol-fed rat showing the appearance of blebs on the cell surface with a diminished number of microvilli. ×4600. (From Yamada *et al.*, 1985a.)

levels observed in response to chronic ethanol feeding may be caused, at least in part, by increased lability of this plasma membrane enzyme (Yamada *et al.*, 1985a). It is well known that ethanol also has striking effects on another plasma membrane enzyme, γ-glutamyltranspeptidase (GGTP): GGTP activity is frequently elevated in alcoholics (Wietholz and Colombo, 1976; Shaw and Lieber, 1978; Horner *et al.*, 1979) and in rats given ethanol (Shaw and Lieber, 1978; Mørland *et al.*, 1977; Ishii *et al.*, 1978; Lahrichi *et al.*, 1982; Nishimura and Teschke, 1982; Yamada *et al.*, 1985b). It has now been found that factors other than a mere increase of liver GGTP may be responsible for the elevated serum levels of this enzyme seen after chronic ethanol consumption: concomitant plasma membrane damage may be necessary for the elevation of serum enzyme level to occur (Yamada *et al.*, 1985b).

ATPase is a third plasma membrane enzyme reported to be affected by ethanol. However, when the effects of ethanol and acetaldehyde on the activities of two hepatic plasma membrane ATPases—Na$^+$, K$^+$)-ATPase and Mg^{2+}-ATPase—were investigated over concentrations ranging from 8 to 90 mM, ethanol did not cause significant inhibition in male rats (Yamada *et al.*, 1985a), although in female rats an eightfold decrease of Na$^+$,K$^+$-ATPase activity was reported (Pascale *et al.*, 1989). Acetaldehyde produced noncompetitive inhibition of Na$^+$,K$^+$-ATPase and Mg^{2+}-ATPase at concentrations of 6 and 56 mM, respectively, and 5′-nucleotidase activity was also inhibited at these con-

centrations. Because the inhibitory concentrations of ethanol and acetaldehyde are higher than are usually found in alcoholic subjects or in experimental animals after alcohol feeding, it seems unlikely that direct suppression of ATPase activity by ethanol or acetaldehyde is responsible for the morphological abnormalities of alcohol-induced liver disease (Gonzalez-Calvin *et al.*, 1983). Similarly, the activation effects of ethanol on liver plasma membrane adenylate cyclase activity require unphysiologically high concentrations (Whetton *et al.*, 1983).

Various studies have demonstrated that the cell surface plays an important role in intracellular homeostasis through activities of membrane enzymes, hormone receptors, and transmembrane transport processes. Many of these dynamic features are influenced, in part, by the fluidity of the lipid bilayer because of its effect on the mobility and/or exposure of membrane proteins. Thee have also been several reports concerning the effect of ethanol on the fluidity of membranes. A currently held view is that *in vitro* ethanol renders membranes more fluid and that chronic ethanol administration alters membrane lipid composition, which results in less fluidity, considered an adaptation to the acute fluidizing effect. Such an adaptation has been reported to occur in brain synaptosomes and erythrocytes (Chin and Goldstein, 1977; Rottenberg *et al.*, 1981) as well as in liver mitochondria and microsomes (Waring *et al.*, 1981; Ponnappa *et al.*, 1982). However, when the effect of chronic ethanol consumption on liver plasma membranes (PM) was studied (Yamada and Lieber, 1984), it was found that in contrast to other membranes, chronic ethanol feeding resulted in an increase in hepatic plasma membrane fluidity as assessed by fluorescence anisotropy. Similar results were observed by Polokoff *et al.* (1985). This alteration was associated with a decrease in membrane vitamin A but an increase in the cholesterol esters content (Kim *et al.*, 1988). Thus, chronic ethanol feeding does not appear to result in a "homeoviscous adaptation" of liver plasma membranes as reported for membranes in other tissues. Rather, an increase in fluidity has been observed. Such changes may relate to the specialized role of the liver in bile excretion, lipoprotein metabolism, and ethanol oxidation.

7.3.2.2c. Promotion of Lipid Peroxidation: Interaction with Cysteine, Glutathione, and Iron.
Aldehydes react quite readily with mercaptans (Schubert, 1936, 1937; Ratner and Clarke, 1937). L-Cysteine could complex with acetaldehyde to form a hemiacetal, which would then transform to L-2-methylthiazolidine-4-carboxylic acid (Schubert, 1937; Debey *et al.*, 1958). It has been suggested that such a complex may be a nontoxic detoxification product, since cysteine was reported to protect against death from acetaldehyde toxicity *in vivo* (Sprince *et al.*, 1974). Cysteine *in vitro* afforded protection against the depression of CO_2 production from palmitate, octanoate, and ketoglutarate by acetaldehyde (Cederbaum and Rubin, 1976). Cysteine is also one of the three amino acids that constitute glutathione (GSH). Binding of acetaldehyde with cysteine and/or glutathione may contribute to a depression of liver glutathione (Shaw *et al.*, 1981). Morton and Mitchell (1985) determined the turnover of GSH in individual animals by measuring the decrease in specific activity of GSH in bile over time after i.v. administration of [35S]cysteine. Rats fed ethanol chronically had significantly increased rates of GSH turnover. Consistent with the results of Vendemiale *et al.* (1984), the increase in turnover of GSH was not caused by an increase in oxidation of GSH. There was, however, an association between GSH turnover and the activity of hepatic GGT in ethanol-fed rats. In contrast to these effects of chronic ethanol consumption, acute ethanol administration inhibited GSH synthesis and produced an increased loss from the liver (Speisky *et al.*, 1985). Glutathione transferase activity (Kocak-Toker *et al.*, 1985) was decreased by acute ethanol administration, and GSH peroxidase after chronic treatment (Morton and Mitchell, 1985).

Glutathione offers one of the mechanisms for the scavenging of toxic free radicals. Although GSH depletion is not necessarily sufficient to cause lipid peroxidation, it is generally agreed that it may favor the peroxidation produced by other factors. Glutathione is important in the protection of cells against electrophilic drug injury in general and against reactive oxygen species in particular. Although it may not be an efficient antioxidant when acting alone, it has been shown to spare and potentiate vitamin E (Barclay, 1988) (see the following). Glutathione depletion also forms the toxicity of xenobiotics such as acetaminophen (see Chapter 6).

A severe reduction in glutathione favors peroxidation (Wendel *et al.*, 1979), and the damage may possibly be compounded by the increased generation of active radicals by the "induced" microsomes following chronic ethanol consumption, as illustrated in Fig. 7.12, (Lieber, 1980). Indeed, it is well known that the microsomal pathway, which requires O_2 and NADPH, is capable of generating lipid peroxides. Theoretically, increased activity of microsomal NADPH oxidase following ethanol consumption (Lieber and DeCarli, 1970; Reitz, 1975) could result in enhanced H_2O_2 and O_2^- production, thereby favoring lipid peroxidation. It has also been claimed that the ethanol-inducible form of rabbit liver microsomal cytochromes P-450 is associated with increased hydroxyl radical formation (Ingelman-Sundberg and Johansson, 1984). However, when the effect of chronic alcohol feeding on lipid peroxidation was studied in rat liver microsomes, no relationship to hydroxyl radical generation was observed (Shaw *et al.*, 1984; Puntarulo and Cederbaum, 1988). Other radicals such as the hydroxyethyl radical (Reinke *et al.*, 1987; Albano *et al.*, 1988), are also produced in liver microsomes, and some of these may initiate lipid peroxidation.

Enhanced lipid peroxidation, possibly mediated by

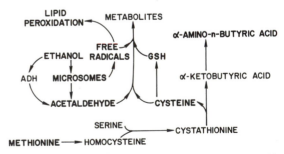

FIGURE 7.12. Hypothetical link among accelerated acetaldehyde production, increased free radical generation by the "induced" microsomes, enhanced lipid peroxidation, and possibly increased α-amino-*n*-butyric acid production. (From Lieber, 1980.)

acetaldehyde (DiLuzio and Stege, 1977), has been proposed as a mechanism for ethanol-induced fatty liver (DiLuzio and Hartman, 1967). The capacity of acetaldehyde to cause lipid peroxidation in the liver has been demonstrated in isolated perfused livers (Muller and Sies, 1982). Whether alcohol administration *in vivo* results in lipid peroxidation and injury has been subject of long-standing debate (Hashimoto and Recknagel, 1968; Scheig and Klatskin, 1969; Bunyan *et al.*, 1969; Comporti *et al.*, 1971), but more recent studies have shown evidence in favor of its occurrence both in non-human primates (Shaw *et al.*, 1981) and humans (Shaw *et al.*, 1983).

Thomas *et al.* (1985) showed that ferritin can provide the iron necessary for initiation of lipid peroxidation. The O_2^- generated by xanthine oxidase reductively releases ferritin-bound iron. Once released, this iron can promote the peroxidation of phospholipid liposomes. It is noteworthy that in alcoholics, the serum ferritin level was elevated in both groups immediately after a drinking bout, significantly more in men with than in those without biochemical signs of liver injury (Välimäki *et al.*, 1983). The serum iron concentration was equally increased but returned to normal during the first week of ethanol withdrawal. Experimentally, iron and ethanol treatments enhanced liver lipid peroxidation; an additive effect in lipid peroxidation was suggested to occur in this condition. In naive rats, very large amounts of ethanol (5–6 g/kg) are required to produce lipid peroxidation (DiLuzio and Hartman, 1967; MacDonald, 1973), whereas a smaller dose (3 g/kg) had no effect (Shaw *et al.*, 1981). By contrast, after chronic ethanol administration to the rat, even the smaller dose of ethanol administered acutely induced liver peroxidation, and this effect could be prevented, at least in part, by the administration of methionine, a precursor of glutathione (Shaw *et al.*, 1981). The ethanol-induced lipid peroxidation was even more striking in the baboon: administration of relatively small doses of ethanol (1–2 g/kg) produced lipid peroxidation and GSH depletion after 5–6 hr. In the baboon chronically fed alcohol (50% of total calories for 1–4 years), alcoholic liver disease, including cirrhosis in some, developed, and such animals exhibit evidence of enhanced hepatic lipid peroxidation and GSH depletion. These changes were observed following an overnight withdrawal from ethanol and were exacerbated by the readministration of ethanol. Evidence for GSH depletion and lipid peroxidation (enhanced diene conjugates) was found in liver biopsies of alcoholics who were withdrawn from alcohol (Shaw *et al.*, 1983).

Experimentally, as discussed before, acute ethanol intake diminished GSH content by 30% and enhanced that of GSSG by 73%. Biliary release of total GSH was reduced in this condition. The combined administration of iron and ethanol further influenced the decrease in hepatic GSH and the increase in GSSG levels elicited by the separate treatments. These data suggest that iron exposure accentuates those changes in lipid peroxidation and in the glutathione status of the liver cell induced by acute ethanol intoxication (Valenzuela *et al.*, 1983). It is also of interest to note that prolonged, heavy alcohol consumption results in the appearance of sialic-acid-deficient transferrin (two residues missing) in human serum. The increased capacity of transferrin deficient in sialic acid to selectively deposit iron in the hepatocyte (Regoeczi *et al.*, 1984) may be of significance for the development of the hepatic siderosis observed in alcoholism and could exacerbate lipid peroxidation (see previous discussion). Variation in iron content of alcoholic beverages is discussed in Section 7.3.3. It is tempting to speculate that the propensity of primates to develop more severe lesions than rats after chronic ethanol consumption may be some way be related, at least in part, to their greater susceptibility to glutathione depletion and the initiation of lipid peroxidation. It is apparent, however, that glutathione depletion *per se* does not suffice to produce liver damage (Siegers *et al.*, 1977), and peroxidation may, under certain conditions, precede glutathione depletion (Sippel, 1983). As mentioned before, concomitant enhanced production of active radicals may be involved, possibly resulting from the microsomal "induction" (see Chapter 1).

NADH inhibits the activity of NAD-dependent xanthine dehydrogenase (XD), thereby favoring that of oxygen-dependent xanthine oxidase (XO) (Kato *et al.*, 1990b; Lieber, 1989) (Fig. 7.10). It has been postulated that because of hypoxia and ethanol, purine metabolites and acetaldehyde accumulate and could be metabolized via XO. This process may lead to the production of oxygen radicals that most probably mediate both inhibition of glycolysis and the direct toxic effects toward liver cells, including peroxidation (Younes and Strubelt, 1987). However, cyanamide (a potent aldehyde dehydrogenase inhibitor) increased acetaldehyde levels fivefold after ethanol, whereas lipid peroxidation was significantly decreased by this treatment, indicating that acetaldehyde was not the substrate involved (Kato *et al.*, 1990b). Physiological substrates for XO—hypoxanthine and xanthine—as well as AMP significantly increased in the liver after ethanol, together with an enhanced urinary output of allantoin (a final product of xanthine metabolism). Allopurinol pretreatment resulted in 90% inhibition of XO activity and also significantly decreased ethanol-induced lipid peroxidation (Kato *et al.*, 1990b). We found a significant inhibition of ethanol-induced accumulation of malondialdehyde *in vivo* by allopurinol, in contrast to the report by Kera *et al.* (1988) but in agreement with observations by Park *et al.* (1988); we have no good explanation for this discrepancy in the effectiveness of allopurinol. This experimental evidence of a protective effect of allopurinol now awaits clinical confirmation.

Antioxidant protective mechanisms involve both enzymatic and nonenzymatic defense systems (Tribble *et al.*, 1987). Impairments in such defense systems have been reported in alcoholics, including alterations of ascorbic acid (Bonjour, 1979), glutathione (see above discussion), selenium (Tanner *et al.*, 1986; Dworkin *et al.*, 1985; Korpera *et al.*, 1985), and vitamin E (Tanner *et al.*, 1986; Yoshikawa *et al.*, 1982; Bjorneboe *et al.*, 1987). These changes could result from direct effects of ethanol or from the malnutrition associated with alcoholism. Furthermore, these defense systems are mutually interrelated; α-tocopherol (αToc), the major antioxidant in the membrane, is viewed as the "last line" of defense against membrane lipid peroxidation (McCay, 1985; Niki, 1987). Bjorneboe *et al.* (1987) reported a reduced hepatic αToc content after chronic ethanol feeding in rats receiving adequate amounts of vitamin E as well as in alcoholics (Bjorneboe *et al.*, 1988). Kawase *et al.* (1989) found that hepatic lipid peroxidation was significantly increased after chronic ethanol feeding in rats receiving a low-vitamin-E diet, indicating that dietary vitamin E is an important determinant of hepatic lipid peroxidation induced by chronic ethanol feeding. Both low dietary vitamin E and ethanol feeding significantly reduced hepatic αToc content, and the lowest hepatic αToc was found in rats receiving a combination of low vitamin E and ethanol. Furthermore, ethanol feeding caused a marked increase of hepatic α-tocopheryl quinone, a metabolite of αToc by free radical reactions. Thus, the combination of ethanol with a low vitamin E intake results in a decrease of hepatic αToc content, which renders the liver more susceptible to free radical attack.

7.3.2.2d. Effects of Acetaldehyde on Free Radical Generation and Enzyme Activities: Cytotoxic Proteins.

Free radical generation by neutrophils may provide a potential mechanism of cellular injury in acute alcoholic liver disease (William and Barry, 1987). This hypothesis is based on the observation that hepatocyte membranes altered by acetaldehyde stimulate neutrophils to produce superoxide anion. However, the concentration of acetaldehyde to which the liver membrane vesicles were exposed was unphysiologically high. In another study, an elevation of chemotactic factor inactivator was shown in alcoholic liver disease (Robbins *et al.*, 1987). This may explain impaired neutrophil chemotaxis in patients with alcoholic hepatitis, possibly favoring both localized infections and overwhelming sepsis (by altering neutrophil influx to sites of inflammation). One mode of acetaldehyde toxicity may involve interference with enzyme activities (Solomon, 1987), possibly secondary to binding with critical functional groups. Minute concentrations of acetaldehyde were found to impair the repair of alkylated nucleoproteins (Espina *et al.*, 1988). Cytotoxic protein molecules generated as a consequence of ethanol metabolism *in vitro* and *in vivo* (Wickramasinghe *et al.*, 1987) have been described and attributed to acetaldehyde–albumin complexes; an interaction of ethanol with macrophages has been incriminated (Wickramasinghe, 1986).

Disulfiram, an inhibitor of acetaldehyde dehydrogenase, raises the acetaldehyde levels after drinking and thereby causes flushing and other adverse effects, which is useful to maintain abstinence. Moderate abnormalities of liver tests are occasionally encountered in patients undergoing treatment with this drug, but they may be more related to drinking than to the taking of disulfiram (Iber *et al.*, 1987).

7.3.3. Role of Congeners

Majchrowicz and Mendelson (1970) showed that acetaldehyde levels in the blood of alcoholic patients were higher after bourbon ingestion than after grain ethyl alcohol and attributed this difference to the acetaldehyde content of the bourbon. However, Freund (1971) could find no measurable amounts of blood acetaldehyde after oral ingestion of an aqueous acetaldehyde solution ten times in excess of the concentration found in most bourbons. Since the amount of acetaldehyde in even the "dirtiest" of alcoholic beverages is about 1/1000 the amount produced by ethanol oxidation in the liver, it is probably not the acetaldehyde in the beverage that leads to higher blood acetaldehyde levels after ingestion but more likely an effect of other congeners of the beverages. These congeners may somehow interfere with acetaldehyde metabolism, initiating a vicious cycle of mitochondrial impairment that further decreases acetaldehyde catabolism. For example, Rubenstein *et al.* (1975) found that pyrogallol (1 and 10 mM concentrations), a metabolic product of gallic acid found in tannins, inhibits rat liver aldehyde dehydrogenase activity *in vitro*. *In vivo*, Collins *et al.* (1974) have shown that pyrogallol (250 mg/kg, i.p.) increases acetaldehyde blood levels in rats when it is given 1 hr before ethanol (3 g/kg, i.p.). Higher aliphatic aldehydes were found to inhibit mitochondrial acetaldehyde metabolism (Hedlund and Kiessling, 1969). Thus, elevated acetaldehyde levels following ingestion of alcoholic beverages, when compared with acetaldehyde levels following an ethanol–acetaldehyde mixture, suggest an effect of congeners on acetaldehyde metabolism.

When taken in relatively moderate amounts for a short period of time, the congener content of alcoholic beverages did not appear to affect the degree of steatosis (DiLuzio, 1962), but the effects of higher amounts of congeners in conjunction with a greater alcohol intake over a longer period of time have not been extensively studied. Some evidence was presented that prolonged consumption of whiskey might exert more striking undesirable effects on the liver than pure ethanol (Jordo and Olsson, 1975) and that certain alcoholic beverages, particularly brandy, were more toxic to liver cell cultures than pure ethanol (Walker *et al.*, 1974). One possible toxic congener is iron. Wine of certain vintages has been shown to contain large amounts of iron (Aron *et al.*, 1961; MacDonald, 1963; MacDonald and Baumslag, 1964), which was found to accumulate in liver and other organs of experimental animals given such wine over long periods of time (Aron *et al.*, 1961); this could also explain, at least in part, iron accumulation in various tissues (including the liver) of wine drinkers with cirrhosis of the liver (Aron *et al.*, 1961; MacDonald, 1963; MacDonald and Baumslag, 1964). By contrast, in non-wine-drinking alcoholics, hepatic iron excess is unusual (Barry, 1973; Jakobovits *et al.*, 1979), although increased iron absorption has been attributed to ethanol itself by some (Charlton *et al.*, 1964) but not others (Celada *et al.*, 1978) (see also Chapters 8 and 10). A unique situation is found in South African blacks, who may develop gross iron overload as a result of drinking home-brewed beer that contains large amounts of iron (Bothwell *et al.*, 1965).

7.3.4. Aggravation of Hepatotoxicity through Microsomal "Induction"

As discussed in Chapter 1, chronic ethanol consumption results in proliferation of the membranes of the smooth endoplasmic reticulum, documented by subfractionation and chemical measurements in micro-somes and by electron microscopy in animals and man (Iseri *et al.*, 1966; Lane and Lieber, 1966). Although these microsomal changes can be interpreted as "adaptive" alterations secondary to "induction" after chronic ingestion of alcohol, some injurious consequences may also ensue. Indeed, accelerated ethanol metabolism results in enhanced production of acetaldehyde and exacerbation of its various toxic manifestations discussed before, including enhanced peroxidation. The latter may also be promoted more directly through enhanced "free radical" formation by the induced microsomes and resulting enzyme inactivations (Dicker and Cederbaum, 1988). Increased microsomal activity may enhance the oxygen requirements (see Chapter 1), thereby aggravating whatever hypoxia may be present for unrelated reasons.

Some compounds acquire hepatotoxicity only after metabolism, or "activation," by the enzymes of the endoplasmic reticulum. One such compound is carbon tetrachloride, the hepatotoxicity of which is greatly increased after chronic alcohol consumption, at least partly because of enhanced activation by microsomes (Hasumura *et al.*, 1974). It is likely that a similar mechanism increases susceptibility in these patients to the hepatotoxic actions of a variety of compounds in the environment, including anesthetics and many commonly used drugs, such as ioniazid and acetaminophen (Sato *et al.*, 1981). As for CCl_4, these compounds are excellent substrates for the ethanol-inducible cytochrome P450IIE1. Selective induction of P450IIE1 in zone 3 (Tsutsumi *et al.*, 1988; Ingelman-Sundberg *et al.*, 1988) contributes to the selective hepatotoxicity of these compounds in that zone after chronic alcohol consumption (see Chapters 1 and 6). Furthermore, glutathione depletion is a major contributory factor (see Section 7.3.2.2c.). In addition, the lower Po_2 in that zone favors peroxidation as demonstrated for CCl_4-mediated hepatotoxicity (De Groot *et al.*, 1988). Not only does ethanol consumption potentiate the hepatotoxicity of these compounds, but the converse also occurs for some of them: CCl_4 pretreatment, for instance, sensitizes the liver to increase collagen deposition following ethanol (Bosma *et al.*, 1988).

Micronutrients such as vitamins may also serve as substrates for the microsomes, and the "induction" of the microsomes may therefore alter vitamin requirements and even affect the integrity of the liver. Indeed, alcoholics have very low vitamin A levels in their livers (Leo and Lieber, 1982). In experimental animals, ethanol administration was shown to depress hepatic vitamin A levels, even when administered with an adequate diet

(Sato and Lieber, 1981). When dietary vitamin A was virtually eliminated, the depletion rate of vitamin A from endogenous hepatic storage was two to three times faster in ethanol-fed rats than in controls. In rats, severe vitamin A depletion was associated with the appearance of multivesicular lysosomes (Leo *et al.*, 1983) (Fig. 7.13). Such lesions were also commonly seen in alcoholic patients with low hepatic vitamin A levels.

Vitamin A supplementation is sometimes used to correct the problems of night blindness and sexual dysfunctions of the alcoholic (see Chapter 17). Such therapy might also be useful with regard to the liver pathology. The therapeutic usage of vitamin A is, however, complicated by the fact that excessive amounts of

vitamin A are known to be hepatotoxic and that the alcoholic has an enhanced susceptibility to this effect (Leo *et al.*, 1982; Leo and Lieber, 1988). In control rats, amounts of vitamin A equivalent to those commonly used for the treatment of the alcoholic were found to be without significant effects on the liver, but in animals chronically fed alcohol signs of toxicity developed, such as striking morphological (Fig. 7.14) and functional alterations of the mitochondria (Leo *et al.*, 1982; Leo and Lieber, 1983). Enhanced toxicity was not associated with an increased vitamin A level in the liver; in fact, because (as mentioned before) alcohol administration tends to decrease vitamin A levels in the liver, even after vitamin A supplementation, alcohol-fed animals had

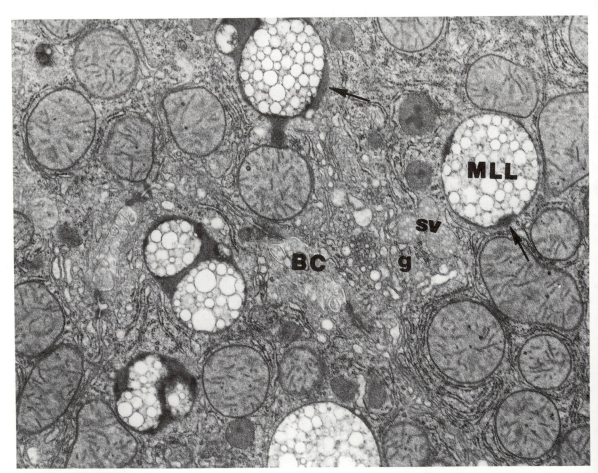

FIGURE 7.13. Multivesicular lipid-containing lysosomes (MLL) surrounding a bile canaliculus (BC) in hepatocytes of rat fed a vitamin-A-deprived diet. Part of the limiting membrane of MLL shows a homogeneous electron-dense crescent (arrows). SV, secretory vesicle; g, Golgi complex. (Uranyl acetate and lead staining, ×24,000). (From Leo *et al.*, 1982.)

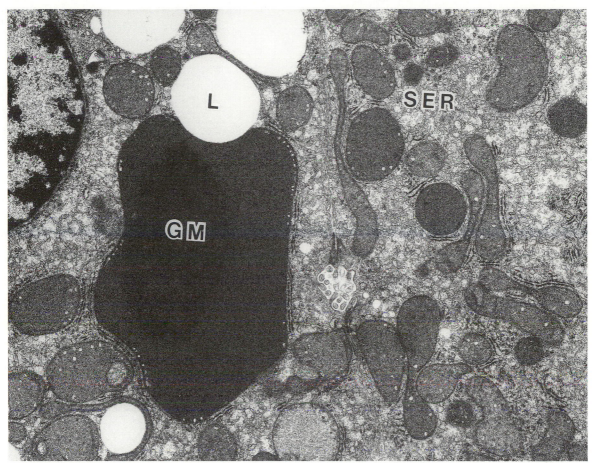

FIGURE 7.14. Electron micrographs of pericentral hepatocytes. High vitamin-A- and ethanol-fed rat. A giant mitochondrion (GM) approaches the size of a nucleus and contains an unusual dense matrix with a large fusiform crystalline inclusion. This as well as adjacent mitochondria display severe disorganization of the cristae. L, lipid droplet; SER, proliferated smooth endoplasmic reticulum (Uranyl acetate and lead staining, ×18,000). (From Leo *et al.*, 1983.)

vitamin A levels in the liver that were not higher than normal values, yet toxicity developed. One possible explanation, still to be proven, is that vitamin A toxicity may be mediated, at least in part, by the enhanced production of a toxic metabolite, as for the case of xenobiotic agents. Thus, the toxicity could be related to the induction of the enzymes of the endoplasmic reticulum after chronic ethanol consumption.

Progression to severe parenchymal injury eventually may offset the microsomal induction, as shown in baboons fed alcohol chronically (Lieber *et al.*, 1975) and in alcoholics with advanced liver disease (Sotaniemi *et al.*, 1977; Farrell *et al.*, 1978), though Hoensch *et al.* (1979) found that alcoholics with liver disease had

higher monooxygenase activity than alcoholics without liver complications.

7.3.5. Role of Alterations of the Immune System, Endotoxins, Ethyl Esters, and Leukotrienes

In the past few years, increasing evidence has suggested that lymphocytes are important in autoimmune tissue-damaging reactions. Accordingly, a cell-mediated mechanism has been implicated in the development of some chronic liver diseases. The pathogenetic role of stimulation alterations of the immune system by ethanol in patients with alcoholic liver disease is discussed in Chapter 9. Several observations suggest that immunologic mechanisms are involved in the perpetuation of hepatic damage by alcohol. Alcoholic hyaline, an abnormal cytokeratin possibly reflecting vitamin A deficiency has ben postulated to trigger an immune response (French, 1981). Cytotoxic activity of mononuclear cells against autologous liver cells was demonstrated in baboons fed ethanol (Lue *et al.*, 1981) and in alcoholics (Kakumu and Leevy, 1977; Poralla *et al.*, 1984), with involvement of both T and non-T cells (Poralla *et al.*, 1984). Alcohol abuse was also reported to lead to transient endotoxemia (Bode *et al.*, 1987), which may contribute to the liver injury (Bhagwandeen *et al.*, 1987). Although primary events have not been differentiated from changes secondary to liver injury, these studies justify further investigation of the role of immunologic factors in the development and progression of alcoholic liver disease.

Ethanol forms ethyl esters *in vivo*. Laposata and Lange (1986) found that concentrations of fatty acid ethyl esters in pancreas, liver, heart, and adipose tissue were significantly higher in acutely intoxicated subjects than in controls. Since this nonoxidative ethanol metabolism occurs in the organs most commonly injured by alcohol abuse, and since some of these organs lack an oxidative ethanol-metabolizing system, it was postulated that fatty acid ethyl esters and their metabolites may play a role in the production of alcohol-induced injury.

Leukotrienes have also been postulated as possible mediators in liver diseases, including the alcoholic variety (Keppler *et al.*, 1988). Indeed, an ethanol-induced inhibition of leukotriene degradation in the liver was demonstrated and provides a novel explanation for some of the inflammatory reactions in acute alcoholic liver disease.

7.3.6. Inhibition of Hepatic Regeneration

There is some debate as to whether ethanol interferes with hepatic regeneration: whereas some found no effect on regeneration (Craig, 1975; Frank *et al.*, 1979), others (Leevy, 1963; Wands *et al.*, 1979) report inhibition of [^{3}H]thymidine incorporation. Although conclusive evidence that ethanol impairs regeneration after partial hepatectomy is still lacking, several results are suggestive of such an effect (Baskin *et al.*, 1988). In liver biopsies, ethanol primarily suppresses mesenchymal cell replication (Leevy, 1966). In the partial hepatectomy model, ethanol decreases replication not only of hepatocytes but also of lipocytes (Tanaka *et al.*, 1991).

Ethanol inhibits the stimulation of ornithine decarboxylase activity resulting from hepatotoxicity (Pösö and Pösö, 1980a) and stabilizes this enzyme (Pösö and Pösö, 1980b). Under these conditions, it also inhibits the synthesis of RNA, probably at the transcriptional level (Pösö and Pösö, 1981). Ethanol also inhibits the repair of DNA, as discussed in Chapter 15. Others found no inhibition by chronic ethanol feedings of the induction of ornithine decarboxylase (after partial hepatectomy) but did find significant inhibition of DNA synthesis (Diehl *et al.*, 1988), an effect attenuated by malotilate (Takada *et al.*, 1987).

7.3.7. Disorders of Collagen Metabolism and Production of Cirrhosis

7.3.7.1. Precursor Lesions

The present prevailing view is that alcoholic cirrhosis develops in response to alcoholic hepatitis. As discussed before, the latter is characterized by ballooning of the hepatocytes, extensive necrosis, and polymorphonuclear inflammation. In addition, there is increased deposition of collagen and glycosaminoglycans (Galambos and Shapiro, 1973). It is understandable that necrosis and inflammation may trigger the scarring process of cirrhosis, but one must question whether this is the only mechanism involved. In some populations in the United States (Nakano *et al.*, 1982; Worner and Lieber, 1985) as well as in Japan (Takada *et al.*, 1982; Minato *et al.*, 1983; Hasumura *et al.*, 1985), cirrhosis commonly develops in alcoholics without an apparent intermediate stage of florid alcoholic hepatitis. This observation raises the question of whether alcohol can promote development of cirrhosis without it being preceded by alcoholic hepatitis. Indeed, in baboons fed alcohol, fatty liver developed, the hepatocytes increased in size, and there was some obvious ballooning. This ballooning was associated with some mononuclear inflammation but very few of the polymorphonuclear cells so characteristic of human alcoholic hepatitis (Popper and Lieber, 1980). Although some clumping was apparent in the cytoplasm by electron microscopy, there was no alcoholic hyalin. Thus, there was no florid picture of alcoholic hepatitis; yet, in one-third of the animals, typical cirrhosis developed (Figs. 7.15 and 7.16).

If one can extrapolate from these baboons to humans, it appears that full-blown alcoholic hepatitis may not be a necessary intermediate step in the development of alcoholic cirrhosis. Sequential clinical observations support this contention (Worner and Lieber, 1985) (Fig.

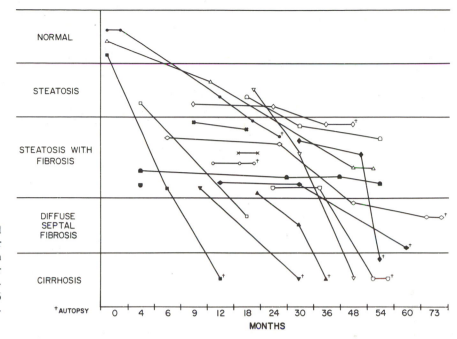

FIGURE 7.15. Sequential development of alcoholic liver injury. The most severe lesion is recorded individually for each time period in 18 baboons fed alcohol up to 6 years. (From Popper and Lieber, 1980.)

7.17). This hypothesis in turn raises the question of which process, in the absence of alcoholic hepatitis, may initiate cirrhosis on chronic consumption of ethanol. It is possible that cirrhosis develops in the presence of minimal inflammation and necrosis, which may suffice to trigger the fibrosis (Lieber, 1977). On the other hand, it is also possible that independent of the necrosis and inflammation, alcohol may have some more direct effects on the metabolism of collagen, the characteristic protein of the fibrous tissues. In alcoholic liver injury there is a great variability in the magnitude of collagen deposition. At the earlier stages, in the so-called simple or uncomplicated fatty liver, collagen is detectable by chemical means only (Feinman and Lieber, 1972; Patek et al., 1976). When collagen deposition is sufficient to become visible by light microscopy, usually it appears first around the central (also called terminal) venules, resulting in so-called "pericentral" or "perivenular" fibrosis or sclerosis, discussed before. Sequential biopsies in alcohol-fed baboons revealed that in some animals by the early fatty liver stage an increased number of mesenchymal cells, particularly myofibroblasts, appear in the perivenular areas (Nakano and Lieber, 1982). Myofibroblasts are present in normal liver and after alcohol feeding (Fig. 7.18; Lieber et al., 1981). After ethanol, proliferation of these cells is eventually accompanied by deposition of abundant collagen bundles, first

in the perivenular areas, leading to perivenular and pericellular fibrosis and ultimately to diffuse fibrosis and cirrhosis.

In the early stages, many filaments with a diameter of 5 nm and fibrils with a diameter of 10 nm were observed around myofibroblasts (Fig. 7.19B). Morphologically, these fibrils are considered to represent type III collagen (Irle et al., 1980). Furthermore, collagen fibrils (with a diameter of about 50 nm) were seen in addition to the other two types of fibrils. They can represent either type III or type I collagen (Hay et al., 1978). Basal laminas were observed between endothelial cells and myofibroblasts or around the myofibroblasts, probably representing type IV collagen. Myofibroblasts have been isolated from surgical baboon liver biopsies by collagenase digestion and Percoll density gradient centrifugation (Savolainen et al., 1984b). By immunofluorescence studies, the cells synthesize collagen types I, III, and IV and laminin. The importance of smooth muscle type cells was also shown by Voss et al. (1982), who cultured tissue specimens from human fibrotic liver. Myofibroblasts' processes also extend into the Disse space. There, however, the most common mesenchymal cell is the lipocyte (also called the fat-storing or Ito cell) (see following discussion). Histochemically procollagens and collagens type I and III are deposited with elastin and fibronectin. Septa were accompanied by

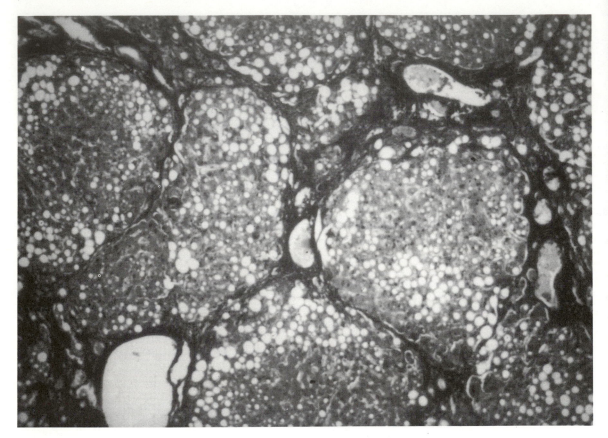

FIGURE 7.16. Cirrhosis in a baboon fed alcohol for 4 years. Fat is regularly distributed through nodules surrounded by connective tissue septa. Chromotrope–aniline blue. ×60. (From Lieber and DeCarli, 1974.)

increases in laminin and collagen type IV in association with the appearance of myofibroblasts (Hahn *et al.*, 1983). There are close similarities of the major structural components and their distribution in human (Hahn *et al.*, 1980) and baboon (Hahn *et al.*, 1983) alcoholic fibrosis.

In cases of alcoholic hepatitis, Kent *et al.* (1976) previously described an increased number of Ito cells. Lipocytes are characterized electron microscopically by abundant lipid droplets (Mak *et al.*, 1984; Mak and Lieber, 1988), microfilament bundles, dense bodies, and pinocytic vesicles (Fig. 7.20). After chronic alcohol consumption, about half of the lipocytes are replaced by transitional cells between lipocytes and fibroblasts (Fig. 7.21); these transitional cells were arbitrarily defined by a lipid droplet volume of less than 20%, measured by morphometry (Mak *et al.*, 1984; Mark and Lieber, 1988). The area of the rough endoplasmic reticulum in

transitional cells is greater than that in lipocytes. A similar increase in rough endoplasmic reticulum of lipocytes was observed in patients with fibrosis in the Disse space (Minato *et al.*, 1983). Transitional cells have abundant microfilaments, dense bodies, and pinocytotic vesicles. They resemble myofibroblasts in the perivenular zones, but they can be differentiated by the lack of surrounding basal laminae. Moreover, myofibroblasts typically have an indented nucleus. Transitional cells are surrounded by abundant collagen fibers and can be seen in the perisinusoidal spaces throughout the lobule associated with a netlike fibrosis, sometimes linking up with the perivenular lesion. There was good correlation among hepatic fibrosis, the percentage of transitional cells, and the area of their rough endoplasmic reticulum (Mak *et al.*, 1984).

Lipocytes, myofibroblasts, and fibroblasts may belong to the same cell family. Transformation of Ito cells

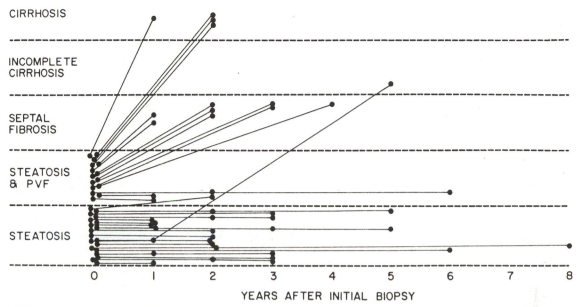

FIGURE 7.17. Progression of fibrosis in alcoholics without hepatitis. Follow-up to 8 years after the initial biopsy. The presence of perivenular fibrosis (PVF) in the initial biopsy is a harbinger of rapid development of fibrosis to more severe stages, including cirrhosis. (From Worner and Lieber, 1985.)

into myofibroblasts has been suggested (Farrell *et al.*, 1977). Many investigators have examined perisinusoidal cells in relation to hepatic fibrogenesis (Wasserman, 1958; Wood, 1963; Schnack *et al.*, 1968; Wisse, 1970; McGee and Patrick, 1972; Galambos and Shapiro, 1973; Galambos *et al.*, 1977; Kent *et al.*, 1977). After inflammatory stimulation or during repair, mesenchymal cells containing fat droplets proliferate in conjunction with collagen deposition; therefore, these cells may play a role in fibrogenesis (Kent *et al.*, 1977). It is likely that some cells previously described in the perisinusoidal space were unrecognized transitional cells. In the experimental model of alcoholic cirrhosis in the baboon, an increased number of lipogranulomas were found in the early stages (Popper and Lieber, 1980). In humans, too, some forms of lipogranulomas have been incriminated in the development of fibrosis (Christoffersen *et al.*, 1971). Thus, by the fatty liver stage, nonparenchymal cells participate in the production of collagen, but we do not know the respective roles of parenchymal and nonparenchymal cells in this process. In addition to the extensive evidence concerning the role of myofibroblasts and Ito cells (with transitional cells) in the production of collagen, several studies reported evidence that hepatocytes can also be contributory. This was originally shown by Diegelmann *et al.* (1983) in hepatocytes

in primary monolayer culture and has been confirmed by Kamegaya *et al.* (1985) and by Clement *et al.* (1986).

7.3.7.2. Mechanisms of Collagen Accumulation

The accumulation of hepatic collagen during the development of cirrhosis could theoretically be accomplished by increased synthesis, decreased degradation, or both. The rate of hepatic fibrous tissue degradation has never been directly measured in human alcoholic liver disease or in any of the animal models of alcoholic liver disease. Therefore, its role in the pathogenesis of hepatic fibrosis is unresolved. Collagen half-life is, however, prolonged in the fibrosis of ethionine intoxication and CCl_4-induced cirrhosis (Hutterer *et al.*, 1964, 1970). The cirrhosis of choline deficiency (to which the rat is highly susceptible and by which primates are much less affected) may be exacerbated in the rodent by administration of ethanol, and in this context collagen degradation may be slowed (Henley *et al.*, 1979).

The mechanisms of collagen degradation in the liver are complex. There appears to be a paradoxical increase in collagenase activity in animals fed ethanol (Okazaki *et al.*, 1977), at least during the early stage of alcoholic liver injury (Okazaki *et al.*, 1977). At that

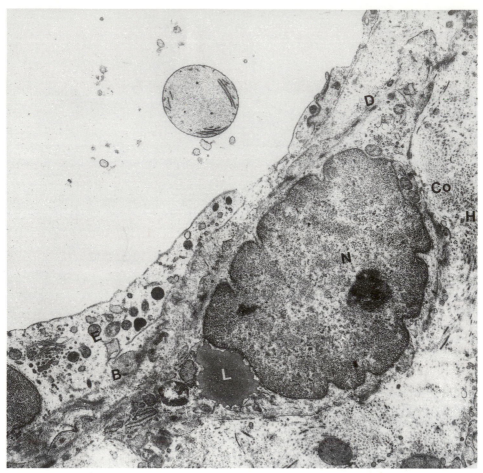

FIGURE 7.18. Nonfibrotic terminal hepatic venule after 4 months of alcohol feeding. Myofibroblasts show characteristic structures, indented nucleus (N), microfilaments with dense bodies (D), basal lamina (B), and lipid droplet (L). Lipid droplet is commonly seen in myofibroblasts. ×10,000. (From Lieber *et al.*, 1981.) E, endothelial cell; Co, collagen fibers; H, hepatocyte.

stage, collagenolytic activity may also be increased, together with the increased collagen synthesis (Kato *et al.*, 1985). Subsequently, collagenase activity may decrease and contribute to the collagen accumulation (Maruyama *et al.*, 1982). Compounds such as polyunsaturated lecithin that stimulate collagenase activity *in*

vitro in lipocytes (Li *et al.*, 1992) also prevent experimental cirrhosis *in vivo* (Lieber *et al.*, 1990b), thereby supporting the hypothesis that cirrhosis might, in part, represent a relative failure of collagen degradation to keep pace with synthesis.

The results observed with lysosomal enzyme activ-

FIGURE 7.19. Liver biopsy of an alcoholic subject. (a) Myofibroblast (mf) around thickened terminal hepatic venule with indented nucleus, microfilaments with dense area (arrow), and many pinocytotic vesicles (v). The myofibroblast is surrounded by numerous collagen fibrils of different diameters and a basal-lamina-like structure (arrowhead). Bleb formation of myofibroblasts beneath endothelial cells is prominent. Lysosomes (L) are observed in endothelial cells and myofibroblasts (uranyl acetate and lead citrate, ×6000). (b) Higher magnification of a part of A. Collagen fibrils of three different sizes are seen around the cell process of a myofibroblast. Intermediate-size (2) and large (3) fibrils are prominent (uranyl acetate and lead citrate, ×22,000). (From Nakano *et al.*, 1982.)

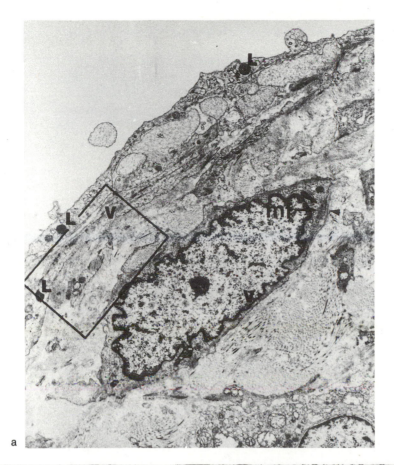

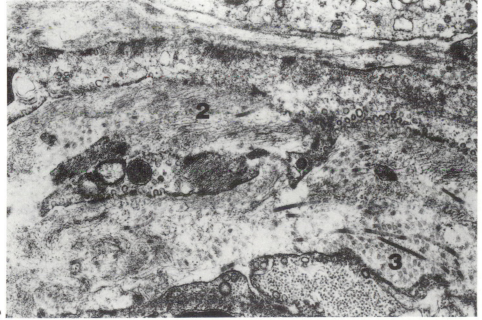

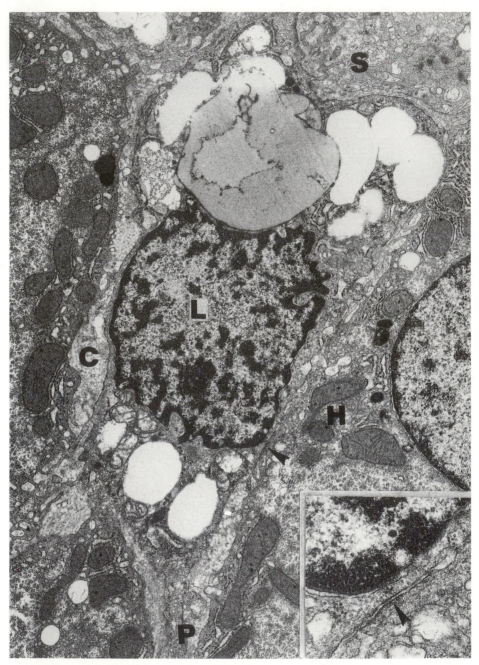

FIGURE 7.20. Lipocyte from a patient with fatty liver. The lipocyte (L) contains lipid droplets of variable sizes with a total lipid volume density of more than 20% of the cell. The rough endoplasmic reticulum is randomly distributed in the cytoplasm as short, narrowed cisternae. The Golgi apparatus is relatively inconspicuous. At the cell periphery and in the cell process (P), there are small aggregates of filaments, dense bodies, and pinocytic vesicles. Cell contacts with increased filamentous density of adjoining plasma membranes (arrowheads) are visible between the lipocyte and a neighboring hepatocyte (H). Collagen fibers (C) are present in the extracellular space. Inset: Cell contacts at higher magnification. S, sinusoid. ×13,000; inset ×50,000. (From Mak and Lieber, 1988.)

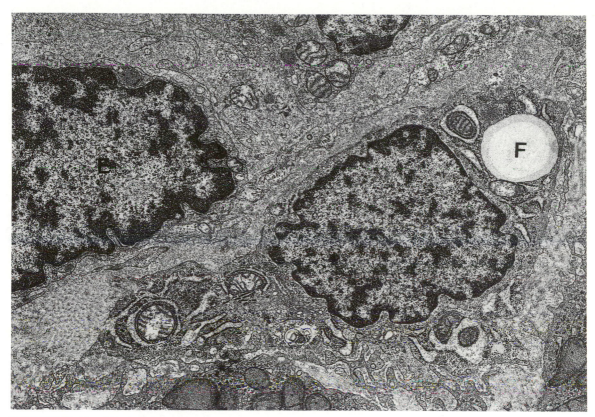

FIGURE 7.21. Transitional cell in the Disse space in a cirrhotic liver. The cell contains a small lipid droplet (F) with a total lipid volume density of less than 20% of the cell volume. The rough endoplasmic reticulum appears conspicuous. Note that the cell profile appears smaller than that of the lipocyte in Fig. 7.20. In the extracellular space, an abundant amount of collagen fibers is present. E, endothelial cell nucleus. ×13,000. (From Mak and Lieber, 1988.)

ity are, in part, contradictory. After 2 weeks of ethanol administration (1 g/kg i.p.), the activities of lysosomal enzymes were decreased (Platt *et al.*, 1971), whereas with larger, oral doses they were unchanged after 1 month and increased after 6 months (Mezey *et al.*, 1976). Henley and Laughrey (1979) reported that in ethanol-treated rats, inhibitors directed toward lysosomal collagenase (tested against radioactive collagen substrate at pH 4) are increased. In patients with cirrhosis, urinary hydroxyproline is usually increased (Emmirch *et al.*, 1967; Kratzsch, 1969; Resnick *et al.*, 1973; Wu and Leevy, 1975; Mezey *et al.*, 1979), possibly reflecting enhanced collagen turnover. It is noteworthy that after alcohol discontinuation, the rate of urinary hydroxyproline excretion was increased, an observation consistent with the possibility that alcohol may inhibit collagen degradation (Wu and Leevy, 1975).

Chronic ethanol consumption with adequate diets led to accumulation of hepatic collagen in rats and baboons, even when fibrosis was not yet histologically detectable (Feinman and Lieber, 1972). In the latter model, the role of increased collagen synthesis is suggested by increased activity of hepatic peptidylproline hydroxylase in rats and primates and increased incorporation of [¹⁴C]proline into hepatic collagen in rat liver slices (Feinman and Lieber, 1972). Increased hepatic peptidylproline hydroxylase activity was also found in patients with alcoholic cirrhosis (Patrick, 1973), hepatitis (Mezey *et al.*, 1979), and in all stages of alcoholic liver disease (Mann *et al.*, 1979). The role of increased collagen synthesis has been confirmed indirectly in man by autoradiographic techniques utilizing liver biopsies (Chen and Leevy, 1975). In baboons given ethanol that developed significant fibrosis, molecular evaluation revealed that the RNA from these livers was more active in *in vitro* protein synthesis, and the type I procollagen mRNA content was significantly higher (per liver RNA) as determined by hybridization analysis (Fig. 7.22). In addition, there was higher mRNA content in the livers of ethanol-fed baboons that developed fibrosis. This increase in type I procollagen mRNA in the baboon model of alcoholic fibrosis may foster fibrogenesis.

Congeners have been incriminated *in vitro* in increased collagen production (Walker and Shand, 1972). Another possible mechanism whereby alcohol consumption may be linked to collagen formation is the increase in tissue lactate secondary to alcohol metabolism (Lieber, 1989; Lieber *et al.*, 1971; Fig. 7.10). Elevated concentrations of lactate have been associated with increased peptidylproline hydroxylase activity both *in vitro* (Green and Goldberg, 1964) and *in vivo* (Lindy *et al.*, 1971). The hepatic free proline pool size, which has been incriminated in the regulation of collagen synthesis (Rojkind and DeLeon, 1970; Chvapil and Ryan, 1973), may be increased by ethanol (Häkkinen and Kulonen, 1975) and is expanded in human portal cirrhosis (Kershenobich *et al.*, 1970). In patients with alcoholic cirrhosis, increased serum free proline and hydroxyproline have been reported (Mata *et al.*, 1975), but high hepatic concentration of free proline did not induce collagen synthesis, at least in rat liver (Forsander *et al.*, 1983). It has also been postulated that lactate may play a role (Kershenobich *et al.*, 1979a) through inhibition of proline oxidase (Kowaloff *et al.*, 1977). This hypothesis illustrates once more the possible impact of ethanol-induced redox changes on intermediary metabolism, including the metabolism of collagen (see Section 7.3.2.1). One of the most promising media-

tors of increased collagen synthesis after ethanol is its metabolite acetaldehyde. In myofibroblast (Savolainen *et al.*, 1984b), fibroblast (Holt *et al.*, 1984), and lipocyte (Moshage *et al.*, 1990) cultures, acetaldehyde increases collagen synthesis and messenger RNA (Brenner and Chojkier, 1987). As discussed before, such messenger RNA is also elevated in the liver of baboons given alcohol (Zern *et al.*, 1985).

7.3.8. Genetic Factors

Although no group is immune to alcoholic cirrhosis, various segments of our society are affected differently. For instance, it is generally recognized that American Jews are relatively spared, whereas American Indians seem to be more prone to develop alcoholic cirrhosis. These differences are commonly attributed to cultural factors leading to alcoholism, although genetic influences favoring either alcoholism or cirrhosis or both have not been ruled out. In addition to the environmental factors discussed before, individual differences in rates of ethanol metabolism appear, in part, to be genetically controlled (Vesell *et al.*, 1971), and the possible role of heredity for the development of alcoholism in humans has been emphasized (Goodwin, 1971). In the last several years, it has been proved beyond doubt that the predisposition for many different diseases is significantly associated with specific histocompatibility antigens (as reviewed in McDevitt and Bodmer, 1972 and Bell and Nordhagen, 1980). In patients with chronic alcoholic liver disease, the HLA-B8 was more prevalent in patients with alcoholic cirrhosis than in controls or in

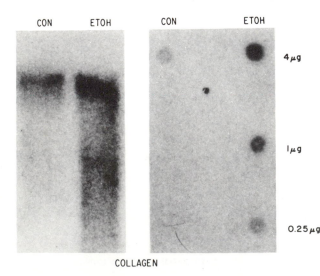

FIGURE 7.22. Northern transfer and dot blot hybridization that identify type I pro-α_1-collagen mRNA in total RNA isolated from livers of control and ethanol-fed baboons that developed fibrotic liver. Both hybridization techniques demonstrate more collagen mRNA in the RNA from ethanol-fed animals. (From Zern *et al.*, 1985.)

patients with fatty liver or minimal fibrosis (Bailey *et al.*, 1976). Another difference was the absence of HLA-A28 in the cirrhotic patients (Morgan *et al.*, 1980). The cirrhotics with HLA-B8 had been drinking for a shorter period of time than patients with comparable clinical and histological diseases but without HLA-B8 (Saunders *et al.*, 1982). These results suggested the hypothesis that genetic factors may be implicated in alcohol-induced liver damage and particularly its progression to cirrhosis. These data, however, have not ben confirmed in other reports (Scott *et al.*, 1977; Bell and Nordhagen, 1978; Gluud *et al.*, 1980). Bell and Nordhagen (1978) found an association with HLA-BW40 and Melendez *et al.* (1979) with HLA-B13, but Scott *et al.* (1977), Simon *et al.* (1977), Shigeta *et al.* (1980), and Gluud *et al.* (1980) did not. These discrepancies in different populations of alcoholic patients with cirrhosis suggest that the HLA locus B is not implicated in alcoholic cirrhosis but that it maintains a close association with nearby genes that are involved. Further studies are obviously required in this emerging field.

7.3.9. Associated Viral Disease

Alcoholism and viral hepatitis can occur in the same patients, but in many subjects with alcoholic cirrhosis there is no evidence for antecedent viral hepatitis; similarly, in the baboons that developed alcoholic cirrhosis, no concomitant virus infection was detectable (C. S. Lieber, unpublished data). Conversely, ethanol infusion did not appear to affect the course of hepatitis B virus infection in chimpanzees (Tabor *et al.*, 1978). Alcoholics have an increased incidence of viral hepatitis, and there has been some question concerning the possible potentiation of alcohol-induced liver damage by the viral disease, but as discussed in Chapter 9, most observations do not substantiate such a mechanism. In particular, there is no evidence that the hepatitis B virus plays a role in alcoholic liver disease (Fong *et al.*, 1988) or in the associated hepatocellular carcinoma (Walters *et al.*, 1988). However, hepatitis C is also commonly associated, and its role is presently being assessed.

7.4. Diagnosis of Alcoholic Liver Injury

At each stage of alcoholic liver disease (fatty liver, alcoholic hepatitis, cirrhosis), the clinician must distinguish between alcoholic and nonalcoholic causes of liver injury. The clinician must also differentiate between the acute, reversible forms of alcoholic liver disease, such as fatty liver and alcoholic hepatitis, and the chronic, mainly irreversible, cirrhosis.

In the evaluation of the patient with either an incidental finding of pathological laboratory values pointing to liver damage or more advanced symptomatology requiring hospitalization, a history of heavy alcohol intake for weeks or months before hospital admission is strongly suggestive of an alcoholic etiology of liver injury. The clinician, however, should realize that estimation of alcohol consumption even in a careful interview may reveal only about 40% of total alcohol consumption (Mäkelä, 1971). Furthermore, the casual estimation of blood ethanol seldom reveals heavy drinkers and does not differentiate between chronic and acute alcohol consumption (Hamlyn *et al.*, 1975). Serum enzyme elevations can be used to detect alcoholics but do not differentiate alcoholic liver injury from other liver diseases (Rollason *et al.*, 1972). At present, there is no definite biochemical marker that can differentiate the alcoholic liver diseases from the nonalcoholic ones, but, as discussed in Chapter 18, several biological markers may help in diagnosing relapse of drinking.

Laboratory tests are not much better for the differentiation of various stages of alcoholic liver disease. This is because of the great variability in the severity of both alcoholic hepatitis and fatty liver. In extreme cases both of them may simulate the clinical picture of liver cirrhosis or even extrahepatic biliary obstruction. The most characteristic clinical signs and laboratory tests or test combinations in the diagnosis and differential diagnosis of alcoholic liver diseases are summarized in Tables 7.1 and 7.2 (Lieber and Salaspuro, 1985). Laparoscopy and endoscopic retrograde cholangiography are indicated in rare cases simulating extrahepatic biliary obstruction. Diagnostic laparotomy should be avoided, since it constitutes a great hazard in acute alcoholic liver disease (Greenwood *et al.*, 1972). The histological confirmation of the diagnosis by liver biopsy obtained either blindly or during laparoscopy is highly desirable.

The aims of the histological interpretation of liver biopsy specimens in alcoholic liver disease are three-fold: (1) to differentiate between nonalcoholic and alcoholic liver diseases; (2) to determine the stage of alcoholic liver disease (fatty liver, perivenular and pericellular fibrosis, alcoholic hepatitis, and cirrhosis); and (3) to evaluate the possible progression of the liver disease in serial biopsies. The first of these goals is clearly emphasized by two studies. Goldberg *et al.* (1977) found ten alcoholic patients whose biopsies demonstrated chronic active or chronic persistent hepatitis. The overlap of

TABLE 7.1.   Clinical Findings in Alcoholic Liver Diseases[a,b]

	Fatty Liver (%)	Alcoholic Hepatitis (%)	Cirrhosis (%)
Anorexia, nausea, fatigue	2–10	70	35
Abdominal pain	2–10	50	34
Fever	2–10	50 (intermittent)	24 (low grade, continuous)
Hepatomegaly	75	95	67
Jaundice	2	55	49
Signs of portal hypertension (ascites splenomegaly)	2	40	50
Spider angiomata	8	30	38

[a]From Lieber and Salaspuro (1985).
[b]The percentages represent mean values. A wide range in the incidence of various symptoms is common in alcoholic hepatitis, depending on the selection of patient material.

laboratory abnormalities was too extensive to allow differentiation. Levin *et al.* (1977) reported that in their population one out of five subjects with alcoholism and chronic abnormalities of liver function had a nonalcoholic liver disease. Furthermore, they concluded that liver biopsy is necessary to establish the diagnosis, since nonalcoholic liver disease could not be detected by the clinical data. The determination of the stage of the alcoholic liver disease is important both for prognosis and treatment (see Section 7.5). Liver biopsy is indicated when the laboratory abnormalities have persisted for 3 to 6 months. The frequency of subsequent biopsies can be decided individually, depending on the initial diagnosis. If the diagnosis of liver cirrhosis is certain,

TABLE 7.2.   Laboratory Findings in Alcoholic Liver Disease[a]

Finding	Comment
	I. Findings suggestive of an alcoholic etiology
AST (SGOT) and ALT (SGPT)	Both enzymes are elevated in about 20% of patients with alcoholic steatosis and cirrhosis and in about 95% of patients with alcoholic hepatitis; SGOT/SGPT ratio increases in alcoholic cirrhosis (59%).
γ-GTP and alkaline phosphatase (ALP)	γ-GTP is greatly elevated in almost all cases and every stage of alcoholic liver disease; ALP is increased in 80% of patients with alcoholic hepatitis and cirrhosis but in only 28% of patients with alcoholic fatty liver; the ratio of γ-GTP/ALP is 1–3 in nonalcoholic liver diseases but 5–35 in alcoholic liver diseases, even in alcoholic steatosis.
Tissue antibodies (smooth muscle, antinuclear, mitochondrial)	Tissue antibodies have been reported to be either increased and normal in alcoholic liver disease. However, the incidence as well as titers are generally low. On this basis the occurrence of tissue antibodies is highly suggestive of nonalcoholic etiology or pathogenetic mechanism of the liver injury.
IgA	Frequently increased in alcoholic liver injury; the magnitude of increase and IgA/IgG ratio can be used to differentiate alcoholic from nonalcoholic liver diseases.
	II. Differentiation between the various stages of alcoholic liver disease
GDH	A mitochondrial enzyme that correlates better with the degree of liver cell necrosis and inflammation than the other enzymes; measurement must follow shortly the alcoholic debauch.
Albumin, prealbumin, elimination tests, NH$_3$	Correlate with the progression and the severity of the liver injury; in fatty liver and clotting factors, abnormal alcoholic hepatitis values normalize after the acute phase of the disease. In cirrhosis abnormal findings are more constant.
IgA	The degree of increase correlates positively with the severity of histological damage.
Lipoproteins	Pre-β-lipoproteins are increased in fatty liver; in severe liver damage (cirrhosis) the pre-β band disappears almost completely, and often lipoproteins migrate as a single wide band.
Nonspecific laboratory findings	Anemia, leukocytosis, and hyperbilirubinemia are common nonspecific findings in alcoholic hepatitis and cirrhosis.

[a]Modified from Lieber and Salaspuro (1985).

further biopsies are not indicated unless a complication such as hepatic carcinoma is suspected. In cases of fatty liver and alcoholic hepatitis, liver biopsy should be repeated once a year if the drinking habits cannot be controlled or the history is unreliable. Changes in laboratory patterns are crucial for making a decision.

7.5. Prognosis and Treatment of Alcoholic Liver Injury

Several studies have demonstrated that the prognosis for acute and chronic alcoholic liver disease is dramatically improved by abstinence from alcohol, whereas continued consumption is associated with increased morbidity and mortality.

7.5.1. Fatty Liver

Alcoholic steatosis is completely reversible in most cases. In some extremely severe cases alcoholic fatty liver may have a fatal outcome (Randall, 1980a,b), but as a rule even those cases needing hospitalization improve within a few days or weeks after cessation of alcohol consumption. In contrast to alcoholic hepatitis, laboratory abnormalities rapidly return to normal levels, and hepatomegaly decreases (Leevy, 1962; Christoffersen and Nielsen, 1972; Devenyi et al., 1970). Fat mobilization may be delayed only if steatosis has been extreme, with fatty cysts, or if it is complicated by obesity or prediabetes.

Other than alcohol withdrawal and the provision of a diet with sufficient calories, vitamins, and protein, generally no treatment is needed. The addition of lipotropes such as choline to the diet has no beneficial value in this condition (Post et al., 1952; Rubin and Lieber, 1968). Anabolic steroids have been found to be efficacious in the treatment of alcoholic fatty liver by Jabbari and Leevy (1967) and Mendenhall (1968) but not by Fenster (1966). On the other hand, these drugs may be somewhat contraindicated (when it comes to the treatment of a relatively benign condition such as fatty liver) because of their putative hepatotoxicity. Various experimental modalities for the treatment of fatty liver are listed in Chapter 4. Thus far they have had no clinical application.

7.5.2. Alcoholic Fibrosis

The treatment is similar to that of alcoholic fatty liver except that efforts at enforcing abstinence should be particularly vigorous in view of the prognostic implications outlined under Section 7.5.5.

7.5.3. Alcoholic Hepatitis

The survival of patients with alcoholic hepatitis is dramatically improved by the discontinuation or reduction of alcohol consumption. Galambos (1974) reported that the 7-year survival is 50% with continued drinking and 80% with reduced drinking. This finding has been confirmed by others (Brunt et al., 1974). Persistence of alcoholic hepatitis is associated with confirmed drinking; progression to cirrhosis is uncommon if abstinence can be maintained (Galambos, 1972). The reported mortality rates for acute alcoholic hepatitis depend on the material selected. Early reports were drawn from autopsy material. Accordingly, an early mortality of two-thirds to three-quarters was reported by Phillips and Davidson (1954), and about 30% by Hardison and Lee (1966). More recently a mortality of from 1.5 to 8% in individuals with alcoholic hepatitis ill enough to require hospitalization but well enough to permit liver biopsy has been found (Harinasuta and Zimmerman, 1971; Green et al., 1963; Christoffersen et al., 1970; Lischner et al., 1971).

Clinical and pathological signs of alcoholic hepatitis regress within weeks or months in most cases. A poorer short-term prognosis can be expected in patients with a prolonged prothrombin time and ascites (Galambos, 1974). The long-term prognosis is unfavorably influenced by the presence of severe histological inflammation and fibrosis in the initial liver biopsy specimen (Galambos, 1974). Elevations of serum enzymes are less valuable prognostic indicators.

The general treatment of the patient with alcoholic hepatitis is the same as that for alcoholic steatosis. The role of adrenocorticosteroid therapy of acute alcoholic hepatitis has been the subject of debate for years. Helman et al. (1971), Lesesne et al. (1978), and Maddrey et al. (1978), in controlled prospective studies, reported significant improvement in survival of steroid-treated patients with encephalopathy but not in those with milder illness. Several other controlled studies, however, have not confirmed these findings (Campra et al., 1973; Depew et al., 1980; Porter et al., 1971; Schlicting et al., 1976; Shumaker et al., 1978). Indeed, in one of the controlled studies an increased risk of fungal infection associated with corticosteroid therapy was reported (Blitzer et al., 1977). In the intervening year, a cooperative study was completed to determine the efficacy of 30 days of treatment with either a glucocorticosteroid (prednisolone) or an anabolic steroid (oxandrolone) in moderate or severe alcoholic hepatitis. Although neither steroid improved short-term survival, oxandrolone therapy was associated with a beneficial effect on long-term survival (Mendenhall *et al.*, 1984).

At present, we do not recommend routine corticosteroid treatment for acute alcoholic hepatitis, though selected cases might be helped.

Propylthiouracil (PTU) has also been suggested for the treatment of alcoholic hepatitis (Orrego *et al.*, 1979), but lack of beneficial therapeutic effect has been reported by another group (Halle *et al.*, 1980). In a review by Szilagyi *et al.* (1983), it was concluded that the current state of knowledge does not allow unequivocal acceptance or rejection of the role of thyroid hormone and antithyroid medication in alcoholic hepatitis. A more recent long-term study reported a lowered mortality in the treated group (Orrego *et al.*, 1987). Curiously, the beneficial effect was not observed in patients with high alcohol consumption: it was restricted to those in whom alcohol intake was moderate, a group known to have a good outcome even without specific treatment. The mechanism of the beneficial effect is not clear. Propylthiouracil has been shown experimentally to protect against alcohol-induced hepatocellular necrosis in hypoxic conditions, but such hypoxic conditions were not documented in the type of patients in whom the beneficial effects were obtained. *d*-Penicillamine has been shown to suppress elevated values of hepatic collagen proline hydroxylase activity while ameliorating urinary excretion of peptide-bound hydroxyproline in patients with alcoholic hepatitis (Mezey *et al.*, 1979), and it has been suggested that it may be useful for the treatment of patients with active alcoholic liver injury (Resnick *et al.*, 1975). These findings await further confirmation. In patients with alcoholic hepatitis, anorexia may require parenteral alimentation, including infusion of amino acids. It is noteworthy that such parenteral alimentation provides no benefit in moderate alcoholic hepatitis; it did more rapidly improve morbidity (liver tests) and probably liver function in severe alcoholic hepatitis but did not improve early mortality (Simon and Galambos, 1988). This issue is discussed in greater detail in Chapter 17. Experimentally, supplementation with *S*-adenosylmethione was shown to correct the ethanol-induced depletion of this compound, the associated decrease in GSH, and general signs of toxicity such as increased circulatory GDH, AST, and ALT activities (Lieber *et al.*, 1990). Clinical applications are now awaited.

7.5.4. Alcoholic Cirrhosis

In one study, dealing presumably with end-stage cirrhosis, withdrawal from alcohol was not found to be beneficial (Soterakis *et al.*, 1973). In most other series, however, dealing with moderately advanced cirrhosis, control of alcohol intake prolonged survival. In patients with histologically proven cirrhosis, the 5-year survival rate for patients without jaundice, ascites, or hematemesis was 88.9% for abstainers and 62.8% for drinkers (Powell and Klatskin, 1968). Similar results were observed by Bode (1979) in Germany and by Orloff *et al.* (1977) in cirrhotics with portacaval shunts. The 5-year survival rates are unfavorably affected by jaundice, ascites, and hematemesis in abstainers (57.5%, 52.4%, and 35%, respectively) and in drinkers (33.3%, 32.7%, and 21%, respectively). The prognosis is still poor during the first year after the discontinuation of drinking. The survival rate of patients who live beyond the first year and who continue to abstain from alcohol is 86.2% for the subsequent 4 years, which is close to the expected survival for individuals of a similar age.

Once cirrhosis is established, the outcome varies widely. The significance of megamitochondria in alcoholic liver disease as a prognosticator of long-term survival has been studied by Chedid *et al.* (1986) in 220 patients. It was concluded that the patients with megamitochondria appeared to represent a subcategory of alcoholic hepatitis with a milder degree of clinical severity. Alternatively, the presence of megamitochondria may represent an earlier stage of the disease. Indeed, megamitochondria appear at the initial fatty liver stage (Lane and Lieber, 1966). Although they persist through early stages of fibrosis and cirrhosis (Floersheim and Bianchi, 1984), they appear to level off and even decrease with more severe disease. This is not to imply that the megamitochondrial lesion in and of itself is benign or even represents a favorable change. Functions of megamitochondria *per se* have not been assessed. Numerous studies, however, have documented that chronic ethanol consumption is associated with alterations of a variety of overall mitochondrial functions (see review by Lieber, 1985).

Corticosteroids did not prolong survival of patients with alcoholic cirrhosis in a controlled study performed by the Copenhagen Study Group for Liver Diseases (Tygstrup *et al.*, 1971; Schlicting *et al.*, 1976).

Colchicine, which inhibits collagen synthesis and procollagen secretion in embryonic tissue (Ehrlich *et al.*, 1974), may provide a new approach to the treatment of alcoholic liver injury (Kershenobich *et al.*, 1979b, 1988). However, as pointed out by Boyer and Ransohoff (1988), the study raises questions regarding differences in severity between treatment and placebo groups and the high dropout rate. Furthermore, the outcome in patients with alcohol cirrhosis was not clearly differentiated from that in an equal number of nonalcoholics. For all these reasons, further results of controlled trials would be desirable.

Prophylactic portasystemic shunting in unselected cirrhotic patients does not improve the overall survival, although the terminal event may be altered from hemorrhage to hepatic failure (Conn and Lindenmuth, 1968; Conn *et al.*, 1972; Jackson *et al.*, 1968; Resnick *et al.*, 1969). Treatment of gastrointestinal bleeding, a major complication in cirrhotics, is discussed in Chapter 10.

Experimentally, various compounds, such as polyunsaturated lecithin (Lieber *et al.*, 1990), prostaglandin derivatives (Ruwart *et al.*, 1988), and proline hydroxylase inhibitors (Fujiwara *et al.*, 1988) have been shown to delay collagen formation. None of these has had controlled clinical applications as yet.

Liver transplantation, originally not applied to alcoholic liver disease, is now increasingly being considered for individuals who have stopped drinking (Kumar *et al.*, 1990).

7.5.5. Prevention of Alcoholic Liver Disease

The main goal in the management of alcoholic liver disease is the withdrawal of alcohol. This is at present the only practical, effective way both to prevent the progression and to improve the prognosis. On this basis a clinician who treats alcoholic liver injury should not limit his efforts to the hepatic symptomatology but should refer his patient to those community organizations willing to take care of alcoholism problems. Furthermore, it should be borne in mind that changes in the overall consumption of alcoholic beverages bear a close relationship to liver cirrhosis mortality and that alcohol control measures can be used to limit consumption; thus, control of alcohol availability is becoming a public health issue (Bruun *et al.*, 1975).

Individual susceptibility to the development of cirrhosis varies. Therefore, it is of practical importance to recognize at an early, still reversible, stage those individuals prone to progress to cirrhosis. In this regard, recognition of perivenular fibrosis on liver biopsy is valuable. As discussed before, this lesion was usually described in association with full-blown alcoholic hepatitis, but it can occur even in the fatty liver stage in the absence of hepatitis (Edmondson *et al.*, 1967; Van Waes and Lieber, 1977a). Experimental studies in alcohol-fed baboons showed that in those animals that progressed to cirrhosis, perivenular fibrosis invariably occurred by the fatty liver stage; by contrast, animals that did not show the lesion did not progress beyond the stage of fatty liver. These experimental data suggest that in the baboon, perivenular fibrosis (PVF) is a common and early warning sign of impending cirrhosis if drinking continues (Van Waes and Lieber, 1977a). These studies have now been expanded by Worner and Lieber (1985) in humans (Fig. 7.17). Patients with PVF at the fatty liver forms were found to be likely to progress to more severe forms of alcoholic liver disease if they continued to consume alcohol. The concept of PVF as a precirrhotic lesion in alcoholics is further strengthened by similar observations in patients with fatty livers after bypass operation for morbid obesity (Marubbio *et al.*, 1976) and in diabetic individuals (Falchuck *et al.*, 1980). Perivenular fibrosis is associated with some interstitial pericellular fibrosis (Fig. 7.4) (Van Waes and Lieber, 1977a); the latter has been subsequently confirmed and emphasized (Nasrallah *et al.*, 1980). From a practical point of view, the observation of PVF in an alcoholic justifies an all-out effort at curtailing drinking. Because of its prognostic value, detection of PVF may be useful in patient management and in the allocation of scarce therapeutic resources.

To detect this precirrhotic lesion, liver biopsy is the only reliable method at present, but it has obvious limitations in its large scale applicability and does not necessarily reflect ongoing fibrogenic activity. As a noninvasive test to assess liver fibrosis, blood measurements of several substances, or enzymes involved in collagen metabolism such as proline, hydroxyproline, procollagen peptides, lactic acid, and propyl hydroxylase activity have been carried out. Among these tests, the radioimmunoassay of the amino-terminal peptides of type III procollagen (P-III-P) either with the original technique of Rohde *et al.* (1983) or some modification (Pierard *et al.*, 1984; Niemelä, 1985) appears to be the most commonly used. However, the assay for P-III-P failed to differentiate patients with fatty liver from those with fatty liver and early fibrosis (Savolainen *et al.*, 1984a). More recently, a modified P-III-P assay using Fab fragments of the antibody (Fab-P-III-P) has been developed and reported to detect degraded fragments of propeptides in addition to the intact propeptide for which the original P-III-P assay has a high affinity (Rohde *et al.*, 1983). Using the test as well as a sensitive radioimmunoassay for fragments of laminin, Sato *et al.* (1986b) found that serum P-III-P, Fab-P-III-P, and laminin concentrations rise significantly with the increasing degree of fibrosis in alcoholics. Values of Fab-P-III-P above 45 ng/ml detected the majority of patients with fibrosis, including 55% of the subjects with PVF, 62% with septal fibrosis, and more than 90% of the patients with cirrhosis. However, with abstinence, Fab-P-III-P levels increased in all alcoholics, and P-III-P values increased in patients with normal P-III-P values on admission. By contrast, the values of laminin decreased during abstinence. Therefore, to interpret serum levels of Fab-P-III-P, P-III-P, and laminin, the duration of abstinence must be taken into consideration: P-III-P, Fab-P-III-P, and laminin measurements in the serum within 1 week of abstinence can contribute to the detection of alcoholic liver disease and the determination of its stage (Nouchi *et al.*, 1987).

Interstitial collagens are synthesized intracellularly as procollagens that contain extension propeptides at the amino and carboxyl ends of their three peptide chains. After secretion, conversion of procollagen to collagen and assembly into fibrils occur. Although the amino-propeptides of type I procollagen are released at an early phase of fibrillogenesis, those of type III procollagen are retained in collagen fibrils in the skin, even at a mature stage (Fleischmajer *et al.*, 1983). Study by Sato *et al.* (1986a) demonstrated that the same difference in the fate of the peptides pertains to the liver. The aminopropeptides were localized along the collagen fibrils around terminal hepatic venules, in the space of Disse, and in portal tracts of control baboons as well as of alcohol-fed animals. There was a characteristic periodicity of 60–70 nm, corresponding to the D (67 nm) collagen fibrils (Fig. 7.23). The presence of propeptides in the fibrils is of relevance for the measurements of their fragments in the serum. As a reflection of the concept that the peptides are cleaved from procollagen on extrusion from the cell, it had been proposed that measurement of circulating peptides may reflect the process of collagen synthesis. The documentation that the peptides are part of the collagen fibers indicates that their appearance in the blood reflects not only synthesis but also breakdown. Such a process might explain why the elevated serum levels observed in patients with liver diseases correlate not only with the degree of hepatic fibrosis but also with the amount of inflammation (Savolainen *et al.*, 1984a; Nouchi *et al.*, 1986).

FIGURE 7.23. Collagen fibrils with aminopropeptides of type III procollagen in perivenular fibrosis of an alcohol-fed baboon. Reaction products are localized periodically along the fibers. ×28,200 (From Sato *et al.*, 1986a.)

REFERENCES

Agostini, A., Vergani, C., Stabilini, R., and Marasini, B.: Determination of seven serum proteins in alcoholic cirrhosis. *Clin. Chim. Acta* **26**:351–355, 1969.

Ainley, C. C., Senapati, A., Brown, I. M. H., Iles, C. A., Slavin, B. M., Mitchell, W. D., Davies, D. R., Keeling, P. W. N., and Thompson, R. P. H.: Is alcohol hepatotoxic in the baboon? *Hepatology* **7**:85–92, 1988.

Albano, E., Tomasi, A., Goria-Gatti, L., and Dianzani, M. U.: Spin trapping of free radical species produced during the microsomal metabolism of ethanol. *Chem. Biol. Interact.* **65**:223–234, 1988.

Alexander, J. F., Lischner, M. W., and Galambos, J. T.: Natural history of alcoholic hepatitis. II. The long term prognosis. *Am. J. Gastroenterol.* **56**:515–525, 1971.

Aron, E., Paoletti, C., Jobard, P., and Gosse, C.: Recherches expérimentales sur la cytosiderose au vin. *Arch. Mal. Appar. Dig.* **50**:745–750, 1961.

Bailey, R. J., Krasner, N., Eddleston, A. L. W. F., Williams, R., Tee, D. E. H., Doniach, D., Kennedy, L. A., and Batchelor, J. R.: Histocompatibility antigens, autoantibodies, and immunoglobulins in alcoholic liver disease. *Br. Med. J.* **2**:727–729, 1976.

Ballard, H., Bernstein, M., and Farrar, J. R.: Fatty liver presenting as obstructive jaundice. *Am. J. Med.* **30**:196–201, 1961.

Baraona, E., Leo, M. A., Borowsky, S. A., and Lieber, C. S.: Alcoholic hepatomegaly: Accumulation of protein in the liver. *Science* **190**:794–795, 1975.

Baraona, E., Leo, M. A., Borowsky, S. A., and Lieber, C. S.: Pathogenesis of alcohol-induced accumulation of protein in the liver. *J. Clin. Invest.* **50**:546–554, 1977.

Barclay, L. R.: The cooperative antioxidant role of glutathione with a lipid-soluble and a water-soluble antioxidant during peroxidation of liposomes initiated in the aqueous phase and in the lipid phase. *J. Biol. Chem.* **263**:16138–16142, 1988.

Barry, M.: Iron and chronic liver disease. *J. R. Coll. Physicians London* **8**:52–62, 1973.

Baskin, G., Henderson, G. I., and Schenker, S.: Ethanol and hepatic regeneration. *Hepatology* **8**:408–411, 1988.

Beckett, A. G., Livingston, A. V., and Hill, K. R.: Acute alcoholic hepatitis. *Br. Med. J.* **ii**:1113–1119, 1961.

Beckett, A. G., Livingston, A. V., and Hill, K. R.: Acute alcoholic hepatitis. *Br. Med. J.* **ii**:675–679, 1962.

Bell, E. T.: Relation of portal cirrhosis to hemochromatosis and to diabetes mellitus. *Diabetes* **4**:435–446, 1955.

Bell, H., and Nordhagen, R.: Association between HLA-BW40 and alcoholic liver disease with cirrhosis. *Br. Med. J.* **1**:822, 1978.

Bell, H., and Nordhagen, R.: HLA antigens in alcoholics, with special reference to alcoholic cirrhosis. *Scand. J. Gastroenterol.* **15**:453–456, 1980.

Bhagwandeen, B. S., Apte, M., Manwarring, L., and Dickeson, J.: Endotoxin induced hepatic necrosis in rats on an alcohol diet. *J. Pathol.* **151**:47–53, 1987.

Biava, C.: Mallory alcoholic hyalin: A heretofore unique lesion of hepatocellular ergastoplasm. *Lab. Invest.* **13**:301–320, 1964.

Birschbach, H. R., Harinasuta, U., and Zimmerman, H. J.: Alcoholic steatonecrosis. II. Prospective study of prevalence of Mallory bodies in biopsy specimens and comparison of severity of hepatic disease in patients with and without this histological feature. *Gastroenterology* **66**:1195–1202, 1974.

Bjorneboe, G. E. A., Bjorneboe, A., Hagen, B. F., Mørland, J., and Drevon, C. A.: Reduced hepatic α-tocopherol content after long-term administration of ethanol to rats. *Biochim. Biophys. Acta* **918**:236–241, 1987.

Bjorneboe, G. E. A., Johnsen, J., Bjorneboe, A., Marklund, S. L., et al.: Some aspects of antioxidant status in blood from alcoholics. *Alcoholism* **12**:806–810, 1988.

Blendis, L. M., Orrego, H., Crossley, I. R., Blake, J. E., Medline, A., and Israel, Y.: The role of hepatocyte enlargement in hepatic pressure in cirrhotic and noncirrhotic alcoholic liver disease. *Hepatology* **2**:539–546, 1982.

Blitzer, B. L., Mutchnick, M. G., Joshi, P. H., Phillips, M. M., Fessell, M., and Conn, H. O.: Adrenocorticosteroid therapy in alcoholic hepatitis: A prospective double-blind randomized study. *Am. J. Dig. Dis.* **22**:477–484, 1977.

Blomhoff, J. P., Skrede, S., and Ritland, S.: Lecithin: Cholesterol acyltransferase and plasma proteins in liver diseases. *Clin. Chim. Acta* **53**:197–207, 1974.

Bode, J. C.: Lebenserwartung bei Leberschäden durch Alkoholabusus. *Lebensversicher. Med.* **6**:159–162, 1979.

Bode, C., Kugler, V., and Bode, J. C.: Endotoxemia in patients with alcoholic and non-alcoholic cirrhosis and in subjects with no evidence of chronic liver disease following acute alcohol excess. *J. Hepatology* **4**:8–14, 1987.

Bonjour, J. P.: Vitamins and alcoholism. *Int. J. Vit. Nutr. Res.* **49**:434–441, 1979.

Borenfreund, E., Higgins, P. J., and Peterson, E.: Intermediate-sized filaments in cultured rat liver tumor cells with Mallory body-like cytoplasm abnormalities. *J. Natl. Cancer Inst.* **64**:232–333, 1980.

Borowsky, S. A., Perlow, W., Baraona, E., and Lieber, C. S.: Relationship of alcoholic hypertriglyceridemia to stage of liver disease and dietary lipid. *Dig. Dis. Sci.* **25**:22–27, 1980.

Bosma, A., Brouwer, A., Seifert, F. W., and Knook, L. D.: Synergism between ethanol and carbon tetrachloride in the generation of liver fibrosis. *J. Pathol.* **156**:15–21, 1988.

Bothwell, T. H., Abrahams, C., Bradlow, B. A., and Charlton, R. W.: Idiopathic and banthu hemochromatosis. Comparative histological study. *Arch. Path.* **79**:163–168, 1965.

Boyer, L. J., and Ransohoff, F. D.: Is colchicine effective therapy for cirrhosis? *N. Engl. J. Med.* **318**:1751–1752, 1988.

Bradley, S. E., Ingelfinger, F. J., and Bradley, G. P.: Hepatic circulation in cirrhosis of the liver. *Circulation* **5**:419–429, 1952.

Bradus, S., Korn, R. J., Chomet, B., and West, M.: Hepatic function and serum enzyme levels in association with fatty metamorphosis of the liver. *Am. J. Med. Sci.* **246**:35–41, 1963.

Brenner, R. R.: Effects of unsaturated fatty acids on membrane structure and enzyme kinetics. *Prog. Lipid Res.* **23**:69–96, 1984.

Brenner, D. A., and Chojkier, M.: Acetaldehyde increases collagen gene transcription in cultured human fibroblasts. *J. Biol. Chem.* **262:**17690–17695, 1987.

Brunt, P. W., Kew, M. C., Scheuer, P. J., and Sherlock, S.: Studies in alcoholic liver disease in Britain. I. Clinical and pathological patterns related to natural history. *Gut* **15:**52–58, 1974.

Bruun, K., Koura, E., Popham, R. E., and Seeley, J. R.: *Liver Cirrhosis Mortality as a Means to Measure the Prevalence of Alcoholism.* Helsinki, The Finnish Foundation for Alcohol Studies, 1960.

Bruun, K., Edwards, G., Lumio, M., Mäkelä, K., Pan, L., Popham, R. E., Room, R., Schmidt, W., Skog, O. J., Sulkunen, P., and Österberg, E.: *Alcohol Control Policies in Public Health Perspective. Report of an International Working Group.* Helsinki, The Finnish Foundation for Alcohol Studies, 1975.

Bunyan, J., Cawthrone, M. A., Diplock, A. T., and Green, J.: Vitamin E and hepatotoxic agents. 2. Lipid peroxidation and poisoning with orotic acid, ethanol and thioacetamide in rats. *Br. J. Nutr.* **23:**309–317, 1969.

Campra, J. L., Hamlin, E. M., Kirschbaum, R. J., Olivier, M., Redecker, A. J., and Reynolds, T. B.: Prednisone therapy of acute alcoholic hepatitis. Report of a controlled trial. *Ann. Intern. Med.* **79:**625–631, 1973.

Carmichael, F. J., Saldivia, V., Israel, Y., McKaigney, J. P., and Orrego, H.: Ethanol-induced increase in portal hepatic blood flow: Interference by anesthetic agents. *Hepatology* **7:**89–94, 1987.

Cederbaum, A. I., and Rubin, E.: Protective effect of cysteine on the inhibition of mitochondrial function by acetaldehyde. *Biochem. Pharmacol.* **25:**963–973, 1976.

Cederbaum, A. I., Lieber, C. S., and Rubin, E.: Effects of chronic ethanol treatment on mitochondrial functions. *Arch. Biochem. Biophys.* **165:**560–569, 1974a.

Cederbaum, A. I., Lieber, C. S., and Rubin, E.: The effect of acetaldehyde on mitochondrial function. *Arch. Biochem. Biophys.* **161:**26–39, 1974b.

Celada, A., Rudolf, H., and Donath, A.: Effect of a single ingestion of alcohol on iron absorption. *Am. J. Hematol.* **5:**225–237, 1978.

Charlton, R. W., Jacobs, P., Sefiel, H., and Bothwell, T. H.: Effect of alcohol on iron absorption. *Br. Med. J.* **2:**1427–1429, 1964.

Chedid, A., Mendenhall, C. L., Tosch, T., Chen, T., Rabin, L., Carcia-Pont, P., Goldberg, S. J., Kiernan, T., Seeff, L. B., Sorrell, M., Tamburro, C., Weesner, R. E., and Zetterman, R.: The Veterans Administration Cooperative Study of Alcoholic Hepatitis: Significance of megamitochondria in alcoholic liver disease. *Gastroenterology* **90:**1858–1864, 1986.

Chen, T. S. N., and Leevy, C. M.: Collagen biosynthesis in liver disease of the alcoholic. *J. Lab. Clin. Med.* **85:**103–112, 1975.

Chin, J. H., and Goldstein, D. B.: Drug tolerance in biomembranes: A spin label study of the effect of ethanol. *Science* **196:**684–685, 1977.

Christoffersen, P., and Juhl, E.: Mallory bodies in liver biopsies with fatty changes but no cirrhosis. *Acta Pathol. Microbiol. Scand.* **79:**201–207, 1971.

Christoffersen, P., and Nielsen, K.: The frequency of Mallory

bodies in liver biopsies from chronic alcoholics. *Acta Pathol. Microbiol. Scand.* **79:**274–278, 1971.

Christoffersen, P., and Nielsen, K.: Histological changes in human liver biopsies from chronic alcoholics. *Acta Pathol. Microbiol. Scand.* **80:**557–565, 1972.

Christoffersen, P., Iversen, K., and Nielsen, K.: A comparative study of two groups of patients with Mallory bodies with and without liver cell necrosis and neutrophilic infiltration. *Scand. J. Gastroenterol.* **5:**633–638, 1970.

Christoffersen, P., Brendstrup, O., Juhl, E., and Poulsen, H.: Lipogranulomas in human liver biopsies with fatty change. *Acta Pathol. Microbiol. Scand.* **79:**150–158, 1971.

Chvapil, M., and Ryan, J. N.: The pool of free proline in acute and chronic liver injury and its effect on the synthesis of collagen-globular protein. *Agents Actions* **3:**38–44, 1973.

Clement, B., Grimaud, J. A., Campion, J. P., Deugnier, Y., and Guillouzo, A.: Cell types involved in collagen and fibronectin production in normal and fibrotic human liver. *Hepatology* **6:**225–234, 1986.

Cohen, J. A., and Kaplan, M. M.: The SGOT/SGPT ratio: An indicator of alcoholic liver disease. *Dig. Dis. Sci.* **24:**835–838, 1979.

Collins, M. A., Gordon, R., Bidgeli, M. G., and Rubenstein, J A.: Pyrogallol potentiates acetaldehyde blood levels during ethanol oxidation in rats. *Chem. Biol. Interact.* **8:**127–130, 1974.

Comporti, M., Burdino, E., and Raja, F.: Fatty acids composition of mitochondrial and microsomal lipids of rat liver after acute ethanol intoxication. *Life Sci.* **10**(Part II):855–866, 1971.

Conn, H. O., and Lindenmuth, W. W.: Prophylactic portacaval anastomosis in cirrhotic patients with esophageal varices. *N. Engl. J. Med.* **279:**725–732, 1968.

Conn, H. O., Schreiber, W. S., and Etkington, S. G.: Cirrhosis and diabetes. II. Association of impaired glucose tolerance with portal systemic shunting in Laennec's cirrhosis. *Am. J. Dig. Dis.* **16:**227–239, 1971.

Conn, H. O., Lindenmuth, W. W., May, C., and Ramsby, G. R.: Prophylactic portacaval anastomosis: A tale of two studies. *Medicine* **51:**27–40, 1972.

Craig, J.: Effects of ethanol and ethionine on DNA synthesis during experimental liver regeneration. *J. Stud. Alcohol* **36:**148–157, 1975.

Craig, R. M., Neumann, T., Jeejeebhoy, K. N., and Yokoo, H.: Severe hepatocellular reaction resembling alcoholic hepatitis with cirrhosis after massive small bowel resection and prolonged total parenteral nutrition. *Gastroenterology* **79:**131–137, 1980.

Debey, H. J., Mackenzie, J. B., and Mackenzie, C. G.: The replacement by thiazolidinecarboxylic acid of exogenous cystine and cysteine. *J. Nutr.* **66:**607–619, 1958.

De Groot, H., Littauer, A., Hugo-Wissemann, D., Wissemann, P., and Noll, T.: Lipid peroxidation and cell viability in isolated hepatocytes in a redesigned oxystat system: Evaluation of the hypothesis that lipid peroxidation preferentially induced at low oxygen partial pressures, is decisive for CCl_4 liver cell injury. *Arch. Biochem. Biophys.* **264:**591–599, 1988.

De la Iglesia, F. A., Porta, E. A., and Hartroft, W. S.: Effects of dietary protein levels on the Saimiri Sciureus. *Exp. Mol. Pathol.* **7:**182–195, 1967.

Denk, H., and Eckerstorfer, R.: Colchicine-induced Mallory body formation in the mouse. *Lab. Invest.* **36:**563–565, 1977.

Denk, H., Franke, W. W., Eckerstorfer, R., Schmid, E., and Kerjascki, D.: Formation and involution of Mallory bodies ("alcoholic hyalin") in murine and human liver revealed by immunofluorescence microscopy with antibodies to pre-keratin. *Proc. Natl. Acad. Sci. USA* **76:**4112–4116, 1979.

Depew, W., Boyer, T., Omata, M., Redeker, A., and Reynolds, T.: Double-blind controlled trial of prednisolone therapy in patients with severe acute alcoholic hepatitis and spontaneous encephalopathy. *Gastroenterology* **78:**524–529, 1980.

Devenyi, P., Rutherdale, J., Sereny, G., and Olin, J. S.: Clinical diagnosis of alcoholic fatty liver. *Am. J. Gastroenterol.* **54:**597–602, 1970.

Dicker, E., and Cederbaum, A. I.: Increased oxygen radical-dependent inactivation of metabolic enzymes by liver microsomes after chronic ethanol consumption. *FASEB J.* **2:**2901–2906, 1988.

Diegelmann, R. F., Guzelian, P. S., Gay, R., and Gay, S.: Collagen formation by the hepatocyte in primary monolayer culture and *in vivo. Science* **219:**1343–1345, 1983.

Diehl, A. M., Chacon, M., and Wagner, P.: The effect of chronic ethanol feeding on ornithine decarboxylase activity and liver regeneration. *Hepatology* **8:**237–242, 1988.

DiLuzio, N. R.: Comparative study of the effect of alcoholic beverages on the development of the acute ethanol-induced fatty liver. *Q. J. Stud. Alcohol* **23:**557–561, 1962.

DiLuzio, N. R., and Hartman, A. D.: Role of lipid peroxidation on the pathogenesis of the ethanol-induced fatty liver. *Fed. Proc.* **26:**1436–1442, 1967.

DiLuzio, N. R., and Stege, T. E.: The role of ethanol metabolites in hepatic lipid peroxidation. In: *Alcohol and the Liver*, Vol. 3 (M. M. Fisher and J. G. Rankin, eds.), New York, Plenum, pp. 45–62, 1977.

Dworkin, B., Rosenthal, W. S., Jankowski, R. H., Gordon, G. G., and Haldea, D.: Low blood selenium levels in alcoholics with and without advanced liver disease. *Dig. Dis. Sci.* **30:**838–844, 1985.

Edmondson, H. A., Peters, R. L., Reynolds, T. B., and Kuzma, O. T.: Sclerosing hyaline necrosis of the liver in the chronic alcoholic. *Ann. Intern. Med.* **59:**646–673, 1963.

Edmondson, H. A., Peters, R. L., Frankel, H. H., and Borowsky, S.: The early stage of liver injury in the alcoholic. *Medicine* **46:**119–129, 1967.

Ehrlich, H. P., Ross, R., and Bornstein, P.: Effects of anti-microtubular agents on the secretion of collagen. *J. Cell Biol.* **62:**390–450, 1974.

Emmrich, R., Haentschel, H. J., and Haentschel, H.: Klinische Bedeutung der Hydroxyprolineauscheidung in Harn. *J. Anat. Med.* **22:**193–196, 1967.

Espina, N., Lima, V., Lieber, C. S., and Garro, A. J.: *In vitro* and *in vivo* inhibitory effect of ethanol and acetaldehyde on O^6 methylguanine transferase. *Carcinogenesis* **9:**761–766, 1988.

Falchuck, K. R., Fiske, S., Federman, C., Trey, C., and Haggitt, R. C.: Pericentral hepatic fibrosis and intracellular hyaline in diabetes mellitus. *Gastroenterology* **78:**535–541, 1980.

Farrell, G. C., Bathal, P. S., and Powell, L. W.: Abnormal liver function in chronic hypervitaminosis A. *Dig. Dis. Sci.* **22:**724–728, 1977.

Farrell, G. C., Cooksley, W. G. E., Hart, P., and Powell, L. W.: Drug metabolism in liver disease. Identification of patients with impaired hepatic drug metabolism. *Gastroenterology* **75:**580–588, 1978.

Farris, W., and Erdman, J.: Protracted hypervitaminosis A following long term low-level intake. *J.A.M.A.* **247:**1317, 1982.

Feinman, L., and Lieber, C. S.: Hepatic collagen metabolism: Effect of alcohol consumption in rats and baboons. *Science* **176:**795, 1972.

Feizi, T.: Immunoglobulins in chronic liver disease. *Gut* **9:**193–198, 1968.

Fenster, L. F.: The nonefficacy of short-term anabolic steroid therapy in alcoholic liver disease. *Ann. Intern. Med.* **65:**738–744, 1966.

Flax, M. H., and Tisdale, W. A.: An electron microscopic study of alcoholic hyalin. *Am. J. Pathol.* **44:**441–453, 1964.

Fleischmajer, R., Olsen, B. R., Timpl, R., Perlish, J. S., and Lovelace, O.: Collagen fibril formation during embryogenesis. *Proc. Natl. Acad. Sci. U.S.A.* **80:**3354–3358, 1983.

Floersheim, G. L., and Bianchi, L.: Ethanol diminishes the toxicity of the mushroom *Amanita phalloides. Experientia* **40:**1268–1270, 1984.

Fong, T. L., Govindarajan, S., Valinluck, B., and Redeker, G. A.: Status of hepatitis B virus DNA in alcoholic disease: A study of a large urban population in the United States. *Hepatology* **8:**1602–1604, 1988.

Forsander, O. A., Pikkarainen, J. A. J., and Salaspuro, M.: A high hepatic concentration of free proline does not induce collagen synthesis in rat liver. *Hepato-Gastroenterology* **30:**6–8, 1983.

Foy, H., Kondi, A., and Mbaya, Y.: Effect of riboflavine deficiency on bone marrow function and protein metabolism in baboons. *Br. J. Nutr.* **18:**307–318, 1964.

Frank, H., Heil, W., and Leodolter, I.: Leber und Bierkonsum vergleichende Untersuchungen an 450 Arbeitern. *Müench. Med. Wochenschr.* **109:**892–897, 1967.

Frank, W. O., Rayes, A. N., Washington, A., and Holt, P. R.: Effect of ethanol administration upon hepatic regeneration. *J. Lab. Clin. Med.* **93:**402–413, 1979.

Franke, W. W., Denk, H., Schmid, E., Osborn, M., and Weber, K.: Ultrastructural, biochemical, and immunologic characterization of Mallory bodies in livers of griseofulvin-treated mice. *Lab. Invest.* **40:**207–220, 1979.

French, S. W.: The Mallory body: Structure, composition and pathogenesis. *Hepatology* **1:**76–83, 1981.

French, S. W., and Davies, P. L.: The Mallory body in the pathogenesis of alcoholic liver disease. In: *Proceedings of the International Symposia on Alcohol and Drug Research, Alcoholic Liver Pathology* (J. M. Khanna, Y. Israel, and H. Kalant, eds.), Toronto, Addiction Research Foundation, pp. 113–144, 1975.

French, S. W., Sim, J. S., Franks, K. E., Burbige, E. J., Denton, T., and Caldwell, M. G.: Alcoholic hepatitis. In: *Alcohol and the Liver* (M. M. Fisher and J. G. Rankin, eds.), New York, Plenum, pp. 261–186, 1976.

French, S. W., Benson, N. C., and Sun, P. S.: Centrilobular liver necrosis induced by hypoxia in chronic ethanol-fed rats. *Hepatology* **4**:912–917, 1984.

Freund, G.: Alcohol, barbiturate, and bromide withdrawal syndrome in mice. In: *Recent Advances in Studies of Alcoholism*. U. S. Dept. of Health, Education and Welfare, Health Services and Mental Health Administration, National Institute of Mental Health, National Institute on Alcohol Abuse and Alcoholism. Washington, D. C., U.S. Government Printing Office, p. 453, 1971.

Fujiwara, K., Ogata, I., Ohta, Y., Hayashi, S., Mishiro, S., Takatsuki, K., Sato, Y., Yamada, S., Hirata, K., Oka, H., Oda, T., Kawaji, H., Matsuda, S., Niyamo, Y., and Tsukudd, R.: Decreased collagen accumulation by a propyl hydroxylase inhibitor in pig serum-induced fibrotic rat liver. *Hepatology* **8**:804–807, 1988.

Gadeholt, G., Aarbakke, J., Dybing, E., Sjoblom, J., and Mørland, J.: Hepatic microsomal drug metabolism, glutamyl transferase activity and *in vitro* antipyrine half-life in rats chronically fed an ethanol diet, a control diet and a chow diet. *J. Pharmacol. Exp. Ther.* **213**:196–203, 1980.

Galambos, J. T.: Natural history of alcoholic hepatitis. III. Histological changes. *Gastroenterology* **63**:1026–1035, 1972.

Galambos, J. T.: Alcoholic hepatitis. In: *The Liver and Its Diseases* (F. Schaffner, S. Sherlock, and C. M. Leevy, eds.), New York, Intercontinental Medical Book, pp. 255–267, 1974.

Galambos, J. T., and Shapiro, R.: Natural history of alcoholic hepatitis. IV. Glycosaminoglycuronans and collagen in the hepatic connective tissue. *J. Clin. Invest.* **52**:2952–2962, 1973.

Galambos, J. T., Hollingsworth, M. A., Falek, A., Warren, W. D., and McCain, J. R.: The rate of synthesis of glycosaminoglycans and collagen by fibroblasts cultured from adult human liver biopsies. *J. Clin. Invest.* **60**:107–114, 1977.

Gall, E. A.: Primary and metastic carcinoma of liver relationship to hepatic cirrhosis. *Arch. Pathol.* **70**:226–232, 1960.

Garagliano, C. F., Mendeloff, A. I., and Lilienfeld, A. M.: Incidence of liver cirrhosis in 16 areas of the U.S. *Gastroenterology* **72**:1060, 1977.

Gerber, M. A., and Popper, H.: Relation between central canals and portal tracts in alcoholic hepatitis: A contribution to the pathogenesis of cirrhosis in alcoholics. *Hum. Pathol.* **3**:199–207, 1972.

Gerber, M. A., Orr, W., Denk, H., Schaffner, F., and Popper, H.: Hepatocellular hyalin in cholestasis and cirrhosis: Its diagnostic significance. *Gastroenterology* **64**:89–98, 1973.

Gluud, C., Aldershvile, J., Dietrichson, O., Hardt, F., Iversen, K., Juhl, E., Nielsen, J. O., Ryder, L. P., Kinhøj, P., and Svejgaard, A.: Human leucocyte antigens in patients with alcoholic liver cirrhosis. *Scand. J. Gastroenterol.* **15**:337–341, 1980.

Goldberg, M, and Thompson, C. M.: Acute fatty metamorphosis of the liver. *Ann. Intern. Med.* **55**:416–432, 1961.

Goldberg, S. J., Mendenhall, C. L., Connell, A., and Chedid, A.: "Nonalcoholic" chronic hepatitis in the alcoholic. *Gastroenterology* **72**:598–604, 1977.

Gonzalez-Calvin, J. L., Saunders, J. B., and Williams, R.: Effects of ethanol and acetaldehyde on hepatitic plasma membrane ATPases. *Biochem. Pharmacol.* **32**:1723–1728, 1983.

Goodwin, D. W.: Is alcoholism hereditary? *Arch. Gen. Psychiatry* **25**:545–549, 1971.

Gordon, E. R.: Mitochondrial functions in an ethanol-induced fatty liver. *J. Biol. Chem.* **248**:8271–8280, 1973.

Gordon, E. R., Rochman, J., Aria, M., and Lieber, C. S.: Lack of correlation between hepatic mitochondrial membrane structure and functions in ethanol-fed rats. *Science* **216**:1319–1321, 1982.

Grace, N. D., and Powell, L. W.: Iron storage disorders of the liver. *Gastroenterology* **64**:1257–1283, 1974.

Grant, B. F., Zobeck, T. S., and Ng, M.-J.-C.: *Liver cirrhosis mortality in the United States, 1971–85*, Surveillance Report #8, USDHHS, PHS, Rockville, Maryland, June 1988.

Green, J., Mistilis, S., and Schiff, L.: Acute alcoholic hepatitis. *Arch. Intern. Med.* **112**:67–78, 1963.

Greenwood, S. M., Leffler, C. T., and Minkowitz, S.: The increased mortality rate of open liver biopsy in alcoholic hepatitis. *Surg. Gyn. Obstet.* **134**:600–604, 1972.

Gregory, D. H., and Levi, D. F.: The clinical pathologic spectrum of alcoholic hepatitis. *Am. J. Dig. Dis.* **17**:479–488, 1972.

Gregory, D. H., Vlahcevic, Z. R., Prugh, M. F., and Swell, L.: Mechanism of secretion of biliary lipids: Role of a microtubular system in hepatocellular transport of biliary lipids in the rat. *Gastroenterology* **74**:93–100, 1978.

Hahn, E., Wick, G., Pencev, D., and Timple, R.: Distribution of basement membrane proteins in normal and fibrotic human liver: Collagen type IV, laminin and fibronectin. *Gut* **21**:63–71, 1980.

Hahn, E. G., Timpl, R., and Lieber, C. S.: Distribution of hepatic collagens, elastin, and structural glycoproteins during the development of alcoholic liver injury in baboons. In: *Biological Approach to Alcoholism: Update* (C. S. Lieber, ed.), Research Monograph-11, DDHS Publications No. (ADM) 83-1261, Washington, D. C., Superintendent of Documents, U.S. Government Printing Office, pp. 403–408, 1983.

Häkkinen, H.-M., and Kulonen, E.: Effect of ethanol on the metabolism of alanine, glutamic acid, and proline in rat liver. *Biochem. Pharmacol.* **24**:199–204, 1975.

Hall, E. M., and Ophuls, W.: Progressive alcoholic cirrhosis: Report of four cases. *Am. J. Pathol.* **1**:477–493, 1925.

Halle, P., Pare, P., Kapstein, E., Kanel, G., Redeker, A. G., and Reynolds, T. B.: Propylthiouracil therapy in severe acute alcoholic hepatitis. *Gastroenterology* **79**:1024, 1980.

Hällen, J., and Krook H.: Follow-up studies on an unselected ten-year material of 360 patients with liver cirrhosis in one community. *Acta Med. Scand.* **173**:479–493, 1963.

Hamalyn, A. N., Brown, A. J., Sherlock, S., and Baron, D. N.: Casual blood ethanol estimations in patients with chronic liver disease. *Lancet* **2**:345–347, 1975.

Hardison, W. G., and Lee, F. I.: Prognosis in acute liver disease of the alcoholic patient. *N. Engl. J. Med.* **275**:61–66, 1966.

Harinasuta, U., and Zimmerman, H. J.: Alcoholic steatonecrosis. I. Relationship between severity of hepatic disease and presence of Mallory bodies in the liver. *Gastroenterology* **60**:1036–1046, 1971.

Harinasuta, U., Chomet, B., Ishak, K., and Zimmerman, H. J.: Steatonecrosis: Mallory body type. *Medicine* **46**:141–162, 1967.

Hashimoto, S., and Recknagel, R. O.: No chemical evidence of hepatic lipid peroxidation in acute ethanol toxicity. *Exp. Mol. Pathol.* **8:** 225–242, 1968.

Hasumura, Y., Teschke, R., and Lieber, C. S.: Increased carbon tetrachloride hepatotoxicity and its mechanism, after chronic ethanol consumption. *Gastroenterology* **66:**415–422, 1974.

Hasumura, Y., Teschke, R., and Lieber, C. S.: Acetaldehyde oxidation by hepatic mitochondria: Its decrease after chronic ethanol consumption. *Science* **189:**727–729, 1975.

Hasumura, Y., Teschke, R., and Lieber, C. S.: Characteristics of acetaldehyde oxidation in rat liver mitochondria. *J. Biol. Chem.* **251:**4908–4913, 1976.

Hasumura, Y., Minato, Y., Nishimura, M., Kaku, Y., Tozuka, S., Koyama, W., Tanaka, Y., and Takeuchi, J.: Hepatic fibrosis in alcoholics: Morphologic characteristics, clinical diagnosis, and natural course. *Pathobiol. Hepat. Fibrosis* **7:**13–24, 1985.

Hay, E. D., Hasty, D. L., and Keihanau, K. L.: Morphological investigation of fibers derived from various types: Fine structure of collagens and their relationship to glycosaminoglycans (GAG). In: *Collagen-Platelet Interaction* (H. Gastpar, K. Kühn, and R. Marx, eds.), New York, Stuttgart-Schattauer Verlag, pp. 129–163, 1978.

Hayashi, N., Kasahara, A., Kurosawa, K., Sasaki, Y., Fusamoto, H., Sato, N., and Kamada, T.: Oxygen supply to the liver in patients with alcoholic liver disease assessed by organ-reflectance spectrophotometry. *Gastroenterology* **88:**881–886, 1985.

Hedlund, S. G., and Kiessling, K.-H.: The physiological mechanism involved in hangover. I. The oxidation of some lower aliphatic fusel alcohols and aldehydes in rat liver and their effect on the mitochondrial oxidation of various substrates. *Acta Pharmacol. Toxicol.* **27:**381, 1969.

Helman, R. A., Temko, M. H., Nye, S. W., and Fallon, H. J.: Alcoholic hepatitis: Natural history and evaluation of predinisolone therapy. *Ann. Intern. Med.* **74:**311–321, 1971.

Henley, K. S., and Laughrey, E. G.: Ethanol induced increase in neutral extractable hydroxyproline containing peptides, (NEHYP), in rat liver. *Alcoholism: Clin. Exp. Res.* **3:**272, 1979.

Henley, K. S., Laughrey, E. G., Appelman, H. D., and Flecker, K.: Effect of ethanol on collagen formation in dietary cirrhosis in the rat. *Gastroenterology* **72:**502–506, 1977.

Hoensch, H., Hartmann, F., Schomerus, H., Bieck, P., and Dölle, W.: Monooxygenase enzyme activity in alcoholics with varying degrees of liver damage. *Gut* **20:**666–672, 1979.

Hoffbauer, F. W., and Zaki, F. G.: Choline deficiency in baboon and rat compared. *Arch. Pathol.* **79:**364–369, 1965.

Holt, K., Bennett, M., and Chojkier, M.: Acetaldehyde stimulates collagen and noncollagen protein production by human fibroblasts. *Hepatology* **4:**843–848, 1984.

Horner, F., Kellen, J. A., Kingston, E., Maharaj, N., and Malkin, A.: Dynamic changes of serum gamma glutamyl transferase in chronic alcoholism. *Enzyme* **24:**217–223, 1979.

Hutterer, F., Rubin, E., and Popper, H.: Mechanism of collagen resorption in reversible hepatic fibrosis. *Exp. Mol. Pathol.* **3:**215–223, 1964.

Hutterer, F., Eisenstadt, M., and Rubin, E.: Turnover of hepatic collagen in reversible and irreversible fibrosis. *Experientia* **26:**244–245, 1970.

Iber, F. L., Lee, K., Lacoursiere, R., and Fuller, R.: Liver toxicity encountered in the Veterans Administration trial of disulfiram in alcoholics. *Alcoholism: Clin. Exp. Res.* **11:**301–304, 1987.

Ingelman-Sundberg, M., and Johansson, I.: Mechanisms of hydroxyl radical formation and ethanol oxidation by ethanol-inducible and other forms of rabbit liver microsomal cytochromes P-450. *J. Biol. Chem.* **259:**6447–6458, 1984.

Ingelman-Sundberg, M. , Johansson, I., Penttilä, E. K., Glaumann, H., and Lindros, O. K.: Centrilobular expression of ethanol-inducible cytochrome P-450 (IIEI) in rat liver. *Biochem. Biophys. Res. Commun.* **157:**55–60, 1988.

International Statistics on Alcoholic Beverages: Production Trade and Consumption 1950–1972. In: *The Finnish Foundation for Alcohol Studies and the World Health Organization Regional Office for Europe.* Forssa, Finland, Aurasen Kirjapaino. Distributed by the Finnish Foundation for Alcohol Studies, Helsinki, Finland, 1977.

Irle, C., Kocher, O., and Gabbiani, G.: Contractility of myofibroblasts during experimental liver cirrhosis. *J. Submicrobiol. Cytol.* **12:**209–217, 1980.

Isaacson, C., Seftel, H. C., Kelley, K. H., and Well, T. H.: Siderosis in the Bantu: The relationship between iron overload and cirrhosis. *J. Lab. Clin. Med.* **58:**845–853, 1961.

Iseri, O. A., and Gottlieb, L. S.: Alcoholic hyaline and mega-mitochondria as separate and distinct entities in liver disease associated with alcoholism. *Gastroenterology* **60:**1027–1035, 1971.

Iseri, O. A., Lieber, C. S., and Gottlieb, L. S.: The ultrastructure of fatty liver induced by prolonged ethanol ingestion. *Am. J. Pathol.* **48:**535–555, 1966.

Ishii, H., Yasuraoka, S., Shigeta, Y., Takagi, S., Kamiya, T., Okuno, F., and Miyamoto, K.: Hepatic and intestinal gamma-glutamyltranspeptidase activity: Its activation by chronic ethanol administration. *Life Sci.* **23:**1393–1397, 1978.

Ishii, H., Okuno, F., Shigeta, Y., Yasuraoka, S., Ebihara, Y., Takagi, T., and Tsuchiya, M.: Significance of serum gamma glutamyl transpeptidase as a marker of alcoholism. *Pharmacol. Biochem. Behav.* **13:**95–99, 1980.

Israel, Y., Videla, L., Fernandes-Videla, V., and Bernstein, J.: Effects of chronic ethanol treatment and thyroxine administration on ethanol metabolism and liver oxidative capacity. *J. Pharmacol. Exp. Ther.* **192:**565–574, 1975.

Iturriaga, H., Pereda, T., Estevez, A., and Ugarte, G.: Serum immunoglobulin changes in alcoholic patients. *Ann. Clin. Res.* **9:**39–43, 1977.

Iturriaga, H., Ugarte, G., and Israel, Y.: Hepatic vein oxygenation, liver blood flow, and the rate of ethanol metabolism in recently abstinent alcoholic patients. *Eur. J. Clin. Invest.* **10:**211–218, 1980.

Jabbari, M., and Leevy, C. M.: Protein anabolism and fatty liver of the alcoholic. *Medicine* **46:**131–139, 1967.

Jackson, F. C., Perin, E. B., Smith, A. G., Dagradi, A. E., and Wadal, H. M.: A clinical investigation of the portacaval shunt. II. Survival analyses of the prophylactic operation. *Am. J. Surg.* **115:**22–42, 1968.

Jakobovits, A. W., Morgan, M. Y., and Sherlock, S.: Hepatic siderosis in alcoholics. *Dig. Dis. Sci.* **24:**305–310, 1979.

Jauhonen, P., Baraona, E., Miyakawa, H., and Lieber, C. S.: Mechanism for selective perivenular hepatotoxicity of ethanol. *Alcoholism: Clin. Exp. Res.* **6:**350–357, 1982.

Jauhonen, P., Baraona, E., Lieber, C. S., and Hassinen, I. E.: Dependence of ethanol-induced redox shift on hepatic oxygen tensions prevailing *in vivo. Alcohol* **2:**163–167, 1985.

Jeghers, H., and Marrao, H.: Hypervitaminosis A: Its broadening spectrum. *Am. J. Clin. Nutr.* **6:**335–339, 1958.

Jolliffe, N., and Jellinek, E. M.: Vitamin deficiencies and liver cirrhosis in alcoholism. Part VII: Cirrhosis of the liver. *Q. J. Stud. Alcohol* **2:**544–583, 1941.

Joly, J. G., Feinman, L., Ishi, H., and Lieber, C. S.: Effect of chronic ethanol feeding on hepatic microsomal glycerophosphate acyltransferase activity. *J. Lipid Res.* **1:**337–343, 1974.

Jones, J. P., Sosa, F. R., Shartsis, J., Shah, P. T., Skromak, E., and Beher, W. T.: Serum cholesterol esterifying and cholesterol ester hydrolyzing activities in liver diseases: Relationships to cholesterol, bilirubin and bile salt concentrations. *J. Clin. Invest.* **50:**259–265, 1971.

Jordo, L., and Olsson, R.: Effect of long-term administration of different hard liquors and red wine on the rat liver. *Acta Pathol. Microbiol. Scand.* **83:**345–354, 1975.

Kakumu, S., and Leevy, C. M.: Lymphocyte cytotoxicity in alcoholic hepatitis. *Gastroenterology* **72:**594–597, 1977.

Kamegaya, K., Okazaki, I., Oda, M., and Tsuchiya, M.: Some aspects of hepatic fibrogenesis in alcoholics—inductive effect of ethanol on liver cells to form collagen fibers. In: *Pathobiology of Hepatic Fibrosis* (C. Hirayama and K. I. Kivirikko, eds.), Elsevier Science Publishers (Biomedical Division), pp. 25–34, 1985.

Kato, S., Murawaki, Y., and Hirayama, C.: Effects of ethanol feeding on hepatic collagen synthesis and degradation in rats. *Res. Commun. Chem. Pathol. Pharmacol.* **47:**163–180, 1985.

Kato, S., Ishii, H., Aiso, S., Yamashita, S., Ito, D., and Tsuchiya, M.: Histochemical and immunohistochemical evidence for hepatic zone 3 distribution of alcohol dehydrogenase in rats. *Hepatology* **12:**66–69, 1990a.

Kato, S., Kawase, T., Alderman, J., Inatomi, N., and Lieber, C. S.: Role of xanthine oxidase in ethanol induced lipid peroxidation in rats. *Gastroenterology* **98:**203–210, 1990b.

Kawase, T., Kato, S., and Lieber, C. S.: Lipid peroxidation and antioxidant defense systems in rat liver after chronic ethanol feeding. *Hepatology* **10:**815–821, 1989.

Keeley, A. F., Iseri, O. A., and Gottlieb, L. S.: Ultrastructure of hyaline cytoplasmic inclusions in a human hepatoma: Relationship to Mallory's alcoholic hyalin. *Gastroenterology* **62:**280–293, 1972.

Kent, G., Inouye, T., Bahu, R., Minick, O. T., and Popper, H.: Vitamin A containing lipocytes and formation of type III collagen in liver injury. *Proc. Natl. Acad. Sci. USA* **73:**3719–3722, 1976.

Kent, G., Inouye, T., Minick, O. T., and Bahu, R. J.: Role of lipocytes (perisinusoidal cells) in fibrogenesis. In: *Kupffer Cells and Other Liver Sinusoidal Cells* (W. Eisse and D. L. Knook, eds.), Amsterdam, Elsevier/North Holland, pp. 73–82, 1977.

Keppler, D., Huber, M., and Baumert, T.: Leukotrienes as mediators in diseases of the liver. *Semin. Liv. Dis.* **8:**357–366, 1988.

Kera, Y., Ohbora, Y., and Komura, S.: The metabolism of acetaldehyde and not acetaldehyde itself is responsible for *in vivo* ethanol-induced lipid peroxidation in rats. *Biochem. Pharmacol.* **37:**3633–3638, 1988.

Kershenobich, D., Fierro, F. J., and Rojkind, M.: The relationship between free pool of proline and collagen content in human liver cirrhosis. *J. Clin. Invest.* **49:**2246–2249, 1970.

Kershenobich, D., Fierro, F. J., and Rojkind, M.: The relationship between serum lactic acid and serum proline in alcoholic liver cirrhosis. *Gastroenterology* **77:**A22, 1979a (abstract).

Kershenobich, D., Uribe, M., Suarez, G. I., Mata, J. M., Perez-Tamayo, R., and Rojkind, M.: Treatment of cirrhosis with colchicine: A double-blind randomized trial. *Gastroenterology* **77:**532–536, 1979b.

Kershenobich, D., Vargas, F., Garcia-Tsao, G., Tomayo, P. R., Gent, M., and Rojkind, M.: Colchicine in the treatment of cirrhosis of the liver. *N. Engl. J. Med.* **318:**1709–1713, 1988.

Kessler, B. J., Liebler, J. B., Bronfin, G. J., and Sass, M.: The hepatic blood flow and splanchnic oxygen consumption in alcoholic fatty liver. *J. Clin. Invest.* **33:**1338–1345, 1954.

Kiessling, K. H., and Pilstrom, L.: Effect of ethanol on rat liver . V. Morphological and functional changes after prolonged consumption of various alcoholic beverages. *Q. J. Stud. Alcohol* **29:**819–827, 1968.

Kim, C.-I., Leo, M. A., Lowe, N., and Lieber, C. S.: Effects of vitamin A and ethanol on liver plasma membrane fluidity. *Hepatology* **8:**735–741, 1988.

Kimura, H., Kako, M., Yo, K., and Oda, T.: Alcoholic hyalins (Mallory bodies) in a case of Weber–Christian disease: Electron microscopic observations of liver involvement. *Gastroenterology* **78:**807–812, 1980.

Klatskin, G., and Yesner, R.: Factors in the treatment of Laennec's cirrhosis. I. Clinical and histological changes observed during a control period of bed-rest, alcohol withdrawal and a minimal basic diet. *J. Clin. Invest.* **28:**723–735, 1949.

Klatskin, G., Krehl, W. A., and Conn, H. O.: The effect of alcohol on the choline requirement. I. Changes in the rat's liver following prolonged ingestion of alcohol. *J. Exp. Med.* **100:**605–614, 1954.

Klion, F. M., and Schaffner, F.: Ultrastructural studies in alcoholic liver disease. *Digestion* **1:**2–14, 1968.

Kocak-Toker, N., Uysal, M., Aykac, G., Sivas, A., Yalcin, S., and Oz, H.: Influence of acute ethanol administration on hepatic glutathione peroxidase and glutathione transferase activities in the rat. *Pharmacol. Res. Commun.* **17:**233–239, 1985.

Koch, O. R., Boveris, A., Sirotzky, De Favelukes, S., Schwarcz, De Tarlovsky, S., and Stoppani, A. O. M.: Biochemical lesions of liver mitochondria from rats after chronic alcohol consumption. *Exp. Mol. Pathol.* **27:**213–220, 1977a.

Koch, O. R., Roatta de Conti, L. L., and Monserrat, A. J.: Effects of alcohol versus sucrose and/or fat-derived calories on the lipotropic requirements. *Nutr. Rep. Int.* **15:**9–17, 1977b.

Konttinen, A., Härtel, G., and Louhija, A.: Multiple serum enzyme analyses in chronic alcoholics. *Acta Med. Scand.* **188:**257–264, 1970.

Korner, W. F., and Vollm, J.: New aspects of the tolerance of retinol in humans. *Int. J. Vit. Nutr. Res.* **45**:362–373, 1975.

Korpera, H., Kumpulainen, J., Luoma, P. V., Arranto, A. J., and Sotaniemi, E. A.: Decreased serum selenium in alcoholics as related to liver structure and function. *Am. J. Clin. Nutr.* **42**: 147–151, 1985.

Kowaloff, E. M., Phang, J. M., Granger, A. S., and Downing, S. J.: Regulation of proline oxidase activity by lactate. *Proc. Natl. Acad. Sci. U.S.A.* **74**:5368–5371, 1977.

Kratzsch, K. H.: Urinary hydroxyproline excretion in cirrhosis of the liver. *Acta Hepatosplenol.* (Stutt.) **16**:2–8, 1969.

Krogsgaard, K., Gluud, C., Henriksen, J. H., and Christoffersen, P.: Correlation between liver morphology and haemodynamics in alcoholic liver disease. *Hepatology* **4**:699–703, 1984.

Krogsgaard, K., Gluud, C., Henriksen, J. H., and Christoffersen, P.: Correlation between liver morphology and haemodynamics in alcoholic liver disease. *Liver* **5**:173–177, 1985.

Kuller, L., Kramer, K., and Fisher, R.: Changing trends in cirrhosis and fatty liver mortality. *Am. J. Publ. Health* **59**:1124–1133, 1969.

Kumar, S., Stauber, R. E., Gavaler, J. S., Basista, M. H., Dindzans, V. J., Schade, R. R., Rabinovitz, M., Tarter, R. E., Gordon, R., Starzl, T. E., and Van Thiel, D. H.: Orthotopic liver transplantation for alcoholic liver disease. *Hepatology* **11**:159–164, 1990.

Kyösola, K., and Salorinne, Y.: Liver biopsy and liver function tests in 28 consecutive long-term alcoholics. *Ann. Clin. Res.* **7**:80–84, 1975.

Lahrichi, M., Ratanasavanh, D., Galteau, M.-M., and Diest, G.: Effect of chronic ethanol administration on gamma-glutamyl-transferase activities in plasma and in hepatic plasma membranes of male and female rats. *Enzyme* **28**:215–257, 1982.

Lancet: Nutrition and liver disease (Editorial). *Lancet* **2**:1132, 1972.

Lane, B. P., and Lieber, C. S.: Ultrastructural alterations in human hepatocytes following ingestion of ethanol with adequate diets. *Am. J. Pathol.* **49**:593–603, 1966.

Laposata, E. A., and Lange, L. G.: Presence of nonoxidative ethanol metabolism in human organs commonly damaged by ethanol abuse. *Science* **231**:497–499, 1986.

Lee, F. I.: Cirrhosis and hepatoma in alcoholics. *Gut* **7**:77–85, 1966.

Leevy, C. M.: Fatty liver: A study of 270 patients with biopsy proven fatty liver and a review of the literature. *Medicine* **41**: 249–276, 1962.

Leevy, C. M.: *In vitro* studies of hepatic DNA synthesis in percutaneous liver biopsy specimens. *J. Lab. Clin. Med.* **61**: 761–779, 1963.

Leevy, C. M.: Abnormalities of hepatic DNA synthesis in man. *Medicine* **45**:423–433, 1966.

Leevy, C. M., Fatty liver: A study of 270 patients with biopsy proven fatty liver and a review of the literature. *Medicine* **41**: 1445–1451, 1968.

Leevy, C. M., Popper, H., and Sherlock, S. (eds.): *Diseases of the Liver and Biliary Tract, Standardization of Nomenclature, Diagnostic Criteria and Diagnostic Methodology, Fogarty International Center Proceedings No. 22*. U.S. Government Printing Office, 1976.

Lelbach, W. K.: Leberschäden bei chronischem Alkoholismuss. I-III. *Acta Hepatosplen.* **13**:321–349, 1966.

Lelbach, W. K.: Leberschäden bei chronischem. Alkoholismuss. *Acta Hepatosplen.* **14**:9–39, 1967.

Lelbach, W. K.: Cirrhosis in the alcoholic and its relation to the volume of alcohol abuse. *Ann. NY Acad. Sci.* **252**:85–105, 1975.

Lelbach, W. K.: Epidemiology of alcoholic liver disease. In: *Progress in Liver Disease* (H. Popper, and F. Schaffner, eds.), New York, Grune & Stratton, pp. 494–515, 1976.

Leo, M. A., and Lieber, C. S.: Hepatic vitamin A depletion in alcoholic liver injury in man. *N. Engl. J. Med.* **307**:597–601, 1982.

Leo, M. A., and Lieber, C. S.: Hepatic fibrosis after long term administration of ethanol and moderate vitamin A supplementation in the rat. *Hepatology* **3**:1–11, 1983.

Leo, M. A., and Lieber, C. S.: New pathway of retinol metabolism in liver microsomes. *J. Biol. Chem.* **260**:5228–5231, 1985.

Leo, M. A., and Lieber, C. S.: Hypervitaminosis A: A liver lover's lament. *Hepatology* **8**:412–417, 1988.

Leo, M. A., Arai, M., Sato, M., and Lieber, C. S.: Hepatotoxicity of moderate vitamin A supplementation in the rat. *Gastroenterology* **82**:194–205, 1982.

Leo, M. A., Sato, M., and Lieber, C. S.: Effect of hepatic vitamin A depletion on the liver in humans and rats. *Gastroenterology* **84**:562–572, 1983.

Leo, M. A., Lowe, N., and Lieber, C. S.: Decreased hepatic vitamin A after drug administration in men and in rats. *Am. J. Clin. Nutr.* **40**:1131–1136, 1984a.

Leo, M. A., Iida, S., and Lieber, C. S.: Retinoic acid metabolism by a system reconstituted with cytochrome P-450. *Arch. Biochem. Biophys.* **234**:305–312, 1984b.

Leo, M. A., Kim, C.-I., and Lieber, C. S.: NAD^+-dependent retinol dehydrogenase in liver microsomes. *Arch. Biochem. Biophys.* **250**:241–249, 1987.

Lesesne, H. R., Bozymski, E. M., and Fallon, H. J.: Liver physiology and disease: Treatment of alcoholic hepatitis with encephalopathy—comparison of prednisolone with caloric supplements. *Gastroenterology* **74**:169–173, 1978.

Levin, D. M., Baker, A. L., Rochman, H., and Boyer, J. L.: Alcoholic patients with chronic abnormalities in liver function harbor unsuspected nonalcoholic liver diseases, detectable only by liver biopsy. *Gastroenterology* **72**:1172, 1977.

Li, J-J., Kim, C-I., Leo, M. A., Mak, K. M., Rojkind, M., and Lieber, C. S.: Polyunsaturated lecithin prevents acetaldehyde-mediated hepatic collagen accumulation by stimulating collagenase activity in cultured lipocytes. *Hepatology* **15**: 373–381, 1992.

Lieber, C. S.: Pathogenesis of alcoholic liver disease: An overview. In: *Alcohol and the Liver*, Vol. 3 (M. M. Fisher and J. G. Rankin, eds.), New York, Plenum, pp. 197–255, 1977.

Lieber, C. S.: Alcohol, liver injury and protein metabolism. *Pharmol. Biochem. Behav.* **13**:17–30, 1980.

Lieber, C. S.: Alcohol and the liver: Metabolism of ethanol, metabolic effects and pathogenesis of injury. *Acta Med. Scand. Suppl.* **703**:11–55, 1985.

Lieber, C. S.: Increased susceptibility of the alcoholic to the toxicity of ethanol, other drugs, industrial solvents and nutrients. *Subst. Abuse* **10:**143–148, 1989.

Lieber, C. S., and DeCarli, L. M.: Reduced nicotinamide-adenine dinucleotide phosphate oxidase: Activity enhanced by ethanol consumption. *Science* **170:**78–80, 1970.

Lieber, C. S., and DeCarli, L. M.: An experimental model of alcohol feeding and liver injury in the baboon. *J. Med. Primatol.* **3:**153–163, 1974.

Lieber, C. S., and Rubin, E.: Alcoholic fatty liver in man on a high protein and low fat diet. *Am. J. Med.* **44:**200–206, 1968.

Lieber, C. S., and Salaspuro, M. P.: Alcoholic liver disease. In: *Liver and Biliary Disease, Second Edition* (R. Wright, H. Millward-Sadler, G. Alberti, and S. Karan, eds.), London, Saunders, pp. 881–947, 1985.

Lieber, C. S., Jones, D. P., and DeCarli, L. M.: Effects of prolonged ethanol intake: Production of fatty liver despite adequate diets. *J. Clin. Invest* **44:**1009–1020, 1965.

Lieber, C. S., Rubin, E., and DeCarli, L. M.: Effects of ethanol on lipid, uric acid, intermediary and drug metabolism including the pathogenesis of the alcoholic fatty liver. In: *The Biology of Alcoholism*, Vol. 1, (B. Kissin and H. Begleiter, eds.), New York, Plenum, pp. 263–305, 1971.

Lieber, C. S., DeCarli, L. M., Gang, H., Walker, G., and Rubin, E.: Hepatic effects of long term ethanol consumption in primates. In: *Medical Primatology—1972*, Part 3 (E, I. Goldsmith and J. Moor-Janowski, eds.), Basel, S. Karger, pp. 270–278, 1972.

Lieber, C. S., DeCarli, L. M., and Rubin, E.: Sequential production of fatty liver, hepatitis and cirrhosis in sub-human primates fed ethanol with adequate diets. *Proc. Natl. Acad. Sci. U.S.A.* **72:**437–441, 1975.

Lieber, C. S., Zimmon, D. S., Kessler, R. E., ad DeCarli, L. M.: Portal hypertension in experimental alcoholic liver injury. *Clin. Res.* **24:**478A, 1976.

Lieber, C. S., Nakano, M., and Worner, T. M.: Ultrastructure of the initial stages of hepatic perivenular fibrosis after alcohol. *Trans. Assoc. Am. Phys.* **94:**292–330, 1981.

Lieber, C. S., Leo, M. A., Mak, K. M., DeCarli, L. M., and Sato, S.: Choline fails to prevent liver fibrosis in ethanol-fed baboons but causes toxicity. *Hepatology* **5:**561–572, 1985.

Lieber, C. S., Baraona, E., Hernandez-Munoz, R., Kubota, S., Sato, N., Kawano, S., and Matsumura, T.: Impaired oxygen utilization: A new mechanism for the hepatotoxicity of ethanol in sub-human primates. *J. Clin. Invest* **83:**1682–1690, 1989.

Lieber, C. S., Casini, A., DeCarli, L. M., Kim, C., Lowe, N., Sasaki, R., and Leo, M. A.: *S*-adenosyl-L-methionine attenuates alcohol-induced liver injury in the baboon. *Hepatology* **11:**165–172, 1990a.

Lieber, C. S., DeCarli, L. M., Mak, K. M., Kim, C. I., and Leo, M. A.: Attenuation of alcohol-induced hepatic fibrosis by polyunsaturated lecithin. *Hepatology* **12:**1390–1398, 1990b.

Lindy, S., Pedersen, F. B., Turto, H., and Uitto, J.: Lactate, lactate dehydrogenase and protocollagen proline hydroxylase in rat skin autograft. *Hoppe–Seyler's Z. Physiol. Chem.* **352:**1113–1118, 1971.

Lischner, M. W., Alexander, J. F., and Galambos, J. T.: Natural history of alcoholic hepatitis. I. The acute disease. *Am. J. Dig. Dis.* **16:**481–494, 1971.

Lue, S. L., Paronetto, F., and Lieber, C. S.: Cytotoxicity of mononuclear cells and vulnerability of hepatocytes in alcoholic fatty liver of baboons. *Liver* **1:**264–267, 1981.

Lundquist, F., Tygstrup, N., Winkler, K. Mellemgaard, K., and Munck-Petersen, S.: Ethanol metabolism and production of free acetate in the human liver. *J. Clin. Invest.* **41:**955–961, 1962.

MacDonald, C. M.: The effect of ethanol on hepatic lipid peroxidation and on the activities of glutathione reductase and peroxidase. *FEBS Lett.* **35:**227–230, 1973.

MacDonald, R. A.: Wines as a source of iron in hemochromatosis. *Nature* **199:**922, 1963.

MacDonald, R. A., and Baumslag, N.: Iron in alcoholic beverages: Possible significance for hemochromatosis. *Am. J. Med. Sci.* **247:**649–654, 1964.

Maddrey, W. C., Weber, F. L., Coulter, A. W., Chura, C. M., Chapanis, N. P., and Walser, M.: Effects of keto analogue essential amino acids in portal-systemic encephalopathy. *Gastroenterology* **71:**190–195, 1978.

Maier, K. P., Haag, S. G., Peskar, B. M., and Gerok, W.: Verlaufsformen alkoholischer Lebererkrankungen. *Klin. Wochenschr.* **57:**311–317, 1979.

Mailliard, M. E., Sorrell, M. F., Volentine, G. D., and Tuma, D. J.: Impaired plasma membrane glycoprotein assembly in the liver following acute ethanol administration. *Biochem. Biophys. Res. Commun.* **123:**951–958, 1984.

Majchrowicz, E., and Mendelson, J H.: Blood concentrations of acetaldehyde and ethanol in chronic alcoholics. *Science* **168:** 1100, 1970.

Mak, K. M., and Lieber, C. S.: Alterations in endothelial fenestrations in liver sinusoids of baboons fed alcohol: A scanning electron microscopic study. *Hepatology* **4:**386–391, 1984.

Mak, K. M., and Lieber, C. S.: Lipocytes and transitional cells in alcoholic liver disease: A morphometric study. *Hepatology* **8:** 1027–1033, 1988.

Mak, K. M., Leo, M. A., and Lieber, C. S.: Alcoholic liver injury in baboons: Transformation of lipocytes to transitional cells. *Gastroenterology* **87:**188–200, 1984.

Mäkelä, K.: Measuring the consumption of alcohol in the 1968–1969 alcohol consumption study. In: *Special Research Institute of Alcohol Studies.* No. 2, Helsinki, Oy Alko Ab., 1971.

Mallory, F. B.: Cirrhosis of the liver. Five different types of lesions from which it may arise. *Bull. Johns Hopkins Hosp.* **22:**69–75, 1911.

Mann, S. W., Fuller, G. C., Rodil, J. V., and Vidins, E. I.: Hepatic prolyl hydroxylase and collagen synthesis in patients with alcoholic liver disease. *Gut* **20:**825–832, 1979.

Martini, G. A., and Bode, C. H.: The epidemiology of cirrhosis of the liver. In: *Alcoholic Cirrhosis and Other Toxic Hepatopathies* (A. Engel and T. Larsson, eds.), Stockholm, Nordiska Bokhandelins forlog, pp. 315–335, 1970.

Marubbio, A. T., Buchwald, H., Schwartz, M. Z., and Varco, R.: Hepatic lesions of central pericellular fibrosis in morbid obesity, and after jejunoileal bypass. *J. Clin. Pathol.* **66:**684–691, 1976.

Maruyama, K., Feinman, L., Fainsilber, Z., Nakano, M., Okazaki, I., and Lieber, C. S.: Mammalian collagenase increases in early alcoholic liver disease ad decreases with cirrhosis. *Life Sci.* **30**:1379–1384, 1982.

Mata, J. M., Kerschenobich, D., Villarreal, E., and Rojkind, M.: Serum free proline and free hydroxyproline in patients with chronic liver disease. *Gastroenterology* **68**:1265–1269, 1975.

Matsuda, Y., Baraona, E., Salaspuro, M., and Lieber, C. S.: Effects of ethanol on liver microtubules and Golgi apparatus. *Lab. Invest.* **41**:455–463, 1979.

Matsuzaki, S., and Lieber, C. S.: Increased susceptibility of hepatic mitochondria to the toxicity of acetaldehyde after chronic ethanol consumption. *Biochem. Biophys. Res. Commun.* **75**:1059–1065, 1977.

McCay, P. B.: Vitamin E.: Interaction with free radicals and ascorbate. *Ann. Rev. Nutr.* **5**:323–340, 1985.

McDevitt, H. O., and Bodmer, W. F.: Histocompatibility antigens, immune responsiveness and susceptibility to disease. *Am. J. Med.* **52**:1–8, 1972.

McGee, J. O., and Patrick, R. S.: The role of perisinusoidal cells in hepatic fibrogenesis: An electron microscopic study of acute carbon tetrachloride liver injury. *Lab. Invest.* **26**:429–440, 1972.

Melendez, M., Vargas-Tank, L., Fuentes, C., Armas-Merino, R., Castillo, D., Wolff, C., Wegmann, M. E., and Soto, J.: Distribution of HLA histocompatibility antigens, ABO blood groups and RH antigens in alcoholic liver disease. *Gut* **20**:288–290, 1979.

Mendeloff, A.: Effect of intravenous infusions of ethanol upon estimated hepatic blood flow in man. *J. Clin. Invest.* **33**:1298–1302, 1954.

Mendenhall, C. L.: Anabolic steroid therapy as an adjunct to diet in alcoholic hepatic steatosis. *Am. J. Dig. Dis.* **13**:783–791, 1968.

Mendenhall, C. L., Anderson, S., Garcia-Pont, P., Goldberg, S., Kiernan, T., Seeff, L. B., Sorrell, M., Tamburro, C., Weesner, R., Zetterman, R., Chedid, A., Chen, T., and Rabin, L.: Short-term and long-term survival in patients with alcoholic hepatitis treated with oxandrolone and prednisolone. *N. Engl. J. Med.* **311**:1464–1470, 1984.

Mezey, E., Potteer, J. J., and Ammon, R. A.: Effect of ethanol administration on the activity of hepatic lysosomal enzymes. *Biochem. Pharmacol.* **25**:2663–2667, 1976.

Mezey, E., Potter, J. J., Iber, F. L., and Maddrey, W. C.: Hepatic collagen proline hydroxylase activity in alcoholic hepatitis effect of d-penicillamine. *J. Lab. Clin. Med.* **93**:92–100, 1979.

Miller, A. I., Moral, M. D., and Schiff, E. R.: Presence of serum α-1-fetoprotein in alcoholic hepatitis. *Gastroenterology* **68**:381–383, 1975.

Milstein, H. J., Bloomer, J. R., and Klatskin, G.: Serum bile acids in alcoholic liver disease. Comparison with histological features of the disease. *Dig. Dis.* **21**:281–285, 1976.

Minato, Y., Hasumura, Y., and Takeuchi, J.: The role of fat-storing cells in Disse space fibrogenesis in alcoholic liver disease. *Hepatology* **3**:559–566, 1983.

Miyakawa, H., Iida, S., Leo, M. A., Greenstein, R. J., Zimmon, D. S., and Lieber, C. S.: Pathogenesis of precirrhotic portal hypertension in alcohol-fed baboons. *Gastroenterology* **88**:143–150, 1985.

Monroe, S., French, S. W., and Zamboni, L.: Mallory bodies in a case of primary biliary cirrhosis: An ultrastructural and morphogenetic study. *Am. J. Clin. Pathol.* **59**:254–262, 1973.

Morgan, M. Y., and Sherlock, S.: Sex-related differences among 100 patients with alcoholic liver disease. *Br. Med. J.* **1**:939–941, 1977.

Morgan, M. Y., Sherlock, S., and Scheuer, P. J.: Acute cholestatis, hepatic failure and fatty liver in the alcoholic. *Scand. J. Gastroenterol.* **31**:299–303, 1978.

Morgan, M. Y., Ross, M. G. R., Ng, C. M., Adams, D. M., Thomas, H. C., and Sherlock, S.: HLA-B8, immunoglobulins, and antibody responses in alcohol-related liver disease. *J. Clin. Pathol.* **3**:488–492, 1980.

Mørland, J., Huseby, N.-E., Sjøblom, M., and Strømme, J. H.: Does chronic alcohol consumption really induce hepatic microsomal gamma-glutamyltransferase activity? *Biochem. Biophys. Res. Commun.* **77**:1060–1066, 1977.

Morton, S., and Mitchell, M. C.: Effects of chronic ethanol feeding on glutathione turnover in the rat. *Biochem. Pharmacol.* **34**:1559–1563, 1985.

Moshage, H., Casini, A., and Lieber, C. S.: Acetaldehyde selectively simulates collagen production in cultured rat liver fat-storing cells but not in hepatocytes. *Hepatology* **12**:511–518, 1990.

Muller, A., and Sies, H.: Role of alcohol dehydrogenase activity and of acetaldehyde in ethanol-induced ethane and pentane production by isolated perfused rat liver. *Biochem. J.* **206**:153–156, 1982.

Nagore, N., and Scheuer, P. J.: The pathology of diabetic hepatitis. *J. Pathol.* **156**:155–160, 1988.

Nakamura, S., Takezawa, Y., Sato, T., Kera, K., and Maeda, T.: Alcoholic liver diseases in women. *Tohoku J. Exp. Med.* **129**:351–355, 1979.

Nakano, M., and Lieber, C. S.: Ultrastructure of initial stages of perivenular fibrosis in alcohol-fed baboons. *Am. J. Path.* **106**:145–155, 1982.

Nakano, M., Worner, T., and Lieber, C. S.: Perivenular fibrosis in alcoholic liver injury: Ultrastructure of histologic progression. *Gastroenterology* **83**:777–785, 1982.

Nasrallah, S. M., Nassar, V. H., and Galambos, J. T.: Importance of terminal hepatic venule thickening. *Arch. Pathol. Lab. Med.* **104**:84–86, 1980.

National Academy of Sciences: *Nutrient Requirements of Nonhuman Primates*, Vol. 14, Washington, D.C., National Research Council, 1978.

Nenci, I.: Identification of actin-like proteins in alcoholic hyaline by immunofluorescence. *Lab. Invest.* **32**:257–260, 1975.

New York City: *Summary of Vital Statistics 1984.* Department of Health, New York, Bureau of Health Statistics and Analysis, 1984.

Niemelä, O.: Radioimmunoassays for type III amino-terminal peptides in humans. *Clin. Chem.* **31**:S1301–1304, 1985.

Niki, E.: Interaction of ascorbate and α-tocopherol. *Ann. N.Y. Acad. Sci.* **493**:186–199, 1987.

Nishimura, M., and Teschke, R.: Effect of chronic alcohol consumption on the activities of liver plasma membrane enzymes: Gamma-glutamyltransferase, alkaline phosphatase and 5'-nucleotidase. *Biochem. Pharmacol.* **31**:377–381, 1982.

Nouchi, T., Lasker, J. M., and Lieber, C. S.: Activation of acetaminophen oxidation in rat liver microsomes by caffeine. *Toxicol. Lett.* **32**:1–8, 1986.

Nouchi, T., Worner, T. M., Sato, S., and Lieber, C. S.: Serum procollagen type III n-terminal propeptides and laminin P_1 peptide in alcoholic liver disease. *Alcoholism: Clin. Exp. Res.* **11**:287–291, 1987.

Novick, D. M., Enlow, R. W., Gelb, A. M., Stenger, R. J., Fotina, M., Winter, J. W., Yancovitz, S. R., Schoenberg, M. D., and Kreek, M. J.: Hepatic cirrhosis in young adults: Association with adolescent onset of alcohol and parenteral heroin abuse. *Gut* **26**:8–13, 1985.

Okamura, K., Harwood, T. R., and Yokoo, H.: Isolation and electrophoretic study on Mallory bodies from the livers of alcoholic cirrhosis. *Lab. Invest.* **33**:193–199, 1975.

Okazaki, I., Feinman, L., and Lieber, C. S.: Hepatic mammalian collagenase: Development of an assay and demonstration of reversal activity after ethanol consumption. *Gastroenterology* **73**:1236, 1977.

Orloff, M. J., Duguay, L. R., and Kosta, L. D.: Criteria for selection of patients for emergency portacaval shunt. *Am. J. Surg.* **134**:146–152, 1977.

Orrego, H., Kalant, H., Israel, Y., Blake, J., Medline, A., Rankin, J. G., Armstrong, A., and Kapur, B.: Effect of short-term therapy with propylthiouracil in patients with alcoholic liver disease. *Gastroenterology* **76**:105–115, 1979.

Orrego, H., Blendis, L. M., Crossley, I. R., Medline, A., MacDonald, A., Ritchie, S., and Israel, Y.: Correlation of intrahepatic pressure with collagen in the Disse space and hepatomegaly in humans and in the rat. *Gastroenterology* **80**:546–566, 1981.

Orrego, H., Blake, J. E., Blendis, I. M., Compton, K. V., and Israel, Y.: Long-term treatment of alcoholic liver disease with propylthiouracil. *N. Engl. J. Med.* **317**:1421–1427, 1987.

Orrego, H., Carmichael, F. J., Saldivia, V., Giles, H. G., Sandrin, S., and Israel, Y.: Ethanol-induced increase in portal blood flow: Role of adenosine. *Am. J. Physiol.* **254**:G495–G501, 1988.

Park, M. K., Rouach, H., Orfanelli, M. H., and Nordmann, R.: Influence of allopurinol and desferrioxamine on the ethanol-induced oxidative stress in rat liver and cerebellum. In: *Alcohol Toxicity and Free Radical Mechanism*, Vol. 71 (R. Nordmann, C. Ribiero, and H. Rouach, eds.), Oxford, Pergamon, pp. 135–139, 1988.

Partin, J. S., Partin, J. C., Schubert, W. K., and McAdams, A. J.: Liver ultrastructure in a betalipoproteinemia: Evolution of micronodular cirrhosis. *Gastroenterology* **67**:107–118, 1974.

Pascale, R., Daino, L., Garcea, R., Frassetto, S., Ruggiu, M. E., Vannini, M. G., Cozzolino, P., and Feo, F.: Inhibition of ethanol of rat liver plasma membrane (Na^+,K^+)ATPase: Protective effect of S-adenosyl-L-methionine, L-methio-nine, and N-acetylcysteine. *Toxicol. Appl. Pharmacol.* **97**:216–229, 1989.

Patek, A. J., Jr., Bowry, S., and Hayes, K. C.: Cirrhosis of choline deficiency in the rhesus monkey: Possible role of dietary cholesterol. *Proc. Soc. Exp. Biol. Med.* **148**:370–374, 1975a.

Patek, A. J., Toth, E. G., Saunders, M. G., Castro, G. A. M., and Engel, J. J.: Alcohol and dietary factors in cirrhosis. *Arch. Intern. Med.* **135**:1053–1057, 1975b.

Patek, A. J., Bowry, S. C., and Sabesin, S. M.: Minimal hepatic changes in rats fed alcohol and a high casein diet. *Arch. Pathol. Lab. Med.* **100**:19–24, 1976.

Patrick, R. S.: Alcohol as a stimulus to hepatic fibrogenesis. *J. Alcoholism* **8**:13–27, 1973.

Pell, S., and D'Alonzo, C. A.: The prevalence of chronic disease among problem drinkers. *Arch. Environ. Health* **16**:679–684, 1968.

Pequignot, G.: Enquete par interrogatoire sur les circonstances diététiques de la cirrhose alcoolique en France. *Bull. Inst. Natl. Hyg.* **13**:719–939, 1958.

Pequignot, G., Chabert, C., Eydoux, H., and Corcowl, M. A.: Increased risk of liver cirrhosis with intake of alcohol. *Rev. Alcoholism* **20**:191–202, 1974.

Pequignot, G., Tuyns, A. J., and Berta, J. L.: Ascitic cirrhosis in relation to alcohol consumption. *Int. J. Epidemiol.* **7**:113–120, 1978.

Perman, G.: Hemochromatosis and red wine. *Acta Med. Scand.* **182**:281–284, 1967.

Pessayre, D., Bichara, M. Geldmann, G., Degott, C., Potet, P., and Benhamou, J.-P.: Perhexiline maleate-induced cirrhosis. *Gastroenterology* **76**:170–177, 1979.

Peters, R. L., Gay, I., and Reynolds, T. B.: Postjejunoileal bypass hepatic disease: Its similarity to alcoholic hepatic disease. *Am. J. Clin. Pathol.* **63**:318–331, 1975.

Petersen, P. U.: Alcoholic hyalin, microfilaments and microtubules in alcoholic hepatitis. *Acta Pathol. Microbiol. Scand.* **85**:384–394, 1977.

Phillips, G. B., and Davidson, C. S.: Acute hepatic insufficiency of the chronic alcoholic. *Arch. Intern. Med.* **94**:585–603, 1954.

Phillips, G. B., and Davidson, C. S.: Liver disease of the chronic alcoholic simulating extrahepatic biliary obstruction. *Gastroenterology* **33**:236–244, 1957.

Pierard, D., Nusgens, B. V., and Lapiere, C. H. M.: Radioimmunoassay for the amino-terminal sequences of type III procollagen in human body fluids measuring fragmented precursor sequences. *Anal. Biochem.* **141**:127–136, 1984.

Pignon, J. P., Bailey, N. C., Baraona, E., and Lieber, C. S.: Fatty acid-binding protein: A major contributor to the ethanol-induced increase in liver cytosolic proteins in the rat. *Hepatology* **7**:865–871, 1987.

Pirola, R., and Lieber, C. S.: Energy wastage in rats given drugs that induce microsomal enzymes. *J. Nutr.* **105**:1544–1548, 1976.

Platt, D., Stein, U., and Heissmeyer, H.: Untersuchungen Zum katabolen Mucopolysaccharid-proteinstoffwechsel der Rattenleber nach Gabe von Athanol. *Z. Klin. Chem. Klin. Biochem.* **9**:126–129, 1971.

Polokoff, M. A., Simon, T. J., Ardon, H. R., Simon, F. R., and

Iwahashi, M.: Chronic ethanol increases liver plasma membrane fluidity. *Biochemistry* **24**:3114–4120, 1985.

Ponnappa, B. C., Waring, A., Hoek, J. B., Rottenberg, H., and Rubin, E.: Chronic ethanol ingestion increases calcium uptake and resistance to molecular disordering by ethanol in liver microsomes. *J. Biol. Chem.* **257**:10141–10146, 1982.

Popham, R. E., Indirect methods of alcoholism prevalence estimation: A critical evaluation. In: *Alcohol and Alcoholism* (R. E. Popham, ed.), Toronto, University of Toronto Press, 1970.

Popham, R. E., Schmidt, W., and DeLint, J. E.: The effect of legal restraint on drinking. In: *Biology of Alcoholism, Vol. 4: Social Biology* (B. Kissin and H. Begleiter, eds.), New York, Plenum, pp. 579–625, 1976.

Popper, H.: The pathogenesis of alcoholic cirrhosis. In: *Alcohol and the Liver* (M. M. Fisher and J. G. Rankin eds.), New York, Plenum, pp. 289–305, 1977.

Popper, H., and Lieber, C. S.: Histogenesis of alcoholic fibrosis and cirrhosis in the baboon. *Am. J. Pathol.* **98**:695–716, 1980.

Popper, H., and Schaffner, F.: The hepatic lesion in Wilson's disease. In: *Wilson's Disease. Some Current Concepts* (J. M. Walshe and J. N. Cummings, eds.), Oxford, Blackwell, pp. 192–197, 1961.

Popper, H., Szanto, P. B., and Parthasarathy, M.: Florid cirrhosis—a review of 35 cases. *Am. J. Clin. Pathol.* **25**:889–901, 1955.

Popper, H., Thung, S. N., and Gerber, M. A.: Pathology of alcoholic liver disease. *Sem. Liv. Dis.* **1**:203, 1981.

Poralla, T., Hutteroth, T. H., Meyer zum Buschenfelde, K. H.: Cellular cytoxocity against autologous hepatocytes in alcoholic liver disease. *Liver* **4**:117–121, 1984.

Porter, H. P., Simon, F. R., Pope, C. E., Volwiler, W., and Fenster, F.: Corticosteroid therapy in severe alcoholic hepatitis: A double-blind drug trial. *N. Engl. J. Med.* **284**:1350–1355, 1971.

Pösö, A. R., and Pösö, H.: Inhibition of ornithine decarboxylase in regenerating rat liver by acute ethanol treatment. *Biochim. Biophys. Acta* **606**:338–346, 1980a.

Pösö, H., and Pösö, A. R.: Stabilization of tyrosine aminotransferase and ornithine decarboxylase in regenerating rat liver by ethanol treatment. *FEBS Lett.* **113**:211–214, 1980b.

Pösö, H., and Pösö, A. R.: Inhibition of RNA and protein synthesis by ethanol in regenerating rat liver: Evidence for transcriptional inhibition of protein synthesis. *Acta Pharmacol. Toxicol.* **49**:125–129, 1981.

Post, J., Benton, J., Breatstone, R., and Hoffman, J.: The effects of diet and choline on fatty infiltration of the human liver. *Gastroenterology* **70**:403–410, 1952.

Powell, L. W.: The role of alcoholism in hepatic iron storage disease. *Ann. N.Y. Acad. Sci.* **252**:124–134, 1975.

Powell, W. J., and Klatskin, G.: Duration of survival in patients with Laennec's cirrhosis. Influence of alcohol withdrawal and possible effects of recent changes in general management of the disease. *Am. J. Med.* **44**:406–420, 1968.

Powell, G. L., Tippett, P. S., Kiorpes, T. C., and Smith, R. H., Fatty acyl-CoA as an effector molecule in metabolism. *Fed. Proc.* **44**:81, 1986.

Puntarulo, S., and Cedarbaum, I.: Effect of oxygen concentration on microsomal oxidation of ethanol and generation of oxygen radicals. *Biochem. J.* 251:787–794, 1988.

Randall, B.: Sudden death and hepatic fatty metamorphosis: A North Carolina survey. *J.A.M.A.* **243**:1723–1725, 1980a.

Randall, B.: Fatty liver and sudden death. *Hum. Pathol.* **11**:147–153, 1980b.

Rankin, J. G.: The natural history and management of the patient with alcoholic liver disease. In: *Alcohol and the Liver* (M. M. Fisher and J. G. Rankin, eds.), New York, Plenum, pp. 365–381, 1977.

Rappaport, A. M.: Liver architecture and microcirculation. In: *Alcoholic Liver Pathology* (J. M. Khanna, Y. Israel, and H. Kalant, eds.), Ontario, The Addiction Research Foundation of Ontario, pp. 43–67, 1975.

Ratner, S., and Clarke, H. T.: The action of formaldehyde upon cysteine. *J. Am. Chem. Soc.* **59**:200–206, 1937.

Ratnoff O. D., and Patek, A. J., Jr.: Natural history of Laennec's cirrhosis of the liver: An analysis of 386 cases. *Medicine* **21**:207–268, 1942.

Regoeczi, E., Chindemi, P. A., and Debanne, M. T.: Transferrine glycans: A possible link between alcoholism and hepatic siderosis. *Alcoholism: Clin. Exp. Res.* **8**:287–292, 1984.

Reinke, L. A., Lai, E. K., DuBose, M., and McCay, P. B.: Reactive free radical generation *in vivo* in heart and liver of ethanol-fed rats: Correlation with radical formation *in vitro*. *Med. Sci.* **7**:9223–9227, 1987.

Reitz, R. C.: A possible mechanism for the peroxidation of lipids due to chronic ethanol ingestion. *Biochim Biophys. Acta* **380**:145–154, 1975.

Resnick, R. H., Chalmer, T. C., Ishihara, A. M., Garceau, A. J., Gallow, A. D., Schimmel, E. M., and O'Hara, E. T.: A controlled study of the prophylactic portacaval shunt: A final report. *Ann. Intern. Med.* **70**:675–688, 1969.

Resnick, R. H., Cerda, J. C., Boitnott, J., Aron, J., and Iber, F. L.: Urinary hydroxyproline excretion in hepatic disorders. *Am. J. Gastroenterol.* **60**:576–584, 1973.

Resnick, R. H., Boitnott, J., Iber, F. L., Makipour, H., and Cerda, J. J.: Penicillamine therapy in acute alcoholic liver disease. In: *Collagen Metabolism in the Liver* (H. Popper and K. Becker, eds.), New York, Stratton Intercontinental, pp. 207–218, 1975.

Reynolds, T. B., Hidemura, R., Michael, H., and Peters, R.: Portal hypertension without cirrhosis in alcoholic liver disease. *Ann. Intern. Med.* **70**:497–506, 1969.

Robbins, R. A., Zetterman, R. K., Kendall, T. J., Gossman, G. L., Monsour, H. P., and Rennard, S. L.: Elevation of chemotactic factor inactivator in alcoholic liver disease. *Hepatology* **7**:872–877, 1987.

Roberts, A. B., Nichola, M. D., Newton, D. L., and Spora, M. B.: *In vitro* metabolism of retinoic acid in hamster intestine and liver. *J. Biol. Chem.* **254**:6296–6302, 1979.

Roberts, A. B., Lamb, L. C., and Sporn, M. B.: Metabolism of all-*trans*-retinoic acid in hamster liver microsomes: Oxidation of 4-hydroxy- to 4-keto-retionoic acid. *Arch. Biochem. Biophys.* **199**:374–383, 1980.

Rohde, H., Langer, I., Krieg, T., and Timpl, R.: Serum and urine analysis of the aminoterminal procollagen peptide type III by radioimmunoassay with antibody Fab fragments. *Collagen Rel. Res.* **3:**371, 1983.

Rojkind, M., and DeLeon, L. D.: Collagen biosynthesis in cirrhotic rat slices: A regulatory mechanism. *Biochim. Biophys. Acta* **217:**512–522, 1970.

Rollason, J. G., Pincherle, G., and Robinson, D.: Serum gamma glutamyl transpeptidase in relation to alcohol consumption. *Clin. Chim. Acta* **39:**75–80, 1972.

Rottenberg, H., Waring, A., and Rubin, E.: Tolerance and cross-tolerance in chronic alcoholics: Reduced membrane binding of ethanol and other drugs. *Science* **213:**583–585, 1981.

Roy, S., Ramalingaswami, V., and Nayak, N. C.: An ultrastructural study of the liver in Indian childhood cirrhosis with particular reference to the structure of cytoplasmic hyaline. *Gut* **12:**693–701, 1971.

Rubenstein, J. A., Collins, M. A., and Tabakoff, B.: Inhibition of liver aldehyde dehydrogenase by pyrogallol and related compounds. *Experientia* **31:**414, 1975.

Rubin, E., and Lieber, C. S.: Experimental alcoholic hepatic injury in man: Ultrastructural changes. *Fed. Proc.* **26:**1458–1467, 1967a.

Rubin, E., and Lieber, C. S.: Early fine structural changes in the human liver induced by alcohol. *Gastroenterology* **52:**1–13, 1967b.

Rubin, E., and Lieber, C. S.: Alcohol induced hepatic injury in nonalcoholic volunteers. *N. Engl. J. Med.* **278:**869–876, 1968.

Rubin, E., Beattie, D. S., and Lieber, C. S.: Effects of ethanol on the biogenesis of mitochondrial membranes and associated mitochondrial functions. *Lab. Invest.* **23:**620–627, 1970.

Rubin, E. Beattie, D. S., Toth, A., and Lieber, C. S.: Structural and functional effects of ethanol on hepatic mitochondria. *Fed. Proc.* **31:**131–140, 1972.

Rutherford, R. B., Boitnott, J. K., Donohoo, J. S., Margolis, S., Sebor, J., and Zuidema, G. D.: Production of nutritional cirrhosis in *Macacca mulatta* monkeys. *Arch. Surg.* **98:**720–730, 1969.

Ruwart, M. J. Rush, B. D., Snyder, K. F., Peters, K. M., Appelman, H. D., and Henley, K. S.: 16,16-Dimethyl prostaglandin E_2 delays collagen formation in nutritional injury in rat liver. *Hepatology* **8:**61–64, 1988.

Sabesin, S. M., Hawkins, H. L., Kuiken, L., and Ragland, J. B.: Abnormal plasma lipoproteins and lecithin–cholesterol acyltransferase deficiency in alcoholic liver disease. *Gastroenterology* **72:**510–518, 1977.

Salaspuro, M., and Sipponen, P.: Demonstration of an intracellular copper-binding protein by orcein staining on long-standing cholestatic liver diseases. *Gut* **17:**787–790, 1976.

Salaspuro, M. P., Sipponen, P., and Makkonen, H.: The occurrence of orcein-positive hepatocellular material in various liver diseases. *Scand. J. Gastroenterol.* **11:**677–681, 1976.

Sato, M., and Lieber, C. S.: Hepatic vitamin A depletion after chronic ethanol consumption in baboons and rats. *J. Nutr.* **111:**2015–2023, 1981.

Sato, M., and Lieber, C. S.: Increased metabolism of retinoic acid after chronic ethanol consumption in rat liver microsomes. *Arch. Biochem. Biophys.* **213:**557–564, 1982.

Sato, C., Matsuda, Y., and Lieber, C. S.: Increased hepatotoxicity of acetaminophen after chronic ethanol consumption in the rat. *Gastroenterology* **80:**140–148, 1981.

Sato, N., Kamada, T., Kawano, S., Hayashi, N., Kishida, Y., Meren, H., Yoshihara, H., and Abe, H.: Effect of acute and chronic ethanol consumption on hepatic tissue oxygen tension in rats. *Pharmacol. Biochem. Behav.* **18:**443–447, 1983.

Sato, S., Leo, M. A., and Lieber, C. S.: Ultrastructural localization of type III procollagen in baboon liver. *Am. J. Pathol.* **122:**212–217, 1986a.

Sato, S., Nouchi, T., Worner, T. M., and Lieber, C. S.: Liver fibrosis in alcoholics: Detection by Fab radioimmunoassay of serum procollagen-III-peptides detects liver fibrosis in alcoholics. *J.A.M.A.* **256:**1471–1473, 1986b.

Saunders, J. B., Haines, A., Portman, B., Wodak, A. D., Powell-Jackson, P. R., Davis, M., and Williams, R.: Accelerated development of alcoholic cirrhosis in patients with HLA-B8. *Lancet* **1:**1381–1384, 1982.

Savolainen, E. R., Goldberg, B., Leo, M. A., Velez, M., and Lieber, C. S.: Diagnostic value of serum procollagen peptide measurements in alcoholic liver disease. *Alcoholism: Clin. Exp. Res.* **8:**384–389, 1984a.

Savolainen, E. R., Leo, M. A., Timpl, R., and Lieber, C. S.: Acetaldehyde and lactate stimulate collagen synthesis of cultured baboon liver myofibroblasts. *Gastroenterology* **87:**777–787, 1984b.

Savolainen, M. J., Baraona, E., Pikkarainen, P., and Lieber, C. S.: Hepatic triacylglycerol synthesizing activity during progression of alcoholic liver injury in the baboon. *J. Lipid Res.* **25:**813–820, 1984.

Savolainen, M. J., Baraona, E., Leo, M. A., and Lieber, C. S.: Pathogenesis of the hypertriglyceridemia at early stages of alcoholic liver injury in the baboon. *J. Lipid Res.* **27:**813–820, 1986.

Schaffer, W. T., Denckla, W. D., Veech, R. L.: Effect of chronic ethanol administration on O_2 consumption in whole body and perfused liver of the rat. *Alcoholism: Clin. Exp. Res.* **5:**192–197, 1981.

Scheig, R., and Klatskin, G.: Some effects of ethanol and carbon tetrachloride on lipoperoxidation in rat liver. *Life Sci.* **8:**855–865, 1969.

Schlicting, P., Juhl, E., Poulsen, H., and Winkel, P.: Alcoholic hepatitis superimposed on cirrhosis: Clinical significance and effect of long-term prednisone treatment. *Scand. J. Gastroenterol.* **11:**304–312, 1976.

Schmidt, W.: Agreement, disagreement in experimental, clinical and epidemiological evidence on the etiology of alcoholic liver cirrhosis: A comment. In: *Alcoholic Liver Pathology* (J. M. Khanna, Y. Israel, and H. Kalant, eds.), Addiction Research Foundation of Ontario, Toronto, Canada, pp. 19–30, 1975.

Schmidt, W.: The epidemiology of cirrhosis of the liver: A statistical analysis of mortality data with special reference to Canada. In: *Alcohol and the Liver* (M. M. Fisher and J. G. Rankin, eds.), New York, Plenum, pp. 1–26, 1977.

Schnack, H., Stockinger, L., and Wewalka, F.: Adventitious connective tissue cells in the space of Disse and their relation to fiber formation. *Rev. Int. Hepatol.* **17:**857–860, 1968.

Schubert, M. P.: Compounds of thiol acids with aldehydes. *J. Biol. Chem.* **114**:341–350, 1936.

Schubert, M. P.: Reactions of semimercaptols with amino compounds. *J. Biol. Chem.* **121**:539–548, 1937.

Scott, B. B., Rajah, S. M., and Losowsky, M. D.: Histocompatibility antigens in chronic liver disease. *Gastroenterology* **72**:122–125, 1977.

Senoo, H., and Wake, K.: Suppression of experimental hepatic fibrosis by administration of vitamin A. *Lab. Invest.* **52**:182–194, 1985.

Shaw, S., and Lieber, C. S.: Mechanism of increased gamma glutamyl transpeptidase after chronic alcohol consumption: Hepatic microsomal induction rather than dietary imbalance. *Subst. Alcohol Actions/Misuse* **1**:423–428, 1978.

Shaw, S., Heller, E. A., Friedman, H. S., Baraona, E., and Lieber, C. S.: Increased hepatic oxygenation following ethanol administration in the baboon. *Proc. Soc. exp. Biol. Med.* **156**:509–513, 1977.

Shaw, S., Jayatilleke, E., and Lieber, C. S.: Hepatic lipid peroxidation: Potentiation by chronic alcohol feeding and attenuation by methionine. *J. Lab. Clin. Med.* **98**:417–425, 1981.

Shaw, S., Rubin, K. P., and Lieber, C. S.: Depressed hepatic glutathione and increased diene conjugates in alcoholic liver disease: Evidence of lipid peroxidation. *Dig. Dis. Sci.* **28**:585–589, 1983.

Shaw, S., Jayatilleke, E., and Lieber, C. S.: The effect of chronic alcohol feeding on lipid peroxidation in microsomes: Lack of relationship to hydroxyl radical generation. *Biochem. Biophys. Res. Commun.* **118**:233–238, 1984.

Shevchuk, O., Baraona, E., Ma, X., Pignon, J-P., and Lieber, C. S.: Gender differences in the response of hepatic fatty acids and cytosolic fatty acid-binding capacity to alcohol consumption in rats. *Proc. Soc. Exp. Biol. Med.* **198**:584–590, 1991.

Shigeta, Y., Ishii, H., Takagi, S., Yoshitake, Y., Hirano, T., Takagi, S., Kohno, H., and Tsuchiya, M.: HLA antigens as immunogenetic markers of alcoholism and alcoholic liver disease. *Pharmacol. Biochem. Behav.* **13**:89–94, 1980.

Shumaker, J. B., Resnick, R. H., Galambos, J. T., Makopour, H., and Iber, F. L.: A controlled trial of 6-methylprednisolone in acute alcoholic hepatitis. *Am. J. Gastroenterol.* **69**:443–449, 1978.

Siegers, C.-P., Schütt, A., and Strubelt, O.: Influence of some hepatotoxic agents on hepatic glutathione levels in mice. *Proc. Eur. Soc. Toxicol.* **18**:160–162, 1977.

Sim, J. S., Franks, K. E., French, S. W., Caldwell, M. G.: Mallory bodies compared with microfilament hyperplasia. *Arch. Pathol. Lab. Med.* **101**:401–404, 1977.

Simon, D., and Galambos, J. T.: A randomized controlled study of peripheral parenteral nutrition in moderate and severe alcoholic hepatitis. *Hepatology* **7**:200–207, 1988.

Simon, J. B., and Scheig, R.: Serum cholesterol esterification in liver disease. *N. Engl. J. Med.* **283**:841–846, 1970.

Simon, M., Bourel, M., Genetet, B., Fauchet, R., Edan, G., and Brissot, P.: Idiopathic hemochromatosis and iron overload in alcoholic liver disease: Differentiation by HLA phenotype. *Gastroenterology* **73**:655–658, 1977.

Sippel, H. W.: Effect of an acute dose of ethanol on lipid peroxidation and on the activity of microsomal glutathione *S*-transferase in rat liver. *Acta Pharmacol. Toxicol.* **53**:135–140, 1983.

Skrede, S., Blomhoff, J. P., Elgjo, K., and Gjone, E.: Serum proteins in diseases of the liver. *Scand. J. Clin. Lab. Invest.* **35**:399–406, 1975.

Skrede, S., Blomhoff, J. P., and Gjone, E.: Biochemical features of acute and chronic hepatitis. *Ann. Clin. Res.* **8**:182–199, 1976.

Smith, F. R., and Goodman, D. S.: Vitamin A transport in human vitamin A toxicity. *N. Engl. J. Med.* **294**:805–808, 1976.

Solomon, L. R.: Evidence for the generation of transaminase inhibitor(s) during ethanol metabolism by rat liver homogenates: A potential mechanism for alcohol toxicity. *Biochem. Med. Metab. Biol.* **38**:9–18, 1987.

Sotaniemi, E. A., Ahlqvist, J., Pelkonen, R. O., Pirttiaho, H., and Luoma, P. V.: Histological changes in the liver and indices of drug metabolism in alcoholics. *Eur. J. Clin. Pharmacol.* **11**:295–303, 1977.

Soterakis, J., Resnick, R. H., and Iber, F. L.: Effect of alcohol abstinence on survival in cirrhotic portal hypertension. *Lancet* **2**:65–67, 1973.

Speisky, H., MacDonald, A., Giles, G., Gunasekara, A., and Israel, Y.: Increased loss and decreased synthesis of hepatic glutathione after acute ethanol administration. *Biochem. J.* **225**:565–572, 1985.

Sprince, H., Parker, C. M., Smith, G., and Gonzales, L. J.: Protection against acetaldehyde toxicity in the rat by L-cysteine, thiamin and L-2-methylthiazolidene-4-carboxylic acid. *Agents Actions* **4**:125–130, 1974.

Stein, S. W., Lieber, C. S., Leevy, C. M., Cherrick, G. R., and Abelmann, W. H.: The effect of ethanol upon systemic and hepatic blood flow in man. *Am. J. Clin. Nutr.* **13**:68–74, 1963.

Stone, W. D., Islam, N. R. K., and Paton, A.: The natural history of cirrhosis. *Q. J. Med.* **37**:119–132, 1968.

Summerskill, W. H. J., Davidson, C. S., Dible, J. H., Mallory, G. K., Sherlock, S., Turner, M. D., and Wolfe, S. J.: Cirrhosis of the liver: A study of alcoholic and nonalcoholic patients in Boston and London. *N. Engl. J. Med.* **262**:1–11, 1960.

Sun, G. Y., and Sun, A. Y.: Ethanol and membrane lipids. *Alcoholism: Clin. Exp. Res.* **9**:164–180, 1985.

Svoboda, D. J., and Manning, R. T.: Chronic alcoholism with fatty metamorphosis of the liver. *Am. J. Pathol.* **44**:645–662, 1964.

Szilagyi, A., Lerman, S., and Resnick, R. H.: Ethanol, thyroid hormones and acute liver injury: Is there a relationship? *Hepatology* **3**:593–600, 1983.

Tabor, E., Gerety, R. J., Barker, L. F., Howard, C. R., and Zuckerman, A. J.: Effect of ethanol during hepatitis B virus infection in chimpanzees. *J. Med. Virol.* **2**:295–303, 1978.

Takada, A., Nei, J., Matsuda, Y., and Kanayama, R.: Clinicopathological study of alcoholic fibrosis. *Am. J. Gastroenterol.* **77**:660–666, 1982.

Takada, A., Nei, J., Tamino, H., and Takase, S.: Effects of malotilate on ethanol-inhibited hepatocyte regeneration in rats. *J. Hepatol.* **5**:336–343, 1987.

Tanaka, Y., Funaki, N., Mak, K. M., Kim, C. I., and Lieber, C. S.: Effects of ethanol and hepatic vitamin A on the proliferation of lipocytes in regenerating rat liver. *J. Hepatology* **12:**344–350, 1991.

Tanner, A. R., Bantock, I., Hinks, L., Lloyd, B., Turner, N. R., and Wright, R.: Depressed selenium and vitamin E levels in an alcoholic population: Possible relationship to hepatic injury through increased lipid peroxidation. *Dig. Dis. Sci.* **31:**1307–1312, 1986.

Targan, S. R., Chow, A. W., and Guze, L. B.: Role of anaerobic bacteria in spontaneous peritonitis of cirrhosis: Report of two cases and review of the literature. *Am. J. Med.* **62:**397–403, 1977.

Teschke, R., Brand, A., and Strohmeyer, G.: Induction of hepatic microsomal gamma-glutamyl transferase activity following chronic alcohol consumption. *Biochem. Biophys. Res. Commun.* **75:**718–724, 1977.

Teschke, R. E., Moreno, F., Heinen, E., Herrmann, J., Kruskemper H.-L., and Strohmeyer, G.: Hepatic thyroid hormone levels following chronic alcohol consumption: Direct experimental evidence in rats against the existence of a hyperthyroid hepatic state. *Hepatology* **3:**469–474, 1983.

Thomas, C. E., Morehouse, L. A., and Aust, S. D.: Ferritin and superoxide-dependent lipid peroxidation. *J. Biol. Chem.* **260:**3275–3280, 1985.

Thompson, J. A., and Reitz, R. C.: Studies on the acute and chronic effects of ethanol ingestion on choline oxidation. *Ann. N.Y. Acad. Sci.* **273:**194–204, 1976.

Tisdale, W. A.: Parenchymal siderosis in patients with cirrhosis after portosystemic shunt surgery. *N. Engl. J. Med.* **265:**928–932, 1961.

Tisdale, W. A., and Klatskin, G.: The fever of Laennec's cirrhosis. *Yale J. Biol. Med.* **33:**94–106, 1960.

Tribble, D. L., Aw, T. Y., and Jones, D. P.: The pathological significance of lipid peroxidation in oxidative cell injury. *Hepatology* **7:**377–387, 1987.

Tsutsumi, M., Shimizu, M., Lasker, J. M., and Lieber, C. S.: Intralobular distribution of ethanol-inducible cytochrome P450IIE1 in liver. *Hepatology* **8:**1237 (abstract), 1988.

Tsutsumi, M., Lasker, J. M., Shimizu, M., Rosman, A. S., and Lieber, C. S.: The intralobular distribution of ethanol-inducible P-450IIE₁ in rat and human liver. *Hepatology* **10:**437–446, 1989.

Tygstrup, N., Juhl, E., and The Copenhagen Study Group for Liver Diseases: The treatment of alcoholic cirrhosis. The effect of continued drinking and prednisone on survival. In: *Alcohol and the Liver* (W. Gerok, K. Sickinger, and H. H. Henneheuser, eds.), New York and Stuttgart, Schattauer Verlag, pp. 519–526, 1971.

Uchida, T., Kao, H., Quispe-Sjogren, M., and Peters, R. L.: Alcoholic foamy degeneration—A pattern of acute alcoholic injury of the liver. *Gastroenterology* **84:**683–692, 1983.

Ugarte, G. R., Iturriaga, H., and Pereda, T.: Possible relationship between the rate of ethanol metabolism and the severity of hepatic damage in chronic alcoholics. *Dig. Dis.* **22:**406–410, 1977.

Valenzuela, A., Fernadez, F., and Videla, L. A.: Hepatic and biliary levels of glutathione and lipid peroxides following iron overload in the rat: Effect of simultaneous ethanol administration. *Toxicol. Appl. Pharmacol.* **70:**87–95, 1983.

Välimäki, M., Härkönen, M., and Ylikahri, R.: Serum ferritin and iron levels in chronic male alcoholics before and after ethanol withdrawal. *Alcohol Alcohol.* **18:**255–260, 1983.

Van Waes, L., and Lieber, C. S.: Early perivenular sclerosis in alcoholic fatty liver, an index of progressive liver injury. *Gastroenterology* **73:**646–650, 1977a.

Van Waes, L., and Lieber, C. S.: Glutamate dehydrogenase, a reliable marker of liver cell necrosis in the alcoholic. *Br. Med. J.* **2:**1508–1510, 1977b.

Vendemiale, G., Jayatilleke, E., Shaw, S., and Lieber, C. S.: Depression of biliary glutathione excretion by chronic ethanol feeding in the rat. *Life Sci.* **34:**1065–1073, 1984.

Vesell, E. S., Page J. G., and Passananti, G. T.: Genetic and environmental factors affecting ethanol metabolism in man. *Clin. Pharmacol. Ther.* **12:**192–201, 1971.

Videla, L. A., and Israel, Y.: Factors that modify the metabolism of ethanol in rat liver and adaptive changes produced by its chronic administration. *Biochem. J.* **118:**275–281, 1970.

Villeneuve, J. P., Pomier, G., and Huet, P. M.: Effect of ethanol on hepatic blood flow in unanesthetized dogs with chronic portal and hepatic vein catheterization. *Can. J. Physiol. Pharmacol.* **59:**598–603, 1981.

Von Olderhausen, H. F.: Alkoholische Leberschäden. *Therapiewoche* **20:**58–73, 1970.

Voss, B., Rauterberg, J., Pott, G., Brehmer, U., Allam, S., Lehmann, R., and Bassewitz, D. V.: Nonparenchymal cells cultivated from explants of fibrotic liver resemble endothelial and smooth muscle cells from blood vessel walls. *Hepatology* **2:**19–28, 1982.

Walker, F., and Shand, J.: Influence of alcohol on collagen synthesis *in vitro. Lancet* **1:**233–234, 1972.

Walker, F., Elmslie, W., Fraser, R. A., Snape, P. E., and Watt, G. C. M.: Cytotoxic effect of alcohol on liver cells and fibroblast *in vitro. Scot. Med. J.* **19:**125–127, 1974.

Walter, E., Blum, H. E., Meier, P., Huonker, M., Schmid, M., Maier, K.-P., Offensperger, W.-B., Offensperger, S., and Gerok, W.: Hepatocellular carcinoma in alcoholic liver disease: No evidence for a pathogenetic role of hepatitis B virus infection. *Hepatology* **8:**745–748, 1988.

Wands, J. R., Carter, E. A., Bucher, N. L. R., and Isselbacher, K. J.: Inhibition of hepatic regeneration in rats by acute and chronic ethanol intoxication. *Gastroenterology* **77:**528–531, 1979.

Waring, A. J., Rottenberg, H., Ohnishi, T., and Rubin, E.: Membranes and phospholipids of liver mitochondria from chronic alcoholic rats are resistant to membrane disordering by alcohol. *Proc. Natl. Acad. Sci. U.S.A.* **78:**2582–2586, 1981.

Waring, A. J., Rottenberg, H., Ohnishi, T., and Rubin, E.: The effect of chronic ethanol consumption on temperature-dependent physical properties of liver mitochondrial membranes. *Arch. Biochem. Biophys.* **216:**51–61, 1982.

Wasserman, F.: The structure of the wall of the hepatic sinusoids in the electron microscope. *Z. Zellforschr.* **49:**13–32, 1958.

Weber, F. L., Mitchell, G. E., Jr., Powell, D. E., Reiser, B. J., and Banwell, J. G.: Reversible hepatotoxicity associated with hepatic vitamin A accumulation in a protein-deficient patient. *Gastroenterology* **82:**118–123, 1982.

Wendel, A., Fenerstein, S., and Konz, K. H.: Acute paracetamol intoxication of starved mice leads to lipid peroxidation *in vivo*. *Biochem. Phys.* **28**:2051–2055, 1979.

Whetton, A. D., Needham, L., Dodd, N. J. F., Heyworth, C. M., and Houslay, M. D.: Forskolin and ethanol both perturb the structure of liver plasma membranes and activate adenylate cyclase activity. *Biochem. Pharmacol.* **32**:1601–1608, 1983.

Wickramasinghe, S. N.: Supernatants from ethanol-containing macrophage cultures have cytotoxic activity. *Alcohol Alcohol.* **21**:263–268, 1986.

Wickramasinghe, S. N., Gardner, B., and Barden, G.: Circulating cytotoxic protein generated after ethanol consumption: Identification and mechanism of reaction with cells. *Lancet* **18**:122–126, 1987.

Wietholz, H., and Colombo, J. P.: Das Verhalten der gamma-glutamyltranspeptidae und anderer Leberenzyme in Plasma während der Alkohol-Entziehungskur. *Schweiz. Med. Wochenschr.* **106**:981–987, 1976.

Wilgram, G. F.: Experimental Laennec type of cirrhosis in monkeys. *Ann. Intern. Med.* **51**:1134–1158, 1959.

Wilkinson, P., Santamaria, J. N., and Rankin, J. G.: Epidemiology of alcoholic cirrhosis. *Aust. Ann. Med.* **18**:222–226, 1969.

William, A. J. K., and Barry, R. E.: Free radical generation by neutrophils: A potential mechanism of cellular injury in acute alcoholic hepatitis. *Gut.* **28**:1157–1161, 1987.

Wisse, E.: An electron microscopic study of the fenestrated endothelial lining of rat liver sinusoids. *J. Ulstruct. Res.* **31**:125–150, 1970.

Wood, R. L.: Evidence of species differences in the ultrastructure of the hepatic sinusoid. *Z. Zellforschr.* **58**:679–692, 1963.

Worner, T. M., and Lieber, C. S.: Plasma glutamate dehydrogenase: Clinical application in patients with alcoholic liver disease. *Alcoholism: Clin. Exp. Res.* **4**:431–434, 1980.

Worner, T. M., and Lieber, C. S.: Perivenular fibrosis as precursor lesion of cirrhosis. *J.A.M.A.* **254**:627–630, 1985.

Wu, A., and Leevy, A. J.: Effect of alcohol on total urinary hydroxy-proline excretion. *Am. J. Gastroenterol.* **64**:217–220, 1975.

Yamada, S., and Lieber, C. S.: Decrease in microviscosity and cholesterol content of rat liver plasma membranes after chronic ethanol feeding. *J. Clin. Invest.* **74**:2278–2289, 1984.

Yamada, S., Mak, K. M., and Lieber, C. S.: Chronic ethanol consumption alters rat liver plasma membranes and potentiates release of alkaline phosphatase. *Gastroenterology* **88**:1799–1806, 1985a.

Yamada, S., Wilson, J. S., and Lieber, C. S.: The effects of alcohol and diet on hepatic and serum gamma-glutamyltranspeptidase activities. *J. Nutr.* **115**:1285–1290, 1985b.

Yokoo, H., Minick, O. T., Batti, F., and Kent, G.: Morphologic variants of alcoholic hyalin. *Am. J. Pathol.* **69**:25–40, 1972.

Yoshikawa, Y., Takemura, S., and Kondo, M.: α-Tocopherol level in liver diseases. *Acta Vitamenol. Enzymol.* **4**:311–318, 1982.

Younes, M., and Strubelt, O.: Enhancement of hypoxic liver damage by ethanol: Involvement of xanthine oxidase and the role of glycolysis. *Biochem. Pharmacol.* **36**:2973–2977, 1987.

Zern, M. A., Leo, M. A., Giambrone, M.-A., and Lieber, C. S.: Increased type I procollagen mRNA levels and *in vitro* protein synthesis in the baboon model of chronic alcoholic liver disease. *Gastroenterology* **89**:1123–1131, 1985.

Zieve, L.: Jaundice, hyperlipemia, and hemolytic anemia: A heretofore unrecognized syndrome associated with alcoholic fatty liver and cirrhosis. *Ann. Intern. Med.* **48**:471–496, 1958.

8

Alcohol and the Hematologic System

John Lindenbaum

The hematologic disorders associated with alcoholism present some of the most challenging (and often confusing) problems in differential diagnosis to the internist and hematologist. Abnormalities involving red cells, white cells, and platelets often occur together but may have differing etiologies. The simultaneous, sometimes synergistic and sometimes opposing effects of acute and chronic ethanol intoxication, liver disease, multiple nutritional deficiencies, infections, and other associated disease states are often difficult to disentangle. The problem is compounded because the alcoholic patient admitted to the hospital is not in a steady state. During the first days or weeks in the hospital, the effects of alcohol on circulating blood cells rapidly recede, liver dysfunction often improves, nutritional depletion states are corrected, and infections are treated. The time at which hematologic observations are made thus becomes crucial: profound thrombocytopenia, for example, may be followed in a few days by striking thrombocytosis; megaloblastic and sideroblastic marrow abnormalities rapidly disappear; an elevated serum iron level during intoxication may be replaced by hypoferremia within 24 to 48 hr of abstinence; and brisk reticulocytosis may quickly succeed reticulocytopenia. The physician unaware of these potential developments will indeed often find the alcoholic patient hematologically baffling.

The clinical investigator interested in the mechanism underlying the hematologic abnormalities must grapple with the same problems. Progress has been made in the last two decades in unraveling some of the major riddles, in good part as the result of planned observations made in human volunteers in whom ethanol was administered under controlled experimental conditions. Our understanding of the disorders of the blood associated with alcoholism has been summarized in two reviews by Eichner (1973) and Lindenbaum (1983). In this chapter, we attempt to bring current concepts up to date, with an emphasis on the clinical syndromes that may be encountered and their underlying pathophysiology. The various disorders affecting red cells, platelets, and white cells are discussed in that order.

8.1. Red Cells

8.1.1. Megaloblastic Anemia and Folate Deficiency

8.1.1.1. Clinical Aspects

Anemia is common in alcoholics. It has been found in 13 to 62% of chronic alcoholics hospitalized for acute or chronic illness (Eichner and Hillman, 1971; Hines and Cowan, 1974; Wu *et al.*, 1975; Heidemann *et al.*, 1981). In two series of consecutively studied patients admitted to hospitals in the United States for complications of chronic alcoholism, megaloblastic anemia resulting from folate deficiency was considered to be the most common cause of anemia, occurring alone or in combination with ringed sideroblasts in approximately 40% of all patients. In addition, several patients in both series were found to have megaloblastic marrow changes in the absence of anemia (Eichner and Hillman, 1971; Hines and Cowan, 1974). Many investigators have found megaloblastic anemia to be common in hospitalized alcoholics (Jarrold and Vilter, 1949; Movitt, 1949, 1950; Jandl, 1955; Jandl and Lear, 1956; Krasnow *et al.*, 1957; Herbert *et al.*, 1960; Herbert *et al.*, 1963; Klipstein and Lindenbaum, 1965; Deller *et al.*, 1965; Kimber *et al.*, 1965b; Jarrold *et al.*, 1967; Hines, 1969; Cowan and Hines, 1971; Wu *et al.*, 1975; Pierce *et al.*, 1976; Heidemann *et al.*, 1981; Savage and Lindenbaum, 1986). In well-nourished alcoholics, however, it is a rare cause of anemia (Eichner *et al.*, 1972).

The megaloblastic marrow changes have usually been attributed to folic acid deficiency, unless an unrelated associated cause of cobalamin lack, such as pernicious anemia (Eichner and Hillman, 1971; Savage and Lindenbaum, 1986), has been present.

The presence of megaloblastic anemia in alcoholics has not been found to correlate well with the severity of hepatic dysfunction (Klipstein and Lindenbaum, 1965; Deller *et al.*, 1965; Wu *et al.*, 1975; Savage and Lindenbaum, 1986). Furthermore, folate deficiency has only been encountered very rarely in cirrhotic patients who were not alcoholics (Klipstein and Lindenbaum, 1965; Kimber *et al.*, 1965b). Megaloblastic anemia develops commonly in imbibers of wine and whiskey, which contain little or no folate (Herbert, 1963), and is less frequently seen in those with a preference for beer, which is rich in the vitamin (Klipstein and Lindenbaum, 1965; Deller *et al.*, 1965; Eichner and Hillman, 1971; Wu *et al.*, 1975).

There is a very strong association between decreased dietary folate intake and the presence of megaloblastic anemia (Herbert *et al.*, 1963; Klipstein and Lindenbaum, 1965; Deller *et al.*, 1965; Jarrold *et al.*, 1967; Hines, 1969; Eichner and Hillman, 1971; Eichner *et al.*, 1972; Wu *et al.*, 1975). In some patients, other factors may contribute to the disturbance in folate balance, including gastrointestinal bleeding, hypersplenism, hemolysis, infection, vomiting, and treatment with drugs such as anticonvulsants or triamterene (Druskin *et al.*, 1962; Klipstein and Lindenbaum, 1965; Kimber *et al.*, 1965b; Lieberman and Bateman, 1968; Corcino *et al.*, 1970; Ballard and Lindenbaum, 1974; Renous *et al.*, 1976; Pillegand *et al.*, 1977; Wilms *et al.*, 1979). Certain patients, for unknown reasons, appear to be especially prone to recurrent episodes of severe folate deficiency (J. Lindenbaum, unpublished observations).

8.1.1.2. Diagnostic Findings

The diagnosis of megaloblastic anemia in alcoholics may be straightforward or difficult. In some patients, all of the "classical" findings associated with florid megaloblastic states are seen, including an elevated mean erythrocyte cell volume (MCV), macroovalocytes and hypersegmented neutrophils on the blood smear, increased serum unconjugated bilirubin, markedly elevated serum lactic dehydrogenase (LDH), and megaloblastic erythroid and myeloid precursors in the bone marrow associated with low serum and erythrocyte folate concentrations (Lindenbaum, 1980a). In many patients, however, particularly those who are only Smildly anemic or who have not yet become anemic, one or more of the classical findings may be absent.

In a recent series of 121 consecutively studied hospitalized alcoholics with anemia, the sensitivity, specificity, and predictive values of various standard diagnostic findings in predicting the presence of megaloblastic change in the bone marrow were evaluated (Savage and Lindenbaum, 1986). Elevation of the MCV (>100 fl) was only present in two-thirds of patients with megaloblastic change; furthermore, the majority of patients with modest MCV elevations (in the range of 100 to 110 fl) did not have megaloblastic anemia. Only an MCV > 110 fl was strongly predictive of megaloblastic marrow morphology; such a high MCV, however, was only seen in one-fourth of the patients. The MCV may be normal in patients with megaloblastic anemia when another disorder, such as iron deficiency, the anemia of chronic disease, or thalassemia trait, is also present (Eichner and Hillman, 1971; Spivak, 1982). The association of iron deficiency and folate deficiency is not uncommon in alcoholics (Eichner and Hillman, 1971; Eichner *et al.*, 1972). In patients with such combined disorders, the characteristic peripheral blood findings of folate deficiency (oval macrocytes and hypersegmented neutrophils) may still be present, often coexisting with microcytosis and hypochromia in a "dimorphic" picture. In some patients with combined deficiencies, however, the blood smear before treatment is only indicative of one deficiency or the other.

The presence of hypersegmented neutrophils on blood smears has a high specificity (95%) but only 78% sensitivity for megaloblastic change (Savage and Lindenbaum, 1986). Neutrophil hypersegmentation may be absent in some patients with megaloblastic anemia with a marked shift to the left in the white blood cell (WBC) differential count or sever granulocytopenia (Lindenbaum and Nath, 1980) or with mild anemia (Savage and

Lindenbaum, 1986). Macroovalocytosis has a higher sensitivity but low specificity. If the presence of more than 3% macroovalocytes is used as a diagnostic criterion, high specificity is obtained at the expense of sensitivity (Savage and Lindenbaum, 1986). Elevations in LDH are of no diagnostic value unless greater than 1000 IU, a finding present in only one in ten patients with megaloblastic anemia. All of these "classical findings" are more likely to be present in severely anemic patients (Savage and Lindenbaum, 1986).

Other "typical" but nonspecific findings usually present in severe megaloblastic anemia include reticulocytopenia, decreased plasma haptoglobin, thrombocytopenia, and (less commonly) granulocytopenia.

The megaloblastic morphological abnormalities on bone marrow examination are frequently more striking in the granulocytic series (Jandl and Lear, 1956; Klipstein and Lindenbaum, 1965; Savage and Lindenbaum 1986). One possible explanation for this is partial reversion of the marrow toward normal after cessation of drinking, since giant bands and metamyelocytes may persist in the marrow for as long as 2 weeks after therapy of a megaloblastic anemia has converted the erythroid series to normoblastic (Nath and Lindenbaum, 1979). In other patients, the coexistence of additional disorders, such as iron deficiency or the anemia of chronic disease, may have "masked" the expression of megaloblastic change in erythroid precursors (Spivak, 1982; Savage and Lindenbaum, 1986). In those with uncomplicated megaloblastic anemia, reticulocytosis develops, and the erythroid marrow reverts to normal during the first 3 to 7 days of therapy with folic acid. These changes occur (but often more gradually) in patients withdrawn from alcohol and treated with hospital diet alone. Hypersegmented neutrophils, however, typically persist unchanged (or increase in numbers) in the peripheral blood during the first 10 days of therapy and may only disappear after about 2 weeks (Figs. 8.1 and 8.2) (Nath and Lindenbaum, 1979).

Reticulocytosis and elevation of the hematocrit often do not ensue with the rapidity normally expected after treatment of megaloblastic anemias in nonalcoholic patients (Savage and Lindenbaum, 1986). This may be the result of another underlying cause of the anemia (Eichner and Hillman, 1971; Savage and Lindenbaum, 1986).

8.1.1.3. Serum and Red Cell Vitamin Levels

In the typical alcoholic patient with megaloblastic anemia, serum and erythrocyte folate concentrations are

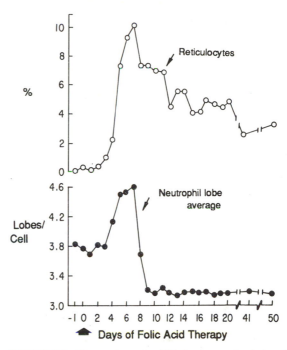

FIGURE 8.1. Changes in neutrophil lobe average (an index of hypersegmentation) in an alcoholic patient with megaloblastic anemia caused by folate deficiency during therapy with large doses of folic acid. During the first 3 days of treatment, the lobe average (shown in bottom half of figure) was unchanged. It then increased at the time of increasing reticulocytosis (upper part of figure) and became normal by day 9.

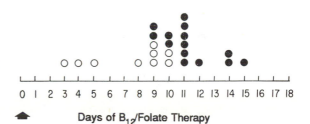

FIGURE 8.2. Time of first fall in peripheral blood neutrophil lobe average in 23 patients with megaloblastic anemia during treatment with large doses of vitamin B_{12} and folic acid. Improvement in hypersegmentation tended to occur more rapidly in patients with associated inflammatory disorders (pneumonia, tuberculosis, or alcoholic hepatitis). (From data reported by Nath and Lindenbaum, 1979.)

low, and serum cobalamin (vitamin B_{12}) levels are normal or elevated (Jandl and Lear, 1956; Herbert et al., 1963; Deller et al., 1965; Kimber et al., 1965b; Klipstein and Lindenbaum, 1965; Jarrold et al., 1967; Hines, 1969; Eichner and Hillman, 1971; Pierce et al., 1976).

In a substantial minority of patients, however, serum or plasma folate concentrations are normal (Herbert et al., 1964; Eichner and Hillman, 1971; Eichner et al., 1972; Herbert et al., 1973; Pierce et al., 1976; Wu et al., 1975; Chalmers et al., 1980; Heidemann et al., 1981), although the anemia subsequently appears to respond to folic acid therapy. In two series of patients from our own experience, serum or plasma folate levels were normal in 25–56% of alcoholics with megaloblastic anemia (Savage and Lindenbaum, 1986). The cause of the "falsely" normal serum folate concentrations was unclear, but it did not correlate with dietary intake of folate before serum levels were obtained or with abnormal liver function (Savage and Lindenbaum, 1986). Erythrocyte folate levels are more reliably depressed in alcoholics with megaloblastic anemia but may be in the lower end of the normal range in 10–30% of patients (Savage and Lindenbaum, 1986).

On the other hand, many alcoholics have low serum folate levels in the absence of morphological evidence of tissue deficiency of the vitamin (Herbert et al., 1963; Klipstein and Lindenbaum, 1965; Jarrold et al., 1967; Hourihane and Weir, 1970; Eichner et al., 1972). Serum folate concentrations fall early in the course of dietary folate deprivation and may be subnormal for weeks or months before stores are depleted to the point of causing disturbances in hematopoiesis (Herbert, 1962; Eichner et al., 1971). In addition, in relatively well-nourished alcoholics, a high serum ethanol levels may be associated with low serum folate concentrations (Eichner et al., 1972; Pierce et al., 1976). Because of the frequency of diminished serum folate values in alcoholic patients in the absence of morphological evidence of folate deficiency, as well as the not uncommon finding of normal levels in the presence of megaloblastic change, the serum folate concentration is of very limited usefulness as a screening test in the diagnostic workup of anemia in alcoholics (Savage and Lindenbaum, 1986). Erythrocyte folate levels are a better indicator of depletion of vitamin stores and are less frequently subnormal in alcoholics. Nonetheless, they are often low in the absence of megaloblastic hematopoiesis. Like the plasma level, the red cell folate lacks sufficient sensitivity and specificity to be a reliable indicator of the presence of megaloblastic change in alcoholics (Savage and Lindenbaum, 1986). Therefore,

if marrow and peripheral blood abnormalities indicative of folate depletion are absent in a patient with low serum and/or red cell folate concentrations, the anemia is a result of some other cause and will not respond to folic acid therapy (Klipstein and Lindenbaum, 1965).

In most alcoholics with megaloblastic anemia, the serum cobalamin concentration is normal or elevated, although a low serum level may be encountered in a patient with primary folate deficiency. The cause of this "secondary" depression in cobalamin, which will return to normal after treatment with folic acid alone, is unknown (Lindenbaum, 1980a). In a minority of alcoholic patients with megaloblastic anemia (as high as 20% in our referral experience), primary deficiency of cobalamin caused by coincidental underlying disorders, such as pernicious anemia, is present (Savage and Lindenbaum, 1986). Because of the importance of promptly diagnosing and treating cobalamin deficiency, we favor measuring serum cobalamin levels in selected alcoholic patients with anemia (see Section 1.10).

8.1.1.4. Pathogenesis

Although a decreased dietary folate intake appears to be a necessary factor in the development of folic acid deficiency in alcoholics, it is now well established that ethanol ingestion also plays an important role. This was first demonstrated by the landmark studies of Sullivan and Herbert (1964), who administered alcohol along with small doses of folic acid to three patients with untreated megaloblastic anemia resulting from folate deficiency (Fig. 8.3). In each case the ingestion of whiskey, wine, or ethanol itself prevented the hematologic response to folic acid therapy. In addition, if alcohol was given after the marrow had become normoblastic, megaloblastic changes promptly recurred. Ethanol also depressed granulocyte and platelet levels. The hematosuppressive effect of alcohol could be overcome with larger doses of folic acid or folinic acid (Sullivan and Herbert, 1964). Subsequently, other investigators were also able to induce megaloblastic marrow abnormalities by the administration of ethanol and a folate-poor diet to human volunteers (Hines, 1969; Eichner and Hillman, 1971, 1973; Hines and Cowan, 1970; Cowan, 1973; Halsted et al., 1973). In addition, Eichner and Hillman (1973) found that when alcohol was given along with a low-folate diet, megaloblastic marrow conversion occurred much more rapidly than when the diet alone was taken. The failure of such morphological changes to occur when alcohol was given with folate supplements to well-nourished volunteers (Lindenbaum and Lieber,

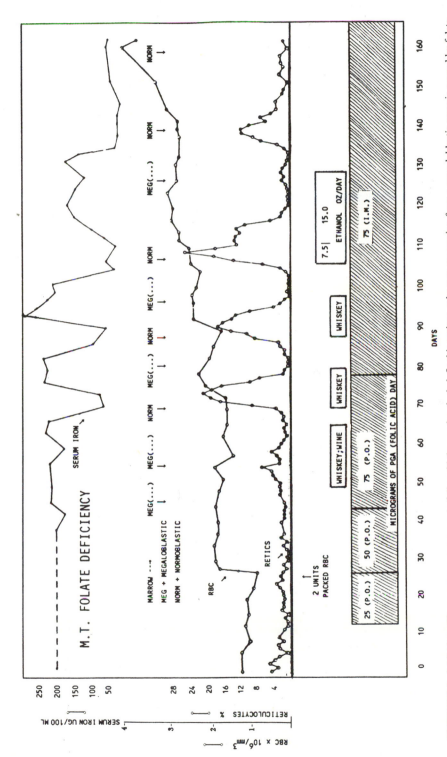

FIGURE 8.3. Inhibition of bone marrow responses to folic acid therapy by alcohol. In this patient, who was recovering from a megaloblastic anemia caused by folate deficiency, reticulocyte responses to doses of 75 μg of folic acid, given orally or parenterally, were reversed by the administration of alcoholic beverages or pure ethanol. The marrow morphology reverted from normoblastic to megaloblastic during each of the four periods of alcohol administration. (From Sullivan and Herbert, 1964.)

1969; Cowan, 1973) indicates that ethanol accelerates the development of megaloblastic erythropoiesis only when folate stores are depleted.

The manner in which ethanol interferes with folate metabolism has not been established. Malabsorption of folic acid may plan a role in some patients with megaloblastic anemia (Klipstein and Lindenbaum, 1965; Halsted *et al.*, 1971). A modest depression in jejunal folate uptake was also noted in chronic beer drinkers who had not developed folate deficiency (Koch, 1980). Alcohol intoxication and folate deficiency appear to act synergistically to depress small bowel function. When ethanol was administered chronically along with an adequate diet with folic acid supplements, human volunteers developed moderate impairment of jejunal salt and water absorption (Halsted *et al.*, 1973; Mekhjian and May, 1977), but folic acid absorption remained normal or minimally impaired (Lindenbaum and Lieber, 1971; Halsted *et al.*, 1973). In contrast, when alcohol was given along with a folate-poor diet, more severe abnormalities of salt and water transport as well as definite folate malabsorption occurred (Halsted *et al.*, 1973; Mekhjian and May, 1977). Folate deficiency alone, in the absence of alcohol, may cause some impairment of jejunal function but has not been shown to interfere with folic acid absorption (Halsted *et al.*, 1973; Cook, 1976; F. Marxer, J. Kaunitz, and J. Lindenbaum, unpublished observations, 1979).

Decreased jejunal folate uptake that was present in alcoholics shortly after admission to the hospital reverted to normal after 1 to 2 weeks of hospital diet despite the continued administration of large amounts of alcohol (Halsted *et al.*, 1971). Similarly, folate malabsorption induced experimentally by the administration of ethanol with a folate-poor diet was corrected by a normal diet with folic acid supplements without reduction of alcohol intake (Halsted *et al.*, 1973). These observations suggest that alcohol only interferes significantly with folic acid absorption when folate deficiency is already present and that deficiency of the vitamin is not primarily the result of impairment of intestinal transport. Furthermore, small bowel dysfunction in alcoholic patients is mainly seen in association with severe folate depletion (F. Marxer, J. Kaunitz, and J. Lindenbaum, unpublished observations, 1979). In monkeys, however, chronic ethanol administration resulted in malabsorption of folic acid that was not associated with decreased serum folate levels, although hepatic concentrations of the vitamin were diminished (Romero *et al.*, 1981).

Other workers have provided evidence that alcohol interferes with folate metabolism at sites other than the gut. In Sullivan and Herbert's (1964) study, alcohol blocked the therapeutic effects of folic acid on megaloblastic anemia regardless of whether the vitamin was given orally or parenterally. When alcohol is administered with a deficient diet to human subjects, serum folate falls to low levels within a few days (Eichner and Hillman, 1973; Hines and Cowan, 1974). Eichner and Hillman (1973) also found that the acute intravenous administration of ethanol caused a marked fall in serum folate levels after 8 hr in fasting volunteers (Fig. 8.4).

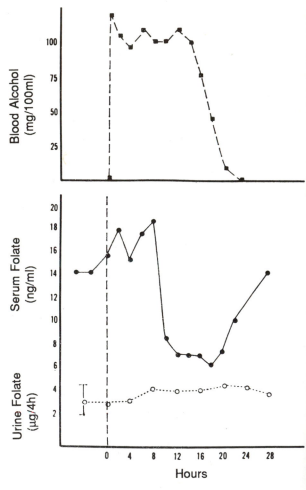

FIGURE 8.4. Effect of the infusion of ethanol intravenously for 13 hr on serum and urinary folate concentrations in a volunteer subject. During the first 45 min, 40 ml of 10% ethanol was given, followed by a steady infusion at a rate of 2 ml/min for the remainder of the period. Serum folate fell abruptly between 8 and 10 hr and promptly returned to normal when blood alcohol levels fell. (From Eichner and Hillman, 1973.)

These workers interpreted the acute effect of alcohol on serum folate levels as most likely caused by a block in the delivery of storage methylfolate from the liver into the circulation or by interruption of an enterohepatic circulation (Eichner and Hillman, 1973; Lane et al., 1976). Alcohol ingestion by human subjects did not alter the rate of clearance of intravenously administered folates from the circulation (Lane et al., 1976).

Recent attempts to define the nature of the alcohol-induced disturbance in folate metabolism in animal models have produced apparently conflicting results. In rats, drainage of the common bile duct caused a fall in serum folate concentrations within 6 hr, suggesting that an enterohepatic circulation was important in maintaining normal serum levels of the vitamin (Hillman et al., 1977; Steinberg et al., 1979). Hillman, Steinberg, and their colleagues have presented evidence, based on experiments in which ethanol was given for 3 days to folate-deficient rats, that the enterohepatic circulation of folate may be blocked by alcohol as the result of increased hepatic retention of the vitamin associated with relatively enhanced formation of folylpolygluta-mates and decreased secretion of folate into bile (Hillman et al., 1977; Steinberg et al., 1981; Hillman and Steinberg, 1982). The interpretation of these findings and the design of the experiments have been challenged, however (Weir et al., 1985). In fact, a bewildering variety of conflicting results has been reported by various investigators, using different animal models and experimental conditions (Weir et al., 1985). Thus, increased, normal, and decreased hepatic folylpolygluta-mate synthesis have been reported in animals fed ethanol (Brown et al., 1973; Hillman et al., 1977; Steinberg et al., 1981, 1982; Wilkinson and Shane, 1982; Keating et al., 1985; Weir et al., 1985). In monkeys fed alcohol, increased excretion of parenterally administered folate in the feces was noted, suggesting that biliary excretion of the vitamin was not impaired (Tamura and Halsted, 1983). It is unclear, therefore, whether alcohol has significant effects on hepatic synthesis or biliary excretion of folates.

A number of studies indicate that alcohol may increase urinary losses of folate. Acute administration of large doses of ethanol to rats (2–4 g/kg) caused a marked increase in urinary folate excretion over an 8-hr period (McMartin, 1984), which could be elicited repeatedly over four successive days by repeated ethanol feeding (McMartin et al., 1986a). The effect was enhanced and prolonged when alcohol metabolism was inhibited with 4-methylpyrazole (McMartin and Collins, 1983). Administration of smaller doses of ethanol

(0.8–1 g/kg) to humans had no effect on total urinary folate excretion (McMartin et al., 1986b). Others also found that ethanol given acutely to human volunteers did not increase urinary folate excretion (Eichner and Hillman, 1973; Lane et al., 1976). Monkeys chronically fed alcohol for 2 to 4 years showed decreased hepatic retention and increased urinary excretion of a parenteral dose of radiolabeled folic acid (Tamura et al., 1981; Tamura and Halsted, 1983). In human volunteers given ethanol for 17 days along with large doses of folic acid, a modest increment in urinary folate excretion was found (Russell et al., 1983). Thus, under certain experimental conditions, alcohol may increase urinary folate losses in animals and humans, but it is uncertain whether this is an important mechanism of the folate deficiency induced by ethanol or whether it is related to the acute reduction in serum folate levels caused by alcohol. It is also unclear whether the increased urinary folate losses are attributable to impairment of folate uptake by the liver or of folate conservation by the kidney. Increased urinary folate levels have also been observed in patients with marked hepatic dysfunction in the absence of alcoholism (Retief and Huskisson, 1969).

Inhibitory effects of ethanol have been described on formate and formaldehyde metabolism (Bertino et al., 1965; Tran et al., 1972; McMartin and Collins, 1983) and on jejunal glycolytic enzymes (Greene et al., 1974). In mice, alcohol does not appear to increase folate catabolism (Kelly et al., 1981). In patients with severe liver disease, decreased hepatic avidity for folate (Cherrick et al., 1965) may contribute to negative folate balance.

Finally, because of the relatively high frequency of normal serum or erythrocyte folate levels in patients with megaloblastic marrow morphology, it has been proposed that in addition to causing folate deficiency, ethanol may have a direct toxic effect on marrow precursor cells, resulting in megaloblastic change (Wu et al., 1975; Heidemann et al., 1981; Chanarin, 1982; Savage and Lindenbaum, 1986). Such an effect remains to be demonstrated experimentally, with the possible exception of a study briefly reported by Sullivan and Liu (1966). Many patients with megaloblastic change associated with normal plasma or red cell folate values will have elevated serum concentrations of homocysteine, which become normal after folic acid treatment, suggesting the presence of a systemic metabolic block in folate utilization despite the normal serum and erythrocyte vitamin levels (D. S. Savage, J. Lindenbaum, S. Stabler, and R. Allen, unpublished observations, 1987). In rats given a low-protein diet and ethanol, hepatic levels of methionine synthetase are decreased (Finkel-stein *et al.*, 1974), an effect that would elevate homo-cysteine levels and might limit the availability of folate coenzymes for DNA synthesis.

In summary, alcohol interferes with folate metabo-lism by a poorly understood mechanism and accelerates the development of megaloblastic anemia in individuals with depleted folate stores.

8.1.1.5. Prevention

Since folate deficiency is a frequent cause of mor-bidity and occasionally results in fatalities in alcoholic patients (Kaunitz and Lindenbaum, 1977), attempts at prevention of this common complication of alcoholism would appear to be worthwhile. Folic acid is soluble in alcoholic beverages, does not alter their taste, and is well absorbed when given in drinks, even in chronic alcoholic patients (Kaunitz and Lindenbaum, 1977) (Fig. 8.5). These workers have recommended that cer-tain alcoholic beverages be fortified with the vitamin as a prophylactic measure.

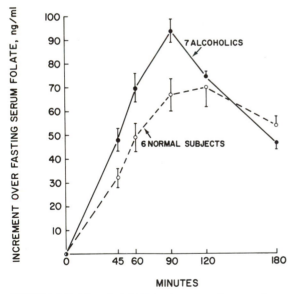

FIGURE 8.5. Mean (±1 S.E.M.) increments in serum *Lacto-bacillus casei* folate concentrations over fasting levels after the administration of folic acid, 40 μg/kg body weight, in red or white wines given to seven chronic alcoholics (solid line) and six normal volunteers (dashed line). (From Kaunitz and Lindenbaum, 1977.)

8.1.2. Sideroblastic Anemia

8.1.2.1. Clinical and Laboratory Findings

A common abnormality of erythropoiesis in hospi-talized alcoholics is the presence in the bone marrow of abnormal sideroblasts (Hines, 1969; Eichner and Hill-man, 1971; Savage and Lindenbaum, 1986). Such cells are red cell precursors that have increased numbers of Prussian-blue-positive cytoplasmic granules containing nonheme iron. In many of the most severely affected cells, the granules from a crown or ring around the nucleus ("ring sideroblast"). Ultrastructurally, the iron is heavily deposited in the matrix between the cristae of mitochondria or in cytoplasmic bodies, which may be lysosomal in nature (Grasso and Hines, 1969; Hines and Grasso, 1970). These morphological abnormalities are associated with all the characteristic features of "in-effective erythropoiesis," that is, erythroid hyperplasia, increased iron uptake by marrow erythroid precursors, subnormal incorporation of iron into the hemoglobin of circulating red cells with reflux of iron into the circula-tion causing hyperferremia, intramedullary destruction of normoblasts, anemia, and reticulocytopenia (Eichner and Hillman, 1971). Since the mitochondria of normo-blasts are typically aligned in a perinuclear location before and during mitosis, the increased proliferative activity of erythroid precursors may be responsible for the "ringed" appearance of these cells.

In two series of consecutively studied patients, significant numbers of ring sideroblasts were found in the bone marrow aspirates of 29 and 31% of hospitalized alcoholics (Eichner and Hillman, 1971; Hines and Cowan, 1974). The percentage of affected cells in such patients varies from less than 10% to 90% (Hines, 1969; Eichner and Hillman, 1971; Hines and Cowan, 1974; Ali and Sweeney, 1974; Pierce *et al.*, 1976). The peripheral blood smear is frequently dimorphic, with a population of normal to macrocytic erythrocytes as well as hypo-chromic and microcytic cells. However, such a dimor-phic smear is neither a sensitive nor a specific predictor of the presence of marrow ring sideroblasts in alcoholics (Savage and Lindenbaum, 1986). The MCV is usually elevated or normal (Eichner and Hillman, 1971; Linden-baum and Roman, 1980; Savage and Lindenbaum, 1986). A more specific finding is the presence of red cells ("siderocytes") containing granules rich in non-heme iron ("Pappenheimer bodies"), which may be de-tected on Wright's-stained peripheral blood smears in one-third to half of cases, particularly during the reticulocytosis that follows alcohol withdrawal (Lindenbaum and Roman, 1980; Savage and Lindenbaum, 1986).

It is difficult to estimate the role of the sideroblastic alterations in causing anemia, because in almost all patients other underling etiologies are present (Savage and Lindenbaum, 1986). There is a strong, although by no means invariable, association with megaloblastic erythropoiesis (Hines, 1969; Eichner and Hillman, 1971; Hines and Cowan, 1974; Pierce *et al.*, 1976; Solomon and Hillman, 1979, Lindenbaum and Roman, 1980; Savage and Lindenbaum, 1986). Many patients have evidence of an underlying anemia of chronic disease (Lindenbaum and Roman, 1980; Savage and Lindenbaum, 1986). The sideroblasts characteristically disappear from the marrow during the first week in hospital but may persist for as long as 12 days (Hines, 1969; Eichner and Hillman, 1971; Hines and Cowan, 1974).

Serum iron concentrations are commonly (but not invariably) in the higher end of the normal range or are elevated in patients with sideroblastic marrows (Hines, 1969; Eichner and Hillman, 1971). One of the factors affecting the detection of hyperferremia is the period elapsing between cessation of alcohol ingestion and the time serum is obtained for iron determination. Serum iron levels fall abruptly after alcohol withdrawal, even in those patients with normal values on admission to hospital (Waters *et al.*, 1966; Hines, 1969; Hourihane and Weri, 1970; Eichner and Hillman, 1971). In many instances, the fall in iron concentration reflects improvement in iron utilization by the previously megaloblastic or sideroblastic marrow, but the decline in serum iron may be seen in the absence of these marrow abnormalities (Fig. 8.6) (Sullivan and Herbert, 1964; Waters *et al.*, 1966; Lindenbaum and Lieber, 1969; Myrhed *et al.*, 1977). The serum ferritin is commonly elevated (Savage and Lindenbaum, 1986).

8.1.2.2. Pathogenesis

Florid sideroblastic changes have only been encountered in alcoholic patients who are also malnourished (Hines, 1969; Eichner and Hillman, 1971; Pierce *et al.*, 1976). Better-nourished alcoholics may show no changes or less impressive accumulations of iron in the cytoplasm of marrow normoblasts (Waters *et al.*, 1966; Hourihane and Weir, 1970; Wu *et al.*, 1975; Pierce *et al.*, 1976). These clinical observations suggest that a nutritional factor is important in the pathogenesis of sideroblastic anemia in alcoholics.

It was initially felt that this factor was most likely to

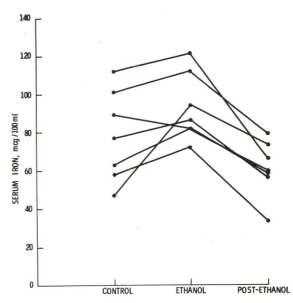

FIGURE 8.6. Changes in serum iron during and after the experimental administration of ethanol constituting 46 to 66% of total caloric intake along with a nutritious diet and vitamin supplements in seven human volunteers. Although none of the subjects developed sideroblastic or megaloblastic marrow morphological alterations on this regimen, serum iron fell in all subjects when ethanol was withdrawn. (From Lindenbaum and Lieber, 1969.)

be pyridoxine deficiency. Chronic alcoholics frequently have low serum concentrations of pyridoxal phosphate (PLP) (Leevy *et al.*, 1965; Walsh *et al.*, 1966; Hines and Cowan, 1974; Davis and Smith, 1974; Lumeng and Li, 1974; Pierce *et al.*, 1976). Hepatic concentrations of vitamin B_6 coenzymes have also been reported to be decreased (Leevy *et al.*, 1965; Frank *et al.*, 1971). Alcoholics with sideroblastic anemia almost always have decreased serum PLP levels (Hines and Cowan, 1974; Pierce *et al.*, 1976). Acetaldehyde has been shown to cause an increased degradation of PLP in erythrocytes and liver cells (Lumeng and Li, 1974; Li *et al.*, 1974; Lumeng, 1978).

The experimental chronic administration of ethanol to human volunteers has resulted in sideroblastic changes in the bone marrow (Fig. 8.7). When alcohol was given with a diet low or borderline in pyridoxine and folate content, ring sideroblasts developed after several weeks in the majority of subjects studied (Hines, 1969; Hines and Cowan, 1970, 1974; Eichner and Hillman, 1971). Serum PLP levels were reported to fall before the development of sideroblastic changes (Hines and Co-

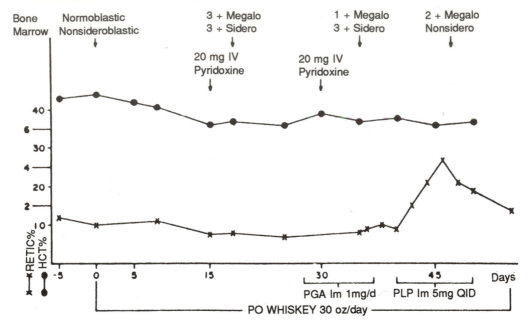

FIGURE 8.7. Development of sideroblastic and megaloblastic marrow morphological changes in a human volunteer during the third week of the administration of whiskey along with a diet low in folate content. The sideroblastic changes did not respond to treatment with intravenous pyridoxine or intramuscular folic acid (PGA) but subsequently cleared after therapy with intramuscular pyridoxal phosphate (PLP), which also elicited an increase in the reticulocyte count. (From Hines and Cowan, 1970.)

wan, 1970, 1974). In contrast, when alcohol was given for comparable periods with an adequate diet plus pyridoxine supplements, ring sideroblasts developed in none of nine subjects of Lindenbaum and Lieber (1969), none of three of Eichner and Hillman (1971), and one of three of Hines and Cowan (1974). In the single volunteer of Hines and Cowan (1974) who developed sideroblastic changes, the morphological abnormalities were milder than those produced by ethanol in conjunction with a deficient diet. The sideroblasts disappeared and reticulocyte responses occurred in three of three subjects given parenteral PLP in large doses for 6 days despite continued alcohol ingestion (Hines, 1969; Hines and Cowan, 1970).

Although this clinical and experimental evidence is suggestive of a relationship between ring sideroblasts and a disturbance in pyridoxine supply or metabolism in alcoholics, many workers have come to doubt this association. Initial reports of depression of erythrocyte pyridoxal kinase (Hines and Cowan, 1970, 1974; Hines, 1975) have not been confirmed by three other groups of investigators (Lumeng and Li, 1974; Chillar et al., 1976; Solomon and Hillman, 1979), and the methodology used by Hines's laboratory to measure erythrocyte pyridoxal

kinase activity has been criticized (Lumeng and Li, 1974; Chillar et al., 1976). Also, red cell PLP concentrations, as measured indirectly by assaying changes in glutamate oxalacetate transaminase activity before and after adding PLP in vitro, have not been found to be depressed in alcoholics with sideroblastic anemia (Chillar et al., 1976; Solomon and Hillman, 1979). Such patients have been found to have markedly elevated erythrocyte coproporphyrin and protoporphyrin levels, a finding that would not be predicted if the primary disturbance were a block in the PLP-dependent ALA-synthetase reaction at an earlier step in porphyrin synthesis (Ali and Sweeney, 1974). Finally, ALA-synthetase activity in bone marrow cells from alcoholics with sideroblastic anemia has been reported to be normal or elevated and not increased by the addition of PLP as compared to controls (Fraser and Schacter, 1980; Tikerpae et al., 1985).

However, ethanol itself may directly inhibit heme synthesis in erythroid cells. Freedman and co-workers have presented evidence that ethanol at 50 to 100 mM concentrations in vitro inhibits heme synthesis in human and rabbit reticulocytes (Freedman et al., 1975; Ibrahim et al., 1979). Alcohol given in vivo inhibited heme synthesis in rabbit reticulocytes (Freedman and Ros-man, 1976). The administration of a single large dose of vodka to normal volunteers caused a marked depression in leukocyte ferrochelatase activity (McColl *et al.*, 1980). If this were also the case in erythroid precursors, it would account for an inhibition of heme synthesis, resulting in the accumulation of protoporphyrin as seen in the red cells of patients with sideroblastic anemia (Ali and Sweeney, 1974). Ethanol may inhibit several other enzymatic steps in heme synthesis, including ALA-dehydratase (Moore *et al.*, 1971; Krasner *et al.*, 1974) and ALA-synthetase (Ibrahim *et al.*, 1979). It is possible that iron deposition in the mitochondrion of erythroid cells is a nonspecific indicator of injury to this organelle by ethanol or acetaldehyde. The administration of alco-hol to pregnant mice was found to result in myelinlike lamellar figures, which were interpreted as evidence of mitochondrial injury, in fetal red cell precursors (Nishi-mura *et al.*, 1981).

If ethanol causes ring sideroblasts by direct inter-ference with heme synthesis, this alone would not ac-count for the strong association of this disorder with nutritional deficiency in alcoholics. One nutritional fac-tor could be folate deficiency, although this is clearly not present in all patients. It is likely, however, that the disturbance in iron metabolism associated with the inef-fective erythropoiesis of folate deficiency may exacer-bate the tendency for iron accumulation in red cell precursors (Pierce *et al.*, 1976).

In summary, alcohol ingestion, usually with an inadequate diet, frequently results in the presence of ring sideroblasts in the bone marrow. Whether the alter-ations in erythroid iron distribution result in anemia is uncertain. Their pathogenesis is not established. Pos-sible factors include (1) a toxic effect of ethanol at one or more steps in heme synthesis and (2) an associated nutritional deficiency state that has not yet been clearly identified.

8.1.3. Vacuolization of Erythroid Precursors

8.1.3.1. Morphology

A third effect of alcohol on erythrocytes, appar-ently unrelated to the megaloblastic and sideroblastic abnormalities, is the induction of vacuolization in mar-row erythroid precursors (Fig. 8.8).

The presence of vacuoles in nucleated red blood cells, first described by McCurdy and colleagues (1962), is a characteristic finding in the majority of recently intoxicated patients (Waters *et al.*, 1966; Jarrold *et al.*, 1967; Hourihane and Weir, 1970; Eichner and Hillman, 1971). Vacuoles may also be seen in white cell precur-sors, although less frequently, unless an associated in-fection is present. Within a week of admission to hospi-tal and withdrawal of alcohol, often within 24 hr, the vacuoles are no longer demonstrable. Vacuolization oc-curs independently of megaloblastic change, folate defi-ciency, sideroblastic change, or thrombocytopenia and is unrelated to nutritional status (Waters *et al.*, 1966; Jarrold *et al.*, 1967; Lindenbaum and Hargrove, 1968; Hines, 1969; Hourihane and Weir, 1970). Similar vac-

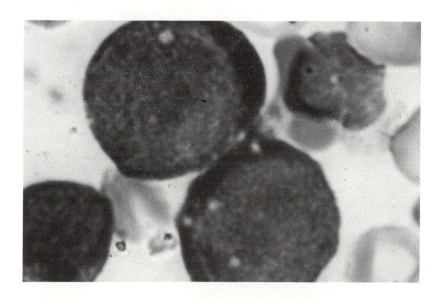

FIGURE 8.8. Vacuolization of bone marrow pronormoblasts in a human volunteer given ethanol along with a nutritious diet for 17 days.

uoles have been reported in the marrow cells of infants born to mothers treated with intravenous alcohol during labor (Lopez and Montoya, 1971).

8.1.3.2. Pathogenesis

The rapid reversibility of the vacuolization after ethanol withdrawal and its morphological resemblance to that seen after chloramphenicol administration suggested a possible toxic depression of erythropoiesis by alcohol (McCurdy et al., 1962; Jarrold et al., 1967). When ethanol was given experimentally (substituted isocalorically for carbohydrate) to well-nourished human volunteers along with excellent protein and vitamin intake, including folic acid supplements, vacuolization of marrow red and white cell precursors resulted (Fig. 8.9) (Lindenbaum and Lieber, 1969). The effect was

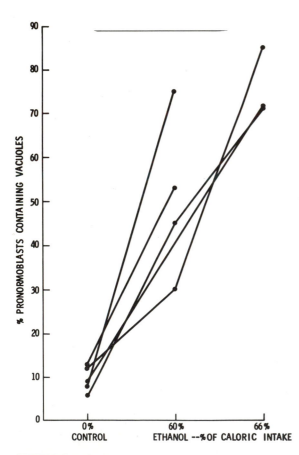

FIGURE 8.9. Increases in percentage of pronormoblasts containing one or more vacuoles in five volunteers receiving ethanol as 60 or 66% of total caloric intake. (From data of Lindenbaum and Lieber, 1969.)

dose related. Vacuole formation was less striking in promyelocytes than in pronormoblasts and was only seen in the former cells after relatively large doses of ethanol. Vacuolization cleared rapidly after cessation of alcohol intake. The vacuoles did not stain with histochemical reactions for fat, mucopolysaccharides, DNA, RNA, peroxidase, or acid or alkaline phosphatases (Lindenbaum and Lieber, 1969). Ultrastructurally, vacuolated red cell precursors in alcoholics show marked membrane convolutions adjacent to the vacuoles, which lack organelles or organized structure, suggesting an effect of alcohol on the cell membrane (Yeung et al., 1973).

The functional importance of the vacuolization is doubtful. When it was produced experimentally, there was no associated fall in hematocrit, reticulocyte count, or granulocyte count, nor was there impairment of the incorporation of radioactive iron into circulating erythrocytes (Lindenbaum and Lieber, 1969). In addition, the vacuoles have frequently been seen clinically in intoxicated patients who were not anemic. The relationship of the vacuolization to changes in serum osmolality secondary to ethanol ingestion requires further study, since similar vacuoles have been reported in a diabetic patient with hyperosmolar coma (Lehane, 1974).

8.1.4. Suppression of Hematopoiesis by Other Mechanisms

Jandl (1955) observed that reticulocytosis often followed abstinence from alcohol in hospitalized patients in the absence of megaloblastic hematopoiesis (Fig. 8.10) and speculated that this represented recovery from ethanol-induced bone marrow suppression. The frequency of reticulocytosis after withdrawal of alcohol is variable (Myrhed et al., 1977; Lundin et al., 1981; Savage and Lindenbaum, 1986). The reticulocytosis may be misinterpreted as evidence of hemolysis (Eichner and Hillman, 1971). The elevation of serum iron induced experimentally by ethanol despite folate supplementation (Sullivan and Herbert, 1964; Lindenbaum and Lieber, 1969) and the fall in serum iron observed in alcoholics without megaloblastic anemia during withdrawal (Waters et al., 1966; Myrhed et al., 1977) could also be considered evidence of suppression of hematopoiesis by ethanol. As summarized in Section 2.1.2, alcohol has been shown to impair marrow production of platelets. Although reticulocytopenia or impairment in iron utilization as measured by ferrokinetic studies was not observed during ethanol administration to vitamin-supplemented subjects (Lindenbaum and Lieber, 1969),

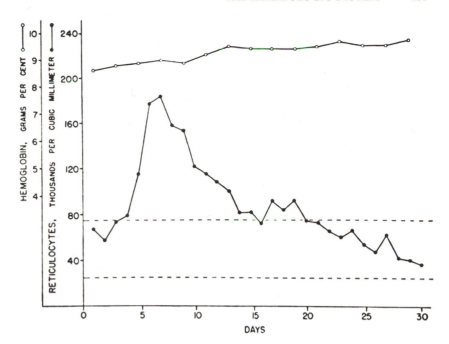

FIGURE 8.10. Mean values for reticulocytes and hemoglobin in five nonmegaloblastic hospitalized patients with alcoholic cirrhosis following hospitalization and abstinence. Dashed lines indicate upper and lower limits of normal for reticulocytes. (From Jandl, 1955.)

it is possible that alcohol ingestion at higher doses or different dose intervals would have suppressed hematopoiesis. The addition of ethanol *in vitro* to murine or human bone marrow cultures at concentrations achievable in the blood of alcoholics suppresses the formation of colonies generated by early and late erythroid progenitor cells, with a more striking effect on the former (erythroid "burst-forming units" or BFU-E) (Meagher *et al.*, 1982; Clark and Krantz, 1986). Acetaldehyde also suppressed erythroid colony formation, but only at concentrations higher than those obtained *in vivo* (Meagher *et al.*, 1982). The relevance of these *in vitro* observations to red cell production in alcoholics is speculative, and it remains an unproven possibility that alcohol suppresses *in vivo* hematopoiesis in the absence of megaloblastic or sideroblastic changes.

In addition to the anemia of chronic disease, a poorly understood, more severe degree of marrow failure as evidenced by ferrokinetic measurements may be encountered in sober patients with alcoholic liver disease (Kimber *et al.*, 1965a). Marrow necrosis may occur in alcoholics (Norgard *et al.*, 1979).

The importance of nutritional deficiency states that cause erythroid marrow failure, such as deficiencies of riboflavin or amino acids, which have been reported to cause anemia in nonalcoholic patients (Adams, 1970; Alfrey and Lane, 1970), has not been evaluated in alcoholics.

8.1.5. Macrocytosis of Alcoholism

8.1.5.1. Clinical Aspects

The presence of macrocytic red cells on blood smears and an elevated erythrocyte MCV are extremely common findings in alcoholic patients. Macrocytosis in alcoholics may be the result of a variety of underlying mechanisms. The finding of enlarged cells or an increased MCV in an alcoholic patient may result from one of at least four causes:

1. Folate deficiency megaloblastic anemia (as discussed earlier in this chapter), in which at least some of the macrocytes are commonly oval in shape and neutrophil hypersegmentation is typically present in addition.
2. Reticulocytosis, usually in response to hemolysis or gastrointestinal bleeding (Kilbridge and Heller, 1969).
3. The macrocytosis of liver disease, caused by the increased deposition of cholesterol and phospholipid on red cell membranes (see Section 1.6.1).
4. The *macrocytosis of alcoholism*, a newly recognized, as yet poorly understood disorder.

In 1974 Wu *et al.* reported that there was a much higher incidence of macrocytosis (as indicated by MCV elevations as measured by the Coulter counter) in alcoholic patients than in a nonalcoholic control group with chronic liver disease. In the same year, Unger and Johnson (1974) reported that during routine screening of the employees of an insurance company, a number of alcoholics were discovered to have an elevated MCV in the absence of folate deficiency, reticulocytosis, or significant hepatic dysfunction. Since then, many reports have appeared confirming these observations (Buffet *et al.*, 1975; Morin and Porte, 1976; Aron *et al.*, 1977; Myrhed *et al.*, 1977; Davidson and Hamilton, 1978; Eschwege *et al.*, 1978; Levi and Chalmers, 1978; Fontan *et al.*, 1978; Khaund, 1978; Whitehead *et al.*, 1978; Wright and Ree, 1978; Chick *et al.*, 1981; Morgan *et al.*, 1981; Cushman *et al.*, 1984; Eriksen *et al.*, 1984; Tonnesen *et al.*, 1986; Bush *et al.*, 1987). The percentage of alcoholics with an elevated MCV has differed widely in various series, from as low as 23% to as high as 96%. The MCV has been utilized as a screening test for alcoholism. Used alone, it has generally lacked sufficient sensitivity, as have other biochemical tests such as the serum γ-glutamyl transpeptidase and the serum aspartate aminotransferase. If various combinations of these tests are used, their sensitivity in detecting alcoholism has been found to be enhanced (Fontan *et al.*, 1978; Chalmers *et al.*, 1981; Eckhardt *et al.*, 1981; Ryback *et al.*, 1982; Cushman *et al.*, 1984; Eriksen *et al.*, 1984; Lumeng, 1986), although still inferior to interviewing tests specifically directed at alcoholism (Bush *et al.*, 1987).

The MCV may be particularly helpful in detecting alcoholism in patients with known liver disease (Levi and Chalmers, 1978; Morgan *et al.*, 1981). Some workers have found a greater prevalence or degree of macrocytosis in alcoholic women than men (Wu *et al.*, 1975; Chaput *et al.*, 1979; Chalmers *et al.*, 1980), although others have not (Whitfield *et al.*, 1978). Alcoholic women with macrocytosis may also have a higher incidence of folate deficiency (Morgan *et al.*, 1981; Chalmers *et al.*, 1980). The MCV may be higher in women with alcoholic liver disease than in men (Levi and Chalmers, 1978; Morgan *et al.*, 1981). The MCV has also been considered to be useful in monitoring abstinence in known alcoholics (Levi and Chalmers, 1978; Chanarin, 1982; Eriksen *et al.*, 1984). In population studies, a correlation between the number of alcoholic drinks taken and MCV has been noted, as well as (possibly) an additional correlation with smoking (Whitehead *et al.*, 1978; Eschwege *et al.*, 1978). The MCV may be better as an indicator of duration and extent of alcohol intake than as a screening test (Shaw *et al.*, 1979; Chick *et al.*, 1981; Tonnesen *et al.*, 1986).

The degree of macrocytosis has usually been mild, with MCVs no higher than 110 fl in the great majority of patients. Anemia is frequently absent or very slight (Unger and Johnson, 1974; Wu *et al.*, 1974; Buffet *et al.*, 1975; Aron *et al.*, 1977; Davidson and Hamilton, 1978; Eschwege *et al.*, 1978; Khaund, 1978; Wright and Ree, 1978). On blood smear, the macrocytes are characteristically round rather than oval, and neutrophil hypersegmentation is absent (Unger and Johnson, 1974; Khaund, 1978; Davidson and Hamilton, 1978). The macrocytosis is uniform; i.e., red cell size distribution is unimodal (Diakhate *et al.*, 1979). The presence of an elevated MCV has correlated poorly with the severity and the presence of liver dysfunction, and it has been seen in association with normal liver biopsies (Unger and Johnson, 1974; Wu *et al.*, 1974, 1975; Buffet *et al.*, 1975; Morin and Porte, 1976; Davidson and Hamilton, 1978).

8.1.5.2. Pathogenesis

Convincing evidence against folate deficiency as the cause of macrocytosis has been provided in many of the reported patients, including normal folate concentrations in serum (Unger and Johnson, 1974; Buffet *et al.*, 1975; Myrhed *et al.*, 1977; Davidson and Hamilton, 1978; Wright and Ree, 1978), erythrocytes (Wu *et al.*, 1975; Belaiche *et al.*, 1978), and liver cells (Wu *et al.*, 1975); absence of *in vitro* evidence of disturbance in the *de novo* synthesis of DNA by bone marrow cells (Wickramasinghe and Longland, 1974; Belaiche *et al.*, 1978); and failure of the macrocytosis to improve after therapy with folic acid and vitamin B_{12} (Unger and Johnson, 1974; Morin and Porte, 1976; Wu *et al.*, 1974). Bone marrow morphology has also been found to be normoblastic in the majority of these patients (Wu *et al.*, 1974; Chalmers *et al.*, 1978, 1980; Belaiche *et al.*, 1978; Diakhate *et al.*, 1979).

The macrocytosis characteristically persists unless the patient abstains from alcohol ingestion. Even with abstention, little or no change in the MCV may occur for many weeks, and it does not usually become normal in less than 1 to 4 months (Unger and Johnson, 1974; Wu *et al.*, 1974; Buffet *et al.*, 1975; Aron *et al.*, 977; Myrhed *et al.*, 1977; Fontan *et al.*, 1978; Chanarin, 1982).

The mechanism leading to the erythrocyte enlargement so common in alcoholics is obscure. Two groups have recently reported studies of red cell membrane lipids in alcoholics with macrocytosis in the absence of anemia or severe liver disease (Alling *et al.*, 1984; Clemens *et al.*, 1986). Although mild abnormalities in liver function were often present in both series, it is likely that most patients had the "macrocytosis of alcoholism." The elevated red cell cholesterol levels typical of the macrocytosis of liver disease (Cooper *et al.*, 1972) were not noted (Clemens *et al.*, 1986), nor was the increased cholesterol : phospholipid ratio characteristic of advanced alcoholic liver disease. However, various abnormalities in the fatty acid composition of erythrocyte phospholipid were present. Although there was some variation in the findings of the two studies, both groups demonstrated increased ratios of saturated and monounsaturated fatty acids to polyunsaturated fatty acids, in part reflecting a decreased concentrations of linoleic acid (18:2) (Alling *et al.*, 1984; Clemens *et al.*, 1986). Similar changes in the proportions of saturated and polyunsaturated fatty acids were also found in plasma (Clemens *et al.*, 1986).

A decreased percentage of polyunsaturated fatty acids has also been reported in the phospholipids of erythrocytes from patients with more severe alcoholic liver disease, with or without associated hemolytic anemia (Goebel *et al.*, 1977, 1979; Salvioli *et al.*, 1978; Owen *et al.*, 1982) and nonalcoholic liver disease (Owen *et al.*, 1982) and in the red cells of rats and mice given ethanol (la Droitte *et al.*, 1984). Increased ratios of saturated and monounsaturated to polyunsaturated fatty acids have also been noted in the platelets of patients with alcoholic and nonalcoholic liver disease (Owen *et al.*, 1981), in hepatic phospholipids and triglycerides of patients with alcoholic fatty liver (Cairns and Peters, 1983), in the plasma cholesterol esters of alcoholics (Warnet *et al.*, 1985), and in phospholipids of brain synaptosomes (Littlejohn and John, 1977) and liver mitochondrial membranes (Waring *et al.*, 1981) after alcohol administration to animals. It has been postulated that such alterations in membrane phospholipids may represent adaptive responses to the tendency of ethanol to render cell membranes more fluid (Littlejohn and John, 1977) or that they reflect increased peroxidation of lipids caused by alcohol (Goebel *et al.*, 1977; Clemens *et al.*, 1986). Increased hepatic concentrations of diene conjugates (evidence of lipid peroxidation) have been observed in baboons and rats after ethanol feeding (Shaw *et al.*, 1981), and increased serum levels of the diene-conjugated nonperoxide isomer of linoleic acid, 9,11-linoleic acid, interpreted as a marker of free radical activity, have recently been observed in the majority of chronic alcoholics (Fink *et al.*, 1985).

It has been suggested that alterations in phospholipid fatty acid ratios or linoleic acid deficiency are in some way responsible for the macrocytosis of alcoholism (Clemens *et al.*, 1986). The relationship of these changes in membrane lipids to macrocytosis, membrane injury, or hemolytic anemia in alcoholics is currently uncertain, however, and will undoubtedly be the subject of further investigations.

In summary, then, a macrocytosis, usually mild in degree and often unassociated with anemia, occurs in the majority of chronic alcoholics; it is not usually associated with folate deficiency or severe liver dysfunction, and it clears slowly several months after withdrawal of alcohol. These observations suggest that alcohol may have induced the macrocytosis, although this has not yet been shown experimentally. The failure of the macrocytosis to improve during many weeks and months of abstention is not characteristic of the other "direct" effects of alcohol on hematopoiesis (e.g., megaloblastic and sideroblastic change, precursor cell vacuolization, thrombocytopenia). The mechanism of the macrocytosis is obscure.

8.1.6. Hemolytic Syndromes and Red Cell Membrane Abnormalities

8.1.6.1. Macrocytosis of Liver Disease

Before discussion of the various hemolytic syndromes encountered in alcoholics, consideration of the more benign erythrocytic morphological abnormalities commonly seen in patients with liver disease, that is, macrocytosis and target cell formation, is in order. It has been recognized by many investigators since the 19th century that macrocytosis of erythrocytes is common in blood smears from patients with both alcoholic and nonalcoholic liver disease (Bingham, 1958).

In the majority of nonalcoholic patients with cirrhosis, the macrocytosis seen on blood smear is not associated with an increased MCV (Bingham, 1958; Werre *et al.*, 1970). Instead, mean cell diameter (rather than volume) is increased ("thin" macrocytosis, "macroplania"). Such flattened cells have an enlarged surface membrane area associated with increased membrane cholesterol and phospholipid content (Neerhout, 1968a; Werre *et al.*, 1970) and are resistant to osmotic lysis (Bingham, 1958; Werre *et al.*, 1970). In some patients with nonalcoholic (and alcoholic) liver disease, the increased deposition of membrane lipid is associated with a true "volume" macrocytosis, that is, with an increased MCV (Bingham, 1958; Neerhout, 1968a; Werre *et al.*, 1970). The macrocytosis of liver disease does not correlate with the presence or absence of anemia, reticulocytosis, or nutritional deficiency (Bingham, 1958). It is said not to occur in patients with fatty liver in the absence of more severe hepatic disease (Bingham, 1958). Normal erythrocytes transfused into

the circulation of cirrhotics acquire the increased osmotic resistance of the recipient's cells (Werre *et al.*,
1970).

8.1.6.2. Target Cells in Liver Disease

The presence of target cells in the blood smears of
some patients with severe hepatic disease appears to be
an exaggerated form of the macrocytosis of liver disease
(Bingham, 1961; Werre *et al.*, 1970; Cooper *et al.*,
1972). The target cells of patients with hepatic disorders
similarly have an increased surface : volume ratio, elevations in both the cholesterol and phospholipid content of
the cell membrane (Neerhout, 1968a; Kilbridge and
Heller, 1969), and increased osmotic resistance (Cooper
and Jandl, 1968). The target cell abnormality also does
not correlate with the presence of anemia (Bingham,
1961; Cooper *et al.*, 1972). The target forms often
disappear when jaundice subsidies (Neerhout, 1968b).
Target erythrocytes from patients with biliary obstruction rapidly lose their osmotic resistance when transfused into normal subjects (Cooper and Jandl, 1968).
This also occurs *in vitro* when the cells are incubated in
normal serum (Cooper and Jandl, 1968). The reverse is
true when normal cells are transfused into patients with
obstructive jaundice or incubated *in vitro* with sera from
jaundiced patients. The changes in osmotic fragility can
be correlated with increases in erythrocyte surface area
and lipid content (Cooper and Jandl, 1968).

8.1.6.3. Low-Grade Hemolysis of Liver Disease

A variety of hemolytic syndromes have been described in patients with alcoholic liver disease. Frequently, it has been suggested that an abnormality in red
cell morphology or in plasma or erythrocyte lipid concentrations is related to the presence of hemolysis. In
order to critically evaluate such inferences, it must be
remembered that a shortened red cell life span has been
reported in many studies in one-third to three-quarters
of patients with alcoholic and nonalcoholic cirrhosis
(Subhiyah and Al-Hindawi, 1967). The degree of shortening of erythrocyte survival has varied from mild to
moderate. Anemia is frequently absent or only mild in
degree because of adequate compensatory production of
red cells by the bone marrow (Hall, 1960). The mechanisms underlying the usually low-grade hemolytic state
are uncertain. Coombs' tests are usually negative, and
the hemolysis is not altered by steroid therapy (Jandl,
1955). Since normal donor cells have a shortened life

span in the cirrhotic patient, an extracorpuscular factor
is present (Jandl, 1955). The frequently associated lowering of the serum iron and iron-binding capacity (Jandl,
1955; Kimber *et al.*, 1965a) are reminiscent of the anemia of chronic inflammatory and neoplastic conditions,
in which there is also a modest impairment of erythrocyte survival (Cartwright, 1966). Increased splenic uptake of red cells has been demonstrated in many cirrhotics using radioactively labeled cells, and this probably
plays a role in some anemic patients (Jandl, 1955;
Subhiyah and Al-Hindawi, 1967; Felsher *et al.*, 1968)
but not in others (Kimber *et al.*, 1965a; Hume *et al.*,
1970).

8.1.6.4. Acanthocytosis ("Spur Cell Anemia")

A clinically important abnormality of the red cell
membrane seen in patients with alcoholic liver disease is
the presence of irregularly spiculated and contracted
erythrocytes, resembling those seen in congenital abetalipoproteinemia ("acanthocytes" or "spur cells"). In
contrast to macrocytosis and targeting, acanthocytosis
has been characteristically observed in patients with
brisk hemolytic anemia. Almost all reported cases have
been in patients with advanced alcoholic cirrhosis who
had hyperbilirubinemia, ascites, splenomegaly, and a
generally poor prognosis (Smith *et al.*, 1964; Silber *et
al.*, 1966; Douglass *et al.*, 1968; Grahn *et al.*, 1968;
Martinez-Maldonado, 1968; Cooper, 1969; McBride
and Jacob, 1970; Cooper *et al.*, 1972, 1974). It has been
estimated to occur in 3% of patients with alcoholic
cirrhosis (Powell *et al.*, 1975). A few cases have been
reported in severe nonalcoholic liver disease, including
cardiac cirrhosis (Grahn *et al.*, 1968), severe viral hepatitis (McBride and Jacob, 1970), neonatal liver disease
(Marie *et al.*, 1967; Tchernia *et al.*, 1968), metastatic
carcinoid tumor (Keller *et al.*, 1971), and childhood
intrahepatic cholestasis (Balistreri *et al.*, 1981).

A marked shortening of the survival of autologous
red cells has been demonstrated, often but not always
associated with increased splenic sequestration (Smith
et al., 1964; Silber *et al.*, 1966; Douglass *et al.*, 1968;
Grahn *et al.*, 1968; Martinez-Maldonado, 1968; Cooper,
1969). When transfused into normal recipients, the patients' erythrocytes have had a shortened survival, and
normal donor cells transfused into the patients also
survive poorly (Smith *et al.*, 1964; Silber *et al.*, 1966;
Douglass *et al.*, 1968; Cooper, 1969). Donor cells have
been reported to acquire membrane spiculations after
transfusion (Turpin *et al.*, 1971). The red cells of a
patient with acanthocytosis survived normally when

transfused into an asplenic recipient (Douglass *et al.*, 1968).

The pathogenesis of acanthocytosis in patients with liver disease remains obscure despite numerous investigations. Studies in which separated cells or plasma from patients and normal controls have been incubated together have yielded conflicting results as to whether the acanthocytosis is reversible or inducible *in vitro* (Smith *et al.*, 1964; Silber *et al.*, 1966; Douglass *et al.*, 1968; Grahn *et al.*, 1968; Martinez-Maldonado, 1968; Cooper, 1969; Keller *et al.*, 1971; Cooper *et al.*, 1974). Some of the reported discrepancies appear to be the result of failure to distinguish between the irregular spiculation and distortion of cell outline of acanthocytes and the more regular and superficial scalloping of crenated cells (burr cells, echinocytes) (Brecher and Bessis, 1972), which may be artifactual and can be induced by incubation of red cells in normal plasma (Douglass *et al.*, 1968; Brecher and Bessis, 1972; Cooper *et al.*, 1974). However, the distinction between spur cells and burr cells may be somewhat arbitrary, since both types of abnormality are frequently encountered in the same smear from a cirrhotic patient (J. Lindenbaum, unpublished observations, 1980).

Cooper *et al.* and a number of other investigators have related abnormalities of red cell morphology in liver disease to alterations in membrane lipids. As previously mentioned, the erythrocyte membranes of patients with target cells have been found to have increased amounts of cholesterol and phospholipid, with a normal or slightly elevated ratio of unesterified cholesterol to phospholipid (FC:PL) (Cooper, 1969; Cooper *et al.*, 1972, 1974; Powell *et al.*, 1975; Salvioli *et al.*, 1978). The increments in both types of membrane lipids result in a benign morphological abnormality (the target cell). In contrast, in patients with spur cells, the increase in membrane lipid is usually confined to free cholesterol, with normal total phospholipids, resulting in a markedly elevated FC:PL ratio (Fig. 8.11) (Cooper, 1969; Cooper *et al.*, 1972, 1974; Powell *et al.*, 1975; Salvioli *et al.*, 1978). Since added cholesterol tends to decrease the fluidity of phospholipid membranes, the imbalance in these two lipids would tend to make spur cell membranes less fluid or more rigid (Cooper, 1977; Balistreri *et al.*, 1981; Owen *et al.*, 1982).

In fact, the acanthocytes of patients with live disease have decreased deformability, which can be demonstrated *in vitro* and which might result *in vivo* in their removal by the spleen (Cooper, 1969; McBride and Jacob, 1970; Keller *et al.*, 1971; Cooper *et al.*, 1974). Another factor that could contribute to diminished membrane fluidity would be a decrease in the ratio of polyunsaturated to saturated acyl chains in the membrane phospholipids (Cooper, 1977). A low linoleic acid (18:2)

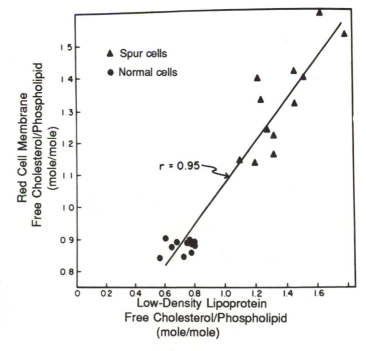

FIGURE 8.11. Relationship between the cholesterol–phospholipid ratio in red cell membranes and in serum low-density lipoproteins in normal subjects and in patients with spur-cell anemia associated with alcoholic cirrhosis. (From Cooper, 1977.)

content of erythrocyte membranes from cirrhotics with spur cell anemia has been reported, along with improvement in the degree of acanthocytosis after intravenous infusions of polyunsaturated lecithin (Salvioli et al., 1978). However, as mentioned earlier in the discussion of the macrocytosis of alcoholism, decreases in linoleic and other polyunsaturated fatty acids and corresponding increases in saturated acyl groups in membrane phospholipids have also been commonly observed in alcoholics without anemia (Alling et al., 1984; Clemens et al., 1986), although it has been argued that subtle, low-grade hemolysis may be detectable in such patients (Kristensson-Aas et al., 1986).

In an attempt to explain the target and spur cell changes, attention has focused on abnormalities in the plasma of patients with liver disease, since circulating red cells are unable to synthesize cholesterol or phospholipids, and erythrocyte membrane lipid composition tends to resemble that in plasma. In fact, elevated membrane cholesterol can be induced in normal red cells by exposure to spur cell plasma in vitro (Cooper, 1969). Cooper et al. (1972) have demonstrated a strong correlation between the FC:PL ratio in serum low-density lipoproteins and that in red cell membranes of patients with liver disease (see Fig. 8.11). A correlation has also been reported with high-density lipoprotein FC:PL ratios by other workers (Salvioli et al., 1978). It has been speculated that changes in serum bile acids, such as the increased concentrations of chenodeoxycholic acid in the serum of patients with spur cells, may contribute to the lipoprotein abnormalities (Cooper and Jandl, 1968; Cooper, 1969; Cooper et al., 1972, 1974; Salvioli et al., 1978). Alterations in the activity of the plasma lecithin:cholesterol acyltransferase (LCAT) activity did not correlate well with membrane lipid composition (Cooper et al., 1972; Powell et al., 1975).

Others, however, have reported cases of acanthocytosis associated with liver disease when red cell cholesterol or cholesterol:phospholipid ratios were normal (Smith et al., 1964; Silber et al., 1966; Douglass et al., 1968; Keller et al., 1971). Dissociation between the changes in red cell cholesterol content and the presence of spur cells has also been reported (McBride and Jacob, 1970; Cooper et al., 1974). In addition, the relationship of either the acanthocytes or the membrane lipid abnormalities to the hemolytic process has not been clearly established (Shohet, 1974).

Splenic destruction of red cells may play an important role in hemolysis in some patients, possibly independent of the erythrocyte membrane abnormalities (Smith et al., 1964; Silber et al., 1966; Grahn et al.,

1968; Cooper et al., 1974; Shohet, 1974). Cooper et al. (1974) have reported an interesting case, however, in which the spur cells were no longer seen following splenectomy, suggesting that splenic "conditioning" may play a role in the genesis of the morphological abnormality. On the other hand, a patient was noted to develop spur cell anemia despite a previous splenectomy (Greenberg and Choi, 1975). More work is needed to unravel the relationship among erythrocyte lipid composition, red cell morphology, and hemolysis in this fascinating disorder.

8.1.6.5. "Zieve's Syndrome"

In 1958, Zieve reported a series of 20 alcoholic patients with hypercholesterolemia, transient hyperbilirubinemia, acute fatty liver, and a mild anemia that was interpreted as hemolytic. The constellation of alcoholic liver disease, hyperlipemia, and hemolysis was felt to represent a distinct syndrome, and it was speculated that the hyperlipemia in some way caused the hemolysis. Subsequently, many additional, more or less similar cases have been reported (Holt and Korst, 1959; Whitcomb and Job, 1960; Myerson, 1968; Kessel, 1962; Strom, 1963; Martini and Dolle, 1965; Blass and Dean, 1966; Zieve, 1966; Gadrat et al., 1967; Albahary et al., 1968; Balcerzak et al., 1968; Westerman et al., 1968; Rudiger et al., 1970; Powell et al., 1972; Goebel et al., 1975, 1977). In some, spherocytosis has been prominent on blood smears (Zieve, 1966; Balcerzak et al., 1968; Powell et al., 1972; Cooper, 1980). The problem has been reviewed critically by Eichner (1973). The evidence that acute hemolytic anemia was actually present in many of the reported cases is equivocal; some may have been recovering from marrow failure or have had chronic low-grade hemolysis associated with liver disease with a fall in hematocrit related to changes in hydration. Even when hemolysis was clearly present, a causal relationship with elevations in serum lipids was not demonstrated (Blass and Dean, 1966; Balcerzak et al., 1968; Eichner, 1973; Cooper, 1980).

It can be argued that hyperlipemia and hemolytic anemia are independent, unrelated complications of alcoholism and that their occasional association in the same patient does not constitute a valid syndrome. On the other hand, the occurrence of cases of acute hemolytic anemia in association with alcoholic liver disease (in the absence of hypophosphatemia or stomatocytosis) cannot be questioned. The association in at least some patients with hyperlipemia could be a useful clue to pathogenesis, suggesting that a disturbance in plasma and red cell lipid metabolism may be responsible for the hemolysis, even if the idea that hyperlipemia causes hemolysis proves to be too simplistic. Red cell cholesterol and phospholipid levels are increased in such patients, as in others with liver disease who do not necessarily have a hemolytic anemia (Westerman *et al.*, 1968; Schubotz *et al.*, 1976; Goebel *et al.*, 1977).

It has been postulated that membrane injury related to peroxidation of fatty acids may be important in the pathogenesis of hemolytic anemia in alcoholics. Subnormal serum and erythrocyte vitamin E concentrations, decreases in the ratio of vitamin E to polyunsaturated fatty acids, and increased *in vitro* sensitivity to hydrogen-peroxide-induced hemolysis were found in alcoholic patients said to have acute hemolytic anemia (Goebel *et al.*, 1977, 1979). Recurrence of hemolysis after resumption of drinking was not prevented by vitamin E administration, however (Goebel *et al.*, 1979).

Abnormalities in red cell enzymatic function have also been reported in such patients, including instability of pyruvate kinase (Goebel *et al.*, 1975, 1977) and impaired hexose monophosphate shunt metabolism, with increased *in vitro* Heinz body formation, possibly as a result of NADP depletion (Smith *et al.*, 1975, 1976).

8.1.6.6. Hemolysis Caused by Hypophosphatemia

In 1971, Jacob and Amsden reported an alcoholic patient with pancreatitis, ketoacidosis, mild cirrhosis, and profound hypophosphatemia (serum P = 0.1 mg/100 ml). A brisk acute hemolytic anemia with spherocytes on blood smear was documented. Markedly diminished levels of erythrocyte ATP and decreased *in vitro* filterability of the patient's red cells were present at the time of hypophosphatemia. The authors speculated that lack of ATP secondary to phosphate depletion resulted in rigidity of the red cell membrane, leading to hemolysis (Jacob and Amsden, 1971). At least two other patients with hemolytic episodes associated with severe hypophosphatemia (and less profoundly depressed red cell ATP levels) have subsequently been reported (Klock *et al.*, 1974; Territo and Tanaka, 1974). Alterations in red cell glycolytic intermediates and increased erythrocyte phospholipid levels were found in one of them (Klock *et al.*, 1974). As Klock *et al.* (1974) indicate, the interpretation of the cause of hemolytic episodes in alcoholic patients requires great caution. This point is illustrated by a report of three patients from a single institution with marked hypophosphatemia associated

with a fall in hemoglobin level (Sarg and Pitchumoni, 1978). In each of these cases, factors other than hypophosphatemia may have been responsible for the drop in hemoglobin, including glucose-6-phosphate dehydrogenase deficiency, folate and iron deficiency, and changes in hydration. Nonetheless, the possibility that hypophosphatemia leads to hemolysis in certain patients seems likely. Profound hypophosphatemia has been induced in dogs by hyperalimentation with glucose and amino acids following a period of starvation. In this model, spherocytosis, decreased red cell filterability and ATP levels, and a brisk hemolytic anemia with increased splenic sequestration of erythrocytes developed in parallel with the fall in serum phosphate (Yawata *et al.*, 1974). All of these abnormalities were prevented or reversed by phosphate administration. The reasons for the not uncommon development of phosphate depletion in alcoholics are not fully understood (Knochel, 1977).

8.1.6.7. Stomatocytes, Knizocytes, and Triangulocytes

Douglass and Twomey (1970) reported four alcoholic patients with transient stomatocytosis. Stomatocytes (red cells with a central slit or mouthlike zone of pallor on Wright's stain) have been reported in association with a wide range of congenital and acquired disorders (Davidson *et al.*, 1977). The four alcoholic patients had mild to moderate hepatic dysfunction associated with fatty liver (Douglass and Twomey, 1970). Varying degrees of anemia and shortening of erythrocyte survival were documented but appeared to be unrelated to the numbers of stomatocytes seen. There was no evidence of increased splenic sequestration. The stomatocytosis recurred in two patients after discharge from hospital and resumption of heavy drinking. The same workers found increased numbers of stomatocytes in 11 of 40 unselected patients with acute alcoholism (Douglass and Twomey, 1970). A number of alcoholics with stomatocytosis have been reported subsequently (Coste *et al.*, 1972; Wisloff and Boman, 1979). We have encountered transient stomatocytosis and hemolytic anemia in an alcoholic patient with fatty liver as well as in a nonalcoholic man with acute viral hepatitis (J. Lindenbaum, unpublished observations, 1980).

Using scanning electron microscopy (SEM), Wisloff and Boman (1979) noted red cells from two alcoholic patients that contained two indentations ("knizocytes"). Cells with three indentations ("triangulocytes") were observed to be invariably present, along with a variety of other morphological abnormalities (including knizocytes, stomatocytes, and target cells), on SEM examination of the blood of 40 alcoholics studied by Homaidan *et al.* (1986). The triangulocytes were still present, although reduced in number, 1 month after withdrawal of alcohol (Homaidan *et al.*, 1986).

The pathogenesis of these interesting abnormalities in red cell morphology, their specificity for alcoholism, and their relationship to membrane injury or hemolytic anemia all remain to be established.

8.1.7. Iron Deficiency

Iron-deficiency anemia, usually the result of gastrointestinal blood loss, is common in alcoholics (Sheehy and Berman, 1954; Kimber *et al.*, 1965a; Eichner *et al.*, 1972). Accumulation of iron in the serum and in bone marrow stores from ineffective erythropoiesis secondary to folate deficiency may sometimes mask the presence of associated iron lack. Deficiency of iron may become apparent only after alcohol withdrawal and correction of the megaloblastic state (Eichner *et al.*, 1972).

The standard laboratory tests useful in the diagnosis of iron-deficiency anemia in nonalcoholic patients often give misleading results in alcoholics. As mentioned previously, serum iron concentrations may fall from levels that are normal or elevated during drinking to low values during the first days of abstinence and may not reflect the adequacy or inadequacy of iron supply. Perhaps because of a coexistent tendency to alcohol-related macrocytosis, the MCV is frequently normal and may even be elevated in alcoholics despite severe iron-deficiency anemia (Savage and Lindenbaum, 1986). The expected changes in the serum iron, serum total iron-binding capacity, and percentage serum transferrin saturation are often absent in alcoholics with iron-deficiency anemia (Savage and Lindenbaum, 1986). Serum ferritin levels are frequently elevated out of proportion to the size of body stores of iron in patients with alcoholic liver disease, possibly as the result of release of ferritin from hepatocytes or the presence of inflammation (Lipschitz *et al.*, 1974; Prieto *et al.*, 1975; Kristenson *et al.*, 1981; Lundin *et al.*, 1981). However, when iron-deficiency anemia is present, serum ferritin levels will not be elevated above the normal range despite severe hepatic dysfunction, although in many patients serum ferritin will be in the lower end of the normal range rather than decreased. The serum ferritin thus appears to be the best noninvasive screening test for iron-deficiency anemia in alcoholics (Savage and Lindenbaum, 1986).

8.1.8. Anemia of Chronic Disease

Anemia secondary to chronic inflammation (e.g., from alcoholic hepatitis) is quite common in alcoholics (Eichner and Hillman, 1971; Savage and Lindenbaum, 1986). The low serum iron typical of the anemia of chronic disease may be absent in alcoholics with this disorder at the time of hospitalization. Although this type of anemia is typically normocytic or slightly microcytic (Cartwright, 1966), in alcoholics the MCV is often elevated, in most cases presumably as the result of an independent effect of ethanol (Savage and Lindenbaum, 1986). Patients with megaloblastic change who fail to respond to folic acid therapy frequently appear to have the anemia of chronic disease (Savage and Lindenbaum, 1986).

8.1.9. Hemodilution

In patients with cirrhosis, expansion of the plasma volume is a common finding (Perera, 1946; Hiller *et al.*, 1949; Bateman *et al.*, 1949). This may result in an apparent anemia, with hematocrit values as low as 28% in the face of a normal red cell mass (Hyde *et al.*, 1952; Sheehy and Berman, 1954; Eisenberg, 1956; Hall, 1960).

8.1.10. Diagnosis and Management of Anemia in Alcoholics

A diagnostic approach to anemia in alcoholics has been proposed (Fig. 8.12) (Savage and Lindenbaum, 1986). Four initial tests are obtained in every patient: the MCV, reticulocyte count, serum ferritin, and careful inspection of the peripheral blood smear. Serum and erythrocyte folate determinations are not recommended, since those tests appear to contribute little of diagnostic value. Serum iron, total iron-binding capacity, cobalamin, and bone marrow aspiration are reserved for selected patients (Fig. 8.12). Patients with neutrophil hypersegmentation, more than 3% macroovalocytes, or MCVs above 110 fl are treated with oral folic acid, 1 mg daily, after a serum cobalamin is obtained to rule out vitamin B_{12} deficiency. The demonstration of iron deficiency (as evidenced by a low serum ferritin, high iron-binding capacity, or absent marrow iron stores) should lead not only to therapy with oral ferrous sulfate but to evaluation for sources of blood loss as well. If reticulocytosis (not accompanied by bleeding or evidence of recovery from megaloblastic anemia) is noted, tests for the presence of hemolysis should be obtained. The

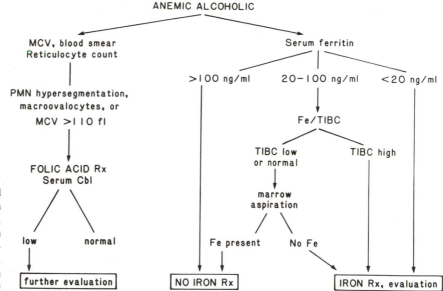

ANEMIC ALCOHOLIC

FIGURE 8.12. A diagnostic and therapeutic approach to anemia in alcoholics. Cbl, cobalamin; Fe, iron; fl, femtoliters; MCV, mean corpuscular volume; PMN, polymorphonuclear neutrophils; TIBC, total iron-binding capacity. (From Savage and Lindenbaum, 1986.)

rationale behind these recommendations is discussed by Savage and Lindenbaum (1986). They are meant to be guidelines rather than rigid algorithms and are not substitutes for a careful history and physical examination, which may lead to modifications in diagnostic approach.

8.2. Platelets

8.2.1. Thrombocytopenia

8.2.1.1. Clinical Aspects

In 1968, two groups of investigators reported an association between alcohol intoxication and thrombocytopenia (Lindenbaum and Hargrove, 1968; Post and Desforges, 1968a). This has subsequently been confirmed by many workers (MacLeod and Michaels, 1969; Cowan and Hines, 1971; Eichner and Hillman, 1971; Coste *et al.*, 1972; Heck and Gehrmann, 1972; Liu, 1973; Myrhed *et al.*, 1977; Wallerstedt and Olsson, 1978; Bjorkholm, 1980; Heidemann *et al.*, 1981; Fink and Hutton, 1983). The degree of thrombocytopenia varies, but it may be profound (Post and Desforges, 1968a). The incidence of thrombocytopenia has been found to vary from 14 to 81% in acutely ill hospitalized alcoholics (Lindenbaum and Hargrove, 1968; MacLeod and Michaels, 1969; Cowan and Hines, 1971; Eichner and Hillman, 1971; Coste *et al.*, 1972; Liu, 1973;

Bjorkholm, 1980; Heidemann *et al.*, 1981). A low platelet count has also been noted in 3 to 43% of chronic alcoholics who were not acutely ill or malnourished (Eichner *et al.*, 1972; Myrhed *et al.*, 1977; Wallerstedt and Olsson, 1978; Heidemann *et al.*, 1981). It is probable that the decrease in circulating platelets associated with alcoholism is the most common cause of thrombocytopenia in the United States (Hines and Cowan, 1974).

In the majority of cases, the thrombocytopenia is not associated with folate deficiency, infection, hypersplenism, or disseminated intravascular coagulation. There has been no correlation with the presence of anemia or abnormalities of liver function (Lindenbaum and Hargrove, 1968; Post and Desforges, 1968a; MacLeod and Michaels, 1969; Cowan and Hines, 1971; Eichner and Hillman, 1971; Coste *et al.*, 1972). Hemorrhagic manifestations are usually absent but may be severe in occasional cases. In some patients there is an associated granulocytopenia (Lindenbaum and Hargrove, 1968; Post and Desforges, 1968a; Eichner and Hillman, 1971; Eichner *et al.*, 1972; Liu, 1973). The platelet count may not change for 1 to 3 days after admission, may even show a fall during this time, or may begin to rise immediately. In any case, a rapid return to or toward normal occurs within a week of alcohol withdrawal (Lindenbaum and Hargrove, 1968; MacLeod and Michaels, 1969; Cowan and Hines, 1971; Coste *et al.*, 1972; Myrhed *et al.*, 1977; Wallerstedt and Olsson, 1978). Failure of the platelet count to rise within

5 to 7 days usually indicates the presence of some other underlying disorder affecting platelets, such as hypersplenism.

After the platelet count returns to normal, a rebound thrombocytosis, which may occasionally exceed levels of 1,000,000 cells/μl, characteristically occurs 5 to 19 days after admission to hospital, usually during the second week (Lindenbaum and Hargrove, 1968; Silvas *et al.*, 1969; Cowan and Hines, 1971; Coste *et al.*, 1972; Davis and Ross, 1973; Moffatt and Schwartz, 1976; Haselager and Vreeken, 1977; Fink and Hutton, 1983). Thrombocytosis after alcohol withdrawal also occurs in the majority of alcoholics whose platelet counts are normal at the time of admission (Fig. 8.13) (Lindenbaum and Hargrove, 1968; Cowan and Hines, 1971). It has been speculated that the period of rebound thrombocytosis in alcoholics may be associated with an increased tendency to thrombotic disorders (Haselager and Vreeken, 1977), but this has not been the observation of most workers in the field.

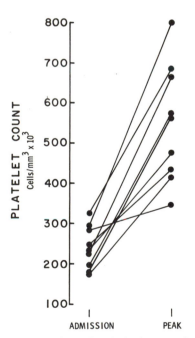

FIGURE 8.13. Admission and peak platelet counts in nine alcoholic patients admitted to hospital and withdrawn from alcohol. Platelet counts were determined three times weekly for at least 2 to 3 weeks. Peak levels were noted 10 to 21 days after admission. (From Lindenbaum and Hargrove, 1968.)

8.2.1.2. Pathogenesis

The clinical finding that thrombocytopenia occurred in the absence of associated folate deficiency and recurred after short periods of binge drinking soon after discharge from the hospital in well-nourished patients suggested that alcohol intoxication itself was the cause of the thrombocytopenia (Lindenbaum and Hargrove, 1968). This hypothesis was subsequently confirmed by several groups, who induced thrombocytopenia experimentally in human subjects by the administration of ethanol or whiskey for 10 days or more (Lindenbaum and Lieber, 1969; Ryback and Desforges, 1970; Cowan and Hines, 1971; Sullivan, 1971; Cowan, 1973; Haut and Cowan, 1974). Ethanol caused a depression in circulating platelets despite the concomitant administration of a nutritious diet and vitamin supplements including large doses of folic acid (Fig. 8.14) (Lindenbaum and Lieber, 1969; Cowan, 1973).

The exact mechanisms whereby alcohol induces thrombocytopenia in the absence of folate deficiency have not been fully elucidated. An increase in platelet size occurs during recovery from thrombocytopenia in alcoholics (Sahud, 1972), suggesting increased production of platelets by the bone marrow after alcohol withdrawal. Megakaryocyte numbers have appeared to be normal or increased at the time of thrombocytopenia in such patients (Lindenbaum and Hargrove, 1968; Post and Desforges, 1968a; MacLeod and Michaels, 1969; Coste *et al.*, 1972), with occasional exceptions (MacLeod and Michaels, 1969; Coste *et al.*, 1972; Ballard, 1980; Gewirtz and Hoffman, 1986). However, it is well recognized that estimation of megakaryocyte numbers from smears of aspirated marrow samples may give a misleading impression of the total megakaryocyte mass. Cowan (1973) induced thrombocytopenia experimentally in two subjects by the administration of alcohol along with a standard hospital diet and large doses of folic acid. Marrow megakaryocyte numbers, as calculated from biopsy specimens that were correlated with quantitative studies of erythropoiesis, were increased 1.4- and 1.9-fold over normal, whereas the life span of ^{51}Cr-tagged autologous platelets was moderately decreased. In these experiments, the induction of thrombocytopenia by ethanol appeared to result from a combination of decreased platelet survival and ineffective thrombocytopoiesis (that is, decreased production of platelets despite a normal to increased marrow megakaryocyte mass). A shortened platelet life span was also reported when thrombocytopenia was produced by alcohol given with a folate-deficient diet (Sullivan and Her-

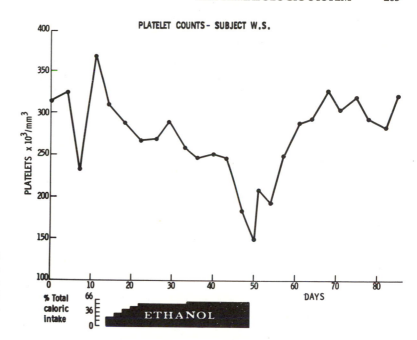

FIGURE 8.14. Serial platelet counts in a volunteer subject who received ethanol in increasing doses up to 66% of daily caloric intake along with a nutritious diet and 1200 μg of folic acid daily. (From Lindenbaum and Lieber, 1969.)

bert, 1964; Cowan, 1973) and in a thrombocytopenic patient studied shortly after admission to hospital (Post and Desforges, 1968a).

In one experimental subject studied by Sullivan (1971), however, platelet life span was found to be normal, and marrow megakaryocyte numbers, as estimated from aspirates rather than biopsy specimens, were reduced. Sullivan *et al.* (1977) subsequently showed that ethanol administration in two human volunteers interfered with the recovery of blood platelet counts after thrombocytopenia had been induced experimentally by removal of circulating platelets. Platelet life span was not affected by ethanol in one of these subjects, and megakaryocyte numbers (estimated from aspirates) appeared normal in both. These workers interpreted their findings as consistent with ineffective thrombocytopoiesis caused by alcohol (Sullivan *et al.*, 1977).

A number of observations *in vivo* and *in vitro* favor the notion that alcohol exerts its inhibitory effect late in megakaryocyte development, in addition to the experimental findings (summarized above) of normal to increased numbers of megakaryocytes despite inadequate platelet production. The rapid increase in the platelet count within 24–96 hr of alcohol withdrawal suggests recovery of a step late in platelet production (Levine *et al.*, 1986). The addition of ethanol at high concentrations *in vitro* did not inhibit the formation of mega-

karyocyte colonies derived from early precursor cells (CFU-Meg) from mouse or human bone marrow (Petursson and Chervenick, 1983; Levine *et al.*, 1986; Clark and Krantz, 1986) or from human peripheral blood (Gewirtz and Hoffman, 1986). It is likely, then, that alcohol acts mainly on differentiated megakaryocytes, although an inhibitory effect on primitive progenitor cells may occur in exceptional cases (Sullivan, 1971; Gewirtz and Hoffman, 1986).

Available data suggest that the thrombocytopenic response to chronic ethanol ingestion may be dose related (Lindenbaum and Lieber, 1969; Sullivan, 1971; Cowan, 1973). In addition, poorly defined host factors may determine the likelihood of occurrence of this complication, since certain patients develop it repeatedly while other heavy drinkers appear to escape it. Also, in experimental studies, the same alcohol dosage regimen will cause thrombocytopenia in some subjects but not in others. In some instances, coexisting folate deficiency may potentiate the effects of ethanol (Sullivan, 1971).

An acute effect of ethanol on the platelet count has also been reported. A transient depression of circulating platelets was demonstrated in three subjects after oral, and in one subject after intravenous, ethanol administration (Ryback and Desforges, 1970; Post and Desforges, 1968b). However, no effect of intravenous alcohol was demonstrable in another subject studied by the same investigators (Post and Desforges, 1968b) or in two other studies (Coste *et al.*, 1972; Cowan, 1973).

In summary, chronic alcohol administration commonly causes thrombocytopenia in the absence of folate deficiency. In most instances, the cause of the thrombocytopenia is probably a combination of a moderately shortened platelet life span and ineffective thrombocytopoiesis, with ethanol acting to inhibit platelet production by differentiated megakaryocytes. The mechanisms underlying these effects on both circulating platelets and their marrow precursors have not been eluciated.

8.2.1.3. Differential Diagnosis of Thrombocytopenia in Alcoholics

Several other entities, in addition to alcohol-induced thrombocytopenia, frequently need to be considered in the differential diagnosis of a low platelet count in the alcoholic. These are listed in Table 8.1 and include folate deficiency (Sullivan and Herbert, 1964), hypersplenism (Aster, 1966), gram-negative or gram-positive septicemia (Riedler *et al.*, 1971), and disseminated intravascular coagulation associated with shock, sepsis, or cirrhosis (Verstraete *et al.*, 1974). Table 8.1 also lists diagnostic findings that may be seen in thrombocytopenia from these causes. Bone marrow aspiration for estimation of megakaryocyte numbers is usually not helpful, since all of the conditions listed (with the exception of folate deficiency in some patients) are associated with normal to increased numbers of platelet precursors. In some instances, observation of the behavior of the platelet count during the first week in the hospital may be necessary in order to clarify the differ-

TABLE 8.1. Common Causes of Thrombocytopenia in Alcoholics

Cause	Helpful Diagnostic Findings
Folate deficiency	MCV > 110 fl, neutrophil hypersegmentation, > 3% macroovalocytes, LDH > 1000 U
Hypersplenism	Enlarged spleen; other evidence of cirrhosis and portal hypertension
Septicemia	Positive blood cultures
Disseminated intravascular coagulation	Decreased fibrinogen, other abnormal clotting tests, increased fibrin degradation products
Alcohol-induced thrombocytopenia	Absence of above findings; increase in platelet count within 5 days

ential diagnosis. In some patients more than one underlying cause may be present. Failure of the platelet count to increase by the seventh hospital day favors the diagnosis of hypersplenism (or some non-alcoholism-related cause of thrombocytopenia, such as autoimmune thrombocytopenia), unless septicemia or hypotension (often associated with intravascular coagulation) have persisted. The diagnosis of alcohol-induced thrombocytopenia is essentially based on exclusion of the other causes listed in the table, followed by the demonstration of spontaneous recovery.

8.2.1.4. Management

Most cases of alcohol-induced thrombocytopenia can be managed by expectant observation of the platelet count after exclusion of the other causes of thrombocytopenia listed in Table 8.1. Corticosteroid therapy does not appear to hasten recovery of the platelet count (Post and Desforges, 1968a). Patients with serious hemorrhage associated with platelet counts of 20,000/μl or less should be given platelet transfusions.

8.2.2. Abnormal Platelet Function

An increasing body of evidence indicates that alcohol administration interferes with platelet function, even in the absence of thrombocytopenia. At the time of admission to hospital, chronic alcoholics have been found to have a variety of abnormalities of platelet function including prolonged bleeding times (Mikhailidis *et al.*, 1986), decreased platelet aggregation tests (Fink and Hutton, 1983; Mikhailidis *et al.*, 1986), and decreased thromboxane A$_2$ release from platelets (Mikhailidis *et al.*, 1986). These findings returned to normal (or may have been followed by evidence of transient hyperfunction of platelets) within 1 to 3 weeks of alcohol withdrawal (Fink and Hutton, 1983; Mikhailidis *et al.*, 1986). In contrast, Arai *et al.* (1986) found evidence of platelet hyperfunction in patients with alcoholic liver disease at the time of hospitalization following heavy drinking.

8.2.2.1. Pathogenesis

Experimentally, alcohol has been shown to interfere with platelet function in human volunteers (Cowan, 1980). Haut and Cowan (1974) found that the chronic administration of large amounts of ethanol with a folate-supplemented hospital diet caused prolongation of the bleeding time, impairment of both primary and secondary aggregation induced by adenosine diphosphate and epinephrine, prolongation of the lag period and diminution in the rate and extent of collagen-induced aggregation, decreased platelet factor 3 availability, and subnormal release of adenine nucleotides (Haut and Cowan, 1974). The abnormalities of platelet function were more severe in subjects who developed thrombocytopenia but were also found in the presence of a normal platelet count. In one subject, abnormalities in platelet ultrastructure and adenine nucleotide metabolism were found (Cowan and Graham, 1975).

Single doses of alcohol given to human volunteers have also been shown to affect *in vitro* platelet function, although the effects have been less striking (Wautier *et al.*, 1981; Kontula *et al.*, 1982; Mikhailidis *et al.*, 1986); one group even reported an enhancement of platelet reactivity (Hillbom *et al.*, 1984). Ethanol in small doses by itself dose not prolong the bleeding time (Deykin *et al.*, 1982; Rosove and Harwig, 1983; Landolfi and Steiner, 1984).

Single modest doses of ethanol that, taken alone, are without effect on the bleeding time have been shown to clearly potentiate the effect of aspirin on this test (Deykin *et al.*, 1982; Rosove and Harwig, 1983). The magnitude and duration of the potentiating effect of alcohol were variable, but were quite striking in some subjects (Deykin *et al.*, 1982; Rosove and Harwig, 1983). Also, when small doses of ethanol were given simultaneously with doses of ibuprofen or indomethacin that, taken alone, had no effect, a more than threefold prolongation of the bleeding time was observed (Deykin *et al.*, 1982). A patient has been reported in whom the combination of aspirin and alcohol ingestion appeared to cause ecchymoses and bleeding into the anterior chamber of an eye associated with transient prolongation of the bleeding time (Kageler *et al.*, 1976).

The mechanisms underlying the impairment in platelet function caused by ethanol have not been clearly elucidated (Cowan, 1980). The addition of ethanol to platelets *in vitro* in concentrations similar to those attained in the blood of patients generally has been found to cause less striking abnormalities than those found after administration *in vivo* (Davis and Phillips, 1970; Stuart, 1979; Cowan, 1980; Quintana *et al.*, 1980; Mikhailidis *et al.*, 1983). In subjects given alcohol in repeated doses, changes in platelet ultrastructure, carbohydrate metabolism, adenine nucleotides, cyclic nucleotides, prostaglandins and thromboxanes, and biogenic amines have been noted, but the significance of any of these changes and their relationship to the platelet

dysfunction *in vivo* caused by ethanol remain to be shown (Cowan, 1980). Platelet damage may be related in part to ethanol-induced hypersomolality (Cowan, 1980). It is possible that alcohol interferes with membrane function in both megakaryocytes and circulating platelets and that the ineffective thrombocytopoiesis, shortened platelet life span, and impaired function are part of a spectrum of varying degrees of membrane damage (Cowan, 1980).

Platelet function is often abnormal *in vitro* and *in vivo* in patients with advanced alcoholic cirrhosis, even in the absence of recent alcohol ingestion (Thomas *et al.*, 1967; Ballard and Marcus, 1976). The cause of the platelet function abnormalities in cirrhosis is not known (Ballard and Marcus, 1976; Owen *et al.*, 1981). It has been postulated that abnormalities in platelet lipids (increased cholesterol/phospholipid ratio, altered phospholipid composition, and decreased arachidonic acid content) may be responsible (Owen *et al.*, 1981).

8.2.2.2. Clinical Significance

It is likely that impaired platelet function caused by alcohol intoxication, as reflected in the prolonged bleeding time, contributes significantly to the bleeding episodes so commonly seen in alcoholics and that cessation of bleeding after hospitalization may in part be related to withdrawal of ethanol. The exact contribution of platelet dysfunction to bleeding phenomena has been difficult to identify, however, in view of the frequent presence of structural lesions (e.g., varices, gastritis).

The more subtle platelet inhibition produced by modest ethanol doses in nonalcoholic patients may also be clinically significant. The potentiating effect of alcohol in interaction with antiinflammatory drugs has already been cited. Impairment in platelet function may possibly also be an important contributory factor to the decreased risk of myocardial infarction observed in light drinkers (Criqui, 1986) as well as the increased risk of hemorrhagic stroke related to all levels of alcohol ingestion (Donohue *et al.*, 1986).

8.3. Granulocytes

Early in this century, Welch, Osler, and others observed that alcoholics had a greatly increased liability to infections, particularly pneumonias, and a greater mortality rate when infected (Welch, 1903; Osler, 1927; Lyons and Saltzman, 1974). Since then, many workers have noted an increased incidence of infections in alcoholics caused by a variety of bacteria, including pneumococci, gram-negative bacilli, anaerobic organisms, *Listeria*, and the tubercle bacillus (Adams and Jordan, 1984). Probably many factors are responsible for the decreased resistance of alcoholics to infections. Several disorders involving impairment of white blood cell function may be contributory.

8.3.1. Granulocytopenia

8.3.1.1. Clinical Aspects

That a paradoxical granulocytopenia may occur in alcoholics with severe bacterial infections has long been recognized (Chomet and Gach, 1967). McFarland and Libre (1963) reported 12 episodes of neutropenia associated with infection (usually bacterial pneumonia) in ten chronic alcoholics. The granulocytopenia, which was present on admission or developed shortly thereafter, was characteristically transient. If the patient survived, the white cell count began to rise within 2 to 4 days of admission, often with a subsequent "rebound" leukocytosis. In such cases, there is frequently an accompanying thrombocytopenia (McFarland and Libre, 1963; Coste *et al.*, 1972). In normal individuals, most of the granulocytes in the body are present in a reserve "compartment" in the bone marrow (Craddock *et al.*, 1960). In infected alcoholic patients with granulocytopenia, bone marrow morphological examination has usually shown decreased cellularity with decreased numbers of mature granulocytes and a paucity of cells beyond the myelocyte or promyelocyte state (McFarland and Libre, 1963). In some patients, megaloblastic changes consistent with folate deficiency may be present, whereas in others there is no morphological or biochemical evidence of folate lack (McFarland and Libre, 1963; Eichner and Hillman, 1971; J. Lindenbaum, unpublished observations, 1980). The marrow findings rapidly return to normal during the first week in hospital.

Episodes of granulocytopenia have also been reported in association with alcohol intoxication in the absence of infection (Lindenbaum and Hargrove, 1968; Eichner *et al.*, 1972; Coste *et al.*, 1972; Liu, 1973). The incidence of leukopenia in two series of noninfected alcoholic patients was reported to be 3.6 and 8.5% (Eichner *et al.*, 1972; Liu, 1973). In both series, the degree of leukopenia was usually mild. There has been a high incidence of associated thrombocytopenia. The granulocytopenia typically returned to normal after a few days in hospital. Only mild hepatic disease, without

splenomegaly, was present. The bone marrow was characteristically normoblastic, red cell folate levels were normal, and serum folate concentrations low, normal, or high. Rebound leukocytosis was not observed during the recovery period (Lindenbaum and Hargrove, 1968; Liu, 1973). Recurrent granulocytopenia in one patient was not prevented by folic acid therapy (Liu, 1973). In two other series of noninfected alcoholics, the incidence of leukopenia was 0 and 5% (Myrhed *et al.*, 1977; Bjorkholm, 1980).

Marrow granulocyte reserves, as estimated morphologically and by the increment in peripheral neutrophil count noted after endotoxin administration, were depleted in alcoholics with and without leukopenia (Fig. 8.15) (McFarland and Libre, 1963; Liu, 1973). The marrow reserves returned to normal after a short period of hospitalization and alcohol withdrawal (McFarland and Libre, 1963; Liu, 1973).

8.3.1.2. Pathogenesis

The reason for the decreased marrow granulocyte reserves in alcoholic patients has not been established. Folate deficiency may play a contributory role in some patients but is probably not the main cause of the granulocytopenia. Unrecognized host factors may be important, in view of the tendency of certain patients to develop recurrent leukopenia with each bout of intoxication (Lindenbaum and Hargrove, 1968; Liu, 1973).

Although Tisman and Herbert (1973) found that the *in vitro* addition of ethanol in concentrations readily achieved *in vivo* to human bone marrow cultures inhibited the growth of granulocyte–macrophage colonies (CFU-GM), most investigators have found no inhibition at ethanol concentrations less than 1 to 3 g/dl (Meagher *et al.*, 1982; Imperia *et al.*, 1984; Clark and Krantz, 1986). Imperia *et al.* (1984) found that ethanol at concentrations of 100 mg/dl and greater inhibited the production of colony-stimulating activity by T cells in culture. Experimental chronic alcohol feeding to dogs has been reported to cause a mild granulocytopenia with decreased marrow cellularity (Beard and Knott, 1966). In experiments in human volunteers, however, ethanol administration has not been shown to result in leukopenia (Lindenbaum and Lieber, 1969; Liu, 1980) except in the presence of severe folate deficiency (Sullivan and Herbert, 1964). Acute alcohol intoxication in rabbits had no effect on their ability to shift granulocytes from marrow reserves into the circulation after endotoxin or glucocorticoid administration (Gluckman and MacGregor, 1978). Similarly, 6 to 47 days of ethanol feeding

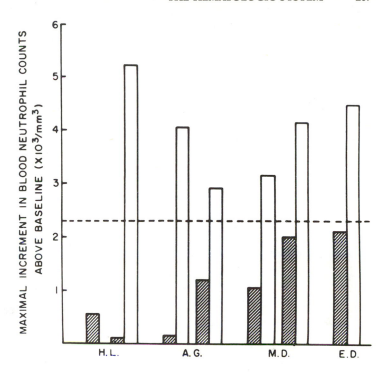

FIGURE 8.15. Bone marrow granulocyte reserves, as measured by the increment in blood neutrophil count after intravenous injection of endotoxin, in four alcoholic patients who were admitted to hospital a total of seven times with neutropenia in the absence of infection, splenomegaly, or megaloblastic erythropoiesis. The test of marrow granulocyte reserves was performed within 48 hr of admission (solid bars) on seven occasions and 6 to 14 days later (open bars) on six occasions. The dotted line represents the lower limit of normal. (From Liu, 1973.)

had no effect on marrow neutrophil reserves in human volunteers (Gluckman *et al.*, 1977; Liu, 1980).

In infected alcoholic patients with granulocytopenia, the decrease in circulating neutrophils is probably a result of a combination of diminished marrow white cell reserves and increased utilization of granulocytes at the site of infection. The neutropenia may well be the result rather than the cause of the infection.

In addition to the above-described entities, the differential diagnosis of neutropenia in alcoholics includes severe folate deficiency as well as hypersplenism in cirrhotics.

8.3.2. Impairment of Granulocyte Adherence and Mobilization

In a series of pioneering experiments, Pickrell (1938) found that severe alcohol intoxication in rabbits markedly decreased the numbers of neutrophils mobilized into sites of inflammation following intradermal, intratracheal, or intrapleural instillation of pneumococci or a nonspecific irritant. Subsequent investigators have made similar observations after the intraperitoneal injection of staphylococci into mice (Louria, 1963), the induction of either a sterile or a pneumococcal peritonitis in rabbits (Buckley *et al.*, 1978), or the pulmon-

ary installation of *Proteus mirabilis* in mice (Astry *et al.*, 1983). The acute oral or intravenous administration of mildly intoxicating doses of ethanol to nutritionally normal human volunteers produced a profound depression of the rate of granulocyte mobilization into areas of traumatized skin (Brayton *et al.*, 1970). The effect appeared to be greater early in the experiments, when blood alcohol levels would have been highest. These findings were confirmed in human subjects given 2 g/kg of ethanol by mouth over a 30-min period (Fig. 8.16) (Gluckman and MacGregor, 1978). On the other hand, the daily administration of 320 ml of 100% ethanol for 6 to 8 days to human volunteers had no effect on granulocyte mobilization studied by the same technique (Gluckman *et al.*, 1977). This was attributed to the relatively low blood alcohol levels at the time of study (Gluckman *et al.*, 1977).

It is likely that the margination and adherence of granulocytes to capillary walls is a necessary preceding step to their diapedesis into sites of inflammation. In Pickrell's (1938) original studies, he noted that the characteristic adherence of leukocytes along the walls of capillaries in infected sites was absent in intoxicated animals. Pickrell suggested that the primary defect induced by ethanol was a failure of capillaries to dilate and increase their permeability, but subsequent workers

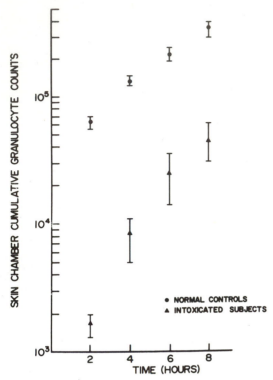

FIGURE 8.16. Cumulative granulocyte delivery into chambers placed on the skin of the forearm over 1-cm abrasions in ten normal controls and six acutely intoxicated subjects. Mean (±1 S.E.M.) values are shown. (From Gluckman and MacGregor, 1978.)

have found that alcohol intoxication does not impair the vasodilation induced in rabbits by injections of histamine or bradykinin (Moses *et al.*, 1968). Recent studies indicate that alcohol directly interferes with the capacity of neutrophils to adhere. The addition of ethanol *in vitro* causes a dose-dependent impairment of the ability of granulocytes to adhere to columns of nylon fiber (MacGregor *et al.*, 1974; Hallengren and Forsgren, 1978). A similar reduction in neutrophil adherence to nylon fiber was found after acute alcohol administration to rabbits (MacGregor *et al.*, 1974; Buckley *et al.*, 1978) and humans (Gluckman and MacGregor, 1978). Like the impairment in granulocyte delivery, the interference with the adherence of human neutrophils appears to be dose related and does not persist into periods of sobriety, even following chronic intoxication (Gluckman *et al.*, 1977; MacGregor *et al.*, 1978). In what may be a related phenomenon, low concentrations of ethanol inhibit the ameboid transformation of human granulocytes that

occurs when they are suspended in plasma *in vitro* (Lichtman *et al.*, 1976).

It is a reasonable, although not conclusively proven, hypothesis that the defect in granulocyte adherence is the cause (or one of the causes) of the failure of neutrophil mobilization (Buckley *et al.*, 1978). As already mentioned, both effects appear to be acute, dose-related actions of ethanol. In the rabbit, both were largely prevented by the prior administration of propranolol (Buckley *et al.*, 1978). These authors speculated that elevation of intracellular cAMP levels (Bryant and Sutcliffe, 1974) might account for the abnormal granulocyte adherence (and its reversal by propranolol). However, exposure to ethanol *in vitro*, even at concentrations of 800 mg/dl, only causes a slight elevation of neutrophil cAMP levels, which can be demonstrated only in the presence of theophylline (Atkinson *et al.*, 1977).

8.3.3. Chemotaxis

The addition of relatively low concentrations of ethanol *in vitro* has been reported to inhibit the chemotactic movement of neutrophils in various systems (Klepser and Nungester, 1939; Phelps and Stanislaw, 1969). Most investigators, however, have only been able to show a detrimental effect of alcohol on *in vitro* chemotaxis at concentrations that are rarely or never achieved *in vivo* (Crowley and Abramson, 1971; Spaguolo and MacGregor, 1975; Hallengren and Forsgren, 1978). The acute administration of ethanol *in vivo* to normal human volunteers had no effect on *in vitro* granulocyte chemotaxis (Spagnuolo and MacGregor, 1975). The daily administration of large doses of alcohol with an adequate diet for 6 to 8 days depressed neutrophil chemotaxis *in vitro* compared to baseline values in three of six chronic alcoholic volunteers. The depression of chemotaxis was demonstrable using the subjects' granulocytes in the presence of normal serum and thus appeared to be at the cellular level (Gluckman *et al.*, 1977). The clinical importance of the impairment in *in vitro* chemotaxis in these volunteers was doubtful, however, in view of their simultaneously normal *in vivo* granulocyte mobilization (Gluckman *et al.*, 1977).

The serum of patients with alcoholic cirrhosis frequently contains inhibitors of neutrophil chemotaxis (DeMeo and Anderson, 1972; Van Epps *et al.*, 1975; Maderazo *et al.*, 1975; Feliu *et al.*, 1977; Campbell, 1981; Rajkovic *et al.*, 1984). These inhibitors inactivate the chemotactic activity of normal human serum and appear to be multiple, with varying molecular weights. Serum-associated inhibition of chemotaxis has also been reported in some alcoholics with mild liver diseases (MacGregor *et al.*, 1978). The presence of the inhibitors may be transient, resembling the behavior of an acute-phase reactant (Van Epps *et al.*, 1975). In addition, cell-localized defects in *in vitro* granulocyte locomotion have also been described in patients with alcoholic liver disease (Rajkovic *et al.*, 1984). The clinical importance of these defects in chemotaxis is uncertain, however, since *in vivo* granulocyte mobilization has been found to be normal in patients with cirrhosis (Brayton *et al.*, 1970; MacGregor, 1986a).

8.3.4. Phagocytosis

Two groups of investigators found that the *in vitro* addition of alcohol at concentrations similar to those attainable during human intoxication had no effect on phagocytosis of particles or bacteria or on bacterial killing by granulocytes (Brayton *et al.*, 1970; Hallengren and Forsgren, 1978), although Stossel *et al.* (1972) found impaired phagocytosis of paraffin oil by guinea pig neutrophils in the presence of 85 mM ethanol. The addition of ethanol *in vitro* caused a dose-dependent inhibition of lysosomal enzyme release from human granulocytes during phagocytosis (Atkinson *et al.*, 1977).

Estimates of the *in vivo* neutrophil phagocytic efficiency of markedly intoxicated animals have been conflicting (Pickrell, 1938; Louria, 1963). The acute or chronic administration of ethanol *in vivo* to human volunteers had no effect on granulocytic phagocytosis or bactericidal ability as studied *in vitro* (Brayton *et al.*, 1970; Spagnuolo and MacGregor, 1975; Gluckman *et al.*, 1977).

Defects in *in vitro* phagocytosis have been described in patients with alcoholic liver disease (Rajkovic *et al.*, 1984; Rajkovic and Williams, 1986).

8.4. Monocytes and Macrophages

Intoxicated animals show decreased clearance of various bacteria instilled into the lungs (Green and Kass, 1964, 1965; Green and Green, 1968; Astry *et al.*, 1983) and peritoneal cavity (Louria, 1963). The mobilization of increased numbers of alveolar macrophages that occurs after aerosol challenge of mice with *Staphylococcus aureus* was decreased by alcohol intoxication (Guarneri and Laurenzi, 1968). Ethanol-treated rats also showed delayed clearance of intravenously injected microaggregated albumin by the reticuloendothelial system (Ali

and Nolan, 1967). Thus, studies in animals indicate that alcohol has a profound effect on the clearance mechanisms of the monocyte–macrophage system, an important line of defense against microorganisms. Decreased plasma clearance of macroaggregated serum albumin, presumably indicating impairment of fixed macrophage function, was found in alcoholic patients on admission to hospital, with normalization within 4 to 7 days of alcohol withdrawal (Liu, 1979).

The mechanism of the macrophage dysfunction induced by ethanol may, at least in part, be similar to that of neutrophils, i.e., a decreased capacity to adhere. The addition of ethanol at concentrations of 100 mg/dl and greater caused a dose-dependent inhibition of the adherence of rabbit alveolar macrophages to tissue culture dishes (Rimland and Hand, 1980). Ethanol added to macrophages causes elevation of intracellular cAMP, which may be responsible for the decreased adherence (Atkinson *et al.*, 1977; Rimland, 1983). The increase in cAMP could, in turn, be the result of increased osmolarity (Rimland, 1983).

Patients with alcoholic cirrhosis often have elevated serum immunoglobulin levels, especially IgA concentrations, as well as high titers of antibodies to the normal enteric flora. High levels of serum IgA may also be seen in alcoholics with normal liver function tests (Drew *et al.*, 1984). These abnormalities may be secondary to decreased clearance of antigens absorbed from the gastrointestinal tract as a result of depression of monocyte–macrophage function by ethanol, abnormal Kupffer cell function caused by cirrhosis (Triger *et al.*, 1972; Bjorneboe *et al.*, 1972), or increased absorption of antigens because of enhanced intestinal permeability from ethanol (Robinson *et al.*, 1981; Bjarnason *et al.*, 1984).

Depression of *in vitro* phagocytosis by alveolar macrophages or monocytes has been found by some investigators to occur after the addition of alcohol (Rimland and Hand, 1980; Rimland, 1983; Morland and Morland, 1984), but not by others (Gee *et al.*, 1974). Inhibitors of *in vitro* monocyte chemotaxis, phagocytosis, fungicidal activity, and metabolism have been found in the serum of patients with alcoholic cirrhosis (Van Epps *et al.*, 1975; Hassner *et al.*, 1981).

The impairment in macrophage–monocyte adherence and clearance of microorganisms induced by alcohol is likely to be of clinical significance in view of the decreased resistance of alcoholics to infections with organisms that are normally eradicated by monocytes and macrophages, such as tubercle bacilli and various bacteria that cause pneumonia (MacGregor, 1986b).

8.5. Lymphocytes

Abnormalities of lymphocyte function are common in alcoholic patients, but the respective etiologic roles of liver disease, malnutrition, various acute illnesses, and direct effects of ethanol itself have not been unraveled. A host of immunologic abnormalities have been described in patients with alcoholic hepatitis and cirrhosis, including hyperimmunoglobulinemia (especially IgA), polyclonal B-cell activation, anergy to various antigens, decreased numbers of total circulating T cells and of T_8 suppressor cells, and diminished blastogenic responses to mitogens (Sorrell and Leevy, 1972; Triger et al., 1972; Bjorneboe et al., 1972; Couzigou et al., 1984; Drew et al., 1984; Watson et al., 1985). This has recently been reviewed by Johnson and Williams (1986).

Lymphocyte abnormalities have been less consistently demonstrated in alcoholic patients with little or no apparent liver disease. Lymphopenia has been reported in some patients, particularly those with an associated granulocytopenia (Myrhed et al., 1977; Bjorkholm, 1980; Liu, 1980). In alcoholics without apparent liver disease, most workers have found normal numbers of circulating T cells and T-cell subsets (Kawamura et al., 1983; Couzigou et al., 1984; Watson et al., 1985). Evidence of in vitro B-cell activation and increased plasma IgA levels have been observed in alcoholics with normal liver function tests (Kawamura et al., 1983; Drew et al., 1984).

Prolonged experimental ethanol administration to human volunteers did not cause reductions in the numbers of circulating lymphocytes (Gluckman et al., 1977; Liu, 1980). Alcohol feeding in rats depressed cutaneous delayed hypersensitivity reactions to dinitrofluorobenzene (Tennenbaum et al., 1969). In somewhat similar studies in human volunteers, alcohol administration for 8 to 28 days had no effect on previously established delayed hypersensitivity responses to various antigens. However, the subjects could not be sensitized to a new antigen (keyhole limpet hemocyanin), which was administered for the first time during the period of ethanol ingestion (Gluckman et al., 1977). On the other hand, chronic alcoholics were readily sensitized to a de novo antigen, dinitrochlorobenzene, when sensitization was attempted during periods of abstinence (Lundy et al., 1975). These findings support the notion that ethanol itself does have subtle effects of T-cell functions in vivo.

A number of effects of alcohol on T-cell function in vitro have been reported. The addition of ethanol, at concentrations achievable in vivo, was found to inhibit mitogen-induced DNA synthesis by many investigators (Tisman and Herbert, 1973; Roselle and Mendenhall, 1982; Kaelin et al., 1984; Kaplan, 1986) but not by others (Sorrell and Leevy, 1972; Atkinson et al., 1977). Exposure to ethanol in vitro has also been found to impair a number of T-cell functions, including migration (Kaelin et al., 1984), elaboration of granulocyte–macrophage colony-stimulating activity (Imperia et al., 1984) and responsiveness to interleukin-2 and phorbol ester (Kaplan, 1986).

A possible explanation for the inhibitory effects of alcohol on T-lymphocyte function would be alterations in cellular cAMP levels. Significant increases in cAMP concentrations of human peripheral blood lymphocytes after exposure to ethanol in vitro, possibly as a result of elevations in adenylate cyclase activity, were noted by Atkinson et al. (1977). Others were only able to demonstrate a potentiating effect of alcohol when added to other agents that cause elevations in cAMP (Hynie et al., 1980).

8.6. Other Factors Favoring Infection

The acute administration of ethanol to human volunteers and to dogs has been reported to cause a transient decrease in the in vitro bactericidal activity of serum against certain organisms (Kaplan and Braude, 1958; Johnson et al., 1969; Marr and Spilberg, 1975). Although this has been attributed to a depression of complement activity in dogs (Marr and Spilberg, 1975), alcohol intoxication in humans did not decrease serum total hemolytic complement levels (Johnson et al., 1969; Spagnuolo and MacGregor, 1975; Gluckman et al., 1977). Diminished serum bactericidal activity against Escherichia coli has been observed in patients with alcoholic cirrhosis (Fierer and Finley, 1979). The clinical relevance of the reported changes in serum bactericidal activity has not been demonstrated.

Other factors undoubtedly contribute to the decreased resistance of alcoholics to infection. These include depression of the cough reflex by ethanol, impairment of reflex closure of the glottis, aspiration and pooling of secretions during periods of deep intoxication, and (possibly) associated cigarette smoking (Berkowitz et al., 1973; Lyons and Saltzman, 1974). Underlying chronic pulmonary dysfunction may also render alcoholics susceptible to infection. Chronic alcoholics have a high incidence of abnormalities in pulmonary function, including evidence of mild obstruction and impairment of diffusing capacity (Banner, 1973; Emirgil *et al.*, 1974). These abnormalities cannot be fully explained on the basis of associated smoking, and it has been argued that they are not the result of previous episodes of pneumonia (Emirgil *et al.*, 1974). It is possible that ethanol itself damages the lungs or affects the pulmonary circulation (Wagner and Heinemann, 1975; Peavy *et al.*, 1980). Hypoxemia associated with intrapulmonary shunting related to underlying cirrhosis and portal hypertension may contribute to pulmonary dysfunction (Emirgil *et al.*, 1974; Heinemann, 1977).

Folate deficiency has been shown to predispose to infection in animals (Haltalin *et al.*, 1970). Evidence of impaired T-lymphocyte function has been reported in women with folic acid deficiency in pregnancy (Gross *et al.*, 1975). In a recent study, however, evidence of depletion of tissue stores of folate was found to be uncommon in alcoholics with infection (D. S. Savage and J. Lindenbaum, unpublished observations, 1980). Although folate deficiency may possibly predispose to infection in individual patients with megaloblastic anemia, it is unlikely to be a major predisposing factor in alcoholics.

Protein–calorie malnutrition may impair lymphocytic cellular immunity (Law *et al.*, 1973) and should be further evaluated as a possible cause of decreased resistance in alcoholics.

In summary, the alcoholic appears to be an immunocompromised host with an increased tendency to become infected and to tolerate infection poorly. The relative roles of the various factors compromising host resistance remain to be established. Systemic factors likely to be significant are alcohol-induced impairment of neutrophil adherence and mobilization, alcohol-induced macrophage dysfunction, diminished marrow granulocyte reserves of unknown etiology, and, possibly, subtle defects in T-cell function. Abnormalities in T-cell responses secondary to protein–calorie malnutrition may be significant in some patients. Local factors, such as diminution of the cough reflex by ethanol, aspiration and pooling of secretions, and (possibly) underlying pulmonary dysfunction, also appear to be important.

8.7. Summary

1. Alcohol intoxication is commonly associated with megaloblastic anemia caused by folate deficiency. Although dietary deficiency is important, alcohol also acts as a weak folate antagonist, lowering the serum level of the vitamin and interfering with folate metabolism via poorly understood mechanisms.

2. Ring sideroblasts are frequently encountered in malnourished alcoholics and have been induced experimentally by ethanol administration, perhaps as the result of inhibition of heme synthesis. Whether the abnormal sideroblasts are related to the development of anemia is uncertain.

3. In the absence of nutritional deficiency, alcohol intoxication results in vacuolization of erythroid precursors but has not been shown to cause anemia.

4. Macrocytosis in the absence of folate deficiency or anemia is very common in alcoholics and is probably caused by chronic intoxication via an obscure mechanism.

5. Alcoholic liver disease often results in alterations in erythrocyte membrane lipids. These may be manifest as benign morphological changes, including macrocytosis and target cell formation, or associated with a profound chronic hemolytic anemia with acanthocytosis when hepatic function is markedly deranged. Transient hemolysis may occur with acute liver injury. The cause of this syndrome is poorly understood. Other morphological curiosities, such as stomatocytes, knizocytes, and triangulocytes, occur commonly in alcoholics for unknown reasons.

6. Marked depression of serum and red cell phosphate levels may lead to acute hemolysis with spherocytosis and decreased erythrocyte membrane fluidity and ATP concentrations.

7. Other causes of anemia in alcoholics include iron deficiency and underlying chronic disease. A modest lowering of the hematocrit may also result solely from an expanded plasma volume.

8. Ethanol administration in the absence of nutritional deficiency commonly leads to thrombocytopenia, most likely as the combined result of a shortened cell life span and ineffective thrombocytopoiesis. In the bone marrow alcohol appears to act primarily at the level of the differentiated megakaryocyte. Thrombocytopenia may also be caused by folate deficiency, hypersplenism, septicemia, and disseminated intravascular coagulation. Ethanol intoxication also interferes with platelet function and prolongs the bleeding time.

9. The alcoholic is an immunocompromised host with an increased susceptibility to and greater morbidity and mortality from various infections.

10. Ethanol intoxication in the presence or absence of infection may occasionally be associated with neutropenia and a diminished marrow granulocyte reserve.

11. Alcohol ingestion also causes decreased granulocyte adherence to the walls of capillaries and reduced mobilization of neutrophils into sites of inflammation. Similar actions of ethanol may affect monocyte–macrophage adherence and clearance functions.

12. Subtle abnormalities in T-cell function may result from alcohol intoxication. Liver disease and malnutrition also cause derangements in T-cell responses. Increased antigen absorption secondary to enhanced intestinal permeability induced by ethanol may contribute to the elevated serum concentrations of IgA immunoglobulins commonly seen in alcoholic liver disease.

13. Other factors, such as depression of the cough reflex by ethanol and aspiration and pooling of secretions, undoubtedly contribute to the predisposition of alcoholics to infection.

REFERENCES

Adams, E. B.: Anemia associated with protein deficiency. *Semin. Hematol.* **7**:55–66, 1970.

Adams, H. G., and Jordan, C.: Infections in the alcoholic. *Med. Clin. North Am.* **68**:179–200, 1984.

Albahary, C., Auffret, M., and Le Gland, J. L.: Syndrome de Zieve, hyperlipidemie alcoolique et anémia hémolytique. *Presse Med.* **76**:371–375, 1968.

Alfrey, C. P., and Lane, M.: The effect of riboflavin deficiency on erythropoiesis. *Semin. Hematol.* **7**:49–54, 1970.

Ali, M. A. M., and Sweeney, G.: Erythrocyte coproporphyrin and protoporphyrin in ethanol-induced sideroblastic erythropoiesis. *Blood* **43**:291–295, 1974.

Ali, M. V., and Nolan, J. P.: Alcohol induced depression of reticulo–endothelial function in the rat. *J. Lab. Clin. Med.* **70**:295–301, 1967.

Alling, C., Gustavsson, L., Kristensson-Aas, A., and Wallerstedt, S.: Changes in fatty acid composition of major glycerophospholipids in erythrocyte membranes from chronic alcoholics during withdrawal. *Scand. J. Clin. Lab. Invest.* **44**:283–289, 1984.

Arai, M., Okuno, F., Nagata, S., Shigeta, Y., Takagi, S., Ebihara, Y., Kobayashi, T., Ishii, H., and Tsuchiya, M.: Platelet

dysfunction and alteration of prostaglandin metabolism after chronic alcohol consumption. *Scand. J. Gastroenterol.* **21**:1091–1097, 1986.

Aron, E., Baglin, M. C., Lamy, J., and Weill, J.: Le dépistage de la consommation alcoolique chronique excessive et le controle du sevrage. Étude comparative des valeurs respective du volume globulaire moyen (VGM) et de l'activité sérique de la gamma-glutamyl-transférase (gamma-GT). *Semin. Hop. Ther.* **53**:1503–1509, 1977.

Aster, R. H.: Pooling of platelets in the spleen: Role in the pathogenesis of hypersplenic thrombocytopenia. *J. Clin. Invest.* **45**:645–657, 1966.

Astry, C. L., Warr, G. A., and Jakab, G. J.: Impairment of polymorphonuclear leukocyte immigration as a mechanism of alcohol-induced suppression of pulmonary antibacterial defenses. *Am. Rev. Respir. Dir.* **128**:113–117, 1983.

Atkinson, J. P., Sullivan, T. J., Kelly, J. P., and Parker, C. W.: Stimulation by alcohol of cyclic AMP metabolism in human leukocytes. Possible role of cyclic AMP in the anti-inflammatory effects of ethanol. *J. Clin. Invest.* **60**:284–294, 1977.

Balcerzak, S. P., Westerman, M. P., and Heinle, E. W.: Mechanism of anemia in Zeive's syndrome. *Am. J. Med. Sci.* **255**:277–287, 1968.

Balistreri, W. F., Leslie, M. H., and Cooper, R. A.: Increased cholesterol and decreased fluidity of red cell membranes (spur cell anemia) in progressive intrahepatic cholestasis. *Pediatrics* **67**:461–466, 1981.

Ballard, H. S.: Alcohol-associated pancytopenia with hypocellular bone marrow. *Am. J. Clin. Pathol.* **73**:830–834, 1980.

Ballard, H. S., and Lindenbaum, J.: Megaloblastic anemia complicating hyperalimentation therapy. *Am. J. Med.* **56**:740–742, 1974.

Ballard, H. S., and Marcus, A. J.: Platelet aggregation in portal cirrhosis. *Arch. Intern. Med.* **136**:316–319, 1976.

Banner, A. S.: Pulmonary function in chronic alcoholism. *Am. Rev. Respir. Dis.* **108**:851–857, 1973.

Bateman, J. C., Shorr, H. M., and Elgvin, T.: Hypervolemic anemia in cirrhosis. *J. Clin. Invest.* **28**:539–547, 1949.

Beard, J. D., and Knott, D. H.: Hematopoietic response to experimental chronic alcoholism. *Am. J. Med. Sci.* **252**:518–525, 1966.

Belaiche, J., Zittoun, J., Marquet, J., and Cattan, D.: La macrocytose de l'alcoolisme chronique est-elle due à un trouble de synthese de l'ADN lié à une carence en folates? *Gastroenterol. Clin. Biol.* **2**:597–602, 1978.

Berkowitz, H., Reichel, J., and Shim, C.: The effect of ethanol on the cough reflex. *Clin. Sci. Mol. Med.* **45**:527–531, 1973.

Bertino, J. R., Ward, J., Sartoreli, A. C., and Silber, R.: An effect of ethanol on folate metabolism. *J. Clin. Invest.* **44**:1028, 1965.

Bingham, J.: The macrocytosis of hepatic disease. I. Thin macrocytosis. *Blood* **14**:694–707, 1958.

Bingham, J. R.: The macrocytosis of hepatic disease: Thin, thick and target macrocytosis. *Can. Med. Assoc. J.* **85**:178–185, 1961.

Bjarnason, I., Ward, K., and Peters, T. J.: The leaky gut of alcoholism: Possible route of entry for toxic compounds. *Lancet* **1**:179–182, 1984.

Bjorkholm, M.: Immunological and hematological abnormalities in chronic alcoholism. *Acta Med. Scand.* **207**:197–200, 1980.

Bjorneboe, M., Prytz, H., and Orskov, F.: Antibodies to intestinal microbes in serum of patients with cirrhosis of the liver. *Lancet* **1**:58–60, 1972.

Blass, J. P., and Dean, H. M.: The relation of hyperlipemia to hemolytic anemia in an alcoholic patient. *Am. J. Med.* **40**:282–289, 1966.

Brayton, R. G., Stokes, P. E., Schwartz, M. S., and Louria, D. B.: Effect of alcohol and various diseases on leukocyte mobilization, phagocytosis, and intracellular bacterial killing. *N. Engl. J. Med.* **282**:123–128, 1970.

Brecher, G., and Bessis, M.: Present status of spiculed red cells and their relationship to the discocyte-echinocyte transformation. A critical review. *Blood* **40**:333–343, 1972.

Brown, J. P., Davidson, G. E., Scott, J. M., and Weir, D. G.: Effect of diphenylhydantoin and ethanol feeding on the synthesis of rat liver folates from exogenous pteroylglutamate (^{3}H). *Biochem. Pharmacol.* **22**:3287–3289, 1973.

Bryant, R. E., and Sutcliffe, M. C.: The effect of 3′,5′-adenosine monophosphate on granulocyte adhesion. *J. Clin. Invest.* **54**:1241–1244, 1974.

Buckley, R. M., Ventura, E. S., and MacGregor, R. R.: Propranolol antagonizes the anti-inflammatory effect of alcohol and improves survival of infected intoxicated rabbits. *J. Clin. Invest.* **62**:554–559, 1978.

Buffet, C., Chaput, J. C., Albuisson, F., Subtil, E., and Etienne, J. P.: La macrocytose dans l'hepatite alcoolique chronique histologiquement prouvée. *Arch. Fr. Mal. App. Dis.* **64**:309–315, 1975.

Bush, B., Shaw, S., Cleary, P., Delbanco, T. L., and Aronson, M. D.: Screening for alcohol abuse using the CAGE questionnaire. *Am. J. Med.* **82**:231–235, 1987.

Cairns, S. R., and Peters, T. J.: Biochemical analysis of hepatic lipid in alcoholic and diabetic and control subjects. *Clin. Sci.* **65**:645–652, 1983.

Campbell, A. C., Dronfield, M. W., Toghill, P. J., and Reeves, W. G.: Neutrophil function in chronic liver disease. *Clin. Exp. Immunol.* **45**:81–89, 1981.

Cartwright, G. E.: Anemia of chronic disorders. *Semin. Hematol.* **3**:351–375, 1966.

Chalmers, D. M., Chanarin, I., Levi, A. J.: Alcohol and the blood. *Br. Med. J.* **3**:203, 1978.

Chalmers, D. M., Chanarin, I., MacDermott, S., Levi, A. J.: Sex-related differences in the haematological effects of excessive alcohol consumption. *J. Clin. Pathol.* **33**:3–7, 1980.

Chalmers, D. M., Rinsler, M. G., MacDermott, S., Spicer, C. C., and Levi, A. J.,: Biochemical and hematological indicators of excessive alcohol consumption. *Gut* **22**:992–996, 1981.

Chanarin, I.: Haemopoiesis and alcohol. *Br. Med. Bull.* **38**:81–86, 1982.

Chaput, J., Lecomte, M., Poynard, T., Buffet, C., Labayle, D., and Etinne, J. P.: Rélations entre le volume globulaire moyen, les consommations alcooliques et tabagiques et l'existence de maladies alcooliques du foie chez des malades hospitalisés. *Gastroenterol. Clin. Biol.* **3**:221–226, 1979.

Cherrick, G. R., Baker, H., Frank, O., and Leevy, C. M.: Observations on hepatic avidity for folate in Laennec's cirrhosis. *J. Lab. Clin. Med.* **66**:446–451, 1965.

Chick, J., Kreitman, N., and Plant, M.: Mean cell volume and gamma-glutamyl-transpeptidase as markers of drinking in working men. *Lancet* **1**:1249–1251, 1981.

Chillar, R. K., Johnson, C. S., and Beutler, E.: Erythrocyte pyridoxine kinase levels in patients with sideroblastic anemia. *N. Engl. J. Med.* **295**:881–883, 1976.

Chomet, B., and Gach, B. M.: Lobar pneumonia and alcoholism: An analysis of thirty-seven cases. *Am. J. Med. Sci.* **253**:300–304, 1967.

Clark, D. A., and Krantz, S. B.: Effects of ethanol on cultured human megakaryocytic progenitors. *Exp. Hematol.* **14**:951–954, 1986.

Clemens, M. R., Kessler, W., Schied, H. W., Schupmann, A., and Waller, H. D.: Plasma and red cell lipids in alcoholics with macrocytosis. *Clin. Chim. Acta* **156**:321–328, 1986.

Cook, G. C.: Absorption of xylose, glucose, glycine, and folic (pteroylglutamic) acid in Zambian Africans with anaemia. *Gut.* **17**:604–611, 1976.

Cooper, R. A.: Anemia with spur cells: A red cell defect acquired in serum and modified in the circulation. *J. Clin. Invest.* **48**:1820–1831, 1969.

Cooper, R. A.: Abnormalities of cell-membrane fluidity in the pathogenesis of disease. *N. Engl. J. Med.* **297**:371–377, 1977.

Cooper, R. A.: Hemolytic syndromes and red cell membrane abnormalities in liver disease. *Semin. Hematol.* **17**:103–112, 1980.

Cooper, R. A., and Jandl, H. J.: Bile salts and cholesterol in the pathogenesis of target cells in obstructive jaundice. *J. Clin. Invest.* **47**:809–822, 1968.

Cooper, R. A., Diloy-Puray, M., Lando, P., and Greenberg, M.S.: An analysis of lipoproteins, bile acids, and red cell membranes associated with target cells and spur cells in patients with liver disease. *J. Clin. Invest.* **51**:3182–3192, 1972.

Cooper, R. A., Kimball, D. B., and Durocher, J. R.: Role of the spleen in membrane conditioning and hemolysis of spur cells in liver disease. *N. Engl. J. Med.* **290**:1279–1283, 1974.

Corcino, J., Waxman, S., and Herbert, V.: Mechanism of triamterene-induced megaloblastosis. *Ann. Intern. Med.* **73**:419–424, 1970.

Coste, T., Gouffier, C. E., and Paraf, A.: Les troubles hématologiques de l'alcoolisme aigu. *Sem. Hop. Paris* **48**:2427–2432, 1972.

Couzigou, P., Vincendeau, P., Fleury, B., Richard-Molard, B., Pierron, A., Bergeron, J. L., Bezian, J. H., Amouretti, M., and Beraud, C.: Étude des modifications des sous-populations lymphocytaires circulantes au cours des hépatopathies alcooliques. Role respectif de l'alcool, de l'insuffisance hépato-cellulaire et de la dénutrition. *Gastroenterol. Clin. Biol.* **8**:915–919, 1984.

Cowan, D. H.: Thrombokinetic studies in alcohol-related thrombocytopenia. *J. Lab. Clin. Med.* **81**:64–76, 1973.

Cowan, D. H.: Effect of alcoholism on hemostasis. *Semin. Hematol.* **17**:137–147, 1980.

Cowan, D. H., and Graham, R. C.: Studies on the platelet defect in alcoholism. *Thromb. Diath. Haemorrh.* **33**:310–327, 1975.

Cowan, D. H., and Hines, J. D.: Thrombocytopenia of severe alcoholism. *Ann. Intern. Med.* **74:**37–43, 1971.

Craddock, C. G., Frei, E., Landy, M., and Smith, W. W.: Quantitative studies of human leukocytes and febrile response to single and repeated doses of purified endotoxin. *Blood* **15:** 840–855, 1960.

Criqui, M. H.: Alcohol consumption, blood pressure, lipids, and cardiovascular mortality. *Alcoholism* **10:**564–569, 1986.

Crowley, J. P., and Abramson, N.: Effect of ethanol on complement-mediated chemotaxis. *Clin. Res.* **19:**415, 1971.

Cushman, P., Jacobson, G., Barboriak, J. J., and Anderson, A. J.: Biochemical markers for alcoholism: Sensitivity problems. *Alcoholism* **8:**253–257, 1984.

Davidson, R. J. L., and Hamilton, P. J.: High mean red cell volume: Its incidence and significance in routine haematology. *J. Clin. Pathol.* **31:**493–498, 1978.

Davidson, R. J., How, J, and Lessels, S.: Acquired stomatocytosis: Its prevalence and significance in routine haematology. *Scand. J. Haematol.* **19:**47–53, 1977.

Davis, J. W., and Phillips, P. E.: The effect of ethanol on human platelet aggregation *in vitro. Atherosclerosis* **11:**473–477, 1970.

Davis, R. E., and Smith, B. K.: Pyridoxal and folate deficiency in alcoholics. *Med. J. Aust.* **2:**257–360, 1974.

Davis, W. M., and Ross, A. O.: Thrombocytosis and thrombocythemia. *Am. J. Clin. Pathol.* **59:**243, 1973.

Deller, D. J., Kimber, C. L., and Ibbotson, R. N.: Folic acid deficiency in cirrhosis of the liver. *Am. J. Dig. Dis.* **10:**35–42, 1965.

DeMeo, A. N., and Anderson, B. R.: Defective chemotaxis associated with a serum inhibitor in cirrhotic patients. *N. Engl. J. Med.* **286:**735–740, 1972.

Deykin, D., Janson, P., and McMahon, L.: Ethanol potentiation of aspirin-induced prolongation of the bleeding time. *N. Engl. J. Med.* **306:**852–854, 1982.

Daikhate, L., Feo, C., Buffet, C., Tchernia, G., Labayle, G., and Etienne, J. P.: Homogénéité des volumes de la population érythrocytaire au cours de la macrocytose de l'éthylisme chronique. *Pathol. Biol.* **27:**99–102, 1979.

Donahue, R. P., Abbott, R. D., Reed, D. M., and Yano, K.: Alcohol and hemorrhagic stroke. *J.A.M.A.* **255:**2311–2314, 1986.

Douglass, C. C., and Twomey, J. J.: Transient stomatocytosis with hemolysis: A previously unrecognized complication of alcoholism. *Ann. Intern. Med.* **72:**159–164, 1970.

Douglass, C. C., McCall, M. S., and Frenkel, E. P.: The acanthocyte in cirrhosis with hemolytic anemia. *Ann. Intern. Med.* **68:**390–397, 1968.

Drew, P. A., Clifton, P. M., LaBrooy, J. T., and Shearman, D. J. C.: Polyclonal B cell activation in alcoholic patients with no evidence of liver dysfunction. *Clin. Exp. Immunol.* **57:**479–486, 1984.

Druskin, M. S., Wallen, M. H., and Bonagura, L.: Anticonvulsant-associated megaloblastic anemia. *N. Engl. J. Med.* **267:** 483–485, 1962.

Eckardt M. J., Ryback, R. S., Rawlings, R. R., and Graubard, B. I.: Biochemical diagnosis of alcoholism: A test of the discriminating capabilities of gamma-glutamyl transpep-

tidase and mean corpuscular volume. *J.A.M.A.* **246:**2707–2710, 1981.

Eichner, E. R.: The hematologic disorders of alcoholism. *Am. J. Med.* **54:**621–630, 1973.

Eichner, E. R., and Hillman, R. S.: The evolution of anemia in alcoholic patients. *Am. J. Med.* **50:**218–232, 1971.

Eichner, E. R., and Hillman, R. S.: Effect of alcohol on serum folate level. *J. Clin. Invest.* **52:**584–591, 1973.

Eichner, E. R., Pierce, H. I., and Hillman, R. S.: Folate balance in dietary-induced megaloblastic anemia. *N. Engl. J. Med.* **284:** 933–938, 1971.

Eichner, E. R., Buchanan, B., Smith, J. W., and Hillman, R. S.: Variations in the hematologic and medical status of alcoholics. *Am. J. Med. Sci.* **273:**35–42, 1972.

Eisenberg, S.: Blood volume in patients with Laennec's cirrhosis of the liver as determined by radioactive chromium-tagged red cells. *Am. J. Med.* **20:**189–195, 1956.

Emirgil, C., Sobol, B. J., Heymann, B., and Shibutani, K.: Pulmonary function in alcoholics. *Am. J. Med.* **57:**69–77, 1974.

Eriksen, J., Olsen, P. S., and Thomsen, Aa. C.: Gamma-glutamyltranspeptidase, aspartate aminotransferase, and erythrocyte mean corpuscular volume as indicators of alcohol consumption in liver disease. *Scand. J. Gastroenterol.* **19:** 813–819, 1984.

Eschwege, E., Papoz, L., Lellouch, J., Claude, J. R., Cubeau, J., Pequignot, G., Richard J. L., and Schwartz, D.: Blood cells and alcohol consumption with special reference to smoking habits. *J. Clin. Pathol.* **31:**654–658, 1978.

Feliu, E., Gougerot, M. A., Hakim, J., Cramer, E., Auclair, C., Rueff, B., and Boivin, P.: Blood polymorphonuclear dysfunction in patients with alcoholic cirrhosis. *Eur. J. Clin. Invest.* **7:**571–577, 1977.

Felsher, B. F., Redeker, A. G., and Reynolds, T. B.: Indirect reacting hyperbilirubinemia in cirrhosis: Its relation to red cell survival. *Am. J. Dig. Dis.* **13:**598–607, 1968.

Fierer, J., and Finley, F.: Deficient serum bactericidal activity against *Escherichia coli* in patients with cirrhosis of the liver. *J. Clin. Invest.* **63:**912–921, 1979.

Fink, R., and Hutton, R. A.: Changes in the blood platelets of alcoholics during alcohol withdrawal. *J. Clin. Pathol.* **36:** 337–340, 1983.

Fink, R., Clemens, M. R., Marjot, D. H., Patsalos, P., Cawood, P., Norden, A. G., Iversen, S. A., and Dormandy, T. L.: Increased free-radical activity in alcoholics. *Lancet* **2:**291–294, 1985.

Finkelstein, J. D., Cello, J. P., and Kyle, W. E.: Ethanol-induced changes in methionine metabolism in rat liver. *Biochem. Biophys. Res. Commun.* **61:**475–481, 1974.

Fontan, M., Goudemand, M., Caridroit, M., Lecompte, L. M., and Tauziede-Castel, C.: Dépistage biologique de l'alcoolisme chronique. Étude combinée de la gamma glutamyl transpeptidase, de l'immunoglobuline A, de l'alpha-1 lipoproteine et du volume globulaire moyen. *Nouv. Presse Med.* **7:**469, 1978.

Frank, O., Luisada-Opper, A., Sorrell, M. F., Thomas, A. D., and Baker, H.: Vitamin deficits in severe alcoholic fatty liver of man calculated from multiple reference units. *Exp. Mol. Pathol.* **15:**191–197, 1971.

Fraser, M. B., and Schacter, B. A.: Increased bone marrow delta-aminolevulinic acid synthetase activity in the acute reversible sideroblastic anemia of alcoholics. *Am. J. Hematol.* **8**:149–156, 1980.

Freedman, M. L., and Rosman, J.: A rabbit reticulocyte model for the role of hemin-controlled repressor in hypochromic anemias. *J. Clin. Invest.* **57**:594–603, 1976.

Freedman, M. L., Cohen, H. S., Rosman, J., and Forte, F. J.: Ethanol inhibition of reticulocyte protein synthesis: The role of haem. *Br. J. Haematol.* **30**:351–363, 1975.

Gadrat, J., Douste-Blazy, I., Ribet, A., Pascal, J. P., and Frexinos, J.: Syndrome de Zieve associé à une pancréatite calcificante. *Presse Med.* **75**:789–792, 1967.

Gee, J. B. L., Kaskin, J., Duncombe, M. P., and Vassallo, C. L. The effects of ethanol on some metabolic features of phagocytosis in the alveolar macrophage. *J. Reticuloendothel. Soc.* **15**:61–68, 1974.

Gewirtz, A. M., and Hoffman, R.: Transitory hypomegakaryocytic thrombocytopenia: Aetiological association with ethanol abuse and implications regarding regulation of human megakaryocytopoiesis. *Br. J. Haematol.* **62**:333–344, 1986.

Gluckman, S. J., and MacGregor, R. R.: Effect of acute alcohol intoxication on granulocyte mobilization and kinetics. *Blood* **52**:551–559, 1978.

Gluckman, S. J., Dvorak, V. C., and MacGregor, R. R.: Host defenses during prolonged alcohol consumption in a controlled environment. *Arch. Intern. Med.* **137**:1539–1543, 1977.

Goebel, K. M., Goebel, F. D., Muhlfellner, G., and Kaffarnik, H.: Red cell metabolism in transient haemolytic anaemia associated with Zeive's syndrome. *Eur. J. Clin. Invest.* **5**:83–91, 1975.

Goebel, K. M., Goebel, F. D., Schubotz, R., and Schneider, J.: Red cell metabolic and membrane features in haemolytic anaemia of alcoholic liver disease (Zieve's syndrome). *Br. J. Haematol.* **35**:573–585, 1977.

Geobel, K. M., Goebel, F. D., Schubotz, R., and Schneider, J.: Hemolytic implications of alcoholism in liver disease. *J. Lab. Clin. Med.* **94**:123–132, 1979.

Grahn, E. P., Dietz, A. A., Stefani, S. S., and Donnelly, W. J.: Burr cells, hemolytic anemia and cirrhosis. *Am. J. Med.* **45**:78–87, 1968.

Grasso, J. A., and Hines, J. D.: A comparative electron microscopic study of refractory and alcoholic sideroblastic anaemia. *Br. J. Haematol.* **17**:35–44, 1969.

Green, G. M., and Kass, E. H.: Factors influencing the clearance of bacteria by the lung *J. Clin. Invest.* **43**:769–776, 1964.

Green, G. M., and Kass, E. H.: The influence of bacterial species on pulmonary resistance to infection in mice subjected to hypoxia, cold stress, and ethanolic intoxication. *Br. J. Exp. Pathol.* **46**:360–366, 1965.

Green, L. H., and Green, G. M.: Differential suppression of pulmonary antibacterial activity as the mechanism of selection of a pathogen in mixed bacterial infection of the lung. *Am. Rev. Respir. Dis.* **98**:819–824, 1968.

Greenberg, M. S., and Choi, E. S. K.: Spur cells in splenectomized patient. *N. Engl. J. Med.* **292**:213–214, 1975.

Greene, H. L., Stifel, F. B., Herman, R. H., Herman, Y. F., and

Rosenweig, N. S.: Ethanol-induced inhibition of human intestinal enzyme activities: Reversal by folic acid. *Gastroenterology* **67**:434–440, 1974.

Gross, R. L., Reid, J. V. O., Newberne, P. M., Burgess, B., Marston, R., and Hift W.: Depressed cell-mediated immunity in megaloblastic anemia due to folic acid deficiency. *Am. J. Clin. Nutr.* **28**:225–232, 1975.

Guarneri, J. J., and Laurenzi, G. A.: Effect of alcohol on the mobilization of alveolar macrophages. *J. Lab. Clin. Med.* **72**:40–51, 1968.

Hall, C. A.: Erythrocyte dynamics in liver disease. *Am. J. Med.* **28**:541–549, 1960.

Hallengren, B., and Forsgren, A.: Effect of alcohol on chemotaxis, adherence and phagocytosis of human polymorphonuclear leucocytes. *Acta Med. Scand.* **204**:43–48, 1978.

Halsted, C. H., Robles, E. A., and Mezey, E.: Decreased jejunal uptake of labeled folic acid (^{3}H-PGA) in alcoholic patients: Roles of alcohol and nutrition. *N. Engl. J. Med.* **285**:701–706, 1971.

Halsted, C. H., Robles, E. A., and Mezey, E.: Intestinal malabsorption in folate-deficient alcoholics. *Gastroenterology* **64**:526–532, 1973.

Haltalin, K. C., Nelson, J. D., Woodman, E. B., and Allen, A. A.: Fatal *Shigella* infection induced by folic acid deficiency in young guinea pigs. *J. Infect. Dis.* **121**:275–287, 1970.

Haselager, E. M., and Vreeken, J.: Rebound thrombocytosis after alcohol abuse: A possible factor in the pathogenesis of thromboembolic disease. *Lancet* **1**:774–775, 1977.

Hassner, A., Kletter, Y., Shlag, D., Yedvab, M., Aronson, M., and Shibolet, S.: Impaired monocyte function in liver cirrhosis. *Br. Med. J.* **282**:1262–1263, 1981.

Haut, M. J., and Cowan, D. H.: The effect of ethanol on hemostatic properties of human blood platelets. *Am. J. Med.* **56**:22–32, 1974.

Heck, J., and Gehrmann, G.: Alkoholische thrombozytendepression. *Med. Wochenschr.* **97**:1088–1092, 1972.

Heidemann, E., Nerke, O., and Waller, H. D.: Alkoholtoxiche veranderungen de hamatopoieses. Eine prospektive studie bei chronischen alkoholikern. *Klin. Wochenschr.* **59**:1303–1312, 1981.

Heinemann, H. O.: Alcohol and the lung. A brief review. *Am. J. Med.* **63**:81–84, 1977.

Herbert, V.: Experimental nutritional folate deficiency in man. *Trans. Assoc. Am. Physicians* **75**:307–320, 1962.

Herbert, V.: A palatable diet for producing experimental folate deficiency in man. *Am. J. Clin. Nutr.* **12**:17–20, 1963.

Herbert, V., Baker, H., Frank, O., Pasher, I., Sobotka, H., and Wasserman, L. R.: The measurement of folic acid activity in serum: A diagnostic aid in the differentiation of the megaloblastic anemias. *Blood* **15**:223–235, 1960.

Herbert, V., Zalusky, R., and Davidson, C. S.: Correlation of folate deficiency with alcoholism and associated macrocytosis, anemia, and liver disease. *Ann. Intern. Med.* **58**:977–988, 1963.

Herbert, V., Streiff, R. R., Sullivan, L. W., and McGeer, P. L.: Deranged purine metabolism manifested by aminoimidazole-carboxamide excretion in megaloblastic anaemias, haemolytic anaemia, and liver disease. *Lancet* **2**:45–46, 1964.

Herbert, V., Tisman, G., Go, L. T., and Brenner, L.: The dU

suppression test using ^{125}I-UdR to define biochemical megaloblastosis. *Br. J. Haematol.* **24:**713–723, 1973.

Hillbom, M., Kaste, M., Kangasaho, M., Numminen, H., and Vapaatalo, H.: Platelet reactivity during ethanol intoxication in healthy men. *Acta Neurol. Scand.* [Suppl. **69**(98):331–332, 1984.

Hiller, G. I., Huffman, E. R., and Levey, S.: Studies in cirrhosis of the liver. I. Relationship between plasma volume, plasma protein concentrations and total circulating proteins. *J. Clin. Invest.* **23:**322–330, 1949.

Hillman, R. S., and Steinberg, S. E.: The effects of alcohol on folate metabolism. *Annu. Rev. Med.* **33:**345–354, 1982.

Hillman, R. S., McGuffin, R., and Campbell, C.: Alcohol interference with the folate enterohepatic cycle. *Trans. Assoc. Am. Physicians* **90:**145–156, 1977.

Hines, J. D.: Reversible megaloblastic and sideroblastic marrow abnormalities in alcoholic patients. *Br. J. Haematol.* **16:**87–101, 1969.

Hines, J. D.: Hematologic abnormalities involving vitamin B_6 and folate metabolism in alcoholic subjects. *Ann. N.Y. Acad. Sci.* **252:**316–327, 1975.

Hines, J. D., and Cowan, D. H.: Studies on the pathogenesis of alcohol-induced sideroblastic bone-marrow abnormalities. *N. Engl. J. Med.* **283:**441–446, 1970.

Hines, J. D., and Cowan, D. H.: Anemia in alcoholism. In: *Drugs and Hematologic Reactions* (N. V. Dimitrov. and J. H. Nodine, eds.), New York, Grune & Stratton, pp. 141–153, 1974.

Hines, J. D., and Grasso, J. A.: The sideroblastic anemias. *Semin. Hematol.* **7:**86–106, 1970.

Holt, F. J., and Korst, D. R.: Transient hemolytic anemia associated with liver disease. *U. Mich. Med. Bull.* **25:**79–84, 1959.

Homaidan, F. R., Kricka, L. J., Bailey, A. R., and Whitehead, T. P.: Red cell morphology in alcoholics: A new test for alcohol abuse. *Blood Cells* **11:**375–385, 1986.

Hourihane, C. O., and Weir, D. G.: Suppression of erythropoiesis by alcohol. *Br. Med. J.* **1:**86–89, 1970.

Hume, R., Williamson, J. M., and Whitelaw, J. W.: Red cell survival in biliary cirrhosis. *J. Clin. Pathol.* **23:**397–401, 1970.

Hyde, G. M., Berlin, N. I., Parsons, R. J., Lawrence, T. H., and Port, S.: The blood volume in portal cirrhosis as determined by P^{32}-labeled red blood cells. *J. Lab. Clin. Med.* **39:**347–353, 1952.

Hynie, S., Lanefelt, F., and Fredholm, B. B.: Effects of ethanol on human lymphocyte levels of cyclic AMP *in vitro*: Potentiation of the response to isoproterenol, prostaglandin E_2 or adenosine stimulation. *Acta Pharmacol. Toxicol.* **47:**58–65, 1980.

Ibrahim, N. G., Spieler, P. J., and Freedman, M. L.: Ethanol inhibition of rabbit reticulocyte haem synthesis at the level of δ-aminolaevulinic acid synthetase. *Br. J. Haematol.* **41:**235–243, 1979.

Imperia, P. S., Chikkappa, G., and Phillips, P. G.: Mechanism of inhibition of granulopoiesis by ethanol. *Proc. Sco. Exp. Biol. Med.* **175:**219–225, 1984.

Jacob, H. S., and Amsden, T.: Acute hemolytic anemia with rigid red cells in hypophosphatemia. *N. Engl. J. Med.* **285:**1446–1450, 1971.

Jandl, J. H.: Anemia of liver disease: Observations on its mechanism. *J. Clin. Invest.* **34:**390–404, 1955.

Jandl, J. H., and Lear, A. A.: The metabolism of folic acid in cirrhosis. *Ann. Inter. Med.* **45:**1027–1044, 1956.

Jarrold, T., and Vilter, R. W.: Hematologic observations in patients with chronic hepatic insufficiency: Sternal bone marrow morphology and bone marrow plasmacytosis. *J. Clin. Invest.* **23:**286–292, 1949.

Jarrold, T., Will, J. J., Davies, A. R., Duffey, P. H., and Bramschreiber, J. L.: Bone marrow-erythroid morphology in alcoholic patients. *Am. J. Clin. Nutr.* **20:**716–722, 1967.

Johnson, R. D., and Williams, R.: Immune responses in alcoholic liver disease. *Alcoholism* **10:**471–486, 1986.

Johnson, W., Stokes, P. E., and Kaye, D.: The effect of intravenous ethanol on the bactericidal activity of human serum. *Yale J. Biol. Med.* **42:**71–85, 1969.

Kaelin, R. M., Semerjian, A., Center, D. M., and Bernardo, J.: Influence of ethanol on human T-lymphocyte migration. *J. Lab. Clin. Med.* **104:**752–760, 1984.

Kageler, W. V., Moake, J. L., and Garcia, C. A.: Spontaneous hyphema associated with ingestion of aspirin and ethanol. *Am. J. Ophthalmol.* **82:**631–634, 1976.

Kaplan, D. R.: A novel mechanism of immunosuppression mediated by ethanol. *Cell. Immunol.* **102:**1–9, 1986.

Kaplan, N. M., and Braude, A. I.: Hemophilus influenza infection in adults: Observations on the immune disturbance. *Arch. Intern. Med.* **101:**515–523, 1958.

Kaunitz, J. D., and Lindenbaum, J.: The bioavailability of folic acid added to wine. *Ann. Intern. Med.* **87:**542–545, 1977.

Kawamura, Y., Tamaki, K., Tamamoto, H., Saito, S., and Aoyagi, T.: Observations on lymphocyte subsets defined by monoclonal antibodies in alcoholics without hepatic dysfunction. *Hepatology* **3:**1046, 1983.

Keating, J. N., Weir, D. G., and Scott, J. M.: The effect of ethanol consumption on folate polyglutamate biosynthesis in the rat. *Biochem. Pharmacol.* **34:**1913–1916, 1985.

Keller, J. W., Majerus, P. W., and Finke, E. H.: An unusual type of spiculated erythrocyte in metastatic liver disease and hemolytic anemia. *Ann. Intern Med.* **74:**732–737, 1971.

Kelly, D., Reed, B., Weir, D., and Scott, J.: Effect of acute and chronic alcohol ingestion on the rate of folate catabolism and hepatic enzyme induction in mice. *Clin. Sci.* **60:**221–224, 1981.

Kessel, L.: Acute transient hyperlipemia due to hepatopancreatic damage in chronic alcoholics (Zieve's syndrome). *Am. J. Med.* **32:**747–757, 1962.

Khaund, R. R.: Macrocytosis and alcoholism. *Lancet* **1:**327, 1978.

Kilbridge, T. M., and Heller, P.: Determinants of erythrocyte size in chronic liver disease. *Blood* **34:**739–746, 1969.

Kimber, C. L., Deller, D. J., Ibbotson, R. N., and Lander, H.: The mechanism of anemia in chronic liver disease. *Q. J. Med.* **34:**33-64, 1965a.

Kimber, C. L., Deller, D. J., and Lander, H.: Megaloblastic and transitional megaloblastic anemia associated with chronic liver disease. *Am. J. Med.* **38:**767-777, 1965b.

Klepser, R. G., and Nungester, W. J.: The effect of alcohol upon the chemotactic response of leukocytes. *J. Infect. Dis.* **65:**196–199, 1939.

Klipstein, F. A., and Lindenbaum, J.: Folate deficiency in chronic liver disease. *Blood* **25**:443–456, 1965.

Klock, J. C., Williams, H. E., and Mentzer, W. C.: Hemolytic anemia and somatic cell dysfunction in severe hypophosphatemia. *Arch. Intern. Med.* **134**:360–364, 1974.

Knochel, J. P.: The pathophysiology and clinical characteristics of severe hypophosphatemia. *Arch. Intern. Med.* **137**:203–220, 1977.

Koch, W.: Intestinal absorption in well-nourished alcoholics. *INSERM* **95**:477–484, 1980.

Kontula, K., Viinikka, L., Ylkiorkala, O., and Ylikahri, R.: Effect of acute ethanol intake on thromboxane and proxtacyclin in human. *Life Sci.* **31**:261–264, 1982.

Krasner, N., Moore, M. R., Thompson, G. G., McIntosh, W., and Goldberg, A.: Depression of erythrocyte δ-aminolaevulinic acid dehydratase activity in alcoholics. *Clin. Sci. Mol. Med.* **46**:415–418, 1974.

Krasnow, S., Walsh, J. R., Zimmerman, H., and Heller, P.: Megaloblastic anemia in "alcoholic" cirrhosis. *Arch. Intern. Med.* **100**:870–880, 1957.

Kirstenson, H., Fex, G., and Trell, E.: Serum ferritin, gamma-glutamyl-transferase and alcohol consumption in healthy middle -aged men. *Drug. Alcohol Dep.* **8**:43–50, 1981.

Kristensson-Aas, A., Wallerstedt, S., Alling, C., Cederblad G., and Magnusson, B.: Haematological findings in chronic alcoholics after heavy drinking with special reference to haemolysis. *Eur. J. Clin. Invest.* **16**:178–183, 1986.

la Droitte, P., Lamboeuf, Y., and de Saint Blanquat, G.: Membrane fatty acid changes and ethanol tolerance in rat and mouse. *Life Sci.* **35**:1221–1229, 1984.

Landolfi, R., and Steiner, M.: Ethanol raises prostacyclin *in vivo* and *in vitro*. *Blood* **64**:679-682, 1984.

Lane, F., Goff, P., McGuffin, R., Eichner, E. R., and Hillman, R. S.: Folic acid metabolism in normal, folate deficient and alcoholic man. *Br. J. Haematol.* **34**:489–500, 1976.

Law, D. K., Dudrick, S. J., and Abdou, N. I.: Immuncompetence of patients with protein-calorie malnutrition: The effects of nutritional repletion. *Ann. Intern. Med.* **79**:545–550, 1973.

Leevy, C. M., Baker, H., TenHove, W., Frank, O., and Cherrick, G.: B-complex vitamins in liver disease of alcoholic. *Am. J. Clin. Nutr.* **16**:339–346, 1965.

Lehane, D. E: Vacuolated erythroblasts in hyperosmolar coma. *Arch. Intern. Med.* **134**:763–765, 1974.

Levi, A. J., and Chalmers, D. M.: Recognition of alcoholic liver disease in a district general hospital. *Gut* **19**:521–525, 1978.

Levine, R. F., Spivak, J. L., Meagher, R. C., and Sieber, F.: Effect of ethanol on thrombopoiesis. *Br. J. Haematol.* **62**:345–354, 1986.

Li, T.-K., Lumeng, L., and Veitch, R. L.: Regulation of pyridoxal 5'-phosphate metabolism in liver. *Biochem. Biophys. Res. Commun.* **61**:677–684, 1974.

Lichtman, M. A., Santillo, P. A., Kearney, E. A., Roberts, G. W., and Weed R. I.: The shape and surface morphology of human leukocytes *in vitro*: Effect of temperature, metabolic inhibitors and agents that influence membrane structure. *Blood Cells* **2**:507–531, 1976.

Lieberman, F. L., and Bateman, J. R.: Megaloblastic anemia possibly induced by triamterene in patients with alcoholic cirrhosis. *Ann. Intern. Med.* **68**:168–173, 1968.

Lindenbaum, J.: Folate and vitamin B_{12} deficiencies in alcoholism. *Semin. Hematol.* **17**:119–129, 1980a.

Lindenbaum, J.: Unpublished observations, 1980b.

Lindenbaum, J.: Drug-induced folate deficiency and the hematologic effects of alcohol. In: *Nutrition in Hematology* (J. Lindenbaum, ed.), New York, Churchill Livingstone, pp. 33–58, 1983.

Lindenbaum, J., and Hargrove, R. L.: Thrombocytopenia in alcoholics. *Ann. Intern. Med.* **68**:526–532, 1968.

Lindenbaum, J., and Lieber, C. S.: Hematologic effects of alcohol in man in the absence of nutritional deficiency. *N. Engl. J. Med.* **281**:333–338, 1969.

Lindenbaum, J., and Lieber, C. S.: Effects of ethanol on the blood, bone marrow and small intestine of man. In: *Biological Aspects of Alcohol* (M. K. Roach, W. M. McIsaac, and P. Creaven, Jr., eds.), Austin, University of Texas Press, pp. 27–53, 1971.

Lindenbaum, J., and Nath, B. J.: Megaloblastic anaemia and neutrophil hypersegmentation. *Br. J. Haematol.* **44**:511–513, 1980.

Lindenbaum, J., and Roman, M. J.: Nutritional anemia in alcoholism. *Am. J. Clin. Nutr.* **33**:2727–2735, 1980.

Lipschitz, D. A., Cook, J. D., and Finch, C. A.: A clinical evaluation of serum ferritin as an index of iron stores. *N. Engl. J. Med.* **290**:1213–1216, 1974.

Littlejohn, J. M., and John, G.: Synaptosomal membrane lipids of mice during continuous exposure to ethanol. *J. Pharm. Pharmacol.* **29**:579–580, 1977.

Liu, Y. K.: Leukopenia in alcoholics. *Am. J. Med.* **54**:605–610, 1973.

Liu, Y. K.: Phagocytic capacity of reticuloendothelial system in alcoholics. *J. Reticuloendothel. Soc.* **25**:605–613, 1979.

Liu, Y. K.: Effects of alcohol on granulocytes and lymphocytes. *Semin. Hematol.* **17**:130–136, 1980.

Lopez, R., and Montoya, M. F.: Abnormal bone marrow morphology in the premature infant associated with maternal alcohol infusion. *J. Pediatr.* **79**:1008–1010, 1971.

Louria, D. B.: Susceptibility to infection during experimental alcohol intoxication. *Trans. Assoc. Am. Physicians* **76**:102–112, 1963.

Lumeng, L.: The role of acetaldehyde in mediating the deleterious effect of ethanol on pyridoxal 5'-phosphate metabolism. *J. Clin. Invest.* **62**:286–293, 1978.

Lumeng, L.: New diagnostic markers of alcohol abuse. *Hepatology* **6**:742–745, 1986.

Lumeng, L., and Li, T.: Vitamin B_6 metabolism in chronic alcohol abuse. *J. Clin. Invest.* **53**:693–704, 1974.

Lundin, L., Hallgren, R., Birgegard, G., and Wide, L.: Serum ferritin in alcoholics and the relation to liver damage, iron state and erythropoietic activity. *Acta Med. Scand.* **209**:327–331, 1981.

Lundy, J., Raaf, J. H., Deakins, S., Wanebo, H. J., Jacobs, D. A., Lee, T., Jacobowitz, D., Spear, C., and Oettgen, H. F.: The acute and chronic effects of alcohol on the human immune system. *Surg. Gynecol. Obstet.* **141**:212–218, 1975.

Lyons, H. A., and Saltzman, A.: Diseases of the respiratory tract in alcoholics. In: *The Biology of Alcoholism, Vol. 3. Clinical Pathology* (B. Kissin, and H. Begleiter, eds.), New York, Plenum, pp. 403–434, 1974.

MacGregor, R. R.: *In vitro* chemotaxis and *in vivo* delivery of PMNS in men with alcoholic cirrhosis. *Clin. Res.* **34**:524, 1986a.

MacGregor, R. R.: Alcohol and immune defense. *J.A.M.A.* **256**:1474–1479, 1986b.

MacGregor, R. R., Gluckman, S. J., Senior, J. R.: Granulocyte function and levels of immunoglobulins and complement in patients admitted for withdrawal from alcohol. *J. Infect. Dis.* **138**:747–753, 1978.

MacGregor, R. R., and Spagnuolo, P. J., and Lentnek, A. L.: Inhibition of granulocyte adherence by ethanol, prednisone, and aspirin, measured with an assay system. *N. Engl. J. Med.* **291**:642–646, 1974.

MacLeod, E. C., and Michaels, L.: Alcohol and the blood. *Lancet* **2**:1198–1199, 1969.

Maderazo, E. G., Ward, P. A., and Quintiliani, R.: Defective regulation of chemotaxis in cirrhosis. *J. Lab. Clin. Med.* **85**: 621–630, 1975.

Marie, J., Fleury, F., Hennequet, A., Desbois, J. C., Cloup, M., and Watchi, J. M.: Ictere grave cirrhogene avec acanthocytose chez le nourrisson. *Arch. Fr. Pediatr.* **24**:585, 1967.

Marr, J. J., and Spilberg, I.: A mechanism for decreased resistance to infection by gram-negative organisms during acute alcoholic intoxication. *J. Lab. Clin. Med.* **86**:253–258, 1975.

Martinez-Maldonado, M.: Role of lipoproteins in the formation of spur cell anaemia. *J. Clin. Pathol.* **21**:620–625, 1968.

Martini, G. A., and Dolle, W.: Krankheitsbilder bei Leberschadigung durch Alkohol. *Dtsch. Med. Wochenschr.* **90**:793–799, 1965.

Marxer, F., and Kaunitz, J., Lindenbaum, J.: Unpublished observations, 1979.

McBride, J. A., and Jacob, H. S.: Abnormal kinetics of red cell membrane cholesterol in acanthocytes: Studies in genetic and experimental abetalipoproteinaemia and in spur cell anaemia. *Bri. J. Haematol.* **18**:383–397, 1970.

McColl, K. E., Thompson, G. G., Moore, M. R., and Goldberg, A.: Acute ethanol ingestion and haem biosynthesis in healthy subjects. *Eur. J. Clin. Invest.* **10**:107–112, 1980.

McCurdy, P. R., Pierce, L. E., and Rath, C. E.: Abnormal bone marrow morphology in acute alcoholism. *N. Engl. J. Med.* **266**:505–507, 1962.

McFarland, E., and Libre, E. P.: Abnormal leucocyte response in alcoholism. *Ann. Intern. Med.* **59**:865–877, 1963.

McMartin, K. E.: Increased urinary folate excretion and decreased plasma folate levels in the rat after acute ethanol treatment. *Alcoholism* **8**:172–178, 1984.

McMartin, K. E., and Collins, T. D.: Role of ethanol metabolism in the alcohol-induced increase in urinary folate excretion in rats. *Biochem. Pharmacol.* **32**:2549–2555, 1983.

McMartin, K. E., Collins, T. D., and Bairnsfather, L.: Cumulative excess urinary excretion of folate in rats after repeated ethanol treatment. *J. Nutr.* **116**:1316–1325, 1986a.

McMartin, K. E., Collins, T. D., Shiao, C. Q., Vidrine, L., and Redetzki, H. M.: Study of dose-dependence and urinary

folate excretion produced by ethanol in humans and rats. *Alcoholism* **10**:419–424, 1986b.

Meagher, R. C., Sieber, F., and Spivak, J. L.: Suppression of hematopoietic-progenitor-cell proliferation by ethanol and acetaldehyde. *N. Engl. J. Med.* **307**:845–849, 1982.

Mekhjian, H. S., and May, E. S.: Acute and chronic effects of ethanol on fluid transport in the human small intestine. *Gastroenterology* **72**:1280–1286, 1977.

Mikhailidis, D. P., Jeremy, J. Y., Barradas, M. A., Green, N., and Dandona, P.: Effect of ethanol on vascular prostacyclin (prostaglandin I_2) synthesis, platelet aggregation, and platelet thromboxane release. *Br. Med. J.* **287**:1495–1498, 1983.

Mikhailidis, D. P., Jenkins, W. J., Barradas, and Jeremy, J. Y.: Platelet function defects in chronic alcoholism. *Br. Med. J.* **293**:715–718, 1986.

Moffatt, T. L., and Schwartz, A.: Rebound thrombocytosis after alcohol withdrawal. *N. Engl. J. Med.* **295**:1322–1323, 1976.

Moore, M. R., Beattie, A. D., Thompson, G. G., and Goldberg, A.: Depression of the δ-aminolevulinic acid dehydratase activity by ethanol in man and rat. *Clin. Sci.* **40**:81–881, 1971.

Morgan, M. Y., Camilo, M. E., Luck, W., Sherlock, S., and Hoffbrand, A. V.: Macrocytosis in alcohol-related liver disease: Its value for screening. *Clin. Lab. Haematol.* **3**:35–44, 1981.

Morin, J., and Porte, P.: Macrocytose érythrocytaire chez les éthyliques. *Nouv. Presse Med.* **5**:273, 1976.

Morland, B., and Moreland, J.: Reduced Fc-receptor function in human monocytes exposed to ethanol *in vitro. Alcohol Alcoholism* **19**:211–217, 1984.

Moses, J. M., and Geschickter, E. H., and Ebert, R. H: Pathogenesis of inflammation. The relationship of enhanced permeability to leukocyte mobilization in delayed inflammation. *Br. J. Exp. Pathol.* **49**:385–394, 1968.

Movitt, E. R.: Megaloblastic bone marrow in liver disease. *Am. J. Med.* **7**:145–149, 1949.

Movitt, E. R.: Megaloblastic erythropoiesis in patients with cirrhosis of the liver. *Blood* **5**:468–477, 1950.

Myerson, R. M.: Acute effects of alcohol on the liver with special reference to the Zieve syndrome. *Am. J. Gastroentrol.* **49**: 304, 1968.

Myrhed, M., Berglund, L., and Bottiger, L. E.: Alcohol consumption and hematology. Acta Med. Scand. **202**:11–15, 1977.

Nath, B. J., and Lindenbaum, J.: Persistence of neutrophil hypersegmentation during recovery from megaloblastic granulopoiesis. *Ann. Intern. Med.* **90**:757–760, 1979.

Neerhout, R. C.: Abnormalities of erythrocyte stromal lipids in hepatic disease. *J. Lab. Clin. Med.* **71**:438–447, 1968a.

Neerhout, R. C.: Reversibility of the erythrocyte lipid abnormalities in hepatic disease. *J. Pediatr.* **73**:364–373, 1968b.

Nishimura, E. T., Beegle, R. G., and Wolf, N. S.: Ultrastructural lesion in fetal hemopoietic cells following ethanol administration to pregnant mice. *Lab. Invest.* **45**:342–436, 1981.

Norgard, M. J., Carpenter, J. T., and Conrad, M. E.: Bone marrow necrosis and degeneration. *Arch. Intern. Med.* **139**:905–911, 1979.

Osler, W.: *Modern Medicine.* Philadelphia, Lea & Febiger, 1927.

Owen, J. S., Hutton, R. A., Day, R. C., Bruckdorfer, K. R., and McIntyre, M.: Platelet lipid composition and platelet aggregation in human liver disease. *J. Lipid Res.* **22**:423–430, 1981.

Owen, J. S., Bruckdorfer, K. R., Day, R. C., and McIntyre, N.: Decreased erythrocyte membrane fluidity and altered lipid composition in human liver disease. *J. Lipid Res.* **23**:124–132, 1982.

Peavy, H. H., Summer, W. R., and Gurtner, G.: The effects of acute ethanol ingestion on pulmonary diffusing capacity. *Chest.* **77**:488–492, 1980.

Perera, G. A.: The plasma volume in Laennec's cirrhosis of the liver. *Ann. Intern. Med.* **24**:643–647, 1946.

Petursson, S. R., and Chervenick, P. A.: Suppression of megakaryocytopoiesis and granulopoiesis by ethanol and acetaldehyde *in vitro*. *Clin. Res.* **31**:320A, 1983.

Phelps, P., and Stanislaw, D.: Polymorphonuclear leukocyte motility *in vitro*. I. Effect of pH, temperature, ethyl alcohol, and caffeine, using a modified Boyden chamber technique. *Arthritis Rheum.* **12**:181–188, 1969.

Pickrell, K. L.: The effect of alcoholic intoxication and ether anesthesia on resistance to pneumococcal infection. *Bull. Johns Hopkins Hosp.* **63**:238–257, 1938.

Pierce, H. I., McGuffin, R. G., and Hillman, R. S.: Clinical studies in alcoholic sideroblastosis. *Arch. Intern. Med.* **136**:283–289, 1976.

Pillegand, B., Tournier, C., Rumeau, J.-M., and Claude, R.: Pancytopenie aigue chez un cirrhotique traité par le triamterene. Un nouveau cas. *Nouv. Presse Med.* **6**:3004–3005, 1977.

Post, R. M., and Desforges, J. F.: Thrombocytopenia and alcoholism. *Ann. Intern. Med.* **68**:1230–1236, 1968a.

Post, R. M., and Desforges, J. F.: Thrombocytopenic effect of ethanol infusion. *Blood* **31**:344–347, 1968b.

Powell, L. W., Roeser, H. P., and Halliday, J. W.: Transient intravascular haemolysis associated with alcoholic liver disease and hyperlipidaemia. *Aust. N.Z. J. Med.* **1**:39–43, 1972.

Powell, L. W., Halliday, J. W., and Knowles, B. R.: The relationship of red cell membrane lipid content to red cell morphology and survival in patients with liver disease. *Aust. N.Z. J. Med.* **5**:101–107, 1975.

Prieto, J., Barry, M., and Sherlock, S.: Serum ferritin in patients with iron overload and with acute and chronic liver diseases. *Gastroenterology* **68**:525–533, 1975.

Quintana, R. P., Lasslo, A., Dugdale, M., Goodin, L. L., and Burkhardt, E. F.: Effects of ethanol and of other factors on ADP-induced aggregation of human blood platelets *in vitro*. *Thromb. Res.* **20**:405–415, 1980.

Rajkovic, I. A, and Williams, R.: Abnormalities of neutrophil phagocytosis, intracellular killing and metabolic activity in alcoholic cirrhosis and hepatitis. *Heaptology* **6**:252–262, 1986.

Rajkovic, I. A., Yousif-Kadaru, A. G. M., Wyke, R. J., and Williams, R.: Polymophonuclear leucocyte locomotion and aggregation in patients with alcoholic liver disease. *Clin. Exp. Immunol.* **58**:654–662, 1984.

Renoux, M., Bernanrd, J.-F., Amar, M., and Boivin, P.: Pancytopenia aigue et mégaloblastose medullaire chez un cirrhotique traité par le triamterene. *Nouv. Presse Med.* **5**:641–642, 1976.

Retief, F. P., and Huskisson, Y. J.: Serum and urinary folate in liver disease. *Br. Med. J.* **2**:150–153, 1969.

Riedler, G. F., Struab, P. W., and Frick, P. G.: Thrombocytopenia in septicemia. *Helv. Med. Acta* **36**:23–38, 1971.

Rimland, D.: Mechanisms of ethanol-induced defects of alveolar macrophage function. *Alcoholism* **8**:73–76, 1983.

Rimland, D., and Hand, W. L.: The effect of ethanol on adherence and phagocytosis by rabbit alveolar macrophages. *J. Lab. Clin. Med.* **95**:918, 1980.

Robinson, G. M., Orrego, Hd., Israel, Y., Devenyi, P., and Kapur, B. M.: Low-molecular-weight polyethylene glycol as a probe of gastrointestinal permeability after alcohol ingestion. *Dig. Dis. Sci.* **26**:971–977, 1981.

Romero, J. J., Tamura, T., and Halsted, C. H.: Intestinal absorption of [³H]folic acid in the chronic alcoholic monkey. *Gastroenterology* **80**:99–102, 1981.

Rosell, G. A., and Mendenhall, C. L.: Alteration of *in vitro* human lymphocyte function by ethanol, acetaldehyde and acetate. *J. Clin. Lab. Immunol.* **9**:33–37, 1982.

Rosove, M. H., and Harwig, S. S. L.: Confirmation that ethanol potentiates aspirin-induced prolongation of the bleeding time. *Thromb. Res.* **31**:525–527, 1983.

Rudiger, H. W., Blume, K. G., Esselborn, H., Glogner, P., Kaffarnik, H., Finke, J., and Lohr, G. W.: Transitorische Hamolyse, Hyperlipamie und unspezifische Leberveranderung bei Alkoholabusus (Zieve-Syndrom). *Blut* **20**:178–184, 1970.

Russell, R. M., Rosenberg, I. H., Wilson, P. D., Iber, F. L., Oaks, E. B., Giovetti, A. C., Otradovec, C. L., Korwoski, P. A., and Press, A. W.: Increased urinary excretion and prolonged turnover time of folic acid during ethanol ingestion. *Am. J. Clin. Nutr.* **38**:64–70, 1983.

Ryback, R., and Desforges, J.: Alcoholic thrombocytopenia in three inpatient drinking alcoholics. *Arch. Intern. Med.* **125**:475–477, 1970.

Ryback, R. S., Eckardt, M J., Felsher, B., and Rawlings, R. R.: Biochemical and hematologic correlates of alcoholism and liver disease. *J.A.M.A.* **248**:2261–2265, 1982.

Sahud, M. A.: Platelet size and number in alcoholic thrombocytopenia. *N. Engl. J. Med.* **286**:355–356, 1972.

Salvioli, G., Rioli, G., Lugli, R., and Salati, R.: Membrane lipid composition of red blood cells in liver disease: Regression of spur cell anaemia after infusion of polyunsaturated phosphatidylcholine. *Gut* **19**:844–850, 1978.

Sarg, M. J., Pitchumoni, C. S.: Hypophospatemic hemolytic syndrome of alcoholics—a common city hospital problem. *Am. J. Med. Sci.* **276**:231–235, 1978.

Savage, D. S., and Lindenbaum, J.: Unpublished observations, 1980.

Savage, D. S., and Lindenbaum, J.: Anemia in alcoholics. *Medicine* **65**:322–338, 1986.

Savage, D. S., Lindenbaum, J., Stabler, S., and Allen, R.: Unpublished observations, 1987.

Schubotz, R., Goebel, K. M., and Kaffarnik, H.: Veranderungen in den Membranlipiden der Erythrocyten bei athanolinduzierter Hyperlipidamie (Zieve-Syndrom). *Klin. Wochenschr.* **54:**827–833, 1976.

Shaw, S., Worner, T. M., Borysow, M. F., Schmitz, R. E., and Lieber, C. S.: Detection of alcoholism relapse: Comparative diagnostic value of MCV, GGTP and AANB. *Alcoholism: Clin. Exp. Res.* **4:**297–301, 1979.

Shaw, S., Jayatilleke, E., Ross, W. A., Gordon, E. R., and Lieber, C. S.: Ethanol-induced lipid peroxidation: Potentiation by long- term alcohol feeding and attenuation by methionine. *J. Lab. Clin. Med* **98:**417, 1981.

Sheehy, T. W., and Berman, A.: The anemia of cirrhosis. *J. Lab. Clin. Med.* **56:**72–82, 1954.

Shohet, S. B.: "Acanthocytogenesis"—or how the red cell won its spurs. *N. Engl. J. Med.* **290:**1316–1317, 1974.

Silber, R., Amorosi, E., Lhowe, J., and Kayden, H. J.: Spur-shaped erythrocytes in Laennec's cirrhosis. *N. Engl. J. Med.* **275:**639–642, 1966.

Silvas, S. E., Turkbas, N., Swaim, W. R., and Doscherholmen, A.: Causes of thrombocytosis. *Minn. Med* **52:**1603–1607, 1969.

Smith, J. A., Lonergan, E. T., and Sterling, K.: Spur-cell anemia: Hemolytic anemia with red cells resembling acanthocytes in alcoholic cirrhosis. *N. Engl. J. Med.* **271:**396–398, 1964.

Smith, J. R., Kay, N. E., Gottlieb, A. J., and Oski, F. A.: Abnormal erythrocyte metabolism in hepatic disease. *Blood* **46:**955–964, 1975.

Smith, J. R., Kay, N. E., Gottlieb, A. J., and Oski, F. A.: Correction of impaired erythrocyte hexose monophosphate shunt metabolism in liver disease by NADP. *Clin. Res.* **24:**320A, 1976.

Solomon, L. R., and Hillman, R. S.: Vitamin B_6 metabolism in anaemic and alcoholic man. *Br. J. Haematol.* **41:**343–356, 1979.

Sorrell, M. F., and Leevy, C. M.: Lymphocyte transformation and alcoholic liver injury. *Gastroenterology* **63:**1020–1025, 1972.

Spagnuolo, P. J., and MacGregor, R. R.: Acute ethanol effect on chemotaxis and other components of host defense. *J. Lab. Clin. Med.* **86:**24–31, 1975.

Spivak, J. L.: Masked megaloblastic anemia. *Arch. Intern. Med.* **142:**2111–2114, 1982.

Steinberg, S. E., Campbell, C. L., and Hillman, R. S.: Kinetics of the normal folate enterohepatic cycle. *J. Clin. Invest.* **64:**83–88, 1979.

Steinberg, S. E., Campbell, C. L., and Hillman, R. S.: Effect of alcohol on hepatic secretion of methylfolate ($CH_3H_4PteGlu_1$) into bile. *Biochem. Pharmacol.* **30:**96–98, 1981.

Steinberg, S. E., Campbell, C. L., and Hillman, R. S.: Effect of alcohol on tumor folate supply. *Biochem. Pharmacol.* **31:**1461–1463, 1982.

Stossel, T. P., Mason, R. J., Hartwig, J., and Vaughan, M.: Quantitative studies of phagocytosis by polymorphonuclear leukocytes: Use of emulsions to measure the initial rate of phagocytosis. *J. Clin. Invest.* **51:**615–623, 1972.

Strom, J.: Zieve's syndrome. Report of a case. *Acta Med. Scand.* **174:**219–222, 1963.

Stuart, M. J.: Ethanol-inhibited platelet prostaglandin synthesis *in vitro. J. Stud. Alcohol.* **40:**1–6, 1979.

Subhiyah, B. W., and Al-Hindawi, A. Y.: Red cell survival and splenic accumulation of radiochromium in liver cirrhosis with splenomegaly. *Br. J. Haematol.* **13:**773–778, 1967.

Sullivan, L. W.: Effect of alcohol on platelet production. In: *Platelet Kinetics: Radioisotopic, Cytological, Mathematical and Clinical Aspects* (J. M. Paulus, ed.), Amsterdam, North Holland Publishing Co., pp. 247–252, 1971.

Sullivan, L. W., and Herbert, V.: Suppression of haematopoiesis by ethanol. *J. Clin. Invest.* **43:**2048–2061, 1964.

Sullivan, L. W., and Liu, Y. K.: Induction of megaloblastic erythropoiesis by ethanol in an individual with normal folate stores. *J. Clin. Invest.* **45:**1078, 1966.

Sullivan, L. W., Adams, W. H., and Liu, Y. K.: Induction of thrombocytopenia by thrombophoresis in man: Patterns of recovery in normal subjects during ethanol ingestion and abstinence. *Blood* **49:**197–207, 1977.

Tamura, T., and Halsted, C. H.: Folate turnover in chronically alcoholic monkeys. *J. Lab. Clin. Med.* **101:**623, 1983.

Tamura, T., Romero, J. J., Watson, J. E., Gong, E. J., and Halsted, C H.: Hepatic folate metabolism in the chronic alcoholic monkey. *J. Lab. Clin. Med.* **97:**654–661, 1981.

Tchernia, G., Navarro, J., Becart, R., and Casasprana, A.: Anémie hémolytique avec acanthocytose et dyslipidemie au cours de deux hépatites néonates. *Arch. Fr. Pediatr.* **25:**729–743, 1968.

Tennenbaum, J. I., Ruppert, R. D., St. Pierre, R. L., and Greenberger, N. J.: The effect of chronic alcohol administration on the immune responsiveness of rats. *J. Allergy* **44:**272–281, 1969.

Territo, M. C., and Tanaka, K. R.: Hypophosphatemia in chronic alcoholism. *Arch. Intern. Med.* **134:**445–447, 1974.

Thomas, D. P., Ream, J., and Stuart, K. R.: Platelet aggregation in patients with Laennec's cirrhosis of the liver. *N. Engl. J. Med.* **276:**1344–1348, 1967.

Tikerpae, J., Samson, D., Lim, C. K., Peters, T. J., and Reynolds, E. H.: The effects of ethanol and anticonvulsants on erythroid δ-aminolaevulinic acid synthase. *Scand. J. Haematol.* **34:**223–227, 1985.

Tisman, G., and Herbert, V.: *In vitro* myelosuppression and immunosuppression by ethanol. *J. Clin. Invest.* **52:**1410–1414, 1973.

Tonnesen, H.,. Hejberg, L., Frobenius, S., and Andersen, J. R.: Erthrocyte mean cell volume—Correlation to drinking pattern in heavy alcoholics. *Acta Med. Scand.* **219:**515–518, 1986.

Tran, N., Laplante, M., and Lebel, E.: Abnormal oxidation of ^{14}C-formaldehyde to $^{14}CO_2$ in erythrocytes of alcoholics and nonalcoholics after consumption of alcoholic beverages. *J. Nucl. Med.* **13:**677–680, 1972.

Triger, D. R., Alp, M. H., and Wright, R.: Bacterial and dietary antibodies in liver disease. *Lancet* **1:**60–63, 1972.

Turpin, F., Vaugier, G., Becart-Michel, R., and Binet, J. L.: Acanthocytose dans un cas de cirrhose alcoolique avec anémie. Transformation en acanthocytes des globules rouges transfusés. *Nouv. Rev. Fr. Hématol.* **11:**791–798, 1971.

Unger, K. W., and Johnson, D.: Red blood cell mean corpuscular volume: A potential indicator of alcohol usage in a working population. *Am. J. Med. Sci.* **267:**281–289, 1974.

Van Epps, D. E., Strickland, R. G., and Williams, R. C.: Inhibitors of leukocyte chemotaxis in alcoholic liver disease. *Am. J. Med.* **59**:200–207, 1975.

Verstraete, M., Verymylen, J., and Collen, D.: Intravascular coagulation in liver disease. *Annu. Rev. Med.* **25**:447–455, 1974.

Wagner, M., and Heinemann, H. O.: Effect of ethanol on phospholipid metabolism by the rat lung. *Am. J. Physiol.* **229**:1316–1320, 1975.

Wallerstedt, S., and Olsson, R.: Changes in "routine laboratory tests" during abstinence after heavy alcohol consumption in chronic alcoholics. *Acta Hepato-Gastroenterol.* **25**:13–19, 1978.

Walsh, M. P., Howorth, P. J. N., and Marks, V.: Pyridoxine deficiency and tryptophan metabolism in chronic alcoholics. *Am. J. Clin. Nutr.* **19**:379–383, 1966.

Waring, A. J. Rottenberg, H, Ohnishi, T., and Rubin, E.: Membranes and phospholipids of liver mitochondria from chronic alcoholic rats are resistant to membrane disordering by alcohol. *Proc. Natl. Acad. Sci. U.S.A.* **78**:2582–2586, 1981.

Warnet, J. M., Cambien, F., Vernier, V., Percoraro, M., Flament, C., Ducimeteire, P., Jacqueson, A., Richard, J. L., and Claude, J. R.: Relation between consumption of alcohol and fatty acids esterifying serum cholesterol in healthy men. *Br. Med. J.* **290**:1859–1861, 1985.

Waters, A. H., Morley, A. A., and Rankin, J. G., Effect of alcohol on haemopoiesis. *Br. Med. J.* **2**:1565–1568, 1966.

Watson, R. R., Jackson, J. C., Hartmann, B., Sampliner, R., Mobley, D., and Eskelson, C.: Cellular immune functions, endorphins, and alcohol consumption in males. *Alcoholism* **9**:248–254, 1985.

Wautier, J. L., Kadeva, H., Deschamps, J. F., Boizard, B., and Pintigny, D.: Inhibition de l'agregation plaquettaire aprés ingestion d'alcool. *Presse Med.* **10**:2040, 1981.

Weir, D. G., McGing, P. G., and Scott, J. M.: Folate metabolism, the enterohepatic circulation and alcohol. *Biochem. Pharmacol.* **34**:1–7, 1985.

Welch, W. H.: The pathological effects of alcohol. In: *Physiological Aspects of the Liquor Problem, Vol. 2* (J. S. Billings, J. S., ed.), Boston, Houghton-Mifflin, p. 351, 1903.

Werre, J. M., Helleman, P. W., Verloop, M. C., and de Gier, J.: Causes of macroplania of erythrocytes in diseases of the liver and biliary tract with special reference to leptocytosis. *Br. J. Haematol.* **19**:223–235, 1970.

Westerman, M. P., Balcerzak, S. P., and Heinle, E. W.: Red cell lipids in Zieve's syndrome: Their relation to hemolysis and to red cell osmotic fragility. *J. Lab. Clin. Med.* **72**:663–669, 1968.

Whitcomb, H. C., and Job, H. J.: Zieve's syndrome: A case report. *Rocky Mt. Med. J.* **57**:48–51, 1960.

Whitehead, T. P., Clark, C. A., and Whitfield, A. G. W.: Biochemical and haematological markers of alcohol intake. *Lancet* **1**:978–981, 1978.

Whitfield, J. B., Hensley, W. J., Bryden, D., and Gallagher, H.: Some laboratory correlates of drinking habits. *Ann. Clin. Biochem.* **15**:297–303, 1978.

Wickramasinghe, S. N., and Longland, J. E.: Assessment of deoxyuridine suppression test in diagnosis of vitamin B_{12} or folate deficiency. *Br. Med. J.* **3**:148–152, 1974.

Wilkinson, J. A., and Shane, B.: Folate metabolism in the ethanol-fed rat. *J. Nutr.* **112**:604–609, 1982.

Wilms, K., Wiedmann, K. H., and Castrillon-Oberndorfer, W. L.: Schwere megaloblastare Anamie durch Triamteren bei einem Patienten mit alkoholischer Leberzirrhose. *Dtsch. Med. Wochenschr.* **104**:814–817, 1979.

Wright, S. G., and Ree, G. H.: Blood in the alcohol stream. *Lancet* **1**:49–50, 1978.

Wu, A., Chanarin, I., and Levi, A. J.: Macrocytosis of chronic alcoholism. *Lancet* **1**:829–831, 1974.

Wu, A., Chanarin, I., Slavin, G., and Levi, A. J.: Folate deficiency in the alcoholic—its relationship to clinical and haematological abnormalities, liver disease and folate stores. *Br. J. Haem.* **29**:469–478, 1975.

Yawata, Y., Hebbel, R. P., Silvis, S., Howe, R., and Jacob, H.: Blood cell abnormalities complicating the hypophosphatemia of hyperalimentation: Erythrocyte and platelet ATP deficiency associated with hemolytic anemia and bleeding in hyperalimented dogs. *J. Lab. Clin. Med.* **84**:643–653, 1974.

Yeung, K. Y., Klug, P. P., Brower, M., and Lessin, L. S.: Mechanism of alcohol induced vacuolization in human bone marrow cells. *Blood* **42**:998, 1973.

Zieve, L.: Jaundice, hyperlipemia and hemolytic anemia: A heretofore unrecognized syndrome associated with alcoholic fatty liver and cirrhosis. *Ann. Intern. Med.* **48**:471–496, 1958.

Zieve, L.: Hemolytic anemia in liver disease. *Medicine* **45**:497–505, 1966.

9

Immunologic Reactions in Alcoholic Liver Disease

Fiorenzo Paronetto

It is now generally accepted that immunologic reactions may contribute to the liver damage seen in patients with alcoholic liver disease (ALD). The exact mechanisms by which immune reactions determine liver damage or whether immunologic alterations are seen in alcoholics without liver damage are, however, still not clear.

Although ethanol has been demonstrated to be a hepatotoxic substance, there are several lines of evidence suggesting that factors other than or in addition to ethanol may be of importance in ALD. Of particular interest in the last 10 years has been the possible involvement of immunologic mechanisms, if not in the initiation, at least in the progression of ALD. It has been claimed that alcohol produces dose- and time-related hepatotoxic damage (Peguignot *et al.*, 1974; Lelbach, 1976; Rankin, 1977). However, several other observations militate against such a tenet: the marked variation of liver damage among patients with similar consumption of ethanol, the lack of cumulative effect of alcohol over time (Sørensen *et al.*, 1984), the lack of correlation between daily alcohol consumption and progression to alcoholic cirrhosis (Marbet *et al.*, 1987), the small percentage of alcoholic patients who develop alcoholic hepatitis or cirrhosis (Galambos, 1974), and the progression of the disease to cirrhosis despite abstinence (Parés *et al.*, 1986).

In the search for explanations, considerable attention has been focused on immune mechanisms, and a host of altered immune reactions have been uncovered. The challenging task of the future is to sort the pathogenetic mechanisms from the concomitant reactions and the inconsequential epiphenomena.

Immunologic aspects of ALD have been the subject of recent reviews (Lieber, 1983, 1984, 1986a, 1987; MacSween and Anthony, 1985; Paronetto, 1985, 1986; Johnson and Williams, 1986; MacGregor, 1986; Zetterman, 1991).

9.1. Morphological Considerations

The early lesions of ALD such as steatosis with lack of inflammatory reaction do not suggest an immunologic pathogenesis. More advanced alterations such as alcoholic hepatitis may be compatible with the involvement of immune reactions. In alcoholic hepatitis, the accumulation of neutrophils is intermixed with a few lymphocytes (French *et al.*, 1979), and there is experimental evidence that lymphocytes sensitized to Mallory bodies may secrete factors chemotactic for neutrophils (Peters *et al.*, 1983). In the more advanced stages abundant accumulation of mononuclear cells is observed, and piecemeal necrosis is a frequent occurrence (Brunt *et al.*, 1974). Cases indistinguishable from chronic active hepatitis have been reported (Goldberg *et al.*, 1977; Crapper *et al.*, 1983; Nei *et al.*, 1983; Montull *et al.*, 1988) and considered related to alcohol because these lesions regressed following abstinence, and no viral markers (including markers for hepatitis C and B viruses) were detected (Takase *et al.*, 1991). Fibrosis that sets in the late stages of ALD may also have an immune mechanism, since activated lymphocytes secrete factors that stimulate fibrogenesis (Johnson and Ziff, 1976; Wahl *et al.*, 1978; Neilson *et al.*, 1980; Wahl and Gatel, 1983), and lymphocytes sensitized to alcoholic hyaline may secrete such factors (Samanta *et al.*, 1985).

9.2. Genetic Factors

Genetic factors have been implicated for a long time in the susceptibility to alcoholism or its sequelae such as ALD, cardiomyopathy, and pancreatitis. Women seem to develop complications of alcoholism more frequently after a shorter period and lower alcoholic intake than men (Krasner *et al.*, 1977; Attali *et al.*, 1981). However, it is not known whether this susceptibility results from genetic factors related to the X chromosome or from differences in body size and hormonal status.

9.2.1. Histocompatibility Antigens (HLA) Class I (HLA A, B, and C)

A number of investigations in various countries have tried to establish an association between alcoholic damage involving the heart, pancreas, and liver and class I (A, B, and C) HLA. Alcoholic cardiomyopathy has been shown to have a higher incidence of HLA Aw30 (Kachru *et al.*, 1980), whereas alcoholic pancreatitis was reported to be linked to HLA B40 (Fauchet *et al.*, 1979). In contrast, a constant association that may point to a common increase in frequency of a particular HLA phenotype was not universally reported in ALD. Different HLA class I phenotypes, however, have been described in association with the same disease in various countries; e.g., patients with alcoholic hepatitis were reported to have an increase of B8 in England, of Cw3 in Japan, and of A2 in Switzerland (Morgan *et al.*, 1980; Shigeta *et al.*, 1980; Bron *et al.*, 1982), suggesting that local subgroups of patients may display increased susceptibility to alcohol and liver damage. It is of interest that when one combines the data from various countries (Eddleston and Davis, 1982), the frequencies of Aw32, B8, B13, B37, and B27 are higher in patients with alcoholic cirrhosis than in control patients, suggesting a weak association between these HLA and cirrhosis. In addition, certain HLA (e.g., B8 and B35) may favor an accelerated development of alcoholic cirrhosis (Saunders *et al.*, 1982; Marbet *et al.*, 1988).

9.2.2. HLA Class II (HLA DR)

The investigations of class II (DR) HLA are of particular importance. These antigens are found on B lymphocytes and macrophages, and associations between certain HLA DR antigens and increased risk of disease involving immunologic abnormalities have been known for several years. Four series of patients with alcoholic cirrhosis from Australia, England, France, and Switzerland had increased frequencies of different HLA (DR1, DR2, DR3, and DR4), underlining again the genetic and possibly environmental heterogeneity of subgroups of patients with ALD. In the only study that included patients with alcoholic hepatitis and cirrhosis, HLA DR3 showed an increase that was, however, not statistically significant (Bron *et al.*, 1982). Increased frequency of HLA DR4 has recently been reported in one study of alcoholic patients who had been admitted to a neuropsychiatric hospital (Córsico *et al.*, 1988).

9.3. Immunoglobulins, Antibodies, and Immune Complexes

9.3.1. Serum Immunoglobulin Elevation

The well-known findings that all immunoglobulins (IgG, IgM, IgE, and particularly IgA) are elevated in the serum of patients with ALD has prompted an investigation of the composition and nature of these immunoglobulins. The characteristic elevation of IgA in ALD is not diagnostic because it is also seen in nonalcoholic liver diseases and in alcoholic patients without liver alterations (Iturriaga *et al.*, 1977; Drew *et al.*, 1984). The serum elevation of IgA is related to a variety of factors (reviewed in Chapter 8), and it is composed both of monomeric (A1) and polymeric (A2) IgA in proportion similar to the serum IgA of normal patients (Kalsi *et al.*, 1983; Seilles *et al.*, 1985; Van de Wiel *et al.*, 1987).

9.3.2. Antibodies to Alcoholic Hyaline

Early studies have described antibodies directed against alcoholic hyalin (Kanagasundaram *et al.*, 1977), a finding not confirmed by another investigation (Kehl *et al.*, 1981). The discrepancy may arise from the use of different methodologies and antisera as well as the variability and complexity of Mallory body preparations.

9.3.3. Autoantibodies

Antibodies to a variety of autoantigens have been described in patients with ALD (Table 9.1). Antinuclear antibodies (ANA), some with specificity against double-stranded DNA (Jain et al., 1976), and anti-smooth-muscle antibodies (SMA), the characteristic antibodies of autoimmune chronic active hepatitis, have been observed in patients with alcoholic cirrhosis, especially in females and in those patients bearing the HLA B8 phenotype (Krasner et al., 1977; Morgan et al., 1980). The antibodies to smooth muscle seem to be directed against G-actin (Cunningham et al., 1985), but a reactivity against cytoskeleton components (intermediate filaments, microfilaments, and microtubules) has been also postulated in patients with ALD (Kurki et al., 1983; Crespi et al., 1986). Antibodies to mitochondria, the characteristic antibodies of primary biliary cirrhosis, are infrequent (Bailey et al., 1976; Morgan et al., 1980; Gluud et al., 1981a; Schuurman et al., 1982; Meliconi et al., 1983), whereas antibodies to "liver-specific protein" (LSP) have been detected frequently in active cirrhosis. It has been claimed that a positive finding of LSP antibodies may indicate those patients with severe ALD (Perperas et al., 1981).

Liver-specific protein is not a single protein but is a macromolecular complex composed of many liver-cell-membrane-derived antigens. The LSP antibodies may not be organ specific because they were observed to react with a kidney antigen that is similar to LSP (Behrens and Paronetto, 1979). Studies in alcoholic patients have indeed indicated that approximately half of the patients have antibodies against the human kidney equivalent of LSP (Manns et al., 1980). Antibodies to LSP have been reported in a large variety of patients with chronic liver disease and are deemed the expression of autoimmune lesions (McFarlane and Williams, 1985). More recently some of these antibodies, especially those seen in patients with autoimmune hepatitis, were demonstrated to react with the asialoglycoprotein receptor protein (McFarlane et al., 1986). It is not known, however, whether these antibodies are also seen in patients with ALD.

Antibodies to liver cell membrane (LMA) have been detected in some patients with alcoholic cirrhosis. These antibodies bind to the surface of normal rabbit hepatocytes isolated by mechanical dispersion, and the antigen involved in this reaction is considered to be distinct from LSP. Anti-LMA antibodies, which were originally reported to be characteristic of autoimmune chronic hepatitis (Hopf et al., 1976), have more recently been considered to be related to piecemeal necrosis rather than to a particular clinical entity (McFarlane and Williams, 1985). When present in patients with ALD, the antibodies are of the IgG and IgA class (Burt et al., 1982), and they may react with LSP and polymerized human albumin (Lee et al., 1985). By a very sensitive radioimmunoassay, serum LMA antibodies of the IgA class were detected in the majority of the patients with alcoholic hepatitis but in only 60% of patients with inactive cirrhosis (Kaku et al., 1988).

9.3.4. Antibodies against Alcohol-Modified Hepatocytes

The antibodies reported to react against liver membrane from alcohol-treated animals are of particular interest. These antibodies are detected in patients with alcoholic hepatitis and active alcoholic cirrhosis and, according to some investigators, also in patients with fatty liver. They have also been observed in the sera of ethanol-treated mice but not in control animals (Goldin and Wickramasinghe, 1987). It is postulated that a new antigen is recognized on the surface of hepatocytes, and since the expression of this antigen is increased when oxidation of acetaldehyde is inhibited by disulfiram, it is suggested that this new determinant is related to the metabolism of ethanol (Crossley et al., 1986), supporting the idea that binding of acetaldehyde to proteins may generate antigenic determinants that trigger an immune response (Israel et al., 1988). Recent studies are attempting to define this antigen. New candidates include a 37-kD protein-acetaldehyde adduct (Lin et al., 1990) and an acetaldehyde-phosphatidylethanolamine adduct (Trudell et al., 1991).

It is of interest that antibodies to acetaldehyde adducts were recently demonstrated not only in animals after immunization with acetaldehyde-modified proteins (Israel et al., 1986) but also in alcoholic patients by immunodiffusion, radioimmunoassay, hemagglutination techniques, and enzyme-linked immunosorbent assays (Hoerner et al., 1986; Niemela et al., 1987; Izumi et al., 1989). Subsequent studies have indicated that these antibodies may also be detected in nonalcoholic patients. Since the highest titers were observed in the more advanced stages of liver disease, the antibodies may be the result rather than the cause of liver damage (Hoerner et al., 1988). However, these antibodies may delineate a

TABLE 9.1. Autoantibodies in Alcoholics and Patients with Alcoholic Liver Disease

Antibody	Methods[a]	Patients/ Controls Numbers	Controls (% positive)	Alcoholics (% positive)	Steatosis (% positive)	Alcoholic Hepatitis (% positive)	Cirrhosis (% positive)[b]	ALD (% positive)	Reference
Alcoholic hyalin	CF, HEM	42 (35)	4.0			76.0	0 Inactive		Kanagasundaram et al., 1977
	CF; HEM, RIA	32				0			Kehl et al., 1981
Antinuclear antibodies (ANA)	RIA	16 (27)	0				75.0		Jain et al., 1976
	IFT	29 (40)	2, 5		0	0	0		Husby et al., 1977
	IFT	105					8.5 M 31.8 F		Krasner et al., 1977
	IFT	69 (69)	15.0				32.0		Gluud et al., 1981a
	IFT	11 (7)	0					0	Meliconi et al., 1983
Ethanol-modified hepatocytes	IFT	40 (20)	5.0		79.0	89.0	59.0		MacSween et al., 1981
	IFT	80 (15)	0		5.0	0	9.0		Krogsgaard et al., 1982
	IFT	65 (30)	5.0	10	74.0	80.0	70.0		Anthony et al., 1983
	IFT, CYT	82 (46)	0		37.0	46.0	44.0		Neuberger et al., 1984
	RIA, IFT	55 (24)	0	0	9.1	62.5	33.3 Inactive 80 Active		Izumi et al., 1985
Liver cell membrane (LMA)	IFT	69 (69)	0				4.3		Gluud et al., 1981a
	IFT	40 (20)	5.0		21.0	22.0	35.0		MacSween et al., 1981
	IFT	210 (50)	4.0		10.6	22.9	66.7 Inactive 27.4 Active		Burt et al., 1982
	IFT	80 (15)	6.7		8.0	11.0	13.0		Krogsgaard et al., 1982
	IFT	15 (31)	0					0	Schuurman et al., 1982
	IFT	65 (30)	5.0	10	22.0	27.0	33.0		Anthony et al., 1983
	RIA	13 (26)				59.2			Frazer et al., 1983
	IFT	11 (7)	0						Meliconi et al., 1983
	IFT	27 (45)	4.0					27.3	Lee et al., 1985
	RIA, IgA	55 (25)	0		40.0	87.5	61.5 Inactive 100 Active	48.0	Kaku et al., 1988
Liver-specific protein (LSP)	RIA	14 (31)	0		12.5	11.2	61.5 Active	29.8	Manns et al., 1980
	RIA	36 (20)	0				0 Inactive		Perperas et al., 1981
Anti-smooth-muscle antibodies (SMA)	RIA	11 (7)	0					36.4	Meliconi et al., 1983
	IFT						23.7 M 45.4 F		Krasner et al., 1977
	IFT	15 (31)						33.0	Schuurman et al., 1982
	IFT	69 (69)	20.0				49.0		Gluud et al., 1981a
	IFT	11 (7)	0					9.1	Meliconi et al., 1983

[a]IFT, immunofluorescence; CF, complement fixation; HEM, hemagglutination; CYT, cytotoxicity; RIA, radioimmunoassay.
[b]M, male; F, female.

group of alcoholic patients with severe liver disease with tendency to progression (Izumi *et al.*, 1989).

9.3.5. Immune Complexes

The observation of high immunoglobulin levels and autoantibodies in patients with ALD has stimulated a search for the presence of immune complexes and the eventual identification of the composition of these complexes. Indeed, immune complexes and aggregated high-molecular-weight immunoglobulins have been demonstrated frequently in alcoholic patients (Penner *et al.*, 1978; Thomas *et al.*, 1978; Abrass *et al.*, 1980; Brown *et al.*, 1983; Stoltenberg and Stoltis, 1984; Spinozzi *et al.*, 1986). Circulating IgA-containing immune complexes have been detected in patients with ALD (Sancho *et al.*, 1981; Coppo *et al.*, 1985). They are not related to alcohol intake but to the severity of liver damage, irrespective of an alcoholic etiology (Van de Wiel *et al.*, 1988). However, increased renal glomerular deposits of IgA, presumably part of immune complexes, have been observed in over 50% of alcoholics with and without liver damage (Smith and Hoy, 1989), suggesting a significant role of IgA in the development of renal damage in alcoholics.

Alcoholic hyalin has been shown to be a component of circulating immune complexes (Govindarajan *et al.*, 1982), and it has been localized in the renal glomeruli of alcoholic patients, often associated with IgG and complement (Burns *et al.*, 1983). The presence of immune complexes or aggregated immunoglobulins may be responsible for a variety of immunologic abnormalities including immunosuppression, an altered modulation of lymphocytes, and activation of monocytes and the complement system, with ensuing tissue-damaging immune reactions (Theophilopoulos and Dixon, 1980; Caulfield and Shaffer, 1987).

9.4. Hepatitis B Virus (HBV)

9.4.1. Antibodies to HBV

In the last decade a number of investigators have focused their attention on whether hepatitis viruses, in addition to ethanol, may contribute to the development of ALD. It has been reported that symptomless patients with HBs antigenemia are at risk of hepatic alteration when drinking an amount of ethanol that is harmless for HBsAg-negative patients (Villa *et al.*, 1982; Nomura *et al.*, 1988). In contrast to the above reports, an epidemiologic study has suggested that HBV infection does not increase the toxic role of alcohol and that alcohol consumption does not increase the risk of HBV infection leading to chronic hepatitis and cirrhosis (Chevillotte *et al.*, 1983).

Most investigators have reported that the prevalence of HBs antigenemia is not or is only minimally increased in patients with ALD. Antibodies directed to HBsAg or to HBcAg may be seen with increased frequencies in patients with ALD, and both anti-HBs and anti-HBc may reach a five- to tenfold increase over controls not only in patients with ALD but also in alcoholic patients without liver damage. Some investigators have also noted that alcoholic patients do not have an increased prevalence of antibodies to hepatitis A virus (Gluud *et al.*, 1982), although an increase of rubella, polio, and cytomegalovirus antibodies has been reported (Vetter *et al.*, 1985). It has been suggested that the high prevalence of HBV markers is connected to a previous HBV infection perhaps related to the lifestyle or the economic status of these patients. It is questionable whether these data are indicative of a concomitant HBV-mediated liver disease because a high prevalence of antibodies was detected in alcoholics without liver damage, and statistical analysis does not indicate that HBV infection increases the influence of ethanol on the production of cirrhosis (Chevillotte *et al.*, 1983). Furthermore, administration of ethanol to chimpanzees did not aggravate the hepatic lesions in the course of experimentally induced HBV infection (Tabor *et al.*, 1978).

9.4.2. HBV DNA in Serum of Alcoholic Patients

A recent investigation in the United States has indicated the inability to detect HBV DNA sequences in the serum of alcoholic patients (Fong *et al.*, 1988). However, early studies from France had indicated that more than 10% of alcoholic patients may have circulating HBV DNA, which is a marker of HBV replication, even in the absence of HBsAg or liver damage (Bréchot *et al.*, 1982, 1985; Nalpas *et al.*, 1985; Pol *et al.*, 1987). It has been postulated that alcohol may interfere with the secretion of viral proteins or that an infection with a particular HBV strain may be encountered more frequently in ALD (Nalpas *et al.*, 1985). Indeed, a virus that shares epitopes with hepatitis B virus and is able to produce hepatitis in chimpanzees has been identified in the blood of an HBsAg-negative alcoholic patient (Wands *et al.*, 1986).

9.4.3. HBV Markers in the Liver of Alcoholic Patients

Hepatitis B core antigen has been demonstrated in the liver in alcoholic cirrhosis even in the absence of serum HBsAg (Omata *et al.*, 1978), and a recent investigation has indicated that 17% of alcoholic patients have tissue HBV markers in the absence of serum markers (Cuccurullo *et al.*, 1987), suggesting a more frequent association of ALD and active HBV infection than was previously suspected by serum analysis. Support for this hypothesis has been sought by searching for HBV DNA in the liver. Although studies from England, Canada, Japan, and the United States on patients with ALD failed to uncover HBV DNA in the liver (Harrison *et al.* 1986; Sherman *et al.*, 1988; Fong *et al.*, 1988; Horiike *et al.*, 1989), studies from France have described HBV DNA in over 10% of HBsAg-negative alcoholic patients with liver disease (Bréchot *et al.*, 1985; Nalpas *et al.*, 1985). These findings have recently been confirmed in another report from England (Fagan *et al.*, 1986).

9.4.4. HBV Vaccination in Alcoholic Patients

The high association of HBV markers and ALD has suggested that alcoholic patients may be at risk of developing HBV infections and hepatocellular carcinoma. In an attempt to prevent HBV infection, some investigators have begun to vaccinate alcoholic patients. It has been found that alcoholic patients, especially those with liver cirrhosis, exhibit a deficient antibody response to HBV vaccination (Degos *et al.*, 1986; Mendenhall *et al.*, 1988) although response to vaccination with different vaccines [e.g., pneumococcal polysaccharide, diphtheria toxoid (Bjorneboe *et al.*, 1970; Smith *et al.*, 1980; Pirovino *et al.*, 1984)] is not impaired. The reason for the specific hyporesponsiveness of alcoholic patients to HBV vaccination is not clear, and further studies are needed to ascertain whether this phenomenon is associated with subclinical HBV infection in alcoholics or with partial tolerance to the virus.

9.5. Hepatitis C Virus (HCV)

A recently developed test for the detection of antibodies to HCV has indicated a major role for HCV in ALD. Prevalence of antibodies to HCV in alcoholics with ALD has been reported to be over 25% in France,

Italy, Spain, and the United States and even higher (over 60%) in Japan, whereas the prevalence of antibodies in alcoholics without ALD was less than 5% (Poynard *et al.*, 1990; Nalpas *et al.*, 1990; Brillanti *et al.*, 1989; Shimizu *et al.*, 1990; Mendenhall *et al.*, 1990; Parés *et al.*, 1990). Antibodies to HCV correlated with clinical and histological severity of the disease. In addition, alcoholics with hepatocellular carcinomas had a prevalence of antibodies to HCV of over 40% in France (Chaput *et al.*, 1990; Poynard *et al.*, 1990, and Nalpas *et al.*, 1990) and 80% in Japan (Shimizu *et al.*, 1990). Although these preliminary data remain to be confirmed, they suggest that alcoholics are a group of patients at high risk for HCV infection and will stimulate future investigations on the role of HCV in the pathogenesis of some ALD.

9.6. Serum Factors and Cytokine Activity

Patients with ALD may harbor, in the serum, factors that interfere with the immune response or with the ability to react to infectious agents. Plasma of ALD patients increases Ig production by normal B cells (Rodriguez *et al.*, 1984) but may contain factors that inhibit lymphocyte proliferation by mitogen (Hsu and Leevy, 1971; Young *et al.*, 1979; Grauer *et al.*, 1984; Roselle *et al.*, 1988). These inhibitory factors are located in a heat-labile high-molecular-weight serum fraction (Behrens *et al.*, 1982). In addition, stimulatory factors are observed after repeated freezing and thawing of plasma of patients with ALD (Behrens *et al.*, 1982).

Activation of mononuclear cells and tissue macrophages may result in altered activity of cytokines, e.g., tumor necrosis factor and interleukin 1 and 2. These proteins may mediate a number of systemic reactions in alcoholics such as febrile response, neutrophilia, anorexia, muscle wasting, hypoalbuminemia, and increased fibroblastic proliferation (Thiele, 1989). Tumor necrosis factor and interleukin 1, after removal of serum inhibitors, are elevated in patients with alcoholic hepatitis (McClain *et al.*, 1986; McClain and Cohen, 1989; Yokota *et al.*, 1987), and the release of tumor necrosis factor is enhanced when circulating mononuclear cells are cultured with lipopolysaccharide. Tumor necrosis factor is enhanced also by solid phase monomeric IgA. The latter is observed along liver sinusoids in ALD suggesting that IgA may be responsible for the increased production of this cytokine (Devière *et al.*, 1991). Tumor necrosis is particularly elevated in the severe cases of alcoholic hepatitis suggesting that it play a role in the pathogenesis (Bird *et al.*, 1990). In contrast, circulating serum interleukin 2 activity, which is produced by normal stimulated lymphocytes and plays a major role in immunomodulation, is decreased in patients with alcoholic hepatitis but not in patients with alcoholic cirrhosis (Saxena *et al.*, 1986). Interleukin-6 (IL-6) is the major mediator of acute phase response and is produced in response to tumor necrosis factor. This cytokine is markedly elevated in patients with alcoholic hepatitis and may mediate hepatic or extrahepatic tissue damage (Devière *et al.*, 1989; Sheron *et al.*, 1991).

9.7. Cell-Mediated Alterations

In addition to derangements of humoral immunity, a number of alterations of cell-mediated immunity have been reported in patients with ALD.

9.7.1. Skin Test Abnormalities

Early investigations have demonstrated a marked decrease in reactivity to a variety of recall antigens (PPD, *Candida*, streptokinase–streptodornase antigens) (Berenyi *et al.*, 1974; Snyder *et al.*, 1978), but it is not certain whether this alteration is related to the alcoholic status or the associated malnutrition (Mills *et al.*, 1983; Mendelhall *et al.*, 1984). Improvement of skin test reactivity is observed following improvement of clinical symptoms and histological alterations of ALD and after peritoneovenous drainage of intractable ascites (Franco *et al.*, 1983).

9.7.2. Phenotypic Expression of Lymphocytes in ALD

9.7.2.1. Phenotypic Expression of Lymphocytes in Peripheral Blood

The advent of monoclonal antibodies has allowed a more precise investigation of the phenotypic expression of circulating lymphocytes. Percentage and absolute numbers of T lymphocytes have been shown to be decreased in patients with cirrhosis, confirming earlier data that used a sheep red cell rosette technique to enumerate T lymphocytes. The alteration of T lymphocytes in noncirrhotic patients with ALD is less clear, and

no constant findings have been reported. Alterations of T-cell subsets were also not constantly reported and, when present, were not related to alcohol consumption, malnutrition, or hepatocellular dysfunction (Couzigu *et al.*, 1984). A decrease of CD8 (suppressor/cytotoxic) lymphocytes with an ensuing increase in the CD4 (helper/inducer)/CD8 cell ratio has been reported by some investigators in alcoholic hepatitis (Ishimaru and Matsuda, 1990). A decrease of a special subset of CD4 cells (CD4$^+$, CD45R$^+$, suppressor-inducer T-cells) has been reported in alcoholic cirrhosis (Müller *et al.*, 1991). Using flow cytometry to study cell populations in the peripheral blood and in the liver no different lymphocyte subset patterns were seen in cirrhotic and noncirrhotic livers and the distribution of lymphocytes in the blood and in the liver did not correlate (Li *et al.*, 1991). Other investigators detected in alcoholic hepatitis and cirrhosis a number of lymphocytes carrying the phenotypic expression of both CD4 and CD8 cells (Jovanovic *et al.*, 1986). In a recent study a decrease of both CD4 and CD8 lymphocytes was reported. Interestingly, after a month of oral nutritional supplementation and abstinence, CD4 but not CD8 cells increased, underscoring the complex, unclear relationship among alcohol intake, malnutrition, and immunologic abnormalities (Roselle *et al.*, 1988). The different findings in various laboratories may be related to the heterogeneity of alcoholic patients studied in different stages of the disease or to difficulties in the performance of phenotypic analysis with monoclonal antibodies. It is hoped that future studies using more precise cell-flow cytometry may reconcile some of the reported differences.

No alteration of B cells has been reported in the peripheral blood of alcoholic patients, whereas an increase of circulating natural killer/killer (NK/K) cells in alcoholic patients without liver disease has suggested that alcohol may be responsible for this alteration. In alcoholic cirrhosis NK cells have been variably reported as normal or increased (Jovanovic *et al.*, 1986; Ruibal-Ares *et al.*, 1987).

It has recently been reported that ethanol and acetaldehyde *in vitro* increase the expression of class II (DR) histocompatibility antigens on human peripheral blood mononuclear cells and activate these cells (Conroy and Allen, 1988). Cells with DR antigens and other markers of lymphocyte activation, such as transferrin receptor and interleukin 2 receptor, are increased in patients with alcoholic cirrhosis (Spinozzi *et al.*, 1986; Devière *et al.*, 1988), suggesting *in vivo* activation of patients' T cells.

Class I histocompatibility antigens are also linked

to the modulation of cellular immunity, including the helper, suppressor, and cytotoxic activity of lymphocytes. Expression of these antigens is increased in the peripheral blood lymphocytes of alcoholics (Kolber *et al.*, 1988) and after *in vitro* incubation of ethanol with lymphoid and nonlymphoid cell lines (Parent *et al.*, 1987), suggesting a direct effect of ethanol on the cellular modulation of class I histocompatibility antigens.

Alteration of lymphocyte phenotype has also been investigated in the ascitic fluid of cirrhotic patients. Interestingly, compared with the percentage in peripheral blood, the proportion of CD8-positive ascitic fluid lymphocytes of alcoholic patients is increased (Pirrone *et al.*, 1983).

9.7.2.2. Phenotypic Expression of Lymphocytes in the Liver

The demonstration that lymphocytes are present in the liver of alcoholic patients (French *et al.*, 1979) and that these lymphocytes may be activated (Sanchez-Tapias, 1977; Bérgroth *et al.*, 1986) and participate in immunologic reactions with liberation of chemotactic factors for neutrophils (Peters *et al.*, 1983) has stimulated investigations to better characterize the lymphocytes located in the portal tracts and in the parenchyma. In mild alcoholic liver disease T lymphocytes were the predominant cells, with a ratio of CD4/CD8 cells of 2, similar to the ratio of normal T cells in the peripheral blood (Bérgroth *et al.*, 1986). T cells are activated because a number of them also bear the DR marker that is considered an indication of cell activation. In alcoholic hepatitis and cirrhosis, lymphocytes with the CD8 (suppressor/cytotoxic) cell phenotype (Fig. 9.1) are observed (Si *et al.*, 1983; Chedid *et al.*, 1990). In areas of piecemeal necrosis, lymphocytes with the CD8 phenotype are abundant, whereas cells bearing the NK/K phenotype are scarce (Si *et al.*, 1984). Thus, these observations suggest that T cells in the liver of alcoholics may be activated and contribute to the liver damage.

9.7.3. Function of Peripheral Blood Lymphocytes

It has been recognized that the phenotypic analysis of mononuclear cells is an inadequate evaluation of the function of lymphocytes because a correlation between surface markers and function of cells is frequently lacking (Alexander *et al.*, 1983). Several investigators have therefore directed their attention to the functional repertoire of lymphocytes in patients with ALD (Table 9.2).

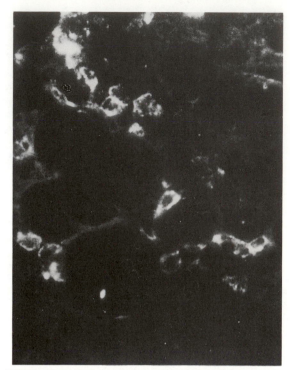

FIGURE 9.1. CD8 (cytotoxic/suppressor) lymphocytes are located in the pericentral zone of a liver biopsy from a patient with alcoholic hepatitis [immunofluorescent staining with Leu-2 monoclonal antibody (Becton Dickinson, Mountain View, CA) ×560].

9.7.3.1. Lymphocyte Stimulation by Mitogens

The degree of lymphocyte stimulation by mitogen has been considered a good parameter of lymphocyte function. Reports have been conflicting. However, dose–response curves compared in patients with ALD using a nonlinear correlation equation showed that patients with ALD display decreased lymphocyte stimulation (Mutchnick and Lee, 1988). These changes were not related to the nutritional status of the patient and were also observed in alcoholic patients without liver disease, suggesting that alcohol is responsible. T-cell activation by anti-CD2 antiserum is reduced in ALD suggesting that the CD2 pathway of lymphocyte activation is markedly defective (Spinozzi *et al.*, 1991).

9.7.3.2. Spontaneous Synthesis of Immunoglobulins by Lymphocytes

Lymphocytes of cirrhotic patients exhibit an increased spontaneous synthesis of all immunoglobulins, especially IgA. This abnormality is also seen in alcoholic patients without liver damage (Drew *et al.*, 1984), suggesting that alcohol induces either an alteration of T-suppressor cells or an increase in the number of primed helper T cells or B cells. The latter possibility may be related to the overstimulation of the immune system by increased antigen absorption or because of inefficient clearance of antigens from the circulation. When lymphocytes of alcoholics are stimulated by a mitogen or α-interferon, the production of immunoglobulins is markedly decreased in comparison to normal controls (Rodriguez *et al.*, 1984; Nouri-Aria *et al.*, 1986), a result explained by the high state of activation of B lymphocytes, which may render these cells unresponsive to further stimulation, or by the presence of a helper defect. This alteration is unlikely to be an epiphenomenon of liver damage because no relationship was found between severity or pattern of liver damage and lymphocyte function (Rodriguez *et al.*, 1984; Nouri-Aria *et al.*, 1986).

9.7.3.3. Suppressor Cell Activity

A number of methodologies have been developed for the study of the suppressor activity of lymphocytes. The majority of investigators have reported a defect of the suppressor cells in patients with alcoholic hepatitis and active, but not inactive, cirrhosis as well as in alcoholic patients without liver damage (Prieto and Mutchnick, 1980). Normal results obtained by some investigators could have been related to inadequate stimulation of suppressor cells.

9.7.3.4. Cytotoxicity of Mononuclear Cells

The observation that mononuclear cells of patients with ALD are directly cytotoxic against not only heterologous hepatocytes or Chang liver cells but also autologous hepatocytes has drawn considerable interest. This activity was primarily ascribed to NK/K cells and T cells (Poralla *et al.*, 1984) and is blocked by liver-specific protein (Cochrane *et al.*, 1977) or alcoholic hyalin (Kakumu and Leevy, 1977). It is seen frequently in patients with cirrhosis and alcoholic hepatitis and, according to one group of investigators, also in patients with steatosis (Poralla *et al.*, 1984). Ethanol preincubation did not increase sensitization of target cells, suggesting that immune reactions are based on metabolic reactions operative *in vivo* (Poralla *et al.*, 1987).

It is of interest that cytotoxicity of lymphocytes has

been confirmed in experimental animals receiving a prolonged administration of alcohol (Lue *et al.*, 1981). In these animals the hepatocytes are particularly vulnerable to the cytotoxic activity of mononuclear cells.

The finding of lymphocytes cytotoxic toward autologous liver cells supports the hypothesis that in ALD, lymphocytes recognize antigens on the surface of hepatocytes and may initiate a direct cytotoxic reaction.

It has been reported that T cells of patients with ALD, especially those with alcoholic hepatitis, are cytotoxic toward autologous acetaldehyde-altered cells, suggesting that acetaldehyde may trigger an immune response against autologous cells (Izumi *et al.*, 1988). It should be pointed out, however, that the experiments mentioned above have been performed *in vitro* utilizing peripheral blood mononuclear cells and that it would be important to ascertain whether the cytotoxic activity is also expressed by lymphocytes that are located in the liver. Future studies on the activity of liver lymphocytes expanded by clonal technology may answer this question.

9.7.3.5. Sensitization of Peripheral Blood Lymphocytes

The technique of inhibition of leukocyte migration has been utilized to detect sensitization of peripheral blood lymphocytes against a number of preparations such as ethanol, acetaldehyde, alcoholic hyalin, and heterologous and autologous liver extracts, which are considered responsible for lymphocyte sensitization. Positive reactions were detected in patients with alcoholic hepatitis, whereas patients with inactive cirrhosis or steatosis or without liver damage failed to show evidence of sensitization in this test.

9.7.3.6. Natural Killer-Cell Activity

Natural killer (NK) cells are leukocytes that are activated by interferon and interleukin 2 (IL-2) and are able to recognize cell surface changes and virally infected or tumorous cells. The activity of NK cells has been variously reported as increased and decreased in patients with cirrhosis (Charpentier *et al.*, 1984; Ruibal-Ares *et al.*, 1987) and as either normal (Charpentier *et al.*, 1984) or increased (Saxena *et al.*, 1980) in alcoholics without evidence of liver damage. Whether alterations of NK cells' activity reflect the individual alteration in reactivity to viral infections and susceptibility to tumors remains to be established.

TABLE 9.2. Alteration of Lymphocyte Functions in Alcoholics and Patients with ALD[a]

Lymphocyte Function	Patients/Controls Numbers	Assay	Method	Alcoholics	Steatosis	Alcoholic Hepatitis	Cirrhosis	ALD	Reference
Proliferation	9 (10)	Mitogen-induced LT					N		Hsu and Leevy, 1971
	24 (40)	Mitogen-induced LT		N			D		Lundy et al., 1975
	11 (12)	Mitogen-induced LT					D	D	Thestrup-Pedersen et al., 1976
	19 (10)	Mitogen-induced LT				D	D		Snyder et al., 1978
	21 (16)	Mitogen-induced LT						N	Watson et al., 1985
	25 (24)	Mitogen-induced LT					D		Ruibal-Ares et al., 1987
	25 (51)	Mitogen-induced LT		D		D	D		Mutchnick and Lee, 1988
Immunoglobulin secretion	10 (43)	IgG	RIA				IgG I		Mutchnick et al., 1981
	9 (23)	IgG	RIA					IgG I	Wands et al., 1981
	34 (61)	IgG, IgA	RIA	IgG, IgA I				IgG, IgA I	Drew et al., 1984
	26 (26)	IgG, IgM	ELISA					IgG, IgM I	Rodriguez et al., 1984
	30 (14)	IgG, IgA, IgM	RIA				IgG, IgA I / IgM I		Rong et al., 1984
	10 (8)	IgG, IgA / IgM	ELISA				IgG, IgA I / IgM N		McKeever et al., 1985
	24 (30)	IgG, IgA / IgM	IFT			IgG, IgA I / IgM N	IgG, IgA I / IgM N	IgG, IgA I / IgM N	Nouri-Aria et al., 1986
	22 (20)	Ig	IFT					IgI	Spinozzi et al., 1986
Suppression	25 (10)	T versus T	LT	A		A	A		Prieto and Mutchnick, 1980
	25 (10)	T versus T	LT				A active / N inactive		Kawanishi et al., 1981
	20 (20)	T versus T	LT					N	Wands et al., 1981
	13 (28)	T versus T	LT				A		Rong et al., 1984
	10 (43)	T versus T	LT					N	McKeever et al., 1985
	15 (11)	T versus T	LT				A		Ruibal-Areas et al., 1987
	25 (24)	T versus T	LT				A		
	13 (28)	T versus B	IgG prod					N	Wands et al., 1981

Function	No. (controls)	Cell/Target	Assay					Reference
Suppression (continued)	14 (20)	T versus B	HEM				A	Alexander et al., 1983
	15 (11)	T versus B	Ig prod				N	McKeever et al., 1985
	24 (30)	T versus B	HEM		A		A	Nouri-Aria et al., 1986
	24 (30)	T versus B	Ig prod		A		A	Spinozzi et al., 1986
	22 (20)	T versus B	Ig prod		A active		A	Kawanishi et al., 1981
	25 (18)	T cells	MLR		N inactive			Scudeletti et al., 1986
	25 (18)	T cells	MLR		A			
	20 (20)	T cells	MLR					
Cytotoxicity	17 (10)	Rabbit hepatocytes			I	I active	N	Cochrane et al., 1977
	63	Chang liver cells		N	I	I active		Kakumu and Leevy, 1977
		Autologous hepatocytes		N	I	I active		
	42 (11)	Autologous hepatocytes		I	I			Poralla et al., 1984
	18 (14)	Acetaldehyde-treated autologous cells			I	I	I	Izumi et al., 1988
Cytokine activity	6 (6)	IL-1			I			McClain et al., 1986
	17 (15)	IL-2			D	N		Saxena et al., 1986
	15 (15)	IL-2				D		Deviere et al., 1988
	16 (16)	Tumor necrosis factor			I			McClain and Cohen, 1989
Sensitization	45 (10)	Autologous liver	LIF		I	N inactive		Sorrell and Leevy, 1972
	45 (10)	Ethanol	LIF		I	N inactive		
		Acetaldehyde	LIF		I	N inactive		
	35 (15)	Liver extract	LIF	N	I			Mihas et al., 1975
	24 (7)	Alcoholic hyalin	LIF		I	N		Zetterman et al., 1976
	33 (15)	Acetaldehyde	LIF	N	I			Actis et al., 1978
	38 (30)	Alcoholic hyalin	LIF		I	I		Triggs et al., 1981
	29 (16)	Alcoholic hyalin	LIF		I (in 30%)	N		Gluud et al., 1981b
NK activity	71 (37)	Cr release	TC	N	D			Charpentier et al., 1984
	32 (15)	Cr release	TC	I				Saxena et al., 1980
	25 (24)	Cr release	TC		I			Ruibal-Ares et al., 1987

[a]N, normal; I, increased; D, decreased; A, altered; HEM, hemolytic plaque assay; LIF, leukocyte migration inhibition; ELISA, enzyme-linked immunosorbent assay; MLR, mixed lymphocyte reaction; LT, lymphocyte proliferation; Prod, production by B cells; TC, K562 target cells.

9.8. Abnormal Antigen Expression on Liver Cells

An abnormal immune reactivity against liver cells suggests an alteration of liver cell constituents, and therefore the search for neoantigens or antigens on the liver cells altered by ethanol or its metabolites is of particular importance (Table 9.3).

9.8.1. Cytokeratins

Mallory bodies, a characteristic feature of alcoholic hepatitis, are composed, at least in part, of aggregated intermediate filaments of cytokeratin type. It has been reported that some of these components have unique antigenic determinants not observed in the keratin normally detected in the hepatocytes (Denk et al., 1979; Morton et al., 1980). However, recent investigations with monoclonal antibodies directed against Mallory bodies have detected only a disorganization but not an antigenic modification of intermediate filament constituents in severe ALD (Barbatis et al., 1986).

Another line of investigation has been directed at "biliary" cytokeratins (cytokeratins 7 and 19 according to the catalogue of Moll et al., 1982) normally located in the bile ducts but not in the hepatocytes. Antisera to these cytokeratins stain the cell membrane of centrilobular hepatocytes in patients with ALD (Fig. 9.2) (Ray, 1987; Van Eyken et al., 1988). "Biliary" cytokeratin reactivity of hepatocytes is seen not only in alcoholic cirrhosis and hepatitis but also in approximately a third of patients with fatty liver, whereas it is observed infrequently in nonalcoholic liver disease (Ray, 1987).

Mallory bodies showed a striking heterogeneity in cytokeratin composition (Kaksuma et al., 1987; Van Eyken et al., 1988) and also contained ubiquitin, a protein produced in response to stress and a signal for degradation of short-lived or abnormal proteins (Lowe et al., 1988; Omar et al., 1989; Manetto et al., 1989).

The recent demonstration of a high-molecular-weight nonkeratin component in all developmental stages of Mallory bodies but not in the normal liver revealed a specific Mallory body antigen that may have a role in its formation (Zatloukal et al., 1990).

A report that tissue polypeptide antigen (TPA)—a protein that is not normally present in hepatocytes but is demonstrable in bile duct epithelium and in a variety of other epithelial cells (Nathrath et al., 1985)—is abnormally expressed in the Mallory bodies and hepatocyte cytoplasm of patients with alcoholic hepatitis and cirrhosis (Burt et al., 1986) can now be explained by the recently reported cross-reactivity between TPA and cytokeratins (Ochi et al., 1985; Hollmann et al., 1985).

9.8.2. Immunoglobulins on the Surface of Hepatocytes

Mechanically isolated hepatocytes display an IgG linear staining in approximately half of the patients with ALD. This staining does not correlate with the severity of liver disease but with elevation of transaminase activity (Trevisan et al., 1983). In contrast, a granular, coarse IgA staining is observed only in cases with alcoholic hepatitis and with alcoholic cirrhosis. The nature of IgA on hepatocytes, whether representing autoantibodies or as part of immune complexes, and its pathogenetic significance are not yet established.

9.8.3. Continuous IgA Staining of Sinusoids

Continuous IgA staining of sinusoids has been repeatedly reported in patients with ALD (Fig. 9.3) (Kater et al., 1979; Swerdlow and Chowdhury, 1983, 1984; Goldin et al., 1986; Van de Wiel et al., 1986, 1987). Staining for IgA does not correlate with severity of liver diseases because it is also observed in patients with fatty liver, but it is seen infrequently in patients with nonalcoholic liver diseases. Sinusoidal IgA, which is composed mainly of monomeric IgA (IgA_1), does not correlate with subclass type and levels of serum IgA, which is mainly of the IgA_2 subclass. It has been suggested that IgA deposition may correlate with the progression of liver disease (Swerdlow and Chowdhury, 1984). Thus, sinusoidal IgA does not seem to reflect a nonspecific deposition of the serum IgA but may represent a distinct effect of alcohol on the liver. IgA_1 is also detected in the skin of patients with ALD and is a major constituent of tissue IgA deposits in a variety of IgA-related disorders, suggesting that this subclass of IgA has an affinity for tissues (Van de Wiel et al., 1986). The pathogenetic significance of sinusoidal IgA deposition in the liver is not as yet established.

9.8.4. HLA Class I in Hepatocytes

Efficient lysis of target cells by CD8 cytotoxic lymphocytes requires that, in addition to specific antigens, target cells carry class I HLA antigens (Zinkernagel and Doherty, 1974). It is thus of particular interest

TABLE 9.3.   Altered Antigen Expression in the Liver of Patients with Alcoholic Liver Disease

Antigen	Location	Patients/Controls Numbers	Controls (% positive)	Steatosis (% positive)	Alcoholic Hepatitis (% positive)	Cirrhosis (% positive)	ALD (% positive)	Reference
IgA	Continuous pattern sinusoids	58 (262)	24.0	85.3	85.7	55.6	76.0	Kater et al., 1979
		59 (21)	0		66.7	75.0	79.7	Swerdlow and Chowdhury, 1983
		40		41.2				Swerdlow and Chowdhury, 1984
		15 (15)	20.0				73.3	Goldin et al., 1986
		41 (41)	12.0	79.0	86.0	78.0	78.0	Van de Wiel et al., 1987
IgA	Hepatocytes	40		81.8	73.3	64.3	72.5	Trevisan et al., 1983
IgG	Hepatocytes	40		0	13.3	78.6	32.7	Trevisan et al., 1983
Bile cytokeratin	Hepatocytes	51 (46)	15.0	36.0	100	100	82.0	Ray, 1987
	Hepatocytes	40	0			Increased		Van Eyken et al., 1988
HLA Class I	Hepatocytes	22 (8)	0		100	100		Barbatis et al., 1981
		11 (2)	0			100	80.0	Van den Oord and Desmet, 1984
		4 (4)	0			75.0		Fukusato et al., 1986
HLA Class II	Hepatocytes	11 (2)	0			0		Van den Oord and Desmet, 1984
		4 (4)	0			0		Fukusato et al., 1986

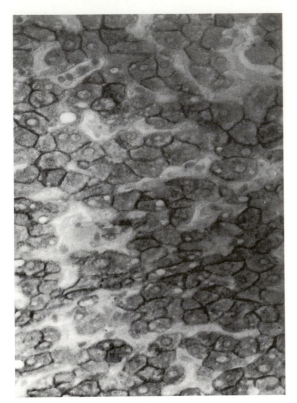

FIGURE 9.2. Honeycomblike membrane hepatocellular staining with "biliary" cytokeratin antibodies is seen in a liver biopsy of a patient with alcoholic hepatitis [monoclonal anticytokeratin antiserum 34 E12 (Enzo, New York); avidin biotin peroxidase staining; ×250].

9.9. Alcohol and Alteration of the Immune Response in Alcoholics without ALD

There is increasing evidence that ethanol may directly affect the immune system, independently of functional alterations of the liver. However, the concomitant effects of malnutrition and environmental factors have not been clearly sorted out.

Much of the evidence of a direct effect of alcohol hinges on the experimental demonstration that administration of alcohol to animals induces numerous immunologic abnormalities, including activation of NK and Kupffer cell function, atrophy of the thymus, and decreased number of T cells, while concomitantly B cells are increased and antibody response may vary according to the nature of the antigen (Ali and Nolan, 1967; Tennenbaum *et al.*, 1969; Grossman *et al.*, 1979; Noland and Camara, 1982; Yamamoto, 1983; Bagasra *et al.*, 1987; Mufti *et al.*, 1988). T-helper cells and ratio of T-helper to T-suppressor cells increased in rats fed etha-

that class I antigens have been demonstrated on the surface of hepatocytes not only of patients with type B hepatitis but also of those with ALD, especially in alcoholic cirrhosis, where a distinct honeycomb pattern along the surface of hepatocytes has been observed (Fig. 9.4). In contrast, in the normal liver, class I HLA are detected primarily in sinusoid lining cells, vascular endothelium, and bile duct epithelium but not in hepatocytes.

The relationship among HLA class I antigens, cytokeratins, and T lymphocytes was not investigated, and future studies in this direction may be rewarding. It is of interest that HLA class II, which are recognized by CD4 lymphocytes, are not abnormally expressed in the hepatocytes of patients with ALD.

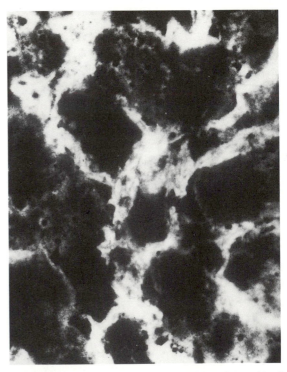

FIGURE 9.3. Immunofluorescence IgA staining of sinusoids of a liver biopsy of a patient with alcoholic hepatitis (fluoresceinated antihuman IgA; ×560).

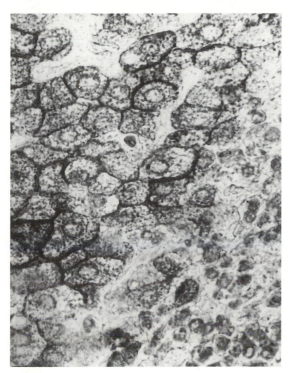

FIGURE 9.4. Staining of hepatocellular membranes of a liver biopsy from a patient with alcoholic hepatitis treated with anti-serum reacting with HLA class I antigens (polyclonal chicken antibody to human β_2 microglobulin, followed by rabbit anti-chicken Ig and peroxidase-conjugated swine antirabbit Ig; ×560).

phagocytosis, and alteration of this function, together with abnormalities of humoral and cell-mediated immunity, are considered responsible for the high incidence in alcoholics of bacterial infections, especially pneumonia and tuberculosis (reviewed in Chapter 8). The high incidence of certain tumors in alcoholics (e.g., tumors of oropharynx, larynx, and esophagus) (Seitz, 1985; Lieber *et al.*, 1986) has suggested an alteration of immune surveillance; however, a cocarcinogenic effect of alcohol has also been considered.

The few studies that focused on immune alterations in alcoholic patients without ALD have reported an increased synthesis of IgA by peripheral blood B cells and an increase in number of NK cells with activity described as normal or increased (Saxena *et al.*, 1980; Charpentier *et al.*, 1984; Jovanovic *et al.*, 1986). Skin reactivity to certain antigens may be altered in alcoholics, but a return to normal reactivity may be related more to amelioration of malnutrition than to alcohol intake (Mendenhall *et al.*, 1984). In contrast, the lymphocyte proliferative response to mitogens is impaired in alcoholic patients without clinical evidence of malnutrition (Mutchnick and Lee, 1988). There is also a suggestion that alcoholics without ALD may develop an alteration of suppressor-cell activity (Prieto and Mutchnick, 1980). Thus, much is to be learned about the direct effect of acute and chronic alcohol intake on the immune response.

nol for more than 1 year (Mufti *et al.*, 1988) and an inherent defect of lymphocyte response to mitogenic stimulation has been described (Jerrells *et al.*, 1989). Natural killer cell activity has been variously reported as normal, increased, and decreased (Saxena *et al.*, 1982; Abdallah *et al.*, 1988; Mufti *et al.*, 1988; Meadow *et al.*, 1989), and changes in NK cell activity have been linked to dietary and nutritional status (Abdallah *et al.*, 1983).

In vitro effects of alcohol include a decreased stimulation of lymphocytes by mitogens and antigens (Tisman and Herbert, 1972; Roselle and Mendenhall, 1982; Glassman *et al.*, 1985), depressed NK-cell activity (Saxena *et al.*, 1980; Ristow *et al.*, 1982; Rice *et al.*, 1983; Mufti *et al.*, 1988), and inhibition of spontaneous and antibody-dependent cell-mediated cytotoxicity (Stacey, 1984; Walia *et al.*, 1987). It is not known, however, whether these *in vitro* alterations parallel the *in vivo* effects of alcohol.

Administration of alcohol has a direct effect on

9.10. Summary

The list of abnormal immune reactions in patients with ALD is quite impressive. Although a direct causal relationship between activation of B lymphocytes and hypergammaglobulinemia is quite conclusive, the pathogenetic significance of humoral, cellular, and hepatic immunologic alterations is still speculative, and so far it cannot be ruled out that the immunologic abnormalities described in patients with alcoholism or ALD are concomitant alterations or inconsequential epiphenomena of liver damage. A major problem has been the unavailability of an experimental model of alcoholic hepatitis and the lack of sequential studies in different stages of the disease. It is expected that future work aimed at a better characterization of the altered antigens and a comprehensive clonal analysis of the functional repertoire of hepatic lymphocytes may clarify the role of autoimmune reactions in the pathogenesis of ALD.

REFERENCES

Abdallah, R. M., Starkey, J. R., and Meadows, G. G.: Alcohol and related dietary effects on mouse natural killer-cell activity. *Immunology* **50**:131–137, 1983.

Abdallah, R. M., Starkey, J. R., and Meadows, G. G.: Toxicity of chronic high alcohol intake on mouse natural killer-cell activity. *Res. Commun. Chem. Pathol. Pharmacol.* **59**:245–258, 1988.

Abrass, C. K., Border, W. A., and Hepner, G.: Non-specificity of circulating immune complexes in patients with acute and chronic liver disease. *Clin. Exp. Immunol.* **40**:292–298, 1980.

Actis, G. C., Ponzetto, A., Rizzetto, M., and Verme, G.: Cell-mediated immunity to acetaldehyde in alcoholic liver disease demonstrated by leukocyte migration test. *Dig. Dis.* **23**:883–886, 1978.

Alexander, G. J. M., Mouri-Aria, K. T., Eddleston, A. L. W. F., and Williams, R.: Contrasting relations between suppressor-cell function and suppressor-cell number in chronic liver disease. *Lancet* **1**:1291, 1983.

Ali, M. V., and Nolan, J. P.: Alcohol induced depression of reticuloendothelial function in the rat. *J. Lab. Clin. Med.* **70**:295–301, 1967.

Anthony, R., Farquharson, M., and MacSween, R. N. M.: Liver membrane antibodies in alcoholic liver disease II. Antibodies to ethanol-altered hepatocytes. *J. Clin. Pathol.* **36**:1302–1308, 1983.

Attali, P., Thibault, N., Buffet, C., Briantais, M. J., Papoz, L., Chaput, J. C., and Etienne, J. P.: Les marqueurs du virus B chex les alcooliques chroniques. *Gastroenterol. Clin. Biol.* **5**:1095–1102, 1981.

Bagasra, O., Howeedy, A., Dorio, R., and Kajdacsy-Balla, A.: Functional analysis of T-cell subsets in chronic experimental alcoholism. *Immunology* **61**:63–69, 1987.

Bailey, R. J., Krasner, N, Eddleston, ALWF, Williams, R, Tee, DEH, Doniach, D, Kennedy, L. A., and Batchelor, J. R.: Histocompatibility antigens, autoantibodies, and immunoglobulins in alcoholic liver diseases with cirrhosis. *Br. Med. J.* **2**:727–729, 1976.

Barbatis, C., Woods, J., Morton, J. A., Fleming, K. A., McMichael, A., and McGee, J. O'D.: Immunohistochemical analysis of HLA (A, B, C) antigens in liver disease using a monoclonal antibody. *Gut* **22**:985–991, 1981.

Barbatis, C., Morton, J., Woods, J. C., Burns, J., Bradley, J., and McGee, J. O'D.: Disorganisation of intermediate filament structure in alcoholic and other liver diseases. *Gut* **27**:765–770, 1986.

Bassendine, M. F., Della Seta, L., Salmeron, J., Thomas, H. C., and Sherlock, S.: Incidence of hepatitis B virus infection in alcoholic liver disease, HBsAg negative chronic active liver disease and primary liver cell cancer in Britain. *Liver* **3**:65–70, 1983.

Behrens, U. J., and Paronetto, F.: Studies on "liver-specific" antigens. I. Evaluation of the liver specificity of "LSP" and "LP-2." *Gastroenterology* **77**:1045–1052, 1979.

Behrens, U., Friedrich, I., Vernace, S., Schaffner, F., and Paro-

netto, F.: Lymphocyte responsiveness per unit volume of blood in patients with chronic nonalcoholic and alcoholic liver disease. Plasma inhibitory factors and functional defects of responder cells. *J. Clin. Lab. Immunol.* **8**:143–152, 1982.

Berenyi, M. R., Straus, B., and Cruz, D.: *In vitro* and *in vivo* studies of cellular immunity in alcoholic cirrhosis. *Am. J. Dig. Dis.* **19**:199–205, 1974.

Bérgroth, V., Konttinen, YT, Segerberg-Konttinen, M., Ollikainen, V., and Salaspuro, M.: Phenotypic *in situ* characterization of lymphocytes in mild alcoholic liver disease. *Acta Pathol. Microbiol. Immunol. Scand. [A]* **94**:337–341, 1986.

Bird, G. L. A., Sheron, N., Goka, A. K. J., Alexander G. J., and Williams, R. S.: Increased plasma tumor necrosis factor in severe alcoholic hepatitis. *Ann. Int. Med.* **112**:917–920, 1990.

Bjorneboe, M., Jensen, K. B., Scheibel, I., Thomsen, A. C., and Bentzon, M. W.: Tetanus antitoxin production and gamma globulin levels in patients with cirrhosis of the liver. *Acta Med. Scand.* **188**:541–546, 1970.

Bréchot, C., Nalpas, B., Couroucé, A. M., Duhamel, G., Callard, P., Carnot, F., Tiollais, P., and Berthelot, P.: Evidence that hepatitis B virus has a role in liver-cell carcinoma in alcoholic liver disease. *N. Engl. J. Med.* **306**:1384–1387, 1982.

Bréchot, C., Pol, S., Thiers, V., Nalpas, B., Degos, F., Carnot, F., Tiollais, P., Wands, J., and Berthelot, P.: Monoclonal anti-HBs antibodies assay and HBV DNA in the serum of HBsAg negative patients. *J. Hepatol.* (Suppl 2.) S200, 1985.

Brillanti, S., Barbara, L., Miglioli, M., and Bonino, F.: Hepatitis C virus: A possible cause of chronic hepatitis in alcoholics. *Lancet* **2**:1390–1391, 1989.

Bron, B., Kubski, D., Widmann, J. J., von Fliedner, V., and Jeannet, M.: Increased frequency of DR3 antigen in alcoholic hepatitis and cirrhosis. *Hepato-gastroenterology* **29**:183–186, 1982.

Brown, S. E., Steward, M. W., Viola, L., Howard, C. R., and Murray-Lyon, I. M.: Chronic liver disease: The detection and characterization of circulating immune complexes. *Immunology* **49**:673–683, 1983.

Brunt, P. W., Key, M. C., Scheuer, P. J., and Sherlock, S.: Studies in alcoholic liver disease in Britain, *Gut* **15**:52–58, 1974.

Burns, J., D'Ardenne, A. J., Morton, J. A., and McGee, J. O. D.: Immune complex nephritis in alcoholic cirrhosis: Detection of Mallory body antigen in complexes by means of monoclonal antibodies to Mallory bodies. *J. Clin. Pathol.* **36**:751–755, 1983.

Burt, A. D., Anthony, R. S., Hislop, W. S., Bouchier, I. A. D., and MacSween, R. N. M.: Liver membrane antibodies in alcoholic liver disease 1. Prevalence and immunoglobulin class. *Gut* **23**:221–225, 1982.

Burt, A. D., Stewart, J., and MacSween, R. N. M.: Immunolocalisation of tissue polypeptide antigen (IPA) in human liver biopsies. *Hepatology* **6**:802, 1986.

Calabrese, E., Gonnelli, E., Ambu, S., Patussi, V., Milani, S., Crispo, A., Masini, R., and Surrenti, C.: Role of hepatitis B virus infection in alcoholic patients. *Ricer. Clin. Lab.* **16**:543–548, 1986.

Caulfield, M. J., and Shaffer, D.: Immunoregulation by antigen/antibody complexes. I. Specific immunosuppression induced *in vivo* with immune complexes formed in antibody excess. *Immunology* **138:**3680–3683, 1987.

Chaput, J. C., Poynard, T., Aubert, A., Lazizi, Y., Hamelin, B., Naveau, S., Bedossa, P., Dubreuil, P., and Pillot, J.: In: *The 1990 International Symposium on Viral Hepatitis and Liver Disease.* Abstract, p. 221, 1990.

Charpentier, B., Franco, D., Paci, L., Charra, M., Martin, B., Vuitton, D., and Fries, D.: Deficient natural killer cell activity in alcoholic cirrhosis. *Clin. Exp. Immunol.* **58:**107–115, 1984.

Chedid, A., and Mendenhall, C. L., *The Veterans Administration Cooperative Study #275: Cell-Mediated Immunity in Alcoholic Liver Disease. Alcohol, Immunomodulation, and AIDS.* 321–332, 1990.

Chevillotte, G., Durbec, J. P., Gerolami, A., Berthezene, P., Bidart, J. M., and Camatte, R.: Interaction between hepatitis B virus and alcohol consumption in liver cirrhosis: An epidemiologic study. *Gastroenterology* **85:**141–145, 1983.

Cochrane, A. M. G., Moussouros, A., Portmann, B., McFarlane, I. G., Thomson, A. D., Eddleston, A. L. W. F., and Williams, R.: Lymphocyte cytotoxicity for isolated hepatocytes in alcoholic liver disease. *Gastroenterology* **72:**918–923, 1977.

Conroy, C., and Allen, J. I.: Ethanol induces expression of class II major histocompatibility antigens on human peripheral blood monocytes. *Hepatology* **8:**1235, 1988.

Coppo, R., Aricó, S., Piccoli, G., Piccoli, G., Basolo, B., Roccatelli, D., Amore, A., Tabone, M., De La Pierre, M., Sessa, A., Delacroix, D. L., and Vaerman, J. P.: Presence and origin of IgA1- and IgA2-containing circulating immune complexes in chronic alcoholic liver diseases with and without glomerulonephritis. *Clin. Immunol. Immunopathol.* **35:**1–8, 1985.

Córsico, R., Pessino, O. L., Morales, V., and Jmelninsky, A.: Association of HLA antigens with alcoholic disease. *J. Stud. Alcohol* **49:**546–550, 1988.

Couzigou, P., Vincendeau, P., Fleury, B., Richard-Molard, B., Pierron, A., Bergeron, J. L., Bezian, J. H., Amouretti, M., and Beraud, C.: Étude des modifications des sous-populations lymphocytaires circulantes au cours des hépatopathies alcooliques. Role respectif de l'alcool, de l'insuffisance hépato-cellulaire et de la dénutrition. *Gastroenterol. Clin. Biol.* **8:**915–919, 1984.

Crapper, R. M., Bhathaland, P. S., and Mackay, I. R.: Chronic active hepatitis in alcoholic patients. *Liver* **3:**327–337, 1983.

Crespi, C., Zauli, D., Bianchi, F. B., and Pisi, E.: Segregation of IgA antibodies to cytoskeleton components in patients with alcoholic liver disease. *Ital. J. Gastroenterol.* **18:**335–337, 1986.

Crossley, I. R., Neuberger, J., Davis, M., Williams, R., and Eddleston, A. L. W. F.: Ethanol metabolism in the generation of new antigenic determinants on liver cells. *Gut* **27:**186–189, 1986.

Cuccurullo, L., Rambaldi, M., Iaquinto, G., Ferraraccio, F., Ambrosone, L., Giardullo, N., and De Vita, A.: Importance of showing HBsAg and HBcAg positivity in the liver for better aetiological definition of chronic liver disease. *J. Clin. Pathol.* **40:**167–171, 1987.

Cunningham, A. L., Mackay, I. R., Frazer, I. H., Brown, C., Pedersen, J. S., Toh, B. H., Tait, B. D., and Clarke, F. M.:

Antibody to G-actin in different categories of alcoholic liver disease. Quantification by an ELISA and significance for alcoholic cirrhosis. *Clin. Immunol. Immunopatol.* **34:**158–164, 1985.

Degos, F., Duhamel, G., Brechot, C., Nalpas, B., Courouce, A. M., Tron, F., and Berthelot, P.: Hepatitis B vaccination in chronic alcoholics. *J. Hepatol.* **2:**402–409, 1986.

Denk, H., Franke, W. W., Eckerstorfer, R., Schmid, E., and Kerjaschki, D.: Formation and involution of Mallory bodies ("alcoholic hyalin") in murine and human liver revealed by immunofluorescence microscopy with antibodies to prekeratin. *Proc. Natl. Acad. Sci. U.S.A.* **76:**4112–4116, 1979.

Devière, J., Denys, C., Schandene, L., Romasco, F., Adler, M., Wybran, J., and Dupont, E.: Decreased proliferative activity associated with activation markers in patients with alcoholic liver cirrhosis. *Clin. Exp. Immunol.* **72:**377–382, 1988.

Devière, J., Content, J., Denys, C., Vandenbussche, P., Schandene, L., Wybran, J., and Dupont, E.: High interleukin-6 serum levels and increased production by leucocytes in alcoholic liver cirrhosis. Correlation with IgA serum levels and lymphokines production. *Clin. Exp. Immunol.* **77:**221–225, 1989.

Devière, J., Vaerman, J. P., Content, J., Denys, C., Schandene, L., Vandenbussche, P., Sibille, Y., and Dupont, E.: IgA triggers tumor necrosis factor α secretion by monocytes: A study in normal subjects and patients with alcoholic cirrhosis. *Hepatology* **13:**670–675, 1991.

Drew, P. A., Clifton, P. M., LaBrooy, J. T., and Shearman, D. J. C.: Polyclonal B cell activation in alcoholic patients with no evidence of liver dysfunction. *Clin. Exp. Immunol.* **57:**479–486, 1984.

Eddleston, A. L. W. F., and Davis, M.: Histocompatibility antigens in alcoholic liver disease. *Br. Med. Bull.* **38:**13–16, 1982.

Fagan, E. A., Trowbridge, R., Davison, F., and Williams, R.: Unexpected detection of HBV-DNA sequences in HBV-seronegative chronic liver diseases. *J. Hepatol.* **3:**(Suppl. 1) S75, 1986.

Fauchet, R., Genetet, B., Gosselin, M., and Gastard, J.: HLA antigens in chronic alcoholic pancreatitis. *Tissue Antigens* **13:**163–166, 1979.

Fong, T. L., Govindarajan, S., Valinluck, B., and Redeker, A. G.: Status of hepatitis B virus DNA in alcoholic liver disease: A study of a large urban population in the United States. *Hepatology* **8:**1602–1604, 1988.

Franco, D., Charra, M., Jeambrun, P., Belghiti, J., Cortesse, A., Sossler, C., and Bismuth, H.: Nutrition and immunity after peritoneovenous drainage of intractable ascites in cirrhotic patients. *Am. J. Surg.* **144:**652–657, 1983.

Frazer, I. H., Kronborg, I. J., and Mackay, I. R.: Antibodies to liver membrane antigens in chronic active hepatitis (CAH). II. Specificity for autoimmune CAH. *Clin. Exp. Immunol.* **54:**213–218, 1983.

French, S. W., Burbige, E. J., Tarder, G., Bourke, E., and Harkin, C. G.: Lymphocyte sequestration by the liver in alcoholic hepatitis. *Arch. Pathol. Lab. Med.* **103:**146–152, 1979.

Fukusato, R., Gerber, M. A., Thung, S. N., Ferrone, S., and Schaffner, F.: Expression of HLA class I antigens on hepatocytes in liver disease. *Am. J. Pathol.* **123:**264–270, 1986.

Galambos, J. T.: Alcoholic hepatitis. In: *The Liver and Its Disease* (F. Schaffner, S. Sherlock, and C. Leevy, eds.), Stuttgart, Thieme, pp. 225–265, 1974.

Glassman, A. B., Bennett, C. E., and Randall, C. L.: Effects of ethyl alcohol on human peripheral lymphocytes. *Arch. Pathol. Lab. Med.* **109**:540–542, 1985.

Gluud, C., Tage-Jensen, U., Bahnsen, M., Dietrichson, O., and Svejgaard, A.: Autoantibodies, histocompatibility antigens and testosterone in males with alcoholic liver cirrhosis. *Clin. Exp. Immunol.* **44**:31–37, 1981a.

Gluud, C., Hardt, F., Aldershvile, J., Christoffersen, P., Lyon, H., and Nielsen, J. O.: Isolation of Mallory bodies and an attempt to demonstrate cell mediated immunity to Mallory body isolate in patients with alcoholic liver disease. *J. Clin. Pathol.* **34**:1010–1016, 1981b.

Gluud, C., Aldershvile, J., Henriksen, J., Kryger, P., and Mathiesen, L.: Hepatitis B and A virus antibodies in alcoholic steatosis and cirrhosis. *J. Clin. Pathol.* **35**:693–697, 1982.

Goldberg, S., Mendenhall, C. L., Connell, A. M., and Chedid, A.: "Nonalcoholic" chronic hepatitis in the alcoholic. *Gastroenterology* **72**:598–604, 1977.

Goldin, R. D., and Wickramasinghe, S. N.: Hepatotoxicity of ethanol in mice. *Br. J. Exp. Pathol.* **68**:815–824, 1987.

Goldin, R. D., Cattle, S., and Boylston, A. W.: IgA deposition in alcoholic liver disease. *J. Clin. Pathol.* **39**:1181–1185, 1986.

Govindarajan, S., Tingberg, H., and Radvan, G.: Circulating immune complexes in alcoholic hepatitis. *Clin. Res.* **30**:995A, 1982.

Grauer, W., Brattig, N. W., Schomerus, H., Frosner, G., and Berg, P. A.: Immunosuppressive serum factors in viral hepatitis. III. Prognostic relevance of rosette inhibitory factor and serum inhibition factor in acute and chronic hepatitis. *Hepatology* **4**:15–19, 1984.

Grossman, C. J., Roselle, G., Sholiton, L. J., and Mendenhall, C. L.: The effect of chronic ethanol (E) on rat thymus: A possible mechanism for the altered immune response of the alcoholic. *Gastroenterology* **77**:A14, 1979.

Harrison, T. J., Anderson, M. G., Murray-Lyon, I. M., and Zuckerman, A. J.: Hepatitis B virus DNA in the hepatocyte: A series of 160 biopsies. *J. Hepatol.* **2**:1–10, 1986.

Hoerner, M., Behrens, U. J., Worner, T., and Lieber, C S.: Humoral immune response to acetaldehyde adducts in alcoholic patients. *Res. Commun. Chem. Pathol. Pharmacol.* **54**:3–12, 1986.

Hoerner, M., Behrens, U. J., Worner, T. M., Blacksberg, I., Braly, L. F., Schaffner, F., and Lieber, C. S.: The role of alcoholism and liver disease in the appearance of serum antibodies against acetaldehyde adducts. *Hepatology* **8**:569–574, 1988.

Hollmann, M., Staab, H. J., Spindler, E., Sproll, M., Anderer, F. A., and Fortmeyer, H. P.: Monoclonal antibody-defined circulating human tumor-associated antigen with epitope shared by cytokeratins. *Biochem. Biophys. Res. Commun.* **128**:34–39, 1985.

Hopf, U., Meyer zum Büschenfelde, K. H., and Arnold, W.: Detection of a liver membrane autoantibody in HBsAg-negative chronic active hepatitis. *N. Engl. J. Med.* **294**:578–582, 1976.

Horiike, H., Michitaka, K., Onji, M., Murota, T., and Ohta, Y.: HBV-DNA hybridization in hepatocellular carcinoma associated with alcohol in Japan. *J. Med. Virology* **28**:189–192, 1989.

Hsu, C. C. S., and Leevy, C. M.: Inhibition of PHA-stimulated lymphocyte transformation by plasma from patients with advanced alcoholic cirrhosis. *Clin. Exp. Immunol.* **8**:749–760, 1971.

Husby, G., Skrede, S., Blomhoff, J. P., Jacobsen, C. D., Berg K., and Gjone, E.: Serum immunoglobulins and organ non-specific antibodies in diseases of the liver. *Scand. J. Gastroenterol.* **12**:297–304, 1977.

Ishimaru, H., and Matsuda, T.: T-cell subsets (Tc, Th, Ts, Tsi) and IL2 receptor-bearing cells in peripheral blood of patients in the acute phase of alcoholic hepatitis. *Alcohol Alcoholism* **25**:353–358, 1990.

Israel, Y., Hurwitz, E., Niemela, O., and Arnon, R.: Monoclonal and polyclonal antibodies against acetaldehyde-containing epitopes in acetaldehyde-protein adducts. *Proc. Natl. Acad. Sci. U.S.A.* **83**:7923–7927, 1986.

Israel, Y., Orrego, H., and Niemela, O.: Immune responses to alcohol metabolites: Pathogenic and diagnostic implications. *Semin. Liver Dis.* **8**:81–90, 1988.

Iturriaga, H., Péreda, T., Estévez, A., and Ugarte, G.: Serum immunoglobulin A changes in alcoholic patients. *Ann. Clin. Res.* **9**:39–43, 1977.

Izumi, N., Sato, C., Hasumura, Y., and Takeuchi, J.: Serum antibodies against alcohol-treated rabbit hepatocytes in patients with alcoholic liver disease. *Clin. Exp. Immunol.* **61**:585–592, 1985.

Izumi, N., Sakai, Y., Hasumura, Y., and Marumo, F.: Specific cytotoxic T-cell response to acetaldehyde-altered cell membrane in alcoholic liver disease. *Hepatology* **8**:1236, 1988.

Izumi, N., Sakai, Y., Koyama, W., and Hasumura, Y.: Clinical significant of serum antibodies against alcohol-altered hepatocyte membrane in alcoholic liver disease. *Alcoholism: Clin. Exp. Res.* **13**:762–765, 1989.

Jain, S., Markham, R., Thomas, H. C., and Sherlock, S.: Double-stranded DNA-binding capacity of serum in acute and chronic liver disease. *Clin. Exp. Immunol.* **26**:35–41, 1976.

Jerrells, T. R., Peritt, D., Marietta, C., and Eckardt, M. J.: Mechanisms of suppression of cellular immunity induced by ethanol. *Alcoholism: Clin. Exp. Res.* **13**:490–493, 1989.

Johnson, R. D., and Williams R.: Immune responses in alcoholic liver disease. *Alcoholism: Clin. Exp. Res.* **10**:471–486, 1986.

Johnson, R. L., and Ziff, M.: Lymphokine stimulation of collagen accumulation. *J. Clin. Invest.* **58**:246–252, 1976.

Jovanovic, R., Worner, T., Lieber, C. S., and Paronetto, F.: Lymphocyte subpopulations in patients with alcoholic liver disease. *Dig. Dis. Sci.* **31**:125–130, 1986.

Kachru, R.B., Proskey, A. J., and Telischi, M.: Histocompatibility antigens and alcoholic cardiomyopathy. *Tissue Antigens* **15**:398–399, 1980.

Kaksuma, Y., Swierenga, S. H. H., Khettry, U., Marceau, N., and French, S. W.: Changes in the cytokeratin intermediate filament cytoskeleton associated with Mallory body formation in mouse and human liver. *Hepatology* **7**:1215–1223, 1987.

Kaku, I., Izumi, N., Hasumura, Y., and Takeuchi, J.: Differences of liver membrane antibody frequency in alcoholic liver disease: Detection of IgG and IgA classes using radio-immunoassay. *Dig. Dis. Sci.* **33:**845–850, 1988.

Kakumu, S., and Leevy, C. M.: Lymphocyte cytotoxicity in alcoholic hepatitis. *Gastroenterology* **72:**594–597, 1977.

Kalsi, J., Delacroix, D. L., and Hodgson, H. J. F.: IgA in alcoholic cirrhosis. *Clin. Exp. Immunol.* **52:**499–504, 1983.

Kanagasundaram, N., Kakumu, S., Chen, T., and Leevy, C. M.: Alcoholic hyaline antigen (AHAg) and antibody (AHAb) in alcoholic hepatitis. *Gastroenterology* **73:**1368–1373, 1977.

Kater, L., Jobsis, A. C., Baart de la Faille-Kuyper, E. H., Bogten, A. J. M., and Grijm, R.: Alcoholic hepatic disease: Specificity of IgA deposits in liver. *Am. J. Clin. Pathol.* **71:**51–57, 1979.

Kawanishi, H., Tavassolie, H., MacDemott, R. P., and Sheagran, J. N.: Impaired concanavalin A-inducible suppressor T-cell activity in active alcoholic liver disease. *Gastroenterology* **80:**510–517, 1981.

Kehl, A., Schober, A., Junge, U., and Winckber, K.: Solid-phase radioimmunoassay for detection of alcoholic hyaline antigen (AHAg) and antibody (anti-AH). *Clin. Exp. Immunol.* **43:**215–221, 1981.

Kolber, M. A., Walls, R. M., Hinners, M. L., and Singer, D. S.: Evidence of increased class I MHC expression on human peripheral blood lymphocytes during acute ethanol intoxication. *Alcoholism: Clin. Exp. Res.* **12:**820–823, 1988.

Krasner, N., Davis, M., Portmann, B., and Williams, R.: Changing pattern of alcoholic liver disease in Great Britain: Relation to sex and signs of autoimmunity. *Br. Med. J.* **1:**1497–1500, 1977.

Krogsgaard, K., Tage-Jensen, U., and Gluud, C.: Liver membrane antibodies in alcoholic liver disease. *Lancet* **1:**1365–1366, 1982.

Kurki, P., Miettinen, A., Salaspuro, M., Virtanen, I., and Stenman, S.: Cytoskeleton antibodies in chronic active hepatitis, primary biliary cirrhosis and alcoholic liver disease. *Hepatology* **3:**297–302, 1983.

Lee, W. M., Martin, K. L., Shelton, L. L., and Galbraith, R. M.: Hepatic membrane antibodies: Studies of prevalence and specificity. *Clin. Exp. Immunol.* **62:**715–723, 1985.

Lelbach, W. K.: Epidemiology of alcoholic liver disease. In: *Progress in Liver Diseases, Vol. V* (H. Popper and F. Schaffner, eds.), New York, Grune & Stratton, pp. 494–515, 1976.

Li, X., Jeffers, L. J., Reddy, K. R., De Medina, M., Silva, M., Villanueva, S., Kilmas, N. G., Esquenazi, V., and Schiff, E. R.: Immunophenotyping of lymphocytes in liver tissue of patients with chronic liver diseases by flow cytometry. *Hepatology* **14:**121–127, 1991.

Lieber, C.: Alcohol and the Liver. In: *The Liver Annual, Vol. III* (I. M. Arias, M. Frenkel, and J. H. P. Wilson, eds.), New York, Elsevier, pp. 107–148, 1983.

Lieber, C.: Alcohol and the liver. In: *The Liver Annual, Vol. IV* (I. M. Arias, M. Frenkel, and J. H. P. Wilson, eds.), New York, Elsevier, pp. 130–186, 1984.

Lieber, C.: Alcohol and the Liver. In: *The Liver Annual, Vol. V* (I. M. Arias, M. Frenkel, and J. H. P. Wilson, eds.), New York, Elsevier, pp. 116–167, 1986a.

Lieber, C. S., Garro, A., Leo, M. A., Mak, K. M., and Worner, T.: Alcohol and cancer. *Hepatology* **6:**1005–1019, 1986b.

Lieber, C.: Alcohol and the Liver. In: *The Liver Annual, Vol. VI* (I. M. Arias, M. Frenkel, and J. H. P. Wilson, eds.), New York, Elsevier, pp. 163–240, 1987.

Lin, R. C., Fillenwarth, M. J., Minter, R., and Lumeng, L.: Formation of the 37-kD protein-acetaldehyde adduct in primary cultured rat hepatocytes exposed to alcohol. *Hepatology* **11:**401–407, 1990.

Lowe, J., Blanchard, A., Morrell, K., Lennox, G., Reynolds, L., Billett, M., Landon, M., and Mayer, R. J.: Ubiquitin is a common factor in intermediate filament inclusion bodies of diverse type in man, including those of Parkinson's disease, Pick's disease, and Alzheimer's disease, as well as Rosenthal fibres in cerebellar astrocytomas, cytoplasmic bodies in muscle, and Mallory bodies in alcoholic liver disease. *J. Pathol.* **155:**9–15, 1988.

Lue, S. L., Paronetto, F., and Lieber, C. S.: Cytotoxicity of mononuclear cells and vulnerability of hepatocytes in alcoholic fatty liver of baboons. *Liver* **1:**264–267, 1981.

Lundy, J, Raaf, J. H., Deakins, S., Wanebo, W. J., Jacobs, D. A., Lee, T.-D., Jacobowitz, D., Spear, C., and Oettgen, H. F.: The acute and chronic effects of alcohol on the human immune system. *Surg. Gynecol. Obstet.* **147:**212–218, 1975.

MacGregor, R. R.: Alcohol and immune defense. *J.A.M.A.* **256:**1474–1479, 1986.

MacSween, R. N. M., and Anthony, R. S.: Immune mechanisms in alcoholic liver disease. In: *Alcoholic Liver Disease* (P. Hall, ed.), New York, John Wiley & Sons, pp. 69–89, 1985.

MacSween, R. N. M., Anthony, R. S., and Farquharson, M.: Antibodies to alcohol-altered hepatocytes in patients with alcohol liver disease. *Lancet* **2:**803–804, 1981.

Manetto, Y., Abdula-Karim, F. W., Perry, G., Tabaton, M., Autilio-Gambetti, L., and Gambetti, P.: Selective presence of ubiquitin in intracellular inclusions. *Am. J. Pathol.* **134:**505–513, 1989.

Manns, M., Meyer zum Büschenfelde, K. H., and Hess, G.: Autoantibodies against liver specific membrane lipoprotein in acute and chronic liver diseases: Studies on organ-, species-, and disease-specificity. *Gut* **21:**955–961, 1980.

Marbet, U. A., Bianchi, L., Meury, U., and Stalder, G. A.: Long-term histological evaluation of the natural history and prognostic factors of alcoholic liver disease. *J. Hepatol.* **4:**364–372, 1987.

Marbet, U. A., Stalder, G. A., Thiel, G., and Bianchi, L.: The influence of HLA antigens on progression of alcoholic liver disease. *Hepatogastroenterol.* **35:**65–68, 1988.

McClain, C. J., and Cohen, D. A.: Increased tumor necrosis factor production by monocytes in alcoholic hepatitis. *Hepatology* **9:**349–351, 1989.

McClain, C. J., Cohen, D. A., Dinarello, C. A., Cannon, J. G., Shedlofsky, S. I., and Kaplan, A. M.: Serum interleukin-1 (IL-1) activity in alcoholic hepatitis. *Life Sci.* **39:**1479–1485, 1986.

McFarlane, B. M., McSorley, C. G., Vergani, D., McFarlane, I. G., and Williams, R.: Serum autoantibodies reacting with the hepatic asialoglycoprotein receptor protein (hepatic lectin) in acute and chronic liver disorders. *J. Hepatol.* **3:**196–205, 1986.

McFarlane, I. G., and Williams R.: Liver membrane antibodies. *J. Hepatol.* **1:**313–319, 1985.

McKeever, U., Mahony, C. O., Whelan, C. A., Weir, D. G., and Feighery, C.: Helper and suppressor T lymphocyte function in severe alcoholic liver disease. *Clin. Exp. Immunol.* **60**:39–48, 1985.

Meadows, G. G., Blank, S. E., and Duncan, D. D.: Influence of ethanol consumption on natural killer cells activity in mice. *Alcoholism: Clin. Exp. Res.* **13**:476–479, 1989.

Meliconi, R., Miglio, F., Stancari, M. V., Baraldini, M., Stefanini, G. F., and Gasbarrini, G.: Hepatocyte membrane-bound IgG and circulating liver-specific autoantibodies in chronic liver disease: Relation to hepatitis B virus serum markers and liver histology. *Hepatology* **3**:155–161, 1983.

Mendenhall, C. L., Anderson, S., Weesner, R. E., Goldberg, S. J., and Crolic, K. A.: Protein-calorie malnutrition associated with alcoholic hepatitis. Veterans Administration Co-operative Study Group on Alcoholic Hepatitis. *Am. J. Med.* **79**:133–135, 1984.

Mendenhall, C., Roselle, G. A., Lybecker, L. A., Marshall, L. E., Grossman, C. J., Myre, S. A., Weesner, R. E., and Morgan, D. D.: Hepatitis B vaccination: Response of alcoholic with and without liver injury. *Dig. Dis. Sci.* **33**:263–269, 1988.

Mendenhall, C. L., Seeff, L., Diehl, A. M., Nelles, M., Agius, C., Ghosn, S., French, S., Gartside, P., Grossman, C. J., Roselle, G. A., and Weesner, R. E.: In: *The 1990 International Symposium on Viral Hepatitis and Liver Disease.* Abstract, 1990, p. 144.

Mihas, A. A., Bull, D. M., and Davidson, C. S.: Cell-mediated immunity to liver in patients with alcoholic hepatitis. *Lancet* **1**:951–953, 1975.

Mills, P. R., Sharkin, A., Anthony, R. S., McLelland, A. S., Main, A. N., MacSween, R. N., and Russell, R. I.: Assessment of nutritional status and *in vivo* immune responses in alcoholic liver disease. *Am. J. Clin. Nutr.* **38**:849–859, 1983.

Moll, R., Franke, W. W., and Schiller, D. L.: The catalog of human cytokeratins: Patterns of expression in normal epithelia, tumors, and cultured cells. *Cell* **31**:11–24, 1982.

Montull, S., Parés, A., Bruguera, M, Caballeria, J., Caballeria, L. L., and Rodés, J.: Chronic active hepatitis in alcoholics. A comparison study with non-A non-B chronic active hepatitis. *J. Hepatol.* **7**:(Suppl. 1) S153, 1988.

Morgan, M. Y., Ross, M. G. R., Ng, C. M., Adams, D. M., Thomas, H. C., and Sherlock, S.: HLA-B8, immunoglobulins, and antibody responses in alcohol-related liver disease. *J. Clin. Pathol.* **33**:488–492, 1980.

Morton, J. A., Fleming, K. A., Trowell, J. M., and McGee, J. O'D.: Mallory bodies—immunohistochemical detection by antisera to unique non-prekeratin components. *Gut* **21**:727–733, 1980.

Mufti, S. I., Prabhala, R., Moriguchi, S., Sipes, I. G., and Watson, R. R.: Functional and numerical alterations induced by ethanol in the cellular immune system. *Immunopharmacology* **15**:85–94, 1988.

Müller, C., Wolf, H., Göttlicher, J., and Eibl, M. M.: Helper-inducer and suppressor-inducer lymphocyte subsets in alcoholic cirrhosis. *Scand. J. Gastroent.* **26**:295–301, 1991.

Mutchnick, M. G., and Lee, H. H.: Impaired lymphocyte proliferative response to mitogen in alcoholic patients. Absence of a relation to liver disease activity. *Alcoholism: Clin. Exp. Res.* **12**:155–158, 1988.

Mutchnick, M. G., Lederman, H. M., Missirian, A., and Johnson, A. G.: *In vitro* synthesis of IgG by peripheral blood lymphocytes in chronic liver disease. *Clin. Exp. Immunol.* **43**:370–375, 1981.

Nalpas, B., Berthelot, P., Thiers, V., Duhamel, G., Courouce, A. M., Tiollais, P., and Bréchot, C.: Hepatitis B virus multiplication in the absence of usual serological markers: A study of 146 chronic alcoholics. *J. Hepatol.* **1**:89–97, 1985.

Nalpas, B., Driss, F., Hamelin, B., Pol, S., and Brèchot, C.: In: *The 1990 International Symposium on Viral Hepatitis and Liver Disease.* Abstract, 1990, p. 220.

Nathrath, W. B. J., Heidenkummer, P., Bjorklund, V., and Bjorklund, B.: Distribution of tissue polypeptide antigen (TPA) in normal human tissues: Immunohistochemical study on unfixed, methanol-, ethanol-, and formalin fixed tissues. *J. Histochem. Cytochem.* **33**:99–109, 1985.

Nei, J., Matsuda, Y., and Takada, A.: Chronic hepatitis induced by alcohol. *Dig. Dis. Sci.* **28**:207–215, 1983.

Neilson, E. G., Jimenez, S. A., and Phillips, S. M.: Cell mediated immunity in interstitial nephritis 3. T-lymphocyte mediated fibroblast proliferation and collagen synthesis, an immune mechanism for renal fibrogenesis. *J. Immunol.* **125**:1708–1714, 1980.

Neuberger, J, Crossley, I. R., Saunders, J. B., Davis, M., Portmann, B., Eddleston, A. L. W. F., and Williams, R.: Antibodies to alcohol altered liver cell determinants in patients with alcoholic liver disease. *Gut* **25**:300–304, 1984.

Niemela, O., Klajner, F., Orrego, H., Vidins, E., Blendis, L., and Israel, Y.: Antibodies against acetaldehyde-modified protein epitopes in human alcoholics. *Hepatology* **7**:1210–1214, 1987.

Nolan, J. P., and Camara, D. S.: Endotoxin, sinusoidal cells and liver injury. In: *Progress in Liver Diseases.* Vol. VII (H. Popper and F. Schaffner, eds.), New York, Grune & Stratton, pp. 361–376, 1982.

Nomura, H., Kashiwagi, S., Hayashi, J., Kajiyama, W., Ikematsu, H., Noguchi, A., Tani, S., and Goto, M.: An epidemiologic study of effects of alcohol in the liver in hepatitis B surface antigen carriers. *Am. J. Epidemiol.* **128**:277–284, 1988.

Nouri-Aria, K. T., Alexander, C. J. M., Portmann, B. C., Hegarty, J. E., Eddleston, A. L. W. F., and Williams, R.: T and B cell function in alcoholic liver disease. *J. Hepatol.* **2**:195–207, 1986.

Ochi, Y., Ura, Y., Hamazu, M., Ishida, M., Kajita, Y., and Nakajima, Y.: Immunological study of tissue polypeptide antigen (TPA)—demonstration of keratin-like sites and blood group antigen-like sites on TPA molecules. *Clin. Chim. Acta* **151**:157–167, 1985.

Omar, R., Saran, B., and Pappolla, M.: Mallory bodies in alcoholic liver disease as evidence of inhibited cytoprotective proteolysis. *Lab. Invest.* **60**:68A, 1989.

Omata, M., Afroudakis, A., Liew, C. T., Ashcavai, M., and Peters, R. L.: Comparison of serum hepatitis B surface antigen (HBsAg) and serum anticore with tissue HBsAg and hepatitis B core antigen (HBcAg). *Gastroenterology* **75**:1003–1009, 1978.

Parent, L. J., Matis, L., and Singer, D. S.: Ethanol: An enhancer of major histocompatibility complex gene expression. *FASEB J.* **1:**469–473, 1987.

Parés, A., Caballeria, J., Bruguera, M., Torres, M., and Rodes, J.: Histological course of alcoholic hepatitis. Influence of abstinence, sex and extent of hepatic damage. *J. Hepatol.* **2:**33–42, 1986.

Parés, A., Barrera, J. M., Caballeria, J., Ercilla, G., Bruguera, M., Caballeria, L., Castillo, R., and Rodes, J.: Hepatitis C virus antibodies in chronic alcoholic patients: Association with severity of liver injury. *Hepatology* **12:**1295–1299, 1990.

Paronetto, F.: Ethanol and the immune system. In: *Alcohol-related Diseases in Gastroenterology* (H. E. Seitz, and B. Kommerell, eds.), Berlin, Springer Verlag, pp. 269–281, 1985.

Paronetto, F.: Cell-mediated immunity in liver disease. *Hum. Pathol.* **17:**168–178, 1986.

Peguignot, G., Chabert, C., Eydoux, H., and Courcol, M. A.: Increased risk of liver cirrhosis with intake of alcohol. *Rev. Alcohol.* **20:**191–202, 1974.

Penner, E., Albini, B., and Milgrom, F.: Detection of circulating immune complexes in alcoholic liver disease. *Clin. Exp. Immunol.* **34:**28–31, 1978.

Perperas, A., Santoulas, T., Portmann, B., Eddleston, A. L. W. F., and Williams, R.: Autoimmunity to a liver membrane lipoprotein and liver damage in alcoholic liver disease. *Gut* **22:**149–152, 1981.

Peters, M., Liebman, H. A., Tong, M. J., and Tinberg, H. M.: Alcoholic hepatitis: Granulocyte chemotactic factor from Mallory body-stimulated human peripheral blood mononuclear cells. *Clin. Immunol. Immunopathol.* **28:**418–430, 1983.

Pirovino, M., Lydick, E., Grob, P. J., Arrenbrecht, S., Altorfer, J., and Schmid, M.: Pneumococcal vaccination: The response of patients with alcoholic liver cirrhosis. *Hepatology* **4:**946–949, 1984.

Pirrone, S., Tosato, F., Rossi, P., Fossaluzza, V., Tonutti, E., and Sala, P. G.: T-cell subsets in peripheral blood and ascitic fluid of patients with alcoholic liver cirrhosis. *Lancet* **2:**518, 1983.

Pol, S., Thiers, V., Nalpas, B., Degos, F., Gazengel, C., Carnot, F., Tiollais, P., Wands, J. R., Berthelot, P., and Brechot, C.: Monoclonal anti-HBs antibodies radioimmunoassay and serum HBV-DNA hybridization as diagnostic tools of HBV infection: Relative prevalence among HBsAg-negative alcoholics, patients with chronic hepatitis or hepatocellular carcinomas and blood donors. *Eur. J. Clin. Invest.* **17:**515–521, 1987.

Poralla, T., Hütteroth, T. H., and Meyer zum Büschenfelde, K. H.: Cellular cytotoxicity against autologous hepatocytes in alcoholic liver disease. *Liver* **4:**117–121, 1984.

Poralla, T., Hütteroth, T. H., Knuth, A., Staritz, M., Dienes, H. P., and Meyer zum Büschenfelde, K. H.: Spontaneous and antibody-dependent cellular immune reactions to ethanol-altered hepatoma cells. *Liver* **7:**50–57, 1987.

Poynard, T., Aubert, A., Lazizi, Y., Naveau, S., Bedossa, P., Dubreuil, P., Pillot, J., and Chaput, J. C.: In: *The 1990 International Symposium on Viral Hepatitis and Liver Disease.* Abstract, 1990, p. 144.

Poynard, T., Aubert, A., Lazizi, Y., Bedossa, P., Hammelin, B., Terris, B., Naveau, S., Dubreuil, P., Pillot, J., and Chaput, J. C.: Independent risk factors for hepatocellular carcinoma in French drinkers. *Hepatology* **13:**896–901, 1991.

Prieto, J. A., and Mutchnick, M. G.: Suppressor T-cell activity (SCA) in acute and chronic alcoholic liver disease. *Gastroenterology* **79:**1046, 1980.

Rankin, J. K.: The natural history and management of the patient with alcoholic liver disease. In: *Alcohol and the Liver* (M. M. Fisher and J. K. Rankin, eds.), New York, Plenum, pp. 365–381, 1977.

Ray, M. B.: Distribution patterns of cytokeratin antigen determinants in alcoholic and nonalcoholic liver diseases. *Hum. Pathol.* **18:**61–66, 1987.

Rice, C., Hudig, D., Lad, P., and Mendelsohn, J.: Ethanol activation of human natural cytoxicity. *Immunopharmacology* **6:**303–316, 1983.

Ristow, S. S., Starkey, J. R., and Hass, G. M.: Inhibition of natural killer cell activity *in vitro* by alcohols. *Biochem. Biophys. Res. Comun.* **105:**1315–1321, 1982.

Rodriguez, M. A., Montano, J. D., and Williams, R. C.: Immunoglobulin production by peripheral blood mononuclear cells in patients with alcoholic liver disease. *Clin. Exp Immunol.* **55:**369–376, 1984.

Rong, P. B., Kalsi, J., and Hodgson, J. F.: Hyperglobulinaemia in chronic liver disease: Relationships between *in vitro* immunoglobulin synthesis, short lived suppressor cell activity and serum immunoglobulin levels. *Clin. Exp. Immunol.* **55:**546–552, 1984.

Roselle, G. A., and Mendenhall, C. L.: Alteration of *in vitro* human lymphocyte function by ethanol, acetaldehyde, and acetate. *J. Clin. Lab. Immunol.* **9:**33–37, 1982.

Roselle, G. A., Mendenhall, C. L., Grossman, C. J., and Weesner, R. E.: Lymphocyte subset alterations in patients with alcoholic hepatitis. *J. Clin. Lab. Immunol.* **26:**169–173, 1988.

Ruibal-Ares, B., de la Barrera, S., Sasiain, M. C., and Colombato, L. A.: Cell-mediated immunity in alcoholic liver cirrhosis. *Medicina (Buenos Aires)* **47:**27–32, 1987.

Samanta, A., Chen, T., and Leevy, C. M.: On the mechanisms of progressive liver injury: Altered DNA and collagen synthesis induced by Mallory bodies. *Gastroenterology* **88:**1692, 1985.

Sanchez-Tapias, J., Thomas, H. C., and Sherlock, S.: Lymphocyte populations in liver biopsy specimens from patients with chronic liver disease. *Gut* **18:**472–475, 1977.

Sancho, J., Egido, J., Sanchez-Crespo, M., and Blasco, R.: Detection of monomeric and polymeric IgA containing immune complexes in serum and kidney from patients with alcoholic liver disease. *Clin. Exp. Immunol.* **47:**327–355, 1981.

Saunders, J. B., Wodak, A. D., Haines, A., Powell-Jackson-P. R., Portmann, B., Davis, M., and Williams, R.: Accelerated development of alcoholic cirrhosis in patients with HLA-B8. *Lancet* **1:**1381–1384, 1982.

Saxena, Q. B., Mezey, E., and Adler, W. H.: Regulation of natural killer activity *in vivo*. I. The effect of alcohol consumption of human peripheral blood natural killer activity. *Int. J. Cancer* **26:**413–417, 1980.

Saxena, Q. B., Saxena, R. K., and Adler, W. H.: Ethanol and

natural killer activity. In: *NK Cells and Other Natural Effector Cells* (R. B. Herberman, ed.), New York, Academic Press, pp. 651–656, 1982.

Saxena, S., Nouri-Aria, K. T., Anderson, M. G., Eddleston, A. L. W. F., and Williams, R.: Interleukin 2 activity in chronic liver disease and the effect of *in vitro* α-interferon. *Clin. Exp. Immunol.* **63:**541–548, 1986.

Schuurman, H. J., Vogten, A. J. M., Schalm, S. W., and Fevery, J.: Clinical evaluation of the liver cell membrane autoantibody assay. *Digestion* **23:**184–193, 1982.

Schudeletti, M., Indiveri, F., Pierri, I., Picciotto, A., and Ferrone, S.: T cells from patients with chronic liver diseases: Abnormalities in PHA-induced expression of HLA class II antigens and in autologous mixed-lymphocyte reactions. *Cell. Immunol.* **102:**227–233, 1986.

Seilles, E., Vuitton, D., Sava, P., Claudé, P., Panouse-Perrin, J., Roche, A., and Delacroix, D. L.: L'IgA et ses différentes formes moléculaires dans le sang veineure mésentérique, portal et périphérique chez l'homme. *Gastroenterol. Clin. Biol.* **9:**607–613, 1985.

Seitz, H. K.: Ethanol and carcinogenesis. In: *Alcohol-Related Diseases in Gastroenterology* (H. K. Seitz and B. Kommerell, eds.), Berlin, Springer-Verlag, pp. 196–212, 1985.

Sherman, M., Campbell, P., Munoz-Calvo, B., Orrego, H., and Compton, K.: Alcoholic liver disease is not associated with occult HBV infection. *Hepatology* **18:**3598, 1988.

Sheron, N., Bird, G., Goka, J., Alexander, G., and Williams, R.: Elevated plasma interleukin-6 and increased severity and mortality in alcoholic hepatitis. *Clin. Exp. Immunol.* **84:**449–453, 1991.

Shigeta, Y., Ishii, H., Takagi, S., Yoshitake, Y., Hirano, T., Takata, H., Kohno, H., and Tsuchiya, M.: HLA antigens as immunogenetic markers of alcoholism and alcoholic liver diseases. *Pharmacol. Biochem. Behav.* **13**(Suppl. 1):89–94, 1980.

Shimizu, S., Kiyosawa, K., Sodeyama, T., Tanaka, E., and Furuta, S.: In: *The 1990 International Symposium on Viral Hepatitis and Liver Disease*. Abstract, 1990, p. 144.

Si, L., Whiteside, T. L., Schade, R. R., and Van Thiel, D. H.: Lymphocyte subsets studied with monoclonal antibodies in liver tissues of patients with alcoholic liver disease. *Alcoholism: Clin. Exp. Res.* **7:**431–435, 1983.

Si, L. Whiteside, T. L., Van Thiel, D. H., and Rabin, B. S.: Lymphocyte subpopulations at the site "piecemeal" necrosis in end stage chronic liver diseases and rejecting liver allografts in cyclosporine-treated patients. *Lab. Invest.* **50:**341–347, 1984.

Smith, S., and Hoy, W. E.: Frequent association of mesangial glomerulonephritis and alcohol abuse: A study of 3 ethnic groups. *Mod. Pathol.* **2:**138–143, 1989.

Smith, W. I., Van Thiel, D. H., Whiteside, T., Janoson, B., Magovern, J., Puet, T., and Rabin, B. S.: Altered immunity in male patients with alcoholic liver disease: Evidence for defective immune regulation. *Alcoholism: Clin. Exp. Res.* **4:**199–206, 1980.

Snyder, N., Bessoff, J., Dwyer, J., and Conn, H.: Depressed delayed cutaneous hypersensitivity in alcoholic hepatitis. *Am. Dig. Dis.* **23:**353–358, 1978.

Sørensen, T. I. A., Orholm, M., Bentsen, K. D., Høybye, G., Eghøje, K., and Christoffersen, P.: Prospective evaluation of alcohol abuse and alcoholic liver injury in men as predictors of development of cirrhosis. *Lancet* **2:**241–244, 1984.

Sorrell, M. F., and Leevy, C. M.: Lymphocyte transformation and alcoholic liver injury. *Gastroenterology* **63:**1020–1025, 1972.

Spinozzi, F., Guerciolini, R., Gerli, R., Gernini, I., Rondoni, F., Frascarelli, A., Rambotti, P., Grignani, F. G., and David S.: Immunological studies in patients with alcoholic liver disease: Evidence for the *in vivo* activation of helper T cells and of the monocyte–macrophage system. *Int. Arch. Allergy Appl. Immunol.* **80:**361–368, 1986.

Spinozzi, F., Bertotto, A., Rondoni, F., Gerli, R., Scalise, F., and Grignani, F.: T-lymphocyte activation pathways in alcoholic liver disease. *Immunology* **73:**140–146, 1991.

Stacey, N. H.: Inhibition of antibody-dependent cell-mediated cytotoxicity by ethanol. *Immunopharmacology* **8:**155–161, 1984.

Stoltenberg, P. H., and Soltis, R. D.: The nature of IgG complexes in alcoholic liver disease. *Hepatology* **4:**101–106, 1984.

Swerdlow, M. A., and Chowdhury, L. N.: IgA subclasses in liver tissues in alcoholic liver disease. *Am. J. Clin. Pathol.* **80:**283–289, 1983.

Swerdlow, M. A., and Chowdhury, L. N.: IgA deposition in liver in alcoholic liver disease. *Arch. Pathol. Lab. Med.* **108:**416–419, 1984.

Tabor, E., Gerety, R. J., Barker, L. F., Howard, C. R., and Zuckerman, A.: Effect of ethanol during hepatitis B virus infection in chimpanzees. *J. Med. Virol.* **2:**295–303, 1978.

Takase, S., Takada, N., Enomoto, N., Yasuhara, M., and Takada, A.: Different types of chronic hepatitis in alcoholic patients: Does chronic hepatitis induced by alcohol exist? *Hepatology* **13:**876–881, 1991.

Tennenbaum, J. I., Ruppert, R. D., St. Pierre, R. L., and Greenberger, N. J.: The effect of chronic alcohol administration on the immune responsiveness of rats. *J. Allergy* **44:**272–281, 1969.

Theofilopoulos, A. N., and Dixon, F. J.: Immune complexes in human diseases. *Am. J. Pathol.* **100:**531–594, 1980.

Thestrup-Pedersen, K., Ladefoged, K., and Andersen, P.: Lymphocyte transformation test with liver-specific protein and phytohaemagglutinin in patients with liver disease. *Clin. Exp. Immunol.* **24:**1–8, 1976.

Thiele, D. L.: Tumor necrosis factor, the acute phase response and the pathogenesis of alcoholic liver disease. *Hepatology* **9:**497–499, 1989.

Thomas, H. C., DeVilliers, D., Potter, B., Hodgson, H., Jain, S., Jewell, D. P., and Sherlock, S.: Immune complexes in acute and chronic liver disease. *Clin. Exp. Immunol.* **31:**150–157, 1978.

Tisman, G., and Herbert, V.: *In vitro* myelosuppression and immunosuppression by ethanol. *J. Clin. Invest.* **52:**1410–1414, 1972.

Trevisan, A., Cavigli, R., Meliconi, R., Stefanini, G. F., Zotti, S., Rugge, M., Noventa, F., Betterle, C., and Realdi, G.: Detection of immunoglobulins G and A on the cell membrane of hepatocytes from patients with alcoholic liver disease. *J. Clin. Pathol.* **36:**530–534, 1983.

Triggs, S. M., Mills, P. R., and MacSween, R. N. M.: Sensitisation of Mallory bodies (alcoholic hyalin) in alcoholic hepatitis. *J. Clin. Pathol.* **34**:21–24, 1981.

Trudell, J. R., Ardies, C. M., Green, C. E., and Allen, K.: Binding of antiacetaldehyde IgG antibodies to hepatocytes with an acetaldehyde-phosphatidylethanolamine adduct on their surface. *Alcoholism: Clin. Exp. Res.* **15**:295–299, 1991.

Van den Oord, J. J., and Desmet, V. J.: Verteilungsmuster der histokompatibilitats-hauptantigene im normalen und pathologischen lebergewebe *Leber Magen Darm* **14**:244–254, 1984.

Van de Wiel, A., Schuurman, H. J., van Riessen, D., Haaijman, J. J., Radl, J., Delacroix, D. L., van Hattum, J., Blok, A. P. R., and Katter, L.: Characteristics of IgA deposits in liver and skin of patients with liver disease. *Am. J. Clin. Pathol.* **86**:724–730, 1986.

Van de Wiel, A., Delacroix, D. L., van Hattum, J., Schuuram, H. J., and Kater, L.: Characteristics of serum IgA and liver IgA deposits in alcoholic liver disease. *Hepatology* **7**:95–99, 1987.

Van de Wiel, A., Valentijn, R. M., Schuurman, H. J., Daha, M. R., Hené, R. J., and Kater, L.: Circulating IgA immune complexes and skin Ig deposits in liver disease relation to liver histopathology. *Dig. Dis. Sci.* **33**:679–684, 1988.

Van Eyken, P., Sciot, R., and Desmet, V. J.: A cytokeratin–immunohistochemical study of alcoholic liver disease: Evidence that hepatocytes can express "bile duct-type" cytokeratins. *Histopathology* **13**:605–617, 1988.

Villa, E., Rubbiani, L., Barchi, T., Ferretti, I., Grisendi, A., De Palma, M., Bellentani, S., and Manenti, F.: Susceptibility of chronic symptomless HBsAg carriers to ethanol-induced hepatic damage. *Lancet* **2**:1243–1244, 1982.

Wahl, S. M., and Gatel, C. L.: Modulation of fibroblast growth by a lymphokine of human T cell and continuous T cell line origin. *J. Immunol.* **130**:1226–1230, 1983.

Wahl, S. M., Wahl, L. M., and McCarthy, J. B.: Lymphocyte mediated activation of fibroblast proliferation and collagen production. *J. Immunol.* **121**:942–946, 1978.

Walia, A. S., Pruitt, K. M., Rodgers, J. D., and Lamon, E. W.: *In vitro* effect of ethanol on cell-mediated cytotoxicity by murine spleen cells. *Immunopharmacology* **13**:11–24, 1987.

Wands, J. R., Dienstag, J. L., Weake, J. R., and Koff, R. S.: *In vitro* studies of enhanced Ig synthesis in severe alcoholic liver disease. *Clin. Exp. Immunol.* **44**:396–404, 1981.

Wands, J. R., Fujita, Y. K., Isselbacher, K. J., Degott, C., Schellekens, H., Dazza, M. C., Thiers, V., Tiollais, P., and Brechot, C.: Identification and transmission of hepatitis B virus-related variants. *Proc. Natl. Adad. Sci. U.S.A.* **83**:6608–662, 1986.

Watson, R. R., Jackson, J. C., Hartmann, B., Sampliner, R., Mobley, D., and Eskelson, C.: Cellular immune functions, endorphins, and alcohol consumption in males. *Alcoholism: Clin. Exp. Res.* **9**:248–254, 1985.

Yamamoto, H.: Chronological changes in the immune response and liver tissue of mice by long-term alcohol administration. *Jpn. J. Gastroenterol.* **80**:2375–2383, 1983.

Yokota, M., Sakamoto, S., Koga, S., and Ibayashi, H.: Decreased interleukin 1 activity in culture supernatant of lipopolysaccharide stimulated monocytes from patients with liver cirrhosis and hepatocellular carcinoma. *Clin. Exp. Immunol.* **67**:335–342, 1987.

Young, G. P., Dudley, F. J., and Van der Weyden, M. B.: Suppressive effect of alcoholic liver disease sera on lymphocyte transformation. *Gut* **20**:833–839, 1979.

Zatloukal, K., Denk, H., Spurej, G., Lackinger, E., Preisegger, K. H., and Franke, W. W.: High molecular weight component of Mallory bodies detected by a monoclonal antibody. *Lab. Invest.* **62**:427–434, 1990.

Zetterman, R. K., Luisada-Opper, A., and Leevy, C. M.: Alcoholic hepatitis: Cell-mediated immunological response to alcoholic hyalin. *Gastroenterology* **70**:382–384, 1976.

Zetterman, R. K.: Autoimmune manifestations of alcoholic liver disease. In: *Autoimmune Liver Diseases* (E. L. Krawitt and R. H. Weisner, eds.), New York, Raven Press, pp. 247–260, 1991.

Zinkernagel, R. M., and Doherty, P. C.: Restriction of *in vitro* T-cell mediated cytotoxicity in lymphocyte choriomeningitis within a syngeneic or semi-allogeneic system. *Nature* **248**:701–702, 1974.

10

Alcohol and the Digestive Tract

Lawrence Feinman, Mark A. Korsten, and Charles S. Lieber

10.1. Introduction

The proximal digestive system is exposed to high concentrations of alcohol. Alcohol influences gastric secretion, and excessive alcohol intake is associated with injury to the mucosa of the stomach. Diarrhea and weight loss are also frequent manifestations of alcoholism. Alterations in the intestinal absorption of nutrients have long been suspected in alcoholics. The present chapter reports on the progress made, emphasizing the past several decades.

We try to distinguish the local effects of alcohol on the mucosa of the digestive tract from its systemic actions. Differences between the acute effects of alcohol and those peculiar to chronic intake are identified. Information drawn from animal models and *in vitro* systems is presented with an attempt to relate it to physiology and disease. Three important topics, alcoholic liver disease, pancreatitis, and the relationship between alcohol and cancer of the digestive system, are dealt with separately in Chapters 7, 10, and 13, respectively.

10.2. Oropharynx and Salivary Glands

Alcohol is excreted via the salivary glands, the concentration in saliva being nearly identical to that in blood (Beazell and Ivey, 1940). Ingestion of dilute alcohol and a number of alcoholic beverages stimulates the flow of saliva (Beazell and Ivey, 1940; Martin and Pangborn, 1971; Winsor and Strongin, 1933). This effect was only observed following the local application of alcohol solutions to the tongue or buccal mucosa. However, after a short period of alcohol-induced stimulation by local administration, the secretory rate is inhibited. On average, salivary secretion is subnormal for 60–90 min and returns to basal levels thereafter (Winsor and Strongin, 1933). The instillation of alcohol into the stomach does not influence the rate of salivary flow.

In alcoholics, enlargement of the parotid gland has been observed in 12% of patients with simple alcohol abuse and in about 50% of patients with alcoholic hepatitis or alcoholic cirrhosis (Bode, 1980). In addition to this morphological alteration, functional changes in saliva production occur. When parotid secretion was measured during stimulation with dilute citric acid, patients with alcoholic cirrhosis exhibited increased rates of flow compared to healthy subjects (Dürr *et al.*, 1975). In the same study, salivary protein excretion was slightly reduced in the alcoholics, but amylase secretion resembled that of control. More recently, Dutta *et al.* (1989) reported patients with alcoholic cirrhosis to have edema, interstromal fat infiltration, and fibrosis of the parotid glands, decreased basal and citric-acid-stimulated flows, and lower salivary concentrations of sodium, bicarbonate, and proteins.

Glossitis and stomatitis have been reported in alcoholics (Beazell and Ivey, 1940; Larato, 1973). These changes are likely the consequences of poor nutrition and respond to treatment with vitamins. Glossitis and stomatitis are not commonly observed in well-nourished subjects despite high alcohol consumption. Pharyngeal cancer is greatly increased in alcoholics, especially when they also smoke (see Chapter 15).

10.3. Esophagus

The ingestion of alcoholic beverages on a chronic basis is associated with peptic esophagitis (Bucher *et al.*, 1978; Martini and Wienbeck, 1974; Messian *et al.*, 1978). Although several clinical observations and experimental reports about the effects of alcohol on the esophagus seem relevant, they do not adequately explain the pathogenesis of esophageal disease. Ethanol causes measurable toxic effects (Chung *et al.*, 1977) in experimental animals at concentrations as low as 5%, abetted by HCl, and histologically obvious damage at concentrations over 30% (Safaie-Shirzai *et al.*, 1974; Salo, 1983; Shirazi and Platz, 1978). Since ethanol has the capacity to injure esophageal mucosa, attention has turned to reasons for possible extended contact of the esophageal mucosa with ethanol and acid. The results are partially conflicting. Abnormalities in esophageal peristalsis were noted in a majority of alcoholics with peripheral neuropathy, although none complained of dysphagia (Fischer *et al.*, 1965; Winship *et al.*, 1968), and in normal subjects after an acute dose of ethanol (Hogan *et al.*, 1972; Mayer *et al.*, 1978). Acute doses of ethanol also interfered with the removal of acid instilled into the lower esophagus (Kjellen and Tibbling, 1979). Acute doses of ethanol decreased the lower esophageal sphincter (LES) pressure (Hogan *et al.*, 1972; Mayer *et al.*, 1978) and promoted gastroesophageal reflux in normal volunteers (Kaufman and Kaye, 1978). Silver found somewhat different results, however, when he restudied this problem in alcoholics without neuropathy during detoxification from alcohol (Silver *et al.*, 1986). The alcoholics initially had increased LES pressures, which reverted to normal after withdrawal from alcohol. These results were subsequently confirmed by Keshavarzian *et al.* (1987). Silver *et al.* (1986) found no abnormalities in acid clearance and only minor abnormalities in peristalsis. The neutralizing capacity of saliva was also investigated. In contrast to control patients, whose salivary bicarbonate content correlated with acid-neutralizing capacity, no such correlation was apparent for the saliva of alcoholic patients. Thus a change in composition of saliva was present in alcoholic subjects.

Chronic ethanol ingestion, especially in conjunction with smoking, is associated with a great increase in esophageal cancer (Breslow and Enstrom, 1974; McCoy *et al.*, 1980; Pottern *et al.*, 1981; Racusen and Krawitt,

1977). Asymptomatic patients who underwent cytological studies of cells exfoliated from their esophagi were found to have increased keratin and fungi, associated "risk factors" for esophageal cancer, in accordance with their history of intensity of alcohol intake (Korsten *et al.*, 1985). This topic receives fuller treatment in Chapter 13.

The backbone of therapy for patients with symptomatic reflux or peptic stricture should be total abstinence from alcohol. Medical therapy also comprises a number of mechanical measures—elevation of the head of the bed, elimination of a bedtime snack, avoidance of restrictive garments. Antacids are prescribed to increase the pH of the refluxed gastric contents; H_2 blockers are useful and should be given more than once per 24 hr to raise gastric pH. Sucralfate is indicated in the presence of erosive lesions. Bethanechol, a cholinergic agent, increases LES pressure and may improve symptoms caused by reflux in some patients. Metaclopramide also may help in treating reflux by increasing LES pressure and enhancing gastric emptying, but neurological side effects limit its usefulness as long-term therapy. Cisapride may be a better alternative. Most peptic strictures can be dilated initially with Eder-Puestow dilators or other devices. In refractory cases surgery may be required, but antireflux operations are reserved for those who remain symptomatic despite maximal medical therapy, which may include omeprazole.

10.4. Stomach

Alcohol ingestion is a cause of acute gastritis and duodenitis. Although alcohol has not been shown to cause chronic peptic ulcer or to retard the healing of peptic ulcer, most physicians usually caution against the use of significant amounts of alcohol by ulcer-prone patients. Drinking alcoholic beverages results in some changes in gastric physiology of unclear significance and in other changes obviously pathogenetic for gastritis. Ethanol influences gastric motility and meal emptying, gastric mucus, the secretion of gastric juice, and the integrity of the gastric mucosa and submucosa. Ethanol is also metabolized in part by the gastrointestinal tract, mainly the stomach, before it reaches the liver, a subject covered in detail in Chapter 16.

10.4.1. Gastric Emptying

The effect of alcohol on gastric emptying in man may depend on the concentration of alcohol in the gastric contents (Pirola, 1978). Hunt and Pathak (1960)

found delayed gastric emptying with increasing ethanol concentrations. Likewise, Cooke (1970, 1972) found that gastric emptying slowed when alcohol was given in concentrations greater than 6 g/100 ml. Others have found that a low dose of ethanol actually speeds gastric emptying (Harichiaux et al., 1970), although higher concentrations cause delay. Pretreatment with alcohol prolongs gastric emptying time in laboratory animals (Barboriak and Meade, 1969) and in humans (Barboriak and Meade, 1970). Barboriak and Meade (1970), for example, administered 120 ml of 100-proof whiskey to normal adults prior to ingestion of a high-fat (60 g) meal; gastric emptying was delayed by 99 ± 32 min.

It was noted that although alcohol might delay gastric emptying of meals, solutions containing ethanol actually were evacuated more quickly than isosmolar glucose solutions (Hunt and Pathak, 1960; Kaufman and Kaye, 1979). Kaufman and Kaye (1979) instilled isovolumetric, isocaloric, and isosmolar test meals to normal volunteers. A solution containing 80 ml ethanol was evacuated more quickly than a solution of 40 ml ethanol and 63 g dextrose, which in turn was emptied faster than a 126-g dextrose meal.

For many years it was postulated that the primary mechanism behind delayed gastric emptying was alcohol-induced "pylorospasm" (Haggard et al., 1941; Tennent, 1941). Although studies could show that diminished gastric emptying occurred above a certain critical blood ethanol level (Haggard et al., 1941; Tennent, 1941), for a long time there was no method to evaluate pyloric function directly. Phaosawasdi et al. (1979) then performed manometry of the pyloric sphincter of normal volunteers following ingestion of a mildly intoxicating dose of alcohol (79 g ethanol; blood levels 100 mg/100 ml). Alcohol caused a decreased pyloric pressure response to duodenal acidification, but there was no change in basal pyloric sphincter pressure and no evidence of pylorospasm. Thus, alcohol most likely delays gastric emptying by altering gastric motility.

The effects of alcohol on gastric emptying of meals are concentration dependent and involve a delay in emptying the solid portions of a meal (Barboriak and Meade, 1970; Tennent, 1941), while enhancing the emptying of liquids (Jian et al., 1986; Kaufman and Kaye, 1979). These effects may be explained by the fact that ethanol retards the activity of the antral gastric muscles, which normally are effective in emptying the solid portions of a meal, and augments the activity of the fundal gastric muscles, which serve to propel the liquid portions (Jian et al., 1986; Sanders and Berry, 1985). The effects of ethanol on the pyloric sphincter, as reviewed above, probably do not retard gastric emptying.

10.4.2. Effect of Ethanol on Gastric Acid Secretion

10.4.2.1. Introduction

Alcohol is described in most textbooks as a powerful stimulant of gastric acid secretion (Pitchumoni and Glass, 1977). In fact, the alcohol test meal (50 ml of 7% ethanol v/v with collections for 1 to 2 hr) was used for many years as a test of gastric acid secretion. Careful review of the literature, however, reveals that oral administration of ethanol actually may cause a diminution in gastric acid output. Early studies in dogs by Dragstedt et al. (1940) suggested that gastric stimulation occurs following alcohol and that the response is similar to that of exogenous histamine. Woodward et al. (1975) found that intravenous administration of 250 ml of 5% ethanol consistently caused acid secretion in the canine Heidenhain pouch, but topical instillation of 5% or 10% ethanol had no effect. Further animal experiments demonstrated actual inhibition of acid secretion because of back-diffusion of hydrogen ion (Davenport, 1967a; Weisbrodt et al., 1973). When ethanol is ingested alone at low concentrations (less than 10%), it stimulates gastric secretion (most clearly in animals) via a combination of direct, gastrin-mediated, and, after absorption, vagus-mediated mechanisms (Becker et al., 1974; Eysselain et al., 1984; Tiscornia et al., 1975; Woodward et al., 1957). When ethanol is ingested alone in concentrations even below 10%, part of its direct effect on parietal cells is to inhibit acid secretion, an effect very likely mediated by cAMP (Puurunen, 1978; Purrunen et al., 1976; Reichstein et al., 1986). Ethanol at 10% inhibited many parameters relevant to gastric secretion (aminopyrine uptake, adenylate cyclase activity, H^+,K^+-ATPase activity) in isolated rabbit gastric glands (Reichstein et al., 1986). In humans, intravenous alcohol causes gastric acid secretion (Hirschowitz et al., 1956; Demol et al., 1985), whereas oral alcohol per se does not (Cooke and Birchall, 1969; Cooke, 1970, 1972). Some wines and beers, however, do have stimulatory properties (see Section 10.4.2.2).

10.4.2.2. Stimulatory Effects

Although the net physiological effect of the acute ingestion of alcohol is to decrease gastric acid secretion, possible mechanisms for the systemic stimulation of the parietal cell have also been studied. Four theories have been proposed: (1) release of endogenous histamine (Dragstedt et al., 1940); (2) vagal stimulation (Hirschowitz *et al.*, 1956); (3) gastrin release (Becker *et al.*, 1974); and (4) changes in mucosal cAMP.

Histamine was originally proposed by Dragstedt *et al.* (1940) as the mediator of alcohol-induced gastric acid secretion. However, Daves *et al.* (1965) failed to demonstrate an increase in histamine concentration in peripheral blood, gastric juice, gastric mucosa, or urine following intravenous infusion of ethanol in the dog, nor could Irvine *et al.* (1960) find a rise in urine histamine. Subsequently, Dinoso *et al.* (1976) showed an increase in the histamine concentration of the gastric mucosa, gastric venous blood, and gastric contents following instillation of 40% v/v ethanol. Thus, there was good evidence that alcohol raised gastric histamine (but without a concomitant rise in gastric acid output). The higher ethanol dose used by Dinoso *et al.* (1976a) may explain why earlier authors (Daves *et al.*, 1965; Irvine *et al.*, 1960) could not demonstrate increased histamine levels. In isolated rabbit gastric glands, low concentrations of ethanol (0.2% and 5%) potentiated the effect of histamine (Reichstein *et al.*, 1986), whereas 10% ethanol had an inhibitory effect.

Alcohol may stimulate parietal cell secretion by a central neural mechanism (Hirschowitz *et al.*, 1956). Alcohol-induced acid secretion was blocked by atropine or vagotomy, indicating that alcohol, like insulin hypoglycemia, stimulates the stomach via the vagus nerve. Kondo and Magee (1977) found no change in acid secretion following bilateral cervical vagal block in the dog, but more recently Kolbel *et al.* (1986) found that atropine partially inhibits gastric secretion by intravenous ethanol, indicating that noncholinergic as well as cholinergic mechanisms are involved in the central effect.

Gastrin release offers an attractive mechanism to explain alcohol-induced gastric acid secretion. Before the advent of radioimmunoassay for measuring gastrin levels, several authors (Woodward *et al.*, 1955, 1957; Irvine *et al.*, 1960) inferred that gastrin was a major factor. Perfusion of isolated antrum with 5% to 10% ethanol solutions caused a rise in volume and acid secretion from fundic Heidenhain pouches; this response could be inhibited by lowering the pH of the antral perfusate to 1.2. Under these circumstances, the gastric juice in the Heidenhain pouch had a low pepsin concentration (Kondo and Magee, 1977) similar to that found on gastrin or pentagastrin stimulation (Dutt and Magee, 1972).

When serum gastrin levels could be measured, the role of gastrin as the cause of alcohol-induced acid secretion was clarified. In dogs, perfusion of the antrum

produced increased antral vein (Becker *et al.*, 1974) and peripheral venous (Bo-Linn and Shanbour, 1977) gastrin. Bo-Linn and Shanbour (1977), however, were unable to demonstrate a concomitant increase in acid output. In humans, Becker *et al.* (1974) found a very slight increase in peripheral gastrin following oral administration of 50% ethanol; gastric acid was not measured. Some authors, using lower ethanol concentrations (up to 16% w/v), were unable to show any change in peripheral gastrin levels (Cooke and Turtle, 1970; Straus *et al.*, 1975). Others (Peterson *et al.*, 1986) also found no increase in serum gastrin or gastric acid secretion when pure ethanol (25 g as 5%, 12%, and 36% solutions) was infused into volunteers by nasogastric tube. Thus, although gastrin may be the mediator of pure ethanol-induced acid stimulation in the dog, its role in humans is doubtful. Interestingly, when ethanol was taken as wine (Lens *et al.*, 1983; Peterson *et al.*, 1986; Singer *et al.*, 1987) or beer (Singer *et al.*, 1987), gastric acid secretion was accompanied by gastrin release, whereas when it was taken as pure ethanol or bourbon whiskey, no evidence for a gastrin mechanism was found.

Finally, it has been proposed that changes in cAMP metabolism might underlie ethanol-induced changes in acid secretion. Ethanol at concentrations of 1–5% increased the content of cAMP in isolated human gastric body mucosa *in vitro* (Puurunen *et al.*, 1976). The ethanol-induced changes in the content of cAMP were attributed to an increase in the activity of adenylate cyclase (Puurunen *et al.*, 1976), which catalyzes the formation of cAMP from ATP. In addition, cAMP phosphodiesterase, the second enzyme controlling the intracellular content of the nucleotide, is competitively inhibited by ethanol concentrations of 5–10% (Puurunen *et al.*, 1976).

The interpretation of measurements of total content of cAMP is complicated by the fact that the ion-transport-related pool of cAMP may be only a small fraction of the total (Field *et al.*, 1976), and the exact site and mechanism of action of cAMP in gastric mucosa and its role in acid secretion are not fully defined.

Thus, the controversy concerning the respective contributions of the many factors that may be involved in the alcohol-induced stimulation of acid secretion continues, complicated by distinct species differences in gastric mucosa response. It is likely that alcohol acts in two (or more) ways to stimulate gastric secretion. When alcohol is absorbed from the gut, or when alcohol is given directly into the bloodstream, it acts on the central nervous system, and the effects are mediated partially by cholinergic mediators of the vagus nerves. In addition, alcohol and/or other constituents of wine and beer exert local effects on the gastric mucosa, presumably both through the release of gastrin and by direct action on the mucosal cell. Such direct effects may be mediated through histamine liberated within the mucosa and/or through the effects of ethanol on cAMP metabolism in the gastric mucosa. However, the stimulatory effects are partially offset by inhibitory ones.

10.4.2.3. Inhibitory Effects

Inhibitory effects are not easy to distinguish from injurious effects. Davenport (1967a) demonstrated that the direct application of ethanol to Heidenhain pouches of dogs resulted in the insorption of hydrogen ion and the exsorption of sodium ion. This observation was confirmed and extended by Dinoso et al. (1970), who found that these ion fluxes and the accompanying mucosal injury could be increased or decreased by the addition of acid or buffer, respectively, to the alcohol perfusate. Using lithium as an ionic analogue of H^+, Smith et al. (1971) reached the same conclusion in a study of humans. On the basis of these results, it was concluded that alcohol alters the permeability of the gastric mucosal "barrier" and that the subsequent ingress of hydrogen is potentially injurious. More recently, others have suggested that the deleterious effects of alcohol are mediated only in part by a change in mucosal permeability (Shanbour et al., 1973; Sernka et al., 1974; Fromm and Robertson, 1976). It was found that topical ethanol increases electrical resistance across the mucosa as well as diminishing the transmucosal electrical potential. Such findings are more consistent with a change in active transport of ions, particularly Cl^-, rather than a simple increase in permeability (which should decrease electrical resistance), although the net ionic changes, namely, an increase in luminal Na^+ and a decrease in luminal H^+, would be the same. A Mg^{2+}- and Mg^{2+}, HCO_3^--ATPase is postulated to link the active transport of Cl^- and HCO_3^-. It has been shown that ethanol may reduce active transport of chloride by inhibiting this ATPase activity (Tague and Shanbour, 1977).

Based on the observation that not only 10% ethanol but also some other hyperosmotic solutions decreased gastric acid secretion in the rat, Sernka and Jackson (1976) assumed that the inhibition of gastric acid secretion by alcohol is a nonspecific hyperosmotic effect. Biggerstaff and Leitch (1977) showed that the effect of 8.5% ethanol on the sodium flux was probably related to

its hyperosmolarity. However, 8.5% ethanol, but not sucrose solution, reduced both transmural potential difference and H^+ secretion, indicating a direct effect of ethanol on active transport (Biggerstaff and Leitch, 1977). Dawson and Cooke (1978) measured the effect of 4% ethanol on transmural fluxes of labeled sodium and chloride across the isolated rat gastric mucosa simultaneously with [^{3}H]mannitol in an attempt to separate transmural ionic movement into cellular and noncellular components. The results of this study support the hypothesis that ethanol inhibits the active transport of sodium and chloride ions and increases the apparent permeability of paracellular pathways for transmural ionic diffusion.

The diminution of acid secretion may also be caused by changes in cAMP content through inhibition of adenylate cyclase (Puurunen and Karppanen, 1975; Puurunen et al., 1977). Puurunen (1978) also reported that ethanol increases the formation of prostaglandins in the gastric mucosa. Prostaglandins of the E and A series inhibit gastric secretion induced by histamine, pentagastrin, and other stimuli in several species (Main and Whittle, 1976; Ramwell and Shaw, 1968). Since there are a number of similarities between the effects of ethanol and prostaglandins on gastric acid secretion, ethanol may inhibit gastric acid secretion not only by lowering the mucosal content of cAMP but also by stimulating the formation of prostaglandins (Puurunen, 1978). This concept was supported by the finding that indomethacin, an inhibitor of prostaglandin synthesis, antagonizes the ethanol-induced inhibition of gastric acid secretion in the rat (Karppanen and Puurunen, 1976).

10.4.2.4. Effect of Chronic Ethanol Administration on Gastric Acid Secretion

As discussed before, alcohol stimulates and inhibits gastric acid production by different mechanisms. On a clinical level, the net effect of these influences undoubtedly depends on the beverage type, alcohol concentration, and chronicity of alcohol abuse.

Chey et al. (1972) investigated the effects of large daily doses (4.4 g/kg) of ethanol on gastric acid secretion in dogs fed a nutritious diet for a period of 12 months. The mean daily secretion of acid from the Heidenhain pouch increased 2.4-fold during the first month of ethanol administration and remained elevated throughout the entire period of observation. The mean maximal acid output (MAO) from the main stomach also increased distinctly within 1 month after the start of alcohol feeding. The MAO remained elevated for 4 months and decreased gradually thereafter, reaching nearly preethanol values after 6 months. It was suggested that the increase in gastric acid secretion induced by chronic excessive intake of alcohol was related to increases in the mean parietal cell mass (Yoshimori et al., 1972) and the increase in the mean parietal cell volume (Lillibridge et al., 1973). Quantitative electron microscopic analysis of the gastric parietal cells demonstrated mitochondrial hypertrophy and a marked increase in the number of vesicotubules in the secretory tubular apparatus (Lillibridge et al., 1973).

In alcohols who drank for more than 5 years, mean basal and maximal acid outputs were significantly less than those of healthy controls of comparable age and sex (Chey et al., 1968). Hyposecretion of acid was also reported in patients with alcohol-induced cirrhosis (Ostrow et al., 1960; Scobie and Summerskill, 1964). However, when gastric secretion was studied in 70 male alcoholics 4 to 8 weeks after the cessation of alcohol, no significant difference in the mean MAO was observed between the alcoholics and the control group of comparable age (Dinoso et al., 1972).

10.4.3. Effect of Alcohol on the Gastric Mucosa

10.4.3.1. Acute Effects

Alcohol has been shown to damage the gastric mucosa of both animals and man. The earliest evidence of mucosal alteration following acute alcohol ingestion was provided by William Beaumont in his classic study of the fistulous stomach of Alexis St. Martin. After alcohol ingestion, Beaumont noted mucosal erythema, superficial ulcerations, and cloudy, viscid secretion in St. Martin's stomach; the mucosa reverted to normal with abstinence. Similar findings were noted in the large endoscopic study of Palmer (1954). More recently, Gottfried et al. (1978) showed by prospective experiments in humans that alcohol caused endoscopically visible antral erythema and friability and microscopic mucosal hemorrhage. Various mechanisms for the pathogenesis of these lesions are discussed below.

10.4.3.1a. Gastric Mucus. Alcohol has been shown experimentally to affect both the gastric mucus barrier as described by Hollander (1954) and the "gastric mucosal barrier" of Davenport (1967b). Following ethanol instillation into canine Heidenhain pouches (10, 20, and 40% v/v), Dinoso et al. (1970) found that mucosubstances of both gastric mucus and surface epithelial cells were significantly reduced. Glowacka et al. (1974) found that ethanol inhibited biosynthesis of galactosamine (a key component of mucus) in human gastric mucosal cells in vitro. Pretreatment with nicotine worsened the alcohol-induced decrease in rat gastric mucus production and the severity of gastric lesions (Wong et al., 1986; Tariq et al., 1986). High concentrations of ethanol ($> 40\%$ v/v) caused dehydration and denaturation of human and pig gastric mucus (Bell et al., 1985). At present, it is not clear whether gastric mucus plays any role in protecting the stomach from alcohol-induced injury. It does not seem to protect against the deep lesions caused by ethanol (Wallace, 1988), although it probably favors restitution of the surface epithelium from the gastric pits (Ito and Lacy, 1985).

10.4.3.1b. Disruption of the Mucosal "Barrier." Alcohol is one of several substances including aspirin and bile salts capable of causing gross and microscopic mucosal alterations, bleeding (with exogenous acid present), changes in ion flux with insorption of H^+ and exsorption of Na^+, decreased electrical resistance, and decreased mucosal-to-serosal potential difference (Davenport, 1967b; Cooke, 1976; Stern et al., 1984). Experiments on the "gastric mucosal barrier" suggest that multiple complex functions rather than passive permeability determine the extent of mucosal damage (Kivilaakso et al., 1978). Therefore, Fromm (1979) has suggested that mucosal changes should be described in physiological terms, and phrases such as "alcohol ruptures the gastric mucosal barrier" be avoided. Despite such changes in semantics, it is clear that alcohol can experimentally produce a number of reversible alterations in mucosal function. In vitro studies with guinea pig gastric mucosa show that some changes in barrier function and morphology induced by 10% ethanol reverse despite continued presence of ethanol (Rutten and Ito, 1986). Disruption of the barrier by ethanol in rats, as measured by fluxes of H^+ and K^+ ions, was reduced by prior ethanol exposure. The protective effect was diminished by indomethacin pretreatment, implicating the necessity for prostaglandin synthesis (Miller and Henagan, 1984).

Changes in mucosal blood flow have been shown to accompany ethanol administration. Augur (1970) used the aminopyrine clearance technique to estimate gastric blood flow in Heidenhain pouches following instillation of ethanol, acetic acid, and aspirin. Blood flow initially increased but then diminished as H^+ back-diffusion occurred and allegedly caused stasis. Using γ-labeled microspheres, considered more accurate than plasma aminopyrine clearance, Cheung *et al.* (1975) found that mucosal blood flow actually increased after alcohol instillation. Gastric vasodilatation protects the mucosa under a variety of circumstances, whereas tissue anoxia may promote ulcerogenesis by preventing mucosal cell buffering of luminal acid (Kivilaakso and Silen, 1979). Gastric mucosal venoconstriction and arteriolar dilatation following alcohol might result from release of histamine (Oates and Hakkinen, 1988).

Other methods besides flux measurements have been used to assess damage in the gastric mucosa. Decreases in both transmucosal potential difference (PD) and electrical resistance have been described in man and experimental animals. Geall *et al.* (1970) and Stern *et al.* (1984) noted an immediate, sustained drop in human gastric PD following 20% alcohol administration. Although PD *per se* does not indicate gastric damage (it is affected by factors such as the gastric secretory state, type of electrode used, and junction potentials), it has been correlated with flux measurements (Davenport, 1964) and ultrastructural changes (Eastwood and Erdmann, 1978).

Several studies suggest that the gastric "barrier" consists of the apical membranes of the mucus-producing surface epithelial cells and the tight junctions between these cells. The effect of alcohol on surface ultrastructure has now been well studied. Eastwood and Kirchner (1974) administered 10% and 25% v/v ethanol with and without exogenous HCl to mice. By light microscopy, surface mucus cells displayed poor staining, nuclear changes, and occasional cellular disruption; extent of damage was related to the alcohol concentration and the presence of exogenous acid. Electron microscopy in severe cases revealed clumping of nuclear chromatin, mitochondrial swelling, and distortion and occasional disruption of the apical cell membrane. Tight junctions were generally intact. Dinoso *et al.* (1976b) found focal separation of tight junctions in the most severely damaged areas following a 40% ethanol dose to dogs. Changes in luminal Na^+, K^+, and protein correlated with both the ethanol concentration and the degree of histological damage. Gottfried *et al.* (1978) gave human volunteers 35% w/v ethanol (1 g/kg dose) and found mottled to diffuse erythema in the distal stomach on endoscopy in all seven patients studied; light microscopy revealed hemorrhage into the subepithelial lamina propria in four of the seven subjects (Fig. 10.1).

Studies with ^{51}Cr-tagged red cells and ^{51}Cr-labeled albumin in humans provide indirect evidence for gastric mucosal damage following alcohol. Although alcohol alone does not cause gastric bleeding in normal subjects, it significantly increases blood loss in persons with atrophic gastritis (Dinoso *et al.*, 1973) or concomitant aspirin ingestion (DeSchepper *et al.*, 1978). Brassinne (1979) was able to demonstrate increased plasma protein shedding following alcohol even in normal subjects. The early events of submucosal vascular injury from alcohol have been nicely documented (Szabo *et al.*, 1985).

10.4.3.1c. Gastritis in the Alcoholic: Pathology. Acute alcohol consumption can cause an erosive/hemorrhagic form of gastritis, whereas chronic consumption of ethanol is more likely to result in a superficial or atrophic type of nonerosive gastritis. In erosive/hemorrhagic gastritis, there are loss of surface epithelial cells and neutrophilic infiltration of the lamina propria and glandular epithelium. The main histological feature of superficial nonerosive gastritis is infiltration of the superficial lamina propria with lymphocytes and plasma cells. Glandular atrophy is the hallmark of atrophic gastritis and is accompanied by dense infiltration of mucosa with mononuclear cells. It should be emphasized that there is poor correlation between the endoscopic and histological appearances of nonerosive gastritis (Sauerbruch *et al.*, 1984; Toukan *et al.*, 1985). What appears like nonerosive hemorrhagic gastritis endoscopically can best be described on biopsy as subepithelial hemorrhage of the foveolar region with surrounding edema and without prominent inflammatory infiltration (Laine and Weinstein, 1988).

10.4.3.1d. Mechanism of the Acute Mucosal Damage: "Cytoprotection." Although inhibition of ion transport (Shanbour *et al.*, 1973) is an important early effect of topical alcohol, altered mucosal permeability also occurs (Kuo *et al.*, 1974). Fromm and Robertson (1976) showed that ouabain, which inhibits Na^+ and Cl^+ transport, did not alter the rate of luminal acid loss in isolated rabbit gastric mucosal chambers. Alcohol, however, increased H^+ insorption in antral mucosa pretreated with ouabain. This demonstrated that alcohol could alter permeability independent of the active transport process. By measuring the transmural fluxes of labeled Na^+ and Cl^- along with [^{3}H]mannitol, Dawson and Cooke (1978) found that alcohol increases the permeability of noncellular pathways for transmural ionic diffusion. Thus, it appears that alcohol directly affects ion transport in the gastric epithelial cell while also causing a "leak" between cells.

It has been suggested that alcohol-induced changes

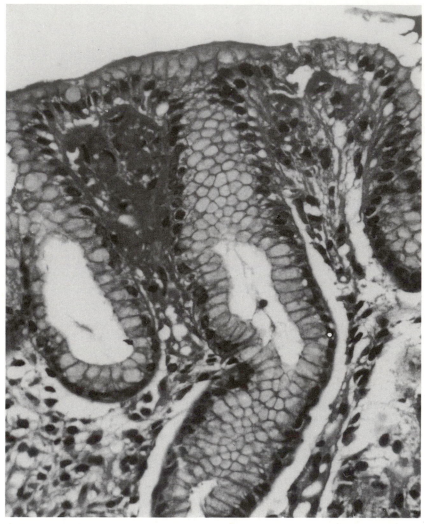

FIGURE 10.1. Subepithelial hemorrhage in the gastric antrum 3 hr after the ingestion of 35% ethanol (1 g/kg). The subepithelial lamina propria is filled with extravasated red blood cells (hematoxylin and eosin, original magnification ×400). (From Gottfried *et al.*, 1978.)

in the gastric mucosa may merely be the result of the hyperosmolarity of ethanol solutions (Sernka and Jackson, 1975, 1976). Sernka and Jackson (1978) found that hyperosmotic NaCl, KCl, dextrose, and $MgCl_2$ produced the same reduction of gastric acid output as ethanol when instilled into the rat stomach. In mice, Kawashima and Glass (1975) showed that a 10% ethanol solution would not produce gastric erosions at any dose, whereas at higher concentrations gross morphological damage was a function of both the concentration and the dose administered. Although hyperosmolarity alone may explain some alcohol-induced changes, ethanol

does have unique features. Biggerstaff and Leitch (1977) found mucosal ulceration in rats following 20% alcohol, although an isosmolar sucrose solution caused no gross change.

Alcohol might render the mucosa more susceptible to the effects of acid or hyperosmolarity by initiating damage by a variety of mechanisms. Partial insight into these mechanisms is afforded by listing the agents that foster or retard the damage and the concomitant measurable changes in mucosal substances. The result, however, is isolated patches of facts without a satisfying overall concept.

Ethanol causes hyperemia and vascular damage to the gastric mucosa within seconds, accompanied by mast cell degranulation (Oates and Hakkinen, 1988). The vascular changes are inhibited by dual lipoxygenase (LO) and cyclooxygenase (CO) inhibitors (BW755C and norhydroguaiaretic acid) and by an H_1 antihistamine (pyrilamine). They are not inhibited by a cyclooxygenase inhibitor (indomethacin) or an H_2 antihistamine (cimetidine). The implication is that histamine and possibly leukotriene C_4 (LTC$_4$) mediate the vascular changes. Superoxide and hydroxyl radicals may be involved in the pathogenesis of ethanol-induced gastric mucosal injury, as superoxide dismutase, allopurinol, and dimethyl sulfoxide protect against deep lesions (not surface epithelial cells) (Terano et al., 1989).

Much emphasis has been placed on the "adaptive cytoprotective" properties of prostaglandins (PG) (Robert, 1979), and surely these are germane to ethanol-induced injury. Exogenous prostaglandins, in doses that do not inhibit acid secretion, prevent gastric necrosis produced by alcohol and a variety of other agents (HCl, hypertonic NaCl, NaOH, boiling water) (Robert et al., 1979). In general, pretreatment with prostaglandins such as 16,16-dimethylprostaglandin E_2 (dmPGE$_2$) protect against the deep necrotic lesions induced by ethanol rather than against damage to surface epithelial cells, although they do seem to enhance restitution of the surface epithelial layer from crypts (Guth et al., 1984; Tarnawski et al., 1985; Wallace and Whittle, 1985). Lipid peroxide levels are increased in gastric mucosa after topical ethanol; the elevated levels are prevented by pretreatment with prostaglandin E_2 and $F_{2\alpha}$ (Miller and Henagan, 1984), which also provided gastric mucosal protection. Other prostaglandins, such as a misoprostol, also afford protection against ethanol-induced gastric damage (Lacy, 1986). Enprostil, a synthetic PGE$_2$, is an exception and potentiates the alcohol damage (Cohen et al., 1990). Sucralfate has afforded protection against alcohol-induced lesions in humans (Cohen et al., 1989), associated with increased levels of 6-ketoprostaglandin $F_{1\alpha}$. The status of sulfhydryl groups appears to be important for the protection against ethanol-induced capillary damage since cysteamine pretreatment affords protection (Trier et al., 1987), as do metals (Dupuy and Szabo, 1986). Intragastric sucralfate (Hollander, 1985) and somatostatin protect against ethanol (in a dose-related fashion: low doses protect and high doses damage) (Diel and Szabo, 1986). Interestingly, the somatostatin effects correlate with mast cell degranulation. Laparotomy protects against alcohol damage in rodents, probably mediated by peptide-containing afferent neu-

rons, which are selectively ablated by capsaicin (from hot red pepper) (Yonei et al., 1990).

In summary, alcohol damages the gastric mucosal "barrier." This may occur because of (1) diminished gastric mucus production, (2) changes in mucosal blood flow, (3) inhibition of active transport, (4) increased intracellular permeability as a result of mast cell release of histamine and release of leukotriene C_4, (5) actual disruption of the cell membrane, (6) hyperosmolarity, (7) changes in PG or cAMP, or (8) via lipoperoxidative mechanisms or sulfhydral group alterations. A combination of effects most likely is involved, some probably nonspecific.

10.4.3.2. Mucosal Changes Caused by Chronic Alcohol Consumption

Much controversy has surrounded the question of whether alcohol causes chronic gastritis in man. Wolff (1970) detected no relationship between alcohol intake and histological evidence of atrophic gastritis (via blind biopsy technique); these findings agreed with earlier observations by Palmer (1954). However, other studies have documented this connection (Joske et al., 1955; Roberts, 1972; Dinoso et al., 1972; Pitchumoni and Glass, 1976). Dinoso et al. (1972) found that only 50% of alcoholics had normal fundic histology compared to 85.7% of the controls; in the antrum, only 15.6% had a normal biopsy, and 65% had antral atrophic gastritis. Parl et al. (1979) also found a significant increase in antral gastritis in alcoholics, not surprising since most of the acute effects of alcohol are directed against the antrum (Gottfried et al., 1978). The incidence of fundal gastritis may be greater too (Parl et al., 1979). Patient selection (nutritional status, definitions of alcoholism), biopsy technique (blind versus directed), and length of abstinence before examination may, at least in part, explain the lack of agreement in the literature regarding the relationship between alcoholism and chronic gastritis. Gender appears unimportant for alcohol-related lesions (Rabinovitz et al., 1989). Blood (Dinoso et al., 1973) and plasma protein loss (Chowdhury et al., 1977) after ethanol intake occurs more readily in patients with chronic gastritis.

Alcoholics with portal hypertension may develop congestion of their gastric mucosa (McCormack et al., 1985). Endoscopically the fundus and body appear reddened and edematous, and pathologically there are dilatation and tortuosity of submucosal veins. This "portal gastropathy" is a cause of bleeding.

Chronic alcohol consumption diminishes the first-pass metabolism of ethanol, which is the amount that gets metabolized before the systemic circulation is reached (DiPadova *et al.*, 1987). The metabolism involves gastric alcohol dehydrogenase activity. The gastric significance of this ethanol metabolism is not yet established.

10.4.4. Upper Gastrointestinal Bleeding in the Alcoholic

10.4.4.1. Bleeding Lesions

Upper gastrointestinal tract (UGI) bleeding is very common in alcoholics. Though it was thought for many years that alcoholics presenting with hematemesis or melena were almost invariably bleeding from peptic ulceration, it is now recognized that alcohol ingestion is not a causative factor in peptic ulcer disease (Cooke, 1978; Langman, 1973; Roth, 1974). Furthermore, large endoscopic series reveal that acute mucosal lesions ("hemorrhagic gastritis") are a preeminent cause (Katz *et al.*, 1976; Belber, 1978) of bleeding in alcoholics. Katz *et al.* (1976) examined the etiology of UGI hemorrhage in 1429 cases at an inner-city hospital with a large alcoholic population. The incidence of acute gastric mucosal lesions was 36.2%, and this did not even include an additional 9% of patients with acute mucosal lesions of the duodenum. Duodenal erosions have been shown to occur even after acute administration of ethanol.

It should be pointed out that several other endoscopic studies both in the United States and abroad have established a lower frequency for acute gastric mucosal lesions as the cause of UGI bleeding than did Katz *et al.* (1976). English studies found acute mucosal lesions in fewer than 8% of patients with hemorrhage (Schiller and Cotton, 1978). Belber (1978) found "gastritis" in 15% of 200 consecutive patients, but the incidence rose to 25% when the alcoholics were examined independently. Lee and Dagradi (1975) found an 18.5% incidence of acute mucosal lesions, with 74% of these patients having recent alcohol and/or salicylate ingestion. A recent survey of 2400 cases of UGI bleeding by the American Society of Gastrointestinal Endoscopy found gastritis to be the cause of bleeding in 23.4% (Silverstein *et al.*, 1981). Although differences in drinking patterns explain, in part, the various rates of acute mucosal lesions, other factors (definition of "significant" hemorrhage, timing of endoscopic examination, number of "undiagnosed" cases) may also be of importance.

The other sources of bleeding found commonly on endoscopy in alcoholics are esophageal varices (15–

19%), chronic peptic ulcer (21–24%), Mallory-Weiss lesions (17%), duodenitis (9%), and esophagitis (7%) (Belber, 1978; Katz *et al.*, 1976). Careful endoscopic examination and scoring of the extent, size, and coloration of esophageal varices enable one to predict which subjects are most likely to have variceal hemorrhage (Snady and·Feinman, 1988). In patients with cirrhosis and known portal hypertension, the incidence of acute mucosal lesions remains high. In one series of 104 patients with known varices and upper gastrointestinal hemorrhage, varices were identified as the bleeding site in 41% of cases, acute mucosal lesions were found in 29%, and peptic ulceration was seen in 16% (Thomas *et al.*, 1979). Other endoscopic studies confirm the fact that only approximately 50% of patients with esophageal varices are actually bleeding from ruptured varices during any single UGI hemorrhage (Waldram *et al.*, 1974; Teres *et al.*, 1976; Novis *et al.*, 1976; Mitchell and Jewell, 1977). Since the clinician cannot tell which patients are bleeding from varices or portal gastropathy, early endoscopy is needed in this group to establish a diagnosis and initiate appropriate therapy. Endoscopy may not alter overall morbidity and mortality of upper gastrointestinal bleeding (Eastwood, 1977), but it is necessary in patients with varices since specific therapy (pitressin infusion, Sengstaken tube, sclerotherapy, later use of propranolol) may be indicated when varices are the cause of bleeding.

The Mallory-Weiss syndrome has long been recognized as a disease of alcoholics. Mucosal tears of the cardioesophageal area are being recognized with increasing frequency because of the increasing use of early endoscopy. Not surprisingly, those centers with large series of Mallory-Weiss lacerations are treating predominantly alcoholic patients. Knauer (1976) reported excessive alcohol intake in 67.5% of his patients, and Graham and Schwartz (1978) reported a 60% incidence of alcoholism. Many patients also have concurrent lesions of the gastroduodenum from their alcohol ingestion. It is important to note that the classic history of vomiting followed by massive hematemesis often cannot be elicited. Graham and Schwartz (1978) found only 29% of 93 patients with such a history. Early endoscopy is the only way to establish the diagnosis, since the lesion is often completely healed within 72 hr and is not visible in radiographic studies.

10.4.4.2. Therapeutic Approach

The therapy required in UGI bleeding depends, of course, on the etiology of the hemorrhage. In patients with acute gastric erosions, bleeding is often not as severe as with peptic ulcer or esophageal varices and tends to halt spontaneously. In Katz's series (Katz *et al.*, 1976), 48% of cases required no transfusions, whereas 20% required over 4 units. Mortality was related to the severity of the underlying alcoholic liver disease. Watson *et al.* (1974) defined a subgroup of alcoholic patients with UGI bleeding from gastric erosions who do not require hospitalization. They suggested that alcoholics under the age of 40 with recent hematemesis can be sent home if there is no fresh blood in the gastric aspirate, no orthostatic hypotension, and a hemoglobin greater than 13 g/dl. Other authors agree that acute gastric erosions caused by alcohol generally represent a benign condition unless complicated by underlying liver disease (Lee and Dagradi, 1975; Roesch, 1978). Outpatient management of alcoholics with UGI bleeding of any subgroup is not the current norm.

The initial approach to the alcoholic patient with UGI bleeding should be directed toward maintaining blood pressure and hemostasis. A nasogastric tube should be placed to assess whether brisk hemorrhage is still occurring. A large-caliber intravenous line should be placed for the immediate delivery of saline to expand the intravascular volume. Packed red blood cells should be administered when available; whole blood may be required for rapid bleeding. If liver disease is present, clotting abnormalities may have to be corrected with fresh frozen plasma and vitamin K. Platelets should be given if thrombocytopenia exists or after multiple transfusions. An electrocardiogram should be performed to evaluate ischemic changes after hypotension and tachycardia have been corrected.

Once the patient is thus stabilized, emergency endoscopy should be carried out. No premedication or topical anesthetic should be used in patients with altered mental status or brisk bleeding unless absolutely necessary because of the danger of pulmonary aspiration. Although the stomach should be cleared of blood prior to endoscopy, the use of ice-water lavage to control bleeding is controversial because it may induce platelet dysfunction and has even been implicated as a cause of acute erosions (Roesch, 1978).

Once the etiology of the hemorrhage has been established, therapy can then be directed against the lesion. As stated above, bleeding from gastric erosions tends to subside spontaneously. Most cases of exsanguination from erosions have occurred because underlying clotting abnormalities could not be controlled. Surgery, usually in the form of a subtotal gastrectomy, is required in a small minority of cases. Bleeding from peptic ulceration in the stomach or duodenum can be very severe but represents no special problem in the alcoholic unless coagulopathy or hepatic failure is present. Special attention should be given to any "visible vessel" seen on endoscopy, since in such cases bleeding is less likely to stop and may prompt therapeutic endoscopic maneuvers or early surgery. In patients with Mallory-Weiss tears, more than 80% of bleeding episodes cease spontaneously, and one-third do not even require transfusion. If hemorrhage continues, surgical repair of the tear may be necessary; mortality is a function of underlying disease (Graham and Schwartz, 1978). Pitressin infusion may help to control hemorrhage in cases of Mallory-Weiss tears or multiple gastric erosions.

Antacids and H_2 blockers are commonly used to treat both acute gastric erosions and peptic ulcer disease, although evidence for efficacy in acute hemorrhage is lacking (LaBrooy *et al.*, 1979; Carstensen *et al.*, 1980).

The use of cimetidine or antacids in established cases of multiple gastric erosions related to alcohol has little justification. Cimetidine has been shown to prevent stress-induced erosions in humans with liver failure (MacDougall *et al.*, 1977). Experimental evidence suggests that H_2 blockade does not prevent erosions caused by alcohol: Robert *et al.* (1979) found that cimetidine pretreatment did not protect the rat gastric mucosa from damage by topical alcohol. Likewise, Dinoso *et al.* (1976a) found extensive damage to canine gastric mucosa despite metiamide pretreatment. This is not surprising, since alcohol is known to exert its deleterious effect on the gastric mucosa even when administered with a buffer (Davenport, 1967a). Hemorrhage often remits spontaneously, and alcohol-induced lesions heal rapidly; therefore, a large controlled trial would be needed to establish the usefulness of antacid or H_2-blocker therapy. Although there is no evidence that antacids or H_2 blockers help in stopping hemorrhage from alcoholic gastritis, the drugs are relatively safe and favored by many clinicians. It is important to point out that antacids may impair cimetidine absorption when the two are taken together (Steinberg and Lewis, 1980).

Preliminary studies (Hosking *et al.*, 1987) indicate that propranolol may be beneficial for bleeding from portal gastropathy.

Algorithms for the treatment of hemorrhage from esophageal varices are available in standard texts. The initial management must be directed toward maintaining blood volume. Endoscopic diagnosis should be made within hours. Some favor initiating sclerotherapy at this time, but the effect on survival is not proved (Snady, 1987; Terblanche *et al.*, 1989). Once the diagnosis is established, the nasogastric tube is taken out since it is usually ineffective in removing clots and may induce gastroesophageal reflux. Coagulopathy is corrected with fresh frozen plasma. If bleeding does not stop, a pitressin infusion is usually effective, starting at a dose of 0.5 U/min. Addition of nitroglycerin decreases vasopressin side effects (Tsai *et al.*, 1986; Gimson *et al.*, 1986). Although at least 75% of cases can be controlled with pitressin, sclerotherapy or a Sengstaken-Blakemore tube is frequently needed. The gastric balloon should be tried alone before inflating the esophageal balloon. A small tube should be inserted above the esophageal balloon to prevent aspiration of oropharyngeal secretions. If sclerotherapy and/or the Sengstaken-Blakemore tube fails to halt the bleeding, or if hemorrhage recurs, there is no widely accepted medical therapy. Transhepatic embolization of the varices with Gelfoam or thrombin may be tried. Emergency surgery is required if hemorrhage cannot be controlled by any other means. The prognosis of such patients is poor. Upper GI bleeding represents a major cause of death in patients with underlying alcoholic cirrhosis.

10.5. Effect of Alcohol on the Small Intestine

10.5.1. Introduction

Disturbances in digestive physiology are common in alcoholics (Mezey, 1975). That diarrhea is frequently associated with alcohol was known since Hippocrates, who prescribed wine as a purgative. Malabsorption has also been documented in approximately 50% of recently hospitalized alcoholics, with and without cirrhosis (Fast *et al.*, 1959; Small *et al.*, 1959; Roggin *et al.*, 1969).

The alcoholic has been shown to have several interacting factors that contribute to altered intestinal function and malabsorption (see Fig. 17.1 in Chapter 17). These factors include a direct effect of alcohol mediated through changes in transport, motility, metabolism, circulation, and cellular structure of the small intestine. In addition to these effects of alcohol *per se*, concomitant nutrient deficiency, hepatic and pancreatic injury, and bacterial overgrowth in the small intestine may aggravate the tendency toward malabsorption and malnutrition in the alcoholic.

10.5.2. Direct Effects of Ethanol on the Intestine

Alcohol is rapidly absorbed from the stomach and upper small intestine. Halsted *et al.* (1973a) administered alcohol orally and intravenously to alcoholic volunteers and measured the concentration of alcohol at various levels in the gastrointestinal tract. After the ingestion of alcohol (0.8 g/kg body wt.), high concentrations of alcohol are obtained in the stomach (7 to 8 g/100 ml) and upper small intestine (1 to 4 g/100 ml). High levels (400 mg/100 ml) are maintained in the upper intestine for more than 1 hr and result in blood levels of 100 to 150 mg/100 ml. After the intravenous administration of alcohol, the concentrations of alcohol in the intestine and blood are similar.

In addition to being absorbed, ethanol is metabolized by the intestine. Alcohol dehydrogenase (ADH) has been found in the stomach and small intestine (Mistilis and Garske, 1969; Carter and Isselbacher, 1971; Mezey, 1975). Also, the mixed-function microsomal oxidases, including the microsomal ethanol-oxidizing system (MEOS) described by Lieber and DeCarli (1970), have been found in the intestine and shown to be induced by ethanol (Seitz *et al.*, 1979) with a prominent increase in the ethanol-inducible cytochrome P450IIE1 (Shimizu *et al.*, 1989). The high concentrations of alcohol found in the intestinal lumen are much greater than those necessary to saturate all known pathways of ethanol oxidation.

Thus, some of the effects of ethanol directly on the intestinal cell may be caused by the high alcohol concentrations or the consequences of its oxidation.

10.5.2.1. Effects of Ethanol on Intestinal Motility

In manometric and cineradiologic studies in normal human subjects, intravenously administered ethanol was reported to cause a significant increase in motility in the second part of the duodenum (Pirola and Davis, 1970). However, in a more recent investigation, i.v. ethanol did not have a significant influence on upper gastrointestinal tract mobility (Kolbel *et al.*, 1986). In contrast, the same groups demonstrated that intragastric ethanol interrupts the phase III interdigestive migrating complex and changes the fasted pattern into a fed pattern (i.e., irregular spiking activity) (Demol *et al.*, 1986). The effects of ethanol on motility in the jejunum and ileum have also been measured. In dogs, intravenous ethanol prolongs the interdigestive cycle in the jejunum from 90 to 200 min (Angel *et al.*, 1980) when electrical activity is measured. Manometric data are not as clear-cut and appear to depend on the anatomic location. In the jejunum, ethanol decreases type I (impeding) waves, whereas in the ileum, it increases type III (propulsion) waves. It is unclear to what extent these various effects contribute to the shortened transit time and diarrhea seen in alcoholics (Keshavarzian *et al.*, 1985). It is also possible that colonic motility may be altered by ethanol, as its intravenous administration is associated with decreased rectosigmoid activity (Berenson and Avner, 1981). This is of interest since a number of diarrheal illnesses have been linked to sigmoid hypomotility (Haddad and Devroede-Bertrand, 1981).

10.5.2.2. Structural Effects of Ethanol on the Intestine

10.5.2.2a. Acute Effects of Ethanol. The concentrations of ethanol to which the upper small intestine is exposed during drinking produce structural alterations of the small intestine both *in vivo* (Maling *et al.*, 1967; Baraona *et al.*, 1974; Hufnagel *et al.*, 1980; Buell *et al.*, 1983) and *in vitro* (Chang *et al.*, 1967; Krawitt, 1974). Intragastric administration of ethanol, on concentrations similar to those of common alcoholic beverages, produced hemorrhagic erosions of rat jejunal villi (Baraona *et al.*, 1974; Krawitt, 1974) (Fig. 10.2). These alterations were seen within 10 min, were well defined at 1 hr, and were less apparent or had even disappeared at 4 and 16 hr. The intensity of the lesions was maximal proximally, decreased toward the ileum, and was proportional to the concentration of ethanol administered (Baraona *et al.*, 1974). A decrease in the number of epithelial cells per crypt persists for several days after an acute intragastric load of ethanol (Hufnagel *et al.*, 1980). Subepithelial intestinal blebs have also been noted in volunteers given a single dose of ethanol (Millan *et al.*, 1980) and during intraluminal ethanol perfusion in the dog (Buell *et al.*, 1983). The morphological lesions and the inhibitory effects of ethanol on calcium absorption and fatty acid oxidation have been reproduced by urea (Baraona *et al.*, 1974, 1975) or mannitol (Krawitt, 1974; Baraona *et al.*, 1975) in hyperosmolar concentrations comparable to those of ethanol.

Similar hemorrhagic erosions of villus tips have been seen in duodenal biopsy specimens from alcoholic volunteers after ingestion of a single dose (1 g/kg body weight) of ethanol (Gottfried *et al.*, 1976). It has been postulated that the caustic effects of these ethanol concentrations are caused, at least in part, by hypertonicity, but this has been questioned (Buell *et al.*, 1983). Instead, changes in epithelial morphology have more recently been linked to disturbances in the mucosal microcirculation.

10.5.2.2b. Effects of Chronic Ethanol Ingestion. Chronic ethanol feeding with nutritionally adequate diets in humans and in the rat produces ultrastructural abnormalities of the intestinal epithelial cells (Rubin *et al.*, 1972). These alterations include mitochondrial abnormalities and dilatation of the endoplasmic reticulum and cisternae of the Golgi apparatus. Biopsies from human subjects also showed focal cytoplasmic degradation. These changes were found in villus and crypt cells both in the jejunum and in the ileum. There were no gross histological alterations (Rubin *et al.*, 1972). However, quantitative studies in the jejunum of the rat showed fewer cells per villus and slightly decreased villus height after ethanol feeding compared to pair-fed controls (Baraona *et al.*, 1974). The number of crypt cells and mitoses were greater in the alcohol-fed rats, suggesting hyperregenerative activity. The shortened jejunal villi were associated with decreased activity of the enzymes typical of mature villus cells, such as lactase, sucrase, and alkaline phosphatase, whereas the morphological changes in the crypt were associated with increased activity of thymidine kinase (an enzyme typical of proliferating crypt cells) and increased incorporation of tritiated thymidine into DNA (Baraona *et al.*, 1974). Thus, chronic ethanol feeding not only altered cell ultrastructure but also the relative numbers of crypt and villus cells, at least in the rat (Baraona *et al.*, 1974). In more recent studies, rat villus height was not altered by ethanol feeding, and enterocyte turnover was decreased rather than increased (Mazzanti and Jenkins, 1987). The two studies are difficult to compare, as the latter results were derived using the metaphase arrest actions of vincristine rather than DNA incorporation, and the feeding pattern prior to sacrifice was not defined.

Decreases in villus height (Hermos *et al.*, 1972) and disaccharide activity (Madzarovova-Nohejlova, 1971) of mucosal biopsy specimens have been found in alcoholics, but the relative importance of alcohol ingestion and nutritional deficiency in the genesis of these abnormalities has not been determined. However, disaccharidase deficiency has been observed in alcoholics without evidence of overt malnutrition or folate or vita-

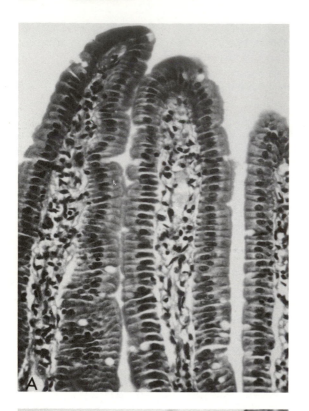

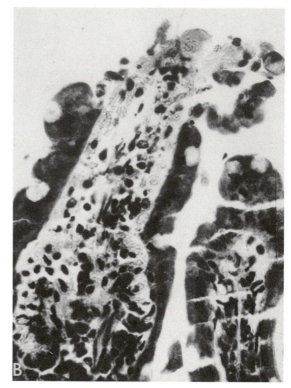

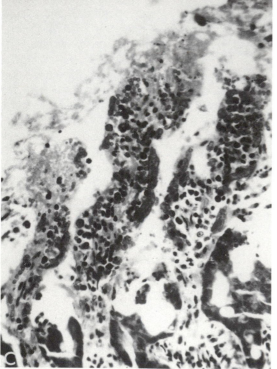

FIGURE 10.2. Microscopic appearance of rat jejunal villi 1 hr after intragastric administration of saline or ethanol solutions. (A) Normal appearance from a control rat after intubation with saline (hematoxylin and eosin, ×160). (B) Typical changes seen after ethanol administration with epithelial cell loss mainly affecting the villus tips and with hemorrhage in the corium (hematoxylin and eosin, ×320). (C) Gross damage to villi in an area of confluent hemorrhage after ethanol administration (hematoxylin and eosin, ×160). (From Baraona *et al.*, 1974.)

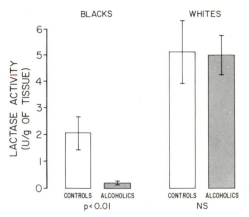

FIGURE 10.3. Comparison of jejunal lactase activity between controls and alcoholics from two racial populations: American blacks and Americans of northern European origin. (From Baraona and Lindenbaum, 1977.)

min B_{12} deficiency (Perlow *et al.*, 1977). In this study, the evidence of lactase deficiency was most striking in the alcoholic black population (Perlow *et al.*, 1977) (Fig. 10.3). Thus, milk intolerance may be unmasked or exaggerated in these populations in which both alcoholism and genetically determined low lactase levels are common (Perlow *et al.*, 1977). Interestingly, a second jejunal biopsy after an additional 2-week period of alcohol abstinence indicated reversibility of the deficiency (Fig. 10.4).

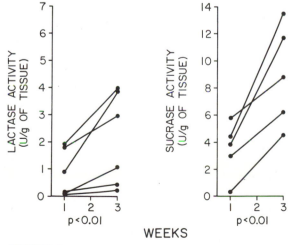

FIGURE 10.4. Jejunal lactase and sucrase activities in successive biopsies done after 1 and 3 weeks of ethanol abstinence in six male alcoholics. (From Perlow *et al.*, 1977.)

10.5.2.3. Effects of Ethanol on Splanchnic Circulation of Blood and Lymph

Ethanol is a well-recognized peripheral vasodilator. It also affects the splanchnic circulation. In humans, hepatic blood flow has been found to be increased after infusion of relatively small doses of ethanol (Mendeloff *et al.*, 1949; Stein *et al.*, 1963), although not all studies have demonstrated this effect (Castenfors *et al.*, 1960; Lundquist *et al.*, 1962). The increased hepatic blood flow was associated with decreased splanchnic vascular resistance and increased cardiac output (Stein *et al.*, 1963). The increased circulation could facilitate the passive diffusion of some substances such as D-xylose (Price *et al.*, 1967). The absorption of actively transported nutrients is unaffected unless extreme circulatory changes are produced (Varro *et al.*, 1967; Bynum and Jacobson, 1971). In addition to an increase in total splanchnic blood flow, intraluminal ethanol causes regional changes of blood flow within the jejunal mucosa (Buell *et al.*, 1983; Buell and Beck, 1984). These alterations in the mucosal microcirculation have been incriminated in the formation of subepithelial blebs noted earlier, a concept (Beck *et al.*, 1986) that has been questioned on both experimental and theoretical grounds (Kvietys *et al.*, 1984).

Mesenteric lymph formation varies with changes in the splanchnic circulation (Yoffrey and Courtice, 1956). It has been shown that ethanol markedly increased the flow of intestinal lymph in rats, most strikingly in animals not previously fed ethanol (Baraona and Lieber, 1975). This effect was smaller or disappeared in rats fed ethanol for several weeks. The lymphagogue effect of ethanol was associated with an increased output of both dietary lipids and plasma proteins into the lymph. The administration of ethanol without exogenous lipids produced a rapid increase in the lymph flow at a time when there were no changes in lymph lipid output (Baraona and Lieber, 1975), indicating that the lymphagogue effect does not depend on increased fat transport. The alterative possibility that the lymphagogue action of ethanol could enhance the transport of dietary lipids is supported by reports that other lymphagogue agents (neostigmine and water overload) (Shepherd and Simmonds, 1959; Baraona and Lieber, 1975) also increased dietary lipid transport into the lymph.

10.5.2.4. Effect of Ethanol on Intestinal Absorption

This section considers the possible mechanisms by which ethanol directly affects transport.

10.5.2.4a. Cell Membrane Effects. The maintenance of the fluidity of biological membranes within close limits is important for normal membrane permeability and for the activity of membrane-bound enzymes (Vessey and Zakim, 1974). Membranes expand (Seeman, 1972) and pass through states of increasing fluidity when exposed to increasing concentrations of various alcohols such as benzyl alcohol and propanol through octanol (Metcalfe *et al.*, 1968; Patterson *et al.*, 1972). The effect of alcohols on biomembrane function and structure depends on the relative hydrophobicity, i.e., the chain length, of the molecule. Thus, the effects of ethanol on membranes is distinctly less pronounced than that of equimolar concentrations of alcohols with greater chain length (Kalant, 1971; Mitjavila *et al.*, 1976). Chin and Goldstein (1977) measured the molecular motion of the spin label, 5-doxylstearic acid, as an index of membrane fluidity of mouse erythrocytes as well as brain mitochondria and synaptosomal membranes. A less restricted molecular motion of this labeled compound indicated a loosening or disordering (increased fluidity) of the membranes. A fluidizing effect was seen after *in vitro* administration of acute doses of ethanol, 0.20–0.35 M. The integrity of small intestinal membranes is also altered by ethanol and its metabolite acetaldehyde. Using vesicles obtained from the brush borders of duodenal and jejunal mucosal cells and a probe (DPH) that localizes in the more hydrophobic interior of the membrane, Ballard *et al.* (1988) demonstrated that ethanol (> 1 M concentrations) decreases the enclosed volume of vesicles (an index of membrane integrity) and that these changes correlated with alterations in the fluorescence anisotropy (r) of the membranes. In contrast, lower concentrations of ethanol have no significant fluidizing effect; only high concentrations of acetaldehyde (44–88 mM), not likely to be found *in vivo*, were found to decrease fluorescence polarization values (Tillotson *et al.*, 1981). The type of probe employed in these studies may be critical; even at lower ethanol concentrations (< 1 M), increased membrane fluidity can be demonstrated using probes (e.g., 5-doxylstearic acid) that concentrate at the surface of the membrane (Hunter *et al.*, 1983).

A fluidizing effect of ethanol on intestinal membranes offers an attractive pathogenetic hypothesis, since disturbances in intestinal function have been observed in the absence of the previously described light and ultrastructural changes.

10.5.2.4b. Effect of Ethanol on Intestinal Permeability. Since ethanol has been found to damage the lining of the small intestine, it can be expected to enhance the permeability of the gut wall. Indeed, Bungert (1973) first demonstrated that large daily doses of ethanol (7 g/kg, administered as a 40% solution by gavage for 3 weeks) increased intestinal permeability for human hemoglobin. Human hemoglobin, applied by gastric tube, was found by immunofluorescence in the epithelial cells of the mucosa and in the liver but not in the mucosa of pair-fed control animals (Bungert, 1973). Ethanol administered to rats by gavage for 4–8 weeks or acutely (Draper *et al.*, 1983) was also shown to increase intestinal permeability to horseradish peroxidase (Worthington *et al.*, 1978). Electron microscopic examination of the tissue in both studies suggested that this macromolecule had entered the intercellular space via the apical junctional complex rather than through the mucosal surface. Similar findings have been reported in humans using ^{51}Cr-labeled EDTA (Bjarnason *et al.*, 1984). Urinary excretion of this usually nonabsorbable compound was significantly increased in alcoholics who had abstained for less than 4 days despite normal duodenal histology. After 2 weeks of abstinence, urinary excretion of [^{51}Cr]EDTA had returned to that of nonalcoholic controls. *In vitro*, jejunal biopsies from alcoholics also demonstrated increased tissue uptake of a number of probes. By whatever mechanism, increased permeability of the absorptive surface to large macromolecules may have clinical significance, especially in terms of antigen exposure and the generation of abnormal immune complexes.

10.5.2.4c. Effect of Ethanol on Intestinal Lipid Absorption. Steatorrhea has been reported in 35% to 56% of alcoholics who have been drinking until a few days prior to testing (Roggin *et al.*, 1969; Mezey *et al.*, 1970). The steatorrhea has usually been mild, though in a few patients a fecal fat excretion greater than 15 g/day has been noted (Roggin *et al.*, 1969; Insunza and Ugarte, 1970; Mezey *et al.*, 1970). The mild steatorrhea of the alcoholic is most often related to impaired pancreatic function and not to a mucosal lesion. Steatorrhea was not seen in five patients given ethanol, 173–253 g/24 hr for 13–37 days (Lindenbaum and Lieber, 1975). The decreased fat absorption seen in many animal studies was most likely related to delayed gastric emptying (Barboriak and Meade, 1969; Bouquillon, 1976; Wilson and Hoyumpa, 1979). Fat absorption is actually increased after a single dose of alcohol in rats (Baraona and Lieber, 1975; Saunders *et al.*, 1982). In the former study, a lipid emulsion containing fatty acids, monoglycerides, and bile salts was administered intraduodenally to rats with ethanol (2.5 g/100 ml), which markedly increased the output of intestinal lymph lipid (Baraona and Lieber, 1975). In the latter study, ethanol increased the transport rate of long-chain fatty acid as assessed in lymph fistula rats. However, the effect was only noted at high rates of oleate absorption (lymph-mediated), suggesting that ethanol affects the lymphatic uptake pathway rather than the portal route (involved in the uptake of low levels of long-chain fatty acids). Chronic ethanol feeding in rats did not affect fecal fat excretion (Rodrigo et al., 1971). It has more recently been appreciated that ethanol may alter in vitro fatty acid uptake by effects on the resistance of the unstirred water layer (Thomson et al., 1984). When this layer is disturbed by mechanical stirring, 4% ethanol failed to alter the rate of uptake of saturated fatty acids and fatty alcohol into rabbit jejunum. In contrast, when the unstirred layer is intact, ethanol decreased the uptake of a number of saturated fatty acids.

It is evident that more studies are required to determine in detail the effect of ethanol on lipid transport across the mucosal membrane. The abnormalities in absorptive function usually return to normal within a few weeks of hospitalization, during which time alcohol has been withdrawn and nutritional status and hepatic and pancreatic function have often improved.

10.5.2.4d. Effect of Ethanol on Sugar Transport.

The intestinal transport of glucose as well as other compounds such as amino acids is believed to depend on binding to a carrier and on the energy supplied by the coupled transport of sodium down its concentration gradient (Schultz, 1977). In order for the cell to maintain a low intracellular sodium concentration necessary for continued operation of the system, cations are pumped out of the cell in exchange for potassium, with energy derived from ATP hydrolysis by Na^+,K^+-ATPase. Despite the effects of ethanol on membrane fluidity (see Section 10.5.2.4a) and the logical assumption that carrier functions would be secondarily impaired (Beck and Dinda, 1981), the latter has not been uniformly demonstrated (Tillotson et al., 1981). Instead, the weight of evidence indicates that ethanol affects glucose transport by altering the sodium concentration gradient, either by inhibiting Na^+,K^+-ATPase activity or by membrane changes that increase Na^+ conductance. Some studies showed no effect (Krasner et al., 1976) on Na^+,K^+-ATPase. But, since others showed that ethanol inhibits Na^+,K^+-ATPase activity in brush border membranes (Mitjavila et al., 1976) and basolateral membranes (Hoyumpa et al., 1977), ethanol could be expected to decrease active transport of glucose.

Indeed, a number of studies have shown that ethanol, at least acutely, inhibits D-glucose transport across the small intestine (Ghirardi et al., 1971; Kuo and Shanbour, 1978; Tillotson et al., 1981; Thomson, 1984). The in vivo perfusion of hamster jejunum with increasing concentrations of ethanol (0.45–1.05 M) caused a concentration-dependent depression of steady-state glucose transport (Fox et al., 1978). The chronic replacement of 36% of dietary calories with ethanol decreased the transport of 200 mM glucose in isolated rat jejunal loops in vivo; however, the absorption of lower concentrations of glucose was unaffected (Lindenbaum et al., 1972). When placed in the mucosal compartment, ethanol reduced glucose and 3-O-methylglucose flux from mucosa to serosa, but when placed in the serosal compartment, it did not affect glucose transport (Chang et al., 1967; Dinda and Beck, 1977). This suggests that ethanol selectively affects uptake across the brush border but not transport across the basolateral membrane. Paradoxically, the uptake of some sugars seems to be increased after chronic ethanol administration. Specifically, it has been shown in rats that galactose uptake is increased after chronic ethanol feeding (Mazzanti et al., 1987) and that glucose absorption is increased by chronic ethanol exposure in humans (Green, 1979).

With regard to D-xylose, decreased absorption has been repeatedly shown in various groups of chronic alcoholics. However, in alcoholics with cirrhosis, the interpretation of the results of the xylose test is difficult, since ascites and/or disturbed renal function may influence the distribution and excretion rate of the pentose. Although D-xylose malabsorption was observed in chronic alcoholics without clinical evidence of severe malnutrition (Roggin et al., 1969; Mezey et al., 1970), xylose excretion normalized after the establishment of an adequate diet (Mezey et al., 1970). With adequate nutrition, D-xylose excretion rates returned to normal independently of continuation of alcohol intake (Mezey et al., 1970). Administration of 173–253 g of ethanol daily to alcoholic subjects for 11–32 days with adequate nutrition and daily vitamin supplements even increased D-xylose absorption (Lindenbaum and Lieber, 1975).

10.5.2.4e. Effect of Ethanol on Amino Acid Absorption.

It has been easy to demonstrate the acute depression of amino acid absorption using high concentrations of ethanol (0.5 to 3.0%) in experimental models of acute ethanol exposure, utilizing short segments of gut perfused in vivo or gut sacs bathed in vitro. It has not been easy to demonstrate depressed absorption acutely using smaller concentrations of ethanol in whole intestines of living animals or in models fed ethanol chronically. Several relevant subjects have been little studied and remain obscure: the effect of ethanol on amino acid absorption from complex mixtures (including peptides) and the overall effect of local changes in amino acid absorption on body nitrogen utilization.

How peptides might be affected by acute or chronic ethanol consumption is not known. Dinda and Beck (1984) found that ethanol depresses hydrolysis of leucylglycine, glycyltyrosine, and phenylalanylglycine in a dose-dependent manner, a reversible change related to a decrease in V_{max}. Hajjar et al. (1981) showed that ethanol decreased intracellular hydrolysis of glycylsarcosine. Although they did not demonstrate any changes in total small intestinal glycylsarcosine absorption after acute or chronic ethanol consumption in rats, their study may have masked differential jejunal versus ileal effects. Also, they examined a single concentration of one dipeptide that is resistant to brush-border peptidases. Others have provided evidence that ethanol affects active transport and passive permeability in the gut. Thus, a depression of active absorption may be underestimated because of an enhanced absorption by diffusion.

Investigation in animals revealed that ethanol administered acutely in vivo or in vitro inhibited the active transport of the amino acids L-phenylalanine, L-leucine, L-glycine, L-alanine, L-methionine, and L-valine. In most in vitro studies a significant inhibition was obtained only by ethanol concentrations of 2% or more. Only in one study did 0.5% ethanol reduce the transport of L-phenylalanine across the jejunal mucosa of rats (Israel et al., 1968). Curiously, an inhibition using 1% ethanol was not observed by others using a similar experimental design (Chang et al., 1967).

Alcohol also inhibits amino acid absorption in vivo in rats (Kuo and Shanbour, 1978) and in humans (Israel et al., 1969). Chang et al. (1967) found a similar increase in L-phenylalanine permeability in the rat small intestine in vitro. These results probably don't indicate a generally increased permeability of the tissue, since the inhibition of amino acid transport was reversible immediately after washing off the ethanol (Chang et al., 1967), and it seems likely that an inhibition of active transport plays a role. The precise nature of this effect is unknown, although inhibition of Na^+,K^+-ATPase activity has been invoked. Fecal nitrogen excretion was found unchanged in ethanol-fed rats (Klatskin, 1961; Rodrigo et al., 1971). However, specific malabsorption of some essential amino acids could result in important alterations of nitrogen balance without detectable changes in total fecal nitrogen excretion. Amino acids escaping

small bowel absorption can be metabolized by colonic bacteria, with the nitrogen reabsorbed as ammonia and eventually excreted in the urine as urea. Indeed, losses of urinary nitrogen have been documented in ethanol-fed rats (Rodrigo et al., 1971).

10.5.2.4f. Effect of Ethanol on Water and Electrolyte Transport. The intestinal absorption of water, sodium, and chloride is reduced in alcoholics in the presence of normal jejunal histology when measured by the triple lumen perfusion technique (Krasner et al., 1976). A similar decrease in electrolyte and water transport occurs in normal human volunteers after administration of ethanol for 2 weeks (Mekhjian and May, 1977). Although changes were more pronounced in subjects fed a folate-deficient diet, they were not prevented fully by dietary folate supplementation. This suggests that the ethanol effect was to some extent independent of folate deficiency. Acute infusion of ethanol into the perfused intestinal segment failed to affect sodium and water absorption in humans (Mekhjian and May, 1977). This latter observation is at variance with results obtained in animals (Dinda et al., 1975; Shanbour and Kuo, 1976; Dinda and Beck, 1977; Fox et al., 1978; Worthington et al., 1978). The effect of 6% ethanol perfusion as a depressant of jejunal water, glucose, and sodium absorption in dogs was not changed by addition of 16,16-dimethylprostaglandin E_2, whereas microvascular injury measured as albumin loss was blunted (Leddin et al., 1988).

The mechanism of the inhibition of water and sodium transport after acute alcohol administration is not clear. As discussed before, it has been postulated that an inhibition of Na^+,K^+-stimulated ATPase may play a role. If this were the case, ethanol should act like ouabain, which reduces sodium transport by inhibiting Na^+,K^+-ATPase of the basolateral membrane only when administered to the serosal side. However, ethanol depressed sodium and water transport in hamster jejunum only when it was placed on the mucosal side and exhibited no effect when added to the serosal solution (Dinda and Beck, 1977). The interpretation of the effect of alcohol on sodium and water absorption is further complicated by data that suggest that ethanol may alter intestinal fluid and electrolyte secretion (Wilson and Hoyumpa, 1979). The in vitro activity of intestinal adenylate cyclase, an enzyme involved in the formation of cAMP, was increased by ethanol (Green et al., 1971). Other data have shown that ethanol administration increased cAMP levels in the small intestine (Wilson and Hoyumpa, 1979). In addition, alcohol concentrations above 5% produced an increase in Na^+ flux from the serosal to mucosal side in rabbit jejunal segments, suggesting an increase in passive permeability.

10.5.2.4g. Effect of Ethanol on Vitamin Transport

Thiamine. Thiamine deficiency is a frequent concomitant of chronic alcoholism (Leevy *et al.*, 1965), and clinical studies indicate that thiamine absorption may be impaired in alcoholic patients (Tomasulo *et al.*, 1968; Thomson *et al.*, 1970). Hoyumpa (1986) found that thiamine transport across the intestinal mucosa of the rat proceeds by a sodium-dependent active process at low concentrations, whereas at higher concentrations thiamine appears to be absorbed by passive diffusion, as confirmed by others (Sklan and Trostler, 1977). Alcohol administration had no effect on the transport of high concentrations of thiamine (Hoyumpa, 1986), suggesting that ethanol inhibits the active transport but not the passive diffusion of the vitamin. Alcohol did not affect thiamine uptake across the brush-border membrane but appears to block thiamine release from the cells across the basolateral membrane (Hoyumpa *et al.*, 1975b). Hoyumpa (1986) found a dose-dependent decrease of Na^+,K^+-ATPase activity in basolateral membranes following addition of ethanol *in vitro*. This decrease in ATPase activity correlated with the reduction of thiamine exit from the intestinal epithelial cell (Hoyumpa *et al.*, 1977). In rats chronically exposed to alcohol, more thiamine was excreted in the feces than in control animals (Balaghi and Neal, 1977), suggesting a decrease in thiamine absorption.

However, chronic administration of ethanol to rats did not affect intestinal thiamine transport or Na^+,K^+-ATPase activity when both were measured at the time ethanol concentrations in the gut lumen or plasma were low (Hoyumpa *et al.*, 1978). When intestinal and plasma ethanol concentrations were consistent with incubation levels, basolateral membrane Na^+,K^+-ATPase activity was reduced, and thiamine exit from the cells was decreased. In humans, studies on the effect of ethanol on thiamine transport gave conflicting results. One study, in which thiamine absorption was measured indirectly by determining the urinary excretion of the radiolabeled vitamin, showed a decrease in thiamine absorption after oral or intravenous ethanol (Thomson *et al.*, 1970). Other studies, in which thiamine absorption was measured by constant intestinal perfusion, showed that thiamine absorption was not affected by intestinal or intravenous ethanol (Desmond *et al.*, 1976; Breen *et al.*, 1985).

Although it is still uncertain that thiamine absorption in humans follows the same mechanisms as in rats, it is important to recognize that in rats, at low concentrations of thiamine, transport has been shown to be an active process, whereas passive transport predominates at high thiamine concentrations (Hoyumpa *et al.*, 1974). Since ethanol has been shown to exert its major effect via inhibition of active transport, theoretically, high concentrations of ethanol at low thiamine concentrations should have the greatest inhibition of absorption. It does appear that inhibition of thiamine transport is dependent more on ethanol concentration than on duration of exposure (Wilson and Hoyumpa, 1979). These findings may be particularly relevant to the periodic binge-drinking alcoholics, who simultaneously may decrease their daily intake of thiamine.

Vitamin B_{12}. Lindenbaum and Lieber (1969) first demonstrated that chronic ingestion of large doses of ethanol causes malabsorption of vitamin B_{12}. In four asymptomatic hematologically normal male alcoholics, who received supplements of folic acid, the isocaloric substitution of ethanol (158–253 g/day) for carbohydrates for 3–8 weeks reduced the mean urinary excretion of [^{57}Co]cyanocobalamin by 42%. Similar results were obtained when additional subjects were studied by the same protocol (Lindenbaum and Lieber, 1975) (Table 10.1). The coadministration of intrinsic factor or pancreatin with cyanocobalamin did not correct the absorptive defect, suggesting interference at the ileal level (Lindenbaum and Lieber, 1969, 1975). The nature of the alcohol effect at the ileal absorption site is unclear (Findlay *et al.*, 1976; Lindenbaum *et al.*, 1973). In the same subjects, however, serum and urinary xylose level after an oral dose were increased during ethanol administration as compared to controls (Lindenbaum and Lieber, 1975). Indeed, vitamin B_{12} deficiency does not seem to be common in alcoholics. In groups of folate-deficient alcoholics with (Herbert *et al.*, 1963; Klipstein and Lindenbaum, 1965) or without (Halsted *et al.*, 1971; Racusen and Krawitt, 1977) cirrhosis, serum vitamin B_{12} concentrations were mostly normal or elevated.

Folic Acid. Among alcoholics, folate deficiency is the most common form of hypovitaminosis (Leevy *et al.*, 1965); malabsorption of the vitamin is one of the etiological factors (Halsted *et al.*, 1971). The uptake of folic acid proceeds by a saturable energy-dependent process at physiological concentrations but is accomplished chiefly by simple diffusion at higher concentrations (Smith, 1973).

By use of an intestinal perfusion technique, alcoholic patients were noted to have a low jejunal uptake of folic acid compared with that of abstinent patients (Hal-

TABLE 10.1. Urinary Excretion of (%) of [57Co] Vitamin B_{12}[a] during Control Periods and Ethanol Administration in Human Volunteers[b]

Subject	Control (pre- and/or post-ethanol) 24 hr	48 hr	Ethanol 24 hr	48 hr	Number of days on ethanol	Maximum ethanol dose (% total caloric intake when tested)
1	42.4	51.4	27.4	33.0	13	46
	40.7[c]	50.6	11.9	20.6	26	
2	12.1	16.2	5.9		21	60
	12.1[c]	16.8	6.7	11.0	31	
	11.2[c]	17.5				
3	17.0	21.0	13.1	17.4	21	60
4	10.5	15.4	5.9	10.2	34	66
5	18.3	18.8	3.1	3.9	21	46
	14.7[c]	19.4	6.1	7.9	28	
6	20.9	30.0	14.5	22.7	23	46
	15.4	25.3	10.2	10.5	37	
7	10.2	14.4	17.3	23.1	25	46
			17.8	25.5	33	
8	7.4	10.6	8.8	11.0	25	46

[a][57Co] Vitamin B_{12} is given with 9 g pancreatin (Viokase®).
[b]From Lindenbaum and Lieber (1975).
[c]Post-ethanol period: 8 to 22 days after cessation of alcohol administration.

sted *et al.*, 1973b). However, this effect was noted to improve with improved nutrition, and further studies in humans revealed that ethanol had no effect on folic acid absorption when the diet was adequate (Halsted *et al.*, 1971). Alcohol does dampen the increase in folate absorption induced by partial starvation (Racusen and Krawitt, 1977). In rats, the combination of a protein-deficient diet and chronic ethanol administration had no effect on folate absorption (Halsted *et al.*, 1974).

Vitamin B_6. Pyridoxine and other nonphosphory-lated forms of the vitamin (pyridoxal and pyridoxamine) are probably absorbed by passive diffusion. When studied in rat inverted jejunal gut sacs or *in vivo* perfused segments, high concentrations of ethanol (4%) enhanced mucosal uptake of pyridoxine, correlating with evidence of mucosal injury; chronic ethanol ingestion produced no further pyridoxine uptake (Middleton *et al.*, 1984). The absorption of phosphorylated species of vitamin B_6 necessitates prior hydrolysis. Increasing concentrations of ethanol (1% to 4%) inhibited pyridoxal 5'-phosphate (PLP) hydrolysis by rat jejunal mucosa. Thus, the absorption of vitamin B_6 could be affected by the concentration of ethanol in the intestine and the state of phosphorylation of ingested vitamin.

Vitamin A. The absorption of vitamin A in rats is saturable but not an energy-dependent process (Hollands and Murachidhara, 1977). The oral administration of ethanol to normal human subjects resulted in decreased blood vitamin A levels (Althausen *et al.*, 1960). However, vitamin A levels were increased in patients with gastrectomy and Bilroth II anastomosis, suggesting that ethanol exerted its effect via delayed gastric emptying.

10.5.2.4h. Effect of Ethanol on Mineral Transport

Calcium. Bone disease is a well-recognized complication of alcoholism (Bikle *et al.*, 1985). Because vitamin D metabolism is normal in alcoholics, most attention has been directed toward the direct effects of ethanol on intestinal calcium absorption. There probably are at least three steps to the transepithelial movement of calcium: calcium entry into the cell, movement through the cell (likely via mitochondria and endoplasmic reticulum), and exit from the cell at the basolateral membrane. *In vitro* studies in rats showed that both acute and chronic oral ethanol reduced calcium transport across the small intestine (Krawitt, 1973). Addition of ethanol to the incubation solution also decreased net calcium flux (Krawitt, 1974). Furthermore, brush-border membranes from chicks ingesting ethanol for several weeks demonstrate decreased uptake of calcium (Bikle *et al.*, 1986). Although calcium uptake by such membranes was stimulated by acute ethanol exposure, the degree of stimulation was considerably less after chronic ethanol feeding. Contrary to expectation, these findings could not be related to the acute and chronic effects of ethanol on membrane fluidity as measured using the fluorescent probe DPH. In non-alcoholic human subjects, acute alcohol ingestion did not alter calcium absorption (Verdy and Caron, 1973).

Magnesium. In alcoholics, magnesium absorption following an oral load of magnesium aspartate was not reduced. However, magnesium depletion occurs in many subjects with chronic alcohol abuse. (Flink *et al.*, 1971). In addition to insufficient dietary magnesium intake (Fisher *et al.*, 1967; Lim and Jakob, 1972) and increased magnesuria (Flink *et al.*, 1971), intestinal loss of magnesium either by vomiting or by diarrhea may contribute to magnesium depletion in alcoholics.

Iron. The duodenal transport of iron in the rat is an active process (Dowdle *et al.*, 1960b). The effect of ethanol on iron absorption remains an unsettled issue. In humans alcohol increased the absorption of ferric chloride in normal subjects but had no effect on ferrous ascorbate or hemoglobin iron (Charlton *et al.*, 1964).

In the rat it has been claimed that ethanol leads to a decrease of iron absorption (Wojcicki *et al.*, 1972). No changes were noted in total body uptake or in the intestinal uptake *in vitro* of radioactive iron after alcohol as a single dose or daily for 4–5 weeks (Murray and Stein, 1965). Others observed that acute ethanol decreased ferrous ion absorption from isolated duodenal loops of rats *in vivo* (Tapper *et al.*, 1968). It appears that the intestinal uptake of the more readily absorbed ferrous form of iron is not affected by acute or chronic alcohol ingestion, whereas the uptake of the less absorbable ferric ion might be stimulated.

Trace Elements. No change in urinary zinc excretion was noted after infusion of 1.09 M ethanol to normal human subjects or alcoholic patients (Sullivan, 1962). However, in chicks, ethanol decreased duodenal absorption of zinc and selenium *in vivo* (Hill, 1960). Chronic ethanol feeding in rats was associated with decreased ileal, but not duodenal, absorption of zinc (Antonson *et al.*, 1978). Absorption of manganese was increased by ethanol (Schafer *et al.*, 1974).

In conclusion, ethanol administration markedly depresses the transport of some substances and stimulates or has little or no effect on that of others (Table 10.2).

Acute Metabolic Effects of Ethanol on Mucosal Epithelial Cells. The cellular effects produced by ethanol could also be caused by metabolic imbalances secondary to its oxidation (see Chapter 7).

Because of its low K_m, ADH should be virtually saturated at ethanol concentrations of about 30 mg/100 ml. Therefore, the effects secondary to the metabolic imbal

TABLE 10.2. Effects of Experimental Alcohol Administration on Intestinal Absorption in the Absence of Nutrition Deficiency[a]

Substance(s)	Single ethanol doses		Chronic ethanol administration	
	Rats	Humans	Rats	Humans
Glucose	↓	NS	↓	0
Xylose	NS	NS	↓	0 or ↑
L-Amino acids	↓	↓	NS	NS
D-Phenylalanine	0	NS	NS	NS
Iron	0 or ↓	0 or ↑	0	NS
Calcium	0 or ↓	0	↓	NS
Sodium, water	NS	0	NS	↓
Manganese	—	NS	NS	NS
Thiamine	—	0 or ↑	NS	NS
Folic acid	NS	0	0	0
Vitamin B$_{12}$	NS	NS	—	↓
Fat	—	NS	—	0 or ↑
Nitrogen	NS	NS	0	NS

[a]NS, not studied; ↑, increased; ↓, decreased; 0, unchanged; —, not measured.

ances produced by the oxidation of ethanol through this pathway should be apparent at this low ethanol concentration. The MEOS has a higher K_m and should be well saturated at about 125 mg/100 ml. Concentrations of ethanol sufficient to saturate both oxidative systems failed to demonstrate significant effects on intestinal lipid metabolism (Baraona *et al.*, 1975).

In support of the possibility that some effects of ethanol could be linked to its oxidation by the small intestine, pyrazole (a known inhibitor of ADH) has been reported to prevent stimulatory effects of ethanol on intestinal triglyceride synthesis (Carter *et al.*, 1971) and transport of manganese (Schafer *et al.*, 1974). However, subsequent studies suggest that pyrazole exerts effects on the intestine that are not linked to inhibition of ethanol oxidation (Carter and Isselbacher, 1973).

The cellular alterations produced by ethanol at the concentrations achieved in the small intestine appear generalized. Concentrations of ethanol of 1.6–3.2 g/100 ml have been shown to inhibit oxygen consumption both in isolated epithelial cells (Van Tuan *et al.*, 1971; Lopez del Pino *et al.*, 1983) and intestinal slices (Israel *et al.*, 1968). It has been argued that the effect on amino acid transport was not by direct interference with oxidative metabolism by ethanol, since cyanide had more profound effects on oxygen consumption for similar decreases in absorption (Israel *et al.*, 1968). Jejunal slices obtained from rats given an acute dose of alcohol prior to sacrifice also showed a depressed oxygen consumption even when incubated without ethanol (Baraona *et al.*, 1974). In addition, previous administration of an ethanol dose *in vivo* diminished the ability of the slices to survive *in vitro* (Baraona *et al.*, 1974), suggesting a more diffuse permanent alteration of the cell.

Intestinal ATP content has been found to be decreased after acute and chronic ethanol administration and after *in vitro* incubation with ethanol (Carter and Isselbacher, 1973). Further studies are needed to determine whether the inhibition of Na^+,K^+-ATPase activity produced by ethanol *in vitro* is of importance for the inhibition of active transport *in vivo*. A stimulation of adenylate cyclase by addition of 2.5% alcohol to jejunal tissue fraction was reported (Greene *et al.*, 1971).

Vesely and Levey (1977) also found an inhibition of guanylate cyclase by ethanol in various rat tissues, including ileum. However, the role of changes in cGMP levels on alterations in intestinal secretion induced by alcohol remains speculative.

The administration of a moderate dose of alcohol to normal volunteers decreased the activity of jejunal glycolytic and gluconeogenic enzymes within 24 hr (Greene *et al.*, 1974). On the other hand, ethanol increased the activity of pyruvate kinase. Folic acid, which is a potent stimulator of jejunal glycolytic enzymes, reversed this inhibitory action of ethanol. Conversely, the response of these activities to folate administration was decreased by the concomitant administration of ethanol (producing a 42–45% increase in the glycolytic activities and no change in the gluconeogenic activity instead of the 1.6- to twofold increase seen after folate alone). Folate and ethanol exerted a synergistic stimulation of pyruvate kinase activity (Greene *et al.*, 1974). It is not clear whether folate activity antagonized a reversible inhibitory effect of ethanol or acted on cell populations that remain unaffected by ethanol.

As in many other tissues, fatty acids are the preferred fuel for the small intestine (Hülsmann, 1971). In rats, intraduodenal infusion of a solution containing ethanol (3.4 g/100 ml) for 15 hr decreased the oxidation of intravenously injected [^{14}C]palmitate to water-soluble metabolites (including $^{14}CO_2$) in the small intestine (Gangl and Ockner, 1975). This effect was attributed to ethanol oxidation by the intestine. In *in vitro* studies, however, the incorporation of palmitate into $^{14}CO_2$ was inhibited by ethanol in concentrations of 2.5 g/100 ml or greater but was unaffected by concentrations of 0.5 and

1 g/100 ml, suggesting that this effect is not linked to ethanol oxidation (Baraona *et al.*, 1975).

A point of controversy concerns the effects of ethanol on intestinal triglyceride synthesis. Intestinal slices incubated with ethanol (2.5–2.6 g/100 ml) were reported to increase triglyceride synthesis by some investigators (Carter *et al.*, 1971) and to decrease it by others (Baraona *et al.*, 1975). In both studies the incorporation of [^{14}C]palmitate into triglyceride fractions isolated by thin-layer chromatography was measured, but in the second study (Baraona *et al.*, 1975) a much higher concentration of fatty acid was used. Under the latter conditions, the incorporation of labeled palmitate into intestinal triglycerides was progressively inhibited by ethanol concentrations from 1 to 5 g/100 ml, while an increasing amount of the fatty acid label was incorporated into a lipid band adjacent to that of triglycerides that was readily visible at high concentrations of both ethanol and fatty acids (Baraona *et al.*, 1975) (Table 10.3). This lipid fraction incorporated not only fatty acids but also equimolar amounts of labeled ethanol and was identified as ethylpalmitate by gas–liquid chromatography and by the splitting off of ethanol during saponification (Baraona *et al.*, 1975). The increased esterification of fatty acids with ethanol disappeared in boiled slices, suggesting that ethyl palmitate formation requires some enzymatic system either in the intestinal cells or in the contaminating luminal contents (Newsome and Rattray, 1965). Indeed, a specific enzyme catalyzing such a reaction has now been demonstrated in a variety of tissues. Contamination of triglycerides with ethyl fatty acid esters could account for as much as 50% apparent increase in triglyceride synthesis in the presence of high ethanol concentration. In any case, it now appears that ethanol has a depressive effect on triglyceride synthesis in humans. Over a range of ethanol concentrations (0.1–5.0%), it was shown that triglyceride synthesis was significantly inhibited when duodenojejunal biopsy specimens were incubated in a micellar lipid solution containing labeled oleate (Zimmerman *et al.*, 1986).

Under other conditions, triglyceride synthesis is increased after ethanol exposure. A number of studies have been performed in rats in the fasting state, 15 to 18 hr after intragastric administration of ethanol in a high dose (7.5 g/kg body weight) (Carter *et al.*, 1971; Middleton *et al.*, 1971) or after prolonged (5–15 hr) jejunal perfusion with high ethanol concentrations (3.4–15 g/100 ml) (Mistilis and Ockner, 1972; Gangl and Ockner, 1975; Rodgers and O'Brien, 1975). In these studies, increased mucosal triglyceride (Carter *et al.*,

TABLE 10.3.   Effects of Ethanol on Palmitate Oxidation
and Esterification by Intestinal Slices[a,b]

Ethanol concentration (g/100 ml)	[14C]Palmitate disappearance (dpm/mg tissue)	14CO2 production (dpm/mg tissue)	[14C]Triglyceride production (dpm/mg tissue)	[14C]Ethylpalmitate production (dpm/mg tissue)
0	1679 ± 98	246 ± 17	343 ± 46	0
0.5	1543 ± 50	202 ± 15	263 ± 34	91 ± 81
1.0	1466 ± 105	220 ± 20	98 ± 10[c]	15 ± 81
2.5	1500 ± 128	88 ± 10[c]	82 ± 14[c]	433 ± 116[c]
5.0	1808 ± 101	34 ± 4[c]	44 ± 9[c]	551 ± 120[c]

[a]From Baraona *et al.* (1975).

[b]Jejunal slices obtained from seven chow-fed rats (fasted overnight). The slices were incubated in Krebs phosphate medium containing a micellar solution of [14C]palmitate, monooleate, and taurocholate for 30 min, without ethanol. Each value represents the mean ± S.E. Identification based on chromatographically characteristic labeling with [14C]ethanol.

[c]$p < 0.01$, paired comparisons with slices incubated without ethanol.

1971; Mistilis and Ockner, 1972, Gangl and Ockner, 1975) and cholesterol (Middleton *et al.*, 1971) synthesis, enhanced activities of jejunal lipid-reesterifying enzymes (microsomal acyl-CoA synthetase and acyl-CoA: monoglyceride acyltransferase) (Rodgers and O'Brien, 1975), and increased lymph output of cholesterol (Middleton *et al.*, 1971) and triglycerides (Mistilis and Ockner, 1972) have been found after ethanol. However, in view of the rapid cell turnover of the small intestine, one wonders whether the observed effects can be attributed not only to direct actions of ethanol on intestinal lipid metabolism but also to cellular necrosis and repair as a consequence of the high ethanol concentrations (Maling *et al.*, 1967; Krawitt, 1974; Baraona *et al.*, 1974) or to increased intraluminal content of lipids and other nutrients derived from sloughed cells.

In summary, the ingestion of alcohol results in intestinal concentrations of ethanol that are much greater than those necessary to saturate all known pathways of ethanol oxidation. The intestinal effects observed at these high concentrations are most likely caused by a direct action of ethanol rather than by the metabolic disturbances associated with its local oxidation. These direct effects include inhibition of active transport and of a variety of metabolic activities, suggestive of diffuse cell damage.

Chronic Metabolic Effects of Ethanol. Striking changes in metabolic activities occur in the small intestine after chronic ethanol feeding along with nutritionally adequate diets. Intestinal slices obtained from rats that have been fed ethanol for several weeks showed increased oxygen consumption (Baraona *et al.*, 1974) and increased ability to oxidize fatty acids (Baraona *et al.*, 1975) and to synthesize triglycerides (Carter *et al.*,

1971; Baraona *et al.*, 1975). The effects of lipid metabolism were associated with increased activity of palmitoyl-CoA synthetase in the mucosal homogenates of ethanol-fed rats (Baraona *et al.*, 1975). This suggests that these rats may develop an increased intestinal capacity to form acyl-CoA esters, a step that is necessary for both fatty acid oxidation and triglyceride synthesis. These metabolic changes are opposite to those found 1 hr after acute administration of a high ethanol dose.

The proliferation of the smooth endoplasmic reticulum seen by electron microscopy after chronic ethanol consumption was found to have a striking biochemical counterpart. A threefold increase in intestinal microsomal cytochrome P-450 content was observed in the ethanol-fed rats compared to their pair-fed controls (Fig. 10.5). Intestinal benzo-α-pyrene hydroxylase activity was also increased threefold in the ethanol-fed rats compared to their pair-fed controls. By histochemical methods with a specific antibody, a significant increase in the ethanol-inducible P450IIE1 was demonstrated in the small intestine (Shimizu *et al.*, 1989). The implications with regard to carcinogenesis are discussed in Chapter 13.

Indirect Effects of Alcohol on Intestinal Function. Malabsorption in alcoholics is usually associated with a history of inadequate diet. Absorptive function improves after admission to the hospital with ingestion of a normal diet despite the continued administration of ethanol, suggesting that malnutrition is an important factor in the pathogenesis of malabsorption in alcoholics (Mezey, 1975; Lindenbaum and Lieber, 1975; Smith, 1973; Halsted *et al.*, 1971).

Multiple nutritional deficiency states occur in alcoholics. Of these, folate and protein deficiency, and

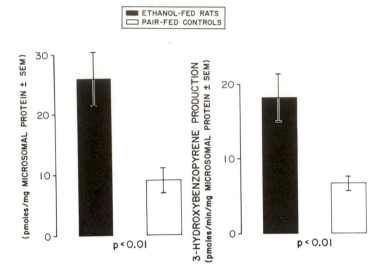

FIGURE 10.5. Microsomal cytochrome P-450 content and benzo-[a]-pyrene hydroxylase activity in the mucosa of the proximal small intestine after 25 days of ethanol administration in seven pairs of rats.

perhaps others, may adversely affect absorptive function.

Severe *folate deficiency* may cause morphological changes in the intestine (Bianchi *et al.*, 1970; Hermos *et al.*, 1972), including villus shortening, decreased mitoses in crypts, macrocytosis, and enlargement of epithelial cell nuclei. Similar changes occur in vitamin B_{12} deficiency. Folate deficiency can alter intestinal function prior to the production of morphological changes.

Studies in several subjects have also demonstrated a synergistic effect between alcohol and folic acid deficiency (Mekhjian and May, 1977; Halsted *et al.*, 1971). A folate-deficient diet augmented the abnormalities in Na^+ and water transport produced by ethanol (Mekhjian and May, 1977).

In alcoholics, there was a strong association between D-xylose, vitamin B_{12}, and folic acid malabsorption and evidence of folate deficiency (Halsted *et al.*, 1973b). Jejunal perfusion studies performed in binge-drinking alcoholics demonstrated reduced absorption of [³H]folic acid. However, this occurred only in subjects who were folate deficient but not when folate levels were normal (Carter and Isselbacher, 1971). In other studies, ethanol ingestion for 2 weeks in seven well-nourished volunteers did not reduce mean folate absorption (Halsted *et al.*, 1971). Importantly, in two subjects induction of dietary folate deficiency, together with ethanol ingestion, resulted in decreased absorption of labeled folic acid, D-xylose, glucose, fluid, and sodium, indicating the important role of folic acid deficiency in the etiology of folate and other nutrient malabsorption. Dietary folate deficiency appears to be a major contributing factor to the diffuse mucosal abnormality seen in alcoholics (Lindenbaum *et al.*, 1974).

Protein malnutrition is not uncommon in alcoholics because of dietary insufficiency and perhaps inhibition by alcohol of intestinal amino acid transport. The contribution of protein malnutrition to the diffuse mucosal abnormality of alcoholics has not been established (Lindenbaum *et al.*, 1974). However, in animals (Glickman and Kirsch, 1973) and in malnourished adults (Mayoral *et al.*, 1967) intestinal function is impaired by protein deficiency.

Severe protein malnutrition can produce a flat mucosa. Studies in protein-deficient children have shown malabsorption of xylose, glucose, and fat (James, 1968; Viteri *et al.*, 1973). The severity of steatorrhea in malnourished children is directly related to the degree of protein deficiency (Viteri *et al.*, 1973).

Protein deficiency also reduced pancreatic exocrine secretion in rats and man (Lemire and Iber, 1967; Tandon *et al.*, 1970), and subclinical protein malnutrition has been cited as a cause of reversible pancreatic insufficiency of alcoholics (Mezey and Potter, 1976). The exact role of protein deficiency is unknown, but it could contribute to malabsorption in the chronic alcoholic by altering both the liminal phase of absorption and mucosal function.

Malabsorption in alcoholics may result from disturbance of the liminal phase of digestion secondary to *pancreatic and hepatic dysfunction*. The presence of pancreatic steatorrhea in alcoholics leads to weight loss, diarrhea, and deficiency of fat-soluble vitamins and correlates with a low pancreatic lipase output following cholecystokinin-pancreozymin or secretin stimulation (Mezey and Potter, 1976). In addition, subclinical pancreatic dysfunction as assessed by an abnormal response to secretin stimulation was present in 44% of a group of alcoholics hospitalized because of their alcoholism (Mezey *et al.*, 1970). In most subjects, the test returned to normal after institution of a normal diet with or without alcohol, suggesting that nutritional factors (especially protein deficiency) may be an important cause of pancreatic dysfunction.

Cirrhosis is associated with decreased excretion of bile salts, subnormal concentration of intraduodenal bile salts, and altered micelles (Vlahcevic *et al.*, 1972; Badley *et al.*, 1970); altered micelles may have secondary effects on lipid absorption and further exacerbate caloric wastage.

In patients with cirrhosis the upper small bowel was said to be more often contaminated with coliforms and *Streptococcus faecalis* than in healthy volunteers (Martini *et al.*, 1957). However, other studies disagreed (Lal *et al.*, 1972; Cornet *et al.*, 1973; Portela-Gomez *et al.*, 1977).

The bacterial microflora in the jejunum of 29 chronic alcoholics was compared to 15 nonalcoholic subjects (Bode *et al.*, 1980). The incidence of positive cultures for several groups of anaerobic bacteria was distinctly higher in the alcoholic groups. The presence or absence of cirrhosis had no effect on the bacterial content in the alcoholics. In this study gastric pH differences between the alcohol and control groups appeared to be the causative factor (Bode *et al.*, 1980); the number of bacteria in the upper small intestine increase in direct proportion to increases in the gastric pH (King and Toskes, 1979).

10.5.3. Summary

Malabsorption is a common finding in alcoholics and cannot be attributed to a single cause. Both ethanol and associated complications of alcoholism contribute to its pathogenesis. The high ethanol concentrations to which the small intestine is exposed during drinking have direct deleterious effects on epithelial cell function and survival. These structural alterations may account for the defective absorption of nutrients that require active transport mechanisms. By contrast, fat absorption is unaffected by ethanol directly. Nutritional, pancreatic, and hepatic complications of alcoholism are the most likely causes of fat malabsorption. Folate deficiency appears to play a major role in the diffuse mucosal dysfunction seen in chronic alcoholics, whereas the role of protein deficiency is not clear.

10.6. Effects of Alcohol on the Colon

It has been stated that the only lesion of the lower gastrointestinal tract known to be related to alcoholism is hemorrhoids (Williams, 1975). This condition may accompany the portal hypertension found in patients with cirrhosis. Although the hemorrhoidal plexuses communicate with the portal system, significant bleeding from hemorrhoids is rarely a problem in portal hypertension (Goldberg *et al.*, 1980; Taylor, 1954). However, varices of the colon may be an unexpected cause of massive rectal bleeding in patients with underlying cirrhosis. Barium x rays are not as useful as would be expected in diagnosing varices of the colon, in part because rarity of the lesion lowers suspicion. The varices can appear as lucent mural filling defects in a linear or cobblestone pattern on barium enema (Brill *et al.*, 1969). These defects may be mistaken for carcinoma or artifact (Fleming and Seaman, 1968). The varices have been reported in the cecum and ascending and descending colon but are more common in the rectum (Levy *et al.*, 1957; Feldman *et al.*, 1962; Fleming and Seaman, 1968; Brill *et al.*, 1969; Doberneck and Jajonski, 1970). The diagnosis can be established by direct intraluminal visualization with sigmoidoscopy or colonoscopy (Geboew *et al.*, 1975; Pickens and Tedesco, 1980) or via the venous phase of mesenteric arteriography.

In addition to the above colonic effects, which are secondary to portal hypertension, alcohol has been shown to have direct effects on colonic motility. Acute alcohol administration (200 ml of 20% alcohol), either by stomach tube or by intravenous infusion, depressed nonpropulsive motility of the colon in dogs, whereas the propulsive motility was increased; a similar effect of alcohol on the propulsive motility of the distal colon was observed in otherwise healthy patients with colostomies. The gastrocolic reflex was augmented following a meal 2.5 hr after alcohol.

The effect of acute alcohol administration on colon morphology in humans has also been studied (Brozinsky *et al.*, 1978). Histological and ultrastructural examination of rectal mucosa was carried out in 11 patients with a history of chronic alcoholism who had also recently consumed ethanol. Goblet cells were noted to be decreased, and a dense mononuclear cell infiltrate was seen on light microscopy. Electron microscopy revealed swollen, distorted mitochondria and dilated and vesicular endoplasmic reticulum. These abnormalities were shown to disappear after a 2-week abstinent period. Serum folate levels were normal in ten of the 11 patients.

Increased incidence of colonic cancer has also been reported in the alcoholic. This is discussed in Chapter 13.

REFERENCES

Althausen, T. L., Uyeyama, K., and Loran, M. R.: Effects of alcohol on absorption of vitamin A in normal and in gastrectomized subjects. *Gastroenterology* **38**:942–945, 1960.

Angel, F., Sava, P., Crenner, F., Lambert A., and Genier, J. F.: Changes in intestinal motility after intravenous injection of alcohol. *C. R. Soc. Biol. (Paris)* **174**:192–198, 1980.

Antonson, D. L., Barak, A. J., and Vanderhood, J. A.: Effect of acute and chronic ethanol ingestion on zinc absorption. *Clin. Res.* **26**:660A, 1978.

Augur, N. A.: Gastric mucosal blood flow following damage by ethanol, acetic acid or aspirin. *Gastroenterology* **58**:311–320, 1970.

Badley, B. W. D., Murphy, G. M., Bouchier, I. A. D., and Sherlock, S.: Diminished micellar phase lipid in patients with chronic nonalcoholic liver disease and steatorrhea. *Gastroenterology* **58**:781–789, 1970.

Balaghi, M., and Neal, R. A.: Effect of chronic ethanol administration on thiamine metabolism in the rat. *J. Nutr.* **107**:2144–2152, 1977.

Ballard, H. J., Wilkes, J. M., and Hirst, B. H.: Effect of alcohols on gastric and small intestinal apical membrane integrity and fluidity. *Gut* **29**:1648–1655, 1988.

Baraona, E., and Lieber, C. S.: Intestinal lymph formation and fat absorption: Stimulation by acute ethanol administration and inhibition by chronic ethanol feeding. *Gastroenterology* **68**: 495, 1975.

Baraona, E., and Lindenbaum, J.: Metabolic aspects of alcohol on the intestine. In: *Metabolic Aspects of Alcoholism* (C. S. Lieber ed.), Baltimore, MD: University Park Press, 1977, pp. 81–116.

Baraona, E., Pirola, R. C., and Lieber, C. S.: Small intestinal damage and changes in cell population produced by ethanol ingestion in the rat. *Gastroenterology* **66**:226–234, 1974.

Baraona, E., Pirola, R. C., and Lieber, C. S.: Acute and chronic effects of ethanol on intestinal lipid metabolism. *Biochim. Biophys. Acta* **388**:19–28, 1975.

Barboriak, J. J., and Meade, R. C.: Impairment of gastrointestinal processing of fat and protein by ethanol in rats. *J. Nutr.* **98**: 373–377, 1969.

Barboriak, J. J., and Meade, R. C.: Effect of alcohol on gastric emptying in man. *Am. J. Clin. Nutr.* **23**:1151–1153, 1970.

Beazell, J. M., and Ivey, A. C.: The influence of alcohol on the digestive tract. *J. Stud. Alcohol* **1**:45–73, 1940.

Beck, I. T., and Dinda, P. K.: Acute exposure of small intestine to ethanol. Effects on morphology and function. *Dig. Dis. Sci.* **26**:817–838, 1981.

Beck, I. T., Morris, G. P., and Buell, M. G.: Ethanol-induced vascular permeability changes in the jejunal mucosa of the dog. *Gastroenterology* **90**:1137–1145, 1986.

Becker, H. D., Reeder, D. D., and Thompson, J. C.: Gastrin release by ethanol in man and in dogs. *Ann. Surg.* **179**:906–909, 1974.

Belber, J. P.: Gastroscopy and duodenoscopy. In: *Gastrointestinal Disease* (M. H. Sleisenger and J. S. Fordtran eds.), Philadelphia, W. B. Saunders, 1978, pp. 691–713.

Bell, A. E., Sellers, L. A., Allen, A., Cunliffe, W. J., Morris, E. R., and Ross-Murphy, S. B.: Properties of gastric and duodenal mucus: Effect of proteolysis, disulfide reduction, bile, acid, ethanol, and hypertonicity on mucus gel structure. *Gastroenterology* **88**:269–280, 1985.

Berenson, M. M., and Avner, D. L.: Alcohol inhibition of rectosigmoid mutability in humans. *Digestion* **22**:210–215, 1981.

Bianchi, A., Chipman, D. W., Dreskin, A., and Rosensweig, N. S.: Nutritional folic acid deficiency with megaloblastic changes in the small-bowel epithelium. *N. Engl. J. Med.* **282**:859–861, 1970.

Biggerstaff, R. J., and Leitch, G. J.: Effects of ethanol on electrical parameters of the *in vivo* rat stomach. *Dig. Dis.* **22**:1064–1068, 1977.

Bikle, D. D., Genant, H. K., Cann, C., Recker, R. R., Halloran, B. P., and Strewler, G. J.: Bone disease in alcohol abuse. *Ann. Intern. Med.* **103**:42–48, 1985.

Bikle, D. D., Gee, E. A., and Munson, S. J.: Effect of ethanol on intestinal calcium transport in chicks. *Gastroenterology* **91**:870–876, 1986.

Bjarnason, I., Ward, K., and Peters, T. J.: The leaky gut of alcoholism: Possible route of entry for toxic compounds. *Lancet* **1**:179–182, 1984.

Bode, J. Ch.: Alcoholhepatitis und Alkohoizirrhose-Klinik, Begleiterkrankungen und Therapie. In: *Leberversagen* (J. Ch. Bode, P. Eckert, M. Fischer, and O. Zelder, eds.), Stuttgart, Thieme, 1980.

Bode, J. Ch. Heidelbach, R., Bode, Ch., Mannheim, W., Durr, H. K., and Martini, C. A.: Bacterial microflora in the jejunum of chronic alcoholics, alcohol and the gastrointestinal tract. *INSERM* **95**:451–456, 1980.

Bo-Linn, G. W., and Shanbour, L. L.: Effects of antral ethanol on gastric acid secretion, potential difference and serum gastrin. *Proc. Soc. Exp. Biol. Med.* **155**:594–598, 1977.

Bouquillon, M.: Effects of acute ethanol ingestion on fat absorption. *Lipids* **11**:848–852, 1976.

Brassinne, A.: Effects of ethanol on plasma protein shedding in the human stomach. *Dig. Dis. Sci.* **24**:44–47, 1979.

Breen, K. J., Buttigieg, R., Iossifidis, S., Lourensz, C., and Wood, B.: Jejunal uptake of thiamine hydrochloride in man: Influence of alcoholism and alcohol. *Am. J. Clin. Nutr.* **42**:121–126, 1985.

Breslow, N. E., and Enstrom, J. E.: Geographic correlations between cancer mortality rates and alcohol-tobacco consumption in the United States. *J. Natl. Cancer Inst.* **53**:631–639, 1974.

Brill, D. R., Bolasny, B., and Vix, V. A.: Colonic varices. *Am. J. Dig. Dis.* **14**:801–804, 1969.

Brozinsky, S., Fani, K., Grosberg, S. J., and Wapnick, S.: Alcohol ingestion-induced changes in the human rectal mucosa: Light and electron microscopic studies. *Dis. Colon Rectum* **21:**329–335, 1978.

Bucher, P., Lepsien, G., Sonnenberg, A., and Blum, A. L.: Verlaufund prognose der Refluxkrankheit bei Konservativer und chirurgischer Behandlung. *Schweiz. Med. Wochenschr.* **108:**2072–2078, 1978.

Buell, M. G., and Beck, I. T.: Effect of ethanol on jejunal regional blood flow in the rabbit. *Gastroenterology* **84:**81–89, 1983.

Buell, M. G., and Beck, I. T.: Ethanol-induced mucosal microvascular stasis and enhanced plasma protein loss in the dog jejunum. *Gastroenterology* **86:**413–420, 1984.

Buell, M. G., Binda, P. K., and Beck, I. T.: Effect of ethanol on morphology and total, capillary and shunted blood flow of different anatomical layers of dog jejunum. *Dig. Dis. Sci.* **28:**1005–1017, 1983.

Bungert, J. H.: Absorption of hemoglobin and hemoglobin iron in alcohol-induced liver injury. *Digestion* **9:**293–308, 1973.

Bynum, T. E., and Jacobson, E. D.: Blood flow and gastrointestinal function. *Gastroenterology* **60:**325–335, 1971.

Carstensen, H. E., Bülow, S., Hansen, O. H., Jakobsen, B. H., Krapup, T., Pedersen, T., Raahave, D., Svendsen, L. B., and Backer, O.: Cimetidine for severe gastroduodenal hemorrhage: A randomized controlled trial. *Scand. J. Gastroenterol.* **15:**103–105, 1980.

Carter, E. A., and Isselbacher, K. J.: The metabolism of ethanol to carbon dioxide by stomach and small intestinal slices. *Proc. Soc. Exp. Biol. Med.* **138:**817–819, 1971.

Carter, E. A., and Isselbacher, K. J.: Effect of ethanol on intestinal adenosine triphosphate (ATP) content. *Proc. Soc. Exp. Biol. Med.* **142:**1171–1173, 1973.

Carter, E. A., Drummey, G. D., and Isselbacher, K. J.: Ethanol stimulates triglyceride synthesis by the intestine. *Science* **174:**1241–1247, 1971.

Castenfors, H., Hultman, E., and Josephson, B.: Effect of intravenous infusions of ethyl alcohol on estimated hepatic blood flow in man. *J. Clin. Invest.* **39:**776–781, 1960.

Chang, T., Lewis, J., and Glazko, A. J.: Effects of ethanol and other alcohols on the transport of amino acids and glucose by everted sacs of rat small intestine. *Biochim. Biophys. Acta* **135:**1000–1007, 1967.

Charlton, R. W., Jacobs, P., Seftel, H., and Bothwell, T. H.: Effect of alcohol on iron absorption. *Br. Med. J.* **2:**1427–1429, 1964.

Cheung, L. Y., Moody, F. G., and Reese, R. S.: Effect of aspirin, bile salt and ethanol on canine mucosal blood flow. *Surgery* **77:**786–792, 1975.

Chey, W. Y., Kusakcioglu, O., Dinoso, V., and Lorber, S. H.: Gastric secretion in patients with chronic pancreatitis and in chronic alcoholics. *Arch. Intern. Med.* **122:**399–403, 1968.

Chey, W. Y., Kosay, S., and Lorber, S. H.: Effects of chronic administration of ethanol on gastric secretion of acid in dogs. *Dig. Dis.* **17:**153–159, 1972.

Chin, J. H., and Goldstein, D. B.: Effects of low concentrations of ethanol on the fluidity of spin-labeled erythrocyte and brain membranes. *Mol. Pharmacol.* **13:**435–441, 1977.

Chowdhury, A. R., Malmud, L. S., and Dinoso, V. P.: Gastrointestinal plasma protein loss during ethanol ingestion. *Gastroenterology* **72:**37–40, 1977.

Chung, R. S. K., Johnson, G. M., and Denbesten, L.: Effect of sodium taurocholate and ethanol on hydrogen ion absorption in rabbit esophagus. *Dig. Dis.* **22:**582–588, 1977.

Cohen, M. M., Bowdler, R., Gervais, P., Morris, G. P., and Wang, H.-R.: Sucralfate protection of human gastric mucosa against acute ethanol injury. *Gastroenterology* **96:**292–298, 1989.

Cooke, A. R., and Turtle, J. R.: Serum gastrins in response to various stimuli. *Clin. Res.* **18:**679, 1970 (abstract).

Cornet, A., Hartmann, L., Courtois, J.-E., Dubrisay, J., Demelier, J.-F., Barbier, J.-Ph., Renault, P., Carnot, F., Debetz, J., and Dadounk, J. P.: Chirrhose du foie et intestin grele confrontations cliniques, biologiques, histologiques, immunologiques et bactériologiques. *Sem. Hop. Paris* **49:**1639–1648, 1973.

Davenport, H. W.: Gastric mucosal injury by fatty and acetylsalicylic acids. *Gastroenterology* **46:**245–253, 1964.

Davenport, H. W.: Ethanol damage to canine oxyntic glandular mucosa. *Proc. Soc. Exp. Biol. Med.* **126:**657–662, 1967a.

Davenport, H. W.: Salicylate damage to the gastric mucosal barrier. *N. Engl. J. Med.* **276:**1307–1312, 1967b.

Daves, I. A., Miller, J. H., Lemmi, C. A., and Thompson, J. C.: Mechanism and inhibition of alcohol-stimulated gastric secretion. *Surg. Forum* **16:**305–307, 1965.

Dawson, D. C., and Cooke, A. R.: Parallel pathways for ion transport across rat gastric mucosa: Effect of ethanol. *Am. J. Physiol.* **235:**E7–15, 1978.

Demol, P., Singer, M. V., Hotz, J., Eysselein, V. E., and Goebell, H.: Different actions of intravenous ethanol on basal (= interdigestive) secretion of gastric acid, pancreatic enzymes and bile acids and gastrointestinal motility in man. *Alcohol Alcoholism* **20:**19–26, 1985.

Demol, P., Singer, M. V., Hotz, J., Hoffmann, U., Hanssen, L. E., Eysselein, V. E., and Goebell, H.: Action of intragastric ethanol on pancreatic exocrine secretion in relation to the interdigestive gastrointestinal motility in humans. *Arch. Int. Physiol. Biochim.* **94:**251–259, 1986.

DeSchepper, P. J., Tjandramaga, T. B., DeRoo, M., Verhaest, L., Daurio, C., Steelman, S. L., and Tempero, K. F.: Gastrointestinal blood loss after diflunisol and after aspirin: Effect of ethanol. *Clin. Pharmacol. Ther.* **23:**669–676, 1978.

Desmond, P. V., Lourenza, C., and Breen, K. J.: Thiamine hydrochloride absorption in man: Normal kinetics and absence of acute effect of ethanol. *Aust. N.Z. J. Med.* **6:**264, 1976.

Diel, F., and Szabo, S.: Dose-dependent effects of linear and cyclic somatostatin on ethanol-induced gastric erosions: The role of mast cells and increased vascular permeability in the rat. *Regul. Peptides* **13:**235–243, 1986.

Dinda, P. K., and Beck, I. T.: On the mechanism of the inhibitory effect of ethanol on intestinal glucose and water absorption. *Am. J. Dig. Dis.* **22:**529–533, 1977.

Dinda, P. K., and Beck, I. T.: Effects of ethanol on cytoplasmic peptidases of the jejunal epithelial cell of the hamster. *Dig. Dis. Sci.* **29:**46–55, 1984.

Dinda, P. K., Beck, I. T., Beck, M., and McElliott, T. F.: Effect of ethanol on sodium dependent glucose transport in the small intestine of the hamster. *Gastroenterology* **68:**1517–1526, 1975.

Dinoso, V. P., Chey, W. Y., Siplet, H., and Lorber, S. H.: Effects of ethanol on the gastric mucosa of the Heidenhain pouch of dogs. *Am. J. Dis. Dis.* **15:**809–817, 1970.

Dinoso, V. P., Chey, W. Y., Braverman, S. P., Rosen, A. P., Ottenberg, D., and Lorber, S. H.: Gastrin secretion and gastric mucosal morphology in chronic alcoholics. *Arch. Intern. Med.* **130**:715–719, 1972.

Dinoso, V. P., Meshkipour, H., and Lorber, S. H.: Gastric mucosal morphology and faecal blood loss during ethanol ingestion. *Gut* **14**:289–292, 1973.

Dinoso, V. P., Chuang, J., and Murthy, S. N. S.: Changes in mucosal and venous histamine concentrations during instillation of ethanol in the canine stomach. *Dig. Dis.* **21**:93–97, 1976a.

Dinoso, V. P., Ming, S.-C., and McNiff, J.: Ultrastructural changes of the canine gastric mucosa after topical application of graded concentrations of ethanol. *Dig. Dis.* **21**:626–632, 1976b.

DiPadova, C., Worner, T. M., Julkunen, R. J. K., and Lieber, C. S.: Effects of fasting and chronic alcohol consumption on the first-pass metabolism of ethanol. *Gastroenterology* **92**:1169–1173, 1987.

Doberneck, R. D., and Janovski, N. A.: Isolated bleeding from colonic varices in patients with liver disease. *Am. J. Dig. Dis.* **15**:834–841, 1970.

Dowdle, E. B., Schachter, D., and Schenker, H.: Active transport of Fe^{59} by everted segments of rat duodenum. *Am. J. Physiol.* **198**:609–613, 1960.

Dragstedt, C. A., Gray, J. S., and Lawton, A. H.: Does alcohol stimulate gastric secretion by liberating histamine? *Proc. Soc. Exp. Biol. Med.* **43**:26–28, 1940.

Draper, L. R., Gyure, L. A., Hall, J. G., and Robertson, D.: Effect of alcohol on the integrity of the intestinal epithelium. *Gut* **24**:399–404, 1983.

Dupuy, D., and Szabo, S.: Protection by metals against ethanol-induced gastric mucosal injury in the rat. *Gastroenterology* **91**:966–974, 1986.

Dürr, H. K., Bode, J. C., Gieseking, R., Haase, R., Arnim, I., and Beckmann, B.: Anderungen der exokrinen Funktion der Glandula parotis und des Pankreas bei Patienten mit Leberzirrhose und chronischem Alkoholismus. *Munch. Med. Wochenschr.* **117**:841, 1975.

Dutt, B., and Magee, D. F.: Pepsin secretion by Heidenhain pouches in dogs. *Am. J. Physiol.* **223**:480–483, 1972.

Dutta, S. K., Dukehart, M., Narang, A., and Latham, P. S.: Functional and structural changes in parotid glands of alcoholic cirrhotic patients. *Gastroenterology* **96**:510–518, 1989.

Eastwood, G. L.: Does early endoscopy benefit the patient with active upper gastrointestinal bleeding? *Gastroenterology* **72**:737–739, 1977.

Eastwood, G. L., and Erdmann, K. R.: Effect of ethanol on canine gastric epithelial ultrastructure and transmucosal potential difference. *Dig. Dis.* **23**:429–435, 1978.

Eastwood, G. L., and Kirchner, J. P.: Changes in the fine structure of mouse gastric epithelium produced by ethanol and urea. *Gastroenterology* **67**:71–84, 1974.

Eysselein, V. E., Singer, M. V., Wentz, H., and Goebell, H.: Action of ethanol on gastrin release in the dog. *Dig. Dis. Sci.* **29**:12–18, 1984.

Fast, B. B., Wolfe, S. J., Stormont, J. M., and Davidson, C. S.: Fat absorption in alcoholics with cirrhosis. *Gastroenterology* **37**:321–324, 1959.

Feldman, M., Smith, J. M., and Warner, C. G.: Varices of the colon: Report of these cases. *J.A.M.A.* **179**:729–730, 1962.

Field, M., Brasitus, T. A., Sheerin, H. E., and Kimberg, D. V.: The role of cyclic nucleotides in the regulation of intestinal ion transport. In: *Stimulus Secretion Coupling in the Gastrointestinal Tract* (R. M. Case and H. Goebell, eds.), Lancaster, MTP Press, 1976, pp. 387–389.

Findlay, J., Sellers, E., and Foster, G.: Lack of effect of alcohol on small intestine binding of the vitamin B$_{12}$-intrinsic factor complex. *Can. J. Physiol. Pharmacol.* **54**:469–476, 1976.

Fischer, R. A., Ellison, C. W., Thayer, W. R., Spiro, H. M., and Glaser, C. W.: Esophageal motility in neuromuscular disorders. *Ann. Intern. Med.* **63**:229–248, 1965.

Fisher, J. M., Mackay, J. R., Taylor, K. B., and Ungar, B.: An immunological study of categories of gastritis. *Lancet* **1**:176–180, 1967.

Fleming, R. J., and Seaman, W. B.: Roentgenographic demonstration of unusual extra esophageal varices. *Am. J. Roentgenol.* **103**:281–290, 1968.

Flink, E. B., Lizarralde, G., and Jacobs, W.: Therapy of delirium tremens with urea (or dexamethasone) and magnesium sulfate. *J. Lab. Clin. Med.* **78**:990, 1971.

Fox, J. E., Bourdages, R., and Beck, I. T.: Effect of ethanol on glucose and water absorption in hamster jejunum *in vivo*. *Am. J. Dig. Dis.* **23**:193–200, 1978.

Fromm, D.: Gastric mucosal "barrier." *Gastroenterology* **77**:396–398, 1979.

Fromm, D., and Robertson, R.: Effects of alcohol on ion transport by isolated gastric and esophageal mucosa. *Gastroenterology* **70**:220–225, 1976.

Gangl, A., and Ockner, R. K.: Intestinal metabolism of plasma free fatty acids. Intracellular compartmentation and mechanism of control. *J. Clin. Invest.* **55**:803, 1975.

Geall, M. G., Phillips, S. F., and Summerskill, W. H. J.: Profile of gastric potential difference in man. *Gastroenterology* **58**:437–443, 1970.

Geboew, K., Broeckaert, L., and Vantrappen, C.: Varices of the colon. Diagnosis by colonoscopy. *Gastrointest. Endosc.* **22**:43–45, 1975.

Ghirardi, P., Marzo, A., Sardini, D., and Marchetti, G.: Changes in intestinal absorption of glucose in rats treated with ethanol. *Experientia* **27**:61–62, 1971.

Gimson, A. E. S., Westaby, D., Hegarty, J., Watson, A., and Williams, R.: A randomized trial of vasopressin and vasopressin plus nitroglycerin in the control of acute variceal hemorrhage. *Hepatology* **6**:410–413, 1986.

Glickman, R. M., and Kirsch, K.: Lymph chylomicron formation during protein synthesis inhibition. Studies of chylomicron apoproteins. *J. Clin. Invest.* **52**:2910–2920, 1973.

Glowacka, D., Kopacz-Jodczyk, T., Mazurkiewicz-Kilczewska, D., and Gindzienski, A.: Effect of ethyl alcohol on the biosynthesis of glucosamine and galactosamine in the human gastric mucosal tissue *in vitro*. *Biochem. Med.* **11**:199–204, 1974.

Goldberg, S. M., Gordon, P. H., and Nivatvongs, S.: *Essentials of Anorectal Surgery*. Philadelphia, J. B. Lippincott, 1980, pp. 83–85.

Gottfried, E. B., Korsten, M. A., and Lieber, C. S.: Gastritis and duodenitis induced by alcohol: An endoscopic and histologic assessment. *Gastroenterology* **70**:890, 1976.

Gottfried, E. B., Korsten, M. A., and Lieber, C. S.: Alcohol-induced gastric and duodenal lesions in man. *Am. J. Gastroenterol.* **70**:587–592, 1978.

Graham, D. Y., and Schwartz, J. T.: The spectrum of the Mallory-Weiss tear. *Medicine* **57**:307–318, 1978.

Green, P. H. R.: Drugs, alcohol and malabsorption. *Am. J. Med.* **67**:1066–1076, 1979.

Greene, H. L., Herman, R. H., and Kraemer, S.: Stimulation of jejunal adenyl cyclase by ethanol. *J. Lab. Clin. Med.* **78**:336–342, 1971.

Greene, H. L., Stifel, F. B., Herman, R. H., Herman, Y. F., and Rosensweig, N. S.: Ethanol-induced inhibition of human intestinal enzymes: Reversal by folic acid. *Gastroenterology* **67**:434–440, 1974.

Guth, P. H., Paulsen, G., and Nagata, H.: Histologic and microcirculatory changes in alcohol-induced gastric lesions in the rat: Effect of prostaglandin cytoprotection. *Gastroenterology* **87**:1083–1090, 1984.

Haddad, H., and Devroede-Bertrand, G.: Large bowel motility disorders. *Med. Clin. North Am.* **65**:1377–1396, 1981.

Haggard, H. W., Greenberg, L. A., and Lolli, G.: The absorption of alcohol with special reference to its influence on the concentration of alcohol appearing in the blood. *Q. J. Stud. Alcohol* **1**:684–726, 1941.

Hajjar, J. J., Krasner, K., Van-Linda, B., and Scheig, R.: Effect of ethanol on glycylsarcosine absorption. *Life Sci.* **29**:1402–1407, 1981.

Halsted, C. H., Robles, E. A., and Mezey, E.: Decreased jejunal uptake of labelled folic acid (H-PCA) in alcoholic patients roles of alcohol and nutrition. *N Engl. J. Med.* **285**:701–706, 1971.

Halsted, C.. H., Robles, E. A., and Mezey, E.: Distribution of ethanol in the human gastrointestinal tract. *Am. J. Clin. Nutr.* **26**:831–834, 1973.

Halsted, C. H., Robles, A., and Mezey, E.: Intestinal malabsorption in folate-deficient alcoholics. *Gastroenterology* **64**:526–532, 1973b.

Halsted, C. H., Bhanthumnavin, K., and Mezey, E.: Jejunal uptake of tritiated folic acid in the rat studied by *in vivo* perfusion. *J. Nutr.* **104**:1674–1680, 1974.

Harichaux, P., Capron, J.-P., Lienard, J., and Freville, M.: Influence de l'ethanol sur l'evacuation gastrique. Etude clinque et experimentale. *Lille Med.* **15**:1059–1065, 1970.

Herbert, V., Zalusky, R., and Davidson, C. S.: Correlation of folate deficiency with alcoholism and associated macrocytosis, anemia, and liver disease. *Ann. Intern. Med.* **58**:977–988, 1963.

Hermos, J. A., Adams, W. H., Liu, Y. K., Sullivan, L. W., and Trier, J. S.: Mucosa of the small intestine in folate-deficient alcoholics. *Ann. Intern. Med.* **76**:957–965, 1972.

Hill, C. H.: The effect of alcohols on absorption and tissue uptake of trace elements. *Fed. Proc.* **28**:300, 1960.

Hirschowitz, B. I., Pollard, H. M., Hartwell, S. W., and London, J.: The action of ethyl alcohol on gastric acid secretion. *Gastroenterology* **30**:244–253, 1956.

Hogan, W. J., Viegas De Andrade, S. R., and Winship, D. H.: Ethanol induced acute esophageal motor dysfunction. *J. Appl. Physiol.* **32**:755–760, 1972.

Hollander, D., Tarnowski, A., Krause, W. J., and Gergely, H.: Protective effect of sucralfate against alcohol-induced gastric mucosal injury in the rat. Macroscopic, histologic, ultrastructural, and functional time sequence analysis. *Gastroenterology* **88**:366–374, 1988.

Hollander, F.: Two-component mucous barrier: Its activity in protecting gastroduodenal mucosa against peptic ulceration. *Arch. Intern. Med.* **93**:107–120, 1954.

Hollands, D., and Murachidhara, K. S.: Vitamin A, intestinal absorption *in vivo*. Influence of luminal factors on transport. *Am. J. Physiol.* **232**:471–477, 1977.

Hosking, S. W., Kennedy, H. J., Seddon, I., and Triger, D. R.: The role of propranolol in congestive gastropathy of portal hypertension. *Hepatology* **7**:437–441, 1987.

Hoyumpa, A. M.: Mechanisms of vitamin deficiencies in alcoholism. *Alcoholism: Clin. Exp. Res.* **10**:573–581, 1986.

Hoyumpa, A. M., Middleton, H., Wilson, F., and Schenker, S.: Dual system of thiamine transport: Characteristics and effect of ethanol. *Gastroenterology* **66**:714, 1974.

Hoyumpa, A. M., Jr., Middleton, H. M., Wilson, F. A., and Schenker, S.: Thiamine transport across the rat intestine. I. Normal characteristics. *Gastroenterology* **65**:1218–1227, 1975.

Hoyumpa, A. M., Jr., Breen, K. J., Schenker, S., and Wilson, F. A.: Thiamine transport across the rat intestine. II. Effect of ethanol. *J. Lab. Clin Med.* **86**:803–816, 1975.

Hoyumpa, A. M., Nichols, S. G., Wilson, F. A., and Schenker, S.: Effect of ethanol on intestinal (Na,K)ATPase and intestinal thiamine transport in rats. *J. Lab. Clin. Med.* **90**:1086–1095, 1977.

Hoyumpa, A. M., Nichols, S. G., and Schenker, S.: Intestinal thiamine transport effect of chronic ethanol administration in rats. *Am. J. Clin. Nutr.* **31**:938–945, 1978.

Hufnagel, H., Bode, C., Bode, J. C., and Lehmann, F.-G.: Damage of rat small intestine induced by ethanol. *Res. Exp. Med. (Berl.)* **178**:65–70, 1980.

Hülsmann, W. C.: Preferential oxidation of fatty acids by rat small intestine. *FEBS Lett.* **17**:35–38, 1971.

Hunt, J. N., and Pathak, J. D.: The osmotic effects of some simple molecules and ions on gastric emptying. *J. Physiol. (Lond.)* **154**:254–269, 1960.

Hunter, C. K., Treanor, L. L., Gray, J. P., Halter, S. A., Hoyumpa, A., and Wilson, F. A.: Effects of ethanol *in vitro* on rat intestinal brush border membranes. *Biochim. Biophys. Acta* **732**:256–266, 1983.

Insunza, I., and Ugarte, G.: Steatorrhea in alcoholics with and without liver cirrhosis. *Rev. Med. Chil.* **98**:669–673, 1970.

Irvine, W. T., Watkin, D. B., and Williams, E. J.: The mechanism by which alcohol stimulates acid secretion. *Gastroenterology* **39**:41–47, 1960.

Israel, Y., Salazar, I., and Rosenmann, E.: Inhibitory effects of alcohol on intestinal amino acid transport *in vivo* and *in vitro*. *J. Nutr.* **96**:499–504, 1968.

Israel, Y., Valenzuela, J. E., Salazar, I., and Ugarte, G.: Alcohol and amino acid transport in the human small intestine. *J. Nutr.* **98**:222, 1969.

Ito, S., and Lacy, E. R.: Morphology of rat gastric mucosal damage, defense, and restitution in the presence of luminal ethanol. *Gastroenterology* **88**:250–260, 1985.

James, W. P. T.: Intestinal absorption in protein caloric malnutrition. *Lancet* **1**:333–335, 1968.

Jian, R., Cortot, A., Ducrot, F., Jobin, G., Chayvialle, J. A., and Modigliani, R.: Effect of ethanol ingestion on postprandial gastric emptying and secretion, biliopancreatic secretions, and duodenal absorption in man. *Dig. Dis. Sci.* **31:**604–614, 1986.

Joske, R. A., Fineki, E. S., and Wood, L. J.: Gastric biopsy: A study of 1000 consecutive biopsies. *Q. J. Med.* **48:**269–294, 1955.

Karppanen, H. O., and Puurunen, J.: Ethanol, indomethacin and gastric acid secretion in the rat. *Eur. J. Pharmacol.* **35:**221–223, 1976.

Katz, D., Pitchumoni, C. S., Thomas, E., and Antonelle, M.: The endoscopic diagnosis of upper-gastrointestinal hemorrhage. *Dig. Dis.* **21:**182–189, 1976.

Kaufman, S. E., and Kaye, M. D.: Induction of gastro-esophageal reflux by alcohol. *Gut* **19:**336–338, 1978.

Kaufman, S. E., and Kaye, M. D.: Effect of ethanol upon gastric emptying. *Gut* **20:**688–692, 1979.

Kawashima, K., and Glass, G. B. J.: Alcohol injury to gastric mucosa in mice and its potentiation by stress. *Dig. Dis.* **20:**162–172, 1975.

Keshavarzian, A., Dangleis, M., Wobbleton, J., Cornish, R., and Iber, F.: Small bowel transit and lactose intolerance in chronic alcoholics. *Gastroenterology* **88:**1444, 1985.

Keshavarzian, A., Iber, F., and Ferguson, Y.: Esophageal manometry and radionuclide emptying in chronic alcoholics. *Gastroenterology* **92:**651–657, 1987.

King, C. E., and Toskes, P.: Small intestine bacterial overgrowth. *Gastroenterology* **76:**1035–1055, 1979.

Kivilaakso, E., and Silen, W.: Pathogenesis of experimental gastric-mucosal injury. *N. Engl. J. Med.* **301:**364–369, 1979.

Kivilaakso, E., Fromm, D., and Silen, W.: Effect of the acid secretory state on intramural pH of rabbit gastric mucosa. *Gastroenterology* **75:**641–648, 1978.

Kjellen, G., and Tibbling, L.: Influence of body position, dry and water swallows, smoking and alcohol on esophageal acid clearing. *Scand. J. Gastroenterol.* **13:**283–288, 1979.

Klatskin, G.: The effect of ethyl alcohol on nitrogen excretion in the rat. *Yal J. Biol. Med.* **34:**124, 1961.

Klipstein, F. A., and Lindenbaum, J.: Folate deficiency in chronic liver disease. *Blood* **25:**443–456, 1965.

Knauer, C. M.: Mallory-Weiss syndrome. Chracterization of 75 Mallory-Weiss lacerations in 528 patients with upper gastrointestinal tract hemorrhage. *Gastroenterology* **71:**5–8, 1976.

Kolbel, C. B., Singer, M. V., Mohle, T., Heinzel, C., Eysselein, V., and Goebell, H.: Action of intravenous ethanol and atropine on the secretion of gastric acid, pancreatic enzymes, and bile acids and the motility of the upper gastrointestinal tract in nonalcoholic humans. *Pacreas* **1:**211–218, 1986.

Kondo, T., and Magee, D. F.: The action of intravenous ethanol on gastric secretion. *Proc. Soc. Exp. Biol. Med.* **156:**299–302, 1977.

Korsten, M. A., Worner, M. D., Feinman, L., Shaw, S., and Federman, Q.: Balloon cytology in screening of asymptomatic alcoholics for esophageal cancer, part I. *Dig. Dis.Sci.* **30:**845–851, 1985.

Krasner, N., Carmichael, R. I., Russell, E. G., Thompson, G. G., and Cochran, K. M.: Alcohol and absorption from the small intestine. *Gut* **17:**249–251, 1976.

Krawitt, E. L.: Ethanol inhibits intestinal calcium transport in rats. *Nature (London)* **243:**88–89, 1973.

Krawitt, E. L.: Effect of acute ethanol administration on duodenal calcium transport. *Proc. Soc. Exp. Biol. Med.* **146:**406–408, 1974.

Kuo, Y. J., and Shanbour, L. L.: Effects of ethanol on sodium 3-O-methyl glucose, and L-alanine transport in the jejunum. *Am. J. Dig. Dis.* **23:**51–56, 1978.

Kuo, Y. J., Shanbour, L. L., and Sernka, T. J.: Effects of ethanol on permeability and ion transport in the isolated dog stomach. *Dig. Dis.* **19:**818–824, 1974.

Kvietys, P.R., Patterson, W. G., Russell, J. M., Barrowman, J. A., and Granger, D. N.: Role of the microcirculation in ethanol-induced mucosal injury in the dog. *Gastroenterology* **87:**562–571, 1984.

LaBrooy, S. J., Misiewicz, J. J., Edwards, J., Smith, P. M., Haggie, S. J., Libman, L., Sarner, M., Wyllie, J. H., Croker, J., and Cotton, P.: Controlled trial of cimetidine in upper gastrointestinal hemorrhage. *Gut* **20:**892–895, 1979.

Lacy, E. R.: Effects of absolute ethanol, misoprostol, cimetidine, and phosphate buffer on the morphology of rat gastric mucosae. *Dig. Dis. Sci.* **31:**101S–107S, 1986.

Laine, L., and Weinstein, W. M.: Histology of alcoholic hemorrhagic "gastritis": A prospective evaluation. *Gastroenterology* **94:**1254–1262, 1988.

Lal, D., Gorbach, S. L., and Levitan, R.: Intestinal micorflora in patients with alcoholic cirrhosis: Urea-splitting bacteria and neomycin resistance. *Gastroenterology* **62:**275–279, 1972.

Langman, M. J. S.: Changing patterns in the epidemiology of peptic ulcer. *Clin. Gastroenterol.* **2:**219–226, 1973.

Larato, D. C.: Gewebeveränderungen in der Mundhöhle bei chronischen Alkoholikern. *Quintessenz* **12:**131, 1973.

Leddin, D. J., Ray, M., Dinda, P. K., Prokopiw, I., and Beck, I. T.: 16,16-Dimethyl prostaglandin E_2 alleviates jejunal microvascular effects of ethanol but not the ethanol-induced inhibition of water, sodium, and glucose absorption. *Gastroenterology* **94:**726–732, 1988.

Lee, E. R., and Dagradi, A. E.: Hemorrhagic erosive gastritis. *Am. J. Gastroenterol.* **63:**201–208, 1975.

Leevy, C. M., Baker, H., Ten Hove, W., Frank, O., and Cherrick, G. R.: B-complex vitamin in liver disease of the alcoholic. *Am. J. Clin. Nutr.* **16:**339–346, 1965.

Lemire, S., and Iber, F. L.: Pancreatic secretion in rats with protein malnutrition. *Johns Hopkins Med. J.* **120:**21–25, 1967.

Lens, H. J., Ferrari-Taylor, J., and Isenberg, J. I.: Wine and five percent ethanol are potent simulants of gastric secretion in human. *Gastroenterology* **85:**1082–1087, 1983.

Levy, J. S., Hardin, J. H., Shipp, H., and Keeling, J. H.: Varices of cecum as an unusual cause of gastrointestinal bleeding. *Gastroenterology* **33:**637–640, 1957.

Lieber, C. S., and DeCarli, L. M.: Hepatic microsomal ethanol-oxidizing system. *In vitro* characteristics and adaptive properties *in vivo*. *J. Biol. Chem.* **245:**2505–2512, 1970.

Lillibridge, C. B., Yoshimori, M., and Chey, W. Y.: Observations on the ultrastructure of oxyntic cells in alcohol-fed dogs. *Am. J. Dig. Dis.* **18:**443–454, 1973.

Lim, P., and Jacob, E.: Magnesium status of alcoholic patients. *Metabolism* **21:**1045–1051, 1972.

Lindenbaum, J., and Lieber, C. S.: Alcohol-induced malabsorption of vitamin B_{12} in man. *Nature* **224:**806, 1969.

Lindenbaum, J., and Lieber, C. S.: Effects of chronic ethanol administration on intestinal absorption in man in the absence of nutritional deficiency. *Ann. N.Y. Acad. Sci.* **252:**228–234, 1975.

Lindenbaum, J., Shea, N., and Saha, J. R.: Alcohol-induced impairment of carbohydrate (CHO) absorption. *Clin. Res.* **20:**459, 1972.

Lindenbaum, J., Saha, R. J., Shea, N., and Lieber, C. S.: Mechanism of alcohol-induced malabsorption of vitamin B_{12}. *Gastroenterology* **64:**762, 1973.

Lindenbaum, J., Pezzimenti, J. F., and Shea, N.: Small-intestinal function in vitamin B_{12} deficiency. *Ann. Intern. Med.* **80:**326, 1974.

Lopez del Pino, V., Hegazy, E., Hauber, G., Remmer, H., and Schwenk, M.: Isolated intestinal cells of guinea pig: a sitable model for assessing direct toxic effects of ethanol on the mucosa of the upper small intestine. *Arch. Toxicol.* **6:**322–326, 1983.

Lundquist, E., Tygstrup, N., Winkler, K., Mellemgaard, K., and Munck-Petersen, S.: Ethanol metabolism and production of free acetate in the human liver. *J. Clin. Invest.* **41:**955–961, 1962.

MacDougall, B. R. D., Bailey, R. J., and Williams, R.: H_2-receptor antagonists and antacids in the prevention of acute gastrointestinal hemorrhage in fulminant hepatic failure. *Lancet* **1:**617–619, 1977.

Madzarovova-Nohejlova, J.: Activite des disacchaaridases intestinales chez l'adulte et chez le buveur chronique de biere de Pilsen. *Biol. Gastreonterol. (Paris)* **4:**325, 1971.

Main, J. H. M., and Whittle, B. J. R.: The role of prostaglandins in gastric acid secretion. In: *Stimulus-Secretion Coupling in the Gastrointestinal Tract* R. M. Case and H. Goebell (eds.), Lancaster, MTP Press, 1976, pp. 147–157.

Maling, H. M., Highman, B., Hunter, J. M., Butler, W. M., and Williams, M. A.: Blood alcohol levels, triglyceride fatty livers, and pathologic changes in rats after single doses of alcohol. In: *Biochemical Factors in Alcoholism* (R. P. Maickel, ed.), New York, Pergamon Press, 1967, pp. 185–199.

Martin, S., and Pangborn, R. M.: Human parotid secretion in response to ethyl alcohol. *J. Dent. Res.* **50:**485–490, 1971.

Martini, G. A., and Wienbeck, M.: Begünstigt Alkohol die Entstehung eines Barrett-Syndroms (Endobrachyösophagus). *Dtsch. Med. Wochenschr.* **99:**434, 1974.

Martini, G. A., Phear, E. A., Ruebner, B., and Sherlock, S.: The bacterial content of the small intestine in normal and cirrhotic subjects. Relation to ethionine toxicity. *Clin. Sci.* **16:**35–51, 1957.

Mayer, E. M., Grabowski, C. J., and Fisher, R. S.: Effects of graded doses of alcohol upon esophageal motor function. *Gastroenterology* **75:**1133–1136, 1978.

Mayoral, L. G., Tripathy, K., Garcia, F. T., Klahr, S., Bolano, S. O., and Ghitis, J.: Malabsorption in the tropics: A second look. *Am. J. Clin Nutr.* **20:**866, 1967.

Mazzanti, R., and Jenkins, W. J.: Effect of chronic ethanol ingestion on enterocyte turnover in rat small intestine. *Gut* **28:**52–55, 1987.

Mazzanti, R., Debham, E. S., and Jenkins, W. J.: Effect of chronic ethanol intake on lactose activity and active galactose absorption in rat small intestine. *Gut* **28:**56–60, 1987.

McCormack, T. T., Sims, J., Eyre-Brook, I., Kennedy, H., Goepel, J., Johnson, A. G., and Triger, D. R.: Gastric lesions in portal hypertension: Inflammatory gastritis or congestive gastropathy? *Gut* **26:**1226–1232, 1985.

McCoy, G. D., Hecht, S. S., and Wynder, E. C.: The roles of tobacco, alcohol and diet in the etiology of upper alimentary and respiratory tract cancers. *Prev. Med.* **9:**622–629, 1980.

Mehkjian, H. S., and May, E. S.: Acute and chronic effects of ethanol on fluid transport in the human small intestine. *Gastroenterology* **72:**1280–1286, 1977.

Mendeloff, A. I., Kramer, P., Ingelfinger, F. J., and Bradley, S. E.: Studies with bromsulfalein, H: Factors altering its disappearance from the blood after a single intravenous injection. *Gastroenterology* **13:**222–234, 1949.

Messian, R. A., Hermos, J. A., Robbins, A. H., Friedlander, D. M., and Schimmel, E. M.: Barrett's esophagus. *Am. J. Gastroenterol.* **69:**458–466, 1978.

Metcalfe, J. C., Seeman, P., and Burgen, A. D. V.: The protein relaxation of benzyl alcohol in erythrocyte membranes. *Mol. Pharmacol.* **4:**87–95, 1968.

Mezey, E.: Intestinal function in chronic alcoholism. *Ann. N.Y. Acad. Sci.* **252:**215–227, 1975.

Mezey, E., and Potter, J. J.: Changes in exocrine pancreatic function produced by altered dietary protein intake in drinking alcoholics. *Johns Hopkins Med. J.* **138:**7–12, 1976.

Mezey, E., Jow, E., Slavin, R. E., and Tobon, F.: Pancreatic function and intestinal absorption in chronic alcoholism. *Gastroenterology* **59:**657–664, 1970.

Middleton, H. M., 3rd, Mills, L. R., and Singh, M.: Effects of ethanol on the uptake of pyridoxine X HCl in the rat jejunum. *Am. J. Clin. Nutr.* **39:**54–61, 1984.

Middleton, W. R. J., Carter, E. A., Drummey, G. D., and Isselbacher, K. J.: Effect of oral ethanol on intestinal cholesterogenesis in the rat. *Gastroenterology* **60:**880–887, 1971.

Millan, M. S., Morris, G. P., Beck, I. T., and Henson, J. T.: Villous damage induced by suction biopsy and by acute ethanol intake in normal human small intestine. *Dig. Dis. Sci.* **25:**513–525, 1980.

Miller, T. A., and Henagan, J. M.: Indomethacin decreases resistance of gastric barrier to disruption by alcohol. *Dig. Dis. Sci.* **29:**141–149, 1984.

Mistilis, S. P., and Garske, A.: Induction of alcohol dehydrogenase in liver and gastrointestinal tract. *Aust. Ann. Med.* **18:**227–231, 1969.

Mistilis, S. P., and Ockner, R. K.: Effects of ethanol on endogenous lipid and lipoprotein metabolism in the small intestine. *J. Lab. Clin. Med.* **80:**34–46, 1972.

Mitchell, C. J., and Jewell, D. P.: The diagnosis of the site of upper gastrointestinal hemorrhage in patients with established portal hypertension. *Endoscopy* **9:**131–135, 1977.

Mitjavila, S., Lacombe, C., and Carrera, G.: Changes in activity of rat brush border enzymes incubated with a homologous series of aliphatic alcohols. *Biochiem. Pharmacol.* **25:**626–630, 1976.

Murray, J., and Stein, N.: Effect of ethanol on absorption of iron in rats. *Proc. Soc. Exp. Biol. Med.* **120**:816–819, 1965.

Newsome, N. H., and Rattray, J. B. M.: The enzymatic esterification of ethanol with fatty acids. *Can. J. Biochem.* **43**:1223–1233, 1965.

Novis, B. H., Duys, P., Barbezat, G. O., Clain, J., Bank, S., and Terblanche, J.: Fireoptic endoscopy and the use of the Sengstaken tube in acute gastrointestinal haemorrhage in patients with portal hypertension and varices. *Gut* **17**:258–263, 1976.

Oates, P. J., and Hakkinen, J. P.: Studies on the mechanism of ethanol-induced gastric damage in rats. *Gastroenterology* **94**:10–21, 1988.

Ostrow, J. D., Timmerman, R. J., and Gray, S. J.: Gastric erosions in human hepatic cirrhosis. *Gastroenterology* **38**:303–313, 1960.

Palmer, E. D.: Gastritis: A reevaluation. *Medicine* **33**:199–290, 1954.

Parl, F. F., Lev, R., Thomas, E., and Pitchumoni, C. S.: Histologic and morphometric study of chronic gastritis in alcohol patients. *Hum. Pathol.* **10**:45–56, 1979.

Patterson, S. J., Butler, K. W., Huang, P., Labelle, Y., Smith, I. C. P., and Schneider, H.: The effects of alcohols on lipid bilayers, a spin label study. *Biochim. Biophys. Acta* **266**:597–602, 1972.

Perlow, W., Baraona, E., and Lieber, C. S.: Symptomatic intestinal disaccharidase deficiency in alcoholics. *Gastroenterology* **72**:680–684, 1977.

Peterson, W. L., Barnett, C., and Walsh, H. J.: Effect of intragastric infusion of ethanol and wine on serum gastrin concentration and gastric secretion. *Gastroenterology* **91**:1390, 1986.

Phaosawasdi, K., Tolin, R., Mayer, E., and Fisher, R. S.: Effects of alcohol on the pyloric sphincter. *Dig. Dis. Sci.* **24**:934–939, 1979.

Pickens, C. A., and Tedesco, F. J.: Colonic varices. *Am. J. Gastroenterology* **73**:73–74, 1980.

Pirola, R. C.: *Drug Metabolism and Alcohol.* Baltimore, University Park Press, 1978.

Pirola, R. C., and Davis, A. E.: Effects of intravenous alcohol on motility of the duodenum and of the sphincter of Oddi. *Aust. Ann. Med.* **19**:24–29, 1970.

Pitchumoni, C. S., and Glass, G. B. J.: Patterns of gastritis in alcoholic. *Biol. Gastroenterol. (Paris)* **9**:11–16, 1976.

Pitchumoni, C. S., and Glass, G. B. J.: Alcohol injury to the gastrointestinal mucosa. In: *Progress in Gastroenterology* (G. B. J. Glass, ed.), New York, Grune & Stratton, 1977, pp. 717–758.

Portela-Gomez, G., Pinto-Correia, J., Danta, S. R., Pires, I. R., Laranjeira, M. R., and Figueiredo, A.: Jejunal microflora in patients with liver cirrhosis. *Arq. Gastroenterol. Sao. Paulo.* **14**:72–75, 1977.

Pottern, L. M., Morris, L. E., Blot, W. J., Ziegler, R. G., and Fraumeni, J. F.: Esophageal cancer among black men in Washington D.C.–I. Alcohol, tobacco and other risk factors. *J. Natl. Cancer Inst.* **67**:777–783, 1981.

Price, J. B., McCullough, W., Peterson, L., Britton, R. C., and Voorhees, A. B.: Effects of portal systemic shunting on intestinal absorption in the dog and in man. *Surg. Gynecol. Obstet.* **125**:305–310, 1967.

Puurunen, J.: Studies on the mechanism of the inhibitory effect of ethanol on the gastric acid output in the rat. *Naunyn Schmiedebergs Arch. Pharmacol.* **303**:87–93, 1978.

Puurunen, J., and Karppanen, H.: Effects of ethanol on gastric acid secretion and gastric mucosal cyclic AMP in the rat. *Life Sci.* **16**:1513–1520, 1975.

Puurunen, J., Karppanen, H., Kairaluoma, M., and Larmi, T.: Effects of ethanol on the cyclic AMP system of the dog gastric mucosa. *Eur. J. Pharmacol.* **38**:275–279, 1976.

Puurunen, J., Hiltunen, K., and Karppanen, H.: Ethanol-induced change in gastric mucosal content of cyclic AMP and ATP in the rat. *Eur. J. Pharmacol.* **42**:85–89, 1977.

Rabinovitz, M., Van Thiel, D. H., Dindzans, V., and Gavaler, J. S.: Endoscopic findings in alcoholic liver disease: Does gender make a difference? *Alcohol* **6**:465–468, 1989.

Racusen, L. C., and Krawitt, E. Z.: Effect of folate deficiency and ethanol ingestion on intestinal folate absorption. *Am. J. Dig. Dis.* **22**:915–920, 1977.

Ramwell, P. W., and Shaw, J. E.: Prostaglandin inhibition of gastric secretion. *J. Physiol (Lond.)* **195**:34P–36P, 1968.

Reichstein, B. J., Okamoto, C., and Forte, J. G.: Effect of ethanol on acid secretion by isolated gastric glands from rabbit. *Gastroenterology* **91**:439–447, 1986.

Robert, A.: Cytoprotection by prostaglandins. *Gastroenterology* **77**:761–767, 1979.

Robert, A., Nezamis, J. E., Lancaster, C., and Hanchar, A. J.: Cytoprotection by prostaglandins in rats. *Gastroenterology* **77**:433–443, 1979.

Roberts, D. M.: Chronic gastritis, alcohol and non-ulcer dyspepsia. *Gut* **13**:768–774, 1972.

Rodgers, J. B., and O'Brien, R. J.: The effect of acute ethanol treatment on lipid-reesterifying enzymes of the rat small bowel. *Am. J. Dig. Dis.* **20**:354–358, 1975.

Rodrigo, C., Antezana, C., and Baraona, E.: Fat and nitrogen balances in rats with alcohol-induced fatty liver. *J. Nutr.* **101**:1307–1310, 1971.

Roesch, W.: Erosions of the upper gastrointestinal tract. *Clin. Gastroenterol.* **7**:623–634, 1978.

Roggin, G. M., Iber, F. L., Kater, R. M. H., and Tobon, F.: Malabsorption in the chronic alcoholic. *Johns Hopkins Med. J.* **125**:321–330, 1969.

Roth, J. L. A.: Drug induced lesions. In: *Gastroenterology, Vol. 1* (3rd ed.) (H. L. Bockus, ed.), Philadelphia, W. B. Saunders, 1974, pp. 487–514.

Rubin, E., Rybak, B. J., Lindenbaum, J., Gerson, C. D., Walker, G., and Lieber, C. S.: Ultrastructural changes in the small intestine induced by ethanol. *Gastroenterology* **63**:801–814, 1972.

Rutten, M. J., and Ito, S.: Structural and functional changes by ethanol on *in vitro* guinea pig gastric mucosa. *Am. J. Physiol.* **251**:G518–G528, 1986.

Safaie-Shirazi, S., Zike, W. L., Brubacher, M., and Debensten, L.: Effect of aspirin, alcohol, and pepsin on mucosal permeability of esophageal mucosa. *Surg. Forum* **25**:335–337, 1974.

Salo, J. A.: Ethanol-induced mucosal injury in rabbit oesophagus. *Scand. J. Gastroenterol.* **18**:713–721, 1983.

Sanders, K. M., and Berry, R. G.: Effects of ethyl alcohol on phasic and tonic contractions of the proximal stomach. *J. Pharmacol. Exp. Ther.* **235**:858–862, 1985.

Sauerbruch, T., Schreiber, M. A., Schussler, P., and Permanetter,

W.: Endoscopy in the diagnosis of gastritis. Diaganostic value of endoscopic criteria in relation to histological diagnosis. *Endoscopy* **16**:101–104, 1984.

Saunders, D. R., Sillery, J., and McDonald, G. B.: Effect of ethanol on transport from rat intestine during high and low rates of oleate absorption. *Lipids* **17**:356–360, 1982.

Schafer, D. F., Stephenson, K. V., Barak, A. J., and Sorrell, M. F.: Effects of ethanol on the transport of manganese by small intestine of the rat. *J. Nutr.* **104**:101–104, 1974.

Schiller, K. F. R., and Cotton, P. B.: Acute upper gastrointestinal hemorrhage. *Clin. Gastroenterol.* **7**:595–604, 1978.

Schultz, S. G.: Sodium-coupled solute transport by small intestine: A status report. *Am. J. Physiol.* **233**:249–254, 1977.

Scobie, B. A., and Summerskill, W. H. J.: Reduced gastric output in cirrhosis: Quantitation and relationships. *Gut* **5**:422, 1964.

Seeman, P.: The membrane actions of anesthetics and tranquilizers. *Pharmacol. Rev.* **24**:583–655, 1972.

Seitz, K., Korsten, M. A., and Lieber, C. S.: Ethanol oxidation by intestinal microsomes: Increased activity after chronic ethanol administration. *Life Sci.* **25**:1443–1448, 1979.

Sernka, T. J., and Jackson, A. F.: Hyperosmotic instillation of rat stomach. *Life Sci.* **17**:435–442, 1975.

Sernka, T. J., and Jackson, A. F.: Responses to ethanol in stomach. *Nutr. Rep. Int.* **14**:415–421, 1976.

Sernka, T. J., and Jackson, A. F.: Gastric acid output in isosmotic ethanol and other isosmotic solutions. *Nutr. Rep. Int.* **17**:437–440, 1978.

Sernka, T. J., Gilleland, C. W., and Shanbour, L. L.: Effects of ethanol on active transport in the dog stomach. *Am. J. Physiol.* **226**:397–400, 1974.

Shanbour, L. L., and Kuo, Y. H.: Effects of ethanol on sodium, chloride and glucose transport. *Fed. Proc.* **35**:707, 1976.

Shanbour, L. L., Miller, J., and Chowdhury, K.: Effects of alcohol on active transport in the rat stomach. *Dig. Dis.* **18**:311–316, 1973.

Shepherd, P., and Simmonds, W. J.: Some conditions affecting the maintenance of a steady lymphatic absorption of fat. *Aust. J. Exp. Biol.* **37**:1–10, 1959.

Shimizu, M., Tsutsumi, M., Lasker, J. M., and Lieber, C. S.: Estraheptic localization of ethanol (E)-inducible P450IIE1. *Alcoholism: Clin. Exp. Res.* **13**:325, 1989.

Shirazi, S. S., and Platz, C. E.: Effect of alcohol on canine esophageal mucosa. *J. Surg. Res.* **25**:373–379, 1978.

Silver, L. S., Worner, T. M., and Korsten, M. A.: Esophageal function in chronic alcoholics. *Am. J. Gastroenterol.* **81**:423–427, 1986.

Silverstein, F. E., Gilbert, D. A., Tedesco, F. J., Buenger, N. K., and Pershing, J.: The national ASGE survey on upper gastrointestinal bleeding. *Gastrointest. Endoscop.* **27**:73–102, 1981.

Singer, M. V., Leffmann, C., Eysselein, V. E., Calden, H., and Goebell, H.: Action of ethanol and some alcoholic beverages on gastric acid secretion and release of gastrin in human. *Gastroenterology* **93**:1247–1254, 1987.

Sklan, D., and Trostler, N.: Site and extent of thiamine absorption in the rat. *J. Nutr.* **107**:353–356, 1977.

Small, M., Longarini, A., and Zamcheck, N.: Disturbances of digestive physiology following acute drinking episodes in 'skid-row' alcoholics. *Am. J. Med.* **27**:575–585, 1959.

Smith, B. M., Skillman, J. J., Edwards, B. G., and Silen, W.: Permeability of the human gastric mucosa: Alteration by acetylsalicylic acid and ethanol. *N. Engl. J. Med.* **285**:716–721, 1971.

Smith, M. E.: The uptake of pteroylglutamic acid by the rat jejunum. *Biochim. Biophys. Acta* **298**:124–129, 1973.

Snady, H.: The role of sclerotherapy in the treatment of esophageal varices: Personal experience and a review of randomized trials. *Am. J. Gastroenterol.* **82**:813–822, 1987.

Snady, H., and Feinman, L.: Prediction of variceal hemorrhage: A prospective study. *Am. J. Gastroenterol.* **83**:519–525, 1988.

Stein, S. W., Lieber, C. S., Leevy, C. M., Cherrick, G. R., and Abelmann, W. H.: The effect of ethanol upon systemic and hepatic blood flow in man. *Am. J. Clin. Nutr.* **13**:68–74, 1963.

Steinberg, W. M., and Lewis, J. H.: Mylanta II inhibits the absorption of cimetidine. *Gastroenterology* **78**:1269, 1980 (abstract).

Stern, A. I., Hogan, D. L., and Isenberg, J. I.: A new method for quantitation of ion fluxes across *in vivo* human gastric mucosa: Effect of aspirin, acetaminophen, ethanol, and hyperosmolar solutions. *Gastroenterology* **86**:60–70, 1984.

Straus, E., Urbach, H. J., and Yalow, R.: Alcohol-stimulated secretion of immunoreactive secretin. *N. Engl. J. Med.* **293**:1031–1032, 1975.

Sullivan, J. F.: Effect of alcohol on urinary zinc excretion. *Q. J. Stud. Alcohol* **23**:216–220, 1962.

Szabo, S., Trier, J. S., Brown, A., and Schnoor, J.: Early vascular injury and increased vasacular permeability in gastric mucosal injury caused by ethanol in the rat. *Gastroenterology* **88**:228–236, 1985.

Tague, L. L., and Shanbour, L. L.: Effects of ethanol on bicarbonate-stimulated ATPase, ATP and cyclic AMP in canine gastric mucosa. *Proc. Soc. Exp. Biol. Med.* **154**:37–40, 1977.

Tandon, B. N., Banks, P. A., George, P. K., Sama, J. K., Ramachandran, K., and Candhi, P. C.: Recovery of exocrine pancreatic funciton in adult protein-calorie malnutrition. *Gastroenterology* **58**:358–362, 1970.

Tapper, S. S., Bushi, S., Ruppert, R. D., and Greenberger, N. J.: Chronic ethanol treatment on the absorption of iron in rats. *Am. J. Med. Sci.* **255**:46–52, 1968.

Tariq, M., Parmar, N. S., and Ageel, A. M.: Effect of nicotine and alcohol pretreatment on the gastric mucosal damage induced by aspirin, phenylbutazone, and reserpine in rats. *Alcoholism* **10**:213–216, 1986.

Tarnawski, A., Hollander, D., Stachura, J., Krause, W. J., and Gergely, H.: Prostaglandin protection of the gastric mucosa against alcohol injury—a dynamic time-related process. Role of the mucosal proliferative zone. *Gastroenterology* **88**:334–352, 1985.

Taylor, F. W.: Portal tension and its dependence on external pressure. *Ann. Surg.* **140**:652–660, 1954.

Tennent, D. M.: The influence of alcohol on the emptying time of the stomach and the absorption of glucose. *Q. J. Stud. Alcohol* **2**:271–276, 1941.

Terano, A., Hiraishi, H., Ota, S.-I., Shiga, J., and Sugimoto, T.: Role of superoxide and hydroxyl radicals in rat gastric mucosal injury induced by ethanol. *Gastroenterol. Jpn.* **24**:488–493, 1989.

Terblanche, J., Burroughs, A. K., and Hobbs, K. E. F.: Controversies in the management of bleeding esophageal varices. *N. Engl. J. Med.* **320:**1392–1398, 1989.

Teres, J., Bordas, J. M., Bru, C., Diaz, F., Bruguera, M., and Rodes, J.: Upper gastrointestinal bleeding in cirrhosis: Clinical and endoscopic correlations. *Gut* **17:**37–40, 1976.

Thomas, E., Rosenthal, W. S., Rymer, W., and Katz, D.: Upper gastrointestinal hemorrhage in patients with alcoholic liver disease and esophageal varices. *Am. J. Gastroenterol.* **72:** 623–629, 1979.

Thomson, A. B. R.: Acute exposure of rabbit jejunum to ethanol: *in vitro* uptake of hexoses. *Dig. Dis. Sci.* **29:**267–274, 1984.

Thomson, A. B. R., Man, S. F. P., and Shnitka, T.: Effect of ethanol on intestinal uptake of fatty acids, fatty alcohols, and cholesterol. *Dig. Dis. Sci.* **29:**631–642, 1984.

Thomson, A. D., Baker, H., and Leevy, C. M.: Patterns of S-thiamine hydrochloride absorption in the malnourished alcoholic patient. *J. Lab. Clin. Med.* **76:**34–45, 1970.

Tillotson, L. G., Carter, E. A., Inui, K.-I., and Isselbacher, K. J.: Inhibition of Na$^+$-stimulated glucose transport function and perturbation of intestinal microvillus membrane vesicles by ethanol and acetaldehyde. *Arch.Biochem. Biophys.* **207:** 360–370, 1981.

Tiscornia, O. M., Palasciano, G., and Dzieniszewski, J.: Simultaneous changes in pancreatic and gastric secretion induced by acute intravenous ethanol infusion. Effect of atropine and reserpine. *Am. J. Gastroenterol.* **63:**389–395, 1975.

Tomasulo, P. A., Kater, R. M. H., and Iber, F. L.: Impairment of thiamine absorption in alcoholism. *Am. J. Clin. Nutr.* **21:** 1341–1344, 1968.

Toukan, A. U., Kamal, M. F., Amr, S. S., Arnaout, M. A., and Abu-Romiyeh, A. S.: Gastroduodenal inflammation in patients with non-ulcer dyspepsia. A controlled endoscopic and morphometric study. *Dig. Dis. Sci.* **30:**313–320, 1985.

Trier, J. S.; Szabo, S., and Allan, C. H.: Ethanol-induced damage to mucosal capillaries of rat stomach. Ultrastructural features and effects of prostaglandin F$_{2\beta}$ and cysteamine. *Gastroenterology* **92:**13–22, 1987.

Tsai, Y.-T., Lay, C.-S., Lai, K.-H., Ng, W.-W., Yeh, Y.-S., Wang, J.-J., Chiang, T.-T., Lee, S.-D., Chiang, B. N., and Lo, K.-J.: Controlled trail of vasopressin plus nitroglycerin vs. vasopressin alone in the treatment of bleeding esophageal varices. *Hepatology* **6:**406–409, 1986.

Van tuan, T., Mitjavila, M. T., and Derache, R.: Effet de l'éthanol sur la respiration des cellules épitheliales isolées de l'intestin de rat. *C.R. Soc. Biol. (Paris)* **165:**2011–2014, 1971.

Varro, V., Jung, I., Szarvas, F., Csernay, L., Savay, G., and Okros, J.: The effect of vasoactive substances on the circulation and glucose absorption of an isolated jejunal loop in dog. *Am. J. Dig. Dis.* **12:**46–59, 1967.

Verdy, M., and Caron, D.: Ethanol et absorption du calcium chez l'humain. *Biol. Gastroenterol. (Paris)* **6:**157–160, 1973.

Vesely, D. L., and Levey, G. S.: Ethanol-induced inhibition of guanylate cyclase in liver pancreas, stomach, and intestine. *Res. Commun. Chem. Pathol. Pharmacol.* **17:**215, 1977.

Vessey, D. A., and Zakim, D.: Membrane fluidity and the regulation of membrane bound enzymes. *Horiz. Biochem. Biophys.* **1:**138–174, 1974.

Viteri, F. E., Flores, J. M., and Alvarado, J.: Intestinal malabsorption in malnourished children before and during recovery.

Relation between severity of protein deficiency and the malabsorption process. *Am. J. Dig. Dis.* **18:**201–211, 1973.

Vlahcevic, Z. R., Juttijudata, P., Bell, C. C., and Swell, L.: Bile acid metabolism in patients with cirrhosis. II. Cholic and chenodeoxycholic acid metabolism. *Gastroenterology* **62:** 1174–1181, 1972.

Waldram, R., Davis, M., Nunnerley, H., and Williams, R.: Emergency endoscopy after gastrointestinal hemorrhage in fifty patients with portal hypertension. *Br. Med. J.* **4:**94–96, 1974.

Wallace, J. L.: Increased resistance of the rat gastric mucosa to hemorrhagic damage after exposure to an irritant. Role of the "mucoid cap" and prostaglandin synthesis. *Gastroenterology* **94:**22–32, 1988.

Wallace, J. L., and Whittle, B. J.: Acceleration of recovery of gastric epithelial integrity by 16,16-dimethyl prostaglandin E$_2$. *Br. J. Pharamcol.* **86:**837–842, 1985.

Watson, W. C., Sallam, M., and Allan, G.: Acute upper gastrointestinal tract bleeding and recent alcohol ingestion. *Can. Med. Assoc. J.* **110:**525–529, 1974.

Weisbrodt, N. W., Kienzle, M., and Cooke, A. R.: Comparative effects of aliphatic alcohols on the gastric mucosa. *Proc. Soc. Exp. Biol. Med.* **142:**450–454, 1973.

Williams, K.: Introduction, alcohol and the gastrointestinal tract. *Ann. N.Y. Acad. Sci.* **252:**214, 1975.

Wilson, F. A., and Hoyumpa, A. M.: Ethanol and small intestinal transport. *Gastroenterology* **76:**388–403, 1979.

Winship, D. H., Carlton, R. C., Zaboralskie, F. F., and Hogan, W. J.: Deterioration of esophageal peristalsis in patients with alcoholic neuropathy. *Gastroenterology* **55:**173–178, 1968.

Winsor, A. L., and Strongin, E. J.: The effect of alcohol on the rate of parotid secretion. *J. Exp. Psychol. [Gen.]* **16:**589–597, 1933.

Wojcicki, J., Samochowiec, L., and Kadykow, M.: Effect of ethanol on iron absorption in the small intestine of rats. *Q. J. Stud. Alcohol* **33:**958–961, 1972.

Wolff, G.: Does alcohol cause chronic gastritis? *Scand J. Gastroenterol.* **4:**289–291, 1970.

Wong, S. H., Ogle, C. W., and Cho, C. H.: The influence of chornic or acute nicotine pretreatment on ethanol-induced gastric ulceration in the rat. *J. Pharm. Pharmacol.* **38:**537–540, 1986.

Woodward, E. R., Robertson, C., Ruttenberg, H. D., and Schapiro, H.: Alcohol as a gastric secretory stimulant. *Gastroenterology* **32:**727–737, 1957.

Woodward, L., Slotten, D. S., and Tillmans, V. C.: Mechanism of alcoholic stimulation of gastric secretion. *Proc. Soc. Exp. Biol. Med.* **89:**428–431, 1955.

Worthington, B. S., Meserole, I., and Syrotuck, J. A.: Effect of daily ethanol ingestion on intestinal permeability to macromolecules. *Am. J. Dig. Dis.* **23:**23–32, 1978.

Yonei, Y., Holzer, P., and Guth, P. H.: Laparotomy-induced gastric protection against ethanol injury is mediated by capsaicin-sensitive sensory neurons. *Gastroenterology* **99:**3–9, 1990.

Yoshimori, M., Chey, N. Y., Escoffery, R., and Lillibridge, C.: Effects of ethanol on gastric secretion of acid and parietal cells in dogs. *Gastroenterology* **62:**732, 1972.

Zimmerman, J., Gati, I., Eisenberg, S., and Rachmilewitz, D.: Ethanol inhibits triglyceride synthesis and secretion by human small intestinal mucosa. *J. Lab. Clin. Med.* **107:**498–501, 1986.

11

Alcohol and the Pancreas

Mark A. Korsten, Romano C. Pirola, and Charles S. Lieber

11.1. Introduction

Chronic pancreatitis exacts a heavy toll in countries with a high prevalence of alcoholism. In fact, of the gastrointestinal complications of alcohol abuse, only hepatic injury has a more significant impact. In recent years, there has been considerable evolution in our understanding of the pathophysiology of chronic pancreatitis and advances in the clinical management of the illness. Much of this newer information has been incorporated into the present chapter. Other sources of information include reviews by DiMagno and Clain (1986), Singh (1986), Bank (1986), and Barkin and Fayne (1986).

11.2. Epidemiology of Alcoholic Pancreatitis

Estimates of the percentage of cases of pancreatitis linked to alcoholism in the United States have ranged from as low as 5% (Thal *et al.*, 1957) to as high as 90% (Albo *et al.*, 1963). Statistics also vary considerably from one country to another in this respect (Fig. 11.1). There are at least two reasons for the wide range of these estimates. First, the diagnosis of alcoholism is not always obvious on clinical grounds, and the prevalence of alcoholism in a study population may be greatly underestimated. Second, there is often a bias in the selection of pancreatic cases. For example, in postmortem studies, severe cases of acute pancreatitis tend to be overrepresented. Death in such patients is often a result of

necrotizing hemorrhagic pancreatitis, a sequela much more common in patients with biliary tract disease.

Nevertheless, true variations in the geographic incidence of alcoholic pancreatitis probably do exist. For example, in England it appeared to be rare up to 1950 and has been diagnosed with increasing frequency since then (Trapnell, 1972). Apart from possible geographic variations in the incidence of alcohol-related pancreatic disease, there may be variations among countries in the clinical presentation. In France, patients with alcoholic pancreatitis present mainly with the features of pancreatic insufficiency, whereas in the United States, Australia, and South Africa, the predominant features are those of acute intermittent inflammation. These apparent differences may be important in assessing the claims for pathogenetic mechanisms from different centers.

Obviously, factors apart from alcohol intake may contribute to the geographic variations in the incidence of alcoholic pancreatitis. Thus, in one English study, it was found that the "hardness" of the drinking water correlated with variations in the incidence of alcoholic pancreatitis in different localities (Bourke *et al.* 1979). Constitutional factors may favor the development of chronic pancreatitis, such as anomalies of the pancreatic duct or sphincter of Oddi, familial hyperlipemia, and the presence of specific genetic markers. The following HLA antigens are reported to show increased incidence in alcoholic pancreatitis: B-40 (Gosselin *et al.*, 1978; Fauchet *et al.*, 1979), Aw23 and Aw24 (Dani *et al.*, 1978), B13 (Gullo *et al.*, 1982), and Bw39 (Wilson *et al.*, 1984). In addition, studies involving other genetic markers have revealed increased incidence of certain

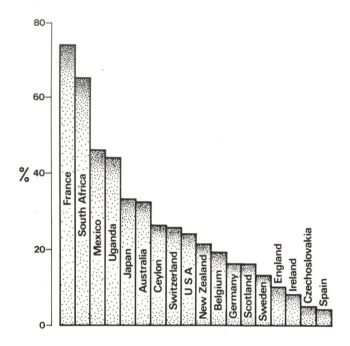

FIGURE 11.1. Geographic variations in the incidence of alcoholic pancreatitis.

blood groups (Marks *et al.*, 1970) and a higher incidence of the MZ α_1-antitrypsin phenotype associated with a reduced incidence of the MM phenotype (Novis *et al.*, 1975).

11.3. The Pathology of Alcoholic Pancreatitis

The pathological features of alcoholic pancreatitis are nonspecific. Even calcification, a finding that strongly suggests an alcoholic etiology, is not pathognomonic.

In the mild forms of alcoholic pancreatitis, Czernobilsky and Mikut (1964) found a high frequency of interstitial edema. Bordalo *et al.* (1977) and Pitchumoni *et al.* (1984) demonstrated that pancreatic injury is common in asymptomatic alcoholics. Fibrosis and atrophy are characteristic features of more severe alcoholic pancreatitis (Fig. 11.2). These lesions are distributed in a patchy fashion throughout the pancreas, often with striking variation in the degree of lobular injury. The irregular distribution of the injury has been attributed to obstruction of peripheral pancreatic ducts by calcified proteinaceous plugs (Sarles *et al.*, 1965). Acinar cells within involved lobules may be replaced by fibrous tissue or goblet cells, and the epithelial cells around the central lumen of the acinus may be flattened (so-called

canalicular dedifferentiation). The number of islet cells within the pancreatic parenchyma has been variably reported to be unchanged (Perrier, 1964), decreased (Gambill *et al.*, 1960), or increased (Sarles *et al.*, 1965).

Pseudocysts often develop after an episode of acute inflammation. Unlike true cysts, they lack a lining epithelium and are surrounded by inflammatory exudate or granulation tissue. Such pseudocysts may or may not communicate with the pancreatic duct (Perrier, 1964).

11.4. Pathogenesis of Alcoholic Pancreatitis

Pancreatitis is an autodigestive process with secondary inflammatory phenomena. Thus, in studying the pathogenesis of pancreatitis, much emphasis has been placed on investigating the synthesis, transport, activation, and inhibition of pancreatic enzymes.

Two key clinical aspects that require explanation are the recurrent nature of acute attacks and the chronic course of the illness. Classical theories often ignore this dual nature of the disease. Thus, hypotheses implicating dysfunction of the sphincter of Oddi tend to ignore the chronic nature of alcoholic pancreatitis. Similarly, the theory of small-duct obstruction by protein plugs tends to overlook the abrupt onset of attacks.

In view of the capacity of the pancreas to resist

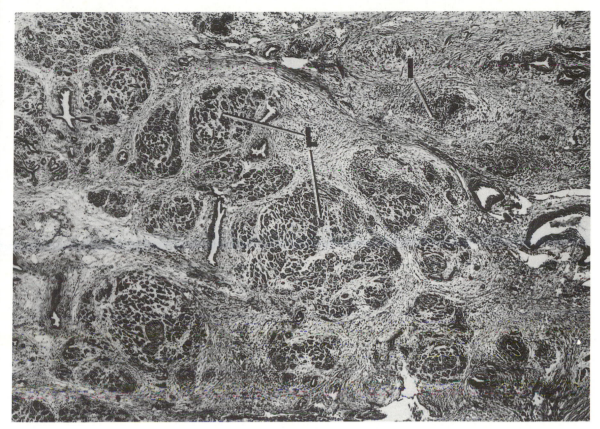

FIGURE 11.2. Moderately advanced chronic pancreatitis with evidence of recent acute inflammation (I). The pancreatic lobules (L) show patchy atrophy and are partly replaced by loose fibrous tissue, which is edematous. (From Pirola and Lieber, 1974.)

autodigestion, it seems plausible to assume that a certain threshold of cell damage must be reached before the process of cell digestion ensues. Whether an actual attack of pancreatitis occurs would depend on the balance between (1) the severity of an acute insult and (2) the capacity of the pancreas to resist autodigestion. Furthermore, it is conceivable that an etiological agent such as ethanol could affect both aspects of this balance, possibly by quite different mechanisms. It is noteworthy that in some countries such as the United States, clinically recognized attacks of pancreatitis occur in only a minority of alcoholics. Yet animal experiments have revealed a number of reproducible effects of alcohol on the pancreas. If these do predispose the pancreas to inflammation, then one needs to postulate the presence of other factors accounting for individual susceptibility to clinical pancreatitis (such as genetic or nutritional factors).

11.4.1. Acute Effects of Ethanol: Sphincter of Oddi

After decades of intense controversy, the various "sphincter theories" have received less attention in recent years. They have been reviewed in detail elsewhere (Pirola and Davis, 1970a). The main objection to these theories is the absence of a "common channel" in many patients and the lack of convincing supportive evidence in animal models.

11.4.2. Acute Changes in Pancreatic Secretion

The oral or intragastric administration of alcohol clearly stimulates the exocrine pancreas in the dog (Walton *et al.*, 1960), the rat (Cavarzan *et al.*, 1975); Korsten *et al.*, 1979), and the pig (Wheatley *et al.*, 1975). It was previously accepted that the pancreatic response to alcohol was mediated by increased gastric acid (and secondary release of secretin from the duodenum). This view, however, has been challenged. For example, when the isolated gastric antrum of the dog is perfused with alcohol, the pancreatic response precedes any increase in gastric acid (Schapiro *et al.*, 1968). In humans, within 5 min of oral ethanol intake, there is a prompt elevation of serum secretin levels. The rapidity of the secretin response suggests that ethanol releases secretin directly from the duodenum. Finally, the pancreatic stimulation after intragastric ethanol is not significantly different in control and achlorhydric (cimetidine-treated) rats (Korsten *et al.*, 1979).

The "direct" effects of ethanol on the exocrine pancreas are more variable. When intravenous alcohol is given to basally secreting rats, there is also a stimulatory effect (Korsten *et al.*, 1981). On the other hand, when the pancreas is stimulated exogenously using secretin or secretin–CCK, depression of pancreatic secretion by alcohol is observed in dogs (Barroso, 1949; Bayer *et al.*, 1972; Tiscornia *et al.*, 1973) and rats (Korsten *et al.*, 1981). In human studies, intravenous ethanol alone has been reported to have no effect on duodenal aspirate volume, but when it is given after secretin stimulation, there is a reduction (Davis and Pirola, 1966; Mott *et al.*, 1972). However, these experiments are difficult to assess in view of the possible effects of ethanol on the sphincter of Oddi. In the dog, intravenous ethanol on a background of secretin or other secretagogues has a dual effect, inhibition followed by stimulation (Kubota *et al.*, 1983). Vagotomy diminished the inhibitory action of intravenous ethanol (Tiscornia *et al.*, 1973), and the authors suggested that this was because intravenous alcohol stimulated inhibitory vagal fibers.

11.4.3. Acute Changes in Pancreatic Metabolism

Little is known about the acute metabolic changes produced by ethanol in the pancreas. Rat pancreatic triglyceride synthesis is stimulated at the expense of phospholipid synthesis by ethanol *in vitro* as well as by acute ethanol administration *in vivo* (Somer *et al.*, 1980a,b). However, this acute effect of ethanol did not occur after chronic ethanol administration (Wilson *et al.*, 1984). The acute effect of ethanol on triglyceride synthesis was also noted by Calderon-Attas *et al.* (1980) in pancreatic fragments. In addition, ethanol enhanced the *in vitro* synthesis of fatty acids from acetate while reducing glycolysis and β-oxidation. Pancreatic phos-

pholipid metabolism is also altered by ethanol. Orrego-Matte *et al.* (1969) demonstrated a decreased appearance of injected ^{32}P in rat pancreas phospholipid after acute (and chronic) ethanol administration. More recently, Chapman and Pattison (1988) reported that ^{32}P incorporation into pancreatic phosphatidylcholine and other phospholipids is acutely inhibited *in vitro* by ethanol. Given the importance of phospholipids in stimulus–response coupling, these results may explain the complex effects of ethanol on the composition of pancreatic secretion. However, these *in vitro* effects are mainly seen at levels of ethanol that are not likely in the clinical setting and, therefore, remain of questionable physiological relevance.

11.4.4. Chronic Effects of Ethanol

11.4.4.1. Animal Models

Lesions comparable to those found in the human pancreas have been produced in dogs (Sarles *et al.*, 1975; Cueto *et al.*, 1976) and rats (Darle *et al.*, 1970; Sarles *et al.*, 1971a; Podol'skii, 1977; Joffe *et al.*, 1979; Kakizaki *et al.*, 1987; Tsukamoto *et al.*, 1988; Grönroos *et al.*, 1988) after chronic ethanol administration. The typical lesion is subcellular (fat deposition, mitochondrial alteration, autophagic vacuoles) and is similar to those observed in the pancreas of chronic alcoholics without clinical pancreatitis (Bordalo *et al.*, 1977). The nature of the lipid that accumulates during the early stages of alcoholic pancreatitis has been investigated. Wilson *et al.* (1982) reported that the only lipid fraction in the rat pancreas increased by chronic ethanol administration was cholesterol ester; this has been confirmed by Rao *et al.* (1984) and Simsek and Singh (1989). More advanced stages of alcohol-induced pancreatic injury appear to require longer periods of ethanol feeding at sustained and high levels (Sarles *et al.*, 1971a; Tsukamoto *et al.*, 1988). The lesions are patchy and characterized by atrophic acini, ductular dilatation and proliferation, intraductal protein precipitates, and fibrosis.

11.4.4.2. Small Duct Obstruction by Proteinaceous Precipitates

Possibly the most widely accepted theory of the pathogenesis of alcoholic pancreatitis is that put forward by Sarles and his colleagues (Sarles, 1974; Sarles *et al.*, 1980). They postulate that a number of factors (including ethanol, hypercalcemia, and genetic factors) favor the precipitation of protein in peripheral pancreatic ducts.

These protein "plugs" could then cause obstructive changes and eventually become calcified and spread to involve the larger ducts. The precipitation and calcification of protein might be favored by an increased concentration of protein in pancreatic juice (Renner *et al.*, 1978; Sahel and Sarles, 1979), by ionic changes (Dreiling *et al.*, 1973; Sarles *et al.*, 1977, 1980; Sahel and Sarles, 1979; Boustiere *et al.*, 1985), and, in part, by decreased levels of a unique protein ("stone protein") that normally prevents formation of calcium crystals (Lohse and Kraemer, 1984; Giorgi *et al.*, 1985; Multigner *et al.*, 1985; Provansal-Cheylan *et al.*, 1986). The latter is a phosphoglycoprotein with a molecular weight of 14,000 da and a high content of acidic amino acids. The complete amino acid sequence of this protein is now known (DeCaro *et al.*, 1987), and it appears unlikely that it simply represents a degradation product of trypsinogen 1 as was previously suggested (Figarella *et al.*, 1984). Alternatively, the intraductal activation of zymogens rather than their increased concentration might lead to these protein precipitates (Allan and White, 1974). Indeed, recent findings in alcoholic subjects support this hypothesis (see Section 11.4.4.3).

The attraction of the protein plug theory is its almost universal applicability to all forms of chronic pancreatitis. Its weakness is the absence of convincing evidence to date that the protein precipitates are the cause and not the result of metabolic derangements in the pancreas. Another problem is that generalized dilatation of the main pancreatic duct is commonly demonstrated by endoscopic retrograde pancreatography in patients with alcoholic pancreatitis. If this were caused by intraductal lesions, such lesions should be readily seen down to the level of the sphincter of Oddi. Such indeed seems to be the case in the patients seen in Marseilles, but is usually only a very late phenomenon in patients in the United States or Australia. Thus, there may be geographic differences in the type of pancreatitis seen.

11.4.4.3. Pancreatic Secretion and Tissue Concentrations of Pancreatic Enzymes

In the dog, the pancreatic response to exogenous cholecystokinin was enhanced after 3 months of alcohol feeding (Sarles *et al.*, 1973) but not after 2 years (Sarles *et al.*, 1977). However, in the latter situation, the pancreatic response to a standard dose of secretin was increased (Sarles *et al.*, 1977). This enhanced sensitivity to secretin may have been caused by changes in the ductular epithelium (reduplication and hyperplasia)

rather than by changes in acinar cells. In humans, chronic alcoholism is also associated with increases in both post-CCK peak protein concentrations (Harada *et al.*, 1979; Sahel and Sarles, 1979) and protein output (Brugge *et al.*, 1985; Rinderknecht *et al.*, 1985). More specifically, chronic intake of ethanol in the human is associated with selective increases in the anionic variant of trypsinogen and the lysosomal enzyme leucine napthylamidase. Because trypsin inhibitor secretion was unaffected, the ratio of trypsinogen to trypsin inhibitor was significantly increased in alcoholics, a finding that might favor intraductal activation of pancreatic zymogen. Indeed, intraductal activation of proenzyme is suggested by the presence of complexes of chymotrypsin and α-proteinase inhibitor in the pancreatic juice of patients with chronic pancreatitis (Miszczuk-Jamska *et al.*, 1983). In addition to protein hypersecretion, it has long been known (Dreiling *et al.*, 1973; Neves *et al.*, 1983) that pancreatic output of bicarbonate is increased in chronic alcoholics prior to the onset of overt pancreatic disease.

Chronic ethanol consumption also alters the pancreatic response to intravenous alcohol. As discussed earlier, intravenous alcohol markedly suppresses secretin-stimulated exocrine secretion in animals and nonalcoholic human subjects. After chronic ethanol ingestion, an acute infusion has the opposite effect, an increase in flow rate and in the outputs of bicarbonate and protein (Sarles *et al.*, 1973).

Variable changes in the tissue concentrations of pancreatic enzymes in animals after chronic ethanol intake have been reported by a number of authors (Tsuzaki *et al.*, 1965; Goslin *et al.*, 1965; Pirola *et al.*, 1970; Sarles *et al.*, 1971b). However, these studies are difficult to assess in view of the absence of pair feeding in some studies or the apparently small amount of ethanol consumed in others. Even when these methodological problems are overcome, both increased and decreased enzyme content have been reported (Singh *et al.*, 1982; Singh, 1983; Bode *et al.*, 1986; Grönroos *et al.*, 1988).

11.5. Nutrition

The role of diet in the pathogenesis of pancreatitis is unsettled. The traditional view is that malnutrition favors the development of pancreatitis in the alcoholic. This is supported by the occurrence of calcifying pancreatitis in regions of the world where kwashiorkor is prevalent (Shaper, 1964) and by the observations by

Pitchumoni *et al*. (1980) and Mezey *et al*. (1988) of poor nutrition in American patients with alcoholic pancreatitis. Furthermore, patients hospitalized after episodes of severe alcoholism and malnutrition have impaired pancreatic secretion (Dinoso *et al*., 1971), with return to normal secretion after dietary restitution even with continued alcohol intake (Mezey and Potter, 1976). It should be noted, however, that such hyposecretion may represent an adaptive response that could even be protective (since a gland containing less enzymes could be less prone to autodigestion).

The possibility that a high nutritional intake favors pancreatitis is raised by the finding that administration of high-fat or -protein diets predisposes animals to experimental pancreatitis (Maki *et al*., 1967; Haig, 1970; Ramo, 1987; Ramo *et al*., 1987), and protein nutrition alters the pancreatic response to chronic ethanol administration. Chronic ethanol intake in rats enhanced overall protein synthesis when the rat diet contained adequate levels of protein, but protein deficiency markedly attenuated these effects (Korsten *et al*., 1990). A number of dietary surveys (Sarles *et al*., 1965; Sarles, 1973; Durbec *et al*., 1978; Goebell and Hotz, 1975; Vantini *et al*., 1977; Uscanga *et al*., 1985) have demonstrated an increased intake of dietary fat and protein in patients with alcoholic pancreatitis. Studies by Pitchumoni *et al*. (1980) also reported that patients with alcoholic pancreatitis consume more fat than patients with alcoholic cirrhosis. However, when the data are adjusted for differences in age and sex (using multivariate analysis), no dietary differences between patients with alcoholic pancreatitis and those with alcoholic cirrhosis could be detected (Wilson *et al*., 1985).

A compromise position on this nutritional controversy has been proposed (Pirola and Lieber, 1974). Prolonged malnutrition could contribute to or aggravate the development of the chronic degenerative features of alcoholic pancreatitis. Good nutrition, by increasing the secretory capacity of the pancreas, would favor acute relapse. In this respect, there is abundant evidence for adaptive increases in pancreatic secretion (Dagorn, 1986) and for its potentiation by ethanol (Goslin *et al*., 1965).

11.6. Summary: Pathogenesis of Alcoholic Pancreatitis

In general, so-called "big duct" theories invoke reflux of injurious agents—bile, activated zymogens, and, most recently (Braganza, 1983; Acheson *et al*., 1985), reactive hepatic metabolites—into the pancreatic

duct from either the biliary system or the duodenum. However, biliopancreatic reflux depends on both a common channel and a favorable pressure gradient between the biliary and pancreatic systems. Neither of these requirements is a constant finding in patients with alcohol-induced disease. Likewise, theories that depend on reflux of noxious duodenal contents into the pancreatic duct overlook the early animal studies that refute this possibility (Pfeffer *et al*., 1957; Robinson and Dunphy, 1963). Finally, although such mechanisms could conceivably evoke a fulminant inflammatory process or even exacerbate an established lesion, they do not adequately address the dual nature of alcoholic pancreatitis, i.e., symptomatic acute attacks and chronic progressive parenchymal destruction.

Sarles *et al*. shifted the investigative focus from "big ducts" to "small ducts." With this, the chronic nature of alcoholic pancreatitis was clearly highlighted. The protein plugs that occlude the ducts are commonly calcified, a process that may be accelerated by a deficiency of "stone protein" in the pancreatic juice (see Section 11.4.4.2). General acceptance of this pathogenetic scheme awaits convincing evidence that protein plugs actually precede the development of pancreatitis and do not merely result from pancreatic inflammation. Moreover, the theory accommodates, only with great difficulty, the occurrence of acute attacks in chronic pancreatitis.

11.7. Clinical Aspects of Alcoholic Pancreatitis

11.7.1. Acute Attack: Symptoms and Signs

Abdominal pain and vomiting are the most common symptoms. Typically, the pain is severe, constant, poorly localized to the upper abdomen with radiation to the back, and often relieved by leaning forward. Abdominal tenderness is almost invariable, but it is generally less than one would expect from the severity of the symptoms.

A mild attack may last 2 to 3 days. Severe attacks may persist for 2 to 4 weeks and are associated with hypotension and a mortality of up to 30%. Mild fever commonly develops in the first few days. A high swinging fever may indicate the development of an abscess. Other complications that may develop within the first week include renal dysfunction, mild jaundice, disturbances of coagulation, hypocalcemia, psychotic reactions, and local spread of inflammation to involve almost any nearby organ.

11.7.2. Clinical Course

The first attack most often occurs after 10 to 15 years of heavy drinking (at least 60–80 g/day, usually more) and is most often seen in males in the fourth decade of life. Recurrent attacks, precipitated by alcohol abuse, typically occur at intervals of weeks or months. With the passage of time, these tend to become more frequent but less severe, and the complications of chronic pancreatitis then become more prominent, notably diabetes and malabsorption. Drug addiction is a common association, possibly because of the relentless pain or because of associated personality disorders. Malabsorption, diabetes, and self-neglect combine to produce a weakened, emaciated state commonly leading to death in middle age.

As discussed before, in some patients, especially in France, the illness is characterized by the insidious onset and steady progression of pancreatic insufficiency with very little in the way of acute inflammatory episodes. In still others, the dominant feature is relentless, almost constant pain. If the patient can remain abstinent, there is some evidence that severity and frequency of painful episodes will decrease (Marks and Bank, 1963), especially in those who have undergone pancreatojejunostomy (Scuro et al., 1983). With time, moreover, progressively more patients remain pain-free for increasing periods of time (Gastard et al., 1973).

11.7.3. Diagnosis

In clinical practice, the diagnosis of alcoholic pancreatitis is often based confidently on the association of compatible clinical features, a history of alcohol abuse, and elevations of the serum amylase level. With time, the diagnosis usually becomes more secure as the relationship of subsequent attacks to alcohol intake becomes apparent, as other features are demonstrated (such as pancreatic calcification and pancreatic insufficiency), and as other causes of abdominal pain and of pancreatitis are excluded. However, the following difficult diagnostic problems may be encountered.

11.7.3.1. How Is the Diagnosis of Pancreatitis Confirmed during an Acute Relapse?

In clinical practice, unequivocal proof of the existence of pancreatitis is rarely obtained. The single most useful diagnostic test of pancreatitis is the serum amylase activity, especially a value more than three times higher than the normal upper limit. However, the serum amylase level can fluctuate markedly within the space of hours, shows poor correlation with clinical severity, is normal in 10% to 20% of patients during an acute attack and may be elevated by many other conditions (Salt and Schenker, 1976) including acute alcohol ingestion (Berk et al., 1979; Dutta et al., 1981). Furthermore, it represents the sum of the activities of isoenzymes of salivary and pancreatic origin, and the isoenzymes of salivary type may originate from a variety of organs, including the pancreas (Salt and Schenker, 1976; Shimamura et al., 1976). Measurement of lipase activity may be helpful because its half-life in the serum is longer than that of amylase. This property may be useful if a patient is first seen some time after the onset of an attack, at a time when the serum amylase has already returned to normal. Furthermore, calculation of the lipase/amylase ratio may allow differentiation between alcoholic and nonalcoholic forms of pancreatitis (Gumaste et al., 1991). Radiologic tests and sonographic examinations are essential during the relapse episode and are discussed subsequently in this chapter.

11.7.3.2. Does Hyperamylasemia in an Alcoholic Indicate Pancreatitis?

This is a common problem in emergency rooms and in patients admitted to alcohol detoxification units. Approximately one-third of alcoholics admitted to alcoholism treatment centers have hyperamylasemia (Dutta et al., 1981), although a single intoxicating dose of alcohol does not cause hyperamylasemia in healthy subjects (Myhre and Nesbett, 1949; Fischer et al., 1976). The cause of hyperamylasemia commonly seen in alcoholics is unknown: it is not related to clinically evident pancreatitis, may be associated with a reduced rather than an elevated amylase clearance, and more frequently involves the salivary rather than the pancreatic type of isoenzyme (Berk et al., 1979; Dutta et al., 1981). In addition, hyperamylasemia may indicate other serious intraabdominal processes such as intestinal ischemia or perforation and biliary tract obstruction.

11.7.3.3. How Is the Alcoholic Etiology Confirmed?

A clinical decision that a case of pancreatitis is the result of alcohol excess is usually made easily and reliably. However, strict validation of the diagnosis is difficult. Furthermore, in a society where alcohol excess is common, there is danger of overlooking other etiological causes. The problem is twofold. The first is the establishment that alcohol intake is excessive. Despite the existence of very useful clinical and laboratory criteria (Chapter 18), unequivocal proof may be impossible in some individual cases. The second problem is proving that the alcohol excess is the cause of the pancreatitis. One has to strike a balance between excluding other etiologies and avoiding overinvestigation. A reasonable approach is to advise total abstention from alcohol, to inquire about abdominal trauma or drugs known to cause pancreatitis, to obtain an abdominal sonogram, and to measure serum lipid and calcium levels. If attacks recur despite abstinence and negative results of tests, then further investigations are necessary. The presence of pancreatic calcification is highly suggestive of an alcoholic etiology, but it should be noted that calcification may spontaneously regress in about one-third of cases over time (Ammann *et al.*, 1988).

11.7.3.4. Is There Significant Primary or Secondary Biliary Tract Disease and What Should Be Done about It?

Some degree of cholestasis is common in pancreatitis and occasionally causes diagnostic confusion. Mild jaundice with tapered narrowing of the common bile duct as it passes through the pancreas is very common in all forms of acute pancreatic inflammation and requires no treatment. Elevation of the serum alkaline phosphatase level is also common. It is usually mild, but occasionally very high levels are encountered in the absence of primary hepatobiliary disease (Hislop and Ryan, 1977). Severe jaundice occasionally develops because of significant biliary tract obstruction from a pancreatic stricture or pseudocyst. It is important in such cases to perform direct cholangiography to exclude the presence of biliary tract calculus or malignancy and to assess the type and severity of the stricture (Shi and Ham, 1980). When the pancreatitis has an alcoholic etiology, the possibility arises that the jaundice is the result of alcoholic liver disease. Curiously, this is an uncommon clinical association (Marks and Bank, 1963), although some of the pathological changes of chronic alcoholic liver disease and alcoholic pancreatitis often coexist (Sobel and Waye, 1963).

A number of studies have defined the clinical course of cholestasis in these patients (Afroundakis and Kaplowitz, 1981; Wisloff *et al.*, 1982; Petrozza *et al.*, 1984). In now appears that there is a high likelihood for the development of biliary cirrhosis in patients with obstruction of the distal common bile duct from alcoholic pancreatitis. To distinguish between fixed and reversible (edematous) stenosis, close clinical observation, including sequential liver biopsies, is imperative. With this approach it has been shown that a significant proportion of patients may spontaneously improve and not require surgery. However, operative decompression should not be delayed if cholangitis, progressive jaundice, or deterioration of liver histology (periportal fibrosis) develops.

11.7.3.5. How Is the Presence of Pancreatic Exocrine Insufficiency Established?

Nonintubative tests are simple and safe but have limited reliability. The simplest is the examination of feces with Sudan III, which stains globules of neutral triglyceride with an orange color. The presence of these globules in a patient with steatorrhea should be a useful indicator of a pancreatic etiology (Luk, 1979), but this differentiation has been questioned (Khouri *et al.*, 1989) inasmuch as fecal triglycerides (by TLC determination) are not increased in patients with pancreatic insufficiency (possibly because lipolysis takes place within the colon). Quantitation of the fecal fat under controlled conditions (72-hr collection, 100 g dietary fat per day) gives a crude indication of progress of the disease or of response to therapy. Furthermore, very high levels (more than double the normal upper limit of 6 g/day) are suggestive of a pancreatic cause of the steatorrhea. The urinary excretion of *para*-aminobenzoic acid (PABA) after an oral dose of N-benzoyl-L-tyrosyl-*p*-aminobenzoic acid (Arvanitakis and Greenberger, 1976; Greenberger, 1976) and the [^{14}C]triolein breath test (TBT) are newer, more rapid tests of pancreatic function. Correction of an abnormal triolein breath test by pancreatic extract is highly suggestive of pancreatic insufficiency (Goff, 1982). The use of the TBT in patients with advanced alcoholic liver injury may produce falsely positive results (Korsten *et al.*, 1987; Absalom *et al.*, 1988) because production of radiolabeled CO_2 from prelabeled fatty acids depends, at least in part, on both normal bile salt production and fatty acid oxidation by the liver. PABA excretion appears to be more independent of liver function and should be employed in this setting (Meyer *et al.*, 1986). Intubative tests provide a more detailed quantitation of pancreatic exocrine function. A standard stimulus such as secretin, cholecystokinin, or both, is given, followed by duodenal aspiration using a double-lumen tube (Dreiling, 1975) or direct cannulation of the pancreatic duct (Robberecht *et al.*, 1975). Probably the best indicators of pancreatic insufficiency are a reduction in flow rate and in bicarbonate concentrations (< 90 meq/liter).

11.7.3.6. What Radiological Tests Should Be Done?

The diagnosis of pancreatitis will be strengthened by plain x-rays showing the radiological features of inflammation in the vicinity of the pancreas. These features include a "sentinel loop" (localized small bowel ileus), a "cut-off" sign (constant constriction of the midtransverse colon with air proximally), pleural effusions (usually left-sided), and basal pulmonary atelectasis. The presence of chronic pancreatic disease is confirmed in a minority of patients by the demonstration of pancreatic calcification (best seen in an oblique or lateral view). The presence of gallstones should be sought, using ultrasonography, oral cholecystography, or both.

Ultrasonography is of particular value in demonstrating the presence of pseudocysts, dilated bile ducts, and gallbladder calculi, but is much less reliable in assessing the size of the pancreas and in demonstrating calculi in the common bile duct. It does not consistently distinguish between the various causes of pancreatic enlargement (malignancy and acute or chronic inflammation) and is less sensitive in obese patients and in the presence of large amounts of bowel gas. Computerized axial tomography (CAT) tends to be more useful than ultrasonography in showing the size, shape, and position of the pancreas, especially in obese patients. It also provides prognostic information: areas of hypovascularity (necrosis), loss of peripancreatic tissue planes, and fluid collections indicate more severe degrees of inflammation and a more guarded outcome (London et al., 1989). Magnetic resonance imaging (MRI) has not yet demonstrated any superiority over CAT in terms of visualizing the pancreas, possibly because pancreatic tissue and adjacent loops of fluid-filled small intestine produce similar intensity signals on both T_1- and T_2-weighted sequences (Smith et al., 1982). In addition, as with CAT, patients lacking retroperitoneal fat are imaged with less clarity. New MRI techniques, however, such as the use of paramagnetic contrast media, may eventually result in improved visualization of the pancreas (Steiner, 1985).

Endoscopic retrograde cholangiopancreatography (ERCP) is the procedure of choice for demonstrating the pancreatic duct system and assessing the presence of obstruction, dilatation, and stricture. With microtransducers, it has been shown that pressures within the pancreatic duct are elevated in chronic alcoholic pancreatitis and that hypermotility may be found within the sphincter of Oddi (Okazaki et al., 1988). In contrast, pressures obtained using pneumohydraulic infusion systems have failed to show these features in other patients with chronic pancreatitis (Novis et al., 1985). The main complications of ERCP are pancreatitis in the presence of inflammation and abscess formation in the presence of a pseudocyst. The former can be minimized by careful injection of contrast during the examination.

11.7.3.7. Has Pancreatic Carcinoma Been Excluded?

Occasionally, the exclusion of pancreatic carcinoma in a patient with chronic alcoholic pancreatitis will represent an extraordinarily difficult diagnostic problem. Both conditions may coexist. Improved methods of diagnosis have consistently failed to improve the 5-year survival rate for those with pancreatic carcinoma. However, proof of the presence of malignancy is helpful not only in planning palliative surgery and in avoiding unnecessary tests but also in the overall care of the patient and his emotional needs. Tests such as angiography, pancreatography, external ultrasonography, and computerized axial tomography are helpful but often lack diagnostic specificity when it is most needed clinically. At this time, endoscopic ultrasonography (EUS) is under intensive study, but its sensitivity and specificity are not yet defined.

Rarely, histological proof may be obtained by endoscopic biopsy. Cytological examination of pancreatic duct contents requires a degree of expertise that appears to be beyond most centers. Against this background, pancreatic biopsy seems attractive, either percutaneously using CAT guidance or at laparotomy.

11.7.4. Prognosis

For the acute attack of pancreatitis the prognosis is considerably better than for gallstone-associated pancreatitis, and a survival rate of greater than 90% can be expected. Prognostic criteria have been proposed by Ranson et al. (1974) and Blamey et al. (1984) but usually are more relevant in acute (biliary) pancreatitis. Poor prognostic signs include an age over 55 years, a blood glucose level greater than 180 mg/100 ml, a white cell count over 16,000/mm³, signs of hemoconcentration, and a serum calcium level of less than 8 mg/100 ml. It remains uncertain whether CAT offers additional predictive value in the acute setting (London et al., 1989).

Chronic pancreatic disease may already be present at the time of the first detectable inflammatory episode

(Strum and Spiro, 1971). Unfortunately, most patients continue to drink heavily, and further episodes, sometimes precipitated by quite small amounts of alcohol, tend to result in severe complications and death in about 10 years. Abstinence from alcohol may be the most important determinant of prognosis. Among patients treated surgically for pancreatic pain, the 10-year survival is 80% in those who stop drinking and 25% to 60% in those who continue to drink (White and Keith, 1973; Leger *et al.*, 1974; Frey *et al.*, 1976).

Carcinoma of the pancreas may occur with increased frequency in alcoholic patients (Burch and Ansari, 1968), but its frequency correlates with cigarette usage rather than with alcohol consumption (Wynder *et al.*, 1973; Lin and Kessler, 1981; Ghadirian *et al.*, 1991).

11.7.5. Therapy

11.7.5.1. Acute Attacks

Traditionally, treatment is aimed at relief of pain, avoidance of pancreatic stimulation, and maintenance of fluid balance and nutrition.

11.7.5.1a. Analgesia. Pentazocine is the pain reliever of choice because it has relatively little effect on the tone of the sphincter of Oddi (Economou and Ward-McQuaid, 1971). For the same reason, meperidine is preferable to morphine (Kjellgren, 1960).

11.7.5.1b. Avoidance of Pancreatic Stimulation. The value of nasogastric aspiration has not been supported by clinical trials (Levant *et al.*, 1974; Naeje *et al.*, 1978; Field *et al.*, 1979; Fuller *et al.*, 1981; Lange and Pederson, 1983; Sarr *et al.*, 1986). However, the results of these studies may be applicable only to certain selected patients with pancreatitis of mild to moderate severity, and it would be premature to abandon this form of treatment routinely. Its use is supported by many anecdotal observations and by physiological considerations. Nasogastric suction often dramatically relieves pain in patients with acute pancreatitis, but this might be a result of relief from gaseous distension and ileus.

11.7.5.1c. Fluid and Nutritional Replacement. Intravenous fluids (dextrose and electrolytes) are given until pain subsides, and oral feeding is then cautiously and progressively introduced. To combat shock and to replace the large volumes of fluid that may be lost retroperitoneally, infusions of plasma, low-molecular-weight dextrans, or blood may be required. Severe cases require monitoring of urinary output and central venous

pressure and special attention to the serum calcium and magnesium levels. In prolonged cases, especially if surgery for a pseudocyst or abscess is contemplated, total parenteral nutrition will be needed. However, it should be borne in mind that parenteral nutrition itself stimulates pancreatic secretion, although less than oral feeding. Furthermore, an unexplained hypercalcemia and pancreatitis can result from parenteral nutrition (Izsak *et al.*, 1980).

11.7.5.1d. Miscellaneous. The search for more effective therapy has been disappointing. Numerous therapeutic agents have been tried and not shown to be of routine value in controlled clinical studies. These include antibiotics (Howes *et al.*, 1975; Finch *et al.*, 1976), aprotinin (Trasylol) (Welbourne *et al.*, 1977), glucagon (Dürr *et al.*, 1976; Olazabal and Fuller, 1978), calcitonin (Goebell, 1976; Martinez and Navarrete, 1984), and cimetidine (Meshkinpour *et al.*, 1979). Somatostatin, an inhibitor of pancreatic exocrine secretion, did not reduce the mortality in pancreatitis but may prevent local complications (Choi *et al.*, 1989).

11.7.5.2. Pseudocysts

Even in the absence of a palpable mass, pancreatic pseudocysts should be suspected in patients with abdominal pain and persistent increases in the serum amylase level. Ultrasonic examination is the primary modality for confirming the diagnosis, since it is non-invasive and has reasonable sensitivity and specificity (Bradley and Clements, 1974; Bradley *et al.*, 1976; Gonzales *et al.*, 1976). ERCP may be helpful in detecting small pseudocysts not detected by ultrasonography (Laxson *et al.*, 1985) and in defining associated abnormalities of the biliary tree. Moreover, when operative intervention is deemed necessary, ERCP may influence the choice of operation in almost one-half of patients (O'Connor *et al.*, 1986). Despite earlier concerns (Wind *et al.*, 1976), ERCP-related sepsis is rare (less than 5% of cases), as is clinically significant acute pancreatitis. Serious complications of untreated pseudocysts include rupture into the free peritoneal cavity or adjacent organs, intracystic hemorrhage, abscess formation, and obstructive jaundice. Nevertheless, the initial treatment of patients with acute pancreatic pseudocysts should be conservative.

As a result of a study that demonstrated that about 25% of pseudocysts will undergo spontaneous resolution during serial ultrasonic examinations, a 3-week period of nonoperative treatment was advised (Bradley and Clements, 1974; Bradley *et al.*, 1976). In addition to allowing resolution, the delay of surgery should improve the encapsulation of the pseudocyst by fibrous tissue. Indefinite delay of elective surgery has been proposed by others (Czaja *et al.*, 1975; Butt, 1977), but this approach has been disputed (Bradley *et al.*, 1979). Thus, it now appears that after 4 to 6 weeks of observation, spontaneous resolution of a pseudocyst is unlikely, and the risk of serious complications exceeds that of elective drainage (Warshaw and Rattner, 1985). For many years, internal cystgastrostomy has been the gold standard of drainage procedures for mature pseudocysts. More recently, percutaneous aspiration of pseudocysts under CAT scan guidance or ultrasonography has been advocated as the initial approach (Barkin *et al.*, 1981). Permanent resolution can be expected, however, in only 25% of mature pseudocysts using this technique. Internal drainage of some pseudocysts can be achieved both percutaneously (Bernardino and Amerson, 1984) and endoscopically (Cremer and Deviere, 1986; Cremer *et al.*, 1989), but these innovative approaches have not yet been prospectively studied in controlled fashion.

11.7.5.3. Chronic Pain

Pain is by far the most common complication of chronic pancreatitis, occurring chronically or intermittently in over half of patients with the disorder. The pain is typically epigastric, radiates to the back, and is often relieved by lying prone with the legs flexed at the hips. As previously noted, the duration and frequency of painful episodes gradually decrease over time (Gastard *et al.*, 1973; Ammann *et al.*, 1984; Miyake *et al.*, 1987; Hayakawa *et al.*, 1989), especially in abstinent patients. The mechanism of pain in chronic pancreatitis is unclear but seems, in part, related to the ability of the pancreas to respond to secretagogues. This would explain the clinical observation that pain in chronic pancreatitis subsides if or when pancreatic insufficiency supervenes. The alternative concept that pain arises from obstruction of the pancreatic duct remains possible. Although the likelihood of finding a pancreatic duct stricture (assessed using ERCP) was found to be no different in patients with and without pain (Bornman *et al.*, 1980), more recent attempts to relieve pain by extracorporeal and endoscopic lithotripsy have been successful.

Treatment of pain in chronic pancreatitis is often discouraging. Basic tenets of management include total abstinence from alcohol, avoidance of large meals, and, if possible, reliance on nonnarcotic analgesics. Chronic use of narcotics for pain relief in this setting is likely to

result in physical dependence. If the above measures prove inadequate and evaluation has excluded other causes of abdominal pain (e.g., pseudocysts, tumor, or ulcers), a trial of pancreatic enzyme therapy is warranted. The therapeutic efficacy of the latter approach has not been uniform (Isaksson and Ihse, 1983; Slaff *et al.*, 1984; Halgreen *et al.*, 1986) but appears most useful in patients with mild pancreatic insufficiency. Other nonoperative modalities include celiac plexus block (Leung *et al.*, 1983; Madsen and Hansen, 1985), acupuncture, and transcutaneous electrical nerve stimulation (TENS). If celiac plexus block is performed, the results are likely to be transient. However, repeated blocks are feasible and might conceivably delay surgery in some patients (DiMagno and Clain, 1986). In the only reported study, the results of acupuncture and TENS were disappointing (Ballegaard *et al.*, 1985). Endoscopic management of pain is a future prospect. Sporadic reports of endoscopic occlusion of the pancreatic duct in patients with intractable pain (Roesch and Demling, 1982) as well as attempts at endoscopic lithotomy (Tsurumi *et al.*, 1984; Cremer *et al.*, 1988) and stent placement (Kozarek *et al.*, 1989) are encouraging but must be validated in controlled studies that include long-term follow-up.

Operative approaches to pain in chronic pancreatitis have traditionally been directed toward drainage of the pancreatic duct. Procedures such as sphincteroplasty, distal pancreaticojejunostomy, and lateral pancreaticojejunostomy (Puestow) are "tailored" to the preoperative pancreatogram and achieve success in a significant (as high as 80%) number of patients. The benefits of these operations do not prove that ductal hypertension *per se* is the cause of pain, since postoperative clinical improvement seems to occur despite endoscopic evidence of anastomotic occlusion (Kugelberg *et al.*, 1976). Alternatively, the benefits of these procedures might derive from pancreatic denervation. This is implied in the success of nondrainage approaches that are directed toward denervation of the pancreas (Warren *et al.*, 1984; Hiraoka *et al.*, 1986).

11.7.5.4. Pancreatic Insufficiency

Steatorrhea is usually improved but not completely corrected by the use of pancreatic extracts (DiMagno *et al.*, 1977; Regan *et al.*, 1979). The choice of extract is influenced not only by the cost and stability of the preparation but also by its acceptability to the patient. Careful explanation and motivation by the physician are important. The stability of the preparation is enhanced by agents that reduce gastric acidity, such as H_2-receptor antagonists or omeprazole, and by enteric coating. At a dose of eight tablets per meal, most preparations will deliver sufficient lipase and trypsin to decrease malabsorption. Moreover, newer preparations with higher lipase content (16,000 units/capsule) may improve compliance. From the clinical standpoint, there is rarely a problem in reversing azotorrhea, since trypsin is more resistant to acid inactivation than lipase (DiMagno and Clain, 1986). However, should steatorrhea persist, gastric acid studies should be performed to assess the adequacy of acid reduction therapy. A reduction in dietary fat (to 50–75 g/day) may also help symptomatically and is usually spontaneously instituted by the patient.

Diabetes mellitus is usually mild but brittle, possibly because of coexistent glucagon deficiency (Banks, 1979). The usual dietary principles of the treatment of diabetes still apply, but reduction of obesity is less of a problem than in most diabetics. Although insulin may be needed, oral agents are less apt to precipitate hypoglycemia. Vascular sequelae of diabetes mellitus are rare in pancreatic diabetes, possibly because the life expectancy of these patients is relatively short.

11.7.5.5. Pancreatic Ascites

The clinical features of pancreatic ascites are relatively nonspecific: increase in abdominal girth, abdominal pain, and weight loss. These symptoms, in the context of known chronic alcoholism, may erroneously suggest cirrhosis with portal hypertension, peritoneal metastases, or tuberculous peritonitis. The most important diagnostic procedure in differentiating these possibilities is paracentesis. Typically, the fluid in pancreatic ascites contains a high amylase concentration (the serum amylase level is usually only minimally elevated) and is exudative (protein concentration >2.5 g/100 ml).

Initial management of pancreatic ascites is medical. Nonoperative therapy usually involves a trial (4 to 6 weeks) of parenteral nutrition (Stephens et al., 1975). If this approach fails to prevent fluid reaccumulation, low-dose pancreatic irradiation may be utilized (Morton et al., 1976). Ascites resistant to these conservative measures requires surgical intervention. If preoperative ERCP demonstrates a pseudocyst, internal drainage (cystgastrostomy or Roux-en-Y cystjejunostomy) is employed. Roux-en-Y pancreatojejunostomy is performed if there is a documented ductal rupture. Surgery should be undertaken within several days of ERCP to avoid septic sequelae from this procedure (Banks, 1979).

11.7.5.6. Summary

We administer intravenous fluids and analgesics during a relapse of pancreatitis and, early in the course, employ ultrasonography to exclude biliary tract disease or detect pancreatic pseudocysts. Unless pain is accompanied by ileus, we do not recommend nasogastric intubation. As the patient's symptoms improve, we slowly and progressively resume oral feeding. After resolution of the acute flare-up, we assess the patient's nutritional status and, if indicated, evaluate the integrity of pancreatic function with measurements of fecal fat, duodenal aspiration (secretin test), or noninvasive tests (e.g., PABA excretion or the triolein breath test). Persistent pain requires a search for complicating processes such as pseudocysts, pancreatic cancer, gastritis, or peptic ulcer disease. ERCP often plays a role in such investigations. If pain is caused by chronic pancreatitis per se, we rely on nonnarcotic analgesics, nonsteroidal antiinflammatory drugs, and pancreatic extracts. We employ repeated celiac plexus blocks to forestall surgery and hope that pain will "burn out" with the passage of time. Pancreatic insufficiency is treated somewhat empirically with varying doses of high-potency pancreatic enzymes and correction of fat-soluble vitamin deficits if present. In intractable cases, we attempt to modify dietary patterns (decrease the ingestion of fat) and eliminate gastric acid either by preventing its production or by its neutralization. In hyperglycemic patients, we use oral agents for as long as possible and administer insulin as a last resort and with great caution. The most important therapeutic goal is abstinence from alcohol, and we encourage the patient to contact a local chapter of Alcoholics Anonymous after inpatient rehabilitation. If heavy drinking recurs, the prognosis is poor, and relapses of the disorder are inevitable.

REFERENCES

Absalom, S. R., Saverymuttu, S. H., Maxwell, J. D., and Levin, G. E.: Triolein breath test of fat absorption in patients with chronic liver disease. Dig. Dis. Sci. **33**:565–569, 1988.

Acheson, D. W. K., Rose, P., Houston, J. B., and Braganza, J. M.: Induction of cytochrome P-450 in pancreatic disease: Consequence, coincidence or cause? Clin. Chim. Acta **153**:73–84, 1985.

Afroudakis, A., and Kaplowitz, N.: Liver histopathology is chronic common bile duct stenosis due to chronic alcoholic pancreatitis. Hepatology **1**:65–72, 1981.

Albo, R., Silen, W., and Goldman, L.: A critical clinical analysis of acute pancreatitis. *Arch. Surg.* **86**:1032–1038, 1963.

Ammann, R. W., Akovbiantz, A., Largiader, F., and Schueler, G.: Course and outcome of chronic pancreatitis: Longitudinal study of a mixed medical–surgical series of 245 patients. *Gastroenterology* **86**:820–828, 1984.

Ammann, R. W., Muench, R., Otto, R., Beuhler, H., Freiburghaus, A. U., and Siegenthaler, W.: Evolution and regression of pancreatic calcification in chronic pancreatitis. *Gastroenterology* **94**:1018–1028, 1988.

Arvanitakis, C., and Greenberger, N. J.: Diagnosis of pancreatic disease by a synthetic peptide: A new test of exocrine pancreatic function. *Lancet* **1**:663–666, 1976.

Ballegaard, S., Christophersen, S., Gamwell Davids, S., Hesse, J., and Vestergaard Olsen, N.: Acupuncture and transcutaneous electric nerve stimulation in the treatment of pain associated with chronic pancreatitis—a randomized study. *Scand. J. Gastroenterol.* **20**:1249–1254, 1985.

Bank, S.: Chronic pancreatitis: Clinical features and medical management. *Am. J. Gastroenterol.* **81**:153–167, 1986.

Banks, P. A.: *Pancreatitis.* New York, Plenum Medical, 1979.

Barkin, J. S., and Fayne, S. D.: Chronic pancreatitis: Update 1986. *Mt. Sinai J. Med. N.Y.* **53**:404–408, 1986.

Barkin, J. S., Smith, F. R., Pereiras, R., Isikoff, M., Levi, J., Livingstone, A., Hill, M., and Rogers, A. I.: Percutaneous aspiration of pancreatic pseudocyst. *Dig. Dis. Sci.* **26**:585–586, 1981.

Barroso, E.: Some effects of alcohol on pancreatic secretion. *Rev. Invest. Clin.* **1**:219–234, 1949.

Bayer, M., Rudick, J., Lieber, C. S., and Janowitz, H. D.: Inhibitory effect of ethanol on canine exocrine pancreatic secretion. *Gastroenterology* **63**:619–626, 1972.

Berk, J. E., Fridhandler, L., and Webb, S. F.: Does hyperamylasemia in a drunken alcoholic signify pancreatitis? *Am. J. Gastroenterol.* **71**:557–562, 1979.

Bernardino, M. E., and Amerson, J. R.: Percutaneous gastrocystostomy: A new approach to pancreatic pseudocyst drainage. *Am. J. Roentgenol.* **143**:1096–1097, 1984.

Blamey, S. L., Imrie, C. W., O'Neill, J., Gilmour, W. A., and Carter, D. C.: Prognostic factors in acute pancreatitis. *Gut* **25**:1340–1346, 1984.

Bode, C., Dürr, H.-K., and Bode, J. C.: Effect of short- and long-term alcohol feeding in rats on pancreatic enzymes content and enzyme secretion in isolated pancreatic lobules *in vitro*. *Int. J. Pancreatol.* **1**:129–139, 1986.

Bordalo, O., Noronha, M., and Dreiling, D. A.: Functional and morphologic studies of the effect of alcohol on the pancreatic parenchyma. *Mt. Sinai J. Med. N.Y.* **44**:481–484, 1977.

Bornman, P. C., Marks, I. N., Girdwood, A. H., Clain, J. E., Narunsky, L., Clain, D. J., and Wright, J. P.: Is pancreatic duct obstruction or stricture a major cause of pain in calcific pancreatitis? *Br. J. Surg.* **67**:425–428, 1980.

Bourke, J. B., Giggs, J. A., and Ebdon, D. S.: Variations in the incidence and spatial distribution of patients with primary acute pancreatitis in Nottingham 1969–1976. *Gut* **20**:366–371, 1979.

Boustiere, C., Sarles, H., Lohse, J., Durbec, J. P., and Sahel, J.: Citrate and calcium secretion in the pure human pancreatic juice of alcoholic and non-alcoholic men and of chronic pancreatitis patients. *Digestion* **32**:1–9, 1985.

Bradley, E.L. III, and Clements, L. J., Jr.: Implications of diagnostic ultrasound in the surgical management of pancreatic pseudocysts. *Am. J. Surg.* **127**:163–173, 1974.

Bradley, E. L., Gonzalez, A. C., and Clements, J. L., Jr.: Acute pancreatic pseudocysts: Incidence and implications. *Ann. Surg.* **184**:734–737, 1976.

Bradley, E. L., Clements, J. L., and Gonzales, A. C.: The natural history of pancreatic pseudocysts: A unified concept. *Am. J. Surg.* **137**:135–141, 1979.

Braganza, J. M.: Pancreatic disease: A casualty of hepatic "detoxification"? *Lancet* **2**:1000–1003, 1983.

Brugge, W. R., Burke, C. A., Brand, D. L., and Chey, W. Y.: Increased interdigestive pancreatic trypsin secretion in alcoholic pancreatic disease. *Dig. Dis. Sci.* **30**:431–439, 1985.

Burch, G. E., and Ansari, A.: Chronic alcoholism and carcinoma of the pancreas. *Arch. Intern. Med.* **122**:273–275, 1968.

Butt, J.: Medical management: Pancreatitis—pancreatic pseudocysts and their complications. *Gastroenterology* **73**:600–601, 1977.

Calderon-Attas, P., Furnelle, J., and Christophe, J.: *In vitro* effects of ethanol and ethanol metabolism in the rat pancreas. *Biochim. Biophys. Acta* **620**:387–399, 1980.

Cavarzan, A., Teixeira, A. S., Sarles, H., Palasciano, G., and Tiscornia, O.: Action of intragastric ethanol on the pancreatic secretion of conscious rats. *Digestion* **13**:145–152, 1975.

Chapman, B. A., and Pattison, N. R.: The effect of ethanol on phospholipid metabolism in rat pancreas. *Biochem. Pharmacol.* **32**:1897–1902, 1988.

Choi, T. K., Mok, F., Zhan, W. H., Fan, S. T., Lai, E. C. S., and Wong, J.: Somatostatin in the treatment of acute pancreatitis: A prospective randomized controlled trial. *Gut* **30**:223–227, 1989.

Cremer, M., and Deviere, J.: Endoscopic management of pancreatic cysts and pseudocysts. *Gastrointest. Endosc.* **32**: 367–368, 1986.

Cremer, M., Vandermeeren, A., and Delhaye, M.: Extracorporeal shock wave lithotripsy (ESWL) for pancreatic stones. *Gastroenterology* **94**:A80, 1988.

Cremer, M., Deviere, J., and Engelholm, L.: Endoscopic management of cysts and pseudocysts in chronic pancreatitis: Long term follow up after 7 years of experience. *Gastrointest. Endosc.* **35**:1–9, 1989.

Cueto, J., Tajen, N., and Zimmerman, B.: Studies of experimental alcoholic pancreatitis in the dog. *Surgery* **62**:159–166, 1976.

Czaja, A. J., Fisher, M., and Marin, G. A.: Spontaneous resolution of pancreatic masses (pseudocysts): Development and disappearance after acute alcoholic pancreatitis. *Arch. Intern. Med.* **135**:558–562, 1975.

Czernobilsky, B., and Mikut, K. W.: The diagnostic significance of interstitial pancreatitis found at autopsy. *Am. J. Clin. Pathol.* **41**:33–34, 1964.

Dani, R., Antunes, L. J., Rocha, W. M., and Nogueira, C. E.: HLA Aw22 and Aw24 associated with chronic alcoholic calcifying pancreatitis. *Arg. Gastroenterol.* **15**:163–166, 1978.

Darle, N., Ekholm, R., and Edlund, Y.: Ultrastructure of the rat exocrine pancreas after long-term intake of ethanol. *Gastroenterology* **58**:62–72, 1970.

Davis, A. E., and Pirola, R. C.: The effects of ethyl alcohol on pancreatic exocrine function. *Med. J. Aust.* **2**:757–760, 1966.

DeCaro, A. M., Bonicel, J. J., Rouimi, P., DeCaro, J. D., Sarles, H., and Rovery, M.: Complete amino acid sequence of an immunoreactive form of human pancreatic stone protein isolated from pancreatic juice. *Eur. J. Biochem.* **168**:201–207, 1987.

DiMagno, E. P., and Clain, J. E.: Chronic pancreatitis. In: *The Exocrine Pancreas, Pathobiology and Diseases* (V. L. W. Go, F. P. Brooks, E. P. DiMagno, J. D. Gardner, E. Lebenthal, and G. A. Scheele, eds.), New York, Raven Press, pp. 541–575, 1986.

DiMagno, E. P., Malagelada, J. R., Go, V. L. W., and Moertel, C. G.: Fate of orally ingested enzymes in pancreatic insufficiency: Comparison of two dosage schedules. *N. Engl. J. Med.* **296**:1318–1322, 1977.

Dinoso, V. P., Chey, W. Y., and Lorber, S. H.: Pancreatic exocrine function in chronic alcoholics. *Gastroenterology* **61**:559–560, 1971.

Dreiling, D. A.: Alcoholism, alcoholic pancreatitis and pancreatic secretion. *Ann. N.Y. Acad. Sci.* **252**:187–199, 1975.

Dreiling, D. A., Greenstein, A. J., and Bordalo, O.: The hypersecretory states of the pancreas. *Am. J. Gastroenterol.* **59**:505–511, 1973.

Durbec, J. P., Sarles, H., and the International Group for the Study of Pancreatic Disease: Relationship between the relative risk of developing chronic pancreatitis and alcohol, protein and lipid consumption. *Digestion* **18**:337–350, 1978.

Dürr, H. K., Zelder, O., Maroske, D., Bode, C., and Bode, J. C.: Zur Behandlung der akuten Pankreatitis nut Glucagon, Bericht uber eine Doppelblindstudie. *Verh. Dtsch. Ges. Inn. Med.* **82**:970–973, 1976.

Dutta, S. K., Doublass, U. A., Nipper, H. C., and Levitt, M. D.: Prevalence and nature of hyperamylasemia in acute alcoholism. *Dig. Dis. Sci.* **26**:136–144, 1981.

Economou, G., and Ward-McQuaid, J. N.: A cross-over comparison of the effect of morphine, pethidine, pentazocine and phenazocine on biliary pressure. *Gut* **12**:218–221, 1971.

Fauchet, R., Genetet, B., Gosselin, J., and Gastrad, J.: HLA antigens in chronic alcoholic pancreatitis. *Tissue Antigens* **13**:163–166, 1979.

Field, B. F., Hepner, G. W., Shabot, M. M., Schwartz, A. A., State, D., Worthen, N., and Wilson, R.: Nasogastric suction in alcoholic pancreatitis. *Dig. Dis. Sci.* **24**:339–344, 1979.

Figarella, C., Amouric, M., and Guz-Crotte, O.: Proteolysis of human trypsinogen 1. Pathogenic implication in chronic pancreatitis. *Biochem. Biophys. Res. Commun.* **118**:154–161, 1984.

Finch, W. T., Sawyers, J. L., and Schenker, S.: A prospective study to determine the efficacy of antibiotics in acute pancreatitis. *Ann. Surg.* **183**:667–671, 1976.

Fischer, J. E., Rosen, H. M., Ebeid, A. M., James, J. H., Keane, J. M., and Soeters, P. B.: The effect of normalization of plasma amino acids on hepatic encephalopathy in man. *Surgery* **80**:77–91, 1976.

Frey, C. F., Child, C. G. III, and Fry, W.: Pancreatectomy for chronic pancreatitis. *Ann. Surg.* **184**:403–413, 1976.

Fuller, R. K., Loveland, J. P., and Frankel, M. H.: An evaluation of the efficacy of nasogastric suction treatment in alcoholic pancreatitis. *Am. J. Gastroenterol.* **75**:349–353, 1981.

Gambill, E. E., Baggenstoss, A. H., and Priestley, J. T.: Chronic relapsing pancreatitis. *Gastroenterology* **39**:404–413, 1960.

Gastard, J., Jouband, F., Farbos, T., Loussouarn, J., Marion, J., Pannier, M., Renaudet, F., Valdazo, R., and Gosselin, M.: Etiology and course of primary chronic pancreatitis in Western France. *Digestion* **9**:416–428, 1973.

Ghadirian, P., Simard, A., Baillargeon, J.: Tobacco, alcohol, and coffee and cancer of the pancreas. A population-based, case-control study in Quebec, Canada. *Cancer* **67**:2664–2670, 1991.

Giorgi, D., Bernard, J. P., deCaro, A., Multigner, L., Lapointe, R., Sarles, H., and Dagorn, J. C.: Pancreatic stone protein. I. Evidence that it is encoded by a pancreatic messenger ribonucleic acid. *Gastroenterology* **89**:381–386, 1985.

Goebell, H.: The role of calcium in pancreatic secretion and disease. *Acta Hepatogastroenterol.* **23**:151-161, 1976.

Goebell, H., and Hotz, J.: Nutritional aspects of chronic pancreatitis in Germany. *Biol. Gastroenterol. (Paris)* **8**:365, 1975.

Goff, J. S.: Two-stage triolein breath test differentiates pancreatic insufficiency from other causes of malabsorption. *Gastroenterology* **83**:44–46, 1982.

Gonzalez, A. C., Bradley, E. L. III, and Clements, J. L., Jr.: Pseudocyst formation in acute pancreatitis: Ultrasonographic evaluation of 99 cases. *Am. J. Roentgenol.* **127**:315–317, 1976.

Goslin, J., Hong, S. S., Magee, D. F., and White, T. T.: Relationship between diet, ethyl alcohol consumption, and some activities of the exocrine pancreas in rats. *Arch. Int. Pharmacodyn. Ther.* **157**:462–469, 1965.

Gosselin, M., Fauchet, R., Genetet, B., and Gastard, J.: Les antigenes HLA dans la pancréatite chronique alcoolique. *Gastroenterol. Clin. Biol.* **2**:883–886, 1978.

Greenberger, A. C.: A tubeless test of exocrine pancreatic function. *Lancet* **1**:663–666, 1976.

Grönroos, J. M., Aho, H. J., Meklin, S. S., Hakala, J., and Nevalainen, T. J.: Pancreatic digestive enzymes and ultrastructure after chronic alcohol intake in the rat. *Exp. Pathol.* **35**:197–208, 1988.

Gullo, L., Tabacchi, P. L., Corazza, G. R., Calanca, F., Campione, O., and Labu, G.: HLA B13 and chronic calcific pancreatitis. *Dig. Dis. Sci.* **27**:214–216, 1982.

Gumaste, V. V., Dave, P. B., Weissman, D., Messer, J.: Lipase/amylase ratio. *Gastroenterology* **101**:1361–1366, 1991.

Haig, T. H. B.: Experimental pancreatitis intensified by a high fat diet. *Surg. Gynecol. Obstet.* **131**:914–918, 1970.

Halgreen, H.: Symptomatic effect of pancreatic enzyme therapy in patients with chronic pancreatitis. *Scand. J. Gastroenterol.* **21**:104–108, 1986.

Harada, H., Yabe, H., Hanafusa, E., Ikuba, I., Takeda, M., Hayashi, T., Negron, A., Ono, A., Yamamoto, N., Mishima, K., and Kimura, I.: Analysis of pure pancreatic juice in patients with chronic alcoholism. *Gastroenterol. Jpn.* **14**:458–466, 1979.

Hayakawa, T., Kondo, T., Shibata, T., Sugimota, Y., and Kitagava, M.: Chronic alcoholism and evolution of pain and prognosis in chronic pancreatitis. *Dig. Dis. Sci.* **34:**33–38, 1989.

Hiraoka, T., Watanabe, E., Katoh, T., Hayashida, N., Mizutani, J., Kanemitsu, K., and Miyauchi, Y.: A new surgical approach for control of pain in chronic pancreatitis: Complete denervation of the pancreas. *Am. J. Surg.* **152:**549–551, 1986.

Hislop, I. G., and Ryan, G. P.: Elevated serum alkaline phosphatase level may aid in the diagnosis of pancreatitis. *Med. J. Aust.* **2:**209–211, 1977.

Howes, R., Zuidema, G. D., and Cameron, J. L.: Evaluation of prophylactic antibiotics in acute pancreatitis. *J. Surg. Res.* **18:**197–200, 1975.

Isaksson, G., and Ihse, I.: Pain reduction by an oral pancreatic enzyme preparation in chronic pancreatitis. *Dig. Dis. Sci.* **28:** 97–1002, 1983.

Izsak, E. M., Shike, M., Roulet, M., and Jeejeebhoy, K. N.: Pancreatitis in association with hypercalcemia in patients receiving total parenteral nutrition. *Gastroenterology* **79:** 555–558, 1980.

Joffe, S. N., Ferrie, M., and O'Hare, R.: The effect of ethanol on plasma biochemistry and the pancreas in the rat. *Acta Hepatogastroenterol.* **26:**323–325, 1979.

Kakizaki, G., Sasahara, M., Aikawa, T., Matsuo, M., Sugawara, Y., Nakamura, K., Endo, S., and Ito, Y.: On the pathogenesis of chronic alcoholic pancreatitis from the viewpoint of experimental results in rats. *Int. J. Pancreatol.* **2:**101–116, 1987.

Khouri, M. R., Ng, S.-N., Huang, G., and Shiau, Y.-F.: Fecal triglyceride excretion is not excessive in pancreatic insufficiency. *Gastroenterology* **99:**848–852, 1989.

Kjellgren, K.: The influence of morphine and pethidine in combination with levallorphan on biliary duct pressure after cholecystectomy. *Br. J. Anaesth.* **32:**2–6, 1960.

Korsten, M. A., Hodes, S. E., Saeli, J. F., Seitz, H. K., and Lieber, C. S.: Effects of ethanol on pancreatic secretion: Roles of gastric acid and exogenous secretin. *Gastroenterology* **76:** 1175, 1979.

Korsten, M. A., Seitz, H., Hodes, S. F., Klingenstein, J., and Lieber, C. S.: The effect of intravenous ethanol on pancreatic secretion in the conscious rat. *Dig. Dis. Sci.* **26:**790–795, 1981.

Korsten, M. A., Klapholz, M. B., Leaf, M. A., and Lieber, C. S.: Use of the triolein breath test in alcoholics with liver damage. *J. Lab. Clin. Med.* **109:**62–66, 1987.

Korsten, M. A., Wilson, J. S., and Lieber, C. S.: Interactive effects of dietary protein and ethanol on rat pancreas: Protein synthesis and enzyme secretion. *Gastroenterology* **99:**229–236, 1990.

Kozarek, R. A., Patterson, D. J., Ball, T. J., and Traverso, L. W.: Endoscopic placement of pancreatic stents and drains in the management of pancreatitis. *Ann. Surg.* **209:**261–266, 1989.

Kubota, K., Magee, D. R., and Sarles, H.: Biphasic action of intravenous ethanol on dog exocrine pancreatic secretion. *Dig. Dis. Sci.* **28:**1116–1120, 1983.

Kugelberg, C. H., Wehlin, L., Arnesjo, B., and Tylen, U.: Endoscopic pancreatography in evaluating results of pancreatico-jejunostomy. *Gut* **17:**267–272, 1976.

Lange, P., and Pederson, T.: Initial treatment of acute pancreatitis. *Surg. Gynecol. Obstet.* **157:**332–334, 1983.

Laxson, L. C., Fromkes, J. J., and Cooperman, M.: Endoscopic retrograde cholangiopancreatography in the management of pancreatic pseudocysts. *Am. J. Surg.* **150:**683–686, 1985.

Leger, L., Lenriot, J. P., and Lemaigre, G.: Five- to twenty-year follow-up after surgery for chronic pancreatitis in 148 patients. *Ann. Surg.* **180:**185–191, 1974.

Leung, J. W. C., Bowen-Wright, M., Aveling, W., Sharvon, P. J., and Cotton, P. B.: Coeliac plexus block for pain in pancreatic cancer and chronic pancreatitis. *Br. J. Surg.* **70:**730–732, 1983.

Levant, J. A., Secrist, D. M., Resein, H., Sturdevant, R. A. L., and Guth, P. H.: Nasogastric suction in the treatment of alcoholic pancreatitis: A controlled study. *J.A.M.A.* **229:**51–53, 1974.

Lin, R. S., and Kessler, I. L.: A multifactorial model for pancreatic cancer in man: Epidemiologic evidence. *J.A.M.A.* **245:**147–152, 1981.

Lohse, J., and Kraemer, R.: Calcium binding to the "stone protein" isolated from pancreatic stones of patients with chronic calcified pancreatitis. *Hoppe Seylers Z. Physiol. Chem.* **365:** 549–555, 1984.

London, N. J. M., Neoptolemos, J. P., Lavella, J., Bailey, I., and James, D.: Contrast-enhanced abdominal computed tomography scanning and prediction of severity of acute pancreatitis: A prospective study. *Br. J. Surg.* **76:**268–271, 1989.

Luk, G. D.: Qualitative fecal fat by light microscopy: A sensitive and specific screening test for steatorrhea and pancreatic insufficiency. *Gastroenterology* **76:**1189, 1979

Madsen, P., and Hansen, E.: Coeliac plexus block versus pancreaticogastrostomy for pain in chronic pancreatitis: A controlled randomized trial. *Scand. J. Gastroenterol.* **20:**1217–1220, 1985.

Maki, T., Kakizaki, G., Sato, T., Saito, Y., Onuma, T., and Nota, N.: Experimental study on alcoholic pancreatitis. *Tohoku J. Exp. Med.* **92:**415–421, 1967.

Marks, I. N., and Bank, S.: Aetiology, clinical features and diagnosis of pancreatitis in the South Western Cape. *S. Afr. Med. J.* **37:**1039–1053, 1963.

Marks, I. N., Bank, S., du Toit, A., Keraan, M. M., Krut, L. H., Mann, J., and Edelstein, I.: Genetic and nutritional factors in calcific pancreatitis. In: *4th World Congress of Gastroenterology Advance Abstracts* (P. Piis, P. Anthonisen, and H. Badeu, eds.), Copenhagen, Danish Medical Association, 1970, p. 220.

Martinez, E., and Navarrete, F.: A controlled trial of synthetic salmon calcitonin in the treatment of severe acute pancreatitis. *World J. Surg.* **8:**354–359, 1984.

Meshkinpour, H., Molinari, M. D., Gardner, L., Berk, J. E., and Hoechler, F. K.: Cimetidine in the treatment of acute alcoholic pancreatitis: A randomized, double-blind study. *Gastroenterology* **77:**687–690, 1979.

Meyer, B., Campbell, D., Curington, C., and Toskes, P.: Liver disease does not affect the bentiromide test. *Gastroenterology* **90:**1551, 1986.

Mezey, E., and Potter, J. J.: Changes in exocrine pancreatic function produced by altered dietary protein intake in drinking alcoholics. *Johns Hopkins Med. J.* **138:**7–12, 1976.

Mezey, E., Kolman, C. J., Diehl, A. M., Mitchell, M. C., and

Herlong, H. F.: Alcohol and dietary intake in the development of chronic pancreatitis and liver disease in alcoholism. *Am. J. Clin. Nutr.* **48:**148–151, 1988.

Miszczuk-Jamska, B., Guy, O., and Figarella, C.: α-Proteinase inhibitor in pure pancreatic juice: Characterization of a complexed form in patients with chronic calcifying pancreatitis and its significance. *Hoppe Seylers Z. Physiol. Chem.* **254:** 1597–1601, 1983.

Miyake, H., Harada, H., Kunichika, K., Ochi, K., and Kimura, I.: Clinical course and prognosis of chronic pancreatitis. *Pancreas* **2:**378–385, 1987.

Morton, R. E., DeLuca, R., Reisman, T. N., Raskin, J. B., and Rogers, A. I.: Pancreatic ascites: Successful treatment with pancreatic radiation. *Dig. Dis. Sci.* **21:**233–336, 1976.

Mott, C. B., Sarles, H., Tiscornia, O., and Gullo, L.: Inhibitory action of alcohol on human exocrine pancreatic secretion. *Am. J. Dig. Dis.* **17:**902–910, 1972.

Multigner, L., Sarles, H., Lombardo, D., and deCaro, A.: Pancreatic stone protein. II. Implication in stone formation during the course of chronic calcifying pancreatitis. *Gastroenterology* **89:**387–391, 1985.

Myhre, J., and Nesbett, S.: Alcohol and pancreatitis: Serum amylase determinations in normal individuals following ingestion of alcohol. *J. Lab. Clin. Med.* **34:**844–845, 1949.

Naeje, R., Salingret, E., Clumeck, N., DeTroyer, A., and Davis, G.: Is nasogastric suction necessary in acute pancreatitis? *Br. Med. J.* **2:**659–660, 1978.

Neves, M. M., Borges, D. R., and Vilela, M. P.: Exocrine pancreatic hypersecretion in Brazilian alcoholics. *Am. J. Gastroenterol.* **78:**513–516, 1983.

Novis, B. H., Young, G. O., Bank, S., and Marks, I. N.: Chronic pancreatitis and α-1-antitrypsin. *Lancet* **2:**748–9, 1975.

Novis, B. H., Bornman, P. C., Girdwood, A. W., and Marks, I. N.: Endoscopic manometry of the pancreatic duct and sphincter zone in patients with chronic pancreatitis. *Dig. Dis. Sci.* **30:** 225–228, 1985.

O'Connor, M., Kolars, J., Ansel, H., Silvis, S., and Vennes, J.: Preoperative endoscopic retrograde cholangiopancreatography in the surgical management of pancreatic pseudocysts. *Am. J. Surg.* **151:**18–24, 1986.

Okazaki, K., Yamamato, Y., Kagiyama, S., Tamura, S., Sakamoto, Y., Nakazawa, Y., Morita, M., and Yamamoto, Y.: Pressure of papillary sphincter zone and pancreatic main duct in patients with chronic pancreatitis in the early stage. *Scand. J. Gastroenterol.* **23:**501–507, 1988.

Olazabal, A., and Fuller, R.: Failure of glucagon in the treatment of alcoholic pancreatitis. *Gastroenterology* **74:**489–491, 1978.

Orrego-Matte, H., Navia, E., Feres, A., and Costamaillere, L.: Ethanol ingestion and incorporation of ^{32}P into phospholipids of pancreas in the rat. *Gastroenterology* **56:**280–285, 1969.

Perrier, C. V.: Symposium on the etiology and pathological anatomy of chronic pancreatitis. Marseilles, 1963. *Am. J. Dig. Dis.* **9:**371–376, 1964.

Petrozza, J. A., Dutta, S. K., Latham, P. S., Iber, F. L., and Gadacz, T. R.: Prevalence and natural history of distal common bile duct stenosis in alcoholic pancreatitis. *Dig. Dis. Sci.* **29:**890–895, 1984.

Pfeffer, R. B., Stasior, O., and Hinton, J. W.: The clinical picture of the sequential development of acute hemorrhagic pancreatitis in the dog. *Surg. Forum* **8:**248–251, 1957.

Pirola, R. C., and Davis, A. E.: The sphincter of Oddi and pancreatitis. *Am. J. Dig. Dis.* **15:**583–587, 1970a.

Pirola, R. C., and Lieber, C. S.: Acute and chronic pancreatitis. In *The Biology of Alcoholism, Vol. 3: Clinical Pathology* (B. Kissin and T. Begleiter, eds.), New York, Plenum Press, pp. 359–402.

Pirola, R. C., Taylor, K. B., Davis, A. E., and Liddelow, A. G.: Effects of ethanol, DL-ethionine, and protein deficiency on rat pancreas. *Am. J. Dig. Dis.* **15:**21–30, 1970.

Pitchumoni, C. S., Sonnenshein, M., Candido, F. M., Pancharam, P., and Cooperman, J. M.: Nutrition in the pathogenesis of alcoholic pancreatitis. *Am. J. Clin. Nutr.* **33:**631–636, 1980.

Pitchumoni, C. S., Glasser, M., Saran, R. M., Panchacharam, P., and Thelmo, W.: Pancreatic fibrosis in chronic alcoholics and nonalcoholics without clinical pancreatitis. *Am. J. Gastroenterol.* **79:**383–388, 1984.

Podol'skii, A. E.: Effect of ethanol on ultrastructure of the pancreatic acini. *Byull. Eksp. Biol. Med.* **83:**377–379, 1977.

Provansal-Cheylan, M., Lusher, M., deCaro, A., Multigner, L., Montalto, G., Sarles, H., and Delaage, M.: Monoclonal antibodies to pancreatic stone protein: Radioimmunoassay and immunological comparison with trypsin 1. *Biochimie* **68:**1109–1113, 1986.

Ramo, O. J.: Antecedent long term ethanol consumption in combination with different diets alters the severity of experimental acute pancreatitis in rats. *Gut* **28:**64–69, 1987.

Ramo, O. J., Apaja-Sarkkinen, M., and Jalovaara, P.: Experimental acute pancreatitis in rats receiving different diets and ethanol. *Res. Exp. Med.* **187:**33–41, 1987.

Ranson, J. H. C., Rifkind, K. M., Roses, D. F., Fink, S. D., Eng, K., and Spencer, F. C.: Prognostic signs and the role of operative management in acute pancreatitis. *Surg. Gynecol. Obstet.* **139:**69–81, 1974.

Rao, G. A., Goheen, S. C., Tsukamoto, H., and Larkin, E. C.: Chronic alcohol consumption may not cause pancreatic steatosis in rats. *IRCS Med. Sci.* **12:**486–487, 1984.

Regan, P. T., Malagelada, J. R., DiMagno, E. P., and Go, V. L. W.: Reduced intraluminal bile acid concentrations and fat maldigestion in pancreatic insufficiency: Correction by treatment. *Gastroenterology* **77:**285–289, 1979.

Renner, I. G., Rinderknecht, H., and Doublas, A. P.: Profile of pure pancreatic secretions in patients with acute pancreatitis: The possible role of proteolytic enzymes in pathogenesis. *Gastroenterology* **75:**1090–1098, 1978.

Rinderknecht, H., Stace, N. H., and Renner, I. G.: Effects of chronic alcohol abuse on exocrine pancreatic secretion in man. *Dig. Dis. Sci.* **30:**65–71, 1985.

Robberecht, P., Cremer, M., Vandermeers-Piret, M.-C., Cotton, P., DeNeef, P., and Christophe, L.: Pancreatic secretion of total protein and of three hydrolases collected in healthy subjects via duodenoscopic cannulation. *Gastroenterology* **69:**374–379, 1975.

Robinson, T. M., and Dunphy, J. E.: Continuous perfusion of bile and protease activators through the pancreas. *J.A.M.A.* **183:** 530–533, 1963.

Roesch, W., and Demling, L.: Endoscopic management of pancreatitis. *Surg. Clin. North Am.* **62**:845–852, 1982.

Sahel, J., and Sarles, H.: Modifications of pure human pancreatic juice induced by chronic alcohol consumption. *Dig. Dis. Sci.* **24**:897–905, 1979.

Salt, W. B. II, and Schenker, S.: Amylase—its clinical significance: A review of the literature. *Medicine (Baltimore)* **55**: 269–289, 1976.

Sarles, H.: An international survey on nutrition and pancreatitis. *Digestion* **9**:289–403, 1973.

Sarles, H.: Chronic calcifying pancreatitis—chronic alcoholic pancreatitis. *Gastroenterology* **66**:604–616, 1974.

Sarles, H., Sarles, J. C., Camatte, R., Muratore, R., Faini, M., Guien, C., Pastro, J., and LeRoy, F.: Observations of 205 confirmed cases of acute pancreatitis, recurring pancreatitis, and chronic pancreatitis. *Gut* **6**:545–559, 1965.

Sarles, H., Lebreuil, G., Tasso, F., Figarella, C., Clemente, F., Devaux, M. A., Fagonde, B., and Payan, H.: A comparison of alcoholic pancreatitis in rat and man. *Gut* **12**:377–388, 1971a.

Sarles, H., Figarella, C., and Clemente, F.: The interaction of ethanol, dietary lipids and proteins on the rat pancreas. I. Digestive enzymes. *Digestion* **4**:13–22, 1971b.

Sarles, H., Tiscornia, O., Palasciano, G., Brasca, A., Hage, G., Devaux, M. A., and Gullo, L.: Effects of chronic intragastric ethanol administration on canine exocrine pancreatic secretion. *Scand. J. Gastroenterol.* **8**:85–96, 1973.

Sarles, H., Tiscornia, O., and Palasciano, G.: Chronic alcoholism and canine exocrine pancreas secretion. A long-term follow-up study. *Gastroenterology* **72**:238–243, 1977.

Sarles, H., Figarella, C., Tiscornia, O., Columb, E., Guy, O., Verine, H., DeCaro, A., Multinger, L., and Lechene, P.: Chronic calcifying pancreatitis (CCP). Mechanism of formation of the lesions. New data and critical study. *In The Pancreas (International Academy of Pathology Monograph)* (P. J. Fitzgerald and A. B. Morrison, eds.), Baltimore, Williams & Wilkins, 1980, pp. 48–66.

Sarles, J. C., Midejean, A., and Devaux, M. A.: Electromyography of the sphincter of Oddi. *Am. J. Gastroenterol.* **63**:221–231, 1975.

Sarr, M. G., Sanfey, H., and Cameron, J. L.: Prospective, randomized trial of nasogastric suction in patients with acute pancreatitis. *Surgery* **100**:500–504, 1986.

Scuro, L. A., Vantini, I., Piubello, W., Micciolo, R., Talamini, G., Benini, L., Benini, P., Pederzole, P., Marzoli, G., Vaona, B., and Cavallini, G.: Evolution of pain in chronic relapsing pancreatitis: A study of operated and non-operated patients. *Am. J. Gastroenterol.* **78**:495–501, 1983.

Shaper, A. G.: Aetiology of chronic pancreatic fibrosis with calcification seen in Uganda. *Br. Med. J.* **1**:1607–1609, 1964.

Shi, E. C. P., and Ham, J. M.: Benign biliary strictures associated with chronic pancreatitis and gallstones. *Aust. N.Z. J. Surg.* **50**:488–492, 1980.

Shimamura, J., Fridhandler, L., and Berk, J. E.: Nonpancreatic-type hyperamylasemia associated with pancreatic cancer. *Am. J. Dig. Dis.* **21**:340–345, 1976.

Simsek, H., and Singh, M.: Effect of chronic ethanol feeding on pancreatic lipid metabolism. *Gastroenterology* **96**:A473, 1989.

Singh, M.: Effect of chronic ethanol feeding on pancreatic enzyme secretion in rats *in vitro*. *Dig. Dis. Sci.* **28**:117–123, 1983.

Singh, M.: Ethanol and the pancreas. In: *The Exocrine Pancreas, Pathobiology and Disease* (V. L. W. Go, F. P. Brooks, E.P. DiMagno, J. D. Gardner, E. Lebenthal, and G. A. Scheele, eds.), New York, Raven Press, pp. 423–442, 1986.

Singh, M., LaSure, M. M., and Bockman, D. E.: Pancreatic acinar cell function and morphology in rats chronically fed an ethanol diet. *Gastroenterology* **82**:425–434, 1982.

Slaff, J., Jacobson, D., Tillman, C. R., Curington, C., and Toskes, P.: Protease-specific suppression of pancreatic exocrine secretion. *Gastroenterology* **87**:44–52, 1984.

Smith, R. W., Reid, A., Hutchinson, J. M. S., and Mallard, J. R.: Nuclear magnetic resonance of the pancreas. *Radiology* **142**: 677, 1982.

Sobel, J. H., and Waye, J. D.: Pancreatic changes in various types of cirrhosis in alcoholics. *Gastroenterology* **45**:341–344, 1963.

Somer, J. B., Thompson, G., and Pirola, R. C.: Influence of ethanol on pancreatic lipid metabolism. *Alcoholism: Clin. Exp. Res.* **4**:341–345, 1980a.

Somer, J. B., Colley, P. W., and Pirola, R. C.: Further studies of the ethanol-induced changes in pancreatic lipid metabolism. *Exp. Mol. Pathol.* **33**:231–239, 1980b.

Steiner, R. E.: Magnetic resonance imaging: Its impact on diagnostic radiology. *Am. J. Roentgenol.* **145**:883–893, 1985.

Stephens, D. B., Martin, C. E., and Sawyers, J. L.: Management of pancreatic ascites. *South. Med. J.* **68**:1234–1238, 1975.

Strum, W. B., and Spiro, H. M.: Chronic pancreatitis. *Ann. Intern. Med.* **74**:264–277, 1971.

Thal, A. P., Perry, J. F., and Egner, O. W.: A clinical and morphologic study of forty-two cases of fatal acute pancreatitis. *Surg. Gynecol. Obstet.* **105**:191–202, 1957.

Tiscornia, O. M., Hage, G., Palasciano, G., Brasaca, A. P., Devaux, M. A., and Sarles, H.: The effects of pentolinium and vagotomy on the inhibition of canine exocrine pancreatic secretion by intravenous ethanol. *Biomedicine* **18**:159–163, 1973.

Trapnell, J.: The natural history and management of acute pancreatitis. *Clin. Gastroenterol.* **1**:147–166, 1972.

Tsukamoto, H., Towner, S. J., Yu, G. S. M., and French, S. W.: Potentiation of ethanol-induced pancreatic injury by dietary fat: Induction of chronic pancreatitis by alcohol in rats. *Am. J. Pathol.* **131**:246–257, 1988.

Tsurumi, T., Fujii, Y., Takeda, M., Tanaka, J., Harada, H., and Oka, H.: A case of chronic pancreatitis successfully treated by endoscopic removal of protein plugs. *Acta Med. Okayama* **38**:169–74, 1984.

Tsuzaki, T., Watanabe, N., and Thal, A. P.: The effect of obstruction of the pancreatic duct and acute alcoholism on pancreatic protein synthesis. *Surgery* **57**:724–729, 1965.

Uscanga, L., Robles-Diaz, G., and Sarles, H.: Nutritional data and etiology of chronic pancreatitis in Mexico. *Dig. Dis. Sci.* **30**: 40–113, 1985.

Vantini, I., Carvallini, G., Angelini, G., Pinbello, W., Bovo, P., Benini, L., Ederle, A., Dobrilla, G., Ballaperta, P., and Scuro, L. A.: Nutritional aspects of chronic pancreatitis in Verona, Italy. *Rendic. Gastroenterol.* **9**:13–17, 1977.

Walton, B., Schapiro, H., and Woodward, E. R.: The effect of alcohol on pancreatic secretion. *Surg. Forum* **11**:365–367, 1960.

Warren, W. D., Millikan, W. J., Henderson, J. M., and Hersh, T.: A denervated pancreatic flap for control of chronic pain in pancreatitis. *Surg. Gynecol. Obstet.* **159**:581–583, 1984.

Warshaw, A. L., and Rattner, D. W.: Timing of surgical drainage for pancreatic pseudocyst: Clinical and chemical criteria. *Ann. Surg.* **202**:720–724, 1985.

Welbourne, R. B., Armitage, P., Gilmore, O. J. A., MacKay, C., Trapnell, J. E., Williamson, R. C. N., and Cox, A. G.: Death from acute pancreatitis. *Lancet* **2**:632–635, 1977.

Wheatley, I. C., Barbezat, G. O., Hickman, R., and Terblanche, J.: The effect of acute ethanol administration on the exocrine pancreatic secretion of the pig. *Br. J. Surg.* **62**:707–712, 1975.

White, T. T., and Keith, R. G.: Long term follow-up study of 50 patients with pancreaticojejunostomy. *Surg. Gynecol. Obstet.* **136**:353–358, 1973.

Wilson, J. S., Colley, P. W., Sosula, L., Pirola, R. C., Chapman, B. A., and Somer, J. B.: Alcohol causes a fatty pancreas: A rat model of ethanol-induced pancreatic steatosis. *Alcoholism: Clin. Exp. Res.* **6**:117–121, 1982.

Wilson, J. S., Gossat, D., Tait, A., Rouse, S., Xu, H., and Pirola, R. C.: Evidence for an inherited predisposition to alcoholic pancreatitis: A controlled HLA typing study. *Dig. Dis. Sci.* **29**:727–30, 1984.

Wilson, J. S., Bernstein, L., McDonald, C., Tait, A., McNeil, D., and Pirola, R. C.: Diet and drinking habits in relation to the development of alcoholic pancreatitis. *Gut* **26**:882–887, 1985.

Wind, G. G., Rubin, P., Waye, J., and Bauer, J. J.: Pancreatic pseudocyst: Is endoscopic retrograde cholangiopancreatography contra-indicated? *Mt. Sinai J. Med. N.Y.* **43**:558–564, 1976.

Wisloff, F., Jakobsen, J., and Osnes, M.: Stenosis of the common bile duct in chronic pancreatitis. *Br. J. Surg.* **69**:52–54, 1982.

Wynder, E. L., Mabuchi, K., Maruchi, N., and Fortner, J. G.: Epidemiology of cancer of the pancreas. *J. Natl. Cancer Inst.* **50**:645–667, 1973.

12

Cardiovascular Effects of Ethanol

Howard S. Friedman

In this chapter the acute and chronic effects of alcohol on cardiac hemodynamics, electrophysiology, and metabolism and regional blood flow are reviewed. Alcoholic cardiomyopathy—the cardiac muscle disorder associated with alcohol abuse—is examined from a clinical and laboratory perspective. The relationship of alcohol use to atherogenesis and hypertension—which has important implications for stroke and heart attack—is also discussed.

12.1. Acute Hemodynamic Effects of Ethanol

12.1.1. Experimental Studies

The acute hemodynamic effects of ethanol have been a subject of some debate, with findings varying to some extent with the experimental conditions of the studies. Investigations performed using isolated heart muscle preparations have demonstrated direct myocardial depressant effects at concentrations of ethanol found with acute alcoholic intoxication in humans. Gimeno *et al.* (1962) observed reversible depressions of peak developed tension in isolated rat atria at concentrations of ethanol as low as 110 mg/dl; Spann *et al.* (1968) found that right ventricular papillary muscles obtained from both normal cats and those with experimentally induced chronic heart failure showed concentration-dependent depression of function at ethanol concentrations ranging from 100 to 500 mg/dl; and Fisher and Kavaler (1975) observed that ethanol at a concentration

of 75 mg/dl or greater depressed the contractile force of frog and cat ventricular muscle in a concentration-dependent manner. By contrast, a concentration of 7.5 mg/dl increased the contractile force of cat but not frog myocardium. Nakano and Moore (1972) also observed a concentration-dependent depression of myocardial force using isolated guinea pig ventricular muscle at concentrations of ethanol as low as 46 mg/dl. These investigators found, in addition, that the depression in muscle function after ethanol could not be explained by the increased osmolality that results from the presence of ethanol in the blood. Hirota *et al.* (1976), using isolated rat ventricular muscle, have observed myocardial depression at concentrations of ethanol as low as 100 mg/dl. The myocardial depressant effects of ethanol are not specific for ethanol but are also evident with other alcohols (Nakano and Moore, 1972). Indeed, the myocardial depressant effects of various alcohols appear to be directly related to the length of their carbon chains (Nakano and Moore, 1972) and therefore to their lipophilic properties.

By contrast to studies using isolated cardiac muscle, investigations that have assessed the effects of ethanol on cardiac "pump" function have yielded more variable findings. At least since 1924, when Sulzer demonstrated in a dog heart–lung preparation that ethanol at a concentration as low as 60 mg/dl caused cardiac dilatation, even before a fall of cardiac output was evident, there has been evidence for a negative inotropic action of alcohol in the beating heart. More recently, however, Webb *et al.* (1967), using an isolated dog heart–lung preparation with cardiac output held constant, found no change in left ventricular function, volumes, or pressures at concentrations of ethanol as high as 900 mg/dl. Using perfused rat hearts, Lochner *et al.* (1969) observed a depression of cardiac function after 15 min of exposure to ethanol at a concentration of 460 mg/dl; a persistent depressant effect was observed only at a concentration of 920 mg/dl. Of interest was their finding that rats that received ethanol in their drinking water for 8 months required a higher concentration (960 mg/dl) to manifest a depression of cardiac performance, and even then there was only a transient effect, suggesting that the heart may develop tolerance to the acute depressant effects of alcohol. Segel *et al.* (1978) also observed in perfused rat hearts a depression of cardiac function: within 3 min of perfusing such hearts with ethanol at a concentration of 300 mg/dl, a sharp decline in cardiac performance occurred. Auffermann *et al.* (1988b) found in perfused hamster hearts a depression of cardiac function after ethanol administration at a concentration of 2 g/dl, characterized by a sharp increase in left ventricular end-diastolic pressure and a slight decrease in systolic pressure, resulting in a decrease in developed pressure. Of interest, pretreatment of these hamsters with verapamil—but not diltiazem—prevented these acute changes by ethanol. Because of the pharmacological concentration of ethanol, the significance of these studies in hamsters is not clear.

Investigations using intact animals to study the cardiac actions of ethanol have also produced conflicting observations. Newman and Valicenti (1971) have observed in open-chest dogs a depression of cardiac force development and cardiac dilatation at an average ethanol level of 273 mg/dl. Webb and Degerli (1965) found in open-chest dogs that an intravenous alcohol infusion of 0.5 g/kg, producing a concentration of 42 to 87 mg/dl, resulted in an increment of cardiac output and stroke volume. At a dosage of 1.5 g/kg, which produced levels of 132 to 207 mg/dl, mean atrial pressure was also increased, but peripheral resistance declined. Despite an apparent enhancement of cardiac function after ethanol, these investigators found that an abnormal myocardial function curve was obtained in response to a volume challenge. In open-chest dogs, following rapid (10 min) intravenous infusion of ethanol (40 and 80 mg/kg), Nakano and Kessinger (1972) also found an increase of heart rate, mean systemic arterial pressure, and myocardial contractile force at an average blood ethanol concentration of 300 mg/dl; at higher blood ethanol concentrations, however, there was a concentration-dependent decline in these parameters.

Mierzwiak *et al.* (1972) examined the effects of intravenous ethanol on contractile function in open-chest dogs using different anesthetics (pentobarbital or chloralose), different types of infusion (continuous or intermittent), and in the presence or absence of autonomic reflexes. At blood ethanol concentrations of 250 mg/dl, irrespective of type of anesthesia or manner of intravenous infusion, left ventricular end-diastolic pressure increased, and left ventricular peak dp/dt declined. In the areflexic animal, these changes were more pronounced, demonstrating the effects of cardiac reflexes in obscuring the direct myocardial depressant actions of ethanol. In a study by Wong (1973) in closed-chest chloralose-anesthetized dogs, the role of the catecholamine-releasing effects of ethanol in masking the direct action of ethanol on the heart was also demonstrated. She found that an intravenous infusion of ethanol produced an elevation of left ventricular end-diastolic pressure and a reduction in stroke work at an estimated blood ethanol concentration of 100 mg/dl in dogs with β-adrenergic and cholinergic blockade, whereas an estimated ethanol concentration of 200 mg/dl was required to produce this effect in neurally intact animals.

Studies of the effect of alcohol in closed-chest anesthetized dogs have generally shown a depression of cardiac function at blood ethanol levels of 110 to 200 mg/dl. Regan *et al.* (1966) found in closed-chest anesthetized dogs that an intravenous infusion of ethanol, producing an average ethanol concentration of 110 mg/dl at 30 min and 201 mg/dl at 120 min, elevated left ventricular end-diastolic pressure and reduced stroke volume and left ventricular peak dp/dt. Moreover, a direct coronary infusion of ethanol at a rate of 68 mg/min also elevated left ventricular end-diastolic pressure and reduced left ventricular peak dp/dt. Friedman *et al.* (1979b), using similar experimental conditions, found that a solution of 2.4 g/kg delivered intravenously over 90 min produced a concentration-dependent depression of cardiac function. At 30 min after beginning ethanol infusion, when blood levels were 118 mg/dl, there was significant cardiac dilatation; at 60 min, when average blood concentration was 199 mg/dl, there was a decline in ejection fraction and left ventricular peak dp/dt normalized for left ventricular end-diastolic volume, an index of contractility. However, there was no change in cardiac output. Mendoza *et al.* (1971) also found that at an average blood ethanol concentration of 190 mg/dl, closed-chest pentobarbital-anesthetized dogs showed a decline in velocity of contraction at a time when cardiac output actually increased, demonstrating the insensitivity of cardiac output in assessing the myocardial response to ethanol.

A study by Horwitz and Atkins (1974) in awake dogs also demonstrated the cardiac depressant effects of ethanol. These investigators found that 0.5 g/kg of ethanol administered rapidly (10 min), resulting in an average blood ethanol concentration of 120 mg/dl, produced a decline in stroke volume and left ventricular peak dp/dt; an additional 1 g/kg of ethanol, resulting in an average concentration of 311 mg/dl, produced further worsening of cardiac function. Horwitz and Atkins (1974) also found that the increment in heart rate observed at high concentrations of ethanol could be inhibited by blockade with propranolol and atropine. Of interest is their observation that with autonomic blockade, at an average ethanol concentration of 120 mg/dl, a bradycardiac response was found. Stratton et al. (1981) also found a decline in left ventricular peak dp/dt in awake dogs after a rapid (3 min) intravenous infusion of 80 mg/kg of ethanol, resulting in an average peak concentration of 160 mg/dl. These researchers also found that at such concentrations of ethanol, left ventricular systolic pressure and heart rate were also reduced. Moreover, in these dogs, after administration of ethanol, a blunted response of heart rate, left ventricular systolic pressure, and left ventricular peak dp/dt to dynamic exercise occurred.

The rate of ethanol administration may account, in part, for some of the differences in the findings of various studies. Jones et al. (1975) found that after a slow intravenous infusion of ethanol, producing an average concentration of 192 mg/dl, no change in cardiac output or left ventricular end-diastolic pressure occurred, although cardiac dilatation and a decline in left ventricular ejection fraction did ensue. By contrast, intracoronary or rapid intravenous infusion of ethanol produced an increment in left ventricular end-diastolic pressure.

Thus, studies of cardiac muscle function demonstrate a direct myocardial depressant effect of ethanol even at concentrations as low as 50 mg/dl. In intact dogs, depending on the rate of infusion of ethanol and other experimental conditions, cardiac "pump" function is also generally found to be depressed at blood ethanol concentrations of 100 to 200 mg/dl. The peripheral circulatory and neurohumoral effects of ethanol tend to obscure the unfavorable action of ethanol on the heart, yet even when cardiac output is normal or actually increased, an abnormality of cardiac function can still be detected.

The ethanol metabolites acetaldehyde and acetate both have circulatory effects. Acetaldehyde, the major metabolite of ethanol, however, actually enhances cardiac function in dogs (Friedman et al., 1979a). At an average blood acetaldehyde level of 40 μM, cardiac output rose, and peripheral resistance declined, but

mean aortic pressure and left ventricular peak dp/dt per left ventricular end-diastolic volume did not change (Friedman et al., 1979a). Thus, the improvement in cardiac performance following acetaldehyde infusion appears to be attributable to the vasodilatory actions of acetaldehyde rather than to a change in the contractile state of the left ventricle.

Vasodilatory properties of acetaldehyde also have been suggested in other studies (Nakano and Prancan, 1972; McCloy et al., 1974; Nakano et al., 1969). Disulfiram, a substance that inhibits acetaldehyde dehydrogenase (Nakano et al., 1969) and depletes norepinephrine and epinephrine stores by inhibiting dopamine β-hydroxylase (Goldstein et al., 1964), produces marked hypotension following ethanol ingestion (Nakano et al., 1969). This hypotensive effect of disulfiram with ethanol has been attributed to the high levels of acetaldehyde produced at a time when the acetaldehyde catecholamine-releasing effect (Perman, 1958; Eade, 1959; Walsh and Truitt, 1968) has been blocked, thereby unmasking a direct arteriolar-dilating effect. A hypotensive effect of acetaldehyde has also been demonstrated following α-adrenergic blockade with phenoxybenzamine (Nakano and Prancan, 1972).

Previous studies using isolated atrium preparations (Kumar and Sheth, 1962; Walsh et al., 1969), isolated perfused hearts (Gailis and Verdy, 1971; Nguyen and Gailis, 1974), and intact, anesthetized (James and Bear, 1967; Nakano and Prancan, 1972; McCloy et al., 1974) and awake (Stratton et al., 1981) animals have demonstrated that acetaldehyde in pharmacological dosages raises systemic blood pressure, augments the cardiac contractile state, and increases heart rate. These sympathomimetic effects of acetaldehyde are related to its catecholamine-releasing properties (Perman, 1958; Eade, 1959; Walsh and Truitt, 1968) and can be prevented by adrenergic blockade (Kumar and Sheth, 1962; James and Bear, 1967; Walsh et al., 1969; Nakano and Prancan, 1972; McCloy et al., 1974). Also, in a preliminary communication a negative inotropic effect has been reported following a prolonged, low-dosage infusion of acetaldehyde (Jesrani et al., 1971). Thus, acetaldehyde's peripheral vasodilatory effects, which would improve cardiac function by reducing cardiac impedance, and indirect sympathomimetic effects, which would improve cardiac contractility, might obscure its suggested or alcohol's known negative inotropic effect. However, because circulating acetaldehyde levels are generally low following ethanol administration, this metabolite probably has little effect on the hemodynamics after alcohol administration.

Acetate, at blood levels found in humans following

administration of ethanol (Lundquist *et al.*, 1962), also has marked hemodynamic effects. In anesthetized and awake dogs, Liang and Lowenstein (1978) found that acetate increases cardiac output and left ventricular peak dp/dt and reduces peripheral resistance in a concentration-dependent fashion. This effect was not blocked by β-adrenergic blockade but was abolished by fluoroacetate, a substance that reduces acetate oxidation. A generalized vasodilatory action was also found. Studies in humans, particularly in the setting of hemodialysis, where the salt had been used as a buffer (Chen *et al.*, 1983), have confirmed these effects. Nitenberg *et al.* (1984) have found that an infusion of sodium acetate, producing an average blood acetate concentration of 3.1 mM, in humans improved various indices of cardiac function, including output, independent of its effects on rate and cardiac "loading" conditions. However, a direct coronary infusion (0.45 mmol/min) did not replicate these changes. In contrast to these effects in the intact animal, Kirkendol *et al.* (1978) have observed a depression of myocardial function with acetate in isolated rabbit papillary muscles. Thus, acetate, like acetaldehyde, could contribute to the acute hemodynamic findings after ethanol administration.

12.1.2. Acute Hemodynamic Effects in Humans

Since ethanol has a complex array of direct and indirect effects, some of which are mutually antagonistic, it should not be surprising that previous studies of the actions of ethanol on cardiac function in humans have yielded conflicting findings. Following ethanol administration, heart rate generally increases (Riff *et al.*, 1969; Juchems and Klobe, 1969; Blomqvist *et al.*, 1970). Yet some studies have shown no change (Dixon, 1907; Ahmed *et al.*, 1973) or merely a transient, inconsistent change in this parameter (Grollman, 1942; Horwitz *et al.*, 1949). In ten subjects—five alcoholics and five nonalcoholics—an infusion of 15% (v/v) ethanol in 5% dextrose administered over 3 hr, resulting in an average peak ethanol concentration of 176 mg/dl, produced a progressive increase of heart rate that was on the average 11 beats/min higher 1 hr after cessation of the infusion than the basal rate, but this change was not statistically significant. Also, the response of heart rate to ethanol was not related to the subject's control rate, the blood ethanol concentration, or a history of alcoholism (Friedman *et al.*, 1974). By contrast, 14 healthy subjects ingesting 1 to 2 g/kg of ethanol showed a slight

but significant increase of rate: heart rate increased from 72 to 77 min^{-1} at a blood ethanol concentration of 39 mg/dl, with a maximal response to 82 min^{-1} occurring when the concentration was 85 mg/dl; however, still higher blood ethanol concentrations did not produce further increments (Juchems and Klobe, 1969). Thus, although heart rate usually increases in response to alcohol, this is not a consistent effect.

Measurements of cardiac output before and after ethanol administration also have not produced uniform findings. In small groups of normal subjects, Grollman (1942) and Stein *et al.* (1963) found that ethanol increased cardiac output. Juchems and Klobe (1969) also observed that cardiac output increased in normal subjects following ingestion of ethanol; the increase in cardiac output, which was related to an increase of both rate and stroke volume, was concentration dependent, with a significant change evident at a blood ethanol concentration of 39 mg/dl and a maximal increment of cardiac output occurring at a peak concentration of 120 mg/dl. Similarly, in ten healthy subjects, Riff *et al.* (1969) found that 6 oz of 90-proof bourbon whiskey, resulting in a blood ethanol level of 85 to 135 mg/dl, produced significant increases in cardiac output and a decline in peripheral resistance. These investigators found no alterations in the hemodynamic response to dynamic exercise after ethanol ingestion.

Blomqvist *et al.* (1970) observed in normal subjects an increase in cardiac output and a lowering of peripheral resistance at rest and with submaximal levels of dynamic exercise at blood ethanol concentrations of 125 to 150 mg/dl following ingestion of 5 oz of 86-proof whiskey, rum, or gin. In this study, the increased cardiac output reflected an increment in heart rate without a change in stroke volume. Other findings reported by these investigators (Blomqvist *et al.*, 1970) include an increase of oxygen consumption at submaximal levels of exercise but no change in peak heart rate or maximal oxygen consumption. In contrast, Gould *et al.* (1972) observed that the ingestion of 2 oz of 87-proof whiskey in less than 10 min produced a decline in cardiac output and an increase in peripheral resistance and arteriovenous oxygen difference in patients with compensated heart disease (normal average resting cardiac output and left ventricular end-diastolic pressure), although normal subjects showed an increase in cardiac output and a lowering of peripheral resistance and arteriovenous oxygen difference. In eight patients with severe heart failure resulting from a dilated cardiomyopathy—four associated with alcohol abuse—Greenberg *et al.* (1982) found that the ingestion of 0.9 g/kg of 80-proof whiskey, producing an average blood ethanol level of 117 mg/dl, resulted in a lowering of mean aortic pressure, peripheral resistance, and pulmonary artery occlusion pressures but no change in cardiac output.

Since cardiac output reflects both heart rate and cardiac "loading conditions," a depression of cardiac pump or muscle function may not be evident if only this parameter is used as a measure of cardiac performance. Left ventricular ejection fraction, which expresses the relationship between stroke volume and left ventricular end-diastolic volume, although also influenced by heart rate, preload, and afterload, is a more sensitive measure of cardiac function than cardiac output. Using echocardiogram-derived measurements, Delgado et al. (1975) observed that ingestion of between 0.7 and 1.15 g/kg of Scotch whiskey, producing blood ethanol levels of 75 to 138 mg/dl, reduced ejection fraction at 30 to 90 min after alcohol was ingested by normal subjects. Kelbaek et al. (1985) also found a reduction of left ventricular ejection fraction assessed by radionuclide angiography after ingestion of alcohol by six healthy young men: a concentration-dependent decrease occurred, with left ventricular ejection fraction falling from 67% to 62% at blood ethanol concentration of 106 mg/dl and to 56% at a blood concentration of 207 mg/dl; alcohol ingestion had, however, no effect on the change in the left ventricular ejection fraction with upright exercise. On a bicycle ergometer subjects showed a progressive increase in this variable with exercise, and only at 75% work capacity, when blood ethanol concentration was 207 mg/dl, did an attenuation of ejection fraction (69% versus 75%) occur.

Using still other indices of cardiac performance, derived from Doppler echocardiography and calibrated carotid pulse tracings and adjusting for "pump" load changes, Lang et al. (1985) observed, in nine nonalcoholic male subjects after ingesting 1.15 g/kg of ethanol (as Scotch whiskey) (producing a blood ethanol concentration of 100 to 108 mg/dl), a depression of cardiac function: the ingestion of alcohol reduced left ventricular end-diastolic dimensions (preload), left ventricular end-systolic wall stress (afterload), systemic blood pressure, and total peripheral resistance. Consequently, the vasodilating actions of ethanol were evident. When adjustments were made for the afterload changes, using methoxamine, a depression of myocardial contractility was also disclosed, as reflected by the Sagawa index, which is the slope of the left ventricular end-systolic pressure versus the end-systolic dimension.

The acute hemodynamic effects of ethanol in hu-mans may also be time dependent. Kupari (1983) compared the effects of ingesting ethanol, 1 g/kg diluted in juice (15%, w/v), to the vehicle by itself. At the peak blood ethanol concentration of 112 mg/dl, which occurred 60 min after the ingestion of alcohol had begun, heart rate and cardiac output (assessed by echocardiography) had increased (above control and greater than the response with vehicle), and peripheral resistance had decreased. Despite persistently high blood ethanol concentrations, these effects became less pronounced during the subsequent 90 min. In contrast to the early effects of alcohol, which resulted in an improvement in cardiac "pump" performance, 120 min after ingestion alcohol was begun there was a worsening of the preejection period/left ventricular ejection time ratio. The lack of an early change in this index might have been related to the confounding effects of the vehicle; indeed, even at 60 min this ratio was significantly worse after ingestion of alcohol than that found after ingesting juice, which by itself had produced an improvement in this variable. In addition to the late abnormality of systolic time intervals, at 120 and 180 min after beginning the alcohol ingestion, left ventricular end-diastolic dimension and systemic blood pressure had also decreased, an effect not observed with the vehicle. Thus, this study demonstrates the complexity of the hemodynamic effects that may follow ingestion of ethanol, the differences in the time constants of those actions, and the importance of the vehicle as a possible confounder.

In addition to Kupari (1983), other investigators (Ahmed et al., 1973; Timmis et al., 1975) have assessed the acute effects of alcohol on systolic time intervals. Ahmed et al. (1973) found in normal subjects that 6 oz of whiskey ingested over 2 hr (giving an average blood ethanol concentration of 74 mg/dl at 60 min) and this same quantity ingested over 1 hr (producing an average level of 50 mg/dl at 30 min) resulted in a worsening of this systolic time ratio even though heart rate did not change and blood pressure response was variable. Timmis et al. (1975) observed a similar response of this index to ethanol ingestion in normal subjects. However, these investigators found that the acute hemodynamic effects of ethanol had an inverse relationship to previous alcohol use; that is, the worse, the control intervals, the less impact ethanol ingestion had on cardiac performance. However, systolic time intervals are influenced not only by changes in myocardial contractility but also by changes in cardiac "load" conditions. Thus, the changes observed by Ahmed et al. (1973) and by Timmis et al. (1975) may be reflective, at least in part, of the changes in cardiac dimensions that may occur following ethanol ingestion.

Sex-specific differences in the acute hemodynamic effects of alcohol have been suggested (Timmis *et al.*, 1979). Ingestion of alcohol in women subjects, producing blood ethanol concentrations averaging 130 mg/dl, had no effect on cardiac output or systolic time intervals, even though an early (at 30 min) increase in heart rate and a late (90 to 120 min) decrease in systemic pressure and left ventricular diastolic dimensions were observed (Timmis *et al.*, 1979). By contrast, Kupari (1983) found that men and women had a similar acute hemodynamic response to alcohol.

Thus, in human beings the cardiac response to ethanol reflects the rate and quantity of alcohol ingested, previous use of alcohol, and the "physiological set" of the subject. In normal subjects, cardiac "pump" function, reflected by cardiac output, will appear to improve following ethanol ingestion, since the effects of ethanol on heart rate and peripheral resistance will obscure its direct myocardial depressant actions. In patients with compensated left ventricular dysfunction, with normal left ventricular end-diastolic pressure and cardiac output, the myocardial depressant effects may predominate, and cardiac output will decline. In patients with decompensated heart failure having a low cardiac output with a high peripheral resistance and heart rate, ethanol's vasodilatory properties may reduce left ventricular end-diastolic pressure and peripheral resistance. Because of concomitant myocardial depression, a fall in aortic pressure without a change in cardiac output may ensue. Even though cardiac output may increase in normal subjects following alcohol ingestion, other indices of myocardial function may still disclose the depressant effects of ethanol. Although these patterns of cardiovascular response have been demonstrated in groups of patients, the rate of ethanol ingestion and the history of alcohol abuse (by affecting the metabolism of ethanol with its attendant effects on arterial pH and serum lactate, acetaldehyde, and acetate) as well as the "psychophysiological" response of the individual may produce atypical responses in some subjects.

12.2. The Acute Metabolic Effects of Ethanol on the Heart

Ethanol has been shown to have an array of acute myocardial metabolic effects, but the biochemical basis for its acute myocardial depressant actions has still not been conclusively determined. Unlike other organs, such as the liver, the heart has at most minimal capacity to oxidize ethanol: the heart lacks alcohol dehydrogenase (Cherrick and Leevy, 1965 and does not appear to have a microsomal ethanol-oxidizing system but, at least in some species, has catalase (Herzog and Fahimi, 1975) and, therefore, in the presence of hydrogen peroxide, may be able to oxidize ethanol (Soffia and Penna, 1987). Soffia and Penna (1987) have found that incubation of rat heart homogenates with 116 mg/dl or 232 mg/dl ethanol generated acetaldehyde and that the addition of glucose oxidase, a hydrogen peroxide producer, to this medium increased the amount of acetaldehyde recovered, whereas the addition of a catalase inhibitor had the opposite effect. However, Lochner *et al.* (1969), using isolated perfused rat hearts, did not observe ethanol oxidation at concentrations of 460 and 920 mg/dl, levels having myocardial depressant effects. These investigators also found no changes in glucose or acetate metabolism, thus making it unlikely that ethanol has an appreciable acute toxic effect on glycolytic or Krebs cycle enzyme systems. Forsyth *et al.* (1973), however, using a similar rat model, found that acetaldehyde is oxidized by the heart and that this effect is blocked by disulfiram, suggesting that aldehyde dehydrogenase is available in rat myocardial tissue.

Recently, nonoxidative metabolism of ethanol with the formation of abnormal alcohol esters has been demonstrated to occur in myocardial tissue (Alling *et al.*, 1984; Lange, 1982; Lange and Sobel, 1983a; Laposata and Lange, 1986). Alling *et al.* (1984) have identified an acidic phospholipid, phosphatidylethanol—previously found only in plants—in various organs, including the heart, following acute and chronic administration of alcohol to rats. Lange's group (Lange, 1982; Lange and Sobel, 1983a,b; Laposata and Lange, 1986) have also identified fatty acid ethyl esters in human organs obtained at autopsy and found that homogenates of those tissues can synthesize these esters at a rate commensurate with the concentrations found. However, the concentrations of fatty acid ethyl esters found in the human heart—even in tissue obtained from acutely intoxicated persons at time of death (<50 mmol/g—and the amounts generated from myocardial tissue (11 ± 1 mmol/g per hr) are small, especially when compared with that found in adipose, pancreatic, and liver homogenates (Laposata and Lange, 1986).

The enzyme pancreatic cholesterol esterase has been found to be a catalyst in the absence of ATP and coenzyme A for the formation of fatty acid ethyl esters from ethanol and nonesterified fatty acids (Lange,

1982); more recently, Mogelson and Lange (1984) have isolated, purified, and characterized the actual enzyme that catalyzes the reaction in rabbit myocardium: fatty acyl ethyl ester synthase. These findings may be more than just a curiosity, since studies by Lange and Sobel (1983b) have demonstrated that a fatty acid ethyl ester, ethyl oleate, binds to rabbit myocardial mitochondria and uncouples mitochondrial oxidative phosphorylation in a time- and concentration-dependent fashion and that such mitochondrial dysfunction occurs at concentrations of the ester found in human hearts. Although fatty acid ethyl esters may exert their mitochondrial inhibitory effect directly on binding to mitochondria, hydrolysis may ensue with the formation of fatty acids, which are potent uncouplers of oxidative phosphorylation (Lange and Sobel, 1983b).

Ethanol has, in fact, been shown to depress mitochondrial respiration acutely (Segel, 1984). Segel and Mason have observed that rat mitochondria demonstrate a depression of respiration with glutamate substrate at concentrations of ethanol of 300–400 mg/dl and a reduction of oxygen consumption at a concentration as low as 300 mg/dl. Acetaldehyde was also found to depress rat mitochondrial respiration in these experiments. However, these effects (reduced ADP/O) were evident at a concentration of acetaldehyde of 1 mM, which is almost 1000 times the levels found circulating in humans after ingestion of alcohol. More recently, Segel (1984) has observed that mitochondria obtained from rat hearts perfused for up to 60 min with 108 mg/dl of ethanol, a concentration that depressed left ventricular function in these animals, also demonstrated reduced respiratory function as evidenced by a reduction in oxygen consumption. Of interest, mitochondria obtained from rat hearts perfused with a concentration of acetaldehyde of 74 μM, a level not influencing left ventricular function, did not manifest depressed respiration.

Despite these demonstrable acute inhibitory actions of ethanol on mitochondrial function, only in pharmacological concentrations are changes in the myocardial concentrations of high-energy phosphates evident. In the isolated rat heart, Aufferman et al. (1988a) found that perfusion with 4 g/dl of ethanol did not change intracellular pH or the concentration of high-energy phosphates measured by [^{31}P]NMR, even though left-ventricular developed pressure decreased. Using isolated perfused golden hamster hearts, however, these investigators (Aufferman et al., 1988b) observed that a lower, but still pharmacological, concentration (2 g/dl) reduced ATP slightly, from 9.8 to 8.8 mmol, but did not

also change phosphocreatine concentration or intracellular pH. Similar observations were made by Lochner et al. (1969): in the isolated perfused rat heart, 920 mg/dl, but not 460 mg/dl, of ethanol reduced ATP but did not change ADP or phosphocreatine. Studies assessing the myocardial redox state after ethanol administration have yielded conflicting findings: Aufferman and co-workers (1988a) have found in the isolated perfused rat that 4 g/dl of ethanol produced a reduction in the NADH redox state, whereas Kikuchi and Kako (1970) found in the isolated perfused rabbit heart that 200 mg/dl changed the ratio of various redox couples toward a more oxidized state.

Because fatty acids are the major energy source for myocardial contractility under normal fasting conditions, the acute myocardial depressant actions of alcohol might be related to changes in lipid metabolism produced by ethanol. Indeed, studies in humans (Wendt et al., 1966; Regan et al., 1969), anesthetized dogs (Regan et al., 1966), rats (Lochner et al., 1969), and rabbits (Kako et al., 1973) have demonstrated that ethanol alters lipid metabolism. Regan et al. (1969) observed in humans that following ingestion of alcohol producing peak serum ethanol concentrations of 150 mg/dl—but not those less 100 mg/dl—myocardial extraction of free fatty acids decreased, and extraction of triglycerides increased; Wendt et al. (1966) also found a tendency for a decrease in myocardial extraction of fatty acids after ingestion of 6 oz of vodka. In anesthetized dogs Regan et al. (1966) found that a 2-hr infusion of ethanol, producing serum ethanol concentrations ranging from 110 mg/dl to 201 mg/dl, reduced myocardial extraction of free fatty acids and increased myocardial extraction of triglycerides; myocardial triglyceride concentration was also found to be increased after this infusion, and the myocardial respiratory quotient was higher than that estimated from myocardial substrate uptake, suggesting that some of the fatty acids extracted were not being oxidized. Studies in the isolated perfused rat tend to support these observations (Lochner et al., 1969).

The changes of myocardial fat metabolism in the anesthetized rabbit following administration of ethanol are similar to those observed in other species (Kikuchi and Kako, 1970; Kako et al., 1973). Following an intravenous ethanol infusion (1.95 g/kg administered over 2–3 hr, an amount estimated to produce blood concentrations of 200 mg/dl) to rabbits, Kako's group (Kikuchi and Kako, 1970; Kako et al., 1973) found an increase in myocardial triglyceride, but not free fatty acid or phospholipid, concentration, a finding not replicated by a norepinephrine infusion or blocked by pretreatment with reserpine. The accumulation of triglycerides was not related to changes in plasma lipids, nor was a change in myocardial lipoprotein lipase detected. In myocardial homogenates obtained after the ethanol infusion, fatty acid oxidation was found to be reduced, and their incorporation into triglycerides, but not into phospholipids, increased. Because such changes were also observed in isolated perfused rabbit hearts, the increases of circulating substrates such as lactate, acetate, or acetoacetate that may occur following ethanol administration is not a necessary condition for the reduction of myocardial fatty acids oxidation. Kako and co-workers (1973) have also observed that the fatty acid composition of the myocardial triglycerides formed during ethanol administration tends to resemble that of plasma fatty acids and that these changes persist for several weeks following the administration of ethanol. Of interest, the addition of carnitine to myocardial homogenates obtained after the ethanol infusion prevented, at least in part, the reduction in fatty acid oxidation and the enhancement of esterification (Kako et al., 1973), demonstrating the reversibility of the acute effects of ethanol on myocardial lipid metabolism and suggesting that the inhibition of fatty acid oxidation after ethanol administration may be related in part to a defect in the carrier system that is responsible for the transfer of fatty-acyl-CoA derivatives from the cytosol to the mitochondria for oxidation.

Ethanol and its metabolites may also affect myocardial protein synthesis. Preedy and Peters (1989) have found that 2½ hr after the intraperitoneal administration of 3.5 g/kg of ethanol to the rat, which produces blood ethanol concentrations ranging from 200 to 300 mg/dl, the rate of protein synthesis and the amount of protein synthesis per unit of RNA in the heart were reduced by 20%. By contrast, Tiernan and Ward (1986), using the same experimental model, did not find significant changes in the heart, even though a reduction of protein synthesis was evident in other organs. Differences in assessing protein synthesis might have accounted for this discrepancy: protein synthesis was measured by Tiernan and Ward (1986) as the incorporation of L-[U-^{14}C]tyrosine delivered as a continuous infusion, whereas Preedy and Peters (1989) used L-[4-^{3}H]phenylalanine delivered by a bolus "flooding dose." Moreover, in the isolated perfused guinea pig heart, Schreiber and co-workers (1972) found no effect of 200 to 300 mg/dl of ethanol on protein synthesis. The reduction of protein synthesis with the systemic administration of ethanol but the lack of change with coronary perfusion suggests that extracardiac factors may be responsible for this finding. Indeed, Schreiber and co-workers (1972)

have found that a 3- to 5-hr perfusion of 800 μM acetaldehyde in the isolated perfused guinea pig heart reduced protein synthesis, and microsomes from normal guinea pig hearts demonstrated 52% of control protein synthesis when incubated for 1 hr with 120 μM acetaldehyde and 65% of control when incubated with 30 to 60 μM acetaldehyde (Schreiber et al., 1974). Acetate, however, is not a likely cause of the inhibition of protein synthesis following ethanol administration, since the addition of 10 mM of this substrate to the isolated perfused rat heart produces an increase in protein synthesis (Smith et al., 1986).

Ethanol might also exert its myocardial depressant actions by a direct effect on the contractile process itself. Because cytosolic concentrations of Ca^{2+} and its accumulation by the sarcoplasmic reticulum will influence myocardial contraction and relaxation, the effects of ethanol on Ca^{2+} fluxes would be of interest. Swartz and co-workers (1974) have found that Ca^{2+} uptake and binding by sarcoplasmic-enriched canine cardiac microsomes is inhibited by ethanol. However, this was not evident at ethanol concentrations less than 2.3 g/dl. Although lower concentrations of ethanol (1.8 g/dl) are required to inhibit Ca^{2+} uptake if such microsomes are phosphorylated by cAMP-dependent protein kinase (Retig et al., 1977), these concentrations are still substantially above relevant levels. Moreover, reduction of Ca^{2+} uptake by the sarcoplasmic reticulum would be expected to inhibit myocardial relaxation, whereas studies that have examined the effects of ethanol on relaxation suggest an enhancement (Hirota et al., 1976).

A study by Williams and co-workers (1975) has also demonstrated a reversible dose-dependent inhibition of Na^+,K^+-activated ATPase activity of plasma membranes obtained from guinea pig hearts with both ethanol (from 460 mg/dl) and acetaldehyde (0.2 to 1 mM). McCall and Ryan (1987) have confirmed these findings in cultured neonatal rat myocytes: 30 min (but not less) after addition of ethanol (100 to 1000 mg/dl) to such cultures $^{42}K^+$ influx, $^{24}Na^+$ efflux, and sodium pump sites per cell ([^{3}H]ouabain-binding sites) were reduced. By contrast, acetaldehyde (1 to 100 μM) did not produce such changes. Of interest, however, when cardiac glycosides produce an inhibition of Na^+,K^+-ATPase, contractility is enhanced rather than depressed.

Also, observations by Puszkin and Rubin (1975) suggest a direct adverse effect of ethanol on contractile proteins. These investigators found that ADP-induced association of human skeletal muscle contractile proteins, actin and myosin, assessed by superprecipitation measured spectrophotometrically, was reversibly inhibited in a dose-dependent fashion by ethanol (from 200 mg/dl) and acetaldehyde (from 50 µM). Whether this phenomenon relates to *in situ* contraction and whether cardiac contractile proteins also demonstrate this finding are not, however, clear.

Ethanol's effects on the biochemical composition and physical properties of the sarcolemma have been proposed as a determinant of the acute myocardial depressant actions of ethanol (Rubin, 1982). Indirect evidence supporting the sarcolemma as a site of acute alcoholic injury is the finding that the lipophilic properties—and therefore the expected predilection for affecting the lipid bilayer of the plasma membrane—of the different alcohols correlate with the magnitude of their myocardial depressant actions (Nakano and Moore, 1972) and, incidentally, with their propensity to shorten the duration of myocyte action potentials (Williams *et al.*, 1980). Although ethanol has been shown to increase the fluidity of the lipid bilayer of various biological membranes acutely (Goldstein, 1984; Taraschi and Rubin, 1985), the relevance of these observations to cardiac myocyte membranes and their importance in explaining myocardial functional changes have still not been determined. Moreover, changes in myocardial sarcolemma integrity, as assessed by changes in transmembrane ionic gradients, did not occur in rats after the intraperitoneal administration of ethanol that produced peak blood ethanol concentrations of 414 mg/dl (Kischuk *et al.*, 1986), even though such changes have been observed after chronic ethanol feeding in such animals (Polimeni *et al.*, 1983).

Thus, substantive alterations in cardiac metabolism occur following ethanol administration at dosages that produce a depression of myocardial function: changes in lipid and protein metabolism, formation of abnormal metabolic products such as fatty acid ethyl esters, depression of mitochondrial function, and a reduction of sodium pump activity have been demonstrated. There is evidence to suggest that changes in the composition and properties of myocyte membranes may also be important. However, a clear definition of the biochemical explanation for the acute myocardial depressant effects of alcohol has still not been determined.

12.3. Changes in Regional Blood Flow Caused by Ethanol

Ethanol has an array of effects on regional circulations independent of its effects on cardiac performance. The feelings of warmth and skin flushing are known to all who use alcohol, and, indeed, studies have shown that ethanol increases cutaneous blood flow (Dixon, 1907; McDowall, 1925; Horwitz *et al.*, 1949; Hughes *et al.*, 1984). Hughes and co-workers (1984) have observed in healthy subjects following ethanol ingestion an increase in skin blood flow at a blood ethanol concentration of approximately 50 mg/dl. By contrast, forearm noncutaneous flow decreases after ethanol administration (Fewings *et al.*, 1966). Fewings and co-workers (1966) found that forearm blood flow decreased in a dose-dependent manner following an intraarterial infusion of 7.5 to 150 mg/dl of ethanol to healthy subjects, an effect that did not appear to be blocked by phenoxybenzamine or sympathectomy. Following ingestion of ethanol, resulting in blood concentrations between 35 and 60 mg/dl, an increase in skin flow could be blocked by sympathetic antagonists. Thus, ethanol ingestion results in an increase in skin flow, an effect that may be neurohumorally mediated, but also in a decrease in limb muscle flow, apparently a direct effect of alcohol.

The effects of ethanol on blood flow to the liver are of particular interest because this organ is one of the major sites of alcohol-induced organ injury. Following intravenous infusions of alcohol in human subjects that produced blood ethanol concentrations between 33 and 103 mg/dl, Stein *et al.* (1963) observed an increase in splanchnic flow. Shaw and co-workers (1977) also found in ketamine-anesthetized baboons, in animals both with and without preexisting liver injury, an increase in splanchnic flow after an intravenous infusion of ethanol that produced an average blood ethanol concentration of 276 mg/dl. Not all studies have found, however, an increase in splanchnic flow following ethanol administration: McSmythe and co-workers (1953) and Horvath and Willard (1962) in anesthetized dogs and Castenfors and co-workers (1960) in humans did not observe a change in splanchnic flow after administration of ethanol.

In studies by Shaw and co-workers (1977) performed on the anesthetized baboon, the increase in splanchnic flow observed after ethanol administration was accompanied by an increment in both hepatic oxygen consumption and hepatic vein oxygen content. In the awake dog, Villeneuve *et al.* (1981) also found that an increase in hepatic flow after intravenous or intragastric administration of 2.0 g/kg of ethanol (which produced blood ethanol concentrations greater than 200 mg/dl) was accompanied by an increase in hepatic vein oxygen, but the change in hepatic oxygen consumption was variable. These findings suggest that hepatic flow increased after ethanol administration more than hepatic oxygen demand required. However, the effects of ethanol on splanchnic blood flow can be inhibited by drugs that affect the metabolism of ethanol (McKaigney *et al.*,

1986; Carmichael *et al.*, 1987). Both 4-methylpyrazone, which inhibits alcohol dehydrogenase (McKaigney *et al.*, 1986), and cyanamide, which inhibits aldehyde dehydrogenase (Carmichael *et al.*, 1987), suppress the increase in splanchnic flow that follows ethanol administration. Despite the high blood acetaldehyde concentrations that occur when cyanamide is administered to the ethanol-treated rat, no change in splanchnic blood flow occurs, suggesting that this metabolite is not an important determinant of the splanchnic flow changes that follow ethanol administration (Carmichael *et al.*, 1987).

Ethanol also appears to have direct actions on splanchnic vessels. With topical application or intravascular administration, ethanol ($\geqslant$50 mg/dl) has been shown to dilate mesenteric vessels of the rat *in vivo* and to antagonize the effects of various vasoconstrictors on these blood vessels (Altura *et al.*, 1979). By contrast, using the same experimental conditions in which ethanol produced vasodilatory actions, Altura and Gebrewold (1981) found that ethanol's metabolites, acetaldehyde (as low as 18 μM) and acetate (0.025–150 μM), had vasoconstrictive actions on splanchnic blood vessels. Because splanchnic flow changes may reflect either hepatic arterial or portal venous flow changes or both, studies measuring selective changes in these circulations are of interest. In both anesthetized and awake dogs, Friedman and co-workers (1983) found an increase in hepatic arterial blood flow after an intravenous infusion of ethanol, resulting in blood ethanol concentrations ranging from 60 to 350 mg/dl. In contrast to these findings, Carmichael and co-workers (1987) found that after oral administration of 1 g/kg of ethanol to the rat, the increment of hepatic blood flow was associated with a decrease in hepatic arterial blood flow.

The effect of ethanol on coronary blood flow has been of clinical interest for more than 200 years (Heberden, 1772). The experimental findings, however, have been a subject of conflicting reports. In the isolated working rat heart (Segel *et al.*, 1978) and in anesthetized dogs, ethanol produced a reduction in both coronary blood flow and cardiac output. Webb and Degerli (1965) also found a decrease in coronary sinus flow in the open-chest anesthetized dog after a rapid (less than 25 min) intravenous infusion of 0.5–1.5 g/kg of ethanol, resulting in blood ethanol concentrations of 42–207 mg/dl; the reduction in flow occurred despite an increase in cardiac output and mean arterial pressure. In addition, apparent coronary vasospastic effects of ethanol have been observed in canine coronary arteries (Altura *et al.*, 1983a; Hayes and Bove, 1988). In the anesthetized

closed-chest dog, Hayes and Bove (1988) found that an intravenous infusion of ethanol, producing blood concentrations ranging from 65 to 255 mg/dl, reduced, in a concentration-dependent relation, the cross-sectional area of the left anterior descending artery, an effect not influenced by α-adrenergic blockade but prevented by a calcium channel antagonist; a reduction in myocardial flow was also evident in these experiments at 2 hr of the infusion, when blood ethanol concentration was 129 mg/dl. Altura and co-workers (1983a) have also found that ethanol produces sustained contractions of helical strips from canine coronary arteries. Although vasospasms were evident at a concentration as low as 39 mg/dl, pronounced changes were only present at 2 g/dl.

In contrast to these investigations, studies in the anesthetized (Lasker *et al.*, 1955; Ganz, 1963; Friedman *et al.*, 1979a, 1981a) and awake dog (Pitt *et al.*, 1970; Friedman *et al.*, 1981b) have shown an increment in coronary flow after intravenous ethanol administration. In these studies an increase in coronary flow after ethanol administration was generally observed when cardiac output had not fallen concomitantly; there were slow rates of infusion, more closely simulating alcohol ingestion; and high blood ethanol levels (greater than 200 mg/dl) were produced. Ethanol has also been found to enhance the increment in coronary blood flow that occurs in the isolated perfused rat heart following norepinephrine administration (Talesnik *et al.*, 1980). In humans, moreover, an intracoronary infusion of 570, but not 285, mg/min of ethanol for 15 min has been found to increase coronary blood flow (Cigarroa *et al.*, 1990).

The mechanisms of the increase in coronary blood flow following ethanol administration are not entirely clear. Although the changes in coronary flow may be caused in part by an increase in myocardial oxygen demand (Friedman *et al.*, 1979a), the fact that the increment in flow occurred irrespective of the change in heart rate (Friedman *et al.*, 1979a, 1981a,b), under controlled hemodynamic conditions (Abel, 1980), and concomitantly with an increment of coronary sinus oxygen content (Friedman *et al.*, 1979a) suggests that the increase in coronary blood flow is not merely a response to increased myocardial oxygen requirements. Ethanol is a direct coronary vasodilator (Abel, 1980), at least as it affects small coronary (R_2) "resistance" arteries and arterioles, and its metabolites, acetaldehyde (Gailis and Verdy, 1971; McCloy *et al.*, 1974; Nguyen and Gailis, 1974; Friedman *et al.*, 1979a; Altura *et al.*, 1983a) and acetate (Liang and Lowenstein, 1978), are also coronary vasodilators. Thus, the increment in coronary blood flow appears to be caused, at least in part, by the vasodilatory actions of ethanol, its metabolites, or both.

The actions of ethanol observed in nonischemic myocardium, however, may be different from those found in ischemic myocardium (Leighninger et al., 1961). In fact, Friedman et al. (1981b) found in anesthetized dogs, at an average blood ethanol concentration of 201 mg/dl, that ethanol appeared to produce a "coronary steal" in ischemic hearts. This phenomenon may occur because the vasculature of ischemic myocardium, which is already nearly maximally dilated, would not respond to the vasodilatory action of ethanol, whereas vessels supplying normal myocardium will dilate, producing a fall in resistance to flow to this area. In effect, blood is drawn away from ischemic myocardium.

The effect of ethanol on brain blood flow is also a subject of conflicting reports. Studies in humans and experimental animals at first suggested that ethanol produces either no change or an increase in brain flow. Thomas (1937) reported that ethanol produces a transitory (generally less than 25 min) vasodilation of pial arteries (100–200 μm outer diameter) in awake and anesthetized cats and rabbits. In contrast, observations by Altura and co-workers (1982; 1983b) in the intact rat indicate that ethanol produces dose-dependent (0.01–10 g/dl) vasoconstriction in pial arterioles (17–58 μm outer diameter).

Studies by Battey and co-workers (1953) in a group of severely intoxicated patients have also been interpreted as supporting the cerebral vasodilatory actions of ethanol. At the time of these brain flow measurements, patients were unresponsive or stuporous; also, hypercapnia and acidemia were present in those patients who had blood gas and pH determinations. Hypercapnia caused by respiratory depression might therefore have obscured a direct action of ethanol on brain flow. Moreover, intravenous infusions of ethanol in humans that result in blood concentrations that do not produce respiratory depression have not been demonstrated to increase total cerebral blood flow (Battey et al., 1953; Fazekas et al., 1955).

In contrast to these earlier studies, Goldman et al. (1973) found in awake rats a reduction of blood flow to the cerebellum and hippocampus (cerebrum) at blood ethanol concentrations ranging between 150 and 250 mg/dl, whereas at concentrations ranging between 30 and 50 mg/dl, no effect was observed. At higher ethanol levels, which resulted in anesthesia and respiratory depression, brain blood flow increased. When respiration was controlled, correcting the blood gas abnormalities, a significant generalized depression of brain blood flow

was observed. Because the fall in brain blood flow paralleled the fall in total cardiac output, the significance of these findings is not clear. In contrast, a decrease in brain blood flow was observed by Friedman et al. (1984) in awake dogs 30 min following the intravenous infusion of ethanol (producing an average blood ethanol concentration of 106 mg/dl), even though arterial gases and pH did not change and cardiac output was not altered significantly. However, in these experiments the effects were generally transient: at 60 min, when blood ethanol concentration was 231 mg/dl, only in the cerebellum was flow still significantly reduced; the effect of ethanol on blood flow was then no longer evident in the cerebrum or brainstem. Hemmingsen and Barry (1979) have also observed the development of tolerance to the effects of ethanol on brain blood flow. Whereas ethanol reduced cerebral blood flow and oxygen consumption in anesthetized rats when blood alcohol concentration was 540 mg/dl and arterial carbon dioxide tension was normal, following 3 to 4 days of administration of 9–12 g/kg per day of ethanol, such changes in cerebral flow and oxygen consumption were no longer evident even at a blood ethanol concentration of 660 mg/ dl. Moreover, these investigators (Hemmingsen and Barry, 1979) confirmed the confounding influence of hypercapnia: in rats made hypercapnic and kept at the same, but elevated, carbon dioxide tension, ethanol reduced cerebral blood flow and oxygen consumption, an effect still evident even after several days of ethanol pretreatment.

Altura and co-workers (1983b) have demonstrated direct cerebrovascular effects of ethanol. Perivascular application, intracarotid infusion, or systemic administration of ethanol produced dose-dependent vasoconstriction of rat cortical arterioles in vivo. Such changes were evident at concentration as low as 10 to 50 mg/dl and lasted for 5 to 40 min. At concentrations greater than 300 mg/dl, intense irreversible spasms were observed. Moreover, these investigators (Altura et al., 1983b) observed that helical strips of rat cerebral arteries developed increased tension in a dose-dependent manner when exposed to alcohol, with effects evident at concentrations as low as 37 mg/dl; these effects were antagonized by calcium channel blockers. In addition to the direct action on the cerebral vasculature, the fall in cerebral blood flow after ethanol administration may also reflect a reduction in brain metabolic demands. Cerebral oxygen consumption has been found to decline with severe intoxication, even when the apparent direct vascular action of ethanol is obscured by hypercapnia (Battey et al., 1953). Also, the cerebrovascular effects of ethanol may be a reflex response to ethanol's action at other sites. Incidentally, depression of cerebral blood flow and oxygen consumption, persisting for at least several weeks, has also been found in patients with Wernicke–Korsakoff syndrome (Shimojyo *et al.*, 1967).

Ethanol has also been shown to influence gastrointestinal blood flow. Following intravenous administration of ethanol, pancreatic flow has been found to decrease at average blood ethanol concentrations ranging from 106 (Friedman *et al.*, 1983) to 298 mg/dl (Horwitz and Myers, 1982). The decrease in pancreatic blood flow appears to be independent of the systemic hemodynamic effects of ethanol, not related to hepatic arterial blood flow (no evidence for "hepatic steal"; Friedman *et al.*, 1983), and not influenced by various pharmacological antagonists (Horwitz and Myers, 1982) but may be attenuated by an intravenous infusion of mannitol (Horwitz and Myers, 1982). Oral administration of 0.5 g/kg of ethanol to awake rats has been demonstrated to increase gastric, small intestine, and colon blood flow, an effect not inhibited by chronic alcohol administration (McKaigney *et al.*, 1986). On the other hand, intravenous administration of ethanol to awake dogs has not been found to influence gastric or small intestine flow (Horwitz and Myers, 1982; Friedman *et al.*, 1983), although Friedman and co-workers (1983) have found an increase of colon blood flow at a blood ethanol concentration 231 mg/dl.

Ethanol has also been found to have other regional vascular effects. Although ethanol has not been shown to have a consistent effect on renal blood flow in dogs (McSmythe *et al.*, 1953; Friedman *et al.*, 1981b), following intravenous administration of ethanol (producing average blood ethanol concentrations ranging from 110 to 230 mg/dl), renal medullary flow has been found to decrease in both anesthetized and awake dogs (Friedman *et al.*, 1981b). Also, in experiments that might have implications for the fetal alcohol syndrome, Altura *et al.* (1982) have found that ethanol produces a concentration-dependent constriction of human umbilical vessels, evident at average concentrations of 52 to 86 mg/dl; this effect was not influenced by pharmacological antagonists.

The metabolite of ethanol, acetate, may contribute to the regional blood flow changes following ethanol administration. In awake dogs an intravenous infusion of 0.075 mmol/kg per min of Tris-acetate for 20 min (producing plasma acetate concentrations of 1 mM) increased myocardial, stomach, small intestine, and colon flow but did not change hepatic, brain, or skin flow; in contrast to the intravenous effects of ethanol,

acetate increased pancreatic flow (Liang and Lowenstein, 1978).

Thus, ethanol and its metabolites have diverse effects on blood vessels and regional blood flow. Experimental conditions—such as route and rate of administration, dosage, duration of study, species differences, use of anesthesia, and whether intact animals or isolated blood vessels are used—have a profound influence on findings. At clinically relevant blood ethanol concentrations in humans or in animals simulating conditions in humans, however, alcohol appears to increase skin, splanchnic, and myocardial blood flow and reduce brain, muscle, limb, and pancreatic flow. Even these effects may be modulated by special conditions; for instance, when total myocardial blood flow increases, flow to ischemic myocardium may still decrease. Although regional blood flow changes may, to a large extent, be merely epiphenomena reflecting local metabolic changes, they may also be factors contributing to organ injury.

12.4. Alcoholic Cardiomyopathy

The concept of myocardial disease associated with chronic abuse of alcohol has become recognized as a separate nosological condition only in the past three decades (Eliazer and Giansiracusa, 1956; Evans, 1959). Before that, the myocardial disorder associated with alcohol abuse was viewed as a form of beriberi heart disease (Weiss and Wilkins, 1936; Blankenhorn, 1946). Because patients with alcoholic heart disease do not conform to the classical features of beriberi heart disease, the concept of "occidental beriberi" was advanced; the criteria used for classifying this disease would, however, satisfy those now used for alcoholic cardiomyopathy. It is now accepted that alcoholics may manifest cardiac dysfunction that is thiamine dependent (alcoholic beriberi) or thiamine independent (alcoholic cardiomyopathy) (Robin and Goldschlager, 1970). Moreover, alcoholic beriberi is now rarely seen.

In 1956, Eliaser and Giansiracusa documented both the lack of a nutritional deficiency and the absence of hepatic disease in a group of chronic alcoholics with cardiac dysfunction. Some investigators nevertheless still question the existence of a unique alcohol-induced cardiomyopathy (Sereny *et al.*, 1978). Recently, however, Urbano-Marquez and co-workers (1989) have demonstrated correlations between total lifetime consumption of ethanol and abnormalities of cardiac function. Although the pathophysiology of alcoholic cardiomyopathy is still not clear, the notion that alcohol abuse can produce a disorder of heart muscle is now widely accepted.

12.4.1. Various Manifestations of Alcohol-Associated Heart Disease

Alcohol abuse can produce a wide spectrum of cardiac disorders. Chronic alcoholics, even those not exhibiting any clinically apparent heart disease, may have subtle abnormalities of cardiac function that can be detected by the use of noninvasive methods (Spodick *et al.*, 1972; Timmis *et al.*, 1975; Wu *et al.*, 1976; Askanas *et al.*, 1980; Kino *et al.*, 1981; Mathews *et al.*,1981; Friedman *et al.*, 1986). Although systolic time intervals are no longer widely used as a measure of cardiac performance, abnormalities of these indices may be evident at rest even when other measurements, such as left ventricular ejection fraction, are normal (Spodick *et al.*, 1972; Timmis *et al.*, 1975; Wu *et al.*, 1976; Askanas *et al.*, 1980; Friedman *et al.*, 1986). Increased left ventricular mass (Kino *et al.*, 1981; Mathews *et al.*, 1981; Ballas *et al.*, 1982; Urbano-Marquez *et al.*, 1989), which has been related to the amount of alcohol consumed over time (Kino *et al.*, 1981; Mathews *et al.*, 1981; Urbano-Marquez *et al.*, 1989), an abnormal ratio of left ventricular internal dimension to left ventricular wall thickness (Mathews *et al.*, 1981), and an abnormal ratio of left ventricular wall stress to left ventricular mass (Friedman *et al.*, 1986) are other subclinical cardiac abnormalities that may be evident in the alcoholic. Of interest, these findings may no longer be present if the alcoholic abstains from alcohol for several years (Reeves *et al.*, 1978). At autopsy, chronic alcoholics have been found to have myocardial hypertrophy and varying degrees of myocardial and perivascular fibrosis (Factor, 1976).

Perhaps more of historic interest than of clinical importance is the disorder characterized by a hyperdynamic circulation, circulatory congestion, and thiamine responsiveness—beriberi heart disease (Weiss and Wilkins, 1936). Although Brigden and Robinson (1964) diagnosed five beer drinkers as having this disorder in a group of 50 patients with alcoholic heart disease, and Alexander (1966a) found 17 cases in 100 patients with idiopathic disease, cardiac beriberi is now a rare cause of heart disease in alcoholics. Nevertheless, this disorder must be kept in mind when treating alcoholics, since even the fulminant form, resembling the Shoshin type, may be seen (Attas *et al.*, 1978). In contrast, the typical patient with heart failure caused by chronic alcoholism has the circulatory features of a congestive cardiomyopathy (Goodwin, 1970): a hypocontractile, dilated heart.

Because the signs of circulatory congestion may not always be present in patients with cardiomyopathy, especially during the early phases or after treatment with diuretics, the term "dilated" cardiomyopathy has been suggested as a generic classification under which alcoholic cardiomyopathy should be placed. However, even this descriptive term does not include all patients with alcoholic cardiomyopathy. Asokan and co-workers (1972) have described a cardiomyopathy in alcoholics with depressed cardiac function characterized by a low cardiac output, a reduced contractile index, and an increased left ventricular end-diastolic pressure, yet having no cardiomegaly. In most studies of alcoholic cardiomyopathy, however, cardiomegaly is found almost invariably in all patients (Brigden and Robinson, 1964; McDonald *et al.*, 1971; Demakis *et al.*, 1974).

A fulminant, variant form of this disorder appeared in epidemic form in habitual beer drinkers in the mid 1960s in Belgium, Quebec City, and Omaha (Morin *et al.*, 1967). Quebec beer drinker's cardiomyopathy—the name given to this condition—was causally related to the presence of trace quantities of cobalt that had been placed in the beer to stabilize the foam. In contrast to patients with typical alcoholic cardiomyopathy, these patients were polycythemic and cyanotic and had large serous cavity effusions and a high early mortality rate (42%).

The cardiac findings in cirrhosis, however, have been a subject of some debate. Although early investigators emphasized the features of a hyperdynamic circulation in this disorder (Claypool *et al.*, 1957; Massumi *et al.*, 1965b), more recent studies (Limas *et al.*, 1974; Kelbaek *et al.*, 1984; Ahmed *et al.*, 1984) have suggested that myocardial disease may also be present in such patients, even if not immediately evident or obscured by other circulatory changes. The circulation of advanced cirrhosis, particularly when portacaval shunts (Murray *et al.*, 1958) or ascites (Fernando and Friedman, 1976) or both are present, is characterized by an increased cardiac output, reduced systemic resistance, increased blood volume, and sometimes by arterial oxyhemoglobin desaturation (Massumi *et al.*, 1965b).

In an echocardiographic study of 20 biopsy-proven cirrhotics, ten of whom had moderate to tense ascites and four hepatic encephalopathy, Fernando and Friedman (1976) found a higher heart rate, left ventricular end-diastolic volume, and left ventricular ejection fraction but a lower systemic resistance than in matched subjects without cardiac, renal, or hepatic disease. Furthermore, when multivariate regression analyses were done in which these cardiovascular findings were related to various clinical features, systemic blood pressure, systemic vascular resistance, and left ventricular diastolic volume were found to be related to estimates of ascites (Friedman and Fernando, in press). By contrast, Murray et al. (1958) were unable to relate the enhanced cardiac output observed in cirrhotics to the presence of ascites or edema but did find a relationship between cardiac output and the concomitant presence of portacaval shunting and advanced hepatic disease. Although these studies did not elucidate the mechanisms for such findings, other studies have disclosed high circulating concentrations of various vasoactive substances that might indeed explain these abnormalities: hyperglucagonemia has been found in alcoholic cirrhosis, particularly when portacaval shunting is present; substance P, a peptide with potent vasodilating actions, has been detected in increased concentrations in the plasma of patients with hepatic coma (Hortnagl et al., 1984); and increased plasma norepinephrine, renin activity, aldosterone, and arginine vasopressin have been observed in cirrhotics with impaired urinary sodium and water excretion, an abnormality associated with the presence of ascites (Bichet et al., 1982).

Despite the presence of a hyperdynamic circulation in some cirrhotics, other studies have demonstrated that such patients also show an abnormal cardiac response to exercise (Gould et al., 1969; Kelbaek et al., 1984) and afterload stress (Regan et al., 1969; Limas et al., 1974; Ahmed et al., 1984); and at autopsy cardiac hypertrophy and myocardial and perivascular fibrosis have been observed (Lunseth et al., 1958).

12.4.2. Alcoholic Cardiomyopathy: Clinical Features

The clinical features of alcoholic cardiomyopathy have been the subject of several surveys and reviews (Evans, 1961; Brigden and Robinson, 1964; Massumi et al., 1965a; Alexander, 1966a; Burch and DePasquale, 1968, 1969; McDonald et al., 1971; Demakis et al., 1974). Alcoholic cardiomyopathy is a primary disorder of heart muscle. By definition, no other cause of heart disease may be present to account for the findings. This disorder is found in patients with a long history of alcohol abuse. Generally, patients have been alcoholics for more than 10 years (Brigden and Robinson, 1964).

Although this disease is found predominantly in men, it is not known whether this sexual predilection is merely a reflection of the prevalence of alcohol abuse or is in fact related to biological factors. On reviewing systolic time intervals in alcoholics, Wu et al. (1976) found that male cirrhotics were more likely to have abnormal systolic intervals than a comparable population of females. From this observation, they concluded that men were more susceptible to the cardiotoxic effects of ethanol than women. In American reviews of alcoholic cardiomyopathy, 85% to 90% of the patients were black (Massumi et al., 1965a; McDonald et al., 1971; Shugoll et al., 1972; Demakis et al., 1974). Patients have been reported to be as young as 21 years old (McDonald et al., 1971) and as old as 72 (Brigden and Robinson, 1964). However, the average patient is generally around 40 years of age when he manifests the disease.

The initial clinical presentation of this disorder may be subtle. Breathlessness is the most common complaint. In the early stages, symptoms may appear to be out of proportion to the signs of heart failure or may be ascribed to a primary pulmonary disease, a not uncommon association in the alcoholic cigarette smoker. The patient may complain of cough, especially occurring at night, and may attribute his illness to persistent upper respiratory infection. In fact, patients who can identify the apparent onset of their illness will frequently describe a "flu-like" illness that does not abate. Whether this reflects a causal event or a precipitating one in a patient with compensated heart failure, or probably most commonly the development of heart failure that is confused with a respiratory infection, is generally not clear. However, the occurrence of a recent viral infection usually cannot be confirmed by serologic methods (McDonald et al., 1971).

Although Evans (1961), Alexander (1966a), and Burch and DePasquale (1969) all comment on the absence of angina pectoris in alcoholic cardiomyopathy, typical ischemic-type chest pain may be a complaint in some patients. Obstructive coronary disease may be coincidentally present, but most patients with this disorder do not have any abnormalities of their coronary circulation. Given the increased left ventricular diastolic wall tension (caused by high left ventricular diastolic pressure and volume), which may impede myocardial flow, and the high systolic wall tension (caused by large systolic dimensions and generally normal systolic pressures) and tachycardia, which will increase myocardial oxygen requirements, patients with advanced alcoholic cardiomyopathy may demonstrate subendocardial myocardial ischemia with exercise. Easy fatigability reflecting a low and fixed cardiac output, palpitations associated with cardiac arrhythmias, anorexia and abdominal discomfort caused by hepatic and intestinal congestion, and swelling resulting from edema are also common symptoms.

12.4.3. Physical Findings

Although many patients with alcoholic cardiomyopathy are not cachectic or malnourished at the onset of their illness, as the disease progresses and anorexia ensues, the wasted appearance of cardiac cachexia is seen. Initially, fluid retention obscures actual weight loss. Between hospitalizations body weight does not change, or it may even increase, but with each admission, as diuresis is accomplished, weight loss becomes apparent. Signs of pulmonary congestion are seen in the early phases of the disease, but as left ventricular dysfunction worsens and pulmonary hypertension and right ventricular failure ensue, venous pressure elevation, hepatomegaly, edema, and even ascites and anasarca are found.

Although elevations in systemic blood pressure may be seen during episodes of acute pulmonary edema or in the early phases of the disease, blood pressure is generally normal or low, even in some patients with a history of hypertension, and pulse pressure is narrow. Cardiomegaly is reflected by displacement of the apical impulse in a caudal and lateral direction. The apical wave form shows (1) prominence of the "a" wave when sinus rhythm is present, reflecting a high left ventricular end-diastolic pressure; (2) a slow rate of systolic impulse development and a sharp contour, indicative of depressed left ventricular wall motion; and (3) a prominent rapid filling wave associated with an increased left ventricular diastolic dimension. The aortic component of the second sound is generally soft, whereas the pulmonic component becomes accentuated as pulmonary hypertension progresses. An S-3 heart sound occurs in 80% to 100% of patients with alcoholic cardiomyopathy (McDonald *et al.*, 1971; Demakis *et al.*, 1974) and reflects left ventricular dilatation. S-4 heart sounds are less common, in part because of the development of atrial fibrillation in 10% to 20% of patients. Systolic murmurs of mitral and triscupid regurgitation caused by papillary muscle dysfunction and annulus enlargement from ventricular dilation are common (40% to 60% of patients) (McDonald *et al.*, 1971; Demakis *et al.*, 1974). In brief, the characteristic auscultatory findings of the dilated cardiomyopathy of chronic alcoholism are a gallop rhythm with loud S-3 and S-4 sounds, an accen-

tuated pulmonic closure sound, a diminished first heart sound, and a soft apical systolic murmur.

12.4.4. Electrocardiogram and Laboratory Findings

Although Evans (1959) has suggested that alcoholic heart disease may have specific electrocardiographic (ECG) features, the findings in this disorder are merely those of a diffuse myocardial disease. The ECG shows broad, enlarged P waves, suggesting an intraatrial conduction abnormality and biatrial enlargement. As atrial dilatation and damage progress, atrial fibrillation or flutter ensues in 10% to 20% of patients (Demakis *et al.*, 1974). Brigden and Robinson (1964) found this arrhythmia some time during the illness in as many as 50% of the patients they reviewed. Ventricular arrhythmias are common and increase in frequency and malignancy as left ventricular function deteriorates. Recognition of their presence is important, since sustained ventricular arrhythmias may account for the high incidence of sudden death in patients with alcoholic cardiomyopathy. Widespread myocardial destruction and generalized fibrosis produce intraventricular conduction abnormalities. Usually these have a nonspecific "left ventricular pattern," but typical left or right bundle branch block patterns are also found in 10% to 15% of patients. Electrocardiographic features of myocardial infarction occur. Pathological "q" waves may be caused by localized areas of myocardial fibrosis but may also be the result of acute myocardial infarction (Regan *et al.*, 1974a, 1975). QRS complex voltage may be reduced, reflecting loss of myocardial mass, presence of pericardial effusion, and generalized edema. More commonly, however, QRS complex voltage is actually increased. The increased QRS complex voltage may reflect actual left ventricular hypertrophy but may also be the result of loss of total body muscle mass, left ventricular dilatation, or even concomitant anemia. Evans (1959) described abnormal T waves, which he has suggested may be specific for alcoholic cardiomyopathy. Although T-wave abnormalities are common in this disorder, the "spinous, cloven, or dimple patterns" that Evans (1959) described appear to be neither sensitive nor specific for alcoholic cardiomyopathy.

Bashour *et al.* (1975) estimated the prevalence of these electrocardiographic abnormalities in a review of electrocardiograms obtained from 65 patients with alcoholic cardiomyopathy: 12% had atrial fibrillation, 34% a first-degree atrioventricular block, 58% an intraventricular conduction abnormality (comprising 31% with a hemiblock pattern, 11% with a left bundle branch block pattern, 9% with a right bundle branch block pattern, and 8% with a nonspecific intraventricular conduction defect), 23% left atrial enlargement, 52% biatrial enlargement, 66% left ventricular hypertrophy, 5% pathological Q waves, and 100% a nonspecific ST-segment and T-wave abnormality. Despite the high prevalence of electrocardiographic intraventricular conduction abnormalities, Bashour et al. (1976) found that none of these patients with cardiomyopathy, including those with electrocardiograms satisfying criteria for a "multifascicular block pattern," progressed to a more advanced form of heart block on follow-up of at least 42 months, consistent with the view that this disorder is not an important cause of complete heart block, although such occurrences have been seen (Leier et al., 1974; H. S. Friedman, unpublished observations).

Luca (1979) performed electrophysiological studies in 20 patients with alcoholic cardiomyopathy. All but one of these patients had abnormal ECGs. Compared with a control group, these patients demonstrated prolonged P–R, intraatrial, and His–Purkinje conduction times. Although atrioventricular conduction time (A–H interval) was not increased, atrial-pacing-induced Wenckebach-type atrioventricular block occurred at a relatively slow heart rate. Moreover, patients with alcoholic cardiomyopathy had lengthening of the atrioventricular refractory period but shortening of atrial and ventricular monophasic action potentials; furthermore, shortening of atrial action potential duration was especially evident in those subject to atrial fibrillation and was associated with an enhanced dispersion of atrial refractoriness.

Chest x-ray films generally show symmetrical cardiomegaly (consistent with four-chamber enlargement), a small aortic silhouette, and absence of cardiac calcifications (Hill et al., 1968). Massive cardiomegaly may be present at times when lung fields are free of congestive changes. Rapid return of heart size to normal in patients responding to treatment in the early phases of alcoholic cardiomyopathy has been referred to as "the accordion heart" (Alexander, 1966a; Hill et al., 1968). Although improvement of left ventricular function with abstinence from alcohol and standard treatment of heart failure may contribute to the dramatic reduction in heart size, it is likely that the clearing of an associated pericardial effusion accounts at least in part for the "accordion heart."

At an advanced stage of the disease, cardiac catheterization will show reduced cardiac output, high ventricular diastolic pressures, and pulmonary hypertension.

The ventricular pressure patterns may show "restrictive" features, but elevation of pulmonary artery pressure and the absence of an equalization of intracardiac diastolic pressures distinguish this order from constrictive pericarditis. Contrast or radionuclide left ventriculography will show increased ventricular volumes with a generalized hypocontractile pattern (Kreulen et al., 1973). In some patients, localized abnormalities of wall motion may also be seen (Kreulen et al., 1973; Regan et al., 1974a, 1975). Left ventriculography may also disclose mild mitral regurgitation, but coronary angiography generally reveals widely patent coronary arteries.

Echocardiograms demonstrate the features of a dilated cardiomyopathy (Mathews et al., 1981). Small pericardial effusions may also be seen, but a large one is unusual. The echocardiographic features and the presence of abnormal systolic time intervals distinguish alcoholic cardiomyopathy from constrictive pericarditis, a condition that might also be suggested by the clinical features of this disorder. Nuclear angiography is another useful technique for the noninvasive diagnosis of alcoholic cardiomyopathy, demonstrating a generalized hypocontractile left ventricle with a low ejection fraction. Thus, noninvasive methods—echocardiography and nuclear angiography—generally obviate the need for cardiac catheterization and contrast angiography for diagnosing alcoholic cardiomyopathy.

12.4.5. Pathology

Alcoholic cardiomyopathy is a dilated cardiomyopathy. There are no pathological features that distinguish this disorder of cardiac muscle from other similar conditions. Alcoholic cardiomyopathy is characterized by dilation of all cardiac chambers (Massumi et al., 1965a; Alexander, 1966a; Burch et al., 1966). Cardiac mass is generally increased, with heart weights averaging 600 g (normal <350 g) (Massumi et al., 1965a; Alexander, 1966a); Alexander (1966a) found that hearts obtained at autopsy from 24 patients with this condition had weights ranging from 375 to 1110 g. The cardiac valves are generally normal, and coronary arteries are patent or minimally atherosclerotic, but this condition does not preclude the coincidental occurrence of significant coronary stenoses (Massumi et al., 1965a; Alexander, 1966a). About one-third of the cases may have mural thrombi and/or evidence of systemic embolism at autopsy (Alexander, 1966a). The myocardium is pale, flabby, and may have small areas of fibrosis (Burch et al., 1966).

The abnormalities of alcoholic cardiomyopathy on

light microscopy are variable, are generally not pronounced, and are not specific. Myocardial fibers may vary in size and degree of abnormality, with myocardial hypertrophy and degeneration both occurring; foci of interstitial and perivascular fibrosis, especially evident in the subendocardium, are also found (Alexander, 1966a; Burch et al., 1966; Tsiplenkova et al., 1986; Vikhert et al., 1986). Fatty tissue infiltration, which becomes more pronounced as myocardial fiber degeneration increases, may also be seen (Vikhert et al., 1986).

An array of abnormalities may be found on electron microscopy, but these findings are also not specific for alcoholic cardiomyopathy (Hibbs et al., 1965; Alexander, 1966a, 1967; Bulloch et al., 1972; Vikhert et al., 1986; Tsiplenkova et al., 1986). Findings include loss of myofibrils, dilatation of the sarcoplasmic reticulum, dehiscence of the intercalated disks, and abnormalities in the number, clustering, and appearance of the mitochondria. Lipid droplet in cells, glycogen particles, and abnormal pigmentation (lipofuscin) have also been described (Hibbs et al., 1965). Histochemical studies have confirmed the presence of larger amounts of lipid in cells, primarily triglyceride (Burch et al., 1966), and a decrease in the amount of mitochondrial enzymes (Burch et al., 1966; Vikhert et al., 1986) consistent with the occurrence mitochondrial injury.

12.4.6. Prognosis and Treatment

The prognosis of patients with alcoholic cardiomyopathy is dependent on the severity of myocardial dysfunction and on whether patients abstain from alcohol (Shugoll et al., 1972; Demakis et al., 1974; Koide et al., 1980; Fuster et al., 1981; Schwarz et al., 1984). Fuster et al. (1981) found in a population of patients with dilated cardiomyopathy (having a median age of 49 years), 20% of whom had a history of alcohol abuse, a 77% mortality rate at 10 years, with two-thirds of the deaths occurring within 2 years of entry into the study. In a prospective study of 69 patients with dilated cardiomyopathy (22% alcohol-related) having a 35% 1-year mortality rate, an apparent alcohol etiology was not a determinant of outcome (40% were dead at 1 year) (Unverferth et al., 1984). By contrast, in Japan, alcoholic cardiomyopathy may be associated with a more favorable prognosis: Koide et al. (1980) found in a small series of patients with this diagnosis that all survived more than 10 years.

Fuster and co-workers (1981) found that age and cardiac dysfunction, reflected by heart size, cardiac output, and left ventricular end-diastolic pressure, were highly predictive of clinical outcome in patients with dilated cardiomyopathy. Similar observations were also reported by Koide and co-workers (1980), and in a multivariate regression analysis Schwarz et al. (1984) found that left ventricular ejection fraction and systolic pressure were independent predictors of outcome, but morphometric findings on myocardial biopsy specimens were not. Mitral regurgitation (Koide et al., 1980) and atrial fibrillation (Koide et al., 1980; Unverferth et al., 1984) are also unfavorable prognostic factors. Unverferth and co-workers (1984) found a 67% 1-year mortality rate in patients with dilated cardiomyopathy and atrial fibrillation. These investigators also found on multivariate analysis that the presence of an electrocardiographic pattern of an intraventricular conduction defect or ventricular arrhythmias was an independent predictor of outcome and confirmed that morphometric findings were not.

Several studies have demonstrated that abstinence from alcohol is a favorable predictor of outcome (Demakis et al., 1974; Koide et al., 1980). In a study of 48 patients with alcoholic cardiomyopathy, the likelihood of clinical improvement and survival was significantly related to abstinence from alcohol (Demakis et al., 1974). Shugoll et al. (1972) found no deaths at follow-up in 12 patients who had stopped drinking, whereas 25% of the 19 who continued to drink died; however, in this study abstinence was not associated with a clinical improvement in survivors. Complete recovery with abstinence has been reported: Schwartz and co-workers (1975) observed in a patient with left ventricular dilatation and hypocontractility, having an ejection fraction of 15%, complete normalization of hemodynamics after 1 year of abstinence from alcohol.

Prolonged bed rest for the treatment of alcoholic cardiomyopathy, as recommended by Burch (Burch and DePasquale, 1968, 1969; Burch et al., 1966), is no longer accepted as being part of the treatment of this disorder, but its rationale, reducing left ventricular wall stress, is probably an important determinant for the efficacy of vasodilator therapy. Venodilators (intermittent nitrate therapy) to reduce left ventricular dimensions and arteriolar vasodilators (hydralazine and especially the angiotensin-converting enzyme inhibitors) to reduce left ventricular impedance are now standard treatment for congestive heart failure with dilated hearts. Because vasodilator therapy not only is beneficial in relieving symptoms but may also improve survival, such therapy may be warranted in the early stages of the disease. Results of ongoing clinical trials may clarify the value of such early intervention. Although digitalis would be expected to improve contractility, its impact on exercise tolerance, especially when compared to vasodilator therapy, is generally minimal, and an increased sensitivity to toxicity has been observed by Burch and DePasquale (1969) in alcoholic cardiomyopathy. Nevertheless, the use of cardiac glycosides as adjunctive therapy in advanced congestive heart failure has now been demonstrated in several recent studies. Low dosage and short-term intravenous treatment with other inotropic agents are frequently helpful in advanced congestive heart failure, especially when hypotension and oliguria are present, but the long-term benefit of such therapy has not been demonstrated, and these drugs may be associated with an adverse proarrhythmic effect. Diuretics are used to treat circulatory congestion. In the advanced stages, large dosages of loop diuretics, sometimes in combination with diuretics that act on the diluting site of the nephron, may be required. In resistant cases intravenous administration of diuretics may be necessary.

The risk of thromboembolism in this disorder is high. Fuster and co-workers (1981), in a retrospective study of patients with dilated cardiomyopathy, found that 18% of those not receiving anticoagulants had a systemic embolism, whereas none who were receiving this therapy had such an event. Moreover, among those with atrial fibrillation, 33% had a systemic embolism during the period of the study (with a median follow-up of 11 years), but only 14% of those not having this arrhythmia had such an occurrence. However, the dangers of anticoagulation in alcoholics preclude the use of this therapy in most patients.

Arrhythmias are treated in the standard fashion. However, identification of patients at risk for sustained ventricular tachycardias is especially important because of the high risk of sudden death in this population. Despite this concern, the risk of proarrhythmia following the administration of antiarrhythmic drugs, especially in those patients with severe left ventricular dysfunction, and the myocardial depressant effects of some of these drugs limit the use of such pharmaceuticals for treating nonsustained ventricular tachyarrhythmias, even when guided by programmed electrical stimulation, in patients with alcoholic cardiomyopathy.

12.5. Effects of Prolonged Alcohol Exposure in Experimental Animals

Although the dilated, hypocontractile heart of alcoholic cardiomyopathy has still not been replicated experimentally, ultrastructural, functional, and bio-chemical abnormalities have been shown to occur in studies in which alcohol has been administered chronically to mice, rats, hamsters, dogs, rhesus monkeys, and perhaps turkey poults. In a study in mice Burch *et al.* (1971) found that animals fed alcohol (without any other dietary control) for 6 to 10 weeks showed mitochondrial swelling, dilatation of the transverse tubular system and intercalated disk, and lipid accumulation in the myocardium, but no gross, light microscopic, or histochemical abnormalities. Subsequently, Berk *et al.* (1975) demonstrated that mice given 20% (v/v) ethanol for 27 weeks showed the same depressed myocardial function as food- and fluid-intake-matched controls, which had become dehydrated and malnourished. In a study that has addressed these concerns, Alexander *et al.* (1977a,b) administered a liquid diet in which ethanol comprised 36% of calories to mice for 15 to 25 weeks. Abnormalities of myocardial structure and metabolism were evident at 15 weeks, and marked changes at 25 weeks: heart size, both absolute and relative to body weight, was increased; interstitial fibrosis was present; abnormalities of mitochondria—including the appearance of giant, fused forms—swelling of the sarcoplasmic reticulum and transverse tubules, and separations of the intercalated disk were found; myocardial triglyceride concentration was elevated; and mitochondrial function, as assessed by inorganic phosphate esterification (P/O), was depressed.

Various investigators have found myocardial contractile abnormalities following chronic ethanol administration in the rat, albeit sometimes without rigorous control of confounders. Maines and Aldinger (1967) observed that rats given 25% ethanol as their only drinking fluid and laboratory chow *ad libitum* demonstrated a deterioration of right ventricular isometric systolic tension beginning at 4 months of such feedings and becoming progressively worse over time. Maruyama *et al.* (1978) have observed that rats given 25% alcohol in their drinking water (which produced an average blood alcohol concentration of 81 mg/dl) for 5 weeks had reduced glycerinated heart muscle function compared to isocaloric-fed controls. Kino *et al.* (1981) have also found depressed myocardial function in rats fed 36% of calories for 5 weeks compared to pair-fed controls. Although the experimental group and controls did not differ in body or heart weight or myocardial cross-sectional area, isolated left ventricular papillary muscle of alcohol-treated rats demonstrated lower peak developed tension and lower maximum rates of rise and fall of tension at all bath calcium concentrations used than pair-fed rats. With a longer period of alcohol feeding, Tepper *et al.* (1986) found even more myocardial function indices abnormal compared to a closely matched control group. Left ventricular papillary muscles from rats fed alcohol as 40% of their caloric intake for 30 weeks, having an average blood ethanol concentration of 40 mg/dl, demonstrated reduced isometric indices—which, in addition to those found by Kino and co-workers (1981), included shortened intervals of contraction and relaxation—and reduced isotonic indices such as lower peak tension, slower maximum velocities of shortening and relengthening, and shortened time to peak shortening. Because the duration of myocardial action potentials was also shortened in the experimental group, these investigators suggested that the observed abnormality of contraction might be related in part to an abbreviation of contraction time. Fisher and Kavaler (1975), however, have found that the acute myocardial depressant effects of ethanol are observed even when shortening of the myocardial action potential by ethanol is prevented by caffeine.

By contrast to the depressed contractile function of isolated myocardium obtained from rats treated with ethanol for prolonged periods, the abnormalities of the intact, isolated perfused heart in such rats may be less pronounced (Whitman *et al.*, 1980; Chan and Sutter, 1982; Segel, 1987; Segel and Dean, 1979; Segel *et al.*, 1981). In rats receiving 55% of their caloric intake from ethanol for 12 to 14 weeks (having an average 2 hr postprandial blood ethanol concentration of 142 mg/dl), Whitman *et al.* (1980) observed no abnormality of baseline cardiac function in an isolated perfused preparation; only when an increased resistance to aortic outflow was produced (an increased afterload) was a difference found: the experimental group showed a lesser, albeit transient, increase in aortic pressure than the pair-fed control group. Chan and Sutter (1982) found that the nonstressed function of isolated perfused hearts of rats given 20% (v/v) ethanol in their drinking water and laboratory chow *ad libitum* for 12 weeks—producing an average blood ethanol concentration of 27 mg/dl—was also not different from a matched control group; however, the response of ventricular pressure to isoproterenol was attenuated in the alcohol-treated rats receiving a perfusate with a normal calcium concentration but relatively augmented when perfusate had a low calcium concentration. Moreover, Segel *et al.* (Segel, 1987; Segel and Dean, 1979; Segel *et al.*, 1981) have found that isolated perfused hearts of rats fed alcohol as 40% of their calories for 40 weeks tended to show normal baseline values and an attenuated functional response to dobutamine, phenylephrine, and glucagon, but not to

ouabain, compared to that of pair-fed controls. Despite the reduced cardiac contractile response to isoproterenol and dobutamine of chronic alcohol-treated rats, Segel and Mason (1982) have not found, however, an abnormality of cardiac β-adrenergic receptors.

Chronic alcohol treatment of rats may also produce visible cardiac abnormalities. Using a pair-fed model in which experimental rats received a diet containing up to 38% of calories as alcohol for as long as 1 year, Segel *et al.* (1975) also observed structural abnormalities in the rat. Myocardial ultrastructural damage was evident after as little as 7 weeks of daily feedings of 10% of calories as alcohol; more pronounced features including functional abnormalities, at 4 to 5 months of 20% to 30% of calories as alcohol; and clear metabolic and mechanical abnormalities and an increased ventricular/body weight ratio at 5 months of ingesting more than 30% of calories as alcohol (Segel *et al.*, 1975). Alcohol-treated rats demonstrated cardiac hypertrophy, myocardial mitochondrial swelling, dilatation of the transverse tubules, sarcoplasmic reticulum, and intercalated disks, and disintegration of myocardial fibrils, findings not evident in food-matched controls. By contrast, Hall and Rowlands (1970) have not detected myocardial ultrastructural changes in rats fed alcohol as 33% of their caloric intake for as long as 14 weeks; however, these rats, which had fatty changes in the liver, also manifested myocardial lipid droplets and pigment deposits.

Chronic ethanol administration to the rat has been shown to produce changes in myocardial metabolism. After 19 weeks of feeding rats 37% of their caloric intake as alcohol, Segel and co-workers (1975) found a depression of myocardial mitochondrial function: respiratory control ratio (oxygen consumption in state 3 respiration/oxygen consumption in state 4 respiration) and respiratory quotient (oxygen consumption in state 3/weight of mitochondrial protein) were reduced. Similar observations have been made by Gvozdjak *et al.* (1973b): rats receiving 250 mg of ethanol/100 g body weight by intraperitoneal injection daily for 10 weeks demonstrated reduced respiration of myocardial homogenates and mitochondria, with mitochondria having a 28% reduction of the respiratory control ratio; myocardial glycogen content and glycolytic activity were also reduced.

Abnormalities of lipid metabolism have also been found after chronic administration of alcohol to rats (Lieber *et al.*, 1966; Reitz *et al.*, 1973; Parker *et al.*, 1974; Williams and Li, 1977). Lieber *et al.* (1966) found that rats receiving 36% of calories as ethanol showed a myocardial accumulation of triglycerides, which could be reduced if long-chain triglycerides were eliminated from the diet. Williams and Li (1977) confirmed those observations. Reitz and co-workers (1973) have also found in rats given 25% (v/v) ethanol in their drinking water and laboratory chow *ad libitum* for 1 month a change in the fatty acid composition of the heart, a finding not evident in weight-matched controls: linoleic acid was elevated, and arachidonic acid reduced. In follow-up studies Reitz' group (Parker *et al.*, 1974) found that the activity of acyl-coenzyme A : carnitine acyltransferase, an enzyme mediating the translocation of fatty acids across the mitochondrial membrane for β-oxidation, was reduced, albeit only for palmitoleate and archidonate and therefore unlikely to account for reduced myocardial fatty acid oxidation.

By contrast, chronic feeding of alcohol to rats does not appear to affect cardiac protein synthesis appreciably. Preedy and Peters (1989) found that chronic alcohol feeding of rats as 36% of calories for 6 weeks did not affect cardiac protein concentration, and the capacity to make protein, as assessed by various RNA (concentration increased) and DNA (concentration not changed) indices, was not adversely changed. An abnormality of myocardial calcium fluxes has also been found in chronic alcohol-treated rat: sarcoplasmic reticulum membrane vesicles from the myocardium of rats treated with ethanol as 39% of their caloric intake for 10 months have been found by Segel and co-workers (1981) to demonstrate a decreased accumulation of calcium both in the absence of oxalate (reduced binding) and in presence of this anion (reduced uptake). The abnormalities of myocardial calcium fluxes and function were found to be reversible and were no longer present after 4 to 6 months of withdrawal of alcohol from the diet (Segel *et al.*, 1981).

Chronic experiments in which inhibitors of ethanol or its metabolites have been administered concurrently with alcohol have demonstrated enhanced myocardial injury. Weishaar and co-workers (1978) have found that rats receiving 36% of their caloric intake as ethanol for 4 to 6 weeks, when given 4-methylpyrazone, an inhibitor of alcohol dehydrogenase, or pargyline, a monoamine oxidase inhibitor associated with increased blood acetaldehyde concentrations following ethanol administration, show more depressed myocardial contractility and greater myocardial metabolic changes than rats receiving only alcohol or just an inhibitor; although all groups receiving alcohol showed depressed myocardial muscle function, reduction of mitochondrial respiration and protein synthesis was observed only in rats also receiving an inhibitor. The significance of this study is not, however, clear because of the concerns that these inhibitors might have had a direct synergistic effect with the alcohol administered and by the disturbingly high values of blood acetaldehyde concentration, even in groups that should not have had an appreciable amount of this substance in the blood. Kino *et al.* (1981) has observed, moreover, that rats receiving aminotriazole, a catalase inhibitor, in addition to ethanol as 36% of caloric intake for 5 to 6 weeks demonstrated marked cardiac morphological abnormalities that were not evident in rats receiving only alcohol or the inhibitor; these findings suggest that catalase, which facilitates the oxidation of ethanol in the presence of hydrogen peroxide, may protect against myocardial injury from alcohol.

Prolonged administration of ethanol to dogs has also been shown to produce abnormalities of cardiac ultrastructure, function, and metabolism; however, the findings in these animals have generally been less pronounced than those in mice and rats. Dogs have been fed alcohol as 36% of calories for as long as 52 months (Thomas *et al.*, 1980) without producing marked changes of cardiac performance. Although a decrease in left ventricular compliance (Thomas *et al.*, 1980) and a reduction of the maximal velocity of myocardial shortening (Sarma *et al.*, 1976) have been observed, minimal if any abnormalities of global left ventricular function have been found, even when the ventricle has been stressed. However, increased Alcian-blue-positive material in the ventricular interstitium (Regan *et al.*, 1974a), dilatation of the intercalated disk (associated with electrical conductance abnormalities: Ettinger *et al.*, 1976), mitochondrial degeneration (Fauvel *et al.*, 1974), and increased left ventricular collagen (Thomas *et al.*, 1980) have been described. Moreover, Regan and co-workers (1974a) have found increased myocardial triglyceride content, increased incorporation of [1-^{14}C]oleic acid into triglycerides and decreased incorporation into phospholipids, particularly in the subendocardium, and increased concentrations of subendocardial sodium but decreased concentrations of subendocardial potassium after ethanol administration for as long as 22 months. Also, Pachinger *et al.* (1973) have found reduced myocardial mitochondrial oxidation, isocitrate dehydrogenase activity, and ATP content after 14 weeks of ethanol administration to the dog, and Bing *et al.* (1974) found diminished calcium uptake and binding by myocardial mitochondria and by sarcoplasmic reticulum after 6 months of such treatment.

Of particular interest are the chronic myocardial effects of ethanol in primates. Vasdev *et al.* (1975) have administered ethanol as 40% of total calories to rhesus monkeys for 3 months. Hearts of the experimental group showed a relative increase in weight. Myocardial fibers showed fatty vacuolization with an increased triglyceride and cholesterol ester content. Evidence of muscle destruction and early fibrosis were also seen. The increased myocardial triglyceride deposits appeared to be caused at least in part by local synthesis, as reflected by the increased incorporation of [1-^{14}C] palmitate into triglyceride.

Two species that spontaneously develop a cardiomyopathy have been used to assess the chronic myocardial effects of ethanol: turkeys and hamsters. Noren and co-workers (1983) gave turkey poults 5% ethanol (25% of their caloric intake), producing blood ethanol concentrations ranging from 80 to 320 mg/dl, as their only fluid intake for the first 56 days of life; control poults received water *ad libitum*. Both groups were given the same caloric intake, but pair feeding was not done. Indeed, even after 1 week, turkey poults receiving alcohol weighed significantly less than the control group. At 28 days of age, poults receiving alcohol had relatively hypertrophied hearts and abnormal systolic time indices, and echocardiograms showed left ventricular dilatation and a reduced ejection fraction; thus, the findings of alcoholic cardiomyopathy were apparently reproduced. Calcium binding and uptake by the sarcoplasmic reticulum were also found to be reduced, and at age 56 days, reduced activity of sarcolemmal Na$^+$,K$^+$-stimulated ATPase, but not of Ca^{2+}-stimulated ATPase activity of sarcoplasmic reticulum, was evident. Although these findings are of interest, because turkey poults are susceptible to developing spontaneously a dilated cardiomyopathy, and nutritional confounders were not controlled, alcohol administration might merely have enhanced the occurrence of turkey cardiomyopathy by producing a malnourished state.

The hamster has also been found to develop myocardial dysfunction after chronic administration of alcohol. Garrett *et al.* (1987) have found that the golden hamster, when given 50% ethanol as its only liquid intake and laboratory chow *ad libitum*, demonstrates cardiac dysfunction and biochemical abnormalities after 7 weeks of such treatment and that these findings are still evident for at least 6 months if ethanol administration is continued (Wu *et al.*, 1987). Although treated and control hamsters had comparable weights, no attempt was made to ensure nutritional adequacy in the experimental groups, and, indeed, the experimental groups had a substantially lower daily total fluid intake. In any case, the isolated perfused heart of the alcohol-treated

hamster showed lower peak and developed left ventricular pressure and a lower rate of development of ventricular pressure; myocardial ATP, ATP/ADP, and adenosine were also decreased, but phosphocreatine was not changed. Of interest, concurrent treatment of experimental animals with verapamil prevented the findings associated with ethanol ingestion. For this reason, the investigators have suggested that chronic alcohol administration may produce myocardial dysfunction by "calcium overload" and/or by membrane injury, abnormalities that might be prevented by verapamil. In addition to the nutritional concerns suggested by the manner of ethanol administration (Berk *et al.*, 1975), the rapidity with which the abnormalities revert after alcohol is stopped and the close resemblance of these findings to the hereditary cardiomyopathy of hamsters, suggesting species peculiarities, may limit the significance of this model.

Several investigators (Rubin, 1979, 1982; Katz *et al.*, 1985) have suggested that ethanol may exert its chronic injurious myocardial effects by first producing sarcolemmal damage. This view is based in part on the observation that ethanol affects the composition and physical properties of biological membranes, changing both the lipid bilayer and membrane proteins (Katz *et al.*, 1985). Such changes increase the "fluidity" of membranes acutely, but chronic ethanol administration results in decreased "fluidity" (Goldstein, 1984). Of interest in this regard are studies by Polimeni and co-workers (1983). These investigators studied the sarcolemmal integrity of rats given 25% (v/v) ethanol as their sole fluid intake and laboratory chow *ad libitum* for as long as 12 weeks; average blood ethanol concentrations of the experimental group ranged from 33 to 67 mg/dl. Although animals receiving alcohol had a lower body weight than the control group, heart-to-body-weight ratio was not different. However, the myocardial extracellular space assessed by morphometry and by tracer (Na$_2$^{35}SO$_4$) distribution and the water distribution were consistent with a shrinkage of the extracellular space and an expansion of the cellular compartment of alcohol-fed rats; moreover, myocardial cellular concentrations of sodium and calcium were increased, and potassium decreased. After 8 weeks of abstinence, these abnormalities tended to revert. Alcohol administration appeared, therefore, to make the sarcolemma more "leaky," resulting in a movement of electrolytes down their electrochemical gradients. Alternatively, inhibition of sarcolemmal Na$^+$,K$^+$-ATPase might have explained these changes. Thus, that primary sarcolemmal injury from chronic administration of ethanol is the pivotal event that leads to alcohol cardiomyopathy is an interesting but still unproven hypothesis.

Chronic alcohol administration might also exert its detrimental cardiac effects by producing vascular injury, and intramyocardial small-vessel disease is, indeed, a common finding of chronic alcoholism (Factor, 1976). Sohal and Burch (1969) have found that mice given 15% ethanol in their drinking water for 3 months demonstrate myocardial capillary injury: focal swelling and necrosis of endothelial cells, endothelial junctional separation, and luminal narrowing. Rabbits given 10 ml of 20% ethanol orally for 3 weeks have also been found to have myocardial capillary "proliferative" changes on morphometric measurements even though they had no myocardial structural abnormalities; these animals also had a normal heart weight but a reduced body weight (Mall et al., 1982). In addition, rats receiving ethanol as 36% of calories for 4 weeks have been shown to have coronary microvascular wall thickening (Herrmann et al., 1984). Although these vascular changes could have an adverse effect on myocardial substrate and oxygen supply, especially during conditions of increased demand, absence of a clear relationship between such findings and myocardial abnormalities suggests that they are not necessary determinants for the development of cardiomyopathy.

Alcohol abuse is associated with an increased concentration of circulating catecholamines (Ogata, 1971). Chronic elevations of these substances have been shown to produce myocardial injury in humans and in experimental animals: myocardial hypertrophy, fatty changes, myofiber degeneration, and interstitial fibrosis have been observed following prolonged exposure to high blood concentrations of catecholamines. Such elevations which have been observed in alcoholics might at least contribute to the development of alcoholic cardiomyopathy. Experimental studies by Rossi et al. (1976) and Morin et al. (1969) support this hypothesis. Rossi and co-workers (1976) have found that the myocardial hypertrophy and injury that occur in rats given 32% (v/v) ethanol as their only fluid intake for 12 weeks were associated with increased myocardial norepinephrine; no myocardial structural changes or elevations in myocardial norepinephrine were evident at 4 weeks of treatment. Morin and co-workers (1969) have observed, moreover, that chronic administration of ethanol aggravates experimental isoproterenol cardiomyopathy. These investigators found that rats receiving 10% ethanol as their only fluid intake, which produced an average blood ethanol concentration of 95 mg/dl, and laboratory chow ad libitum for 3 months developed myocardial

injury more frequently and of greater severity after high-dosage isoproterenol administration than weight-matched controls.

Concurrent infection and hypertension in alcoholics with an enhanced susceptibility to myocardial injury have also been suggested as causes for alcoholic cardiomyopathy. Exaggerated myocardial damage from trypanosomal (Miller and Abelman, 1967) and Cocksackie B-3 (Morin et al., 1969) infections has been observed in rats treated chronically with alcohol. Hypertension, a condition that is prevalent in alcoholics, especially during acute withdrawal (see Section 12.7), has been suggested by several investigators (Dickenson, 1972; Friedman et al., 1986) as a cause of dilated cardiomyopathy. Friedman et al. (1986) have observed that alcoholics who manifested hypertension during detoxification had worse systolic time indices than alcoholics not having this finding. Moreover, this abnormality was evident even when blood pressure and heart rate had returned to normal values. Hypertensive alcoholics also demonstrated features of left ventricular "hyperfunction," which has been suggested as a prodromal finding in hypertensives who develop congestive heart failure. Also, Chan et al. (1985) have observed that rats developing hypertension after alcohol administration have a high incidence of circulatory congestion and death when exposed to additional stress. Thus, clinical and experimental evidence suggests that alcohol-associated hypertension might at least contribute to the development of congestive heart failure in alcoholics.

Thus, chronic administration of alcohol to experimental animals has resulted in many of the cardiac abnormalities found in humans with alcohol-associated myocardial injury; however, a dilated, hypocontractile heart—the hallmark of alcoholic cardiomyopathy—has not been convincingly reproduced. Myocardial hypertrophy and ultrastructural changes have been found in various experimental models: mitochondrial structural abnormalities; dilatation of the transverse tubular system, sarcoplasmic reticulum, and intercalated disk; interstitial fibrosis; myofibril disruption; and microvascular changes. An increased accumulation of triglycerides, at least in part because of their increased formation in the heart, and changes in myocardial fatty acid composition are consistent findings. Mitochondrial respiration has been found to be decreased—a finding associated with reduced isocitrate dehydrogenase activity and reduced myocardial ATP content. After chronic alcohol administration myocardial mitochondria and sarcoplasmic reticula show a reduced ability to take up and bind calcium. Cardiac muscle function is generally depressed after prolonged ethanol administration, but "pump" function is, at most, mildly disturbed. Indeed, the inability to replicate alcoholic cardiomyopathy suggests that alcohol alone may not be sufficiently toxic to heart muscle to produce such damage. The presence of an additional factor, such as infection, nutritional deficiency, hypertension, or cobalt, may therefore be necessary.

12.6. Holiday Heart

Cardiac arrhythmias have been related to alcohol abuse even in subjects not having evidence of heart disease or an electrolyte imbalance, such as hypokalemia (Scherf et al., 1967) or hypomagnesemia. This association has been popularized under the rubric of "holiday heart," a term coined by Ettinger et al. (1978) because such paroxysms of arrhythmia occur especially after binge drinking, sometimes in subjects without a history of habitual use of alcohol. Although subjects with "holiday heart" may present with ventricular arrhythmias, characteristically, atrial arrhythmias are observed, especially atrial fibrillation.

The association of paroxysmal atrial fibrillation with alcohol use has been the subject of several reports. Thornton (1984) reported on four subjects who usually drank little or no alcohol but developed atrial fibrillation after an alcohol binge, which reverted within 24 hr of their hospitalization. Rich et al. (1985), in a case-control study, found a twofold greater prevalence of heavy alcohol use in patients with idiopathic atrial fibrillation than in a matched hospital population; of interest, patients with alcohol-related atrial fibrillation were more likely to develop alcohol withdrawal symptoms than other hospitalized patients with a history of alcohol abuse. The only clinical features that distinguished patients with alcohol-related atrial fibrillation from those with this arrhythmia unrelated to alcohol use were the greater number of men and the lesser awareness of their tachycardia in the alcoholic population (Rich et al., 1985).

The electrophysiological basis for the proarrhythmic actions of ethanol is not entirely clear. Studies performed on isolated myocardium and Purkinje fibers have generally demonstrated that ethanol shortens action potential duration. Gimeno et al. (1962) observed a 10% shortening in the duration of the action potential of isolated rat atrium at an ethanol concentration of 440 mg/dl; Fisher and Kavaler (1975) also found a shortening of the action potential in isolated frog and cat ventricular muscle at similar ethanol concentrations. In isolated canine Purkinje fibers, Snoy et al. (1980) observed a decrease in the action potential duration at concentrations of 183–350 mg/dl; these investigators found, however, that at ethanol concentrations of 244 mg/dl and 350 mg/dl, the rate of depolarization and the action potential amplitude also decreased. Snoy et al. (1980) observed, in addition, that action potential duration shortening was related to an enhanced rate of early repolarization, whereas the rate of late repolarization was actually reduced. Also in isolated canine Purkinje fibers, Williams et al. (1980) confirmed that ethanol shortens (by 5–10%) action potential duration in a concentration-dependent manner (100–300 mg/dl). In contrast, no change in the rate of depolarization or in the action potential amplitude was observed. These differences may be explained by methodological problems in the Snoy et al. (1980) study: the tissue bath was acidotic, lacked magnesium, and had a low potassium concentration.

Williams et al. (1980) also found that other alcohols besides ethanol shortened action potential duration and that such changes could be directly related to their lipophilic properties. Ethanol's metabolites were not found, however, to shorten the Purkinje fiber action potential: acetaldehyde in pharmacological concentrations lengthened action potential duration, an effect that was antagonized by β-blockers; acetate produced no changes on the action potential (Williams et al., 1980).

Williams et al. (1980) also did not observe an effect of ethanol on spontaneous sinoatrial activity in guinea pig right atria. The absence of a direct enhancing effect of alcohol on sinus activity has also been reported by James and Bear (1967) in canine sinus artery perfusion studies; these investigators observed an increase of sinus rate only when acetaldehyde was infused at a concentration at which it releases myocardial norepinephrine. Carpentier and Gallardo-Carpentier (1981) have found effects of ethanol at a low concentration (5–40 mg/dl) on isolated guinea pig ventricular muscle action potentials that could also be related to a catecholamine-mediated action.

Thus, electrophysiological studies performed on isolated atrial and ventricular muscle and Purkinje fibers in several species have consistently demonstrated that ethanol, at concentrations observed in the blood of humans after the ingestion of alcohol, produces a shortening of action potential duration, an effect that appears to be related to its lipophilic properties. Neither alcohol nor its metabolites appear to affect directly sinoatrial node activity; sinus tachycardia following ingestion of alcohol would therefore require a neurohumoral-mediated mechanism.

There are no electrocardiographic features in humans specifically associated with acute alcohol ingestion or alcohol abuse. Sereny (1971) reviewed the electrocardiograms of 1000 male alcoholics admitted to a medical unit: 82% of records were interpreted as being normal. Sinus tachycardia observed in 8.6% of the patients and nonspecific T-wave findings in 4% were the two most common abnormalities found. Of interest, three of nine patients given alcohol after their electrocardiographic T-wave disturbance had normalized with abstinence again showed this finding. However, the T-wave morphology did not have any characteristic features, as previously suggested by Evans (1959).

The acute cardiac electrophysiological effects of alcohol have been assessed in humans with nonalcoholic heart disease (Gould et al., 1974, 1978) and in humans with "holiday heart," both with some patients having evidence of myocardial disease (Greenspon and Schaal, 1983) and with none having such findings (Engel and Luck, 1983). Gould et al. (1974) indicated in one report that 2 oz of whiskey in 12 patients with heart disease had no effect on intraatrial (P–A interval) or atrioventricular nodal (A–H interval) conduction but prolonged His–Purkinje (H–V interval) conduction time, whereas in another report of 14 patients who also had heart disease and were given 2 oz of whiskey, Gould and his co-workers (1978) described a prolongation of intraatrial conduction but no change in His–Purkinje conduction time. In the latter study, programmed electrical stimulation was also performed, which demonstrated a shortening of both atrioventricular nodal and ventricular refractory periods.

Greenspon and Schaal (1983) performed electrophysiological studies in 14 chronic alcohol users, 12 of whom averaged at least 50 ml of ethanol each day, who had symptoms suggesting the occurrence of cardiac arrhythmias, with four demonstrating a sustained arrhythmia at the time of admission. In no patient was an arrhythmia induced by programmed electrical stimulation when the subject was abstinent; however, following ethanol administration, which produced a blood ethanol concentration averaging 66 mg/dl (30–100 mg/dl), sustained atrial flutter or fibrillation was elicited in five. In one of the patients, without clinical or laboratory findings of heart disease, a sustained ventricular tachycardia was evoked after alcohol administration (Greenspon et al., 1979). Greenspon et al. also found in these 14 patients that alcohol lengthened the H–V interval, markedly in a patient who had Mobitz type II

atrioventricular block at time of admission, increased heart rate, and shortened the corrected sinoatrial conduction time, a measure of sinoatrial node activity. In contrast to the findings of Gould et al. (1978), these investigators did not observe a change in atrioventricular nodal or ventricular refractoriness.

Engel and Luck (1983) performed similar studies in 14 men, 11 of whom were "habitual alcohol abusers"—four with episodes of atrial fibrillation or flutter when drinking alcohol, but none with other features of heart disease. In four alcohol abusers atrial tachyarrhythmias were elicited in the absence of alcohol; in five of the ten subjects such arrhythmias were evoked only after the ingestion of alcohol, 2 to 4 oz of whiskey producing a blood ethanol concentration averaging 77 mg/dl. Of interest, two of the subjects who developed atrial flutter after ingesting alcohol had no history of alcohol abuse. These studies, like those of Gould and co-workers (1978) and Greenspon and Schaal (1983), failed to show a change in atrial refractoriness or dispersion of atrial recovery time. Thus, alcohol-related arrhythmias can be elicited in some patients by programmed electrical stimulation, especially following the administration of alcohol. The findings disclosed by electrophysiological studies in humans have varied between reports, perhaps as a consequence of differences in the populations studied. However, the electrophysiological conditions that might explain "holiday heart"—such as alterations in atrial refractoriness, conduction velocity, or dispersion of recovery time—have not been demonstrated by these methods.

Alcohol is associated not only with proarrhythmic effects but also with antiarrhythmic actions, at least experimentally (Paradise and Stoelting, 1965; Madan and Gupta, 1967). In anesthetized dogs ethanol has been found to slow myocardial electrical conduction (Goodkind et al., 1975; Gilmour et al., 1981). Goodkind et al. (1975) found that an infusion of 5% to 40% (v/v) ethanol into the left coronary artery of the dog or an intravenous bolus of ethanol that produced blood ethanol concentrations averaging 239–341 mg/dl prolonged atrioventricular and intraventricular conduction in a dose-dependent fashion but not His–Purkinje conduction; these findings, moreover, could not be explained merely by the increase in blood osmolality that occurs after such perturbations. Gilmour et al. (1981) also observed that an intracoronary infusion of ethanol at concentrations of 10% and 30% (v/v) or an intravenous infusion producing an average concentration of 245 mg/dl (but not one averaging 130 mg/dl) in the dog delayed myocardial electrical conduction; the amplitude of myocardial electrograms was also attenuated by these infusions. By contrast, these investigators (Gilmour *et al.*, 1981) found that such intravenous (but not intraarterial) infusions *lessened* both the slowing of conduction and the reduction of amplitude of myocardial electrograms resulting from acute myocardial ischemia. Moreover, this effect was associated with a protective action against inducibility of ventricular fibrillation. Also, Kostis *et al.* (1973) observed that ethanol, at a blood concentration averaging 148–238 mg/dl, increased ventricular fibrillation threshold in the intact dog; Paradise and Stoelting (1965) found that ethanol inhibits digitalis-induced ventricular tachycardia under similar experimental conditions; and Madan and Gupta (1967) reported that ethanol, at a blood concentration averaging 125 mg/dl, suppresses myocardial-infarction-induced ventricular tachycardia, at least transiently, in the awake dog. In addition, Bernauer (1986) found that pretreatment of anesthetized intact and isolated perfused rats with ethanol reduced in a dose-dependent fashion the occurrence of ischemic, but not reperfusion, ventricular arrhythmias.

Alcohol has also been found to have a favorable effect on the duration of experimental atrial fibrillation (Madan and Gupta, 1967; Nguyen *et al.*, 1987). Madan and Gupta (1967) reported that a rapid intravenous infusion of 0.5 g/kg of ethanol, which produced an average blood concentration of 125 mg/dl, shortened the duration of acetylcholine-induced atrial fibrillation in the anesthetized dog by 45%. Similar effects on electrically induced atrial fibrillation have been reported by Nguyen *et al.* (1987) at an average blood ethanol concentration of 208 mg/dl; shortening of the duration of atrial fibrillation ensued in these experiments despite the concomitant occurrence of arterial acidosis, hypotension, and myocardial depression.

The differences between studies that have demonstrated proarrhythmic actions of alcohol and those showing antiarrhythmic effects are not readily apparent but might, perhaps, be related to experimental considerations: the proarrhythmic actions of ethanol may result from elevations of catecholamines or from focal release of norepinephrine by ethanol or acetaldehyde. Under conditions in which baseline catecholamines are already elevated, such as in the acute open-chest dog, the neurohumoral effects of ethanol might be obscured. Another possibility is that some humans who manifest "holiday heart" may have subclinical alcoholic cardiomyopathy. If ethanol indeed has antiarrhythmic actions in humans, however, this might be one of the mechanisms for its protective effect in coronary artery disease.

12.7. Hypertension

Since 1915, when Lian reported alcoholism to be a cause of hypertension in French servicemen, alcohol use has been related to elevations of systemic blood pressure. Numerous studies (Arkwright *et al.*, 1982a; Cairns *et al.*, 1984; Cooke *et al.*, 1982; Fortmann *et al.*, 1983; Gyntelberg and Meyer, 1974; Harburg *et al.*, 1980; Gordon and Kannel, 1983; Jackson *et al.*, 1985; Klatsky *et al.*, 1986; Kondo and Ebihara, 1984; Kromhout *et al.*, 1985; MacMahon *et al.*, 1984; Milon *et al.*, 1982; Paulin *et al.*, 1985; Savdie *et al.*, 1984; Ueshima *et al.*, 1984; Wallace *et al.*, 1982) have now confirmed in diverse populations that alcohol use elevates systemic blood pressure independent of the confounding influence of age, body weight, or cigarette smoking. Moreover, systemic hypertension has been found to be more prevalent with use, and especially with abuse, of alcohol.

Although alcohol use has been related to both systolic and diastolic pressures, the influence of alcohol appears to be greater on the systolic value. However, in contrast to the independent effects of age and adiposity on blood pressure, the contribution of alcohol use seems to be relatively small. Even so, the impact is such that alcohol use can be related to an increased prevalence of hypertension (Arkwright *et al.*, 1982a; Barboriak *et al.*, 1982a; Cairns *et al.*, 1984; Gruchow *et al.*, 1985; MacMahon *et al.*, 1984; Milon *et al.*, 1982), and, moreover, in a longitudinal study, heavy drinking has been shown to increase the risk of developing hypertension (Dyer *et al.*, 1981).

A controversy, still unsettled, is whether a "threshold" exists at which alcohol consumption leads to an elevation of blood pressure. Some studies have even suggested a dip in blood pressure for light alcohol usage, producing U- or J-shaped relationships. In the Tecumseh Community Health Study, Harburg *et al.* (1980) found lower blood pressures in light drinkers than in abstainers. Similar findings were obtained in the Framingham Study: the trough for adjusted systolic blood pressure occurred at about one drink per day in men and four to five drinks per day in women, with more than five drinks per day required for an increased adjusted systolic blood pressure. In the Lipid Research Clinical Prevalence Study (Wallace *et al.*, 1982), the lowest blood pressure in women was observed at less than two drinks per week, and an increased value was not evident at less than three drinks per day. In the Kaiser–Permanente Study, Klatsky *et al.* (1977) found that a fall in blood pressure was evident in women who drank two or fewer drinks per day; in 1986, Klatsky *et al.* reported, however, that as little as one to two drinks per day elevates blood pressure in both men and women.

The discrepancy between these studies and those that failed to identify a dip or a threshold might be related to differences in group categorization and to the impact of confounding factors. In the Tecumseh Community Health Study (Harburg *et al.*, 1980), adjustment for age and body weight was shown to lessen the apparent dip in blood pressure observed in women who were light drinkers. Studies by Cooke *et al.* (1982), Mac-Mahon *et al.* (1984), and Milon *et al.* (1982) have confirmed that correction for age and adiposity makes the dip disappear. Also, it is unlikely that one to two drinks per week would have a long-term physiological effect on blood pressure, supporting the perception that light drinkers may have other characteristics that influence blood pressure.

Not only has there been controversy regarding the presence of a threshold for the hypertensive effects of alcohol, but there are also data that suggest a plateau. Several studies (Cooke *et al.*, 1982; Klatsky *et al.*, 1986; Savdie *et al.*, 1984) have shown that the blood-pressure-elevating effects of alcohol level off at approximately nine drinks per day. Whether this is merely a statistical artifact, as suggested by Klatsky *et al.* (1986), is not clear. Certainly, the myriad of biomedical and psychosocial factors present in the alcoholic are likely to confound a simple linear relationship.

Other factors that might influence the impact of alcohol consumption on blood pressure include type of beverage used, race, and age. The association of alcohol use and blood pressure is observed in countries where the preferred beverage may be beer, wine, sake, or whiskey. Also, there is evidence that older persons—those over 50 years—may have an increased sensitivity to the blood-pressure-elevating effects of alcohol (Barboriak *et al.*, 1982a; Fortmann *et al.*, 1983; Klatsky *et al.*, 1986; Milon *et al.*, 1982). Fortmann *et al.* (1983) observed that the increased sensitivity of older women to the hypertensive effects of alcohol was less evident in those receiving supplemental estrogens, suggesting a modulating influence of hormones on this effect. Despite the limitations inherent in epidemiologic studies and the problem of confounding influences, the body of evidence from population studies supports a positive relationship of alcohol use to systemic blood pressure and an increased prevalence of systemic hypertension in alcohol abusers.

Alcohol withdrawal is characterized by an intense psychomotor and sympathoadrenal reaction. Not sur-

prisingly, hypertension is frequently observed in this setting. Beevers *et al.* (1982) found that 50% of alcoholics undergoing detoxification had a systolic pressure greater than 140 mm Hg or a diastolic pressure greater than 90 mm Hg or both; Mehta and Sereny (1979) observed that 40% of such patients had a systolic pressure greater than 160 mm Hg or a diastolic pressure greater than 95 mm Hg or both; and Clark and Friedman (1985) found, in a study of alcoholics undergoing detoxification but with those having delirium tremens excluded, that 33% had the blood pressure elevations reported by Mehta and Sereny (1979). Although the peak blood pressures in such patients were generally observed on the day of admission for detoxification, in 20% of those with hypertension, the peak occurred on the second or third day (Clark and Friedman, 1985). In most of the hypertensives blood pressure gradually declined to normotensive values. However, even after all features of withdrawal had abated, 9% of the alcoholics still had systemic hypertension (Saunders *et al.*, 1979; Clark and Friedman, 1985), a value that probably reflects the prevalence of non-alcoholic-associated hypertension in this population. Also, alcoholics whose blood pressure returned to normal values still showed an exaggerated blood pressure response to a cold pressor test administered on the fourth or fifth hospital day (Clark and Friedman, 1985).

The blood pressure response during alcoholic detoxification has been related to withdrawal symptoms and to estimated recent use of alcohol or biochemical markers of such use. Beevers *et al.* (1982) found that age- and sex-adjusted blood pressures had a linear relationship to severity of withdrawal symptoms. These investigators also found a correlation of blood pressure with estimated daily intake of alcohol during the 3 months preceding detoxification; moreover, after exclusion of cirrhotics, blood pressure correlated with markers of recent alcohol use such as γ-glutamyltransferase and mean corpuscular volume. In an early study Saunders *et al.* (1979) found that alcoholics with fatty liver and well-compensated cirrhosis had higher blood pressures than those with no histological evidence of liver damage. In contrast, cirrhotics with ascites, jaundice, or varices had the lowest blood pressures, both during and after detoxification (Saunders *et al.*, 1979), an effect probably related to the hemodynamic alterations produced by advanced liver disease such as the appearance of arteriovenous shunts and high levels of some circulating vasoactive substances.

Peak blood pressure and blood pressure response during withdrawal have been related to the age of the alcoholic. Alcoholics with hypertension were observed to be older than those not showing this finding (Beckman et al., 1981; Clark and Friedman, 1985). Moreover, persistence of hypertension after withdrawal was most commonly found in older alcoholics (more than 50 years of age), whereas the lowest blood pressure and the pattern of a progressive decline in blood pressures were found in the youngest alcoholics (Beckman et al., 1981).

Alcohol detoxification is associated with elevation of various hormones. Plasma renin activity, cortisol, and aldosterone have all been correlated with the severity of withdrawal (Bannan et al., 1984). However, only plasma cortisol has been related, and that weakly, to blood pressure after alcohol withdrawal (Bannan et al., 1984). Bannan et al. (1984) also observed that urine volume and 24-hr sodium excretion were diminished during alcohol detoxification and that the 24-hr urine volume varied inversely with diastolic pressure.

Increased blood and urinary catecholamines have been found on acute ingestion (Ireland et al., 1984), with chronic use (Ogata et al., 1971), and after withdrawal of alcohol (Ogata et al., 1971; Clark and Friedman, 1985). Levels of epinephrine and its metabolites are especially increased during alcohol detoxification (Ogata et al., 1971; Clark and Friedman, 1985). Clark and Friedman (1985) found high average blood levels of epinephrine in both hypertensive and normotensive alcoholics undergoing detoxification; however, elevated blood epinephrine levels were more frequently observed in the hypertensives, with 86% having this finding versus only 44% of the normotensives. Moreover, hypertensives, at a time when blood pressure was normal and withdrawal symptoms had abated, demonstrated an exaggerated and atypical response to a cold pressor test. Not only did blood levels of norepinephrine increase after a cold pressor test, as observed in nonalcoholics (Arkwright et al., 1982b), but plasma epinephrine also increased after this stress, a finding not observed in nonalcoholics. Thus, hypertension during alcohol detoxification appears to be associated with enhanced adrenal cortical and medullary function. However, whether the increased adrenal activity is sufficient to explain hypertension during alcohol withdrawal is not entirely clear.

During alcohol detoxification, plasma electrolytes, including magnesium (Clark and Friedman, 1985), and atrial natriuretic factor (Clark et al., 1987) have not been found to be different in hypertensives. Although increased plasma levels of certain prostaglandin metabolites have also been observed in alcoholic hypertensives, the relationship of these substances to blood pressure is not clear (Clark et al., 1987). Changes in vascular and baroreceptor responsiveness, important determinants of blood pressure, have not been reported in hypertensive alcoholics. By contrast, a comprehensive study of such responses, in which drinkers (one or two drinks per day) with higher blood pressures were compared to nondrinkers, failed to disclose any differences (Arkwright et al., 1982b).

Since 1984, clinical reports have appeared that have provided a conclusive link between alcohol use and elevation of systemic blood pressure. Potter and Beavers (1984) studied the effects of continued drinking, withdrawal, and then resumption of alcohol in hypertensives who used on the average 60 g, but less than 80 g, per day. Those given their estimated usual amount to drink showed no fall in blood pressure until alcohol was stopped, with a significant fall in systolic and diastolic blood pressure evident within 72 hr thereafter. Drinking a comparable amount of alcohol after 3 days of abstinence resulted, within 48 hr, in a significant elevation of blood pressure. These changes were evident in most, but not all, of the participants. Also, blood pressure effects were found to be accompanied by changes in plasma cortisol concentration but were not related to plasma renin activity, 24-hr urine volume, or 24-hr urinary sodium, potassium, or metanephrine.

A similar blood pressure response has been reported in normotensive users of alcohol (averaging approximately three drinks per day) who merely curtailed or resumed their normal use (Puddey et al., 1985). Male volunteers, averaging 35 years of age, were assigned either to a group that maintained a normal drinking pattern for 6 weeks followed by 6 weeks in which only a low-alcohol-content beer (0.9% v/v) was permitted or to a group in which the same protocol was used but alcohol was first curtailed. By the second week of follow-up, a significant change in blood pressure was evident. Changes of systolic and diastolic pressure were found to correlate with both estimated use of alcohol and objective markers of alcohol consumption. However, a multivariate regression analysis, with weight change considered, only the relationship of alcohol use with supine systolic blood pressure remained significant. Moreover, one-third of the participants did not show a fall in blood pressure on the low-alcohol regimen. Despite the modest changes in blood pressure associated with alcohol use (1.1. mm Hg of systolic pressure per drink per day), this study (Puddey et al., 1985) demonstrated a direct effect of alcohol and blood pressure in a normotensive, nonalcoholic population that merely curtailed their use of alcohol.

Following the same protocol, Puddey et al. (1987)

examined the effects of alcohol intake on blood pressure in hypertensive, moderate to heavy drinkers (averaging four to five drinks per day). An estimated 85% reduction of alcohol use during the curtailment period produced significant decreases of systolic and diastolic pressures. These changes in blood pressure correlated with biochemical markers of alcohol use, and the findings were still evident after adjustment for weight changes. As this group observed in normotensive subjects (Puddey *et al.*, 1985), the independent effects of alcohol were modest, with less than 1 mm Hg of blood pressure related to each drink per day.

Predilection for hypertension and recentness of exposure to alcohol may influence the effects of alcohol intake on blood pressure. Malholtra *et al.* (1985) compared the effects of 5 days of 1 g/kg of alcohol per day (approximately five to six drinks per day) to abstinence in volunteers. Normotensive light drinkers who had not used alcohol for the 2 weeks preceding the study showed no differences in blood pressure; hypertensive light drinkers who had not taken alcohol or antihypertensive treatment for the 2 weeks preceding the study showed only higher standing blood pressures during the drinking period; whereas hypertensive, moderate drinkers who continued to drink their usual amount but had not taken antihypertensive treatment for the 2 weeks preceding the study showed higher standing and supine blood pressures during the drinking period. Thus, hypertensives who use alcohol regularly might derive an especially beneficial effect from abstinence or even from curtailment of alcohol consumption.

Recently, Chan and Sutter (1983) have reported that male Wistar rats given alcohol, as 20% (v/v) of drinking water, developed an increased blood pressure after 4 weeks of treatment. The increment of blood pressure was enhanced by chronic stress produced by 30 min per day of exposure to heat radiation (Chan *et al.*, 1985). Measurements obtained at 12 weeks showed that alcohol-treated animals had an increased plasma volume and a reduced 24-hr urine volume. Although plasma electrolytes were not different—except perhaps for a trend toward a lower plasma potassium concentration in the experimental group, alcohol-treated rats had a reduced urinary sodium concentration and an increase in both urinary calcium and magnesium concentrations. However, vascular responsiveness to catecholamines was not enhanced in aortic and portal vein strips obtained from these rats, and, in fact, portal veins showed a reduced contractile response to norepinephrine in a bathing medium with a low calcium concentration. By contrast, alcohol treatment increased plasma norepi-

nephrine levels. Moreover, the enhanced blood pressure response produced by chronic heat stress was associated with further elevation of plasma norepinephrine concentration but with no additional expansion in plasma volume.

Altura's group (1980, 1982a) has demonstrated, however, that changes in vascular responsiveness might indeed occur in rats after chronic ethanol treatment. Male Wistar rats given a liquid diet containing 6.8% (v/v) alcohol (control group received an equicaloric volume of a sucrose solution) for up to 24 months showed tolerance to the inhibitory effects of pharmacological concentrations of ethanol on spontaneous and evoked contractile responses of aortic and portal vein strips. Moreover, in a review, Altura and Altura (1982) have reported that alcohol-treated rats showed enhanced evoked vascular contractions at 12 and 24 weeks and that blood vessels obtained from these rats demonstrated dose–response curves consistent with the development of "supersensitivity." These changes in blood vessel reactivity were accompanied by an increased calcium and a decreased magnesium vascular content. Because these divalent cations are known modulators of contraction, such findings could explain the changes in vascular reactivity produced by chronic alcohol ingestion.

In brief, systemic hypertension has been produced in rats following chronic alcohol administration. Increased sympathetic activity reflected by elevated plasma norepinephrine, sodium and water retention, and an enhanced excretion of calcium and magnesium accompany the elevation of blood pressure. Chronic stress, which can by itself lead to an increased blood pressure in rats, enhances the hypertensive and noradrenergic effects of alcohol. Other studies in rats given alcohol for up to 24 months have also demonstrated changes in blood vessel responsiveness and vascular divalent cation content. Further investigation is required, however, to determine whether these experimental models are indeed relevant to alcohol-associated hypertension in humans.

Thus, clinical and experimental studies have demonstrated that alcohol consumption influences blood pressure. Because of the general tendency to underestimate alcohol intake, particular care must be taken in obtaining an accurate history of alcohol use in hypertensives. When alcohol abuse is suspected but denied, biochemical markers of alcohol abuse may be helpful. Because even one to two drinks per day may elevate blood pressure, especially in those with hypertension, hypertensives should be advised to use alcohol infrequently and then only in small amounts. Not only may alcoholism be the cause of hypertension, but the alcoholic is not likely to comply with recommendations for treatment. Alcohol-associated hypertension should therefore be regarded as a form of secondary hypertension that can be diagnosed without sophisticated testing and treated without surgery or drugs.

12.8. Stroke

Alcohol abuse has been associated with acute ischemic (Lee, 1979; Hillbom and Kaste, 1978, 1981) and hemorrhagic (Donahue et al., 1986; Gill et al., 1986; Stampfer et al., 1988) stroke. Moreover, relations between even moderate (15–39 ml of ethanol per day) use of alcohol and hemorrhagic stroke have been found (Donahue et al., 1986).

An association of alcohol abuse and thrombotic stroke has been reported in young adults. In a retrospective case review of 49 patients under age 50 years with a cerebral infarction but no history of hypertension, Lee (1979) found that 79% of the men had a history of alcohol abuse; the presence of such a history was twofold greater than in the case controls. In both a retrospective survey (1978) and a prospective study (1981) Hillbom and Kaste observed that alcohol intoxication frequently preceded thrombotic stroke occurring in men under 40 years of age. In a chart review of 76 patients with thrombotic stroke, ethanol-related cases comprised 40% of those aged 16–19 years, 25% between 20 and 29 years, but only 13% between 30 and 39 years. Hypertension did not appear to be a causal determinant in this study. In a prospective study of 23 consecutive patients under 40 years of age with an acute ischemic brain infarction, 43% had been ethanol intoxicated within 24 hr of the event; alcohol use was two- to fivefold greater in this cohort than in matched subjects, and hypertension was not an important causal factor, having been present in only four patients (Hillbom and Kaste, 1981).

In the Yugoslavia Cardiovascular Disease Study, Kozararevic et al. (1980) found a relationship between alcohol consumption and stroke at 7-year follow-up in male subjects having ranged in age from 35 to 62 years. However, on multivariate analysis in which systolic blood pressure, a covariate, was included as an independent variable, the relationship between stroke and alcohol consumption was no longer evident. By contrast, other studies have demonstrated that the relationship between alcohol use and stroke may be evident even apart from its effects on blood pressure. In a 12-year follow-up of the Honolulu Heart Program, comprising men ranging in age from 45 to 69, alcohol use was found to be related to hemorrhagic stroke even when adjustments were made for age and various risk factors, including hypertension: relative risk of a hemorrhagic stroke increased from 2.3 that of nondrinkers in light drinkers (1–14 ml of ethanol per day) to 2.9 in heavy drinkers (40 ml per day); specifically, this added risk was related to the increased occurrence of subarachnoid hemorrhage, with moderate drinkers having a relative risk of such an event of 3.5 and heavy drinkers one of 3.8 (Donahue et al., 1986).

The increased risk of subarachnoid hemorrhage in drinkers has also been observed in women. Stampfer et al. (1988) reported in a 4-year follow-up of nurses, ranging in age from 35 to 59 years, an incremental change in the relative risk of subarachnoid hemorrhage with increasing use of alcohol that was independent of age and other cardiovascular risk factors. By contrast, the relative risk of ischemic stroke, and stroke in general, was actually decreased in women drinkers; no added risk was evident even in those women who drank more than two drinks each day. However, the impact of alcohol abuse on stroke could not be determined in this survey because fewer than 2% of these women drank more than three drinks per day. In a retrospective case-control study, Gill et al. (1986) found that reported light use of alcohol (one or fewer drinks per day), with adjustment made for other risk factors, was also associated with an overall reduction in the relative risk of stroke in men; however, men who reported three or more drinks per day had a relative risk of stroke of 4.2. The relations of various biochemical markers of alcohol use to the occurrence of stroke also followed this J-shaped trend.

Alcohol use has an array of effects that, depending on conditions, might exert either a favorable or an unfavorable influence on the occurrence of cerebrovascular accidents. When used in moderation, alcohol generally has a favorable effect on the lipid profile that is associated with the occurrence of atherogenic vascular occlusive events (see Section 12.9). By contrast, even when used in moderation, alcohol tends to elevate systemic blood pressure (see Section 12.7). Because hypertension is the leading risk factor for stroke, this relationship would favor the occurrence of cerebrovascular events, especially hemorrhagic stroke. Alcohol abuse may produce cardiac arrhythmias and cardiomyopathy, which are associated with the development of cardiac thrombus and the attendant risk of embolic stroke. In addition, ethanol has been found to reduce cerebral blood flow in awake experimental animals (Altura *et al.*, 1983a; Friedman *et al.*, 1984); Altura and co-workers (1983a) have found, moreover, that in rats at a blood ethanol concentration of 300 mg/dl, cerebral arteriolar spasms are observed, often followed by vascular rupture. The acute effects of alcohol on rheology would favor conditions for thrombotic stroke, whereas its effects on platelets, coagulation, and thrombolysis might lessen the risk of thrombotic stroke but might enhance that of a hemorrhagic stroke (see Section 12.9).

Thus, alcohol abuse has been associated with the occurrence of cerebral ischemic infarction in young adults. Alcohol users, even those having no more than one or two drinks each day, have a higher risk of a subarachnoid hemorrhage than nondrinkers, and the risk increases with use. The added risk may be related in part of the effects of alcohol use on blood pressure, but this mechanism would not be sufficient, because an unfavorable relationship is still present when an adjustment is made for blood pressure change. Alcohol use in moderation, especially in women, appears also to have a protective effect on cerebrovascular disease, related, perhaps, to those mechanisms that also favorably influence coronary artery disease.

12.9. Relationship of Alcohol Use to Coronary Artery Disease

The effects of alcohol use on coronary artery disease have been a subject of controversy. The suggestion that alcoholism protects against the development of coronary artery disease has been refuted by critical examination of autopsy findings (Wilens, 1947; Viel *et al.*, 1966). Wilens (1947) concluded on the basis of a review of 519 autopsies in alcoholics, which were compared with 600 in nonalcoholics, that the apparent low incidence of obstructive coronary disease in alcoholics could be explained by the relatively young age at which they died and by the reduced incidence of hypertension, diabetes mellitus, and obesity in this population; Viel *et al.* (1966) found, moreover, no relationship between a history of alcoholism, which was associated with the occurrence of cirrhosis, and the presence of atherosclerosis in a review of 777 autopsies in men. Autopsy studies of cirrhotics *per se*, however, have disclosed that moderate to severe coronary atherosclerosis is less common in cirrhotics, and their incidence of myocardial infarction is only about 25% (Howell and Manion, 1960) of that found in individuals dying of other diseases, even

when matched for age, sex, and race (Ruebner *et al.*, 1961). Ruebner *et al.* (1961) have pointed out a statistical bias inherent in such studies: two lethal diseases, even when unrelated, will appear to be related inversely in autopsy studies, such as that evidenced also by the negative association of cirrhosis with extrahepatic malignant tumors. Parrish and Eberly (1961) have demonstrated, furthermore, the importance of the control group in arriving at conclusions: when compared to the findings in victims of accidents, neither chronic alcoholics nor cirrhotics were found to have more severe coronary artery disease; when compared to their respective null groups, however, a difference was evident. Because accident victims, in particular, tend to be more representative of the general population than the overall autopsy population, biases not present in accident victims, such as a relatively high prevalence of diabetes mellitus and hypertension, did not confound the comparison.

By contrast, epidemiologic studies have generally demonstrated a favorable influence of alcohol use, at least in moderation, on the occurrence of coronary events. Both longitudinal investigations such as the Chicago Western Electric Study (Dyer *et al.*, 1980; 1981), the Honolulu Heart Study (Yano *et al.*, 1977) and the Yugoslavia Cardiovascular Disease Study (Kozararevic *et al.*, 1980) and case-control investigations such as the Kaiser–Permanente Study (Klatsky *et al.*, 1974, 1981a) and a study by Hennekens *et al.* (1978) have demonstrated favorable relationships between alcohol use and the occurrence of coronary events. In a 17-year follow-up of 1899 white men, ranging in age from 40 to 55 years at entry, employed by the Chicago Western Electric Company, Dyer and co-workers (1981) observed that the incidence of death from coronary heart disease decreased progressively from the occasional drinker to those who drank up to four to five drinks each day. In a 10-year follow-up of 7705 Japanese–American men living in Hawaii, Yano and co-workers (1977) found a favorable influence of alcohol use on coronary events, which followed a linear trend up to 60 ml of alcohol each day, even when confounding risk factors were considered.

In a 7-year follow-up of 11,121 Yugoslavian men, ranging in age from 35 to 62 years at entry, Kozararevic *et al.* (1980) found an inverse relationship between alcohol use as an independent risk factor and nonfatal myocardial infarction and fatal coronary heart disease. Klatsky and co-workers (1981a) observed in members of the Kaiser Foundation Health Plan that the percentage of "nondrinkers" with fatal and nonfatal coronary events was higher than the percentage of drinkers in a control group, and Hennekens *et al.* (1978) observed a risk ratio of 0.6 in drinkers compared to nondrinkers in a study in which 568 married men dying of coronary heart disease were contrasted with a matched group. A favorable influence of alcohol use on coronary events has also been observed in women: in a study based on a diet questionnaire completed by 98,462 nurses, Stampfer and co-workers (1988) found, at a 2- to 4-year follow-up of 98% of these women, that age-adjusted relative risk of nonfatal myocardial infarction and fatal heart disease was related inversely to alcohol use. The relationship of alcohol use to reduction of coronary events appears, moreover, to be dependent on the quantity of alcohol used rather than on the type of beverage imbibed (Yano *et al.*, 1977; Hennekens *et al.*, 1979). A puzzling observation, however, has been made by Nanji (1985): he found that although coronary heart disease mortality rate for 27 countries varied inversely with the percentage of alcohol consumed in these countries as wine, it varied directly with the percentage consumed as beer. Yano *et al.* (1977) have found, in contrast, a favorable relationship between alcohol use and coronary events even in beer drinkers.

Not all studies have observed a favorable influence of alcohol use on coronary events. In the British regional heart study of 7729 men, ranging in age from 40 to 59 years at entry, Shaper *et al.* (1987) found no significant benefit of alcohol use on coronary events after an average follow-up of 6.2 years. In this study, unlike other investigations, biochemical markers were used to validate reported alcohol intake. Light drinkers (one or two drinks daily) had the fewest coronary events on follow-up, but they also had the most favorable risk profile. As expected, however, those who reported no alcohol use had the highest prevalence of coronary heart disease on entering the study; when only individuals without evidence of coronary disease at first examination were analyzed, the apparent advantage in light drinkers was no longer evident. A curious finding was the consistently worst outcome in the occasional drinker (one or two drinks each month or on special occasions) at follow-up.

Even if light or moderate use of alcohol reduces nonfatal and fatal coronary events, it is not clear whether such use has a benefit on overall mortality rates. Klatsky *et al.* (1981a), reporting on a 10-year follow-up of the Kaiser–Permanente study, found fewer deaths overall in individuals imbibing two or fewer drinks each day when compared to all other groups, including those groups comprising individuals who drank more, for whom the death rate was increased; when the moderate drinkers were compared to abstainers, however, no statistically significant advantage was evident. Despite finding a favorable independent influence of alcohol consumption on coronary events, Kozararevic *et al.* (1980) did not find a relationship between alcohol use and overall risk of dying. A similar dissociation between the effects of light or moderate alcohol use on coronary heart disease mortality rates (a reduction) and those on total mortality rates (no effect) has been observed by Dyer *et al.* (1981).

In this regard, alcohol use may be similar to taking aspirin or lipid-lowering drugs for the primary prevention of coronary heart disease; those interventions have also been shown to reduce coronary events without influencing total mortality rates. Moreover, alcohol abuse has been associated with an increase in coronary heart disease. Pell and D'Alonzo (1968) compared 922 alcoholics employed by the Dupont Company to a matched group of nonalcoholics stratified by blood pressure; at 5-year follow-up both male and female alcoholics had a higher coronary heart disease mortality rate than nonalcoholics. Dyer *et al.* (1981) found that white male "problem drinkers" employed by the Chicago Peoples Gas Company and those white males who drank more than six drinks each day employed by the Chicago Western Electric Company had an increased mortality rate in general and an increased coronary heart disease mortality rate in particular after a 15-year follow-up; however, when the effects of hypertension were considered, this relationship was no longer apparent. In a longitudinal study of Swedish men aged 50 years, registration at a Temperance Board was an independent determinant (blood pressure considered) of heart disease mortality at 9-year follow-up (Wilhelmsen *et al.*, 1973). Thus, alcohol use appears to have a bimodal influence on coronary events, an apparent favorable effect in light or moderate drinkers and an adverse effect in problem drinkers; however, alcohol consumption has never been shown, at any level of use, to have a beneficial effect in reducing death rates when all causes are considered.

Coronary angiographic studies also suggest that alcohol use may have a favorable influence on coronary artery disease. Barboriak *et al.* (1977, 1979, 1982a) have related estimates of coronary stenoses, assessed by coronary arteriography, to alcohol use in individuals undergoing diagnostic investigations. When compared to subjects who reported drinking less than 1 oz of alcohol weekly, those who reported greater use, especially individuals drinking more than 6 oz weekly, had less occlusive disease and fewer myocardial infarctions.

These observations applied to both men and women and appeared to be independent of other coronary risk factors, such as smoking (Barboriak *et al.*, 1982a). Moreover, in individuals undergoing diagnostic coronary angiography and not found to have obstructive disease, Fried and co-workers (1986) have observed that those consuming one to three drinks daily had coronary arteries with larger internal diameters than those imbibing less than one drink daily.

Because coronary events are generally the consequences of atherosclerotic and thrombotic occlusions, alcohol use would be expected to influence such occurrences by affecting atherogenesis or clotting or both. Epidemiologic and experimental evidence suggests that alcohol use may indeed affect both processes. The effects of alcohol use on plasma lipids may be of particular importance in the development of atherosclerosis. A comprehensive review of the relationship of alcohol to lipid metabolism is presented in Chapter 4.

In brief, alcohol use is believed to exert its "protective" effects on coronary artery atherosclerosis by favorably affecting the plasma lipoprotein profile. The Cooperative Lipoprotein Phenotyping Study has demonstrated that alcohol use is positively associated with high-density lipoprotein (HDL) and negatively associated with low-density lipoprotein (LDL) and that these findings relate—LDL directly and HDL inversely—to the prevalence of coronary heart disease, myocardial infarction in particular, in the population studied, the Honolulu Heart Study cohort (Kagan *et al.*, 1981). A recent investigation (Haskell *et al.*, 1984) has disputed the importance of this observation. In a study in which moderate alcohol users stopped drinking and then resumed their normal drinking patterns, Haskell *et al.* (1984) found that whereas plasma HDL-cholesterol concentrations did, in fact, relate to alcohol use, the changes were confined to the HDL_3 subfraction, which has not been related to coronary disease, and not to the subfraction that has been related to coronary artery disease protection, HDL_2, which did not change. By contrast, in a similar study in which changes of the protein contents of HDL, the apolipoproteins, were measured, Camargo *et al.* (1985) found that apolipoproteins AI, which relates inversely to risk of coronary events, and AII, which has no clear relationship, both increased with alcohol use. Moreover, Masarei and co-workers (1986) have observed in a crossover study in which moderate alcohol users received either light (0.9% v/v) or normal (5%) beer for 6 weeks that serum lipoprotein lipid and apolipoprotein changes could be related to alcohol consumption independent of the ef-

fects of various confounders such as weight change. During periods of increased alcohol consumption, HDL-cholesterol, its subfractions HDL_2 and HDL_3, and its major apolipoproteins, AI and AII, all increased.

Experimental studies in primates have also demonstrated favorable effects of alcohol administration on plasma lipoproteins (Hojnacki *et al.*, 1988) and atherosclerosis (Rudel *et al.*, 1981). Hojnacki and co-workers (1988) have found that squirrel monkeys fed up to 36% of their caloric intake as ethanol showed a dose-related linear increase in HDL-cholesterol that was primarily related to an increment of HDL_2. Increases of LDL-cholesterol and HDL_3 became evident only when the diet contained 24% of calories as alcohol, and the most favorable lipoprotein relations were visible at 12%. The effects of alcohol on lipoproteins appeared to be time dependent, with animals receiving low-dosage ethanol showing higher dosage responses after several months of low-dosage treatment. In this species an 18% dosage appears to be pivotal; at this percentage of calories as alcohol, which is associated with a blood ethanol concentration exceeding 100 mg/dl, changes in lipid metabolism occur that would favor atherogenesis: LDL-cholesterol and apoliprotein B increase, and lecithin : cholesterol acyltransferase (LCAT) decreases (Hojnacki *et al.*, 1988). Changes in lipoproteins and atherosclerosis have been observed in another primate model (*Macaca nemestrina*): Rudel and co-workers (1981) have observed that a diet containing 36% of calories as ethanol protected against the atherogenicity of a high-cholesterol diet (1.0 mg/kcal), particularly in the coronary circulation; this protective effect was related to changes in the molecular weight and composition of plasma lipoproteins.

Alcohol use has important effects on platelets and coagulation that might contribute to its coronary "protective" actions. Kagan and co-workers (1981) have observed in the Honolulu Heart Study that even though alcohol use appeared to have a favorable influence in preventing myocardial infarction, it is less valuable in obviating the occurrence of angina pectoris; this suggested to these investigators a nonatherogenic mechanism for alcohol's coronary "protective" actions. Studies in healthy volunteers have demonstrated that following ingestion of alcohol sufficient to produce blood ethanol concentrations less than 100 mg/dl, the prolongation of bleeding time by aspirin is enhanced (Deykin *et al.*, 1982), and the blood concentration of the metabolite of prostacyclin, 6-keto-$PGF_{1\alpha}$ is increased (Landolfi and Steiner, 1984). With blood ethanol concentrations greater than 200 mg/dl for sustained periods, which occur frequently in alcoholics, thrombocytopenia and direct effects of alcohol on platelet function may also be found (Haut and Cowan, 1980). In the Northwick Park Heart Study, Meade and co-workers (1987) found that alcohol use is related inversely to plasma fibrinogen concentration, a risk factor for coronary events. Alcohol may also enhance fibrinolysis: ethanol has been found to increase the secretion of plasminogen activator by bovine endothelial cells (Laug, 1983). By contrast, rebound thrombocytosis has been observed in alcoholic withdrawal; such occurrences could increase the risk for thromboembolic events in the alcoholic (Haselager and Vreeken, 1977).

The effects of alcohol on angina pectoris has been a subject of controversy. In his classic description of angina pectoris, Heberden (1772) had advised that alcohol be used in its treatment. From the acute effects of ethanol on cardiac hemodynamics, which might increase myocardial oxygen demand, and on myocardial blood flow, which might reduce flow to ischemic myocardium (Friedman, 1981), there is little to suggest a beneficial effect of alcohol in this condition. Clinical studies, moreover, have failed to demonstrate any benefit of alcohol in alleviating or preventing angina pectoris. Stearns et al. (1946) found that 1 oz of whiskey failed to shorten the duration of angina pectoris induced by exercise; these investigators also did not find a change in the exercise tolerance of patients with angina pectoris who imbibed 1 oz of whiskey before they exercised, even when they had taken a similar amount on the four days preceding testing. Russek et al. (1950) also observed that 1–2 oz of whiskey imbibed 5 to 30 min before an exercise test did not prevent the electrocardiographic ST-segment depressions that occurred in patients with angina pectoris, but it seemed to mask the actual chest discomfort; nitroglycerin, by contrast, prevented the occurrence of both chest pains and ST-segment depressions in these patients.

Orlando and co-workers (1976) have found, moreover, in a double-blind, randomized study, in which the effects of 2 and 5 oz of ethanol on exercise stress testing were contrasted with those of a noncaloric beverage, that the mean exercise time until angina was shortened and the extent of ST-segment depression was exaggerated at both dosages of ethanol. Moreover, several reports (Fernandez et al., 1973; Kashima et al., 1982; Matsuguchi et al., 1984; Takizawa et al., 1984) of alcohol-provoked Prinzmetal variant angina have appeared. In these reports of vasospastic angina, generally with patent coronary arteries, but not always (Fernandez et al., 1973; Matsuguchi et al., 1984), episodes have

usually occurred several hours after alcohol had been ingested (sometimes more than 10 hr afterwards) (Matsuguchi et al., 1984; Takizawa et al., 1984), suggesting a reactive withdrawal phenomenon to the vasodilating actions of ethanol, its metabolites, or both in susceptible individuals. A recent report (Matsuguchi et al., 1988) of alcohol administration preventing such attacks is consistent with this view. Of interest, ethanol has not been found to influence the myocardial blood flow responses to ergonovine, a substance used to provoke vasospastic angina, in the dog (H. S. Friedman, unpublished observations).

Regan and co-workers (1974a; 1975) have suggested that some alcoholics might have an increased susceptibility to myocardial infarction in the presence of patent or minimally stenotic coronary arteries. Because of their finding of increased myocardial glycoprotein and coronary perivascular fibrosis in alcoholics who succumbed from an acute myocardial infarction with patent or near-patent coronary arteries, these investigators concluded that a toxic cardiomyopathy had been the cause of the occurrence. Moreyra et al. (1982) have reported also on three men who had acute myocardial infarction after binge drinking and were subsequently found to have patent coronary arteries; two were young men (18 and 22 years). These reports, however, appeared before performing coronary arteriography in the presence of acute myocardial infarction had become a common practice and before cocaine was widely recognized as a cause of acute myocardial infarction. Because spontaneous coronary thrombolysis may occur in a substantial number of typical thrombotic coronary occlusions, resulting in even patent coronary arteries on angiography, and no mention was made of cocaine use in these reports, it is still not clear whether alcohol-induced myocardial infarction is, indeed, a distinct nosological disorder.

Alcohol abusers both with and without coronary disease appear to be at increased risk of sudden death. Epidemiologic studies have disclosed an increased incidence of sudden, unexpected death in alcoholics (Kramer et al., 1968; Randall, 1980; Sarkioja and Hirvonen, 1984). The occurrence of sudden death has been described in young individuals, generally in the third to the fifth decade of life, in whom the only remarkable finding at autopsy is hepatic fatty metamorphosis (Kramer et al., 1968; Randall, 1980); generally, the blood ethanol concentration found in these alcoholics at autopsy is less than 50 mg/dl (Randall, 1980). The cause of these events is not known, although ventricular fibrillation evoked by an intense sympathoadrenal response to withdrawal or by an electrolyte imbalance such as profound hypokalemia or hypomagnesemia would probably be the best explanation. Also, in a multivariate regression analysis in which individuals dying suddenly of coronary events were compared with those having nonfatal myocardial infarction, being a heavy alcohol consumer (average alcohol consumption of 167 g/week for sudden death and 96 g/week for nonfatal events) was found to be an independent predictor of sudden death; of interest, the analysis did not support an arrhythmogenic explanation for this relationship (Fraser and Upsdell, 1981).

Thus, alcohol use appears to have a protective effect in coronary artery disease, albeit without altering overall mortality rate. In contrast, alcohol abuse is associated with an increased risk of dying of a coronary event. Alcoholics without coronary artery disease, moreover, may also be at risk of dying suddenly, although the explanation for such occurrences is not clear.

REFERENCES

Abel, F. L.: Direct effects of ethanol on myocardial performance and coronary resistance. *J. Pharmacol. Exp. Ther.* **212**:28–33, 1980.

Ahmed, S. S., Levinson, G. E., and Regan, T. J.: Depression of myocardial contractility with low doses of ethanol in normal man. *Circulation* **48**:378–385, 1973.

Ahmed, S. S., Howard, M., Hove, W., Leevy, C. M., and Regan, T. J.: Cardiac function in alcoholics with cirrhosis: Absence of overt cardiomyopathy—myth or fact? *J. Am. Coll. Cardiol.* **3**:696–702, 1984.

Alexander, C. S.: Idiopathic heart disease. *Am. J. Med.* **41**:213–228, 1966a.

Alexander, C. S.: Electron microscopic observations in alcoholic heart disease. *Br. Heart J.* **29**:200–206, 1967.

Alexander, C. S., Forsyth, G. W., Nagasawa, H. T., and Kohloff, J. G.: Alcoholic cardiomyopathy in mice: Myocardial glycogen, lipids and certain enzymes. *J. Mol. Cell. Cardiol.* **9**:235–245, 1977a.

Alexander, C. S., Sekhri, K. K., and Nagasawa, H. T.: Alcoholic cardiomyopathy in mice electron microscopic observations. *J. Mol. Cell. Cardiol.* **9**:247–254, 1977b.

Alling, C., Gustavsson, L., Mansson, J., Benthin, G., and Anggard, E.: Phosphatidylethanol formation in rat organs after ethanol treatment. *Biochim. Biophys. Acta* **793**:119–122, 1984.

Altura, B. M., and Altura B. T.: Microvascular and vascular smooth muscle actions of ethanol, acetaldehyde, and acetate. *Fed. Proc.* **41**:2447–2451, 1982.

Altura, B. M., and Gebrewold, A.: Failure of acetaldehyde or acetate to mimic the splanchnic arteriolar or venular dilator actions of ethanol: Direct *in situ* studies on the microcirculation. *Br. J. Pharmacol.* **73**:580–582, 1981.

Altura, B. M., Ogunkoya, A., Gebrewold, A., and Altura, B. T.: Effects of ethanol on terminal arterioles and muscular venules: Direct observations on the microcirculation. *J. Cardiovas. Pharmacol.* **1**:97–113, 1979.

Altura, B. T., Pohorecky, L. A., and Altura, B. M.: Demonstration of tolerance to ethanol in noninvasive tissue: Effects of vascular smooth muscle. *Alcoholism: Clin. Exp. Res.* **4**:62–69, 1980.

Altura, B. M., Altura, B. T., Carella, A., Chatterjee, M., Halevy, S., and Tejani, N.: Alcohol produces spasms of human umbilical blood vessels: Relationship to fetal alcohol syndrome (FAS). *Eur. J. Pharmacol.* **86**:311–312, 1982.

Altura, B. M., Altura, B. T., and Gebrewold, A.: Alcohol-induced spasms of cerebral blood vessels: Relation to cerebrovascular accidents and sudden death. *Science* **220**:331–333, 1983a.

Altura, B. M., Altura, B. T., and Carella, A.: Ethanol produces coronary vasospasm: Evidence for a direct action of ethanol on vascular muscle. *Br. J. Pharmacol.* **78**:260–262, 1983b.

Andren, L., and Hansson, L.: Circulatory effects of stress in essential hypertension. *Acta Med. Scand.* **646**:69–72, 1981.

Arkwright, P. D., Beilin, L. J., Rouise, I., Armstrong, B. K., and Vandongen, R.: Effects of alcohol use and other aspects of lifestyle on blood pressure levels and prevalence of hypertension in a working population. *Circulation* **66**:60–66, 1982a.

Arkwright, P. D., Beilin, L. J., Vandongen, R., Rouse, I. A., and Lalor, C.: The pressor effect of moderate alcohol consumption in man: A search for mechanisms. *Circulation* **66**:515–518, 1982b.

Askanas, A., Udoshi, M., and Sadjadi, S.: The heart in chronic alcoholism: A noninvasive study. *Am. Heart J.* **99**:9–16, 1980.

Asokan, S. K., Frank, M. J., and Witham, A. C.: Cardiomyopathy without cardiomegaly in alcoholics. *Am. Heart J.* **84**:13–18, 1972.

Attas, M., Hanley, H. G., Stultz, D., Jones, M. R., and McAllister, R. G.: Fulminant beriberi heart disease with lactic acidosis: Presentation of a case with evaluation of left ventricular function and review of pathophysiologic mechanisms. *Circulation* **58**:566–572, 1978.

Auffermann, W., Camacho, S. A., Wu, S., Litt, L., Parmley, W. W., Higgins, C. B., and Wikman-Coffelt, J.: Acute alcohol cardiac depression is associated with a decrease in intracellular water and a reduction of the redox state. *J. Am. Coll. Cardiol.* **2**:224A, 1988a.

Auffermann, W., Wu, S., Parmley, W. W., Higgins, C. B., Sievers, R., and Wikman-Coffelt, J.: Reversibility of acute alcohol cardiac depression: ^{31}P NMR in hamsters. *FASEB J.* **2**:256–263, 1988b.

Ballas, M., Zoneraich, S., Yunis, M., Zoneraich, O., and Rosner, F.: Noninvasive cardiac evaluation in chronic alcoholic patients with alcohol withdrawal syndrome. *Chest* **82**:148–153, 1982.

Bannan, L. T., Potter, J. F., Beevers, D. G., Saunders, J. B., Walters, J. R. F., and Ingram, M. C.: Effect of alcohol withdrawal on blood pressure, plasma renin activity, aldosterone, cortisol and dopamine β-hydroxylase. *Clin. Sci.* **66**:659–663, 1984.

Barboriak, J. J., Rimm, A. A., Anderson, A. J., Schmidhoffer, M., and Tristani, F. E.: Coronary artery occlusion and alcohol intake. *Br. Heart J.* **39:**289–293, 1977.

Barboriak, J. J., Anderson, A. J., Rimm, A. A., and Tristani, F. E.: Alcohol and coronary arteries. *Alcoholism: Clin. Exp. Res.* **3:**29–32, 1979.

Barboriak, P. N., Anderson, A. J., Hoffmann, R. G., and Barboriak, J. J.: Blood pressure and alcohol intake in heart patients. *Alcoholism: Clin. Exp. Res.* **6:**234–238, 1982a.

Barboriak, J. J., Anderson, A. J., and Hoffmann, R. G.: Smoking, alcohol and coronary artery occlusion. *Atherosclerosis* **43:**277–282, 1982b.

Bashour, T. T., Fahdul, H., and Cheng, T. O.: Electrocardiographic abnormalities in alcoholic cardiomyopathy. *Chest* **68:**24–27, 1975.

Bashour, T. T., Fahdul, H., and Cheng, T. O.: Multifascicular block in cardiomyopathy. *Cardiology* **61:**89–97, 1976.

Battey, L. L., Heyman, A., and Patterson, J. L.: Effects of ethyl alcohol on cerebral blood flow and metabolism. *J.A.M.A.* **152:**6–10, 1953.

Beckman, H., Frank, R. R., Robertson, R. S., Brady, K. A., and Coin, E. J.: Evaluation of blood pressure during early alcohol withdrawal. *Ann. Emerg. Med.* **10:**32–34, 1981.

Beevers, D. B., Bannan, L. T., Saunders, J. B., Paton, A., and Walters, T. R. F.: Alcohol and hypertension. *Contrib. Nephrol.* **30:**92–97, 1982.

Berk, S. L., Block, P. J., Toselli, P. A., and Ullrick, W. C.: The effects of chronic alcohol ingestion in mice on contractile properties of cardiac and skeletal muscle: A comparison with normal and dehydrated–malnourished controls. *Experientia* **31:**1302–1303, 1975.

Bernauer, W.: The effect of ethanol on arrhythmias and myocardial necrosis in rats with coronary occlusion and reperfusion. *Eur. J. Pharmacol.* **126:**179–187, 1986.

Bichet, D. G., Van Putten, V. J., and Schrier, R. W.: Potential role of increased sympathetic activity in impaired sodium and water excretion in cirrhosis. *N. Engl. J. Med.* **25:**1552–1557, 1982.

Bing, R. J., Tillmanns, H., Fauvel, J. M., Seeler, K., and Mao, J. C.: Effect of prolonged alcohol administration on calcium transport in heart muscle of the dog. *Circ. Res.* **35:**33–38, 1974.

Blomqvist, G., Saltin, B., and Mitchell, J. H.: Acute effects of ethanol ingestion on the response to submaximal and maximal exercise in man. *Circulation* **42:**463–470, 1970.

Brigden, W., and Robinson, J.: Alcoholic heart disease. *Br. Med. J.* **2:**1293–1289, 1964.

Bulloch, R. T., Pearce, M. B., Murphy, M. L., Jenkins, B. J., and Davis, J. L.: Myocardial lesions in idiopathic and alcoholic cardiomyopathy. *Am. J. Cardiol.* **29:**15–25, 1972.

Burch, G. E., and DePasquale, N. P.: Alcoholic cardiomyopathy. *Cardiologia* **52:**48–56, 1968.

Burch, G. E., and DePasquale, N. P.: Alcoholic cardiomyopathy. *Am. J. Cardiol.* **29:**723–731, 1969.

Burch, G. E., Phillips, J. H., and Ferran, V. J.: Alcoholic cardiomyopathy. *Am. J. Med. Sci.* **251:**89–104, 1966.

Burch, G. E., Colcolough, H. L., Harb, J. M., and Tsui, C. Y.: The effect of ingestion of ethyl alcohol, wine and beer on the myocardium of mice. *Am. J. Cardiol.* **27:**522–528, 1971.

Cairns, V., Keil, U., Kleinbaum, D., Doering, A., and Stieber, J.: Alcohol consumption as a risk factor for high blood pressure: Munich blood pressure study. *Hypertension* **6:**122–131, 1984.

Camargo, C. A., Williams, P. T., Vranizan, K. M., Albers, J. J., and Wood, P. D.: The effect of moderate alcohol intake on serum apolipoproteins A-1 and A-11. *J.A.M.A.* **19:**2854–2857, 1985.

Carmichael, F. J., Israel, Y., Saldivia, V., Giles, H. G., Meggiorini, S., and Orega, H.: Blood acetaldehyde and the ethanol-induced increase in splanchnic circulation. *Biochem. Pharmacol.* **36:**2673–2678, 1987.

Carpentier, R. G., and Gallardo-Carpentier, A.: Effect of ethanol on guinea pig ventricular action potentials. *J. Electrocardiol. (San Diego)* **14:**333–340, 1981.

Castenfors, H., Hultman, E., and Josephson, B.: Effect of intravenous infusions of ethyl alcohol on estimated hepatic blood flow in man. *J. Clin. Invest.* **39:**776–781, 1960.

Chan, T. C. K., and Sutter, M. C.: The effects of chronic ethanol consumption on cardiac function in rats. *Can. J. Physiol. Pharmacol.* **60:**777–782, 1982.

Chan, T. C. K., and Sutter, M. C.: Ethanol consumption and blood pressure. *Life Sci.* **33:**1965–1973, 1983.

Chan, T. C. K., Wall, R. W., and Sutter, M. D.: Chronic ethanol consumption, stress, and hypertension. *Hypertension* **7:**519–524, 1985.

Chen, T. S., Friedman, H. S., Smith, A. J., and Del Monte, M. L.: Hemodynamic changes during hemodialysis: Role of dialyzate. *Clin. Nephrol.* **20:**190–196, 1983.

Cherrick, G. R., and Leevy, C. M.: The effect of ethanol metabolism on levels of oxidized and reduced nicotinamide–adenine dinucleotide in liver, kidney, and heart. *Biochim. Biophys. Acta* **107:**29–37, 1965.

Cigarroa, G., Lange, R. A., Popma, J. J., Yurow, G., Sills, M. N., Firth, B. G., Phil, D., and Hillis, L. D.: Ethanol-induced coronary vasodilatation in patients with and without coronary artery disease. *Am. Heart J.* **119:**254–259, 1990.

Clark, L. T., and Friedman, H. S.: Hypertension associated with alcohol withdrawal: Assessment of mechanisms and complications. *Alcoholism: Clin. Exp. Res.* **9:**125–130, 1985.

Clark, L. T., Hoover, E. L., Crandall, D. L., Cervoni, P., Haynes, S., and El Sherif, N.: Atrial natriuretic peptide and prostaglandin responses to alcohol induced hypertension. *Clin. Res.* **35:**440A, 1987.

Claypool, J. G., Delp, M., and Lin, T. K.: Hemodynamic studies in patients with Laennec's cirrhosis. *Am. J. Med. Sci.* **234:**48–55, 1957.

Cooke, K. M., Frost, G. W., Thornell, I. R., and Stokes, G. S.: Alcohol consumption and blood pressure. *Med. J. Aust.* **1:**65–69, 1982.

Delgado, C. E., Fortuin, N. J., and Ross, R. S.: Acute effects of low doses of alcohol on left ventricular function by echocardiography. *Circulation* **31:**535–540, 1975.

Demakis, J. G., Proskey, A., Rahimtoola, S. H., Jamil, M., Sutton, S. C., Rosen, K. M., Gunnar, R. M., and Tobin, J. R.: The natural course of alcoholic cardiomyopathy. *Ann. Intern. Med.* **80:**293–297, 1974.

Deykin, D., Janson, P., and McMahon, L.: Ethanol potentiation of aspirin-induced prolongation of the bleeding-time. *N. Engl. J. Med.* **306:**852–854, 1982.

Dickinson, J.: Debate: That congestive cardiomyopathy is really hypertensive heart disease in disguise. *Postgrad. Med. J.* **48:** 777–789, 1972.

Dixon, W. E.: The action of alcohol on the circulation. *J. Physiol. (Lond.)* **35:**346–366, 1907.

Donahue, R. P., Abbott, R. D., Reed, D. M., and Yano, K.: Alcohol and hemorrhagic stroke. *J.A.M.A.* **255:**2311–2314, 1986.

Dyer, A. R., Stamler, J., Paul, O., Lepper, M., Shekelle, R. B., McKean, H., and Garside, D.: Alcohol consumption and 17-year mortality in the Chicago Western Electric Company Study. *Prev. Med.* **9:**78–90, 1980.

Dyer, A. R., Stamler, J., Paul, O., Berkson, D. M., Shekelle, R. B., Lepper, M. H., McKean, H., Lindberg, H. A., Garside, D., and Tokich, T.: Alcohol, cardiovascular risk factors and mortality: The Chicago experience. *Circulation* **64**(Suppl. III)**:**20–27, 1981.

Eade, N. R.: Mechanism of sympathomimetic action of aldehydes. *J. Pharmacol. Exp. Ther.* **127:**29–34, 1959.

Eliaser, M., and Giansiracusa, F. J.: The heart and alcohol. *Calif. Med.* **84:**234–236, 1956.

Engel, T. R., and Luck, J. C.: Effect of whiskey on atrial vulnerability and "holiday heart." *J. Am. Coll. Cardiol.* **1:**816–818, 1983.

Ettinger, P. O., Lyons, M., Oldewurtel, H. A., and Regan, T. J.: Cardiac conduction abnormalities produced by chronic alcoholism. *Am. Heart J.* **91:**66–78, 1976.

Ettinger, P. O., Wu, C. F., De La Cruz, C., Weisse, A. B., Ahmed, S. S., and Regan, T. J.: Arrhythmias and the "holiday heart": Alcohol-associated cardiac rhythm disorders. *Am. Heart. J.* **95:**555–562, 1978.

Evans, W.: The electrocardiogram of alcoholic cardiomyopathy. *Br. Heart J.* **21:**445–456, 1959.

Evans, W.: Alcoholic cardiomyopathy. *Am. Heart J.* **61:**556–567, 1961.

Factor, S.: Intramyocardial small-vessel disease in chronic alcoholism. *Am. Heart J.* **92:**561–575, 1976.

Fauvel, J. M., Tillmanns, H. T., Pachinger, O., Mao, J. C., and Bing, R. J.: Étude expérimentale de l'administration prolongée d'alcool sur les lipides artériels, l'ultrastructure et la performance cardiaques. *Arch. Mal. Coeur. Vaiss.* **67:**837–846, 1974.

Fazekas, J. F., Albert, S. M., and Alman, R. W.: Influence of chlorpromazine and alcohol on cerebral hemodynamics and metabolism. *Am. J. Med. Sci.* **230:**128–132, 1955.

Fernandez, D., Rosenthal, J. E., Cohen, L. S., Hammond, G., and Wolfson, S.: Alcohol-induced Prinzmetal variant angina. *Am. J. Cardiol.* **32:**238–239, 1973.

Fernando, H. A., and Friedman, H. S.: Demonstration of the hyperdynamic heart of cirrhosis by echocardiography. *Clin. Res.* **24:**613A, 1976.

Fewings, J. D., Hanna, J. D., Walsh, J. A., and Whelan, R. F.: The effects of ethyl alcohol on blood vessels of the hand and forearm in man. *Br. J. Pharmacol. Chemother.* **27:**93–106, 1966.

Fisher, V. J., and Kavaler, F.: The action of ethanol upon the action potential and contraction of ventricular muscle. In: *Recent Advances in Studies on Cardiac Structure and Metabolism, Vol. 5* (A. Fleckenstein and N. S. Dalla, eds.), Baltimore, University Park Press, 1975, pp. 415–422.

Forsyth, G. W., Nagasawa, H. T., and Alexander, C. S.: Acetaldehyde metabolism by the rat heart (37624). *Proc. Soc. Exp. Biol. Med.* **144:**498–500, 1973.

Fortmann, S. P., Haskell, W. L., Vranizan, K., Brown, B. W., and Farquhar, J. W.: The association of blood pressure and dietary alcohol: Differences by age, sex, and estrogen use. *Am. J. Epidemiol.* **118:**497–507, 1983.

Fraser, G. E., and Upsdell, M.: Alcohol and other discriminants between cases of sudden death and myocardial infarction. *Am. J. Epidemiol.* **114:**462–476, 1981.

Friedman, H. S.: Acute effects of ethanol on myocardial blood flow in the nonischemic and ischemic heart. *Am. J. Cardiol.* **47:**61–67, 1981.

Friedman, H. S., Matsuzaki, S., Choe, S. S., Fernando, H. A., Celis, A., Zaman, Q., and Lieber, C.: Demonstration of dissimilar acute haemodynamic effects of ethanol and acetaldehyde. *Cardiovasc. Res.* **13:**477–487, 1979a.

Friedman, H. S., Dowd, A., and Neal, C.: Effects of ethanol on myocardial blood flow in the ischemic heart. *Clin. Res.* **27:** 562A, 1979b.

Friedman, H. S., Lowery, R., Archer, M., and Scorza, J.: The effect of ethanol on coronary blood flow in awake dogs. *Clin. Res.* **29:**193A, 1981a.

Friedman, H. S., Lowery, R., Archer, M., and Scorza, J.: The effects of ethanol on regional blood flow in awake dogs. *Clin. Res.* **29:**650A, 1981b.

Friedman, H. S., Lowery, R., Shaughnessy, E., and Scorza, J.: The effects of ethanol on pancreatic blood flow in awake and anesthetized dogs. *Proc. Soc. Exp. Biol. Med.* **174:**377–382, 1983.

Friedman, H. S., Lowery, R., Archer, M., Shaughnessy, E., and Scorza, J.: The effects of ethanol on brain blood flow in awake dogs. *J. Cardiovasc. Pharmacol.* **6:**344–348, 1984.

Friedman, H. S., Vasavada, B. C., Malec, A. M., Hassan, K. K., Shah, A., and Siddiqui, S.: Cardiac function in alcohol-associated systemic hypertension. *Am. J. Cardiol.* **57:**227–231, 1986.

Friedman, H. S., and Fernando, H. A.: Ascites as a marker for the hyperdynamic heart of Laennec's cirrhosis. *Alcoholism Clin. Exp. Res.* (in press).

Fuster, V., Gersh, B. J., Giuliani, E. R., Tajik, A. J., Brandenburg, R. O., and Frye, R. L.: The natural history of idiopathic dilated cardiomyopathy. *Am. J. Cardiol.* **47:**525–531, 1981.

Gailis, L., and Verdy, M.: The effect of ethanol and acetaldehyde on the metabolism and vascular resistance of the perfused heart. *Can. J. Biochem.* **49:**227–233, 1971.

Ganz, V.: The acute effect of alcohol on the circulation and on the oxygen metabolism of the heart. *Am. Heart* **66:**494–497, 1963.

Garrett, J. S., Coffelt-Wikman, J., Sievers, R., Finkbeiner, W. E., and Parmley, W. W.: Verapamil prevents the development of alcoholic dysfunction in hamster myocardium. *J. Am. Coll. Cardiol.* **9:**1326–1331, 1987.

Gill, J. S., Zezulka, A. V., Shipley, M. J., Gill, S. K., and Beevers, D. G.: Stroke and alcohol consumption. *N. Engl. J. Med.* **315:**1041–1051, 1986.

Gilmour, R. F., Jr., Ruffy, R., Lovelace, D. E., Mueller, T. M., and Zipes, D. P.: Effect of ethanol on electrogram changes and regional myocardial blood flow during acute myocardial ischaemia. *Cardiovasc. Res.* **15:**47–58, 1981.

Gimeno, A. L., Gimeno, M. F., and Webb, J. L.: Effects of ethanol on cellular membrane potentials and contractility of isolated rat atrium. *Am. J. Physiol.* **203**:194–196, 1962.

Goldman, H., Sapirstein, L. A., Murphy, S., and Moore, J.: Alcohol and regional blood flow in brains of rats. *Proc. Soc. Exp. Biol. Med.* **144**:983–988, 1973.

Goldstein, D. B.: The effects of drugs on membrane fluidity. *Am. Rev. Pharmacol. Toxicol.* **24**:43–64, 1984.

Goldstein, M., Anagnoste, B., Lauber, E., and McKereghan, M. R.: Inhibition of dopamine β-hydroxylase by disulfiram. *Life Sci.* **3**:763–767, 1964.

Goodkind, M. J., Gerber, N. H., Jr., Mellen, J. R., and Kostis, J. B.: Altered intracardiac conduction after acute administration. *J. Pharmacol. Exp. Ther.* **194**:633–638, 1975.

Goodwin, J. F.: Congestive and hypertrophic cardiomyopathy. *Lancet* **1**:731–739, 1970.

Gordon, T., and Kannel, W. B.: Drinking and its relation to smoking, BP, blood lipids and uric acid. *Arch. Intern. Med.* **143**:1366–1374, 1983.

Gould, L., Shariff, M., Zahir, M., and Di Lieto, M.: Cardiac hemodynamics in alcoholic patients with chronic liver disease and presystolic gallop. *J. Clin. Invest.* **48**:860–868, 1969.

Gould, L., Zahir, M., DeMartino, A., and Gomprecht, R. F.: Cardiac effects of a cocktail. *J.A.M.A.* **218**:1799–1802, 1972.

Gould, L., Reddy, C. V. R., Patel, N., and Gomprecht, R. F.: Effects of whiskey on heart conduction in cardiac patients. *Q. J. Stud. Alcohol* **35**:26–33, 1974.

Gould, L., Reddy, C. V. R., Becker, W., Keun-Chang Oh, K. C., and Kim, S. G.: Electrophysiologic properties of alcohol in man. *J. Electrocardiol.* **11**:219–226, 1978.

Greenberg, B. H., Schutz, R., Grunkemeier, G. L., and Griswold, H.: Acute effects of alcohol in patients with congestive heart failure. *Ann. Intern. Med.* **97**:171–174, 1982.

Greenspon, A. J., and Schaal, S. F.: The "holiday heart": Electrophysiologic studies of alcohol effects in alcoholics. *Ann. Intern. Med.* **98**:135–139, 1983.

Greenspon, A. J., Stang, J. M., Lewis, R. P., and Schaal, S. F.: Provocation of ventricular tachycardia after consumption of alcohol. *N. Engl. J. Med.* **301**:1049–1050, 1979.

Grollman, A.: The influence of alcohol on the circulation. *Q. J. Stud. Alcohol* **3**:5–14, 1942.

Gruchow, H. W., Sobocinski, K. A., and Barboriak, J. J.: Alcohol, nutrient intake, and hypertension in US adults. *J.A.M.A.* **253**:1567–1570, 1985.

Gvozdjak, A., Bada, V., Kruty, F., Niederland, T. R., and Gvozdjak, J.: Chronic effect of ethanol on the metabolism of myocardium. *Biochem. Pharmacol.* **22**:1807–1811, 1973a.

Gvozdjak, A., Bada, V., Kruty, F., Niederland, T. R., and Gvozdjak, J.: Effect of ethanol on the metabolism of the myocardium and its relationship to development of alcoholic myocardiopathy. *Cardiology* **58**:290–297, 1973b.

Gyntelberg, F., and Meyer, J.: Relationship between blood pressure and physical fitness, smoking and alcohol consumption in Copenhagen males aged 40–59, *Acta Med. Scand.* **195**:375–380, 1974.

Hall, J. L., and Rowlands, D. T.: Cardiotoxicity of alcohol. An electron microscopic study in the rat. *Am. J. Pathol.* **60**:153–160, 1970.

Harburg, E., Ozgoren, F., Hawthorne, V. M., and Schork, M. A.: Community norms of alcohol usage and blood pressure: Tecumseh, Michigan. *Am. J. Public Health* **70**:813–820, 1980.

Haselager, E. M., and Vreekem, J.: Rebound thrombocytosis after alcohol abuse: A possible factor in the pathogenesis of thromboembolic disease. *Lancet* **1**:774–775, 1977.

Haskell, W. L., Camargo, C., Williams, P. T., Vranizan, K. M., Krauss, R. M., Lindgren, F. T., and Wood, P. D.: The effect of cessation and resumption of moderate alcohol intake on serum high-density-lipoprotein subfractions. *N. Engl. J. Med.* **310**:806–810, 1984.

Hayes, S. N., and Bove, A.: Ethanol causes epicardial coronary artery vasoconstriction in the dog. *Circulation* **78**:165–170, 1988.

Heberden, W.: Some account of a disorder of the breast. *Medical Transactions* **2**:59–67, 1772.

Hemmingsen, R., and Barry, D. I.: Adaptive changes in cerebral blood flow and oxygen consumption during ethanol intoxication in the rat. *Acta Physiol. Scand.* **106**:249–255, 1979.

Hennekens, C. H., Rosner, B., and Cole, D. S.: Daily alcohol consumption and fatal coronary heart disease. *Am. J. Epidemiol.* **107**:196–200, 1978.

Hennekens, C. H., Willett, W., Rosner, B., Cole, D. S., and Mayrent, S. L.: Effects of beer, wine, and liquor in coronary deaths. *J.A.M.A.* **242**:1973–1974, 1979.

Herrmann, H.-J., Morvai, V., Ungvary, N. C., and Muhlig, P.: Long term effects of ethanol on coronary microvessels of rats. *Microcirculation, Endothelium, and Lymphatics* **1**:589–610, 1984.

Herzog, V., and Fahimi, H. D.: Identification of peroxisomes (microbodies) in mouse myocardium. *J. Mol. Cell. Cardiol.* **8**:271–281, 1975.

Hibbs, R. G., Ferrans, V. J., Black, W. C., Weilbaecher, D. G., Walsh, J. J., and Burch, G. E.: Alcoholic cardiomyopathy: An electron microscopic study. *Am. Heart J.* **69**:766–779, 1965.

Hill, C. A., Harle, T. S., and Gaston, W.: Cardiomyopathy: A review of 59 patients with emphasis on the plain chest roentgenogram. *Am. J. Roentgenol. Radium Ther. Nucl. Med.* **104**:433–439, 1968.

Hillbom, M., and Kaste, M.: Does ethanol intoxication promote brain infarction in young adults? *Lancet* **2**:1181–1184, 1978.

Hillbom, M., and Kaste, M.: Ethanol intoxication: A risk factor for ischemic brain infarction in adolescents and young adults. *Stroke* **12**:422–425, 1981.

Hirota, Y., Bing, O. H. L., and Abelmann, W. H.: Effect of ethanol on contraction and relaxation of isolated rat ventricular muscle. *J. Mol. Cell. Cardiol.* **8**:727–732, 1976.

Hojnacki, J. L., Cluette-Brown, J. E., Mulligan, J. J., Hagan, S. M., Mahony, K. E., Witzgall, S. K., Osmoilski, T. V., and Barboriak, J. J.: Effect of ethanol dose on low density lipoproteins and high density lipoprotein subfractions. *Alcoholism: Clin. Exp. Res.* **12**:149–154, 1988.

Hortnagl, H., Singer, E. A., Lenz, K., Kleinberger, G., and Lochs, H.: Substance P is markedly increased in plasma of patients with hepatic coma. *Lancet* **1**:480–483, 1984.

Horvath, S. M., and Willard, P. W.: Effect of ethyl alcohol upon splanchnic hemodynamics. *Proc. Soc. Exp. Biol. Med.* **3**:295–300, 1962.

Horwitz, L. D., and Atkins, J. M.: Acute effects of ethanol on left ventricular performance. *Circulation* **49:**124–128, 1974.

Horwitz, L. D., and Myers, J. H.: Ethanol-induced alterations in pancreatic blood flow in conscious dogs. *Circ. Res.* **50:**250–256, 1982.

Horwitz, O., Montgomery, H., Longaker, E. D., and Sayen, A.: Effects of vasodilator drugs and other procedures on digital cutaneous blood flow, cardiac output, blood pressure, pulse rate, body temperature, and metabolic rate. *Am. J. Med. Sci.* **218:**669–682, 1949.

Howell, W. L., and Manion, W. C.: The low incidence of myocardial infarction in patients with portal cirrhosis of liver: A review of 639 cases of cirrhosis of the liver from 17,731 autopsies. *Am. Heart J.* **60:**341–343, 1960.

Hughes, J. H., Henry, R. E., and Daly, M. J.: Influence of ethanol and ambient temperature on skin blood flow. *Ann. Emerg. Med.* **13:**597–600, 1984.

Ireland, M. D., Vandongen, R., Davidson, L., Beilin, L. J., and Rouse, I. L.: Acute effects of moderate alcohol consumption on blood pressure and plasma catecholamines. *Clin. Sci.* **66:**643–648, 1984.

Jackson, R., Stewart, A., Beaglehole, R., and Scragg, R.: Alcohol consumption and blood pressure. *Am. J. Epidemiol.* **122:**1037–1044, 1985.

James, T. N., and Bear, E. S.: Effects of ethanol and acetaldehyde on the heart. *Am. Heart. J.* **74:**243–254, 1967.

Jesrani, M. U., Gopinathan, K., Khan, M. T., Oldewurtel, H. A., and Regan, T. J.: Acetaldehyde and the myocardial depressant effects of ethanol. *Circulation* **43**(Suppl. II):II–127, 1971.

Jones, R. D., Kleinerman, J. I., and Luria, M. H.: Observations on left ventricular failure induced by ethanol. *Cardiovasc. Res.* **9:**286–294, 1975.

Juchems, R., and Klobe, R.: Hemodynamic effects of ethyl alcohol in man. *Am. Heart J.* **78:**133–135, 1969.

Kagan, A., Yano, K., Rhoads, G. G., and McGee, D. L.: Alcohol and cardiovascular disease: The Hawaiian experience. *Circulation* **64**(Suppl. III):27–30, 1981.

Kako, K. J., Liu, M. S., and Thornton, M. J.: Changes in fatty acid composition of myocardial triglyceride following a single administration of ethanol to rabbits. *J. Mol. Cell. Cardiol.* **5:**473-489, 1973.

Kashima, T., Tanaka, H., Arikawa, K., and Ariyama, T.: Variant angina induced by alcohol ingestion. *Angiology* **33:**137–139, 1982.

Katz, A. M., Freston, J. W., Massines, F. C., and Herbette, L. G.: Membrane damage and the pathogenesis of cardiomyopathies. *J. Mol. Cell. Cardiol.* **17**(Suppl. 2):11–20, 1985.

Kelbaek, H., Eriksen, J., Brynjolf, I., Raboel, A., Lund, J. O., Munck, O., Bonnevie, O., and Godtfredsen, J.: Cardiac performance in patients with asymptomatic alcohol cirrhosis of the liver. *Am. J. Cardiol.* **54:**852–855, 1984.

Kelbaek, H., Gjorup, T., Bynjolf, I., Christensen, N. J., and Godtfredsen, J.: Acute effects of alcohol on left ventricular function in healthy subjects at rest and during upright exercise. *Am. J. Cardiol.* **55:**164–167, 1985.

Kikuchi, T., and Kako, K. J.: Metabolic effects of ethanol on the rabbit heart. *Circ. Res.* **26:**625–634, 1970.

Kino, M., Imamitchi, H., Morigutchi, M., Kawamura, K., and

Takatsu, T.: Cardiovascular status in asymptomatic alcoholics, with reference to the level of ethanol consumption. *Br. Heart. J.* **46:**545–551, 1981.

Kirkendol, P. L., Pearson, J. E., Bower, J. D., and Holbert, R. D.: Myocardial depressant effects of sodium acetate. *Cardiovasc. Res.* **12:**127–136, 1978.

Kischuk, R. P., Otten, M. D., and Polimeni, P. I.: Effect of acute alcoholic intoxication on myocardial electrolyte and water distributions. *J. Mol. Cell. Cardiol.* **18:**197–205, 1986.

Klatsky, A. L., Friedman, G. D., and Siegelaub, A. B.: Alcohol consumption before myocardial infarction. *Ann. Intern. Med.* **81:**294–301, 1974.

Klatsky, A. L., Friedman, G. D., Siegelaub, A. B., and Gerard, M. J.: Alcohol consumption and blood pressure. *N. Engl. J. Med.* **296:**1194–1200, 1977.

Klatsky, A. L., Friedman, G. D., and Siegelaub, A. B.: Alcohol and mortality: A ten-year Kaiser–Permanente experience. *Ann. Intern. Med.* **95:**139–145, 1981a.

Klatsky, A. L., Friedman, G. D., and Armstrong, M. A.: The relationships between alcohol beverage use and other traits to blood pressure: A new Kaiser–Permanente study. *Circulation* **73:**628–636, 1986.

Koide, T., Kato, A., Takabatake, Y., Itzuka, M., Uchida, Y., Ozeki, K., Morooka, S., Kakihana, M., Serizawa, T., Tanaka, S., Ohya, T., Momomura, S., and Murao, S.: Variable prognosis in congestive cardiomyopathy: Role of left ventricular function, alcoholism, and pulmonary thrombosis. *Jpn. Heart J.* **21:**451–463, 1980.

Kondo, K., and Ebihara, A.: Alcohol consumption and blood pressure in a rural community of Japan. In: *Nutritional Prevention of Cardiovascular Disease* (W. Lovenberg and Y. Yamori, eds.), Orlando, FL, Academic Press, 1984, pp. 217–220.

Kostis, J. B., Horstmann, E., Mavrogeorgis, E., Radzius, A., and Goodkind, M. J.: Effect of alcohol on the ventricular fibrillation threshold in dogs. *Q. J. Stud. Alcohol* **34:**1315–1322, 1973.

Kozararevic, D., McGee, D., Vojvodic, N., Dawber, T., Racic, Z., and Gordon, T.: Frequency of alcohol consumption and morbidity and mortality. *Lancet* **1:**613–616, 1980.

Kramer, K., Kuller, L., and Fisher, R.: The increasing mortality attributed to cirrhosis and fatty liver in Baltimore (1957–1966). *Ann. Intern. Med.* **69:**273–282, 1968.

Kreulen, T. H., Gorlin, R., and Herman, M. V.: Ventriculographic patterns and hemodynamics in primary myocardial disease. *Circulation* **47:**299–308, 1973.

Kromhout, D., Bosschieter, E. B., and Coulander, C. L.: Potassium, calcium, alcohol intake and blood pressure: The Zytphen study. *Am. J. Clin. Nutr.* **41:**1299–1304, 1985.

Kumar, M. A., and Sheth, U. K.: The sympathomimetic action of acetaldehyde on isolated atria. *Arch. Int. Pharmacodyn. Ther.* **137:**188–198, 1962.

Kupari, M.: Acute cardiovascular effects of ethanol: A controlled non-invasive study. *Br. Heart J.* **49:**174–182, 1983.

Landolfi, R., and Steiner, M.: Ethanol raises prostacyclin *in vivo* and *in vitro*. *Blood* **64:**679–682, 1984.

Lang, R. M., Borrow, K. M., Neumann, A., and Feldman, T.: Adverse cardiac effects of acute alcohol ingestion in young adults. *Ann. Intern. Med.* **102:**742–747, 1985.

Lange, L. G.: Nonoxidative ethanol metabolism: Formation of fatty acid ethyl esters by cholesterol esterase. *Proc. Natl. Acad. Sci. U.S.A.* **79**:3954–3957, 1982.

Lange, L. G., and Sobel, B. E.: Myocardial metabolites of ethanol. *Circ. Res.* **52**:479–482, 1983a.

Lange, L. G., and Sobel, B. E.: Mitochondrial dysfunction induced by fatty acid ethyl esters, myocardial metabolites of ethanol. *J. Clin. Invest.* **72**:724–731, 1983b.

Laposata, E. A., and Lange, L. G.: Presence of nonoxidative ethanol metabolism in human organs commonly damaged by ethanol abuse. *Science* **231**:497–499, 1986.

Lasker, N., Sherrod, T. R., and Killalm, F.: Alcohol on the coronary circulation of the dog. *J. Pharmacol. Exp. Ther.* **113**:441–420, 1955.

Laug, W. E.: Ethyl alcohol enhances plasminogen activator secretion by endothelial cells. *J.A.M.A.* **250**:772–776, 1983.

Lee, K.: Alcoholism and cerebrovascular thrombosis in the young. *Acta Neurol. Scand.* **59**:270–274, 1979.

Leier, C. V., Schaal, S. F., Leighton, R. F., and Whayne, T. F.: Heart block in alcoholic cardiomyopathy. *Arch. Intern. Med.* **134**:766–768, 1974.

Leighninger, D. S., Rueger, R., and Beck, C. S.: Effect of pentaerythritol tetranitrate, amyl nitrite and alcohol on arterial blood supply to ischemic myocardium. *Am. J. Cardiol.* **7**:533–557, 1961.

Liang, C. S., and Lowenstein, J. M.: Metabolic control of the circulation: Effects of acetate and pyranide. *J. Clin. Invest.* **62**:1029–1038, 1978.

Lieber, C. S., Spritz, N., and DeCarli, L. M.: Accumulation of triglycerides in heart and kidney after alcohol ingestion. *J. Clin. Invest.* **45**:1041, 1966.

Limas, C. J., Guiha, N. H., Lekagul, O., and Cohn, J. N.: Impaired left ventricular function in alcoholic cirrhosis with ascites. *Circulation* **49**:755–760, 1974.

Lochner, A., Cowley, R., and Brink, A. J.: Effect of ethanol on metabolism and function of perfused rat heart. *Am. Heart J.* **78**:709–789, 1969.

Luca, C.: Electrophysiological properties of right heart and atrioventricular conducting system in patients with alcoholic cardiomyopathy. *Br. Heart J.* **42**:274–281, 1979.

Lundquist, F., Tygstrup, N., Winkler, K., Mellemgaard, K., and Munck-Petersen, S.: Ethanol metabolism and production of free acetate in the human liver. *J. Clin. Invest.* **41**:955–961, 1962.

Lunseth, J. H., Olmstead, E. G., Forks, G., and Abboud, F.: A study of heart disease in one hundred eight hospitalized patients dying with portal cirrhosis. *Arch. Intern. Med.* **102**:405–413, 1958.

MacMahon, S. W., Black, R. B., Macdonald, G. J., and Hall, W.: Obesity, alcohol consumption and blood pressure in Australian men and women: The National Heart Foundation of Australia risk factor prevalence study. *J. Hypertens.* **2**:85–91, 1984.

Madan, B. R., and Gupta, R. S.: Effect of ethanol in experimental auricular and ventricular arrhythmias. *Jpn. J. Pharmacol.* **17**:683–684, 1967.

Maines, J. E., and Aldinger, E. E.: Myocardial depression accompanying chronic consumption of alcohol. *Am. Heart J.* **73**:55–63, 1967.

Malhotra, H., Mehta, S. R., Mathur, D., and Khandelwal, P. D.: Pressor effects of alcohol in normotensive and hypertensive subjects. *Lancet* **2**:584–586, 1985.

Mall, G., Mattfeldt, T., Rieger, P., Volk, B., and Frolov, V. A.: Morphometric analysis of the rabbit myocardium after chronic ethanol feeding—early capillary changes. *Basic Res. Cardiol.* **77**:57–67, 1982.

Maruyama, Y., Sarma, J. S. M., Fischer, R., Bertuglia, S., and Bing, R. J.: Effect of alcohol on the contractile properties. *Curr. Alcoholism.* **3**:393–404, 1978.

Massumi, R. A., Rios, J. C., Gooch, A. S., Nutter, D., De Vita, T., and Datlow, D. W.: Primary myocardial disease: Report of fifty cases and review of the subject. *Circulation* **31**:19–41, 1965a.

Massumi, R. A., Rios, J. C., and Ticktin, H. E.: Hemodynamic abnormalities and venous admixture in portal cirrhosis. *Am. J. Med. Sci.* **250**:67/275–75/283, 1965b.

Mathews, E., Gardin, J. M., Henry, W., DelNegro, A. A., Fletcher, R. D., Snow, J. A., and Epstein, S. E.: Echocardiographic abnormalities in chronic alcoholics with and without overt congestive heart failure. *Am. J. Cardiol.* **47**:570–578, 1981.

Matsuguchi, T., Araki, H., Anan, T., Hayata, N., Nakagaki, O., Takeshita, A., and Nakamura, M.: Provocation of variant angina by alcohol ingestion. *Eur. Heart J.* **5**:906–912, 1984.

Matsuguchi, T., Araki, H., Nakamura, N., Etoh, Y., Okamatsu, S., Takeshita, A., and Nakamura, M.: Prevention of vasospastic angina by alcohol ingestion: Report of two cases. *Angiology* **38**:394–400, 1988.

McCall, D., and Ryan, K.: The effects of ethanol and acetaldehyde on Na pump function in cultured rat heart cells. *J. Mol. Cell. Cardiol.* **19**:453–463, 1987.

McCloy, R. B., Prancan, A. V., and Nakano, J.: Effects of acetaldehyde on the systemic pulmonary, and regional circulations. *Cardiovasc. Res.* **8**:216–226, 1974.

McDonald, C. D., Burch, G. E., and Walsh, J. J.: Alcoholic cardiomyopathy managed with prolonged bed rest. *Ann. Intern. Med.* **74**:681–691, 1971.

McDowall, R. J. S.: Action of alcohol on the circulation. *J. Pharmacol. Exp. Ther.* **25**:289–295, 1925.

McKaigney, J. P., Carmichael, F. J., Saldivia, V., Israel, Y., and Orrego, H.: Role of ethanol metabolism in the ethanol-induced increase in splanchnic circulation. *Am. J. Physiol.* **250**:G518–G523, 1986.

McSmythe, C., Heinemann, H. O., and Bradley, S. E.: Estimated hepatic blood flow in the dog: Effect of ethyl alcohol on its renal blood flow, cardiac output and arterial pressure. *Am. J. Physiol.* **172**:737–742, 1953.

Meade, T. W., Imeson, J., and Stirling, Y.: Effects of changes in smoking and other characteristics on clotting factors and the risk of ischaemic heart disease. *Lancet* **2**:986–988, 1987.

Mehehta, B., and Sereny, G.: Cardiovascular manifestations during alcohol withdrawal. *Mt. Sinai J. Med.* **46**:484–485, 1979.

Mendoza, L. C., Hellberg, K., Rickart, A., Tillich, G., and Bing, R. J.: The effect of intravenous ethyl alcohol on the coronary circulation and myocardial contractility of the human and canine heart. *J. Clin. Pharmacol.* **7**:165–176, 1971.

Mierzwiak, D. S., Wildenthal, K., and Mitchell, J. H.: Acute effects of ethanol on the left ventricle in dogs. *Arch. Int. Pharmacodyn. Ther.* **199:**43–53, 1972.

Miller, H., and Abelman,   : Effects of dietary ethanol upon experimental trypanosomal (*T. Cruzi*) myocarditis. *Proc. Soc. Exp. Biol.* **126:**193–198, 1967.

Milon, H., Froment, A., Gaspard, P., Guidollet, J., and Ripoll, J. P.: Alcohol consumption and blood pressure in a French epidemiological study. *Eur. Heart J.* **3**(Suppl. C):59–64, 1982.

Mogelson, S., and Lange, L. G.: Nonoxidative ethanol metabolism in rabbit myocardium: Purification to homogeneity of fatty acyl ethyl ester synthase. *Biochemistry* **23:**4075–4081, 1984.

Moreyra, A. E., Kostis, J. B., Passannante, A. J., and Juo, P. T.: Acute myocardial infarction in patients with normal coronary arteries after acute ethanol intoxication. *Clin. Cardiol.* **5:**425–430, 1982.

Morin, Y. L., Foley, A. R., Martineau, G., and Roussel, J.: Quebec beer-drinkers' cardiomyopathy: Forty-eight cases. *Can. Med. Assoc. J.* **97:**881–904, 1967.

Morin, Y., Roy, P. E., Mohiuddin, S. M., and Tasker, P. K.: The influence of alcohol on viral and isoproterenol cardiomyopathy. *Cardiovasc. Res.* **3:**363–368, 1969.

Murray, J. F., Dawson, A. M., and Sherlock, S.: Circulatory changes in chronic liver disease. *Am. J. Med.* **24:**358–367, 1958.

Nakano, J., and Kessinger, M.: Cardiovascular effects of ethanol, its congeners and synthetic bourbon in dogs. *Eur. J. Pharmacol.* **17:**195–201, 1972.

Nakano, J., and Moore, S. E.: Effect of different alcohols on the contractile force of the isolated guinea-pig myocardium. *Eur. J. Pharmacol.* **20:**266–270, 1972.

Nakano, J., and Prancan, A. V.: Effects of adrenergic blockade on cardiovascular responses to ethanol and acetaldehyde. *Arch. Int. Pharmacodyn. Ther.* **196:**259–268, 1972.

Nakano, J., Holloway, J. E., and Schackford, J. S.: Effects of disulfiram on the cardiovascular responses to ethanol in dogs and guinea pigs. *Toxicol. Appl. Pharmacol.* **14:**439–446, 1969.

Nanji, A. A.: Alcohol and ischemic heart disease: Wine, beer, or both. *Int. J. Cardiol.* **8:**487–490, 1985.

Newman, W. H., and Valicenti, J. F.: Ventricular function following acute alcohol administration: A strain-gauge analysis of depressed ventricular dynamics. *Am. Heart J.* **81:**61–68, 1971.

Nguyen, M. H., and Gailis, L.: Effect of acetaldehyde on the isolated, non-working guinea-pig heart: Independence of the coronary flow increase from changes in heart rate and oxygen consumption. *Can. J. Physiol. Pharmacol.* **52:**602–612, 1974.

Nguyen, T. N., Friedman, H. S., and Mokraoui, A. M.: Effects of alcohol on experimental atrial fibrillation. *Alcoholism: Clin. Exp. Res.* **11:**474–476, 1987.

Nitenberg, A., Huyghebaert, M. F., Blanchet, F., and Amiel, C.: Analysis of increased myocardial contractility during sodium acetate infusion in humans. *Kidney Int.* **26:**744–751, 1984.

Noren, G. R., Staley, N. A., Einzig, S., Mikell, F. L., and Asinger, R. W.: Alcohol-induced congestive cardiomyopathy: An animal model. *Cardiovasc. Res.* **17:**81–87, 1983.

Ogata, M., Mendelson, J. H., Mello, N. K., and Majchrowicz, E.: Adrenal function and alcoholism. *Psychosom. Med.* **33:**159–180, 1971.

Orlando, J., Aronow, W. S., Cassidy, J., and Prakash, R.: Effect of ethanol on angina pectoris. *Ann. Intern. Med.* **84:**652–655, 1976.

Pachinger, O., Tillmanns, H., Mao, J. C., Fauvel, J. M., and Bing, R. J.: The effect of prolonged administration of ethanol and cardiac metabolism and performance in the dog. *J. Clin. Invest.* **52:**2690–2696, 1973.

Paradise, R. R., and Stoelting, V.: Conversion of acetyl strophanthidin-induced ventricular tachycardia to sinus rhythm by ethyl alcohol. *Arch. Int. Pharmacodyn.* **157:**312–321, 1965.

Parker, S. L., Thompson, J. A., and Reitz, R. C.: Effects of chronic ethanol ingestion upon acyl-CoA : Carnitine acyltransferase in liver and heart. *Lipids* **9:**520–525, 1974.

Parrish, H. M., and Eberly, A. L., Jr.: Negative association of coronary atherosclerosis with liver cirrhosis and chronic alcoholism: A statistical fallacy. *J. Indiana State Med. Assoc.* **54:**340–347, 1961.

Paulin, J. M., Simpson, F. O., and Waal-Manning, J. H.: Alcohol consumption and blood pressure in a New Zealand community study. *N.Z. Med. J.* **98:**425–428, 1985.

Perman, E. S.: The effect of acetaldehyde on the secretion of adrenaline and noradrenaline from the suprarenal gland of the cat. *Acta Physiol. Scand.* **13:**71–76, 1958.

Pitt, B., Sugishita, Y., Green, H. L., and Gottlieb, C. F.: Coronary hemodynamic effects of ethyl alcohol in the conscious dog. *Am. J. Physiol.* **219:**175–177, 1970.

Polimeni, P. I., Otten, M.D ., and Hoeschen, L. E.: *In vivo* effects of ethanol on the rat myocardium: Evidence for a reversible, non-specific increase of sarcolemmal permeability. *J. Mol. Cell. Cardiol.* **15:**113–122, 1983.

Potter, J. F., and Beevers, D. G.: Pressor effect of alcohol in hypertension. *Lancet* **1:**119–122, 1984.

Preedy, V. R., and Peters, T. J.: The acute and chronic effects of ethanol on cardiac muscle protein synthesis in the rat *in vivo*. *Alcohol* **7:**97–102, 1990.

Puddey, I. B., Beilin, L. J., Vandongen, R., Rouse, I. L., and Rogers, P.: Evidence for a direct effect of alcohol consumption on blood pressure in normotensive men: A randomized controlled trial. *Hypertension* **7:**707–713, 1985.

Puddey, I. B., Beilin, L. J., and Vandongen, R.: Regular alcohol use raises blood pressure in treated hypertensive subjects. *Lancet* **1:**647–651, 1987.

Puszkin, S., and Rubin, E.: Adenosine diphosphate effect of contractility of human muscle actomyosin: Inhibition by ethanol and acetaldehyde. *Science* **188:**1319–1320, 1975.

Randall, B.: Sudden death and hepatic fatty metamorphosis. *J.A.M.A.* **243:**1723–1725, 1980.

Reeves, W. C., Nanda, N. C., and Gramiak, R.: Echocardiography in chronic alcoholics following prolonged periods of abstinence. *Am. Heart J.* **95:**578–583, 1978.

Regan, R. J., Koroxenidis, G., Moschos, C. B., Oldewurtel, H. A., Lehan, P. H., and Hellems, H. K.: The acute metabolic and hemodynamic responses of the left ventricle to ethanol. *J. Clin. Invest.* **45:**270–278, 1966.

Regan, T. J., Levinson, G. E., Oldewurtel, H. A., Frank, M. J., Weisse, A. B., and Moschos, C. B.: Ventricular function in noncardiacs with alcoholic fatty liver: Role of ethanol in the production of cardiomyopathy. *J. Clin. Invest.* **48**:397–406, 1969.

Regan, T. J., Chia, F. W., Weisse, A. B., Haider, B., and Ahmed, S. S.: Acute myocardial infarction in toxic cardiomyopathy without coronary obstruction. *Tran. Assoc. Am. Physicians* **86**:193–199, 1974a.

Regan, T. J., Khan, M. I., Ettinger, P. O., Haider, B., Lyons, M. M., and Oldewurtel, H. A.: Myocardial function and lipid metabolism in the chronic alcoholic animal. *J. Clin. Invest.* **54**:740–752, 1974b.

Regan, T. J., Wu, C. F., Weisse, A. B., Moschos, C. B., Ahmed, S. S., and Lyons, M. M.: Acute myocardial infarction in toxic cardiomyopathy without coronary obstruction. *Circulation* **51**:453–460, 1975.

Reitz, R. C., Helsabeck, E., and Mason, D. P.: Effects of chronic alcohol ingestion on the fatty acid composition of the heart. *Lipid* **8**:80–84, 1973.

Retig, J. N., Kirchberger, M. A., Rubin, E., and Katz, A. M.: Effects of ethanol on calcium transport by microsomes phosphorylated by cyclic AMP-dependent protein kinase. *Biochem. Pharmacol.* **26**:393–396, 1977.

Rich, E. C., Siebold, C., and Campion, B.: Alcohol-related acute atrial fibrillation: A case-control study and review of 40 patients. *Arch. Intern. Med.* **145**:830–833, 1985.

Riff, D. P., Jain, A. C., and Doyle, J. T.: Acute hemodynamic effects of ethanol on normal human volunteers. *Am. Heart J.* **78**:592–597, 1969.

Robin, E., and Goldschlager, N.: Persistence of low cardiac output after relief of high output by thiamine in a case of alcoholic beriberi and cardiac myopathy. *Am. Heart J.* **80**:103–108, 1970.

Rossi, M. A., Oliveira, J. S. M., Zucoloto, S., and Becker, P. F. L.: Norepinephrine levels and morphologic alterations of myocardium in chronic alcoholic rats. *Beitr. Pathol.* **159**:51–60, 1976.

Rubin, E.: Alcoholic myopathy in heart and skeletal muscle. *N. Engl. J. Med.* **301**:28–33, 1979.

Rubin, E.: Alcohol and the heart: Theoretical considerations. *Fed. Proc.* **41**:2460–2464, 1982.

Rudel, L. L., Leathes, C. W., Bond, M. G., and Bullock, B. C.: Dietary ethanol-induced modifications in hyperlipoproteinemia and atherosclerosis in nonhuman primates (*Macaca nemestrina*). *Atheriosclerosis* **1**:144–155, 1981.

Ruebner, B. H., Miyai, K., and Abbey, H.: The low incidence of myocardial infarction in hepatic cirrhosis: A statistical artefact. *Lancet* **2**:1435–1436, 1961.

Sarkioja, T., and Hirvonen, J.: Causes of sudden unexpected deaths in young and middle-aged persons. *Forensic Sci.* **24**:247–261, 1984.

Sarma, J. S. M., Ikeda, S., Fischer, R., Maruyama, Y., Weishaar, R., and Bing, R. J.: Biochemical and contractile properties of heart muscle after prolonged alcohol administration. *J. Mol. Cell. Cardiol.* **8**:951–972, 1976.

Saunders, J. B., Beevers, D. G., and Paton, A.: Factors influencing blood pressure in chronic alcoholics. *Clin. Sci.* **57**:295–298, 1979.

Savdie, E., Grosslight, G. M., and Adena, M. A.: Relation of alcohol and cigarette consumption to blood pressure and serum creatinine levels. *J. Chronic. Dis.* **37**:617–623, 1984.

Scherf, D., Cohen, J., and Shafiia, H.: Ectopic ventricular tachycardia, hypokalemia and convulsions in alcoholics. *Cardiologia* **50**:129–139, 1967.

Schreiber, S. S., Briden, K., Oratz, M., and Rothschild, M. A.: Ethanol, acetaldehyde, and myocardial protein synthesis. *J. Clin. Invest.* **51**:2820–2826, 1972.

Schreiber, S. S., Oratz, M., Rothschild, M. A., Reff, F., and Evans, C.: Alcoholic cardiomyopathy II: The inhibition of cardiac microsomal protein synthesis by acetaldehyde. *J. Mol. Cell. Cardiol.* **6**:207–213, 1974.

Schwartz, L., Sample, B., and Wigle, D.: Severe alcoholic cardiomyopathy reversed with abstention from alcohol. *Am. J. Cardiol.* **36**:963–966, 1975.

Schwarz, F., Mall, G., Zebe, H., Schmitzer, E., Manthey, J., Scheurlen, H., and Kubler, W.: Determinants of survival in patients with congestive cardiomyopathy: Quantitative morphologic findings and left ventricular hemodynamics. *Circulation* **70**:923–928, 1984.

Segel, L. D.: Mitochondrial respiration after cardiac perfusion with ethanol or acetaldehyde. *Alcoholism: Clin. Exp. Res.* **8**:560–563, 1984.

Segel, L. D.: Alcoholic cardiomyopathy in rats: Inotropic responses to phenylephrine, glucagon, ouabain, and dobutamine. *J. Mol. Cell. Cardiol.* **19**:1061–1072, 1987.

Segel, L. D., and Dean, T. M.: Acute effects of acetaldehyde and ethanol on rat heart mitochondria. *Res. Commun. Chem. Pathol. Pharmacol.* **25**:461–474, 1979.

Segel, L. D., and Mason, D. T.: Beta-adrenergic receptors in chronic alcoholic rat hearts. *Cardiovasc. Res.* **16**:34–39, 1982.

Segel, L. D., Rendig, S. V., Choquet, Y., Chacko, K., Amsterdam, E. A., and Mason, D. T.: Effects of chronic graded ethanol consumption on the metabolism, ultrastructure, and mechanical function of the rat heart. *Cardiovasc. Res.* **9**:649–663, 1975.

Segel, L. D., Miller, R. R., and Mason, D. T.: Depressive effects of acute ethanol exposure on function of isolated working rat hearts. *Cardiovasc. Med.* **13**:211–213, 1978.

Segel, L. D., Rendig, S. G., and Mason, D. T.: Alcohol-induced cardiac hemodynamic and Ca^{2+} flux dysfunction are reversible. *J. Mol. Cell. Cardiol.* **13**:443–455, 1981.

Sereny, G.: Effects of alcohol on the electrocardiogram. *Circulation* **44**:558–564, 1971.

Sereny, S., Mehta, B., and Sethna, D.: Chronic alcoholic cardiomyopathy fact or fiction? *Drug Alcohol Depend.* **3**:331–343, 1978.

Shaper, A. G., Phillips, A. N., Pocock, S. J., and Walker, M.: Alcohol and ischemic heart disease in middle aged British men. *Br. Med. J.* **294**:733–737, 1987.

Shaw, S., Heller, E. A., Friedman, H. S., Baraona, E., and Lieber, C. S.: Increased hepatic oxygenation following ethanol administration in the baboon. *Proc. Soc. Exp. Biol. Med.* **156**:509–513, 1977.

Shimojyo, S., Scheinberg, P., and Reinmuth, O.: Cerebral blood flow and metabolism in Wernicke–Korsakoff syndrome. *J. Clin. Invest.* **46**:849–854, 1967.

Shugoll, G. I., Bowen, P. J., Moore, P. J., and Lenkin, L. M.: Follow-up observations and prognosis in primary myocardial disease. *Arch. Intern. Med.* **129**:67–72, 1972.

Smith, D. M., Fuller, S. J., and Sugden, P. H.: The effects of lactate, acetate, glucose, insulin, starvation and alloxan-diabetes on protein synthesis in perfused rat hearts. *Biochem. J.* **236**:543–547, 1986.

Snoy, F. J., Harker, R. J., Thies, W., and Greenspan, K.: Ethanol-induced electrophysiological alteration in canine cardiac purkinje fibers. *J. Stud. Alcohol* **41**:1023–1030, 1980.

Soffia, F., and Penna, M.: Ethanol metabolism by rat heart homogenates. *Alcohol* **4**:45–48, 1987.

Sohal, R. S., and Burch, G. E.: Effect of ethanol ingestion on the myocardial capillaries of mice. *Cardiovasc. Res.* **3**:369–372, 1969.

Spann, J. F., Jr., Mason, D. T., Beiser, G. D., and Gold, H. K.: Actions of ethanol on the contractile state of the normal and failing cat papillary muscle. *Clin. Res.* **16**:249, 1968.

Spodick, D. H., Pigott, V. M., and Chirife, R.: Preclinical cardiac malfunction in chronic alcoholism. *N. Engl. J. Med.* **287**:677–680, 1972.

Stampfer, M. J., Colditz, G. A., Willett, W. C., Speizer, F. E., and Hennekens, C. H.: A prospective study of moderate alcohol consumption and the risk of coronary disease and stroke in women. *N. Engl. J. Med.* **319**:267–273, 1988.

Stearns, S., Riseman, J. E. F., and Gray, W.: Alcohol in the treatment of angina pectoris. *N. Engl. J. Med.* **234**:578–583, 1946.

Stein, W. S., Lieber, C. S., Leevy, M. C., Cherrick, G. R., and Abelmann, W. H.: The effect of ethanol upon systemic and hepatic blood flow in man. *Am. J. Clin. Nutr.* **13**:68–74, 1963.

Stratton, R., Dormer, K. J., and Zeiner, A. R.: The cardiovascular effects of ethanol and acetaldehyde in exercising dogs. *Alcoholism: Clin. Exp. Res.* **5**:56–63, 1981.

Sulzer, R.: The influence of alcohol on the isolated mammalian heart. *Heart* **11**:141–150, 1924.

Swartz, M. H., Repke, D. I., Katz, A. M., and Rubin, E.: Effects of ethanol on calcium binding and calcium uptake by cardiac microsomes. *Biochem. Pharmacol.* **23**:2369–2376, 1974.

Takizawa, A., Yasue, H., Omote, S., Nagao, M., Hyon, H., Nishida, S., and Horie, M.: Variant angina induced by alcohol ingestion. *Am. Heart J.* **107**:25–27, 1984.

Talesnik, J., Belo, S., and Israel, Y.: Enhancement of noradrenaline-induced metabolic coronary dilatation by ethanol. *Eur. J. Pharmacol.* **61**:279–286, 1980.

Taraschi, T. F., and Rubin, E.: Effects of ethanol on the chemical and structural properties of biologic membranes. *Lab. Invest.* **52**:120–131, 1985.

Tepper, D., Capasso, J. M., and Sonnenblick, E. H.: Excitation-contraction coupling in rat myocardium: Alternations with long term ethanol consumption. *Cardiovasc. Res.* **20**:369–374, 1986.

Thomas, C. B.: The cerebral circulation. Effect of alcohol on cerebral vessels. *Arch. Neurol. Psychiatry* **38**:331–339, 1937.

Thomas, G., Haider, B., Oldewurtel, H. A., Lyons, M. M., Yeh, C. K., and Regan, T. J.: Progression of myocardial abnormalities in experimental alcoholism. *Am. J. Cardiol.* **46**:233–241, 1980.

Thornton, J. R.: Atrial fibrillation in healthy non-alcoholic people after an alcoholic binge. *Lancet* **2**:1013–1014, 1984.

Tiernan, J. M., and Ward, L. C.: Acute effects of ethanol on protein synthesis in the rat. *Alcohol and Alcoholism* **21**:171–179, 1986.

Timmis, G. C., Ramos, C. R., Gordon, S., and Gangaharan, V.: The basis for differences in ethanol-induced myocardial depression in normal subjects. *Circulation* **51**:1144–1148, 1975.

Timmis, C. G., Gordon, S., Ramos, G. R., and Gangaharan, V.: The relative resistance of normal young women to ethanol-induced myocardial depression. *Angiology* **30**:733–743, 1979.

Tsiplenkova, V. G., Vikhert, A. A., and Cherpachenko, N. M.: Ultrastructural and histochemical observations in human and experimental alcoholic cardiomyopathy. *J. Am. Coll. Cardiol.* **8**:22A–32A, 1986.

Ueshima, H., Shimamoto, T., Iida, M., Konishi, M., Tanigaki, M., Doi, M., Tsujioka, K., Nagano, E., Tsuda, C., Ozawa, H., Kojima, S., and Komachi, Y.: Alcohol intake and hypertension among urban and rural Japanese populations. *J. Chron. Dis.* **37**:585–592, 1984.

Unverferth, D. V., Magorien, D. R., Moeschberger, L. M., Baker, B. P., Fetters, K. J., and Leier, V. C.: Factors influencing the one-year mortality of dilated cardiomyopathy. *Am. J. Cardiol.* **54**:147–152, 1984.

Urbano-Marquez, A., Estruch, R., Navarro-Lopez, F., Grau, M. J., Mont, L., and Rubin, E.: The effects of alcoholism on skeletal and cardiac muscle. *N. Engl. J. Med.* **320**:409–415, 1989.

Vasdev, S. C., Chakravati, R. N., Subrahmanyam, D., Jain, A. C., and Wahi, P. L.: Myocardial lesions induced by prolonged alcohol feeding in rhesus monkeys. *Cardiovasc. Res.* **9**:134–140, 1975.

Viel, B., Donoso, S., Salcedo, D., Rojas, P., Varela, A., and Alessandri, R.: Alcoholism and socioeconomic status hepatic damage and arteriosclerosis. *Arch. Intern. Med.* **117**:84–91, 1966.

Vikhert, M. A., Tsiplenkova, G.V., and Cherpachenko, M. N.: Alcoholic cardiomyopathy and sudden cardiac death. *J. Am. Coll. Cardiol.* **8**:3A–11A, 1986.

Villeneuve, J. P., Pomier, G., and Huet, P. M.: Effects of ethanol on hepatic blood flow in unanesthetized dogs with chronic portal and hepatic vein catheterization. *Can. J. Physiol. Pharmacol.* **59**:598–603, 1981.

Wallace, B. R., Barrett-Connor, E., Criqui, M., Wahl, P., Hoover, J., Hunninghake, D., and Heiss, G.: Alterations in blood pressures associated with combined alcohol and oral contraceptive use—the lipid research clinics prevalence study. *J. Chronic Dis.* **35**:251–257, 1982.

Walsh, J. M., and Truitt, B. E., Jr.: Release of 7-H³ norepinephrine in plasma and urine by acetaldehyde and ethanol in cats and rabbits. *Fed. Proc.* **27**:601, 1968.

Walsh, J. M., Hollander, B. P., and Truitt, B. E., Jr.: Sympathomimetic effects of acetaldehyde on the electrical and contractile characteristics of isolated left atria of guinea pigs. *J. Pharmacol. Exp. Ther.* **167**:173–186, 1969.

Webb, W. R., and Degerli, I. U.: Ethyl alcohol and the cardiovascular system. *J.A.M.A.* **191**:1055–1058, 1965.

Webb, W. R., Gupta, D. N., Cook, W. A., Sugg, W. L., Bashour, F. A., and Unal, M. O.: Effects of alcohol on myocardial contractility. *Dis. Chest* **52**:602–605, 1967.

Weishaar, R., Bettuglia, M. A. S., Ashikawa, K., Sarma, J. S. M., and Bing, R. J.: Comparative effects of chronic ethanol and acetaldehyde exposure on myocardial function in rats. *J. Clin. Pharmacol.* **18**:377–387, 1978.

Weiss, S., and Wilkins, R. W.: The nature of the cardiovascular disturbances in vitamin deficiency states. *Trans. Assoc. Am. Physicians* **51**:339–372, 1936.

Whitman, V., Musselman, J., Schuler, H. Y., and Fuller, E. O.: Metabolic and functional consequences of chronic alcoholism and the rat heart. *J. Mol. Cell. Cardiol.* **12**:1249–1260, 1980.

Wilens, S. L.: The relationship of chronic alcoholism to atherosclerosis. *J.A.M.A.* **135**:1136–1139, 1947.

Wilhelmsen, L., Wedel, H., and Tibblin, G.: Multivariate analysis of risk factors for coronary heart disease. *Circulation* **48**:950–958, 1973.

Williams, E. S., and Li, T. K.: The effect of chronic alcohol administration of fatty acid metabolism and pyruvate oxidation of heart michondria. *J. Mol. Cell. Cardiol.* **9**:1003–1011, 1977.

Williams, E. S., Mirro, M. J., and Bailey, J. C.: Electrophysiological effects of ethanol, acetaldehyde, and acetate on cardiac tissues from dog and guinea pig. *Circ. Res.* **47**:473–478, 1980.

Williams, J. W., Tada, M., Katz, A. M., and Rubin, E.: Effect of ethanol and acetaldehyde on the $(Na^+ + K^+)$-activated adenosine triphosphatase activity of cardiac plasma membranes. *Biochem. Pharmacol.* **24**:27–32, 1975.

Wong, M.: Depression of cardiac performance by ethanol unmasked during autonomic blockage. *Am. Heart J.* **86**:508–515, 1973.

Wu, C. F., Sudhakar, M., Jaferi, G., Ahmed, S. S., and Regan, T. J.: Preclinical cardiomyopathy in chronic alcoholics: A sex difference. *Am. Heart J.* **91**:281–286, 1976.

Yano, K., Rhoads, G. G., and Kagan, A.: Coffee, alcohol and risk of coronary heart disease among Japanese men living in Hawaii. *N. Engl. J. Med.* **297**:405–409, 1977.

13

Effects of Alcohol Abuse on Skeletal Muscle

Finbarr C. Martin and Timothy J. Peters

13.1. Introduction

The last few decades have seen a resurgence of interest in alcoholic myopathy, but its recognition is far from new. As long ago as in 1822, James Jackson, Harvard Professor of Medicine, described painful neuritis and profound muscle weakness in a chronic alcoholic (Jackson, 1822). Jackson presented clinical evidence to emphasize the muscular origin of the weakness in chronic alcoholics. In 1849, Huss in Sweden described muscle weakness in a chronic alcoholic without clinical neuropathy (Huss, 1849, cited in Ekbom *et al.*, 1964). The relationship of muscle wasting to alcoholic peripheral neuropathy continues to be controversial. Toward the end of the 19th century, Gudden reported the first histological studies of muscles of five chronic alcoholics examined at postmortem (Gudden, 1895). Three had abnormal muscles, and his description of the acute myopathic changes requires little alteration today.

Recent interest followed a case report by Hed *et al.* in 1955 of a 55-year-old chronic alcoholic man with swollen muscles, myoglobinuria, and acute renal failure. At postmortem, the subject had hepatic steatosis and edema, necrosis, and hyaline degeneration of the muscle fibers. These workers later described another alcoholic male with a similar but nonfatal myopathy that recurred after bouts of heavy drinking (Fahlgren *et al.*, 1957), and subsequently they suggested a causal relationship between alcohol ingestion and acute myo-

necrosis in a series of 12 patients (Hed *et al.*, 1962). Two years later, in 1964, the same group (Ekbom *et al.*, 1964) described the less dramatic clinical syndrome of progressive proximal wasting and weakness in a series of 16 alcoholics, some of whom had no muscle pain or tenderness. They found that in all cases there were elevated serum activities of creatine kinase and aldolase. In 1967, Nygren reported elevated serum kinase activities in 30 of 79 chronic alcoholics without any muscular symptoms or delirium tremens, and this finding was confirmed by others (Velez-Garcia *et al.*, 1966; Perkoff *et al.*, 1967; Lafair and Myerson, 1968). At the same time the first systematic neurophysiological study of chronic alcoholics found that of a series of 24 patients without clinically evident muscle weakness or elevation of serum creatine kinase activity, 18 had myopathic features on electromyography (Faris *et al.*, 1967).

13.2. Clinicopathological Classification

From the above brief historical account, it can be seen that the categories of acute myopathy, chronic myopathy, and subclinical myopathy emerged. Indeed, this was the classification adopted in 1971 (Perkoff, 1971), although the acute form was further subdivided on the basis of three clinical presentations. Unfortunately, this classification was not completely satisfactory, since there were a number of histopathological and electrophysiological features common to all forms. Spargo (1981) has developed the classification emphasizing focal necrosis as the salient feature of acute myopathy, but he included, as characteristics of chronic myopathy, histopathological features that would be considered by others (Dubowitz and Brooke, 1973) to be features of neuropathy as it affects muscle. Spargo's subclinical myopathy included histopathological features of both the acute and chronic forms of myopathy. Developments in the histochemical study of muscle biopsies (Dubowitz, 1981) has allowed a more precise characterization of the muscle wasting of chronic myopathy (Slavin et al., 1983), so that the histopathological distinction between the acute and chronic forms is more satisfactory. In practice they may coexist, and each may be clinically obvious or quite subclinical (Martin and Peters, 1985a,b). Table 13.1 summarizes the principal features of acute and chronic alcoholic myopathy.

13.3. Clinicopathological Features of Acute Myopathy

Acute alcoholic myopathy is a toxic rhabdomyolysis responsible for the dramatic clinical picture described by Hed et al. (1962) of an acute onset of painful, tender, swollen, firm, but weak muscles of one or more groups. Any muscles may be involved, including the calf, where deep venous thrombosis may be simulated. It is now clear that typical but milder histopathological features of acute myopathy may be detected in muscle biopsies from patients with much less florid clinical features, more typical of chronic myopathy (Martin et al., 1985).

Acute myopathy is associated with elevated serum levels of creatine kinase (Velez-Garcia et al., 1966; Per-

koff et al., 1967; Lafair and Myerson, 1968; Myerson and Lafair, 1970) and lactate dehydrogenase (Pittman and Decker, 1971; Spector et al., 1979). A preceding heavy alcoholic bout is common, and repeated bouts may cause recurring myopathy (Hed et al., 1962). Muscle cramps may be extremely frequent and painful in severe cases of acute myopathy (Perkoff et al., 1967) but also occur in less florid cases and in drinkers without evident myopathy or electrolyte disturbances (Martin et al., 1985).

Myoglobinuria is a well-recognized complication (Hed et al., 1955; Fahlgren et al., 1957); Hed et al., 1962; Perkoff et al., 1967) and may lead to acute tubular necrosis with renal failure (Valaitis et al., 1960; Koffler et al., 1976). It may be overlooked or mistaken for hematuria or bilirubinuria. Hyperkalemia has been reported in acute alcoholic rhabdomyolysis (Fahlgren et al., 1957; Hed et al., 1962; Kahn and Meyer, 1970) and is probably caused by the combination of abnormal release of potassium from necrotic fibers and renal insufficiency. A form of acute myopathic weakness with intracellular edema has been reported in chronic alcoholics with hypokalemia (Lynch, 1969) but is distinct from alcoholic myopathy.

The histological picture is of rhabdomyolysis with fiber swelling, vacuoles, fragmentation with loss of myofibrillary striations, and hyaline and granular degeneration (Hed et al., 1955; Valaitis et al., 1960; Klinkerfuss et al., 1967; Kahn and Meyer, 1970; Pittman and Decker, 1971). An acute inflammatory reaction may be present (Lafair and Myerson, 1968; Kahn and Meyer, 1970). Excess fat both within and surrounding the fibers may be a characteristic feature of alcoholic acute myopathy (Ekbom et al., 1964; Hughes, 1974). In our series of seven patients with definite, albeit mild, acute myopathic features (Martin et al., 1985), four showed an abnormally high (>3%) proportion of nuclei situated

TABLE 13.1. Comparison of Principal Features of Acute and Chronic Skeletal Muscle Myopathy in Alcohol Abuse

Feature	Acute Myopathy	Chronic Myopathy
Alcohol intake	Bout drinking	Long-standing abuse
Muscles affected	Patchy, often distal	Proximal, symmetrical
Weakness	Variable	Frequently
Tenderness	Marked	No
Associated neuropathy	Sometimes	Frequently
Acute tubular necrosis	In severe cases	No
Serum creatine kinase	Raised, often very high	Occasional moderate rise
Fiber types affected	All types	Type II (especially IIb)

internally rather than, as normal, adjacent to the sarcolemma (Dubowitz and Brooke, 1973).

Basophilic cytoplasm, a feature of muscle fiber regeneration (Dubowitz and Brooke, 1973), has also been reported (Klinkerfuss et al., 1967). Reduced activities of myosin Mg^{2+}-ATPase and nicotinamide adenine dinucleotide dehydrogenase have been demonstrated histochemically (Martinez et al., 1973). Electron microscopic studies have shown extensive disintegration and separation of myofilaments (Douglas et al., 1966; Klinkerfuss et al., 1967), swollen mitochondria with blurred cristae (Douglas et al., 1966; Fisher et al., 1971), dilatation of the sarcoplasmic reticula (Klinkerfuss et al., 1967), and tubular aggregates adjacent to the sarcolemma (Chui et al., 1975).

13.4. Clinicopathological Features of Chronic Myopathy

Chronic alcoholic myopathy presents with a gradual onset of usually painless symmetrical proximal muscle wasting and weakness (Ekbom et al., 1964; Perkoff et al., 1967; Martin et al., 1985), most obviously in the legs (Mancall et al., 1966; Perkoff et al., 1967). Severe cases may result in difficulty in rising from a chair or in climbing stairs. It may be present in a patient who has also developed acute myopathy. Some accounts describe tenderness in affected muscles. Only histopathological study of such tender muscles can exclude coexisting acute myopathy as the explanation for this feature.

Chronic alcoholic myopathy is similar to that seen in osteomalacia (Young et al., 1978), in which muscles may also be tender, in iatrogenic or spontaneous Cushing's syndrome (Pleasure et al., 1970), and in hypothyroidism (McKeran et al., 1975), and specific investigation is required to exclude these disorders.

Profound weight loss may be seen in chronic alcoholics (Morgan, 1982), and severe chronic myopathy will certainly contribute to this. The cachexia of advanced decompensated cirrhosis, however, affects the muscles of hands, face, and trunk as well as the limbs. It has not been established whether histopathological changes in the muscles of these cachectic patients are similar to those in the proximal muscles of myopathic individuals without liver disease. Neurological abnormalities are also seen in up to 50% of chronic alcoholics (Perkoff, 1971). Signs of peripheral neuropathy and ataxia may thus complicate the clinical picture but do not explain the proximal localization of the weakness (Perkoff, 1971; Martin et al., 1985).

Routine hematoxylin and eosin-stained muscle sections may appear superficially normal, with variation of fiber size as the sole abnormality. The development of histochemical techniques (Dubowitz and Brooke, 1973), particularly myosin Mg^{2+}-ATPase staining, has allowed the characterization of skeletal muscle fibers as types I, IIa, and IIb (Brooke and Engel, 1969). The major histopathological change in chronic alcoholic myopathy, as in the other chronic myopathies mentioned above, is atrophy of the type IIb fibers (Slavin et al., 1983). The observation of changes such as a mild increase in fibrous tissue and regenerative features (Klinkerfuss et al., 1967) in patients with clinical signs of chronic myopathy is difficult to evaluate, since those studies did not include staining to differentiate fiber types, and doubt remains whether in fact the histology was more typical of chronic myopathy or muscle recovering from an episode of acute myopathy or both.

Changes of neuropathic atrophy, isolated atrophic angular fibers, or fiber-type grouping typical of the denervation–renervation process may be seen in biopsies on a background of generalized type II fiber atrophy (Martin et al., 1985).

Focal infiltration with fat is an occasional feature (Ekbom et al., 1964; Lynch, 1969; Hughes, 1974). Our study of hospital inpatients selected only on the basis of their having consumed over 100 g ethanol daily for over 3 years revealed excessive lipid staining with oil Red-O in 18 of 90 (20%) quadriceps muscle biopsies (Martin et al., 1985). The lipid droplets were of variable size and distributed throughout the fibers. Both fiber types were affected, though often to differing extents. All the affected biopsies had type II atrophy, and two also showed changes of acute myopathy.

Detailed biochemical study of eight of the biopsies revealed that the fatty accumulation was caused entirely by an excess of triglyceride with some minor differences in fatty acid profiles compared to the triglyceride in normal muscle biopsies (Sunnasy et al., 1983). The amount of triglyceride correlated with stated alcohol consumption.

As was reported above for acute myopathy, an increased proportion of centrally placed nuclei was also seen in five biopsies showing chronic myopathy but without acute myopathy (Martin et al., 1985). The significance of this is not clear, but it may represent evidence of regeneration following acute myopathy. Chui et al. (1975) noted that regenerative sarcoplasmic tubules are often seen surrounding a group of nuclei. On the other hand, it has also been noted in muscle with type II atrophy in a patient with hypothyroidism, a disease not associated with acute rhabdomyolysis (McKeran *et al.*, 1975).

Histological staining for glycogen phosphorylase clearly distinguished fiber types I and II in 16 chronic myopathic biopsies but was reduced in intensity in two, particularly in the type II fibers of one severely atrophic biopsy (Martin, 1984). Phosphorylase staining was normal in ten biopsies of patients with acute alcoholic myopathy (Perkoff *et al.*, 1966).

Electron microscopic reports have emphasized increased glycogen granules and lipid droplets (Hughes, 1974; Chui *et al.*, 1975), thickening of Z lines in regenerating foci (Klinkerfuss *et al.*, 1967; Kahn and Meyer, 1970; Mair and Tome, 1972), and reduced numbers (Douglas *et al.*, 1966; Fisher *et al.*, 1971) and volume (Kiessling *et al.*, 1975) of mitochondria. Mitochondrial changes were not seen in biopsies that at the light microscopic level showed severe type II atrophy but no evidence of acute myopathy (Martin, 1984). A convoluted appearance of the sarcolemma was noted in biopsies with type II atrophy (Martin *et al.*, 1982), but the exact relationship of ultrastructural abnormalities to fiber atrophy is uncertain, since fiber types cannot be unequivocally distinguished on electron microscopy.

13.5. Prevalence of Alcoholic Myopathy

Widely differing prevalence rates have been reported for the various forms of alcoholic myopathy. Variability results partly from the different populations of patients studied and partly from different diagnostic criteria. In a postmortem study of 100 alcoholics with alcoholic liver disease, 8% had muscle fiber necrosis or degenerative changes sufficient to be called acute myopathy (Kahn and Meyer, 1970). In a critical review of previous case reports of alcoholic myopathy, Oh (1972) suggested that a diagnosis of acute myopathy required definite histopathological evidence of muscle necrosis or elevated serum creatine kinase activity in an established chronic alcoholic with clinical symptoms or signs of muscle damage. He identified only seven such patients in a consecutive series of 938 hospitalized alcoholics.

In our recent survey of 151 hospital inpatients, each of whom had consumed over 100 g alcohol for over 3 years, but many of whom were not regarded by themselves or their doctors as alcoholics, seven (5%) had evidence on muscle biopsy of acute myopathy (Martin *et al.*, 1985). Only one presented with the classical picture of acute rhabdomyolysis with renal impairment, and five of the patients had normal serum activities of creatine kinase. A further 21 patients (15% of total series) had raised serum creatine kinase activities without evidence of acute myopathy on muscle biopsy. This finding conflicts with the general view that serum enzyme changes are sensitive markers of acute myopathy but may be explained by false-negative biopsy findings from sampling error.

Other series of hospitalized alcoholics, also unselected for muscle problems, included 38% of 79 patients (Nygren, 1967), 78% of 27 patients (Velez-Garcia *et al.*, 1966), and 61% of 59 patients (Perkoff *et al.*, 1967) with raised creatine kinase activities. Half of the above mentioned 151 patients (Martin *et al.*, 1985) had objective evidence of muscle weakness, but type II fiber atrophy was even more widespread, being present in 98 (65%).

13.6. Pathogenesis of Alcoholic Myopathy

Because there is little, if any, oxidation of ethanol by skeletal muscle, alterations to its metabolic or physical integrity either result from the direct impact of ethanol itself or are secondary to metabolic derangement elsewhere, acting through nervous or humoral influences in muscle. The role of acetaldehyde is uncertain in view of the controversy about the blood concentrations achieved (Knop *et al.*, 1981; Eriksson *et al.*, 1983; Lynch *et al.*, 1983).

Rubin and colleagues have studied the effect of ethanol on normal volunteers or animals in an attempt to distinguish its direct effect from those of the common sequelae of chronic alcoholism, malnutrition, neuropathy, and liver disease (Song and Rubin, 1972; Rubin *et al.*, 1976; Rubin, 1979). The pathogenic relationship between acute and chronic myopathy is by no means clear, although it has been suggested that the acute form occurs under appropriate conditions on the background of mild chronic damage (Spargo, 1981). A number of possible mechanisms of damage have been suggested, and these are discussed in turn.

13.6.1. Ethanol and Acute Myopathy

Ethanol alters the fluidity of cell membranes by entering into an ordered arrangement within the lipid bilayer (Wilson and Hoyumpa, 1979). Such alterations, if they occur *in vivo*, might be responsible for the intracellular edema and mitochondrial swelling seen in human acute myopathy and experimental animals. Muscle preparations from human subjects showed reduced contractility of actomyosin, and isolated sarcoplasmic reticulum membranes showed decreased calcium uptake. These abnormalities were not accompanied by changes evident at the light-microscopic level, but serum creatine kinase activities were raised, returning to normal within 2 weeks of abstention.

Further evidence of a direct pathological effect on muscle membranes has come from the demonstration of a rise in serum creatine kinase activities in normal well-fed volunteers in response to an infusion of ethanol (Lafair and Myerson, 1968). Transient ultrastructural changes have also been observed in the muscles of adequately fed rats following a single inebriating dose of alcohol (Munsat et al., 1973). These changes were dose dependent and consisted of mitochondrial swelling with distortion of the cristae. Histochemically, the mitochondria were abnormal. All these changes resolved within 24 hr of the alcohol challenge.

The relevance of these acute changes to the development of a myopathy is unclear. Reduced activity of Mg^{2+}-ATPase was reported in one alcoholic patient with acute myopathy, but the activities of Ca^{2+}-ATPase and of the mitochondrial enzymes succinate dehydrogenase, glutamate dehydrogenase, and malate dehydrogenase were normal in this and another similar patient (Martin et al., 1984a).

The only animal model to simulate human rhabdomyolysis has been the rat, in which it is produced by the combination of 2 to 4 weeks of exposure to ethanol and a brief period of food deprivation (Haller and Drachman, 1980). The pathological changes were accompanied by reduced tissue phosphate concentrations. Anderson et al. (1980) showed that the muscle contents of phosphorus and magnesium were reduced in acute alcoholic myopathy, and they proposed a pathogenic role for phosphate deficiency. This suggestion is consistent with the observation of reversible pathological alterations in muscle cells in experimental phosphorus deficiency (Fuller et al., 1976).

13.6.2. Peripheral Neuropathy and Chronic Myopathy

Mancall et al. (1966) reported on ten nutritionally deprived chronic alcoholics with symmetrical proximal weakness and less marked distal weakness. Electromyography of the proximal muscles showed myopathic features, whereas the distal muscles showed evidence of neurogenic lesions. Proximal muscle biopsies showed fiber necrosis and atrophy. They postulated a primary

chronic alcoholic myopathy but conceded that most of their patients also had a peripheral neuropathy.

Faris and Reyes (1971) claimed that chronic myopathic wasting does not occur without demonstrable peripheral neuropathy, which they considered to be the primary problem. However, a later large neurophysiological study reported myopathic changes without neuropathy in 6% of patients (Worden, 1976). Although the predominant clinical form of peripheral neuropathy in alcoholics is symmetrical distal sensory neuropathy, there is substantial evidence for a distal motor component with slowed motor nerve conduction (Ekbom et al., 1964). More significantly, reduced H-reflex response amplitudes and reduced conduction velocities in the proximal segments of the peripheral nerves of chronic alcoholics have also been reported (Guineneuc and Bathien, 1976; Liberson et al., 1979).

More recently, Langohr et al. (1983) studied 13 chronic alcoholics with generalized muscle wasting; 11 of the patients had an axonal sensory polyneuropathy. Muscle biopsies were taken from tibialis anterior and revealed fiber-type grouping, angular fibers, and target fibers, indicating neurogenic damage in addition to type II fiber atrophy. They could find no evidence to support a primary myopathic process. Unfortunately, tibialis anterior is known to be affected commonly in alcoholic neuropathy, and thus their studies cannot exclude a primary myopathy affecting proximal muscles.

In our series of 151 hospitalized heavy drinkers mentioned above (Martin et al., 1985) 72% of the 98 patients with type II fiber atrophy and 59% of the 59 patients with completely normal muscle biopsies had clinical evidence of neuropathy. Nineteen patients were studied in more detail (Mills et al., 1986). Twelve had electrophysiological evidence of sensory neuropathy, which was subclinical in five. There were no electrophysiological or histological features of denervation. One patient had severe neuropathy and normal muscle fibers, and three patients with muscle atrophy had normal electrophysiological findings, thus prompting the conclusion that the two pathologies may coexist but are pathogenetically distinct.

13.6.3. Cirrhosis, Malnutrition, and Protein Metabolism

Muscle wasting is a prominent clinical feature of decompensated cirrhosis, and evidence has been reported suggesting that the rate of myofibrillar protein breakdown is accelerated in such patients (Zoli et al., 1982). A recent study of chronic alcoholic patients without liver failure or fever that incorporated certain methodological improvements found no evidence of increased muscle breakdown rates in subjects with or without severe muscle wasting (Martin and Peters, 1985b). Severe muscle wasting can certainly occur in chronic alcoholics without cirrhosis (Martin *et al.*, 1985).

The relative roles of altered rates of protein synthesis and breakdown in the pathogenesis of muscle-wasting disorders is uncertain (Millward *et al.*, 1976; Rennie *et al.*, 1982). Several studies have indicated that ethanol impairs protein metabolism of a variety of organs *in vitro*, including liver (Perin *et al.*, 1979), brain (Noble and Tewari, 1979), and heart (Schreiber *et al.*, 1974). The effect of ethanol on protein synthesis *in vivo* has been reported recently (Tiernan and Ward, 1986). Whole-body protein synthesis was reduced by 41%, and that of skeletal muscle by 75%. These figures are in effect the averages of differing rates for individual protein constituents of the various organs.

It seems likely that muscle wasting in chronic alcoholics occurs as a result of a reduced myofibrillary protein synthesis rate, but whether this is caused by limited substrate supply or a derangement of intracellular control factors is not known. No relationship could be established between the severity of proximal muscle fiber atrophy and the protein nutrition status of chronic alcoholics as judged by serum alkaline ribonuclease measurements (Duane and Peters, 1988).

Specific deficiencies of folic acid, iron, thiamine, pyridoxine, riboflavin, and ascorbic acid are all seen in chronic alcoholics, but none has been associated with specific muscle pathology. The cortisol status of alcoholics has been studied by a variety of workers following reports of a pseudo-Cushingoid syndrome induced by alcohol (Smals *et al.*, 1976). There is, however, no evidence of a general abnormality in plasma or urinary cortisol levels in alcoholic subjects with proximal muscle wasting (Duane and Peters, 1987).

13.6.4. Mineral Metabolism and Muscle Wasting

The proximal myopathy of osteomalacia consists of type II fiber atrophy (Young *et al.*, 1978). Ethanol ingestion results in increased urinary losses of calcium and magnesium (Kalbfleisch *et al.*, 1963) and phosphate (McDonald and Margen, 1979), and there is considerable evidence of reduced tissue concentrations of magnesium in chronic alcoholics (Fuller *et al.*, 1976; Ward *et al.*, 1983). In humans, hypomagnesemia may result in impairment of parathyroid hormone secretion and con-

sequent hypocalcemia. Could minor disturbances of mineral metabolism, not severe enough to result in hypocalcemia or radiologically evident metabolic bone disease, nevertheless result in chronic myopathy? Hickish *et al.* (1986) studied the vitamin D status of 25 male alcoholics and found a quarter to be deficient, but there was no relationship between plasma vitamin D levels and quadriceps muscle strength.

Magnesium also plays a key role in the activity of some enzymes of the glycogenolytic pathway discussed later, and thus, magnesium deficiency could impair muscle metabolism with possible pathogenic implications.

We have reported reduced muscle concentrations of magnesium proportional to the degree of muscle atrophy in chronic alcoholics (Ward *et al.*, 1983), but this may be a reflection of the loss of relatively magnesium-rich type II fiber tissue rather than its cause.

13.6.5. Ethanol and Disordered Energy Metabolism

A number of acute alterations in skeletal muscle carbohydrate metabolism have been demonstrated in both nonalcoholic volunteers (Juhlin-Dannfelt *et al.*, 1977) and chronic alcoholics (Hed *et al.*, 1977). Spargo (1981) postulated that the pathogenesis of alcoholic myopathy was multifactorial but that binge drinking or abrupt food withdrawal might cause sufficient acute metabolic impairment to critically reduce the amino acid and glucose supply to muscle cells, so that in a chronic alcoholic with subclinical myopathy the effect might be to precipitate acute myopathy.

Hepatic oxidation of acetaldehyde forms acetate. There is increased acetate uptake and reduced glucose uptake by the skeletal muscle of normal volunteers during an infusion of ethanol (Juhlin-Dannfelt *et al.*, 1977). Reduced forearm muscle glucose uptake has been noted in chronic alcoholics after a period of heavy drinking (Hed *et al.*, 1977). A substantial increase was noted after 2 weeks of abstinence. Further, ethanol appears to potentiate reactive hypoglycemia (O'Keefe and Marks, 1977). Thus, there may be a shortfall of substrate supply to skeletal muscle during periods of intense activity.

Glycogen is the most important immediate source of substrate for glycolysis during such intense activity. Adenosine triphosphate (ATP) is produced by substrate-level phosphorylation via the glycolytic pathway. Continued flux through the pathway at times of high ATP demand is maintained by the conversion of pyruvate to lactate in the cytosol, thus regenerating NAD^+. This reaction is catalyzed by lactate dehydrogenase activity. Reduced lactate output from normal muscles exposed to ethanol (Chui *et al.*, 1978) and from muscles of chronic alcoholics (Nygren, 1971) has been noted during exercise, although the latter has been contested recently (Ward and Peters, 1983). Kiessling *et al.* (1975) reported reduced activities of both lactate and triose phosphate dehydrogenases in biopsies from chronic alcoholics with type II fiber atrophy, and Suominen *et al.* (1974) reported reduced activities of lactate dehydrogenase and hexokinase. We were unable to confirm reduced activities of lactate dehydrogenase activity in biopsies with histological features of either acute myopathy or chronic myopathy (Martin *et al.*, 1984b). It seems unlikely that moderate reduction of activity of an enzyme catalyzing a reaction that is generally at equilibrium would have pathophysiological implications.

The amount of glycogen in myopathic tissue appears increased on electron microscopy (Hughes, 1974; Martin *et al.*, 1982), but chemical determination has not confirmed this impression (Suominen *et al.*, 1974; Martin *et al.*, 1984b). Activities of glycogen phosphorylase in myopathic tissue was reduced in proportion to the severity of fiber atrophy (Martin *et al.*, 1984b). This is of uncertain pathogenic significance, since it may simply reflect loss of fiber that is rich in this activity compared with type I fiber (Harris *et al.*, 1976). The activity of phosphofructokinase, which catalyzes the rate-regulating step for glycolysis, may be more crucial. Assay of this activity has also shown reduction in proportion to the severity of type II fiber atrophy (Martin *et al.*, 1984b), but similar considerations about the interpretation apply for this as were discussed for phosphorylase.

Interesting though these alterations in enzyme activity may be, their relationship to events *in vivo* is uncertain. Magnetic resonance imaging using ^{31}P has been employed by P. Duane, G. Rada, and T. J. Peters (unpublished data) to investigate whether there is significant derangement of the glycogenolytic pathway activity in the forearm muscles of chronic alcoholics. Severely affected patients showed the changes seen in McArdle syndrome (congenital glycogen phosphorylase deficiency), but other less severely affected subjects were normal.

In summary, a variety of abnormalities have been reported in normal muscle under the influence of ethanol and in affected muscle of chronic alcoholics, but no satisfactory explanation has yet been established for the pathogenesis of either the acute or chronic form of alcoholic myopathy or the relationship between them.

13.7. Treatment and Prognosis

The treatment of mild acute myopathy will take place as part of the general supportive management of the alcoholic patient, with attention to hydration, repletion of vitamins and electrolyes, and analgesia or sedation as required. The treatment of severe rhabdomyolysis is a medical emergency requiring intravenous rehydration with careful monitoring of fluid and electrolyte balance and urine output. Acute tubular necrosis resulting in renal failure is likely to resolve satisfactorily, so renal dialysis is worthwhile during the acute episode. Subsequent abstention from alcohol and attention to nutrition and physical rehabilitation will be required.

The treatment of chronic myopathy consists of abstention from alcohol and adoption of a nutritionally adequate diet. Physical exercise will assist the return to normal of muscle strength as bulk increases. Isokinetic exercises with periods of intense activity are theoretically more likely to build up type II fibers, but damaged muscles may be more sensitive to this form of exercise (Edwards *et al.*, 1984).

The initial account by James Jackson described a reasonable clinical recovery after prolonged hospital stay with abstention from alcohol. This optimistic prognosis was reemphasized from clinical observation by Perkoff (1971). We studied 25 patients after variable periods of follow-up, up to 15 months (Fig. 13.1). There was a marked improvement with reduction of atrophy in the 16 patients who abstained from alcohol. The nine

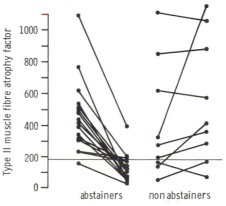

FIGURE 13.1. Sequential measurements of type II fiber areas in 16 abstainers studied on admission and 3–18 months later and in nine nonabstainers studied 3–10 months later. Upper limit of normal range for atrophy factor shown by horizontal line. (From Martin *et al.*, 1985.)

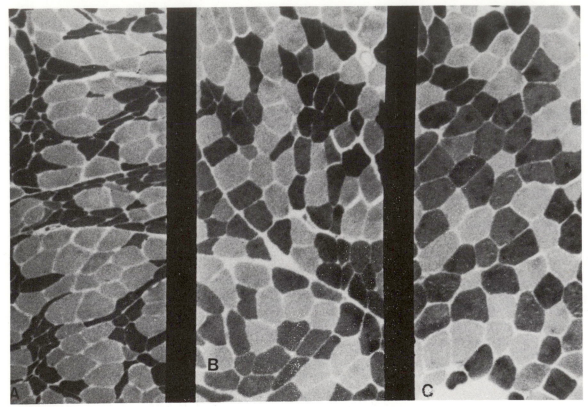

FIGURE 13.2. Myosin Mg^{2+}-ATPase-stained section of quadriceps biopsy from male alcoholic on admission (A) and after 3 (B) and 6 (C) months of abstention. Note the progressive increase in the size of the type II (dark-stained) fibers ($\times 150$). (From Peters *et al.*, 1985.)

patients who continued to drink either remained the same or worsened (Peters *et al.*, 1985). It is of interest that the recovery of the 16 patients was not associated with a consistent change in the size of their type I muscle fibers or an increase in body weight or clinical recovery from peripheral neuropathy. Figure 13.2 shows the sequential studies on one patient, illustrating that his recovery was incomplete after 3 months but complete after 6 months of abstention.

ACKNOWLEDGMENT. Our thanks go to Jean Ponder for secretarial assistance.

REFERENCES

Anderson, R., Cohen, M., Haller, R., Elms, J., Carter, N. W., and Knochel, J. P.: Skeletal muscle phosphorus and magnesium deficiency in alcoholic myopathy. *Miner. Electrolyte Metab.* **4**:106–112, 1980.

Brooke, M. H., and Engel, W. K.: The histographic analysis of human muscle biopsies with regard to fiber types. 2. Disease of the upper and lower motor neurone. *Neurology (Minneap.)* **19**:378–393, 1969.

Chui, L. A., Neustein, H., and Munsat, T. L.: Tubular aggregates in subclinical alcoholic myopathy. *Neurology (Minneap.)* **25**:405–412.

Chui, L. A., Munsat, T. L., and Craig, J. R.: Effect of ethanol on lactic acid production by exercised normal muscle. *Muscle Nerve* **1**:57–61, 1978.

Douglas, R. M., Fewings, J. D., Casley-Smith, J. R., and West, R. F.: Recurrent rhabdomyolysis precipitated by alcohol. A case report with physiological and electron microscopic studies of skeletal muscle. *Aust. Ann. Med.* **15**:251–256, 1966.

Duane, P., and Peters, T. J.: Glucocorticoid status in chronic alcoholics with and without skeletal muscle myopathy. *Clin. Sci.* **76**:601–603, 1987.

Duane, P., and Peters, T. J.: Nutritional status in alcoholics with and without chronic skeletal muscle myopathy. *Alcohol Alcoholism* **23**:271–277, 1988.

Duane, P., Symes, A., Raja, K., and Peters, T. J.: Alcoholic myopathy and nutritional status as determined by serum alkaline ribonuclease activity (EC 3.1.4.22). *Clin. Sci.* **70**(Suppl. 13):84P, 1986.

Dubowitz, V.: Histochemistry of Muscle Disease. In: *Disorders of Voluntary Muscle* (J. N. Walton, ed.), Edinburgh, Churchill Livingstone, 1981, pp. 261–295.

Dubowitz, V., and Brooke, M. H.: *Muscle Biopsy: A Modern Approach.* London, W. B. Saunders, 1973.

Edwards. R. H. T., Jones, D. A., Newham, D. J., and Chapman, S. J.: Role of mechanical damage in pathogenesis of proximal myopathy in man. *Lancet* **1**:548–551, 1984.

Ekbom, K., Hed, R., Kirstein, L., and Astrom, K. E.: Muscle affections in chronical alcoholism. *Arch. Neurol.* **10**:449–458, 1964.

Eriksson, C. J. P.: Human blood acetaldehyde concentration during ethanol oxidation (update 1982). *Pharmacol. Biochem. Behav.* **18**(Suppl. 1):141–150, 1983.

Fahlgren, H., Hed, R., and Lundmark, C.: Myonecrosis and myoglobinuria in alcohol and barbiturate intoxication. *Acta Med. Scand.* **158**:405–412, 1957.

Faris, A. A., and Reyes, M. G.: Reappraisal of alcoholic myopathy, clinical and biopsy study on chronic alcoholics without muscle weakness or wasting. *J. Neurol. Neurosurg. Psychiatry* **34**:86–92, 1971.

Faris, A. A., Reyes, M. G., and Abrams, B. M.: Subclinical alcoholic myopathy: Electromyographic and biopsy study. *Trans. Am. Neurol. Assoc.* **92**:102–106, 1967.

Fisher, E. R., Punterri, A. J., Jung, Y., Corredor, D. G., and Donowski, T. S.: Alcoholism and other concomitants of mitochondrial inclusions in skeletal muscles. *Am. J. Med. Sci.* **214**:97–106, 1971.

Fuller, T. J., Carter, N. W., Barcenas, C., and Knochel, J. P.: Reversible changes of the muscle cell in experimental phosphorus deficiency. *J. Clin. Invest.* **57**:1019–1024, 1976.

Gudden, H.: Klinische und anatomische Beitrage fur Kenntniss der multiplen Alkoholneuritis nebst Bemerkungen uber die Regenerations Vorange in periperhen Nervensystem. *Arch. Psychiatr. Nervenkr.* **28**:642–741, 1895.

Guineneuc, P., and Bathien, N.: Two patterns of results in polyneuropathies investigated with the H reflex. Correlation between proximal and distal conduction velocities. *J. Neurol. Sci.* **30**:83–94, 1976.

Haller, R. G., and Drachman, D. B.: Alcoholic rhabdomyolysis: An experimental model in the rat. *Science* **208**:412–415, 1980.

Hed, R., Larsson, H., and Wahlgren, F.: Acute myoglobinuria. Report of a case with a fatal outcome. *Acta Med. Scand.* **152**:459–462, 1955.

Hed, R., Lundmark, C., Fahlgren, H., and Orell, S.: Acute muscular syndrome in chronic alcoholism. *Acta Med. Scand.* **171**:585–599, 1962.

Hed, R., Lindblad, L. E., Nygren, A., and Sundblad, L.: Forearm glucose uptake during glucose tolerance tests in chronic alcoholics. *Scand. J. Clin. Lab. Invest.* **37**:229–233, 1977.

Hickish, T., Johnson, M., Colston, K., and Maxwell, J. D.: Muscle strength and vitamin D status in male alcoholics. *Clin. Sci.* **70**(Suppl. 13):86P, 1986.

Hughes, J. T.: *Pathology of Muscle*. Philadelphia, W. B. Saunders, 1974, pp. 144–149.

Jackson, J.: On a peculiar disease resulting from the use of ardent spirits. *N. Engl. J. Med. Surg.* **11**:351–353, 1822.

Juhlin-Dannfelt, A., Ahlborg, G., Hagenfeldt, L., Jorfeldt, L.,

and Felig, P.: Influence of ethanol on splanchnic and skeletal muscle substrate turnover during prolonged exercise in man. *Am. J. Physiol.* **233**:E195–E202, 1977.

Kahn, L. B., and Meyer, J. S.: Acute myopathy in chronic alcoholism. A study of 22 autopsy cases with ultrastructural observations. *Am. J. Clin. Pathol.* **53**:516–530, 1970.

Kalbfleisch, J. B., Lindeman, R. D., Ginn, H. E., and Smith, W. D.: Effects of ethanol administration on urinary excretion of magnesium and other electrolytes in alcoholic and normal subjects. *J. Clin. Invest.* **42**:1471–1475, 1963.

Kiessling, K. H., Pilstrom, L., Bylund, A. C., Piehl, K., and Saltin, B.: Effects of chronic ethanol abuse on structure and enzyme activities of skeletal muscle in man. *Scand. J. Clin. Lab. Invest.* **35**:601–607, 1975.

Klinkerfuss, G., Bleisch, V., Dioso, M. M., and Perkoff, G. T.: A spectrum of myopathy associated with alcoholism: 11 light and electron observations. *Ann. Intern. Med.* **67**:493–510, 1967.

Knop, J., Angelo, H., and Christensen, J. M.: Is role of acetaldehyde in alcoholism based on an analytical artefact? *Lancet* **2**:102, 1981.

Koffler, A., Friedler, R. M., and Massry, S. G.: Acute renal failure due to non-traumatic rhabdomyolysis. *Ann. Intern. Med.* **85**:23–28, 1976.

Lafair, J. S., and Myerson, R. M.: Alcoholic myopathy. *Arch. Intern. Med.* **122**:417–422, 1968.

Langohr, H. D., Weitholter, H., and Peiffer, J.: Muscle wasting in chronic alcoholics: Comparative histochemical and biochemical studies. *J. Neurol. Neurosurg. Psychiatry* **46**:248–254, 1983.

Liberson, W. T., Chong, Y. C., and Fried, P.: EMG studies in alcoholism 11. *Electromyogr. Clin. Neurophysiol.* **19**:15–26, 1979.

Lynch, C., Lim, C. K., Thomas, M., and Peters, T. J.: Assay of blood and tissue aldehydes by the HPLC analysis of their 2,4-dinitrophenylhydrazine derivatives. *Clin. Chim. Acta* **130**:117–122, 1983.

Lynch, P. G.: Alcoholic myopathy. *J. Neurol. Sci.* **9**:449–462, 1969.

Mair, W. G. P., and Tome, F. M. S.: *Atlas of the Ultrastructure of the Diseased Human Muscle*. Edinburgh, Churchill Livingstone, 1972, p. 176.

Mancall, E. L., McEntee, W. J., Hirschborn, A., and Gonyea, E.: Proximal muscle weakness and atrophy in the chronic alcoholic. *Neurology (Minneap.)* **16**:301–305, 1966.

Martin, F. C.: *Alcoholic Skeletal Myopathy. A Clinical, Histopathological and Biochemical Study*. MD Thesis, University of London, 1984.

Martin, F., and Peters, T. J.: Alcoholic muscle disease. *Alcohol Alcoholism* **20**:125–136, 1985a.

Martin, F. C., and Peters, T. J.: Assessment *in vitro* and *in vivo* of muscle degradation in chronic skeletal muscle myopathy of alcoholism. *Clin. Sci.* **68**:693–700, 1985b.

Martin, F. C., Slavin, G., and Levi, A. J.: Alcoholic muscle disease. *Br. Med. Bull.* **38**:53–56, 1982.

Martin, F. C., Slavin, G., Levi, A. J., and Peters, T. J.: Investigation of the organelle pathology of skeletal muscle in chronic alcoholism. *J. Clin. Pathol.* **37**:448–454, 1984a.

Martin, F. C., Levi, A. J., Slavin, G., and Peters, T. J.: Glycogen content and activities of key glycolytic enzymes in muscle biopsies from control subjects and patients with chronic alcoholic skeletal myopathy. *Clin. Sci.* **66:**69–78, 1984b.

Martin, F.C., Ward, K., Slavin, G., Levi, J., and Peters, T. J.: Alcoholic skeletal myopathy, a clinical and pathological study. *Q. J. Med.* **55:**233–251, 1985.

Martinez, A. J., Hooshnand, H., and Faris, A. A.: Acute alcoholic myopathy. Enzyme histochemistry and electron microscopic findings. *J. Neurol. Sci.* **20:**245–252, 1973.

McDonald, J. J., and Margen, C. S.: Wine versus ethanol in human nutrition. III Calcium, phosphorus and magnesium balance. *Amer. J. Clin. Nutr.* **32:**823–833, 1979.

McKeran, R. O., Slavin, G., Andrews, T. M., Ward, P., and Mair, W. G. P.: Muscle fiber type changes in hypothyroid myopathy. *J. Clin. Pathol.* **28:**659–663, 1975.

Mills, K. R., Ward, K., Martin, F., and Peters, T. J.: Peripheral neuropathy and myopathy in chronic alcoholism. *Alcohol Alcoholism* **21:**357–362, 1986.

Millward, D. G., Garlick, P. J., Nnanyelugo, D. O., and Waterlow, J. C.: The relative importance of muscle protein synthesis and breakdown in the regulation of muscle mass. *Biochem. J.* **156:**185–188, 1976.

Morgan, M. Y.: Alcohol and nutrition. *Br. Med. Bull.* **38:**21–29, 1982.

Munsat, T. L., Neustein, H., Higgins, T., Chui, L. A.: Experimental acute alcoholic myopathy. *Neurology (Minneap.)* **23:**407, 1973.

Myerson, R. M., and Lafair, J. S.: Alcoholic muscle disease. *Med. Clin. North Am.* **54:**723–730, 1970.

Noble, E. P., and Tewari, S.: Metabolic aspects of alcoholism in the brain. In: *Metabolic Effects of Alcohol* (P. Avogaro, C. R. Sirtori, and E. Tremoli, eds.), Amsterdam, Elsevier/North Holland, 1979, pp. 149–185.

Nygren, A.: Serum creatine phosphokinase in chronic alcoholism. *Acta Med. Scand.* **182:**383–387, 1967.

Nygren, A.: The ischaemic lactic acid response and the muscle LDH-isoenzyme pattern in alcoholics. *Acta Med. Scand.* **190:**283–285, 1971.

Oh, S. J.: Alcoholic myopathy, a critical review. *Alabama J. Med. Sci.* **9:**79–95, 1972.

O'Keefe, S. J. D., and Marks, V.: Lunchtime gin and tonic, a cause of reactive hypoglycaemia. *Lancet* **1:**1286–1288, 1977.

Perin, A., Sessa, A., and Desiderio, M. A.: Ethanol and liver protein synthesis. In: *Metabolic Effects of Alcohol* (P. Avogaro, C. R. Sirtori, and E. Tremoli, eds.), Amsterdam, Elsevier/North Holland, 1979, pp. 281–292.

Perkoff, G. T.: Alcoholic myopathy. *Annu. Rev. Med.* **22:**125–132, 1971.

Perkoff, G. T., Hardy, P., and Velez-Garcia, E.: Reversible acute muscular syndrome in chronic alcoholism. *N. Engl. J. Med.* **274:**1277–1284, 1966.

Perkoff, G. T., Dioso, M. M., Bleisch, V., and Klinkerfuss, G.: A spectrum of myopathy associated with alcoholism. I. Clinical and laboratory features. *Ann. Intern. Med.* **67:**481–492, 1967.

Peters, T. J., Martin, F., and Ward, K.: Chronic alcoholic skeletal myopathy—common and reversible. *Alcohol* **2:**485–489, 1985.

Pittman, J. F., and Decker, J. P.: Acute and chronic myopathy associated with alcoholism. *Neurology (Minneap.)* **21:**293–296, 1971.

Pleasure, D. E., Walsh, G. O., and Engel, W. K.: Atrophy of skeletal muscle in patients with Cushing's syndrome. *Arch. Neurol.* **22:**118–125, 1970.

Rennie, M. J., Edwards, R. H. T., and Millward, D. J.: Increased protein degradation in muscle disease: Cause or effect. *Muscle Nerve* **5:**85–86, 1982.

Rubin, E.: Alcoholic myopathy in heart and skeletal muscle. *N. Engl. J. Med.* **301:**28–33, 1979.

Rubin, E., Katz, A. M., Lieber, C. S., Stein, E. P., and Puskin, S.: Muscle damage produced by chronic alcohol consumption. *Am. J. Pathol.* **83:**499–516, 1976.

Schreiber, S. S., Oratz, M., Rothschild, M. A., Reff, F., and Evans, C.: Alcoholic cardiomyopathy. II. The inhibition of cardiac microsomal protein synthesis by acetaldehyde. *J. Mol. Cell. Cardiol.* **6:**207–213, 1974.

Slavin, G., Martin, F., Ward, P., Levi, J., and Peters, T. J., : Chronic alcohol excess is associated with selective but reversible injury to type 2B muscle fibers. *J. Clin. Pathol.* **36:**772–777, 1983.

Smals, A. G., Kloppenborg, P. W., Njo, K. T., Knoben, J. M., and Ruland, C. M.: Alcohol-induced Cushinoid syndrome. *Br. Med. J.* **2:**1298, 1976.

Song, S. K., and Rubin, E. : Ethanol produces muscle damage in human volunteers. *Science* **175:**327–328, 1972.

Spargo, E.: Alcohol and muscle disease. *Br. J. Alcohol Alcoholism* **16:**124–134, 1981.

Spector, R., Choudhury, A., Cancilla, P., and Lakin, R.: Alcoholic myopathy diagnosis by alcohol challenge. *J.A.M.A.* **242:**1648–1649, 1979.

Sunnasy, D., Cairns, S. C., Martin, F., Slavin, G., and Peters, T. J.: Chronic alcoholic skeletal muscle myopathy: A clinical, histological and biochemical assessment of muscle lipid. *J. Clin. Pathol.* **36:**778–784, 1983.

Suominen, H., Forsberg, S., Heikkinen, E., and Osterback, L.: Enzyme activities and glycogen concentration in skeletal muscle in alcoholism. *Acta Med. Scand.* **196:**199–202, 1974.

Tiernan, J., and Ward, L. C.: Acute effects of ethanol on protein synthesis in the rat. *Alcohol Alcoholism* **21:**171–179, 1986.

Valaitis, J., Pilz, C. G., Oliver, H., and Chomet, B.: Myoglobinuria, myoglobinuric nephrosis and alcoholism. *Arch. Pathol.* **70:**195–202, 1960.

Velez-Garcia, E., Hardy, P., Dioso, M., and Perkoff, G. T.: Cysteine-stimulated serum creatine phosphokinase: Unexpected results. *J. Lab. Clin. Med.* **68:**636–645, 1966.

Ward, K., and Peters, T. J.: Ischaemic lactate response in alcoholism—a reappraisal. *Clin. Sci.* **65:**21P, 1983.

Ward, K., Martin, F., Raja, K., and Peters, T. J.: Muscle magnesium in alcoholic muscle and liver disease. *Clin. Sci.* **65:**67P–68P, 1983.

Wilson, F. A., and Hoyumpa, A. M.: Ethanol and small intestinal transport. *Gastroenterology* **76:**388–403, 1979.

Worden, R. E.: Pattern of muscle and nerve pathology in alcoholism. *Ann. N.Y. Acad. Sci.* **273:**351-359, 1976.

Young, A., Brenton, D. P., and Edwards, R. H. T.: Analysis of muscle weakness in osteomalacia. *Clin. Sci. Mol. Med.* **54:**31P, 1978.

Zoli, M., Marchesini, G., Dondi, C., Bianchi, G. P., and Pisi, E.: Myofibrillar protein catabolic rates in cirrhotic patients with and without muscle wasting. *Clin. Sci.* **62:**683–686, 1982.

14

The Effects of Alcohol on the Nervous System

Clinical Features, Pathogenesis, and Treatment

Maurice Victor

Virtually every organ system is affected adversely by the excessive ingestion of alcohol,* but none more frequently or more seriously than the nervous system. Unlike the effects of alcohol on nonneural organs and structures, the clinical effects on the nervous system are remarkably diverse. This diversity derives from the fact that alcohol or alcoholism induces its clinical effects by several different mechanisms, namely, by a direct toxic effect on nerve cells, by the withdrawal of alcohol after a period of chronic intoxication, by inducing a state of nutritional deficiency that may affect the central or peripheral nervous system or both, and by crossing the placenta and damaging the fetal brain. The varied mechanisms by which alcohol affects the nervous system serve as the basis for the classification shown in Table 14.1.

14.1. Alcoholic Intoxication

The usual forms of alcoholic intoxication, or drunkenness, are so familiar as to require little elabora-

tion. Varying degrees of exhilaration and excitement, loss of restraint, irregularities of behavior, loquacity, slurred speech, incoordination of movement and gait, irritability and combativeness, drowsiness, stupor, and coma are the common features. These are regularly accompanied by a variety of nonneurological effects (tachycardia, vasodilation, vertigo, urinary frequency, and so on), which are described fully in other chapters of this volume.

It is also a commonplace observation that the behavioral abnormalities induced by a given amount of alcohol vary considerably from one person to another—depending on whether the subject is a man or woman, an abstainer, an occasional drinker, or an alcoholic addict. The effects of acute ingestion of alcohol on nonaddicted persons have been the subject of many studies, some dating from the turn of the century (McDougall, 1904; Rivers, 1908; Dodge and Benedict, 1915; Miles, 1918, 1932). All manner of motor performance—whether the simple maintenance of a standing posture, the control of speech, or highly organized and complex motor skills—is affected adversely by alcohol. The movements involved in these acts not only are made more slowly than normal but also are more inaccurate and random in character and therefore less well adapted to the accomplishment of specific ends. The same is true of eye

---

*The term alcohol is used throughout this chapter to designate only ethyl alcohol or ethanol, the active ingredient in beer, wine, whiskey, gin, brandy, and other common alcoholic beverages.

TABLE 14.1. Classification of the Alcoholic Neurological Diseases

Alcoholic intoxication (including coma, pathological intoxication, and blackouts)

The abstinence or withdrawal syndrome (tremulousness, hallucinosis, seizures, delirium tremens)

Nutritional diseases of the nervous system secondary to alcoholism

    Wernicke disease and Korsakoff psychosis

    Alcoholic cerebellar degeneration

    Peripheral neuropathy (neuropathic beriberi)

    Optic neuropathy (tobacco–alcohol amblyopia)

    Pellagra

Diseases of uncertain pathogenesis most often observed in alcoholics

    Central pontine myelinolysis

    Marchiafava–Bignami disease

    Fetal alcohol syndrome (see Chapter 15)

    Alcohol myopathy (see Chapter 13)

    Alcoholic dementia

    Alcoholic cerebral atrophy

Hepatic encephalopathy

---

movements. Following the acute ingestion of alcohol by healthy young persons, the peak saccadic velocity (the speed with which the eyes move from one visual target to another) is reduced, and the smooth pursuit movements become fragmented as a result of the appearance of saccadic and dysmetric elements (Wilkinson and Kime, 1974). These observations explain why even mild degrees of intoxication impair the performance of skilled tasks that require eye-hand coordination.

Alcohol also impairs the efficiency of cognitive functions—learning, remembering, thinking, and making judgments. With increasing blood alcohol concentrations, there is a progressive impairment in the ability to form associations, whether of words or figures; in the powers of attention and concentration; and in judgment and discrimination. Paradoxically, at low blood alcohol concentrations (after 2 oz of whiskey), intelligent young men actually solved problems in symbolic logic better than they did without alcohol; after 4 oz their performance was about normal, and with larger amounts their performance deteriorated (Keller, 1959).

Miles (1932) constructed a scale relating increasing degrees of clinical intoxication to blood alcohol levels: at levels of 30 mg/dl, mild euphoria was detectable; at 50 mg/dl, mild incoordination; at 100 mg/dl,

ataxia; at 200 mg/dl, confusion and a reduced level of mental activity; at 300 mg/dl, stupor; and at 400 mg/dl, deep anesthesia. The value of such a scale is limited to subjects similar to those observed by Miles, i.e., nonhabituated men in whom the blood alcohol level rises steadily over a 2-hr period. These relationships have little validity in chronic drinkers, in whom the phenomenon of tolerance is operative. Also it appears that nonaddicted women, after consuming comparable amounts of alcohol, have higher blood alcohol levels than men, even with allowance for differences in size—the result perhaps of a decreased gastric oxidation of alcohol (Frezza et al., 1990).

The physiological effects of ethanol on the nervous system have been reviewed in detail by Kalant (1979) and Kalant and Woo (1981). They make the point, as do others (Pohorecky, 1977), that a biphasic action of alcohol is evident at all levels of organization of the nervous system. In the case of single nerve cells, the administration of low concentrations of ethanol, acting directly on the neuronal membranes, increases their electrical excitability (increase in frequency and regularity of spontaneous spikes); progressively higher concentrations depress this excitability and finally abolish it. In multineuronal pathways, the biphasic effect depends more on different dose-response functions of modulatory inhibitory and excitatory neurons than on the direct biphasic actions of alcohol on individual cells. This is true of the simplest neuronal loops (e.g., monosynaptic spinal reflexes) as well as highly complex systems of the brain (EEG activity, evoked potentials). Thus, the monosynaptic reflex is subject to modulation at the level of the sensory receptor, the afferent terminal, and the efferent motoneuron, and a particular reflex response will depend on the balance between the facilitatory modulating effects of ethanol and its inhibitory action (Kalant, 1979). Low concentrations of ethanol produce an increase in low-amplitude fast (β) activity in the EEG, associated with arousal and desynchronization; increasingly higher doses cause progressive slowing and synchronization of the EEG in conjunction with depression of the state of consciousness.

14.1.1. Alcoholic Coma

The precise incidence of this form of intoxication cannot be stated. In the series of Plum and Posner (1980), comprising 500 patients who presented with nontraumatic coma, approximately 30% proved to be caused by "exogenous toxins," of which alcohol was presumably the most frequent.

The diagnosis of alcoholic coma is not always immediately evident. The usual features—flushed face, stupor, odor of alcohol—are not invariably present; moreover, they may be found in patients with coma of other types. In contrast to the more common and lesser manifestations of alcoholic intoxication, alcoholic coma may prove lethal. The measurement of blood alcohol is particularly helpful in diagnosis. Levels of 500 mg/dl are lethal in 50% of patients (Sellers and Kalant, 1976). However, severe chronic drinkers with levels in this range may sometimes show few signs of intoxication, and survival has been observed with much higher concentrations. Profound stupor and coma in the presence of relatively low blood alcohol concentrations should always raise the suspicion of some complicating illness (particularly meningitis, pneumonia, or other infection, head injury, or hepatic encephalopathy) or the concurrent ingestion of another sedative drug.

14.1.2. Pathological Intoxication (Alcohol Idiosyncratic Intoxication)

In a small proportion of individuals, alcohol has an extreme excitatory effect rather than a sedative one. In the past, the terms "pathological intoxication," "complicated intoxication," and "acute alcoholic paranoid state" were applied to this state, and indiscriminately included under these titles were variant forms of delirium tremens, all variety of epileptic phenomena, and psychopathic and criminal behavior (see reviews of Banay, 1944; Perr, 1986). Gradually there has emerged the clinical concept of an unusual reaction to alcohol characterized by an outburst of blind fury, combativeness, and destructive behavior occurring in persons without a history of alcoholism or mental or neurological disease. Often the patient can be subdued only with difficulty. The attack terminates with deep sleep, which occurs spontaneously or in response to the administration of sedatives. On awakening, the patient claims to have no memory of the episode. Allegedly, such a reaction may follow the ingestion of small amounts of alcohol, but in our experience the amount of ingested alcohol has always been substantial. Unlike the usual forms of alcohol intoxication or alcohol withdrawal, the atypical syndrome has not been produced in experimental subjects, and the diagnosis depends entirely on the aforementioned rather arbitrary anecdotal criteria.

The main disorders to be distinguished from the idiosyncratic reaction are the outbursts of rage and violence that characterize the behavior of certain sociopaths and occur occasionally as part of temporal lobe seizures. Diagnosis may be difficult and depends on eliciting other manifestations of temporal lobe epilepsy or sociopathy.

Vagaries of nomenclature have contributed to the ambiguity that surrounds this subject. The traditional term—pathological intoxication—is obviously inappropriate, insofar as all forms of intoxication are pathological. Unusual, atypical, or idiosyncratic intoxication is more suitable, indicating, as these terms do, a form of intoxication that differs from the usual or typical ones. Alcohol idiosyncratic reaction is the term suggested by the *Diagnostic and Statistical Manual of Mental Disorders* (DSM-III R).

The idiosyncratic reaction to alcohol has been ascribed to many factors (previous craniocerebral trauma, "underlying hysterical or epileptoid temperament," sociopathy), but there are few factual data to support these beliefs. An analogy may be drawn between pathological intoxication and the paradoxical reaction that occasionally follows the administration of barbiturates.

The diagnosis of pathological or idiosyncratic intoxication may have important legal implications. Alcoholism and the usual forms of drunkenness are no longer considered crimes, but intoxication is not an acceptable defense for criminal acts that are committed while intoxicated. However, a person with idiosyncratic intoxication is considered temporarily insane and therefore not responsible for any harmful actions during the period of intoxication (Perr, 1986).

14.1.3. Blackouts

This term, in the language of the alcoholic, refers to transient episodes of amnesia that occur during periods of intoxication. With sobriety, the patient cannot recall events that occurred while he or she was intoxicated, even though the state of consciousness (as judged by others) appeared little altered at that time. However, a systematic assessment of mental function during the amnesic period has not been made.

The nature and significance of such amnesic episodes is unclear. Some psychiatrists deny that a loss of memory occurs and view such "blackouts" as a form of malingering; others speak of "repression," which prevents the conscious awareness of painful memories. These views are purely speculative. It has been proposed that alcohol has a deleterious effect on short-term memory function in alcoholics who experience "blackouts" (Goodwin *et al.*, 1970; Tamerin *et al.*, 1971), but this notion has been discredited by Mello (1973). She showed that tests of short-term memory (up to 6 min) were performed equally well by men with a history of blackouts and those without such a history at blood alcohol levels above 200 mg/dl; furthermore, there was little difference in the accuracy of performance during intoxication and sobriety. It is widely held that "blackouts," as defined above, are a distinctive feature of alcohol addiction or an early predictor of alcohol addiction. However, the same phenomenon may occur in persons who never become addicted to alcohol. The salient facts are that there has been an episode of intoxication of a degree that prevented the formation of memories, and rarely will the amount of alcohol consumed in the course of moderate social drinking produce this effect.

14.1.4. Pathogenesis

The symptoms of alcoholic intoxication result from the depressant action of alcohol on neurons of the central nervous system (CNS), the primary site of action being the cell membrane and membrane-associated proteins, which function as receptors and ion channels. Alcohol increases the proportion of membrane lipid that is in a fluid rather than a gel state ("fluidization"), a process that disrupts the function of the protein components—particularly the GABA receptors and their associated chloride-ion channels and the voltage-dependent calcium channels. The disruption of membrane protein function underlies alcohol intoxication; the adaptation of cell membranes exposed chronically to alcohol, by which the membranes are kept in a suitable state of fluidity, may account for tolerance (Goldstein, 1983). The effects of alcohol on the nerve cell and its membrane have been the subject of intensive study in recent years (Porter *et al.*, 1990; Charness, 1992).

From a clinical point of view, small amounts of alcohol may have a direct stimulating effect on medullary neurons that govern respiration, but the effect is slight and transient; with increasing amounts of alcohol, the effect is wholly a depressant one. The hyperactivity of tendon reflexes in the initial stages of intoxication probably represents a transitory escape of spinal motor neurons from higher inhibitory control. In these respects, alcohol acts in a manner akin to the inhalation anesthetics. In distinction to the latter agents, the margin between the dose of alcohol that produces surgical anesthesia and that which dangerously depresses respiration is very narrow. This fact inserts an element of urgency in the diagnosis and treatment of alcoholic narcosis.

14.1.5. Treatment

14.1.5.1. Mild to Moderate Intoxication

Mild to moderate intoxication requires no special treatment. Certain time-honored remedies such as a warm shower followed by a cold one, strong coffee, forced activity, or induction of vomiting may temporarily counteract the effect of alcoholic intoxication, but there is no evidence that any of these methods or any of the "sobering-up" medicines (analeptics) significantly influence the rate of disappearance of alcohol from the blood. Similarly, alcoholic stupor is usually of short duration, and if the vital signs are normal, no special therapeutic measures are necessary. The idiosyncratic reaction to alcohol ("pathological intoxication") usually requires the use of restraints and the intramuscular administration of sedatives in substantial doses, e.g., phenobarbital (Luminal®) 200 mg, amobarbital (Amytal®) 500 mg, or haloperidol (Haldol®) 5 to 10 mg; each may be repeated once in 30 or 40 min if necessary.

14.1.5.2. Management of Alcoholic Coma

This represents a medical emergency. The goal of treatment is to tide the patient over the crisis in respiration and its complications. One should make certain, by insertion of an endotracheal tube, that the patient has a clear airway, and an automatic positive-pressure respiratory should be available in case of respiratory paralysis. If no injury to the head or neck is obvious, the patient should be placed in a semiprone position to prevent aspiration of secretions and vomitus. If shock has supervened, immediate treatment with fluids, vasopressor drugs, and steroids must be instituted; in any event, an intravenous infusion of 5% glucose in water should be started. Since alcohol is absorbed rapidly from the stomach, gastric lavage is usually not necessary; moreover, this procedure carries the risk of aspiration of gastric contents. However, if concurrent ingestion of other drugs is known or suspected, gastric aspiration may be helpful. The bladder should be emptied and drainage instituted.

The vital signs should be measured frequently, and accumulated secretions removed by suctioning and frequent turning of the patient. Disorders to which alcoholics are particularly vulnerable—subdural hematoma, pneumonia, meningitis, hepatic failure, and gastrointestinal bleeding—must be systematically excluded. Chest films, CT scan, lumbar puncture, and examination of the blood for the presence of alcohol and other drugs ("toxic screen") are routine procedures in these circumstances.

Hemodialysis or peritoneal dialysis is an effective means of lowering the blood alcohol concentration but is necessary only in patients with extremely high blood alcohol concentrations, particularly if accompanied by acidosis, and in those who have concurrently ingested methanol or ethylene glycol or some other dialyzable drug.

14.2. Alcohol Withdrawal Syndrome

Included under this title are several closely related groups of symptoms—tremulousness, hallucinations, seizures, and delirium. Each of these manifestations can be observed in relatively pure form and is so described, but usually they occur in various combinations. These symptoms are seen mainly in spree or periodical drinkers but also in steady drinkers in whom alcohol has been withdrawn for some reason. Invariably these symptoms occur on a background of sustained excessive drinking, but they become manifest only after a period of relative or absolute abstinence from alcohol, hence the designation abstinence or withdrawal syndrome.

14.2.1. Incidence

The frequencies of the various withdrawal states relative to one another and to other major complications of alcoholism are indicated in Table 14.2. Also evident from the table is the fact that the abstinence syndrome in one form or another represents the most common cause for admission of an alcoholic to a hospital. However, the figures do not reflect the true incidence of alcoholic intoxication, stupor, and coma, since only a small proportion of such patients who reach the emergency room are admitted to the hospital (Victor and Adams, 1953); as a corollary, intoxicated patients who do gain admission represent the most serious forms of intoxication (combative, stuporous, and comatose patients).

14.2.2. Clinical Features

14.2.2.1. Alcoholic Tremulousness

The most frequent manifestations of the abstinence syndrome are tremulousness, commonly referred to as "the shakes" or "the jitters," combined with general

TABLE 14.2. Analysis of 266 Consecutive Patients with Complications of Alcohol Abuse Admitted to the Boston City Hospital over a 60-Day Period[a]

Complication	Number of patients admitted	Percentage of total
Acute intoxication	56	21.0
Stupor or coma (27 patients)		
Combative state (15 patients)		
Acute alcoholic tremulousness	92	34.6
Tremor and transitory hallucinations	30	11.3
Pure auditory hallucinosis	6	2.3
Typical delirium tremens	14	5.3
Atypical delirious–hallucinatory states	11	4.1
Nutritional diseases of the nervous system	8	3.0
Others[b]	49	18.4
Rum fits[c]	32	12.0

[a]From Victor and Adams (1953).
[b]Included are patients in whom the alcoholic complications could not be categorized precisely because of the presence of head injury or overwhelming medical and/or surgical disease
[c]Alcohol withdrawal seizures occurring alone or with other withdrawal symptoms.

irritability, nausea, and vomiting. These symptoms first appear after a few days of sustained drinking, in the morning, following the relatively brief period of abstinence represented by a night's sleep. The symptoms subside under the influence of a few drinks, after which the patient is able to drink for the remainder of the day without distress. This cycle repeats itself for days or weeks on end; in most instances, a spree that culminates in admission to the hospital lasts about 2 weeks. Drinking is eventually terminated, most often because of increasing severity of the recurring tremor, vomiting, and weakness but also because of intercurrent injury or infection or lack of funds, and sometimes by dint of will.

Tremulousness and associated symptoms reach their peak of intensity 24 to 36 hr after the complete cessation of drinking, at which time the clinical picture is distinctive: the patient is alert and startles easily, the face is flushed, the conjunctivae are injected, and tachycardia is usually present. Anorexia, nausea, retching, general muscular weakness, and insomnia are equally prominent. The patient tends to be preoccupied, inattentive, and disinclined to answer questions, or he may respond in a rude or perfunctory manner. He may be mildly disoriented in time and unable to reconstruct the events of the final days of his drinking spree, but a significant degree of confusion is lacking, and the patient is usually aware of his surroundings and the nature of his illness.

A fast-frequency tremor is a prominent feature. It is variable in severity, tending to diminish when the patient is inactive and in quiet surroundings and to increase with attempted motor activity or emotional stress. The tremor may be of such severity that the patient cannot stand without help, speak clearly, or feed himself. More often, there is only a fine rapid tremor of the outstretched hands. Or there may be little objective evidence of tremor, and the patient complains only of being "shaky inside." Physiological analysis, using surface electrodes on agonist and antagonist muscles of the outstretched arms and hands, has disclosed two types of alcohol withdrawal tremor (LeFebvre-D'Amour *et al.*, 1978): (1) a tremor frequency greater than 8 Hz (mean 8.3 Hz) in which there is continuous activity in antagonistic muscles (as in physiological tremor, but of greater amplitude); and (2) a tremor frequency less than 8 Hz (mean 7.2 Hz), characterized by discrete bursts of EMG activity occurring synchronously in antagonistic muscles, similar to one form of familial tremor. Both types of alcoholic tremor differ from tremors of slower frequency (e.g., parkinsonian tremor) in which opposing muscles contract alternately and rhythmically.

Within a few days, the florid appearance, anorexia, weakness, tachycardia, and tremor subside to a large extent, but the overalertness, tendency to startle easily, jerkiness of movement, and general feeling of uneasiness may not subside completely for 10 to 14 days. According to Projesz and Begleiter (1983), certain electrophysiological changes (increased amplitudes of somatosensory evoked potentials and decreased latencies of auditory brainstem potentials) may persist for as long as 3 weeks after withdrawal and well after the clinical abnormalities have subsided.

14.2.2.2. Alcoholic Hallucinosis

Symptoms of disordered sense perception occur in approximately one-quarter of patients with the alcohol withdrawal syndrome. The patient may complain of "bad dreams," i.e., disturbed sleep associated with nightmarish episodes, which he finds difficult to separate from real experience. Sounds and shadows may be misinterpreted, or familiar objects may appear distorted and assume unreal forms (illusions). True hallucinations are less common; most often they are purely visual or auditory or mixed visual and auditory, and less often tactile or olfactory.

14.2.2.3. Acute and Chronic Auditory Hallucinosis

For many years it has been recognized that there is a special type of alcoholic psychosis, the central feature of which is an auditory hallucinosis (Kraepelin, 1946; Wernicke, 1900). Kraepelin believed that alcoholic auditory hallucinosis was related to delirium tremens, but Bleuler (1951) considered it to be a long-standing form of schizophrenia. Subsequent writings have tended to support one or the other of these contentions. These writings have been reviewed by Victor and Hope (1957), who also analyzed the clinical features of 76 personally studies cases of this disorder.

The dominant feature of the illness is the occurrence of auditory hallucinations in a patient whose sensorium is otherwise little affected; i.e., confusion, disorientation, and obtundation are minimal or absent, and memory is not significantly impaired. The hallucinations may take the form of a low-pitched hum or chant or of unstructured sounds such as buzzing, ringing, shots, clicking, and so on ("the elementary hallucinations" of Bleuler); in most cases vocal hallucinations predominate. The voices may be unidentifiable, or they may be attributed to the patient's family, friends, or neighbors, and rarely to God, radio, or radar. The voices may address the patient directly, but more frequently they discuss him in the third person. In the majority of cases the voices are maligning, reproachful, or threatening and are disturbing to the patient; but sometimes they are not unpleasant, in which case they leave the patient undisturbed. The voices may be very vivid and tend to be exteriorized; i.e., they come from behind the door, from the corridor, or through the wall. In response, the patient may call the police for protection or barricade himself against invaders; he may even attempt suicide to avoid what the voices threaten.

In most patients the disorder of sense perception becomes manifest during the first night after the cessation of drinking and becomes more prominent during the nights that follow. The hallucinations may intrude intermittently or may be more or less continuous. As a rule, the hallucinatory episode is short-lived; in 85% of our patients the duration was 6 days or less. A few patients, while actively hallucinating, are apparently aware that their hallucinations are unreal, but most of them do not appreciate the unreality of the hallucinations. Insight tends to be regained gradually; the patient goes through a stage in which he begins to doubt the reality of his hallucinations and is reluctant to talk about them, or he may question his sanity. Full recovery is characterized by the realization that the voices were imaginary and by the ability to recall, often with remarkable clarity, the abnormal thought content of the psychotic episode. Finally, there are a few patients who insist on the reality of their hallucinatory experience, even though they are no longer hallucinating.

In a small proportion of patients with auditory hallucinosis, the disease assumes a chronic form. In such patients, after a short period of time (as early as 1 week) the symptomatology begins to change. The patient becomes quiet and resigned, despite the fact that the hallucinations remain unchanged in content. Ideas of reference and influence and other poorly systematized paranoid delusions become prominent. The disorder resembles schizophrenia—illogical thinking, vagueness, tangential associations, and a dissociation of affect and thought content. Nevertheless, there are important diagnostic differences: the alcoholic illness has a considerably later age of onset than classic schizophrenia; the past history does not disclose schizoid traits; and the symptoms develop in close temporal relationship to a drinking bout. Moreover, alcoholic patients with hallucinosis are not distinguished by a high familial incidence of schizophrenia (Schuckit and Winokur, 1971; Scott, 1967).

14.2.2.4. Withdrawal Seizures ("Rum Fits")

In the clinical setting under discussion, i.e., where relative or absolute abstinence follows a period of sustained chronic intoxication, patients are particularly vulnerable to the development of convulsive seizures.

Characteristically, the seizures are grand mal in type, i.e., major generalized convulsions with loss of consciousness. The occurrence of focal seizures in this setting indicates the presence of focal cerebral disease (caused most often by craniocerebral trauma) in addition to the factor of alcohol withdrawal. There may be only a single seizure or, more often, two to six seizures, or sometimes more, occurring over a span of several hours. In about 3% of cases the seizures take the form of status epilepticus.

The most distinctive attribute of these seizures is their relationship to the cessation of drinking. This is depicted in Fig. 14.1, which is based on the study of 162 patients in whom the interval between the cessation of drinking and the onset of seizures could be determined accurately; over 90% of the seizures occurred during the 7- to 48-hr period following cessation of drinking, with a peak incidence between 13 and 24 hr (Victor and Brausch, 1967).

Seizures that occur during the withdrawal period are often associated with varying degrees of tremor, anorexia, insomnia, weakness, and hallucinosis. In our series, delirium tremens (i.e., the syndrome of gross confusion, disordered perception, and psychomotor and autonomic nervous system overactivity) developed in 28% of patients with seizures; in these cases the seizures invariably preceded the delirium. The postictal confusional state may blend with the onset of delirium tremens, or the confusion may clear to a varying extent, sometimes for 24 hr or even longer, before the delirium becomes manifest (Victor and Adams, 1953).

The attributes described above serve to identify

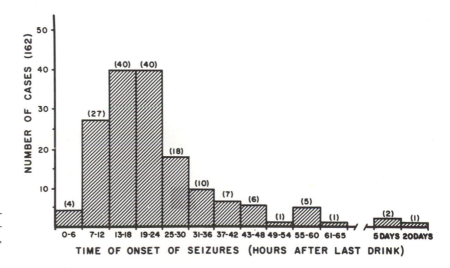

FIGURE 14.1. Relationship of onset of seizures to cessation of drinking. (From Victor and Brausch, 1967.)

seizures that occur only in relation to alcohol withdrawal ("rum fits") and to distinguish them from idiopathic or posttraumatic seizures, which may be precipitated by drinking but also occur in the interdrinking period. Seizures of the latter type are made worse and more frequent by drinking (Lennox, 1941; Victor and Brausch, 1967). Such seizures, unlike "rum fits," may be precipitated by a relatively short period of intoxication, e.g., an evening or weekend of heavy social drinking; nevertheless, the factor of withdrawal is still operative, insofar as the seizures tend to occur not when the patient is intoxicated but the morning after, in the "sobering-up" period.

Except for a transient dysrhythmia in the withdrawal period, EEG abnormalities in patients with "rum fits" are no more frequent than in the normal adult population. By contrast, persistent EEG abnormalities are found in 50% to 75% of nonalcoholic patients with idiopathic epilepsy (Greenblatt et al., 1944; Victor, 1968). Another distinctive feature of patients with "rum fits" is their sensitivity to photic stimulation (Victor and Brausch, 1967). This may take the form of a tonic-clonic convulsion with loss of consciousness or of coarse myoclonic movements of the face, neck, and arms that spread to involve the trunk and legs (photoparoxysmal or photomyogenic responses). Photic stimulation practically never provokes such responses in normal individuals and only rarely does so in patients with idiopathic epilepsy (Hughes et al., 1960; Victor, 1968). Contrariwise, almost half of the patients admitted to a hospital because of withdrawal seizures show a positive response to photic stimulation; a similarly high incidence of positive responses has been found in patients without spontaneous seizures but with severe tremulousness and other withdrawal symptoms (Victor and Brausch, 1967).

The foregoing observations discredit the notion that seizures in the alcoholic patient merely represent "latent epilepsy" made manifest by alcohol or that alcohol precipitates convulsive seizures in patients who are subject or constitutionally predisposed to them. Instead, the EEG reflects a sequence of changes engendered by alcohol itself: a decrease in the frequency of brain waves during the period of chronic intoxication; a rapid return of the EEG to normal immediately after the cessation of drinking; a brief period of dysrhythmia (sharp waves and paroxysmal changes) coinciding with the period of vulnerability to convulsive seizures, and again a rapid return of the EEG to normal (Wikler et al., 1956).

14.2.2.5. Delirium Tremens

This is the most serious form of the alcohol withdrawal syndrome. It is characterized by a state of profound confusion, delusions, vivid hallucinations, tremor, agitation, and sleeplessness as well as by signs of increased autonomic nervous system activity such as fever, tachycardia, and profuse sweating. Defined in this way, delirium tremens is a relatively rare disease, constituting only 5% of the alcoholic neurological disorders and 9% of the withdrawal syndromes at the Boston City Hospital (Table 14.2). Delirium tremens develops in one of several ways. The patient, an excessive and steady drinker for many months or years, may have been admitted to the hospital for some unrelated infectious illness, accident, or operation. In these circumstances, the delirium develops on the second or third day after admission to the hospital, sometimes even later. More often the problem is clearly one of alcoholism from the beginning. Following a protracted drinking spree, the symptoms may already have evolved through a stage of tremulousness, hallucinosis, and/or seizures, and the patient may even appear to be recovering from these symptoms when delirium tremens supervenes.

In severe cases, the clinical picture is one of the most colorful in medicine. The patient is constantly agitated, looking suspiciously about him, picking at and disarranging the bedclothes, or tugging at his restraints. More complex restless movements are usually related to some delusion or hallucination; characteristic are the turning and retraction of the head and eyes to engage some imaginary person in conversation, persistent searching movements in which the patient ransacks his bedclothes for something he supposes to be concealed there, and reenacting of complex habitual acts having to do with the patient's occupation. Some of the patient's incessant activity may be explained by the misbelief that he is imprisoned in some building; he strives to escape, and although he may be persuaded briefly to return to bed, he is soon attempting to get up again.

A coarse, irregular tremor and jactitations affect the extremities, face, and tongue. Picking and fumbling movements are typical. Speech is often slurred and garbled to the point of unintelligibility. The patient may shout and scream for hours on end, or speech may be reduced to an almost inaudible muttering. Neologisms, paraphasias, and elements of echolalia and palilalia can sometimes be recognized, indicating the presence of a mild dysphasia as well as a disturbance of articulation.

Signs of overactivity of the autonomic nervous

system are characteristic of delirium tremens and serve, more than any other feature, to distinguish it from nonalcoholic forms of delirium. Dilated pupils, tachycardia, fever, hypotension or hypertension, and hyperhidrosis are the usual manifestations; excessive pilomotor responses, pallor or flushing, nausea, constipation, and diarrhea may also be present.

Confusion is usually profound in degree. The patient may be unaware of the most obvious facts about his immediate situation, e.g., whether he is standing or lying, what clothing he is wearing, and whether he is indoors or on the street. Powers of attention and concentration are markedly impaired. If the patient's attention can be gained for a brief period, momentary flashes of clear insight and accurate responses may be obtained. He may even refer to bits of information from a previous examination that had seemingly left him unimpressed at that time. These observations indicate that the confusion is caused mainly by inattention and inaccurate interpretation of sensory experiences and that other aspects of intellectual function, particularly memory, are relative intact. In the older psychiatric literature this state was called clouding of consciousness.

Aberrations of perception may represent false interpretations of sensory impressions, i.e., illusions, or they may be unfounded in reality, i.e., hallucinations, although no one can ever be certain that some sensory stimulus is not involved in their genesis. Another aspect of the perceptual disorder is the patient's ready response to suggestion. He may, for example, be provoked to go through the motions of opening and then drinking a bottle of beer or lighting a cigar simply by pretending to hand him such objects. Poorly systematized delusions may be added; frequently these have a paranoid coloring—the patient believes that he is being stalked or pursued and that he is in imminent danger of being shot or castrated.

Most cases of delirium tremens are benign and short-lived, ending either abruptly or gradually. Less frequently there may be one or more relapses, several episodes of delirium of varying severity being separated by intervals of relative or complete lucidity, the entire process lasting for several days or, exceptionally, for several weeks. The course and outcome of delirium tremens, as we observed them in 101 patients, are summarized in Table 14.3. Most of the fatal cases were associated with an infectious illness or injury; in some, no complicating illnesses could be found, the patients dying in a state of hyperthermia or peripheral circulatory collapse. In others, death came so suddenly that the

TABLE 14.3. Course and Termination of 101 Cases of Delirium Tremens[a]

Nonfatal outcome (86 cases)		
Single episode		
Abrupt ending (49 cases)		
Gradual ending (27 cases)		
Duration (69 cases)		
24 h or less	(10 cases; 14.5%)	⎫
25–48 h	(17 cases; 24.6%)	⎬ 82.6%
49–72 h	(30 cases; 43.5%)	⎭
73–96 h	(6 cases; 8.7%)	
More than 4 days	(6 cases; 8.7%)	
Recurrent episodes lasting 3–31 days (10 cases)		
Fatal outcome (15 cases)		

[a]From Victor and Adams (1953).

terminal events could not be discerned, even after postmortem examination.

14.2.2.6. Atypical Delirious-Hallucinatory States

Reference is made here to a number of diagnostically confusing variants of delirium tremens in which one facet of the delirium tremens complex assumes prominence to the practical exclusion of other symptoms. Although referred to as "atypical," they occur about as frequently as classic delirium tremens (see Table 14.2). In some patients there is a transient state of quiet confusion, agitation, or peculiar behavior, or the patient may become disturbed to the point of violence. In another group, a vivid delusional state dominates the clinical picture. The patient may relate a loosely connected and fantastic tale, such as fighting pitched battles, participating in the Indian wars or bank robberies, or the like; moreover, the patient may not be completely disoriented and later may be able to recall some of the abnormal ideas. Or several of these symptoms may be combined so that the condition is difficult to distinguish from delirium tremens. These confusional-hallucinatory-delusional states are like typical delirium tremens in that a similar interval of time separates their onset from the cessation of drinking, and the duration of both illnesses is much the same. Unlike typical delirium tremens, the atypical states always present as single circumscribed episodes, are only rarely preceded by seizures, and do not end fatally. This may simply indicate that the atypical state is a partial or less severe form of the disease.

The pathology of delirium tremens has been the subject of numerous writings, most of them highly uncritical. Cerebral edema or "wet brain," pial congestion and opacity of the meninges, and diffuse pyknosis and "acute swelling" of cerebral neurons are the abnormalities that have been described and quoted repeatedly (Courville, 1955; Marchand, 1932). Pial congestion and meningeal opacity are ubiquitous neuropathological findings and by no means characteristic of delirium tremens, and the aforementioned neuronal changes are probably artifactual. Pathological examination of our own extensive material has been unrevealing, as one might expect of a disease that is essentially reversible. Edema and brain swelling were absent except to a mild degree in cases in which shock or anoxia had occurred terminally, and there were no significant microscopic changes in the brain. In fact, the brains of alcoholics are somewhat reduced in weight (Harper, 1983).

14.2.2.7. Pathogenesis of Delirium Tremens and Related Disorders

The one indispensable factor in the genesis of the tremulous-hallucinatory-convulsive-delirious states is the relative or absolute withdrawal of alcohol following a period of chronic intoxication. This idea is far from new, but only relatively recently has it received general acceptance. In 1848, Magnus Huss (quoted by Jellinek, 1943) drew a distinction between convulsions in epileptics who drank and convulsions brought about by the cessation of drinking, a relationship that was also recognized by other writers in the last century (Bratz, 1899; Féré, 1896). Bonhoeffer (1901) stated that the sudden withdrawal of alcohol was an important factor in the precipitation of delirium tremens, a concept subsequently reaffirmed by Hare (1915), Kalinowsky (1942), Lambert (1934), and Osler (1917). Despite these authoritative statements, this point of view failed to prevail. In the years that followed, statements denying the causative role of alcohol withdrawal came to carry as much weight as statements affirming it (Jelliffe and White, 1929). One notion in particular gained credence in the 1930s, namely, that "abstinence is in itself an expression of the beginning of delirium" (Bumke and Kant, 1936). In the United States this view was elaborated by Noyes (1939) and by Bowman et al. (1939). The latter authors regarded "disgust for alcohol" as well as nausea and vomiting, consequent to gastritis and hepatitis, as the initial symptoms of delirium tremens rather than as possible precipitating factors. (Conveniently overlooked was the development of delirium tremens and kindred disorders in other settings—following the abrupt imposition of abstinence by injury, infection, admission to the hospital for surgery, etc.) Some authors rejected the withdrawal theory on the grounds that only a small proportion of alcoholics developed delirium tremens after being admitted to jail or a hospital (Bostock, 1939; Bowman et al., 1939; Piker, 1937). Evidence such as this prompted Bowman and Jellinek to write, in 1941, that the alcohol withdrawal theory had been virtually discarded in this country. In Europe, also, the theory was relegated to obsolescence, judging from the writings of Bleuler (1951) and of Bumke and Kant (1936).

The idea that delirium tremens and related disorders simply represent the extreme effects of alcoholic intoxication is untenable for several reasons. The symptoms of toxicity (slurred speech, staggering, drowsiness, stupor, etc.) are in themselves distinctive and quite different from the symptom complex of tremor, hallucinosis, fits, and delirium. The former symptoms are invariably associated with an elevated blood alcohol concentration, and the latter with a reduction in blood alcohol from a previously higher level (Isbell et al., 1955). It should be emphasized that blood concentrations need not be reduced to zero in order to precipitate withdrawal symptoms; the latter may become manifest when the concentrations have fallen to 50 to 100 mg/dl (Freund, 1969; Mendelson and LaDou, 1964). Finally, the symptoms of intoxication worsen with the continued ingestion of alcohol (the drowsy patient who continues to drink becomes stuporous and comatose); conversely, the administration of alcohol may nullify symptoms such as tremor and hallucinations, and even delirium tremens can be suppressed if the patient is given a pint or more of whiskey daily (Victor and Adams, 1961).

Our observations, which strongly support the withdrawal theory, are summarized in Fig. 14.2. Represented in this figure are a group of patients with alcoholic tremulousness, seizures, hallucinosis, and delirium tremens, in whom the time of onset of symptoms could be determined. Since the number of patients in each group was unequal, the number in each group was raised to 100, and the onset of symptoms, in percentage of the group, was plotted against the time that drinking ceased. The period of drinking in this figure is greatly preshortened; the notching of the base line represents the recurring morning nausea and tremor, which were controlled by a few drinks. These observations, made on open medical wards, were corroborated by Isbell et al. (1955) and later by Mendelson and LaDou (1964), who studied the effect of alcoholic intoxication and withdrawal in human volunteers under controlled conditions.

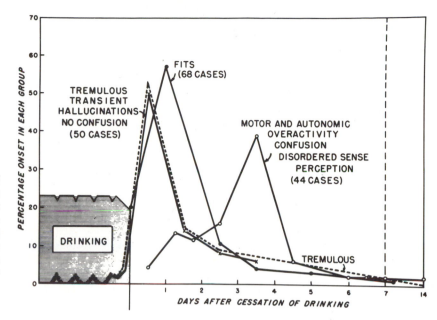

FIGURE 14.2. Relationship of acute neurological disturbances to cessation of drinking. The section marked "drinking" is greatly preshortened. The periodic notching of the base line represents the nausea and tremor that follows a night's sleep and can be controlled by a few drinks. (From Victor and Adams, 1953.)

Finally, unmistakable alcohol withdrawal phenomena have been produced in mice (McQuarrie and Fingl, 1958; Freund, 1969; Goldstein, 1972; Griffiths et al., 1973); in dogs (Essig and Lam, 1968); in rats (French and Morris, 1972; Lieber and DeCarli, 1973), and in rhesus monkeys (Ellis and Pick, 1969). The inhalation model, developed by Goldstein (1972), made it possible to examine the effects of alcohol dosage and duration of alcohol exposure on both the severity of withdrawal signs and the persistence of physical dependence. Various animal models of alcoholism that fulfill the pharmacological criteria of tolerance and physical dependence have been reviewed by Mello (1976) and by Mendelson and Mello (1978).

It is noteworthy that some chronic alcoholics do not develop an overt withdrawal syndrome on cessation of drinking. The reasons for this are unclear. It has been suggested that the cessation of drinking, though a necessary condition for the initiation of delirium and related disorders, may not be a sufficient condition and that other factors (nutritional and metabolic), in addition to the cessation of drinking, constitute the necessary complex of events that culminate in the withdrawal syndrome (Mendelson and Mello, 1978). Nutritional deficiency is probably not an essential factor in the genesis of withdrawal symptoms insofar as the latter may occur in patients who are adequately nourished and may subside uneventfully in patients denied all foods and vita-

mins (Isbell et al., 1955; Mendelson and LaDou, 1964; Victor and Adams, 1961). The claim that exhaustion of the adrenal cortex is a factor in the genesis of delirium tremens (Smith, 1950) has not been substantiated (Czaja and Kalant, 1961; Krusius et al., 1958).

The activation of the sympathetic nervous system on withdrawal of alcohol is reflected in an increase of urinary and plasma catechol levels of a magnitude that is observed after moderate physical exercise (Carlsson and Haggendal, 1967). Elevated concentrations of spinal fluid norepinephrine and its major metabolite (MHPG) have also been demonstrated in the alcohol withdrawal period (Hawley et al., 1985).

Low serum concentrations of potassium, sodium, and calcium are observed frequently but not consistently in patients with chronic alcoholism, during both intoxication and withdrawal. The presence of these abnormalities may simply reflect the poor nutritional state of many alcoholics. On the other hand, the manifestations of alcohol withdrawal are regularly associated with low serum magnesium levels (Mendelson et al., 1959; Randall et al., 1959; Suter and Klingman, 1955; Wacker and Vallee, 1958). Moreover, a depression of serum magnesium in both animals and humans may be associated with neuromuscular irritability and convulsions (Klingman et al., 1955; MacIntyre, 1967; Vallee et al., 1960), and some of these phenomena can be reversed by the administration of magnesium (Hirschfelder and Haury,

1934; Vallee *et al.*, 1960). Also, there is evidence that chronic alcoholism may induce true magnesium deficiency (see review of Flink, 1986).

Wolfe and Victor (1969) demonstrated that the extent to which a patient becomes hypomagnesemic during alcohol withdrawal correlates with a vulnerability to seizures and a heightened sensitivity to photic stimulation; moreover, the intravenous administration of magnesium sulfate during this period decreased the susceptibility to seizures. The concentration of serum magnesium may return to normal spontaneously by the time the patient develops delirium tremens, indicating that hypomagnesemia cannot be incriminated in the genesis of the latter syndrome (Wacker and Vallee, 1958; Wolfe and Victor, 1969).

Probably of greater significance than hypomagnesemia is the rise in arterial pH that accompanies the withdrawal state (Sereny *et al.*, 1966; Wolfe and Victor, 1969). The relationships among hypomagnesemia, rise in arterial pH, and increased sensitivity of the patient to seizures are illustrated in Fig. 14.3.

The alkalosis that characterizes the withdrawal state has proved to be respiratory in type, the result of tachypnea and increased depth of respiration (Wolfe *et al.*, 1969). In two groups of human volunteers, one of which was subjected to sustained alcoholic intoxication for 14 days (five subjects) and the other for 60 days (four subjects), it was observed that elevation of arterial pH, lowering of PCO_2, and abstinence symptoms (sweating, tremor, hyperreflexia, and convulsions) began as early as 8 hr after the cessation of drinking and reached a peak intensity between 15 and 30 hr (Wolfe *et al.*, 1969). These data, which are illustrated in Fig. 14.4, were confirmed in a study of 31 alcoholic subjects on an open neurological ward (Wolfe and Victor, 1971; Victor, 1973). On the basis of these findings we have suggested that the compounded effects of acute hypomagnesemia and respiratory alkalosis, each of which is known to be associated with hyperexcitability of the nervous system, are responsible for alcohol withdrawal symptoms.

14.2.3. Treatment of the Withdrawal Syndrome

The following remarks are directed to the management of delirium tremens, but the same principles are

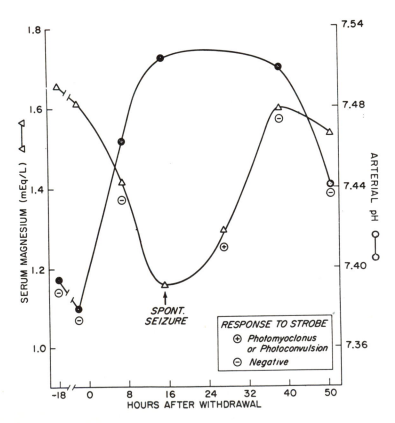

FIGURE 14.3. Relationship between hypomagnesemia, rise in arterial pH, and photic sensitivity in the alcohol withdrawal period. (From Wolfe and Victor, 1969.)

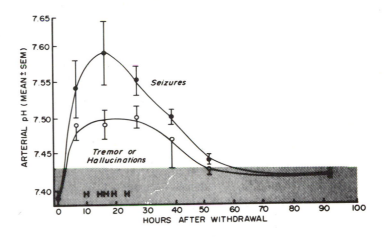

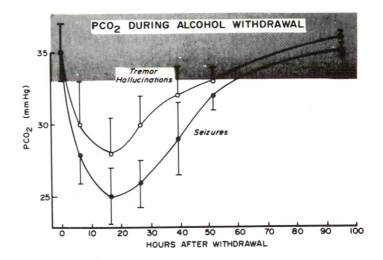

FIGURE 14.4. Rise in arterial pH and drop in PCO_2 during alcohol withdrawal and their relationship to withdrawal symptoms. (From Victor, 1973.)

applicable to the less serious forms of the withdrawal syndrome.

Treatment begins with a careful search for an associated injury or infection, particularly cerebral laceration, subdural hematoma, pneumonia, or meningitis. To this end, chest films, lumbar puncture, and CT scans or MRI should be obtained routinely, and liver function should be assessed. The noisy and agitated patient should be placed apart from other patients; if an attendant or family can be with the patient, it is preferable to allow the patient to move about the room rather than to tie him in bed. Such arrangements are rarely possible in

a general hospital, in which case mechanical restraints are necessary. The most important aspects of treatment are the administration of fluids and the correction of electrolyte depletion. Severe diaphoresis requires the administration of large amounts of fluid (up to 10 liters daily), of which about one-quarter should be normal saline. The addition of other electrolytes is governed by the laboratory findings. The correction of low sodium concentrations must be carried out with great caution, lest a central pontine myelinolysis be induced (see Section 14.8).

Occasionally, the abuse of alcohol induces a severe

degree of hypoglycemia (Field *et al.*, 1963; Freinkel *et al.*, 1963). This complication requires a background of fasting or starvation of at least several days' duration, with severe depletion of liver glycogen. In this setting the ingestion of alcohol may inhibit gluconeogenesis and the output of hepatic glucose. Such patients require parenterally administered glucose in large quantities, always supplemented by B vitamins.

Alcohol-associated ketoacidosis is an uncommon occurrence in severe chronic alcoholics, in whom gastritis or pancreatitis leads to severe vomiting and inability to take food or alcohol. During the drinking period, high blood alcohol levels actually suppress ketogenesis; with abstinence, the block in ketogenesis is removed, and the liver is flooded with fatty acids, the conversion of which leads to production of ketone bodies. The ketoacidosis is usually corrected by instituting the general measures described above.

In patients with delirium tremens, the temperature, pulse, and blood pressure should be recorded at frequent intervals in anticipation of peripheral circulatory collapse or hyperthermia, the two most common lethal complications of this disease. Should shock supervene, one must act quickly, utilizing whole-blood transfusion, fluids, and vasopressor drugs. The occurrence of hyperthermia demands the use of a cooling mattress in addition to specific treatment for any infection that may be present.

Also, one must not neglect the many small measures that allay the patient's fear and suspicion and reduce the tendency to hallucinate. The room should be kept lighted, and the patient moved as little as possible. Every procedure should be explained in detail, and presence of a member of the family may help the patient maintain contact with reality.

The use of drugs for the treatment of delirium tremens and related disorders has been the subject of a vast literature (see reviews of Victor, 1966; Gessner, 1979). Extravagant claims, unsupported by meaningful data, have been made for innumerable drugs. Many authors have failed to define what is being treated and to distinguish between the common, early (minor) withdrawal symptoms, from which patients always recover and which are responsive to practically all sedative drugs, and the late and potentially lethal delirium tremens, which is relatively unresponsive to drugs.

Practically all sedative drugs are effective in controlling the minor withdrawal symptoms. Of these, chlordiazepoxide is the most popular. Shorter-acting benzodiazepines, such as oxazepam and chlorazepam, are probably more suitable in patients with liver disease.

Propranolol, an effective agent in the treatment of essential tremor, may also reduce the tremor of alcohol withdrawal (Zilm *et al.*, 1975). Other autonomic blocking agents that effectively suppress withdrawal symptoms are lofexidine, an α_2 agonist that blocks autonomic outflow centrally (Cushman *et al.*, 1985), and the β blocker atenolol (Kraus *et al.*, 1985). The tricyclic anticonvulsant carbamazepine may also be useful in this respect (Butler and Messiha, 1986).

It is questionable whether any of the aforementioned drugs, with the possible exception of chlordiazepoxide (Kaim *et al.*, 1969; Sellers *et al.*, 1983) and chlormethiazole, a drug that is cross-tolerant with ethanol (Palsson, 1986), are effective in preventing delirium tremens. In general, the phenothiazines should be avoided in the treatment of withdrawal symptoms because of their epileptogenic properties. Moreover, the therapeutic advantages of these drugs over paraldehyde have not been proved by controlled studies (Gessner, 1979). The main advantage of paraldehyde is its safety, and this can be assured by using ampules of the drug and administering it by mouth, in doses of 8 to 12 ml in orange juice, or rectally, diluted in water. If parenteral medication is necessary, then diazepam, 10 mg intravenously, may be given and repeated once or twice at 20- to 30-min intervals until the patient is calm but awake; or phenobarbital sodium or amobarbital sodium, in doses of 120 mg, or haloperidol (10 mg) may be given intramuscularly, provided that there is no serious liver disease.

It needs to be emphasized that the object of drug therapy in delirium tremens is not the absolute suppression of agitation and tremor; to accomplish this requires amounts of drug that could seriously depress respiration. One's aim is merely to blunt the agitation and prevent exhaustion and to facilitate the administration of parenteral fluids and nursing care. In the milder forms of the withdrawal syndrome, drugs are useful in assuring rest and sleep.

Corticosteroids have no place in the treatment of the withdrawal syndrome. Certain adverse effects of these hormones, namely, the masking of infection, the deleterious effect on tuberculosis and peptic ulcer, and the tendency to produce a negative nitrogen balance and excessive excretion of potassium, pose special dangers to the alcoholic patient.

Finally, it needs to be reemphasized that most alcoholics who develop withdrawal symptoms are also nutritionally depleted. It is good practice, particularly *if the patient is receiving glucose intravenously, to administer B vitamins*, even though withdrawal symptoms are not primarily caused by vitamin deficiency.

14.2.4. Treatment of Withdrawal Seizures

In most cases, withdrawal seizures ("rum fits") do not require the use of anticonvulsant drugs. Characteristically, as the patient withdraws from alcohol, there is only a single seizure or a brief run of seizures, which often have ceased by the time the patient is seen by a physician or by the time certain medicines, such as phenytoin, become effective. The parenteral administration of phenobarbital or the use of oral loading doses of chlordiazepoxide or diazepam (Sellers *et al.*, 1983; Devenyi and Harrison, 1985) in the early stages of the withdrawal period may prevent "rum fits," particularly in patients with a previous history of this disorder. The administration of phenytoin is not effective in these circumstances (Alldredge and Simon, 1990). Of course, in patients who have been taking phenytoin regularly, the drug should not be discontinued abruptly. In about 3% of cases, withdrawal seizures take the form of status epilepticus, and this should be treated like status in nonalcoholics.

The prescription of anticonvulsant medicines for alcoholics who have recovered from withdrawal seizures, with the idea of preventing future seizures, is impractical; if the patient remains abstinent, he will be free of seizures, and if he resumes drinking, he usually abandons his medications. Of course, this does not apply to alcoholics with idiopathic or posttraumatic epilepsy, who require continued treatment with anticonvulsants in the interdrinking period.

14.3. Nutritional Diseases of the Nervous System Secondary to Alcoholism

This category comprises the Wernicke-Korsakoff syndrome, "alcoholic" cerebellar degeneration, peripheral neuropathy, optic neuropathy, and pellagra. In the genesis of these disorders, the role of alcohol is secondary. Alcohol acts mainly by displacing food in the diet but also by increasing the demand for B vitamins, particularly thiamine, which are necessary to metabolize the carbohydrate furnished by alcohol itself. Also, there is evidence that alcoholics have a decreased capacity to absorb thiamine and folic acid, among other nutrients (Halstead *et al.*, 1971, 1973; Lindenbaum and Lieber, 1969, 1975; Thomson *et al.*, 1968; Tomasulo *et al.*, 1968), and to digest fat because of impaired pancreatic function (Halstead *et al.*, 1973; Mezey *et al.*, 1970). The disorders of absorption and digestion are probably caused by disturbances of nutrition rather than by the

action of alcohol itself and, once established, serve to exaggerate the deficiencies that result from dietary deprivation. Hoyumpa (1986) has discussed other mechanisms that may be involved in the absorption (as well as storage and metabolism) of vitamins in the alcoholic.

In economically developed countries, nutritional diseases of the nervous system are observed almost exclusively among alcoholics; such an association is less frequent in undeveloped countries, but significant nonetheless. The incidence of alcoholic nutritional diseases cannot be stated precisely. We found signs of nutritional disease in only 3% of the alcoholics at the Boston City Hospital (Table 14.2). Nevertheless, because of the prevalence of alcoholism, the number of alcohol-induced deficiency diseases is far from insignificant. Notable also is the high incidence of mortality and morbidity in patients with these diseases. Although chronic alcoholism is three to six times more common in men than in women (Keller and Gurioli, 1976), the incidence of deficiency diseases is almost equal in the two sexes (Victor and Laureno, 1978).

In general, the alcoholic and dietary habits of patients with nutritional disease are to be distinguished from those of the spree drinker. The latter drinks excessively for a period of days or weeks, during which time he eats little; between sprees, however, he rapidly replenishes his nutritional state. In the patient with nutritional disease, the period of drinking and undernutrition is sustained for months and often for years.

The general effects of malnutrition—muscular wasting, mucocutaneous lesions, circulatory abnormalities, scorbutic lesions, and anemia—often accompany the neurological abnormalities. All of the nutritional disorders that occur in the alcoholic population have been encountered under conditions in which alcohol plays no part, e.g., in prisoner-of-war camps and communities in which polished rice is the staple of the diet.

14.3.1. The Wernicke–Korsakoff Syndrome

14.3.1.1. Definitions

Wernicke encephalopathy is a disease of acute or subacute onset, characterized by nystagmus, abducens and conjugate gaze palsies, ataxia of gait, and mental disturbance (global confusion, apathy, drowsiness). This triad of clinical signs was first described in 1881 by Carl Wernicke, who named the disease *polioencephalitis hemorrhagica superioris*. It is now known to be caused by nutritional deficiency, more specifically by a deficiency of thiamine, and is observed predominantly though not exclusively in alcoholics.

Korsakoff psychosis (amnestic or amnesic-confabulatory syndrome; psychosis polyneuritica) refers to a unique mental disorder in which retentive memory is impaired out of proportion to other cognitive functions in an otherwise alert and responsive patient. The first comprehensive account of this disorder was given by the Russian psychiatrist S. S. Korsakoff in a series of articles published from 1887 to 1891 (see Victor and Yakovlev, 1955, for a historical review). This disorder, like Wernicke disease, is most often associated with alcoholism and malnutrition, but it may be a symptom of various other diseases that have their basis in lesions of the diencephalon or the inferomedial portions of the temporal lobes.

In the alcoholic, nutritionally deficient patient, Korsakoff psychosis is usually associated with Wernicke disease. Stated in another way, Korsakoff psychosis is the psychic manifestation of Wernicke disease. For this reason, and others to be elaborated further on, this symptom complex should be called the Wernicke–Korsakoff syndrome if both the Wernicke and Korsakoff components are evident and Wernicke disease if only the ocular-ataxic-confusional symptoms are present and the amnesic disorder is lacking.

The Wernicke–Korsakoff syndrome is not uncommon. It was found in 2.2% of 3548 consecutive adult autopsies (Victor and Laureno, 1978), and about the same prevalence has been recorded by Harper (1983).

The following account of the clinical and pathological features is based largely on our own writings on this disease (Victor *et al.*, 1989).

14.3.1.2. Clinical Features

The triad of clinical abnormalities originally described by Wernicke—ophthalmoplegia, ataxia, and disturbances of consciousness and mentation—are still diagnostically useful. The disease may present with any one of these abnormalities, but more often they occur in combination.

Disturbances of consciousness and mentation are present in practically all cases. Three fairly distinct derangements of function can be recognized. Most often the patient presents with a characteristic disorder of behavior and mental function that we have referred to as a "global confusional state." The latter is compounded of several elements: apathy and an incapacity to sustain physical or mental activity; impairment of awareness and responsiveness; disorientation, inattention, and failure of concentration; derangements of perceptual function and memory; and lethargy and drowsiness progressing to stupor and coma if the disorder remains unrecognized and untreated.

Some patients, at the time of admission to the hospital, show the signs of alcohol withdrawal. These may vary from a mild state of agitation, tremor, and confusion to a full-blown delirium tremens.

The unique disorder of retentive memory, Korsakoff psychosis, may already be evident at the time of admission to the hospital. More often, the memory disorder is obscured by the profound general confusion, and only later, as this subsides, does the Korsakoff psychosis stand out as the most prominent feature of the illness.

Two essential abnormalities characterize the Korsakoff amnesic state: (1) an impaired ability to recall information that had been acquired over a period of years before the onset of the illness (retrograde amnesia) and (2) an impaired ability to make new memories, i.e., to acquire new information or to learn (anterograde amnesia). Confabulation, an ill-defined symptom that has come to be regarded as a specific feature of Korsakoff psychosis, is neither consistently present nor a requisite for diagnosis. Other cognitive functions, particularly those of attention and concentration, and certain perceptual and conceptual functions are also impaired, but to a relatively minor degree. As a rule, patients have only limited insight into their disability and tend to be apathetic, inert, and indifferent to persons and events around them.

Ocular abnormalities are a consistent clinical manifestation of the illness. The most common ocular abnormality is nystagmus (both horizontal and vertical), to which may be added abducens palsies and palsies of conjugate gaze. The abducens palsy is always bilateral but need not be symmetrical. In approximately half the patients, there is evidence of internuclear ophthalmoplegia or other palsies of conjugate gaze—most frequently of horizontal and upward gaze and sometimes of downward gaze. In patients who present with complete ophthalmoplegia, nystagmus may be absent but becomes evident after treatment has been instituted and some degree of recovery of ocular movement has occurred. Such is the sensitivity of the ocular palsies to thiamine that even a meal or two, prior to admission to the hospital, may ablate them or leave only subtle evidence of their presence, most often in the form of a gaze-evoked horizontal nystagmus.

Pupillary abnormalities are uncommon presenting signs; in patients with severe ophthalmoplegia, the pupils may be small and nonreactive. Funduscopic examination occasionally discloses small retinal hemorrhages, but we have never observed papilledema in this disease. Pallor of the optic disks, also a rare finding, is usually attributable to an associated optic neuropathy. So-called ocular bobbing has been reported in a patient with Wernicke disease (Luda, 1980).

Ataxia is the third major presenting symptom. In its most severe form the patient literally cannot stand or walk without support. Less severe forms are characterized by a wide-based stance and a slow, short-stepped, unsteady gait. The mildest degrees are disclosed only on tandem walking. By contrast, individual movements of the limbs are relatively little affected. An ataxic tremor of cerebellar type, brought out by heel-to-knee-to-shin testing, occurs in a small proportion of patients (20% in our series), and intention tremor of the arms is even less frequent.

Signs of *peripheral neuropathy*, usually mild in degree, are found in more than 80% of cases with the Wernicke–Korsakoff syndrome (Victor *et al.*, 1989). In the acute stage of the disease, vestibular function, as measured by standard caloric tests, is usually impaired; this has been referred to as vestibular paresis (Ghez, 1969). An isolated case of bilateral facial palsy of lower motor neuron type has been reported, presumably from involvement of the facial nerve nuclei or their intramedullary fibers (Rice *et al.*, 1984).

A number of systemic abnormalities are common in patients with Wernicke–Korsakoff syndrome—tachycardia (> 100/min), redness and papillary atrophy of the tongue, cheilosis, angular stomatitis, and abnormalities of the skin. *Impairment of liver function* was present in two-thirds of our patients, and in 17% there was clinical evidence of cirrhosis—jaundice, ascites, spider angiomata, and the signs of portal-systemic shunting. *Hypothermia* is a relatively uncommon but well-recognized finding in the acute stage of the disease.

Although the overt signs of beriberi heart disease are rare, disorders of cardiovascular function such as tachycardia, exertional dyspnea, postural hypotension, and minor ECG abnormalities are frequent (Gravallese and Victor, 1957). Cardiac output, measured by the Evans blue dye-dilution technique, is elevated out of proportion to oxygen consumption, in association with low peripheral vascular resistance; these abnormalities can be reversed by treatment with thiamine (Gravallese and Victor, 1957). *Postural hypotension* is probably due to impaired function of the autonomic nervous system, more specifically a defect in the sympathetic outflow (Birchfield, 1984).

14.3.1.3. Ancillary Findings

In the initial phase of the illness, about half the patients show electroencephalographic (EEG) abnormalities consisting of a diffuse slowing, mild to moderate in degree. More remarkable is the fact that an equal number show no EEG abnormalities at all. Despite the localization of lesions to discrete portions of the diencephalon and brainstem, total cerebral blood flow and cerebral oxygen and glucose consumption may be greatly reduced in the acute stages of the disease, and these defects may still be present after several weeks of treatment (Shimojyo *et al.*, 1967). These observations make the point that profound reductions in brain metabolism need not be reflected in EEG abnormalities or in depression of the state of consciousness and that the latter is a function of the location of the lesion, not the degree of the overall metabolic defect.

Hemorrhagic lesions of the diencephalon and brainstem (which are found in about 10% of cases of Wernicke–Korsakoff disease) are readily disclosed by CT scans (Escobar *et al.*, 1983; McDowell and LeBlanc, 1984; Roche *et al.*, 1988). The CT scans may also disclose the nonhemorrhagic lesions of this diseases (Mensing *et al.*, 1984), but MRI is far more sensitive in this respect (Gallucci *et al.*, 1990; Donnal *et al.*, 1990; Squire *et al.*, 1990). Undoubtedly MRI will prove to be useful in diagnosis, particularly when eye signs are absent or stupor and coma have supervened (Victor, 1991).

The cerebrospinal fluid (CSF) is normal or shows only a modest elevation of protein content. Protein values above 100 mg/dl or a pleocytosis should suggest the presence of a subdural hematoma, meningitis, or some other complicating illness.

The blood pyruvate levels are consistently elevated in untreated cases of Wernicke disease, but this abnormality is insufficiently specific to be of practical value. The concentration of blood transketolase, an enzyme of the hexose monophosphate shunt that requires thiamine pyrophosphate (TPP) as a cofactor, more accurately estimates the state of thiamine nutrition. Before treatment with thiamine, patients with Wernicke disease show a marked reduction in their transketolase activity (as low as one-third of normal values) and a striking TPP effect (up to 50%). Restoration of these values toward normal occurs within a few hours of the administration of thiamine, and completely normal values are usually attained within 24 hr (Dreyfus, 1962).

14.3.2. Course and Prognosis

The manifestations of the Wernicke–Korsakoff syndrome respond to specific treatment, i.e., the administration of thiamine, in a fairly uniform manner. The most predictable effects are on the ocular abnormalities, which often begin to recover within hours after the administration of thiamine and practically always within a day or two. Recovery is usually complete within several weeks, but a vertical nystagmus may persist for a month or longer, and, in about 60% of cases, a fine horizontal nystagmus remains as a permanent stigma of the disease. In respect to its persistence, horizontal nystagmus is unique among the ocular signs.

Improvement of ataxia is somewhat delayed, and recovery tends to be incomplete. More than half of the patients are left with a somewhat slow, shuffling, wide-based gait and an inability to walk tandem. The residual disturbances of gait, like horizontal nystagmus, help to identify obscure and chronic cases of dementia as alcoholic-nutritional in origin. Vestibular function recovers at about the same rate as the ataxia of stance and gait, and recovery is usually complete.

In about 15% of patients the symptoms of apathy, drowsiness, and global confusion subside completely in a week or two. In the remainder, the symptoms recede more slowly, and as they do, the defects in memory and learning (Korsakoff psychosis) stand out more clearly. Once the amnesic state is established, complete or almost complete recovery occurs in only a small proportion of patients (somewhat less than 20% in our series). In an equal number of patients, no meaningful improvement in memory or learning function can be discerned. The majority of patients fall between these two extremes: some improve only slightly; others are eventually able to carry out routine household or institutional tasks under supervision.

It is apparent from the foregoing account that Wernicke encephalopathy and Korsakoff psychosis are not separate diseases but that the changing ocular and ataxic signs and the transformation from a global confusional state to an amnesic syndrome are simply successive stages in the process of recovery of a single disease. Of 186 patients in our series who presented with Wernicke disease and survived the acute illness, 157 (84%) showed this sequence of recovery.

The mortality rate among patients with the Wer-nicke–Korsakoff syndrome is high—20% in the early stages of the disease and 17% in the chronic stages (Table 14.4).

14.3.3. Pathological Changes

14.3.3.1. Causes of Death

The most common causes of death in the early stages of the disease were decompensated liver disease and some form of infection—pneumonia, pulmonary tuberculosis, pancreatitis, bacterial meningitis, and septicemia, in that order of frequency. Myocardial infarction, carcinoma, intercurrent infection, and liver failure were the major causes of death in the chronic stages. Cirrhosis of the Laennec type was found in 36 of 81 of our autopsied patients; many of these patients died in hepatic coma. In the acute group, 11 patients showed only fatty metamorphosis of liver, and three showed hepatitis. Noteworthy also is the fact that the liver was entirely normal in 31 patients.

A mild to moderate degree of sulcal widening in the frontal regions and dilatation of the lateral and third ventricles were observed in approximately one-fourth of the cases in our series; histologically, however, there were no changes to account for the shrinkage of the frontal cortex. Enlargement of the third ventricle may be associated with symmetrical crescent-shaped zones of degeneration, gray-brown in color, occupying the paraventricular zones but sparing the immediate subependymal tissue (Fig. 14.5).

The mammillary bodies were affected in virtually every case (Fig. 14.6). Symmetrically distributed le-

TABLE 14.4. General Pathological Findings in 81 Cases of Wernicke–Korsakoff Syndrome[a]

Complication	Group 1 (N = 44)[b]	Group 2 (N = 37)[c]
Infection	34 (77%)	15 (41%)
Cirrhosis	23 (52%)	13 (35%)
Carcinoma	2 (5%)	10 (27%)
Pulmonary embolism	3 (7%)	5 (14%)
Acute myocardial infarction	0	2 (5%)
Pulmonary edema	2 (5%)	3 (8%)
Delirium tremens (clinical)	3 (7%)	0
Miscellaneous	2 (5%)	3 (8%)

[a]From Victor et al. (1989).
[b]Patients in whom the average time of death was 8 days after the onset of the Wernicke–Korsakoff syndrome.
[c]Patients in whom the average time of death was 3.2 years after the onset.

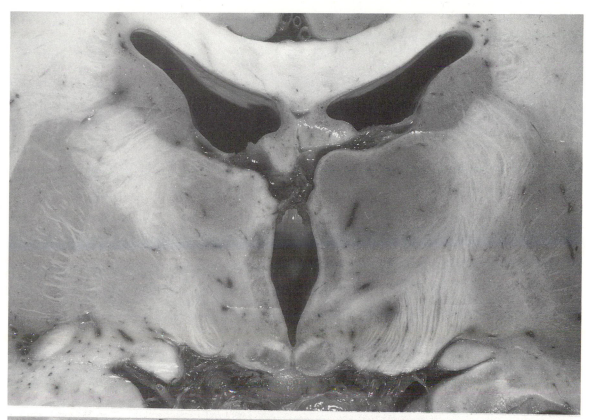

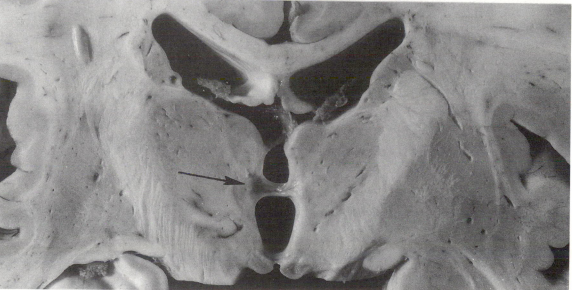

FIGURE 14.5. Wernicke–Korsakoff syndrome: coronal sections of formalin-fixed brain. Above: Symmetrical bands of necrosis in the walls of the third ventricle and in the center of the mammillary bodies. The patient died 10 days after the acute onset of Wernicke disease. Below: Similar lesions in a patient who died in the chronic (amnesic) stage of the disease. Arrow points to a lesion in the massa intermedia. In both cases, the third and lateral ventricles are moderately dilated. (From Victor *et al.*, 1989.)

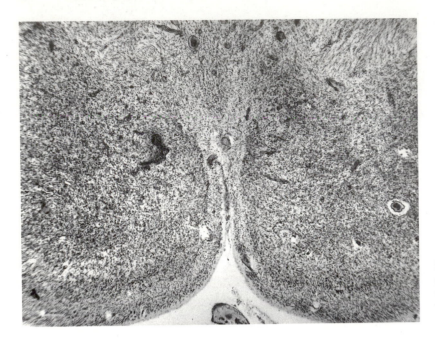

FIGURE 14.6. Wernicke–Korsakoff syndrome. Nissl-stained section of the mammillary bodies showing severe loss of neurons and replacement with histocytes and astrocytes.

sions of similar appearance may be observed in the periaqueductal gray matter, inferior and superior colliculi, fornices, and in the floor of the fourth ventricle, in the region of the dorsal motor nuclei of the vagi (Fig. 14.7). Hemorrhages were observed in only a small proportion of cases (10%) in the same general areas as the necrotic lesions. Usually they are scattered asymmetrically and are petechial in size, but occasionally they may be larger, and rarely they are profuse and confluent, outlining the necrotic lesions (Tashiro and Nelson, 1972).

Midsagittal section of the cerebellum frequently disclosed a characteristic abnormality of the vermis consisting of atrophy of some or all of the folia anterior and superior to the primary fissure (Fig. 14.8). The distribution and approximate extent of these lesions are illustrated in Fig. 14.9.

Histologically, the most advanced lesions are characterized by virtually complete necrosis of tissue, although some of the glial elements always survive; hence, cavitation does not result. In somewhat less severe lesions, nerve cells, axis cylinders, and myelin sheaths are variably destroyed, with an appropriate increase in reactive glial cells. Still less severe lesions consist of slight to moderate nerve cell loss with a greater loss of myelinated fibers; this type of change is most often found in the mammillary bodies. In the thalamus, the medial parts of which contain few myeli-

nated fibers, the reverse obtains; that is, nerve cells are lost, and myelin is relatively spared. The blood vessels in and around the lesions are prominent but not definitely increased in number; these effects are probably secondary to the tissue changes, since the latter are observed in many cases without any apparent vascular alterations. In the cerebellar cortex, the most prominent abnormality is a loss of Purkinje cells, accompanied by an increase in the Bergmann glia. Severe lesions are characterized, in addition, by a patchy loss of nerve cells in the granular layer, thinning of the molecular layer, and a proliferation of small astrocytes in the latter zone (Fig. 14.10).

14.3.3.2. Clinical-Anatomic Correlations

Our observations indicate that the lesions of the medial dorsal nuclei of the thalamus (and not those of the mammillary bodies) are the crucial ones with respect to the memory disorder. Five patients in our series showed severe changes in the mammillary bodies despite the fact that they had betrayed no memory defect during life; on the other hand, all of our patients known to have had Korsakoff psychosis during life showed lesions of the medial dorsal nuclei of the thalamus. The failure of the mammillary body lesions to produce a memory defect is supported by a number of experimental studies in animals (Victor *et al.*, 1989).

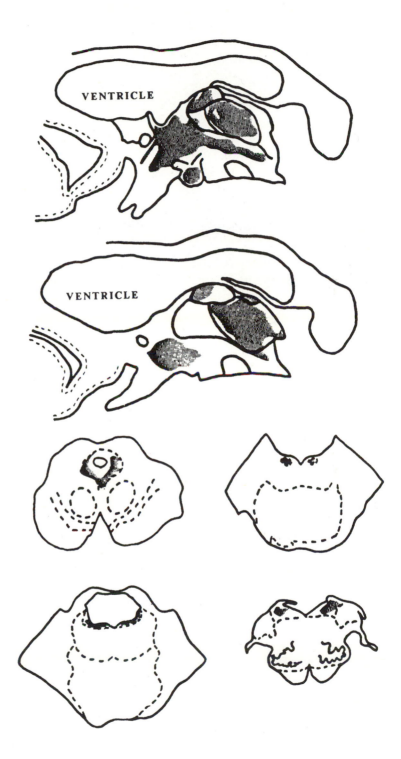

FIGURE 14.7. Topography of the lesions in the Wernicke–Korsakoff syndrome. Above: Diagrams of the lesions in the wall and floor of the third ventricle in the sagittal and parasagittal planes. Shaded areas show the pattern of involvement of the thalamus (mainly medial-dorsal nucleus) and hypothalamus, including the mammillary body. Below: Horizontal sections of the brainstem showing (clockwise, from the upper left) lesions of the periaqueductal gray matter, abducens nuclei, vestibular nuclei, and floor of the fourth ventricle. (From Victor *et al.*, 1971.)

FIGURE 14.8. Alcoholic cerebellar degeneration. Sagittal section through the middle of the vermis, showing the typical atrophy of the anterior-superior lobules. (From Victor *et al.*, 1959).

With respect to the ocular signs, our data suggest that vertical nystagmus is related to lesions in the upper pons or midbrain, and horizontal nystagmus to lesions of the medial vestibular nuclei. Lateral gaze palsy results most likely from lesions near the paraabducens complex, and vertical gaze palsy from the pretectal and periaqueductal lesions. The lateral rectus palsies relate to lesions of the sixth nerve nuclei, and ptosis and pupillary abnormalities to lesions of the third nerve nuclei.

It is likely that the disorder of vestibular function ("vestibular paresis") is responsible for the severe disturbance of equilibrium in the acute phase of Wernicke disease. However, the persistent ataxia of stance and gait is related to the lesion of the superior vermis; ataxia of the legs is associated with extension of the lesion to the anterior parts of the anterior lobes, and ataxia of the arms with a lesion of the more posterior parts of the anterior lobes (Victor *et al.*, 1959).

The lesion responsible for the olfactory deficit in Korsakoff patients is uncertain. It may be in the mammillary nuclei, ventromedial nuclei of the hypothalamus, or the thalamic dorsomedial nuclei, all of which are thought to function as relay nuclei in the central connections mediating olfactory sensation (Mair *et al.*, 1980).

14.3.4. Etiology of the Wernicke–Korsakoff Syndrome

Most if not all of the symptoms of the Wernicke–Korsakoff syndrome result from a deficiency of thiamine (Jolliffe *et al.*, 1941; de Wardener and Lennox, 1947; Phillips *et al.*, 1952; Victor and Adams, 1961). As already indicated, ophthalmoplegia recovers quickly and completely under the influence of this vitamin alone; nystagmus and ataxia recover somewhat more slowly and often incompletely. Symptoms of apathy and drowsiness worsen rapidly if thiamine is withheld and clear rapidly after the administration of this vitamin. Patients who present only with ocular and ataxic signs will develop an apathetic-confusional state if they are denied thiamine and can be protected from developing this state by the administration of thiamine alone (Victor and Adams, 1961). Lesions resembling those of Wer-

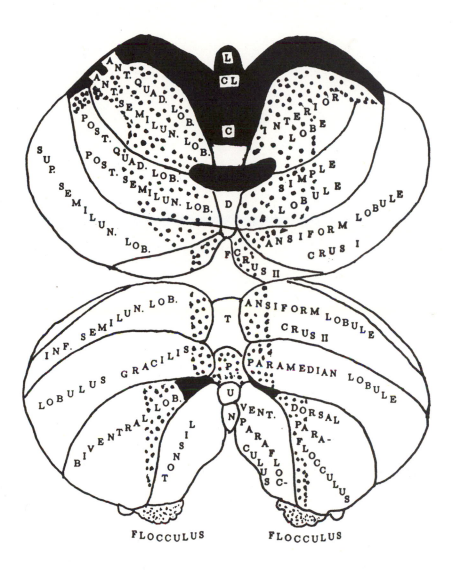

FIGURE 14.9. Alcoholic cerebellar degeneration. Diagram illustrates the topography and graded severity of the cerebellar cortical lesions. Two commonly used terminologies are indicated, one on each side of the diagram. Vermian divisions are indicated by initials. (From Victor *et al.*, 1959.)

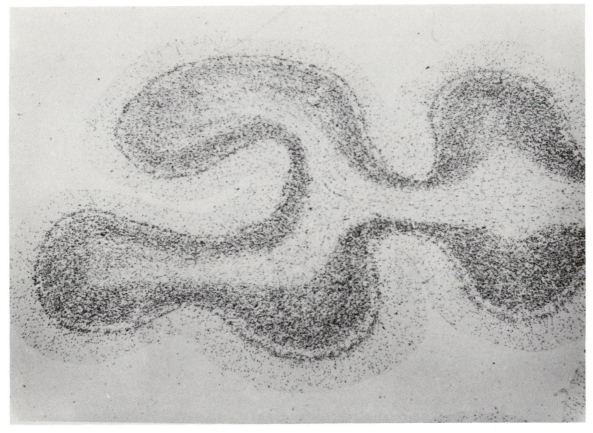

FIGURE 14.10. Alcoholic cerebellar degeneration. Low-power microscopic section demonstrating severe involvement of the vermian (culmen) cortex. The molecular layer is narrow; Purkinje cells are absent, and the Bergmann glia are greatly increased; the granule cell layer is thinned. Gliosis of the molecular layer is apparent. Cresyl violet stain.

nicke disease in humans in both distribution and histo-logical character have been produced in a variety of avian and mammalian species (Dreyfus and Victor, 1961).

The precise mechanisms by which thiamine defi-ciency produces nervous system lesions and the selec-tive vulnerability of certain parts of the nervous system to such a deficiency remain to be clarified. McEntee *et al.* (1984) have pointed out that the paraventricular lesions lie in monoamine-containing pathways. In 25 patients with the Wernicke–Korsakoff syndrome they found that levels of MHPG, the primary brain metabo-lite of norepinephrine, were decreased in the CSF. Also, the administration of clonidine, a putative α-noradren-ergic agonist, seemed to improve the memory disorder

in these patients. These authors suggested that damage to the ascending norepinephrine containing neurons in the brainstem and diencephalon accounts for the amne-sia in the Wernicke–Korsakoff syndrome.

Blass and Gibson (1977) found that transketolase in fibroblasts from four alcoholics with Wernicke–Korsa-koff disease bound TPP less avidly than did the trans-ketolase from control lines. Presumably this defect in transketolase would be insignificant if the diet were adequate but would be deleterious if the diet were low in thiamine. These findings, which have been corroborated by Mukherjee *et al.* (1987), implicate a hereditary factor in the genesis of the Wernicke–Korsakoff syndrome and would explain why only a small proportion of nutri-tionally deficient alcoholics develop this disease.

14.3.5. Treatment

The recognition of the symptoms and signs of the Wernicke–Korsakoff syndrome, or even the suspicion of their presence, demands the immediate administration of thiamine. Although 2–3 mg of this vitamin may sometimes modify the ocular signs, much larger doses are needed to assure sustained recovery and repletion of tissue stores of thiamine. An initial dose of 50 mg of fresh thiamine intravenously plus 50 mg intramuscularly should be given, and the latter dose repeated each day until the patient resumes a normal diet. The prompt use of thiamine in large doses prevents progression of the disease and reverses those lesions, or portions thereof, in which permanent damage has not yet occurred, i.e., the biochemical lesions. In patients who show only ocular and ataxic signs, the prompt administration of thiamine is crucial in preventing the development of an amnesic psychosis. Strict bed rest is mandatory in the acute stage of the illness. Signs of congestive failure call for rapid digitalization as well as specific treatment with thiamine. The further management of the disease involves the use of a balanced diet and all the B vitamins, since the patient is usually deficient in more than thiamine alone.

A particular danger attends the treatment of severely malnourished alcoholics with glucose solutions, namely, precipitation of the Wernicke–Korsakoff syndrome. The occurrence of this syndrome, sometimes with the rapid evolution of stupor, has been observed to follow prolonged intravenous administration of fluids or hyperalimentation (Baughman and Papp, 1976; Novak and Victor, 1974), and the refeeding of prisoners of war (Cruickshank, 1950) or patients who had been on a starvation diet (Drenick et al., 1966). A similar mechanism is probably operative in each of these circumstances: an increment of carbohydrate intake and a decrease in the thiamine-to-calorie ratio.

In respect to long-term management of the chronic amnesic state, disposition of the patient to his family, nursing home, or mental institution is undertaken on the basis of the severity of the Korsakoff psychosis as well as the existing social circumstances, but only after one is certain that no further improvement in memory function is possible.

14.4. Alcoholic Cerebellar Degeneration

14.4.1. Definition

This title designates a stereotyped cerebellar syndrome in alcoholics characterized clinically by an ataxia of gait and of the legs and, to a lesser degree, of the arms, and pathologically by a degeneration of the cerebellar cortex, more or less confined to the anterior portions of the vermis and the anterior lobes (Victor et al., 1959). This complication of alcoholism is relatively common; in a series of 3548 postmortem examinations in adults, the typical lesions were found in 4% of cases, with a male : female ratio of 3.6 : 1 (Victor and Laureno, 1978).

14.4.2. Clinical Features

This disorder occurs on a background of excessive drinking and impaired nutrition of many years' duration; about half of the patients have signs of peripheral neuropathy, usually mild in degree, and a similar proportion have liver disease of varying severity. The history may disclose the occurrence of weight loss at the onset of the ataxia, although signs of deficiency may not be evident by the time the patient is seen by a physician.

On examination, the most prominent abnormalities are those of stance and gait. The feet are placed widely apart, the trunk is tilted slightly forward with the arms held somewhat away from the sides, and the patient walks slowly with small irregular steps and with eyes fixed on the ground. The ataxia of gait is accentuated on turning or making rapid postural adjustments, as when coming to a halt and sitting down. As the disease stabilizes, the steps become more regular, and the gait assumes a more rhythmic quality. On standing there is an instability of the trunk and often a rhythmic tremor of the head as well. The instability of the trunk takes the form of a specific 3-Hz rhythmic swaying in the anteroposterior direction (Mauritz et al., 1979). In most patients, there is also an ataxic tremor of the legs, brought out on heel-to-shin testing. The arms are affected to a lesser extent and sometimes not at all. Nystagmus, dysarthria, and postural tremor are observed infrequently.

In most cases, the cerebellar syndrome evolves over a period of several weeks or months, after which it remains unchanged for many years. In some the onset is more abrupt, and in others the symptoms worsen in relation to an infectious illness or an attack of delirium tremens. Occasionally the cerebellar syndrome appears after a prolonged drinking bout and then resolves completely in a matter of a few days or weeks. Presumably, in such cases, the derangement is purely a functional one ("biochemical lesion") that has not progressed to the point of structural change.

In its mildest form the clinical abnormality is limited to one of station and gait and may be evident only when the patient attempts to walk a narrow line or heel to toe; in such cases, the pathological changes are restricted to the anterior superior portions of the vermis (Victor *et al.*, 1959; Victor and Ferrendelli, 1970). In the more severe forms of the disease, the lesion extends into the anterior parts of the anterior lobes, giving rise to an ataxia of the legs and, less frequently, into the more posterior folia of the anterior lobe (the experimentally demonstrated arm area). The pathological changes and their topography are illustrated in Figs. 14.8 and 14.9.

So-called alcoholic cerebellar degeneration does not differ from the cerebellar manifestations of Wernicke disease, either clinically or pathologically. Probably they represent the same disease process, one term being used when the cerebellar syndrome occurs alone and the other when the cerebellar abnormalities are associated with ocular and mental signs. Most patients come to medical attention long after the onset of cerebellar symptoms and improve only slightly (mainly through recovery of peripheral neuropathy). Acute alcoholic cerebellar degeneration may respond to large doses of thiamine (Graham *et al.*, 1971).

14.5. Alcoholic Neuropathy

14.5.1. Clinical Features

The prevalence of this condition can only be stated in general terms. In a consecutive series of 1030 alcoholics admitted to the Boston City Hospital, 92 (9%) proved, on clinical examination, to have peripheral nerve disease (Victor and Adams, 1953). Men and women were affected in equal numbers, a feature that has been noted repeatedly in writings on this subject.

Two features were common to all of our patients with alcoholic neuropathy: (1) abuse of alcohol, usually severe in degree and of many years' duration, and (2) nutritional deficiency. The diets of these patients consisted mainly of bread, margarine, coffee, sugar, doughnuts, pie, canned soups, beans, and spaghetti and were conspicuously low in meat and fish, cereals, fresh fruit, and vegetables. A loss of 30 to 40 lb from optimum weight was recorded in more than half of our patients. Some beer drinkers were actually obese, but their dietary intake was unbalanced and inadequate. These observations are in close agreement with those of Wechsler (1933), Minot *et al.* (1933), Jolliffe *et al.* (1936), and Wortis *et al.* (1938), who originally documented the dietary inadequacies of alcoholics with neuropathy and recognized this disorder for what it was,

neuropathic beriberi (for full discussion, see Victor, 1984).

The symptoms and signs of alcoholic neuropathy are quite diverse. In its mildest form, the neuropathy is virtually asymptomatic and is detected only on neurological examination, which discloses thinness and tenderness of the leg muscles, loss or depression of the Achilles reflexes and occasionally of the patellar reflexes as well, and an inconstant impairment of superficial sensation over the feet and shins. If alcoholic patients are examined with physiological techniques (conduction velocities in sensory and motor fibers, latencies of the H reflex), a significant proportion show impairment of function before any clinical symptoms of neuropathy have developed (Blackstock *et al.*, 1972; Mayer, 1966).

The majority of alcoholics with peripheral neuropathy are symptomatic, however. Weakness, paresthesias, and pain, usually insidious in onset and slowly progressive, are the common complaints. Some patients describe an evolution of symptoms over a period of several days, but usually this proves to be a worsening of symptoms that had been present before. The legs are affected before the arms, and often the symptoms are limited to the legs. In about one-quarter of the patients, pain and paresthesias are the main symptoms. The pain is described as a dull, constant ache in the feet or legs, less often as sharp and lancinating. Complaints of coldness (subjective only) of the feet are common. More distressing are "burning" feelings affecting the soles mainly, less often the dorsal aspects of the feet, and rarely the hands. The dysesthesias are made worse by contactual stimuli and may be so severe that the patient finds it intolerable to stand or walk (despite the relative preservation of strength) or to bear the weight of bedclothes on his feet or to pick up a utensil. The term "burning feet" has been applied to this syndrome, but "painful extremities" would be more appropriate, since the burning feeling is accompanied by other types of pain and the hands may be involved as well as the feet.

Examination discloses varying degrees of motor, sensory, and reflex loss. As the symptoms suggest, the signs are symmetrical in distribution and often confined to the legs. Usually, the loss of strength is more readily demonstrated in the distal than in the proximal musculature of the limbs and may be striking in degree, taking the form of complete paralysis of dorsiflexion of the feet and hands, i.e., a foot and wrist drop (Fig. 14.11). However, proximal muscles of the limbs are also affected, as indicated by difficulty in arising from a squatting position. Occasionally the weakness is more prominent in the muscles of the hips and thighs than in

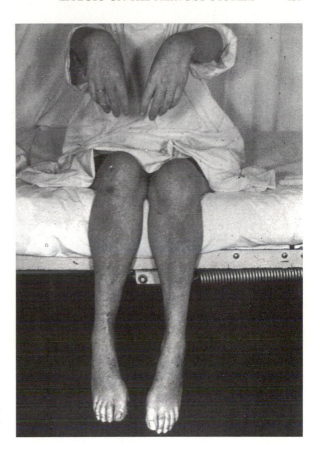

FIGURE 14.11. Alcoholic polyneuropathy. Note the distal symmetrical wasting of the limbs and the weakness illustrated by wrist drop and foot drop. (From Victor, 1986.)

the distal muscles. Absolute paralysis of the legs is observed only rarely; immobilization by contractures at the knees and ankles is a more common occurrence. Tenderness of muscles, particularly of the calves, is a characteristic finding.

Weakness of the legs, even of slight degree, is practically always accompanied by a loss of tendon reflexes. Marked depression or loss of the Achilles reflex is a sensitive index of alcoholic polyneuropathy and often persists even after motor and sensory function has returned. In the arms also, weakness is usually accompanied by a decrease or loss of deep tendon reflexes; in contrast to the legs, deep tendon reflexes are sometimes retained in the arms despite marked weakness of the hands. In a small proportion of patients, particularly those in whom pain and burning paresthesias are the major complaints, the deep tendon reflexes are retained and may even be of greater than average briskness.

Excessive sweating of the feet and hands is a common manifestation, indicative of involvement of the peripheral (postganglionic) sympathetic nerve fibers. Postural hypotension is observed only rarely, probably because of sparing of the splanchnic nerves in all but the most advanced forms of the disease (see below).

Sensory loss or impairment usually involves both superficial and deep modalities, although one modality may seemingly be affected out of proportion to others. The most consistent attribute of the sensory changes is their distal and symmetrical distribution, shading off over a considerable vertical extent of the limbs. Rarely the sensory loss extends to the lower abdomen. "Hyperesthesia," a heightened response to tactile and painful stimuli, usually indicates an underlying sensory deficit; once the stimulus is perceived, however, it may have a painful or unpleasant quality (hyperpathia).

The incidence of motor, reflex, and sensory abnormalities in a series of our patients with alcoholic neuropathy and the combinations in which they occurred are summarized in Table 14.5.

Stasis edema and pigmentation, glossiness, and

TABLE 14.5. Incidence of Motor, Reflex, and Sensory Abnormalities in Alcoholic Neuropathy[a]

Neuropathic abnormality	Legs (189 cases)	Arms (57 cases)
Loss of reflexes alone	45 (24%)[b]	6 (11%)[c]
Loss of sensation alone	10 (5%)	10 (18%)
Weakness alone	0	5 (9%)
Weakness and sensory loss	2 (1%)	10 (18%)
Reflex and sensory loss	40 (21%)	2 (4%)
Sensory, motor, and reflex loss	66 (35%)	17 (30%)
Data incomplete	26 (14%)	7 (12%)

[a]Source: Victor et al. (1989).
[b]Figures in parentheses indicate percentage of 189 cases.
[c]Figures in parentheses indicate percentage of 57 cases.

thinness of the skin of the lower legs and feet are common findings. Major dystrophic changes, in the form of perforating plantar ulcers and painless destruction of the bones and joints of the feet, are observed occasionally (Miller and Hunt, 1978; Said, 1980; Sandri et al., 1976; Thornhill et al., 1973).

The cranial nerves are usually spared, but in some patients with severe affection of the peripheral nerves, dysphagia, hoarseness and weakness of the voice (from vocal cord paralysis), hypotension, and other derangements of autonomic nervous function are observed (Duncan et al., 1980; Novak and Victor, 1974). In such patients, the responses in heart rate to the Valsalva maneuver, deep breathing, tilting, baroreceptor stimulation, and administration of atropine are significantly decreased. These symptoms have their basis in lesions of the vagus nerves and paravertebral sympathetic chains (Novak and Victor, 1974). The disorder of esophageal peristalsis that occurs in some patients with alcohol neuropathy is also best explained by vagal dysfunction (Winship et al., 1968).

The CSF is usually normal, although a modest elevation of protein content is found in a small number of patients.

14.5.2. Pathological Findings

The essential pathological change is a noninflammatory degeneration of the peripheral nerves and, in advanced cases, of the anterior and posterior nerve roots. The degenerative process is more intense in the distal than in the proximal segments. Both the myelin and axis cylinders are destroyed, the former probably earlier and to a greater extent than the latter (axonal degeneration). Denny-Brown (1958) and Bischoff (1971)

considered segmental demyelination to be the essential peripheral nerve lesion, but subsequent studies indicated that segmental demyelination is a rare finding and that axonal degeneration is the basic histological abnormality (Behse and Buchtal, 1977; Said and Landrieu, 1978). That the major abnormality in alcoholic polyneuropathy is one of axonal degeneration, initially affecting the distal segments of the peripheral nerves ("dying-back neuropathy"), has been amply corroborated by electrophysiological studies (see review by Victor, 1984).

Dorsal root ganglion cells may be lost to a variable extent, and the anterior horn cells of the spinal cord show a secondary or "axonal" reaction. The degeneration of the posterior columns, particularly of the columns of Goll, is secondary to the degeneration of dorsal root ganglion cells and posterior roots.

14.5.3. Prognosis and Treatment

Recovery from alcoholic polyneuropathy is a slow process. In the mildest cases there may be a considerable restoration of motor function in a few weeks; in the severe forms of the disease, several weeks may pass before the first signs of recovery become manifest and a year before the patient is able to walk unaided.

The foremost consideration is to supply adequate nutrition in the form of a balanced diet and supplemental B vitamins in large daily doses. If the patient will not or cannot eat, parenteral feeding becomes necessary; the vitamins may be given intramuscularly or added to intravenous fluids. A suitable parenteral preparation is Berroca-C or Parenterovite, one ampule daily.

Pressure of bedclothes may be avoided by placing a cradle support over the legs. Aching from immobility is best managed by gentle massage and passive movement of the limbs. Aspirin or acetaminophen, in a dosage of 0.3 to 1.0 g (5 to 15 grains) every 4 hr, is usually sufficient to control hyperpathia; occasionally codeine in doses of 15 to 30 mg needs to be added. Stronger opioids should be avoided, particularly if the pain is chronic in nature.

The successful regeneration of peripheral nerve may be of no avail if the denervated muscles have been allowed to undergo contracture and the joints to become fixed. During the day, the patient's legs should be positioned so that his soles rest firmly against a footboard in order to prevent shortening of the heel cords. In cases of severe paralysis, molded splints should be applied to the arms and hands and legs and feet during periods of rest. Pressure on the heels and elbows can be avoided by padding the splints and by turning the patient frequently if he is unable to do so himself. As soon as the patient's general condition permits, the limbs should be passively manipulated through a full range of movement several times daily. As function returns, more vigorous physiotherapeutic measures can be undertaken.

The tendency of the patient to resume his alcoholic habits when he leaves the hospital constitutes a special problem. Suitable arrangements must be made for close supervision of the patient during the long and tedious convalescence.

14.6. Alcohol (Tobacco–Alcohol) Amblyopia

This title refers to a now-uncommon alcohol-induced disorder of vision that is not attributable to a lesion of the cornea or other parts of the refractive mechanism. Insofar as the primary lesion is probably in the optic nerve and the condition is caused by nutritional deficiency, the terms *nutritional optic neuropathy* and *nutritional or deficiency amblyopia* are preferable.

Typically the patient complains of a blurring of vision for near and distant objects, developing gradually over a period of several days or weeks. Examination discloses a reduction in visual acuity and the presence of central or centrocecal scotomas, larger for colored than for white test objects. These changes are always bilateral and roughly symmetrical (Fig. 14.12). Funduscopic abnormalities are observed occasionally; these vary from a mild papillitis with slight hyperemia and blurring of the disk margins to pallor of the optic disks in the most advanced cases. Retinal hemorrhages occur rarely. Untreated, the disease progresses to irreversible optic atrophy. Under the influence of a nutritious diet and B-vitamin supplements, improvement occurs in all but the most chronic cases, the degree of recovery depending on the duration of the amblyopia before therapy is instituted.

The pathological changes are illustrated in Fig. 14.13. They consist of degeneration of the optic nerves, chiasm, and tracts, more or less confined to the papillomacular bundles, associated with a loss of retinal ganglion cells, particularly in the region of the macula (Victor *et al.*, 1960; Victor and Dreyfus, 1965). The primary lesion is more likely in the optic nerve than in the retinal ganglion cells.

The evidence is overwhelming that this type of amblyopia is caused by nutritional deficiency and not by

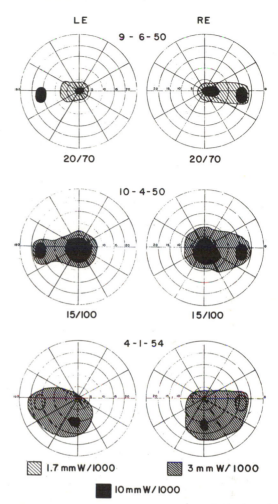

FIGURE 14.12. "Alcohol" amblyopia. Visual fields show the characteristic bilateral centrocecal scotomas involving the fixation point and blind spot. The patient was a 57-year-old malnourished alcoholic who continued to abuse alcohol until his death, 10 years after the onset of visual disturbance. The pathological changes in the optic nerves are shown in Fig. 14.13. (From Victor and Dreyfus, 1965.)

the toxic effects of alcohol and/or tobacco (Victor, 1963, 1970). Nor is there any critical evidence to support the view that the amblyopia results from the combined effects of vitamin B_{12} deficiency and chronic poisoning with cyanide generated in tobacco smoke (Potts, 1973; Victor, 1970). The specific nutritional factor that is responsible for alcohol amblyopia, if there is a single one, has not been determined. It appears that under certain conditions a deficiency of one of several B vitamins, or perhaps a combination of them, may pro-

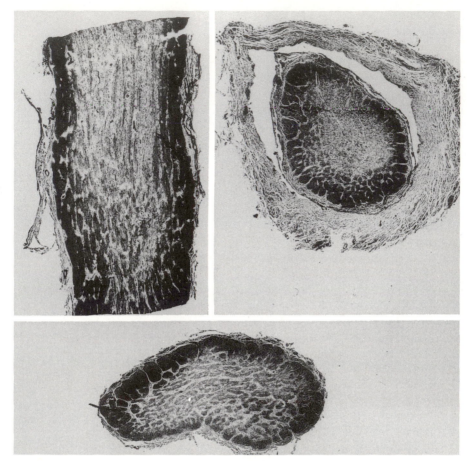

FIGURE 14.13. "Alcohol" amblyopia. Sections of the optic nerve stained for myelin. Upper left: Longitudinal section. Upper right: Cross section through retrobulbar portion. Lower: Cross section anterior to chiasm. There is extensive destruction in the area of the papillomacular bundle. (From Victor and Dreyfus, 1965.)

duce retrobulbar neuropathy. Such a causal relationship also appears to be operative in the nutritional polyneuropathies.

14.7. Pellagra

In the early part of this century pellagra attained epidemic proportions in the United States, particularly in the southern states and in the alcoholic population of large urban centers. Since 1940 the prevalence of pellagra has diminished greatly, attributable no doubt to the general practice of fortifying bread and cereals with niacin. Nevertheless, occasional cases are still observed in the alcoholic population (Bomb *et al.*, 1977; Ishii and Nishihara, 1981; Ronthal and Adler, 1969; Shah *et al.*, 1971; Spivak and Jackson, 1977).

The major neurological manifestations are the cerebral ones. Initially, symptoms are mainly mental and may be mistaken for those of neurosis: insomnia, fatigue, nervousness, irritability, and feelings of depression. Examination discloses mental dullness, apathy, impairment of memory and sometimes an acute confusional psychosis based on cerebral cortical changes (Fig. 14.14). The pellagrous manifestations of spinal cord involvement are referable to both the posterior and lateral columns, predominantly the former. The signs of peripheral nerve affection are indistinguishable from those of neuropathic beriberi. In the past, a syndrome comprising tremor, extrapyramidal rigidity, suck and grasp reflexes, and coma and referred to as "nicotinic acid deficiency encephalopathy" was thought to be a manifestation of pellagra (Jolliffe *et al.*, 1940, Spillane, 1947). Recently, this notion has been resurrected by

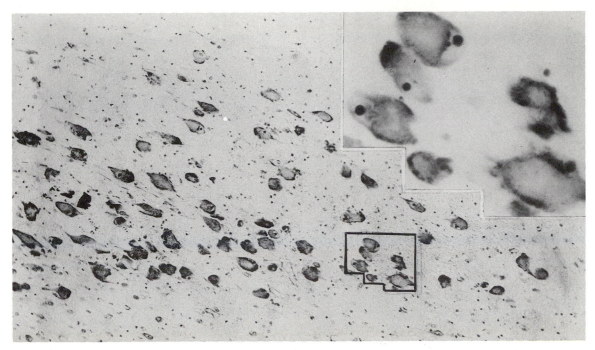

FIGURE 14.14. Pellagra. A Nissl-stained section from the motor cortex of a patient with pellagra showing the characteristic rounding, chromatolysis, and eccentricity of the nuclei of the Betz cells (inset).

Serdaru *et al.* (1988). The syndrome has never been clearly defined, clinically or pathologically.

The entity of alcoholic pellagra was discredited many years ago by Spies and DeWolfe (1933). They demonstrated clearly that alcoholic pellagra and endemic pellagra are identical and that the role of alcohol in the genesis of pellagra is simply one of substitution of drink for food; this is the prevailing view today. The cutaneous, gastrointestinal, and neurasthenic symptoms result from a deficiency of either nicotinic acid or tryptophan, the amino acid precursor of nicotinic acid (Goldsmith *et al.*, 1952). These manifestations respond to treatment with niacin and tryptophan, whereas the neurological disturbances are resistant to such treatment probably because they are caused by a deficiency of pyridoxine (Victor and Adams, 1956) rather than by niacin-tryptophan deficiency.

14.8. Central Pontine Myelinolysis

In 1959, R. D. Adams and his colleagues first drew attention to a unique disease occurring in alcoholic and undernourished patients, the most remarkable feature of

which was the presence of a large, symmetrical focus of demyelination in the center of the basis pontis (Fig. 14.15). Clinically, the illness had manifested itself in two of the original patients by a spastic bulbar paralysis and quadriplegia that led to death in 13 and 26 days, respectively. In two other patients there were no symptoms referable to the lesions, presumably because of their small size. Once attention was focused on this disorder, many cases were recognized, and sizable series were collected in several centers, including our own (Burcar *et al.*, 1977; Endo and Oda, 1981; Forno and Rivera, 1975; Mathews and Moosy, 1975; Wright *et al.*, 1979).

14.8.1. Clinical Features

Central pontine myelinolysis (CPM) is not a common disease. We found only nine cases in a series of 3548 consecutive autopsies in adults (0.25%); the two sexes were affected almost equally, and the cases did not fall into any one age period. About one-fifth of the cases now being recorded are in children.

The outstanding clinical characteristic of CPM is its invariable association with some other serious, often life-threatening, disease(s). In more than half of the

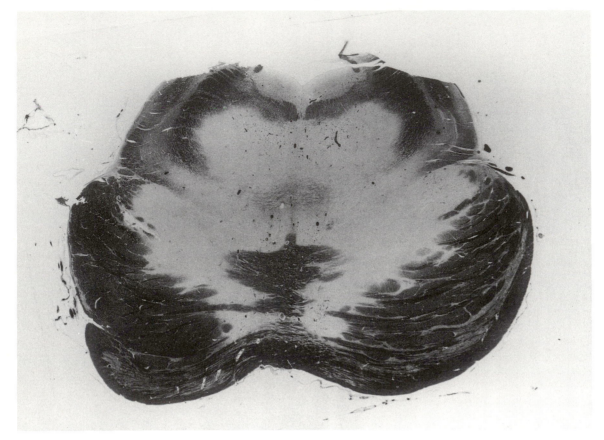

FIGURE 14.15. A severe instance of central pontine myelinolysis. A sharply demarcated focus of demyelination occupies not only the basis pontis but also most of the ventral portion of the pontine tegmentum.

reported cases it has appeared in the late stages of chronic alcoholism, often in association with the Wernicke–Korsakoff syndrome and polyneuropathy, which is the reason for including it in this chapter. Whenever a gravely ill patient, particularly an alcoholic with a general medical disease, develops a quadriplegia, pseudobulbar palsy, and the "locked-in" syndrome over a period of several days, one is justified in making a clinical diagnosis of CPM. This will be possible in fewer than one-third of cases in which the disease is demonstrated at autopsy. However, the capacity of the CT scan and MRI to visualize the pontine lesion has greatly increased the possibility of making a premortem diagnosis and has made possible the recognition of variants of the disease. We have observed several patients who recovered completely or partially over months, with disappearance of the lesion in CT scans and MRI.

14.8.2. Etiology and Pathogenesis

There are no data to support the idea that CPM is caused by the toxic effect of alcohol, by a specific nutritional deficiency, or by ischemia, hypoxia, or the compressive effect of temporal lobe herniation. Attention is now focused on the pathogenetic effects of hyponatremia and more particularly on the rapid correction of the serum sodium. In the cases of Burcar *et al.* (1977), Wright *et al.* (1979), and Tomlinson *et al.* (1976), hyponatremia was initially severe and had been rapidly corrected; the latter authors suggested that the vigorous correction of electrolyte imbalance was responsible for the clinical worsening. That profound hyponatremia and its rapid correction are important in the pathogenesis of this disease has been demonstrated in dogs by Laureno (1983) and in humans by MRI studies (Brunner *et al.*,

1990). McKee *et al.* (1988) have adduced evidence that extreme serum hyperosmolality, independent of profound hyponatremia and its correction, is an important factor in the causation of CPM.

The rate at which profound hyponatremia should be corrected is not entirely certain. According to the best available data, the serum sodium should not be raised by more than 12 mmol/liter in the first 24 hr or by more than 20 mmol/liter in the first 48 hr.

14.9. Marchiafava–Bignami Disease (Primary Degeneration of the Corpus Callosum)

14.9.1. Clinical-Pathological Features

Marchiafava–Bignami disease is more readily defined by its pathological than its clinical features. The principal alteration is in the middle portion of the corpus callosum, which on gross examination appears somewhat rarefied and sunken and reddish or gray-yellow in color, depending on its age. Microscopically, one observes clearly demarcated zones of demyelination with relative sparing of the axis cylinders and an abundance of fatty macrophages. Inflammatory changes are absent. Less consistently, lesions of a similar nature are found in the central portions of the anterior and posterior commissures and the brachia pontis (Fig. 14.16).

Symmetrically placed lesions have also been observed in the columns of Goll, superior cerebellar peduncles, and cerebral hemispheres, involving the centrum semiovale and the convolutional white matter (King and Meehan, 1936). In several cases, the lesions of deficiency amblyopia, Wernicke disease, and Marchiafava–Bignami disease have been conjoined (Delay *et al.*, 1958; Garde *et al.*, 1952; Girard *et al.*, 1953).

Many of the reported cases, as first pointed out by Jequier and Wildi (1956), have had cortical lesions of a special type: the neurons in the third layer of the frontal and temporal lobe cortices had disappeared and were replaced by a fibrous gliosis. A similar lesion has been observed in patients with chronic hepatocerebral disease (Victor *et al.*, 1965). Morel (1939), who gave the first description of this cortical laminar sclerosis, did not observe an association with Marchiafava–Bignami or chronic liver disease; in subsequent reports, the cortical lesion was consistently associated with interruption of the corpus callosum (Delay *et al.*, 1959).

Marchiafava–Bignami disease is decidedly rare. The disease affects persons in middle and late adult life. With rare exceptions, all of the reported cases have been in chronic alcoholic males.

The clinical features are variable, and no syndrome of uniform type has emerged. Some patients have presented in a state of terminal stupor or coma, and others with manifestations of chronic inebriation and alcohol withdrawal. In yet another group, a slowly progressive dementia has been described; dysarthria, slowing and unsteadiness of movement, transient sphincteric incontinence, hemiparesis, and apractic or aphasic disorders were superimposed. In two patients who came to our attention, the clinical manifestations were essentially those of bilateral frontal lobe disorder—motor and mental slowness, apathy, prominent grasping and sucking reflexes, gegenhalten, incontinence, and a slow, hesitant, wide-based gait.

In view of the great variability of the clinical picture, the diagnosis of Marchiafava–Bignami disease is understandably difficult. In fact, the diagnosis is rarely made during life. The occurrence, in a chronic alcoholic, of a frontal lobe syndrome or a symptom complex that points to a diagnosis of Alzheimer disease or frontal-corpus callosum tumor but in whom the symptoms remit should suggest the diagnosis of Marchiafava–Bignami disease. Both CT scans and MRI should prove helpful in identifying the lesion in such patients.

The pathogenesis and etiology of this disease are not known.

14.10. Alcoholic Dementia

The term "alcoholic dementia" is long entrenched in psychiatric writings, as are several other vague designations that are used more or less interchangeably—"alcoholic deteriorated state," "organic or chronic brain syndrome due to alcohol," and "alcoholic pseudoparesis." Implicit in these terms is the concept of a special form of alcoholic dementia that is (1) distinctive and different from the well-known dementias secondary to alcoholism, i.e., Korsakoff psychosis, "alcoholic" pellagra, Marchiafava–Bignami disease, and chronic hepatocerebral degeneration, and (2) attributable to the direct toxic effects of alcohol on the brain. Such a syndrome has never been delineated with any precision, as is abundantly clear from a sampling of writings on this subject (see references to Chafetz, 1975; Lewis, 1952; Strecker *et al.*, 1951; Bleuler, 1951; Bowman and Jellinek, 1941; Keller and McCormick, 1982).

It has been proposed that the term alcoholic dementia be used to designate an abnormal mental state

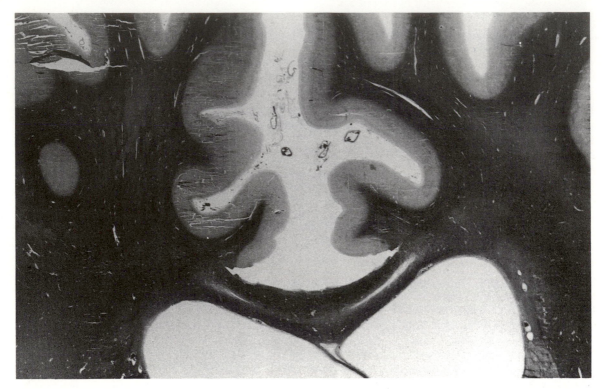

FIGURE 14.16. Marchiafava–Bignami disease. Coronal section of the formalin-fixed brain showing a slitlike area of degeneration in the central portion of the corpus callosum (myelin stain). The patient was a 62-year-old alcoholic man with a 3-month history of mental confusion, slowness of gait, and urinary incontinence. Examination disclosed prominent frontal lobe signs (grasping and sucking reflexes, incontinence, short-stepped, wide-based gait) and apraxias (he was unable to dress himself or tie his shoes despite preservation of strength and sensation). The signs improved almost to normal over a 4-month period, and he died two months later following a head injury. (From Victor, 1986).

that resembles Korsakoff psychosis but is separable from it on clinical grounds (Horvath, 1975; Lishman, 1981; Seltzer and Sherwin, 1978; Cutting, 1978). This proposition is based on the premise that Korsakoff psychosis is a fairly pure disorder of memory of acute onset and that patients with more global symptoms of gradual onset must therefore represent a separate disorder. This premise is flawed on both clinical and pathological grounds. In a considerable proportion of alcoholics, Korsakoff psychosis evolves subacutely or chronically; moreover, Korsakoff patients invariably show a disturbance of cognitive functions and certain behavioral abnormalities that depend little or not at all on memory. More importantly, so-called alcoholic dementia lacks a distinctive, defined pathology. The neuropathological changes described by Courville (1955) and quoted repeatedly as characteristic of alcoholic dementia (as well as delirium tremens) are simply not acceptable, for reasons alluded

to earlier in this chapter. In our experience, the majority of patients who have come to autopsy with a diagnosis of alcoholic dementia have shown the lesions of the Wernicke–Korsakoff syndrome, the clinical features of which have not been recognized during life. Traumatic lesions of varying type and severity were sometimes added. Isolated cases have shown the lesions of anoxic encephalopathy, acute and chronic hepatic encephalopathy, communicating hydrocephalus, Alzheimer disease, Marchiafava–Bignami disease, ischemic infarction, or some other disease quite unrelated to alcoholism. Practically always, in our material, the clinical state could be accounted for by one or a combination of these disease processes, and there has been no need to invoke a separate entity due to the toxic effect of alcohol on the brain. This theme has been fully developed in a publication by Victor and Adams (1985), to which the interested reader is directed for a detailed discussion.

14.11. Alcoholic Cerebral Atrophy

This disorder, like alcoholic dementia, does not constitute a clinical-pathological entity. The concept of an alcoholic cerebral atrophy was essentially the product of pneumoencephalography (PEG), which demonstrated that mild to moderate enlargement of the lateral and third ventricles and diffuse widening of the cerebral sulci, particularly of the frontal lobes, were common findings in hospitalized alcoholics. These radiological abnormalities were referred to as "brain damage" and "cerebral atrophy," although pathological evidence of cerebral atrophy was entirely lacking. Nor could these abnormalities be related to a distinctive clinical syndrome; enlarged ventricles and widened sulci were observed as frequently in alcoholics without neuropsychiatric symptoms as in those with such symptoms (Brewer and Perrett, 1971; Haug, 1968). Furthermore, in the symptomatic alcoholics with PEG evidence of "atrophy," there was no consistent relationship between the degree of atrophy and the severity of the neuropsychiatric symptoms (see review by Victor and Adams, 1985, for pertinent references).

Computed tomographic scans have yielded much the same information. The ventricular and sulcal enlargement in alcoholic subjects is generally mild to moderate in degree (this was also true of the PEG studies); only in exceptional instances does the degree of enlargement approach that observed in Alzheimer disease or other *bona fide* dementias. In many of the alcoholics with ventricular and sulcal enlargement there is no evidence of impaired cerebral function, either clinically or on psychometric testing, and among the impaired patients there is no consistent correlation between the severity of ventricular and sulcal enlargement and mental impairment. In fact, in the series of Epstein *et al.* (1977) and of Hill and Mikhael (1979), many alcoholics with overt neuropsychiatric symptoms showed no CT abnormalities at all. Of particular interest in this respect are the findings of Wilkinson (1982). He demonstrated that in clinically normal alcoholics, the radiological measures of brain atrophy are age related, and once the age factor is partialled out, the CT findings in these patients do not differ significantly from those of nonalcoholic controls. Not surprisingly, the CT abnormalities in clinically impaired alcoholics (mainly patients with varying degrees of the Wernicke–Korsakoff syndrome) were not age related.

The use of multiple CT scans, made initially in close relation to a period of chronic drinking and repeated at a later date, after several months of abstinence, has indicated that so-called alcoholic cerebral atrophy is potentially reversible (Carlen *et al.*, 1978; Artmann *et al.*, 1981; Ron *et al.*, 1982; Zipursky *et al.*, 1989). Moreover, ventricular and sulcal enlargement, disclosed by CT scanning, and the reversibility of these abnormalities, are known to occur in a number of metabolic and nutritional disorders other than alcoholism, namely, in anorexia nervosa (Enzmann and Lane, 1977; Heinz *et al.*, 1977), Cushing syndrome (Heinz *et al.*, 1977), kwashiorkor (Dublin and Dublin, 1978), and in patients who had been treated for prolonged periods with steroids (Bentson *et al.*, 1978; Langenstein *et al.*, 1979; Okuno *et al.*, 1980). The nature of these reversible abnormalities is not fully understood (Tewari and Noble, 1979; Carlen and Wilkinson, 1980).

A pathological basis for the radiological abnormalities has not been established. Courville (1955) has described a series of changes that purportedly underlie cerebral atrophy in alcoholics, but his illustrations reveal no definite histological abnormalities; only artifacts are shown. As indicated earlier, about 25% of our patients with Wernicke–Korsakoff disease showed dilatation of the lateral ventricles and widening of the frontal sulci, but we could not identify any definite changes in the cortex or white matter by standard histological techniques. Nor have experimental studies settled the problem. Studies in mice and rats have purportedly shown that protracted ingestion of alcohol in the absence of malnutrition results in a partial loss of cells in the hippocampal and dentate gyri (Riley and Walker, 1978). An immense amount of alcohol was required to produce these changes (corresponding, in a 70-kg man, to between 2.5 and 3 liters of 86-proof whiskey daily for 10 years). On the other hand, Phillips *et al.* (1981) have demonstrated that cerebral cortical neurons of rats are remarkably resistant to the effects of alcohol, showing no signs of degeneration even after the cortex was superfused by a concentration of ethanol three times greater than the concentration that caused death by paralysis of the respiratory center.

In summary, the occurrence of ventricular and sulcal enlargement in alcoholics who show no clinical evidence of cerebral dysfunction, and the potential reversibility of these abnormalities, are difficult to reconcile with the concept of cerebral atrophy, a term conventionally used by neuropathologists to designate disturbances of neurons that lead to their gradual degeneration and death (Greenfield, 1958). To speak of these CT abnormalities as "reversible atrophy" takes unwarranted license with the term and does nothing to enhance our understanding of the process. In the present state of our knowledge it would be preferable to refer to ventricular enlargement sulcal widening as such, rather than as cerebral atrophy.

14.12. Hepatic Encephalopathy

Embraced by this title are the cerebral abnormalities consequent to liver disease and the portal–systemic shunting of blood. All forms of liver disease may give rise to an encephalopathy, but most often the latter complicates Laennec cirrhosis of the alcoholic type. This was the underlying type of liver disease in 41 of 60 cases of hepatic coma originally reported by Adams and Foley (1953). In a consecutive series of 3548 autopsies in adults, we found the lesions of hepatic encephalopathy in 273 (7.7%) and in 70% of the latter the liver disease proved to be of the alcoholic type (Victor and Laureno, 1978).

In the alcoholic, two characteristic but overlapping forms of hepatic encephalopathy can be recognized: (1) a subacute neurological syndrome that complicates (or terminates) liver failure and is manifest as hepatic stupor or coma and (2) a chronic (non-Wilsonian) syndrome of hepatocerebral degeneration.

14.12.1. Subacute Hepatic Encephalopathy (Hepatic Coma)

The clinical syndrome, as delineated originally by Adams and Foley (1953), is characterized by a derangement of consciousness, presenting first as mental confusion with increased or decreased psychomotor activity, followed by progressive drowsiness, stupor, and coma. The confusional state is frequently combined with a characteristic intermittency of sustained muscle contraction, imparting an irregular "flapping" movement to the outstretched hands (asterixis). The electroencephalogram (EEG) becomes abnormal early in the course of the disordered mental state. It consists of bilaterally synchronous slow waves, in the δ range, which at first are interspersed with α activity and later, as the coma deepens, displace all normal activity. Some patients show only random high-voltage asynchronous slow waves. The variable occurrence of a fluctuating rigidity of the trunk and limbs, grimacing, suck and grasp reflexes, exaggeration or asymmetry of tendon reflexes, Babinski signs, and focal or generalized seizures round out the clinical picture. This state evolves over a period of days to weeks and often terminates

fatally; or, after reaching a certain stage, the symptoms may regress completely or partially.

14.12.2. Chronic Hepatic Encephalopathy

Patients who survive an episode of hepatic coma are occasionally left with residual neurological abnormalities, such as tremor of the head or arms, asterixis, grimacing, choreatic twitching of the limbs, dysarthria, ataxia of gait, or impairment of intellectual function, and these symptoms may worsen with repeated episodes of coma (Victor et al., 1965). In other patients with chronic liver disease, fixed neurological abnormalities may become manifest in the absence of discrete episodes of hepatic coma. In either circumstance, there ensues a progressive dementia of variable severity, combined with dysarthria, ataxia, intention tremor, and choreoathetosis that is most prominent in the cranial musculature. A coarse rhythmic tremor of the arms, corticospinal tract signs, and EEG abnormalities are frequently added. Other less frequent signs are muscular rigidity, grasp reflexes, tremor in repose, nystagmus, asterixis, and, in isolated cases, an ataxic tremor that has been referred to as action or intention myoclonus (Lance and Adams, 1963). In essence, each of the neurological abnormalities that occurs in hepatic coma may also be observed in patients with chronic hepatocerebral degeneration, the only difference being that the abnormalities are evanescent in the former and chronic and irreversible in the latter.

Sherlock et al. (1954) have described a chronic form of hepatic encephalopathy in which a disorder of mood, personality, and intellect persists for months or even years. Despite their chronicity, the neuropsychiatric symptoms fluctuate widely in severity or are intermittent in nature and are essentially reversible if proper therapeutic measures are instituted. For this reason, the chronic syndrome of Sherlock et al. (1954) accords more closely with hepatic coma in its many variations than with the chronic hepatocerebral syndrome described above.

14.12.3. Pathology

The neuropathological changes take two forms, corresponding to the clinical syndromes outlined above. (1) Patients who die in hepatic coma show a diffuse increase in the number and size of protoplasmic astrocytes in the deep layers of the cerebral cortex, in the lenticular nucleus, thalamus, substantia nigra, and red, dentate, and pontine nuclei, with little or no visible alteration in the nerve cells or other parenchymal elements (Fig. 14.15); (2) in patients with chronic irreversible neurological abnormalities, the cerebral lesions take the form of a patchy but diffuse cortical laminar or pseudolaminar necrosis and polymicrocavitation at the corticomedullary junction, in the striatum (particularly in the superior pole of the putamen), and in the cerebellar white matter (Figs. 14.16 and 14.17). Other changes are a diffuse increase in the size and number of protoplasmic astrocytes; the presence of periodic acid-Schiff (PAS)-positive inclusions, consisting mainly of glycogen, in the astrocytic nuclei; and degeneration of nerve cells and medullated fibers in the cerebral and cerebellar cortex, dentate, and lenticular nuclei (Figs. 14.18 and 14.19).

14.12.4. Pathogenesis

Gabuzda *et al.* (1952), while investigating the diuretic properties of ammonium cation exchange resins in alcoholics with cirrhosis, noted that these resins (which bind sodium and liberate ammonium) induced not only a diuresis but also certain neurological and EEG abnormalities, which they recognized as the early manifestations of hepatic coma ("impending hepatic coma"). They proposed that, because of hepatic disease, the patients were unable to remove the ammonia that had been absorbed from the bowel and that the neurological abnormalities were caused by ammonia intoxication. In a series of subsequent studies, it was shown that a syndrome identical to hepatic coma could be provoked in cirrhotic patients by the administration of ammonium

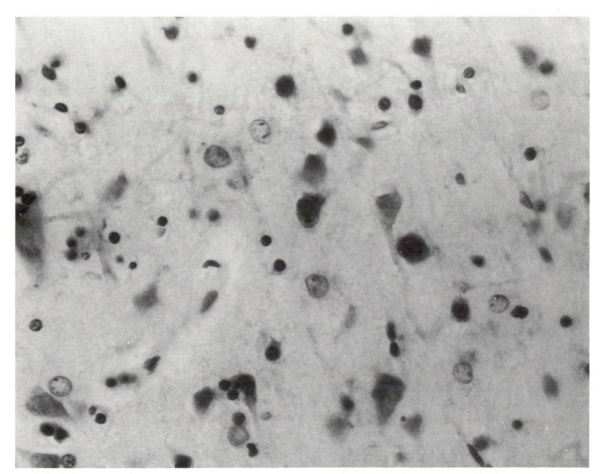

FIGURE 14.17. Nissl-stained section of the deep layers of the cerebral cortex from an alcoholic patient who died in hepatic coma. The astrocytic (Alzheimer type 2) nuclei are increased in size and number, and some occur in groups. (From Victor *et al.*, 1965.)

FIGURE 14.18. Chronic hepatic encephalopathy. Section of the motor-sensory cortex showing laminar or "band necrosis" in layers 4 and 5, with interruption of the medullated fibers. Myelin (Spielmeyer) stain. (From Victor *et al.*, 1965.)

chloride, diammonium citrate, urea, or dietary protein and that withdrawal of these agents or restriction of protein intake resulted in remission of the signs of hepatic coma and reversal of the EEG pattern to normal (Phillips *et al.*, 1952; Schwartz *et al.*, 1954). McDermott and Adams (1954) observed the same sequence of events in a patient whose superior mesenteric vein had been joined to the inferior vena cava (Eck fistula), the liver being free of disease. Thus, it was established that an abnormality of nitrogen metabolism and, more specifically, an elevated blood ammonia level was a fundamental factor in the genesis of hepatic coma.

These early observations were followed by a spate of experimental studies on the role of ammonia and other chemical factors in the genesis of hepatic encephalopathy. Even a listing of these studies is beyond the scope of this chapter. The interested reader is referred to a number of excellent reviews of this subject: Zieve (1987), Lockwood (1987), Cooper and Plum (1987), and Horrocks (1987).

It is noteworthy that despite the implication of many factors other than hyperammonemia in the pathogenesis of hepatic encephalopathy, the theory of ammonia intoxication has provided the few effective means of treatment of this condition: restriction of dietary protein; mechanical cleansing of the bowel; oral administration of neomycin, which eliminates urease-producing organisms from the bowel; surgical exclusion of the colon; and the use of lactulose, an inert sugar that effectively acidifies the colonic contents. As a corollary, the salutary therapeutic effects of these measures, the common attribute of which is a lowering of blood ammonia, lend strong support to the theory of "ammonia intoxication."

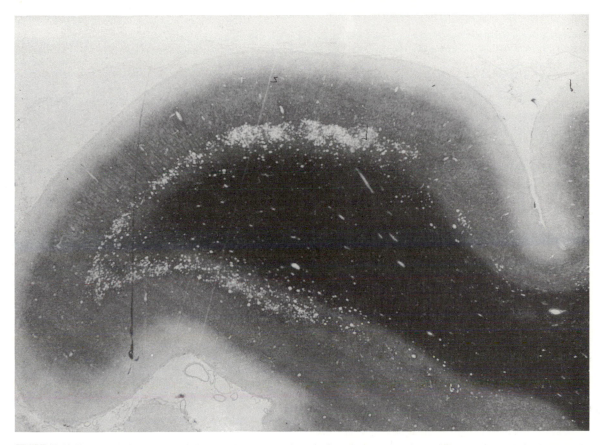

FIGURE 14.19. Chronic hepatic encephalopathy. Low-power view of a frontal gyrus. An advanced degree of vacuolation involves the deep layers of the cortex and to a lesser degree the underlying white matter. Loyez myelin stain. (From Victor *et al.*, 1965.)

REFERENCES

Adams, R. D., and Foley, J. M.: The neurological disorder associated with liver disease. *Res. Publ. Assoc. Res. Nerv. Ment. Dis.* **32:**198, 1953.

Adams, R. D., Victor, M., and Mancall, E. L.: Central pontine myelinolysis. *Arch. Neurol. Psychiatry* **81:**154, 1959.

Alldredge, B. K., and Simon, R. P.: Treatment of alcohol withdrawal seizures with phenytoin. In: *Alcohol and Seizures.* (R. J. Porter, R. H. Mattson, J. A. Cramer, and I. Diamond, eds.), Philadelphia, F. A. Davis, 1990, pp. 290–297.

Artmann, H., Gall, M. V., Hacker, H., and Herrlich, J.: Reversible enlargement of cerebral spinal fluid space in chronic alcoholics. *Am. J. Neuroradiol.* **2:**23–27, 1981.

Banay, R. S.: Pathologic reaction to alcohol: 1. Review of the literature and original case reports. *Q. J. Stud. Alcohol* **4:**580, 1944.

Baughman, F. A., Jr., and Papp, J. P.: Wernicke's encephalopathy with intravenous hyperalimentation: Remarks on similarities between Wernicke's encephalopathy and the phosphate depletion syndrome. *Mt. Sinai J. Med. N.Y.* **43:**48–52, 1976.

Behse, F., and Buchthal, F.: Alcoholic neuropathy: Clinical, electrophysiological and biopsy findings. *Ann. Neurol.* **2:**95, 1977.

Bentson, J., Reza, M., Winter, J., and Wilson, G.: Steroids and apparent cerebral atrophy on computed tomography scans. *J. Comput. Assist. Tomogr.* **2:**16–23, 1978.

Birchfield, R. I.: Postural hypotension in Wernicke's disease: A manifestation of autonomic nervous system involvement. *Am. J. Med.* **36:**404, 1984.

Bischoff, A.: Die alkoholische Polyneuropathie. *Dtsch. Med. Wochenschr.* **96:**317–322, 1971.

Blackstock, E., Ruschworth, G., and Gath, D.: Electrophysiological studies in alcoholism. *J. Neurol. Neurosurg. Psychiatry* **35:**326–334, 1972.

Blass, J. P., and Gibson, G. E.: Abnormality of a thiamine-requiring enzyme in patients with Wernicke-Korsakoff syndrome. *N. Engl. J. Med.* **297:**1367–1370, 1977.

Bleuler, E.: *Textbook of Psychiatry* (trans. A. A. Brill), New York, Macmillan, 1951, p. 327.

Bomb, B. S., Bedi, H. K., and Bhatnagar, K.: Post-ischemic paresthesia in pellagrins. *J. Neurol. Neurosurg. Psychiatry* **40:**265–267, 1977.

Bonhoeffer, K.: *Die akuten Geisteskrankheiten der Gewohnheit-strinker. Eine klinische Studie VIII*, Jena, Fisher, 1901, p. 226.

Bostock, J.: Alcoholism and its treatment. *Med. J. Aust.* **26**:136, 1939.

Bowman, K. M., and Jellinek, E. M.: Alcoholic mental disorders. *Q. J. Stud. Alcohol* **2**:312, 1941.

Bowman, K. M., Wortis, H., and Keiser, S.: The treatment of delirium tremens. *J.A.M.A.* **112**:1217, 1939.

Bratz, D.: Alkohol und Epilepsie. *Allg. Z. Psychiatrie* **56**:334, 1899.

Brewer, C., and Perrett, L.: Brain damage due to alcohol consumption: An air-encephalographic, psychometric and electro-encephalographic study. *Br. J. Addict.* **66**:170–182, 1971.

Brunner, J. E., Redmond, J. M., Haggar, A. M., Kruger, D. F., and Elias, S. B.: Central pontine myelinolysis and pontine lesions after rapid correction of hyponatremia: A prospective magnetic resonance imaging study. *Ann. Neurol.* **27**:61–66, 1990.

Bumke, O., and Kant, F.: Trunksucht und chronisher Alko-holismus. In: *Handbuck der Neurologie*, Vol. 13 (O. Bumke and O. Foerster, eds.), Berlin, Springer, 1936, p. 828.

Burcar, P. J., Notenberg, M. D., and Yarnell, P. R.: Hyponatremia and central pontine myelinolysis. *Neurology (Minneap.)* **27**:223–226, 1977.

Butler, D., and Messiha, F. S.: Alcohol and withdrawal and carbamazepine. *Alcohol* **3**:113–129, 1986.

Carlen, P. L., and Wilkinson, D. A.: Alcoholic brain damage and reversible deficits. *Acta Psychiatr. Scand. [Suppl.]* **286**:89–101, 1980.

Carlen, P. L., Wortzman, G., Holgate, R. C., Wilkinson, D. A., and Rankin, J. C.: Reversible cerebral atrophy in recently abstinent chronic alcoholics measured by computed tomography scans. *Science* **200**:1076–1078, 1978.

Carlsson, C., and Haggendal, J.: Arterial noradrenaline levels after ethanol withdrawal. *Lancet* **2**:889, 1967.

Chafetz, M. E.: Alcoholism and alcoholic psychoses. In: *Comprehensive Textbook of Psychiatry, 2nd ed.* (A. M. Freeman, H. I. Kaplan, and J. B. Sadock, eds.), Baltimore, Williams & Wilkins, 1975, p. 343.

Charness, M. E.: Molecular mechanisms of ethanol intoxication, tolerance, and physical dependence. Chap. 5. In: *Medical Diagnosis and Treatment of Alcoholism* (Mendelson, J. H., and Mello, N. K., eds.), New York, McGraw-Hill, 1992, pp. 155–199.

Cooper, A. J. L., and Plum, F.: Biochemistry and physiology of brain ammonia. *Physiol. Rev.* **67**:440–519, 1987.

Courville, C. B.: *Effects of Alcohol on the Nervous System of Man.* Los Angeles, San Lucas, 1955.

Cruickshank, E. K.: Wernicke's encephalopathy. *Q. J. Med.* **19**:327, 1950.

Cushman, P. Jr., Forbes, R., Lerner, W., and Stewart, M.: Alcohol withdrawal syndromes: Clinical management with lofex-idine. *Alcoholism* **9**:103–108, 1985.

Cutting, J.: The relationship between Korsakoff's syndrome and "alcoholic dementia." *Br. J. Psychiatry* **132**:240, 1978.

Czaja, C., and Kalant, H.: The effect of acute alcoholic intoxication on adrenal ascorbic acid cholesterol in the rat. *Can. J. Biochem. Physiol.* **39**:327, 1961.

Delay, J., Boudin, G., Brion, S., and Barbizet, J.: Étude anatamo-clinique de huit encephalopathies alcooliques: Encepha-lopathie de Wernicke et syndromes voisins. *Encephale* **47**:99, 1958.

Delay, J., Brion, S., Escourelle, R., and Sanchez, A.: Rapports entre la dégénérescence du corpos calleux de Marchiafava-Bignami et la sclerose laminaire corticale de Morel. *Ence-phale* **49**:281, 1959.

Denny-Brown, D. E.: The neurological aspects of thiamine defi-ciency. *Fed. Proc.* **17**(Suppl. 2):35, 1958.

Denvenyi, P., and Harrison, M. J.: Prevention of alcohol with-drawal seizures with oral diazepam loading. *Can. Med. Assoc. J.* **135**:798–800, 1985.

de Wardener, H. E., and Lennox, B.: Cerebral beriberi (Wer-nicke's encephalopathy). *Lancet* **1**:11, 1947.

Dodge, R., and Benedict, F. G.: *Psychological Effects of Alcohol* (Publication 232), Washington, D.C., Carnegie Institution of Washington, 1915.

Donnal, J. F., Heinz, E. R., and Burger, P. C.: MR of reversible thalamic lesions in Wernicke syndrome. *Am. J. Neuroradiol.* **11**:893–894, 1990.

Drenick, E. J., Joven, C. B., and Swenseid, M. E.: Occurrence of acute Wernicke's encephalopathy during prolonged starvation for the treatment of obesity. *N. Engl. J. Med.* **274**:937–939, 1966.

Dreyfus, P. M.: Clinical application of blood transketolase deter-minations. *N. Engl. J. Med.* **267**:596, 1962.

Dreyfus, P. M., and Victor, M.: Effects of thiamine deficiency on the central nervous system. *Am. J. Clin. Nutr.* **9**:414, 1961.

Dublin, A. B., and Dublin, W. A.: Cerebral pseudoatrophy and computed tomography: Two illustrative case reports. *Surg. Neurol.* **10**:209, 1978.

Duncan, G., Johnson, R. H., Lambie, D. G., and Whiteside, E. A.: Evidence of vagal neuropathy in chronic alcoholics. *Lancet* **2**:1053–1057, 1980.

Ellis, F. W., and Pick, J. R.: Ethanol-induced withdrawal reactions in rhesus monkeys. *Pharmacologist* **11**:256, 1969.

Endo, Y., and Oda, M.: Central pontine myelinolysis. *Acta Neuro-pathol.* **53**:145–153, 1981.

Enzmann, D. R., and Lane, B.: Cranial computed tomography findings in anorexia nervosa. *J. Comput. Assist. Tomogr.* **1**:410–414, 1977.

Epstein, P. S., Pisani, V. D., and Fawcett, J. A.: Alcoholism and cerebral atrophy. *Clin. Exp. Res.* **1**:61–65, 1977.

Escobar, A., Aruffo, C., and Rodriguez-Carbajal, J.: Wernicke's encephalopathy. A case report with neurophysiologic and CT scan studies. *Acta Vitaminol. Enzymol.* **5**:125–131, 1983.

Essig, C. F., and Lam, R. C.: Convulsions and hallucinatory behavior following alcohol withdrawal in the dog. *Arch. Neurol.* **18**:626–632, 1968.

Fere, E.: *L'Epilepsie.* Paris, Gauthier-Willars et Fils, 1896, p. 227.

Field, J. B., Williams, H. E., and Mortimore, G. E.: Studies on the mechanism of ethanol-induced hypoglycemia. *J. Clin. Invest.* **42**:497, 1963.

Flink, E. B.: Magnesium deficiency in alcoholism. *Alcoholism: Clin. Exp. Res.* **10**:590–594, 1986.

Forno, L. S., and Rivera, L.: Central pontine myelinolysis. *J. Neuropathol. Exp. Neurol.* **34**:77, 1975.

Freinkel, N., Singer, D. L., Arky, R. A., Bleicher, S. J., Anderson, J. B., and Silbert, C. K.: Alcohol hypoglycemia: I. Carbohydrate metabolism of patients with clinical alcohol hypoglycemia and the experimental reproduction of the syndrome with pure ethanol. *J. Clin. Invest.* **42:**1112, 1963.

French, S. W., and Morris, J. R.: Ethanol dependence in the rat induced by nonintoxicating levels of ethanol. *Res. Commun. Chem. Pharmacol.* **4:**221–233, 1972.

Freund, G.: Alcohol withdrawal syndrome in mice. *Arch. Neurol.* **21:**315–320, 1969.

Frezza, M., DiPadova, C., Pozzato, G., Terpin, M., Baraona, E., and Lieber, C.S.: High blood alcohol levels in women. The role of decreased gastric alcohol dehydrogenase activity and first-pass metabolism. *N. Engl. J. Med.* **322:**95–99, 1990.

Gabuzda, G. J., Phillips, G. B., and Davidson, C. S.: Reversible toxic manifestations in patients with cirrhosis of the liver given cation exchange resins. *N. Engl. J. Med.* **246:**124, 1952.

Gallucci, M., Bozzao, A., Splendiani, A., Masciocchic, C., and Passariello, R.: Wernicke encephalopathy: MR findings in five patients. *Am. J. Neuroradiol.* **11:**887–892, 1990.

Garde, A., Girard, P. F., and Pallasse, A.: L'encephalopathie hemorrhagique de Gayet-Wernicke et la degenerescence subaique du corps calleux (Marchiafava-Bignami): A propos d'une observation anatomoclinique. *Rev. Lyon Med.* **1:**99, 1952.

Gessner, P. K.: Drug therapy of the alcohol withdrawal syndrome. In: *Biochemistry and Pharmacology of Ethanol, Vol. II*, (E. Majehrowicz and E. P. Noble, eds.), New York, Plenum, 1979, pp. 375–435.

Ghez, C.: Vestibular paresis: A clinical feature of Wernicke's disease. *J. Neurol. Neurosurg. Psychiatry* **32:**134–139, 1969.

Girard, P. F., Garde, A., and Devic, M.: Considérations terminologiques, étiologiques, anatomiques et cliniques concernant l'encephalopathie de Gayet-Wernicke: Ses rapports avec le syndrome de Marchiafava-Bignami et la psychose de Korsakow. *Rev. Neurol.* **88:**236, 1953.

Goldsmith, G. A., Sarrett, H. P., Register, U. D., and Gibbens, J.: Studies of niacin requirement in man: Experimental pellagra in subjects on corn diets low in niacin and tryptophan. *J. Clin. Invest.* **31:**533, 1952.

Goldstein, D. B.: Relationship of alcohol dose to intensity of withdrawal signs in mice. *J. Pharmacol. Exp. Ther.* **180:**203–215, 1972.

Goldstein, D. B.: *Pharmacology of Alcohol.* New York, Oxford University Press, 1983.

Goodwin, D. W., Othmer, E., Halikas, J. A., and Freemon, F.: Loss of short-term memory as a predictor of the alcoholic "blackout." *Nature* **227:**201–202, 1970.

Graham, J. R., Woodhouse, D., and Read, F. H.: Massive thiamine dosage in an alcoholic with cerebellar cortical degeneration. *Lancet* **2:**107, 1971.

Gravallese, M. A., and Victor, M.: Circulatory studies in Wernicke's encephalopathy. *Circulation* **15:**836, 1957.

Greenblatt, M., Levin, S., and DeCorli, F.: The electroencephalogram associated with chronic alcoholism, alcoholic psychosis, and alcoholic convulsions. *Arch. Neurol. Psychiatry* **52:**290, 1944.

Greenfield, J. G.: *Neuropathology* (J. G. Greenfield, ed.), London, Edward Arnold, 1958, p. 26.

Griffiths, P. J., Littleton, J. M., and Ortiz, A.: A method for the induction of dependence to ethanol in mice. *Br. J. Pharmacol.* **47:**669–670, 1973.

Halstead, C. H., Robles, E. A., and Mezey, E.: Decreased jejunal uptake of labeled folic acid (^{3}H-PGA) in alcoholic patients: Roles of alcohol and nutrition. *N. Engl. J. Med.* **285:**701, 1971.

Halstead, C. H., Robles, E. A., and Mezey, E.: Intestinal malabsorption in folate-deficient alcoholics. *Gastroenterology* **64:**526, 1973.

Hare, F.: Alcohol and delirium tremens. *Br. Med. J.* **1:**446, 1915.

Harper, C. G.: The incidence of Wernicke's encephalopathy in Australia—a neuropathological study of 131 cases. *J. Neurol. Neurosurg. Psychiatry* **46:**593, 1983.

Haug, J. O.: Pneumoencephalographic evidence of brain damage in chronic alcoholics: A preliminary report. *Acta Psychiatry. Scand. [Suppl.]* **203:**135–143, 1968.

Hawley, R. J., Major, L. F., Schulman, E. A., and Linnoila, M.: Cerebrospinal fluid 3-methoxy-4-hydroxyphenylglycol and norepinephrine levels in alcohol withdrawal. *Arch. Gen. Psychiatry* **42:**1056–1062, 1985.

Heinz, E. R., Martinez, J., and Haenggeli, A.: Reversibility of cerebral atrophy in anorexia nervosa and Cushing's syndrome. *J. Comput. Assist. Tomogr.* **1:**415–418, 1977.

Hill, S. Y., and Mikhael, M. A.: Computerized transaxial tomographic and neuropsychological evaluations in chronic alcoholics and heroin abusers. *Am. J. Psychiatry* **136:**598, 1979.

Hirschfelder, A. D., and Haury, V. G.: Clinical manifestations of high and low plasma magnesium. *J.A.M.A.* **102:**1138, 1934.

Horrocks, L. A. (ed.): Symposium proceedings on "Neurochemical Consequences of Hepatic Disease," *Neurochemical Pathology* 6:1/2, pp. 1–166, April, 1987.

Horvath, T. B.: Clinical spectrum and epidemiological features of alcoholic dementia. In: *Alcohol, Drugs and Brain Damage* (J. G. Rankin, ed.), Toronto, Alcoholism and Drug Addiction Research Foundation of Ontario, 1975, p. 1.

Hoyumpa, A. M.: Mechanisms of vitamin deficiencies in alcoholism. *Alcoholism: Clin. Exp. Res.* **10:**573–581, 1986.

Hughes, J. R., Curtin, M. J., and Brown, V. P.: Usefulness of photic stimulation in routine clinical electroencephalopathy. *Neurology (Minneap.)* **10:**777, 1960.

Isbell, H., Fraser, H. F., Wikler, A., Belleville, R. E., and Eisenman, A. S.: An experimental study of the etiology of "rum fits" and delirium tremens. *Q. J. Stud. Alcohol* **16:**1, 1955.

Ishii, N., and Nishihara, Y.: Pellagra among chronic alcoholics: Clinical and pathological study of 20 necropsy cases. *J. Neurol. Neurosurg. Psychiatry* **44:**209–215, 1981.

Jelliffe, S. E., and White, W. A.: *Disease of the Nervous System.* Philadelphia, Lea & Febiger, 1929, p. 1066.

Jellinek, E. M.: Classics of the alcohol literature: Magnus Huss' *Alcoholismus Chronicus. Q. J. Stud. Alcohol* **4:**85, 1943.

Jequier, M., and Wildi, E.: Le syndrome de Marchiafava-Bignami. *Schweiz. Arch. Neurol. Neurochir. Psychiatr.* **77:**393, 1956.

Jolliffe, N., Colbert, C. N., and Joffe, P. M.: Observations on the etiologic relationship of vitamin B (B_1) to polyneuritis in the alcohol addict. *Am. J. Med. Sci.* **107:**515, 1936.

Jolliffe, N., Bowman, K. M., Rosenblum, L. A., and Fein, H. D.: Nicotinic acid deficiency encephalopathy. *J.A.M.A.* **114:**307, 1940.

Jolliffe, N., Wortis, H., and Fein, H. D.: The Wernicke syndrome. *Arch. Neurol. Psychiatry* **46**:569, 1941.

Kaim, S. C., Klett, C. S., and Rothfeld, B.: Treatment of the acute alcohol withdrawal state: A comparison of four drugs. *Am. J. Psychiatry* **125**:1640–1646, 1969.

Kalant, H.: Alcohol and electrophysiology of the central nervous system. In: *Advances in Pharmacology and Therapeutics, Drug-Action Modification-Comparative Pharmacology, Vol. 8* (G. Olive, ed.), New York, Pergamon Press, 1979, p. 199.

Kalant, H., and Woo, N.: Electrophysiological effects of ethanol on the nervous system. *Pharmacol. Ther.* **14**:431–457, 1981.

Kalinowsky, L. B.: Convulsions in non-epileptic patients on withdrawal of barbiturates, alcohol, and other drugs. *Arch. Neurol. Psychiatry* **48**:946, 1942.

Keller, M.: Other effects of alcohol. In: *Drinking and Intoxication: I. Physiological and Psychological Effects of Alcohol* (R. G. McCarthy, ed.), Glencoe, IL, Free Press, 1959, p. 13.

Keller, M., and Gurioli, C.: *Statistics on Consumption of Alcohol and on Alcoholism.* New Brunswick, NJ, Rutgers Center for Alcohol Studies, 1976.

Keller, M., and McCormick, M.: *A Dictionary of Words about Alcohol.* New Brunswick, NJ, Rutgers Center for Alcohol Studies, 1982.

King, L. S., and Meehan, M. C.: Primary degeneration of the corpus callosum (Marchiafava's disease). *Arch Neurol. Psychiatry* **36**:547, 1936.

Klingman, W. O., Suter, C., Green, R., and Robinson, I.: Role of alcoholism and magnesium deficiency in convulsions. *Trans. Am. Neurol. Assoc.* **80**:162, 1955.

Kraepelin, E.: Alcoholic mental disturbances. In: *Lectures on Clinical Psychiatry* (authorized translation from second German edition, revised) (T. Johnstone, ed.), Baltimore, William Wood, 1946, p. 102.

Kraus, M. L., Gottlieb, L. D., Horwitz, R. I., and Anscher, M.: Randomized clinical trial of atenolol in patients with alcohol withdrawal. *N. Engl. J. Med.* **313**:905–909, 1985.

Krusius, F. E., Vartia, K. O., and Forsander, O.: Experimentelle Studien uber die biologische Wirkung von Alkohol: II. Alkohol und Nebennierenrinden-Funktion. *Ann. Med. Exp. Biol. Fenn.* **36**:424, 1958.

Lambert, A.: The intoxicants. In: *Textbook of Medicine* (3rd edition) (R. Cecil, ed.), Philadelphia, W. B. Saunders, 1934, p. 568.

Lance, J. W., and Adams, R. D.: The syndrome of intention or action myoclonus as a sequel to hypoxic encephalopathy. *Brain* **86**:111, 1963.

Langenstein, I., Willig, R. P., and Kuhne, D.: Reversible cerebral atrophy caused by corticotrophin. *Lancet* **1**:1246, 1979.

Laureno, R.: Central pontine myelinolysis following rapid correction of hyponatremia. *Ann. Neurol.* **13**:232–242, 1983.

LeFebvre-D'Amour, M., Shahani, B. T., and Young, R. R.: Tremor in alcoholic patients. In: *Physiological Tremor, Pathological Tremors and Clonus.* (J. E. Desmedt, ed.), Basel, Karger, 1978, pp. 106–164.

Lennox, W. G.: Alcohol and epilepsy. *Q. J. Stud. Alcohol* **2**:1, 1941.

Lewis, A.: Psychoses—alcoholic psychoses. In: *British Encyclo-*

pedia of Medical Practice, Vol. 10 (2nd ed.) London, Butterworth & Co., 1952, p. 394.

Lieber, C. S., and DeCarli, L. M.: Ethanol dependence and tolerance: A nutritionally controlled experimental model in the rat. *Res. Commun. Chem. Pathol. Pharmacol.* **6**:983–991, 1973.

Lindenbaum, J., and Lieber, C. S.: Alcohol-induced malabsorption of vitamin B_{12} in man. *Nature* **224**:806, 1969.

Lindenbaum, J., and Lieber, C. S.: Effects of chronic ethanol administration on intestinal absorption in man in the absence of nutritional deficiency. *Ann. N.Y. Acad. Sci.* **252**:228–234, 1975.

Lishman, W. A.: Cerebral disorder in alcoholism, syndromes of impairment. *Brain* **104**:1–20, 1981.

Lockwood, A. H.: Hepatic encephalopathy: Experimental approaches to human metabolic encephalopathy. *CRC Crit. Rev. Neurobiol.* **3**:105–133, 1987.

Luda, E.: Le signe du bobbing oculaire dans l'encephalopathie de Wernicke. *Rev. Otoneuroophtalmol.* **52**:123–125, 1980.

MacIntyre, I.: Magnesium metabolism. *Adv. Intern. Med.* **13**:143–154, 1967.

Mair, R. G., Capra, C., McEntee, W. J., and Engen, T.: Odor discrimination and memory in Korsakoff's psychosis. *J. Exp. Psychol.* **6**:455, 1980.

Marchand, L.: Les lesions du systeme nerveux, du foie, des reins et de la rate dans le "delirium tremens" des alcooliques. *Ann. Anat. Pathol.* **9**:1026, 1932.

Mathews, T., and Moosy, J.: Central pontine myelinolysis: Lesion evolution and pathogenesis. *J. Neuropathol. Exp. Neurol.* **34**:77, 1975.

Mauritz, K. H., Dichgans, J., and Hufschmidt, A.: Quantitative analysis of stance in late cortical cerebellar atrophy of the anterior lobe and other forms of cerebellar ataxia. *Brain* **102**:461–482, 1979.

Mayer, R. F: Peripheral nerve conduction in alcoholics. *Psychosom. Med.* **28**:475, 1966.

McDermott, W. V., and Adams, R. D.: Episodic stupor associated with an Eck fistula in the human with particular reference to the metabolism of ammonia. *J. Clin. Invest.* **33**:1, 1954.

McDougall, W.: On a new method for the study of concurrent mental operations and of mental fatigue. *Br. J. Psychol.* **1**:435, 1904.

McDowell, J. R., and LeBlanc, H. J.: Computed tomographic findings in Wernicke-Korsakoff syndrome. *Arch. Neurol.* **41**:453–454, 1984.

McEntee, W. J., Mair, R. G., and Langlais, P. J.: Neurochemical pathology in Korsakoff's psychosis: Implications for other cognitive disorders. *Neurology (N.Y.)* **34**:648–652, 1984.

McKee, A. C., Winkelman, M. D., and Banker, B. Q.: Central pontine myelinolysis in severely burned patients: Relationship to serum hyperosmolality. *Neurology* **38**:1221, 1988.

McQuarrie, D. G., and Fingl, E.: Effects of single doses and chronic administration of ethanol on experimental seizures in mice. *J. Pharmacol. Exp. Ther.* **124**:264, 1958.

Mello, N. K.: Short-term memory function in alcohol addicts during intoxication. In: *Alcohol Intoxication and Withdrawal: Experimental Studies, Proceedings 39th International Con-*

gress on Alcoholism and Drug Dependence (M. M. Gross, ed.), New York, Plenum Press, 1973. p. 333.

Mello, N. K.: Animal models for the study of alcohol addiction. *Psychoneuroendocrinology* **1**:347, 1976.

Mendelson, J. H., and LaDou, J.: Experimentally induced chronic intoxication and withdrawal in alcoholics: II. Psychophysiological findings. *Q. J. Stud. Alcohol (Suppl.)* **2**:14, 1964.

Mendelson, J. H., and Mello, N. K.: Basic mechanisms underlying physical dependence upon alcohol. *Ann. N.Y. Acad. Sci.* **311**:69, 1978.

Mendelson, J. H., Wexler, D., Kubzansky, P., Leiderman, P. H., and Solomon, P.: Serum magnesium in delirium tremens and alcoholic hallucinosis. *J. Nerv. Ment. Dis.* **128**:352, 1959.

Mensing, J. W. A., Hoogland, P. H., and Slooff, J. L.: Computer tomography in the diagnosis of Wernicke's encephalopathy: A radiologic-neuropathologic correlation. *Ann. Neurol.* **16**: 363, 1984.

Mezey, E., Jow, E., Slavin, R. E., and Tobon, F.: Pancreatic function and intestinal absorption in chronic alcoholism. *Gastroenterology* **59**:657, 1970.

Miles, W. R.: *Effect of Alcohol on Psychophysiological Functions* (Publication 266). Washington, D.C., Carnegie Institute of Washington, 1918.

Miles, W. R.: Psychological effects of alcohol on man. In: *Alcohol and Man* (H. Emerson, ed.), New York, Macmillan, 1932, p. 224.

Miller, R. M., and Hunt, J. A.: The radiologic features of alcoholic ulcer-osteolytic neuropathy in blacks. *S. Afr. Med. J.* **54**:159–161, 1978.

Minot, G. R., Strauss, M. B., and Cobb, S.: "Alcoholic" polyneuritis: Dietary deficiency as a factor in its production. *N. Engl. J. Med.* **208**:1244, 1933.

Morel, F.: Une forme anatomoclinique particuliere de l'alcoolisme chronique cliniquement rappelant la pseudo-paralysie des anciens auteurs, anatomiquement présentant une sclérose corticale laminaire. *Schweiz. Arch. Neurol. Psychiatr.* **44**: 305, 1939.

Mukherjee, A. B., Svoronos, S., Ghazanfari, A., Martin, P. R., Fisher, A., Roecklein, B., Rodbard, D., Staton, R., Behar, D., and Berg, C. J.: Transketolase abnormality in cultured fibroblasts from familial chronic alcoholic men and their male offspring. *J. Clin. Invest.* **79**:1039–1043, 1987.

Novak, D. J., and Victor, M.: Affection of the vagus and sympathetic nerves in alcoholic polyneuropathy. *Arch. Neurol.* **30**: 273–284, 1974.

Noyes, A. P.: *Modern Clinical Psychiatry (2nd ed.).* Philadelphia, W. B. Saunders, 1939.

Okuno, T., Ito, M., Konishi, Y., Yoshioka, M., and Nakano, Y.: Cerebral atrophy following ACTH therapy. *J. Comput. Assist. Tomogr.* **4**:20–23, 1980.

Osler, W.: *Principles and Practice of Medicine (8th ed.).* New York, Appleton, 1917, p. 398.

Palsson, A.: The efficacy of early chlormethiazole medication in the prevention of delirium tremens. A retrospective study of the outcome of different drug treatment strategies at the Helsingborg psychiatric clinics, 1975–1980. *Acta Psychiatr. Scand. [Suppl.]* **329** (73):140–145, 1986.

Perr, I. N.: Pathological intoxication and alcohol idiosyncratic intoxication. I. Diagnostic and clinical aspects. *J. Forensic. Sci.* **31**:806–811, 1986.

Phillips, G. B., Schwartz, R., Gabuzda, G. J., and Davidson, C. S.: The syndrome of impeding hepatic coma in patients with cirrhosis of the liver given certain nitrogenous substances. *N. Engl. J. Med.* **247**:239, 1952.

Phillips, S. C., Cragg, B. G., and Singh, S. C.: The short-term toxicity of ethanol to neurons in rat cerebral cortex tested by topical application *in vivo* and a note on a problem in estimating ethanol concentrations in tissue. *J. Neurol. Sci.* **49**:353–361, 1981.

Piker, P.: On the relationship of sudden withdrawal of alcohol to delirium tremens. *Am. J. Psychiatry* **93**:1387, 1937.

Plum, F., and Posner, J. B.: *The Diagnosis of Stupor and Coma, 3rd ed.*, Philadelphia, F. A. Davis, 1980, p. 2.

Pohorecky, L. A.: Biphasic action of alcohol. *Biobehav. Rev.* **1**: 231, 1977.

Porter, R. J., Mattson, R. H., Cramer, J. A., and Diamond, I. (eds.): *Alcohol and Seizures*, Philadelphia, F. A. Davis, 1990, pp. 25–139.

Potts, A. M.: Tobacco amblyopia. *Surv. Ophthalmol.* **17**:313–319, 1973.

Projesz, B., and Begleiter, H.: Brain dysfunction and alcohol. In: *The Biology of Alcoholism, Vol. 7* (B. Kissen and H. Begleiter, eds.), New York, Plenum Press, 1983, p. 449.

Randall, R. E., Rossmeisl, E. C., and Bleifer, K. H.: Magnesium depletion in man. *Ann. Intern. Med.* **50**:257, 1959.

Rice, J. P., Horowitz, M., and Chin, D.: Wernicke-Korsakoff syndrome with bilateral facial nerve palsies. *J. Neurol. Neurosurg. Psychiatry* **47**:1356–1357, 1984.

Riley, J. N., and Walker, D. W.: Morphological alterations in hippocampus after long-term alcohol consumption in mice. *Science* **201**:646–648, 1978.

Rivers, W. H. R.: *The Influence of Alcohol and Other Drugs on Fatigue.* London, Edward Arnold, 1908.

Roche, S. W., Lane, R. J. M., and Wade, J. P. H.: Thalamic hemorrhages in Wernicke-Korsakoff syndrome demonstrated by computed tomography. *Ann. Neurol.* **23**:312, 1988.

Ron, M. A., Acker, W., Shaw, G. K., and Lishman, W. A.: Computerized tomography of the brain in chronic alcoholism. A survey and follow-up study. *Brain* **105**:497–514, 1982.

Ronthal, M., and Adler, H.: Motor nerve conduction velocity and the electromyograph in pellagra. *S. Afr. Med. J.* **43**:642–644, 1969.

Said, G.: A clinicopathologic study of acrodystrophic neuropathies. *Muscle Nerve* **3**:491–501, 1980.

Said, G. and Landrieu, P.: Etude quantitative des fibres nerveuses isolees dans les polynevrites alcooliques. *J. Neurol. Sci.* **35**:317–330, 1978.

Sandri, R., Bruschi, M., and Conte, L.: Le ulcere perforanti plantari nelle polyneuriti alcooliche. *Minerva Med.* **67**:2919–2926, 1976.

Schuckit, M. A., and Winokur, G.: Alcoholic hallucinosis and schizophrenia: A negative study. *Br. J. Psychiatry* **119**:549–550, 1971.

Schwartz, R., Phillips, G. B., Seegmiller, E., Gabuzda, G. J., and Davidson, C. S.: Dietary protein in the genesis of hepatic coma. *N. Engl. J. Med.* **251:**685, 1954.

Scott, D. F.: Alcoholic hallucinosis: An aetiological study. *Br. J. Addict.* **62:**113–125, 1967.

Sellers, E. M., and Kalant, H.: Alcohol intoxication and withdrawal. *N. Engl. J. Med.* **294:**757–762, 1976.

Sellers, E. M., Naranjo, C. A., Harrison, M., Devenyi, P., Roach, C., and Sykora, K.: Diazepam loading: Simplified treatment of alcohol withdrawal. *Clin. Pharmacol. Ther.* **34:**822–826, 1983.

Seltzer, B., and Sherwin, I.: "Organic brain syndromes": An empirical study and critical review. *Am. J. Psychiatry* **135:** 13–21, 1978.

Serdaru, M., Hausser-Hauw, C., Laplane, D., Buge, A., Castaigne, P., Goulon, M., Lhermitte, F., and Hauw, J. J.: The clinical spectrum of alcoholic pellagra encephalopathy. *Brain* **111:**829–842, 1988.

Sereny, G., Rapoport, A., and Husdan, H.: The effect of alcohol withdrawal on electrolyte and acid-base balance. *Metabolism* **15:**896–904, 1966.

Shah, D. R., Singh, S. V., and Jain, I. L.: Neurological manifestations in pellagra. *J. Assoc. Physicians India* **19:**443–446, 1971.

Sherlock, S., Summerskill, W. H. J., White, L. P., and Phear, E. A.: Portal-systemic encephalopathy: Neurological complications of liver disease. *Lancet* **2:**453, 1954.

Shimojyo, S., Scheinberg, P., and Reinmuth, O.: Cerebral blood flow and metabolism in the Wernicke-Korsakoff syndrome. *J. Clin. Invest.* **46:**849–854, 1967.

Smith, J. J.: The treatment of acute alcoholic states with ACTH and adrenocortical hormones. *Q. J. Stud. Alcohol* **11:**190, 1950.

Spies, T. D., and DeWolfe, H. F.: Observation on etiological relationship of severe alcoholism to pellagra. *Am. J. Med. Sci.* **186:**521, 1933.

Spillane, J. D.: *Nutritional Disorders of the Nervous System.* Baltimore, Williams & Wilkins, 1947, p. 48.

Spivak, J. L., and Jackson, D. L.: Pellagra: An analysis of 18 patients and a review of the literature. *Johns Hopkins Med. J.* **140:**295–309, 1977.

Squire, L. R., Amaral, D. G., and Press, G. A.: Magnetic resonance imaging of the hippocampal formation and mammillary nuclei distinguish medial temporal lobe and diencephalic amnesia. *J. Neurosci.* **10:**3106–3117, 1990.

Strecker, E. A., Ebaugh, F. G., and Ewalt, J. R.: *Practical Clinical Psychiatry.* New York, Blakiston, 1951, pp. 150, 155, and 169.

Suter, C., and Klingman, W.: Neurologic manifestations of magnesium depletion states. *Neurology (Minneap.)* **4:**691, 1955.

Tamerin, J. S., Weiner, S., Poppen, R., Steinglass, P., and Mendelson, J. H.: Alcohol and memory: Amnesia and short-term memory function during experimentally-induced intoxication. *Am. J. Psychiatry* **127:**1659–1664, 1971.

Tashiro, K., and Nelson, J. S.: Hemorrhagic necrosis of superior cerebellar vermis associated with acute Wernicke's encephalopathy. *J. Neuropathol. Exp. Neurol.* **31:**185, 1972.

Tewari, S., and Noble, E. P.: Effect of ethanol on cerebral protein and ribonucleic acid synthesis. In: *Biochemistry and Phar-*

macology of Ethanol (E. Majchrowicz and E. P. Noble, eds.), New York, Plenum Press, 1979, p. 541.

Thomson, A., Baker, H., and Leevy, C. M.: Thiamine absorption in alcoholism. *Am. J. Clin. Nutr.* **21:**537, 1968.

Thornhill, H. L., Richter, R. W., Shelton, M. L., and Johnson, C. A.: Neuropathic arthropathy (Charcot forefeet) in alcoholics. *Orthop. Clin. North Am.* **4:**7–20, 1973.

Tomasulo, P. A., Kater, R. M. H., and Iber, F. L.: Impairment of thiamine absorption in alcoholism. *Am. J. Clin. Nutr.* **21:** 1341–1344, 1968.

Tomlinson, B., Pierides, A., and Bradley, W.: Central pontine myelinolysis: Two cases associated with electrolyte disturbance. *Q. J. Med. (N. Ser.)* **45:**373, 1976.

Vallee, B. L., Wacker, E. C., and Ulmer, D. D.: The magnesium deficiency tetany syndrome in man. *N. Engl. J. Med.* **262:** 155, 1960.

Victor, M.: Tobacco-alcohol amblyopia: A critique of current concepts of this disorder, with special reference to the role of nutritional deficiency in its causation. *Arch. Ophthalmol.* **70:** 313, 1963.

Victor, M.: Treatment of alcoholic intoxication and the withdrawal syndrome: A critical analysis of the use of drugs and other forms of therapy. *Psychosom. Med.* **28:**636, 1966.

Victor, M.: The pathophysiology of alcoholic epilepsy. *Res. Publ. Assoc. Res. Nerv. Ment. Dis.* **46:**431–454, 1968.

Victor, M: Tobacco, amblyopia, cyanide poisoning and vitamin B_{12} deficiency: A critique of current concepts. In: *Miami Neuro-Ophthalmology Symposium, Vol. 5* (J. L. Smith, ed.), Hallandale, FL, Huffman, 1970, pp. 33–48.

Victor, M.: The role of hypomagnesemia and respiratory alkalosis in the genesis of alcohol-withdrawal symptoms. *Ann. N.Y. Acad. Sci.* **215:**235–248, 1973.

Victor, M: Polyneuropathy due to nutritional deficiency and alcoholism. In: *Peripheral Neuropathy* (2nd ed.) (P. J. Dyck, P. K. Thomas, E. H. Lambert, and R. Bunge, eds.), Philadelphia, W. B. Saunders, 1984, p. 1899.

Victor, M.: Neurologic disorders due to alcoholism and malnutrition. In: *Clinical Neurology* (A. B. Baker, and R. J. Joynt, eds.), Philadelphia, Harper and Row, vol. 4, Chap. 61, 1986.

Victor, M.: MR in the diagnosis of the Wernicke–Korsakoff syndrome. *Am. J. Neuroradiol.* **11:**895–896, 1990.

Victor, M., and Adams, R. D.: The effect of alcohol upon the nervous system. *Res. Publ. Assoc. Res. Nerv. Ment. Dis.* **32:** 526, 1953.

Victor, M., and Adams, R. D.: Neuropathy of experimental vitamin B_6 deficiency in monkeys. *Am. J. Clin. Nutr.* **4:**346, 1956.

Victor, M., and Adams, R. D.: On the etiology of the alcoholic neurologic diseases: With special references to the role of nutrition. *Am. J. Clin. Nutr.* **9:**379, 1961.

Victor, M., and Adams, R. D.: The alcoholic dementias. In: *Handbook of Clinical Neurology, Neurobehavioral Disorders, Vol. 2* (P. J. Vinken, G. W. Bruyn, and H. L. Klawans, eds.), Amsterdam, Elsevier, 1985, pp. 335–352.

Victor, M., and Brausch, C. C.: The role of abstinence in the genesis of alcoholic epilepsy. *Epilepsia* **8:**1–20, 1967.

Victor, M., and Dreyfus, P. M.: Tobacco-alcohol amblyopia: Further comments on its pathology. *Arch. Ophthalmol.* **74:**649, 1965.

Victor, M., and Ferrendelli, J. A.: The nutritional and metabolic diseases of the cerebellum: Clinical and pathological aspects. In: *The Cerebellum in Health and Diseases* (W. S. Fields and W. D. Willis, eds.), St. Louis, Warren H. Green, 1970, p. 412.

Victor, M., and Hope, J. M.: The phenomenon of auditory hallucinations in chronic alcoholism. *J. Nerv. Ment. Dis.* **126**:451, 1957.

Victor, M., and Laureno, R.: The neurologic complications of alcohol abuse: Epidemiologic aspects. *Adv. Neurol.* **19**:603–617, 1978.

Victor, M., and Yakovlev, P. I.: S. S. Korsakoff's psychic disorder in conjunction with peripheral neuritis: A translation of Korsakoff's original article with brief comments on the author and his contribution to clinical medicine. *Neurology (Minneap.)* **5**:394, 1955.

Victor, M., Adams, R. D., and Mancall, E. L.: A restricted form of cerebellar degeneration occurring in alcoholic patients. *Arch. Neurol.* **1**:577, 1959.

Victor, M. Mancall, E. L. and Dreyfus, P. M.: Deficiency amblyopia in the alcoholic patient: A clinicopathologic study. *Arch. Opthtahalmol.* **64**:1, 1960.

Victor, M., Adams, R. D., and Cole, M.: The acquired (non-Wilsonian) type of chronic hepatocerebral degeneration. *Medicine* **44**:345, 1965.

Victor, M., Adams, R. D., and Collins, G. H.: *The Wernicke-Korsakoff Syndrome. A Clinical and Pathological Study of 245 Patients, 82 with Postmortem Examinations (2nd ed., revised.)* Philadelphia, F. A. Davis, 1989.

Wacker, W. E. C., and Vallee, B. L.: Magnesium metabolism. *N. Engl. J. Med.* **259**:431, 1958.

Wechsler, I. S.: Etiology of polyneuritis. *Arch. Neurol. Psychiatry* **29**:813, 1933.

Wernicke, C.: *Grundriss der Psychiatrie.* Leipzig, Thieme, 1900.

Wikler, A., Pescor, F. T., Fraser, H. F., and Isbell, H.: Electroencephalographic changes associated with chronic alcoholic intoxication and the alcohol abstinence syndrome. *Am. J. Psychiatry* **113**:106, 1956.

Wilkinson, D. A.: Accelerated mental aging in alcoholism: Working hypothesis or uncontrolled variable? In: *Cerebral Defects in Alcoholism* (D. A. Wilkinson, ed.), Toronto, Addiction Research Foundation, 1982, p. 61.

Wilkinson, I. M. S., and Kime, R.: Alcohol and human eye movement. *Trans. Am. Neurol. Assoc.* **99**:38–41, 1974.

Winship, D. H., Caffisch, C. R., Zboralske, F. F., and Hogan, W. J.: Deterioration of esophageal peristalis in patients with alcoholic neuropathy. *Gastroenterology* **55**:173–178, 1968.

Wolfe, S. M., and Victor, M.: The relationship of hypomagnesemia and alkalosis to alcohol withdrawal symptoms. *Ann. N.Y. Acad. Sci.* **162**:973–984, 1969.

Wolfe, S. M., and Victor, M.: The physiological basis of the alcohol withdrawal syndrome. In: *Recent Advances in Studies of Alcoholism* (N. K. Mello and J. H. Mendelson, eds.), Washington, U.S. Government Printing Office, 1971, p. 188.

Wolfe, S. M., Mendelson, J. H., Ogato, M., Victor, M., Marshall, W., and Mello, N.: Respiratory alkalosis and alcohol withdrawal. *Trans. Assoc. Am. Physicians* **82**:344–352, 1969.

Wortis, H., Wortis, S. B., and Marsh, F. I.: Vitamin C studies in alcoholics. *Am. J. Psychiatry* **94**:891, 1938.

Wright, D., Laureno, R., and Victor, M.: Pontine and extrapontine myelinolysis. *Brain* **102**:361, 1979.

Zieve, L.: Pathogenesis of hepatic encephalopathy. *Metab. Brain Dis.* **2**:147–165, 1987.

Zilm, D. H., Sellers, E. M., MacLeod, S. M., and Degani, N.: Letter: Propranolol effect on tremor in alcoholic withdrawal. *Ann. Intern. Med.* **83**:234–236, 1975.

Zipursky, R. B., Lim, K. O., and Pfefferbaum, A.: MRI study of brain changes with short-term abstinence from alcohol. *Alcohol Clin. Exp. Res.* **13**:664–666, 1989.

15

Alcohol Abuse: Carcinogenic Effects and Fetal Alcohol Syndrome

Description, Diagnosis, and Prevention

Anthony J. Garro, Barbara H. J. Gordon, and Charles S. Lieber

Alcohol abuse is associated with an increased risk of cancer and birth defects, two disease processes that at some stage involve alterations in gene expression and cell development. The biochemical, molecular, and physiological mechanisms through which alcohol consumption produces these pathological effects are not thoroughly understood and seem to be multifactorial in nature. This chapter focuses largely on the experimental systems that have provided insights into potential mechanisms underlying the carcinogenic and teratogenic effects of alcohol abuse. The epidemiologic studies that established the association between alcohol abuse and cancer and birth defects have recently been reviewed (Garro and Lieber, 1990; Abel and Sokol, 1989) and are not discussed in detail here.

15.1. Alcohol and Cancer

Alcohol abuse has long been recognized as a major risk factor for cancers of the upper alimentary and upper respiratory tract. Cancers at several sites, including the large bowel, breast, pancreas and stomach, also have been correlated with alcohol consumption. In addition, alcohol abuse has been associated with an increased risk

of hepatocellular carcinoma, but in contrast to the aforementioned cancers, this has been attributed not to alcohol consumption *per se* but to the ensuing cirrhosis. There is some evidence, however, that risk of hepatocellular carcinoma is increased in alcoholics even in the absence of cirrhosis (Sakurai, 1969; Lieber *et al.*, 1979).

15.2. Ethanol as a Cocarcinogen

The earliest and most consistently recognized effect of alcohol consumption on cancer risk has been the increased incidence of upper alimentary and upper respiratory tract cancers produced by the combination of alcohol and tobacco use (see Garro and Lieber, 1990, for a review). In a recent case-control study, for example, the increase in oropharyngeal cancer risk, expressed as odds ratios for the combination of these two factors, was approximately 38 times background; heavy drinking in the absence of smoking yielded an odds ratio of 5.8, and heavy smoking without drinking resulted in a 7.4-fold increased risk (Blot *et al.*, 1988). Some studies have indicated a differential risk for upper alimentary/upper respiratory tract (UAT/URT) cancers depending on the type of alcoholic drink, it being greater for spirits than for beer and wine (Wynder and Bross, 1961; Martinez, 1969; Hirayama, 1979; Yu *et al.*, 1988), whereas others related risk to the total amount of ethanol consumed rather than to any particular type of beverage (Williams and Horm, 1977; Jensen, 1979; Tuyns *et al.*, 1979). Even though drinking *per se* produces an increased risk of UAT/URT cancer, the synergistic interaction seen between drinking and smoking has led to the hypothesis that the principal effect of ethanol on the carcinogenic process is that of a cocarcinogen. This hypothesis is supported by the animal studies described below.

15.3. Animal Studies

Although there is no experimental evidence that ethanol *per se* is a carcinogen, numerous studies conducted over the last 25 years have shown that ethanol is capable of acting as a cocarcinogen at several different body sites with a variety of chemical carcinogens.

Because of the relationship between smoking and drinking in cancer, many of the early investigations employed polycyclic aromatic hydrocarbons (PAHs), which are common constituents of tobacco smoke, as the inducing carcinogens. In one of the first such studies, Protzel *et al.* (1964) found that rats fed ethanol in their drinking water exhibited a decreased latent period and increased frequency of buccal tumors induced by topically applied benzo[a]pyrene (BP). In a similar study, ethanol (as a 50% solution) painted over areas that had been pretreated with dimethylbenzanthracene (DMBA) was reported to act as a promoter of DMBA-induced neoplastic transformation in hamster cheek pouches (Elzay, 1966, 1969a). Although ethanol did not affect tumor latency or incidence in these experiments, it did increase the frequency of parabasilar budding and dyskeratoses in exposed animals. Ethanol, used as a solvent, also was reported to act as a cocarcinogen for DMBA in a mouse skin-painting study (Stenback, 1969) and, in a similar application, to enhance the induction of esophageal tumors in mice (Horie *et al.*, 1965). Intraperitoneal injection of ethanol (1.5 mg/kg) for 7 days prior to injection of BP also was reported to increase the frequency of muscle tumors in mice (Capel *et al.*, 1978).

In more recent studies of tobacco-associated carcinogens, McCoy *et al.* (1981, 1986) reported that ethanol administered either as part of an isocaloric pair-feeding regimen or in drinking water increased the incidence of nasal and tracheal tumors induced in hamsters by i.p. N-nitrosopyrrolidine (NPY) but had no effect on tumor induction by N'-nitrosonornicotine

(NNN). Similarly, Griciute *et al.* (1986) did not observe an effect of ethanol on the incidence of NNN-induced tumors when the NNN was administered as an alcoholic solution but did find a decreased latent period in the alcohol group. On the other hand, ethanol consumption has been reported to enhance the frequency of NNN-induced nasal tumors in rats, but this effect was seen only in rats that were fed an ethanol-containing diet 4 weeks prior to and during the NNN exposure (Castonguay *et al.*, 1984). Isocaloric pair-feeding of ethanol also has been shown to promote the progression of esophageal tumors initiated by N-nitrosomethylbenzylamine (MBN) in rats (Mufti *et al.*, 1989a). Interestingly, in this latter study, feeding ethanol either prior to or during carcinogen exposure decreased the incidence of esophageal lesions and tumors. In contrast, prefeeding ethanol (4% in drinking water) in combination with a zinc-deficient diet enhanced MBN induction of esophageal tumors (Gabrial *et al.*, 1982; Newberne *et al.*, 1986). Dietary ethanol administered as a 5% solution in drinking water also has been reported to promote the development of esophageal tumors induced in rats by N-methyl-N-amylnitrosamine (Shimizu, 1986).

Conflicting results have been obtained with the use of dimethylnitrosamine (DMN) and diethylnitrosamine (DEN) in combination with ethanol. Simultaneous administration of ethanol and DMN increased the number of esophageal papillomas and epidermoid carcinomas in rats (Gibel, 1967) and brain tumors in mice (Griciute *et al.*, 1981), but chronic feeding of ethanol had no observable effect on DMN hepatic tumor induction (Teschke *et al.*, 1983). Similarly, the reported effects of ethanol on DEN-induced hepatocarcinogenesis have ranged from enhancement (Porta *et al.*, 1985) to no effect (Schmahl *et al.*, 1965) to a protective effect (Habs and Schmahl, 1981). Finally, with respect to another hepatocarcinogen, vinyl chloride (VC), Radike *et al.* (1977, 1981) reported that its tumorigenic effect was enhanced in rats by the administration of ethanol in drinking water.

Experimentally, ethanol also has been reported to enhance tumor induction at two other sites with which it has been associated as a risk factor in epidemiological studies, namely, the large bowel (see Seitz and Simonowski, 1989, for a recent review) and breast. Seitz *et al.* (1984) showed that prefeeding an ethanol-containing liquid diet (36% total caloric intake) enhanced the induction of rectal tumors in rats by 1,2-dimethylhydrazine. This effect was also seen with the direct-acting carcinogen azoxymethylmethylnitrosamine (AMMN) (Garzon *et al.*, 1986). Conflicting results have been seen with the experimental induction, by azoxymethane (AOM), of colonic cancer in rats. Prefeeding either a low-alcohol beer (12% of caloric intake) or a low dose of ethanol (9% of calories) for 3 weeks prior to AOM administration increased the incidence and proportion of tumors in the left but not the right colon, whereas high-alcohol beer and a high-dose ethanol diet, 23% and 18% of total calories, respectively, reduced tumors in the right colon and had no effect in the left (Hamilton *et al.*, 1987a). Similar inhibiting effects of ethanol prefeeding on AOM-induced colonic cancer have been observed in studies over a range of ethanol and AOM concentrations (Hamilton *et al.*, 1987b, 1988).

With respect to ethanol's effect on breast cancer, two experimental models have shown an enhancement of tumors with ethanol feeding. Ethanol administered as a 12% solution in drinking water, commencing immediately after weaning, significantly reduced the latent period in the genesis of spontaneous mammary adenocarcinoma in C3H/St mice (Schrauzer *et al.*, 1979). Of particular interest in this latter study was the finding that serum prolactin levels were actually lower in the alcohol-fed mice relative to the controls, as this observation is incompatible with the hypothesis that ethanol promotes breast cancer by stimulating prolactin secretion (Williams, 1976). Ethanol administered either by gavage or in a pair-feeding regimen for 3 to 5 weeks prior to carcinogen administration also was reported to enhance the initiation of mammary cancers induced either with DMBA or methylnitrosourea (Grubbs *et al.*, 1988; Singletary *et al.*, 1990).

In summary, the weight of the animal data indicates that ethanol consumption can indeed enhance the carcinogenic activity of a broad spectrum of organ-specific carcinogens in several animal species. This cocarcinogenic effect may be influenced, however, by the dose and timing of the ethanol exposure.

15.4. Biochemical Bases for Ethanol's Activity as a Cocarcinogen

In the context of a multistage theory of chemical carcinogenesis, initiation generally refers to the DNA-damaging effects of chemicals that either are direct-acting electrophiles, capable of reacting with nucleophilic centers on DNA, or are converted to electrophilic derivatives during the course of their metabolism (Miller, 1970; Miller and Miller, 1977). The principal enzyme system that is involved in this metabolism is the microsomal cytochrome P-450-dependent mixed-function oxidases. The inductive effect of ethanol on this enzyme system has focused attention on ethanol's ca-

pacity to influence carcinogenesis through its effects on P-450-mediated carcinogen and retinoid metabolism. Postinitiation, promotion, and progression stages of carcinogenesis have been identified through the effects of carcinogens that neither are DNA-reactive nor give rise to DNA-reactive metabolites and that exert their effects after initiation has been completed. Examples of such compounds include the phorbol esters, which promote the development of skin tumors at sites that had previously been initiated with polycyclic hydrocarbons such as BP (for reviews, see Weinstein, 1978; Weinstein *et al.*, 1979; Berenblum, 1979) and phenobarbital, which promotes hepatic tumor development in animals that had previously been treated with hepatocarcinogens (Peraino *et al.*, 1973; Kitagawa *et al.*, 1979).

In most of the animal studies reviewed above, no attempt was made to determine whether ethanol's cocarcinogenic activity was related to effects on initiation, promotion, or progression. Furthermore, in a number of studies that considered the phase at which ethanol was acting, conflicting results have been reported, particularly with respect to ethanol's capacity to influence the postinitiation phase of carcinogenesis. For example, Stenback (1969) indicated that ethanol did not act as a promoter when applied to mouse skin that had been pretreated with 9,10-DMBA, whereas Elzay (1966, 1969a) reported that a topically applied 50% solution of ethanol promoted carcinogenesis by 7,12-DMBA that had been painted on hamster cheek pouches. More recently, Mufti *et al.* (1989a), reported that ethanol, administered in the diet as part of a pair-feeding study, affected the promotion but not the initiation of esophageal tumors initiated in rats by MBN.

The mechanisms whereby ethanol may act directly as a cocarcinogen in either the initiation or promotion phase of chemical carcinogenesis include (1) cytotoxic effects of ethanol and its metabolites, (2) the induction of microsomal enzymes that in turn affect the metabolism of carcinogens and retinoids and also affect lipid peroxidation, (3) a diminished capacity to detoxify electrophiles, (4) inhibition of repair of carcinogen-DNA adducts, and (5) suppression of immune responses. Indirect consequences of alcohol abuse, which may increase cancer risk, include increased risk of hepatitis B virus infection and specific dietary deficiencies.

15.4.1. Cytotoxic and Mitogenic Effects

Replicating DNA, because of its partially single-stranded nature, is more reactive with many chemical carcinogens than resting DNA (Singer and Fraenkel-Conrat, 1969). Repeated cell injury and repair in the presence of carcinogens would be expected, therefore, to sensitize tissues to chemical carcinogens. Liver cancer, for example, is chemically induced more readily when carcinogen exposure is superimposed on a regenerating liver (Craddock, 1978). Moreover, cell population and chemicals which induce cell proliferation increase the risk of cancer (Cohen and Ellwein, 1990, 1991). In support of the idea that some direct cytotoxic effect of ethanol on the tissues it contacts plays a significant role in alcohol-associated cancers, Williams and Horm (1977) noted that the Third National Cancer Survey data indicated a gradient of decreasing risk that paralleled the successive dilution of alcohol in the alimentary tract and portal circulation: highest in the oral cavity, lower in the larynx and esophagus, and lowest in the stomach, pancreas, and rectum. Local effects of drinking on the alimentary tract have been demonstrated in a number of studies. In an oral cytology survey Anderson (1972) found that diskaryotic cells occurred with higher frequency in heavy drinkers relative to other patients, and Winship *et al.* (1968) noted the occurrence of alcohol-associated functional abnormalities in the esophagus that may represent either direct myopathic or neuropathic effects of ethanol. There is, however, a lack of convincing experimental evidence to support the assumption that ethanol facilitates carcinogen penetration in the UAT by affecting the permeability of the mucosal barrier. Fromm and Robertson (1976), for example, did not find evidence for significant changes in mucosal permeability following ethanol. Also, the epidemiologic association of UAT and URT cancers with the consumption of weak alcoholic beverages such as beer does not support the idea that the cocarcinogenic effect of ethanol resides in its capacity to damage the esophageal mucosa. There are reports, discussed in more detail in Chapter 10, that ethanol does alter gastric permeability and active transport of various ions in the stomach (Davenport, 1967; Dinoso *et al.*, 1972; Smith *et al.*, 1971), but even here it has been suggested that the deleterious effects of ethanol are mediated only in part by a change in mucosal permeability (Shanbour *et al.*, 1973; Sernka *et al.*, 1974; Fromm and Robertson, 1976).

Acute ethanol administration has been associated with gastritis, and this effect has been confirmed in binge drinkers (Gottfried *et al.*, 1978), but the incidence of chronic gastritis in heavy drinkers is a more controversial issue. Whereas some investigators have suggested a connection between this condition and alcohol abuse (Joske *et al.*, 1955; Dinoso *et al.*, 1972; Roberts, 1972; Pitchumoni *et al.*, 1976), other studies failed to detect a relationship between alcohol intake and histo-

logical evidence of atrophic gastritis (Palmer, 1954; Wolff, 1970). The reason for this inconsistency is not known, but it may result, at least in part, from patient selection (nutritional status, definitions of alcoholism), biopsy techniques (blind versus directed), and length of abstinence before examination. A possible increase in the incidence of atrophic gastritis in heavy alcohol users would be of particular interest, since this lesion appears to be a precursor of human gastric carcinoma.

In the liver, large numbers of autopsy studies have shown that the occurrence of hepatoma is closely associated with cirrhosis, which is generally related to alcohol consumption. The incidence of cirrhosis in patients with hepatoma varies from 60% to 90% in different studies (Tuyns, 1978). It should also be noted that a few studies have indicated that in alcoholics hepatomas may occur even in the absence of cirrhosis (Sakurai, 1969; Lieber *et al.*, 1979). This may indicate that alcohol-induced haptic cell injury below the level of identifiable cirrhosis could act as a predisposing condition to carcinogenesis.

Ethanol also may stimulate cell proliferation in the absence of any marked antecedent cytotoxic effect. Chronic ethanol consumption has been reported to stimulate rectal cell proliferation in the rat (Simanowski *et al.*, 1986; Seitz and Simanowski, 1989), and the ethanol-associated increase in dimethylhydrazine-induced cancer in this species was suggested to be related to this cell proliferative effect (Seitz *et al.*, 1984). This proposal is further supported by the observation that chronic ethanol consumption is also cocarcinogenic in the rectum with the direct-acting carcinogen AMMN (Garzon *et al.*, 1986). Ethanol also has been shown to be mitogenic for esophageal epithelium (Mak *et al.*, 1987a,b; Haentjens *et al.*, 1987) and also to potentiate tracheal squamous metaplasia caused by vitamin A deficiency in rats (Mak *et al.*, 1984).

15.4.2. Microsomal Enzyme Induction

15.4.2.1. Carcinogen Metabolism

The association of alcohol consumption with cancers at sites that do not come into contact with high concentrations of alcohol suggests that mechanisms other than, or in addition to, the direct cytotoxic effects of ethanol play a role in carcinogenesis. One possible explanation for ethanol's ability to act as a cocarcinogen at remote sites as well as at ethanol contact sites resides in ethanol's capacity to act as an inducer of the microsomal cytochrome P-450-dependent biotransformation system (Rubin and Lieber, 1968; Joly *et al.*, 1973;

Lieber, 1973). It is well known that this enzyme system is involved in the metabolic conversion of many structurally diverse chemical carcinogens to highly reactive electrophilic intermediates capable of reacting with critical macromolecules including nucleic acids and proteins (Miller, 1970; Gillette, 1976; Miller and Miller, 1977). Furthermore, it has been suggested that there is an association between the levels and distribution of various P-450 isozymes and some cancers (Kellerman *et al.*, 1973; Emery *et al.*, 1978; Ayesh *et al.*, 1984; Nebert, 1988; Kadlubar *et al.*, 1988; Caporaso *et al.*, 1989). Work conducted in this laboratory and by others has shown that dietary ethanol does indeed result in the induction of carcinogen-activating enzymes not only in the liver, the major site of xenobiotic metabolism, but also in a number of other tissues in which alcohol-associated cancers are observed. These tissues include the lungs and intestines, which are major portals of entry for tobacco smoke and dietary carcinogens, and the esophagus (for a recent review see Lieber *et al.*, 1987). Induction of P-450 in the esophagus may be particularly relevant to carcinogenesis at this site because of both the specificity of P-450 isozymes (Swann *et al.*, 1984) and the low concentrations of other detoxifying enzyme systems in this tissue (Farinati *et al.*, 1989).

The general approach used in these studies has been to isolate microsomal preparations from tissues of alcohol-fed and control-diet animals, generally rats, hamsters, or mice, and then assay these preparations for

their ability to metabolically convert procarcinogens either to mutagens detectable in the Ames *Salmonella* mutagenesis assay (Ames *et al.*, 1975; Maron and Ames, 1983) or to other detectable end products.

Enhanced microsomal conversion of many structurally diverse carcinogens has been observed after an inductive pretreatment with ethanol. The carcinogens utilized in these studies have included compounds and mixtures found in tobacco smoke such as BP, NPY, NNN, and tobacco pyrolyzate (Seitz *et al.*, 1978, 1979, 1981a; Lieber *et al.*, 1979; McCoy and Wynder, 1979; Farinati *et al.*, 1985), models of dietary carcinogens such as DMN and tryptophan pyrolyzate (Maling *et al.*, 1975; Garro *et al.*, 1981; Seitz *et al.*, 1981a), and other hepatotoxins and carcinogens such as carbon tetrachloride, 2-aminofluorene, 2-acetylaminofluorene, benzene, 4-aminobiphenyl, benzidine, and methylazoxymethanol (Maling *et al.*, 1975; Lieber *et al.*, 1979; Sato *et al.*, 1980, 1981; Driscoll and Snyder, 1984; Ioannides and Steele, 1986; Fiala *et al.*, 1987). In some instances these inductive effects have exhibited tissue, substrate, gender, and species specificities. For example, in the intestine ethanol increased microsomal activation of BP and tryptophan pyrolyzate but not tobacco pyrolyzate, whereas, as seen in Table 15.1, lung microsomes from ethanol-fed rats exhibited an enhanced capacity to activate the promutagens in tobacco pyrolyzate but did not exhibit any increased activity toward BP or tryptophan pyrolyzate (Seitz *et al.*, 1981a). Although the mutagen or mutagens being activated in the tobacco pyrolyzate

TABLE 15.1.   Effect of Chronic Ethanol Consumption on the Capacity of Lung Microsomes to Activate Tobacco Pyrolyzate, Benzo[a]pyrene, and Tryptophan Pyrolyzate to Mutagens[a]

Treatment	Tobacco pyrolyzate[b]		Benzo[a]pyrene		Tryptophan pyrolyzate[c]	
	Amount/plate (mg tobacco)	His+ revert./plate[d]	Amount/plate (μg)	His+ revert./plate[d]	Amount/plate (μg tryptophan)	His+ revert./plate[d]
Ethanol-fed rats	0.5	110	1.56	55	83	29
	1.0	240	3.13	78	167	98
	2.0	483	6.25	101	333	160
			12.5	110	666	168
			25.0	113		
Pair-fed controls	0.5	62	1.56	53	83	26
	1.0	118	5.13	75	167	100
	2.0	181	6.25	99	333	157
			12.5	109	666	165
			25.0	114		

[a]Mutagenesis assays were performed with microsomes from pools of three animals each. From Seitz *et al.* (1982a).
[b]Microsomal protein/plate was 0.7 mg. Similar results were obtained with 0.9 and 1.1 mg microsomal protein/plate.
[c]Microsomal protein/plate was 1.1 mg.
[d]Numbers are averages of His+ colonies of *Salmonella typhimurium* TA 100 on duplicate plates. The number of spontaneous His+ revertants have been subtracted.

are not known, it is of interest that lung microsomes from alcohol-fed rats also exhibit an enhanced capacity to activate the tobacco mutagen NPY (Farinati *et al.*, 1985; Fig. 15.1). Ioannides and Steele (1986) also reported both inductive and inhibitory effects toward different PAHs; Seitz *et al.* (1981b) demonstrated a gender-specific effect with respect to induction of BP metabolism in the rat, and Anderson *et al.* (1986) have reported that in contrast to rats, in which ethanol induces DMN demethylase activity, no induction is seen in mice.

Ethanol's ability to induce DMN demethylase activity (Fig. 15.2) is of particular interest, as it is detectable over a DMN concentration range of 0.3–100 mM (Garro *et al.*, 1981). This is in contrast to other microsomal enzyme inducers such as phenobarbital, 3-methylcholanthrene, and polychlorinated biphenyls, which increase the activity of DMN demethylase isozymes whose activity is detectable only at relatively high DMN concentration (> 40 mM) and repress the activity of low-K_m DMN demethylases (see Garro and Lieber, 1990, for a review). This effect of ethanol appears to be caused by the induction of a unique species of cytochrome P-450 (Ohnishi and Lieber, 1977; Koop *et al.*, 1982) that differentially affects the activation of various carcinogens; a selective affinity for DMN has indeed been demonstrated with the ethanol-induced cytochrome P-450 (Yang *et al.*, 1985). Ethanol is also an effective competitive inhibitor of DMN demethylase activity (see Driver and Swann, 1987, for a review). This capacity to act as both an inducer and an inhibitor may explain the conflicting reports of ethanol's influence on DMN-mediated carcinogenicity alluded to in Section 15.3, particularly when the route of exposure and the presence or absence of ethanol at the time of exposure are taken into account. As pointed out by Swann *et al.* (1984), when DMN is administered orally the liver can effect a "first-pass clearance" up to a DMN dose of 30 μg/kg. At higher doses the hepatic enzymes become saturated, and methylation of the kidneys and other organs occurs. Ethanol, when given to rats in relatively low amounts, equivalent to a person drinking 0.5 liter of beer, prevents this "first-pass" clearance and can produce a fivefold increase in the methylation of kidney DNA.

In summary, ethanol consumption increases the capacity for microsomal activation of many classes of chemical carcinogens in different tissues. The significance of this effect of ethanol vis-à-vis actual cancer risk will be influenced by a variety of other factors operating *in vivo*, including the carcinogen-detoxifying capacity of various tissues, the route of carcinogen exposure, and, in the alcohol abuser in particular, whether or not ethanol is present in the circulation at the time of carcinogen exposure.

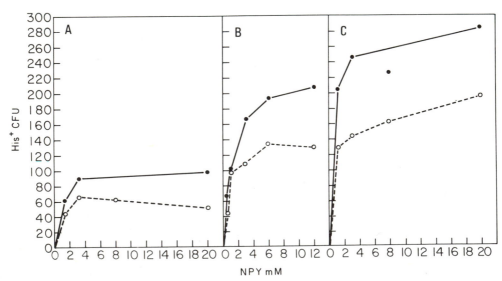

FIGURE 15.1. His+ revertants of TA1535 induced by NPY activation by lung microsomes from ethanol-fed (●) and control animals (○). Each point represents the average number of induced His+. The microsomal protein concentrations used in A, B, and C were 1.7, 2.9, and 3.8 mg/ml, respectively. The experiments reported in each panel were carried out using different pooled lung samples from six pairs of pair-fed animals. (From Farinati *et al.*, 1985.)

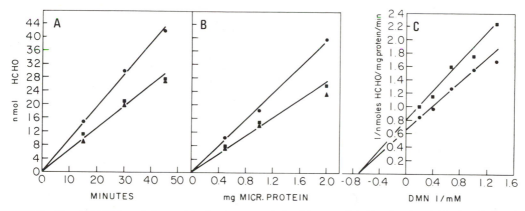

FIGURE 15.2. Hepatic DMN demethylase activity in microsomes isolated from rats fed either the ethanol (●)-, isocaloric carbohydrate (■)-, or isocaloric fat (▲)-containing diets. Formaldehyde was measured as [^{14}C]formaldehyde produced from di-[^{14}C]methylnitrosamine by incubation with the microsomes in the presence of 5 mM NADPH at 37°C in a 0.1 M sodium phosphate, 2 mM EDTA (pH 6.4) buffer. (A) Time course of formaldehyde production in a 0.5-ml reaction mixture containing 1 mM DMN (specific activity, 1300 dpm/nmol) and 1 mg microsomal protein. (B) A 0.5-ml reaction mixture containing 1 mM DMN (specific activity, 2100 dpm/nmol) after a 25-min incubation. (C) Lineweaver–Burk plot of DMN demethylase activity at a microsomal protein concentration at 2 mg/ml, DMN specific activity at 370 dpm/nmol, and incubation time of 25 min. (From Garro *et al.*, 1981.)

15.4.2.2 Effects on Retinol (Vitamin A) Metabolism

Ethanol consumption results in a severe depression in hepatic vitamin A levels through at least two mechanisms. (1) It increases mobilization of vitamin A from the liver to other organs (Sato and Lieber, 1981; Leo *et al.*, 1986a). (2) It induces a cytochrome P-450-mediated breakdown of both retinol and retinoic acid (Sato and Lieber, 1982; Leo and Lieber, 1985). These effects of ethanol may be of importance in carcinogenesis, as vitamin A plays an essential role in the maintenance of normal growth and control of cell differentiation in a variety of epithelial and mesenchymal tissues (see Sporn and Roberts, 1983; Sporn *et al.*, 1984, for reviews). In addition, there is both epidemiologic and experimental evidence that vitamin A has anticarcinogenic properties affecting both the initiation and promotion stages of carcinogenesis (for a recent review, see Ziegler, 1989).

Dietary carotenoids and retinyl esters are the major sources of vitamin A, which is stored in the liver in the form of retinyl esters (Rasmussen *et al.*, 1985). In animals, retinoic acid is just as effective as retinal as a dietary supplement (Zile and DeLuca, 1968), and retinoic acid is more effective than either retinol or retinal as an anticarcinogen or inducer of cellular differentiation *in vitro* (Strickland and Mahdavi, 1978; Breitman *et al.*, 1980; Lotan, 1980). Epidemiologic studies involving different geographic locales have associated dietary retinoid deficiency and low serum vitamin A levels with increased cancer risk, particularly of the esophagus and lung (Kmet and Mahboubi, 1972; MacLennan *et al.*, 1977; Atukorola *et al.*, 1979; Mettlin *et al.*, 1980; Kark *et al.*, 1981; Kvale *et al.*, 1983). These epidemiologic studies have been supported by animal studies, which have demonstrated the efficacy of retinoids in prevention of cancers at different body sites (Moon *et al.*, 1977; Becci *et al.*, 1978; Sporn and Newton, 1981; Hill and Grubbs, 1982).

Ethanol consumption has been observed to interact synergistically with vitamin A deficiency in increasing the incidence of tracheal squamous metaplasia (Fig. 15.3; Mak *et al.*, 1984, 1987a). This is of particular interest with respect to UAT/URT cancers, since squamous metaplasia is one of the earliest stages preceding the development of carcinoma *in situ*, with the latter often being found in association with invasive carcinomas (Umiker and Storey, 1952; Williams, 1952; Weller, 1953; Valentine, 1956; Auerbach *et al.*, 1961). In addition, in the same ethanol-consuming, vitamin-A-deficient rats, the tracheal epithelium, which was not as yet involved in the formation of squamous metaplasia, exhibited a number of morphological abnormalities. The ciliated cells contained an increased number of lysosomes and had compound cilia (Mak *et al.*, 1984, 1987a). Increased

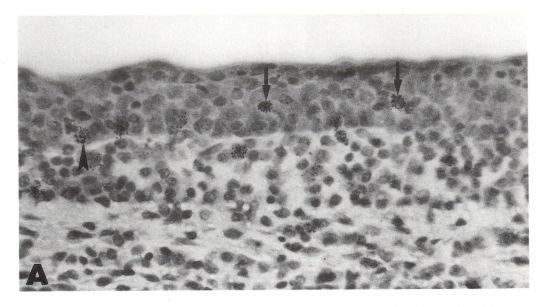

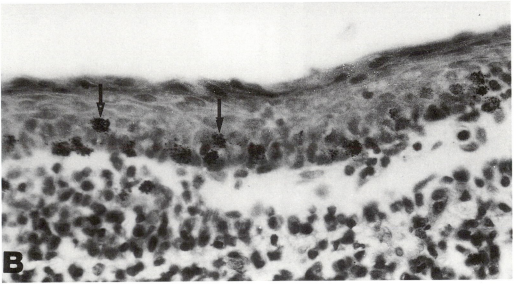

FIGURE 15.3. Autoradiographs illustrating cell proliferation in squamous metaplasia. (A) From a rat fed the vitamin-A-deficient diet. This autoradiograph shows four labeled basal cells in the basal layer at the bottom of the epithelium; one of them is marked by an arrowhead. In the suprabasal layers, labeled cells are also revealed (arrows). (B) From a rat fed the vitamin-A-deficient diet plus ethanol. More labeled basal cells are visible than in A. Labeled suprabasal cells are also seen (arrows). Hematoxylin and eosin, × 500. (From Mak *et al.*, 1987b.)

numbers of lysosomes also have been observed in ciliated tracheal cells following exposure to carcinogens (Harris *et al.*, 1971), and compound cilia have been observed in animals exposed to carcinogens and in humans with bronchial cancer (Reznik-Schuller, 1975; McDowell *et al.*, 1976). More recently, in a study of the effects of vitamin A deficiency and ethanol on esophageal mucosa, it was found that in contrast to the trachea, vitamin A deficiency altered cellular differentiation, but this alteration was not influenced by ethanol, and ethanol, independently of the vitamin A deficiency, stimulated basal cell proliferation (Mak *et al.*, 1987b).

In light of the effect of ethanol on vitamin A metabolism, it may prove useful in formulating hypotheses for the cocarcinogenic effect of ethanol in the UAT/URT to take into account what is known of the effects of vitamin A on differentiation and carcinogenesis. On a molecular level retinoids appear to control the expression of genes that are involved in the cytoskeleton matrix as well as some oncogenes (Sporn and Roberts, 1983). Retinoids also may influence carcinogen metabolism through the induction of specific P-450 isozymes (Leo *et al.*, 1984; Qin and Huang, 1986; McCarthy *et al.*, 1987) and by directly interfering with the P-450-mediated activation of procarcinogens (Fig. 15.4; Leo *et al.*, 1986b). Finally, retinoids also may inhibit tumor development by stimulating various aspects of cell-mediated immune responses (see Bendich, 1989, for a review).

15.4.2.3. Effect on Lipid Peroxidation

Lipid peroxidation has been implicated as having a promotional effect on the carcinogenic process (King and McCay, 1983; Perera *et al.*, 1987). This suggestion is based in large part on the antagonistic effect of dietary polyunsaturated fats and dietary antioxidants in carcinogenesis. In general, dietary fats enhance tumorigenesis (Carroll, 1980; Reddy *et al.*, 1980), whereas antioxidants inhibit the process (Wattenberg, 1978; Ip, 1981; King and McCay, 1983). Although there are a number of explanations for the effects of dietary fat on carcinogenesis, including a nonspecific caloric effect, an increase in the levels of membrane peroxidation, which is inhibited by antioxidants, is another possibility under active consideration (Rogers, 1983).

Experimentally, microsomes from ethanol-fed rats have been shown to generate reactive oxygen intermediates such as superoxide, peroxide, and hydroxyl radical at elevated rates compared with controls (Lieber and DeCarli, 1970; Thurman, 1973; Klein *et al.*, 1983; Dicker and Cederbaum, 1987, 1988). This is associated with more lipid peroxidation in ethanol-fed animals (Dianzani, 1985; Videla and Valenzuela, 1985; Szebeni *et al.*, 1986; Nordmann *et al.*, 1987), and furthermore, there is evidence for ethanol-associated lipid peroxidation in humans (Suematsu *et al.*, 1981; Shaw *et al.*, 1983).

15.4.3. Diminished Capacity to Detoxify Electrophiles

Glutathione (GSH) plays a key role in the detoxification of electrophiles and in the reduction of lipid peroxides. Acute ethanol consumption has been reported to produce a marked decrease in hepatic GSH levels (Estler and Ammon, 1966; MacDonald *et al.*, 1977; Guerri and Grisolia, 1980; Fernandez and Videla, 1981). This effect of ethanol could therefore contribute both to an increase in the number of carcinogen-DNA adducts produced as a result of electrophile production from carcinogens and to increased levels of lipid peroxidation. Several mechanisms contribute to the decreased hepatic GSH levels, with the most significant apparently being an increased efflux of GSH from hepatocytes

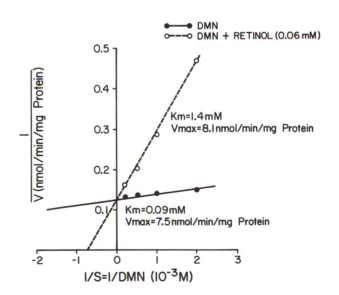

FIGURE 15.4. Effect of retinol on hepatic microsomal DMN demethylation. Liver microsomes of rats fed ethanol were incubated with various concentrations of DMN, with or without retinol. (From Leo *et al.*, 1986b.)

(Speisky *et al.*, 1985). Other contributing factors include the reaction of acetaldehyde both with GSH itself (Vina *et al.*, 1980; Speisky *et al.*, 1985) and with cysteine, a GSH precursor (Lieber, 1980). There also is a decrease in hepatic synthesis of GSH following acute ethanol treatment (Speisky *et al.*, 1985). In contrast, chronic ethanol consumption produces a transient increase in hepatic GSH (Hetu *et al.*, 1982) and does not affect esophageal GSH levels in the rat (Farinati *et al.*, 1989).

15.4.4. Inhibition of DNA-Alkylation Repair

DNA repair processes are important in protecting cells from chemical carcinogens that alter DNA structure and sequences. Such alterations result in either somatic mutations or the expression of oncogenes and ultimately lead to the uncontrolled cellular growth characteristic of tumors. Cells possess a number of enzyme systems capable of repairing different types of DNA damage, and patients born with DNA repair deficiencies are at greater risk of developing cancer (Cleaver, 1980; Becker, 1986). Chronic alcohol consumption may increase cancer risk by inhibition of the DNA repair protein O^6-methylguanine transferase (O^6-MeGT), which removes alklyl groups (methyl and ethyl) from the O^6 position of guanine (Bogden *et al.*, 1981; Craddock *et al.*, 1982; Lemaitre *et al.*, 1982). In rats, chronic and acute alcohol consumption causes an increased persistence of DMN-induced hepatic O^6-MeG DNA adducts and acetaldehyde has been shown to inhibit both rat and human O^6-MeGT activity (Garro *et al.*, 1986; Mufti *et al.*, 1988b; Espina *et al.*, 1988).

The major DNA base alkylation products generated by exposure to alkylating nitroso compounds such as DMN are, in order of frequency of occurrence, N^7-MeG, O^6-MeG, and O^4-MeT (Newbold *et al.*, 1980; Eadie *et al.*, 1984). Persistence of O^6-MeG in DNA of various organs has been associated with carcinogenicity of several alkylating agents (Kleihues and Cooper, 1976; Kleihues *et al.*, 1979; Lindamood, 1984), and alkylation of the O^4 position of thymine also may be significant (Swenberg *et al.*, 1984; Dyroff *et al.*, 1986). O^6-MeGT is an alkyl transferase that transfers methyl or ethyl groups from the O^6 position of guanine to a cysteine residue located in the protein, which in turn inactivates the transferase (Pegg and Perry, 1981; Pegg *et al.*, 1983; Harris *et al.*, 1983).

The first indication that chronic alcohol consumption interfered with the repair of O^6-MeG adducts came

from studies of the effects of ethanol consumption on DMN-induced hepatic DNA alkylation. In these experiments it was observed that O^6-MeG adducts, but not N^7-MeG adducts, persisted for longer periods in ethanol-fed rats than in controls (Garro *et al.*, 1986; Mufti *et al.*, 1988b). Moreover, this effect appeared to be specific for O^6-MeG repair, as removal of acetylaminofluorene adducts, which are repaired by a separate excision pathway (Kriek, 1972), was not affected (Garro *et al.*, 1986). Isolation of O^6-MeGT from ethanol-fed and control-diet animals showed a loss of hepatic O^6-MeGT activity following ethanol consumption. This *in vivo* decrease in O^6-MeGT activity appears to result primarily from acetaldehyde generated by ethanol metabolism. Pretreatment of animals with disulfiram (Antabuse®), which inhibits acetaldehyde dehydrogenase activity and leads to higher and more prolonged levels of acetaldehyde following ethanol administration, exacerbated the loss of O^6-MeGT activity following ethanol administration (Fig. 15.5; Espina *et al.*, 1988). Furthermore, both rat and human O^6-MeGT were shown to be significantly inhibited *in vitro* by acetaldehyde at concentrations as low as 0.1 µM (Fig. 15.6). Ethanol also was observed to inhibit O^6-MeGT *in vitro*, but at concentrations in the range of 10–50 mM, and this inhibition appeared to be by trace levels of acetaldehyde that were generated spontaneously or produced by residual alcohol dehydrogenase activity in the O^6-MeGT preparations (Espina *et al.*, 1988). It must be noted that some studies failed to detect an effect of ethanol on the repair of DMN-induced O-^{6}-MeG adducts (Schwarz *et al.*, 1982; Belinsky *et al.*, 1982) or an inhibition of O^6-MeGT activity by acetaldehyde at concentrations up to 300 µM (Krokan *et al.*, 1985; Grafstrom *et al.*, 1985). The reasons for these conflicting results are not known at the present time but may reflect methodological differences in the various studies (Garro *et al.*, 1986; Espina *et al.*, 1988).

15.4.5. Induction of Sister Chromatid Exchange

Acetaldehyde, the first metabolite of ethanol, induces sister chromatid exchanges (SCEs) in cells grown in tissue culture (Obe and Ristow, 1977; Veghelyi and Osztovics, 1978; Ristow and Obe, 1978; Alvarez *et al.*, 1980). Daily treatment of Chinese hamster cell cultures with concentrations of acetaldehyde ranging from 0.25 $\times 10^{-3}$ to 1.5 $\times 10^{-3}$ % (v/v) produced a dose-dependent increase in SCEs (Obe and Beck, 1979). Acetaldehyde also has been shown to induce SCEs in human lympho-

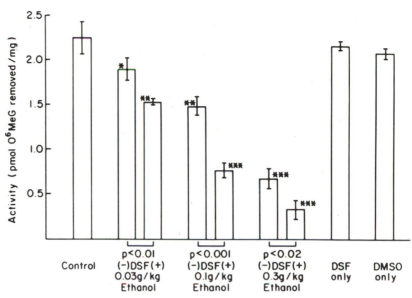

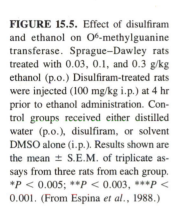

FIGURE 15.5. Effect of disulfiram and ethanol on O^6-methylguanine transferase. Sprague–Dawley rats treated with 0.03, 0.1, and 0.3 g/kg ethanol (p.o.) Disulfiram-treated rats were injected (100 mg/kg i.p.) at 4 hr prior to ethanol administration. Control groups received either distilled water (p.o.), disulfiram, or solvent DMSO alone (i.p.). Results shown are the mean ± S.E.M. of triplicate assays from three rats from each group. *$P < 0.005$; **$P < 0.003$, ***$P < 0.001$. (From Espina *et al.*, 1988.)

cytes exposed *in vitro* (He and Lambert, 1985; Lambert and He, 1988). Along these same lines, Obe and Ristow (1979), from a comparison between their own data on chromosomal aberrations in peripheral blood lymphocytes from alcoholics and literature values, concluded that there is an elevation of chromosomal aberrations in alcoholics. The potential significance of these observations with respect to tumor promotion is related to the hypothesis that compounds with SCE-inducing activity could theoretically act as promoters (Kinsella and Radman, 1978). By increasing the frequency of SCEs, such compounds could enhance the possibility that recessive mutations are expressed. In addition, stimulation of chromosome damage and rearrangement could foster the expression of latent oncogenes. Acetaldehyde has been shown to enhance the tumorigenicity of BP in

FIGURE 15.6. Inhibition of O^6-MeGT activity by acetaldehyde. The acetaldehyde inhibition values are expressed as percentages of enzyme activities observed in the absence of acetaldehyde: (●) isocitrate dehydrogenase (IDH); (○) glutathione-S-transferase (GSHT); (▼) glyceraldehyde-3-phosphate dehydrogenase (GAPD); (△) yeast alcohol dehydrogenase (YADH). The O^6-MeGT activities of rat liver extract (■) and human liver extract (×) were 4.89 pmol O^6-MeG/mg and 14.09 pmol O^6-MeG/mg, respectively. The activities of IDH, GSHT, GAPD, and YADH were 302, 875, 7.30, and 1.65 U/mg, respectively, where 1 unit (U) equals the conversion of 1 nmol substrate per minute. (From Espina *et al.*, 1988.)

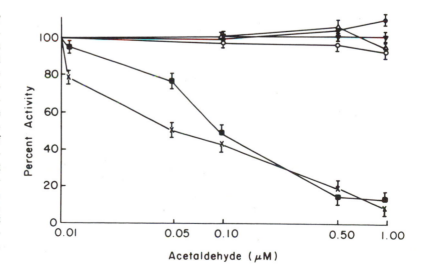

hamster lung (Feron, 1979) and itself induces laryngeal tumors in hamsters and nasal tumors in rats (Woutersen *et al.*, 1984; Feron *et al.*, 1982). Consistent with these observations, acetaldehyde, in the presence of ethanol, has been shown to form mixed acetal-nucleoside DNA adducts (Fraenkel-Conrat and Singer, 1988) and to be mutagenic in a number of test systems (see Dellarco, 1988, for a recent review).

15.4.6. Immunosuppression

Many tumor cells display novel surface antigens that, theoretically, should lead to the recognition and elimination of these cells by the immune system. Because there is both epidemiologic and experimental evidence linking alcohol abuse and ethanol or its metabolites to the suppression of immune responses (see Chapter 8; Mufti *et al.*, 1989b for recent reviews), alcohol-associated immunosuppression has been considered for some time to be a possible contributing factor in the increased cancer incidence seen in alcohol abusers (Vitale *et al.*, 1977). For the most part, however, the epidemiologic studies in which decreased immune responses have been associated with alcohol abuse have involved patients who already had alcoholic liver disease (see Chapter 8). It is difficult to assess, therefore, whether the immunologic defects observed were caused directly by ethanol or reflected other aspects of the disease process including malnutrition, which is known to affect immune responses (Petro *et al.*, 1984). Nevertheless, there are animal and *in vitro* studies involving isolated immunocompetent cells that have shown that both acute and chronic exposure to either ethanol or some of its metabolites impairs cell-mediated immune functions in the absence of marked liver dysfunctions. For example, Roselle and Mendenhall (1984) reported a significant decrease in lymphocyte response to mitogens after chronic ethanol treatment in guinea pigs. Jerrells *et al.* (1986) demonstrated that acute ethanol administration in rats resulted in a rapid loss of lymphocytes from spleen and thymus. Mufti *et al.* (1988a) showed a similar depletion of splenic lymphocytes following chronic ethanol consumption in rats and a change in T-helper/T-suppressor cell ratios. Aldo-Bensen (1989) has reported that ethanol, at concentrations of 100–150 mg/dl directly suppresses B-cell responses. There are conflicting reports on the effects of ethanol on the activities of NK and K cells, which are believed to play a role in defense against tumors. Saxena *et al.* (1980, 1981) reported an increase in NK activity following ethanol in both hu-

mans and rodent. *In vitro* studies of NK activity were consistent with the *in vivo* observations showing a moderate increase at ethanol concentrations up to 0.2% followed by a decline in activity above 2% ethanol (Mufti *et al.*, 1988a). Meadows *et al.* (1989), on the other hand, reported that baseline and IL-2-stimulated NK activity was decreased in mice after 2 weeks on a diet containing ethanol at approximately 40% of total calories.

Even though there are clear immunologic defects associated with alcohol abuse, some of which may be the direct consequences of pathological effects of ethanol or its metabolites, there is reason to question the significance of these effects regarding general chemical carcinogenesis. Although immunosuppressed patients or animals do exhibit increased cancer incidences, the cancers observed are mostly lymphoreticular neoplasms, i.e., cancers of the immune system itself (Frizzera *et al.*, 1980; Baird *et al.*, 1982). Furthermore, nude mice, which are genetically defective with respect to T-cell-mediated immune responses, do not exhibit an increased incidence of spontaneous tumors in organs other than those of the immune system, nor are they more susceptible than normal mice with respect to chemically induced cancers (Rygaard and Povlson, 1974; Stutman, 1979). Nevertheless, the immune system may play a vital role in the defense against virally induced tumors, particularly in hepatitis-B-virus-associated hepatocellular carcinoma.

15.5. Hepatitis B Virus and Hepatocellular Carcinoma

Hepatitis B virus (HBV) infection is associated with increased risk of hepatocellular carcinoma (Beasley *et al.*, 1981; Beasley, 1982), and alcoholics have an increased incidence of HBV infection (Mills *et al.*, 1979; Hislop *et al.*, 1981; Orholm *et al.*, 1981; Gluud *et al.*, 1982; Chevilotte *et al.*, 1983). The HBV DNA is capable of integrating into host genomic DNA, particularly in chronic HBV carriers (Brechot *et al.*, 1981; Shafritz *et al.*, 1981), and then may induce chromosomal alterations important in hepatocarcinogenesis (Henderson *et al.*, 1988). Brechot *et al.* (1982), in a study of patients with various stages of alcohol liver disease and alcohol liver disease with hepatocellular carcinoma, found that whereas eight out of 51 subjects with alcohol liver disease without carcinoma had integrated HBV DNA in their livers, all 20 subjects with cirrhosis and hepatocellular carcinoma had integrated HBV DNA sequences. Further support for the interaction among alcohol, HBV infection, and hepatocellular carcinogenesis has been provided by Ohnishi *et al.* (1982), who noted that hepatocarcinogenesis was hastened significantly in HBsAg carriers who continued to drink.

15.6. Dietary Deficiencies and Alcohol Abuse

In addition to the direct effects of alcohol consumption on vitamin A metabolism, which, as discussed earlier, may influence both the initiation and promotion of chemically induced cancers, other alcohol-associated dietary deficiencies also may contribute to cancer risk. In cases of chronic alcoholic abuse, ethanol may account on the average for as much as 50% of an individual's daily caloric intake. In this regard, a case-control study among black males in Washington, D.C. noted that poor nutritional status was an important risk factor in alcohol-associated esophageal cancer (Pottern *et al.*, 1981; Ziegler *et al.*, 1981).

15.6.1. Iron and Zinc Deficiencies

Chronic iron deficiency, which is seen in alcohol abusers, also has been associated with an increased risk of UAT cancer both in women with Plummer–Vinson syndrome, which is characterized by dysphagia and UAT atrophy (Wynder *et al.*, 1957), and in inhabitants of Central Asia (Kmet and Mahboubi, 1972). Chronic iron deficiency also may play a role in the etiology of gastric cancer (Vitale and Gottlieb, 1975) and may influence cell-mediated immune responses (Chandra *et al.*, 1977; Beisel *et al.*, 1981). Alcoholics also have hyperzincuria and reduced zinc levels (Vallee *et al.*, 1956; Sullivan and Lankford, 1962), and experimentally zinc-deficient diets have shown to enhance esophageal tumor induction in rats treated with MBN (Gabrial *et al.*, 1982). The zinc-deficient diet presumably either enhances the metabolic activation of MBN by microsomal P-450 or stimulates cell proliferation in the esophagus.

15.6.2. Riboflavin (Vitamin B$_2$)

Riboflavin deficiency is common among alcoholics, especially in lower socioeconomic groups, and riboflavin deficiency also has been implicated in the Plummer-Vinson syndrome (Wynder and Fryer, 1958). Experimentally, Wynder and Chan (1970) reported that in mice, riboflavin deficiency is associated with epithelial hyperplasia and increased susceptibility of skin to chemically induced cancer.

15.6.3. Pyridoxine (Vitamin B$_6$)

Pyridoxine deficiency also has been associated with alcohol abuse and appears to be related to the acetaldehyde derived from ethanol metabolism (Lumeng and Li, 1974; Lumeng, 1978). The decreased hepatic transaminase activity that is associated with alcohol abuse (Matloff *et al.*, 1980) is apparently a result of acetaldehyde's effect on pyridoxine (Solomon, 1987), which acts as a cofactor for these enzymes. Pyridoxine also plays a key role in hematopoiesis and in both cell-mediated and humoral immune responses (Axelrod, 1971; Beisel *et al.*, 1981), which ultimately may affect viruses such as HBV. Wynder (1976) also has suggested that pyridoxine deficiency is associated with enhanced hepatocarcinogenesis.

15.6.4. Vitamin E

Alcoholics have been reported to have abnormally low blood levels of vitamin E (Losowsky and Leonard, 1967; Myerson, 1968), and in at least one experimental system, vitamin E has been shown to interact synergistically with another antioxidant, selenium, in preventing mammary carcinogenesis in rats (Horvath and Ip, 1983).

15.6.5. Lipotrope Deficiency

Chronic ethanol consumption in rats increases requirements for lipotropic factors such as choline and methionine (Best *et al.*, 1949; Finkelstein *et al.*, 1974). Lipotrope-deficient diets also have been shown to enhance the hepatocarcinogenicity of chemical carcinogens in rats (Rogers and Newberne, 1980; Rogers, 1983), and ethanol further enhances this effect of a lipotropic diet (Porta *et al.*, 1985). The relevance of these observations to humans is questionable, however, since the hepatic injury induced by lipotrope deficiency appears to be primarily a disease of rats (see Chapter 4 for a full discussion of this topic).

15.7. Fetal Alcohol Syndrome: Description, Diagnosis, and Prevention

Chronic alcohol consumption during pregnancy has been associated with a cluster of abnormalities in fetal development that is referred to as the fetal alcohol syndrome (FAS), which includes decreased pre- and postnatal growth, craniofacial abnormalities, and impaired central nervous system development (see Table 15.2; Sokol *et al.*, 1986) (Lemoine *et al.*, 1968; Jones and Smith, 1973; Jones *et al.*, 1973). Behavioral problems such as hyperactivity and learning disabilities that occur in the absence of physical dysmorphology but may be related to abnormal brain function also have been associated with drinking during pregnancy and have been referred to as alcohol-related birth defects (ARBD) or fetal alcohol effects (FAE). Clinical studies suggest that 32–50% of alcoholic women who drink heavily during pregnancy give birth to offspring affected by alcohol (Ouellette *et al.*, 1977; Erb and Andersen, 1978; Lancet, 1983). Others have reported the association between social drinking, i.e., one to two drinks per week during the first trimester or one to two drinks per week during the second trimester, and an increased risk of miscarriage (Kline *et al.*, 1980; Harlap and Shiono, 1980). More than 2 oz of absolute alcohol per day is generally considered heavy consumption of alcohol.

In the U.S. the incidence of FAS is estimated to be 0.3 to 0.5 per 1000 live births (Abel and Sokol, 1991).

This translates into approximately 1400 FAS children born in the United States per annum (Abel and Sokol, 1991). Some FAE (excluding FAS) are seen in about 3–4 children per 1000 live births, a possible 30,000 children born each year in the United States (Lancet, 1983; Little *et al.*, 1983, 1986). These births result in an annual estimated cost of over $1 million for residential care of the mentally retarded alone caused by FAS. The estimated cost for treatment of FAS-related growth retardation, anatomic abnormalities, and mental retardation combined is $74.6 million per year (Abel and Sokol, 1991). Alcohol consumption during pregnancy has been suggested to be one of the most prevalent causes of preventable mental retardation, learning disabilities, and behavior problems (Clarren and Smith, 1978; Abel, 1984).

Several books have fully reviewed the FAS, including human case reports and experimental studies (Abel, 1984; Rosett and Weiner, 1984). There are also extensive publications on alcohol effects on brain development and function (Abel, 1981; Colangelo and Jones, 1982; Ciba Foundation, 1983; West, 1986, 1987).

15.7.1. Description of the Fetal Alcohol Syndrome

The three most striking and prevalent abnormalities that define the FAS as first described (Jones *et al.*, 1973; Jones and Smith, 1973) and later named by the Fetal Alcohol Syndrome Study Group of the Research Society on Alcoholism in the United States (Rosett, 1980) are (1) generalized prenatal growth deficiency without postnatal catch-up, (2) craniofacial anomalies, and (3) central nervous system abnormalities including mental retardation (see Table 15.2; Sokol *et al.*, 1986). Many other organ systems, however, are also affected by prenatal alcohol exposure.

The decrement in size seen at birth in FAS children includes decreased birth length, birth weight, and head circumference in approximately 95% of cases (Jones and Smith, 1975; Mulvihill and Yeager, 1976; Clarren and Smith, 1978).

The craniofacial anomalies include ocular abnormalities such as short palpebral fissures, ptosis, and strabismus (Jones *et al.*, 1973), hypognathia, broad nasal bridge, and curved upper lip with deficient philtrum development (Jones and Smith, 1973; Streissguth *et al.*, 1980; Neri *et al.*, 1988). As shown in Fig. 15.7 (Schardein, 1985), similar anomalies have been demonstrated in mice following an acute dose of ethanol on day 7 of gestation, the time of murine neural plate and

---

TABLE 15.2.   Minimal Criteria for the Diagnosis of Fetal Alcohol Syndrome (FAS) as Proposed by the Fetal Alcohol Study Group of the Research Society on Alcoholism[a]

Prenatal and/or postnatal growth retardation in weight and/or length for gestational age (below tenth percentile)
CNS involvement, including:
   Signs of neurological abnormality
   Development delay
   Intellectual impairment
   Morphological abnormalities, e.g., head circumference below third percentile or brain malformation
Characteristic facial dysmorphology, qualitatively described as including:
   Microphthalmia and/or short palpebral fissures
   Elongated midface
   Poorly developed philtrum, thin upper lip, and flattening of the maxillary area

[a]Adapted from Sokol *et al.* (1986).

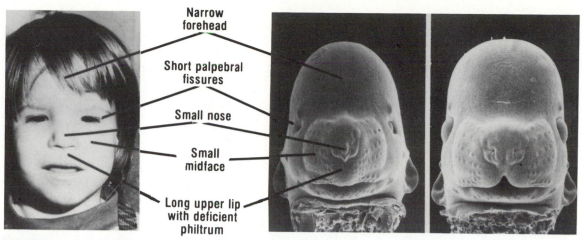

FIGURE 15.7. Face of a FAS child at 2 years 6 months compared to day-14 mouse treated with 7.9 mg/kg ethanol on day 7, shown on the left, compared to day-15 control, on the right. Courtesy of K. Sulik. (From Schardein, 1985.)

midline facial development (Sulik *et al.*, 1981). Many of the craniofacial and structural abnormalities as well as the intellectual impairment seen in humans persist throughout development (Figs. 15.8A, B, and C and 15.9A, B, and C; Streissguth, 1985).

The central nervous system abnormalities involve a myriad of structural and functional defects (for reviews see Abel, 1981; Colangelo and Jones, 1982; West, 1986, 1987). The consequences of this effect, i.e., abnormal brain development, decreased intelligence and learning ability, and abnormal behavior, constitute the overriding social concerns resulting from excessive exposure to ethanol during pregnancy. Bonthius *et al.* (1988) have reported that the incidence of microcephaly, which is indicative of a decrement in brain growth, is related in a dose-dependent fashion to maternal blood alcohol levels. Abnormal cellular development and cell migration are also associated with prenatal alcohol exposure (Clarren *et al.*, 1978; Clarren and Smith, 1978; Diaz and Samson, 1980; Samson, 1986; Samson and Grant, 1984). Recent work has refined the descriptions of developmental abnormalities relating them to specific brain regions and subcellular organelles (Stibler *et al.*, 1983; West *et al.*, 1986; West, 1987; Dow and Riopelle, 1985; Savage *et al.*, 1989). These may be related to abnormalities of brain biochemistry and neuronal function (Druse Manteuffel, 1990; U.S. Public Health Service, 1990). Specific behavioral alterations in humans as well as in animal models include tremulousness, poor grasp, poor sucking response (Ouellette *et al.*, 1977; Martin *et al.*, 1978; Riley and Rockwood, 1984), poor eye–hand

coordination, abnormal sleep patterns (Streissguth *et al.*, 1978; Rosett and Sandler, 1979), attention deficits, delayed reaction time, delayed language development, and short-term memory deficits (Iosub *et al.*, 1981; Streissguth, 1986) as well as other learning and behavior deficits in children and animal models (Riley *et al.*, 1979; Randall *et al.*, 1986; Streissguth and LaDue, 1987; Rockwood and Riley, 1986; Bond, 1986).

Mental retardation, operationally defined as an IQ below 70, is observed in approximately 50% of FAS children (Mulvihill and Yeager, 1976). The severity of intellectual deficits is positively correlated with the severity of anatomic defects in these children (Streissguth *et al.*, 1978). The classic FAS child, however, represents only a portion of a greater number of FAE children who exhibit some behavioral and learning abnormalities without necessarily exhibiting the entire spectrum of craniofacial and other teratogenic effects of prenatal alcohol exposure.

Hypoplasia of the optic nerve has been demonstrated in 46% of the FAS children studied, with excessively tortuous retinal arteries seen frequently in these subjects (Strömland, 1981, 1985). A significant percentage of children with FAS also have been reported to have an astigmatic ocular abnormality caused by altered corneal shape (Garber, 1984). The overriding clinical importance of these malformations is their association with decreased visual acuity. In one study of FAS children, only 23% had normal vision (Strömland, 1981, 1985). Some researchers have indicated a higher percentage of subnormal auditory acuity in children diagnosed with

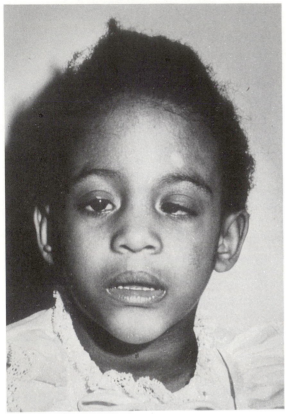
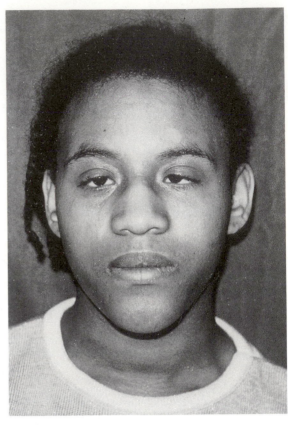

A   B

FIGURE 15.8. Female at ages (A) 3 years 9 months and (B, C) 14 years 2 months, originally reported as patient 4 in 1973 (Jones and Smith, 1973; Jones *et al.*, 1973). Short palpebral fissures, hypoplastic philtrum, strabismus, and ptosis persist. Increased growth of the nose and mandible and the short, stocky stature are often associated with puberty in girls with FAS. Courtesy of A. P. Streissguth. (From Streissguth *et al.*, 1985.)

the FAS. Others report normal hearing in the FAS children tested, although some of these children have delayed language development and short-term memory deficits, which could be related to hearing loss (Streissguth, 1986). Finally, pinnal anomalies such as rotation of the helix and long helical roots have been reported in about 50% of FAS cases examined (Mulvihill and Yeager, 1976; Neri *et al.*, 1988).

Other anatomic defects associated with prenatal alcohol exposure include cardiac ventral septal defects in children and animals (Lemoine *et al.*, 1968; Randall *et al.*, 1977a), urogenital abnormalities (Randall *et al.*, 1977a; Randall and Taylor, 1979), and muscular abnormalities including hypotonia with abnormal muscle fiber development (Lemoine *et al.*, 1968; Adickes and Shuman, 1983; Kronick, 1976; Chernoff, 1977; Ihemelandu, 1984). Delayed and/or abnormal skeletal devel-

opment including radioulnar synostosis, flexion contractures, and bilateral bone fusion in the hands are reported (Cremin and Jaffer, 1981; Jaffer *et al.*, 1981; Streissguth *et al.*, 1985). Alterations in vitamin D metabolism have been reported (Milne and Baran, 1985; Turner *et al.*, 1987). Other reports suggest alterations in calcium, phosphorus, and magnesium nutriture rather than direct effects on vitamin D metabolism in rats (Mezey *et al.*, 1979). These findings, together with the skeletal abnormalities described in children, do suggest altered formation and mineralization of the bone but require further study to determine whether they are, in fact, related to prenatal alcohol exposure.

Several cases with hepatic abnormalities including hepatomegaly, increased hepatic lipids, fibrotic tissue, and bile duct proliferation as well as elevated SGOT, LDH, or serum alkaline phosphatase levels, and high or

FIGURE 15.8. *(Continued)*

c

low bilirubin have been described in the offspring of alcoholics (Habbick *et al.*, 1979; Newman *et al.*, 1979; Møller, 1979; Lefkowitch *et al.*, 1983). It is questionable, however, whether these changes were caused by alcohol because they were discovered years after the cessation of the exposure. Animal studies (in which ethanol exposure, food intake, and environmental conditions were controlled) do suggest, however, that some hepatic biochemical and morphological abnormalities may be causally related to ethanol exposure *in utero*. Chronic prenatal exposure to ethanol increases hepatic fat deposits, but only after the appearance of measurable fetal hepatic ADH activity (Rovinski *et al.*, 1987). A description of hepatic morphology at birth in alcohol-preferring mice suggests that chronic maternal alcohol consumption increases glycogen and fat with altered, but not enlarged, mitochondria and hepatic cell necrosis (Amankwah *et al.*, 1984). Delayed development of fetal hepatic gluconeogenic enzymes may also play a role

(Canny and Roe, 1983). The significance of this delay is unclear, however, since other effects of prenatal ethanol exposure in this rat model were not replicated.

15.7.2. Possible Causal Mechanisms for FAS

Although there is no animal model for "full" FAS, many of the teratogenic effects of prenatal ethanol exposure that are seen in humans have been replicated in animals, which have served as useful models for studying FAS. The species studied include chicks exposed chronically to ethanol vapor (Sandor and Elias, 1968) and mice, rats, dogs, and nonhuman primates consuming ethanol-containing diets (Streissguth *et al.*, 1980).

15.7.2.1. Maternal Diet

Inadequacy of the maternal diet had been thought to play a role in the development of the FAS, but several lines of evidence indicate that its role is probably minor (Abel and Dintcheff, 1978; Weinberg, 1984; Goad, 1984). Although few reports in humans give any information regarding maternal nutritional status, those that do, do not demonstrate a specific, consistent pattern of malnutrition. One of the early descriptions of children affected prenatally by ethanol suggests that eight of the 11 alcoholic mothers who gave birth to FAS offspring consumed diets low in both protein and calories (Ulleland, 1972). Another study reported that both maternal and fetal umbilical plasma zinc levels were lower in the alcoholic women than in nonalcoholics. Dysmorphic features in these offspring were also inversely related to plasma zinc concentrations at delivery (Flynn, 1981). Zinc deficiency alone can cause similar dysmorphic anomalies, and an alcohol–zinc interaction is suggested by other data in rats and mice (Tanaka *et al.*, 1982, 1983; Miller *et al.*, 1983; Keppen *et al.*, 1985) but not in nonhuman primates (Fisher *et al.*, 1988). The role of zinc in humans with FAS remains to be fully explored.

Another line of evidence suggesting that alcohol-induced malnutrition is not the cause of impaired fetal development is that a neonate born to a malnourished pregnant woman would be proportionally small in terms of weight, length, brain weight, and head circumference (Stein and Susser, 1975; Winick *et al.*, 1973), whereas the FAS newborn or animal exhibits decreased brain size and head circumference, but this decrement is proportionally less than the reduction in body weight (Lemoine *et al.*, 1968; Jones and Smith, 1973; Ouellette *et al.*, 1977; Henderson *et al.*, 1979; Weinberg, 1984). Further

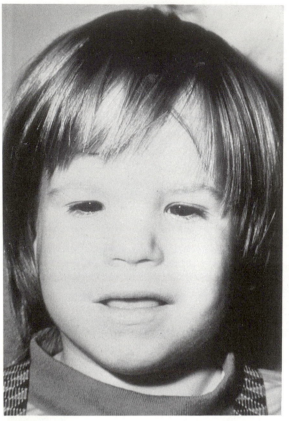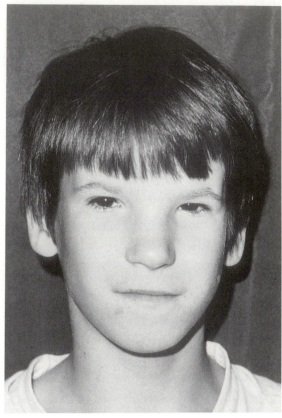

A   B

FIGURE 15.9. Male at ages (A) 2 years 6 months and (B, C) 12 years 2 months, originally reported as patient 8 in 1973 (Jones and Smith, 1973; Jones *et al.*, 1973). Short palpebral fissures, epicanthal folds, flat midface, hypoplastic philtrum, and thin upper vermilion border can be seen at all ages. This young adolescent boy with FAS has the characteristic features of short, lean prepubertal stature with FAS. Courtesy of A. P. Streissguth. (From Streissguth *et al.*, 1985.)

more, in animal models, despite an adequate diet, disproportionate decrements in body and brain weights at birth result when alcohol-derived calories equal or exceed 30% (Weinberg, 1984). Others suggest that when ethanol-derived calories provide fewer than 30% of total calories for animals fetal birth weights are not decreased significantly, but placental weights are increased (Weiner *et al.*, 1981; Gordon *et al.*, 1985a). In contrast, when ethanol provides 30% or more of the total caloric intake, both placental growth and function as well as fetal growth are affected (Henderson *et al.*, 1982; Fisher *et al.*, 1985; Ghishan *et al.*, 1982; Weinberg, 1984, 1985).

15.7.2.2. Teratogenicity of Ethanol and Acetaldehyde

Ethanol has been implicated as a primary teratogen, and a dose–response relationship between ethanol consumption and fetal anomalies has been described (Chernoff, 1977). In support of these findings, ethanol is present in fetal tissues minutes after an acute dose of radiolabeled ethanol is administered to the pregnant dam (Idanpaan-Heikkila *et al.*, 1971, 1972). Decreased DNA and protein synthesis with delayed development and decreased crown–rump length have been reported in rat

FIGURE 15.9. (*Continued*)

embryos exposed to ethanol while grown *in vitro* (Brown *et al.*, 1979; Priscott, 1982), and soft tissue abnormalities induced by prenatal alcohol consumption have been associated with decreased DNA and protein synthesis (Lochry *et al.*, 1981). Animals treated acutely with ethanol in a "binge-drinking" paradigm have offspring with specific facial, limb, and central nervous system abnormalities, which are correlated with the dose and day of exposure (Sulik *et al.*, 1981; Webster *et al.*, 1980, 1983).

Acetaldehyde, the first metabolite of ethanol, is a highly reactive compound capable of binding to biologically important substances. It is also a potential mutagen and potent vasoconstrictor and, therefore, has been incriminated as a potential teratogen. In humans, higher maternal blood acetaldehyde levels occur after one drink in alcoholic women whose offspring were FAS children compared to alcoholics and teetotaling women whose offspring were normal children (Veghelyi, 1983). Despite one early study in which the presence of acet-

aldehyde could not be demonstrated in rat fetuses (Kesaniemi and Sipple, 1975), acetaldehyde has been reported in the blood of rats, monkeys, and ewes after administration of an acute dose of ethanol (Mukherjee and Hodgen, 1982; Brien *et al.*, 1983, 1985; Gordon *et al.*, 1985b; Guerri and Sanchis, 1985) and in mice and guinea pigs after chronic ethanol treatment (Randall *et al.*, 1977b; Clarke *et al.*, 1986). Unlike ethanol, fetal acetaldehyde levels are significantly lower than those in the maternal circulation. On day 20 of pregnancy, 90 min after an acute gavage of ethanol in liquid diet, fetal and umbilical blood acetaldehyde concentrations were one-quarter of maternal levels in rats (Gordon *et al.*, 1985b). Similar data have been reported in mice, monkeys, and rats (Randall *et al.*, 1977b; Mukherjee and Hodgen, 1982; Guerri and Sanchis, 1985).

The human placenta metabolizes ethanol to acetaldehyde and also transfers acetaldehyde from the maternal to the fetal circulation (Karl *et al.*, 1988). These data indicate that acetaldehyde crosses from the maternal circulation to that of the fetus despite a partial placental barrier. Unfortunately, no safe maternal blood ethanol or acetaldehyde concentrations are known. Acetaldehyde injection or oral administration to pregnant mice or rats increases resorptions, decreases birth weight and crown-rump length, and alters neurological development (O'Shea and Kaufman, 1979, 1981; Sreenathan *et al.*, 1982; Webster *et al.*, 1983). In rats, delayed skeletal ossification, shortened umbilical cord length, and altered placental development also occur (Sreenathan *et al.*, 1982). However, acetaldehyde itself may not be the responsible agent. In most of these studies no inhibitors were used to prevent acetaldehyde reduction to ethanol via alcohol dehydrogenase (ADH), the preferred reaction at physiological pH. In several species, however, fetal hepatic ADH activity is undetectable or very low during gestation (Pikkarainen and Raiha, 1967; Clarke *et al.*, 1986; Sanchis and Guerri, 1986), suggesting that this back reaction is probably inconsequential in addition to the fact that only micromolar concentrations of acetaldehyde are effective as a teratogen. The acetaldehyde that reaches the fetus, therefore, is probably derived from maternal blood or placental ethanol metabolism.

The effects of acetaldehyde on fetal weight and resorptions were reported to be exacerbated by treatment with disulfiram, an inhibitor of acetaldehyde oxidation (Dreosti *et al.*, 1981). Greater retardation in skeletal development was also reported with this treatment (Veghelyi *et al.*, 1978). However, one animal study found no difference in the teratogenesis of alcohol after disulfiram treatment (Webster *et al.*, 1983). There is a case report of chromosomal abnormalities in a FAS child born to a woman who drank while being treated with disulfiram (Qazi *et al.*, 1979). At concentrations known to occur in human alcoholics, acetaldehyde can cause sister chromatid exchanges in tissue culture (Veghelyi *et al.*, 1978). An increased incidence of the coexistence of FAS and Down syndrome (trisomy 21) in the offspring of second-generation alcoholic women seems to implicate acetaldehyde, although no blood levels of this compound were reported (Bingol *et al.*, 1987). *In vitro* studies have shown that acetaldehyde can cause decreased thymidine incorporation into DNA (Dreosti *et al.*, 1981), concentration-dependent decreases in overall growth, decreases in accumulation of DNA and protein in embryo explants (Campbell and Fantel, 1983), and prolonged cell-culture doubling times (Wickramasinghe and Malik, 1986). Acetaldehyde also has recently been shown to inhibit fetal DNA methylation, an effect that may alter the regulation of fetal gene expression (Garro *et al.*, 1990).

15.7.2.3. Selective Fetal Malnutrition: Placental Role

Placental functions include gas, nutrient, drug, and waste product transfer in addition to hormone production and metabolism. Thus placental functions are analogous to those of the respiratory tract, kidney, gastrointestinal tract, and endocrine organs, making it a unique organ essential to normal embryonic and fetal development. Fetal exposure to ethanol and acetaldehyde are dependent on placental transfer of these substances. Acetaldehyde levels are determined by its transfer and production versus its disappearance via placental, maternal, and/or fetal liver metabolism. Earlier, evidence was given to indicate that the fetus can be exposed to both ethanol and acetaldehyde, suggesting that most net transfer of both ethanol and acetaldehyde is probably from the maternal to the fetal circulation. As a secondary mechanism, in addition to the possible primary toxicity of ethanol and acetaldehyde, the effects of these compounds on placental growth and function are also potential contributory mechanisms for the prenatal effects of alcohol.

To pass from maternal blood to fetal blood in the human hemochorial placenta, substances must traverse two cellular layers of fetal origin (Beck, 1981). In contrast, the rodent hemotrichorial placenta contains three cellular layers of fetal origin, and maternal blood circulates through the placenta in "tubules" unlike the maternal "lacunae" in the human placenta. The rodent placenta also differs from the human because it does not produce estrogens and progesterone. Despite these differences there are important morphological parallels. The most striking placental effect of ethanol exposure is its effect on overall size and weight. Both smaller placentas (Kaminski *et al.*, 1978) and placentas of normal size (Baldwin *et al.*, 1982) have been reported in humans born with FAS. In the rat, enlarged (Patwardhan *et al.*, 1981; Weiner *et al.*, 1981; Ghishan *et al.*, 1982; Ghishan and Green, 1983; Gordon *et al.*, 1985a; Fisher *et al.*, 1985) and normal-sized (Henderson *et al.*, 1979; Lin, 1981b) placentae have been documented when animals are fed an alcohol-containing diet before and during or throughout pregnancy but not when the alcohol diet was begun on day 6 (Lin, 1981a). In rats, the enlarged placentae contain increased DNA and RNA content with no change in DNA-to-RNA or protein-to-DNA ratios (Gordon *et al.*, 1985a; Fisher *et al.*, 1985), suggesting an overall increase in the number of nuclei with comparable increases in other cellular components.

Ethanol consumption during pregnancy affects placental function as well as development in animals. Rats fed alcohol at 30–36% of total caloric intake have reduced rates of *in vivo* placental transfer of glucose, zinc, valine, and other amino acids (Ghishan *et al.*, 1982; Patwardhan *et al.*, 1981; Lin, 1981a; Snyder *et al.*, 1986). Zinc supplementation does not reverse the deficit in placental transfer of this mineral (Ghishan and Greene, 1983), suggesting that it is alcohol or its metabolism that causes the abnormality and not decreased nutrient transfer alone. Chronic or acute ethanol consumption also reduces *in vitro* transport of various branched- and straight-chain amino acids in rats and nonhuman primates (Lin, 1981a,b; Henderson *et al.*, 1982; Fisher *et al.*, 1982; Gordon *et al.*, 1982). Chronic ethanol treatment during pregnancy in rats also reduces placental folic acid receptor binding (Fisher *et al.*, 1985), suggesting less transfer of this nutrient. When the total calories derived from ethanol are reduced to 27% and protein is increased from 18% to 25%, no statistical difference is demonstrated for *in vivo* placental uptake of zinc, glucose, or α-aminoisobutyric acid in rats (Jones *et al.*, 1981; Gordon *et al.*, 1985a). Even when ethanol represented 27% of the available calories and protein accounted for 25%, total fetal amino acid accumulation was significantly decreased, suggesting a possible fetal abnormality preventing normal amino acid uptake (Gordon *et al.*, 1985a). In isolated human placentae, only ethanol concentrations in excess of those known to exist in heavily drinking alcoholics significantly inhibited the uptake of actively transported non-metabolizable amino acids in one study (Fisher et al., 1981) but not in another (Schenker et al., 1989). Perhaps chronic ethanol consumption changes in vivo amino acid transport by decreasing placental energy production needed for active transport or by altering the membrane itself, a phenomenon known to occur in liver after chronic ethanol exposure but that remains unstudied in the human placenta.

Other factors affecting nutrient transfer to the fetal circulation include blood flow and placental morphology. Diminished blood flow to the placenta in rats has been reported after chronic ethanol feeding (Jones et al., 1981). Placental umbilical vascular spasm as a possible mechanism for a period of anoxia has been reported in monkey (Mukherjee and Hodgen, 1982). More recently, the same was seen in human umbilical artery with physiological concentrations of alcohol (Savoy-Moore et al., 1989). Morphological alterations that could influence transport include thickening of the villous basement membrane from hyperplasia in addition to a decrease in fetal arterial vessels returning fetal blood to the human placenta. Both of these changes occur after chronic alcohol consumption (Amankwah and Kaufmann, 1984). Villitis in human placentae with non-specific morphological changes have been reported, though no definitive correlation was made with FAS offspring in these women (Baldwin et al., 1982). Thus, ethanol, acetaldehyde, or their metabolism can alter placental morphology, which may alter transport. To fully understand the effects of ethanol on the placental growth and transport and thereby fetal development, the role of the accumulation and transfer of toxic waste products and of placental hormone synthesis affecting maternal plasma volume and nutrient transport itself require further study.

15.7.2.4. Prostaglandins

Prostaglandins may play a role in the decreased growth and teratogenic abnormalities seen in FAS. Prostaglandin synthesis inhibitors such as aspirin and indomethacin lessen the detrimental effects of prenatal alcohol exposure on growth in chicks and mice (Pennington, 1988a; Pennington et al., 1985; Randall and Anton, 1984; Randall et al., 1987). Chicks injected with various doses of ethanol show a dose-related decrease in prostaglandin catabolism resulting in increased prostaglandin and cAMP concentrations and cAMP binding activity without changes in either adenylate cyclase or

protein kinase activities (Pennington et al., 1983; Pennington, 1988b). These may contribute to the depressed brain DNA content and cell number with resultant decreases in brain weight, as protein kinase itself may directly regulate DNA synthesis. Urinary excretion of some vasoactive prostaglandins increases in chronic alcohol-consuming women and their neonates (Ylikorkala et al., 1988). Whether the mechanism for this effect is altered prostaglandin metabolism or is secondary to altered kidney function is not known, nor is the impact of these changes on fetal development.

15.7.2.5. Other Factors Influencing FAS

In addition to the total amount of ethanol consumed during pregnancy, other factors that may influence the risk of bearing a FAS child include the timing, duration, and intensity of ethanol consumption, the mother's capacity to metabolize ethanol and acetaldehyde, and smoking history.

The timing, intensity (levels of fetal blood alcohol), and duration of prenatal alcohol exposure are major factors determining the severity and magnitude of the abnormalities detected at birth or later in life (West, 1986, 1987; Bonthius et al., 1988; Clarren et al., 1988; Hoyseth and Jones, 1989). Although the precise effects of exposure during "critical periods" of development are not fully understood, in part because of difficulties of extrapolation across species and the incomplete nature of available data, it is clear that different effects on embryonic development are seen relative to time of exposure. Exposure during the first trimester in humans and other primates (the equivalent of the first and second trimester in rodents) results in major organ abnormalities with facial dysmorphology (Sulik et al., 1981, 1986; Ernhart et al., 1987; West, 1986; West et al., 1987; Neri et al., 1988; Hoyseth and Jones, 1989). Ethanol exposure of rats and mice during the period corresponding to the first and second trimesters in humans can cause teratologic abnormalities, brain dysmorphology, exencephaly, hydrocephaly, ancephaly, neural tube defects, and altered behavioral and learning patterns (Randall et al., 1977b; Sulik et al., 1986; Samson, 1986; West and Purce, 1986; West, 1987). The third trimester in humans is a major determinant of birth weight, length, and head circumference and is the period during which a "brain growth spurt" occurs. Ethanol exposure during this period is associated with decreased brain and overall birth weights (Little et al., 1980, 1984; Weiner and Larson, 1987). Exposure of rodents to ethanol during the analogous period also causes decreased brain size, microcephaly, learning deficits, and altered behavior (West, 1986, 1987; Bonthius et al., 1988).

Attempts have been made, through both epidemiologic and experimental studies, to assess the effects of different types of drinking patterns, i.e., chronic versus binge, on FAS. The results of a study conducted by Jones et al. (1984), in which women were divided into three groups (binge, heavy, and light drinkers), suggest that it is the chronic heavy drinker who is at the greatest risk of bearing a FAS child. These results have been supported by rodent (Randall et al., 1977b) and primate (Clarren and Bowden, 1982; Inouye et al., 1985; Clarren et al., 1987) studies, which indicate that the teratogenic effects of ethanol are seen primarily under conditions of persistent ethanol exposure with associated high blood alcohol levels.

Since both ethanol and acetaldehyde have been implicated as proximal teratogens in the FAS, the mother's capacity to metabolize these compounds would be expected to influence the risk of FAS. Many factors, including genetic background, previous drinking history, stage of alcoholism, liver function, other drug exposures, and nutritional status, all may influence these metabolic capacities (see Chapters 1 and 2 for a detailed review).

Smoking is known to have a deleterious effect on pregnancy outcome, and since a high percentage of alcoholics also smoke, with older women drinking and smoking more than younger women (Ernhart et al., 1985; Shino et al., 1986; Sokol et al., 1986; Plant and Plant, 1988; Rubin et al., 1988; Heller et al., 1988), smoking may contribute to the severity of birth defects seen in FAS children. For example, decreased birth weights and lengths and an increased incidence of low-birth-weight babies who later exhibit behavioral and learning disabilities have been noted in the offspring of women who smoke throughout pregnancy (Abel, 1984).

In conclusion, there are multiple mechanisms through which abusive ethanol consumption can result in the congenital abnormalities seen in the FAS. These mechanisms include maternal nutritional status and direct toxicity of ethanol and acetaldehyde to the placenta and fetus and may involve vasoactive substances such as prostaglandins and be exacerbated by the toxic effects of smoking with decreased placental blood perfusion.

15.7.3. Preventing Fetal Alcohol Syndrome

At present the only remedy for the FAS lies in the area of prevention. Although an occasional drink during pregnancy may not be harmful to the fetus, it is perhaps prudent to heed the advertisement of one public awareness campaign: ". . . the best drink is no drink at all . . ." Nevertheless, it is a reality that in the United States between 5% and 7% of pregnant women are heavy drinkers (Rosett et al., 1981, 1983a; Larson, 1983; Little et al., 1984). All of the physical, social, and environmental stresses that hinder these women from controlling their alcohol intake need to be addressed. Several studies indicate that prenatal care plus counseling and education for addicted women in a supportive, nonjudgmental environment have the greatest potential for lowering alcohol consumption, thereby decreasing the risk of the adverse effects of prenatal alcohol exposure (Little et al., 1983, 1984; Weiner and Larson, 1987; Weiner et al., 1989).

Educational programs for health care workers including obstetricians have been shown to increase awareness of the FAS and to result in increased professional advice to abstain from alcohol during pregnancy (Little et al., 1983). An educational approach that includes programs to increase the awareness and understanding the role of alcohol in fetal development, training of physicians to evaluate pregnant women's alcohol consumption, and counseling for inner-city alcoholic subjects have been used (Weiner et al., 1982; Rosett et al., 1983a). A teaching packet with text and slides for health professionals and community health workers has also been developed (Rosett and Weiner, 1985). Two short films are also currently available for use to stimulate group discussions and increase self-awareness training among health professionals (Documentations for Learning, 1986). These medical and public health specialists report a positive impact of their programs. Similar efforts in Sweden also indicate that pregnant women are highly motivated and are willing to decrease their alcohol intake during pregnancy. The offspring in these cases exhibit a more normal birth weight and a decrease in other alcohol-associated effects (Weiner and Larson, 1987).

There is hesitancy among some concerning the recommendation that all pregnant women should refrain from consuming any alcoholic beverages during pregnancy. Nevertheless, it seems clear that for women at high risk, the benefits of decreased alcohol intake during pregnancy include the normalization of fetal birth weight and neonatal sleep patterns when heavy drinking is curtailed before or even during the third trimester (Little et al., 1980; Weiner and Larson, 1987). However, a similar incidence of morphological abnormalities in the newborns has been reported for 18 pregnant women who reduced their intake compared to those seen in 25 heavy drinkers who continued to drink heavily (Rosett *et al.*, 1983b; Weiner and Larson, 1987), supporting the evidence from animal studies that morphological abnormalities result when interference with normal early development occurs. Therefore, the earlier problem drinking can be identified and treated, the less severe is the risk of abnormal development.

In summary, clearly, the most effective means of preventing FAS is the avoidance of heavy alcohol consumption before and during pregnancy. Several recommendations have been issued suggesting abstinence from alcohol consumption during pregnancy (American Surgeon General's Advisory Report on Alcohol and Pregnancy, 1981; American Medical Association, 1982; Royal College of Psychiatrists, 1982). Public health efforts do increase awareness among some segments of the population but do not affect the drinking of female alcoholics who drink heavily during pregnancy. It is important to add that on the positive side, there is general agreement that prenatal alcohol effects can be prevented or ameliorated by women who decrease alcohol intake during pregnancy. Normalization of birth weight and behavioral and mental abnormalities have been documented in women who began pregnancy consuming moderate to heavy amounts of ethanol (30–125 g per day) but decreased their consumption some time during the first trimester of pregnancy (Larson *et al.*, 1985). These findings have been replicated in different geographic locations with women of different genetic backgrounds (Weiner and Larson, 1987). Others have reported similar findings with a variety of different initial ethanol intakes (Little *et al.*, 1980; Rosett *et al.*, 1981), though Jones reports different results (1986). Thus, educational programs and supportive counseling focused on heavily drinking women at high risk for delivering FAS or FAE babies can have a positive impact on a pregnant woman's alcohol consumption during her pregnancy. This in turn has positive long-term consequences for fetal outcome and later development.

ACKNOWLEDGMENTS. Work in the authors' laboratories was supported by The March of Dimes Birth Defects Foundation grants #15-155 and #15-0443 and by the Alcohol Research Center Grant #AA 03508.

REFERENCES

Abel, E. L.: Fetal Alcohol Syndrome, Vol. I: An Annotated Comprehensive Bibliography, Boca Raton, CRC Press, 1981.

Abel, E. L.: Smoking and pregnancy. *J. Psychoact. Drugs* **16**:327–338, 1984.

Abel, E. L., and Dincheff, B. A.: Effects of prenatal alcohol exposure on growth and development in rats. *J. Pharmacol. Exp. Ther.* **207**:916–921, 1978.

Abel, E. L., and Sokol, R. J.: A revised conservative estimate of the incidence of FAS and its economic impact. *Alcoholism: Clin. Exp. Res.* **15**:514–524, 1991.

Abel, E. L., and Sokol, R. J.: Alcohol consumption during pregnancy: The dangers of moderate drinking. In: *Alcoholism, Biochemical and Genetic Aspects* (W. H. Goedde, and D. P. Agarwal, eds.), New York, Pergamon Press, pp. 228–237, 1989.

Adickes, E. D., and Shuman, R. M.: Fetal alcohol myopathy. *Pediatr. Pathol.* **1**:369–384, 1983.

Aldo-Benson, M.: Mechanisms of alcohol-induced suppression of B-cell response. *Alcoholism: Clin. Exp. Res.* **13**:469–475, 1989.

Alvarez, M., Cimino, L., Cory, M., and Gordon, R.: Ethanol induction of sister chromatid exchanges in human cells *in vitro. Cell. Genet.* **27**:66–69, 1980.

Amankwah, K. S., and Kaufmann, R. C.: Ultrastructure of human placenta: Effects of maternal drinking. *Gynecol. Obstet. Invest.* **18**:311–316, 1984.

Amankwah, K. S., Kaufmann, R. C., and Weberg, A. D.: Ultrastructure changes in neonatal liver tissue: Effects of maternal drinking. *Gynecol. Obstet. Invest.* **17**:213–218, 1984.

American Medical Association Council on Scientific Affairs: Fetal Effects of Maternal Alcohol Use. *Proc. House of Delegates Report #4*. Chicago, AMA, 1982, pp. 184–193.

American Surgeon General's Advisory Report on Alcohol and Pregnancy: *FDA Drug Bull.* **11**:9–10, 1981.

Ames, B. M., McCann, J., and Yamaski, E.: Methods for detecting carcinogens and mutagens with the Salmonella/mammalian-microsome mutagenicity test. *Mutat. Res.* **31**:347–364, 1975.

Anderson, D. L.: Intraoral site distribution of malignancy and preinvasive malignant cell transformation in dental patients and alcoholism. *Acta Cytol.* **16**:322–326, 1972.

Anderson, L. M., Harrington, G. W., Pylypiw, H. M., Jr., Hagiwara, A., and Magee, P.N.: Tissue levels and biological effects of *N*-nitrosodimethylamine in mice during chronic low- or high-dose exposure with or without ethanol. *Drug Metab. Dispos.* **14**:733–739, 1986.

Atukorola, S., Bawsu, T. K., Dickerson, W. T., Donaldson, D., and Sakula, A.: Vitamin A, zinc and lung cancer. *Br. J. Cancer* **40**:927–931, 1979.

Auerbach, O., Stout, A. P., Hammond, E. C., and Garfinkel, L.: Changes in bronchial epithelium in relation to cigarette smoking and in relation to lung cancer. *N. Engl. J. Med.* **265**:253–287, 1961.

Axelrod, A. E.: Immune processes in vitamin deficiency states. *Am. J. Clin. Nutr.* **24**:265–271, 1971.

Ayesh, R., Idle, J. R., Ritchie, J. C., Crothers, M. J., and Hetzel, M. R.: Metabolic oxidation phenotypes as markers for susceptibility to lung cancer. *Nature* **312**:169–170, 1984.

Baird, S. M., Beattie, G. M., Lennon, R. A., Lipsick, J. S., Jensen, F. C., and Kaplan, N. O.: Induction of lymphoma in antigenically stimulated athymic mice. *Cancer Res.* **42**:198–206, 1982.

Baldwin, V. J., MacLeod, P. M., and Benirschke, K.: *Placental findings in alcohol abuse in pregnancy in birth defects.* Original article series, No. 18 (3A):89–94, March of Dimes Birth Defect Foundation, 1982.

Beasley, R. P.: Hepatitis B virus as the etiologic agent in hepatocellular carcinoma. *Hepatology* 2:215–265, 1982.

Beasley, R. P., Hwang, L., Lin, C., and Chien, C. S.: Hepatocellular carcinoma and hepatitis B virus. *Lancet* II:1129–1133, 1981.

Becci, P. J., Thompson, H. J., Grubbs, C. J., Squire, R. A., Brown, C. C., Sporn, M. B., and Moon, R. C.: Inhibitory effect of 13-*cis*-retinoic acid on urinary bladder carcinogenesis induced in C57BL/6 mice by *N*-butyl-*N*-(4-hydroxybutyl)nitrosamine. *Cancer Res.* 38:4463–4466, 1978.

Beck, F.: Comparative placental morphology and function. In: *Developmental Toxicology* (C. A. Kimmel, and J. Buelke-Sam, eds.), New York, Raven Press, pp. 35–54, 1981.

Becker, Y.: Cancer in ataxia-telangiectasia patients: Analysis of factors leading to radiation-induced and spontaneous tumors. *Anticancer Res.* 6:1021–1032, 1986.

Beisel, W. R., Edelman, R., Nauss, K., and Suskind, R. M.: Single-nutrient effects of immunologic functions. Report of a workshop sponsored by the Department of Food and Nutrition and its Nutrition Advisory Group of the American Medical Association. *J.A.M.A.* 245:53–58, 1981.

Belinsky, S. A., Bedell, M. A., and Swenberg, J. A.: Effect of chronic ethanol diet on the replication, alkylation and repair of DNA from hepatocytes and nonparenchymal cells following dimethylnitrosamine administration. *Carcinogenesis* 3:1293–1297, 1982.

Bendich, A.: Carotenoids and the immune response. *J. Nutr.* 119:112–115, 1989.

Berenblum, I.: Sequential aspects of chemical carcinogens. In: *Skin in Cancer* (F. F. Becker, ed.), New York, Plenum Press, 1979.

Best, C. H., Hartroft, W. S., Lucas, C. S., and Ridout, J. H.: Liver damage produced by feeding alcohol or sugar and its prevention by choline. *Br. Med. J.* II:1001–1006, 1949.

Bingol, N., Fuchs, M., Iosub, S., Kumar, S., Stone, R. K., and Gromisch, D. S.: Fetal alcohol syndrome associated with trisomy 21. *Alcoholism: Clin. Exp. Res.* 11:42–44, 1987.

Blot, W. A., McLaughlin, J. K., Winn, D. M., Austin, D. F., Greenberg, R. S., Preston-Martin, S., Bernstein, L., Schoenberg, J. B., Stemhagen, A., and Fraumeni, J. F.: Smoking and drinking in relation to oral and pharyngeal cancer. *Cancer Res.* 48:3282–3287, 1988.

Bogden, M. M., Eastman, A., and Bresnick, E.: A system in mouse liver for the repair of O^6-methylguanine lesions in methylated DNA. *Nucleic Acids Res.* 9:3089–3103, 1981.

Bond, N. W.: Fetal alcohol exposure and hyperactivity in rats: The role of the neurotransmitter systems involved in arousal and inhibition. In: *Alcohol and the Developing Brain* (J. R. West, ed.), New York, Oxford University Press, pp. 45–70, 1986.

Bonthius, D. J., Goodlett, C. R., and West, J. R.: Blood alcohol concentration and severity of microcephaly in neonatal rats depend on the pattern of alcohol administration. *Alcohol* 5:209–214, 1988.

Brechot, C., Hadchouel, M., Scotto, J., Degos, F., Charnay, P.,

Trepo, C., and Tiollas, P.: Detection of hepatitis B virus DNA in liver and serum: a direct appraisal of the chronic carrier state. *Lancet* II:765–768, 1981.

Brechot, C., Nalpas, B., and Courouce, A.: Evidence that hepatitis B virus has a role in liver cell carcinoma in alcoholic liver disease. *N. Engl. J. Med.* 306:1384–1387, 1982.

Breitman, T. R., Selonick, S. E., and Collins, S. J.: Induction of differentiation of the human promyelocytic leukemia cell line (HL-60) by retinoic acid. *Proc. Natl. Acad. Sci. U.S.A.* 77:2936–2940, 1980.

Brien, J. F., Loomis, C. W., and Tranmer, J.: Disposition of ethanol in human maternal venous blood and amniotic fluid. *Am. J. Obstet. Gynecol.* 146:181–186, 1983.

Brien, J. F., Clarke, D. W., Richardson, B., and Patrick, J.: Disposition of ethanol in maternal blood, fetal blood and amniotic fluid of third-trimester pregnant ewes. *Am. J. Obstet. Gynecol* 152:583–590, 1985.

Brown, N. A., Goulding, E. H., and Fabro, S.: Ethanol embryotoxicity: Direct effects on mammalian embryos *in vitro*. *Science* 206:573–575, 1979.

Campbell, M. A., and Fantel, A. G.: Teratogenicity of acetaldehyde *in vitro*: Relevance to the fetal alcohol syndrome. *Life Sci.* 132:2641–2647, 1983.

Canny, N. E., and Roe, D. A.: Effects of maternal ethanol consumption on the ontogeny of glucogenesis in the perinatal period. *Drug Nutrient Int.* 2:193–202, 1983.

Capel, I. D., Jenner, M., Pinnock, M. H., and Williams, D. C.: The effect of chronic alcohol intake upon the hepatic microsomal carcinogen-activation system. *Oncology* 35:168–170, 1978.

Caporaso, N., Hayes, R. B., Dosemeci, M., Hoover, R., and Ayesh, R.: Lung cancer risk, occupational exposure, and the debrisoquine metabolic phenotype. *Cancer Res.* 49:3675–3679, 1989.

Carroll, K. K.: Lipids and carcinogenesis. *J. Environ. Pathol. Toxicol.* 3:253–271, 1980.

Castonguay, A., Rivenson, A., Trushin, N., Reinhart, J., Spathopoulos, S., Weiss, C. J., Reiss, B., and Hecht, S. S.: Effects of chronic ethanol consumption on the metabolism and carcinogenicity of N-nitrosonornicotine in F344 rats. *Cancer Res.* 44:2285–2290, 1984.

Chandra, R. K., Au, B., Woodford, G., and Hyam, P.: Iron status, immune response and susceptibility to infection. In: *Iron Metabolism* (H. Krebs, ed.), (Ciba Foundation Symposium No. 51) Amsterdam, Elsevier/Excerpta/North-Holland, pp. 249–268, 1977.

Chernoff, G. F.: The fetal alcohol syndrome in mice: An animal model. *Teratology* 15:223–229, 1977.

Chevilotte, G., Durbec, J. P., Gerolami, A., Berthezene, R., Bidart, J. M., and Camatte, R.: Interaction between hepatitis B virus and alcohol consumption in liver cirrhosis. *Gastroenterology* 85:141–145, 1983.

Ciba Foundation: *Mechanisms of Alcohol Damage in Utero* (R. Porter, M. O'Connor, and J. Whelan, eds.), (CIBA Foundation Symposium No. 105) London, Pittman, 1983.

Clarke, D. W., Steenaart, N. A. E., and Brien, J. F.: Disposition of ethanol and activity of hepatic and placental alcohol dehydrogenase and aldehyde dehydrogenases in the third-trimester pregnant guinea pig for single and short-term oral ethanol administration. *Alcoholism: Clin. Exp. Res.* **10:**330–336, 1986.

Clarren, S. K., and Bowden, D. M.: Fetal alcohol syndrome: A new primate model for binge drinking and its relevance to human ethanol teratogenesis. *J. Pediatr.* **101:**819–824, 1982.

Clarren, S. K., and Smith, D. W.: The fetal alcohol syndrome. *N. Engl. J. Med.* **298:**1063–1067, 1978.

Clarren, S. K., Alward, E. C., Somi, S. M., Streissguth, A. P., and Smith, D. W.: Brain malformations related to prenatal exposure to ethanol. *J. Pediatr.* **92:**64–67, 1978.

Clarren, S. K., Bowden, D. M., and Astley, S. J.: Pregnancy outcomes after weekly oral administration of ethanol during gestation in the pig-tailed macaque (*Macaca nemestrina*). *Teratology* **35:**345–354, 1987.

Clarren, S. K., Astley, S. J., and Bowden, D. M.: Physical anomalies and developmental delays in non-human primate infants exposed to weekly doses of ethanol during gestation. *Teratology* **37:**561–570, 1988.

Cleaver, J. E.: DNA damage, repair systems and human hypersensitivity diseases. *J. Environ. Pathol. Toxicol.* **3:**53–68, 1980.

Cohen, S. M., and Ellwein, L. B.: Cell proliferation in carcinogenesis. *Science* **249:**1007–1011, 1990.

Cohen, S. M., and Ellwein, L. B.: Genetic errors, cell proliferation and carcinogenesis. *Cancer Res.* **51:**6493–6505, 1991.

Colangelo, W., and Jones, D. G.: The fetal alcohol syndrome: a review and assessment of the syndrome and its neurological sequelae. *Prog. Neurobiol.* **19:**271–314, 1982.

Craddock, V. M.: Cell proliferation and induction of liver cancer. In: *Primary Liver Tumors* (H. Remmer, H. M. Bolt, P. Bannaschi, and H. Popper, eds.), Lancaster England, MTP Press, pp. 377–383, 1978.

Craddock, V. M., Henderson, A. R., and Gash, S.: Nature of constitutive and induced mammalian O6-methylguanine DNA repair enzymes. *Biochem. Biophys. Res. Commun.* **107:**546–553, 1982.

Cremin, B. J., and Jaffer, Z.: Radiological aspects of the fetal alcohol syndrome. *Pediatr. Radiol.* **1:**151–153, 1981.

Davenport, H. W.: Ethanol damage to canine oxyntic glandular mucosa. *Proc. Soc. Exp. Biol. Med.* **126:**657–662, 1967.

Dellarco, V. L.: A mutagenicity assessment of acetaldehyde. *Mutat. Res.* **195:**1–20, 1988.

Dianzani, M. U.: Lipid peroxidation in ethanol poisoning: A critical reconsideration. *Alcohol Alcoholism* **20:**161–173, 1985.

Diaz, J., and Samson, H. H.: Impaired brain growth in neonatal rats exposed to ethanol. *Science* **208:**751–753, 1980.

Dicker, E., and Cederbaum, A. I.: Hydroxyl radical generation by microsomes after chronic ethanol consumption. *Alcoholism: Clin. Exp. Res.* **11:**309–314, 1987.

Dicker, E., and Cederbaum, A. I.: Increased oxygen radical-dependent inactivation of metabolic enzymes by liver microsomes after chronic ethanol consumption. *FASEB J.* **2:**2901–2906, 1988.

Dinoso, V. P., Chey, W. Y., Braverman, S. P., Rosen, A. P., Ottenberg, D., and Lorber, S. H.: Gastric secretion and gastric mucosal morphology in chronic alcoholics. *Arch. Intern. Med.* **130:**715–719, 1972.

Documentations for Learning: *Taking a Drinking History: "Counseling and Referral."* Boston, Fetal Alcohol Education Program, 1986.

Dow, K. E., and Riopelle, M. J.: Ethanol neurotoxicity: Effects on neurite formation and neurotrophic factor production *in vitro*. *Science* **228:**591–593, 1985.

Dreosti, L. E., Ballard, J., Belling, G. B., Record, I. R., Manuel, S. J., and Hetzel, B. S.: The effect of ethanol and acetaldehyde on DNA synthesis in growing cells and on fetal development in the rat. *Alcoholism: Clin. Exp. Res.* **5:**357–362, 1981.

Driscoll, K. E., and Snyder, C. A.: The effects of ethanol and repeated benzene exposures on benzene pharmacokinetics. *Toxicol. Appl. Pharmacol.* **73:**525–532, 1984.

Driver, E. H., and Swann, P. F.: Alcohol and human cancer. Review. *Anticancer Res.* **7:**309–320, 1987.

Druse, M. J.: Effects of *in utero* ethanol exposure on the development of neurotransmitter systems. In: *Development of the Central Nervous System: Effect of Alcohol and Opiates* (M. Miller, ed.), New York, Alan R. Liss, 1990.

Dyroff, M. C., Richardson, F. C., Popp, J. A., Bedell, M. A., and Swenberg, J. A.: Correlation of O4-ethyldeoxythymidine accumulation, hepatic initiation and hepatocellular carcinoma induction in rats continuously administered dimethylnitrosamine. *Carcinogenesis* **7:**241–248, 1986.

Eadie, J. S., Conrad, M., Toorchen, D., and Topal, M. D.: Mechanism of mutagenesis by O6-methylguanine. *Nature* **308:**201–203, 1984.

Elzay, R. P.: Local effect of ethanol in combination with DMBA in hamster cheek pouch. *J. Dent. Res.* **45:**1788–1795, 1966.

Elzay, R. P.: Local effect of ethanol in combination with DMBA on hamster cheek pouch. *J. Dent. Res.* **45:**1788–1795, 1969a.

Elzay, R. P.: Effects of alcohol and cigarette smoke as promoting agents in hamster pouch carcinogenesis. *J. Dent. Res.* **48:**1200–1205, 1969b.

Emery, A. E., Danford, N., Anand, R., Duncan, W., and Paton, L.: Arylhydrocarbon-hydroxylase inducibility in patients with cancer. *Lancet* **I:**470–472, 1978.

Erb, L., and Andersen, B. D.: The fetal alcohol syndrome (FAS). *Clin. Pediatr.* **17:**644–649, 1978.

Ernhart, C. B., Wolf, A. W., Linn, P. L., Sokol, R. J., Kennard, M. J., and Filipovich, H. F.: Alcohol-related birth defects: Syndromal anomalies, intrauterine growth retardation, and neonatal behavior assessment. *Alcoholism: Clin. Exp. Res.* **9:**447–453, 1985.

Ernhart, C. B., Sokol, R. J., Martier, S., Moron, P., Nadier, D., Ager, J. W., and Wolf, A.: Alcohol teratogenicity in the human: A detailed assessment of specificity, critical period and threshold. *Am. J. Obstet. Gynecol.* **156:**33–39, 1987.

Espina, N., Lima, V., Lieber, C. S., and Garro, A. J.: *In vitro* and *in vivo* inhibitory effect of ethanol and acetaldehyde on O6-methylguanine transferase. *Carcinogenesis* **9:**761–766, 1988.

Estler, C. J., and Ammon, H. P. T.: Glutathion und SH-Gruppenhaltige Enzyme in der Leber Weisser Mause nach Einmaliger Alkoholgabe. *Med. Pharmacol. Exp.* **15:**299–306, 1966.

Farinati, F., Zhou, Z. C., Bellah, J., Lieber, C. S., and Garro, A. J.: Effect of chronic ethanol consumption on activation of nitrosopyrrolidine to a mutagen by rat upper alimentary tract, lung and hepatic tissue. *Drug Metab. Dispos.* **13:**210–214, 1985.

Farinati, F., Lieber, C. S., and Garro, A. J.: The effects of chronic ethanol consumption on carcinogen activating and detoxifying systems in rat upper alimentary tract tissue. *Alcoholism: Clin. Exp. Res.* **13**:357–360, 1989.

Fernandez, V., and Videla, L. A.: Effect of acute and chronic ethanol ingestion on the content of reduced glutathione of various tissues of the rat. *Experientia* **37**:392–294, 1981.

Feron, V. J.: Effects of exposure to acetaldehyde in Syrian hamsters simultaneously treated with benzo(a)pyrene or dimethylnitrosamine. *Prog. Exp. Tumor Res.* **24**:162–176, 1979.

Feron, V. J., Kruysse, A., and Woutersen, R. A.: Respiratory tract tumors in hamsters exposed to acetaldehyde vapor alone or simultaneously to benzo(a)pyrene or diethylnitrosamine. *Eur. J. Cancer Clin. Oncol.* **18**:13–31, 1982.

Fiala, E. S., Sohn, O. S., and Hamilton, S. R.: Effects of chronic dietary ethanol on *in vivo* and *in vitro* metabolism of methylazoxymethanol and on methylazoxymethanol-induced DNA methylation in rat colon and liver. *Cancer Res.* **47**:5939–5943, 1987.

Finkelstein, J. D., Cello, J. P., and Kyle, W. E.: Ethanol-induced changes in methionine metabolism in rat livers. *Biochem. Biophys. Res. Commun.* **61**:525–531, 1974.

Fisher, S. E., Atkinson, M., Van Thiel, D. H., Rosenblum, E., David, R., and Holzman, I.: Selective fetal malnutrition: The effect of ethanol and acetaldehyde upon *in vitro* uptake of alpha amino isobutyric acid by human placenta. *Life Sci.* **29**:1283–1288, 1981.

Fisher, S. E., Atkinson, M., Burnap, J. K., Jacobson, S., Sehgal, P. K., Scott, W., and Van Thiel, D. H.: Ethanol-associated selective fetal malnutrition: A contributing factor in the fetal alcohol syndrome. *Alcoholism: Clin. Exp. Res.* **6**:197–201, 1982.

Fisher, S. E., Inselman, L., Doffy, L., Atkinson, M., Spencer, H., and Change, B.: Ethanol and fetal nutrition: Effect of chronic ethanol exposure on rat placental growth and membrane associated folic acid receptor binding activity. *J. Pediatr. Gastro. Nutr.* **4**:645–649, 1985.

Fisher, S. E., Alcock, N. W., Amirian, J., and Altshuler, H. L.: Neonatal and maternal hair zinc levels in a nonhuman primate model of the fetal alcohol syndrome. *Alcoholism: Clin. Exp. Res.* **12**:417–421, 1988.

Flynn, A.: Zinc status of pregnant alcoholic women: A determinant of fetal outcome. *Lancet* **I**:572–576, 1981.

Fraenkel-Conrat, H., and Singer, B.: Nucleoside adducts are formed by cooperative reaction of acetaldehyde and alcohols: possible mechanism for the role of ethanol in carcinogenesis. *Proc. Natl. Acad. Sci. U.S.A.* **85**:3758–3761, 1988.

Frizzera, G., Rosai, J., Dehner, L., Spector, B. D., and Kersey, J. H.: Lymphoreticular disorders in primary immunodeficiencies: New findings based on up-to-date histologic classification of 35 cases. *Cancer* **96**:692–699, 1980.

Fromm, D., and Robertson, R.: Effects of alcohol on ion transport by esophageal mucosa. *Gastroenterology* **70**:220–225, 1976.

Gabrial, G. N., Schrager, T. F., Newberne, P. M.: Zinc deficiency, alcohol, and a retinoid: Association with esophageal cancer in rats. *J. Natl. Cancer Inst.* **68**:785–789, 1982.

Garber, J. M.: Steep corneal curvature: A fetal alcohol syndrome landmark. *Journal Amer. Optometric Ass.* **55**:595–598, 1984.

Garro, A. J., and Lieber, C. S.: Alcohol and cancer. *Annu. Rev. Pharmacol. Toxicol.* **30**:219–249, 1990.

Garro, A. J., Seitz, H. K., and Lieber, C. S.: Enhancement of dimethylnitrosamine metabolism and activation to a mutagen following chronic ethanol consumption. *Cancer Res.* **41**:120–124, 1981.

Garro, A. J., Espina, N., Farinati, F., and Salvagnini, M.: The effects of ethanol consumption on carcinogen metabolism and on O⁶-methylguanine transferase-mediated repair of alkylated DNA. *Alcoholism: Clin. Exp. Res.* **10**:73–77, 1986.

Garro, A. J., McBeth, D., Lima, V., and Lieber, C. S.: Dietary ethanol inhibits fetal DNA methylation: Implications for the fetal alcohol syndrome. *FASEB J.* **4**:A64, 1990.

Garzon, F. T., Seitz, H. K., Simanowski, U. A., Berger, M. R., and Schmahl, D.: Alcohol as a modifying agent in experimental carcinogenesis. *Dig. Dis. Sci.* **31**:99S, 1986.

Ghishan, F. K., and Greene, M. L.: Fetal alcohol syndrome: Failure of zinc supplementation to reverse the effect of ethanol on placental transport of zinc. *Pediatr. Res.* **17**:529–531, 1983.

Ghishan, F. K., Patwardhan, R., and Greene, H. L.: Fetal alcohol syndrome: Inhibition of placental zinc transport as a potential mechanism for fetal growth retardation in the rat. *J. Lab. Clin. Med.* **100**:45–52, 1982.

Gibel, W.: Experimentelle Unterauchungen zur Synkarzinogenese beim Esophaguskarginom. *Arch. Geschwulstforsch.* **30**:181–189, 1967.

Gillette, J. R.: Environmental factors in drug metabolism. *Fed. Proc.* **35**:1142–1147, 1976.

Gluud, C., Aldershvile, J., Henriksen, J., Kryger, J., and Mathiesen, L.: Hepatitis B and A virus antibodies in alcoholic steatosis and cirrhosis. *J. Clin. Pathol.* **35**:693–697, 1982.

Goad, P. T.: The role of maternal diet in the development toxicology of ethanol. *Toxicol. Appl. Pharmacol.* **73**:256–267, 1984.

Gordon, B. H. J., Durandin, R. M., Rosso, P., and Winick, M.: Placental amino acid transport in alcohol fed rats. *Fed. Proc.* **41**:496, 1982.

Gordon, B. H. J., Streeter, M. L., Rosso, P., and Winick, M.: Prenatal alcohol exposure: Abnormalities in placental growth and fetal amino acid uptake in the rat. *Biol. Neonate* **47**:113–119, 1985a.

Gordon, B. H. J., Baraona, E., Miyakawa, H., Finkelman, F., and Lieber, C. S.: Exaggerated acetaldehyde response after ethanol administration during pregnancy and lactation in rats. *Alcoholism: Clin. Exp. Res.* **9**:17–22, 1985b.

Gottfried, E. B., Korsten, M. A., and Lieber, C. S.: Alcohol induced gastric and duodenal lesions in man. *Am. J. Gastroenterol.* **70**:587–592, 1978.

Grafstrom, R. C., Curren, R. D., Yang, L. L., and Harris, C. C.: Genotoxicity of formaldehyde in cultured human fibroblasts. *Science* **228**:89–91, 1985.

Griciute, L., Castegnaro, M., and Bereziat, J.: Influence of ethyl alcohol on carcinogenesis with *N*-nitrosodimethylamine. *Cancer Lett.* **13**:345, 1981.

Griciute, L., Castegnaro, M., Bereziat, J. C., and Cabral, J. R. P.: Influence of ethyl alcohol on the carcinogenic activity of *N*-nitrosonornicotine. *Cancer Lett.* **31**:267, 1986.

Grubbs, C. J., Juliana, M. M., and Whitaker, L. M.: Effect of ethanol on initiation of methylnitrosourea (MNU) and di-ethylbenzanthracene (DMBA)-induced mammary cancers. *Proc. Ann. Meet. Am. Assoc. Cancer Res.* **29:**A590, 1988.

Guerri, C., and Grisolia, S.: Changes in glutathione in acute and chronic alcohol intoxication. *Pharmacol. Biochem. Behav.* **13** (Suppl. 1)**:**53–61, 1980.

Guerri, C., and Sanchis, R.: Acetaldehyde and alcohol levels in pregnant rats and their fetuses. *Alcohol* **2:**267–270, 1985.

Habbick, B. F., Casey, R., Zaleski, W. A., and Murphy, F.: Liver abnormalities in three patients with fetal alcohol syndrome. *Lancet* **I:**580–581, 1979.

Habs, M., and Schmahl, D.: Inhibition of the hepatocarcinogenic activity of dimethylnitrosamine (DENA) by ethanol in rats. *Hepatogastroenterology* **28:**242–244, 1981.

Haentjens, P., De Backer, A., and Willems, G.: Effect of an apple brandy from Normandy and of ethanol on epithelial cell proliferation in the esophagus of rats. *Digestion* **37:**184–192, 1987.

Hamilton, S. R., Hyland, J., McAvinchy, D., Chaudhry, Y., Hartka, L., Kim, H. T., Cichon, P., Floyd, J., Turjman, N., Kessie, G., Nair, P. P., and Dick, J.: Effects of chronic dietary beer and ethanol consumption on experimental colonic carcinogenesis by azoxymethane in rats. *Cancer Res.* **47:**1551–1559, 1987a.

Hamilton, S. R., Sohn, O. S., and Fiala, E. S.: Effects of timing and quantity of chronic dietary ethanol consumption on azoxymethane-induced colonic carcinogenesis and azoxymethane metabolism in Fischer 344 rats. *Cancer Res.* **47:**4305–4311, 1987b.

Hamilton, S. R., Sohn, O. S., and Fiala, E. S.: Inhibition by dietary ethanol of experimental colonic carcinogenesis induced by high-dose azoxymethane in F344 rats. *Cancer Res.* **48:**3313–3318, 1988.

Harris, A. L., Karran, P., and Lindahl, T.: O^6-methylquanine-DNA methyltransferase of human lymphoid cells: Structural and kinetic properties and absence in repair-deficient cells. *Cancer Res.* **43:**3247–3252, 1983.

Harris, C. C., Sporn, M. B., Kaufman, D. G., Smith, J. M., Baker, M. S., and Saffiotti, C.: Acute ultrastructural effects of benzo(a)pyrene and ferric oxide on the hamster tracheobronchial epithelium. *Cancer Res.* **31:**1977–1989, 1971.

He, S.-M., and Lambert, B.: Induction and persistence of SCE-inducing damage in human lymphocytes exposed to vinyl acetate and acetaldehyde *in vitro. Mutat. Res.* **158:**201–208, 1985.

Heller, J., Anderson, H. R., Bland, J. M., Brooke, O. G., Peacock, J. L., and Stewart, C. M.: Alcohol in pregnancy: Patterns and association with socio-economic psychological and behavioral factors. *Br. J. Addict.* **83:**541–551, 1988.

Henderson, A. S., Ripley, S., Hino, O., and Rogler, C. E.: Identification of a chromosomal aberration associated with a hepatitis B DNA integration in human cells. *Cancer Genet. Cytogenet.* **30:**269–275, 1988.

Henderson, G. I., Hoyumpa, A. M., McClain, C., and Schenker, S.: The effects of chronic and acute alcohol administration on fetal development in the rat. *Alcoholism: Clin. Exp. Res.* **3:** 99–105, 1979.

Henderson, G. I., Rashmi, V., Patwardhan, R. V., McLeroy, S.,

and Schenker, S.: Inhibition of placental amino acid uptake in rats following acute and chronic ethanol exposure. *Alcoholism: Clin. Exp. Res.* **6:**495–505, 1982.

Hetu, C., Yelle, L., and Joly, J. G.: Influence of ethanol on hepatic glutathione content and on the activity of glutathione S-transferases and epoxide hydrase in the rat. *Drug Metab. Dispos.* **10:**246–250, 1982.

Hill, D., and Grubbs, C. J.: Retinoids as chemopreventive and anticancer agents intact animals. *Anticancer Res.* **2:**111–124, 1982.

Hirayama, T.: Diet and cancer. *Nutr. Cancer* **1:**67–81, 1979.

Hislop, W. S., Follet, E. A. C., Bouchier, I. A. D., and MacSween, R. N. M.: Serological markers of hepatitis B in patients with alcoholic liver disease: A multiple centre survey. *J. Clin. Pathol.* **34:**1017–1019, 1981.

Horie, A., Kohchi, S., Kuratsune, M.: Carcinogenesis in the esophagus. II: Experimental production of esophageal cancer by administration of ethanolic solutions of carcinogens. *Gann* **56:**429–441, 1965.

Horvath, P. M., and Ip, C.: Synergistic effect of vitamin E and selenium in the chemoprevention of mammary carcinogenesis in rats. *Cancer Res.* **43:**5335–5341, 1983.

Hoyseth, K. S., and Jones, P. J. H.: Mini review: Ethanol induced teratogenesis—characterization, mechanisms and diagnostic approaches. *Life Sci.* **44:**643–649, 1989.

Idanpaan-Heikkila, J. E., Fritchie, G. E., and McIsaac, W. M.: Placental transfer of C^{14}-ethanol. *Am. J. Obstet. Gynecol.* **110:**426–428, 1971.

Idanpaan-Heikkila, J., Jouppila, P., Akerblom, H. K., Isoaho, R., Kauppila, E., and Koivisto, M.: Elimination and metabolic effects of ethanol in mother, fetus and newborn infant. *Am. J. Obstet. Gynecol.* **112:**387–393, 1972.

Ihemelandu, E. C.: Effect of maternal alcohol consumption on pre- and postnatal muscle development of mice. *Growth* **48:** 35–43, 1984.

Inouye, R. N., Kokich, V. G., Clarren, S. K., and Bowden, D. M.: Fetal alcohol syndrome: An examination of craniofacial dysmorphology in *Macca nemestriana. J. Med. Primatol.* **14:**35–48, 1985.

Ioannides, C., and Steele, C. M.: Hepatic microsomal mixed-function oxidase activity in ethanol-treated hamsters and its consequences on the bioactivation of aromatic amines to mutagens. *Chem. Biol. Interact.* **59:**129–139, 1986.

Iosub, S., Fuchs, M., Bingol, N., and Gromisch, D. S.: Fetal alcohol syndrome revisited. *Pediatrics* **68:**475–479, 1981.

Ip, C.: Prophylaxis of mammary neoplasia by selenium supplementation in the initiation and promotion phases of chemical carcinogenesis. *Cancer Res.* **41:**4386–4390, 1981.

Jaffer, Z., Nelson, M., and Beighton, P.: Bone fusion in the fetal alcohol syndrome. *J. Bone Joint Surg.* **63:**569–571, 1981.

Jensen, O. M.: Cancer morbidity and causes of death among Danish brewery workers. *Int. J. Cancer* **23:**454–465, 1979.

Jerrells, T. R., Marietta, C. A., Eckhardt, M. J., Majchrowicz, E., and Weight, F. F.: Effects of ethanol administration on parameter of immunocompetency in rats. *J. Leukocyte Biol.* **39:**499, 1986.

Joly, J. G., Ishii, H., Teschke, R., Hasumura, Y., and Lieber, C. S.: Effect of chronic ethanol feeding on the activities and submicrosomal distribution of reduced nicotinamide adenine dinucleotide phosphate (NADPH)-cytochrome P-450 reductase and the demethylase for aminopyrine and ethylmorphine. *Biochem. Pharmacol.* **22:**1532–1535, 1973.

Jones, K. L.: Fetal alcohol syndrome. *Pediatr. Rev.* **8:**122–126, 1986.

Jones, K. L., and Smith, D. W.: Recognition of the FAS in early infancy. *Lancet* **II:**999–1001, 1973.

Jones, K. L., and Smith, D. W.: The fetal alcohol syndrome. *Teratology* **12:**1–10, 1975.

Jones, K. L., Smith, D. W., Ulleland, C. N., and Streissguth, A. P.: Pattern of malformation in offspring of chronic alcoholic mothers. *Lancet* **I:**1267–1271, 1973.

Jones, K. L., Polletti, A., and Garner, H. H.: The Italian experience in mental health care. *Hosp. Commun. Psychiatry* **37:** 795–802, 1984.

Jones, P. J. M., Leichter, J., and Lee, M.: Placental blood flow in rats fed alcohol before and during gestation. *Life Sci.* **29:** 1153–1159, 1981.

Joske, R. A., Finkel, E. S., Wood, L. J.: Gastric biopsy: A study of 1000 consecutive biopsies. *Q. J. Med.* **48:**269–294, 1955.

Kadlubar, F. F., Talaska, G., Lang, N. P., Benson, R. W., and Roberts, D.: Assessment of exposure and susceptibility to aromatic amine carcinogens. In: *Methods for Detecting DNA Damaging Agents in Humans: Applications in Cancer Epidemiology and Prevention* (H. Bartsch, K. Hemminiki, and I. K. O'Neill, eds.) Lyons IARC Scientific Publications, No. 89, pp. 166–174, 1988.

Kaminski, M., Rumeau, C., and Schwartz, D.: Alcohol consumption in pregnant women and the outcome of pregnancy. *Alcoholism: Clin. Exp. Res.* **2:**155–162, 1978.

Kark, J. D., Smith, A. H., Switzer, B. R., Hames, N., and Curtis, G.: Serum vitamin A (retinol) and cancer incidence in Evans County, Georgia. *J. Natl. Cancer Inst.* **66:**7–16, 1981.

Karl, P. I., Gordon, B. H. J., Lieber, C. S., and Fisher, S. E.: Acetaldehyde production and transfer by the perfused human placental cotyledon. *Science* **242:**273–275, 1988.

Kellerman, G., Luyten-Kellerman, M., and Shaw, C. R.: Aryl-hydrocarbon-hydroxylase inducibility and bronchogenic carcinoma. *N. Engl. J. Med.* **289:**934–937, 1973.

Keppen, L. D., Pysher, T., Rennert, O. M.: Zinc deficiency acts as a co-teratogen with alcohol in fetal alcohol syndrome. *Pediatr. Res.* **19:**944–947, 1985.

Kesaniemi, Y. A., and Sipple, H. W.: Placental and fetal metabolism of acetaldehyde in rat. I. Contents of ethanol and acetaldehyde in placenta and fetus of the pregnant rat during ethanol oxidation. *Acta Pharmacol. Toxicol.* **37:**43–48, 1975.

King, M. M., and McCay, P. B.: Modulation of tumor incidence and possible mechanisms of inhibition of mammary carcinogenesis by dietary antioxidants. *Cancer Res.* **43**(Suppl.): 2485a–2490s, 1983.

Kinsella, A. R., and Radman, M.: Tumor promoter induces sister chromatid exchanges: Relevance to mechanisms of carcinogenesis. *Proc. Natl. Acad. Sci. U.S.A.* **75:**6149–6153, 1978.

Kitagawa, T., Pitot, H. C., Miller, E. C., and Miller, J.: Promotion by dietary phenobarbital of hepatocarcinogenesis by 2-methyl-*N,N*-dimethyl-4-aminoazobenzene in the rat. *Cancer Res.* **39:**112, 1979.

Kleihues, P., and Cooper, H. K.: Repair excision of alkylated bases from DNA *in vivo. Oncology* **33:**86–88, 1976.

Kleihues, P., Doejer, G., Keefer, L. K., Rice, J. M., Roller, P. P., and Hodgson, R. M.: Correlation of DNA methylation by methyl (acetoxmethyl) nitrosamine with organ-specific carcinogenicity in rats. *Cancer Res.* **39:**5136–5140, 1979.

Klein, S. M., Cohen, G., Lieber, C. S., and Cederbaum, A. I.: Increased microsomal oxidation of hydroxyl radical scavenging agents and ethanol after chronic consumption of ethanol. *Arch. Biochem. Biophys.* **223:**425–433, 1983.

Kline, J., Stein, Z., Shrout, P., Susser, M., and Warburton, D.: Drinking during pregnancy and spontaneous abortion. *Lancet* **II:**176–180, 1980.

Kmet, J., and Mahboubi, E.: Esophageal cancer in the Caspian Littoral of Iran. *Science* **175:**846–853, 1972.

Koop, D. R., Morgan, E. T., Tarr, G. E., and Coon, M. J.: Purification and characterization of a unique isozyme of cytochrome P-450 from liver microsomes of ethanol-treated rabbits. *J. Biol. Chem.* **257:**8472–8480, 1982.

Kriek, E.: Persistent binding of a new reaction product of the carcinogen *N*-hydroxyl-*N*-2-acetylaminofluorene with guanine in rat liver DNA *in vivo. Cancer Res.* **32:**2042–2048, 1972.

Krokan, H., Grafstrom, R. C., Lundquist, K., Esterbauer, H., and Harris, C. C.: Cytotoxicity, thiol depletion and inhibition of O^6-methylguanine-DNA methyltransferase by various aldehydes in cultured human bronchial fibroblasts. *Carcinogenesis* **6:**1755–1759, 1985.

Kronick, J. B.: Teratogenic effects of ethyl alcohol administered to pregnant mice. *Am. J. Obstet. Gynecol.* **124:**676–680, 1976.

Kvale, G., Bjelke, E., and Gart, J.: Dietary habits and lung cancer risk. *Inst. J. Cancer* **31:**397–405, 1983.

Lambert, B., and He, S.-M.: DNA and chromosomal damage induced by acetaldehyde in human lymphocytes *in vitro. Ann. N.Y. Acad. Sci.* **534:**369–376, 1988.

Lancet: Alcohol and the fetus—is zero the only option? *Lancet* **I:**682–683, 1983.

Larson, G.: Prevention of fetal alcohol effects: An antenatal program for early detection of pregnancies at risk. *Acta Obstet. Gynecol. Scand.* **62:**171–178, 1983.

Larson, G., Bohlin, A. B., and Tunell, R.: Prospective study of children exposed to variable amounts of alcohol *in utero. Arch. Dis. Child.* **60:**316–321, 1985.

Lefkowitch, J. H., Roshton, A. R., and Feng-Chen, K. C.: Hepatic fibrosis in fetal alcohol syndrome. *Gastroenterology* **85:**951–957, 1983.

Lemaitre, M., Renard, A., and Verly, W. G.: A common chromatin factor involved in the repair of O^6 methylguanine and O^6 ethyl guanine lesions in DNA. *FEBS Lett.* **144:**242–256, 1982.

Lemoine, P., Haurrousseau, M., Borteyru, J. P., and Menuet, J. C.: les enfants de parents alkooliques. Anomalies observées: A propos de 127 cas. *Quest Med.* **21:**476–482, 1968.

Leo, M. A., and Lieber, C. S.: New pathway for retinol metabolism in liver microsomes. *J. Biol. Chem.* **260:**5228–5231, 1985.

Leo, M. A., Iida, S., and Lieber, C. S.: Retinoic acid metabolism by a system reconstituted with cytochrome P-450. *Arch. Biochem. Biophys.* **234:**305–312, 1984.

Leo, M. A., Kim, C., and Lieber, C. S.: Increased vitamin A in esophagus and other extrahepatic tissues after chronic ethanol consumption in the rat. *Alcoholism: Clin. Exp. Res.* **6:**487–492, 1986a.

Leo, M. A., Lowe, N., and Lieber, C. S.: Interaction of drugs and retinol. *Biochem. Pharmacol.* **35:**3949–3953, 1986b.

Lieber, C. S.: Hepatic and metabolic effects of alcohol. *Gastroenterology* **65:**821–846, 1973.

Lieber, C. S.: Alcohol-liver injury and protein metabolism. *Pharmacol. Biochem. Behav.* **13** (Suppl. 1):17–30, 1980.

Lieber, C. S., and DeCarli, L. M.: Quantitative relationship between amount of dietary fat and severity of alcoholic fatty liver. *Am. J. Clin. Nutr.* **23:**474–478, 1970.

Lieber, C. S., Seitz, H. K., Garro, A. J., and Worner, T. M.: Alcohol-related diseases and carcinogenesis. *Cancer Res.* **39:**2863–2886, 1979.

Lieber, C. S., Baraona, E., Leo, M. A., and Garro, A. J.: Metabolism and metabolic effects of ethanol, including interaction with drugs, carcinogens and nutrition. *Mutat. Res.* **186:**201–233, 1987.

Lin, G. W. J.: Effect of ethanol feeding during pregnancy on placental transfer of alpha-amino isobutyric acid in the rat. *Life Sci.* **28:**595–601, 1981a.

Lin, G. W. J.: Fetal malnutrition: A possible cause of the Fetal Alcohol Syndrome. *Prog. Biochem. Pharmacol.* **18:**115–121, 1981b.

Lindamood, C. III, Bedell, M. A., Billings, K. C., Dyroff, M. C., and Swenberg, J. A.: Dose response for DNA alkylation. [^{3}H]thymidine uptake into DNA, and O^6-methylguanine-DNA methyltransferase activity in hepatocytes of rats and mice continuously exposed to dimethylnitrosamine. *Cancer Res.* **44:**196–200, 1984.

Little, R. E., Streissguth, A. P., Barr, H. M., and Herman, C. S.: Decreased birth weight in infants of alcoholic women who abstained during pregnancy. *J. Pediatr.* **96:**974–977, 1980.

Little, R. E., Streissguth, A. P., Guzinski, G. M., Grathwohl, H. L., Blumhagen, J. M., and McIntyre, C. E.: Change in obstetrician advice following a two-year community education program on alcohol use and pregnancy. *Am. J. Obstet. Gynecol.* **146:**23–28, 1983.

Little, R. E., Young, A., and Streissguth, A. P.: Preventing fetal alcohol effects: Effectiveness of a demonstration project. In: *Mechanisms of Alcohol Damage in Utero* (CIBA Foundation Symposium No. 105), London, Pittman, pp. 254–274, 1984.

Little, R. E., Asker, R. L., Sampson, P. D., and Renwick, J. H.: Fetal growth and moderate drinking in early pregnancy. *Am. J. Epidemiol.* **123:**270–278, 1986.

Lochry, E. A., Randall, C. L., Goldsmith, A. A., and Sutker, P. B.: Effects of acute alcohol exposure during selected days of gestation in C$_3$H mice. *Neurobehav. Toxicol. Teratol.* **4:**15–19, 1981.

Losowsky, M. S., and Leonard, P. J.: Evidence of vitamin E deficiency in patients with malabsorption or alcoholism and the effects of therapy. *Gut* **8:**539–543, 1967.

Lotan, R.: Effects of vitamin A and its analogs (retinoids) on normal and neoplastic cells. *Biochim. Biophys. Acta* **605:**33–91, 1980.

Lumeng, L.: The role of acetaldehyde in mediating the deleterious effect of ethanol on pyridoxal 5'-phosphate metabolism. *J. Clin. Invest.* **62:**286–293, 1978.

Lumeng, L., and Li, T.: Vitamin B$_6$ metabolism in chronic alcohol abuse. *J. Clin. Invest.* **53:**693–704, 1974.

MacDonald, C. M., Dow, J., and Moore, M. R.: A possible protective role for sulphydryl compounds in acute alcoholic liver injury. *Biochem. Pharmacol.* **26:**1529–1531, 1977.

MacLennan, R., DaCosta, J., Day, N. E., Law, C. H., Ng, Y. K., and Shanmugaratnam, K.: Risk factors for lung cancer in Singapore Chinese, population with high female incidence rates. *Int. J. Cancer* **20:**854–860, 1977.

Mak, K. M., Leo, M. A., and Lieber, C. S.: Ethanol potentiates squamous metaplasia of the rat trachea caused by vitamin A deficiency. *Trans. Assoc. Am. Physicians* **97:**210–221, 1984.

Mak, K. M., Leo, M. A., and Lieber, C. S.: Potentiation by ethanol consumption of tracheal squamous metaplasia caused by vitamin A deficiency in rats. *J. Natl. Cancer Inst.* **79:**1001–1010, 1987a.

Mak, K. M., Leo, M. A., and Lieber, C. S.: Effect of ethanol and vitamin A deficiency on epithelial cell proliferation and structure in the rat esophagus. *Gastroenterology* **93:**362–367, 1987b.

Maling, H. M., Stripp, B., Sipes, I. G., Highman, B., Saul, W., and Williams, M. A.: Enhanced hepatotoxicity of carbon tetrachloride and dimethylnitrosamine by pretreatment of rats with ethanol and some comparisons with potentiation by isopropanol. *Toxicol. Appl. Pharmacol.* **33:**291–308, 1975.

Maron, D. M., and Ames, B. N.: Revised methods for the Salmonella mutagenicity test. *Mutat. Res.* **113:**173–215, 1983.

Martin, D. C., Martin, J. C., Streissguth, A. P., and Lund, C. A.: Sucking frequency and amplitude in newborns as a function of maternal drinking and smoking. *Curr. Alcohol* **5:**359–366, 1978.

Martinez, I.: Factors associated with cancer of the esophagus, mouth and pharynx in Puerto Rico. *J. Natl. Cancer Inst.* **42:**1069–1094, 1969.

Matloff, D. S., Selinger, M. J., and Kaplan, M. M.: Hepatic transaminase activity in alcoholic liver disease. *Gastroenterology* **78:**1380–1382, 1980.

McCarthy, D. J., Lindamood, C. III, and Hill, D. L.: Effects of retinoids on metabolizing enzymes and on binding of benzo(a)pyrene to rat tissue DNA$_1$. *Cancer Res.* **41:**2849–2854, 1987.

McCoy, G. D., and Wynder, E.: Etiological and preventive implications in alcohol carcinogenesis. *Cancer Res.* **39:**2844–2850, 1979.

McCoy, G. D., Hecht, S. S., Katayama, S., and Wynder, E. L.: Differential effect of chronic ethanol consumption on the carcinogenicity of *N*-nitrosopyrrolidine and *N'*-nitrosonornicotine in male Syrian golden hamsters. *Cancer Res.* **41:**2849–2854, 1981.

McCoy, G. D., Hecht, S. S., and Furuya, K.: The effect of chronic ethanol consumption on the tumorigenicity of *N*-nitrosopyrrolidine in male Syrian golden hamsters. *Cancer Lett.* **33:**151–159, 1986.

McDowell, E. M., Barrett, L. A., Harris, C. C., and Trump, B. F.: Abnormal cilia in human bronchial epithelium. *Arch. Pathol. Lab. Med.* **100:**429–436, 1976.

Meadows, G. G., Blank, S. E., Duncan, D. D.: Influence of ethanol consumption on natural killer cell activity in mice. *Alcoholism: Clin. Exp. Res.* **13:**476–479, 1989.

Mettlin, C., Graham, S., Priore, R., Marshall, J., and Swanson, M.: Diet and cancer of the esophagus. *Nutr. Cancer* **2:**143–147, 1980.

Mezey, E., Patter, J. J., and Merchant, C. R.: Effect of ethanol feeding on bone composition in the rat. *Am. J. Clin. Nutr.* **32:**25–29, 1979.

Miller, J. A.: Carcinogenesis by chemicals: An overview—G.H.A. Clowes Memorial Lecture. *Cancer Res.* **30:**559–576, 1970.

Miller, J. A., and Miller, E. C.: Ultimate chemical carcinogens as reactive mutagenic electrophiles. In: *Origins of Human Cancer* (H. H. Hiatt, J. D. Watson, and J. A. Winsten, eds.), New York, Cold Springs Harbor Laboratory Press, pp. 605–627, 1977.

Miller, S., Del Villano, B. C., Flynn, A., and Krumhanst, M.: Interaction of alcohol and zinc in fetal dysmorphogenesis. *Pharmacol. Biochem. Behav.* **18:**311–315, 1983.

Mills, P. R., Rennington, T. H., Kay, P., MacSween, R. N. M., and Watkinson, G.: Hepatitis B antibody in alcoholic cirrhosis. *J. Clin. Pathol.* **32:**778–782, 1979.

Milne, M., and Baran, D. T.: Inhibitory effect of maternal alcohol ingestion on rat pup 25-hydroxy vitamin D production. *Pediatr. Res.* **19:**102–104, 1985.

Møller, J.: Hepatic dysfunction in patient with fetal alcohol syndrome. *Lancet* **I:**605–606, 1979.

Moon, R. C., Grubbs, C. J., Sporn, M. B., and Goodman, D. G.: Retinyl acetate inhibits mammary carcinogenesis induced by *N*-methyl-*N*-nitrosourea. *Nature* **267:**620–621, 1977.

Mufti, S. I., Prabhala, R., Moriguchi, S., Sipes, I. G., and Watson, R. R.: Functional and numerical alterations induced by ethanol in the cellular immune system. *Immunopharmacology* **15:**85–94, 1988a.

Mufti, S., Salvagnini, M., Lieber, C. S., and Garro, A. J.: Chronic ethanol consumption inhibits repair of dimethylnitrosamine-induced DNA alkylation. *Biochem. Biophys. Res. Commun.* **152:**425–431, 1988b.

Mufti, S. I., Becker, G., and Sipes, I. G.: Effect of chronic dietary ethanol consumption on the initiation and promotion of chemically-induced esophageal carcinogenesis in experimental rats. *Carcinogenesis* **10:**303–309, 1989a.

Mufti, S. I., Darban, H. R., and Watson, R. R.: Alcohol, Cancer and Immunomodulation. *CRC Crit. Rev. Hematol. Oncol.* **9:**243–261, 1989b.

Mukherjee, A. B., and Hodgen, C. D.: Maternal ethanol exposure induces transient impairment of umbilical circulation and fetal hypoxia in monkeys. *Science* **218:**700–702, 1982.

Mulvihill, J. J., and Yeager, A. M.: Fetal alcohol syndrome. *Teratology* **13:**345–348, 1976.

Myerson, R. M.: Acute effects of alcohol on the liver with special reference to the Zieve syndrome. *Am. J. Gastroenterol.* **49:**304–311, 1968.

Nebert, D. W.: Genes encoding drug-metabolizing enzymes: possible role in human disease. In: *Phenotypic Variation in Populations: Relevance to Risk Assessment* (A. D. Woodhead, M. A. Bender, and R. C. Leonard, eds.), New York, Plenum Press, 1988.

Neri, G., Sammito, V., Romano, C., Sanfilippo, S., Opitz, J. M.: Facial midline defect in the Fetal Alcohol Syndrome: Embryogenetic considerations in two clinical cases. *Am. J. Med. Genet.* **29:**477–482, 1988.

Newberne, P. M., Charnley, G., Adams, K., Cantor, M., Roth, D., and Supharkarn, V.: Gastric and oesophageal carcinogenesis: models for the identification of risk and protective factors. *Fed. Chem. Toxicol.* **24:**1111–1119, 1986.

Newbold, R. F., Warren, W., Medcalf, A. S. C., and Amos, J.: Mutagenicity of carcinogenic methylating agents is associated with a specific DNA modification. *Nature* **283:**596–599, 1980.

Newman, S. L., Flannery, D. B., and Caplan, D. B.: Simultaneous occurrence of extra hepatic biliary atresia and fetal alcohol syndrome. *Am. J. Dis. Child.* **133:**101, 1979.

Nordmann, R., Ribiere, C., and Rovach, H.: Involvement of iron and iron-catalyzed free radical production in ethanol metabolism and toxicity. *Enzyme* **37:**57–69, 1987.

O'Shea, K. S., and Kaufman, M. H.: The teratogenic effect of acetaldehyde: implications for the study of the fetal alcohol syndrome. *J. Anat.* **126:**65–76, 1979.

O'Shea, K. S., and Kaufman, M. H.: Effect of acetaldehyde on the neuroepithelium of early mouse embryos. *J. Anat.* **132:**107–118, 1981.

Obe, G., and Beck, B.: Mutagenic activity of aldehydes. *Drug Alcohol Depend.* **4:**91–94, 1979.

Obe, G., and Ristow, H.: Acetaldehyde but not alcohol induces sister chromatid exchanges in Chinese hamster cells *in vitro*. *Mutat. Res.* **56:**211–213, 1977.

Obe, G., and Ristow, H.: Mutagenic, cancerogenic and teratogenic effects of alcohol. *Mutat. Res.* **65:**229–259, 1979.

Ohnishi, K., and Lieber, C. S.: Reconstitution of the microsomal ethanol-oxidizing system: Qualitative and quantitative changes of cytochrome P-450 after chronic ethanol consumption. *J. Biol. Chem.* **252:**7124–7131, 1977.

Ohnishi, K., Iida, S., Iwama, S., Goto, M., Nomura, F., Takashi, M., Mishima, A., Kuno, K., Kimura, K., Musha, H., Kotota, K., and Okuda, K.: The effect of chronic habitual alcohol intake on the development of liver cirrhosis and hepatocellular carcinoma: relation to hepatitis B surface antigen carriers. *Cancer* **49:**672–677, 1982.

Orholm, M., Alderschvile, J., Tage-Jensen, U., Schlichting, P., Nielsen, J., Hardt, F., and Christoffersen, P.: Prevalence of hepatitis B virus infection among alcoholic patients with liver disease. *J. Clin. Pathol.* **34:**1378–1380, 1981.

Ouellette, E. M., Rosett, M. L., Rosman, N. P., and Weiner, L.: Adverse effects on offspring of maternal alcohol abuse during pregnancy. *N. Engl. J. Med.* **297:**528–530, 1977.

Palmer, E. D.: Gastritis: A re-evaluation. *Medicine* **33:**199–290, 1954.

Patwardhan, R. V., Schenker, S., Henderson, G. I., Abou-Mourad, N. N., and Hoyumpa, A. M.: Short-term and long-term ethanol administration inhibits the placental uptake and transport of valine in rats. *J. Lab. Clin. Med.* **98:**251–262, 1981.

Pegg, A. E., and Perry, W.: Alkylation of nuclei acids and metabolism of small doses of dimethylnitrosamine in the rat. *Cancer Res.* **41:**3128–3132, 1981.

Pegg, A. E., Weist, L., Foote, R. S., Mitra, S., and Perry, W.: Purification and properties of O^6-methylguanine-DNA transmethylase from rat liver. *J. Biol. Chem.* **258**:2327–2333, 1983.

Pennington, S. N.: Alcohol metabolism and fetal hypoplasia in chick brain. *Alcohol* **5**:91–94, 1988a.

Pennington, S. N.: Ethanol-induced growth inhibition: The role of cyclic AMP-dependent protein kinase. *Alcoholism: Clin. Exp. Res.* **12**:125–129, 1988b.

Pennington, S. N., Boyd, J. W., Kalmus, G. W., and Wilson, R. W.: The molecular mechanism of fetal alcohol syndrome (FAS) I. Ethanol induced growth suppression. *Neurobehav. Toxicol. Teratol.* **5**:259–262, 1983.

Pennington, S., Allen, Z., Runion, J., Farmer, P., Rowland, L., and–Kalmus, G.: Prostaglandin synthesis inhibitors block alcohol-induced fetal hypoplasia. *Alcoholism: Clin Exp. Res.* **9**:433–439, 1985.

Peraino, C., Fry, R. J. M., Staffeldt, E., and Kisilleshi, W. E.: Effect of varying the exposure to phenobarbital on its enhancement of 2-acetylaminofluorene-induced hepatic tumorigenesis in the rat. *Cancer Res.* **33**:2701–2705, 1973.

Perera, M. I., Katyal, S. L., and Shinozuka, H.: Choline-deficient diet enhances the initiating and promoting effects of methapyrilene hydrochloride in rat liver as assayed by the induction of gamma-glutamyltranspeptidase-positive hepatocyte foci. *Br. J. Cancer* **56**:774–778, 1987.

Petro, T. M., Watson, R. R., and Bhattacharjee, J. K.: Immunity to bacterial pathogens in the protein-malnourished host. In: *Nutrition, Disease Resistance and Immune Function.* New York, Marcel Dekker, 1984.

Pikkarainen, P., and Raiha, N. C. R.: Development of alcohol dehydrogenase activity in the human liver. *Pediatr. Res.* **1**:165–168, 1967.

Pitchumoni, C. S., and Jerzy-Glass, G. B.: Patterns of gastritis in alcoholics. *Biol. Gastro–Enterol.* **9**:11–16, 1976.

Plant, M. L., and Plant, M. A.: Maternal use of alcohol and other drugs during pregnancy and birth abnormalities: Further results from a prospective study. *Alcohol* **23**:229–223, 1988.

Porta, E. A., Markell, N., and Dorado, R. D.: Chronic alcoholism enhances hepatocarcinogenicity of diethylnitrosamine in rats fed a marginally methyl-deficient diet. *Hepatology* **5**:1120–1125, 1985.

Pottern, L. M., Morris, L. E., Blot, W. J., Ziegler, R. G., and Fraumeni, J. F.: Esophageal cancer among black men in Washington D.C. I. Alcohol, tobacco and other risk factors. *J. Natl. Cancer Inst.* **67**:777–783, 1981.

Priscott, P. K.: The effects of ethanol on rat embryos developing *in vitro. Biochem. Pharmacol.* **31**:3641–3643, 1982.

Protzel, M., Gardina, A. C., and Albano, U. H.: The effect of liver imbalance on the development of oral tumors in mice following the applications of benzpyrene or tobacco tar. *Oral Surg.* **18**:622, 1964.

Qazi, Q. M., Madahar, C., Masakawa, A., and McGann, B.: Chromosome abnormality in a patient with fetal alcohol syndrome. *Curr. Alcohol* **5**:155–161, 1979.

Qin, S., and Huang, C. C.: Influence of mouse liver stored vitamin A on the induction of mutations (Ames Tests) and SCE of bone marrow cells by aflatoxin B$_1$, benzo(a)pyrene, or cyclophosphamide. *Environ. Mutagen.* **8**:839–847, 1986.

Radike, M. J., Stemmer, K. L., Brown, P. G., Larson, E., and Bingham, E.: Effect of ethanol and vinyl chloride on the induction of liver tumors: Preliminary report. *Environ. Health Perspect.* **21**:153–155, 1977.

Radike, M. J., Stemmer, K. L., and Bingham, E.: Effect of ethanol on vinyl chloride carcinogenesis. *Environ. Health Perspect.* **41**:59–62, 1981.

Randall, C. L., and Anton, R. F.: Aspirin reduces alcohol-induced prenatal mortality and malformations in mice. *Alcoholism: Clin. Exp. Res.* **8**:513–515, 1984.

Randall, C. L., and Taylor, W. J.: Prenatal ethanol exposure in mice: Teratogenic effects. *Teratology* **19**:305–312, 1979.

Randall, C. L., Taylor, W. J., and Walker, D. W.: Ethanol-induced malformations in mice. *Alcoholism: Clin. Exp. Res.* **1**:219–224, 1977a.

Randall, C. L., Taylor, W. J., Tabakoff, B., and Walker, D. W.: Ethanol as a teratogen. In: *Alcohol and Aldehyde Metabolizing Systems, Vol. III* (R. G. Thurman, et al., eds.), New York, Academic Press, pp. 659–670, 1977b.

Randall, C. L., Becker, H. C., and Middaugh, L. D.: Effect of prenatal ethanol exposure on activity and shuttle avoidance behavior in adult C-57 mice. *Alcohol Drug Res.* **6**:351–360, 1986.

Randall, C. L., Anton, R. F., and Becker, H. C.: Alcohol, pregnancy and prostaglandins. *Alcoholism: Clin. Exp. Res.* **11**:26–32, 1987.

Rasmussen, M., Blomhoff, R., Helgerud, P., Solberg, L. A., Berg, T., and Norum, K. R.: Retinol and retinyl esters in parenchymal and nonparenchymal rat liver cell fractions after long-term administration of ethanol. *J. Lipid Res.* **26**:1112–1119, 1985.

Reddy, B. S., Cohen, L. A., McCoy, G. D., Hill, P., Weisburger, J. H., and Wynder, E. L.: Nutrition and its relationship to cancer. *Adv. Cancer Res.* **32**:238–245, 1980.

Reznik-Schuller, H.: Ciliary alterations in hamster respiratory tract epithelium after exposure to carcinogens and cigarette smoke. *Cancer Lett.* **1**:7–13, 1975.

Riley, E. P., and Rockwood, G. A.: Alterations in suckling behavior in preweanling rats exposed to alcohol prenatally. *Nutr. Behav.* **1**:289–299, 1984.

Riley, E. P., Lochry, E. A., and Shapiro, N. R.: Lack of response inhibition in rats prenatally exposed to alcohol. *Psychopharmacology* **62**:47–52, 1979.

Riley, E. P., Barron, S., and Hannigan, J. H.: Response inhibition deficits following prenatal alcohol exposure: A comparison to the effects of hippocampal lesions in rats. In: *Alcohol and Brain Development* (J. West, ed.), New York, Oxford University Press, pp. 71–102, 1986.

Ristow, H., and Obe, G.: Acetaldehyde induces cross-links in DNA and causes sister-chromatid exchanges in human cells. *Mutat. Res.* **58**:115–119, 1978.

Roberts, D. M.: Chronic gastritis, alcohol and non-ulcer dyspepsia. *Gut.* **13**:768–774, 1972.

Rockwood, G. A., and Riley, E. P.: Suckling deficits in rat pups exposed to alcohol *in utero. Teratology* **33**:145–151, 1986.

Rogers, A. E.: Influence of dietary content of lipids and lipotropic nutrients on chemical carcinogens in rats. *Cancer Res.* **43**(Suppl):2477s–2484s, 1983.

Rogers, A. E., and Newberne, P. M.: Lipotrope deficiency in experimental carcinogenesis. *Nutr. Cancer* **2**:104–112, 1980.

Roselle, G. A., and Mendenhall, C. L.: Ethanol-induced alterations in lymphocyte function in the guinea pig. *Alcoholism: Clin. Exp. Res.* **8**:62–67, 1984.

Roselle, G. A., Mendenhall, C. L., and Grossberg, C. J.: Ethanol and soluble mediators of host response. *Alcoholism: Clin. Exp. Res.* **13**:494–498, 1989.

Rosett, H.: A clinical perspective of the fetal alcohol syndrome. *Alcoholism: Clin. Exp. Res.* **4**:119–122, 1980.

Rosett, H. L., and Sandler, L. W.: Effects of maternal drinking on neonatal morphology and state regulations. In: *The Handbook of Infant Development* (J. D. Osofsky, ed.), John Wiley & Sons, pp. 809–836, 1979.

Rosett, H. L., and Weiner, L.: *Alcohol and the Fetus*. New York, Oxford Press, 1984.

Rosett, H. L., and Weiner, L.: *Alcohol and the Fetus: A Teaching Package*. Boston, Fetal Alcohol Education Program, pp. 1–83, 1985.

Rosett, H. L., Weiner, L., and Edelin, E. C.: Strategies for prevention of fetal alcohol effects. *Obstet. Gynecol.* **57**:1–7, 1981.

Rosett, H. L., Weiner, L., Lee, A., Zuckerman, B., Dooling, E., and Oppenheimer, E.: Patterns of alcohol consumption and fetal development. *Obstet. Gynecol.* **61**:539–546, 1983a.

Rosett, H. L., Weiner, L., and Edelin, K. C.: Treatment experience with pregnant problem drinkers. *J.A.M.A.* **249**:2029–2033, 1983b.

Rovinski, B., Hosein, E. A., Lee, H., Lau-You-Hin, G., and Rastogi, N. K.: Hepatology of maternal ethanol consumption in rat offspring: An assessment with a study of the ontogenic development of ethanol-oxidizing systems. *Alcohol Drug Res.* **7**:195–205, 1987.

Royal College of Psychiatrists: *R. Coll. Psychiatr. Bull.* **6**:69, 1982.

Rubin, E., and Lieber, C. S.: Hepatic microsomal enzymes in man and rat, induction and inhibition by ethanol. *Science* **162**:690–691, 1968.

Rubin, D. H., Krasilnikoff, P. A., and Leventhal, J. M.: Cigarette smoking and alcohol consumption during pregnancy by Danish women and their spouses—a potential source of fetal mortality. *J. Drug Alcohol Abuse* **14**:405–417, 1988.

Rygaard, J., and Povlson, C. O.: The mouse mutant nude does not develop spontaneous tumors. An argument against immunological surveillance. *Acta Pathol. Microbiol. Immunol. Scand.* **82**:62–67, 1974.

Sakurai, M. A.: A histopathologic study of the effect of alcohol on cirrhosis and hepatoma of autopsy cases in Japan. *Acta Pathol. Jpn.* **19**:283–314, 1969.

Samson, H. H.: Microcephaly and Fetal Alcohol Syndrome: human and animal studies. In: *Alcohol and Brain Development* (J. R. West, ed.), New York, Oxford University Press, pp. 167–183, 1986.

Samson, H. H,. and Grant, K. A.: Ethanol-induced microcephaly in neonatal rats: Relation to dose. *Alcoholism: Clin. Exp. Res.* **8**:201–203, 1984.

Sanchis, R,. and Guerri, C.: Alcohol-metabolizing enzymes in placenta and fetal liver: Effect of chronic ethanol intake. *Alcoholism: Clin. Exp. Res.* **10**:39–44, 1986.

Sandor, S. T., and Elias, S. T.: The influence of aethyl-alcohol on the development of the chick embryo. *Rev. Roum. Embryol.* **5**:51–76, 1968.

Sato, A., Nakagima, T., Koyama, Y.: Effects of chronic ethanol consumption of hepatic metabolism of aromatic and chlorinated hydrocarbons. *Br. J. Ind. Med.* **37**:382–386. 1980.

Sato, A., Nakagima, T., and Koyama, Y.: Dose related effects of a single dose of ethanol on the metabolism in rat liver of some aromatic and chlorinated hydrocarbons. *Toxicol. Appl. Pharmacol.* **60**:8–15, 1981.

Sato, M., and Lieber, C. S.: Hepatic vitamin A depletion after chronic ethanol consumption in baboons and rats. *J. Nutr.* **111**:2015–2023, 1981.

Sato, M., and Lieber, C. S.: Increased metabolism of retinoic acid after chronic ethanol consumption in rat liver microsomes. *Arch. Biochem. Biophys.* **213**:557–564, 1982.

Savage, D. D., Montano, C. Y., Paxton, L. L., and Kasarskis, E. J.: Prenatal ethanol exposure decreases hippocampal mossy fiber zinc in 45 day old rats. *Alcoholism: Clin. Exp. Res.* **13**:588–593, 1989.

Savoy-Moore, R. T., Dombrowski, M. P., Cheng, E. A., and Sokol, R. J.: Low dose alcohol contracts the human umbilical artery *in vitro*. *Alcoholism: Clin. Exp. Res.* **13**:40–42, 1989.

Saxena, Q. B., Mezey, E., and Adler, W. H.: Regulation of natural killer activity *in vivo*. II. The effect of alcohol consumption on human peripheral blood natural killer activity. *Int. J. Cancer* **26**:413, 1980.

Saxena, Q. B., Saxena, R. K., and Adler, W. H.: Regulation of natural killer activity in vivo. IV. High natural killer cell activity in alcohol drinking mice. *Ind. J. Exp. Biol.* **19**:1001, 1981.

Schardein, J. L.: Current status of drugs as teratogens in man. *Prog. Clin. Biol. Res.* **163**:181–190, 1985.

Schenker, S., Dicke, J. M., Johnson, R. F., Hays, S. E., and Henderson, G. I. K.: Effect of ethanol on human placental transport of model amino acids and glucose. *Alcoholism: Clin. Exp. Res.* **1**:112–119, 1989.

Schmahl, D., Thomas, C., Sattler, W., and Scheld, G. F.: Expermentelle Untersuchunger zur Synkarginogenese III Mittleung Versuche zur Krebserzeugun bei Ratten bei gleichzeit!ger Gabe von Diethylnitrosamin und Tetrachlorkohylenstoff bzw. Aethylalkohol Zugleich ein experimenteller Beitrag zur Frage du Alkoholcirrhose. *Z. Krebsforsch* **66**:526–532, 1965.

Schrauzer, G. N., McGinness, J. E., Ishmael, D., and Bell, L. J.: Alcoholism and cancer. I. Effects of long-term exposure to alcohol on spontaneous mammary adenocarcinoma and prolactin levels in C3H/st mice. *J. Stud. Alcohol* **40**:240–246, 1979.

Schwarz, M., Weisbeck, G., Hummel, J., and Krunz, W.: Effect of ethanol on dimethylnitrosamine activation and DNA synthesis in rat liver. *Carcinogenesis* **3**:1071–1075, 1982.

Seitz, H. K., and Simonowski, U. A.: Ethanol and colorectal carcinogenesis. In: *Colorectal Cancer: From Pathogenesis to Prevention* (H. K. Seitz, U. A. Simanowski, and N. A. Wright, eds.), Berlin, Springer Verlag, pp. 177–189, 1989.

Seitz, H. K., Garro, A. J., and Lieber, C. S.: Effect of chronic ethanol ingestion on intestinal metabolism and mutagenicity of benzo(a)pyrene. *Biochem. Biophys. Res. Commun.* **85**:1061–1066, 1978.

Seitz, H. K., Garro, A. J., and Lieber, C. S.: Enhanced hepatic activation of procarcinogens after chronic ethanol consumption. *Gastroenterology* **77**:40, 1979.

Seitz, H. K., Garro, A. J., and Lieber, C. S.: Enhanced pulmonary and intestinal activation of procarcinogens and mutagens after chronic ethanol consumption in the rat. *Eur. J. Clin. Invest.* **11**:33–38, 1981a.

Seitz, H. K., Garro, A. J., and Lieber, C. S.: Sex dependent effect of chronic ethanol consumption in rats on hepatic microsome-mediated mutagenicity of benzo(a)pyrene. *Cancer Lett.* **13**:97–102, 1981b.

Seitz, H. K., Czygan, P., Waldherr, R., Veith, S., Raedsch, R., Kassmodel, H., and Kommerell, B.: Enhancement of 1,2-dimethylhyrazine-induced rectal carcinogenesis following chronic ethanol consumption in the rat. *Gastroenterology* **86**:886–891, 1984.

Sernka, T. J., Gilleland, C. W., and Shanbour, L. L.: Effects of ethanol on active transport in the dog stomach. *Am. J. Physiol.* **226**:397–400, 1974.

Shafritz, D. A., Shouval, D., Sherman, H. I., Hadziyannis, S. J., and Kew, M. C.: Integration of hepatitis B virus DNA into the genome of liver cells in chronic liver disease and hepatocellular carcinomas. *N. Engl. J. Med.* **305**:1067–1073, 1981.

Shanbour, L. L., Miller, J., and Chowdhury, T. K.: Effects of alcohol on active transport in the rat stomach. *Am. J. Dig. Dis.* **18**:311–316, 1973.

Shaw, S., Rubin, K. P., and Lieber, C. S.: Depressed hepatic glutathione and increased diene conjugates in alcoholic liver disease: Evidence of lipid peroxidation. *Dig. Dis. Sci.* **28**:585–589, 1983.

Shimizu, T.: Experimental study of esophageal cancer-effect of alcohol, vitamin C, prostaglandin E2 and tegafur on carcinogenesis by N-methyl-N-amylnitrosamine and the development of esophageal carcinoma. *Nippon Gan Chiryo Gakkai Zashi* **21**:1232–1243, 1986.

Shino, P. H., Klebanoff, M. A., and Rhoads, G. G.: Smoking and drinking during pregnancy. *J.A.M.A.* **255**:82–84, 1986.

Simanowski, U. A., Seitz, H. K., Baier, B., Kommerell, B., Schmidt, Gayk, H., and Wright, N. A.: Chronic ethanol consumption selectively stimulates rectal cell proliferation in the rat. *Gut* **27**:278–282, 1986.

Singer, B., and Fraenkel-Conrat, H.: The role of conformation in chemical mutagenesis. In: *Progress in Nuclei Acid Research and Molecular Biology* (J. N. Davidson and W. E. Cohn), New York, Academic Press, **9**:1–29, 1969.

Singletary, K., Odoms, A., Nelshoppen, J., and McNary, M.: Enhancement by ethanol of DMBA-induced mammary tumorigenesis. *FASEB J.* **4**:A1171, 1990.

Smith, B. M., Skillman, J. J., Edwards, B. C., and Silen, W.: Permeability of the human gastric mucosa: Alteration by acetylsalicylic acid and ethanol. *N. Engl. J. Med.* **285**:716–721, 1971.

Snyder, A. K., Singh, S. P., and Pullen, G. L.: Ethanol-induced intrauterine growth retardation: Correlation with placental glucose transfer. *Alcoholism: Clin. Exp. Res.* **10**:167–170, 1986.

Sokol, R. J., Ager, J., Martier, S., Debanne, S., Ernhart, C., Kuzman, J., and Miller, S. I.: Significant determinants of susceptibility to alcohol teratogenicity. *Ann. N.Y. Acad. Sci.* **477**:87–102, 1986.

Solomon, L. R.: Studies on the mechanism of acetaldehyde-mediated inhibition of rat liver transaminases. *Clin. Chim. Acta* **168**:207–217, 1987.

Speisky, H., MacDonald, A., Giles, G., Orrega, H., and Israel, Y.: Increased loss and decreased synthesis of hepatic glutathione after acute ethanol administration. *Biochem. J.* **225**:565–572, 1985.

Sporn, M. B., and Newton, D. L.: Retinoids and chemoprevention of cancer. In: *Inhibition of Tumor Induction and Development* (M. S. Zedeck and M. Lipkin, eds.), New York, Plenum Press, pp. 71–100, 1981.

Sporn, M. B., and Roberts, A. B.: Role of retinoids in differentiation and carcinogenesis. *Cancer Res.* **43**:3034–3040, 1983.

Sporn, M. B., Roberts, A. B. and Goodman, D. S. (eds.): *The Retinoids*. New York, Academic, 1984.

Sreenathan, R. N., Padmanabhan, R., and Singh, S.: Teratogenic effects of acetaldehyde in the rat. *Drug Alcohol Depend.* **9**:339–350, 1982.

Stein, A., and Susser, M.: The Dutch famine, 1944–1945, and the reproductive process. I. Effects on six indices at birth. *Pediatr. Res.* **9**:70–76, 1975.

Stenback, F.: The tumorigenic effect of ethanol. *Acta Pathol. Microbiol. Scand.* **77**:325–326, 1969.

Stibler, H., Burns, E., Kruckeborg, T., Gaetano, P., Cerven, E., Borg, S., and Tabakoff, B.: Effect of ethanol on synaptosomal sialic acid metabolism in the developing rat brain. *J. Neurol. Sci.* **59**:21–35, 1983.

Straus, B., Berenyi, M. R., Huang, J. M., and Straus, E.: Delayed hypersensitivity in alcoholic cirrhosis. *Am. J. Dig. Dis.* **16**:509–516, 1971.

Streissguth, A. P.: The behavioral teratology of alcohol: Performance behavioral and intellectual deficits in prenatally exposed children. In: *Alcohol and Brain Development* (J. R. West, ed.), New York, Oxford University Press, pp. 3–44, 1986.

Streissguth, A. P., and LaDue, R. A.: Fetal alcohol syndrome and fetal alcohol effects: Teratogenic causes of mental retardation and development disabilities: In: *Toxic Substances and Mental Retardation: Neurobehavioral Toxicology and Teratology* (S. R. Schroeder, ed.), Washington, D.C., American Association of Mental Retardation, 1987, pp. 1–32.

Streissguth, A. P., Herman, C. S., and Smith, D. W.: Intelligence behavior, and dysmorphogenesis in the fetal alcohol syndrome: A report on 20 patients. *J. Pediatr.* **92**:363–367, 1978.

Streissguth, A. P., Landesman-Dwyer, S., Martin, J. C., and Smith, D. W.: Teratogenic effects of alcohol in humans and laboratory animals. *Science* **209**:353–361, 1980.

Streissguth, A. P., Clarren, S. K., and Jones, K. L.: Natural history of the fetal alcohol syndrome: A ten-year follow-up of eleven patients. *Lancet* **I**:85–91, 1985.

Strickland, S., and Mahdavi, V.: The induction of differentiation in teratocarcinoma stem cells by retinoic acid. *Cell* **15**:393–403, 1978.

Strömland, K.: Eyeground malformations in the fetal alcohol syndrome. *Neuropediatrics* **12**:97–98, 1981.

Strömland, K.: Ocular abnormalities in the fetal alcohol syndrome. *Acta Ophthalmol. [Suppl.]* **63**:5–49, 1985.

Stutman, O.: Chemical carcinogenesis in nude mice: Comparison between nude mice from homozygous matings and effect of age and carcinogen dose. *J. Natl. Cancer Inst.* **62:**353–358, 1979.

Suematsu, T., Matsumura, T., Sato, N., and Kumano, M.: Lipid peroxidation in alcoholic liver disease in humans. *Alcoholism: Clin. Exp. Res.* **5:**427–430, 1981.

Sulik, K.: Chemically Induced Birth Defects. In: *Drug and Chemical Toxicology* (J. L. Schardein, ed.), New York, Marcel Dekker, 1985, pp. 785.

Sulik, K. K., Johnston, M. C., and Webb, M. A.: Fetal alcohol syndrome: Embryogenesis in a mouse model. *Science* **214:**936–938, 1981.

Sulik, K. K., Johnston, M. C., Daft, P. A., Russell, W. E., and Dehart, D. B.: Fetal alcohol syndrome and Di George Anomaly: Critical exposure periods for craniofacial malformations as illustrated in an animal model. *Am. J. Med. Genet. Suppl.* **2:**97–112, 1986.

Sullivan, J. F., and Lankford, H. G.: Urinary excretion of zinc in alcoholism and postalcoholic cirrhosis. *Am. J. Clin. Nutr.* **10:**153–157, 1962.

Swann, P. F., Coe, A. M., and Mace, R.: Ethanol and dimethylnitrosamine metabolism and disposition in the rat. Possible relevance to the influence of ethanol on human cancer incidence. *Carcinogenesis* **5:**1337–1343, 1984.

Swenberg, J. A., Dyroff, M. C., Bedell, M. A., Popp, J. A., Huh, M., Kirstein, A., and Rajewsky, M. F.: O^4-ethyldeoxythymidine, but not O^6-ethyldeoxyguanosine accumulates in DNA of hepatocytes of rats exposed continuously to diethylnitrosamine. *Proc. Natl. Acad. Sci. U.S.A.* **81:**1692–1695, 1984.

Szebeni, J., Eskelson, C. D., Mufti, S. I., Watson, R., and Sipes, I. G.: Inhibition of ethanol induce ethane exhalation by carcinogenic pretreatment of rats 12 months earlier. *Life Sci.* **39:**3587–3591, 1986.

Tanaka, H., Nakazawa, K., and Suzukin, A. M.: Prevention possibility for brain dysfunction in rat with the fetal alcohol syndrome—low-zinc—status and hypoglycemia. *Brain Dev.* **4:**429–438, 1982.

Tanaka, H., Inomata, K., and Arima, M.: Zinc supplementation in ethanol-treated pregnant rats increases the metabolic activity in the fetal hippocampus. *Brain Dev.* **5:**549–554, 1983.

Teschke, R., Minzlaff, M., Oldiges, H., and Frenzel, H.: Effect of chronic alcohol consumption on tumor incidence due to dimethylnitrosamine administration. *J. Cancer Res. Clin. Oncol.* **106:**58–64, 1983.

Thurman, R. G.: Induction of hepatic microsomal NADPH-dependent production of hydrogen peroxide by chronic prior treatment with ethanol. *Mol. Pharmacol.* **9:**670–675, 1973.

Turner, R. T., Greene, V. S., and Bell, D. M.: Demonstration that ethanol inhibits bone matrix synthesis and mineralization in the rat. *Bone Min. Res.* **2:**61–66, 1987.

Tuyns, A. J.: Alcohol and cancer. *Alcohol Health Res. World* **2:**20–31, 1978.

Tuyns, A. J., Pequignot, G., and Abbatucci, J. S.: Oesophageal cancer and alcohol consumption: Importance of type of beverage. *Int. J. Cancer* **23:**443–447, 1979.

Ulleland, C. N.: The offspring of alcoholic mothers. *Ann. N.Y. Acad. Sci.* **197:**167–169, 1972.

Umiker, W., and Storey, C.: Bronchogenic carcinoma *in situ.* Report of a case with positive biopsy, cytological examination, and lobectomy. *Cancer* **5:**369–374, 1952.

U.S. Public Health Service: *Seventh Special Report to U.S. Congress on Alcohol and Health* (ADM #281880002), Rockville, MD, U.S. Dept. HHS, PHS, ADAMHA, NIAAA, pp. 144–161, 1990.

Valentine, E. H.: Squamous metaplasia of the bronchus. A study of metaplastic change occurring in the epithelium of the major bronchus in cancerous and noncancerous cases. *Cancer* **10:**272–279, 1957.

Vallee, B. C., Wacher, W. E., Bartholomay, A. F., and Robin, E. D.: Zinc metabolism in hepatic dysfunction I. Serum zinc concentration in Laennec's cirrhosis and their validation by sequential analysis. *N. Engl. J. Med.* **255:**403–408, 1956.

Veghelyi, P. V.: Fetal abnormality and maternal ethanol metabolism. *Lancet* **II:**53–54, 1983.

Veghelyi, P., and Osztovics, M.: The alcohol syndromes: The intrarecombigenic effect of acetaldehyde. *Experientia* **34:**195–196, 1978.

Veghelyi, P. V., Osztovics, M., and Szaszovszky, E.: Maternal alcohol consumption and birth weight. *Br. Med. J.* **2:**1365–1366, 1978.

Videla, L. A., and Valenzuela, A.: Alcohol ingestion, liver glutathione and lipoperoxidation: Metabolic interrelations and pathological implications. *Life Sci.* **31:**2395–2407, 1985.

Vina, J., Estrella, J. M., Guerri, C., and Romero, F. J.: Effect of ethanol on glutathione concentration in isolation hepatocytes. *Biochem. J.* **188:**549–552, 1980.

Vitale, J. J., and Gottlieb, L. S.: Alcohol and alcohol-related deficiencies as carcinogens. *Cancer Res.* **35:**336–338, 1975.

Vitale, J. J., Broitman, S. A., Vavrousek-Jakuba, E., Rodday, P. W., and Gottlieb, L. S.: The effects of iron deficiency and the quality and quantity of fat on chemically induced cancer. *Adv. Exp. Med. Biol.* **91:**229–242, 1977.

Wattenberg, L. W.: Inhibition of chemical carcinogenesis. *J. Natl. Cancer Inst.* **60:**11–18, 1978.

Webster, W. S., Walsh, D. A., Lipson, A. M., and McEwen, S. E.: Teratogenesis after acute alcohol exposure in inbred and outbred mice. *Neurobehav. Toxicol* **2:**227–234, 1980.

Webster, W. S., Walsh, D. A., McEwen, S. E., and Lipson, A. H.: Some teratogenic properties of ethanol and acetaldehyde in C57BL/65: Implications for the study of the Fetal Alcohol Syndrome. *Teratology* **27:**231–243, 1983.

Weinberg, J.: Nutritional issues in prenatal alcohol exposure. *Neurobehav. Toxicol. Teratol.* **6:**261–269, 1984.

Weinberg, J.: Effects of ethanol and maternal nutritional status on fetal development. *Alcoholism: Clin. Exp. Res.* **9:**261–269, 1985.

Weiner, C. P., Sabbagha, R. E., Tamura, R. K., and DalCompo, S.: Sonographic abdominal circumference: dynamic versus static imaging. *Am. J. Obstet. Gynecol.* **139:**953–955, 1981.

Weiner, L., and Larson, G.: Clinical prevention of fetal alcohol effects: A reality. *Alcohol Health Res. World* **11:**60–94, 1987.

Weiner, L., Rosett, H. L., and Edelia, K. C.: Behavioral evaluation of fetal alcohol education for physicians. *Alcoholism: Clin. Exp. Res.* **6:**230–233, 1982.

Weiner, L., Morse, B. A., and Garrido, P.: FAS/FAE: Focusing prevention on women at risk. *Int. J. Addict.* **24**:385–395, 1989.

Weinstein, I. B.: Current concepts on mechanisms of chemical carcinogens. *Bull. N.Y. Acad. Med.* **54**:366–383, 1978.

Weinstein, I. B., Yamasaki, H., Wigler, M., Lee, L., Fisher, P. B., Jeffrey, A., and Greenberger, D.: Molecular and cellular events associated with the action of initiating carcinogens and tumor promoters. In: *Carcinogens: Identification and Mechanisms of Action* (A. C. Griffin and C. R. Shaw, eds.), New York, Raven Press, 1979, pp. 339–418.

Weller, R. W.: Metaplasia of bronchial epithelium: A postmortem study. *Am. J. Clin. Pathol.* **23**:768–774, 1953.

West, J. R. (ed.): *Alcohol and Brain Development.* New York, Oxford University Press, 1986.

West, J. R.: Fetal alcohol-induced brain damage and the problem of determining temporal vulnerability: A review. *Alcohol Drug Res.* **7**:423–441, 1987.

West, J. R., and Pierce, D. R.: Perinatal alcohol exposure and neuronal damage. In: *Alcohol and Brain Damage* (J. R. West, ed.), New York, Oxford University Press, 1986, pp. 120–157.

West, J. R., Hamre, K. M., and Cassell, M. D.: Effects of ethanol exposure during the third trimester equivalent on neuron number in rat hippocampus and dentate gyrus. *Alcoholism: Clin. Exp. Res.* **10**:190–196, 1986.

Wickramasinghe, S. N., and Malik, F.: Acetaldehyde causes a prolongation of the doubling time and an increase in the model volume of cells in culture. *Alcoholism: Clin. Exp. Res.* **10**:350–354, 1986.

Williams, M. J.: Extensive carcinoma *in situ* in the bronchial mucosa associated with two invasive bronchogenic carcinomas. Report of case. *Cancer* **5**:740–747, 1952.

Williams, R. R.: Breast and thyroid cancer and malignant melanoma promoted by alcohol-induced pituitary secretion of prolactin, T.S.H and M.S.H. *Lancet* **I**:996–999, 1976.

Williams, R. R., and Horm, J. W.: Association of cancer sites with tobacco and alcohol consumption and socioeconomic studies of patients: Interview study from Third National Cancer Survey. *J. Natl. Cancer Inst.* **58**:525–547, 1977.

Winick, M., Brasel, J. A., and Velasco, E. G.: Effects of prenatal nutrition upon pregnancy risk. *Clin. Obstet. Gynecol.* **16**:184–198, 1973.

Winship, D. H., Carlton, R. C., Zaboralskie, F. F., and Hogan, W. J.: Deterioration of esophageal peristalsis in patients with alcoholic neuropathy. *Gastroenterology* **55**:173–178, 1968.

Wolff, G.: Does alcohol cause chronic gastritis? *Scan. J. Gastroenterol.* **4**:289–291, 1970.

Woutersen, R. A., Appelman, L. M., Feron, V. J., and Van Der Heijden, C. A.: Inhalation toxicity of acetaldehyde in rats. II. Carcinogenicity study: Interim results after 15 months. *Toxicology* **31**:123–133, 1984.

Wynder, E. L.: Nutrition and cancer. *Fed. Proc.* **35**:1309–1315, 1976.

Wynder, E. L., and Bross, I. J.: A study of etiological factors in cancer of the esophagus. *Cancer* **14**:389–413, 1961.

Wynder, E. L., and Chan, P. C.: The possible role of riboflavin deficiency in epithelial neoplasia. II. Effect on skin tumor development. *Cancer* **26**:1221–1224, 1970.

Wynder, E. L., and Fryer, J. H.: Etiologic considerations of Plummer–Vinson (Paterson–Kelly) syndrome. *Ann. Intern. Med.* **49**:1106–1128, 1958.

Wynder, E. L., Hultberg, S., Jacobsson, F., and Bross, I. J.: Environmental factors in cancer of upper alimentary tract: Swedish study with special reference to Plummer–Vinson (Paterson–Kelly) syndrome. *Cancer* **10**:470–487, 1957.

Yang, C. S., Tu, Y. Y., Koop, D. R., and Coon, M. J.: Metabolism of nitrosamines by purified rabbit liver cytochrome P-450 isozymes. *Cancer Res.* **45**:1140–1145, 1985.

Ylikorkala, O., Halmesmaki, E., and Viinikka, L.: Urinary prostacyclin and thromboxane metabolites in drinking pregnant women and in their infants: Relational to the fetal alcohol effects. *Obstet. Gynecol.* **71**:61–66, 1988.

Yu, M. C., Garabrant, D. H., Peters, J. M., and Mack, T. M.: Tobacco, alcohol, diet, occupation and carcinoma of the esophagus. *Cancer Res.* **48**:3843–3848, 1988.

Ziegler, R. G.: A review of epidemiologic evidence that carotenoids reduce the risk of cancer. *J. Nutr.* **119**:116–122, 1989.

Ziegler, R. G., Morris, L. E., Blot, W. J., Pottern, L. M., Hoover, R., and Fraumeni, J. R.: Esophageal cancer among black men in Washington D.C. II. Role of nutrition. *J. Natl. Cancer Inst.* **67**:1199–1206, 1981.

Zile, M., and DeLuca, H. F.: Retinoic acid: some aspects of growth-promoting activity in the albino rat. *J. Nutr.* **94**:302–308, 1968.

16

Alcohol and the Kidney

Murray Epstein

When the liver is full of fluid and this overflows into the peritoneal cavity, so that the belly becomes full of water, death follows.
Hippocrates, ca. 400 B.C.

The interrelationship of liver disease and simultaneous kidney dysfunction has been appreciated for thousands of years. More than 2400 years ago, Hippocrates recorded the association of ascites and liver disease (cited in Atkinson, 1956). During the subsequent two millennia, the scope of liver–kidney interrelationships has been amplified and extended; currently, such interrelationships are voluminous and exceedingly complex. Table 16.1 comprises a proposed framework for considering those conditions that involve both the liver and the kidney and in which relationships have been reasonably well defined. In brief, three major categories of disease states subtend such interrelationships: (1) disorders involving both the liver and the kidney directly, (2) primary disorders of the kidney with secondary hepatic involvement, and (3) primary disorders of the liver with secondary renal dysfunction.

Derangements of renal function that secondarily complicate primary disorders of the liver are the most clinically significant. Table 16.2 summarizes some of the alterations in renal function and electrolyte metabolism that frequently accompany liver disease. These complications are diverse and comprise a wide continuum, varying from complications with little clinical significance to others that are serious and require therapeutic intervention (Epstein, 1985a, 1988a).

It is generally assumed that the observed alterations in renal function are attributable solely to an impairment of hepatic function. Since the majority of patients with liver disease and associated renal dysfunction in the United States are alcoholics, it is also possible that the alterations in renal function may be attributable to effects of alcohol on the kidney. This chapter examines the relationship of alcohol itself on renal function.

16.1. Changes in Renal Anatomy

The structural changes of the kidney associated with alcohol administration have been studied. Chaikoff *et al.* (1948), were among the first to examine the effect of alcohol on the kidney. They administered alcohol by gastric tube to dogs maintained on a high-protein diet and then observed striking alterations in renal morphology, including glomerular changes characterized by hypercellularity and thickening of the basement membrane. Concomitantly, there were tubular changes including enlarged and altered tubular epithelial cells.

Lieber *et al.* (1966) observed that in ethanol-fed rats there is increased lipid accumulation. For years it has been noted that chronic alcoholics with fatty nutritional cirrhosis had significant bilateral nephromegaly. To further characterize this abnormality, Laube *et al.* (1967) studied the kidneys of 89 chronic alcoholics in various stages of liver disease and compared them with the kidneys of patients with other forms of cirrhosis and with various control groups. They observed that a significant nephromegaly was present in all cases of cirrhosis. Among the chronic alcoholics, the greatest nephro-

TABLE 16.1.   Abnormalities Affecting the Liver and the Kidney Simultaneously[a]

Disorders involving both the liver and kidney directly
Primary disorders of the kidney with secondary hepatic involvement
Primary disorders of the liver with secondary renal dysfunction
  Extrahepatic biliary obstruction with secondary impairment of renal function
  Parenchymal liver disease with secondary renal dysfunction

[a]Reproduced with permission from Epstein, 1988b.

megaly was associated with fatty metamorphosis of the liver without cirrhosis. The degree of nephromegaly was directly proportional to the amount of hepatomegaly. To ascertain which tissue component was responsible for the increased weight, assays of kidney protein, lipid, carbohydrate, and water content were performed. Assay of renal tissue fluid, protein, lipid, and carbohydrate revealed no accumulation of any one of these tissue components. They suggested that nephromegaly was the result of added tissue mass, best explained on the basis of cellular hypertrophy and hyperplasia.

Van Thiel *et al.* (1977) systematically assessed the effects of ethanol feeding on renal function and gross and microscopic morphology in alcohol-fed rats as compared to similar studies in isocalorically fed animals ingesting the same diet except that dextrimaltose was isocalorically substituted for ethanol. Alcohol-fed animals had significantly reduced renal function and interstitial edema compared to their isocaloric controls. When

renal mass and renal constituent analysis were normalized for body weight, alcohol-fed animals were found to have renal hypertrophy characterized by significantly increased absolute amounts of protein, fat, and water.

In summary, it is clear that alcohol ingestion induces gross and microscopic changes in the kidney.

16.2. Renal Hemodynamics

The alterations in renal hemodynamics that occur in the setting of established liver disease, and particularly decompensated cirrhosis, have been well documented (Epstein, 1988a, c, 1990a; Klingler *et al.*, 1970; Papper, 1958). In patients with decompensated cirrhosis, GFR and RPF vary over a very broad spectrum, ranging at the extremes from supernormal values to severe renal failure. Although it is noted that the lowest values of GFR and RPF tend to be observed in the sickest patients, such an association is by no means invariable, and there is great variation in renal hemodynamics in all phases of liver disease. The mechanisms responsible for the abnormal values are not readily apparent. Supernormal values have received relatively little investigative attention and have not been explained.

The mechanisms responsible for reduction in renal hemodynamics have not been fully elucidated. As I have detailed in several recent reviews (Epstein, 1988b, 1990a), several hormonal, neural, and hemodynamic mechanisms have been implicated or suggested; these are enumerated in Table 16.3. Prominent among these are activation of the renin–angiotensin system, alter-

TABLE 16.2.   Renal Abnormalities in Liver Disease[a]

Parenchymal liver disease with secondary impairment of renal function:
  Deranged renal sodium handling
  Impaired renal water excretion
  Impaired renal concentrating ability
  Hepatorenal syndrome
  Acute renal failure
  Glomerulopathies associated with
    Cirrhosis
    Acute viral hepatitis
    Chronic viral hepatitis
  Impaired renal acidification
Extrahepatic biliary obstruction with secondary impairment of renal function
  Acute renal failure

[a]Reproduced with permission from Epstein, 1986.

TABLE 16.3.   Mechanisms that May Participate in the Renal Failure of Liver Disease

Hormonal
  Activation of the renin–angiotensin system
  Alterations in renal prostaglandins
    Diminished vasodilatory protaglandins
    Increased vasoconstrictor thromboxanes
  Alterations in kallikrein–kinin system
  Atrial natriuretic peptides
  Vasoactive intestinal peptide
  Endotoxemia
  Glomerulopressin deficiency
Neural and hemodynamic
  Alterations in intrarenal blood flow distribution
  Increase in sympathetic nervous system activity

ations in renal prostaglandins, increased activity of the sympathetic nervous system, and chronic endotoxemia.

Despite the obvious importance of the problem, relatively few studies have examined the effect of ethanol administration on renal hemodynamics in humans. The majority of studies have been carried out in animal models, particularly the dog. Studies by Sargent et al. (1974, 1975, 1979) suggest that acute alcohol ingestion does not alter either renal hemodynamics or the excretion of sodium (for a review, see Beard and Knott, 1968). These studies, however, have not extended beyond 6 hr post-ingestion even though some pathophysiological sequelae of acute alcohol ingestion have been shown to be consequential to secondary effects of alcohol, such as catecholamine release (Nakano and Prancan, 1972) and acetaldehyde production.

Indeed, it has been reported that a single dose of 3 g/kg of ethyl alcohol to the dog produces isosmotic elevations in the plasma volume, but the increases were evident only during the period 10 to 26 hr after ingestion (Nicholson and Taylor, 1940). Furthermore, chronic administration of 3 g/kg of ethyl alcohol to the dog produced isosmotic elevations in the plasma volume and the extracellular fluid volume after 1 week, and the volume expansions were maintained for the remaining 7 weeks of the study (Beard et al., 1965). In humans, chronic alcoholism has been reported to cause similar alterations (Beard and Knott, 1968).

In order to determine whether the effects documented acutely were sustained during the latter stages of acute ethanol ingestion, Sargent et al. (1974) studied the time-dependent effects of acute alcohol administration (3 g/kg) and the cumulative effects of a 7-day administration of alcohol (3 g/kg per day) on renal function in the dog. In comparison to an equivolumetric administration of water, the alcohol treatment produced elevations in glomerular filtration rate and effective renal plasma flow at 10 and 18 hr after the acute dosage and after the chronic treatment. At 26 hr after the acute and after the chronic administration, the filtration fraction was significantly elevated. The percentage tubular reabsorption of sodium showed an increasing trend during the acute and chronic administrations, attaining significance at 26 hr after the acute treatment. The results suggest that acute or chronic alcohol administration produces renal vasodilation.

Relatively few studies have been conducted in humans. The available studies suggest that alcohol produces only minor fluctuations in renal blood flow and glomerular filtration rate even at high blood levels (Rubini et al., 1955; Kalbfleisch et al., 1963). The effects of acute administration of ethanol in one dose on plasma volume, extracellular fluid volume, and blood sodium, potassium, and chloride concentrations seem to vary with the species examined, the route and dose of ethanol administration, and the period after ethanol during which observations are made.

In summary, although the relatively few available studies in humans suggest that alcohol does not induce major changes in renal hemodynamics, additional systematic studies are necessary to substantiate this formulation.

16.3. Renal Sodium Handling

In contrast to the well-established impairment of sodium excretion that complicates the course of most patients with advanced liver disease (Epstein, 1986, 1987a, 1988b, 1990b; Papper, 1958), data relating to the effects of ethanol on sodium excretion are fragmentary and not wholly consistent. Theoretical considerations have suggested that ethanol may alter renal sodium handling. Israel and Kalant (1963) demonstrated that ethanol in concentrations well below the lethal range exert a definite inhibitory effect on the active transport of sodium by the isolated frog skin. Subsequently, Israel-Jacard and Kalant (1965) reported that ethanol significantly inhibited the active reaccumulation of K^+ by precooled rabbit kidney cortex. In light of these findings suggesting that ethanol inhibits the active transport of cations by many types of cells including renal epithelial cells, it is not altogether surprising that ethanol has been proposed to modulate renal sodium handling.

A review of the available literature discloses that ethanol has generally been found to reduce the urinary excretion of sodium and potassium (Nicholson and Taylor, 1938; Rubini et al., 1955; Ogata et al., 1968; Kalbfleisch et al., 1963; Klingman and Haag, 1958) and to lower urine osmolality (Ogata et al., 1968; Kleeman et al., 1955; Roberts, 1963). However, some exceptions have been reported.

The first thorough investigation of electrolyte excretion in response to ethanol intake was reported by Nicholson and Taylor (1938). They observed that alcohol ingestion (180–212 ml of 95% alcohol/8 hr period) resulted in the renal retention of sodium, potassium, and chloride.

In 1955, Rubini et al. (1955) administered 120 ml of 100-proof bourbon whiskey to human subjects and found that alcohol caused an increase in urine flow and free water clearance as well as a consistent decrease in the excretion of sodium, potassium, and chloride. This finding strongly suggested that the rise in the free water clearance was caused by inhibition of the release of the antidiuretic hormone. In addition, these same investigators in the same year found that alcohol diuresis may be blocked by a prior increase in the circulation of endogenous antidiuretic hormone as well as by the administration of exogenous vasopressin. Most investigators now agree that alcohol diuresis results from suppression of the release of antidiuretic hormone.

Huang et al. (1957) observed an increased salt and water excretion with an oral dose of 1 gr alcohol per kilogram body weight and salt retention with an increased extracellular volume when 4 gr per kilogram was orally administered in dogs. Klingman and Haag (1958) found that a lethal dose of alcohol administered to dogs produced a decrease in plasma chloride, transient hypokalemia, oliguria, and, in most cases, anuria with a decreased concentration of urinary electrolytes. Barlow and Wooten (1959) studied plasma and erythrocyte alterations in 64 human chronic alcoholics prior to treatment. They found that plasma sodium, potassium, and chloride and red-cell chloride concentrations were significantly decreased, while red-cell sodium remained unaltered.

Huang et al. (1957) found the usual decrease after large amounts of alcohol but increased excretion after small doses. Kleeman et al. (1955) did not observe any changes in solute excretion when ethanol was given at the height of water diuresis. Based on these reports it was suggested that the hydration state might be a factor determining whether ethanol increases or decreases the excretion of sodium and potassium. Linkola (1974) examined the possibility that the state of hydration moderates the natriuretic response. He investigated the effects of ethanol on urine sodium, potassium, and osmolality in rats maintained in a positive water balance produced by prior consecutive water loadings. Ethanol loading was followed by rapid rises in urine output, sodium concentration, and osmolality and a somewhat slower rise in potassium concentration. Linkola (1975) reported that when ethanol is ingested with excessive amounts of fluid or in a prehydrated state, natriuresis occurs instead of sodium retention.

In summary, the available data suggest that alcohol does not markedly alter renal sodium handling. Although ethanol has generally been noted to reduce sodium excretion modestly, the ultimate effect appears to be modified by the state of hydration.

16.4. Hypophosphatemia in the Alcoholic

It is now widely recognized that acute hypophosphatemia occurs commonly in hospitalized alcoholics (Stein et al., 1966; Fankushen et al., 1964; Knochel, 1977, 1980; Knochel et al., 1975; Ryback et al., 1980). In those whose alcoholism is not severe, it may not occur at all. It is very clear that hypophosphatemia can occur in normal persons by several mechanisms, and thus, severe phosphorus deficiency does not necessarily cause hypophosphatemia. In severe alcoholics, hypophosphatemia can be anticipated in more than 50%. To demonstrate its occurrence requires daily blood sampling for the first 3–5 hospital days. When it does not appear in the anticipated setting, one should suspect a concomitant event that confounds its recognition such as ongoing rhabdomyolysis, acidosis, an inadequate circulatory volume, or renal insufficiency. A typical pattern of serum phosphorus concentration is one that is normal or slightly low when a patient is admitted to the hospital, followed by a progressive decline over the following 24 to 72 hr and a return toward normal thereafter.

Hypophosphatemia in the alcoholic could occur by several mechanisms (Table 16.4). Phosphorus deficiency consequent to an inadequate intake per se can cause hypophosphatemia (Lotz et al., 1968). Many severe alcoholics consume a poor diet, and undoubtedly, this is an important factor underlying phosphorus deficiency in many of them. However, although dietary deficiency may be important, it does not always play a critical role.

Hyperventilation with respiratory alkalosis can cause profound hypophosphatemia, even in normal subjects (Mostellar and Tuttle, 1964). Reduction of CO_2 tension in extracellular fluid will cause an equal reduction inside the cell. As intracellular pH rises, phosphofructokinase, the rate-regulating enzyme of glycolysis, is activated. As the rate of phosphorylation of glucose

TABLE 16.4.   Possible Causes of Severe Hypophosphatemia in the Severe Alcoholic

Phosphorus deficiency
Respiratory alkalosis
Insulin
Hyperalimentation
Renal losses
Magnesium deficiency

and its metabolites increases, free phosphate ions are consumed, and hypophosphatemia results.

Hypophosphatemia of modest degree occurs in normal subjects during infusion of insulin. Since, in normal subjects, insulin administration does not induce severe hypophosphatemia, the role of insulin in hypophosphatemia in alcoholics appears to be contributory rather than a principal factor. Since insulin stimulates hydrogen ion movement from the cell, the cell becomes alkalinized, thereby increasing activity of phosphofructokinase. Therefore, phosphate ions are utilized in glucose phosphorylation, and hypophosphatemia results (Bakris, 1992; Ram *et al.*, 1992).

Hospitalized alcoholics often receive fluids intravenously containing glucose or amino acids. Both are potent stimuli for insulin release. Phosphorus uptake into cells during refeeding or hyperalimentation is another important cause of hypophosphatemia. However, the pattern of hypophosphatemia is considerably different in the hyperalimented patient. As a rule, these patients are wasted but remain capable of an anabolic response to administered nutrients. Over a period of 8 days or more, their serum phosphorus falls progressively to very low levels as a result of phosphate incorporation into new protoplasm. Characteristically, this pattern is seen most often in the nonalcoholic patient with a severe wasting illness. In contrast, the alcoholic usually becomes hypophosphatemic during the first few hospital days (Knochel, 1988).

Another cause for hypophosphatemia in the alcoholic is phosphorus wasting into the urine. Most chronic alcoholics excrete about 0.5–1.5 g phosphorus into their urine on the day of admission to the hospital, if measurements are obtained before their serum phosphorus concentration declines. Phosphaturia of such magnitude in persons not consuming phosphorus points to cellular release incident to catabolism or acidosis. Most severe alcoholics are starving when they become sufficiently sick to accept hospitalization. However, as their serum phosphorus falls to low levels, phosphorus virtually disappears from the urine. Hypophosphatemia in conjunction with diminished phosphorus excretion, a pattern similar to that seen during treatment for diabetic ketoacidosis or during respiratory alkalosis, indicates that phosphorus has been shifted into cells. This pattern argues against renal tubular injury as a major cause of phosphaturia.

In summary, acute hypophosphatemia commonly occurs in alcoholic patients. There are multiple mechanisms that contribute to the hypophosphatemia.

16.5. Effects on Renal Calcium Handling

In the severely ill, hospitalized alcoholic, hypocalcemia occurs with about the same frequency as hypophosphatemia and hypomagnesemia. It can be sufficiently severe to cause tetany (Estep *et al.*, 1969). One possible cause is hypoalbuminemia (Table 16.5). Since albumin binds approximately 1 mg calcium/g, a patient whose albumin concentration is 3.0 g/100 ml would be expected to show a total serum calcium concentration of about 8 mg/100 ml, of which 5 mg/100 ml would be unbound. Alcoholics are often hypoalbuminemic as a result of protein deficiency or liver disease; however, hypoalbuminemia cannot always explain hypocalcemia in such patients (Knochel, 1988). In these patients, serum calcium clearly falls below the predicted value for their prevailing albumin concentration. Thus, hypoalbuminemia does not consistently explain hypocalcemia in alcoholics.

Vitamin D deficiency may also be implicated as a possible cause of hypocalcemia in alcoholics, especially if significant liver disease coexists (Velentzas *et al.*, 1977). Hydroxylation of cholecalciferol in the 25 position occurs in the liver, and this may be impaired in liver disease (Meyer *et al.*, 1978). Since vitamin D is fat soluble, it may not be absorbed properly in those alcoholics who have steatorrhea. Thus, endogenously produced cholecalciferol may not undergo appropriate metabolism in the liver to become activated, and active vitamin D ingested in the diet may not be absorbed. Calcium absorption would be impaired, and hypocalcemia might result (Massry and Goldstein, 1983).

Rhabdomyolysis can cause hypocalcemia in alco-

TABLE 16.5.   **Possible Causes of Hypocalcemia in the Alcoholic**[a]

Hypoalbuminemia
Magnesium deficiency
Blunts calcemic response to PTH[b]
Blunts hydroxyproline response to PTH
Reduces serum PTH levels
Decreased absorption of calcium in the gut
Hypoparathyroidism with magnesium deficiency
Vitamin D deficiency
Steatorrhea
Rhabdomyolysis
Ethanol

[a]Adapted from Knochel (1983).
[b]Parathyroid hormone.

holics as a result of calcium precipitation in injured muscle as $CaHPO_4$ or $CaCO_3$.

Finally, hypocalcemia could result from the action of ethanol. In experimental studies by Peng *et al.* (1972), animals given very large quantities of ethanol became hypocalcemic. In addition, rats with fatal concentrations of blood ethanol showed no hypercalcemic response to injected parathyroid hormone (PTH) (Peng *et al.*, 1972).

Several investigators have demonstrated that the administration of ethanol to healthy subjects increases urinary calcium excretion. Kalbfleisch *et al.* (1963) assessed the effects of ethanol administration on urinary excretion of calcium in nine normal and four alcoholic subjects. Following appropriate control periods, 30 ml of 100% ethanol was given orally, followed by 5% ethanol intravenously at a rate of 5 ml/min. All ethanol-loaded subjects manifested a marked increase in calcium excretion (mean of 89%). The increase in calcium excretion was invariably observed within 20 min after ethanol administration, with the maximal effect generally observed at 60 to 80 min. The changes in calcium excretion were independent of change in GFR, renal plasma flow, or urinary flow rate. Subsequently, Markkanen and Nanto (1966) examined the effect of ethanol infusion (50 ml of 94% ethanol infused in 500 ml of normal saline over 3 hr) to nine normal subjects. They observed an increase in serum calcium and inorganic phosphorus levels during ethanol infusion, with a subsequent return to preinfusion levels. Concomitantly, urinary excretion of calcium and inorganic phosphorus increased.

16.6. Effects on Renal Magnesium Handling

16.6.1. Description

Chronic alcoholism is the most common cause of hypomagnesemia in the United States. Similar to hypophosphatemia and hypokalemia, it is by far most common in sick, hospitalized alcoholics. Daily blood sampling is necessary to reveal its high incidence. Recognition of this abnormality is important, since it responds readily to treatment. During the phase of recovery from chronic alcoholism there is a positive magnesium balance. The manifestations of chronic alcoholism resemble the cutaneous and neuromuscular manifestations of magnesium deficiency as they are seen in animals.

The mechanisms mediating the hypomagnesemia are multiple (Table 16.6). There is some evidence in experimental animals that ethanol interferes with magnesium absorption from the gut in that rats given alcohol require significant magnesium supplementation of their diet. In addition, it is clear that the ingestion of ethanol causes a profound increase in urinary magnesium excretion (Kalbfleisch *et al.*, 1963; McCollister *et al.*, 1963).

Kalbfleisch *et al.* (1963) assessed the effects of ethanol on urinary excretion of magnesium in nine normal and four alcoholic subjects. Following appropriate control periods, 30 ml of 100% ethanol was given orally, followed by 5% ethanol intravenously at a rate of 5 ml/min. All ethanol-loaded subjects manifested a marked increase in magnesium excretion (mean of 167%; range of 90% to 357%). Of interest, there was no significant difference between the magnesiuria of normal and alcoholic subjects. An increase in magnesium excretion was invariably observed within 20 min after ethanol administration, with the maximal effect generally observed at 60 to 80 min. An examination of the magnesiuria disclosed that the changes in magnesium excretion were independent of GFR, renal plasma flow, or urinary flow rate.

It has been postulated that ethanol may exert its magnesiuric effects through lactate production. Barker *et al.* (1959) have previously demonstrated that lactate administration results in an increased urinary excretion of magnesium. Since the administration of ethanol results in increased levels of blood lactate, it is possible that the magnesiuria of ethanol might be related to an increased lactate production and excretion. It has also been postulated that ethanol or one of its intermediate metabolites exerts a direct renal tubular effect on magnesium exchange at the level of the tubule. Such a mechanism, however, remains speculative.

TABLE 16.6.   Possible Causes of Hypomagnesemia in the Alcoholic[a]

Magnesium deficiency
Inadequate diet
Steatorrhea
Diarrhea
Ethanol
Phosphorus deficiency
ATP–magnesium complex formation
Fatty-acid–magnesium complex formation

[a]Adapted from Knochel (1983).

16.6.2. Therapy

The therapy for magnesium depletion in cirrhotic patients does not differ importantly from that of other magnesium-depleted patients. Although prevention is generally the linchpin of therapy, this is usually not feasible in the alcohol-abusing patient with cirrhosis, who frequently comes to the hospital in a depleted state. I recommend the following schema (adapted from Oster, 1985) as a general formula for therapy.

It is notoriously difficult to estimate the size of a magnesium deficit, since the serum [Mg] only variably represents body stores. Nevertheless, in a significantly depleted individual the deficit is believed to be in the range of 1.0 to 2.0 mEq/kg of body weight. If this is the case, and considering the inevitable urinary loss of a portion of the administered magnesium (see below), it should be obvious that the usual dose of 2 ml of 50% magnesium sulfate given intramuscularly in each buttock, often used by house staff physicians for patients with alcohol withdrawal syndrome, may not nearly suffice if the patient happens to be severely magnesium depleted.

Mild depletion of magnesium does not necessarily require specific therapy. Whenever possible, and certainly for mild depletion, the oral route of replacement is preferable (Oster, 1985). Excessive ingestion of magnesium salts, however, may produce diarrhea and prevent the desired positive balance of magnesium. Nevertheless, in some patients with underlying gastrointestinal disease, oral administration of magnesium may not only correct magnesium depletion but sometimes also reduce the diarrhea. One may begin by using magnesium-containing antacids; aluminum hydroxide gel can be given to minimize magnesium-induced diarrhea.

It has been suggested that the percentage of magnesium absorbed from a dosage of 30 to 90 ml of magnesium antacid four times daily is 6% to 15%. Thus, a 30-ml q.i.d. dose of Maalox Therapeutic Concentrate, which contains about 10 mEq of magnesium per 5 ml, might provide as much as 36 mEq of absorbable magnesium per day.

Milk of magnesia USP (Phillips) is a suspension of magnesium hydroxide containing approximately 36 mg (3 mEq) of elemental magnesium per milliliter. Initially 5 ml can be given once a day and subsequently increased to four times a day. As an alternative, 250 or 500 mg of magnesium oxide [containing 150 mg (12.3 mEq) or 300 mg (24.7 mEq) of elemental magnesium, respectively] may be given four times daily if diarrhea does not occur. Overdosage of magnesium oxide has been reported to result in acidosis and hyperkalemia.

Because the percentage of intestinal reabsorption is quite variable, the serum [Mg] must be followed closely during replacement therapy, and several days should be allowed to achieve total body replenishment.

When hypomagnesemia is severe and symptomatic, the parenteral route is mandatory. The standard preparation is $MgSO_4 \cdot 7H_2O$) (usually available in either 10% or 50% solutions). Of these, only the 10% solution is sufficiently dilute to be given safely by vein. One gram of magnesium sulfate provides 98 mg (8.1 mEq) of magnesium. The 50% solution can be given intramuscularly or intravenously after appropriate dilution (see below). It has been suggested that since sulfate ions bind calcium, magnesium chloride might be preferred to magnesium sulfate for treating patients with coexisting hypocalcemia.

It can be expected that about 50% of an intravenous load will be lost in the urine, and 50 mEq or more of magnesium may be needed within the first 12 hr. It has been suggested that the slower the infusion of magnesium, the less will be lost in the urine, since the plasma [Mg] determining the renal threshold will less likely be exceeded. Generally, no more than 100 mEq of magnesium should be given in a 12-hr period, and considerable caution is needed with this dose. Rarely, much larger amounts have been given safely to patients with normal renal function. During rapid infusion the patient should remain supine in order to avoid hypotension. In order to avoid overdose, close attention should be given to alteration of deep tendon reflexes, respiration, and serial serum magnesium levels. Furthermore, calcium chloride, the antidote for severe hypermagnesemia, should be readily available.

There are many schemata for the parenteral administration of magnesium. A reasonably simple one is provided in Table 16.7. The basic idea is to give magnesium at the rate of approximately 1 mEq/kg per day on the first day, followed by a reduction in dosage to the range of 0.3 to 0.5 mEq/kg per day for 3–5 days. A simple way to accomplish this is to place 40.5 mEq (5 g of $MgSO_4 \cdot 7H_2O$) in 1 liter of solution, which is infused over 3 hr and repeated every 12 hr until the appropriate total amount has been given. Whether the magnesium is diluted in saline or glucose solutions is a matter of clinical judgment, but one should remember that glucose administration may worsen hypokalemia. If the GFR is considerably depressed, one should be much more cautious. Rates of administration should be re-

TABLE 16.7.   Schemata for Magnesium Therapy in Adults with Normal Glomerular Filtration Rate[a]

Intramuscular 50% $MgSO_4 \cdot 7H_2O$

Day 1:	4 ml (2 ml in each site, 16.2 mEq total) every 2 hr for three doses, then every 4 hr for four doses
Day 2:	2 ml (8.1 mEq) every 4 hr for six doses
Day 3:	2 ml every 6 hr

Intravenous 50% $MgSO_4 \cdot 7H_2O$

Day 1:	12 ml (48.6 mEq) in 1000 ml of solution over 3 hr followed by 2 liters of solution, each containing 12 ml (48.6 mEq) of magnesium, over next 20 hr
Days 2 to 5:	12 ml distributed equally in the day's total intravenous fluids

[a]Reproduced with permission from Oster (1985).

duced by about 75% and the serum [Mg] monitored frequently (perhaps twice daily). Likewise in patients with atrioventricular heart block of severe grades or bifascicular blocks, the intravenous infusion of magnesium must be done with particular caution; pharmacologically, magnesium has many similarities to the calcium channel-blocking agents.

In summary, it is apparent that alcohol abuse contributes to multiple electrolyte disturbances. Table 16.8 summarizes several of these disturbances and contributing mechanisms.

16.7.   Renal Water Handling

The serum sodium level is determined by the balance of water in relation to that of sodium—too much water causes hyponatremia; too little causes hypernatremia. Hyponatremia connotes a serum sodium con-

TABLE 16.8.   How Alcohol Abuse Contributes to Electrolyte Disturbances

Disturbance	Cause(s)
Hyponatremia	Massive intake of solute-free fluid (for example, beer)
Hypokalemia	Dietary deficiency or gastrointestinal losses
	Leaky membranes
	Extracellular-to-intracellular shifts
Hypophosphatemia	Dietary deficiency or malabsorption
	Increased cellular uptake
Hypomagnesemia	Dietary deficiency or malabsorption
	Phosphorus deficiency

centration below the lower limit of normal, i.e., < 135 mEq/liter. In the cirrhotic patient, exchangeable sodium is increased. Therefore, if the patient is hyponatremic, the amount of water retained must be disproportionally greater than that of sodium. That is, there must be a specific defect in renal water excretion (i.e., diluting capacity) (Epstein, 1985b).

Impairment of renal diluting capacity occurs frequently in patients with cirrhosis (Vaamonde, 1988; Epstein, 1985b, 1987a,b; Aldersberg and Fox, 1943). Hyponatremia, the expression of this impaired capacity to excrete water, is a commonly encountered clinical problem in cirrhotic patients. Indeed, hyponatremia probably represents the single most common electrolyte abnormality that confronts the physician treating patients with cirrhosis of the liver. The pathogenesis of this abnormality and its clinical manifestations have recently been reviewed (Epstein, 1985b).

In contrast to the well-established and widely recognized impairment of renal water handling in established cirrhosis, the question of whether renal water handling is perturbed in noncirrhotic alcoholic patients has not been extensively investigated. On the one hand, the effects of ethanol administration *per se* are well established. Anyone who has ever imbibed has very soon had cause to reflect on a possible relationship between alcohol and urine formation. Nicholson and Taylor (1938) observed a biphasic response to alcohol ingestion. Following the ingestion of 180–212 ml of 95% alcohol during an 8-hr period, there was a water diuresis during the initial 4 hr. During the second 4-hr period, only small amounts of urine were voided despite the continued ingestion of alcohol during this period. In 1940, the same authors (Nicholson and Taylor, 1940) administered 4 ml of 95% ethanol per kilogram body weight to five human subjects and eight dogs. They reported a slight increase in plasma sodium concentration and a diuresis occurring within the first hour after giving alcohol; during the remaining 24 hr, practically no urine was voided. Furthermore, it was observed that no appreciable change occurred in plasma hematocrit in humans. In contrast, the hematocrit decreased in the dogs.

This cause-and-effect relationship was first put on a sound scientific basis by Eggleton (1942), who showed that acute ethanol ingestion provoked a diuresis leading to a negative water balance. She suggested that a suppression of antidiuretic hormone release from the posterior pituitary gland was responsible for this sequence of events. Ethanol acutely inhibits the release of vasopressin, resulting in an increase in free water excretion.

However, this effect persists only while the blood ethanol level is rising. In subjects under 50 years of age, a rising blood ethanol concentration suppresses plasma immunoreactive arginine vasopressin levels despite a concomitant increase in the serum osmolality (Helderman *et al.*, 1978). In contrast, subjects older than 50 years quickly escape the suppressive effect of the ethanol infusion and manifest sharp increases in their plasma arginine vasopressin levels. This escape occurs even though the increase in serum osmolality observed in the older subjects is similar to that encountered in the younger group. The difference in vasopressin response to ethanol between younger and older individuals may relate, in part, to an increase in osmoreceptor sensitivity with advancing age (Helderman *et al.*, 1978). Whether a similar variation of the arginine vasopressin response to ethanol infusion occurs in chronic alcoholics is not known.

16.8. Chronic Effects of Ethanol

When normal nutrition is maintained, chronic ethanol administration usually produces no alteration in the serum sodium concentration. However, there is some evidence to suggest that water and solute retention may occur in alcoholism. Chronic ethanol administration (2–4 g/kg, given enterally) to dogs over an 8-week period increases their total body water, extracellular volume, and plasma volume through an isosmotic retention of water and solutes by the kidneys (Beard *et al.*, 1965). In humans, withdrawal from chronic alcohol use is associated with a decline in total body water and in intra- and extracellular volume that are accompanied by a solute diuresis, suggesting that chronic alcoholism may induce both solute and water retention. These changes are not associated with alterations in the serum electrolytes and are not usually apparent. However, chronic volume expansion induced by ethanol use may contribute to the development of hypertension that is common in chronic alcoholism (Beevers *et al.*, 1982; Klatsky *et al.*, 1977).

16.9. Hyponatremia in Alcoholics without Cirrhosis

16.9.1. Occurrence

Chronic alcoholic patients can develop hyponatremia for any of the reasons that produce this abnormality in nonalcoholic subjects. However, from the preceding discussion it is apparent that acute or chronic ethanol use *per se* does not usually alter the serum sodium concentration. Therefore, hyponatremia in a noncirrhotic alcoholic patient is most often multifactorial in origin.

A hyponatremic syndrome that is peculiar to beer drinkers has been described (Demanet *et al.*, 1971; Hilden and Svendsen, 1975). Although the frequency of this syndrome is not clear, it may comprise a significant percentage of cases of symptomatic hyponatremia. Demanet *et al.* (1971) noted that 12 of 67 patients admitted to the hospital with severe hyponatremia (serum sodium concentration $\leq$ 120 mEq/liter) had neurological symptoms attributable solely to the electrolyte disorder. Seven of these 12 individuals were heavy beer drinkers who consumed at least 4 liters of beer daily. Although there was no evidence of overt volume depletion in these alcoholics, vomiting and diarrhea or prehospital diuretic therapy was present in all and contributed to the sodium deficit, as evidenced by a marked sodium retention during the 4 days subsequent to hospitalization. The excessive consumption of free water (as beer), however, was felt to have been the major factor leading to the development of the hyponatremia in these subjects.

Hyponatremia also may occur in beer drinkers in the absence of any evidence of sodium depletion. Hilden and Svendsen (1975) observed hyponatremia in five patients who consumed at least 5 liters of beer daily with no additional form of nourishment. The degree of hyponatremia and the severity of the associated symptoms were less prominent than in the patients reported by Demanet *et al.* (1971). Urinary osmolality was appropriately low in those patients in whom it was measured, confirming that vasopressin secretion was suppressed. The genesis of the hyponatremia in these patients appeared to be the consumption of free water in excess of solute sufficient to allow for excretion of the water. Because the minimum urinary osmolality is approximately 50–60 mOsm/kg water, roughly 300 mOsm of solute must be excreted to allow the elimination of 5 liters of water. Because of their poor protein intake and the low solute content of beer, these patients probably excrete less than 300 mOsm/day. Therefore, "beer-drinkers' hyponatremia" is a disorder of water overload *per se* and is not a direct consequence of ethanol ingestion.

16.9.2. Therapeutic Implications

The impaired diluting ability afflicting many patients with advanced liver disease may have important implications for management. As has been emphasized, hyponatremia connotes a dilutional state secondary to an impaired capacity to excrete water. A *sine qua non* of this observation is that the rational basis for treating hyponatremia is fluid restriction. Regardless of the degree of dilutional impairment, appropriate fluid restriction will eventually repair this abnormality (Epstein, 1985b).

As a caveat, it must be emphasized that in many hospitals it is quite difficult to adhere to the physician's orders for fluid restriction. As examples, paramedical and nursing personnel may not consider the fluids administered with medication as part of the fluid restriction. We are often dismayed to observe pitchers of water and several containers of juice inadvertently placed at the bedside of patients thought to be severely fluid-restricted. Thus, patients who ostensibly are on rigid fluid restriction may indeed be receiving 1500 or even 2000 ml of fluid per day. It is not surprising that the serum sodium concentration of such patients does not increase or may even decrease. Because of this problem, it is frequently necessary to resort to absolute fluid restriction with levels approaching 0–200 ml/day for the first few days in order to initiate the return of the serum sodium concentration toward normal.

Because many patients with advanced liver disease manifest an impairment of renal diluting ability, one may inquire when therapy should be instituted to normalize this disturbance. Some insight into this question may be afforded by recent experimental evidence suggesting that hyponatremia of even a moderate degree (≤ 125 mEq/liter) may result in irreversible neurological deficits if allowed to persist. Thus, it seems reasonable to correct hyponatremia in all patients with liver disease in whom the serum sodium is < 130 mEq/liter.

Because of the potential interactions of potassium and renal water handling, hypokalemia should be corrected. It has been proposed that potassium depletion may exacerbate dilutional hyponatremia by enhancing the effects of ADH at the level of the renal tubule, by increasing ADH release, or by stimulating angiotensin II generation with a resultant increase in thirst.

Another major consideration sometimes not given sufficient attention is the inappropriate use of drugs that may adversely affect renal diluting capacity. Specifically, there may be numerous drugs administered to cirrhotic patients that could impair renal diluting capacity. Examples include nonsteroidal antiinflammatory drugs and chlorpropamide. For a more complete listing the reader is referred to the review by Epstein (1985b).

Finally, in rare circumstances, one may be forced to consider the use of dialysis for the correction of severe hyponatremia in cirrhosis. Ring-Larsen *et al.* (1973)

treated 16 patients with decompensated cirrhosis, hepatic coma, renal failure (eight patients), and hyponatremia with peritoneal dialysis. Serum sodium averaged 119 mEq/liter before dialysis and remained above 130 mEq/liter for the week after dialysis. Such heroic measures, however, rarely are necessary, even in the most severely decompensated cirrhotic patient.

16.10. Acid–Base Disturbances Induced by Alcohol

The initial observations indicating that ingestion of alcohol alters acid–base balance were those of Thomas (1898). He reported that the carbon dioxide content and capacity of the blood were diminished. Himwich *et al.* (1933) made observations on the acid–base balance in dogs and humans after feeding alcohol. Their studies were made over short periods of time, and no attempt was made to study the electrolyte balance; however, they observed no change in the total base or chlorides of the serum. They found that there was a reduction in the carbon dioxide content and carbon dioxide capacity of the blood, which was accompanied by a fall in the blood pH. There was also an increase in the blood lactic acid and sugar. It was their belief that the increase in lactic acid was brought about by the conversion of muscle glycogen to lactic acid by the action of alcohol.

Both chronic and acute alcohol abuse may cause or contribute to a wide range of acid–base and electrolyte disturbances (Table 16.9). In each patient, however, a variety of factors can interact with alcohol to modify the clinical course and severity of the disorder. Among the more important factors that determine the clinical expression of alcohol-induced disturbances are the amount and duration of drinking, the nutritional status of the patient, and the coexistence of other diseases.

16.10.1. Acute Disorders

The acute ingestion of excessive amounts of alcohol may lead to any one of the fundamental acid–base disturbances (Oster and Perez, 1986). Because alcohol (ethanol) is a central nervous system depressant, extreme intoxication may depress the sensitivity of the respiratory center to carbon dioxide; the results are hypoventilation and respiratory acidosis. In contrast, alcohol withdrawal in patients with severe chronic alcoholism may produce both central nervous system hyperexcitability and increased respiratory carbon dioxide sensitivity, leading to hyperventilation and alkalosis.

TABLE 16.9.   How Alcohol Abuse Contributes to Acid–Base Disturbances

Disturbance	Cause(s)
Metabolic acidosis	Accidental or intentional ingestion of ethylene glycol (automotive antifreeze), methanol (wood alcohol), paraldehyde (sedative used for alcohol withdrawal), or aspirin (overdose)
	Lactic acidosis (associated with sepsis or shock)
	Ketoacidosis (starvation combined with alcohol ingestion)
Respiratory acidosis	Profound intoxication with respiratory depression
Metabolic alkalosis	Volume depletion (from vomiting, prolonged gastric suction, or profuse diarrhea)
	Alkali administration (antacid abuse)
Respiratory alkalosis	Alcohol withdrawal (sympathetic nervous system hyperactivity)
	Cirrhosis of the liver (extensive arteriovenous shunting; endogenous toxins that stimulate respiration)

The development of metabolic acid–base disturbances in alcoholic patients is somewhat less straightforward. These disturbances generally occur only when a number of conditions are met.

16.10.2. Metabolic Alkalosis

Aside from hangover, gastritis and vomiting are probably the most common consequences of an acute alcoholic binge. If vomiting is copious or prolonged, the loss of volume, salt, and acid may cause metabolic alkalosis. At times, this alkalosis may be severe.

The pathophysiology of metabolic alkalosis has been clearly elucidated. Vomiting or gastric suction causes losses of water, sodium, chloride, and acid equivalents. The resultant contraction of plasma volume leads to decreased renal perfusion and to increased renal renin secretion, which, in turn, stimulates aldosterone secretion. Aldosterone acts to promote sodium–potassium exchange in the distal nephron; increased aldosterone production contributes to the hypokalemia characteristic of this condition (Oster, 1983).

16.10.3. Alcoholic Ketoacidosis

Ethanol can induce lactic acidosis (Kreisberg *et al.*, 1971), ketoacidosis (Cahill, 1981; Cooperman *et al.*, 1974; Fulop and Hoberman, 1975; Soffer and Hamburger, 1982; Levy *et al.*, 1973), mild acetic acidosis (Halperin *et al.*, 1983), and vomiting-induced metabolic

alkalosis. Because of the potential for a complex acid–base disturbance, Halperin *et al.* (1983) prefer the term alcoholic acidosis rather than alcoholic ketoacidosis.

Alcoholic ketoacidosis is a not-uncommon, potentially serious complication of severe and usually chronic alcohol abuse. It can occur at any age, and despite earlier evidence (Jenkins *et al.*, 1971) there is no unusual predominance in women. For unknown reasons certain individuals appear to be particularly prone to develop this syndrome and may do so repeatedly, even though others drink just as heavily. Although the syndrome typically occurs in nondiabetic patients, it is not surprising that alcohol abuse can also be responsible for ketoacidosis in diabetics, and it has been suggested that these patients are more subject to severe acidemia (Kreisberg, 1978).

The clinical features of this condition have been detailed previously (Oster and Epstein, 1984). The fact that alcoholic ketoacidosis was once considered rare might be explained, at least in part, by a masking of the degree of ketosis (Cooperman *et al.*, 1974; Fulop and Hoberman, 1975; Soffer and Hamburger, 1982; Levy *et al.*, 1973) by an increase in the β-hydroxybutyrate/acetoacetate (β-OHB/AcAc) ratio (Kreisberg, 1978; Cahill, 1981; Levy *et al.*, 1973; Lefevre *et al.*, 1970). The syndrome typically occurs following an alcoholic binge in chronic alcohol abusers. Often there has been no caloric intake other than ethanol for several days (Cooperman *et al.*, 1974). Blood alcohol levels are frequently negative (Jenkins *et al.*, 1971) or low when patients are first seen. Nausea, vomiting, and abdominal pain (sometimes accounting for cessation of alcohol ingestion) and tachypnea, hypothermia, and volume contraction are common (Halperin *et al.*, 1983). Not infrequently, other intercurrent problems complicate the clinical picture. These include not only lactic acidosis but gastrointestinal bleeding, bacterial infection, acute pancreatitis, and delirium tremens. It seems likely that these associated conditions, rather than alcoholic ketosis *per se*, are the principal cause of mortality. Prior to therapy, hyperuricemia (Fulop and Hoberman, 1975; Soffer and Hamburger, 1982; Levy *et al.*, 1973) and hyperphosphatemia (Miller *et al.*, 1978) are common.

16.11.   Acidosis Caused by Toxin Ingestion

Ethylene glycol, methanol, and isopropanol are intoxicants that alcoholics occasionally substitute for ethanol when the latter agent is unavailable. Each is metabolized by alcohol dehydrogenase. The common features of intoxication with these drugs are central nervous system depression and the presence of a metabolic acidosis most often associated with a high anion gap. However, because the products of metabolism of these agents differ, the clinical syndromes produced by them have unique and characteristic features.

16.12. Abnormalities of Acid–Base Status in Liver Disease

Aside from the effects of alcohol on acid–base homeostasis, alcohol may also exert indirect effects through its ability to produce chronic liver disease. Once established, chronic liver disease can bring about a wide spectrum of acid–base abnormalities.

There are several important associations between liver function and acid–base balance (Oster, 1983). First, primarily because of its metabolism of organic acid anions (Halperin and Jungas, 1983; Halperin *et al.*, 1985), particularly lactate and certain amino acids, the liver has an important influence on normal acid–base homeostasis. Second, under certain conditions such as ketoacidosis, the liver may become a major net producer of hydrogen ions (H^+). Third, patients with various types of liver disease often develop complicating acid-base disturbances. Finally, the disturbance may be a complication of therapy, as when diuretic therapy results in metabolic alkalosis (Oster and Vaamonde, 1983).

16.12.1. Prevalence

Alkalosis is the most common general disturbance. In fact, of 884 separate measurements in 631 cirrhotic patients (11 studies), alkalosis was present in 71% (Oster, 1983). Isolated respiratory alkalosis was approximately three times more common than isolated metabolic alkalosis (38% versus 13%) and was the most frequent abnormality in seven of the 11 studies. The degree of hyperventilation and resulting alkalemia is usually modest. There is a very strong association between the presence of hypokalemia and metabolic alkalosis.

Acidosis is far less prevalent than alkalosis (13% versus 71% of the measurements reviewed), and respiratory acidosis is rare (only 2% of measurements) and is often an ominous preterminal finding (Oster, 1983). Acidosis occurs typically in the setting of hypotension or renal insufficiency and often is caused by lactic acidosis. Except in chronic autoimmune liver disease (see Section 16.13.1), the complete form of distal renal tubular acidosis (RTA) is rare.

Respiratory alkalosis occurs more frequently than

metabolic alkalosis, but diuretic therapy resulting in potassium depletion, vomiting, and massive blood transfusion may cause metabolic alkalosis. Metabolic acidosis is considerably less common and is usually attributable to lactic acidosis, to renal insufficiency when the hepatorenal syndrome supervenes, and, rarely, to RTA. Respiratory acidosis is extremely uncommon in patients with liver disease but can occur in the presence of extreme ascites and pleural effusions, when either severe hypokalemia or hypophosphatemia impairs respiratory muscle activity, or as a terminal event.

Mixed acid–base disturbances occur frequently in patients with chronic decompensated severe liver disease. Indeed, such disorders were observed in 10 of 14 studies reviewed (Oster, 1983). By far the most common complex disturbance is respiratory plus metabolic alkalosis.

16.12.2. Pathophysiology of Acid–Base Disorders

16.12.2.1. Respiratory Alkalosis

As mentioned, respiratory alkalosis is commonly found in patients with severe clinical manifestations of liver disease; its pathogenesis remains uncertain. Hypoxemia, attributed to ventilation–perfusion imbalance secondary to ascites and to venous admixture from portopulmonary or intrapulmonary shunts has been suggested as a leading cause. Nevertheless, the hypoxemia is frequently mild, and there may be no correlation between the degree of hypoxemia and the arterial PCO_2.

Hyperammonemia has been considered a possible cause of hyperventilation. Although a lack of correlation in some studies between plasma ammonia levels and the degree of alveolar ventilation casts some doubt on the importance of the association, both hyperammonemia and hyperventilation are common clinical manifestations in patients with Reye's syndrome. Finally, increased plasma progesterone levels, hyponatremia, intracellular acidosis, and physical stimuli are other possible causes of hyperventilation (Kaehny, 1983).

16.12.2.2. Metabolic Alkalosis

The metabolic alkalosis that occurs in patients with chronic liver disease can almost always be attributed either to extrarenal or iatrogenic factors (Oster and Vaamonde, 1983). Secondary aldosteronism probably only rarely causes metabolic alkalosis unless diuretics are administered, because in the patient with peripheral edema and ascites, the delivery of sodium to the collecting tubule is limited. Therefore, even though the avidity for sodium reabsorption in the cortical collecting tubule is high, limited availability of sodium attenuates the voltage-related enhancement of potassium and H^+ secretion resulting from sodium reabsorption (Oster and Vaamonde, 1983). Thus, alkalosis [and potassium depletion (Perez and Oster, 1983)] of renal origin is rare in the untreated patient with decompensated cirrhosis. Once diuretics are given, however, a higher rate of sodium delivery to the distalmost nephron is assured, and renal potassium wasting, hypokalemia, and hypochloremic alkalosis often ensue.

In addition to diuretic administration, the cause of metabolic alkalosis in decompensated cirrhosis may be of extrarenal origin (e.g., vomiting). Alkalosis of any origin will tend to be maintained (bicarbonaturia prevented) not only by a decrease in effective plasma volume but also by mineralocorticoid excess, hypokalemia, and possibly inhibition of tubular chloride reabsorption or reduction in glomerular filtration rate (Oster and Vaamonde, 1983).

16.12.2.3. Lactic Acidosis

Despite the importance of the liver in lactate metabolism, frank lactic acidosis is not common in patients with chronic liver disease. Basal lactate levels are usually either within normal limits or only modestly increased, that is, usually less than 4 mmol/liter. When lactic acidosis occurs, it usually is in association with factors, such as sepsis, shock, and bleeding, that either further depress hepatic or renal lactate utilization or increase lactate generation (Kreisberg, 1980). Other potential factors include ethanol abuse, medications such as phenformin, fructose, or sorbitol, hypotension, and hyperventilation.

Although severe lactic acidosis complicates the terminal course of only a small percentage of patients with liver disease, it is important because of its ominous prognosis. Lactic acidosis generally occurs in the presence of severe hepatic damage and in conjunction with other metabolic insults.

16.13. Renal Tubular Acidosis

16.13.1. Autoimmune Liver Disease

Renal tubular acidosis has been described in association with both autoimmune liver disease (ALD) and ethanol-induced chronic liver disease; it is a common and sometimes important feature of patients with various types of chronic ALD, e.g., chronic active hepatitis

(CAH), primary biliary cirrhosis (PBC), and cryptogenic cirrhosis (Puig et al., 1980).

Most patients with RTA and ALD have either chronic PBC or CAH. The association of liver disease and RTA appears to be less common in cryptogenic cirrhosis and, for the complete variety, rare in alcoholic cirrhosis. The association of RTA and ALD seems to be strongest in patients with PBC (Pares et al., 1981). Almost always, the acidification defect is of the distal type; nevertheless, "maximal" tubular bicarbonate reabsorption may also be decreased, implying an additional abnormality in the proximal tubule. Multiple proximal defects characteristic of the Fanconi syndrome do not occur.

The clinical manifestations of patients with RTA and ALD are similar to those of patients with RTA in the absence of liver disease. Evidence for hepatic dysfunction and vasculitic manifestations is common. No correlation has been found between the severity of hepatic disease and the presence or absence of RTA. As expected, many of the patients are young, and most are female. Because of hypokalemia, acidemia, and, less commonly, osteomalacia, patients may complain of polydipsia, polyuria, muscular weakness, and bone pain. Nephrocalcinosis and nephrolithiasis are infrequent. The incomplete syndrome of distal RTA (inability to lower urine pH normally in response to acid loading without spontaneous acidosis) is far more common than spontaneous hyperchloremic acidosis (Oster, 1983).

Although the pathogenesis of RTA in ALD is uncertain, cell-mediated immunologic perturbations appear to play a major role in mediating not only the hepatic lesions but the renal tubular defects as well (Oster, 1983). In PBC as well as Wilson's disease, copper-induced tubular lesions appear to play an important pathogenic role (Pares et al., 1981). Significantly higher plasma and urinary copper levels have been observed in CAH patients with RTA, as well as a correlation between these variables and the minimal urinary pH achieved after an ammonium chloride load. Perhaps failure of biliary excretion of copper leads to abnormal urine and plasma levels and to deposition of the metal in the distal tubule (Pares et al., 1981).

16.13.2. Alcoholic Cirrhosis

The frequency and pathogenesis of RTA in alcoholic cirrhosis differ to a great extent from those in patients with ALD. Of 16 reports totaling 255 patients, there were 98 instances (38%) of incomplete distal RTA but only eight instances of the complete syndrome (Oster, 1983). Thus, the finding of a low plasma HCO_3^-

together with an elevated plasma Cl^- should raise the suspicion of an acid–base disorder other than RTA, such as acute diarrhea or respiratory alkalosis. Some investigators believe that RTA is somewhat more likely to occur in patients with severe liver disease. There is no evidence of an association between hyperglobulinemia and RTA in cirrhotic patients.

The important question of whether RTA increases the risk of encephalopathy remains unanswered. Either the acidosis or the hypokalemia associated with RTA might facilitate renal ammoniagenesis; the increased urine pH might divert some of this "extra" ammonia into the renal venous blood. To my knowledge, however, only one group of investigators has reported a correlation between RTA and a history of repeated episodes of hepatic encephalopathy.

Despite some opinion to the contrary, decreased delivery of sodium to the distalmost nephron appears to play an important pathogenic role in the impaired acidification of patients with alcoholic cirrhosis. Several investigators have observed intense avidity for sodium reabsorption in their subjects with RTA. Furthermore, the administration of diuretics, sodium sulfate, or sodium phosphate, which increase distal sodium delivery, often results in a marked lowering of urine pH.

Several lines of evidence link distal sodium delivery and reabsorption to H^+ secretion (Kurtzman, 1983). Renal acidification is less efficient in the presence of sodium depletion, and a modest sodium deficiency impairs the ability to lower urine pH maximally. In several varieties of incomplete distal RTA, increasing sodium delivery with sodium sulfate leads to reduction of urine pH.

In summary, an incomplete syndrome of distal RTA is a common complication of alcoholic liver disease. A low distal delivery of sodium, possibly because of a reduction in the transtubular potential difference, is probably partially responsible. The clinical importance of this incomplete RTA remains unknown. A small number of patients appear to have clinically important renal potassium wasting as a related or associated defect, and this, perhaps by facilitating renal ammoniagenesis, might predispose to the development of hepatic encephalopathy.

16.14. Effects on the Renin–Aldosterone System and Adrenocortical Function

Ethanol changes the hydration state of the body by its diuretic action and exerts varying effects on blood and urine electrolytes. Ethanol is also known to produce hypoglycemia and increased activity of the sympathetic nervous system. The renin–angiotensin–aldosterone system is affected by changes in the blood volume, the electrolyte state of the body, hypoglycemia, and the activity of the sympathetic nervous system (Epstein and Norsk, 1988). Linkola *et al.* (1976, 1979) investigated the state of the renin–aldosterone axis during ethanol intoxication to assess its role in the pathogenesis of hangover.

It is clear that ethanol exerts significant effects on the renin–aldosterone system and that the effects vary with the extent of ethanol administration. It has been demonstrated that plasma renin activity (PRA) is increased in humans by the drinking of intoxicating doses of alcohol. Puddey *et al.* (1985) evaluated the effects of moderate alcohol consumption on the renin-aldosterone system in 20 normal men aged 20–24 years. The subjects consumed either 750 ml nonalcoholic beer as a control or the same beverage with 1 ml/kg alcohol added, which increased the plasma alcohol concentration to 16.7 ± 1.0 ($\pm$S.E.) mM within 70 min.

The PRA increased more than twofold 90 min after the ingestion of alcohol. This was accompanied by a decrease in diastolic blood pressure and a fall in plasma potassium, both possible stimuli to the rise in PRA. A late increase in plasma sodium, also occurring 90 min after alcohol ingestion, was attributed to plasma volume contraction after an alcohol–induced diuresis. This may have been an additional factor in stimulating renin release.

Puddey *et al.* (1985) concluded that the acute increase in PRA associated with moderate alcohol consumption is predominantly a secondary response to changes in fluid and electrolyte balance and blood pressure. Although a direct action of alcohol on renin release was not excluded, the possibility that repeated activation of the renin-angiotensin system mediates the pressor effect of regular moderate alcohol consumption is, therefore, diminished.

There is relatively little information on the adrenocortical response to ethanol in humans. Jenkins and Connolly (1968) reported that the acute administration of ethanol increased plasma cortisol in those subjects whose plasma ethanol concentration was greater than 100 mg/100 ml.

16.14.1. Possible Role of Atrial Natriuretic Factor (ANF)

Another vasoactive hormone that has attained increasing importance is atrial natriuretic factor (ANF).

Recent studies have demonstrated that ANF participates in the regulation of volume homeostasis both in animals and humans and is a major determinant of renal sodium handling (Epstein *et al.*, 1987). Investigators in several laboratories have assessed its possible role in the pathogenesis of deranged sodium and water homeostasis in patients with decompensated cirrhosis (Atlas and Epstein, 1988; Epstein, 1989; Epstein *et al.*, 1989). In contrast, data are not available regarding the effects of alcohol consumption on ANF release or the possible role of ANF in mediating the fluid and electrolyte alterations of acute alcoholism.

16.15. Effects on Blood Pressure

Several lines of evidence have demonstrated an association between regular moderate alcohol consumption and an increase in blood pressure (Beevers *et al.*, 1982; Kaysen and Noth, 1984), leading to the concept of alcohol-related hypertension. This concept is not new, but it has largely been forgotten. In fact, in 1915, Lian published a report of the prevalence of hypertension in relation to drinking habits in French army officers. The highest prevalence was in "*très grands buveurs*," with intermediate levels in "*grands buveurs*" and "*moyens buveurs*," and the lowest prevalence found in "*les sobres*."

Many population studies have demonstrated a close relation between blood pressure and alcohol intake. It is noteworthy that this relationship is observed within the range of blood pressures conventionally considered to be normal and within the range of alcohol intakes normally regarded as moderate (Klatsky *et al.*, 1971).

The most important population study to show a relation between alcohol and blood pressure is the Kaiser-Permanente project (Klatsky *et al.*, 1977). In this study, reliable blood pressure measurements and careful alcohol histories were obtained in over 80,000 men and women, and the association was found to be independent of age, sex, ethnic origin, obesity, smoking habit, and social class.

In a subsequent epidemiological study, Maheswaran *et al.* (1987) confirmed a correlation between blood pressure and alcohol intake measured by a 7-day alcohol diary.

Beevers (1977; Beevers *et al.*, 1982) demonstrated that the elevation in the blood pressure was closely related to biochemical and hematological indices of liver disease. The increase in blood pressure was also closely related to the severity of withdrawal symptoms,

and it fell with recovery. Patients who remained abstinent remained normotensive, whereas in those who resumed drinking blood pressure rose again.

There are now four clinical studies of hypertensive patients which demonstrated that reducing alcohol intake lowers blood pressure and restarting drinking raises pressure (Potter and Beevers, 1984; Malhotra *et al.*, 1985; Puddey *et al.*, 1987; Ueshima *et al.*, 1987) The mechanisms responsible for this pressor effect of alcohol have not been established. They may relate to sympathetic nervous activity, as judged by both the nature of the withdrawal symptoms and the correlation between change in blood pressure and change in dopamine β-hydroxylase (Beevers *et al.*, 1982). The high plasma renin levels are most probably accounted for by the low urinary sodium excretion and mild clinical dehydration. Changes in renin and aldosterone did not correlate with blood pressure. The few high plasma cortisol concentrations and the correlation between systolic blood pressure and plasma cortisol levels probably reflect a degree of alcoholic pseudo-Cushing syndrome (Beevers *et al.*, 1982). Arkwright *et al.* (1982b) studied the effect of moderate habitual alcohol consumption on blood pressure and demonstrated that alterations in plasma catecholamines did not appear to mediate the hypertension. In a more recent study, Ireland *et al.* (1984) examined the acute effects of moderate alcohol consumption on blood pressure, plasma catecholamines, and cortisol. They observed that alcohol resulted in a relative rise in epinephrine and a delayed increase in norepinephrine concentration. Additional studies are necessary to further characterize and elucidate the intriguing association between alcohol consumption and hypertension and to delineate its pathophysiology.

16.16. Summary

It is evident that in addition to the alterations in renal function attributable solely to an impairment of liver disease, alcohol *per se* induces a wide array of renal functional abnormalities. The major electrolyte disturbances attributable to alcohol abuse include dilutional hyponatremia, hypokalemia, hypophosphatemia, and hypomagnesemia. When alcohol abuse persists with resultant established liver disease, sodium retention supervenes, manifested by ascites and edema.

Finally, it is evident that both chronic and acute alcohol abuse may cause or contribute to a wide range of acid–base and electrolyte disturbances. The determinants of the clinical expression of alcohol-induced disturbances include the amount and duration of drinking, the nutritional status of the patient, and the coexistence of other diseases.

Despite the serious nature of many of these abnormalities, only a few have been the subject of systematic, rigorous investigations. In light of the profound clinical implications of these disturbances, it is hoped that some readers will use this review as a basis for formulating and undertaking future investigations on this subject. I believe that the next several years in this field will be exciting and that we will witness additional studies further characterizing some of the toxic renal–fluid–electrolyte effects of ethanol and delineating their pathophysiology. The next review of the renal functional abnormalities of alcoholism should make interesting reading.

REFERENCES

Aldersberg, D., and Fox, C. L.: Changes of the water tolerance tests in hepatic disease. *Ann. Intern. Med.* **19:**642–650, 1943.

Arkwright, P. D., Beilin, L. J., Rouse, I., Armstrong, B. K., and Vandongen, R.: Effects of alcohol use and other aspects of lifestyle on blood pressure levels and prevalence of hypertension in a working population. *Circulation* **66:**60, 1982a.

Arkwright, P. D., Beilin, L. J., Vandongen, R., Rouse, I., and Lalor, D.: The pressor effect of moderate alcohol consumption in man: A search for mechanisms. *Circulation* **66:**515, 1982b.

Atkinson, M.: Ascites in liver disease. *Postgrad. Med. J.* **32:**482–485, 1956.

Atlas, S. A., and Epstein, M.: Atrial natriuretic factor: Implications in cirrhosis and other edematous disorders. In: *The Kidney in Liver Disease* (3rd ed.), (M. Epstein, ed.), Baltimore: Williams & Wilkins, 1988, pp. 429–455.

Bakris, G. L.: Abnormalities of calcium and the diabetic, hypertensive patient: implications for renal preservation. In: *Calcium Antagonists in Clinical Medicine*, (M. Epstein, ed.), Philadelphia, Hanley & Belfus, 1992.

Barker, E. S., Elkinton, J. R., and Clark, J. K.: Studies of the renal excretion of magnesium in man. *J. Clin. Invest.* **38:**1733, 1959.

Barlow, G., and Wooten, R. L.: Electrolyte alterations in human plasma and erythrocytes associated with chronic alcoholism. *Proc. Soc. Exp. Biol. Med.* **101:**44, 1959.

Beard, J. D., and Knott, D. H.: Fluid and electrolyte balance during acute withdrawal in chronic alcoholic patients. *J.A.M.A.* **204:**133–139, 1968.

Beard, J. D., Barlow, G., and Overman, R. R.: Body fluids and blood electrolytes in dogs subjected to chronic ethanol administration. *J. Pharmacol. Exp. Ther.* **148:**348–355, 1965.

Beevers, D. G.: Alcohol and hypertension. *Lancet* **2:**114–115, 1977.

Beevers, D. B., Bannan, L. T., Saunders, J. B., Paton, A., and

Walters, J. R. F.: Alcohol and hypertension. *Contrib. Nephrol.* **30:**92, 1982.

Cahill, G. F., Jr.: Ketosis. *Kidney Int.* **20:**416–425, 1981.

Chaikoff, I. L., Entenman, C., Gillman, T., and Connor, C. L.: Pathologic reactions in the livers and kidneys of dogs fed alcohol while maintained on a high protein diet. *Arch. Pathol.* **45:**435–446, 1948.

Cooperman, M. T., Davidoff, F., Spark, R., and Pallota, J.: Clinical studies of alcoholic ketoacidosis. *Diabetes* **23:**433–439, 1974.

Demanet, J. C., Bonnyns, M., Bleiberg, H., and Stevens-Rocmans, C.: Coma due to water intoxication in beer drinkers. *Lancet* **2:**1115–1117, 1971.

Eggleton, M. G.: The diuretic action of alcohol in man. *J. Physiol.* **101:**172–191, 1942.

Epstein, M.: Hepatorenal syndrome. In: *Bockus Gastroenterology* (4th ed.) (J. E. Berk, ed.), Philadelphia, W. B. Saunders, 1985a, pp. 3138–3149.

Epstein, M.: Derangements of renal water handling in liver disease. *Gastroenterology* **89:**1415–1425, 1985b.

Epstein, M.: The sodium retention of cirrhosis: A reappraisal. *Hepatology* **6:**312–315, 1986.

Epstein, M.: Pathogenesis of sodium retention in liver disease. In: *Body Fluid Homeostasis.* (B. Brenner and J. Stein, eds.), New York, Churchill-Livingstone, 1987a, pp. 299–333.

Epstein, M.: Renal complications in liver disease. In: *Diseases of the Liver* (6th ed.), (L. Schiff and E. Schiff, eds.), Philadelphia, J. B. Lippincott, 1987b, pp. 903–923.

Epstein, M.: The kidney in liver disease. In: *The Liver: Biology and Pathobiology* (I. M. Arias, H. Jakoby, D. Popper, D. Schachter, and D. A. Shafritz, eds.), New York, Raven Press, 1988a, pp. 1043–1063.

Epstein, M.: Renal sodium handling in liver disease. In: *The Kidney in Liver Disease* (3rd ed.) (M. Epstein, ed.), Baltimore, Williams & Wilkins, 1988b, pp. 3–30.

Epstein, M.: Liver disease. In: *Textbook of Nephrology* (2nd ed.), (S. G. Massry, and R. J. Glossock, eds.), Baltimore, Williams & Wilkins, 1988c, pp. 966–977.

Epstein, M.: Atrial natriuretic factor in patients with liver disease. *Am. J. Nephrol.* **9:**89–100, 1989.

Epstein, M.: Functional renal abnormalities in cirrhosis: Pathophysiology and management. In: *Hepatology. A Textbook of Liver Disease* (2nd ed.), Philadelphia, W. B. Saunders, 1990a, pp. 493–512.

Epstein, M.: Disorders of sodium balance. In: (J. H. Stein, ed.), Boston, Little, Brown, 1990b, pp. 844–853.

Epstein, M., and Norsk, P.: Aldosterone in liver disease. In: *The Kidney in Liver Disease* (3rd ed.), (M. Epstein, ed.) Baltimore, Williams & Wilkins, 1988, pp. 356–373.

Epstein, M., Loutzenhiser, R., Friedland, E., Aceto, R. M., Camargo, M. J. F., and Atlas, S. A.: Relationship of increased plasma ANF and renal sodium handling during immersion-induced central hypervolemia in normal humans. *J. Clin. Invest.* **79:**738–745, 1987.

Epstein, M., Loutzenhiser, R., Norsk, P., and Atlas, S.: Relationship between plasma ANF responsiveness and renal sodium handling in cirrhotic humans. *Am. J. Nephrol.* **9:**133–143, 1989.

Estep, H., Shaw, W. A., Watlington, C., Hobe, R., Holland, W., and Tucker, S. G.: Hypocalcemia due to hypomagnesemia and reversible parathyroid hormone unresponsiveness. *J. Clin. Endocrinol. Metab.* **29:**842–848, 1969.

Fankushen, D., Raskin, D., Dimich, A., and Wallach, S: The significance of hypomagnesemia in alcoholic patients. *Am. J. Med.* **37:**802–812, 1964.

Fulop, M., and Hoberman, H. D.: Alcoholic ketosis. *Diabetes* **24:** 785–790, 1975.

Halperin, M. L., and Jungas, R. L.: The metabolic production and renal disposal of hydrogen ions—an examination of the biochemical processes. *Kidney Int.* **24:**709–713, 1983.

Halperin, M. L., Hammeke, M., Josse, R. G., and Jungas, R. L.: Metabolic acidosis in the alcoholic: A pathophysiological approach. *Metabolism* **32:**308–315, 1983.

Halperin, M. L., Goldstein, M. B., Stinebaugh, B. J., and Jungas, R. L.: Biochemistry and physiology of ammonium excretion. In: *The Kidney—Physiology and Pathophysiology* (D. W. Seldin and G. Giebisch, eds.), New York, Raven Press, 1985, pp. 1471–1490.

Helderman, J. H., Vestal, R. E., Rowe, J. W., Tobin, J. D., Andres, R., and Robertson, G. L.: The response of arginine vasopressin to intravenous ethanol and hypertonic saline in man: The impact of aging. *J. Gerontol.* **33:**39–47, 1978.

Hilden, T., and Svendsen, T. L.: Electrolyte disturbances in beer drinkers. A specific "hypo-osmolality syndrome." *Lancet* **2:** 245–246, 1975.

Himwich, H. E., Nahum, L., Rakieten, N., Fazikas, J. F., Dubois, D., and Gilder, E. F.: The metabolism of alcohol. *J.A.M.A.* **100:**651–654, 1933.

Huang, K. C., Knoefel, P. K., Shimomura, L., and King Buren, N.: Some effects of alcohol on water and electrolytes in the dog. *Arch. Int. Pharmacodyn.* **109:**90–98, 1957.

Ireland, M. A., Vandongen, R., Davidson, L., Beilin, L. J., and Rouse, I. L.: Acute effects of moderate alcohol consumption on blood pressure and plasma catecholamines. *Clin. Sci.* **66:** 643, 1984.

Israel, Y., and Kalant, H.: Effect of ethanol on the transport of sodium in frog skin. *Nature* **200:**476–478, 1963.

Israel-Jacard, Y., and Kalant, H.: Effect of ethanol on electrolyte transport and electrogenesis in animal tissues. *J. Cell. Comp. Physiol.* **65:**127–132, 1965.

Jenkins, J. S., and Connolly, J.: Adrenocortical response to ethanol in man. *Br. Med. J.* **2:**804–805, 1968.

Jenkins, D. W., Eckel, R. E., and Craig, J. W.: Alcoholic ketoacidosis. *J.A.M.A.* **217:**177–183, 1971.

Kaehny, W. D.: Respiratory acid–base disorders. *Med. Clin. North Am.* **67:**915–928, 1983.

Kalbfleisch, J. M., Lindemann, R. D., Ginn, H. E., and Smith, W. O.: Effect of ethanol administration on urinary excretion of magnesium and other electrolytes in alcoholic and normal subjects. *J. Clin. Invest.* **42:**1471–1475, 1963.

Kaysen, G., and Noth, R. H.: The effects of alcohol on blood pressure and electrolytes. *Med. Clin. North Am.* **68:**221–246, 1984.

Klatsky, A. L., Friedman, G. D., Siegelaub, A. B., and Gerard, M. J.: Alcohol consumption and blood pressure. *N. Engl. J. Med.* **296:**1194–1200, 1977.

Kleeman, C. R., Rubini, M. E., Lamdin, E., and Epstein, F. H.: Studies on alcohol diuresis. II. The evaluation of ethyl alcohol as an inhibitor of the neurohypophysis. *J. Clin. Invest.* **34:**448–455, 1955.

Klingler, E. L., Vaamonde, C. A., Vaamonde, L. S., Lancestremere, R. G., Morosi, H. J., Frish, E., and Papper, S.: Renal function changes in cirrhosis of the liver. *Arch. Intern. Med.* **125:**1010–1015, 1970.

Klingman, G. I., and Haag, H. B.: Studies on severe alcohol intoxication in dogs. I. Blood and urinary changes in lethal intoxication. *Q. J. Stud. Alcohol* **19:**203–225, 1958.

Knochel, J. P.: The pathophysiology and clinical characteristics of severe hypophosphatemia. *Arch. Intern. Med.* **137:**203–220, 1977.

Knochel, J. P.: Hypophosphatemia in the alcoholic. *Arch. Intern. Med.* **140:**613–615, 1980.

Knochel, J. P.: Hypophosphatemia. *West. J. Med.* **134:**15–26, 1981.

Knochel, J. P.: Derangements of phosphate magnesium and calcium homeostasis in chronic alcoholism. In: *The Kidney in Liver Disease* (3rd ed.), (M. Epstein, ed.), Baltimore: Williams & Wilkins, 1988.

Knochel, J. P., Bilbrey, G. L., Fuller, T. J., and Carter, N. W.: The muscle cell in chronic alcoholism: The possible role of phosphate depletion in alcoholic myopathy. *Ann. N.Y. Acad. Sci.* **252:**274–286, 1975.

Kreisberg, R. A.: Diabetic ketoacidosis: New concepts and trends in pathogenesis and treatment. *Ann. Intern. Med.* **88:**681–695, 1978.

Kreisberg, R. A.: Lactate homeostasis and lactic acidosis. *Ann. Intern. Med.* **92:**227–237, 1980.

Kreisberg, R. A., Owen, W. C., and Siegal, A. M.: Ethanol-induced hyperlacticacidemia: Inhibition of lactate utilisation. *J. Clin. Invest.* **50:**166–174, 1971.

Kurtzman, N. A. (principal discussant): Acquired distal renal tubular acidosis. *Kidney Int.* **24:**807–819, 1983.

Laube, H., Norris, H. T., and Robbins, S. L.: The nephromegaly of chronic alcoholics with liver disease. *Arch. Pathol.* **84:** 290–294, 1967.

Lefevre, A., Adler, H., and Lieber, C. S.: Effect of ethanol on ketone metabolism. *J. Clin. Invest.* **49:**1775–1782, 1970.

Levy, L. J., Duga, J., Girgis, M., and Gordon, E. E.: Ketoacidosis associated with alcoholism in nondiabetic subjects. *Ann. Intern. Med.* **78:**213–219, 1973.

Lian, C.: L'alcoolisme, cause d'hypertension artérielle. *Bull. Acad. Med.* **74:**525–528, 1915.

Lieber, C. S., Spritz, N., and DeCarli, L. M.: Accumulation of triglycerides in heart and kidney after alcohol ingestion. *J. Clin. Invest.* **45:**1041, 1966.

Linkola, J.: Effects of ethanol on urine sodium and potassium concentrations and osmolality in water-loaded rats. *Acta Physiol. Scand.* **92:**212–216, 1974.

Linkola, J.: Natriuresis after diluted ethanol solutions. *Lancet* **2:** 1157, 1975.

Linkola, J., Fyhrquist, F., Nieminen, M. M., Weber, T. H., and Tontti, K.: Renin–aldosterone axis in ethanol intoxication and hangover. *Eur. J. Clin. Invest.* **6:**191–194, 1976.

Linkola, J., Fyhrquist, F., and Ylikahri, R.: Renin, aldosterone and cortisol during ethanol intoxication and hangover. *Acta Physiol. Scand.* **106:**75–82, 1979.

Lotz, M., Zisman, E., and Bartter, F. C.: Evidence for a phosphorus depletion syndrome in man. *N. Engl. J. Med.* **278:** 409–413, 1968.

Maheswaran, R., Potter, J. F., and Beevers, D. G.: The role of alcohol in hypertension.*J. Clin. Hypertens.* **2:**172–178. 1986.

Malhotra, H., Mehta, S. R., Mathur, D., and Khandelwal, P. D.: Pressor effects of alcohol in normotensive and hypertensive subjects. *Lancet* **2:**584–586, 1985.

Markkanen, T., and Nanto, V.: The effect of ethanol infusion on the calcium–phosphorus balance in man. *Experientia* **22:**753–754, 1966.

Massry, S. G., and Goldstein, D. A.: Vitamin D and liver disease. In: *The Kidney in Liver Disease* (2nd ed.), (M. Epstein, ed.), New York, Elsevier, 1983, pp. 229–242.

McCollister, R. J., Flink, E. B., and Lewis, M. D.: Urinary excretion of magnesium in man following ingestion of alcohol. *Am. J. Clin. Nutr.* **12:**415–420, 1963.

Meyer, M., Wexler, S., Jedvab, M., Spirer, Z., Keysar, N., and Shibolet, A.: Malabsorption of vitamin D in man and rat with liver cirrhosis. *J. Mol. Med.* **3:**29–37, 1978.

Miller, P. D., Heinig, R. E., and Waterhouse, C.: Treatment of alcoholic acidosis. *Arch. Intern. Med.* **138:**67–72, 1978.

Mori, T. A., Puddey, I. B., Wilkinson, S. P., Beilin, L. J., and Vandongen R.: Urinary steroid profiles and alcohol-related blood pressure elevation. *Clin. Exp. Pharmacol. Physiol.* **18:** 287–290, 1991.

Mostellar, M. E., and Tuttle, E. P., Jr.: The effects of alkalosis on plasma concentration and urinary excretion of inorganic phosphate in man. *J. Clin. Invest.* **43:**138–149, 1964.

Nakano, J., and Prancan, A. V.: Effects of adrenergic blockade on cardiovascular responses to ethanol and acetaldehyde. *Arch. Int. Pharmacodynamie et de thérapie* **196:**259–268, 1972.

Nicholson, W. M., and Taylor, H. M.: The effect of alcohol on the water and electrolyte balance in man. *J. Clin. Invest.* **17:**279–285, 1938.

Nicholson, W. M., and Taylor, H. M.: Blood volume studies in acute alcoholism. *Q. J. Stud. Alcohol* **1:**472, 1940.

Ogata, M., Mendelson, J. H., and Mello, N. K.: Electrolytes and osmolality in alcoholics during experimentally induced intoxication. *Psychosom. Med.* **30:**463, 1968.

Okel, B. B., and Hurst, J. W.: Prolonged hyperventilation in man: associated electrolyte changes and subjective symptoms. *Arch. Intern. Med.* **108:**757–762, 1961.

Oster, J. R.: Acid–base homeostasis and liver disease. In: *The Kidney in Liver Disease* (2nd ed.) (M. Epstein, ed.), New York, Elsevier, 1983, pp. 147–182.

Oster, J. R.: Magnesium. *South. Med. J.* **78:**1111–1113, 1116–1126, 1985.

Oster, J. R., and Epstein, M.: Acid–base aspects of ketoacidosis. *Am. J. Nephrol.* **4:**137–151, 1984.

Oster, J. R., and Perez, G. O.: Acid–base disturbances in liver disease. *J. Hepatol.* **2:**299–306, 1986.

Oster, J. R., and Vaamonde, C. A.: Metabolic alkalosis. In: *Textbook of Nephrology* (S. G. Massry and R. J. Glassock, eds.), Baltimore, Williams & Wilkins, 1983, pp. 3.130–3.138.

Papper, S.: The role of the kidney in Laennec's cirrhosis of the liver. *Medicine* (Baltimore) **37:**299–316, 1958.

Pares, A., Rimola, A., Bruguera, M., Mas, E., and Rodes, J.: Renal tubular acidosis in primary biliary cirrhosis. *Gastroenterology* **80:**681–686, 1981.

Peng, T., Cooper, C. W., and Munson, P. L.: The hypocalcemic effect of ethyl alcohol in rats and dogs. *Endocrinology* **91:** 586–593, 1972.

Perez, G. O., and Oster, J. R.: Altered potassium metabolism in liver disease. In: *The Kidney in Liver Disease* (2nd ed.), (M. Epstein, ed.), New York, Elsevier, 1983, pp. 183–201.

Potter, J. F., and Beevers, D. G.: Pressor effect of alcohol in hypertension. *Lancet* **1:**119–122, 1984.

Puddey, I. B., Vandongen, R., Beilin, L. J., and Rouse, I. L.: Alcohol stimulation of renin release in man: Its relation to the hemodynamic, electrolyte, and sympatho-adrenal responses to drinking. *J. Clin. Endocrinol. Metab.* **61:**37–42, 1985.

Puddey, I. B., Beilin, L. J., and Vandongen, R.: Regular alcohol use raises blood pressure in treated hypertensive subjects. A randomised controlled trial. *Lancet* **1:**647–651, 1987.

Puig, J. G., Anton, F. M., Gomez, M. E., Aguado, A. G., Barbado, J., Arnalich, F., Vazquez, J. J., Vazquez, J. O., and Montero, A.: Complete proximal tubular acidosis (type 2, RTA) in chronic active hepatitis. *Clin. Nephrol.* **13:**287–292, 1980.

Ram, J. L., Standley, M. S., and Sowers, J. R.: Calcium function in vascular smooth muscle and its relationship to hypertension. In: *Calcium Antagonists in Clinical Medicine*, (M. Epstein, ed.), Philadelphia, Hanley & Belfus, 1992, pp. 29–42.

Ring-Larsen, H., Clausen, E., and Ranck, L.: Peritoneal dialysis in hyponatremia due to liver failure. *Scand. J. Gastroenterol.* **8:**33–40, 1973.

Roberts, K. E.: Mechanism of dehydration following alcohol ingestion. *Arch. Intern. Med.* **112:**154, 1963.

Rubini, M. E., Kleeman, C. R., and Lamdin, E.: Effect of ethyl alcohol ingestion on water, electrolyte and acid base metabolism. *J. Clin. Invest.* **34:**439–447, 1955.

Ryback, R. S., Eckardt, M. J., and Pautler, C. P.: Clinical relationships between serum phosphorus and other blood chemistry values in alcoholics. *Arch. Intern. Med.* **140:**673–677, 1980.

Sargent, W. Q., Simpson, J. R., and Beard, J. D.: The effect of acute and chronic alcohol administration on renal hemodynamics and monovalent ion excretion. *J. Pharmacol. Exp. Ther.* **188:**461–474, 1974.

Sargent, W. Q., Simpson, J. R., and Beard, J. D.: Renal haemodynamics and electrolyte excretion after reserpine and ethanol. *J. Pharmacol. Exp. Ther.* **193:**356, 1975.

Sargent, W. Q., Beard, J. D., and Knott, D. H.: Acid–base balance following ethanol intake and during acute withdrawal from ethanol. In: *Biochemistry and Pharmacology of Ethanol* Vol. 2. (J. Majchrowicz, and E. P. Nobel, eds.), New York, Plenum Press, 1979, p. 17.

Soffer, A., and Hamburger, S.: Alcoholic ketoacidosis: A review of 30 cases. *J. Am. Med. Wom. Assoc.* **37:**106–110, 1982.

Stein, J. H., Smith, W. O., and Ginn, H. E.: Hypophosphatemia in acute alcoholism. *Am. J. Med. Sci.* **252:**78–83, 1966.

Thomas, I.: Ueber die Wirkung einiger narkotischer Stoff auf die

Blutgase, die Blutalkalescens und die rother Blutkorperchen. *Arch. Exp. Pathol. Pharmakol.* **41:**1–18, 1898.

U.S. Department of Health, Education and Welfare: *Mortality Trends for Leading Causes of Death, United States—1950–1969. Vital and Health Statistics*, Ser. 20–26, DHEW Publication No. (HRA) 74–1853. Washington, DHEW, 1970, p. 4.

Ueshima, H., Ogihara, T., Baba, S., Tabuchi, Y., Mikawa, K., Hashizume, K., Mandai, T., Ozawa, H., Kumahara, Y., Asakura, S. *et al.* The effect of reduced alcohol consumption on blood pressure: A randomised, controlled, single blind study. *J. Human Hypertension.* **1:**113–119, 1987.

Vaamonde, C. A.: Renal water handling in liver disease. In: *The Kidney in Liver Disease* (3rd ed.) (M. Epstein, ed.), Baltimore: Williams & Wilkins, 1988.

Van Dyke, H. B., and Ames, R. G.: Alcohol diuresis. *Acta Endocrinol.* **7:**110–121, 1951.

Van Thiel, D. H., William, W. D., Gavaler, J. S., Little, J. M., Estes, L. W., and Rabin, B. S.: Ethanol—its nephrotoxic effect in the rat. *Am. J. Pathol.* **89:**67–84, 1977.

Velentzas, C., Oreopoulos, D. G., Brances, L., Wilson, D. R., and Marquez-Julio, A.: Abnormal vitamin-D levels in alcoholics. *Ann. Intern. Med.* **86:**198, 1977.

17

Nutrition

Medical Problems of Alcoholism

Lawrence Feinman and Charles S. Lieber

17.1. General Nutritional Status in the Alcoholic Including Disorders of Minerals and Vitamins

Alcoholism remains one of the major causes of nutritional deficiency syndromes in the United States. The relationship between alcoholism and nutrition is complicated by many factors (Fig. 17.1). Alcohol is directly toxic to many tissues and thereby alters the nutritional requirement for various foodstuffs by interfering with their intake, absorption, metabolism, and loss from the body. The nutritional content of alcoholic beverages is meager in the extreme, providing mostly "empty calories." Nutritional therapy requires an assessment of nutritional deficit and extent of organ damage, which may dictate increased needs or limited tolerance. For example, the alcoholic with vitamin C or thiamin deficiency is diagnosed with ease, and replacement is straightforward. The diagnosis of vitamin A deficiency is much harder to establish, and therapy with vitamin A must take into account the potential hepatotoxicity of the vitamin during concomitant alcoholism. As another example, patients with alcoholic liver disease can develop encephalopathy with dietary protein intakes below the daily requirement. Nutritional planning for these patients must include consideration of special sensitivity to avoid iatrogenic complications. This chapter reviews the pathophysiology and diagnosis of malnutrition in the alcoholic with regard to various dietary constituents

and outlines specific recommendations for nutritional therapy.

17.2. Nutritive Aspects of Alcohol and Alcoholic Beverages

Alcoholic beverages are almost devoid of nutritive value except for their water content, variable amounts of carbohydrate, and ethanol, an energy source whose biological impact is not exactly predicted from the 7.1 kcal/g value obtained on its combustion in the bomb calorimeter. The carbohydrate content of whiskey, cognac, or vodka is near nothing; red and dry white wines contain from 2 to 10 g/liter; beer and sherry may contain 30 g/liter; sweetened white wines and port may contain as much as 120 g/liter (Pekkanen and Forsander, 1977). Although harmful levels of iron, lead, or cobalt may be present in some beverages, the remainder of the nonalcoholic constituents found in alcoholic beverages have not for the most part been shown to contribute to nutrition or to damage (Feinman and Lieber, 1988).

Americans consume 4.5% of total calories as alcohol (Scheig, 1970). Heavy drinkers may derive more than one-half of their daily calories as ethanol. Although the combustion of ethanol in a bomb calorimeter indicates a value of 7.1 kcal/g, we know that its biological value is less when compared to carbohydrates on a calorie basis. To illustrate, subjects on a metabolic ward

515

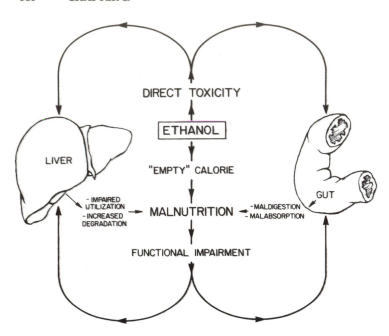

FIGURE 17.1. Interaction of direct toxicity of ethanol with malnutrition from primary or secondary deficiencies. Secondary malnutrition may be caused by either maldigestion and malabsorption or impaired utilization (decreased activation and/or increased inactivation) of nutrients. Both "direct toxicity" of ethanol and malnutrition (whether primary or secondary) may affect function and structure of liver and gut. (From Lieber, 1982.)

given additional calories as ethanol either failed to gain weight (Lieber *et al.*, 1965) or did not gain as much weight as when given additional calories as carbohydrate or fat (Pirola and Lieber, 1972) (Figs. 17.2 and 17.3). Patients in an open hospital ward also failed to gain weight when given additional calories as ethanol

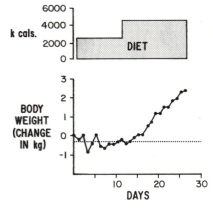

FIGURE 17.3. Effect on body weight of adding 2000 kcal/day as chocolate to the diet of the same subject as in Fig. 17.2. The dotted line represents the mean change during the control period. (From Pirola and Lieber, 1972.)

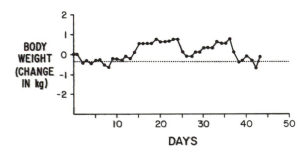

FIGURE 17.2. Effect on body weight of adding 2000 kcal/day as ethanol to the diet of one subject. The dotted line represents the mean change during the control period. (From Pirola and Lieber, 1972.)

(Mezey and Faillace, 1971). When ethanol was substituted for carbohydrate as 50% of calories in a balanced diet under metabolic ward conditions, weight loss occurred (Pirola and Lieber, 1972). Others have reported variable responses when additional calories were given as ethanol (Crouse and Grundy, 1984); lean individuals did not gain weight, but half of the obese individuals gained some weight. Body composition studies were not reported: they might have helped to interpret the results.

Ethanol probably alters metabolic rate, which may explain its failure to promote weight gain. It increases oxygen consumption in normal subjects and does so to a greater extent in alcoholics (Tremolieres and Carre, 1961). Its substitution for carbohydrate increases the metabolic rates of both humans and rodents (Stock and Stuart, 1974; Stock *et al.*, 1973). A 15% increase in thermogenesis occurred in rats after only 10 days of ethanol diet (Stock and Stuart, 1974). Only a small part of the energy wastage was attributable to metabolism by brown fat (Rothwell and Stock, 1984). Diet-induced thermogenesis was increased by ethanol in humans. Energy wastage on consumption of ethanol may occur via oxidation without phosphorylation by the microsomal ethanol-oxidizing system (Pirola and Lieber, 1972). Chronic ethanol consumption induces this pathway, after which energy wastage is aggravated (Pirola and Lieber, 1975, 1976). Others have implicated (Israel *et al.*, 1975) and argued against (Teschke *et al.*, 1983) an uncoupling of mitochondrial NADH reoxidation, abetted by a hyperthyroid state or catecholamine release. (Further details concerning the energy contributions of ethanol may be found in Chapter 1.)

In summary, alcoholic beverages have little of nutritive value other than calories; as a source of calories, alcohol is not as adequate as equivalent carbohydrate. The lower body weight associated with alcohol consumption (Williamson *et al.*, 1987) is probably caused in part by the above factors.

17.3. Nutritional Status of Alcoholics

Alcoholism can undermine nutritional status. At one time it was estimated that 20,000 alcoholics were suffering major illnesses from malnutrition in the United States each year, accounting for 7½ million days of hospitalization (Iber, 1971). The subgroup of alcoholics hospitalized for medical complications of alcoholism has the most severe malnutrition. They have had inadequate dietary protein (Patek *et al.*, 1975), have signs of protein malnutrition (Iber, 1971; Mendenhall *et al.*, 1985), and have anthropomorphic measurements indicative of impaired nutrition: height/weight ratio is lower (Morgan, 1981), muscle mass estimated by creatinine/height index is reduced (Mendenhall *et al.*, 1984; Morgan, 1981), and triceps skin folds are thinner (Mendenhall *et al.*, 1984; Morgan, 1981; Simko *et al.*, 1982).

Continued drinking results in weight loss, whereas abstinence results in weight gain (World *et al.*, 1984a,b) in patients with and without liver disease (Simko *et al.*, 1982).

Many patients who drink to excess are either not malnourished, or are less malnourished than the group hospitalized for medical problems. Those with moderate alcohol intake (Bebb *et al.*, 1971), even those admitted to hospital for alcohol rehabilitation (Neville *et al.*, 1968) rather than for medical problems, hardly differ nutritionally from controls (matched for socioeconomic and health history) except that the females have a lower level of thiamin saturation (Neville *et al.*, 1968).

The wide range in nutritional status of our alcoholic population may reflect differences in what they eat. Moderate alcohol intake, alcohol accounting for 16% of total calories (alcohol included), is associated with slightly increased total energy intake (Gruchow *et al.*, 1985). Perhaps because of the energy considerations already discussed, this group with higher total caloric intake has no weight gain despite physical activity levels comparable to those of the non-alcohol-consuming population. This level of alcohol intake, and even slightly higher levels (23%) (Hillers and Massey, 1985), are associated with a substitution of alcohol for carbohydrate in the diet. In those individuals consuming more than 30% of total calories as alcohol, significant decreases in protein and fat intake occur too, and their intake of vitamins A, C, and thiamin may descend below the recommended daily allowances (Gruchow *et al.*, 1985). Calcium, iron, and fiber intake are also lowered (Hillers and Massey, 1985).

The mechanisms underlying the altered pattern of food intake are not exactly known. Suppression of appetite was postulated (Westerfeld and Shulman, 1959) but has not been studied much. Depressed consciousness during inebriation, hangover, and gastroduodenitis caused by ethanol partly explain the decreased food intake.

In summary, alcohol intake is associated with a wide spectrum of nutritional states; the vast majority of the alcohol-consuming public has slight if any detectable impairment, whereas those hospitalized for medical complications may be severely malnourished. The contribution of subtle nutritional alterations produced by ethanol to the pathogenesis of ethanol-induced or other disease states is largely undetermined. There is no proof that nutritional deficiency causes alcoholism, and nutrition therapy *per se* for alcoholism has not been successful (Hillman, 1974).

17.4. Water-Soluble Vitamins

17.4.1. Thiamin

Thiamin deficiency is most often present in alcoholics in our society and is responsible for Wernicke–Korsakoff syndrome and beriberi heart disease and possibly contributes to polyneuropathy. When prolonged thiamin deprivation was induced in the rhesus monkey, neuroanatomical lesions comparable to the human syndrome were produced without the need for concomitant alcohol intake (Witt and Goldman-Rakic, 1983a,b). There has been no confirmation that Wernicke–Korsakoff syndrome is associated with an inborn abnormality in transketolase affinity for coenzyme (Blass and Gibson, 1977).

Thiamin intake will surely be insufficient for those alcoholics who rely on alcoholic beverages for most of their energy needs. However, when obvious deficiency symptoms are not present, it is difficult for the clinician to assess thiamin status. Blood thiamin concentration is not often used clinically; erythrocyte transketolase activity is most often measured, followed by the increased activity on addition *in vitro* of its cofactor, thiamin pyrophosphate (TPP). An exaggerated increase in transketolase activity has been considered useful for recognizing thiamin deficiency by some but not all workers. For example, when subjects were refed thiamin, a clear increase in urinary thiamin was noted, and the TPP effect diminished (Somogyi, 1976). However, some found that a great proportion of obviously thiamin-deficient subjects did not demonstrate a TPP effect (Camilo *et al.*, 1981). They postulate that profound thiamin deficiency, alcohol (Bitsch *et al.*, 1982), or liver disease (Fennelly, *et al.*, 1967) may affect the levels of apoenzyme transketolase or its binding to cofactor and thus prevent the TPP effect. Others found that decreased erythrocyte transketolase activity alone correlated best with thiamin deficiency in patients with Wernicke's encephalopathy and was more frequently abnormal than the TPP effect when the entire group of alcoholics was considered (Wood *et al.*, 1977). The apparent abnormality of apotransketolase present in those with an absent TPP effect was infrequently linked to liver disease; it was postulated to reflect thiamin deficiency *per se* (Wood *et al.*, 1977). In any case, in experimental animals (Shaw *et al.*, 1981) and in well-nourished alcoholics who take in normal or greater amounts of thiamin, thiamin levels in the organs are maintained (Hoyumpa, 1983), and there is no abnormality in the relative amounts of phosphorylated species of thiamin (Dancy *et al.*, 1984).

The ability of alcoholics to absorb thiamin, when admitted to hospital, is impaired relative to control patients (Tomasulo *et al.*, 1968) when measured by radioactive thiamin excretion, an admittedly crude test affected by steps other than absorption. Even though stool weights were normal in those studies, folic acid deficiency (Hoyumpa, 1983) could have contributed to malabsorption. Using refined testing, Thomson and Majumdar (1981) still saw reduced thiamin absorption with alcohol, but only in a minority of subjects. Jejunal perfusion studies (Breen *et al.*, 1985) could not show an effect of 5% alcohol on thiamin absorption in humans. Interestingly, it was observed that thiamin absorption is lower in middle-aged subjects (alcoholics and controls) than in younger individuals. Studies in rats have progressed further, but in view of the above they may not be completely relevant to human disease. In rodents, thiamin absorption is accomplished by an active system with a low K_m and a passive system with a high K_m (Hoyumpa, 1983). Alcohol interferes with thiamin absorption by the low-concentration active pathway and presumably (if humans are shown to be similarly constituted) could be an important factor for individuals with marginal or low thiamin intake. The effects of alcohol on thiamin activation and storage in the liver are controversial (Hoyumpa, 1983).

Thiamin should be provided to all alcoholics because there is an appreciable incidence of thiamin deficiency in alcoholics; it is difficult to assess clinically any but the most glaring thiamin deficiency syndromes; it is important to reverse early neurological disease; and thiamin replacement is easy and safe. Fifty milligrams per day of thiamin should be given parenterally until oral intake can be established, followed by 50 mg/day orally for weeks, or longer if neurological problems persist. Some have proposed adding thiamin to alcoholic beverages. The feasibility of this approach is indicated by the absorption of thiamin from Sorghum beer despite its 3% alcohol and liver yeast content (Katz *et al.*, 1985).

17.4.2. Riboflavin

When there is a general lack of B-vitamin intake, riboflavin deficiency may be encountered. Deficiency was found in 50% of a small group of patients with medical complications severe enough to warrant hospital admission (Rosenthal *et al.*, 1973). Although none exhibited the classic signs of riboflavin deficiency, they had an abnormal activity coefficient (AC), which returned to normal 2 to 7 days after intramuscular replacement with 5 mg riboflavin daily. Activity coefficient is measured as the ratio of erythrocyte glutathione reductase activity on addition of flavin adenine dinucleotide (FAD) divided by the activity with no additions.

Riboflavin is readily absorbed, excreted in the urine when ingested in excess, and has no described toxicity. It is usually given to alcoholic patients as part of a multivitamin preparation.

17.4.3. Pyridoxine

Neurological, hematological, and dermatological disorders can be caused in part by pyridoxine deficiency. Pyridoxine deficiency, measured as low plasma pyridoxal-5′-phosphate (PLP), was reported in over 50% of alcoholics without hematological findings or abnormal liver function tests (Lumeng and Li, 1974). Inadequate intake may partly explain low plasma PLP, but increased destruction and reduced formation may be important: PLP is more rapidly destroyed in erythrocytes in the presence of acetaldehyde, the first product of ethanol oxidation, perhaps by displacement of PLP from protein and consequent exposure to phosphatase (Lumeng and Li, 1974; Lumeng, 1978). Fairly high levels of acetaldehyde were used, so the significance of the proposed mechanism is uncertain. Studies in dogs that were given pyridoxine and ethanol to achieve blood ethanol levels of 300 mg/dl showed decreased PLP from pyridoxine (Parker et al., 1979). The decreased plasma levels of PLP were ascribed to decreased formation, since removal of PLP from plasma is largely carried out by tissues other than liver, kidney, intestines, and spleen (Lumeng et al., 1984).

Clinical management generally involves provision of pyridoxine in the usual multivitamin dosage unless neuropathy or pyridoxine-responsive anemia has been diagnosed. Since ataxia from sensory neuropathy has been ascribed to toxicity from as little as 200 mg of pyridoxine per day, the indiscriminate use of large doses of this vitamin must be avoided (Schaumberg et al., 1983; Parry and Bredesen, 1985).

17.4.4. Folic Acid

Alcoholics tend to have a low folic acid status when they are drinking heavily and their folic acid intake is reduced. When a group of unselected alcoholics was studied, 37.5% had low serum folate levels, and 17.6% had low red blood cell folate levels (World et al., 1984a). In monkeys folate deficiency can be created by feeding ethanol up to 50% of total calories for over 2 years despite provision of an otherwise adequate diet: hepatic folate is low, and there is evidence for decreased folate absorption (Romero et al., 1981). Malnourished alcoholics without liver disease also absorb folic acid less well than their better-nourished counterparts (Halsted et al., 1971). Folic acid absorption, usually increased by partial starvation, is less increased in rats when alcohol is ingested (Racusen and Krawitt, 1977). It has not been clearly shown, however, that either protein deficiency or alcohol (Halsted et al., 1971; Lindenbaum and Lieber, 1971) decreases folate absorption. Thus, it is still unclear what aspects of malnutrition adversely affect folate absorption and under what clinical circumstances alcohol may interfere with folate absorption.

Alcohol accelerates the production of megaloblastic anemia in patients with depleted folate stores (Lindenbaum, 1982) and suppresses the hematological response to folic acid in folic-acid-depleted patients (Sullivan and Herbert, 1964). Alcohol also has other effects on folate metabolism, but their significance is not clear: alcohol given acutely causes a decrease in serum folate, partly explained by increased urinary excretion (Russell et al., 1983); alcohol given chronically to monkeys decreases hepatic folate levels, partly through the inability of the liver to retain folate (Tamura et al., 1981) and perhaps partly from increased urinary and fecal losses (Tamura and Halsted, 1983).

The clinical approach to folate deficiency without anemia is straightforward. A diet providing adequate folate, perhaps with additional folate, will replete stores in a matter of weeks. If malabsorption persists after this period, evaluation for causes other than folate deficiency should be instituted. When the patient is anemic, the diagnostic evaluation is more complex (Lindenbaum, 1982). In addition to folate deficiency, the direct effect of alcohol on the bone marrow, liver disease, hypersplenism, bleeding, iron deficiency, infection, and the use of anticonvulsants, are all commonly encountered and will exert separate and combined influences on the hematological picture. It should be kept in mind, first, that in well-nourished alcoholics folic acid deficiency is a rare cause of anemia (Eichner et al., 1972), and second, that a search for folic acid deficiency (serum or red cell folate levels) as an explanation for anemia is unwarranted unless some or all of the morphological features of the vitamin deficiency are present (macroovalocytes, hypersegmentation of polymorphonuclear leukocytes, megaloblastosis of the bone marrow). The following sequence has been proposed for the development of folic acid deficiency: negative folate balance (serum folate less than 3 ng/ml); folate depletion (red blood cell folate less than 160 ng/ml); folate-deficient erythropoiesis (red blood cell folate less than 120 ng/ml, neutrophil lobe average greater than 3.5, liver folate less than 1.2 g/g); folate-deficient anemia (low hemoglobin, elevated mean corpuscular volume, macroovalocytosis) (Herbert, 1987). When there is combined iron and folate deficiency, the expression of macrocytosis will be modified, or a dimorphic red blood cell population may occur. Hypersegmentation of leukocyte nuclei and macroovalocytosis may persist for several weeks after folate replacement has been started (Lindenbaum, 1982).

Some have proposed adding folate to alcoholic beverages (Kaunitz and Lindenbaum, 1977), since the taste of the beverage is not altered and vitamin absorption is adequate.

17.4.5. Vitamin B$_{12}$

Alcoholics do not commonly get vitamin B$_{12}$ deficiency. Serum vitamin B$_{12}$ levels usually are normal in patients with folate deficiency, whether they have cirrhosis (Herbert et al., 1963; Klipstein and Lindenbaum, 1965) or not (Halsted et al., 1971; Racusen and Krawitt, 1977). The low prevalence of clinically significant deficiency probably results from the large body stores of vitamin B$_{12}$ and the reserve capacity for absorption, since there are several factors in the context of alcoholism that would favor depletion. Pancreatic insufficiency, for example, results in decreased vitamin B$_{12}$ absorption as measured by the Schilling test. In this situation there is insufficient luminal protease activity and alkalinity, which normally serve to release vitamin B$_{12}$ from r protein, secreted by salivary glands, intestines, and possibly stomach (Herzlich and Herbert, 1986). Alcohol ingestion has also been shown to decrease vitamin B$_{12}$ absorption in volunteers after several weeks of intake (Lindenbaum and Lieber, 1975). The alcohol effect may be in the ileum, since coadministration of intrinsic factor or pancreatin does not correct the Schilling test results. Whether or not the binding of intrinsic factor–vitamin B$_{12}$ complex to ileal sites is abnormal remains controversial (Findlay et al., 1976; Lindenbaum et al., 1973).

17.4.6. Vitamin C

The vitamin C status of alcoholic patients (mostly admitted to hospital) is lower than those of non-alcoholics as measured by serum ascorbic acid, peripheral leukocyte ascorbic acid, or urinary ascorbic acid after an oral challenge (Bonjour, 1979). In addition to a lower mean ascorbic acid level for the alcoholic group compared to controls, some 25% of patients with Laennec's cirrhosis had serum ascorbic acid levels below the range of healthy controls (Bonjour, 1979). Ascorbic acid status is low in alcoholic patients with and without liver disease. When alcohol exceeds 30% of total calories, vitamin C intake generally falls below recommended dietary allowances (Gruchow et al., 1985). Inadequate ascorbic acid intake provided only a partial explanation for low ascorbic acid status. For patients who are not clearly scorbutic, it is unknown if their low ascorbic acid levels have clinical significance. Daily supplementation with 175 to 500 mg of ascorbic acid may be necessary for weeks to months to restore plasma ascorbate and urinary ascorbate to normal (Bonjour, 1979).

17.5. Fat-Soluble Vitamins

17.5.1. Vitamin A

The interaction of alcoholism with vitamin A is especially interesting because it involves the intake and possibly the absorption of the vitamin and its metabolism, and there is evidence that alcohol may modulate the role of vitamin A in hepatotoxicity and carcinogenesis.

Ingestion of vitamin A is not significantly below normal for Americans imbibing up to a mean of 400 calories of alcohol per day (or less than 20% of total calories) Gruchow et al., 1985), since the vitamin A density of the nonalcoholic portion of the diet is similar to that ingested by control populations. Americans consuming 24% of calories as alcohol, however, ingest 75% of the RDA for vitamin A (Hillers and Massey, 1985). Probably those with intense alcohol intake, 50% of calories or more as alcohol, eat even less vitamin A. Chilean wine drinkers eat only 25% of the RDA for vitamin A when they drink 150 g of alcohol daily (Bunout et al., 1983). Regular alcohol consumption in elderly American men is associated with lower vitamin A intake (Barboriak et al., 1978). In a single study (Althausen et al., 1960) vitamin A absorption was inhibited (17% reduction) by 120 ml of wine. Vitamin A absorption may be substantially reduced as a result of chronic alcoholic pancreatitis.

The effect of acute alcohol ingestion on blood vitamin A levels has been variously reported as unchanged in humans (Russell et al., 1979), increased in the dog (Lee and Lucia, 1965), and increased as retinol-bound lipoproteins in the rat (Sato and Lieber, 1980). The effect of chronic ethanol consumption on hepatic vitamin A has been consistent and profound: hepatic vitamin A stores are decreased whether dietary vitamin A is low or high. Rodents fed alcohol chronically 5 g/kg per day had a 20% decrease in the liver (Nadkarni *et al.*, 1979). Higher alcohol intakes, 36% of calories or about 14 g/kg per day, decreased hepatic vitamin A by 60% in 4 to 6 weeks and by 72% in 7 to 9 weeks with no change in serum vitamin A or retinol-binding protein (RBP) and an increase in hepatic retinol-binding protein (Sato and Lieber, 1980). A fivefold increase in dietary vitamin A did not prevent hepatic depletion by alcohol. When baboons were fed 50% of calories as alcohol, they showed a 60% decrease in hepatic vitamin A after 4 months and a 95% decrease in 24 to 84 months (Sato and Lieber, 1980). Hepatic vitamin A levels show progressive decrease with increasing severity of lesions to include cirrhosis (Leo and Lieber, 1982) (Fig. 17.4) in humans.

Enhancement of hepatic vitamin A degradation by alcohol consumption is one likely explanation for vitamin A depletion. Vitamin A degradation via metabolism

of retinoic acid to 4-hydroxy- and 4-oxoretinoic acid and other polar metabolites is catalyzed by microsomal enzymes inducible by ethanol consumption (Sato and Lieber, 1982b), but these enzymes are not active enough to deplete vitamin A stores. A newly discovered microsomal pathway for oxidation of retinol to polar metabolites (Leo and Lieber, 1985) also is inducible by alcohol (Leo *et al.*, 1986) and is a more likely candidate for vitamin A depletion. In addition, alcohol promotes vitamin A mobilization from the liver (Leo *et al.*, 1986).

One of the clinical consequences of altered vitamin A status is night blindness arising from low tissue vitamin A. Abnormal dark adaptation occurred in 15% of alcoholics without cirrhosis and 50% with cirrhosis (Bonjour, 1981). A serum vitamin A level of 1.4 µmol/liter or more seemed to exclude retinal dysfunction with 95% confidence (Carney and Russell, 1980). The correlation of serum vitamin A with tissue stores is complicated by the presence of liver disease and protein deficiency.

The effects of vitamin A status and its related pathology are widespread throughout the body and are affected by ethanol consumption. Hepatotoxicity from diminished vitamin A includes the presence of multivesicular lysosomes and is potentiated by concomitant alcohol intake (Leo *et al.*, 1983). Hepatotoxicity of increased vitamin A also includes fibrosis (Leo and Lieber, 1983a,b) and is potentiated by concomitant alcohol (Leo *et al.*, 1982). Ethanol increases squamous metaplasia from vitamin A depletion in rodent trachea (Mak *et al.*, 1984) and increases vitamin A in lung and esophagus (Lieber, 1987). The role of alcohol in increasing vitamin A content of some tissues while having the opposite effect in other tissues, speeding or altering the conversion of vitamin A to metabolites, may have important consequences for the hepatotoxicity, fibrosis, and squamous metaplasia described above. Alcohol-mediated changes in vitamin A status may be of importance for the association of vitamin A or carotene levels with malignancies (Anonymous, 1985a; Lieber, 1987).

Several factors make therapy a complicated affair in the setting of alcoholism: assessment of tissue stores of vitamin A is difficult; vitamin A in high doses is toxic; even usual doses of vitamin A are potentially toxic concomitant with continued intake of alcohol and monitoring vitamin A hepatotoxicity is difficult in the presence of continued alcohol intake. Vitamin A replacement should therefore be modest for patients who cannot be assured an alcohol- and drug-free environment. Replacement of vitamin A should be considered for those who are confirmed as deficient and who can be assured abstinence from alcohol. Night blindness (or abnormal

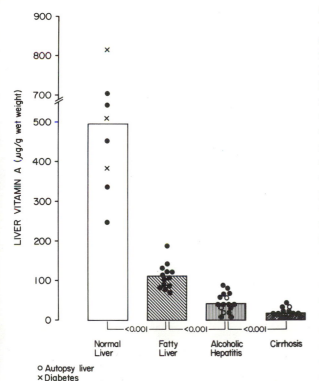

FIGURE 17.4. Hepatic vitamin A levels in subjects with normal livers and various stages of alcoholic liver injury. To convert vitamin A values to micromoles/g, multiply by 0.003491. Figures below the graph denote *P* values. (From Leo and Lieber, 1982.)

dark adaptation) with low serum vitamin A (<30 µg/dl or 1.4 µM) is evidence of deficiency. Vitamin A at dosage of 10,000 µg/day for several weeks should provide an adequate trial. A low serum zinc (<80 µg/dl) can be treated with 660 µg $ZnSO_4$ per day (*vide infra*). Zinc therapy might also be tried subsequent to failure of vitamin A treatment. Documented fat malabsorption should prompt parenteral replacement of vitamin A. These recommendations seem reasonable but are not based on extensive clinical trials.

17.5.2. Vitamin D

It is clear that alcoholics have illness related to abnormalities of calcium, phosphorus, and vitamin D homeostasis. They have decreases in bone density (Saville, 1965) and bone mass (Gascon-Barré, 1985), increased susceptibility to fractures (Nilsson, 1970), and increased osteonecrosis (Solomon, 1973). Low blood calcium, phosphorus, magnesium, and 25-OH-vitamin D have been reported, indicating disturbed calcium metabolism (Gascon-Barré, 1985).

In patients with alcoholic liver disease, vitamin D deficiency probably derives from too little vitamin D substrate, which results from poor dietary intake, malabsorption because of cholestasis and/or pancreatic insufficiency, and insufficient sunlight. The physiological sequence is conversion of 7-dehydrocholesterol of animal origin to cholecalciferol (vitamin D_3) by photolysis, followed by hydroxylation to 25-hydroxycholecalciferol (25-OH-vitamin D_3) by the liver and finally hydroxylation to the most active compound, 1,25-dihydroxycholecalciferol [1,25-$(OH)_2$-vitamin D_3] by kidney, bone, and placenta. Cholecalciferol absorption depends on intact absorption of fat. The status of 25-hydroxylation is adequate even in advanced liver disease (Gascon-Barré, 1985; Long, 1983; Posner *et al.*, 1978), although some disagree (Jung *et al.*, 1978). There is a lower concentration of vitamin-D-binding globulin, a protein synthesized in the liver, in alcoholic liver disease (Long, 1983), but because of its excess of binding sites its decrease is an unlikely cause of 25-OH-vitamin D_3 deficiency.

Osteomalacia in patients with liver disease should be treated by increasing intake of vitamin D_3, ultraviolet irradiation therapy, and correction of fat malabsorption to keep plasma calcium, phosphorus, and 25-OH-D normal.

17.5.3. Vitamin K

Vitamin K deficiency in alcoholism arises when there is an interruption of fat absorption as a result of pancreatic insufficiency, biliary obstruction, or intestinal mucosal abnormality secondary to folic acid deficiency. Inadequacy of dietary vitamin K is not a likely cause of clinical deficiency unless there is concomitant sterilization of the large gut, a reliable source of the vitamin. Alcohol-induced hepatocyte injury interferes with utilization of available vitamin K with a consequent drop in blood levels of clotting factors whose syntheses depend on this vitamin: II, VII, IX, X. Vitamin K serves as a cofactor for the microsomal carboxylase reaction that results in posttranslational modification of these proteins, the conversion of glutamic acid (Glu) residues to γ-carboxyglutamic acid (Gla) residues, necessary for function. Abnormally high levels of inactive factor II (prothrombin) are found in the plasma (Blanchard *et al.*, 1981) in the presence of cirrhosis or vitamin K deficiency. If there is doubt as to whether hepatocellular dysfunction or lack of availability of vitamin K to the liver is responsible for low vitamin-K-dependent clotting factors in the blood, a trial of intramuscular vitamin K is helpful (Roberts and Cederbaum, 1972).

17.5.4. Vitamin E and Selenium

Vitamin E and selenium serve protective roles as antioxidants and interact physiologically (Bieri *et al.*, 1983; Scott, 1980; Martin, 1985; Levander and Burk, 1986). Vitamin E is a powerful antioxidant that prevents peroxidation of cellular and subcellular membrane phospholipids. Selenium is also involved in antioxidant functions and is a component of red blood cell glutathione peroxidase. Vitamin E and selenium function synergistically: vitamin E reduces selenium requirement, prevents its loss from the body, and maintains it in an active form; selenium spares vitamin E and reduces the requirement for the vitamin (Martin, 1985).

Thus far, vitamin E deficiency is not a recognized complication of alcoholism but has occurred in adults with diverse causes of fat malabsorption (Bieri *et al.*, 1983) and primary biliary cirrhosis (Knight *et al.*, 1986). The clinical manifestations included decreased erythrocyte survival, neurological disturbances (areflexia, gait disturbance, decreased proprioception and vibratory sensation, ophthalmoplegia). It would not be surprising if vitamin E deficiency were to be found in patients with severe steatorrhea from pancreatic insufficiency or prolonged cholestatic states of alcoholic liver injury. Indeed, in alcoholic patients with pancreatic insufficiency and fat malabsorption, low serum vitamin E levels were recently reported (Kalvaria *et al.*, 1986) and alcohol consumption was shown to be associated with a lowering of hepatic vitamin E in rats (Kawase *et al.*, 1989).

Selenium metabolism is of great theoretical interest to hepatologists in view of the proposed lipoperoxidative mechanisms of drug- and alcohol-induced liver injury (Lieber, 1987). Interestingly, serum selenium levels have been recorded as low in the alcoholic, especially in the presence of liver disease, but this may be a consequence of liver injury (Dworkin *et al.*, 1985; Korpela *et al.*, 1985a,b; Lieber, 1987; Shah *et al.*, 1985; Valimaki *et al.*, 1983), since other, nonalcoholic patients with liver disease also have low levels. No recommendations for dietary modifications of vitamin E or selenium intake in alcoholism are warranted.

17.6. Minerals and Electrolytes

17.6.1. Salt and Water Retention of Cirrhosis

Alcoholics with chronic liver disease often have disorders of water and electrolyte balance. Sodium and water retention are clinically apparent as peripheral edema, ascites, and pleural effusions. Patients may have weight gain, increasing abdominal girth, ankle swelling, and respiratory difficulties or umbilical herniation as further complications. Not only is sodium retained avidly, but a water load cannot be excreted normally (Gabuzda, 1970a) (see Chapter 16). Low body potassium may result from vomiting, diarrhea, hyperaldosteronism, muscle wasting, renal tubular acidosis, and diuretic therapy. Potassium depletion may contribute to the appearance of renal vein ammonia and worsen hepatic encephalopathy (Shear and Gabuzda, 1970).

The pathogenesis of fluid retention and ascites is complex. At the hepatic level, portal hypertension, hypoalbuminemia, and alterations of lymph flow are important factors for ascites formation (Summerskill *et al.*, 1970). Many of the endocrine accompaniments suggest that the body is reacting to a diminished "effective circulating volume" (total blood volume is normal or elevated,but a disproportionately large fraction is sequestered in the splanchnic region) (Epstein, 1982): hyperreninemia, hyperaldosteronemia, increased blood norepinephrine, reversal of salt and water retention by restoration of nonsplanchnic volume via head-out body immersion in water or via peritoneal–venous (LeVeen) shunting of ascites. Of additional pathogenetic significance may be an insufficiency of factors that promote salt loss by the kidneys, such as atrial natriuretic factor (Epstein, 1986), and an abnormality of renal hemo-

dynamics based in part on alterations in renal prostaglandins (Epstein, 1986).

Patients with cirrhosis and fluid overload may require urgent relief, as when ascites and pleural effusion are causing respiratory difficulties or when imminent rupture of an umbilical hernia may result in lethal peritonitis. Thoracentesis and/or paracentesis should be performed promptly. Usually, however, there is no rush to diurese patients once a diagnostic tap has shown the fluid to be a noninfected transudate. Treatment is eventually aimed at preventing recurrence of fluid retention. Dietary management combines sodium and water restriction. It is difficult to provide a palatable diet on a long-term basis with less than 0.5 to 1 g of sodium and 1500 ml of total fluid daily. At least these amounts are recommended with addition of spironolactone, followed, if necessary, by small doses of diuretics (hydrochlorthiazide or furosemide) to achieve an initial daily weight loss of no more than 0.5 kg (Boyer, 1986). More rapid weight loss is probably safe when the patient has mobilizable peripheral edema and when the patient can be observed carefully (Boyer, 1986). Accelerated diuresis risks renal failure. Use of prostaglandin inhibitors such as nonsteroidal antiinflammatory drugs (NSAID) carries the potential risk of altering renal hemodynamics and precipitating renal failure. Careful monitoring for the development of hypokalemia (or hyperkalemia), hyponatremia, and renal failure must be undertaken. For patients in whom a reasonable program of salt and water restriction and diuretic therapy is not successful, a peritoneal–venous LeVeen shunt may be useful (LeVeen, 1985; Smajdja and Franco, 1985; Wapnic *et al.*, 1979). Best results have been obtained for patients without encephalopathy, coagulopathy, or severe jaundice.

17.6.2. Zinc

Zinc is an essential element for humans. Deficiency has resulted in growth retardation, male hypogonadism, rough dry skin, disordered taste, poor appetite, and mental lethargy. Zinc is absorbed from the small bowel. In the rat the ileum has the greatest capacity for absorption (Antonson *et al.*, 1979) and appears to increase absorption when zinc deficiency is produced experimentally (Smith *et al.*, 1978). Among the enzymes that contain zinc are hepatic alcohol dehydrogenase (which converts ethanol to acetaldehyde), ocular retinol dehydrogenase (which converts retinol to retinal), and hepatic and erythrocyte superoxide dismutases (which serve to protect against oxidative damage). It is not surprising that zinc deficiency induced in rats results in poor growth and lowered hepatic alcohol dehydrogenase, with diminished rate of elimination of a test dose of ethanol (Anonymous, 1985b; Huber and Gershoff, 1975). A diminished capacity for conversion of retinol to retinal was found in the retinas of zinc-deficient rats (Huber and Gershoff, 1975). Hypogonadism in zinc-deficient rats results from Leydig cell failure rather than from pituitary changes (McClain et al., 1984).

Alcoholic cirrhosis is associated with abnormalities of zinc homeostasis, although the clinical implications are uncertain. Patients have low plasma zinc (Vallee et al., 1956), low liver zinc (Vallee et al., 1957), and an increase in urinary zinc (Sullivan, 1962b; Vallee et al., 1957). Acute ethanol ingestion, however, does not cause zincuria (Sullivan, 1962a). The low zinc content of chronic alcoholics with cirrhosis is thought to be caused by decreased intake and decreased absorption as well as increased urinary excretion. Many Americans have a diet that is marginal in zinc (Sandstead, 1973). Alcoholics fall into several of those groups with marginal intake. It is interesting that zinc absorption was shown to be low in alcoholic cirrhotics but not in cirrhosis of other etiologies (Valberg et al., 1985). Some instances of night blindness not fully responsive to vitamin A replacement (see Section 17.5.1) have responded to zinc replacement. The possibility that human hypogonadism of alcoholism may involve perturbations of vitamin A and zinc interactions has not been studied.

Currently the therapeutic use of zinc in alcoholism is restricted to the treatment of night blindness not responsive to vitamin A. Sullivan and his colleagues (Sullivan, 1962a,b; Sullivan et al., 1979) could not raise serum zinc in patients with alcoholic cirrhosis: zinc was increased in the urine.

17.6.3. Magnesium

Chronic alcoholism is associated with magnesium deficiency (Flink, 1986): alcoholics have low blood magnesium and low body exchangeable magnesium; symptoms in alcoholics resemble those in patients with magnesium deficiency of other etiologies; alcohol ingestion causes magnesium excretion; on withdrawal from alcohol magnesium balance is positive; hypocalcemia in alcoholics may be responsive only to magnesium repletion. Transient hypoparathyroidism may account for magnesiuria during acute alcoholism (Laitinen et al., 1991). The correlation of magnesium content of blood with that of other tissues and with the severity of clinical symptoms in individual cases is imperfect,

although it is statistically obvious among groups. Magnesium replacement should be seriously considered for symptomatic patients with measurably low serum magnesium, for anorectic patients with low serum magnesium, and for hypocalcemic alcoholics not responsive to calcium replacement. The majority of alcoholics will replete body stores of magnesium readily from normal dietary sources.

17.6.4. Iron

Alcoholics may be iron-deficient as a result of the several gastrointestinal lesions to which they are prone, which may bleed (esophagitis, esophageal varices, gastritis, duodenitis). Iron overload of the liver was described in Bantus who consumed alcoholic beverages prepared in iron containers, which thereby contributed a large amount of elemental iron to their diet. In most alcoholics, the iron content of the liver is normal or only modestly elevated, although there may be stainable iron in reticuloendothelial cells, possibly from hepatic necrosis or bouts of hemolysis. It is unclear whether increased intestinal absorption of iron caused by alcohol (Chapman et al., 1983a) or hepatic uptake of iron from serum in established alcoholic liver disease (Chapman et al., 1983b) contributes significantly to an increase in hepatic iron. Iron therapy should be restricted to clear states of deficiency.

Alcoholism has been reported to result in qualitative changes in transferrin (Stibler et al., 1978, 1980), the serum transport protein for iron: a higher fraction of molecules bear a reduced sialic acid content. This provides a useful test for chronic alcohol consumption (Behrens et al., 1988a,b). The synthesis of transferrin is decreased at the stage of alcoholic cirrhosis, and so is serum transferrin concentration (Potter et al., 1985). At the stage of alcoholic fatty liver the serum transferrin concentration is normal, although both catabolic rate and presumably synthesis are increased (Potter et al., 1985). The significance of these changes in transferrin are not yet apparent.

17.7. Nutritional Therapy

The nutritional therapy of alcoholism is directed at the prevention of illness caused by alcoholism, the treatment of documented or presumed deficiencies, and the management of complications of alcoholism. As has been discussed above, individuals consuming over 30% of total calories as alcohol have a high probability of ingesting less than the recommended daily amounts of carbohydrate, protein, fat, vitamins A, C, and B (especially thiamin), and minerals such as calcium and iron. It is sensible to recommend a complete diet comparable to that of nonalcoholics to forestall deficiency syndromes, although we do not know if organ damage from direct toxicity of alcohol (e.g., alcoholic liver disease) can thereby be prevented. The feasibility and desirability of adding thiamin and perhaps folic acid to alcoholic beverages has been discussed above but not yet done.

The management of observed deficiencies of protein and calories is straightforward in the absence of organ damage. The treatment of gross malnutrition of proteins and calories in the context of severe acute and chronic liver disease is discussed below. Nervous system damage from thiamin lack is serious and treatable with a great margin of safety; therefore, thiamin deficiency should be presumed if not definitely disproved. Parenteral therapy with 50 mg of thiamin per day should be given until similar doses can be taken by mouth. Riboflavin and pyridoxine should be routinely given at the dosages usually contained in standard multivitamin preparations. Adequate folic acid replacement can be accomplished with the usual hospital diet. Additional replacement is optional unless deficiency is severe. Vitamin A replacement should only be given for well-documented deficiency and to patients whose abstinence from alcohol is assured (see discussion in Section 17.5.1 of hepatotoxicity of hypervitaminosis A with alcohol). Vitamin A at dosages of 10,000 µg/day may then be given. Zinc replacement should be given for night blindness unresponsive to vitamin A replacement. Magnesium replacement is recommended for symptomatic patients with low serum magnesium. Iron deficiency that has been clearly diagnosed may be replaced in the usual manner orally.

The treatment of the complications of alcoholism is given in full detail in appropriate other chapters of this book. Wernicke–Korsakoff syndrome requires at least 50 mg of thiamin daily (parenterally if necessary) for prolonged periods. Beriberi heart failure responds quickly to thiamin. Peripheral nerve damage will necessitate months or years of B-vitamin therapy. Acute pancreatitis may require withholding oral feeding for prolonged periods, during which time central venous alimentation must be given. Chronic pancreatic exocrine deficiency is treated by dietary manipulation (often decreases in fat) with oral pancreatic enzymes at mealtime.

The nutritional management of acute and chronic liver disease caused by alcoholism is being actively investigated with the aim of defining feeding programs to reverse malnutrition, ameliorate liver disease, and decrease mortality without promoting hepatic encephalopathy. A diet providing 25 to 35 kcal and 1 to 1.25 g of protein per kilogram of ideal body weight, not unlike that advocated for healthy adults, should be attempted (Gabuzda, 1970b; Soberon et al., 1987). Lipid should constitute 30% to 35% of calories, and carbohydrate 40%. If patients with alcoholic hepatitis cannot achieve these intakes from the trays offered to them, it is uncertain whether force feeding by stomach tube or parenteral routes is desirable. From the evidence reviewed below, it seems likely that the severely anorectic patient will benefit from nasogastric feeding with an energy-dense mixture such as Isocal-HCN or from intravenous supplementation with a standard amino acid mixture and glucose. The usefulness of proteins or amino acids of special composition is much less certain.

Soberon et al. (1987) treated a small group of anorectic patients with alcoholic hepatitis by giving nasogastric feedings containing Isocal-HCN to provide 35 kcal/kg ideal body weight per day. Values exceeding baseline intakes of energy and protein were thus accomplished, nitrogen balance was more positive, and percentage intestinal absorption of energy, fat, and protein shifted toward normal. Encephalopathy was not affected. The study was too small to reveal anything about changes in liver function or mortality. These findings extend the observations of others (Smith et al., 1982; Mendenhall et al., 1985) that patients with acute and chronic liver disease can often tolerate large amounts of protein (80 to 140 g/day) and show improvement of nutritional parameters.

Parenteral therapy using conventional mixtures of amino acids to provide at least 80 g of protein equivalent per day have been fairly safe as far as promotion of hepatic encephalopathy is concerned. Nasrallah and Galambos (1980) showed an improvement in survival, but others could show only improved nitrogen balance (Diehl et al., 1985) or some aspects of liver function (Naveau et al., 1986). For example, Naveau et al. (1986) found only an improved serum bilirubin after 28 days in a group receiving 40 kcal/g and 200 mg N/kg by central vein in addition to the 40 kcal/kg and 40 g protein they were offered by mouth, compared to a prospectively studied group of patients offered only oral intake. Calvey et al. (1985) randomized patients, regardless of encephalopathy, to receive a controlled diet alone or supplements consisting of 2000 kcal and 10 g N per day orally, nasogastrically or intravenously as necessary. Although patients receiving 10 g N or more per day achieved positive nitrogen balance, in the absence of renal failure, the supplemented group was not benefited in terms of nutritional parameters or survival.

The treatment of hepatic encephalopathy is discussed in detail elsewhere in this book. Dietary aspects of management consist of diminishing protein intake to about 20 g of protein per day for a patient who can eat but to 0 g in the comatose. Such drastic reductions can only be maintained for a day or two. Every attempt must be made to provide sufficient calories to prevent catabolism of endogenous protein, which would contribute to encephalopathy and is deleterious in its own right. Dietary protein is raised in a stepwise fashion, with continual monitoring for return of encephalopathy, until a satisfactory protein intake is achieved (0.5 to 1.2 g/kg ideal body weight per day). As discussed above, if anorexia precludes adequate intake of energy and nitrogen, nasogastric feeding or parenteral infusions should be given.

Provision of dietary protein of unusual quality, such as vegetable protein, which is lower in aromatic amino acids than casein, for example, or the use of keto analogues of essential amino acids to lower the overall nitrogen load while providing the carbon skeletons of these amino acids, or the use of oral (*Hepatic Aid*) or intravenous (*F080*) amino acid mixtures high in branched-chain amino acids and low in aromatic amino acids, have not been consistently more successful in well-controlled studies than standard mixtures. Although Greenberger *et al.* (1977) claimed an advantage for vegetable protein, others (De Bruin *et al.*, 1974) could not confirm this. Similarly, in patients given lactulose, Shaw *et al.* (1983) found no advantage (in terms of encephalopathy) for dietary proteins of vegetable origin; in fact, the vegetable proteins were less well tolerated and led to more negative nitrogen balance. Some studies (Maddrey *et al.*, 1976) showed improved mental status with keto analogues of essential amino acids, but did not show improved nitrogen balance. Of the better-controlled studies utilizing mixtures high in branched chain and low in aromatic amino acids, beneficial effects have been claimed (Egberts *et al.*, 1985; Horst *et al.*, 1984) and denied (Errikson *et al.*, 1982; Mendenhall *et al.*, 1985; McGhee *et al.*, 1983). Thus, mixtures such as *Hepatic Aid* and *F080*, although theoretically attractive because they represent a partial reversal of the blood amino acid profile seen in hepatic encephalopathy (increased aromatic amino acid concentrations and lowered branched-chain amino acids), are, however, more expensive, inadequately tested against appropriate amino

acid or protein controls in the presence of comparable energy sources, and are deficient in cystine, cysteine, and glutathione (Chawla *et al.*, 1984). They have induced decrease in plasma tyrosine and cystine of unknown significance (Millikan *et al.*, 1983).

REFERENCES

Althausen, T. L., Uyeyama, K., and Loran, K.: Effects of alcohol on absorption of vitamin A in normal and in gastrectomized subjects. *Gastroenterology* **38**:942–945, 1960.

Anonymous: Vitamin A and cancer. *Lancet* **2**:325–326, 1985a.

Anonymous: Zinc deficiency impairs ethanol metabolism. *Nutr. Rev.* **43**:158–159, 1985b.

Antonson, D. L., Barak, A. J., and Vanderhoof, J. A.: Determination of the site of zinc absorption in rat small intestine. *J. Nutr.* **109**:142–147, 1979.

Barboriak, J. J., Rooney, C. B., Leitschuh, T. H., and Anderson, A. J.: Alcohol and nutrient intake of elderly men. *J. Am. Diet. Assoc.* **72**:493–495, 1978.

Bebb, H. T., Houser, H. B., Witschi, J. C., Littell, A. S., and Fuller, R. K.: Calorie and nutrient contribution of alcoholic beverages to the usual diets of 155 adults. *Am. J. Clin. Nutr.* **24**:1042–1052, 1971.

Behrens, U. J., Worner, T. M., Braly, L. F., Schaffner, F., and Lieber, C. S.: Carbohydrate-deficient transferrin (CDT), a marker for chronic alcohol consumption in different ethnic populations. *Alcoholism: Clin. Exp. Res.* **12**:427–432, 1988a.

Behrens, U. J., Worner, T. M., and Lieber, C. S.: Changes in carbohydrate deficient transferrin (CDT) levels after alcohol withdrawal. *Alcoholism: Clin. Exp. Res.* **12**:539–544, 1988b.

Bieri, J. G., Corash, L., and Hubbard, V. S.: Medical progress: Medical uses of vitamin E. *N. Engl. J. Med.* **308**:1063–1071, 1983.

Bitsch, R., Hansen, J., and Hötzel, D.: Thiamin metabolism in the rat during long-term alcohol administration. *Int. J. Vitam. Nutr. Res.* **52**:126–133, 1982.

Blanchard, R. A., Furie, B. C., Jorgensen, M., Kruger, S. F., and Furie, B.: Acquired vitamin K-dependent carboxylation deficiency in liver disease. *N. Engl. J. Med.* **305**:242–248, 1981.

Blass, J. P., and Gibson, G. E.: Abnormality of a thiamin-requiring enzyme in patients with Wernicke–Korsakoff syndrome. *N. Engl. J. Med.* **297**:1367–1370, 1977.

Blomstrand, R., Löf, A., and Osterling, H.: Studies on the metabolic effects of long-term administration of ethanol and 4-methylpyrazole in the rat. Vitamin A depletion in the liver. *Nutr. Metab.* **21**(Suppl. 1):148–151, 1977.

Bonjour, J. P.: Vitamins and alcoholism 1. Ascorbic acid. *Int. J. Vitam. Nutr. Res.* **49**:434–441, 1979.

Bonjour, J. P.: Vitamin and alcoholism. *Int. J. Vitam. Nutr. Res.* **51**:166–177, 1981.

Boyer, T. D.: Editorial. Removal of ascites: What's the rush? *Gastroenterology* **90**:2022–2023, 1986.

Breen, K. J., Buttigieg, R., Iossifidis, S., Lourensz, C., and Wood, B.: Jejunal uptake of thiamin hydrochloride in man: Influence of alcoholism and alcohol. *Am. J. Clin. Nutr.* **42:**121–126, 1985.

Bunout, D., Gattás, V., Iturriaga, H., Perez, C., Pereda, T., and Ugarte, G.: Nutritional status of alcoholic patients: Its possible relationship to alcoholic liver damage. *Am. J. Clin. Nutr.* **38:**469–473, 1983.

Calvey, H., Davis, M., and Williams, R.: Controlled trial of nutritional supplementation, with and without branched chain amino acid enrichment, in treatment of acute alcoholic hepatitis. *J. Hepatol.* **1:**141–151, 1985.

Camilo, M. E., Morgan, M. Y., and Sherlock, S.: Erythrocyte transketolase activity in alcoholic liver. *Scand. J. Gastroenterol.* **16:**273–279, 1981.

Carney, E. A., and Russell, R. M.: Correlation of dark adaptation test results with serum vitamin A levels in diseased adults. *J. Nutr.* **110:**552–557, 1980.

Chapman, R. W., Morgan, M. Y., Bell, R., and Sherlock, S.: Hepatic iron uptake in alcoholic liver disease. *Gastroenterology* **84:**143–147, 1983a.

Chapman, R. W., Morgan, M. Y., Boss, A. M., and Sherlock, S.: Acute and chronic effects of alcohol on iron absorption. *Dig. Dis. Sci.* **28:**321–327, 1983b.

Chawla, R. K., Lewis, F. W., Kutner, M. H., Bate, D. M., Roy, R. G., and Rudman, D.: Plasma cysteine, cystine, and glutathione in cirrhosis: *Gastroenterology* **87:**770–776, 1984.

Crouse, J. R., and Grundy, S. M.: Effects of alcohol on plasma lipoproteins and cholesterol and triglyceride metabolism in man. *J. Lipid Res.* **25:**486–496, 1984.

Dancy, M., Evans, G., Gaitonde, M. K., and Maxwell, J. D.: Blood thiamin and thiamin phosphate ester concentrations in alcoholic and non-alcoholic liver diseases. *Br. Med. J.* **289:**79–82, 1984.

De Bruijn, K. M., Blendis, L. M., Zilm, D. H., Carl, P. L., and Andersen, Y. H.: Effect of dietary protein manipulations in subclinical portal-systemic encephalopathy. *Gut* **24:**53–60, 1983.

Diehl, A. M., Boitnott, J. K., Herlong, H. F., Potter, J. J., Van Duyn, M. A. Chandler, E., and Mezey, E.: Effect of parenteral amino acid supplementation in alcoholic hepatitis. *Hepatology* **5:**57–63, 1985.

Dworkin, B., Rosenthal, W. S., and Jankowski, R. H.: Low blood selenium levels in alcoholics with and without advanced liver disease. *Dig. Dis. Sci.* **30:**838–844, 1985.

Egberts, E.-H., Schomerus, H., Hamster, W., and Jurgens, P.: Branched chain amino acids in the treatment of latent portosystemic encephalopathy: A double placebo-controlled crossover study. *Gastroenterology* **88:**887–895, 1985.

Eichner, E. R., Buchanan, B., Smith, J. W., and Hillman, R. S.: Variations in the hematologic and medical status of alcoholics. *Am. J. Med. Sci.* **273:**35–42, 1972.

Epstein, F. H.: Underfilling versus overflow in hepatic ascites. *N. Engl. J. Med.* **307:**1577–1578, 1982.

Epstein, M.: Editorials. The sodium retention of cirrhosis; A reappraisal. *Hepatology* **6:**312–315, 1986.

Errikson, L. S., Persson, A., and Wahre, J.: Branched-chain amino acids in the treatment of hepatic encephalopathy. *Gut* **23:**801–806, 1982.

Feinman, L., and Lieber, C. S.: Toxicity of ethanol and other

components of alcoholic beverages. *Alcoholism: Clin. Exp. Res.* **12:**2–6, 1988.

Fennelly, J., Frank, O., Baker, H., and Leevy, C. M.: Red blood cell-transketolase activity in malnourished alcoholics with cirrhosis. *Am. J. Clin. Nutr.* **20:**946–949, 1967.

Findlay, J., Sellers, E., and Forstner, G.: Lack of effect of alcohol on small intestine binding of the vitamin B_{12}–intrinsic factor complex. *Can. J. Physiol. Pharmacol.* **54:**469–476, 1976.

Flink, E. B.: Magnesium deficiency in alcoholism. *Alcoholism: Clin. Exp. Res.* **10:**590–594, 1986.

Gabuzda, G. J.: Nutrition and liver disease. *Med. Clin. North Am.* **54:**1455–1472, 1970a.

Gabuzda, G. J.: Cirrhosis, ascites and edema: Clinical course related to management. *Gastroenterology* **58:**546–553, 1970b.

Gascon-Barré, M.: Hypothesis and review. Influence of chronic ethanol consumption on the metabolism and action of vitamin D. *J. Am. Coll. Nutr.* **4:**565–574, 1985.

Greenberger, N. J., Carey, J., Schenker, S., Bettinger, L., Stamnes, C., and Beyer, P.: Effect of vegetable and animal protein diets in chronic hepatic encephalopathy. *Am. J. Dig. Dis.* **22:**845–855, 1977.

Gruchow, H. W., Sobocinski, K. A., Barboriak, J. J., and Scheller, J. G.: Alcohol consumption, nutrient intake and relative body weight among U.S. adults. *Am. J. Clin. Nutr.* **42:**289–295, 1985.

Halsted, C. H., Robles, E. Z., and Mezey, E.: Decreased jejunal uptake of labeled folic acid (^{3}H-PGA) in alcoholic patients: role of alcohol and nutrition. *N. Engl. J. Med.* **285:**701–706, 1971.

Herbert, V.: Folate deficiency. In: *Abstracts XXI Congress, International Society of Haematology*, Sydney, International Society of Haematology, 1986, p. 216.

Herbert, V., Zalusky, R., and Davidson, C. S.: Correlation of folate deficiency with alcoholism and associated macrocytosis, anemia, and liver disease. *Ann. Intern. Med.* **58:**977–988, 1963.

Herzlich, B., and Herbert, V.: Rapid collection of human intrinsic factor uncontaminated with cobalophilin (R binder). *Am. J. Gastroenterol.* **81:**678–680, 1986.

Hillers, V. N., and Massey, I. K.: Interrelationships of moderate and high alcohol consumption with diet and health status. *Am. J. Clin. Nutr.* **41:**356–362, 1985.

Hillman, R. W.: Alcoholism and malnutrition. In: *Biology of Alcoholism* Vol. III, (B. Kissin and H. Begleiter (eds.), New York, Plenum Press, 1974, pp. 513–560.

Horst, D., Grace, N. D., Conn, H. O., Schiff, E., Schenker, S., Viteri, A., Law, D., and Atterbury, C. E.: Comparison of dietary protein with an oral, branched chain-enriched amino acid supplement in chronic portal–systemic encephalopathy: A randomized controlled trial. *Hepatology* **4:**279–287, 1984.

Hoyumpa, A. M.: Alcohol and thiamin metabolism. *Alcoholism: Clin. Exp. Res.* **7:**11–14, 1983.

Huber, A. M., and Gershoff, S. N.: Effects of zinc deficiency on the oxidation of retinol and ethanol in rats. *J. Nutr.* **105:**1486–1490, 1975.

Iber, F. L.: In alcoholism, the liver sets the pace. *Nutr. Today* **6:**2–9, 1971.

Israel, Y., Videla, L., and Bernstein, L.: Liver hypermetabolic state after chronic alcohol consumption: Hormonal interrelations and pathologic implications. *Fed. Proc.* **34:**2052–2059, 1975.

Jung, R. T., Davie, M., Hunter, J. O., Chalmer, T. M., and Lawson, D. E. M.: Abnormal vitamin D metabolism in cirrhosis. *Gut* **19:**290–293, 1978.

Kalvaria, I., Labadarios, D., Shephard, G. S., Visser, L., and Marks, I. N.: Biochemical vitamin E deficiency in chronic pancreatitis. *Int. J. Pancreatol.* **1:**119–128, 1986.

Katz, D., Metz, J., and van der Westhuyzen, J.: Intestinal absorption of thiamin from yeast containing sorghum beer. *Am. J. Clin. Nutr.* **42:**666–670, 1985.

Kaunitz, J. D., and Lindenbaum, J.: The bioavailability of folic acid added to wine. *Ann. Intern. Med.* **87:**542–545, 1977.

Kawase, T., Kato, S., and Lieber, C. S.: Lipid perioxidation and antioxidant defense systems in rat liver after chronic ethanol feeding. *Hepatology* **10:**815–821, 1989.

Klipstein, F. A., and Lindenbaum, J.: Folate deficiency in chronic liver disease. *Blood* **25:**443–456, 1965.

Knight, R. E., Bourne, A. J., Newton, M., Black, A., Wilson, P., and Lawson, M. J.: Neurologic syndrome associated with low levels of vitamin E in primary biliary cirrhosis. *Gastroenterology* **91:**209–211, 1986.

Korpela, H., Kumpulainen, J., and Luoma, P.: Decreased serum selenium in alcoholics as related to liver structure and function. *Am. J. Clin. Nutr.* **42:**147–151, 1985a.

Korpela, H., Kumpulainen, J. T., and Sotaniemi, E. A.: The role of selenium deficiency in the pathogenesis of alcoholic liver disease. *Nutr. Res.* **40**(Suppl. I):424–425, 1985b.

Laitinen, K., Lamberg-Allardt, C., Tunninen, R., Karonen, S.-L., Tahtela, R., Ylikahri, R., and Valimaki, M.: Transient hypoparathyroidism during acute alcohol intoxication. *N. Engl. J. Med.* **324:**721–727, 1991.

Lee, M., and Lucia, S. P.: Effect of ethanol on the mobilization of vitamin A in the dog and in the rat. *Stud. Alcohol* **26:**1–8, 1965.

Leo, M. A., and Lieber, C. S.: Hepatic vitamin A depletion in alcoholic liver injury. *N. Engl. J. Med.* **307:**597–601, 1982.

Leo, M. A., and Lieber, C. S.: Interaction of ethanol with vitamin A. *Alcoholism: Clin. Exp. Res.* **7:**15–21, 1983a.

Leo, M. A., and Lieber, C. S.: Hepatic fibrosis after long-term administration of ethanol and moderate vitamin A supplementation in the rat. *Hepatology* **3:**1–11, 1983b.

Leo, M. A., and Lieber, C. S.: New pathway for retinol metabolism in liver microsomes. *J. Biol. Chem.* **260:**5228–5231, 1985.

Leo, M. A., Arai, M., Sato, M., and Lieber, C. S.: Hepatotoxicity of moderate vitamin A supplementation in the rat. *Gastroenterology* **82:**194–205, 1982.

Leo, M. A., Sato, M., and Lieber, C. S.: Effect of hepatic vitamin A depletion on the liver in humans and rats. *Gastroenterology* **84:**562–572, 1983.

Leo, M. A., Kim, C., and Lieber, C. S.: Increased vitamin A in esophagus and other extrahepatic tissues after chronic ethanol consumption in the rat. *Alcoholism: Clin. Exp. Res.* **10:**487–492, 1986.

Levander, O. A., and Burk, R. F.: Report on the 1986 A.S.P.E.N. research workshop on selenium in clinical nutrition. *Parent. Ent. Nutr.* **10:**545–549, 1986.

LeVeen, H. H.: The LeVeen shunt. *Ann. Rev. Med.* **36:**453–469, 1985.

Lieber, C. S.: *Medical Disorders of Alcoholism: Pathogenesis and Treatment.* W. B. Saunders, Co., Philadelphia, 1982.

Lieber, C. S.: Alcohol and the liver. In: *Liver Annual—VI* (I. M. Arias, M. S. Frenkel, and J. H. P. Wilson, eds.), Amsterdam, Excerpta Medica, 1987, pp. 163–240.

Lieber, C. S., Jones, D. P., and DeCarli, L. M.: Effects of prolonged ethanol intake: Production of fatty liver despite adequate diets. *J. Clin. Invest.* **44:**1009–1021, 1965.

Lindenbaum, J.: Alcohol and the hematologic system. In: *Medical Disorders of Alcoholism. Pathogenesis and Treatment,* Vol. 22, *Major Problems in Internal Medicine* (C. S. Lieber, ed.), Philadelphia, W. B. Saunders, 1982, pp. 313–362.

Lindenbaum, J., and Lieber, C. S.: Effects of ethanol on the blood, bone marrow, and small intestine of man. In: *Biological Aspects of Alcohol,* Vol. III (M. K. Roach, W. M. McIssac, and P. J. Creaven, eds.), Austin, University of Texas Press, 1971, pp. 27–45.

Lindenbaum, J., and Lieber, C. S.: The effects of chronic ethanol administration on intestinal absorption in man in the absence of nutritional deficiency. *Ann. N. Y. Acad. Sci.* **252:**228–234, 1975.

Lindenbaum, J., Saha, J. R., Shea, N., and Lieber, C. S.: Mechanism of alcohol-induced malabsorption of vitamin B_{12}. *Gastroenterology* **64:**762, 1973.

Long, R. G.: Vitamin D in chronic liver disease. In: *Liver in Metabolic Diseases* (L. Bianchi, W. Gerok, L. Landmann, K. Sickinger, and G. A. Stadler, eds.), Lancaster, MTP Press, 1983, pp. 421–427.

Lumeng, L.: The role of acetaldehyde in mediating the deleterious effect of ethanol on pyridoxal 5′-phosphate metabolism. *J. Clin. Invest.* **62:**286–293, 1978.

Lumeng, L., and Li, T.-K.: Vitamin B_6 metabolism in chronic alcohol. Pyridoxal phosphate levels in plasma and the effects of acetaldehyde on pyridoxal phosphate synthesis and degradation in human erythrocytes. *J. Clin. Invest.* **53:**693–704, 1974.

Lumeng, L., Schenker, S., Li, T.-K., Brashear, R. E., and Compton, M. C.: Clearance and metabolism of plasma pyridoxal 5′phosphate in the dog. *J. Lab. Clin. Med.* **103:**59–69, 1984.

Maddrey, W. C., Weber, F. L., Coulter, A. W., Chura, C. H., Chapanis, N. P., and Waiser, M.: Effects of keto analogues of essential amino acids in portal–systemic encephalopathy. *Gastroenterology* **71:**190–195, 1976.

Mak, K. M., Leo, M. A., and Lieber, C. S.: Ethanol potentiates squamous metaplasia of the rat trachea caused by vitamin A deficiency. *Trans. Assoc. Am. Physicians* **98:**210–221, 1984.

Martin, D. W., Jr.: Fat soluble vitamins. In: *Harper's Review of Biochemistry* (20th ed.) (D. W. Martin, Jr., P. A. Mayes, V. W. Rodwell, and D. K. Granner, eds.), Los Altos, CA, Lange Medical Publications, 1985, pp. 118–127.

McClain, C. J., Gavaler, J. S., and Van Thiel, D. H.: Hypogonadism in the zinc-deficient rat: Localization of the functional abnormalities. *J. Lab. Clin. Med.* **104:**1007–1015, 1984.

McGhee, A., Henderson, M., Millikan, W. J., Bleier, J. C., Vogel, R., Kassouny, M., and Rudman, D.: Comparison of the effects of hepatic-aid and a casein modular diet on encephalopathy, plasma amino acids, and nitrogen balance in cirrhotic patients. *Ann. Surg.* **197:**288–293, 1983.

Mendenhall, C. L., Anderson, S., Weesner, R. E., Goldberg, S. T., and Crolic, K. A.: Protein–calorie malnutrition associated with alcoholic hepatitis. *Am. J. Med.* **76:**211–222, 1984.

Mendenhall, C., Bongiovanni, G., Goldberg, S., Miller, B., Moore, J., Rouster, S., Schneider, D., Tamburro, C., Tosch, T., Weesner, R., and the VA Cooperative Study Group: VA cooperative study on alcoholic hepatitis III: Changes in protein–calorie malnutrition associated with 30 days of hospitalization with and without enteral nutritional therapy. *Parent. Ent. Nutr.* **9:**590–596, 1985.

Mezey, E., and Faillace, L. A.: Metabolic impairment and recovery time in acute ethanol intoxication. *J. Nerv. Ment. Dis.* **153:**445–452, 1971.

Millikan, W. J., Jr., Henderson, J. M., Warren, W. D., Riepe, S. P., Kutner, M. H., Wright-Bacon, L., Epstein, C., and Parks, R. B.: Total parenteral nutrition with F080 in cirrhotics with subclinical encephalopathy. *Ann. Surg.* **197:**294–304, 1983.

Morgan, M. Y.: Enteral nutrition in chronic liver disease. *Acta Chir. Scand. [Suppl.]* **507:**81–90, 1981.

Nadkarni, G. D., Deshpande, U. R., and Pahuja, D. N.: Liver vitamin A stores in chronic alcoholism in rats: Effect of propylthiouracil treatments. *Experientia* **35:**1059–1060, 1979.

Nasrallah, A., and Galambos, J. T.: Amino acid therapy of alcoholic hepatitis. *Lancet* **2:**1276–1277, 1980.

Naveau, S., Pelletier, G., Poynard, T., Attali, P., Poitrine, A., Buffet, C., Etienne, J. P., and Chaput, J. C.: A randomized clinical trial of supplementary parenteral nutrition in jaundiced alcoholic cirrhotic patients. *Hepatology* **6:**270–274, 1986.

Neville, J. N., Eagles, J. A., Samson, G., and Olson, R. E.: Nutritional status of alcoholics. *Am. J. Clin. Nutr.* **21:**1329–1340, 1968.

Nilsson, B. E.: Conditions contributing to fracture of the femoral neck. *Acta Chir. Scand.* **136:**383–384, 1970.

Parker, T. H., Marshall, J. P., Roberts, R. K., Wang, S., Schiff, E. R., Wilkinson, G. R., and Schenker, S.: Effect of acute alcohol ingestion on plasma pyridoxal 5′-phosphate. *Am. J. Clin. Nutr.* **32:**1246–1252, 1979.

Parry, G., and Bredesen, D. E.: Sensory neuropathy with low-dose pyridoxine. *Neurology (N.Y.)* **35:**1466–1468, 1985.

Patek, A. J., Toth, E. G., Saunder, M. G., Castro, G. A. M., and Engel, J. J: Alcohol and dietary factors in cirrhosis. *Arch. Intern. Med.* **135:**1053–1057, 1975.

Pekkanen, L., and Forsander, O.: Nutritional implications of alcoholism. *Nutr. Bull.* **4:**91–102, 1977.

Pirola, R. C., and Lieber, C. S.: The energy cost of the metabolism of drugs including alcohol. *Pharmacology* **7:**185–196, 1972.

Pirola, R. C., and Lieber, C. S.: Energy wastage in rats given drugs that induce microsomal enzymes. *J. Nutr.* **105:**1544–1548, 1975.

Pirola, R. C., and Lieber, C. S.: Hypothesis: Energy wastage in alcoholism and drug abuse: Possible role of hepatic microsomal enzymes. *Am. J. Clin. Nutr.* **29:**90–93, 1976.

Posner, D. B., Russell, R. M., Absood, S., Connor, T. B., Davis, C., Martin, L., Williams, J. B., Norris, A. H., and Merchant, C.: Effective 25-hydroxylation of vitamin D_2 in alcoholic cirrhosis. *Gastroenterology* **74:**866–870, 1978.

Potter, G. J., Chapman, R. W. G., Nunes, R. M., Sorrentino, D., and Sherlock, S.: Transferrin metabolism in alcoholic liver disease. *Hepatology* **5:**714–721, 1985.

Racusen, L. C., and Krawitt, E. L.: Effect of folate deficiency and ethanol ingestion on intestinal folate absorption. *Am. J. Dig. Dis.* **22:**915–920, 1977.

Roberts, H. R., and Cederbaum, A.: The liver and blood coagulation: Physiology and pathology. *Gastroenterology* **63:**297–320, 1972.

Romero, J. J., Tamura, T., and Halsted, C. H.: Intestinal absorption of ^3H folic acid in the chronic alcoholic monkey. *Gastroenterology* **80:**99–102, 1981.

Rosenthal, W. S., Adham, N. F., Lopez, R., and Cooperman, J. M.: Riboflavin deficiency in complicated chronic alcoholism. *Am. J. Clin. Nutr.* **26:**858–860, 1973.

Rothwell, N. J., and Stock, M. J.: Influence of alcohol and sucrose consumption on energy balance and brown fat activity in rat. *Metabolism* **33:**768–771, 1984.

Russell, R. M., Giovetti, A., Garrett, M., Thompson, J. N., and Mackey, E.: Lack of direct ethanol effect on hepatic vitamin A mobilization. *Gastroenterology* **77:**A36, 1979.

Russell, R. M., Rosenberg, I. H., Wilson, P. D., Iber, F. L., Oaks, E. B., Glovetti, A. C., Otradovec, C. L., Karwoski, B. S., and Press, A. W.: Increased urinary excretion and prolonged turnover time of folic acid during ethanol ingestion. *Am. J. Clin. Nutr.* **38:**64–70, 1983.

Sandstead, H. H.: Zinc nutrition in the United States. *Am. J. Clin. Nutr.* **26:**1251–1260, 1973.

Sato, M., and Lieber, C. S.: Hepatic vitamin A depletion after chronic ethanol consumption. *Gastroenterology* **79:**1123, 1980.

Sato, M., and Lieber, C. S.: Increased metabolism of retinoic acid after chronic ethanol consumption in rat liver microsomes. *Arch. Biochem. Biophys.* **213:**557, 1982.

Saville, P. D.: Changes in bone mass with age and alcoholism. *J. Bone Joint Surg.* **47:**492–499, 1965.

Schaumberg, H., Kaplan, J., Windebank, A., Vick, N., Rasmus, S., Pleasure, D., and Brown, M. J.: Sensory neuropathy from pyridoxine abuse. A new megavitamin syndrome. *N. Engl. J. Med.* **309:**445–448, 1983.

Scheig, R.: Effects of ethanol on the liver. *Am. J. Clin. Nutr.* **23:**467–473, 1970.

Scott, M. L.: Advances in our understanding of vitamin E. *Fed. Proc.* **39:**2736–2739, 1980.

Shah, N., Smith, A., and Picciano, M. F.: Plasma selenium levels in alcoholic liver disease and primary biliary cirrhosis. *Nutr. Res.* **40**(Suppl I)**:**385–387, 1985.

Shaw, S., Gorkin, B. D., and Lieber, C. S.: Effect of chronic alcohol feeding on thiamin status: Biochemical and behavioral correlates. *Am. J. Clin. Nutr.* **34:**856–860, 1981.

Shaw, S. T., Worner, T., and Lieber, C. S.: Comparison of animal and vegetable protein sources in the dietary management of hepatic encephalopathy. *Am. J. Clin. Nutr.* **28:**59–63, 1983.

Shear, L., and Gabuzda, G. J.: Potassium deficiency and endogenous ammonium overload from kidney. *Am. J. Clin. Nutr.* **23:**614–618, 1970.

Simko, V., Connell, A. M., and Banks, B.: Nutritional status in alcoholics with and liver disease. *Am. J. Clin. Nutr.* **35:**197–203, 1982.

Smajdja, C., and Franco, D.: The LeVeen shunt in the effective treatment of intractable ascites in cirrhosis. A study on 140 patients. *Ann. Surg.* **201:**488–493, 1985.

Smith, J., Horowitz, J., Henderson, J. M., and Heymsfield, S.: Enteral hyperalimentation in undernourished patients with cirrhosis and ascites. *Am. J. Clin. Nutr.* **35:**56–72, 1982.

Smith, K. T., Cousins, R. J., Silbon, B. L., and Faille, M. L.: Zinc absorption and metabolism by isolated, vascularly perfused rat intestine. *J. Nutr.* **108:**1849–1857, 1978.

Soberon, S., Pauley, M. P., Duplantier, R., Fan, A., and Halsted, C. H.: Metabolic effects of enteral formula feeding in alcoholic hepatitis. *Hepatology* **7:**1204–1209, 1987.

Solomon, L.: Drug induced arthropathy and necrosis on the femoral head. *J. Bone Joint Surg. [Br.]* **55:**246–261, 1973.

Somogyi, J. C.: Early signs of thiamin deficiency. *Bibl. Nutr. Dieta.* **23:**78–85, 1976.

Stibler, H., Allgulander, C., Borg, S., and Kjellin, K. G.: Abnormal microheterogeneity of transferrin in serum and cerebrospinal fluid in alcoholism. *Acta Med. Scand.* **204:**49–56, 1978.

Stibler, H., Sydow, O., and Borg, S.: Quantitative estimation of abnormal microheterogeneity of transferrin in alcoholics. *Pharmacol. Biochem. Behav.* **13**(Suppl. 1)**:**47–51, 1980.

Stock, A. L., Stock, M. J., and Stuart, J. A.: The effect of alcohol (ethanol) on the oxygen consumption of fed and fasting subjects. *Proc. Nutr. Soc.* **32:**40A–41A, 1973.

Stock, M. J., and Stuart, J. A.: Thermic effects of ethanol in the rat and man. *Nutr. Metabol.* **17:**297–305, 1974.

Sullivan, J. F.: Effect of alcohol on urinary zinc excretion. *Q. J. Stud. Alcohol* **23:**216–220, 1962a.

Sullivan, J. F.: The relation of zincuria to water and electrolyte excretion in patients with hepatic cirrhosis. *Gastroenterology* **42:**439–442, 1962b.

Sullivan, J. F., Williams, R. V., and Burch, R. E.: The metabolism of zinc and selenium in cirrhotic patients during six weeks of zinc ingestion. *Alcoholism: Clin. Exp. Res.* **3:**235–239, 1979.

Sullivan, L. W., and Herbert, V.: Suppression of hematopoiesis by ethanol. *J. Clin. Invest.* **43:**2048–2062, 1964.

Summerskill, W. H. J., Barnardo, D. E., and Baldus, W. P.: Disorders of water and electrolyte metabolism in liver disease. *Am. J. Clin. Nutr.* **23:**499–507, 1970.

Tamura, T., and Halsted, C. H.: Folate turnover in chronically alcoholic monkeys. *J. Lab. Clin. Med.* **101:**623–628, 1983.

Tamura, T., Romero, J. J., Watson, J. E., Gong, E. J., and Halsted, C. H.: Hepatic folate metabolism in the chronic alcoholic monkey. *J. Lab. Clin. Med.* **97:**654–661, 1981.

Teschke, R., Moreno, F., Heinen, E., Herrmann, J., Kruskemper, H. L., and Strohmeyer, G.: Is there any evidence of a hyperthyroid hepatic state following chronic alcohol intake? *Alcohol Alcoholism* **18:**151–155, 1983.

Thomson, A. D., and Majumdar, S. K.: The influence of ethanol on intestinal absorption and utilization of nutrients. *Clin. Gastroenterol.* **10:**263–293, 1981.

Tomasula, P. A., Kater, R. M. H., and Iber, F. L.: Impairment of thiamin absorption in alcoholism. *Am. J. Clin. Nutr.* **21:**1341–1344, 1968.

Tremolieres, J., and Carre, L.: Etudes sur les modalites d'oxydation de l'alcool chez l'homme normal et alcoolique. *Rev. Alcool.* **7:**202–227, 1961.

Valberg, L. S., Flanagan, P. R., Ghent, C. N., and Chamberlain, M. J.: Zinc absorption and leukocyte zinc in alcoholic and nonalcoholic cirrhosis. *Dig. Dis. Sci.* **30:**329–333, 1985.

Valimaki, M. J., Harju, K. J., and Ylikahri, R. H.: Decreased serum selenium in alcoholics—a consequence of liver dysfunction. *Clin. Chim. Acta* **130:**291–296, 1983.

Vallee, B. L., Wacker, W. E. C., Bartholomay, A. F., and Robin, E. D.: Zinc metabolism in hepatic dysfunction. I. Serum zinc concentrations in Laënnec's cirrhosis and their validation by sequential analysis. *N. Engl. J. Med.* **255:**403–408, 1956.

Vallee, B. L., Wacker, E. C., Bartholomay, A. F., and Hock, F. L.: Zinc metabolism in hepatic dysfunction. *N. Engl. J. Med.* **257:**1055–1065, 1957.

Wapnic, S., Grossberg, S. J., and Evans, M. I.: Randomized prospective matched pair study comparing peritoneovenous shunt and conventional therapy in massive ascites. *Br. J. Surg.* **66:**667–670, 1979.

Westerfeld, W. W., and Schulman, M. P.: Metabolism and caloric value of alcohol. *J.A.M.A.* **170:**197–203, 1959.

Williamson, D. F., Forman, M. R., Binkin, N. J., Gentry, E. M., Reminton, P. L., and Trowbridge, F. L.: Alcohol and body weight in United States adults. *Am. J. Public Health* **77:**1324–1330, 1987.

Witt, E. D., and Goldman-Rakic, P. S.: Intermittent thiamin deficiency in the rhesus monkey. I. Progression of neurological signs and neuroanatomical lesions. *Ann. Neurol.* **13:**376–395, 1983a.

Witt, E. D., and Goldman-Rakic, P. S.: Intermittent thiamin deficiency in the rhesus monkey. II. Evidence for memory loss. *Ann. Neurol.* **13:**396–401, 1983b.

Wood, B., Breen, K. J., and Penington, D. G.: Thiamin status in alcoholism. *Aust. N.Z. J. Med.* **7:**475-484, 1977.

World, M. J., Ryle, P. R., Jones, D., Shaw, G. K., and Thompson, A. D.: Differential effect of chronic alcoholic intake and poor nutrition on body weight and fat stores. *Alcohol Alcoholism* **19:**281–290, 1984a.

World, M. J., Ryle, P. R., Pratt, O. E., and Thompson, A. D.: Alcohol and body weight. *Alcohol Alcoholism* **19:**1–6, 1984b.

18

Biological Markers of Alcoholism

Alan S. Rosman and Charles S. Lieber

18.1. Utility of a Biological Marker of Alcoholism

18.1.1. Practical Application of a Marker of Alcoholism

A biological marker of ethanol consumption would be a valuable tool in the diagnosis and treatment of alcoholism. There are several areas of potential clinical application. First, a biological marker would be useful for the early detection of alcoholism, as many patients do not seek medical attention at an early stage (Clark, 1981). Previous studies have demonstrated the medical and social advantages of early intervention in the treatment of alcoholism (Reiff, 1985; Paredes, 1985). In addition, a biological marker would improve the monitoring of sobriety in alcoholics undergoing treatment. Furthermore, many patients who are suffering from a disease in which alcoholism is the suspected etiological agent (e.g., chronic pancreatitis, cirrhosis) fail to give an accurate history of their alcohol consumption. These patients are then subjected to an exhaustive diagnostic workup to rule out other etiological factors. A reliable biological marker of alcoholism would expedite the diagnostic workup and perhaps avoid invasive procedures (e.g., endoscopic retrograde pancreatography, liver biopsy). Finally, selected patients with alcohol-induced liver disease may become candidates for liver transplantation if there is a proven record of abstinence (Maddrey and Van Thiel, 1988); a reliable biological marker would provide objective verification of abstinence and thus improve the selection process.

A biological marker of alcoholism would also be valuable in clinical research. Currently, clinical trials evaluating the efficacy of various treatment modalities of alcoholism usually rely on the patient's self-report of alcohol consumption as the therapeutic endpoint. Because the patient's self-report may not always be reliable (Sobell *et al.*, 1979; Orrego *et al.*, 1979; Watson *et al.*, 1984; Peachey and Kapur, 1986; Fuller *et al.*, 1988), a biological marker of alcohol consumption may provide a more objective assessment of abstinence. In addition, a reliable biological marker would be a valuable tool for epidemiologic studies investigating the incidences of alcoholism in different communities.

A marker of alcohol consumption would have useful forensic applications. Individuals arrested for driving while intoxicated could be screened for chronic alcohol consumption and then referred to an appropriate treatment program (Luchi *et al.*, 1978; Gjerde and Morland, 1987). In addition, workers involved in areas of public safety could be effectively monitored for alcohol abuse.

18.1.2. Limitations of Patient History and Screening Instruments

Although chronic ethanol consumption produces diverse biochemical and physiological changes, a reliable and simple marker is still not available. Presently, the diagnosis of alcoholism by physicians is primarily based on a patient's history. However, because denial may play an important role in alcoholism, patients may be reluctant to seek medical care (Moore and Murphy, 1961; Clark, 1981; Spickard, 1986). In a prospective

531

study of patients admitted to the medical and orthopedic services of a community hospital, only 29% of patients with a history of alcohol dependence or alcohol abuse considered themselves to be alcoholic (Bush *et al.*, 1987). Even when the alcoholic patient comes to medical attention, the diagnosis of alcoholism is frequently overlooked in a variety of clinical settings (Barrison *et al.*, 1980). Thus, physicians were only able to identify correctly 25% of alcoholics attending a general medical clinic (Persson and Magnusson, 1988).

Even when the alcoholic patient is identified and referred for treatment, the long-term monitoring of sobriety is suboptimal. Physicians mainly rely on a patient's self-report of alcohol consumption for identifying relapse. However, various studies have demonstrated the lack of reliability of a patient's self-report of alcohol consumption. In clinical studies in which alcohol consumption was monitored by both patients' self-report and daily urine specimens for ethanol (Orrego *et al.*, 1979; Peachey and Kapur, 1986), more than half of the alcoholic patients who drank while undergoing therapy denied any alcohol use. In the Veterans Administration Cooperative Study, which evaluated the efficacy of disulfiram (Fuller *et al.*, 1988), drinking behavior was assessed using patients' self-report, collateral history, and serial urine or blood tests for ethanol. Approximately 35% of patients who claimed to be abstinent were indeed drinking during the study period, as confirmed by collateral history or laboratory tests. Furthermore, the number of drinking days was underestimated by at least 28% when compared to collateral history. Thus, a patient's self-report is insensitive in the detection of relapse and will underestimate the patient's alcohol consumption.

Since the previous studies used blood or urine ethanol determination as a gold standard, the drinking behavior may be further underestimated. Because the rate of ethanol metabolism is about 10 g/hr (Rowland and Tozer, 1989), blood or urine ethanol testing will only identify recent intake. Furthermore, blood ethanol determination cannot discriminate between sporadic drinking and excessive alcohol consumption.

A number of questionnaires have been developed in order to improve the detection of alcoholism. Most of the questions inquire about the social consequences related to drinking behavior rather than directly asking the patient to quantify his alcohol consumption. One of the most investigated questionnaires is the 25-item Michigan Alcoholism Screening Test (MAST), dealing with social, legal, and medical effects of alcoholism (Selzer, 1971). The MAST has been demonstrated to have clinical value (Moore, 1971, 1972), but its format and length have limited its application. Various modifications of the original MAST have included the ten-item brief MAST (Pokorny *et al.*, 1972), the 35-item Self-Administered Alcohol Screening Test (SAAST) (Swenson and Morse, 1975), and the nine-item SAAST (Davis *et al.*, 1987). These modifications have generally improved the simplicity of administration without sacrificing diagnostic accuracy (Kristenson *et al.*, 1983; Davis *et al.*, 1987). Two other screening items, the four-item CAGE (Mayfield *et al.*, 1974) and the five-item Skinner Trauma Questionnaire (Skinner *et al.*, 1984), are easy to administer and have greater sensitivity and specificity than standard laboratory markers of alcoholism (Kristenson and Trell, 1982; Bernadt *et al.*, 1982; Skinner *et al.*, 1984), but, nevertheless, they have several limitations. First, these tests are dependent on patient cooperation and veracity. In addition, their accuracy may be affected by socioeconomic and cultural factors (Walters *et al.*, 1983; Monteiro *et al.*, 1986; Bilal *et al.*, 1987). Finally, the utility of these questionnaires in monitoring alcoholic patients undergoing treatment has not been validated.

18.1.3. Clinical Utility of Laboratory Tests

When clinical evaluation results in a diagnostic uncertainty, physicians may turn to laboratory tests for further assistance. The principles for evaluating diagnostic tests have been discussed in several review articles (McNeil *et al.*, 1975; Gottfried and Wagar, 1983; Sox, 1986). These principles are summarized below to serve as a basis for evaluating the utility of laboratory markers of alcohol consumption.

The usefulness of a diagnostic test lies in its ability to discriminate accurately between patients with a disease and patients without a disease. The diagnostic characteristics used to assess the discriminating ability of a laboratory test include sensitivity and specificity. Sensitivity is the proportion of patients with a disease who have a positive (i.e., abnormal) test. Specificity is the proportion of patients without the disease who have a negative (i.e., normal) test. Both the sensitivity and specificity will vary with the cutoff point used to define a positive test (McNeil *et al.*, 1975). In addition, sensitivity is dependent on the population being studied. Several laboratory markers of alcoholism are only abnormal in patients at advanced stages of their disease. Because many of these markers may be less sensitive in detecting alcoholics at an early stage, the sensitivity of these markers determined in hospitalized alcoholics should not be extrapolated to ambulatory alcoholics without direct verification. Specificity is also affected by the nature of the control group selected. Laboratory markers that are normal in healthy, nonalcoholic volunteers (i.e., laboratory workers) can be affected by a variety of nonalcoholic conditions. Thus, laboratory markers of alcoholism may become less specific when the control group incudes patients afflicted with a variety of nonalcoholic medical conditions.

In utilizing laboratory tests for diagnostic purposes, physicians tend to use the predictive values of a given test. The predictive value of a positive test is the probability of a disease being present when a test is positive. The predictive value of a negative test is the probability of the disease being absent when a test is negative. By applying Bayes's theorem, the predictive values can be calculated from the sensitivity, specificity, and the pretest probability. The pretest probability is equivalent to the prevalence of the disease when the laboratory test is used to screen a population. When the test is applied in the diagnostic evaluation of a particular patient, the pretest probability is the physician's estimation of the probability of disease prior to laboratory testing. In Fig. 18.1, the predictive value of a positive test of a hypothetical marker with a sensitivity and specificity of 90% is plotted as a function of the prevalence of the disease. At a prevalence of 10%, the predictive value of a positive test is only 50%. This predictive value may not be high enough for clinical utility in many situations. However, if clinical evaluation suggested that a patient had a 50% probability of alcoholism prior to laboratory testing (i.e., the pretest probability), the predictive value of a positive test would be 90%. In this setting, the test would be useful in corroborating a clinical suspicion of alcoholism. Thus, the predictive values of a test are influenced by the pretest probability.

18.1.4. Characteristics of the Ideal Marker

For maximal clinical application, an ideal marker of alcohol consumption should adhere to the following criteria:

1. The test procedure should be noninvasive and result in only minimal inconvenience to the patient (e.g., a blood, urine, or breath test).
2. The test should achieve a high-enough sensitivity and specificity to be a useful screening tool.
3. A screening marker of alcoholism should be able to discriminate accurately between social drinking and excessive alcohol consumption.
4. In contrast, a marker of relapse in alcoholics who are undergoing treatment should be sensitive to even low levels of drinking, as the goal of most treatment programs is complete abstinence (Peele, 1984).
5. The possible causes of false-positive tests should be readily identified and easy to exclude in light of the social stigma associated with the diagnosis of alcoholism.
6. The marker should not be significantly affected by either nonalcoholic liver disease or by the patient's nutritional status.
7. The marker should correlate with either the quantity of alcohol consumed or the duration of alcohol consumption.
8. The marker should persist for at least a few days after abstinence but then normalize after a reasonable abstinence period. In addition, the half-life of the marker during abstinence should be predictable. Finally, the marker should revert to its abnormal state during a relapse episode.
9. There should be no interference with the test by the various agents used in the treatment of alcoholism (benzodiazepines, folate, disulfiram).

No currently available marker meets all the criteria (Salaspuro, 1986). However, there have been recent advances in the understanding of the biochemical and physiological consequences of chronic alcohol use. Thus, the possible metabolic alterations resulting from excessive ethanol consumption that could be applied to the development of a biological marker include:

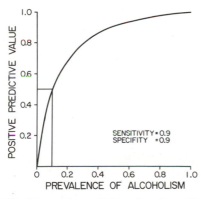

FIGURE 18.1. The positive predictive value of a positive test for a theoretical marker is plotted as a function of the prevalence of alcoholism using Bayes's theorem. The theoretical marker is assumed to have a sensitivity and specificity of 90%. If the marker was used to screen a population with a 10% prevalence of alcoholism, the predictive value of a positive test would be only 50%.

1. Lasting toxic effects on a specific organ system (e.g., liver, red blood cells).
2. Induction of ethanol-inducible cytochrome P-450.
3. Alterations in posttranslational modifications of various proteins.
4. Stable adducts of ethanol and acetaldehyde.

18.1.5. Defining Alcoholism

In order to evaluate the accuracy of a laboratory test, a gold standard that ascertains the true disease state is needed. However, a universally accepted definition of alcoholism is lacking. The revised third edition of the *Diagnostic and Statistical Manual of Mental Disorders* (American Psychiatric Association, 1987) (also known as DSM-III-R) subdivides alcoholism into two categories, alcohol abuse and alcohol dependence. The DSM-III-R criteria for alcohol abuse include the following features:

1. Continued use of alcohol despite awareness that further use may cause or exacerbate occupational, social, psychological, or medical problems.
2. Recurrent use of alcohol in situations where its use may be physically hazardous.
3. Duration of some these symptoms for at least 1 month.

The DSM-III-R diagnostic features of alcohol dependence further require that patients have impaired control of alcohol consumption and may be associated with symptoms of tolerance and withdrawal. The DSM-III-R lists specific diagnostic criteria for alcohol dependence. However, there may be some limitations in using the DSM-III-R criteria as a screening test of alcoholism. First, many of the adverse physical and psychological consequences of alcohol use may not be apparent to either the patient or physician, resulting in the underdiagnosis of alcoholism. In addition, application of these criteria may be dependent on patient cooperation and veracity and may be influenced by social and cultural factors.

An alternative approach to defining alcoholism relates to the quantity of alcohol consumed. Several studies have suggested that the risk of organ damage is related to cumulative level of alcohol consumption (Lelbach, 1975; Pequignot *et al.*, 1978; Tuyns and Pequignot, 1984). Thus, the risk of organ damage is increased when chronic alcohol consumption exceeds 40 g/day for males and 20 g/day for females (Pequignot *et al.*, 1978; Tuyns and Pequignot, 1984). Although this approach may be more objective than the DSM-III-R criteria, many patients fail to give an accurate history of their alcohol consumption (Sobell *et al.*, 1979; Orrego *et al.*, 1979; Watson *et al.*, 1984; Peachey and Kapur, 1986; Fuller *et al.*, 1988).

18.2. Currently Available Markers

Excessive alcohol consumption can produce abnormalities in a variety of currently available laboratory tests (Salaspuro, 1986). However, none of these tests has adequate sensitivity and specificity for the screening of alcoholism in the general population. Despite their limitations, clinicians currently utilize these tests for corroborating their clinical suspicion of alcoholism. Commonly used markers include GGT, MCV, liver enzymes, and high-density lipoprotein cholesterol.

18.2.1. γ-Glutamyl Transpeptidase

γ-Glutamyl transpeptidase (GGT) is an enzyme primarily involved in the hydrolysis of γ-glutamyl peptides and plays an important role in glutathione metabolism (Tate and Meister, 1981). The enzyme is found in a variety of organs, including the liver, biliary tract, pancreas, and kidneys (Tate and Meister, 1981). In the hepatocyte, GGT is localized in the plasma membrane, endoplasmic reticulum, and microvilli of the bile canalicular membrane (Ishii *et al.*, 1986).

Most studies have demonstrated that an acute dose of ethanol in normal human volunteers does not result in a significant increase in GGT (Luchi *et al.*, 1978; Gill *et al.*, 1982; Devgun *et al.*, 1985; Clark *et al.*, 1982; Dunbar *et al.*, 1982). However, chronic ethanol administration has been shown to increase serum GGT in experimental animals (Ishii *et al.*, 1978; Gadeholt *et al.*, 1980; Teschke *et al.*, 1977; Shaw and Lieber, 1980a), in healthy human volunteers (Belfrage *et al.*, 1973), and in alcoholics (Shaw and Lieber, 1980a). Possible mechanisms for the increase in serum GGT by chronic ethanol administration include:

1. Induction of hepatic GGT by ethanol (Ishii *et al.*, 1978; Gadeholt *et al.*, 1980; Teschke *et al.*, 1977; Shaw and Lieber, 1980a).
2. Increased synthesis of hepatic GGT secondary to cholestasis (Kryszewski *et al.*, 1973).
3. Damage to hepatocytes resulting in the release of enzymes (Wu *et al.*, 1976).

4. Increased lability of hepatic plasma membranes, resulting in increased release of GGT (Yamada *et al.*, 1985a).

Administration of ethanol (75 g/day) in nonalcoholic volunteers over a 5-week period resulted in a significant increase in GGT in most of the subjects (Belfrage *et al.*, 1973). However, there was a high degree of intersubject variability in the elevation of GGT after chronic ethanol administration. Numerous clinical studies have also shown that serum GGT is not uniformly elevated in all alcoholics. Many of the studies investigating the sensitivity of serum GGT as a marker of alcoholism are summarized in Tables 18.1 and 18.2. Table 18.1 lists the studies performed in hospitalized alcoholics and Table 18.2 incudes studies evaluating ambulatory alcoholic patients. These studies selected alcoholic patients who were drinking actively and also included alcoholic patients without apparent liver disease. The pooled sensitivity of GGT for alcoholics (combining both hospitalized and outpatients) can be calculated to be 52%. The pooled sensitivity for hospitalized alcoholics (Table 18.1) is 62% as compared to 43% in outpatients (Table 18.2). The difference may be secondary to the greater severity of alcoholism in hospitalized patients. Furthermore, serum GGT is a more sensitive marker of alcoholism in patients with suspected liver disease than in patients without apparent liver disease (Moussavian *et al.*, 1985). Finally, this test is more sensitive in patients who have recently been drinking (Gluud *et al.*, 1981; Sanchez-Craig and Annis, 1981).

There is only a weak correlation between the serum GGT and the daily alcohol consumption in alcoholics (Chick *et al.*, 1981; Rollason *et al.*, 1972; Garvin *et al.*, 1981; Sanchez-Craig and Annis, 1981; Gluud *et al.*, 1981; Papoz *et al.*, 1981; Whitfield *et al.*, 1981; Bernadt *et al.*, 1982; Poikolainen *et al.*, 1985; Moussavian *et al.*, 1985; Latcham, 1986). Furthermore, there is no significant correlation between the serum GGT and the duration of alcoholism (Wadstein and Skude, 1979b; Garvin *et al.*, 1981; Gluud *et al.*, 1981; Moussavian *et al.*, 1985).

During abstinence, serum GGT will decrease according to its average half-life of 26 days (Orrego *et al.*, 1985). However, the half-life may be prolonged in alcoholics suffering from chronic liver disease (Moussavian *et al.*, 1985). Monitoring serial serum GGT levels may be of value during the treatment of alcoholism, as increases of sequential serum GGT levels above the remission baseline level are suggestive of a relapse. Two studies reported that the sensitivity for sequential GGT measurements in the detection of relapse is 45–80% (Shaw *et al.*, 1979; Irwin *et al.*, 1988).

The specificity of the serum GGT is dependent on the nature of the control group selected. A variety of other microsomal inducing agents (e.g., anticonvulsants, anticoagulants, and oral contraceptives) can in-

TABLE 18.1.   Studies in Inpatients Evaluating Serum GGT as a Marker of Alcoholism

Author (year)	Number of alcoholics	Percentage with increased GGT	Selection of patients
Rosalki and Rau (1972)	32	72%	Alcoholics admitted for psychiatric therapy
Danielsson *et al.* (1978)	38	47%	Alcoholics admitted for detoxification
Shaw *et al* (1978)	100	74%	Alcoholics admitted for detoxification
Wadstein and Skude (1979a)	156	66%	Alcoholics admitted for detoxification
Garvin *et al.* (1981)	69	55%	Alcoholics admitted for detoxification
Chick *et al.* (1981)	30	60%	Alcoholics admitted for detoxification
Bernadt *et al.* (1982)	42	36%	Psychiatric inpatients screened for alcoholism
Moussavian *et al.* (1985)	97	52%	Alcoholics without liver disease
Moussavian *et al.* (1985)	15	100%	Alcoholics with liver disease
Bell *et al.* (1985)	101	70%	Alcoholics admitted for detoxification
Latcham (1986)	181	54%	Alcoholics admitted for detoxification
Stibler *et al.* (1986)	77	76%	Alcoholics admitted for detoxification
Bush *et al.* (1987)	104	54%	Hospital patients screened for alcoholism
Storey *et al.* (1987)	20	55%	Alcoholics admitted for detoxification
Schellenberg and Weill (1987a)	50	86%	Alcoholics admitted for detoxification
Roine *et al.* (1987a)	49	55%	Alcoholics admitted for detoxification
Behrens *et al.* (1988a)	72	61%	Alcoholics admitted for detoxification

TABLE 18.2.   Studies in Outpatients Evaluating Serum GGT as a Marker of Alcoholism

Author (year)	Number of alcoholics	Percentage with increased GGT	Selection of patients
Rosalki and Rau (1972)	44	77%	Outpatient alcoholics
Baxter et al. (1980)	52	44%	Surgical clinic patients screened for alcoholism
Gluud et al. (1981)	159	45%	Outpatient alcoholics who are actively drinking
Sanchez-Craig and Annis (1981)	14	57%	Outpatients who are heavy consumers of alcohol
Chick et al. (1981)	77	52%	Company employees screened for alcoholism
Kristenson and Trell (1982)	223	38%	Male residents of Malmö, Sweden screened for alcoholism
Clark et al. (1983)	87	51%	London government employees screened for alcoholism
Skinner et al. (1984)	68	39%	Alcoholics seeking outpatient treatment
Skinner et al. (1984)	9	33%	Family-practice patients screened for alcoholism
Puchois et al (1984)	78	59%	Outpatients screened for alcoholism
Korri et al. (1985)	39	51%	Intoxicated emergency room patients screened for alcoholism
Hollstedt and Dahlgren (1987)	100	42%	Female alcoholics attending outpatient rehabilitation
Roine et al. (1987a)	31	26%	Skid-row alcoholics in a homeless shelter
Barrison et al. (1987)	124	18%	Alcoholic patients referred for outpatient treatment
Persson and Magnusson (1988)	208	40%	Outpatients screened for alcoholism

crease the serum GGT (Whitfield *et al.*, 1973; Martin *et al.*, 1976; Salaspuro, 1986). Furthermore, serum GGT can be elevated by a variety of medical conditions including nonalcoholic liver disease, biliary tract disease, hyperlipidemia, and hyperthyroidism (Rutenburg *et al.*, 1963; Whitfield *et al.*, 1972; Azizi, 1982; Salaspuro, 1986). Thus, the use of serum GGT as a screening test for alcoholism in hospitalized patients will result in a large number of false positives (Chick *et al.*, 1982). However, the specificity of serum GGT in the screening of ambulatory patients for alcoholism is approximately 85% (Baxter *et al.*, 1980; Chick *et al.*, 1982; Skinner *et al.*, 1984).

In conclusion, serum GGT is not an ideal screening test of alcoholism because of the lack of sensitivity and moderate specificity. However, this test is useful in confirming a clinical suspicion of alcoholism and may be of value in monitoring the treatment of alcoholics.

18.2.2. Mean Corpuscular Volume

The mean corpuscular volume (MCV) can be readily derived from the complete blood count by calculating the ratio of the hematocrit to the red cell count. If an increased MCV is associated with megaloblastic changes, the peripheral smear may reveal oval-shaped macrocytes and hypersegmented polymorphonuclear leukocytes. However, macrocytosis can also occur without megaloblastic changes, resulting in round macrocytes. Chronic alcohol consumption can result in macrocytosis by several mechanisms:

1. Megaloblastic changes secondary to folate deficiency.
2. Reticulocytosis secondary to hemolysis or bleeding.
3. Macrocytosis secondary to liver disease.
4. Macrocytosis of alcoholism.

The last mechanism, "the macrocytosis of alcoholism," appears to be the most common cause of macrocytosis in alcoholics (Wu *et al.*, 1975). Although the pathogenesis of this entity is poorly understood, it is felt to be a direct toxic effect of ethanol rather than secondary to folate deficiency or liver dysfunction (Wu *et al.*, 1975; Carney and Sheffield, 1978; Okany *et al.*, 1983). Furthermore, specific megaloblastic changes (e.g., hypersegmented polymorphonuclear cells, oval-shaped cells) are not commonly found in this condition. The pathogenesis of this entity is further discussed in Chapter 8.

Wu and co-workers (1974) initially reported that the majority of alcoholics hospitalized for gastrointestinal or neurological problems had macrocytosis. Table 18.3 lists the studies evaluating the MCV as a marker of alcoholism utilizing a threshold MCV greater than 96–99. If the studies that included alcoholics without apparent liver disease are pooled, the calculated sensitivity of MCV is 37%. In addition, the MCV may be elevated in 60–80% of patients with alcoholic liver disease (Morgan *et al.*, 1981; Eriksen *et al.*, 1984). However, approximately 20% of patients with nonalcoholic liver disease will also have this finding (Morgan *et al.*, 1981). Most

TABLE 18.3. Studies Evaluating MCV as a Marker of Alcoholism

Author (year)	Number of alcoholics	Percentage with increased MCV	Selection of patients
Baxter *et al.* (1980)	52	16%	Surgical clinic patients screened for alcoholism
Chick *et al.* (1981)	30	40%	Alcoholics admitted for detoxification
Chick *et al.* (1981)	77	27%	Company employees screened for alcoholism
Morgan *et al.* (1981)	20	80%	Alcoholics with suspected liver disease
Bernadt *et al.* (1982)	42	0%	Psychiatric inpatients screened for alcoholism
Clark *et al.* (1983)	87	12%	London government employees screened for alcoholism
Skinner *et al.* (1984)	68	49%	Alcoholics seeking outpatient treatment
Skinner *et al.* (1984)	8	25%	Family-practice patients screened for alcoholism
Eriksen *et al.* (1984)	28	61%	Alcoholics with suspected liver disease
Cushman *et al.* (1984)	543	47%	Alcoholics referred for treatment
Tonnesen *et al.* (1986)	64	45%	Outpatient alcoholics who are actively drinking
Latcham (1986)	224	35%	Alcoholics admitted for detoxification
Hollstedt and Dahlgren (1987)	100	48%	Female alcoholics attending outpatient rehabilitation
Barrison *et al.* (1987)	124	37%	Alcoholics referred for outpatient treatment
Behrens *et al.* (1988a)	72	28%	Alcoholics admitted for detoxification

studies report only a weak correlation between the MCV and daily ethanol consumption in alcoholics (Bagrel *et al.*, 1979; Chick *et al.*, 1981; Papoz *et al.*, 1981; Morgan *et al.*, 1981; Whitfield *et al.*, 1981; Bernadt *et al.*, 1982; Poikolainen *et al.*, 1985; Tonnesen *et al.*, 1986; Latcham, 1986). However, one study reported a significant correlation between the MCV and the alcohol intake during the recent drinking period (Tonnesen *et al.*, 1986).

The duration of macrocytosis in alcoholics who have abstained is usually prolonged and variable. Thus, most studies have reported than an increased MCV can persist for several months in many alcoholics despite abstinence and folate replacement (Unger and Johnson, 1974; Myrhed *et al.*, 1977; Shaw *et al.*, 1979; Morgan *et al.*, 1981). Because of the prolonged half-life, the MCV has a low sensitivity in the detection of relapse in alcoholic patients (Shaw *et al.*, 1979).

In screening a population for alcoholism using the MCV, other causes of macrocytosis should also be considered. Thus, alcoholism was the underlying cause of macrocytosis in only 15% of patients in a tertiary care hospital (Davidson and Hamilton, 1978). Other etiological factors included nutritional deficiencies (B_{12}, folic acid), hypothyroidism, nonalcoholic liver disease, reticulocytosis, hematological malignancies, and drugs (Davidson and Hamilton, 1978). Furthermore, both smoking and age have been demonstrated to increase the MCV (Eschwege *et al.*, 1978; Chalmers *et al.*, 1979). However, the specificity of MCV when screening ambulatory patients for alcoholism will be at least 90–95%

because of the relatively low incidence of folate and B_{12} deficiencies, nonalcoholic liver diseases, and hematological malignancies in outpatients (Baxter *et al.*, 1980; Chick *et al.*, 1981; Skinner *et al.*, 1984).

In conclusion, the MCV is not an ideal screening test of alcoholism, primarily because of its low sensitivity. It is also of limited value in monitoring alcoholics during treatment. However, the MCV may be useful in conjunction with other tests in corroborating a clinical suspicion of alcoholism.

18.2.3. Liver Enzymes

A variety of "liver" enzymes, including the transaminases, alkaline phosphatase, and glutamate dehydrogenase, are commonly utilized to evaluate the alcoholic patient (Lieber, 1988). The transaminases and alkaline phosphatase are commonly included in screening panels, and the glutamate dehydrogenase can be readily performed by commercial laboratories using a spectrophotometric enzymatic assay.

18.2.3.1. Transaminases

The transaminases are a group of enzymes that catalyze the transfer of an amino group from an amino acids to an α-ketoacid. Two transaminases, aspartate aminotransferase (AST) (also referred to as glutamic–oxaloacetic transaminase, GOT) and alanine aminotransferase (ALT) (also referred to as glutamic–pyruvic transaminase, GPT) are commonly performed as routine screening tests of liver function. However, AST is also present in a variety of extrahepatic tissues including muscle and myocardium and thus may be significantly elevated in muscular disorders and in myocardial injury (Zimmerman and West, 1963). Serum ALT is more specific for liver disease, as the extrahepatic contribution is usually insignificant (Zimmerman and West, 1963). However, severe muscle injury can also produce moderate elevations of the serum ALT (Dreyfus et al., 1959; Zimmerman and West, 1963). In the hepatocyte, AST is present in both the mitochondria and the cytosol, whereas ALT is primarily localized to the cytosol (Williams and Hoofnagle, 1988). Hepatocellular injury can result in elevation of the serum transaminases either by increasing the cell membrane permeability or by cell necrosis (Clermont and Chalmers, 1967).

A variety of studies have demonstrated that chronic ethanol administration can elevate the serum transaminases levels in nonhuman primates (Lieber et al., 1975), in human volunteers (Belfrage et al., 1973), and in alcoholics admitted to a metabolic ward (Shaw and Lieber, 1980a). However, in the clinical studies (Belfrage et al., 1973; Shaw and Lieber, 1980a), the elevations in the serum transaminases were modest and only infrequently exceeded the reference range. A further

study suggested that the magnitude of the increase in transaminases in alcoholics during ethanol administration is related to the severity of the liver disease (Nishimura et al., 1980).

Thus, the variable effect of ethanol on the transaminases theoretically limits the application of transaminases as a marker of alcoholism. A variety of clinical studies shown in Table 18.4 confirm that only a minority of alcoholics without apparent liver disease have elevated transaminases. Thus, the pooled sensitivity of AST as a marker of alcoholism is 35%, significantly less than the serum GGT. Furthermore, a lower sensitivity is observed in outpatients screened for alcoholism. Serum ALT is even less sensitive than serum AST. However, 3–5% of alcoholics will have an elevated serum AST despite a normal serum GGT (Rosalki and Rau, 1972; Skude and Wadstein, 1977; Cushman et al., 1984). Finally, the correlation between alcohol consumption and serum AST is poor (Rollason et al., 1972; Bagrel et al., 1979; Whitfield et al., 1981).

18.2.3.2. AST : ALT Ratio

A variety of clinical studies (Harinasuta et al., 1967; Cohen and Kaplan, 1979; Correia et al., 1981)

TABLE 18.4.   Studies Evaluating the Transaminases as a Marker of Alcoholism

Author (year)	Number of alcoholics studied	Percentage with increased AST	Percentage with increased ALT	Selection of patients
Rosalki and Rau (1972)	31	35%	23%	Hospitalized alcoholics
Rosalki and Rau (1972)	33	27%	15%	Outpatient alcoholics
Wadstein and Skude (1979a)	156	70%	—	Alcoholics admitted for detoxification
Baxter et al. (1980)	52	10%	—	Surgical clinic patients screened for alcoholism
Gluud et al. (1981)	159	30%	—	Outpatient alcoholics who are actively drinking
Bernadt et al. (1982)	42	21%	—	Psychiatric inpatients screened for alcoholism
Clark et al. (1983)	87	29%	—	Government employees screened for alcoholism
Cushman et al. (1984)	543	31%	—	Alcoholics referred for treatment
Korri et al. (1985)	39	26%	—	Emergency room patients screened for alcoholism
Moussavian et al. (1985)	97	40%	33%	Hospitalized alcoholics without liver disease
Moussavian et al. (1985)	15	87%	60%	Hospitalized alcoholics with liver disease
Bell et al. (1985)	104	53%	49%	Alcoholics admitted for detoxification
Stibler et al. (1986)	77	68%	55%	Alcoholics admitted for detoxification
Roine et al. (1987a)	31	23%	13%	Skid-row alcoholics in a homeless shelter
Roine et al. (1987a)	49	59%	22%	Alcoholics admitted for detoxification
Hollstedt and Dahlgren (1987)	100	18%	—	Outpatient female alcoholics
Behrens et al. (1988a)	72	43%	38%	Alcoholics admitted for detoxification
Persson and Magnusson (1988)	208	10%	—	Outpatients screened for alcoholism

have suggested that the serum AST : ALT ratio may help discriminate between alcoholic and nonalcoholic liver disease. As shown in Fig. 18.2, most cases of alcoholic liver disease have a serum AST : ALT ratio greater than 2, whereas most cases of nonalcoholic liver disease have an AST : ALT ratio less than 1 (Cohen and Kaplan, 1979; Correia *et al.*, 1981). However, patients with advanced forms of chronic viral hepatitis commonly have a serum AST : ALT ratio greater than 1 (Cohen and Kaplan, 1979; Williams and Hoofnagle, 1988).

A variety of mechanisms have been proposed to explain the increased AST : ALT ratio in alcoholic liver disease. Hepatocyte mitochondria are abundant in AST but lack ALT. Ethanol intoxication can produce a selective injury to the mitochondria, resulting in release of mitochondrial AST and elevation of the serum AST : ALT ratio (Williams and Hoofnagle, 1988). A second possible mechanism for the increased AST : ALT ratio is related to pyridoxal phosphate deficiency, which commonly occurs in alcoholics (Lumeng and Li, 1974; Rossouw *et al.*, 1978). Both laboratory and clinical studies have suggested that ALT activity can be depressed because of pyridoxine deficiency, resulting in an increased serum AST : ALT ratio (Ning *et al.*, 1966; Ludwig and Kaplowtiz, 1980; Matloff *et al.*, 1980; Diehl *et al.*, 1984).

18.2.3.3. Alkaline Phosphatase

The term alkaline phosphatase covers a family of enzymes that catalyze the hydrolysis of phosphate esters and have an optimal pH at alkaline conditions. Alkaline phosphatase is found in a variety of tissues including liver, bone, intestine, placenta, kidney, and leukocytes (Fishman, 1974). In the hepatocyte, alkaline phosphatase is located in the sinusoidal and canalicular membranes and in the cytosol (Hagerstrand, 1975). Elevations of alkaline phosphatase can result from cholestasis, cholangitis, infiltrative diseases of the liver, cirrhosis, hepatomas, and metastatic neoplasms. Proposed mechanisms for the elevation of alkaline phosphatase in hepatic diseases include (Kaplowitz *et al.*, 1982; Yamada *et al.*, 1985b):

1. Failure to excrete circulating alkaline phosphatase.
2. Increased enzyme synthesis (as seen in cholestasis).
3. Appearance of an abnormal form of alkaline phosphatase.
4. Failure of the liver to clear intestinal alkaline phosphatase.
5. Increased release of alkaline phosphatase from hepatic plasma membrane secondary to increased membrane lability.

Various studies have reported that alkaline phosphatase is not a sensitive marker of alcoholism. In alcoholic patients without apparent liver disease, the reported sensitivity is only 5–30% (Rosalki and Rau, 1972; Gluud *et al.*, 1981; Moussavian *et al.*, 1985). Even in alcoholics with liver disease, the reported sensitivity is approximately 50% (Wu *et al.*, 1976; Moussavian *et al.*, 1985). Furthermore, alkaline phosphatase elevation is not specific for alcoholic liver disease, as it may be elevated in a variety of hepatic and extrahepatic disorders.

The ratio of serum γ-glutamyl transpeptidase to alkaline phosphatase may be of value in discriminating between alcoholic and nonalcoholic liver disease. A ratio greater than 1.4 is suggestive of alcoholic liver disease, and a ratio greater than 3.5 is usually diagnostic (Lai *et al.*, 1982).

18.2.3.4. Glutamate Dehydrogenase

Glutamate dehydrogenase (GDH) is a mitochondrial enzyme that catalyzes the oxidative deamination of glutamate. The enzyme is primarily found in the liver

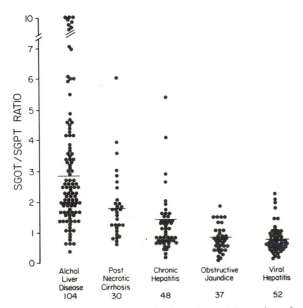

FIGURE 18.2. SGOT/SGPT ratios in patients with a variety of liver diseases. Most patients with alcoholic liver disease had a SGOT/SGPT ratio greater than 2. (Modified with permission from Cohen and Kaplan, 1979.)

and is concentrated in the perivenular area of the liver acinus (Van Waes and Lieber, 1977). In addition, GDH has been reported to reflect hepatic cell necrosis in alcoholic liver disease (Van Waes and Lieber, 1977; Worner and Lieber, 1980).

The reported sensitivity of serum GDH for alcoholism is 24–63% (Bernadt et al., 1982; Nalpas et al., 1984). However, serum GDH rapidly returns to normal after 5–10 days of cessation from drinking (Worner and Lieber, 1980). Thus, serum GDH should be measured within 48 hr of drinking for optimal clinical utility (Lieber, 1988). Finally, Worner and Lieber (1980) reported that there was no significant correlation between serum GDH and alcohol consumption.

The specificity of serum GDH as marker of alcoholic liver disease is disputed. Takase et al. (1985) reported that serum GDH was useful in discriminating between alcoholic and nonalcoholic liver disease. In contrast, two other studies reported significant overlap in GDH activity between alcoholic liver disease and a variety of nonalcoholic liver disorders (Nalpas et al., 1984; Ellis et al., 1978).

18.2.4. High-Density-Lipoprotein Cholesterol

High-density lipoprotein (HDL) functions in the transport of excess cholesterol from various tissues to the liver and may have a protective role against atherosclerosis. Furthermore, HDL is involved in the exchange of apolipoproteins, cholesterol, and phospholipids with other lipoproteins. HDL particles contain about 50% protein and 50% lipid by weight. The major protein components are apolipoproteins AI and AII, and the major lipid components include cholesterol and phospholipids. In addition, HDL can be separated by analytic ultracentrifugation into two subfractions, HDL_2 and HDL_3 (Naito, 1987a). Because direct measurement of total HDL mass and its subfractions is cumbersome, it is usually performed by specialized research laboratories (Naito, 1987a). However, HDL-cholesterol can be readily measured by precipitating the low-density (LDL) and very-low-density lipoproteins (VLDL) with heparin–manganese and measuring the remaining cholesterol in solution (primarily of HDL origin) by an enzymatic colorimetric method (Naito, 1987b). Thus, HDL-cholesterol can be routinely performed by most clinical laboratories.

The metabolism of HDL and its interaction with chronic ethanol consumption is complex and is further described in Chapter 4. Nonhuman primate models have been developed that demonstrated that chronic ethanol

administration increased serum HDL-cholesterol (Karsenty et al., 1985; Hojnacki et al., 1988). In addition, numerous clinical studies have been performed in human volunteers. Acute administration of ethanol did not result in a significant increase in serum HDL-cholesterol (Schneider et al., 1985; Taskinen et al., 1985). In contrast, chronic administration of either moderate doses of alcohol (20–40 g/day) or high doses of alcohol (greater than 70 g/day) significantly elevated serum HDL-cholesterol in nonalcoholic volunteers (Belfrage et al., 1977; Hulley and Gordon, 1981; Hartung et al., 1983; Fraser et al., 1983; Thornton et al., 1983; Crouse and Grundy, 1984; Haskell et al., 1984; Schneider et al., 1985; Camargo et al., 1985; Burr et al., 1986; Bertiere et al., 1986; Masarei et al., 1986). In addition, numerous studies in alcoholic patients hospitalized for detoxification have reported that the serum HDL-cholesterol was elevated at the time of admission and decreased to control values after 1–2 weeks of abstinence (Danielsson et al., 1978; Barboriak et al., 1980; Ekman et al., 1981; LaPorte et al., 1981; Devenyi et al., 1981; Taskinen et al., 1982; Bell et al., 1985; Tateossian et al., 1985; Desai et al., 1986). In studies investigating the effects of moderate ethanol consumption, the HDL_3 subfraction accounted for most of the increase in HDL-cholesterol, with only minimal change in the HDL_2 subfraction (Haskell et al., 1984; Bertiere et al., 1986). In contrast, the HDL_2 subfraction accounted for the major part of the increases in HDL-cholesterol in chronic alcoholics (Ekman et al., 1981; Valimaki et al., 1986).

Two of the mechanisms that have been postulated to explain the increase in HDL-cholesterol resulting from ethanol consumption are these:

1. Induction of microsomal enzymes resulting in increased synthesis of apolipoproteins AI and AII and secretion of nascent HDL (which can be then converted to HDL_3) (Cushman et al., 1982; Luoma et al., 1982). This mechanism has been used to explain the increase in the HDL_3 subfraction resulting from moderate alcohol consumption (Taskinen et al., 1987).
2. Increased plasma lipoprotein lipase activity resulting in an accelerated turnover of VLDL and an increased conversion of HDL_3 to HDL_2 (Schneider et al., 1985; Ekman et al., 1981; Taskinen et al., 1982; Sane et al., 1984). This mechanism has been used to explain the increase in HDL_2 subfraction resulting from heavy alcohol consumption (Taskinen et al., 1987).

However, the presence of severe liver disease appears to attenuate the effects of ethanol consumption on HDL-cholesterol. Several studies have reported that alcoholics with advanced liver disease had either normal or depressed levels of serum HDL-cholesterol at the time of admission (Devenyi *et al.*, 1981; Tateossian *et al.*, 1985; Malmendier *et al.*, 1983; Okamoto *et al.*, 1988). Furthermore, these levels did not change with abstinence in patients with advanced liver disease (Devenyi *et al.*, 1981; Tateossian *et al.*, 1985).

The high variability of serum HDL-cholesterol in the normal population (Heiss *et al.*, 1980) and the complexity of the effects of alcohol on HDL metabolism limit the application of HDL-cholesterol as a marker of alcohol consumption. Table 18.5 summarizes the studies investigating the sensitivity of HDL-cholesterol in a variety of alcoholic populations. The pooled sensitivity of HDL-cholesterol is only 31%. Various studies have indicated that serum HDL-cholesterol may be more sensitive in alcoholics without apparent liver disease (Devenyi *et al.*, 1981; Malmendier *et al.*, 1983; Tateossian *et al.*, 1985; Okamoto *et al.*, 1988), in contrast to most of the other markers, which are more sensitive in alcoholics with liver disease. Thus, the addition of serum HDL-cholesterol in combination with other markers will improve the sensitivity for detecting excessive alcohol consumption (Danielsson *et al.*, 1978; Puchois *et al.*, 1984; Cushman *et al.*, 1984; Bell *et al.*, 1985).

Most studies have reported a poor correlation between serum HDL-cholesterol and daily alcohol consumption in the alcoholic population (LaPorte *et al.*, 1981; Sanchez-Craig and Annis, 1981; Bernadt *et al.* 1982; Bell *et al.*, 1985; Dai *et al.*, 1985; Poikolainen *et al.*, 1985). In addition, correlations between the HDL subclasses (HDL$_2$ and HDL$_3$) and alcohol consumption were not significant (Dai *et al.*, 1985). Furthermore, Dai *et al.* (1985) noted that the relationship between HDL-

cholesterol and alcohol consumption was not linear; serum HDL-cholesterol tended to increase with ethanol consumption up to 350 g/day and then declined with heavier drinking.

A variety of factors other than alcohol can also affect HDL metabolism. Thus, elevations in serum HDL-cholesterol may be caused by genetic factors, diet, exercise, and a variety of medications (e.g., bile-acid-binding resins, estrogens, nicotinic acid, and gemfibrozil) (Glueck, 1985; Glueck *et al.*, 1985). In the adult population, females tend to have higher HDL-cholesterol levels than males, requiring the use of appropriate reference values for screening for alcohol use (Cushman *et al.*, 1984). Finally, as moderate alcohol consumption can elevate serum HDL-cholesterol, this test may not be able to discriminate accurately between social drinkers and alcoholics (Skinner *et al.*, 1984).

In conclusion, serum HDL-cholesterol is a poor screening test of alcoholism because of its lack of sensitivity and specificity. However, HDL-cholesterol may be of value when used in combination with other markers of alcoholism. Further studies are needed to determine whether utilizing various subfractions or components of HDL will result in a clinically useful marker of alcohol consumption.

18.3. Special Laboratory Markers

A number of special laboratory markers of alcoholism have been developed. Some of these markers show much promise for clinical application in the near future.

18.3.1. α-Amino-*n*-Butyric Acid

α-Amino-*n*-butyric acid (AANB) is a nonessential amino acid derived from the catabolism of methionine, serine, and threonine (Shaw and Lieber, 1983). Plasma

TABLE 18.5.   Studies Evaluating HDL-C as a Marker of Alcoholism

Author (year)	Number of alcoholics	Percentage with increased HDL-C	Selection of patients
Danielsson *et al.* (1978)	38	66%	Alcoholics admitted for detoxification
Skinner *et al.* (1984)	68	26%	Alcoholics seeking outpatient treatment
Skinner *et al.* (1984)	9	0%	Family practice patients screened for alcoholism
Cushman *et al.* (1984)	543	32%	Alcoholics referred for treatment
Sanchez-Craig and Annis (1981)	14	28%	Outpatients who are heavy consumers of alcohol
Puchois *et al.* (1984)	78	19%	Outpatients screened for alcoholism
Bell *et al.* (1985)	104	32%	Alcoholics admitted for detoxification

and tissue levels of AANB can be measured either by an amino acid analyzer (Shaw et al., 1976) or by gas–liquid chromatography (Ellingboe et al., 1978). Studies have demonstrated that chronic ethanol administration resulted in a threefold increase in plasma AANB levels in rats (Shaw and Lieber, 1980b) and a sevenfold increase in plasma AANB levels in baboons (Shaw and Lieber, 1978) as compared to pair-fed controls. Furthermore, a clinical investigation in which alcohol was reinstituted to alcoholics after a 6-week abstinent period resulted in a twofold increase in plasma AANB from control levels (Shaw and Lieber, 1978). Plasma AANB then returned to control levels following cessation of alcohol administration. The mechanism for the increase in AANB by ethanol consumption is unknown. It has been postulated that since chronic ethanol administration results in induction of the hepatic enzymes involved in methionine catabolism (Finkelstein et al., 1974), increased hepatic production of AANB may ensue (Shaw and Lieber, 1980b).

However, a variety of nutritional and metabolic factors can also affect plasma levels of AANB and other animo acids. Thus, low-protein diets can depress plasma AANB levels (Swendseid et al., 1968), whereas keto-acidosis can elevate serum AANB (Swendseid et al., 1967; Felig et al., 1970). However, these nutritional and metabolic factors tend to affect both AANB and the branched-chain amino acids in a similar degree, whereas chronic ethanol consumption has a greater effect on AANB than on the branched-chain amino acids (Shaw and Lieber, 1978).

Studies have attempted to use the ratio of plasma AANB to leucine as a marker of alcohol consumption in order to correct for the various nutritional and metabolic influences on AANB. Three studies reported that alcoholic patients have a higher plasma AANB-to-leucine ratio that nonalcoholic controls (Shaw et al., 1976; Morgan et al., 1977; Chick et al., 1982). However, studies have also noted that the plasma AANB-to-leucine ratio could not accurately discriminate between alcoholic patients and controls (Morgan et al.,1977; Dienstag et al., 1978; Eriksson et al., 1979; Hilderbrand et al., 1979; Jones et al., 1981; Chick et al., 1982). Furthermore, the plasma AANB-to-leucine ratio may be elevated in patients with nonalcoholic liver disease (Morgan et al., 1977; Dienstag et al., 1978; Eriksson et al., 1979; Yudkoff et al., 1979). The discriminating ability of plasma AANB could be improved by using a nonlinear model to describe the relationship between AANB and leucine (Shaw et al., 1978).

Despite the limited success of AANB as a screen-

ing marker of alcoholism, Shaw et al. (1979) reported that sequential AANB measurements may be helpful in identifying relapses in alcoholics undergoing treatment. In their study, alcoholics admitted for detoxification and rehabilitation were monitored with sequential markers (MCV, GGT, and AANB) obtained during admission, after 3 weeks of inpatient treatment, and during follow-up outpatient visits. Elevations of AANB above the base line obtained at discharge were seen in 50% of patients who had relapsed, as shown in Fig. 18.3. In contrast, the sensitivities of sequential serum GGT and MCV for relapse detection were 45% and 11%, respectively. The combination of plasma AANB and serum GGT had a 79% sensitivity for detecting relapses in alcoholic patients.

In conclusion, neither the plasma AANB nor the plasma AANB-to-leucine ratio serves as a good screening marker of alcoholism because of the many variables that affect amino acid metabolism. Nevertheless, sequential measurements of AANB may be useful for detecting relapses in patients undergoing treatment.

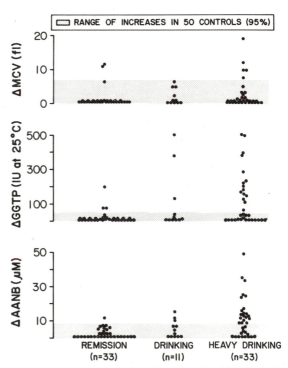

FIGURE 18.3. Increases in MCV, GGT, and AANB above the discharge value in alcoholics who relapsed as compared to alcoholics in remission. A majority of alcoholics who relapsed had an increase in either the GGT or the AANB. (From Shaw et al., 1979.)

18.3.2. Carbohydrate-Deficient Transferrin

Recent studies have suggested that ethanol may have important effects on transferrin metabolism. Transferrin is a plasma protein that is synthesized and secreted by the liver and is primarily involved in the transport of iron. Transferrin exhibits polymorphism in the human population. There are three major genetic variants known as transferrin B, C, and D (Giblett, 1962). The most frequent allele in most populations is the C type (Giblett, 1962). The pattern of inheritance appears to be both autosomal and codominant (Giblett, 1962). Transferrin also possesses a variety of carbohydrate moieties including sialic acid, galactose, N-acetylglucosamine, and mannose (Spik et al., 1975). When analyzed by isoelectric focusing, the normal forms of transferrin have an isoelectric point (pI) of 5.2–5.7 (Stibler and Borg, 1986). However, because sialic acid is a charged saccharide, alterations in the sialic acid content of transferrin will change the pI (Stibler and Borg, 1986).

Stibler et al. (1978) initially reported heterogeneity of transferrin in sera obtained from alcoholics when analyzed by isoelectric focusing. In addition to the main transferrin components with a pI of 5.2–5.7, transferrin variants with a pI of 5.7–5.9 were identified in most alcoholics (Stibler et al., 1978, 1979). The abnormal variant forms accounted for 5–10% of transferrin in alcoholics (Stibler et al., 1980; Vesterberg et al., 1984; Storey et al., 1985; Schellenberg and Weil, 1987b). Further studies revealed that the altered pI of these transferrin variants was caused by a reduction in the sialic acid component (Stibler and Borg, 1981; Petren et al., 1987). Thus, whereas the normals form of transferrin possess at least four molecules of sialic acid, the variant forms possess three or fewer molecules of sialic acid (Stibler and Borg, 1981). Stibler and Borg (1986) also demonstrated that these abnormal forms of transferrin were lacking in other carbohydrate moieties (e.g., galactose and N-acetylglucosamine).

The mechanism for the alteration in transferrin is unclear. However, ethanol exposure can inhibit hepatic synthesis of glycoproteins (Sorrell and Tuma, 1978; Sorrell et al., 1983). Thus, the impairment of hepatic glycosylation may result in the secretion of a carbohydrate-deficient transferrin. An alternative mechanism may be impaired hepatic elimination of the carbohydrate-deficient transferrin by ethanol exposure (Stibler and Hultcrantz, 1987).

Isoelectric focusing of serum is not suitable for routine clinical laboratories, so a simplified method of quantifying the carbohydrate-deficient transferrin (CDT) was developed (Stibler et al., 1986; Storey et al., 1987). This technique involves separating the CDT from the main transferrin component by ion-exchange chromatography and then quantifying the CDT by a radioimmune assay. Clinical studies have investigated the utility of CDT as a screening marker of alcoholism. Stibler et al. (1986) and Behrens et al. (1988a,b) utilized the absolute value of CDT in their studies, whereas Storey et al. (1987) used the ratio of CDT to total transferrin. Both of these methods resulted in a relatively sensitive marker of excessive alcohol consumption. Both Stibler et al. (1986) and Storey et al. (1987) reported sensitivities of at least 90%. Behrens et al. (1988b), studying a racially mixed population at the Bronx Veterans Affairs Medical Center, reported a sensitivity of 81% as shown in Fig. 18.4. In all three studies, CDT was superior to currently used markers of alcoholism (GGT, MCV, transaminases) (Stibler et al., 1986; Storey et al., 1987; Behrens et al., 1988b). Similar results were obtained using a newly developed method consisting of isoelectric focusing coupled with Western blotting (Xin et al., 1991), whereas a commerically available CDT kit measures only one quarter of the CDT present and has a lower sensitivity and specificity especially in populations including patients with liver disease (Xin et al., 1992).

Elevations of CDT appear to be specific for excessive alcohol consumption. The CDT was not elevated by moderate alcohol consumption (approximately 40 g/day of ethanol) (Stibler et al., 1986). Furthermore, the test had a specificity of 98% when applied to a population of patients with a variety of nonalcoholic medical disorders (Stibler et al., 1986). In addition, various medications that can elevate the serum GGT and transaminases (e.g., anticonvulsants, anticoagulants, antipsychotics) did not increase the CDT above the reference value (Stibler et al., 1986). Finally, CDT may be of value in discriminating between alcoholic and nonalcoholic liver disease. Two studies reported that none of their patients with nonalcoholic liver disease had an elevated CDT (Storey et al., 1987; Stibler and Hultcrantz, 1987). However, Behrens et al. (1988b) reported occasional false positives in nonalcoholic liver disease. It is noteworthy that 38% of patients with primary biliary cirrhosis had elevated CDT values (Fig. 18.5).

Stibler et al. (1986) reported a significant correlation between CDT concentration and daily alcohol consumption in a population that included normal consumers and alcoholics ($r = 0.64$). When the alcoholics were analyzed separately, the correlation was still signif-

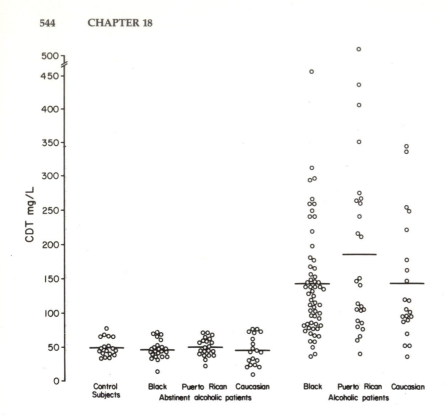

FIGURE 18.4. Levels of CDT in 18 nonalcoholic controls, 62 abstinent alcoholics, and 107 alcoholic patients who were actively drinking. The sensitivity of CDT in detecting actively drinking alcoholics was 81% at the Bronx VA. (From Behrens *et al.*, 1988b.)

icant but lower ($r = 0.22$). However, Behrens *et al.* (1988b) reported no significant correlation between these variables in their alcoholic population.

In most alcoholic patients, CDT levels decreased progressively with cessation of drinking and reached normal levels after 1–2 months of abstinence, as shown in Fig. 18.6 (Stibler *et al.*, 1986; Behrens *et al.*, 1988a). The biological half-life of CDT levels during the abstinent period is 16–17 days (Stibler *et al.*, 1986; Beh-

rens *et al.*, 1988a). However, some alcoholics may have CDT levels that fluctuate during the abstinent period rather than progressively decline (Behrens *et al.*, 1988a). Further studies are therefore needed to determine whether CDT can be used to monitor alcoholics for relapse.

In conclusion, initial studies suggest that CDT may be a useful marker of alcoholism, CDT has a higher sensitivity and specificity than the currently available

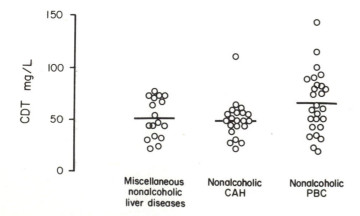

FIGURE 18.5. Levels of CDT in patients with nonalcoholic liver disease. In this study performed at the Bronx VA and Mount Sinai Medical Centers in New York, most patients with nonalcoholic liver disease had normal CDT levels. However 38% of patients with primary biliary cirrhosis had elevated CDT values. (From Behrens *et al.*, 1988b.)

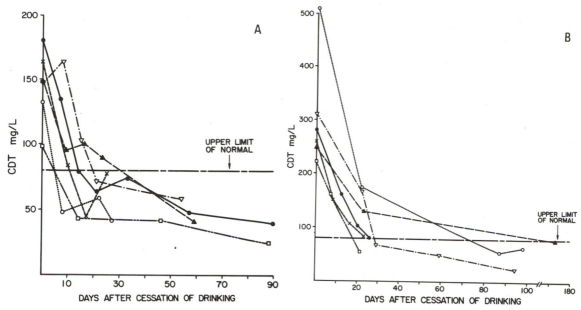

FIGURE 18.6. Sequential CDT levels in alcoholics with an initial CDT value below 200 mg/liter (A), or greater than 200 mg/liter (B). (From Behrens *et al.*, 1988a.)

markers. However, its utility in monitoring alcoholics for relapse requires further investigation.

18.3.3. Apolipoproteins AI and AII

Apolipoproteins AI and AII are the major components of HDL. A variety of immunologic techniques for measuring apolipoproteins have been developed (Karlin *et al.*, 1976; Curry *et al.*, 1976; Cheung and Albers, 1977; Goldberg *et al.*, 1980). Moderate alcohol consumption (14–40 g/day) can elevate serum apolipoproteins AI and AII in nonalcoholic volunteers (Fraser *et al.*, 1983; Camargo *et al.*, 1985; Masarei *et al.*, 1986; Moore *et al.*, 1988). Furthermore, alcoholics who were admitted for detoxification had increased apolipoprotein AI and AII levels, which normalized after 2–4 weeks of abstinence (Barboriak *et al.*, 1981). However, alcoholics with moderate to severe liver disease had depressed levels of these apolipoproteins (Malmendier *et al.*, 1983; Okamoto *et al.*, 1988), thus limiting the sensitivity of these tests. Puchois *et al.* (1984) reported that the sensitivities of serum apolipoproteins AI and AII were 28% and 45%, respectively. The combination of these apolipoprotein measurements with other conventional markers improved the composite sensitivity (Puchois *et al.*, 1984).

18.3.4. Mitochondrial Aspartate Aminotransferase

Total aminotransferase (tAST) of hepatic origin consists of two isoenzymes: mitochondrial aminotransferase (mAST) and cytosolic aminotransferase (cAST) (Morino *et al.*, 1964; Schmidt *et al.*, 1967a). Alcohol consumption can result in selective injury to the mitochondria, preferentially releasing mAST into the serum (Ishii *et al.*, 1979). Thus, the serum mAST has been suggested as a possible marker of alcohol consumption (Nalpas *et al.*, 1984). The mAST can be quantitated by both chromatographic (Schmidt *et al.*, 1967b) and immunochemical methods (Rej, 1980).

Ishii *et al.* (1979) initially reported that chronic ethanol consumption can elevate both the serum mAST and the ratio of mAST to the total serum AST (mAST : tAST). A study by Nalpas *et al.* (1986) reported that 84% of alcoholics with normal liver function tests and 95% of alcoholics with liver disease have elevated serum mAST levels (greater than 1.2 units/liter). However, most patients with nonalcoholic liver disease also have elevated serum mAST levels (Nalpas *et al.*, 1984, 1986; Okuno *et al.*, 1988). Nalpas *et al.* (1986) improved the specificity by using the mAST : tAST ratio. Thus, over 80% of actively drinking alcoholics had an elevated mAST:tAST ratio (greater than 7%) as compared to 11% of patients with nonalcoholic liver disease.

In conclusion, serum mAST may be a sensitive marker of alcoholism but is also elevated in non-alcoholic liver disease. The use of the serum mAST:tAST ratio may improve the specificity for alcoholic liver disease.

18.3.5. Acetaldehyde

As discussed in Chapter 2, chronic alcohol consumption can have significant effects on the blood levels of the ethanol metabolites. A variety of studies have reported that an ethanol challenge will result in increased plasma acetaldehyde levels in alcoholics as compared to controls (Korsten *et al.*, 1973; Lindros *et al.*, 1980; Maring *et al.*, 1983; DiPadova *et al.*, 1987). The increased acetaldehyde levels in chronic alcoholics have been attributed to both accelerated oxidation of ethanol and reduced metabolism of acetaldehyde (Lindros *et al.*, 1980; Pikkarainen *et al.*, 1981; Palmer and Jenkins, 1982; Nuutinen *et al.*, 1983). However, plasma acetaldehyde is not a reliable marker of ethanol consumption because acetaldehyde will be metabolized within a few hours after ethanol ingestion (Korsten *et al.*, 1975; DiPadova *et al.*, 1987). Furthermore, as discussed in Chapter 2, methods for plasma acetaldehyde determinations have been difficult to standardize.

With improved methods for acetaldehyde deter-minations, two studies have reported that a significant amount of blood acetaldehyde may be reversibly bound to erythrocytes (Baraona *et al.*, 1987; Peterson and Polizzi, 1987; Peterson *et al.*, 1988). After an ethanol challenge, acetaldehyde concentrations were three- to tenfold higher in red cells than in plasma (Baraona *et al.*, 1987). Furthermore, as shown in Fig. 18.7, elevations in red blood cell acetaldehyde may persist for several days or weeks following ethanol ingestion in alcoholic patients (Hernandez-Munoz *et al.*, 1989). In addition, chronic alcohol consumption may increase the binding capacity of erythrocytes for acetaldehyde (Hernandez-Munoz *et al.*, 1989). This increased binding capacity may reflect elevations in red blood cell cysteine, which occur after chronic alcohol consumption (Hernandez-Munoz *et al.*, 1989). Furthermore, as shown in Fig. 18.8, elevated red blood cell cysteine levels in alcoholics may persist for several days to weeks following alcohol withdrawal (Hernandez-Munoz *et al.*, 1989). Further investigations of the binding properties of erythrocytes for acetalde-hyde may provide a useful marker of alcoholism.

18.3.6. Ethanol and Acetaldehyde Adducts

Both ethanol and its metabolite, acetaldehyde may form stable chemical adducts with a variety of biological substances. Phospholipid–ethanol adducts have been identified in various tissues of laboratory animals administered ethanol (Alling *et al.*, 1983, 1984). Because

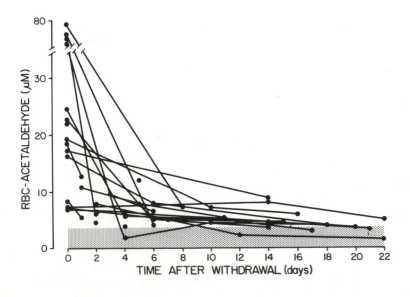

FIGURE 18.7. Evolution of RBC acetaldehyde in alcoholics following alcohol withdrawal. Elevations in RBC acetaldehyde may persist for several days after cessation of drinking. Normal values ± 2 S.D. are indicated by the shaded area. (From Hernandez-Munoz *et al.*, 1989.)

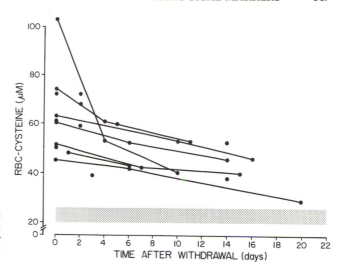

FIGURE 18.8. Evolution of RBC cysteine in alcoholics following alcohol withdrawal. Elevations in RBC cysteine may persist for several days or weeks after cessation of drinking. Normal values ± 2 S.D. are indicated by the shaded area. (From Hernandez-Munoz *et al.*, 1989.)

acetaldehyde possesses a carbonyl group, it may react with the nucleophilic groups in amino acids and proteins (Sorrell and Tuma, 1987; Collins, 1988). Thus, acetaldehyde may initially form a Schiff base with various amino groups (Sorrell and Tuma, 1987). Because Schiff bases are usually chemically unstable, a second reaction (e.g., reduction, rearrangement) is required in order to generate a stable adduct (Sorrell and Tuma, 1987); however, it is not certain whether these stabilizing reactions commonly occur *in vivo*.

The characterization of biologically stable acetaldehyde adducts *in vivo* may possibly yield markers of prior acetaldehyde levels analogous to the use of glycosylated hemoglobin (Hb_{Alc}) as an indicator of glucose control in diabetics (Jovanovic and Peterson, 1981). A variety of biological compounds that may form adducts with acetaldehyde include plasma amines, plasma proteins, hemoglobin, and intracellular proteins such as microsomal P450IIE1 (Behrens *et al.*, 1988c) and collagen (Behrens *et al.*, 1990). Furthermore, some of these acetaldehyde adducts may act as neoantigens, triggering the formation of humoral antibodies.

18.3.6.1. Acetaldehyde–Amine Adducts

Acetaldehyde can chemically form condensation products with a variety of biogenic amines. The condensation of acetaldehyde with catecholamines (e.g., norepinephrine, dopamine) will yield tetrahydroisoquinolines (TIQs), and the condensation of acetaldehyde with indolamines (e.g., tryptamine, serotonin) will generate tetrahydro-β-carbolines (THBCs). Various studies have reported that alcoholics have increased plasma or urinary levels of TIQ compounds (Collins *et al.*, 1979; Sjoquist *et al.*, 1981a,b; Adachi *et al.*, 1986; Faraj *et al.*, 1989) and THBC compounds (Beck *et al.*, 1982; Rommelspacher *et al.*, 1985). However, at least two factors limit the application of these tests as markers of ethanol consumption. First, some of these studies report a moderate degree of overlap between control and alcoholic patients (Collins *et al.*, 1979; Adachi *et al.*, 1986; Faraj *et al.*, 1989). In addition, many of the acetaldehyde–amine condensation products are present in a variety of foods and alcoholic beverages (Collins, 1988).

18.3.6.2. Acetaldehyde–Hemoglobin Adducts

Stevens *et al.* (1981) demonstrated that acetaldehyde–hemoglobin adducts may form *in vitro* in the presence of chemical reducing agents. Other studies have further confirmed that acetaldehyde–hemoglobin adducts can form *in vitro* with acetaldehyde concentrations comparable to levels seen after chronic alcohol consumption (Lumeng and Durant, 1985; Nguyen and Peterson, 1984).

The identification of acetaldehyde–hemoglobin adducts formed *in vivo* has been elusive. Several studies *in vitro* have reported that incubation of hemoglobin with acetaldehyde will increase the fast-migrating fraction when analyzed by cation-exchange chromatography

(Stevens *et al.*, 1981; Tsuboi *et al.*, 1981; Hoberman, 1983; Homaidan *et al.*, 1984). However, the application of column chromatography for the identification *in vivo* of acetaldehyde–hemoglobin adducts has yielded conflicting results. Several studies noted an increase in the fast-migrating hemoglobin fractions in alcoholics (Stevens *et al.*, 1981; Hoberman and Chiodo, 1982; Hoberman, 1983; Stockham and Blanke, 1988), but other studies reported no significant difference from nonalcoholic controls (Homaidan *et al.*, 1984; Mobley and Huisman, 1982; Peterson *et al.*, 1986). Other analytical techniques, including isoelectric focusing (Gordis and Herschkopf, 1986), measurement of acetaldehyde released by acid hydrolysis (Lucas *et al.*, 1988), and HPLC analysis of tryptic digests (Stockham and Blanke, 1988), have been inconsistent in identifying hemoglobin adducts *in vivo*. Thus, at the present time, acetaldehyde–hemoglobin adducts only remain a potential marker of alcohol consumption.

18.3.6.3. Antibodies against Acetaldehyde Adducts

A study by Israel *et al.* (1986) demonstrated that acetaldehyde adducts may act as neoantigens, generating antibodies specific for acetaldehyde epitopes in laboratory mice. In addition, the study reported that chronic administration of ethanol resulted in the generation of antibodies specific for protein–acetaldehyde conjugates. However, clinical studies evaluating these antibodies as a marker of alcoholism have not been consistent. An initial study by Hoerner *et al.* (1986) reported that alcoholics had significantly increased titers of antibodies against acetaldehyde adducts as compared to controls, as shown in Fig. 18.9. A further study by Hoerner *et al.* (1988) revealed that the titers were increased in most alcoholic patients in various stages of liver disease. In contrast, Niemela *et al.* (1987) reported that only patients with alcoholic hepatitis had significantly elevated titers of antibodies against acetaldehyde adducts. Finally, Hoerner *et al.* (1988) reported that patients with nonalcoholic liver disease also had elevated titers (shown in Fig. 18.9), limiting the clinical application of this test as a marker of alcoholism.

18.3.7. Erythrocyte Aldehyde Dehydrogenase

Several studies have reported that chronic ethanol consumption is associated with reduced hepatic aldehyde dehydrogenase (ALDH) activity, which results in impaired metabolism of acetaldehyde (Jenkins

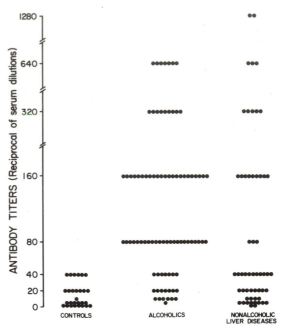

FIGURE 18.9. Serum antibody titers against acetaldehyde adducts in healthy controls, alcoholics, and patients with nonalcoholic liver disease. (From Hoerner *et al.*, 1988.)

and Peters, 1980; Lebsack *et al.*, 1981; Nuutinen *et al.*, 1983; Zorzano *et al.*, 1989). Erythrocytes also possess ALDH with kinetic properties similar to hepatic cytosolic ALDH (Inoue *et al.*, 1979; Rawles *et al.*, 1987). In addition, erythrocyte ALDH activity may reflect hepatic erythrocyte ALDH activity (Matthewson and Record, 1986) and thus may be similarly affected by chronic ethanol consumption. Several studies have reported that erythrocyte ALDH activity is reduced in alcoholics relative to controls (Maring *et al.*, 1983; Agarwal *et al.*, 1983a; Lin *et al.*, 1984; Takase *et al.*, 1985; Matthewson and Record, 1986; Towell *et al.*, 1986). Furthermore, erythrocyte ALDH is not significantly affected by nonalcoholic liver disease (Matthewson and Record, 1986; Zorzano *et al.*, 1989). However, a variety of factors limit the use of erythrocyte ALDH as a marker of active drinking. First, most studies reported at least a moderate degree of overlap in values between actively drinking alcoholics and controls (Maring *et al.*, 1983; Agarwal *et al.*, 1983a; Takase *et al.*, 1985; Matthewson and Record, 1986; Towell *et al.*, 1986). In addition, a variety of medications such as nitrates and oral hypoglycemics can depress erythrocyte ALDH activity (Maring *et al.*, 1983; Hellstrom *et al.*, 1983; Towell *et al.*, 1985).

18.3.8. β-Hexosaminidase

Chronic ethanol ingestion has been demonstrated to increase hepatic lysosomal enzyme activity in laboratory rats (Mezey *et al.*, 1976). Further studies in human volunteers suggested that administration of 60 g/day of ethanol over a 10-day period could increase serum β-hexosaminidase, a lysosomal enzyme (Isaksson *et al.*, 1985). Furthermore, Hultberg *et al.* (1980) reported that 95% of alcoholics admitted for detoxification in their study had elevated serum β-hexosaminidase. However, β-hexosaminidase can also be elevated in non-alcoholic liver diseases (Hultberg *et al.*, 1981a,b, 1983) and pregnancy (Hultberg and Isaksson, 1981). Thus, further studies are needed to evaluate the clinical utility of this laboratory test.

18.3.9. Urinary Dolichols

Dolichols are long-chain polyisoprenoid alcohols that function as carrier lipids in the biosynthesis of glycoproteins (Waechter and Lennarz, 1976) and may also influence the fluidity of cell membranes (Chojnacki and Dallner, 1988). Although almost all tissues are capable of synthesizing dolichols, the highest rates of biosynthesis in rats occur in the kidney, liver, and spleen (Elmberger *et al.*, 1987). At the subcellular level, dolichols are primarily synthesized in the microsomes (Adair and Keller, 1982; Wong and Lennarz, 1982; Ekstrom *et al.*, 1982).

High-performance liquid chromatographic (HPLC) methods have been developed for determination of dolichols in tissues, plasma, and urine (Yamada *et al.*, 1985c; Turpeinen, 1986). Several studies have reported that a variety of physiological and pathological states can affect dolichol metabolism in humans (Chojnacki and Dallner, 1988).

Several studies have investigated the effect of alcohol consumption on urinary dolichols. An acute dose of ethanol (5.5 g/kg body weight) can increase urinary dolichols in healthy nonalcoholic volunteers (Roine *et al.*, 1987b). Studies have further reported that urinary dolichols were significantly elevated in most hospitalized alcoholic patients (Pullarkat and Raguthu, 1985; Roine *et al.*, 1987a). However, the clinical utility of urinary dolichols as a marker of alcoholism may be limited because of their short decay half-life (3 days) (Roine, 1988) and lack of specificity for alcoholism (Roine *et al.*, 1989a). However, they may be potentially useful when used in conjunction with other markers of alcohol consumption.

18.3.10. Acetate

As discussed in Chapter 2, acetaldehyde is further metabolized to acetate in the liver. Nuutinen *et al.* (1985) reported that an acute dose of ethanol resulted in higher blood acetate levels in alcoholics than in controls. The elevation in acetate levels was attributed to the enhanced elimination of ethanol secondary to chronic ethanol consumption (Nuutinen *et al.*, 1985). Korri *et al.* (1985) evaluated the utility of blood acetate in screening intoxicated patients in the emergency room for chronic alcohol abuse. In their study, 65% of alcoholic patients had elevated acetate levels while intoxicated as compared to 8% of controls and occasional drinkers. Because acetate is rapidly metabolized after ethanol ingestion, blood acetate levels would be of value only in differentiating alcoholics from sporadic drinkers in the setting of acute intoxication (e.g., patients arrested for drunk driving).

18.3.11. Methanol

Methanol is a congener of many alcoholic beverages (Carroll, 1970) and is also formed endogenously in humans (Eriksen and Kulkarni, 1963). The oxidation of methanol is catalyzed by alcohol dehydrogenase (Mani *et al.*, 1970). Chronic alcohol consumption may inhibit the metabolism of methanol (Majchrowicz and Mendelson, 1971). Roine *et al.* (1989b) reported that serum methanol was significantly elevated in alcoholics as compared to sporadic drinkers arrested for drunk driving. Because methanol is rapidly metabolized after ethanol ingestion, serum methanol shares the limitations of serum acetate as a marker.

18.4. Biological Markers of a Predisposition to Alcoholism

Research during the last two decades has suggested a genetic predisposition to alcoholism (Crabb, 1990). Two forms of alcoholism have been identified (Cloninger, 1980). Type 1 is usually adult onset, has fewer social problems, occurs in both sexes, and is environmentally determined. Type 2 is usually early onset (before age 25), is associated with antisocial behavior, occurs mainly in males, and has a stronger genetic predisposition. In light of these findings, a biological marker of a predisposition to alcoholism would provide further insight into the pathogenesis of alcoholism.

18.4.1. Platelet Markers

Recent studies have suggested that various platelet enzymes, particularly monoamine oxidase (MAO), may reflect a predisposition to alcoholism. Monoamine oxidase is a mitochondrial enzyme that plays an important role in the regulation of neuronal stores of catecholamines and indoleamines (Kopin, 1964). Autopsy studies have suggested that alcoholics may have decreased brain MAO activity as compared to controls (Gottfries et al., 1975; Oreland et al., 1983). Platelets also possess MAO activity that may possibly reflect brain levels (Oreland et al., 1981).

Numerous studies have investigated the utility of platelet MAO activity as a biological marker of alcoholism. Several studies have reported decreased platelet MAO activity in alcoholics as compared to controls (Wiberg et al., 1977; Major and Murphy,1978; Sullivan et al., 1979; Alexopoulos et al., 1983; Agarwal et al., 1983b; Giller et al., 1984; Dolinsky et al., 1985; Faraj et al., 1987; Pandey et al., 1988). In contrast, a few studies were unable to detect a significant difference between alcoholics and controls (Takahashi et al., 1976; Lykouras et al., 1987; Tabakoff et al., 1988). These inconsistencies can be partially explained by the lack of standardization of the platelet MAO assay. Tsuji (1985) reported that the type of substrate used for the assay can affect the observed differences between alcoholics and controls. Furthermore, two studies reported that platelet MAO activity was only decreased in patients with the familial form of alcoholism (i.e., type 2) but not in patients with the type 1 form (von Knorring et al., 1985; Pandey et al., 1988). Studies have also reported that the depression in platelet MAO may persist for several years in alcoholics despite abstinence (Sullivan et al., 1978; Faraj et al., 1987), further suggesting that platelet MAO may reflect a predisposition for alcoholism. Finally, Tabakoff et al. (1988) reported that platelet adenylate cyclase activity is also altered in both acutely drinking and abstinent alcoholics.

18.4.2. Lymphocyte Markers

Lymphocytes have adenosine receptors, and their activation results in increases in cyclic adenosine monophosphate (cAMP) levels (Marone et al., 1978). Studies have reported that lymphocytes from alcoholics have altered basal and adenosine-receptor-stimulated cAMP levels as compared to controls (Diamond et al., 1987; Nagy et al., 1988). The alterations associated with alcoholism may, in part, result from genetic factors (Nagy et al., 1988).

18.4.3. Recombinant DNA Approaches

The use of DNA probes and endonuclease mapping could potentially identify restriction-fragment-length polymorphisms associated with alcoholism (Devor et al., 1988). Using this approach, Blum et al. (1990) identified an allelic form of the dopamine D_2 receptor gene that was associated with alcoholism. However, a study by Bolos et al. (1990) failed to confirm this association. While further studies are needed to resolve this conflict, the full potential of molecular biology in the field of alcoholism has yet to be realized.

18.5. Combination of Markers

18.5.1. Types of Combinations

At the present time, no single marker has been shown to have sufficient diagnostic accuracy to be useful for the screening of ambulatory patients for alcoholism (Salaspuro, 1986). Therefore, several investigators have attempted to combine markers in order to improve diagnostic accuracy. One can combine the markers qualitatively in order to improve the sensitivity or specificity. Alternatively, one can use multivariate analysis techniques (e.g., discriminant analysis) to mathematically combine the laboratory markers and generate a formula that may be used for diagnostic classification.

18.5.2. Qualitative Combinations

A qualitative combination of markers can be used to improve the sensitivity or specificity. In order to optimize the sensitivity,one can use a battery of laboratory tests and require that at least one of the tests be positive for the diagnosis of alcoholism. For example, in the study by Cushman et al. (1984), the sensitivities for the GGT, AST, and MCV were 48%, 47%, and 31%, respectively. However, 61% of the alcoholics had at least one abnormal test. Table 18.6 summarizes the improvement in sensitivity resulting from the combination of markers in this manner. However, this form of combination (where only one of the tests has to be positive) will decrease the specificity. For example, in a study where company directors were carefully screened for alcoholism (Chick et al., 1981), GGT and MCV had specificities of 78% and 94%, respectively. However, 26% of the nonalcoholics had either an elevated MCV or GGT, resulting in a specificity of 74%.

One can optimize specificity by using a battery of laboratory markers and require that all of the tests be

TABLE 18.6.   Studies Evaluating the Sensitivity of the Combination
of Markers in the Detection of Alcohol

Author (year)	Study population	GGT	MCV	AST	GGT or MCV	GGT or MCV or AST
Cushman et al.(1984)	Alcoholics attending rehabilitation program	48%	47%	31%	—	61%
Chick et al. (1981)	Manual workers screened for alcoholism	53%	23%	—	53%	—
Chick et al. (1981)	Company directors screened for alcoholism	50%	32%	—	63%	—
Chick et al. (1981)	Hospitalized male alcoholics	60%	40%	—	73%	—
Hollstedt and Dahlgren (1987)	Outpatient female alcoholics	42%	48%	18%	67%	68%
Latcham (1986)	Hospitalized alcoholics	52%	35%	—	64%	—

positive for a diagnosis of alcoholism. In the study of screened company directors by Chick et al. (1981), the diagnostic criterion that both MCV and GGT be positive for a diagnosis of alcoholism had a specificity of 98%, higher than the specificity of the individual tests. However, this form of combination will decrease the sensitivity to 17%.

A statistical technique has been developed that can qualitatively combine markers to optimize both the sensitivity and specificity and has been applied to the classification of alcoholic patients (Stamm et al., 1984). However, the diagnostic criteria developed by this technique require further validation.

18.5.3. Linear Discriminant Analysis

Linear discriminant analysis (LDA) is a statistical technique that results in a linear combination of variables to create a discriminant score or function. For example, if one uses a battery of laboratory markers (MCV, GGT, and AST) to classify patients as either alcoholics or nonalcoholics, one can generate the following discriminant function:

$$Y = A + B_1 \times MCV + B_2 \times GGT + B_3 \times AST$$

In this example, Y is the discriminant score; MCV, GGT, and AST are the laboratory values and serve as the independent variables; A is a constant provided by the analysis; and B_1, B_2, and B_3 are discriminant coefficients provided by the analysis. Linear discriminant analysis calculates the values for the discriminant coefficients B_1, B_2, and B_3 and the constant A such that the generated discriminant scores will optimally classify patients. In this example, patients with a discriminant score above the threshold score will be classified as alcoholics, whereas patients with a discriminant score below the threshold will be classified as nonalcoholics. In addition, in dealing with a large number of laboratory tests,

discriminant analysis can determine the relative contribution of a laboratory test to the discriminant function. Those tests that fail to contribute significantly to the discriminant function can be eliminated from the function. Finally, the ability of the discriminant function to classify patients accurately as alcoholics or nonalcoholics (e.g., sensitivity, specificity) needs to be determined. Ideally, the accuracy of the discriminant function should be determined by evaluating a population that is different from the one used for the derivation of the function ("cross-validation"). However, in cases of limited numbers of patients, various statistical techniques have been developed that use the study population for validating the discriminant function (Dolinsky and Schnitt, 1988).

Several studies have used linear discriminant analysis to classify alcoholics from nonalcoholics using conventional laboratory tests (Chalmers et al., 1981; Bliding et al., 1982; Shaper et al., 1985; Freedland et al., 1985; Chan et al., 1987). As shown in Table 18.7, these studies used different populations to develop the discriminant function. Thus, the discriminant analysis selected different laboratory tests as contributing to the discriminant function. For example, Chalmers et al. (1981) attempted to discriminate alcoholics from nonalcoholic patients with gastroenterology disorders. Because many of the nonalcoholics had liver disease, the transaminases were not useful in discriminating alcoholics from nonalcoholics. However, the transaminases may be useful as a marker of alcoholism when applied to a population with a low incidence of nonalcoholic liver disease. Furthermore, discriminant analysis may select laboratory tests normally not considered markers of alcoholism (e.g., potassium, phosphorus). Many hospitalized alcoholics are malnourished, resulting in decreased serum potassium and phosphorus. Thus, if the study population contains a large number of malnourished alcoholics, the discriminant analysis will then

TABLE 18.7.   Studies Using Linear Discriminant Analysis to Identify Alcoholics

Author (year)	Population studied	Laboratory values selected for discriminant function	Sensitivity	Specificity
Chalmers et al. (1981)[a]	Alcoholics, gastroenterology clinic patients, and patients with nonalcoholic liver disease	MCV, GGT, alkaline phospatase	87%	94%
Bliding et al. (1982)	Alcoholic and nonalcoholic Swedish military recruits	MCV, GGT, ALT, AST, uric acid, apolipoprotein AI, ferritin	54%	94%
Shaper et al. (1985)	Alcoholic and nonalcoholic patients participating in the British Regional Heart Study	GGT, HDL-C, uric acid, lead, mean corpuscular hemoglobin (MCH)	41%	98%
Chan et al. (1987)	Young adult alcoholics and nonalcoholic college students visiting student health clinic	BUN, potassium, MCV	89%	92%
Freedland et al. (1985)[a]	Hospitalized alcoholics and nonalcoholic psychiatric inpatients	BUN, AST, calcium, phosphorus, albumin	59%	72%

[a]The sensitivity and specificity were determined by cross-validation.

utilize laboratory tests associated with malnutrition for the discriminant function. However, the discriminant function may not be reliable when applied to a population of well-nourished alcoholics.

Table 18.7 also lists the sensitivities and specificities of the studies using linear discriminant analysis. However, it is difficult to compare these results as the studies used different populations for their analysis. Furthermore, the studies used different statistical methods to derive and validate the discriminant function. In general, the use of linear discriminate analysis was more accurate than individual laboratory tests in identifying alcoholics. However, none of the discriminant functions developed by these studies achieved a sensitivity greater than 90%. At the present, linear discriminant analysis may be helpful in corroborating a clinical diagnosis but does not appear to be accurate enough to be used as a screening test.

18.5.4. Quadratic Discriminant Analysis

Quadratic discriminant analysis uses a nonlinear combination of laboratory tests to develop a discriminant function. Linear discriminant analysis relies on mean differences in the laboratory values between groups to generate the discriminant function. However, quadratic discriminant analysis can also utilize the difference in relationships between tests to help differentiate alcoholics from nonalcoholics (Ryback et al., 1982a).

Four studies that have utilized quadratic discriminant analysis in alcohol research are summarized in Table 18.8. The sensitivities and specificities derived from validating samples for each study are also listed in that table. Three studies (Eckardt et al., 1981; Ryback et al., 1982b; Hillers et al., 1986) reported that quadratic discriminant analysis was very accurate in correctly classifying alcoholics and nonalcoholics. In contrast, Freedland et al. (1985) concluded that quadratic discriminant analysis was not better than linear discriminate analysis in classifying alcoholics when the discriminant function was tested by cross-validation. Furthermore, there are potential limitations to the use of quadratic discriminant analysis. The computations may be quite complex and thus not easily amenable to clinical application. In addition, the development of an accurate quadratic discriminant function using a large number of tests may require a very large sample size (Brown, 1984). Finally, the accuracy of quadratic discriminant function may decrease when applied to a different sample of patients (Freedland et al., 1985).

18.6.   Summary

Currently used markers of alcohol consumption can be helpful in corroborating a clinical suspicion of alcoholism. The combination of markers in a qualitative or quantitative fashion can further improve the diagnostic accuracy. At present, no conventional marker or combination of markers has sufficient diagnostic accuracy to be useful for screening low-risk populations. Because of its high sensitivity and specificity, carbohydrate-deficient transferrin has potential clinical use

TABLE 18.8.   Studies Using Quadratic Discriminant Analysisis to Identify Alcoholics

Author (year)	Population studied	Sensitivity	Specificity
Eckardt *et al.* (1981)	Hospitalized alcoholics and ambulatory nonalcoholics attending venereal disease clinic	95%	94%
Ryback *et al.* (1982b)	Alcoholics with liver disease and nonalcoholics with liver disease at Long Beach Veterans Affairs Hospital	92%	80%
Hillers *et al.* (1986)	Hospitalized alcoholics, men sentenced to driving school, college students, and university employees	91%	90%
Freedland *et al.* (1985)[a]	Hospitalized alcoholics and nonalcoholic psychiatric inpatients	32%	88%

[a]The sensitivity and specificity were determined by cross-validation.

TABLE 18.9.   Characteristics of Useful Laboratory Markers of Alcoholism

Marker	Estimated sensitivity	Half-life for decay	Causes of false positives
GGT	50%	26 days (variable)	Nonalcoholic liver disease, biliary tract disease, microsomal inducing agents, hyperlipidemia, obesity, hyperthyroidism
MCV	37%	Months (variable)	B_{12} deficiency, folate deficiency, hypothyroidism, malignancies, nonalcoholic liver disease
AST	35%	Weeks (variable)	Nonalcoholic liver disease, muscle disorders, myocardial infarction
ALT	30–35%	Weeks (variable)	Nonalcoholic liver disease
HDL-C	30%	5–10 days	Genetic factors, diet, exercise, medications, social drinking
CDT	80–90%	17–19 days	Primary biliary cirrhosis

when it becomes available as a routine test. Because patient self-report may not be reliable, the use of markers in monitoring alcoholics for relapse may lead to more objective assessment of treatment. However, only a few investigators have used these markers for monitoring relapses, and further studies are needed. Table 18.9 summarizes the diagnostic features of the currently available markers (as well as CDT). Future research on the biological and metabolic effects of alcohol consumption may lead to the establishment of more accurate markers or may make current markers more practical. The development of reliable biological markers would lead to significant improvements in the diagnosis and treatment of alcoholism.

REFERENCES

Adachi, J., Mizoi, Y., Fukunaga, T., Ueno, Y., Imamichi, H., Ninomiya, I., and Naito, T.: Individual difference in urinary excretion of salsolinol in alcoholic patients. *Alcohol* 3:371–375, 1986.

Adair, W. L., and Keller, R. K.: Dolichol metabolism in rat liver. Determination of the subcellular distribution of dolichyl phosphate and its site and rate of *de novo* biosynthesis. *J. Biol. Chem.* **257**:8990–8996, 1982.

Agarwal, D. P., Tobar-Rojas, L., Harada, S., and Goedde, H. W.: Comparative study of erythrocyte aldehyde dehydrogenase in alcoholics and control subjects. *Pharmacol. Biochem. Behav.* **18**:(Suppl. 1):89–95, 1983a.

Agarwal, D. P., Philippu, G., Milech, U., Ziemsen, B., Schrappe, O., and Goedde, H. W.: Platelet monoamine oxidase and erythrocyte catechol-O-methyltransferase activity in alcoholism and controlled abstinence. *Drug Alcohol Depend.* **12**:85–91, 1983b.

Alexopoulos, G. S., Lieberman, K. W., and Frances, R. J.: Platelet MAO activity in alcoholic patients and their first-degree relatives. *Am. J. Psychiatry* **140**:1501-1504, 1983.

Alling, C., Gustavsson, L., and Änggard, E.: An abnormal phospholipid in rat organs after ethanol treatment. *FEBS Lett.* **152**:24–28, 1983.

Alling, C., Gustavsson, L., Mansson, J. E., Benthin, G., and Änggard, E.: Phosphatidylethanol formation in rat organs after ethanol treatment. *Biochim. Biophys. Acta* **793**:119–122, 1984.

American Psychiatry Association: *Diagnostic and Statistical Manual of Mental Disorders* (3rd ed., rev.). Washington, APA, 1987, pp. 165–185.

Azizi, F.: γ-Glutamyl transpeptidase levels in thyroid disease. *Arch. Intern. Med.* **142**:79–81, 1982.

Bagrel, A., d'Houtaud, A., Gueguen, R., and Siest, G.: Relations between reported alcohol consumption and certain biological variables in an "unselected" population. *Clin. Chem.* **25:** 1242–1246, 1979.

Baraona, E., Di Padova, C., Tabasco, J., and Lieber, C. S.: Red blood cells: A new major modality for acetaldehyde transport from liver to other tissues. *Life Sci.* **40:**253–258, 1987.

Barboriak, J. J., Jacobson, G. R., Cushman, P., Herrington, R. E., Lipo, R. F., Daley, M. E., and Anderson, A. J.: Chronic alcohol abuse and high density lipoprotein cholesterol. *Alcoholism: Clin. Exp. Res.* **4:**346–349, 1980.

Barboriak, J. J., Alaupovic, P., and Cushman, P.: Abstinence-induced changes in plasma apolipoprotein levels of alcoholics. *Drug Alcohol Depend.* **8:**337–343, 1981.

Barrison, I. G., Viola, L., and Murray-Lyon, I. M.: Do housemen take an adequate drinking history? *Br. Med. J.* **281:**1040, 1980.

Barrison, I. G., Ruzek, J., and Murray-Lyon, I. M.: Drink-watchers—description of subjects and evaluation of laboratory markers of heavy drinking. *Alcohol Alcoholism* **22:**147–154, 1987.

Baxter, S., Fink, R., Leader, A. R., and Rosalki, S. B.: Laboratory tests for excessive alcohol consumption evaluated in general practice. *Br. J. Alcohol Alcoholism* **15:**164–166, 1980.

Beck, O., Bosin, T. R., Lundman, A., and Borg, S.: Identification and measurement of 6-hydroxy-1-methyl-1,2,3,4-tetrahydro-β-carboline by gas chromatography–mass spectrometry. *Biochem. Pharmacol.* **31:**2517–2521, 1982.

Behrens, U. J., Worner, T. M., and Lieber, C. S.: Changes in carbohydrate-deficient transferrin levels after alcohol withdrawal. *Alcoholism: Clin. Exp. Res.* **12:**539–544, 1988a.

Behrens, U. J., Worner, T. M., Braly, L. F., Schaffner, F., and Lieber, C. S.: Carbohydrate-deficient transferrin, a marker for chronic alcohol consumption in different ethnic populations. *Alcoholism: Clin. Exp. Res.* **12:**427–432, 1988b.

Behrens, U. J., Hoerner, M., Lasker, J. M., and Lieber, C. S.: Formation of acetaldehyde adducts with ethanol-inducible P450IIE1 *in vivo*. *Biochem. Biophys. Res. Commun.* **154:** 584–590, 1988c.

Behrens, U. J., Ma, X. L., Bychenok, S., Baraona, E., and Lieber, C. S.: Acetaldehyde–collagen adducts in CCL_4-induced liver injury in rats. *Biochem. Biophys. Res. Commun.* **173:**111–119, 1990.

Belfrage, P., Berg, B., Cronholm, T., Elmqvist, D., Hagerstrand, I., Johansson, B., Nilsson-Ehle, P., Norden, G., Sjovall, J., and Wiebe, T.: Prolonged administration of ethanol to young, healthy volunteers: Effects on biochemical, morphological and neurophysiological parameters. *Acta Med. Scand. [Suppl.]* **552:**1–44, 1973.

Belfrage, P., Berg, B., Hagerstrand, I., Nilsson-Ehle, P., Tornqvist, H., and Wiebe, T.: Alterations of lipid metabolism in healthy volunteers during long-term ethanol intake. *Eur. J. Clin. Invest.* **7:**127–131, 1977.

Bell, H., Stromme, J. H., Steensland, H., and Bache-Wiig, J. E.: Plasma HDL-cholesterol and estimated ethanol consumption in 104 patients with alcohol dependence syndrome. *Alcohol Alcoholism* **20:**35–40, 1985.

Bernadt, M. W., Mumford, J., Taylor, C., Smith, B., and Murray, R. M.: Comparison of questionnaire and laboratory tests in

the detection of excessive drinking and alcoholism. *Lancet* **1:**325–328, 1982.

Bertiere, M. C., Betoulle, D., Apfelbaum, M., and Girard-Globa, A.: Time-course, magnitude and nature of the changes induced in HDL by moderate alcohol intake in young non-drinking males. *Atherosclerosis* **61:**7–14, 1986.

Bilal, A. M., Kristof, J., and El-Islam, M. F.: A cross-cultural application of a drinking behaviour questionnaire. *Addict. Behav.* **12:**95–101, 1987.

Bliding, G., Bliding, A., Fex, G., and Tornqvist, C.: The appropriateness of laboratory tests in tracing young heavy drinkers. *Drug Alcohol Depend.* **10:**153–158, 1982.

Blum, K., Noble, E. P., Sheridan, P. J., Montgomery, A., Ritchie, T., Jagadeeswaran, P., Nogami, H., Briggs, A. H., and Cohn, J. B.: Allelic association of human dopamine D_2 receptor gene in alcoholism. *J.A.M.A.* **263:**2055–2060, 1990.

Bolos, A. M., Dean, M., Lucas-Derse, S., Ramsburg, M., Brown, G. L., and Goldman, D.: Population and pedigree studies reveal a lack of association between the dopamine D_2 receptor gene and alcoholism. *J.A.M.A.* **264:**3156–3160, 1990.

Brown, G. W.: Discriminant analysis. *Am. J. Dis. Child.* **138:**395–400, 1984.

Burr, M. L., Fehily, A. M., Butland, B. K., Bolton, C. H., and Eastham, R. D.: Alcohol and high-density-lipoprotein cholesterol: A randomized controlled trial. *Br. J. Nutr.* **56:**81–86, 1986.

Bush, B., Shaw, S., Cleary, P., Delbanco, T. L., and Aronson, M. D.: Screening for alcohol abuse using the CAGE questionnaire. *Am. J. Med.* **82:**231–235, 1987.

Camargo, C. A., Williams, P. T., Vranizan, K. M., Albers, J. J., and Wood, P. D.: The effect of moderate alcohol intake on serum apolipoproteins A-I and A-II: A controlled study. *J.A.M.A.* **253:**2854–2857, 1985.

Carney, M. W. P., and Sheffield, B.: Serum folate and B_{12} and haematological status of in-patient alcoholics. *Br. J. Addict.* **73:**3–7, 1978.

Carroll, R. B.: Analysis of alcohol beverages by gas–liquid chromatography. *Q. J. Stud. Alcohol* (Suppl. 5):6–19, 1970.

Chalmers, D. M., Levi, A. J., Chanarin, I., North, W. R. S., and Meade, T. W.: Mean cell volume in a working population: The effects of age, smoking, alcohol and oral contraception. *Br. J. Haematol.* **43:**631–636, 1979.

Chalmers, D. M., Rinsler, M. G., MacDermott, S., Spicer, C. C., and Levi, A. J.: Biochemical and haematological indicators of excessive alcohol consumption. *Gut* **22:**992–996, 1981.

Chan, A. W. K., Welte, J. W., and Whitney, R. B.: Identification of alcoholism in young adults by blood chemistries. *Alcohol* **4:**175–179, 1987.

Cheung, M. C., and Albers, J. J.: The measurement of apolipoprotein A-I and A-II levels in men and women by immunoassay. *J. Clin. Invest.* **60:**43–50, 1977.

Chick, J., Kreitman, N., and Plant, M.: Mean cell volume and gamma-glutamyl-transpeptidase as markers of drinking in working men. *Lancet* **1:**1249–1251, 1981.

Chick, J., Longstaff, M., Kreitman, N., Plant, M., Thatcher, D., and Waite, J.: Plasma α-amino-*n*-butyric acid : leucine ratio and alcohol consumption in working men and in alcoholics. *J. Stud. Alcohol* **43:**583–587, 1982.

Chojnacki, T., and Dallner, G.: The biological role of dolichol. *Biochem. J.* **251**:1–9, 1988.

Clark, P. M. S., Kricka, L. J., and Zaman, S.: Drivers, binge drinking, and gamma-glutamyltranspeptidase. *Br. Med. J.* **285**:1656–1657, 1982.

Clark, P. M. S., Holder, R., Mullet, M., and Whitehead, T. P.: Sensitivity and specificity of laboratory tests for alcohol abuse. *Alcohol Alcoholism* **18**:261–269, 1983.

Clark, W. D.: Alcoholism: Blocks to diagnosis and treatment. *Am. J. Med.* **71**:275–286, 1981.

Clermont, R. J., and Chalmers, T. C.: The transaminase tests in liver disease. *Medicine* **46**:197–207, 1967.

Cloninger, C. R.: Neurogenetic adaptive mechanisms in alcoholism. *Science* **236**:410–416, 1980.

Cohen, J. A., and Kaplan, M. M.: The SGOT/SGPT ratio—an indicator of alcoholic liver disease. *Dig. Dis. Sci.* **24**:835–838, 1979.

Collins, M. A.: Acetaldehyde and its condensation products as markers in alcoholism. *Recent Dev. Alcohol* **6**:387–403, 1988.

Collins, M. A., Num, W. P., Borge, G. F., Teas, G., and Goldfarb, C.: Dopamine-related tetrahydroisoquinolines: Significant urinary excretion by alcoholics after alcohol consumption. *Science* **206**:1184–1186, 1979.

Correia, J. P., Alves, P. S., and Camilo, E. A.: SGOT–SGPT ratios. *Dig. Dis. Sci.* **26**:284, 1981.

Crabb, D. W.: Biological markers for increased risk of alcoholism and for quantitation of alcohol consumption. *J. Clin. Invest.* **85**:311–315, 1990.

Crouse, J. R., and Grundy, S. M.: Effects of alcohol on plasma lipoproteins and cholesterol and triglyceride metabolism in man. *J. Lipid Res.* **25**:486–496, 1984.

Curry, M. D., Alaupovic, P., and Suenram, C. A.: Determination of apolipoprotein A and its constitutive A-I and A-II polypeptides by separate electroimmunoassays. *Clin. Chem.* **22**:315–322, 1976.

Cushman, P., Barboriak, J. J., Liao, A., and Hoffman, N. E.: Association between plasma high density lipoprotein cholesterol and antipyrine metabolism in alcoholics. *Life Sci.* **30**:1721–1724, 1982.

Cushman, P., Jacobson, G., Barboriak, J. J., and Anderson, A. J.: Biochemical markers for alcoholism: Sensitivity problems. *Alcoholism: Clin. Exp. Res.* **8**:253–257, 1984.

Dai, W. S., LaPorte, R. E., Hom, D. L., Kuller, L. H., D'Antonio, J. A., Gutai, J. P., Wozniczak, M., and Wohlfahrt, B.: Alcohol consumption and high density lipoprotein cholesterol concentration among alcoholics. *Am. J. Epidemiol.* **122**:620–627, 1985.

Danielsson, B., Ekman, R., Fex, G., Johansson, B. G., Kristensson, H., Nilsson-Ehle, P., and Wadstein, J.: Changes in plasma high density lipoproteins in chronic male alcoholics during and after abuse. *Scand. J. Clin. Lab. Invest.* **38**:113–119, 1978.

Davidson, R. J. L., and Hamilton, P. J.: High mean red cell volume: Its incidence and significance in routine haematology. *J. Clin. Pathol.* **31**:493–498, 1978.

Davis, L. J., Hurt, R. D., Morse, R. M., and O'Brien, P. C.: Discriminant analysis of Self-Administered Alcoholism

Screening Test. *Alcoholism: Clin. Exp. Res.* **11**:269–273, 1987.

Desai, K., Owen, J. S., Wilson, D. T., and Hutton, R. A.: Platelet aggregation and plasma lipoproteins in alcoholics during alcohol withdrawal. *Thromb. Haemostas.* **55**:173–177, 1986.

Devenyi, P., Robinson, G. M., Kapur, B. M., and Roncari, D. A. K.: High-density lipoprotein cholesterol in male alcoholics with and without severe liver disease. *Am. J. Med.* **71**:589–594, 1981.

Devgun, M. S., Dunbar, J. A., Hagart, J., Martin, B. T., and Ogston, S. A.: Effects of acute and varying amounts of alcohol consumption on alkaline phosphatase, aspartate transaminase, and γ-glutamyltransferase. *Alcoholism: Clin. Exp. Res.* **9**:235–237, 1985.

Devor, E. J., Reich, T., and Cloninger, C. R.: Genetics of alcoholism and related end-organ damage. *Semin. Liver Dis.* **8**:1–11, 1988.

Diamond, I., Wrubel, B., Estrin, W., and Gordon, A.: Basal and adenosine receptor-stimulated levels of cAMP are reduced in lymphocytes from alcoholic patients. *Proc. Natl. Acad. Sci. U.S.A.* **84**:1413–1416, 1987.

Diehl, A. M., Potter, J., Boitnott, J., Van Duyn, M. A., Herlong, H. F., and Mezey, E.: Relationship between pyridoxal 5′-phosphate deficiency and aminotransferase levels in alcoholic hepatitis. *Gastroenterology* **86**:632–636, 1984.

Dienstag, J. L., Carter, E. A., Wands, J. R., Isselbacher, K. J., and Fischer, J. E.: Plasma α-amino-*n*-butyric acid to leucine ratio: Nonspecificity as a marker for alcoholism. *Gastroenterology* **75**:561–565, 1978.

DiPadova, C., Worner, T. M., and Lieber, C. S.: Effect of abstinence on the blood acetaldehyde response to a test dose of alcohol in alcoholics. *Alcoholism: Clin. Exp. Res.* **11**:559–561, 1987.

Dolinsky, Z. S., and Schnitt, J. M.: Discriminant function analysis of clinical laboratory data. Use in alcohol research. *Recent Dev. Alcohol* **6**:367–385, 1988.

Dolinsky, Z. S., Shaskan, E. G., and Hesselbrock, M. N.: Basic aspects of blood platelet monoamine oxidase activity in hospitalized men alcoholics. *J. Stud. Alcohol* **46**:81–85, 1985.

Dreyfus, J. C., Schapira, G., and Schapira, F.: Serum enzymes in the physiopathology of muscle. *Ann. N.Y. Acad. Sci.* **75**:235–249, 1959.

Dunbar, J. A., Hagart, J., Martin, B., Ogston, S., and Devgun, M. S.: Drivers, binge drinking, and gammaglutamyltranspeptidase. *Br. Med. J.* **285**:1083, 1982.

Eckardt, M. J., Ryback, R. S., Rawlings, R. R., and Graubard, B. I.: Biochemical diagnosis of alcoholism. A test of the discriminating capabilities of γ-glutamyl transpeptidase and mean corpuscular volume. *J.A.M.A.* **246**:2707–2710, 1981.

Ekman, R., Fex, G., Johansson, B. G., Nilsson-Ehle, P., and Wadstein, J.: Changes in plasma high density lipoproteins and lipolytic enzymes after long-term, heavy ethanol consumption. *Scand. J. Clin. Lab. Invest.* **41**:709–715, 1981.

Ekstrom, T., Eggens, I., and Dallner, G.: Biosynthesis of dolichol and dolichylphosphate in rat hepatocytes *in vivo*. *FEBS Lett.* **150**:133–136, 1982.

Ellingboe, J., Mendelson, J. H., Varanelli, C. C., Neuberger, O., and Borysow, M.: Plasma alpha amino-*n*-butyric acid:leucine ratio. Normal values in alcoholics. *J. Stud. Alcohol* **39:**1467–1476, 1978.

Ellis, G., Goldberg, D. M., Spooner, R. J., and Ward, A. M.: Serum enzyme tests in diseases of the liver and biliary tree. *Am. J. Clin. Pathol.* **70:**248–258, 1978.

Elmberger, P. G., Kalen, A., Appelkvist, E. L., and Dallner, G.: *In vitro* and *in vivo* synthesis of dolichol and other main mevalonate products in various organs of the rat. *Eur. J. Biochem.* **168:**1–11, 1987.

Eriksen, J., Olsen, P. S., and Thomsen, A. C.: Gamma-glutamyl-transpeptidase, aspartate aminotransferase, and erythrocyte mean corpuscular volume as indicators of alcohol consumption in liver disease. *Scand. J. Gastroenterol.* **19:**813–819, 1984.

Eriksen, S. P., and Kulkarni, A. B.: Methanol in normal human breath. *Science* **141:**639–640, 1963.

Eriksson, S., Fex, G., Johansson, B., and Nilsson, B.: Plasma alpha-amino-*n*-butyric acid/leucine ratio in alcoholism. *N. Engl. J. Med.* **300:**93–94, 1979.

Eschwege, E., Papoz, L., Lellouch, J., Claude, J. R., Cubeau, J., Pequignot, G., Richard, J. L., and Schwartz, D.: Blood cells and alcohol consumption with special reference to smoking habits. *J. Clin. Pathol.* **31:**654–658, 1978.

Faraj, B. A., Lenton, J. D., Kutner, M., Camp, V. M., Stammers, T. W., Lee, S. R., Lolies, P. A., and Chandora, D.: Prevalence of low monoamine oxidase function in alcoholism. *Alcoholism: Clin. Exp. Res.* **11:**464–467, 1987.

Faraj, B. A., Camp, V. M., Davis, D. C., Lenton, J. D., and Kutner, M.: Elevation of plasma salsolinol sulfate in chronic alcoholics as compared to nonalcoholics. *Alcoholism: Clin. Exp. Res.* **13:**155–163, 1989.

Felig, P., Marliss, E., Ohman, J. L., and Cahill, G. F.: Plasma amino acid levels in diabetic ketoacidosis. *Diabetes* **19:**727–729, 1970.

Finkelstein, J. D., Cello, J. P., and Kyle, W. E.: Ethanol-induced changes in methionine metabolism in rat liver. *Biochem. Biophys. Res. Commun.* **61:**525–531, 1974.

Fishman, W. H.: Perspectives on alkaline phosphatase isoenzymes. *Am. J. Med.* **56:**617–650, 1974.

Fraser, G. E., Anderson, J. T., Foster, N., Goldberg, R., Jacobs, D., and Blackburn, H.: The effect of alcohol on serum high density lipoprotein (HDL). A controlled experiment. *Atherosclerosis* **46:**275–286, 1983.

Freedland, K. E., Frankel, M. T., and Evenson, R. C.: Biochemical diagnosis of alcoholism in men psychiatry patients. *J. Stud. Alcohol* **46:**103–106, 1985.

Fuller, R. K., Lee, K. K., and Gordis, E.: Validity of self-report in alcoholism research: Results of a Veterans Administration cooperative study. *Alcoholism: Clin. Exp. Res.* **12:**201–205, 1988.

Gadeholt, G., Aarbakke, J., Dybing, E., Sjoblom, M., and Morland, J.: Hepatic microsomal drug metabolism, glutamyl transferase activity and *in vivo* antipyrine half-life in rats chronically fed an ethanol diet, a control diet and a chow diet. *J. Pharmacol. Exp. Ther.* **213:**196–203, 1980.

Garvin, R. B., Foy, D. W., and Alford, G. S.: A critical examin-

ation of gamma-glutamyl transpeptidase as a biochemical marker for alcohol abuse. *Addict. Behav.* **6:**377–383, 1981.

Giblett, E. R.: The plasma transferrins. *Prog. Med. Genet.* **2:**34–63, 1962.

Gill, G. V., Baylis, P. H., Flear, C. T. G., Skillen, A. W., and Diggle, P. H.: Acute biochemical responses to moderate beer drinking. *Br. Med. J.* **285:**1770–1773, 1982.

Giller, E., Nocks, J., Hall, H., Stewart, C., Schnitt, J., and Sherman, B.: Platelet and fibroblast monoamine oxidase in alcoholism. *Psychiatry Res.* **12:**339–347, 1984.

Gjerde, H., and Morland, J.: Concentrations of carbohydrate-deficient transferrin in dialysed plasma from drunken drivers. *Alcohol Alcoholism* **22:**271–276, 1987.

Glueck, C. J.: Nonpharmacologic and pharmacologic alteration of high-density lipoprotein cholesterol: Therapeutic approaches to prevention of atherosclerosis. *Am. Heart J.* **110:**1107–1115, 1985.

Glueck, C. J., Laskarzewski, P. M., Rao, D. C., and Morrison, J. A.: Familial aggregation of coronary risk factors. In: *Coronary Artery Disease: Prevention, Complication, and Treatment* (W. E. Connor, and J. D. Bristow, eds.), Philadelphia, J.B. Lippincott, 1985. pp. 173–192.

Gluud, C., Andersen, I., Deitrichson, O., Gluud, B., Jacobsen, A., and Juhl, E.: Gamma-glutamyltransferase, aspartate aminotransferase and alkaline phosphatase as markers of alcohol consumption in out-patient alcoholics. *Eur. J. Clin. Invest.* **11:**171–176, 1981.

Goldberg, R. B., Karlin, J. B., Juhn, D. J., Scanu, A. M., Edelstein, C., and Rubenstein, A. H.: Characterization and measurement of human apolipoprotein A-II radioimmunoassay. *J. Lipid Res.* **21:**902–912, 1980.

Gordis, E., and Herschkopf, S.: Application of isoelectric focusing in immobilized pH gradients to the study of acetaldehyde-modified hemoglobin. *Alcoholism: Clin. Exp. Res.* **10:**311–319, 1986.

Gottfried, E. L., and Wagar, E. A.: Laboratory testing: A practical guide. *DM* **29:**1–41, 1983.

Gottfries, C. G., Oreland, L., Wiberg, A., and Winblad, B.: Lowered monoamine oxidase activity in brains from alcoholic suicides. *J. Neurochem.* **25:**667–673, 1975.

Hagerstrand, I.: Distribution of alkaline phosphatase activity in healthy and diseased human liver tissue. *Acta. Pathol. Microbiol. Scand. (A)* **83:**519–526, 1975.

Harinasuta, U., Chomet, B., Ishak, K., and Zimmerman, H. J.: Steatonecrosis—Mallory body type. *Medicine* **46:**141–167, 1967.

Hartung, G. H., Foreyt, J. P., Mitchell, R. E., Mitchell, J. G., Reeves, R. S., and Gotto, A. M.: Effect of alcohol intake on high-density lipoprotein cholesterol levels in runners and inactive men. *J.A.M.A.* **249:**747–750, 1983.

Haskell, W. L., Camargo, C., Williams, P. T., Vranizan, K. M., Krauss, R. M., Lindgren, F. T., and Wood, P. D.: The effect of cessation and resumption of moderate alcohol intake on serum high-density-lipoprotein subfractions. A controlled study. *N. Engl. J. Med.* **310:**805–810, 1984.

Heiss, G., Johnson, N. J., Reiland, S., Davis, C. E., and Tyroler,

H. A.: The epidemiology of plasma high-density lipoprotein cholesterol levels. The Lipid Research Clinics Program Prevalence Study summary. *Circulation* **62**(Suppl. IV):116–136, 1980.

Hellstrom, E., Tottmar, O., and Widerlov, E.: Effects of oral administration or implantation of disulfiram on aldehyde dehydrogenase activity in human blood. *Alcoholism: Clin. Exp. Res.* **7**:231–236, 1983.

Hernandez-Munoz, R., Baraona, E., Blacksberg, I., and Lieber, C. S.: Characterization of the increased binding of acetaldehyde to red blood cells in alcoholics. *Alcoholism: Clin. Exp. Res.* **13**:654–659, 1989.

Hilderbrand, R. L., Hervig, L. K., Conway, T. L., Ward, H. W., and Markland, F. S.: Alcohol intake, ratio of plasma alpha-amino-*n*-butyric acid to leucine, and gamma-glutamyl transpeptidase in nonalcoholics. *J. Stud. Alcohol* **40**:902–905, 1979.

Hillers,, V. N., Alldredge, J. R., and Massey, L. K.: Determination of habitual alcohol intake from a panel of blood chemistries. *Alcohol Alcoholism* **21**:199–205, 1986.

Hoberman, H. D.: Post-translational modification of hemoglobin in alcoholism. *Biochem. Biophys. Res. Commun.* **113**:1004–1009, 1983.

Hoberman, H. D., and Chiodo, S. M.: Elevation of the hemoglobin A$_1$ fraction in alcoholism. *Alcoholism: Clin. Exp. Res.* **6**:260–266, 1982.

Hoerner, M., Behrens, U. J., Worner, T. M., and Lieber, C. S.: Humoral immune response to acetaldehyde adducts in alcoholic patients. *Res. Commun. Chem. Pathol. Pharmacol.* **54**:3–12, 1986.

Hoerner, M., Behrens, U. J., Worner, T. M., Blacksberg, I., Braly, L. F., Schaffner, F., and Lieber, C. S.: The role of alcoholism and liver disease in the appearance of serum antibodies against acetaldehyde adducts. *Hepatology* **8**:569–574, 1988.

Hojnacki, J. L., Cluette-Brown, J. E., Mulligan, J. J., Hagan, S. M., Mahony, K. E., Witzgall, S. K., Osmolski, T. V., and Barboriak, J. J.: Effect of ethanol dose on low density lipoproteins and high density lipoprotein subfractions. *Alcoholism: Clin. Exp. Res.* **12**:149–154, 1988.

Hollstedt, C., and Dahlgren, L.: Peripheral markers in the female "hidden alcoholic." *Acta Psychiatr. Scand.* **75**:591–596, 1987.

Homaidan, F. R., Kricka, L. J., Clark, P. M. S., Jones, S. R., and Whitehead, T. P.: Acetaldehyde–hemoglobin adducts: An unreliable marker of alcohol abuse. *Clin. Chem.* **30**:480–482, 1984.

Hulley, S. B., and Gordon, S.: Alcohol and high-density lipoprotein cholesterol. Causal inference from diverse study designs. *Circulation* **64**(Suppl. III):57–63, 1981.

Hultberg, B., and Isaksson, A.: A possible explanation for the occurrence of increased β-hexosaminidase activity in pregnancy serum. *Clin. Chim. Acta* **113**:135–140, 1981.

Hultberg, B., Isaksson, A., and Tiderstrom, G.: β-Hexosaminidase, leucine aminopeptidase, cystidyl aminopeptidase, hepatic enzymes and bilirubin in serum of chronic alcoholics with acute ethanol intoxication. *Clin. Chim. Acta* **105**:317–323, 1980.

Hultberg, B., Braconier, J. H., Isaksson, A., and Jansson, L.:

β-Hexosaminidase level in serum from patients with viral hepatitis as a measure of reticuloendothelial function. *Scand. J. Infect. Dis.* **13**:241–245, 1981a.

Hultberg, B., Isaksson, A., and Jansson, L.: β-Hexosaminidase in serum from patients with cirrhosis and cholestasis. *Enzyme* **26**:296–300, 1981b.

Hultberg, B., Isaksson, A., Joelsson, B., Alwmark, A., Gullstrand, P., and Bengmark, S.: Pattern of serum beta-hexosaminidase in liver cirrhosis. *Scand. J. Gastroenterol.* **18**:877–880, 1983.

Inoue, K., Nishimukai, H., and Yamasawa, K.: Purification and partial characterization of aldehyde dehydrogenase from human erythrocytes. *Biochim. Biophys. Acta* **569**:117–123, 1979.

Irwin, M., Baird, S., Smith, T. L., and Schuckit, M.: Use of laboratory tests to monitor heavy drinking by alcoholic men discharged from a treatment program. *Am. J. Psychiatry* **145**:595–599, 1988.

Isaksson, A., Blanche, C., Hultberg, B., and Joelsson, B.: Influence of ethanol on the human serum level of beta-hexosaminidase. *Enzyme* **33**:162–166, 1985.

Ishii, H., Yasuraoka, S., Shigeta, Y., Takagi, S., Kamiya, T., Okuno, F., Miyamoto, K., and Tsuchiya, M.: Hepatic and intestinal gamma-glutamyltranspeptidase activity: Its activation by chronic ethanol administration. *Life Sci.* **23**:1393–1398, 1978.

Ishii, H., Okuno, F., Shigeta, Y., and Tsuchiya, M.: Enhanced serum glutamic oxaloacetic transaminase activity of mitochondrial origin in chronic alcoholics. In: *Currents in Alcoholism*, Vol. V (M. Galanter, ed.), New York, Grune & Stratton, 1979, pp. 101–108.

Ishii, H., Ebihara, Y., Okuno, F., Munakata, Y., Takagi, T., Arai, M., Shigeta, S., and Tsuchiya, M.: γ-Glutamyl transpeptidase activity in liver of alcoholics and its histochemical localization. *Alcoholism: Clin. Exp. Res.* **10**:81–85, 1986.

Israel, Y., Hurwitz, E., Niemela, O., and Arnon, R.: Monoclonal and polyclonal antibodies against acetaldehyde-containing epitopes in acetaldehyde–protein adducts. *Proc. Natl. Acad. Sci. U.S.A.* **83**:7923–7927, 1986.

Jenkins, W. J., and Peters, T. J.: Selectively reduced hepatic acetaldehyde dehydrogenase in alcoholics. *Lancet* **1**:628–629, 1980.

Jones, J. D., Morse, R. M., and Hurt, R. D.: Plasma α-amino-*n*-butyric acid/leucine ratio in alcoholics. *Alcoholism: Clin. Exp. Res.* **5**:363–365, 1981.

Jovanovic, L., and Peterson, C. M.: The clinical utility of glycosylated hemoglobin. *Am. J. Med.* **70**:331–338, 1981.

Kaplowitz, N., Eberle, D., and Yamada, T.: Biochemical tests for liver disease. In: *Hepatology: A Textbook of Liver Disease* (D. Zakim and T. D. Boyer, eds.), Philadelphia, W. B. Saunders, 1982, pp. 583–612.

Karlin, J. B., Juhn, D. J., Starr, J. I., Scanu, A. M., and Rubenstein, A. H.: Measurement of human high density lipoprotein apolipoprotein A-I in serum by radioimmunoassay. *J. Lipid Res.* **17**:30–37, 1976.

Karsenty, C., Baraona, E., Savolainen, M. J., and Lieber, C. S.: Effects of chronic ethanol intake on mobilization and excretion of cholesterol in baboons. *J. Clin. Invest.* **75:**976–986, 1985.

Kopin, I. J.: Storage and metabolism of catecholamines: The role of monoamine oxidase. *Pharmacol. Rev.* **16:**179–191, 1964.

Korri, U. M., Nuutinen, H., and Salaspuro, M.: Increased blood acetate: A new laboratory marker of alcoholism and heavy drinking. *Alcoholism: Clin. Exp. Res.* **9:**468–471, 1985.

Korsten, M. A., Matsuzaki, S., Feinman, L., and Lieber, C. S.: High blood acetaldehyde levels after ethanol administration. Difference between alcoholic and nonalcoholic subjects. *N. Engl. J. Med.* **292:**386–389, 1975.

Kristenson, H., and Trell, E.: Indicators of alcohol consumption: Comparisons between a questionnaire (Mm-MAST), interviews and serum γ-glutamyl transferase (GGT) in a health survey of middle-aged males. *Br. J. Addict.* **77:**297–304, 1982.

Kristenson, H., Ohlin, H., Hulten-Nosslin, M. B., Trell, E., and Hood, B.: Identification and intervention of heavy drinking in middle-aged men: Results and follow-up of 24–60 months of long term study with randomized controls. *Alcoholism: Clin. Exp. Res.* **7:**203–209, 1983.

Kryszewski, A. J., Neale, G., Whitfield, J. B., and Moss, D. W.: Enzyme changes in experimental biliary obstruction. *Clin. Chim. Acta* **47:**175–182, 1973.

Lai, C. L., Ng, R. P., and Lok, A. S. F.: The diagnostic value of the ratio of serum gamma-glutamyl transpeptidase to alkaline phosphatase in alcoholic liver disease. *Scand. J. Gastroenterol.* **17:**41–47, 1982.

LaPorte, R., Valvo-Gerard, L., Kuller, L., Dai, W., Bates, M., Cresanta, J., Williams, K., and Palkin, D.: The relationship between alcohol consumption, liver enzymes and high-density lipoprotein cholesterol. *Circulation* **64**(Suppl. III): 67–72, 1981.

Latcham, R. W.: Gamma glutamyl transpeptidase and mean corpuscular volume: Their usefulness in the assessment of inpatient alcoholics. *Br. J. Psychiatry* **149:**353–356, 1986.

Lebsack, M. E., Gordon, E. R., and Lieber, C. S.: Effect of chronic ethanol consumption on aldehyde dehydrogenase activity in the baboon. *Biochem. Pharmacol.* **30:**2273–2277, 1981.

Lelbach, W. K.: Cirrhosis in the alcoholic and its relation to the volume of alcohol abuse. *Ann. N.Y. Acad. Sci.* **252:**85–105, 1975.

Lieber, C. S.: Blood markers of alcoholic liver disease. *Recent Dev. Alcoholism* **6:**351–365, 1988.

Lieber, C. S., DeCarli, L. M., and Rubin, E.: Sequential production of fatty liver, hepatitis and cirrhosis in sub-human primates fed ethanol with adequate diets. *Proc. Natl. Acad. Sci. U.S.A.* **72:**437–441, 1975.

Lin, C. C., Potter, J. J., and Mezey, E.: Erythrocyte aldehyde dehydrogenase activity in alcoholism. *Alcoholism: Clin. Exp. Res.* **8:**539–541, 1984.

Lindros, K. O., Stowell, A., Pikkarainen, P., and Salaspuro, M.: Elevated blood acetaldehyde in alcoholics with accelerated ethanol elimination. *Pharmacol Biochem. Behav.* **13**(Suppl. 1):119–124, 1980.

Lucas, D., Menez, J. F., Bodenez, P., Baccino, E., Bardou, L. G., and Floch, H. H.: Acetaldehyde adducts with haemoglobin: Determination of acetaldehyde released from haemoglobin by acid hydrolysis. *Alcohol Alcoholism* **23:**23–31, 1988.

Luchi, P., Cortis, G., and Bucarelli, A.: Forensic considerations on the comparison of "serum γ-glutamyltranspeptidase" ("γ-GT") activity in experimental acute alcoholic intoxication and in alcoholic car drivers who caused road accidents. *Forens. Sci.* **11:**33–39, 1978.

Ludwig, S., and Kaplowitz, N.: Effect of pyridoxine deficiency on serum and liver transaminases in experimental liver injury in the rat. *Gastroenterology* **79:**545–549, 1980.

Lumeng, L., and Durant, P. J.: Regulation of the formation of stable adducts between acetaldehyde and blood proteins. *Alcohol* **2:**397–400, 1985.

Lumeng, L., and Li, T. K.: Vitamin B_6 metabolism in chronic alcohol abuse. Pyridoxal phosphate levels in plasma and the effects of acetaldehyde on pyridoxal phosphate synthesis and degradation in human erythrocytes. *J. Clin. Invest.* **53:**693–704, 1974.

Luoma, P. V., Sotaniemi, E. A., Pelkonen, R. O., and Ehnholm, C.: High-density lipoproteins and hepatic microsomal enzyme induction in alcohol consumers. *Res. Commun. Chem. Pathol. Pharmacol.* **37:**91–96, 1982.

Lykouras, E., Moussas, G., and Markianos, M.: Platelet monoamine oxidase and plasma dopamine-β-hydroxylase activities in non-abstinent chronic alcoholics. Relation to clinical parameters. *Drug Alcohol Depend.* **19:**363–368, 1987.

Maddrey, W. C., and Van Thiel, D. H.: Liver transplantation: An overview. *Hepatology* **8:**948–959, 1988.

Majchrowicz, E., and Mendelson, J. H.: Blood methanol concentrations during experimentally induced ethanol intoxication in alcoholics. *J. Pharmacol. Exp. Ther.* **179:**293–300, 1971.

Major, L. F., and Murphy, D. L.: Platelet and plasma amine oxidase activity in alcoholic individuals. *Br. J. Psychiatry* **132:**548–554, 1978.

Malmendier, C. L., Mailier, E. L., Amerijckx, J. P., and Fischer, M. L.: Plasma levels of apolipoproteins A-I, A-II and B in alcoholism: Relation to the degree of histological liver damage, and to liver function tests. *Hepatogastroenterology* **30:** 236–239, 1983.

Mani, J. C., Pietruszko, R., and Theorell, H.: Methanol activity of alcohol dehydrogenases from human liver, horse liver, and yeast. *Arch. Biochem. Biophys.* **140:**52–59, 1970.

Maring, J. A., Weigand, K., Brenner, H. D., and Von Wartburg, J. P.: Aldehyde oxidizing capacity of erythrocytes in normal and alcoholic individuals. *Pharmacol. Biochem. Behav.* **18**(Suppl. 1):135–138, 1983.

Marone, G., Plaut, M., and Lichtenstein, L. M.: Characterization of a specific adenosine receptor on human lymphocytes. *J. Immunol.* **11:**2153–2159, 1978.

Martin, J. V., Martin, P. J., and Goldberg, D. M.: Enzyme induction as a possible cause of increased serum-triglycerides after oral contraceptives. *Lancet* **1:**1107–1108, 1976.

Masarei, J. R.L., Puddey, I. B., Rouse, I. L., Lynch, W. J., Vandongen, R., and Beilin, L. J.: Effects of alcohol consumption on serum lipoprotein-lipid and apolipoprotein concentrations. Results from an intervention study in healthy subjects. *Atherosclerosis* **60:**79–87, 1986.

Matloff, D. S., Selinger, M. J., and Kaplan, M. M.: Hepatic

transaminase activity in alcoholic liver disease. *Gastroenterology* **78:**1389–1392, 1980.

Matthewson, K., and Record, C. O.: Erythrocyte aldehyde dehydrogenase activity in alcoholic subjects and its value as a marker for hepatic aldehyde dehydrogenase in subjects with and without liver disease. *Clin. Sci.* **70:**295–299, 1986.

Mayfield, D., McLeod, G., and Hall, P.: The CAGE questionnaire: Validation of a new alcoholism screening instrument. *Am. J. Psychiatry* **131:**1121–1123, 1974.

McNeil, B. J., Keeler, E., and Adelstein, S. J.: Primer on certain elements of medical decision making. *N. Engl. J. Med.* **293:**211–215, 1975.

Mezey, E., Potter, J. J., and Ammon, R. A.: Effect of ethanol administration on the activity of hepatic lysosomal enzymes. *Biochem. Pharmacol.* **25:**2663–2667, 1976.

Mobley, R., and Huisman, T. H. J.: Minor hemoglobins (Hb A_1) in chronic alcoholic patients. *Hemoglobin* **6:**79–81, 1982.

Monteiro, M. G., Pires, M. L. N., and Masur, J.: The trauma questionnaire for detecting alcohol abuse: Limiting factors. *Alcohol* **3:**287–289, 1986.

Moore, R. A.: The prevalence of alcoholism in a community general hospital. *Am. J. Psychiatry* **128:**638–639, 1971.

Moore, R. A.: The diagnosis of alcoholism in a psychiatric hospital: A trial of the Michigan Alcoholism Screening Test (MAST). *Am. J. Psychiatry* **128:**1565–1569, 1972.

Moore, R. A., and Murphy, T. C.: Denial of alcoholism as an obstacle to recovery. *Q. J. Stud. Alcohol* **22:**597–609, 1961.

Moore, R. D., Smith, C. R., Kwiterovich, P. O., and Pearson, T. A.: Effect of low-dose alcohol use versus abstention on apolipoproteins A-I and B. *Am. J. Med.* **84:**884–890, 1988.

Morgan, M. Y., Milson, J. P., and Sherlock, S.: Ratio of plasma alpha-amino-*n*-butyric acid to leucine as an empirical marker of alcoholism: Diagnostic value. *Science* **197:**1183–1185, 1977.

Morgan, M. Y., Camilo, M. E., Luck, W., Sherlock, S., and Hoffbrand, A. V.: Macrocytosis in alcohol-related liver disease: Its value for screening. *Clin. Lab. Haematol.* **3:**35–44, 1981.

Morino, Y., Kagamiyama, H., and Wada, H.: Immunochemical distinction between glutamic–oxaloacetic transaminases from the soluble and mitochondrial fractions of mammalian tissues. *J. Biol. Chem.* **239:**943–944, 1964.

Moussavian, S. N., Becker, R. C., Piepmeyer, J. L., Mezey, E., and Bozian, R. C.: Serum gamma-glutamyl transpeptidase and chronic alcoholism. Influence of alcohol ingestion and liver disease. *Dig. Dis. Sci.* **30:**211–214, 1985.

Myrhed, M., Berglund, L., and Bottiger, L. E.: Alcohol consumption and hematology. *Acta Med. Scand.* **202:**11–15, 1977.

Nagy, L. E., Diamond, I., and Gordon, A.: Cultured lymphocytes from alcoholic subjects have altered cAMP signal transduction. *Proc. Natl. Acad. Sci. U.S.A.* **85:**6973–6976, 1988.

Naito, H. K.: High-density lipoprotein (HDL) cholesterol. In: *Methods in Clinical Chemistry* (A. J. Pesce and L. A. Kaplan, eds.), St. Louis, C. V. Mosby, 1987a, pp. 1179–1194.

Naito, N. K.: Cholesterol. In: *Methods in Clinical Chemistry* (A. J. Pesce and L. A. Kaplan, eds.), St. Louis, C. V. Mosby, 1987b, pp. 1156–1178.

Nalpas, B., Vassault, A., Le Guillou, A., Lesgourgues, B., Ferry, N., Lacour, B., and Berthelot, P.: Serum activity of mitochondrial aspartate aminotransferase: A sensitive marker of alcoholism with or without alcoholic hepatitis. *Hepatology* **4:**893–896, 1984.

Nalpas, B., Vassault, A., Charpin, S., Lacour, B., and Berthelot, P.: Serum mitochondrial aspartate aminotransferase as a marker of chronic alcoholism: Diagnostic value and interpretation in a liver unit. *Hepatology* **6:**608–614, 1986.

Nguyen, L. B., and Peterson, C. M.: The effect of acetaldehyde concentrations on the relative rates of formation of acetaldehyde-modified hemoglobins. *Proc. Soc. Exp. Biol. Med.* **177:**226–233, 1984.

Niemela, O., Klajner, F., Orrego, H., Vidins, E., Blendiŝ, L., and Israel, Y.: Antibodies against acetaldehyde-modified protein epitopes in human alcoholics. *Hepatology* **7:**1210–1214, 1987.

Ning, M., Baker, H., and Leevy, C. M.: Reduction of glutamic pyruvic transaminase in pyridoxine deficiency in liver disease. *Proc. Soc. Exp. Biol. Med.* **121:**27–30, 1966.

Nishimura, M., Hasumura, Y., and Takeuchi, J.: Effect of an intravenous infusion of ethanol on serum enzymes and lipids in patients with alcoholic liver disease. *Gastroenterology* **78:**691–695, 1980.

Nuutinen, H., Lindros, K. O., and Salaspuro, M.: Determinants of blood acetaldehyde level during ethanol oxidation in chronic alcoholics. *Alcoholism: Clin. Exp. Res.* **7:**163–168, 1983.

Nuutinen, H., Lindros, K., Hekali, P., and Salaspuro, M.: Elevated blood acetate as indicator of fast ethanol elimination in chronic alcoholics. *Alcohol* **2:**623–626, 1985.

Okamoto, Y., Fujimori, Y., Nakano, H., and Tsujii, T.: Role of the liver in alcohol-induced alteration of high-density lipoprotein metabolism. *J. Lab. Clin. Med.* **111:**482–485, 1988.

Okany, C. C., Bond, A. N., and Wickramasinghe, S. N.: Effects of ethanol on cell volume and protein synthesis in a human lymphoblastoid cell line (Raji). *Acta Haematol.* **70:**24–34, 1983.

Okuno, F., Ishii, H., Kashiwazaki, K., Takagi, S., Shigeta, Y., Arai, M., Takagi, T., Ebihara, Y., and Tsuchiya, M.: Increase in mitochondrial GOT (m-GOT) activity after chronic alcohol consumption: Clinical and experimental observations. *Alcohol* **5:**49–53, 1988.

Oreland, L., Wiberg, A., Asberg, M., Traskman, L., Sjostrand, L., Thoren, P., Bertilsson, L., and Tybring, G.: Platelet MAO activity and monoamine metabolites in cerebrospinal fluid in depressed and suicidal patients and in healthy controls. *Psychiatry Res.* **4:**21–29, 1981.

Oreland, L., Wiberg, A., Winblad, B., Fowler, C. J., Gottfries, C. G., and Kiianmaa, K.: The activity of monoamine oxidase-A and -B in brains from chronic alcoholics. *J. Neural Transm.* **56:**73–83, 1983.

Orrego, H., Blendis, L. M., Blake, J. E., Kapur, B. M., and Israel, Y.: Reliability of assessment of alcohol intake based on personal interviews in a liver clinic. *Lancet* **2:**1354–1356, 1979.

Orrego, H., Blake, J. E., and Israel, Y.: Relationship between gamma-glutamyl transpeptidase and mean urinary alcohol levels in alcoholics while drinking and after alcohol withdrawal. *Alcoholism: Clin. Exp. Res.* **9:**10–13, 1985.

Palmer, K. R., and Jenkins, W. J.: Impaired acetaldehyde oxidation in alcoholics. *Gut* **23:**729–733, 1982.

Pandey, G. N., Fawcett, J., Gibbons, R., Clark, D. C., and Davis, J. M.: Platelet monoamine oxidase in alcoholism. *Biol. Psychiatry* **24:**15–24, 1988.

Papoz, L., Warnet, J. M., Pequignot, G., Eschwege, E., Claude, J. R., and Schwartz, D.: Alcohol consumption in a healthy population. Relationship to γ-glutamyl transferase activity and mean corpuscular volume. *J.A.M.A.* **245:**1748–1751, 1981.

Paredes, A.: Cost effectiveness in alcoholism treatment. *Subst. Abuse* **6:**29–37, 1985.

Peachey, J. E., and Kapur, B. M.: Monitoring drinking behavior with alcohol dipstick during treatment. *Alcoholism: Clin. Exp. Res.* **10:**663–666, 1986.

Peele, S.: The cultural context of psychological approaches to alcoholism. Can we control the effects of alcohol? *Am. Psychol.* **39:**1337–1351, 1984.

Pequignot, G., Tuyns, A. J., and Berta, J. L.: Ascitic cirrhosis in relation to alcohol consumption. *Int. J. Epidemiol.* **7:**113–120, 1978.

Persson, J., and Magnusson, P. H.: Comparison between different methods of detecting patients with excessive consumption of alcohol. *Acta Med. Scand.* **223:**101–109, 1988.

Peterson, C. M., and Polizzi, C. M.: Improved method for acetaldehyde in plasma and hemoglobin-associated acetaldehyde: Results in teetotalers and alcoholics reporting for treatment. *Alcohol* **4:**477–480, 1987.

Peterson, C. M., Polizzi, C. M., and Frawley, P. J.: Artefactual increase in hemoglobins A_{1a+b} in blood from alcoholic subjects. *Alcoholism: Clin. Exp. Res.* **10:**219–220, 1986.

Peterson, C. M., Jovanovic-Peterson, L., and Schmid-Formby, F.: Rapid association of acetaldehyde with hemoglobin in human volunteers after low dose ethanol. *Alcohol* **5:**371–374, 1988.

Petren, S., Vesterberg, O., and Jornvall, H.: Differences among five main forms of serum transferrin. *Alcoholism: Clin. Exp. Res.* **11:**453–456, 1987.

Pikkarainen, P. H., Gordon, E. R., Lebsack, M. E., and Lieber, C. S.: Determinants of plasma free acetaldehyde levels during the oxidation of ethanol. Effects of chronic ethanol feeding. *Biochem. Pharmacol.* **30:**799–802, 1981.

Poikolainen, K., Karkkainen, P., and Pikkarainen, J.: Correlations between biological markers and alcohol intake as measured by diary and questionnaire in men. *J. Stud. Alcohol* **46:**383–387, 1985.

Pokorny, A. D., Miller, B. A., and Kaplan, H. B.: The brief MAST: A shortened version of the Michigan Alcoholism Screening Test. *Am. J. Psychiatry* **129:**342–345, 1972.

Puchois, P., Fontan, M., Gentilini, J. L., Gelez, P., and Fruchart, J. C.: Serum apolipoprotein A-II, a biochemical indicator of alcohol abuse. *Clin. Chim. Acta* **144:**185–189, 1984.

Pullarkat, R. K., and Raguthu, S.: Elevated urinary dolichol levels in chronic alcoholics. *Alcoholism: Clin. Exp. Res.* **9:**28–30, 1985.

Rawles, J. W., Rhodes, D. L., Potter, J. J., and Mezey, E.: Characterization of human erythrocyte aldehyde dehydrogenase. *Biochem. Pharmacol.* **36:**3715–3722, 1987.

Reiff, S.: A cost-effectiveness study of alcoholism treatment in a health maintenance organization. *Subst. Abuse* **6:**24–28, 1985.

Rej, R.: An immunochemical procedure for determination of mitochondrial aspartate aminotransferase in human serum. *Clin. Chem.* **26:**1694–1700, 1980.

Roine, R. P.: Effects of moderate drinking and alcohol abstinence on urinary dolichol levels. *Alcohol* **5:**229–231, 1988.

Roine, R. P. Turpeinen, U., Ylikahri, R., and Salaspuro, M.: Urinary dolichol—a new marker of alcoholism. *Alcoholism: Clin. Exp. Res.* **11:**525–527, 1987a.

Roine, R. P., Ylikahri, R., Koskinen, P., Suokas, A., Hamalainen, J., and Salaspuro, M.: Effect of heavy weekend drinking on urinary dolichol levels. *Alcohol* **4:**509–511, 1987b.

Roine, R., Humaloja, K., Hamalainen, J., Nykanen, I., Ylikahri, R., and Salaspuro, M.: Significant increases in urinary dolichol levels in bacterial infections, malignancies and pregnancy but not in other clinical conditions. *Ann. Med.* **21:**13–16, 1989a.

Roine, R. P., Eriksson, C. J. P., Ylikahri, R., Penttila, A., and Salaspuro, M.: Methanol as a marker of alcohol abuse. *Alcoholism: Clin. Exp. Res.* **13:**172–175, 1989b.

Rollason, J. G., Pincherle, G., and Robinson, D.: Serum gamma glutamyl transpeptidase in relation to alcohol consumption. *Clin. Chim. Acta* **39:**75–80, 1972.

Rommelspacher, H., Damm, H., Schmidt, L., and Schmidt, G.: Increased excretion of harman by alcoholics depends on events of their life history and the state of the liver. *Psychopharmacology* **87:**64–68, 1985.

Rosalki, S. B., and Rau, D.: Serum γ-glutamyl transpeptidase activity in alcoholism. *Clin. Chim. Acta* **39:**41–47, 1972.

Rossouw, J. E., Labadarios, D., Davis, M., and Williams, R.: Vitamin B_6 and aspartate aminotransferase activity in chronic liver disease. *S. Afr. Med. J.* **53:**436–438, 1978.

Rowland, M., and Tozer, T. N.: *Clinical Pharmacokinetics: Concepts and Applications*. Philadelphia, Lea & Febiger, 1989, p. 385.

Rutenburg, A. M., Goldbarg, J. A., and Pineda, E. P.: Serum γ-glutamyl transpeptidase activity in hepatobiliary pancreatic disease. *Gastroenterology* **45:**43–48, 1963.

Ryback, R. S., Eckardt, M. J., Rawlings, R. R., and Rosenthal, L. S.: Quadratic discriminant analysis as an aid to interpretive reporting of clinical laboratory tests. *J.A.M.A.* **248:**2342–2345, 1982a.

Ryback, R. S., Eckardt, M. J., Felsher, B., and Rawlings, R. R.: Biochemical and hematologic correlates of alcoholism and liver disease. *J.A.M.A.* **248:**2261–2265, 1982b.

Salaspuro, M.: Conventional and coming laboratory markers of alcoholism and heavy drinking. *Alcoholism: Clin. Exp. Res.* **10:**5S–12S, 1986.

Sanchez-Craig, M., and Annis, H. M.: Gamma-glutamyl transpeptidase and high-density lipoproteins cholesterol in male problem drinkers: Advantages of a composite index for predicting alcohol consumption. *Alcoholism: Clin. Exp. Res.* **5:**540–544, 1981.

Sane, T., Nikkila, E. A., Taskinen, M. R., Valimaki, M., and Ylikahri, R.: Accelerated turnover of very low density lipoprotein triglycerides in chronic alcohol users. A possible mechanism for the up-regulation of high density lipoprotein by ethanol. *Atherosclerosis* **53:**185–193, 1984.

Schellenberg, F., and Weill, J.: Serum desialotransferrin in the detection of alcohol abuse. *Alcohol Alcoholism* (Suppl. 1): 625–629, 1987a.

Schellenberg, F., and Weill, J.: Serum desialotransferrin in the detection of alcohol abuse. Definition of a Tf index. *Drug Alcohol Depend.* **19**:181–191, 1987b.

Schmidt, E., Schmidt, W., and Otto, P.: Isoenzymes of malic dehydrogenase, glutamic oxaloacetic transaminase and lactic dehydrogenase in serum in diseases of the liver. *Clin. Chim. Acta* **15**:283–289, 1967a.

Schmidt, E., Schmidt, F. W., and Mohr, J.: An improved simple chromatographic method for separating the isoenzymes of malic dehydrogenase and glutamic oxaloacetic transaminase. *Clin. Chim. Acta* **15**:337–342, 1967b.

Schneider, J., Liesenfeld, A., Mordasini, R., Schubotz, R., Zofel, P., Kubel, F., Vandre-Plozzitzka, C., and Kaffarnik, H.: Lipoprotein fractions, lipoprotein lipase and hepatic triglyceride lipase during short-term and long-term uptake of ethanol in healthy subjects. *Atherosclerosis* **57**:281–291, 1985.

Selzer, M. L.: The Michigan Alcoholism Screening Test: The quest for a new diagnostic instrument. *Am. J. Psychiatry* **127**:1653–1658, 1971.

Shaper, A. G., Pocock, S. J., Ashby, D., Walker, M., and Whitehead, T. P.: Biochemical and haematological response to alcohol intake. *Ann. Clin. Biochem.* **22**:50–61, 1985.

Shaw, S., and Lieber, C. S.: Plasma amino acid abnormalities in the alcoholic. Respective role of alcohol, nutrition, and liver injury. *Gastroenterology* **74**:677–682, 1978.

Shaw, S., and Lieber, C. S.: Mechanism of increased gamma glutamyl transpeptidase after chronic alcohol consumption. Hepatic microsomal induction rather than dietary imbalance. *Subst. Alcohol Actions Misuse* **1**:423–428, 1980a.

Shaw, S., and Lieber, C. S.: Increased hepatic production of alpha-amino-*n*-butyric acid after chronic alcohol consumption in rats and baboons. *Gastroenterology* **78**:108–113, 1980b.

Shaw, S., and Lieber, C. S.: Plasma amino acids in the alcoholic: Nutritional aspects. *Alcoholism: Clin. Exp. Res.* **7**:22–27, 1983.

Shaw, S., Stimmel, B., and Lieber, C. S.: Plasma alpha-amino-*n*-butyric acid to leucine ratio: An empirical biochemical marker of alcoholism. *Science* **194**:1057–1058, 1976.

Shaw, S., Lue, S. L., and Lieber, C. S.: Biochemical tests for the detection of alcoholism: Comparison of plasma alpha-amino-*n*-butyric acid with other available tests. *Alcoholism: Clin. Exp. Res.* **2**:3–7, 1978.

Shaw, S., Worner, T. M., Borysow, M. F., Schmitz, R. E., and Lieber, C. S.: Detection of alcoholism relapse: Comparative diagnostic value of MCV, GGTP, and AANB. *Alcoholism: Clin. Exp. Res.* **3**:297–301, 1979.

Sjoquist, B., Borg, S., and Kvande, H.: Catecholamine derived compounds in urine and cerebrospinal fluid from alcoholics during and after long-standing intoxication. *Subst. Alcohol Actions Misuse* **2**:63–72, 1981a.

Sjoquist, B., Borg, S., and Kvande, H.: Salsolinol and methylated salsolinol in urine and cerebrospinal fluid from healthy volunteers. *Subst. Alcohol Actions Misuse* **2**:73–77, 1981b.

Skinner, H. A., Holt, S., Schuller, R., Roy, J., and Israel, Y.: Identification of alcohol abuse using laboratory tests and a history of trauma. *Ann. Intern. Med.* **101**:847–851, 1984.

Skude, G., and Wadstein, J.: Amylase, hepatic enzymes and bilirubin in serum of chronic alcoholics. *Acta Med. Scand.* **201**:53–58, 1977.

Sobell, M. B., Sobell, L. C., and VanderSpek, R.: Relationships among clinical judgment, self-report, and breath-analysis measures of intoxication in alcoholics. *J. Consult. Clin. Psychol.* **47**:204–206, 1979.

Sorrell, M. F., and Tuma, D. J.: Selective impairment of glycoprotein metabolism by ethanol and acetaldehyde in rat liver slices. *Gastroenterology* **75**:200–205, 1978.

Sorrell, M. F., and Tuma, D. J.: The functional implications of acetaldehyde binding to cell constituents. *Ann. N.Y. Acad. Sci.* **492**:50–62, 1987.

Sorrell, M. F., Nauss, J. M., Donohue, T. M., and Tuma, D. J.: Effects of chronic ethanol administration on hepatic glycoprotein secretion in the rat. *Gastroenterology* **84**:580–586, 1983.

Sox, H. C.: Probability theory in the use of diagnostic tests. An introduction to critical study of the literature. *Ann. Intern. Med.* **104**:60–66, 1986.

Spickard, A.: Alcoholism: The missed diagnosis. *South. Med. J.* **79**:1489–1492, 1986.

Spik, G., Bayard, B., Fournet, B., Strecker, G., Bouquelet, S., and Montreuil, J.: Studies on glycoconjugates. LXIV. Complete structure of two carbohydrate units of human serotransferrin. *FEBS Lett.* **50**:296–299, 1975.

Stamm, D., Hansert, E., and Feuerlein, W.: Detection and exclusion of alcoholism in men on the basis of clinical laboratory findings. *J. Clin. Chem. Clin. Biochem.* **22**:79–96, 1984.

Stevens, V. J., Fantl, W. J., Newman, C. B., Sims, R. V., Cerami, A., and Peterson, C. M.: Acetaldehyde adducts with hemoglobin. *J. Clin. Invest.* **67**:361–369, 1981.

Stibler, H., and Borg, S.: Evidence of a reduced sialic acid content in serum transferrin in male alcoholics. *Alcoholism: Clin. Exp. Res.* **5**:545–549, 1981.

Stibler, H., and Borg, S.: Carbohydrate composition of serum transferrin in alcoholic patients. *Alcoholism: Clin. Exp. Res.* **10**:61–64, 1986.

Stibler, H., and Hultcrantz, R.: Carbohydrate-deficient transferrin serum in patients with liver diseases. *Alcoholism: Clin. Exp. Res.* **11**:468–473, 1987.

Stibler, H., Allgulander, C., Borg, S., and Kjellin, K. G.: Abnormal microheterogeneity of transferrin in serum and cerebrospinal fluid in alcoholism. *Acta Med. Scand.* **204**:49–56, 1978.

Stibler, H., Borg, S., and Allgulander, C.: Clinical significance of abnormal heterogeneity of transferrin in relation to alcohol consumption. *Acta Med. Scand.* **206**:275–281, 1979.

Stibler, H., Sydow, O., and Borg, S.: Quantitative estimation of abnormal microheterogeneity of serum transferrin in alcoholics. *Pharmacol. Biochem. Behav.* **13**(Suppl. 1):47–51, 1980.

Stibler, H., Borg, S., and Joustra, M.: Micro anion exchange chromatography of carbohydrate-deficient transferrin in serum in relation to alcohol consumption (Swedish Patent 8400587-5). *Alcoholism: Clin. Exp. Res.* **10**:535–544, 1986.

Stockham, T. L., and Blanke, R. V.: Investigation of an acetaldehyde–hemoglobin adduct in alcoholics. *Alcoholism: Clin. Exp. Res.* **12**:748–754, 1988.

Storey, E. L., Mack, U., Powell, L. W., and Halliday, J. W.: Use of chromatofocusing to detect a transferrin variant in serum of alcoholic subjects. *Clin. Chem.* **31:**1543–1545, 1985.

Storey, E. L., Anderson, G. J., Mack, U., Powell, L. W., and Halliday, J. W.: Desialylated transferrin as a serological marker of chronic excessive alcohol ingestion. *Lancet* **1:** 1292–1294, 1987.

Sullivan, J. L., Stanfield, C. N., Schanberg, S., and Cavenar, J.: Platelet monoamine oxidase and serum dopamine-β-hydroxylase activity in chronic alcoholics. *Arch. Gen. Psychiatry* **35:**1209–1212, 1978.

Sullivan, J. L., Cavenar, J. O., Maltbie, A. A., Lister, P., and Zung, W. W. K.: Familial biochemical and clinical correlates of alcoholics with low platelet monoamine oxidase activity. *Biol. Psychiatry* **14:**385–394, 1979.

Swendseid, M. E., Yamada, C., Vinyard, E., Figueroa, W. G., and Drenick, E. J.: Plasma amino acid levels in subjects fed isonitrogenous diets containing different proportions of fat and carbohydrate.. *Am. J. Clin. Nutr.* **20:**52–55, 1967.

Swendseid, M. E., Yamada, C., Vinyard, E., and Figueroa, W. G.: Plasma amino acid levels in young subjects receiving diets containing 14 or 3.5 g nitrogen per day. *Am. J. Clin. Nutr.* **21:** 1381–1383, 1968.

Swenson, W. M., and Morse, R. M.: The use of a Self-Administered Alcoholism Screening Test (SAAST) in a medical center. *Mayo Clin. Proc.* **50:**204–208, 1975.

Tabakoff, B., Hoffman, P. L., Lee, J. M., Saito, T., Willard, B., and De Leon-Jones, F.: Differences in platelet enzyme activity between alcoholics and nonalcoholics. *N. Engl. J. Med.* **318:**134–139, 1988.

Takahashi, S., Tani, N., and Yamane, H.: Monoamine oxidase activity in blood platelets in alcoholism. *Folia Psychiatr. Neurol. Jpn.* **30:**455–462, 1976.

Takase, S., Takada, A., Tsutsumi, M., and Matsuda, Y.: Biochemical markers of chronic alcoholism. *Alcohol* **2:**405–410, 1985.

Taskinen, M. R., Valimaki, M., Nikkila, E. A., Kuusi, T., Ehnholm, C., and Ylikahri, R.: High density lipoprotein subfractions and postheparin plasma lipases in alcoholic men before and after ethanol withdrawal. *Metabolism* **31:**1168–1174, 1982.

Taskinen, M. R., Valimaki, M., Nikkila, E. A., Kuusi, T., and Ylikahri, R.: Sequence of alcohol-induced initial changes in plasma lipoproteins (VLDL and HDL) and lipolytic enzymes in humans. *Metabolism* **34:**112–119, 1985.

Taskinen, M. R., Nikkila, E. A., Valimaki, M., Sane, T., Kuusi, T., Kesaniemi, Y. A., and Ylikahri, R.: Alcohol-induced changes in serum lipoproteins and in their metabolism. *Am. Heart J.* **113:**458–464, 1987.

Tate, S. S., and Meister, A.: γ-Glutamyl transpeptidase: Catalytic, structural and functional aspects. *Mol. Cell. Biochem.* **39:** 357–368, 1981.

Tateossian, S., Peynet, J. G., Legrand, A. G., Collet, B., Rossignol, J. A., Delattre, J. J., and Rousselet, F. J.: Variations in HDL and VLDL levels chronic alcoholics. Influence of the degree of liver damage and of withdrawal of alcohol. *Clin. Chim. Acta* **148:**211–219, 1985.

Teschke, R., Brand, A., and Strohmeyer, G.: Induction of hepatic microsomal gamma-glutamyltransferase activity following chronic alcohol consumption. *Biochem. Biophys. Res. Commun.* **75:**718–724, 1977.

Thornton, J., Symes, C., and Heaton, K.: Moderate alcohol intake reduces bile cholesterol saturation and raises HDL cholesterol. *Lancet* **2:**819–821, 1983.

Tonnesen, H., Hejberg, L., Frobenius, S., and Andersen, J. R.: Erythrocyte mean cell volume—correlation to drinking pattern in heavy alcoholics. *Acta Med. Scand.* **219:**515–518, 1986.

Towell, J., Garthwaite, T., and Wang, R.: Erythrocyte aldehyde dehydrogenase and disulfiram-like side effects of hypoglycemics and antianginals. *Alcoholism: Clin. Exp. Res.* **9:** 438–442, 1985.

Towell, J. F., Barboriak, J. J., Townsend, W. F., Kalbfleisch, J. H., and Wang, R. I. H.: Erythrocyte aldehyde dehydrogenase: Assay of a potential biochemical marker of alcohol abuse. *Clin. Chem.* **32:**734–738, 1986.

Tsuboi, K. K., Thompson, D. J., Rush, E. M., and Schwartz, H. C.: Acetaldehyde-dependent changes in hemoglobin and oxygen affinity of human erythrocytes. *Hemoglobin* **5:**241–250, 1981.

Tsuji, M.: Measurement of platelet monoamine oxidase using three different substrates in patients with alcoholism and schizophrenia. *Folia Psychiatr. Neurol. Jpn.* **39:**521–530, 1985.

Turpeinen, U.: Liquid-chromatographic determination of dolichols in urine. *Clin. Chem.* **32:**2026–2029, 1986.

Tuyns, A. J., and Pequignot, G.: Greater risk of ascitic cirrhosis in females in relation to alcohol consumption. *Int. J. Epidemiol.* **13:**53–57, 1984.

Unger, K. W., and Johnson, D.: Red blood cell mean corpuscular volume: A potential indicator of alcohol usage in a working population. *Am. J. Med. Sci.* **267:**281–289, 1974.

Valimaki, M., Nikkila, E. A., Taskinen, M. R., and Ylikahri, R.: Rapid decrease in high density lipoprotein subfractions and postheparin plasma lipase activities after cessation of chronic alcohol intake. *Atherosclerosis* **59:**147–153, 1986.

Van Waes, L., and Lieber, C. S.: Glutamate dehydrogenase: A reliable marker of liver cell necrosis in the alcoholic. *Br. Med. J.* **2:**1508–1510, 1977.

Vesterberg, O., Petren, S., and Schmidt, D.: Increased concentrations of a transferrin variant after alcohol abuse. *Clin. Chim. Acta* **141:**33–39, 1984.

von Knorring, A. L., Bohman, M., von Knorring, L., and Oreland, L.: Platelet MAO activity as a biological marker in subgroups of alcoholism. *Acta Psychiatr. Scand.* **72:**51–58, 1985.

Wadstein, J., and Skude, G.: Changes in amylase, hepatic enzymes and bilirubin serum upon initiation of alcohol abstinence. *Acta Med. Scand.* **205:**313–316, 1979a.

Wadstein, J., and Skude, G.: Serum ethanol, hepatic enzymes and length of debauch in chronic alcoholics. *Acta Med. Scand.* **205:**317–318, 1979b.

Waechter, C. J., and Lennarz, W. J.: The role of polyprenol-linked sugars in glycoprotein synthesis. *Annu. Rev. Biochem.* **45:** 95–110, 1976.

Walters, G. D., Jeffrey, T. B., Kruzich, D. J., Greene, R. L., and

Haskin, J. J.: Racial variations on the MacAndrew Alcoholism Scale of the MMPI. *J. Consult. Clin. Psychol.* **51**:947–948, 1983.

Watson, C. G., Tilleskjor, C., Hoodecheck-Schow, E. A., Pucel, J., and Jacobs, L.: Do alcoholics give valid self-reports? *J. Stud. Alcohol* **45**:344–348, 1984.

Whitfield, J. B., Pounder, R. E., Neale, G., and Moss, D. W.: Serum γ-glutamyl transpeptidase activity in liver disease. *Gut* **13**:702–708, 1972.

Whitfield, J. B., Moss, D. W., Neale, G., Orme, M., and Breckenridge, A.: Changes in plasma γ-glutamyl transpeptidase activity associated with alterations in drug metabolism in man. *Br. Med. J.* **1**:316–318, 1973.

Whitfield, J. B., Allen, J. K., Adena, M., Gallagher, H. G., and Hensley, W. J.: A multivariate assessment of alcohol consumption. *Int. J. Epidemiol.* **10**:281–288, 1981.

Wiberg, A., Gottfries, C. G., and Oreland, L.: Low platelet monoamine oxidase activity in human alcoholics. *Med. Biol.* **55**:181–186, 1977.

Williams, A. L. B., and Hoofnagle, J. H.: Ratio of serum aspartate to alanine aminotransferase in chronic hepatitis. Relationship to cirrhosis. *Gastroenterology* **95**:734–739, 1988.

Wong, T. K., and Lennarz, W. J.: The site of biosynthesis and intracellular deposition of dolichol in rat liver. *J. Biol. Chem.* **257**:6619–6624, 1982.

Worner, T. M., and Lieber, C. S.: Plasma glutamate dehydrogenase: Clinical application in patients with alcoholic liver disease. *Alcoholism: Clin. Exp. Res.* **4**:431–434, 1980.

Wu, A., Chanarin, I., and Levi, A. J.: Macrocytosis of chronic alcoholism. *Lancet* **1**:829–830, 1974.

Wu, A., Chanarin, I., Slavin, G., and Levi, A. J.: Folate deficiency in the alcoholic—its relationship to clinical and hematological abnormalities, liver disease and folate stores. *Br. J. Haematol.* **29**:469–478, 1975.

Wu, A., Slavin, G., and Levi, A. J.: Elevated serum gamma-glutamyl-transferase (transpeptidase) and histological liver damage in alcoholism. *Am. J. Gastroenterol.* **65**:318–323, 1976.

Xin, Y., Lasker, J. M., Rosman, A. S., and Lieber, C. S.: Isoelectric focusing/Western blotting: A novel and practical method for quantitation of carbohydrate-deficient transferrin in alcoholics. *Alcoholism: Clin. Exp. Res.* **15**:814–821, 1991.

Xin, Y., Rosman, A. S., Lasker, J. M., and Lieber, C. S.: Measurement of carbohydrate-deficient transferrin by isoelectric focusing/Western blotting and by micro anion-exchange chromatography/radioimmunoassay: Comparison of diagnostic accuracy. *Alcohol Alcohol.* (in press), 1992.

Yamada, S., Wilson, J. S., and Lieber, C. S.: The effects of ethanol and diet on hepatic and serum γ-glutamyltransferase activities in rats. *J. Nutr.* **115**:1285–1290, 1985a.

Yamada, S., Mak, K. M., and Lieber, C. S.: Chronic ethanol consumption alters rat liver plasma membranes and potentiates release of alkaline phosphatase. *Gastroenterology* **88**:1799–1806, 1985b.

Yamada, K., Yokohama, H., Abe, S., Katayama, K., and Sato, T.: High-performance liquid chromatographic method for the determination of dolichols in tissues and plasma. *Anal. Biochem.* **150**:26–31, 1985c.

Yudkoff, M., Blazer-Yost, B., Cohn, R., and Segal, S.: On the clinical significance of the plasma α-amino-n-butyric acid: leucine ratio. *Am. J. Clin. Nutr.* **32**:282–285, 1979.

Zimmerman, H. J., and West, M.: Serum enzyme levels in the diagnosis of hepatic disease. *Am. J. Gastroenterol.* **40**:387–402, 1963.

Zorzano, A., Ruiz del Arbol, L., and Herrera, E.: Effect of liver disorders on ethanol elimination and alcohol and aldehyde dehydrogenase activities in liver and erythrocytes. *Clin. Sci.* **76**:51–57, 1989.

Index